C0-CCP-091

The Oxford Handbook of Interdisciplinarity

THE OXFORD HANDBOOK OF **Interdisciplinarity**

Editor-in-Chief

Robert Frodeman

University of North Texas

Associate Editors

Julie Thompson Klein

Wayne State University

Carl Mitcham

Colorado School of Mines

Managing Editor

J. Britt Holbrook

University of North Texas

OXFORD
UNIVERSITY PRESS

Great Clarendon Street, Oxford OX2 6DP

Oxford University Press is a department of the University of Oxford.
It furthers the University's objective of excellence in research, scholarship,
and education by publishing worldwide in

Oxford New York

Auckland Cape Town Dar es Salaam Hong Kong Karachi
Kuala Lumpur Madrid Melbourne Mexico City Nairobi
New Delhi Shanghai Taipei Toronto

With offices in

Argentina Austria Brazil Chile Czech Republic France Greece
Guatemala Hungary Italy Japan Poland Portugal Singapore
South Korea Switzerland Thailand Turkey Ukraine Vietnam

Published in the United States
by Oxford University Press Inc., New York

First edition published 2010

British Library Cataloguing in Publication Data

Data available

Library of Congress Cataloging in Publication Data

The oxford handbook of interdisciplinarity / editor in chief, Robert Frodeman... [et al.].
p. cm.
ISBN 978-0-19-923691-6 (hardback)
1. Interdisciplinary approach to knowledge. 2. Interdisciplinary research.
3. Congnitive science—Research. I. Frodeman, Robert.
BD255.094 2010
001—dc22 2010014847

Typeset by SPI Publisher Services, Pondicherry, India
Printed in Great Britain
on acid-free paper by
CPI Antony Rowe, Chippenham, Wiltshire

ISBN 978-0-19-923691-6

1 3 5 7 9 10 8 6 4 2

Editorial Board

Preface

The Oxford handbook of interdisciplinarity (HOI) has been nearly 10 years in the making. By way of preface, it is useful to tell part of this history.

While the editors of this volume have been involved in interdisciplinary research—and research into interdisciplinarity—for decades, our active collaboration dates from 2001. During the 2001–02 academic year Frodeman served as the Hennebach Visiting Professor in the Humanities at the Colorado School of Mines (CSM), where Mitcham was (and remains) a professor within the Division of Liberal Arts and International Studies. Their common interests in interdisciplinarity led to the creation of a project entitled 'New Directions in the Earth Sciences and the Humanities', launched with seed money from CSM. Soon thereafter, familiarity with her work led to an invitation to Klein to join these early efforts. Together with HOI advisory board member Nancy Tuana, in 2002 New Directions went live with the stated goal of conducting 'experiments in interdisciplinarity'.

New Directions began by issuing a request for interdisciplinary teams to propose projects at the intersection of the earth sciences and the humanities. Projects were to be focused on environmental questions relating to the theme of water. After receiving 31 proposals, six were chosen for funding of $10,000 each, contingent on the raising of a 1:1 match. Over the next few years New Directions attracted several hundred thousand dollars of funding from a number of entities—most prominently the National Science Foundation (NSF), but also the National Endowment for the Humanities (NEH), the National Aeronautics and Space Administration (NASA), the Environmental Protection Agency (EPA), the Cooperative Institute for Research in Environmental Sciences (CIRES), the National Center for Atmospheric Research (NCAR), the Geological Survey of Canada (GSC), and two universities, the Columbia University Earth Institute and the Pennsylvania State University Rock Ethics Institute.

The six teams also agreed to meet regularly to exchange insights arising from their projects. The first workshop was held at Biosphere 2 near Tucson, Arizona in the spring of 2002, at the site of a failed idealistic interdisciplinary and transdisciplinary experiment—that is, Biosphere 2. The lessons recounted there highlighted the need for some type of summary account of interdisciplinary research. A second workshop took place at CSM in the fall of 2002. New Directions continued its case-based approach to interdisciplinarity by including a field trip to the nearby Rocky Flats nuclear weapons production facility. This meeting also led to the 2003 publication of a special issue of the *CSM Quarterly* collecting papers on the theory and practices in interdisciplinarity, multidisciplinarity, cross-disciplinarity, and more.

New Directions researchers continued to sort out meanings and opportunities in interdisciplinary experience, for instance at a third workshop hosted by Tuana at Pennsylvania State University in the fall of 2003. With Frodeman's move to the University of North Texas (UNT) in the fall of 2004, New Directions expanded its remit to 'New Directions: Science, Humanities, Policy'. New Directions 2.0 received substantial sustaining funding from UNT, which served as the occasion for expanding our base concerns to reflect on larger issues. Indeed, in different combinations we were now reaching out to the fields of science, technology, and society studies as well as science policy to promote discussions on the differences between narrow and broader forms of interdisciplinarity, interdisciplinarity as critical assessment of knowledge production, and the needs for a humanities policy that might complement science policy.

With this rebranding New Directions 2.0 turned its focus to a series of larger, thematic workshops that drew in specific groups of researchers and scholars. The first of these was held in St Petersburg, Russia in the summer 2004. Entitled 'Cities and Rivers: St Petersburg and the Neva River', this week-long NSF-funded research looked at the inter- and transdisciplinary challenges faced by St Petersburg in addressing its water quality and quantity issues (<http://enspire.syr.edu/nevaworkshop/>). Then, in the aftermath of hurricane Katrina, New Directions 2.0 received NSF funding for 'Cities and Rivers 2: New Orleans, the Mississippi Delta, and Katrina', held in March 2006 (<http://www.ndsciencehumanitiespolicy.org/katrina/>). This workshop focused on the breakdown between knowledge producers and users that clearly contributed to this disaster. Other workshops where held at NASA Ames (on environmental ethics and space policy, spring 2007) and in southern Chile (on the challenges facing frontier ecosystems, also spring 2007). Information about all of these workshops can be found at <http://www.csid.unt.edu/>.

About this time we approached Oxford University Press (OUP) about the creation of a handbook that would pull together disparate strands of insight concerning inter- and transdisciplinarity. Once the proposal was accepted by OUP, workshop meetings centered on efforts not to simply explore interdisciplinary in particular case studies and projects but to take interdisciplinarity itself as a project. Klein argued for expanding our links to Europe and for contacting a network developed around the concept of transdisciplinarity. One result was a meeting hosted by Peter Weingart and Wolfgang Krohn at the Center for Interdisciplinary Research (ZIF) in the fall of 2006. Efforts were also made to engage the leading group for the study of interdisciplinarity in the United States, the Association for Integrated Studies.

In our proposal to OUP we described the goal as:

> to introduce a greater degree of order into the field of interdisciplinary research, education, and practice by creating a work that will become a basic reference for all future attempts at interdisciplinarity. ... This handbook will offer a historical survey of attempts at interdisciplinarity, a review of successes and failures within both research and education and across the sciences and the humanities, and identify a set of best practices that will serve as the launching point for future explorations of interdisciplinarity.

Finally, institutional support of New Directions 2.0 at the University of North Texas increased by an order of magnitude in the fall of 2008, when New Directions 2.0 was absorbed by the Center for the Study of Interdisciplinarity (CSID, <http://www.csid.unt.edu/>). Interdisciplinary centers for the study of one or another issue are common throughout higher education. But in the Anglophone world at least there have been few, if any, that focused on interdisciplinarity itself. In association with CSID (directed by Frodeman and J. Britt Holbrook, managing editor of HOI, with Klein and Mitcham as senior fellows), the Oxford Handbook project outgrew its concern for bridging the geosciences and the humanities and was transformed into broader field of university teaching, research, and public outreach.

Interdisciplinarity takes place at multiple sites and on multiple levels, and in multiple types and forms. Ironically, interdisciplinarity is divided into scientific, humanistic, social scientific, and forms which not even its most ardent practitioners and proponents can easily transcend. But by creating a base understanding of interdisciplinarity across its many forms and articulating the issues that repeatedly arise on different levels, we hope that this handbook can contribute to critical assessment of a vibrant new dimension of knowledge production and use.

Robert Frodeman
Julie Thompson Klein
Carl Mitcham

Acknowledgements

Crucial early support for this work was given by the Hennebach Program in the Humanities at the Colorado School of Mines. Thanks are due to Arthur Sacks, then director of the Division of Liberal Arts and International Studies, for supporting New Directions early on. Warren Burggren, Dean of Arts and Sciences, the University of North Texas, also offered continuing and stalwart support. Special thanks also to Wendy Wilkins, Provost, University of North Texas, for supporting the creation of CSID which provided the resources necessary to bring this to fruition. In addition, we offer our thanks to Oxford University Press (OUP): Ian Sherman of OUP first championed the idea of this volume at OUP, and Helen Eaton brought it to completion.

We acknowledge the fine work of Jonathon Parker and Nathan Bell, editorial assistants at the University of North Texas, who proofread the final versions of the chapters and of Steven Hrotic, postdoctoral fellow at CSID. Finally, J. Britt Holbrook deserves special recognition for the central role he played as managing editor for the volume.

Robert Frodeman
Julie Thompson Klein
Carl Mitcham

Contents

Contributor biographies xv

Introduction xxix
Robert Frodeman

PART 1: **THE TERRAIN OF KNOWLEDGE**

1 **A short history of knowledge formations** 3
Peter Weingart

2 **A taxonomy of interdisciplinarity** 15
Julie Thompson Klein

3 **Interdisciplinary cases and disciplinary knowledge** 31
Wolfgang Krohn

Box: Prospects for a philosophy of interdisciplinarity 39
Jan C. Schmidt

4 **Deviant interdisciplinarity** 50
Steve Fuller

5 **Against holism** 65
Daniel Sarewitz

PART 2: **INTERDISCIPLINARITY IN THE DISCIPLINES**

6 **Physical sciences** 79
Robert P. Crease

Box: **Mathematics and root interdisciplinarity** 88
Erik Fisher and David Beltran-del-Rio

7 **Integrating the social sciences: theoretical knowledge, methodological tools, and practical applications** 103
Craig Calhoun and Diana Rhoten

8 **Biological sciences** 119
Warren Burggren, Kent Chapman, Bradley Keller, Michael Monticino, and John Torday

9 **Art and music research** 133
Julie Thompson Klein and Richard Parncutt

10 **Engineering** 147
Patricia J. Culligan and Feniosky Peña-Mora

11 **Religious studies** 161
Sarah E. Fredericks

12 **Information research on interdisciplinarity** 174
Carole L. Palmer

Box: **Transcending discipline-based library classifications** 180
Richard Szostak

PART 3: **KNOWLEDGE INTERDISCIPLINED**

13 **A field of its own: the emergence of science and technology studies** 191
Sheila Jasanoff

14 **Humanities and technology in the Information Age** 206
Cathy N. Davidson

Box: ***Vectors*: an interdisciplinary digital journal** 210
Tara McPherson

15 **Media and communication** 220
Adam Briggle and Clifford G. Christians

16 **Cognitive science** 234
Paul Thagard

17 **Computation and simulation** 246
Johannes Lenhard

18 **Interdisciplinarity in ethics and the ethics of interdisciplinarity** 259
Anne Balsamo and Carl Mitcham

19 **Design as problem solving** 273
Prasad Boradkar

Box: **InnovationSpace at Arizona State University** 276
Prasad Boradkar

20 **Learning to synthesize: the development of interdisciplinary understanding** 288
Veronica Boix Mansilla

Box: **ZiF** 292
Ipke Wachsmuth

Box: **Creativity and interdisciplinarity** 296
Thomas Kowall

PART 4: **INSTITUTIONALIZING INTERDISCIPLINARITY**

21 **Evaluating interdisciplinary research** 309
Katri Huutoniemi

22 **Peer review** 321
J. Britt Holbrook
23 **Policy challenges and university reform** 333
Clark A. Miller
Box: **Transdisciplinary efforts at public science agencies: NSF's SciSIP program** 339
Erin Christine Moore
24 **Administering interdisciplinary programs** 345
Beth A. Casey
25 **Undergraduate general education** 360
William H. Newell
Box: **The Association for Integrative Studies** 364
William H. Newell
26 **Interdisciplinary pedagogies in higher education** 372
Deborah DeZure
27 **Facilitating interdisciplinary scholars** 387
Stephanie Pfirman and Paula J. S. Martin
28 **Doctoral student and early career academic perspectives** 404
Jessica K. Graybill and Vivek Shandas
29 **A memoir of an interdisciplinary career** 419
Daniel Callahan

PART 5: **KNOWLEDGE TRANSDISCIPLINED**

30 **Solving problems through transdisciplinary research** 431
Gertrude Hirsch Hadorn, Christian Pohl, and Gabriele Bammer
Box: **td-net—the Swiss Academies of Arts and Sciences' forum for transdisciplinary research** 434
Christian Pohl and Gertrude Hirsch Hadorn
Box: **Sustainability foresight: participative approaches to sustainable utility sectors** 441
Bernhard Truffer
31 **Systems thinking** 453
Sytse Strijbos
Box: **Mapping interdisciplinary research** 457
Katy Börner and Kevin W. Boyack
32 **Cross-disciplinary team science initiatives: research, training, and translation** 471
Daniel Stokols, Kara L. Hall, Richard P. Moser, Annie X. Feng, Shalini Misra, and Brandie K. Taylor
Box: **Managing consensus in interdisciplinary teams** 482
Rico Defila and Antonietta Di Giulio

33 **The environment** 494
J. Baird Callicott
Box: **Biocultural conservation in Cape Horn: the Magellanic woodpecker as a charismatic species** 499
Ximena Arango, Ricardo Rozzi, Francisca Massardo, and José Tomás Ibarra
34 **Health sciences and health services** 508
Jennifer L. Terpstra, Allan Best, David B. Abrams, and Gregg Moor
Box: **Telehospice: a case study in healthcare intervention research** 511
Elaine M. Wittenberg-Lyles, Debra Parker Oliver, and George Demiris
35 **Law** 522
Marilyn Averill
36 **Risk** 536
Sven Ove Hansson
37 **Corporate innovation** 546
Bruce A. Vojak, Raymond L. Price, and Abbie Griffin

Index 561

Contributor biographies

David B. Abrams, PhD, is executive director of the Steven A. Schroeder Institute for Tobacco Research and Policy Studies at the American Legacy Foundation. He directed the Office of Behavioral and Social Sciences Research (OBSSR) at the National Institutes of Health (NIH). Previously, he was Professor of Psychiatry and Human Behavior and of Community Health at Brown University, and founding director of Brown's Transdisciplinary Centers for Behavioral and Preventive Medicine. He has published widely and received 65 NIH grant awards as principal investigator or co-principal investigator. He is a fellow of the American Psychological Association and the Academy of Behavioral Medicine Research.

Ximena Arango has a MSc in management and conservation of sub-Antarctic natural resources from the Universidad de Magallanes and her research has focused on the definition and implementation of the Magellanic woodpecker as a flagship species for the Cape Horn Biosphere Reserve (CHBR) and the exploring of other potential flagship species at the regional scale. She is the education coordinator for the Omora Ethnobotanical Park at Puerto Williams (IEB-UMAG) where she carries out local environmental education programs with the school in Puerto Williams at the biosphere reserve.

Marilyn Averill is an attorney who is currently a doctoral student in environmental studies at the University of Colorado at Boulder, where she is affiliated with the Center for Science and Technology Policy Research. Her research and teaching interests focus on international environmental governance and the politics of science, particularly within the context of climate change. Her publications include 'Climate litigation: shaping public policy and stimulating debate' in Moser and Dilling, *Creating a climate for change: communicating climate change and facilitating social change* (2007) and 'Climate litigation: ethical implications and social impacts' in the *Denver University Law Review.*

Anne Balsamo focuses on the relationship between culture and technology. She is Professor of Interactive Media and Gender Studies in the School of Cinematic Arts, and co-founded Onomy Labs, Inc. a Silicon Valley design and fabrication company that builds cultural technologies. Previously she was a member of Research on Experimental Documents which created experimental reading devices and new media genres. Her books include *Technologies of the gendered body: reading cyborg women* (Duke University Press, 1996) on the implications of emergent biotechnologies, and *Designing culture: the technological imagination at work* which examines the relationship between cultural theory, interdisciplinary collaboration, and technological innovation.

Gabriele Bammer (BSc, BA, PhD) is a professor at the National Centre for Epidemiology and Population Health, ANU College of Medicine, Biology and Environment at The Australian National University and research fellow at the Program in Criminal Justice Policy and Management. John F. Kennedy School of Government, Harvard University. She is developing the new discipline of integration and implementation sciences and testing its application in collaborative projects on illicit drug policy, natural resource management, and policing and security in Australia, and the impact of global environmental change on food systems in the Indo-Gangetic plain. Her publications include *Uncertainty and risk: multidisciplinary perspectives*, edited with Michael Smithson (Earthscan, 2008).

David Beltran-del-Rio earned his MS in applied mathematics from the University of Colorado at Boulder in 2003 and his BA from St. John's College in 1993. He has also worked with the Fiske Planetarium, educating the public on the cosmological and mathematical origins of calendars and time keeping.

Allan Best, PhD, is a health systems researcher with the Vancouver Coastal Health Research Institute, and a clinical professor at the University of British Columbia. His research applies systems thinking to partnerships between interdisciplinary/intersectoral teams of academics and health-system decision makers in the co-production of problem-based knowledge to guide systems change, with a particular interest in alternative models for putting knowledge into action. Key publications include a 2003 *American Journal of Health Promotion* article on the use of theory, and a 2007 article in the *American Journal of Preventive Medicine* focusing on systems thinking for population health.

Veronica Boix Mansilla (EdD in human development and psychology, MS in education) is principal investigator at the Harvard Interdisciplinary Studies Project, which she co-directs with Howard Gardner. Standing at the cross-roads of cognitive/developmental psychology, epistemology, and pedagogy as applied to disciplines such as history, biology, and the visual arts, her research and publications focus on the nature of disciplinary and interdisciplinary understanding. Over more than a decade, Veronica's work at Harvard Project Zero has advanced knowledge and practical tools in areas such as the development of epistemological beliefs, learning and cognition in history and science, performance-based assessment, and education for understanding.

Prasad Boradkar is an associate professor in the School of Design Innovation at Arizona State University. He is the director of InnovationSpace, a transdisciplinary laboratory at Arizona State University where students and faculty partner with researchers and businesses to explore human-centered product design concepts which improve society and the environment. Prasad is the author of *Designing things: a critical introduction to the culture of objects*, (Berg Press, 2010).

Katy Börner is the Victor H. Yngve Professor of Information Science at the School of Library and Information Science at Indiana University, with additional affiliations with Informatics, Cognitive Science, and the Biocomplexity Institute. She is also founding director of the Cyberinfrastructure for Network Science Center. Her research focuses on the development of data analysis and visualization techniques for information management; she is particularly interested in the evolution of scientific disciplines, the analysis of online activity, and the development of collaborative cyberinfrastructures. Her latest book is *Atlas of science: visualizing what we know* (MIT Press, expected 2010).

Adam Briggle, with a PhD in environmental studies, is an assistant professor in the Department of Philosophy and Religion Studies at the University of North Texas. His research interests lie at the intersections of science, technology, ethics, and politics. He served as a postdoctoral fellow at the University of Twente, the Netherlands, where he researched the implications of new media technologies for quality of life. He is author of *A rich bioethics: public policy, biotechnology, and the Kass Council (2010)* with the University of Notre Dame Press.

Kevin W. Boyack joined SciTech Strategies, Inc. in 2007 after working at Sandia National Laboratories, where he spent several years in Computation, Computers, Information and Mathematics Center. He holds a PhD in Chemical Engineering from Brigham Young University. His current interests and work are related to information visualization, knowledge domains, science mapping with associated metrics and indicators, network analysis, and the intergration and analysis of multiple data types.

Warren W. Burggren is professor in the Department of Biological Sciences and dean of the College of Arts and Sciences at the University of North Texas. His research focuses upon developmental and evolutionary physiology at the tissue, organ system, and organismal levels. Additionally, he is interested in interdisciplinary activities surrounding physiological complexity and quantitative methods for describing how complexity changes during the developmental processes of animals. Burggren has published approximately 190 articles, chapters, monographs, and textbooks in comparative animal physiology.

Craig Calhoun is president of the Social Science Research Council and university professor at NYU where he also directs the Institute for Public Knowledge. Interdisciplinary throughout his career, Calhoun studied anthropology and history as well as sociology at Columbia, Manchester, and Oxford, and has held joint appointments in those fields and in communications. He edited the *Oxford dictionary of social sciences* (2002). His most recent books are *Nations matter* (Routledge 2007) and *The roots of radicalism* (Chicago University Press 2008).

Daniel Callahan (PhD, philosophy, Harvard) is senior researcher and president emertius at the Hastings Center, which he cofounded, and an elected member of the Institute of Medicine of the National Academy of Sciences. He has been a Senior Lecturer at the Harvard Medical School, and is now a Senior Scholar, Institute of Politics and Policy Studies, Yale University. Dr. Callahan is the author or editor of 41 books, including *Taming the Beloved Beast: How Medical Technology Costs are Destroying Our Health Care System* (Princeton University Press, 2009) and *Medicine and the Market: Equity v. Choice* (Johns Hopkins University Press, 2006).

J. Baird Callicott is Regents Professor of Philosophy and chair of the department of philosophy and religion studies at the University of North Texas. He is the co-editor-in-chief of the *Encyclopedia of environmental ethics and philosophy* and author or editor of a score of books and author of dozens of journal articles, encyclopedia articles, and book chapters in environmental philosophy and ethics. Callicott has served the International Society for Environmental Ethics as president and Yale University as bioethicist-in-residence. Callicott is perhaps best known as the leading modern exponent of Aldo Leopold's land ethic.

Beth A. Casey (PhD, English and comparative literature) is associate professor emeritus of English and Canadian Studies at Bowling Green State University, where she also served as director of general education for 27 years. She taught Canadian studies, English Canadian literature, and American literature and frequently serves as a consultant on development for interdisciplinary studies and general education in colleges and universities across the United States. Her publications include: *The quiet revolution: the transformation and reintegration of the humanities*, *Developing and administering interdisciplinary programs*, and *Administering interdisciplinary programs: creating climates for change*.

Kent Chapman (PhD Arizona State at Tempe) is a botanist and professor of biology, University of North Texas, where he serves as director of the Center for Plant Lipid Research. In addition to his own research and publishing activities, he coordinates diverse research activities focusing on basic and applied aspects of research in the regulation of plant lipid metabolism.

Clifford G. Christians is professor and director of the Institute of Communications Research at the University of Illinois, Urbana-Champaign, with joint appointments in journalism and media studies. His research focuses on communication ethics, interpretation analysis, and dialogic theory. His co-edited book on *Media ethics: cases and moral reasoning* (first published, 1983; 7th edition, 2005) helped establish this interdisciplinary field. Other publications include *Jacques Ellul: interpretative essays* (1981), *Communication ethics and universal values* (1997), and *Moral engagement in public life* (2002).

Robert P. Crease is chairman of the Department of Philosophy at Stony Brook University, New York. He has written, edited, or translated a dozen books in history and philosophy of science. He is the organizer of the Trust Institute at Stony Brook, which presents interdisciplinary programs about contemporary issues. His articles and reviews have appeared in *The Atlantic Monthly*, *The New York Times Magazine*, *The Wall Street Journal*, *Newsday*, and elsewhere. He writes a monthly column, 'Critical point', on the social dimensions of science for *Physics World* magazine.

Patricia Culligan, (BSc in civil engineering, MPhil and PhD in soil mechanics) is professor of civil engineering and engineering mechanics at Columbia University, New York and Vice Dean of Academic Affairs for Columbia's School of Engineering and Applied Science. Her interdisciplinary teaching and research focuses on geo-environmental engineering, sustainable urban development, and engineering for developing communities, with an emphasis on community engagement. Dr Culligan is the author or co-author of five books, three book chapters, and over 70 publications in refereed journals and conference proceedings. Her recent book, *Ecogowanus: urban remediation by design* (Plunz and Culligan, 2007), explores intersections of engineering and architectural design in the reclamation of contaminated urban land.

Cathy N. Davidson is the Ruth F. DeVarney Professor of English and the John Hope Franklin Humanities Institute Professor of Interdisciplinary Studies at Duke University. From 1999 to 2006, she was Duke's vice provost for interdisciplinary studies, a position that spanned all nine schools of the university. She is the co-founder of HASTAC (pronounced 'haystack', an acronym for humanities, arts, science, and technology advanced collaboratory) and writes on subjects ranging from eighteenth-century literacy and to contemporary digital learning.

Rico Defila is scientific secretary at the Interdisciplinary Center for General Ecology (IKAÖ) at the University of Bern (Switzerland) and a member of its management board. He studied law at the University of Bern. His primary research interests are the theory and methodology of inter- and transdisciplinary teaching and research and the institutionalization of interdisciplinary academic fields in universities. He conducts research projects and teaches courses in these fields, and he advises inter- and transdisciplinary research projects, programs, and institutes.

George Demiris (MSc, University of Heidelberg, PhD, University of Minnesota) is an associate professor and program director at the University of Washington in both the schools of nursing and medicine. His research interests include the application of home-based technologies for adults with chronic conditions and disabilities, smart homes and ambient assisted living applications, and the use of telehealth in home care and hospices. As an active member of the Telehospice Project, Demiris strongly advocates interdisciplinary approaches in his work, and has presented and published widely in a number of disciplines.

Deborah DeZure (PhD, interdisciplinary humanities and education) is assistant provost for faculty and organizational development at Michigan State University. Deborah is on the editorial board of *Change Magazine* and four journals on college teaching and learning. She served on the board of directors for the Association for Integrative Studies and *About Campus*. As a senior fellow at the Association of American Colleges and Universities (AAC&U), she was co-principal investigator on a project based on the Integrative Learning Project: Opportunities to Connect. Publications edited by her include: *Learning from change: landmarks on teaching and learning in higher education from* Change Magazine *(1969–99)* (2000) and *To improve the academy* (1997).

Annie X. Feng is a behavioral scientist, a SAIC Frederick Contractor, supporting the behavioral research program of the Division of Cancer Control and Population Sciences at the National Cancer Institute (NCI). Annie is a member of the NCI evaluation team of large initiatives, engaging in the evaluation of transdisciplinary initiatives and the study of team science. Her research interests include examining the talent development trajectory of scientists, team science, and transdisciplinary training. She received her BA and MA in English language and literature in China and her doctorate in education from St John's University in New York.

Erik Fisher (PhD in Environmental Studies, MA in Classics) is assistant professor with a joint appointment in the School of Politics and Global Studies and in the Consortium for Science, Policy and Outcomes at Arizona State University. He is also assistant director for international activities in the Center for Nanotechnology in Society and principal investigator of the STIR (Socio-Technical Integration Research) Project. He studies the multilevel governance of emerging technologies, spanning the nested chains of agency 'from lab to legislature'. He is co-editor of *The yearbook of nanotechnology in society, volume 1: presenting futures* (Springer 2008).

Sarah E. Fredericks, with a PhD in science, philosophy, and religion, is an assistant professor in the Department of Philosophy and Religion Studies at the University of North Texas (Denton, TX). Her interdisciplinary teaching and research focus on religion and ecology, religion and science, and energy ethics.

Robert Frodeman (PhD in philosophy, MS in geology) is a professor in the Department of Philosophy and Religion Studies at the University of North Texas. He is also director of the Center for the Study of Interdisciplinarity (<http://www.csid.unt.edu>), which seeks to develop the theory and practice of interdisciplinarity broadly construed. Publications include *Earth matters: the earth sciences, philosophy, and the claims of community* (Prentice Hall, 2000),

Rethinking nature (Indiana, 2004), *Geo-logic: breaking ground between philosophy and the earth sciences* (SUNY, 2003), and the *Encyclopedia of environmental ethics and philosophy* (co-edited with Baird Callicott).

Steve Fuller, originally trained in history and philosophy of science, is professor of sociology at the University of Warwick, UK. Active in the interdisciplinary field of science and technology studies for the past quarter century, he is best known for the 'social epistemology' research program, which is the name of a journal he founded in 1987 and the title of his first book. Recent books include *The knowledge book: key concepts in philosophy, science and culture* and *New frontiers in science and technology studies*. In 2007 Fuller was awarded a 'higher doctorate' (DLitt) by Warwick for his contributions to scholarship.

Antonietta Di Giulio is university lecturer at the Interdisciplinary Center for General Ecology (IKAÖ) at the University of Bern (Switzerland). She studied philosophy, and her doctoral thesis deals with the notion of sustainability as it emerges from the documents of the UN. Her primary research and teaching interests are the theory and methodology of inter- and transdisciplinary teaching and research, as well as education for sustainable development. She conducts research projects and teaches courses in these fields, and she advises inter- and transdisciplinary research projects, programs, and institutes.

Jessica K. Graybill, with a PhD in geography and urban ecology, is an Assistant professor in the geography department at Colgate University. Her interdisciplinary pedagogy and research is in urban ecology, urban political ecology, and urban and environmental issues of the Former Soviet Union. Publications include 'The rough guide to interdisciplinarity: graduate student experiences' (2006) and 'Continuity and change: (re)constructing geographies of the environment in late Soviet and post Soviet Russia' (2007).

Abbie Griffin BS, chemical engineering (Purdue University), MBA (Harvard University), PhD (Massachusetts Institute of Technology) holds the Royal L. Garff Presidential Chair in Marketing at the David Eccles School of Business at the University of Utah. Her research investigates means for measuring and improving the process of new product development. She is a member of the board of directors of Navistar International, and was editor of the *Journal of Product Innovation Management*. Previously, she worked as a plant engineer for Polaroid Corporation, in product and technology commercialization for Corning Glass Works, and in technology consulting for Booz, Allen and Hamilton.

Kara L. Hall is a health scientist in the Office of the Associate Director of the Behavioral Research Program in the Division of Cancer Control and Population Sciences at the National Cancer Institute (NCI). During her career, Dr Hall has participated in a variety of interdisciplinary clinical and research endeavors. Since arriving at NCI, Dr Hall has focused on advancing dissemination and implementation research and the science of team science as well as promoting the use, testing, and development of health behavior theory in cancer control research.

Sven Ove Hansson is professor of philosophy and head of the Department of Philosophy and the History of Technology, Royal Institute of Technology, Stockholm. He has led several interdisciplinary research programs in the areas of risk and environmental protection. He has also served on the board of the Swedish Natural Science Foundation and on the government's advisory board of researchers. His books include *Setting the limit: occupational health standards* *and the limits of science* (Oxford University Press 1998) and *The structures of values and norms* (Cambridge University Press 2001). He is editor-in-chief of *Theoria*. Homepage: <http://www.infra.kth.se/~soh>.

Gertrude Hirsch Hadorn, with a PhD in education and a habilitation in philosophy, is a professor at the Department of Environmental Sciences at ETH Zurich, and has been President of the transdisciplinarity-net of the Swiss Academies of Arts and Sciences, and is a member of the editorial board of the transdisciplinary journal *GAIA*. Her research interests actually include the philosophy of environmental and sustainability science, concepts and methodology of transdisciplinary research, environmental ethics and ethics of science. Her publications include the *Principles for designing transdisciplinary research* (2007) together with Christian Pohl and she has been a co-editor of the *Handbook of transdisciplinary research* (2008).

J. Britt Holbrook (PhD in philosophy, Emory University) is assistant director of the Center for the Study of Interdisciplinarity (<http://www.csid.unt.edu/>) and research assistant professor within the Department of Philosophy and Religion Studies at the University of North Texas. His research focuses on the relations between science, technology, and society, as well as interdisciplinarity in general, and on inter- and transdisciplinary pressures on peer review in particular. He is guest editor of a special edition of *Social Epistemology* devoted to the integration of societal impacts criteria into the process of peer review at public science funding agencies.

Katri Huutoniemi is a doctoral candidate in environmental policy at the University of Helsinki, Finland, and was a visiting fellow at Harvard Graduate School of Arts and Sciences for the year 2007–08. Her doctoral thesis investigates the role of interdisciplinarity in knowledge production and the challenges it poses to traditional research evaluation. Her publications include a co-authored report *Promoting interdisciplinary research: the case of the Academy of Finland* (2005) and a co-authored article *Analyzing interdisciplinarity: typology and indicators* (2010).

José Tomás Ibarra has an MSc in conservation and wildlife management from the Pontificia Universidad Católica de Chile. He is teacher of the course 'Natural history of Chilean wildlife' at the same university. Since 2006, he has been working within the Omora Ethnobotanical Park in the Cape Horn Biosphere Reserve (CHBR). His applied research tries to link wildlife ecology with the field of biocultural diversity.

Sheila Jasanoff is Pforzheimer Professor of Science and Technology Studies at Harvard University's John F. Kennedy School of Government. Previously, she was founding chair of the Department of Science and Technology Studies at Cornell University. Trained in law at Harvard Law School, she has published widely on the role of science and technology in modern democratic societies, with a particular focus on the use of science in legal and political decision making. Her books include *Controlling chemicals* (co-authored, 1985), *The fifth branch* (1990), *Science at the bar* (1995), and *Designs on nature* (2005).

Bradley Keller (PhD University of Kansas) is the director of pediatric cardiovascular research at the Cardiovascular Innovation Institute at the University of Louisville, where he has also joined the Department of Pediatrics faculty and the medical staff at Kosair Children's Hospital. Keller is principal investigator or co-principal investigator on five National Institutes of Health grants, and has been widely recognized for his research into the development of three-dimensional tissues for heart repair and regeneration. As the US medical director for

Touching Hearts in Tibet, he is also interested in using telemedicine and computer technology for screening, monitoring, and treatment of children in remote locations.

Julie Thompson Klein is professor of humanities at Wayne State University, an internationally recognized expert on interdisciplinary research, education, and problem solving, and a recipient of the Kenneth Boulding Award for outstanding scholarship on interdisciplinarity. She has lectured and consulted internationally, served on national task forces, and advised public and private agencies. Publications include *Interdisciplinarity: history, theory, and practice* (1990), *Interdisciplinary studies today* (1994), *Crossing boundaries: knowledge, disciplinarities, and interdisciplinarities* (1996), *Transdisciplinarity: joint problem solving among science, technology, and society* (2001), *Interdisciplinary education in K-12 and college* (2002), *Mapping interdisciplinary studies* (1999), and *Humanities, culture, and interdisciplinarity* (2005) and *Creating interdisciplinary campus cultures* (2010).

Thomas Kowall, PhD in education, is research professor with the international MBA program, Ecole Nationale des Ponts et Chaussées (ENPC), Paris. Courses and lectures on aesthetics, communication, management, philosophy, and strategy to audiences in Berkeley, Boston/Cambridge, Buenos Aires, Casablanca, Chicago, Nagoya, Orlando, Muscat, Paris, Shanghai, Tokyo, Toronto, Tozeur (Tunisia), Vancouver, and Victoria. His doctoral thesis was 'Distinguishing disciplines: a philosophical analysis of identity conditions for academic disciplines'. Tom was a student of Sue Larson and Donald Davidson (Stanford) and wrote his thesis under Ian Winchester (Toronto).

Wolfgang Krohn, philosopher and social scientist, is professor emeritus of science and technology studies at the University of Bielefeld. His research interests comprise studies on the relationships between science and technology, both historical and contemporary; the aesthetic dimensions of science and technology; and the spread of 'real world' experimentation outside the laboratories in social, ecological, and technological innovation projects. Recent publications include: *Society as experiment: sociological foundations for a self-experimental society* (with M. Gross, 2005), *Realexperimente* (with M. Gross and H. Hoffmann-Riem, 2005), and *Ästhetik in der Wissenschaft* (2006).

Johannes Lenhard received his PhD in mathematics. He works as a member of the scientific staff at the Center for Interdisciplinary Research (ZiF) and at the philosophy department of Bielefeld University. A main focus of his research is on computer instrumentation and the diverse modifications of scientific and societal practice coming with it. Recent publications include *Simulation: pragmatic construction of reality*, edited with G. Küppers and T. Shinn (Springer 2007).

Paula J. S. Martin has chaired interdisciplinary environmental studies programs at Emory University and Juniata College where she was assistant provost and professor of environmental science. She served on the Pennsylvania Consortium for Interdisciplinary Environmental Policy, on the Executive Board of the Council for Environmental Deans and Directors (CEDD) and co-chaired CEDD's Interdisciplinary Scholars Career Development Committee. She is now assistant director of Kenai Peninsula College, University of Alaska, Anchorage.

Francisca Massardo, PhD in plant physiology, is an ethnobotanist and plant physiologist. Her research includes ethnobotany, nectar production and the reproductive biology of Magellanic flora. She was involved in the creation of the Omora Ethnobotanical Park in Navarino Island and the Cape Horn Biosphere Reserve. She is a founding member of the UNT-Chile Field Station and is currently a professor at Universidad de Magallanes, Chile.

Tara McPherson is an associate professor at the School of Cinematic Arts, University of Southern California. Her *Reconstructing Dixie: race, gender and nostalgia in the imagined South* received the 2004 John G. Cawelti Award and was a finalist for the Kovacs Book Award. She is co-editor of the anthologies *Hop on pop: the politics and pleasures of popular culture*, and the forthcoming *Interactive frictions: the politics and pleasures of popular culture*, and is the editor of *Digital Youth, Innovation and the Unexpected*. She also edits *Vectors* (www.vectorsjournal.org) and *The International Jounal of Learning and Media* (ijlm.net). Her new media research focuses on issues of convergence, gender, and race, as well as upon the development of new tools and paradigms for digital publishing, learning, and authorship.

Clark A. Miller is associate professor in the Consortium for Science, Policy & Outcomes and School of Politics and Global Studies at Arizona State University. He holds a PhD in electrical engineering from Cornell and has held several positions at the intersection of science and technology studies, government, political science, international public affairs, environmental studies, and science and technology policy. His research and teaching focus on science, technology, and global governance, with an emphasis on the construction of policy reasoning in transnational policy debates. He is the editor of *Changing the atmosphere: expert knowledge and environmental governance* (MIT Press, 2001).

Shalini Misra is a PhD candidate in planning, policy, and design in the School of Social Ecology at the University of California, Irvine. She has a BS in civil engineering (Gujarat University, India) and an MS in sustainable resource management (Technical University of Munich, Germany). Her interdisciplinary research interests include the health, interpersonal, and community-level impacts of the Internet and other digital communication technologies. She is also interested in understanding the factors that influence the success of transdisciplinary collaboration, training, and action research initiatives.

Carl Mitcham is professor of liberal arts and international studies at the Colorado School of Mines as well as a faculty affiliate of the Center for Science and Technology Policy Research (University of Colorado, Boulder) and the European Graduate School (Saas Fee, Switzerland). His interdisciplinary teaching and research focus is on ethics and policy issues related to science, technology, and society studies. Publications include *Thinking through technology* (1994) and the *Encyclopedia of science, technology, and ethics* (2005).

Michael Monticino is a professor in the Department of Mathematics and dean of Toulouse School of Graduate Studies at the University of North Texas. Working with organizations like IBM, ARGO Data Resources, and the US Navy, he has applied statistics, probability models and stochastic control techniques to solve real-world problems. He was also founding president of the Mathematical Association of America Special Interest Group for Business Industry and Government. Current projects include: modeling ecosystem dynamics in relation to human land-use decision making, and applying ideas from complex system theory for the improvement of physiological research.

Gregg Moor is a project manager and partner in the InSource research group. Since 1990, he has worked extensively in the area of interorganizational partnerships, project administration, and research, focusing on population and public health, and health services provision and support. He takes a systems approach, applying methodologies such as social network analysis, system dynamics modeling, and concept mapping to health system challenges. He played an integral role in the development of the InSource rapid review methodology, which integrates published evidence with tacit knowledge from content experts and local stakeholders, to produce concrete recommendations for decision makers, within a short timeframe.

Erin Christine Moore is a PhD candidate in environmental studies at the University of Colorado, Boulder, where she also works with the Center for Science and Technology Policy Research. Erin is interested in issues related to science, technology, and society. Her research is inherently interdisciplinary, focusing especially on a philosophical understanding of the ways science and technology are understood, appropriated, and exploited in the world. Erin has published articles in environmental ethics and science policy. She has a master's degree in philosophy from the University of North Texas.

Richard P. Moser, PhD in clinical psychology (Pacific Graduate School of Psychology) is a research psychologist within the Office of the Associate Director, Behavioral Research Program, at the National Cancer Institute (NCI). He is the data coordinator for the Health Information National Trends Survey (HINTS), provides analytic support for in-house research projects, helps lead evaluation activities for the division, and performs his own research. His research interests include statistical methodology, health cognitions, and end-of-life issues. Previously, he performed alcoholism research at the Palo Alto VA hospital, taught statistics at several psychology graduate schools and consulted for the statistical software company SPSS.

William H. Newell is a professor of interdisciplinary studies at Miami University in Oxford, OH, where he has taught since 1974. He has served as executive director of the Association for Integrative Studies for 40 years and was its founding president in 1979. He has published three books and 40 articles and chapters on interdisciplinary studies and/or complex systems in higher education, public administration, and social science history. He has served as consultant, external evaluator, or public lecturer over a hundred times at colleges and universities in the United States, Canada, and New Zealand.

Carole L. Palmer is an associate professor at the Graduate School of Library and Information Science at the University of Illinois at Urbana-Champaign. Her research investigates how information systems and services can best support the research work of scientists and scholars. Her focus is on information technologies to improve interdisciplinary inquiry and scientific discovery, and use-based development of digital resources and tools. Her publications include *Work at the boundaries of science: information and the interdisciplinary research process* (1991) and numerous papers on scholarly information work in the digital environment.

Debra Parker Oliver (MSW, PhD in rural sociology, University of Missouri) is an associate professor of family and community medicine at the University of Missouri. Her goals are improving gerontology and end-of-life care, and, as part of the Telehospice Project, relies on interdisciplinary approaches. Her mixed-methods intervention studies and applications of telehealth technologies have been supported by several grants (including from the National

Cancer Institute), and have been widely published. She has also been a president of the Missouri End of Life Coalition and the Missouri Hospice and Palliative Care Organization, and been hospice director for Community Hospices of America.

Richard Parncutt is professor of systematic musicology at the University of Graz. His research focuses on music cognition (structure, performance, modeling), the origins of music, and musicological interdisciplinarity. He holds undergraduate degrees in music and physics and a PhD in music, physics, and psychology from Australian universities, and was a postdoc in Germany, Sweden, Canada and the UK. He is or was a board member of all leading international journals in music psychology and several other journals in systematic musicology. He founded the series 'Conference on interdisciplinary musicology' and is academic editor of the *Journal of Interdisciplinary Music Studies.*

Feniosky Peña-Mora (ScD civil engineering systems) is Dean of the School of Engineering and Applied Science at Columbia University. His interdisciplinary teaching and research is in engineering, information technology, and management. Author of more than 100 publications in refereed journals, conference proceedings, book chapters, and textbooks, his publications include *Introduction to construction dispute resolution* (2002) and *Ad hoc distributed shared memory system for disaster relief situations* (2006). His research work has also resulted in three patents and two pending patents.

Stephanie Pfirman is Alena Wels Hirschorn '58 and Martin Hirschorn Professor and chair of the Barnard College, Columbia University, Department of Environmental Science, past president of the Council of Environmental Deans and Directors (CEDD), co-chair of CEDD's Interdisciplinary Scholars Career Development Committee, and co-principal investigator of the National Science Foundation-sponsored Advancing Women in the Sciences Initiative of the Columbia Earth Institute. Current interests include understanding changes in Arctic sea ice, and development of women scientists and interdisciplinary scholars.

Christian Pohl, with a PhD in environmental sciences, is co-director of the transdisciplinarity-net of the Swiss Academies of Arts and Sciences and lecturer at ETH Zurich. His research interest is in the analysis and design of transdisciplinary research in the field of sustainable development. Publications include *Principles for designing transdisciplinary research* (2007) together with Gertrude Hirsch Hadorn and he was a co-editor of the *Handbook of transdisciplinary research* (2008).

Raymond L. Price, BS, psychology, MA, organizational behavior (Brigham Young University), and PhD in organizational behavior (Stanford Graduate School of Business) is the Severns Chair for Human Behavior, Professor in Industrial and Enterprise Systems Engineering at the University of Illinois at Urbana-Champaign, with an appointment as professor of human resource education. Previously, he was vice-president of human resources at Allergan, Inc, director of employee training and development for Boeing Commercial Airplane Group, and held various management positions with Hewlett-Packard. He is the founding director of the College of Engineering's Technology Entrepreneurship Center and the campus' Leadership Center.

Diana Rhoten Ph.D. inecology, MA in environmental philosophy is the program director of Knowledge Institutions at the Social Science Research Council and a program director in the

Office of Cyberinfrastructure at the National Science Foundation. Her research focuses on the processes and outcomes of interdisciplinary knowledge production and innovation. She is particularly interested in how the emergence of collaborative research strategies and the growing significance of virtual communities are changing the structure and culture of science. Recent publications can be found in *Annual Review of Law and Social Science* (2007), *Research Policy* (2007), and *Science* (2004).

Ricardo Rozzi, PhD in ecology, MA in enviromental philosophy is an ecologist and environmental philosopher whose main research focuses on environmental ethics and the conservation of biocultural diversity in the Cape Horn region. He was involved in the creation of the Omora Ethnobotanical Park in Navarino Island and led the efforts to establish the Cape Horn Biosphere Reserve. He is a founding member of the UNT-Chile field station and an adjunct scientist at the Institute of Ecology and Biodiversity (IEB-Chile). He is currently an associate professor at Universidad de Magallanes (Chile) and at the Department of Philosophy and Religion Studies at the University of North Texas.

Daniel Sarewitz received a PhD in geosciences but since 1989 has been a practitioner and researcher in the interdisciplinary area of science policy, with particular interests in science and decision making, and the governance of science and technology. He currently directs the Consortium for Science, Policy, and Outcomes at Arizona State University, where he is a professor in the School of Life Sciences and School of Sustainability. He is the author of *Frontiers of illusion: science, technology, and the politics of progress* (1996) as well as numerous scholarly and general-interest articles about science and society.

Jan C. Schmidt is professor of philosophy in the Unit of Social, Culture and Technology Studies at Darmstadt University of Applied Sciences, Germany. He has been associate professor for philosophy of science and technology at Georgia Tech and assistant professor of physics at the University of Mainz. He received a PhD in theoretical physics at the University of Mainz and a habilitation in philosophy at Darmstadt University of Technology. Dr Schmidt worked for the Wuppertal Institute for Climate, Energy, and Environment, and as scientific consultant in industry projects. He is a founding member of the interdisciplinary Network of Technology Assessment (NTA) in Europe.

Vivek Shandas is Associate Professor in the Toulan School of Urban Studies and Planning, and a research associate in the Center for Urban Studies at Portland State University. His research and teaching interests include innovative models for interdisciplinary education, geographic information systems, and urban ecology. Dr Shandas has an undergraduate degree in biology, masters degrees in economics and environmental policy, and completed his PhD in urban design and planning at the University of Washington (Seattle). He publishes widely in social and natural science journals, and serves as a technical advisor on several local and state organizations.

Daniel Stokols is professor of planning, policy, and design and psychology and social behavior in the School of Social Ecology at the University of California, Irvine (UCI). He served as director and dean of social ecology at UCI from 1988–98. His recent research has examined factors that influence the success of transdisciplinary research and training programs. He is past president of the Division of Population and Environmental Psychology of the American

Psychological Association (APA). He is currently serving as scientific consultant to the National Cancer Institute, Division of Cancer Control and Population Sciences, on transdisciplinary research initiatives.

Sytse Strijbos, with a PhD in philosophy, was affiliated with the Vrije Universiteit, Amsterdam and North West University, Potchefstroom Campus, South Africa as a special professor. He is now chairperson of the European Branch of the International Institute of Development and Ethics, an independent institute established in 2004 (<http://www.iide-online.org>). His interdisciplinary teaching and research is in systems thinking, philosophy, and the ethical issues of our technological world. Recent publications are *In search of an integrative vision for technology: interdisciplinary studies in information systems* (2006) and *From our side: emerging perspectives of development and ethics* (2008).

Richard Szostak is professor of economics at the University of Alberta. Szostak's research interests span the fields of economic history, methodology, history of technology, ethics, study of science, information science, and especially the theory and practice of interdisciplinarity. He has served on the board of the Association for Integrative Studies for most of the last decade. He has also served on the governing councils of the interdisciplinary programs in humanities computing, science technology and society, and religious studies at the University of Alberta.

Brandie K. Taylor is an evaluation specialist in the Strategic Planning and Evaluation Branch, Office of Strategic Planning and Financial Management, National Institute of Allergy and Infectious Diseases, National Institutes of Health. Her current work encompasses the various aspects of program evaluation and her interdisciplinary research interests are in health behavioral research and the science of team science. Ms Taylor has an undergraduate degree in honors psychology, with minors in biological science and political science and a masters degree in clinical psychology from West Virginia University. She is currently working toward the completion of her PhD in clinical psychology.

Jennifer L. Terpstra, MPH, is a PhD candidate at the University of British Columbia in the interdisciplinary studies graduate program. Jennifer's PhD studies are structured around the implementation sciences, including the study of complex systems, knowledge management, participatory research, and evaluation. Her research focuses on integrating disciplinary and stakeholder knowledge, and providing research support for policy and practice change. Jennifer utilizes a mixed-methods approach and is particularly interested in bringing systems methods, such as system dynamics modeling and agent-based modeling, to population health research. Jennifer is funded through a doctoral research award in public health through the Canadian Institute for Health Research.

Paul Thagard is professor of philosophy, psychology, and computer science at the University of Waterloo, where he directs the cognitive science program. His research is in two interdisciplinary fields, cognitive science and philosophy and history of science. His most recent books *are Hot thought: mechanisms and applications of emotional cognition* and *Mind: introduction to cognitive science* (2nd edn).

John Torday (MSc, PhD) is a professor in the Division of Obstetrics/Maternal Fetal Medicine at the Harbor-UCLA Medical Center. As co-director of the Torday-Rehan Lab, Torday argues that we cannot take advantage of the underlying patterns in human biology and medicine without an understanding of evolution and descent-with-modification, and is critical of contemporary descriptive paradigms that may dismiss observations thought to be 'counterintuitive'. Reciprocally, a research paradigm focused on evolution within the cellular context would not only benefit evolutionary theory, but also lend itself to the application of genomics to medicine.

Bernhard Truffer, PhD (University of Fribourg, Switzerland) is currently head of Comprehensive Innovation Research in Utility Sectors (CIRUS), which investigates urban water management and energy supply with a goal of sustainable transformation of infrastructure sectors. His interests in interdisciplinarity have led him to study the relationship between research groups and parent institutions, multilevel perspectives on technological innovations, as well as the developing water management systems in China, Australia, and Japan. Truffer is also currently associated with the Science Policy Group at the Social Science Research Centre (WZB), Berlin, and lectures on geography at the University of Berne.

Bruce A. Vojak, BS, MS, PhD in electrical engineering (University of Illinois at Urbana-Champaign) and MBA (University of Chicago's Booth School of Business) is associate dean for administration in the College of Engineering at the University of Illinois at Urbana-Champaign. He is adjunct professor of electrical and computer engineering and industrial and enterprise systems engineering. Formerly, he was director of advanced technology for Motorola, and held various positions at Amoco Corporation. In his early career, he was a research staff member at the MIT Lincoln Laboratory. He currently works in strategic technology management, and serves on the board of directors for Midtronics, Inc.

Ipke Wachsmuth is Professor of Artificial Intelligence at Bielefeld University, where he was also director of the Center for Interdisciplinary Research (Zentrum für interdisziplinäre Forschung, Zif) for seven years. He is a former president of the German Cognitive Science Society and currently coordinator of the Collaborative Research Center 'Alignment in Communication' at Bielefeld. His more recent research interests are in virtual reality and artificial agents, including gestural and multimodal interaction as well as emotion and intentionality.

Peter Weingart studied sociology, economics, and constitutional law at the universities of Freiburg, Berlin (Free University) and Princeton. He is professor emeritus of sociology, sociology of science and science policy at the University of Bielefeld, Germany and was, until 2009, director of the Institute for Science and Technology Studies (IWT). He was director of the Center for Interdisciplinary Research (ZiF) from 1989 to 1994, and is a member of the Berlin-Brandenburg Academy of Sciences as well as the Academy of Engineering (acatech). His current research interests are science advice to politics, science–media interrelation, and science communication. He assumed the editorship of *Minerva* in 2007.

Elaine M. Wittenberg-Lyles (PhD, University of Oklahoma) recently joined the University of North Texas as an assistant professor of communication studies. Her research program entails an examination of the interpersonal processes occurring in the context of death and dying. She is an active member of the Telehospice Project, an interdisciplinary team researching the use of technology to support caregivers of hospice patients. In addition to her prolific publishing in journals like *Qualitative Health Research* and the *Journal of Pain and Symptom Management*, she has co-authored *Communication as comfort: multiple voices in palliative care* on the complexities of communication in end-of-life care.

Introduction

In an introduction to a volume such as this it is standard to offer a synoptic account of the contents, with a general statement of the goals of the volume and summaries of the material contained within. The tone taken is scholarly and impersonal, emodying accepted academic standards.

A book on interdisciplinarity raises the possibility of a different approach. Examining the current methods and goals of knowledge production—at its core, the remit of a volume on interdisciplinarity—challenges the academic status quo. Conventions concerning the proper depth of inquiry or degree of academic rigor tend to marginalize unconventional claims or approaches, which are often hard to document or measure. And so interdisciplinary work is often accused of dilettantism and shoddy standards. These dangers are real enough. But at its best, interdisciplinarity represents an innovation in knowledge production—making knowledge more relevant, balancing incommensurable claims and perspectives, and raising questions concerning the nature and viability of expertise.

Rather than simply a summary of what is contained within the volume, this introduction also constitutes a reflection on interdisciplinarity and the future of knowledge.

* * *

Most people think of knowledge like they think of money—that they can never have enough of it. But the beneficial nature of continued, indeed infinite, knowledge production is the great unexamined premise of our justly named 'knowledge society'. It is a belief that has guided the progress of Western civilization since the Enlightenment.

This despite the fact that it is evident that knowledge can sometimes do more harm than good. After all, even robust knowledge can be misapplied: its introduction ill-timed, or its variant forms need the counter-weight of other kinds of knowledge in order to answer a question or solve a problem. The pursuit of knowledge also carries a variety of personal or moral dangers. Seeking knowledge to exercise greater control over the world can, for instance, betray a lack of control over oneself. Or it can become dysfunctional, obsessive, or escapist, hindering effective action. These are obvious dangers, familiar to all lovers of knowledge.

But our fundamental assumption is rarely questioned. Knowledge—rather than, say, moderating our desires—is seen as the answer to all of our quandaries. And so we are awash in all the cognates of knowledge: data, facts, information, statistics, and records. Has this cornucopia of knowledge led to an increase in wisdom? Has it led to happier, more fulfilled lives? Research on happiness (e.g. Lane 2000) indicates that beyond a certain point additional wealth leads to no increase in felt satisfaction with one's life. Could the same be true for knowledge? Should knowledge also be subject to an Aristotelian mean?

Still the data keep pouring in, from university think-tanks, biotech labs, space probes, and the reaches of the Internet. NASA's latest project is called the Lunar Reconnaissance Orbiter (LRO). In round numbers it will return 200 megabytes of data a day of 'spectral image cubes'. A book may total 1 megabyte of data, so the LRO will return the equivalent of 200 books a day of data. This is, of course, only a drop in our ocean of data: a recent IBM commercial announced that each day we generate eight times the knowledge contained in all the world's libraries—as if this were something to celebrate rather than be concerned about.

To one degree or another, the contributors to this volume share the intuition that the solution to our social, political, intellectual, and economic problems does not simply lie in the accumulation of more and more knowledge. What is needed today is a better understanding of the relations between fields of knowledge, a better grasp of the ways knowledge produced in the academy moves into society, and a better sense of the dangers as well as the opportunities of continued knowledge production.

As a whole, then, this volume focuses on the question of what is *pertinent* knowledge. It also implicitly raises the question of whether, in a given situation, knowledge is pertinent at all.

* * *

The *Oxford handbook of interdisciplinarity* (HOI) surveys the state of interdisciplinary knowledge today—knowledge that spans the disciplines and interdisciplinary fields and crosses the space between the academy and society at large. Its 37 chapters and 14 boxes provide both a snapshot and a critique of the state of knowledge integration as interdisciplinarity approaches its century mark. Despite the limitations inherent to such a project we hope that it fulfills the goal of all handbooks: to supply a ready and concise compendium of information about a topic of increasing importance.

Of course, when the subject is interdisciplinarity the very idea of a compendium becomes problematic. One may ask whether the editors are latter-day Encyclopedists—or unreconstructed positivists—who propose to offer a unified account of all knowledge. Or perhaps to provide a *mathesis universalis* of postmodern culture, summarizing all learning within the bounds of a single volume. The Oxford HOI harbors no such ambitions. It does not offer a synthesis of the disciplines, an overarching theory of interdisciplinary education, or a universal methodology of inter- or transdisciplinary research—although some of its individual authors may harbor such aspirations.

What, then, are the goals of this handbook? First, it provides a picture of current efforts of knowledge production that cross or bridge disciplinary boundaries ('interdisciplinarity'), and of the growing effort to make knowledge products more pertinent to non-academic actors ('transdisciplinarity').[1] Building from previous such efforts,[2] HOI offers the most synoptic and

[1] In the title of this volume, following US generic usage, 'interdisciplinarity' covers both the integration of knowledge across disciplines, narrow and wide, and the intercourse between (inter)disciplines and society. The latter often goes by the name of transdisciplinarity, particularly in Europe. Where further distinctions are needed they will be made.

[2] Among those who have preceded us, there have been monographs (e.g., Lattuca, 2001; Stehr, 2006; Klein, 1990, and 2005, and 2010; Fuller, S. and J. Collier, 2004; Repko, 2008); anthologies of reprinted articles (Newell, 1998); and collections of original essays (e.g., Weingart and Stehr, 2002; Hirsch Hadorn, et al., 2008).

broad-based account of interdisciplinarity to date. Its original essays bring together many of the leading thinkers on interdisciplinarity and its resonances in particular subject domains, including—crucially—accounts of the institutional and administrative aspects of interdisciplinarity.

Second, and more to the point here, HOI heralds the centrality of philosophic reflection for twenty-first century society. Not, it must be immediately stated, philosophy as it was done across the last century. From the perspective of the history of philosophy, twentieth-century philosophy was an aberration—a field disciplined, and a specialist's domain, one more regional expression of knowledge in principle no different from other fields such as geology or chemistry. (The point applies generally across the humanities, which embraced specialization rather than seeing their role as at heart integrative in nature.) While philosophers have always had a fraught relationship with their community, in the last half of the twentieth century the connection was broken: academic philosophy was characterized by great technical acumen wedded to societal irrelevance (Kuklick 2000).

Interdisciplinarity represents the resurgence of interest in a larger view of things. As such, interdisciplinarity is inherently philosophical, in the non-professionalized and non-disciplined sense of the term. The impetus for this was in the first instance extra-academic in origin. As knowledge production expanded, with much of it since World War II funded by public funds, demands for accountability have grown. The assumption of a linear or automatic connection between knowledge and social benefit has given way to sharp questions about the usefulness of knowledge. The power of knowledge to constantly overturn society (Marx: all that is solid melts into air) calls for a field of study, or an antidiscipline, devoted to the examination of knowledge in the largest possible compass. An antidiscipline, because it is crucial that such a study resist being once again drawn in by the gravitational pull of disciplinary approaches and standards.

The fields of social epistemology, science and technology studies, and science policy have each made important efforts in this direction. However, philosophy and the humanities—before they became exercises in logic chopping and nook-dwelling expertise—had the best claim and pedigree to being broad and incisive studies of the relation between knowledge and the good life. It is a twentieth-century irony that just when antidisciplines were most needed the humanities withdrew into specialization. To be clear: this is not to place the discipline of philosophy over other disciplines. It is rather to state that, in an *ex post facto* manner, the very search for and challenging of disciplinary standards *is* (or at least, *was*) philosophy. The corollary is that insofar as a field becomes disciplined it cannot offer the peculiar kind of insights that our times require.

To state it again: this volume does not seek to somehow constitute a unified field theory or methodology of all knowledge. Such dreams are chimerical: there never will be *the* interdisciplinary method any more than there exists *the* scientific method. Interdisciplinarity represents a new word for a perennial challenge which will never be fully answered. Experienced hands can offer hints and rules of thumb constituting a rough theory and practice of interdisciplinarity. The chapters presented here provide a number of such insights. But success at integrating different perspectives and types of knowledge—whether for increased insight, or for greater purchase on a societal problem—is a matter of manner rather than of method, requiring a sensitivity to nuance and context, a flexibility of mind, and an adeptness at navigating and translating concepts.

In aggregate, the 51 essays presented here represent a snapshot of the current and evolving state of academia-based knowledge production in the early years of the twenty-first century.

* * *

In her chapter Julie Thompson Klein offers a masterful set of definitions of 'interdisciplinarity' and other cognate terms. But the term may be approached in another, more psychological and archeological manner, where we are alive to more obscurely felt resonances. 'Interdisciplinarity' often functions apophatically: it announces an absence, expressing our dissatisfaction with current modes of knowledge production. It contains a collective unconscious of worries about the changing place of knowledge in society, and expresses a feeling that the academy has lost its way. Excessive specialization, the lack of societal relevance, and the loss of the sense of the larger purpose of things are tokens of these concerns.

The assumptions that we have relied on—that knowledge is inherently beneficial, or that scientists and scholars can justify the pursuit of knowledge in terms of 'curiosity' or the innate love of knowledge—now have the faint, yet unmistakable, scent of anachronism. In a similar manner, academic research programs are often badly out of step with our needs on the ground. Climate research within the United States continues to be funded at $2 billion a year, though it is unclear what further insights are likely to result. Climate change is almost certainly occurring, and it is almost certainly caused by human activities; but greater certainty or greater specificity than that is unlikely. Similarly, within the humanities, the philosophy of science has been 'pure' for decades, built on the assumption that the epistemological aspects of scientific research can be separated from the social, ethical, political, economic, and religious causes and consequences of science. Only the former counted as 'the philosophy of science'—while all around us, science and technology were transforming our lives. The philosophy of science was disciplined, when it needed to be interdisciplinary.[3]

'Interdisciplinarity' should not be treated as a shibboleth or a sign of one's advanced thinking. Neither is it an incantation that will magically solve our problems. Interdisciplinarity is simply a means. But to what end? Pragmatically put, toward the ends of greater insight and greater success at problem solving. More fundamentally, however, interdisciplinarity is a means toward the end of preserving or achieving the good life in a complex, global, rapidly innovating society. That is, interdisciplinarity constitutes an implicit philosophy of knowledge—not an 'epistemology', but rather a general reflection on whether and to what degree knowledge can help us achieve the perennial goal of living the good life. It is the newest expression of a very old question.

This point needs to be stated squarely, because in a global age when pluralism and relativism have become default positions, means (such as new technologies, or new techniques of knowledge production) have a tendency to become ends. Despite the riches that it has brought us, disciplinary knowledge has tacitly functioned as an abdication. By focusing on

[3] One can find tentative moves in a new direction, for instance in the 2006 formation of the Society for the Philosophy of Science in Practice.

standards of excellence internal to a discipline academics have been able to avoid larger responsibilities of how knowledge contributes to the creation of a good and just society.

The worst abdication has been by the humanities. Intimidated by success across the sciences, the humanities also embraced an analytic model of knowledge production. A few protested: at the beginning of the twentieth century William James spoke of the 'plaster-grey temperament of our balding young PhDs boring each other in seminaries, and writing those direful reports of the literature in the "Philosophical Review"'. But most interpreted the change in positive terms as an increase in rigor. Making a point so obvious that he places it in rhetorical parentheses, University of Chicago philosopher Brian Leiter stated in 2006 that we must aim for the highest possible pitch of philosophical rigor: '(Which "camp" of philosophy could possibly be committed to less careful analysis, less thorough argumentation?)'. But rigor should not be our paramount value. It must be balanced with other virtues such as timeliness, cost, and pertinence to one's audience. Rigor of argumentation—like knowledge production itself—should be subject to a mean.

One discerns a growing movement to reflect on such matters, which sometimes goes by the name of the philosophy of interdisciplinarity. This is a positive development, but we should be alive to the dangers of disciplinary capture, where new questions become just one more regional study or specialist's nook, as has happened with most new attempts to make connections across varied domains. At the very least, if we are going to have a philosophy of interdisciplinarity, it should be complemented by philosophy *as* interdisciplinarity.

The former would be given over to philosophical specialists who address questions such as whether 'interdisciplinarity' carries any distinctive epistemic content and whether there are specifically interdisciplinary objects or methodologies. It would be another 'philosophy of...'—another species of the philosophic enterprise. Rigor would be a paramount intellectual virtue. The philosophy of interdisciplinarity would treat reflections on interdisciplinarity in a *disciplinary* manner, as a discrete domain of reflection. In time this study would result in a scholarly, peer-reviewed literature, conferences, and journals.

In contrast, philosophy as interdisciplinarity points toward something closer to what Heidegger called fundamental ontology, or in his later writings, *Denken*. It strikes a balance between breadth, depth, timeliness, and societal relevance. Moreover, it constitutes a philosophical *practice* where philosophers and humanists work as much outside as within the study. Call it field philosophy, in analogy with field rather than lab science: philosophical spirits (with or without a PhD in philosophy) participate at the project level with others such as scientists, engineers, and policy makers, community groups or NGOs, helping to draw out the philosophic dimensions of controversies that stymie progress (Frodeman 2008). Philosophy as interdisciplinarity would not eschew theoretical questions; quite the opposite. But its theory would be rooted in and always return to extra-philosophic practices.

More particularly, the nature and possibility of expertise would be central to its concerns. The literature on expertise has grown significantly in recent years,[4] but it has not connected its points to questions of interdisciplinarity. It is noteworthy, but rarely noted, that the pursuit of specialization today lacks epistemological warrant. Specialization and

[4] See, for instance, Crease and Selinger, 2006, Ericsson, et al, 2006, and Collins and Evans, 2007.

expertise are built upon two assumptions—that it is possible to get down to the bottom of things, and that it is possible to study parts of the world in isolation from the world at large. The first of these is undercut by the work of modern physics, which suggests that there are no Newtonian 'simples' or irreducible pieces of matter to be found. (And if there is no bottom to things, how deep is deep enough?) More generally, however, this is the everyday experience of those in academic life. Every question raises another question, every argument a counter-argument, ad infinitum.

Secondly, our assumptions concerning the viability of expertise have been guided by the metaphorics of the laboratory, where the separation of a bench experiment from the world at large has been thought to be relatively epistemologically unproblematic. Certainty becomes possible when we make conditions and results replicable by controlling the materials used and constraining the parameters of the experiment. The upshot: robust results, but within a self-contained bubble. When those results enter the larger world we lose our controls, with often quite unexpected results.

It turns out that the world is more ecological than we had hoped. The epistemological pretensions of laboratory science have been dashed: rather than it being possible to study phenomena in isolation, everything is implicated with everything else, at least potentially. We can go deeper into a given subject only by passing over examination of the lateral connections between that subject and the rest of the universe of thought and action. But this bias for the deep rather than for the broad is rarely defended. It is in fact indefensible. Nonetheless, specialization and expertise remain the coin of the academic realm, for reasons of ease of measurement rather than any inherent virtue to the approach.

Although greater clarity and depth of insight are always possible, life is lived *in media res*. Every topic of research is infinite; there is no final or unimpeachable answer that does not give rise to another question. Current standards for what counts as expertise in a given domain are as much a reflection of political and sociological factors such as societal relevance, funding streams, or intellectual fashion as they are of inherent epistemological standards. If this point is not yet reflected in disciplinary peer review and standards for tenure and promotion, it has been recognized by institutional bodies such as the National Research Council, which recently has depicted a profusion of interdisciplinary activity that challenges conventional notions of the static nature of expertise (Committee on Facilitating Interdisciplinary Research 2004).

Reality does not allow us to control its parameters. There is no beginning or end to thinking, no straightforward path to scholarly relevance. It is as if, in our drilling down into the bedrock of knowledge, our drill bit strikes open air—revealing a cavern with a variety of wonders, but with no imperative concerning which direction we should head. Of course it is possible to know things: airplanes fly because they are competently designed; certain interpretations of Plato are better than others; we trust our physician's interpretation more than our own. But these cases are proximate. The densely imbricated nature of existence means that expertise has limits—and that these limits cannot be defined beforehand.

Confucius claimed: 'to know that you know what you know, and to know that you don't know what you don't know, is true wisdom'. The problem with this dictum is that it is very hard to draw the boundary between one and the other. Any knowledge that we possess—with the obvious exception of those domains we construct ourselves, such as the deductive world of geometry—is intrinsically fallible, proximate, and unbounded. Attempts to understand the world or any part of it need to be inter- and transdisciplinary in nature—even if this means that we lose the comfort of disciplinary guarantees of expertise.

* * *

Identifying the optimal structure for a handbook of interdisciplinarity raises a number of ontological and taxonomic issues. Disciplines are not simply mirrors held up to reality. They are economic devices and psychological supports as much as reflections of the way things are. Their boundaries are permeable and subject to movement. From the outside disciplines can appear conceptually unified, but those within often find themselves in internecine conflict. Analytic and continental philosophers argue over who is a 'real' philosopher; geology and biology departments house both systematic and historical approaches; the social sciences struggle over qualitative versus quantitative methods. In response, an account of interdisciplinarity needs to be tentative and flexible in nature.

The sections and chapters included here try to mark out the major boundary crossings between disciplines and between academia and society. Nonetheless, difficult judgments abound. Should there be a single article on interdisciplinary social science, or individual chapters on geography, sociology, and economics? A chapter on the transdisciplinary nature of environmental concerns, or one more tightly focused on, say, climate change? Individual authors were chosen through a combination of stature, the editors' knowledge and contacts, and availability. The result is a collection of authors—many of whom are scholars of international reputation—whose views on interdisciplinarity are often quite at variance with one another.

The chapters themselves are ordered in terms of a fivefold division. Part 1, 'The terrain of knowledge', offers a set of historical, taxonomic, and philosophic accounts of the genesis and development of disciplinary knowledge. The chapters of this section constitute the most synoptic of the five divisions. In 'A short history of knowledge formations' Peter Weingart provides a summary account of the development of the disciplines. Julie Thompson Klein's 'A taxonomy of interdisciplinarity' provides the definitions of terms relied on by the rest of the authors of this volume. And Wolfgang Krohn's 'Interdisciplinary cases and disciplinary knowledge' frames interdisciplinary problem solving in terms of the complexity and contingency of case work.

At turns descriptive and evaluative, the chapters of this section explore the overall landscape of interdisciplinarity, providing critical commentaries on the possibility, use, and desirability of interdisciplinary knowledge. Thus Steve Fuller challenges the 'Whig' sense of intellectual history by offering his own account of 'Deviant interdisciplinarity', while in 'Against holism' Daniel Sarewitz points out the limits of interdisciplinary approaches to knowledge. All told, this section is likely to be of greatest interest to readers concerned with the historical and theoretical dimensions of interdisciplinarity.

Part 2, 'Interdisciplinarity in the disciplines', begins from knowledge as we find it today in the academy and explores the distinctive manifestations of interdisciplinarity from specific disciplinary perspectives. The seven chapters of this section make it clear that interdisciplinarity manifests itself differently in different disciplinary contexts—that 'Interdisciplinary work by an art historian looks markedly different from that by a sociologist of art'.[5] Early reviewers of this work pointed to the paradoxical nature of pro-

[5] Ken Wissoker, "Negotiating a Passage between Disciplinary Borders," *Items and Issues* [Social Science Research Council], vol. 1 (Fall 2000), p. 1.

viding an account of interdisciplinarity in terms of the disciplines. But the paradox is only apparent: disciplinarity is the precondition for interdisciplinarity. As a self-conscious movement interdisciplinarity only arose in the face of academic specialization that so markedly accelerated in the late nineteenth century.

While the chapters of this section overlap to some degree, they largely break down along discrete disciplinary lines. In 'Physical sciences' Robert Crease treats physics as an opportunity to explore the challenges of coordination, quality assessment, and communication across different academic cultures—issues that bedevil all interdisciplinary work. In 'Integrating the social sciences: theoretical knowledge, methodological tools, and practical applications' Craig Calhoun and Diana Rhoten examine the two great interdisciplinary engagements of the postwar era within the social sciences, the development of area studies and of quantitative research methods. In 'Biological sciences' Warren Burggren and colleagues take on the daunting task of summarizing the varieties of interdisciplinarity within biology, ranging from biochemistry to medicine to mathematics to bioengineering. And in their chapter on 'Art and music research' Julie Thompson Klein and Richard Parncutt emphasize the universality of art and music across cultures while reviewing new critical approaches to art and music scholarship and teaching.

In their chapter 'Engineering' Patricia J. Culligan and Feniosky Peña-Mora use civil engineering to focus their analysis of the future responsibilities of engineers within society. Sarah E. Fredericks' examination of 'Religious studies' treats religion as an academic discipline distinct from religious practice and explores the play of anthropology, sociology, philosophy, and other disciplines in the field. In the final chapter of this section, Carole Palmer's 'Information research on interdisciplinarity' provides an overview of research on interdisciplinarity in library and information science and discusses ways to manage the problem of information scatter.

Part 3, 'Knowledge interdisciplined', examines the development of regions of knowledge that have grown up in the spaces between established disciplines (e.g. Sheila Jasanoff on science and technology studies in 'A field of its own: the emergence of science and technology studies'), or have formed as both their own specialty and as a methodology or perspective for other domains of knowledge (e.g. Johannes Lenhard on 'Computation and simulation'). Some of these regions are new—see for instance Paul Thagard's account of 'Cognitive science'—while others are as ancient as the tradition of Western thought itself (Anne Balsamo and Carl Mitcham's account of 'Ethics'). It remains an open question whether these interdisciplines will evolve into disciplines, or whether they embody different models for intellectual activity.

The effects of cultural and technological innovation on knowledge also receive attention here. Cathy Davidson looks at cross-fertilization effects in 'Humanities and technology in the Information Age', while in their chapter on 'Media and communication' Adam Briggle and Clifford Christians offer an account of the consequences of new media on knowledge production, dissemination, and consumption. Prasad Boradkar's chapter on 'Design as problem solving' examines the transdisciplinary developments of material culture from spoons to cities. Finally, in 'Learning to synthesize: the development of interdisciplinary understanding' Veronica Boix Mansilla utilizes psychological studies of cognition in order to develop an account of interdisciplinary learning.

The nine chapters of Part 4, 'Institutionalizing interdisciplinarity', look at the opportunities and challenges of making higher education more open and supportive of the inter- and transdisciplinary dimensions of knowledge. Antiquated administrative forms can strangle innovation; but despite this fact the institutional expressions and challenges of interdisciplinarity are often neglected. To have a realistic chance of success the reformation of knowledge must operate simultaneously at the levels of teaching, research, and administration.

The chapters of this section therefore address a wide range of issues relating to assessing, administering, and institutionalizing interdisciplinary work. In 'Evaluating interdisciplinary research' Katri Huutoniemi proposes three different perspectives on how to appraise interdisciplinary research. J. Britt Holbrook's chapter on 'Peer review' discusses the institution of peer review and the pressures driving it toward inter- and transdisciplinarity. Beth Casey's 'Administering interdisciplinary programs' examines the development of innovative policies across the United States in support of interdisciplinary programs, schools, or colleges. In his chapter on 'Undergraduate general education' William H. Newell offers an overview of the evolving role of interdisciplinary studies in undergraduate education in the United States. The section also includes Deborah DeZure's 'Interdisciplinary pedagogies in higher education', which surveys the array of instructional methods now available for promoting interdisciplinary learning outcomes.

Other chapters in this section look at these institutional questions from the perspective of individual researchers. Stephanie Pfirman and Paula Martin's 'Facilitating interdisciplinary scholars' identifies means for better facilitating interdisciplinary scholarship within the discipline-based academy. And in their chapter 'Doctoral student and early career academic perspectives' Jessica Graybill and Vivek Shandas discuss the unique and sometimes disconcerting challenges faced by graduate students trained in interdisciplinary approaches once they graduate.

Finally, two essays fall into categories all their own. In 'Policy challenges and university reform', Clark Miller uses science policy as a prism for thinking about the future of the university. And in 'A memoir of an interdisciplinary career', Dan Callahan gives a personal account of the development of an interdisciplinary institute devoted to bioethics, the Hastings Center.

Part 5, 'Knowledge transdisciplined', presents knowledge integration from an extra-academic perspective—as starting from societal needs and perspectives. The section begins with 'Solving problems through transdisciplinary research', where Gertrude Hirsch Hadorn, Christian Pohl, and Gabriele Bammer explore two current approaches to integrative research—transdisciplinary research from Europe, and integration and implementation sciences from Australia—in order to prevent or mitigate problems such as violence, disease or environmental pollution. In 'Systems thinking', Sytse Strijbos explicates this now classic term encompassing postwar developments in fields such as cybernetics, information theory, game and decision theory, automaton theory, systems engineering, and operations research. And in 'Cross-disciplinary team science initiatives: research, training, and translation', Dan Stokols and a team of researchers at the National Cancer Institute look at the unique challenges of large-scale team research projects that involve hundreds of scientists from different fields and locations working together on a common problem.

Another set of chapters in this section are topically based. In 'The environment', J. Baird Callicott frames the discussion of interdisciplinarity and the environment in terms of the recent development of the transdiscipline of conservation biology. Jennifer Terpstra and colleagues show how health outcomes result from the interplay of factors from the cellular to the socio-political level in 'Health science and health services'. Similarly, in her chapter 'Law', Marilyn Averill shows how the development of law has been inherently interdisciplinary, transdisciplinary, and problem based, shaping individual and group behavior and the distribution of social costs and benefits. In his chapter on 'Risk' Sven Ove Hansson discusses the mix of disciplines (e.g. psychology, epidemiology, statistics) used to access risk in areas as wide-ranging as air pollution to airbag regulation. Finally, in 'Corporate innovation', Bruce A. Vojak and colleagues examine the role of 'serial innovators' who are adept in repeatedly developing successful corporate innovations through the combination of technical skills, insight into customer needs, and political savvy necessary for getting their projects accepted for commercialization.

Of course, even with this range of perspectives significant gaps remain. Some of the gaps were inadvertent. A planned chapter on interdisciplinarity in economics did not quite come to fruition. The same is true for a proposed chapter on literature, history, and philosophy (even though the humanities are reasonably well represented), and for specific chapters within area studies, e.g. gender studies. And the 14 boxes placed throughout the volume help to mitigate these gaps. Nonetheless, a volume such as this will always remain a work in progress.

Robert Frodeman

References

Collins, H. and Evans, R. (2007). *Rethinking expertise.* Chicago: University of Chicago Press.

Committee on Facilitating Interdisciplinary Research (2004). *Facilitating interdisciplinary research.* Washington, DC: National Academies Press,

Crease, R. and Selinger, E. (2006). *The philosophy of expertise.* New York: Columbia University Press.

Ericsson, K. A. *et al.* (2006). *The Cambridge handbook of expertise and expert performance.* Cambridge: Cambridge University Press.

Frodeman, R. (2008). Philosophy unbound: environmental thinking at the end of the earth. *Environmental Ethics*, **30**(3), 313–24.

Fuller, S. and Collier, J. (2004). *Philosophy, rhetoric, and the end of knowledge*, 2nd edn [orig. edn Fuller, 1993] Hillsdale NJ: Lawrence Erlbaum Associates.

Hirsch Hadorn, G., Hoffmann-Riem, H., Biber-Klemm, S. *et al.* (2008). *Handbook of transdisciplinary research.* Dordrecht: Springer.

Klein, J.T. (1990). *Interdisciplinarity: history, theory, and practice.* Detroit, MI: Wayne State University Press.

Klein, J.T. (2005). *Humanities, culture, and interdisciplinarity: the changing American academy.* New York: SUNY Press.

Klein, J.T. (2010). *Creating interdisciplinary campus cultures.* San Francisco: Jossey Bass. Co-published by the Association of American Colleges and Universities.

Kuklick, B. (2000). *A history of philosophy in America, 1720–2000.* Oxford: Oxford University Press.

Lane, R.E. (2000). *The loss of happiness in market democracies.* New Haven, CT: Yale University Press.

Lattuca, L. (2001). *Creating interdisciplinarity: interdisciplinary research and teaching among college and university faculty.* University Park, PA: Penn State University Press.

Leiter, B. (2006). *The future for philosophy.* Oxford: Oxford University Press.

Newell, W.H. (ed.) (1998) *Interdisciplinarity: essays from the literature.* New York: College Entrance Exam Board.

Repko, A.F. (2008). *Interdisciplinary research: process and theory.* Thousand Oaks, CA: Sage Publishing.

Stehr, N. (2005). *Knowledge politics: governing the consequences of science and technology.* Boulder, CO: Paradigm Books.

Weingart, P. and Stehr, N. (eds) (2000). *Practising interdisciplinarity.* Toronto: University of Toronto Press.

PART 1
The terrain of knowledge

CHAPTER 1

A short history of knowledge formations

PETER WEINGART

The scientific disciplines such as physics, chemistry, biology, and, in the social sciences, psychology, sociology, and economics shape not only our perception of the sciences proper but also of the world around us as if they were the given structure of the world. A look back in history reveals that they are a fairly recent phenomenon, barely 200 years old in their present form (with precursors going back further), and are well on their way to yet another transformation.

Long before disciplines emerged, philosophers from Plato onward have sought to categorize human knowledge and, thus, to understand how that knowledge was best gained and ordered. A superficial look at the Greek philosophers' epistemology best illustrates the historical nature of the classifications. Aristotle differentiated '*scientia*' (*episteme*) as the knowledge about causes and reasons from mere opinions (*doxa*) that are often subjective, and from technology (*techne*) and the arts (*ars*) as the knowledge requisite to create or construct. Only scientific knowledge can claim to be universally valid. Science is, thus, distinct from practical orientation, a theoretically oriented activity. Theoretical knowledge is gained by observation and contemplation and comprises three areas (or disciplines in the modern sense): mathematics, physics, and (first) philosophy. Mathematics consists of geometry, arithmetic optics, and harmonics. Physics is the knowledge of the material world and all forms of life (i.e. today's biology). Philosophy includes knowledge of the cosmos and theology. The ancient concept of science excluded the crafts and arts as knowledge controlling actions. The Roman *Stoa* (*c.* 300 BCE) subsequently developed a classification of knowledge in opposition to Aristotle's that included practical knowledge and distinguished logics, physics, and ethics. Subsequently Aristotelian and Stoic classifications overlapped and merged with medieval concepts of the '*artes liberales*' that constituted what was then considered the comprehensive system of knowledge: grammar, rhetoric, logic, arithmetic, music, geometry, and astronomy (Kambartel 1996).

Since then the 'classifications of the sciences' as well as the theories of science and knowledge production have continued to evolve.[1] Such classifications, of which modern

[1] One pertinent example of the historical nature of disciplines is the distinction between 'science' and the 'humanities' in the Anglo Saxon world as opposed to the term '*Wissenschaft*' in the German speaking countries. Here, 'scientific disciplines' refers to the embracing notion of '*Wissenschaft*' unless stated otherwise.

academic disciplines are but one example, both reflect and, in turn, structure the production as well as the distribution of knowledge, i.e. research and teaching. Beyond that, they also shape the application of knowledge. However, they do so only if they become institutionalized, which is an important difference between the classification of knowledge preceding the emergence of disciplines and disciplines proper.

In other words, whether in terms of disciplines or any other type of categorization, providing a taxonomy of knowledge is an essential element in the ordering of knowledge. Like any other social fact, disciplines are subject to change, albeit usually only gradually and at long intervals. Their relative stability is the precondition for societies to be able to accumulate knowledge and at the same time to select and forget what is no longer relevant when conditions have changed. Disciplines, like any other classificatory principle of knowledge, therefore have the function of mediating and directing social change.

By tracing the development of modern disciplines from when they first emerged to the present, when as some observers claim they seem to be disappearing, the nature of knowledge formations may be captured best.

1.1 Knowledge production in the seventeenth and eighteenth centuries

Since the beginning of modern science and knowledge production is generally associated with Francis Bacon, his survey of the sciences may be taken as a starting point. It achieved widespread influence and was accepted by the French encyclopedists (Diderot, d'Alembert). Bacon differentiated, on the one hand, between Natural and Civil History dealing with works of nature and man, respectively. On the other hand, the sciences were distinguished into Theology and Philosophy, and Philosophy into a Doctrine of the Deity (Natural Theology), a Doctrine of Nature and a Doctrine of Man. These branches of philosophy, while further differentiated, were, according to Bacon, joined in a common trunk, the Primary Philosophy (Flint 1904, p. 106). The details of this and subsequent classifications are not important in this context. The point highlighted here is that the underlying principle of the classifications is the respective approach or method of gaining knowledge rather than—as in the case of modern disciplines—an ordering of subject matter. Philosophy is concerned with reasons and causes of (all) things, while history—in Bacon's scheme—is not science but the basis of science, describing and ordering singular things. Thus, classifications ordered human knowledge in hierarchies, in which mathematics and philosophy were considered to be at the top. The (hierarchical) differentiation of natural history, philosophy, and mathematics as modes of generating knowledge was primary.

Up to the end of the eighteenth century disciplinary differentiation is only secondary (e.g. Medicine as part of Bacon's Human Philosophy). These disciplines did not have a social function of their own but only served as repositories of certified knowledge. Disciplines were relatively unimportant until the end of the eighteenth century (Stichweh 1984, pp. 14–15). In particular, there was little relation between the classifications of knowledge and the structure of universities (in terms of 'faculties'). The hierarchy of faculties as organizational structures of the universities, institutionalized since the Middle Ages, placed philosophy at the bottom, and medicine, law, and theology above it. Although this hierarchy of the faculties had also represented a classification of knowledge, it subsequently lost acceptance. At the end of the eighteenth century the notion of a 'lower' faculty and 'higher faculties' counted as past. Nonetheless it continued to have an impact on students and professors alike. Philosophy was considered a propaedeutic subject. Students wanted to study one of the sciences, instead. Professors perceived their careers as ascending through the hierarchy of faculties, accumulating teaching positions in different fields (Stichweh 1984, p. 33).

What led to the change of this premodern order of knowledge? What were the chief properties of the 'new', disciplinary order of knowledge that emerged around the end of the eighteenth century? One chief reason for the change is seen in the growing pressure that data collection had on the disciplines. Until this point, data had been collected and ordered in ever-growing spatial classification systems. Linnaeus' *Systema naturae* contained 549 species in its first edition of 1735 and 7000 in the last edition in 1766–8, having grown from 10 to 2300 pages. The growing and unmanageable complexity of these systems was felt by contemporary scholars, motivating them to develop new methods in order to limit the realm of possible experience. The traditional methods of information processing, the classification and spatial ordering of knowledge, had to be given up. At the turn from the eighteenth to the nineteenth century the temporalization of complex sets of information replaced the spatially conceived classification systems of natural history (Lepenies 1976, pp. 16–18). Thinking in terms of development became a new technique of systematization.

1.2 The differentiation of knowledge into disciplines

Beginning in the latter half of the seventeenth century and increasingly throughout the eighteenth, science became the particular activity of collecting and ordering all available knowledge, the delineating and systematic arranging of topics, and the ever more intense interaction between participants in scientific communities. This resulted in the dramatic growth of science in terms of the amount of information produced and communicated. Thus, problems of overload and integration arose. This is seen as the antecedent condition of disciplinary (i.e. internal) differentiation. The number of experiential data and theories of science grew to a critical mass which generated innovations and motivated further research. At the same time, growth increased the pressure to treat data selectively according to criteria specific to science. That is the essence of differentiation (Stichweh 1984, p. 42).

The internal differentiation of scholarly activity into disciplines began gaining momentum through two developments. One was increasing abstraction, for example through the mathematical conceptualization of objects. This means that science to a decreasing degree gained its information about the world directly from its environment. Instead, ever more objects were reconstructed and organized under independent (typically mathematical) criteria. A growing stock of concepts, theories, and instruments mediated the experiences gathered, i.e. experience was no longer grasped immediately, but rather constructed on the level of the concept. At the same time, and this is the second development, the modern scientific mode of gaining knowledge was expanded to new subject matters. The perception of concrete objects as the initial step of discipline formation was replaced by the constitution of problems by a discipline, i.e. ultimately by a group of scholars who shared common concepts and methods, forming a community. The problems thus defined were then applied to new objects. This self-referential augmentation of objects of scientific analysis is the mechanism of *specialization*. It represents a centrifugal force for science in that the scientific methods and instruments of analysis become more generalizable, and thus more effective in being applied to new objects and phenomena. A pertinent example is molecular biology, which emerged from a systematic application of the methods of physics to the problem of explaining life (Cairns *et al.*1966). The result has been the creation of a new (sub-)discipline devoted to the study of formerly 'biological' phenomena on the molecular level, requiring new methods and instruments and using new concepts.

The emergence of disciplines in the modern sense, which took place around 1800, implied the shift from occasions arising externally to science for the collection of experience and data to problems for research generated 'within' science itself. This meant that the judgment of relevance also became subject to control by the respective groups of scholars. Their language became gradually more specialized and removed from everyday language. Thus, the public reception of scholarly research changed.

Previously, throughout the eighteenth century, books, articles, and even experiments were still addressed to the general public. The more specialized communication among scholars became, the more it was addressed to themselves. This 'closure' of disciplinary communication communities was expressed through specialized journals and in the organization of scholarly associations. As the communication turned inward and became self-referential the disciplinary community became the relevant public. At the same time this process generated a division between specialists and laypersons that has become increasingly pronounced since then. Popularization emerged as a separate activity not intended to contribute new knowledge but limited to the translation and mediation of scientific knowledge to the broader (educated) public. The increasingly esoteric nature of knowledge production led to a growing distance from practical concerns and increased resistance to commercial and technical applications that had previously legitimized the utility of the sciences. By 1830 the process of discipline formation, notably with respect to their establishment at German universities, had brought about physics and chemistry as fully fledged disciplines (Nye 1993, p. 4). At universities in the United States the differentiation of disciplines took a little longer, not least because of the pragmatic orientation of scientists and engineers concerned with the building of a new country. Many of today's universities which have their roots in the Land Grant (Morrill Act) of 1862 started out as agricultural schools and testing stations.

The change in the self-perception of science and scientists is apparent in contemporary statements deploring the loss of the unity of science. Especially in Germany, the unity of science remained an ideal associated with the unity of the academy, particularly the Royal Prussian Academy in Berlin, whose founding president, Gottfried Wilhelm Leibniz (died 1716), counted as the personification of that unity. Nye cites Lothar Meyer in 1864 pleading for the need to reunite 'the now severed sciences' (Nye 1993, p. 4). At about the same time Hermann von Helmholtz observed that no one 'could oversee the whole of science and keep the threads in one hand and find orientation. The natural result is that each individual researcher is forced to choose an ever smaller area as his workplace and can only maintain incomplete knowledge of neighboring areas' (Helmholtz 1896, p. 162).

The ongoing specialization into disciplines caused a fundamental change in the orientation of scientists. Scholars of the eighteenth century wrote textbooks and compendia. Their aspirations and concept of a career was to become knowledgeable in several fields of science and thereby advance in the hierarchy of sciences. In contrast, in the context of the new disciplinary order of science, *originality*, the discovery of new phenomena and explanations, became the primary objective of science. Research became organized on the basis of a division of labor into numerous highly specialized activities. Emil Du Bois-Reymond commented with nostalgia in 1882 that 'a thousand busy ants are producing daily countless details... only concerned to attract attention for a moment and obtain the best price for their goods', that the 'stream of discovery is split into ever more and ever more unimportant trickles' (Du Bois-Reymond 1886, p. 450). Three decades earlier the philologist August Boeckh, as cited by Daston, had already characterized the new dynamics of specialization when he declared that 'no problem was too small not to be worthy of a serious scientific analysis' (Daston 1999, p. 74).

The differentiation into disciplines also had a profound institutional impact. Throughout the eighteenth century the (royal) academies had been the organizational framework within which knowledge was accumulated. The academies represented the undifferentiated state of science insofar as they were assemblies of scholars who deliberated on questions of all fields of knowledge. But as a type of institution they became organizationally incapable of coping with the growing specialization. By the close of the century research had moved out of the academies into the universities. The universities proved to be the form of organization better suited for the accommodation of different, quite heterogeneous, disciplines with their specific 'cultures' and the pursuit of research in the modern sense. Thus, the two institutions switched roles, the academies becoming the institutional place for the collection and conservation of knowledge, while new knowledge was produced and disseminated at the universities (Stichweh 1984, p. 73).

The decline of the academies was exacerbated by the changes which disciplinary differentiation triggered in communication among scientists. The exchange among scholars in the academy began to suffer from the increasing distance between the disciplines. Actual investigations, and reports and discussions about new developments, were taken to the new specialized scientific associations that began to be established at the end of the eighteenth century. Moreover, the academies proved to be too slow for the communicative needs of the specialized communities of scholars. The scientific associations assumed the function of publishing specialized journals which accelerated the speed of communication relative to the more cumbersome transactions of the academies.

After the turn from the eighteenth to the nineteenth century, knowledge formation and dissemination that had been united in the one academy differentiated into a system of different interconnected functions. The universities combined the production of new knowledge (i.e. research) and teaching, producing young researchers and professionals. The academies became increasingly marginalized as honorific associations. The actual needs of specialized communication were fulfilled by the new scientific associations that formed outside the universities. Although at first they comprised all disciplines, like the German Gesellschaft Deutscher Naturforscher und Ärzte (GDNÄ) and subsequently the British Association for the Advancement of Science (BAAS) and the American Association for the Advancement of Science (AAAS), they soon began to differentiate internally along disciplinary lines (see below). They organized conferences and journals and thus became the core institutions for disciplinary communication and the construction of disciplinary 'identities' or 'cultures.' Disciplinary associations thereby became the structuring principle of knowledge formation in the nineteenth and twentieth centuries.

1.3 The nature of disciplines

The essence of discipline formation and evolution is self-referential communication. Self-referentiality is given when the communication is 'closed' towards the environment and the evaluation of relevance and quality of research is limited to the members of the respective disciplinary community. As self-referential communication communities, disciplines have a dual identity. Their social identity is constituted by the rules of membership, i.e. teaching, examinations, certificates, careers, the attribution of reputation, and, thus, the formation of a hierarchical social structure. Their factual identity is constituted by the contents of communication. It concerns the delineation of a subject matter, a common set of problems and theories, concepts and specific methods to study it, the criteria of quality of achievement which are the basis for the evaluation and attribution of reputation by peer review. The procedure of peer review, where the members of the particular disciplinary community are judged competent to make an evaluation, also constitutes the borderline between experts and laymen with reference to the communicated knowledge.

Academic disciplines are not formal organizations but social communities bonded by communication. However, they take on different organizational forms with respect to different functions. The most important of these is education, which in universities is organized in terms of faculties or departments. These are the central structural elements of universities, and although they differ somewhat from one system of higher education to another they have more in common. The faculties or departments, constituted by chairs or professorial positions, represent disciplinary knowledge, and on that basis determine the contents of teaching curricula and of formal degrees which certify successful completion of studies.

As young students become socialized into a respective disciplinary culture, they become educated in the disciplinary contents, and they become accredited when they finish their studies successfully. At the end of this process, and given a disciplinary labor market, the interests of a discipline are defended by their members, both within universities and outside. The interest at stake is influence, i.e. the power of defining a field of investigation *vis-à-vis* other competing disciplines, since influence can be translated into career opportunities (Turner 2000).

Disciplines are not only institutionalized in university faculties but also in scholarly associations. These, in fact, have functions not only internal to the disciplines but also with regard to their economic, political, and social environment. Internally they coordinate the communication process by staging conferences and running disciplinary journals. To the outside they represent the interests of the disciplinary communities in various ways. The certification of disciplinary training and formal accreditation are attempts to secure a monopoly for a certain sector of the professional or semiprofessional job market. Particularly in Europe, the associations have had state support in controlling access to the job market by setting standards, thereby keeping competitors out. The virtual monopoly of lawyers in state bureaucracies or the exclusion of psychologists from the health system by the medical profession, although now past, are cases in point. In this sense, disciplines act like guilds.

Disciplinary associations also represent the disciplines' interests to politics (and politics for science, in particular). This concerns, first, mostly funding programs when decisions about large investments into certain research areas are at stake. Second, access to advisory bodies which may entail not only political influence *per se* but future funding opportunities may also be on the agenda. Thus, disciplinary associations also act as lobbying groups.

Even though the disciplines have a large degree of autonomy in determining their own development, and a virtual monopoly of expertise on their respective subject matter, they depend on external resources. Thus, they are dependent on how they are perceived by funders, whether these be the state, research councils, or private foundations. Most funding organizations reserve a certain percentage of their resources to respond to bottom-up proposals from the research community, the rationale being that researchers know best which problems to tackle. This is funding *basic research* as it responds to research priorities determined by and within the disciplines themselves. The newly founded European Research Council was established explicitly with the intention of supporting 'frontier research'. Under such funding schemes the perception of disciplines by the funding organizations is largely consonant with that in the universities.

But research councils and foundations (let alone government departments and industry) have their own priorities, their own political or economic goals which they want to realize by distributing funds among research fields. For this purpose they formulate research programs that are not identical with the disciplinary nomenclature of the universities but that may extend across disciplinary boundaries. 'Nanotechnology' and 'climate research' are such priority areas which encompass several of the traditional disciplines such as physics and chemistry as well as more specialized subdisciplines (e.g. biochemistry, atmospheric chemistry).These organizations have perceptions of the disciplines which may differ considerably from the disciplines' self-conceptions.

However, the formulation of such funding programs, no matter how focused or how general, depends on the advice from (disciplinary) scientists who pass judgment about the state of research in the respective fields, about the feasibility of a research program, and about the delineation of an intended problem area. In other words, funding programs of state and private funders have to be connected to the disciplinary structure (i.e. the reality of research), but they may set different priorities from those that would emanate from the disciplines themselves.

The second half of the twentieth century has seen a notable shift in science policy. As resources for research have grown continuously, not only has the share of industrial expenditures for R&D increased relative to public funds, but also the relative impact of political funding programs has become stronger relative to the funding of basic research. The willingness of political and economic players to let the disciplines determine their own research priorities has evidently decreased.

1.4 Growth and disciplinary specialization

Disciplines were a new organizational mode for the production and ordering of knowledge that responded to the limitations of the classificatory systems of knowledge at the end of the eighteenth century. By extending scientific analysis to every subject matter and by opening up a potentially unlimited succession of abstractions, disciplines seemed to be a social mechanism for knowledge production without foreseeable constraints. The dramatic growth of science that had been unleashed by the emergence of disciplines in the early nineteenth century continued at an exponential rate for almost 200 years and only started to level off somewhat in the 1980s. Growth rates (usually measured in number of publications) differ somewhat between different disciplines and from country to country and time period to time period, indicating that both disciplinary methodological conditions as well as the external infusion of resources determine the dynamics (Schummer 1997, p. 116; Weingart 2003, p. 186). Thus, for example the geosciences have a doubling time of about 8 years and mathematics of 20 years. It is obvious that exponential growth of any system cannot continue forever. In the case of science the ultimate limit would be the respective population (Price 1971). Either the growth rate declines, the system under observation collapses, or the identity of the unit concerned (here 'science') changes. But before these signs of crisis occur, the system has two ways to react to growth: (1) structuring, i.e. hierarchization of attention, and (2) internal differentiation, i.e. specialization (Weingart 2003, p. 188).

Selective attention is a well-known phenomenon in communication in general and in scientific communication in particular. The general rule (the 80/20 rule) is that 20% of scientific publications draw the attention of 80% of readers (measured in number of citations). This merely underscores the fact that there are fairly stable reputation structures in all disciplines. The second mechanism is connected to the first, however. Although disciplines are, in one sense, surprisingly stable, they are subject to continuous change as well, namely to internal specialization.

This may be illustrated by one of the very few quantitative studies of the growth of a discipline and subsequent specialization. The study of the development of the humanities in Germany from 1954–85 showed that in the discipline of English Language and Literature in 1954, 24 professors published 12 books and a smaller number of articles, all of which could be easily read by the community including students in the course of a year. In 1985, 300 professors published 60 books and 600 articles, too many publications to be read by any one member of the community. The consequence of growth: 'The disciplinary denomination as a frame of reference for identification and orientation must have become more abstract and been replaced by internal, narrower limits' (Weingart *et al.* 1991, p. 288). The field which in the mid-1950s could be considered a discipline had within three decades grown roughly tenfold and differentiated internally into a multitude of specialties. This is reflected *inter alia* in the number of journals and the dispersion of articles among them. In 1954, ten articles were published in six journals, five of them in the same journal (*Anglia*). Thirty years later that journal attracted only 11.8% of all articles published in the field, whereas 314 articles were published in 135 more specialized journals (Weingart *et al.* 1991, p. 293). It is evident that the practitioners of what used to be the discipline 'English Language and Literature' (*Anglistik*) no longer know each other personally nor do they even know all of each others' publications.

The former discipline is no longer a whole community of scholars but is fractured. As can be expected, the same process of internal specialization can be observed at the level of scholarly associations. The first general scientific association encompassing all disciplines was the German GDNÄ, founded in 1822, which served as a model for the BAAS, founded in 1831. In America the AAAS was founded in 1848 as a re-formation of the Association of American Geologists and Naturalists. These associations subsequently differentiated into specialized sections which follow largely disciplinary lines. As these associations are interdisciplinary and their membership is not limited to scientists—following the model of the former academies—the disciplines emigrated and formed specialized associations. It has been shown for Germany that their number increased between 1900 and 1999 from 35 to 275. But the process of specialization does not stop there. These associations differentiate further into even more specialized sections. For example the German Physics Association (the Deutsche Physikalische Gesellschaft (DPS)) has formed 29 subsections since 1951, seven of which were founded after 1974. Likewise, the Society of German Chemists (Gesellschaft Deutscher Chemiker (GDC)) has split into 21 special groups (Schwechheimer and Weingart 2007, p. 188).

Not all such specializations are the result of developments internal to the discipline. External motivations and opportunities, changes in contexts of application, economic developments, competition between disciplines, demand for expertise, etc., can also play a crucial role. So far the process of specialization has not come to an end even if one can observe a slowing down. For that reason the original 'disciplines' such as physics, chemistry, biology, psychology, etc. have long lost their function as communities of communication. However, they are still the common framework for various subdisciplines. Depending on the specific function, either disciplines or subdisciplines or even so-called 'specialties' become the relevant reference, be it for the organization of research institutes, the structuring of university departments, the labeling of a journal, or the demarcation of a funding program. But the essential criterion on which such demarcations are based is still the same: the boundary of a network of meaningful communication of scientific substance.

1.5 Interdisciplinarity and transdisciplinarity: new modes of knowledge production?

Uneasiness about the loss of unity of science goes back to the very time when that unity was lost, i.e. in the early nineteenth century. Since then the call for a reunification or for interdisciplinarity has been persistent. In the 1930s the 'unity of science' movement was initiated by philosophers of science and natural scientists but remained without impact (Neurath, *et al.* 1938). In the late 1960s and into the 1970s in the contexts of debates about technology gaps, technology forecasting, and protection of the environment, the Organisation for Economic Co-operation and Development (OECD) triggered a new debate on interdisciplinarity (Apostel *et al.* 1972). In that publication Erich Jantsch first coined the term 'transdisciplinarity' which was taken up two decades later by Gibbons *et al.* (1994) to diagnose the emergence of a new mode of knowledge production termed 'Mode 2'. The thesis that the traditional disciplinary 'Mode 1' of knowledge production has given way to a new transdisciplinary mode of knowledge production has since then initiated animated discussions among analysts and the mobilization of conflicting evidence. Together with a series of similar pronouncements of a fundamental change in knowledge production, these analyses beg the question of whether they truly signal the advent of a new order of knowledge formation or if they only describe surface phenomena (Funtowicz and Ravetz 1993; Ziman 1994). Has the disciplinary organizational mode of science really come to an end?

The claims of change (insofar as they are relevant here) can be summarized as follows: the university has lost its monopoly as the institution of knowledge production since many other organizations are also performing that function. Transitory networks and contexts are formed which replace traditional disciplines. Knowledge production outside disciplines is no longer the search for basic laws (fundamental research) but takes place in contexts of application (Funtowicz and Ravetz 1993, p. 121; Gibbons *et al.* 1994, p. 4). Disciplines are no longer the crucial frames of orientation for the delineation of subject matters and the formulation of research problems. Research is, instead, characterized by transdisciplinarity: solutions to problems appear in contexts of application and research results are no longer communicated in journals. The criteria of quality are no longer determined by disciplines alone but additional criteria, social, political, and economical, are applied to determine quality (Funtowicz and Ravetz 1993, p. 90; Gibbons *et al.* 1994, p. 8).

The thesis of a new mode of knowledge production has been based on impressionistic evidence only, motivated perhaps by the emergence of a 'knowledge market' in which a multitude of think-tanks and specialized commercial research institutes offer their services for governments and industry. The concept of a 'knowledge society' which has found its way from scholarly discourse into the mass media has probably contributed to the plausibility of the thesis. It has not been supported by theoretical considerations or by systematic empirical evidence. Theoretically, two reasons could account for the emergence of inter- and transdisciplinary structures that would replace traditional disciplines. First, with the continuously growing number of specialties (i.e. research fields below the level of disciplines) the probability increases that, due to the proximity of such fields, new recombinations will occur which will result in new 'interdisciplinary' research fields. The organizational status of these fields, however, still follows the mode of 'internal' specialization. After a period of emergence they form into another specialized field. Second, inter- and transdisciplinary research fields are promoted by funding agencies in the interest of directing research to politically desired goals. This process is conditioned by the fact that the 'externally' defined subject matters, research problems, and values or interests can trigger sustained research. Examples for the first reason are physical chemistry and molecular biology, examples for the second are climate research and gender studies. The latter are combinations of disciplines or subdisciplines that are joined in research centers, journals, and funding programs but that remain intellectually independent and continue to develop individually.

The replacement of a discipline-based mode of knowledge production by a new mode is not corroborated by empirical data either. The commanding role of the universities as the core institution of knowledge production, and by implication the role of disciplines as their organizational structures, is unfettered. At the same time the expansion and differentiation of courses and degrees has continued unabated (Frank and Meyer 2007, p. 20). Part and parcel of this process is the incorporation of ever more observable phenomena and social and technical activities into the knowledge producing and diffusing portfolio of the university. However, this does not result in parochial curricula and research programs. 'By the end of the twentieth century, science curricula...are more differentiated and specialized, but the specialization involved can easily be followed and understood by specialists anywhere in the world' (Frank and Meyer 2007, p. 30).

Another indicator of structural changes in the organization of knowledge is scholarly associations. Systematic study shows an increase of inter- or multidisciplinary associations, mostly in the broader fields of medicine and biology. Apart from the fact that this observation is based on self-descriptions which may reflect an adaptation to political expectations, the emergence of these encompassing associations has primarily occurred in applied fields (Schwechheimer and Weingart 2007, pp. 194–5).

Thus, disciplines and their derivatives, specialties, and research fields, remain the principal organizational unit for the production and diffusion of knowledge. However, the process of differentiation and the concomitant scientification (i.e. the expansion of the perception of the world in terms of scientific methods and concepts) soften the once rigid boundaries of the disciplines and allow for the emergence of interdisciplinary fields and the effect of 'external' occasions to initiate cross boundary research activities. To postulate that disciplines lose their function amounts to claiming that the development of scientific knowledge is exclusively directed by 'external' societal and political interests. It would actually imply a reversal of the differentiation process that has been under way for more than two centuries. Since that is unlikely to happen, traditional disciplines and inter-, multi-, and transdisciplinary research fields will exist side by side.

References

Apostel, L., Berger, G., Briggs, A., and Michaud, G. (eds) (1972). *Interdisciplinarity: problems of teaching and research in universities*. Paris: Organization for Economic Cooperation and Development.

Cairns, J. *et al.* (ed.) (1966). *Phage and the origins of molecular biology*. Cold Spring Harbor: Cold Spring Harbor Laboratory of Molecular Quantitative Biology.

Daston, L. (1999). Die Akademien und die Einheit der Wissenschaften: die Disziplinierung der Disziplinen. In: J. Kocka, R. Hohlfeld, and P.T. Walther (eds) *Die Königlich Preußische Akademie der Wissenschaften zu Berlin im Kaiserreich*, pp. 61–84. Berlin: Akademie Verlag.

Du Bois-Reymond, E. (1886). Über die wissenschaftlichen Zustände der Gegenwart. *Reden*, Vol. 2. Leipzig: Veit & Co.

Flint, R. (1904). *Philosophy as Scientia Scientiarum and a history of classifications of the sciences.* High Wycombe: University Microfilms Ltd for the College of Librarianship, Wales.

Frank, D.J. and Meyer, J.W. (2007). Worldwide expansion and change in the university. In: G. Krücken, A. Kosmützky, and M. Torka (eds) *Towards a multiversity?*, pp. 19–44. Bielefeld: Transcript Verlag.

Funtowicz, S.O. and Ravetz, J.R. (1993). The emergence of post-normal science. In: R. von Schomberg (ed.) *Science, politics, and morality: scientific uncertainty and decision making*, pp. 95–123. Dordrecht: Kluwer Academic Publishers.

Gibbons, M. *et al.* (1994). *The new production of knowledge.* London: Sage.

von Helmholtz, H. (1896). Über das Verhältnis der Naturwissenschaften zur Gesamtheit der Wissenschaften. *Vorträge und Reden*, 4th edn, Vol. 1, pp. 157–86. Braunschweig: Vieweg und Sohn.

Kambartel, F. (1996). Wissenschaft. In: J. Mittelstraß (ed.) *Enzyklopädie Philosophie und Wissenschaftstheorie*, Vol. 4, pp. 719–21. Stuttgart: Metzler.

Lepenies, W. (1976). *Das Ende der Naturgeschichte: Wandel Kultureller Selbstverständlichkeiten in den Wissenschaften des 18. und 19. Jahrhunderts.* Munich: Hanser Verlag.

Neurath, O., Carnap, R., and Morris, C.W. (eds) (1938). *Foundations of the unity of science: toward an international encyclopedia of unified science*, Vol. 1. Chicago: University of Chicago Press.

Nye, M.J. (1993). *From chemical philosophy to theoretical chemistry: dynamics of matter and dynamics of disciplines, 1800–1950.* Berkeley: University of California Press.

Price, D. de Solla (1971). *Little science, big science*, paperback edn. New York: Columbia University Press.

Schummer, J. (1997). Scientometric studies on chemistry I: the exponential growth of chemical substances, 1880–1995. *Scientometrics* **39**(1), 107–23.

Schwechheimer, H. and Weingart, P. (2007). Dimensionen der Veränderung der Disziplinenlandschaft. In: P. Weingart, M. Carrier and W. Krohn (eds) *Nachrichten aus der Wissensgesellschaft: Analysen zur Veränderung der Wissenschaft*, pp. 182–228. Weilerswist: Velbrück Wissenschaft.

Stichweh, R. (1984). *Zur Entstehung des modernen Systems wissenschaftlicher Disziplinen: Physik in Deutschland 1740–1890.* Frankfurt am Main: Suhrkamp.

Turner, S. (2000). What are disciplines? And how is interdisciplinarity different? In: P. Weingart and N. Stehr (eds) *Practising interdisciplinarity*, pp. 46–65. Toronto: University of Toronto Press.

Weingart, P. (2003). Growth, differentiation, expansion and change of identity – the future of science. In: B. Jörges and H. Nowotny (eds) *Social studies of science and technology: looking back ahead*, Sociology of the Sciences Yearbook, Vol. 23, pp. 183–200 Dordrecht: Kluwer Academic Publishers.

Weingart, P. *et al.* (1991). *Die sog. Geisteswissenschaften: Außenansichten.* Frankfurt am Main: Suhrkamp.

Ziman, J. (1994). *Prometheus bound. Science in a dynamic steady state.* Cambridge: Cambridge University Press.

CHAPTER 2

A taxonomy of interdisciplinarity

JULIE THOMPSON KLEIN

Taxonomies classify entities according to similarities and differences, whether they are animal species, artistic genres, or medical symptoms. Since the late nineteenth century, taxonomies of knowledge in the Western intellectual tradition have been dominated by a system of disciplinarity that demarcates domains of specialized inquiry. Over the latter half of the last century, though, the system was supplemented and challenged by an increasing number of interdisciplinary activities. This proliferation gave rise, in turn, to new taxonomies that registered expansion of the genus *Interdisciplinarity*, propelled by new species of integration, collaboration, complexity, critique, and problem solving. The new classification schemes differentiated forms of disciplinary interaction, motivations for teaching and research, degrees of integration and scope, modes of interaction, and organizational structures. The first major interdisciplinary typology was published in 1972, created for an international conference held in France in 1970 and co-sponsored by the Organisation for Economic Co-operation and Development (OECD) (Apostel *et al.* 1972). Other labels soon followed, producing a sometimes confusing array of jargon. However, the three most widely used terms in the OECD typology—'multidisciplinary', 'interdisciplinary', and 'transdisciplinary'—constitute a core vocabulary for understanding both the genus of *Interdisciplinarity* and individual species within the general classification.

This chapter distinguishes *Multidisciplinarity* and *Interdisciplinarity* (*ID*) then describes species of *Methodological ID* and *Theoretical ID*, *Bridge Building* and *Restructuring*, *Instrumental ID* and *Critical ID*. After that, it defines major trendlines in the current heightened momentum for *Transdisciplinarity* and closes with the most recent typologies and reflections on the problem of taxonomy. Taxonomies construct the ways in which we organize knowledge and education. However, they are neither permanent nor complete and their boundaries change. A comparative picture of defining characteristics is an important index of patterns of practice and change. Table 2.1 is an overview of key terms in the chapter and the literature it cites.

Table 2.1 Defining characteristics in typologies of interdisciplinarity

Multidisciplinarity	Interdisciplinarity	Transdisciplinarity
• juxtaposing	• integrating	• transcending
• sequencing	• interacting	• transgressing
• coordinating	• linking	• transforming
	• focusing	
	• blending	
• complementing		• hybridizing
• Encyclopedic ID		Systematic Integration
• Indiscriminate ID		Transsector Interaction
• Pseudo ID		

Partial Integration ←-----------------------------→ Full Integration

Contextualizing ID		ConceptualID
Auxiliary ID	Supplementary ID	Structural ID/Unifying ID
Composite ID	Generalizing ID	Integrative ID

Degrees of Collaboration

Shared ID ←-------------→ Cooperative ID

- Narrow versus Broad or Wide ID
- Methodological versus Theoretical ID
- Bridge Building versus Restructuring
- Instrumental versus Critical ID
- Endogenous versus Exogenous ID

2.1 Multidisciplinary juxtaposition and alignment

In a comparative study of taxonomies, Lisa Lattuca found that most definitions treat the *integration* of disciplines as the 'litmus test' of interdisciplinarity. In fields that prioritize critique of knowledge over synthesizing existing disciplinary components, the premise is disputed, along with the view that disciplinary grounding is the necessary basis for interdisciplinary work. Nonetheless, integration is the most common benchmark and, combined with degrees of disciplinary *interaction*, provides a comparative framework for understanding differences in types of interdisciplinary work (Lattuca 2001, pp. 78, 109).

In the OECD classification, *Multidisciplinarity* was defined as an approach that juxtaposes disciplines. Juxtaposition fosters wider knowledge, information, and methods. Yet, disciplines remain separate, disciplinary elements retain their original identity, and the existing structure of knowledge is not questioned. This tendency is evident in conferences, publications, and research projects that present different views of the same topic or problem in serial order. Similarly, many so-called 'interdisciplinary' curricula are actually a multidisciplinary assemblage of disciplinary courses, including programs of general education and interdisciplinary fields that ask students to take a selection of department-based courses. The keywords in Rebecca Crawford Burns' typology of integrative education provide everyday images of multidisciplinary juxtaposition. When disciplines and school subjects are aligned in parallel fashion, they are in a *Sequencing* mode and, when intentionally aligned, in a *Coordinating* mode (Burns 1999, pp. 8–9). In either case, however, integration and interaction are lacking. Several technical terms shed further light on the nature of *Multidisciplinarity* in both education and research.

2.1.1 Encyclopedic, indiscriminate, and pseudo forms

Multidisciplinarity is encyclopedic in character. In a six-part typology, Margaret Boden defined *Encyclopedic ID* as a 'false' or at best a 'weak' form. It is an expansive enterprise typically lacking intercommunication, a trait embodied in joint degrees, the journals *Science* and *Nature*, and collocated information on the World Wide Web (Boden 1999, pp. 14–15). Comparably, in the OECD conference Heinz Heckhausen defined *Indiscriminate ID* as an encyclopedic form, citing the *studium generale* of German education, vocational training that prepares workers to handle a variety of problems with 'enlightened common sense', and exposure to multiple disciplines in professional education. A second form, *Pseudo ID*, is embodied in the erroneous proposition that sharing analytical tools such as mathematical models of computer simulation constitutes 'intrinsic' interdisciplinarity (Heckhausen 1972, p. 87). A number of disciplines have also been described as 'inherently interdisciplinary' because of their broad scope. Philosophy, literary studies, and religious studies were early examples, followed by anthropology, geography, and many interdisciplinary fields. A wide compass alone, however, does not constitute interdisciplinarity.

2.1.2 Contextualizing, informed, and composite relationships

The loose and restricted relationship of disciplines in multidisciplinarity is illustrated by the familiar practice of applying knowledge from one discipline in order to contextualize another. For instance, a scholar might use the discipline of history to inform readers about a particular movement in philosophy or use philosophy to provide an epistemological context for interpreting a particular event. In *Contextualizing ID*, Boden stipulates, other disciplines are taken into account without active cooperation. She cites the engineering profession's effort to include social contexts of practice, and the Academy of Finland Integrative Research (AFIR) team mentions the example of a research proposal for an extensive reference book on Scandinavian history. Authors from multiple disciplines were to be involved, but their chapters would be arrayed in encyclopedic sequence (Boden 1999, pp. 15–16; Bruun *et al.* 2005, pp. 112–13).

The label *Composite ID* names another familiar practice—applying complementary skills to address complex problems or to achieve a shared goal. Heckhausen cited major societal problems such as war, hunger, delinquency, and pollution. He deemed peace research and city planning 'interdisciplinarities in the making', because they simulate exploring interdependences among a 'jigsaw puzzle-like composition' of adjacent fields. He also noted the Apollo space project (Heckhausen 1972, p. 88). In *Composite ID*, the AFIR team found, production of knowledge retains a strong disciplinary thrust. However, results are integrated within a common framework. In the biosciences, for instance, technical knowledge from many fields and expensive instruments are often shared. For example, a research proposal for a forest technology project included a large array of approaches in the forest sciences. The approaches were dissimilar but did not cause conceptual barriers because of their historical coexistence within forestry (Bruun *et al.* 2005, p. 114).

2.2 Interdisciplinary integration, interaction, and collaboration

When integration and interaction become proactive, the line between multidisciplinarity and interdisciplinarity is crossed. Integrated designs, Burns indicates, restructure existing approaches through explicit *focusing* and *blending* (Burns 1999, pp. 11–12). Lattuca adds the image of *linking* issues and questions that are not specific to individual disciplines. In education, for example, courses achieve a more holistic understanding of a cross-cutting question or problem, such as historical and legal perspectives on public education or biological and psychological aspects of human communication (Lattuca 2001, pp. 81–3). Purposes differ, however. A course on the environment does not have the same motivation as building the infrastructure of a new interdiscipline such as clinical and translational science or borrowing the concept of imagery from art history in a political science research project on visual symbols in election campaigns. Scope varies as well. William Newell depicts a spectrum moving from partial to full integration, and the focus may be narrow or wide. *Narrow ID* occurs between disciplines with compatible methods, paradigms, and epistemologies, such as history and literature and the AFIR example of forest sciences. Fewer disciplines are typically involved as well, simplifying communication. *Broad* or *Wide ID* is more complex. It occurs between disciplines with little or no compatibility, such as sciences and humanities. They have different paradigms or methods and more disciplines and social sectors may be involved (Newell 1998, p. 533).

Many believe that interdisciplinarity is synonymous with collaboration. It is not. However, heightened interest in teamwork to solve complex intellectual and social problems has reinforced the connection, especially in team teaching and research management. Here too, degrees of integration and interaction differ. In *Shared ID*, Boden designates, different aspects of a complex problem are tackled by different groups. They possess complementary skills, communicate results, and monitor overall progress. Yet, daily cooperation does not necessarily occur. In contrast, *Cooperative ID* requires teamwork, exemplified by the collaboration of physicists, chemists, engineers, and mathematicians in the Manhattan Project to build an atomic bomb and in research on public policy issues such as energy and law and order (Boden 1999, pp. 17–19). In a four-level typology, Simon and Goode (1989, pp. 220–1) sketched the range of interactions that occurs in both research and teaching. The least degree is the reductive role of supplying background or contextual information to other disciplines. Elaboration or explanation of findings is the next level, but is still limited. At higher levels of interaction, joint definition of variables or categories occurs and, in the greatest degree, fundamental questions are refined by integrating the approaches of all the participants into the research design. Differing degrees of integration and interaction are further evident in *Methodological ID* versus *Theoretical ID*.

2.2.1 Methodological ID

Methodological ID and *Theoretical ID* are often differentiated in taxonomies. The typical motivation in *Methodological ID* is to improve the quality of results. The typical activity is borrowing a method or concept from another discipline in order to test a hypothesis, to answer a research question, or to help develop a theory (Bruun *et al.* 2005, p. 84). Here, as well, degrees of integration and interaction differ. If a borrowing does not result in a significant change in practice, Heckhausen stipulated, the relationship of disciplines is *Auxiliary*. If the borrowing becomes more sophisticated and an enduring dependence develops, the relationship becomes *Supplementary,* exemplified by incorporation of psychological testing in pedagogy and neurophysiological measures in psychology (Heckhausen 1972, pp. 87–89). When new laws become the basis for an original discipline, such as electromagnetics or cybernetics, a new *Structural* relationship emerges (Boisot 1972, pp. 94–5). Some methodologies have also formed the foundation for recognized specialties such as statistics, oral history, and econometrics (Becher 1989, p. 49).

The history of interdisciplinary approaches in the social sciences yields an extended illustration. In a six-part typology, Raymond Miller identified two kinds of *Methodological ID*. The first, Shared Components, includes research methods that are shared across disciplines, such as statistical inference. The second, Cross-Cutting Organizing Principles, are focal concepts or fundamental social processes used to organize ideas and findings across disciplines, such as 'role' and 'exchange' (Miller 1982, pp. 15–19). New engineering and technological methods that were developed during World War II stimulated postwar borrowings of cybernetics, systems theory, information theory, game theory, and new conceptual tools of communication theory and decision theory. In addition, the roster of shared methods includes techniques of surveying, interviewing, sampling, polling, case studies, cross-cultural analysis, and ethnography. In the latter decades of the twentieth century, a 'third methodological movement' also emerged, marked by new borrowings that combine quantitative and qualitative traditions (Mahan 1970; Tashakkori and Teddlie 2003; Smelser 2004, p. 60). For a four-part typology of ID social sciences classified as cross-fertilizations, formal collaborations, topics that catalyze new fields, and problem-oriented research beyond the academy, see Calhoun and Rhoten in Chapter 7 of this volume.

Relations with the humanities changed as well. In 1980, Clifford Geertz identified a broad shift within intellectual life in general and the social sciences in particular. The model of physical sciences and a laws-and-instances explanation was being supplanted by a case-and-interpretation model and symbolic form analogies borrowed from the humanities. Social scientists were increasingly representing society as a game, drama, or text, rather than a machine or quasi-organism. They were also borrowing methods of speech-act analysis, discourse models, and cognitive aesthetics, crossing the traditional boundary of explanation and interpretation. Conventional rubrics remain, but they are often jerry-built to accommodate a situation Geertz dubbed increasingly 'fluid, plural, uncentered, and ineradicably untidy'. Postpositivist, poststructural, constructivist, interpretive, and critical paradigms also stimulated new interactions in the interdisciplinary study of culture within and across humanities and social sciences.

2.2.2 Theoretical interdisciplinarity

Theoretical ID connotes a more comprehensive general view and epistemological form. The outcomes include conceptual frameworks for analysis of particular problems, integration of propositions across disciplines, and new syntheses based on continuities between models and analogies. Individual projects also exhibit theoretical imperatives. One research proposal the AFIR team examined sought to develop a model of mechanisms that mediate mental stress experiences into physiological reactions and eventually coronary heart disease. Previous studies emphasized correlation of single stress factors or separate personal traits associated with the disease. In contrast, the project aimed to develop an interdisciplinary theory based on integration of psychological and medical elements and testing the conceptual tool of inherited 'temperament' (Bruun *et al.* 2005, p. 86).

For Boden, the highest levels of the genus *Interdisciplinarity* are Generalizing ID and Integrated ID. In Generalizing ID, a single theoretical perspective is applied to a wide range of disciplines, such as cybernetics or complexity theory. In Integrated ID, which Boden pronounces 'the only true interdisciplinarity', the concepts and insights of one discipline contribute to the problems and theories of another, manifested in computational neuroscience and the philosophy of cognitive science. Individuals may find their original disciplinary methods and theoretical concepts modified as a result of cooperation, fostering new conceptual categories and methodological unification (Boden 1999, pp. 19–22). Comparably, Lattuca judges Conceptual ID to be a '[t]rue or full' form of interdisciplinarity. The core issues and questions lack a compelling disciplinary basis, and a critique of disciplinary understanding is often implied (Lattuca 2001, p. 117). Talk of 'true' or 'full' interdisciplinarity leads to a further distinction—between motivations of Bridge Building and Restructuring.

2.3 Bridge building versus restructuring

The Nuffield Foundation in London identified two basic metaphors of interdisciplinarity—bridge building and restructuring. Bridge building occurs between complete and firm disciplines. Restructuring detaches parts of several disciplines to form a new coherent whole. The Foundation also noted a third possibility that occurs when a new overarching concept or theory subsumes the theories and concepts of several existing disciplines, akin to the notion of transdisciplinarity (Apostel *et al.* 1972, pp. 42–5). The difference between bridge building and restructuring is illustrated by Landau, Proshansky, and Ittelson's classification of two phases in the history of interdisciplinary approaches in social sciences. The first phase, dating from the close of World War I to the 1930s, was embodied in the founding of the Social Science Research Council and the University of Chicago school of social science. The interactionist framework at Chicago fostered integration, and members of the Chicago school were active in efforts to construct a unified philosophy of natural and social sciences. The impacts were widely felt, and on occasion disciplinary 'spillage' led to the formation of hybrid disciplines such as social psychology and political sociology. However, traditional categories of knowledge and academic structures remained intact.

The second phase, dating from the close of World War II, was embodied in 'integrated' social science courses, a growing tendency for interdisciplinary programs to become 'integrated' departments, and the concept of behavioral science. The traditional categories that anchored the disciplines were questioned, and lines between them began to blur, paving the way toward a new theoretical coherence and alternative divisions of labor. The behavioral science movement sought an alternative method of organizing social inquiry, rather than tacking imported methods and concepts onto traditional categories. The field of area studies is another prominent case. In contrast to earlier 'interdisciplinary' borrowing, it was a new 'integrative' conceptual category with greater analytic power, stimulating a degree of theoretical convergence in the concepts of role, status, exchange, information, communication, and decision-making (Landau *et al.* 1962, pp. 8, 12–17).

2.4 Interdisciplinary fields and hybrid specializations

The formation of new interdisciplinary domains is a major instance of restructuring. Miller identified four pertinent categories in his typology. *Topics* are associated with problem areas. 'Crime', for instance, is a social concern that appears in multiple social science disciplines and in criminal justice and criminology. 'Area', 'labor', 'urban', and 'environment' also led to new academic programs, and study of the 'aged' produced the field of gerontology. *Life experience* became prominent in the late 1960s and 1970s with the emergence of ethnic studies and women's studies. *Hybrids* are 'interstitial cross-disciplines' such as social psychology, economic anthropology, political sociology, biogeography, culture and personality, and economic history. *Professional preparation* also led to new fields with a vocational focus, such as social work and nursing and, Neil Smelser adds, fields of application to problem areas such as organization and management studies, media studies and commercial applications, and planning and public policy (Smelser 2004, p. 61).

Ursula Hübenthal's keyword for the formal intersection of topics and objects is 'Intermeshing', in contrast to 'Complementing' interests among disciplines that remain apart (Hübenthal 1994, p. 63). Heckhausen called the higher level of formality *Unifying ID*, an outcome that occurred when biology reached the subject matter level of physics, forming biophysics (Heckhausen 1972, pp. 88–9). Within the field of science, technology, and society studies, Susan Cozzens also noted a specialized interdisciplinary bridge formed by alliances of economists of scientific research and technological development with historians and sociologists of technology interested in technological innovations (Cozzens 2001, p. 57). Observing a historical increase in hybrids, Dogan and Pahre identified two stages in the process. The first stage is specialization, and the second is continuous reintegration of fragments of specialties across disciplines. There are two types of hybrids. The first kind becomes institutionalized as a subfield of a discipline or a permanent cross-disciplinary program. The second kind remains informal. Hybrids often form in the gaps between subfields. Child development, for example, incorporates developmental psychology, language acquisition, and socialization (Dogan and Pahre 1990, pp. 63, 66, 72).

One of the myths of interdisciplinarity is that the 'inter-discipline' of today is the 'discipline' of tomorrow (Apostel *et al.* 1972, p. 9). Their trajectories vary greatly, however. Some fields remain embryonic, while others develop epistemological strength anchored by shared thematic principles, unifying core concepts, and a new community of knowers with a common interlanguage. Economic and social capital are powerful determinants in the political economy of knowledge. The growth of area studies was enabled by significant amounts of funding from the Ford Foundation. Molecular biology also enjoyed a level of funding lacking in social psychology, and the same discrepancy is evident today in the differing status of biomedicine and cultural studies.

Labels are not absolute states of being, either. Richard Lambert (1991) describes area studies as a 'highly variegated, fragmented phenomenon, not a relatively homogeneous intellectual tradition'. Much of what may be called 'genuinely interdisciplinary' work occurred at the juncture of four disciplines that provided the initial bulk of area specialists: history, literature and language, anthropology, and political science. In that hybrid intellectual space, a historically informed political anthropology developed using material in local languages. Blending of disciplinary perspectives occurred most often at professional meetings and in research by individual specialists. Broadly defined themes have been the dominant pattern in scholarly papers, creating a collective 'multidisciplinary' perspective, and the topic of any one event 'drives the disciplinary mix'. At the same time, area studies is 'subdisciplinary' in the sense that research by individuals has tended to concentrate on particular subdomains, while the field at large is 'transdisciplinary' in the broad scope of its endeavors.

2.4.1 Instrumental ID versus critical ID

The difference between Instrumental ID and Critical ID is a major faultline in the discourse of interdisciplinarity. In an analysis of forms of interdisciplinary explanation, Mark Kann identified three political positions. Conservative elites want to solve social and economic problems, without concern for epistemological questions. Liberal academics demand accommodation but maintain a base in the existing structure. Radical dissidents challenge the existing structure of knowledge, demanding that interdisciplinarity respond to the needs and problems of oppressed and marginalized groups (Kann 1979, pp. 187–8). Methodological ID is 'instrumental' in serving the needs of a discipline. During the 1980s, though, another kind of Instrumental ID gained visibility in science-based areas of economic competition, such as computers, biotechnology and biomedicine, manufacturing, and high-technology industries. Peter Weingart (2000, p. 39) treats this type of activity as *Strategic* or *Opportunistic ID*. In this instance, *Interdisciplinarity* serves the market and national needs.

In contrast, *Critical ID* interrogates the dominant structures of knowledge and education with the aim of transforming them, raising questions of value and purpose silent in *Instrumental ID*. New fields in Miller's *Life experience* category were often imbued with a critical imperative, prompting Douglas Bennett to call them a 'sacred edge' in the reopened battle over inclusion and exclusion (Bennett 1997, p. 144). Older fields, such as American studies, also took a 'critical turn' in the 1960s and 1970s, and a 'new interdisciplinarity' emerged in the humanities (Klein 2005, pp. 153–75). Salter and Hearn (1996) call interdisciplinarity the necessary 'churn in the system', aligning it with a dynamic striving for change that disturbs continuity and routine. This imperative is signified in a new rhetoric of 'anti', 'post', 'non', and 'de-disciplinary' that is prominent in cultural studies, women's and ethnic studies, literary studies, and postmodern approaches across disciplines. An increasing number of faculty in humanities and social sciences, Lattuca reports, do interdisciplinary work with an explicit intent to deconstruct disciplinary knowledge and boundaries, blurring the boundaries of the epistemological and the political (Lattuca 2001, pp. 15–16, 100).

The disciplines are also implicated in Critical ID. Giles Gunn's typology of interdisciplinary approaches to literary studies identifies multiple approaches to mapping. The simplest strategy is on disciplinary grounds, tracing the relationship of one discipline to another, such as 'literature and...' philosophy or psychology, and so forth. The map changes, though, if another question is asked. What new subjects and topics have emerged? New examples appear, including the history of the book, psychoanalysis of the reader, and the ideology of gender, race, and class. Each topic, in turn, projected further lines of investigation. 'The threading of disciplinary principles and procedures', Gunn found, 'is frequently doubled, tripled, and quadrupled in ways that are not only mixed but, from a conventional disciplinary perspective, somewhat off center'. They do not develop in linear fashion but are characterized by overlapping, underlayered, interlaced, cross-hatched affiliations, collations, and alliances with ill-understood and unpredictable feedbacks. The final and most difficult approach to mapping is rarely acknowledged. Correlate fields and disciplines have changed, challenging assumptions about the strength of boundaries while working to erode them. 'The inevitable result of much interdisciplinary study, if not its ostensible purpose', Gunn concluded, 'is to dispute and disorder conventional understandings of relations between such things as origin and terminus, center and periphery, focus and margin, inside and outside' (Gunn 1992, pp. 241–3, 248–9).

The distinction between Instrumental and Critical forms is not absolute. Research on problems of the environment and health often combines critique and problem solving.

Nonetheless, a clear division appears in the classification of motivations. Observing trends in the medical curriculum, Bryan Turner (1990) affirms that when interdisciplinarity is conceived as a short-term solution to economic and technological problems, pragmatic questions of reliability, efficiency, and commercial value take center stage. In social medicine and sociology of health, in contrast, interdisciplinarity emerged as an epistemological goal. Researchers focused on the complex causality of illness and disease that factors in psychological, social, and ethical factors missing from the hierarchical biomedical model.

2.4.2 Transdisciplinarity

In the OECD typology, Transdisciplinarity (TD) was defined as a common system of axioms that transcends the narrow scope of disciplinary worldviews through an overarching synthesis, such as anthropology construed as the science of humans. Conference participants Jean Piaget and Andre Lichnerowicz regarded TD as a conceptual tool capable of producing interlanguages. Piaget treated it as a higher stage in the epistemology of interdisciplinary relationships based on reciprocal assimilations, and Lichnerowicz promoted 'the mathematic' as a universal interlanguage. Erich Jantsch embued TD with a social purpose in a hierarchical model of the system of science, education, and innovation (Jantsch 1972; Lichnerowicz 1972; Piaget 1972). The intellectual climate of the times was evident in the organizing languages of the OECD seminar—logic, cybernetics, general systems theory, structuralism, and organization theory. Since then, the term has proliferated, becoming a descriptor of broad fields and synoptic disciplines, a team-based holistic approach to health care, and a comprehensive integrative curriculum design driven by the keyword 'transcending'. A defining essay on the website td-net notes that TD research has developed in different contexts, fostering different types with different goals (Transdisciplinarity Net 2009). Four major trendlines define the current heightened momentum.

2.5 Current TD trendlines

One trendline is the contemporary version of the historical quest for systematic integration of knowledge. This quest spans ancient Greek philosophy, the medieval Christian *summa*, the Enlightenment ambition of universal reason, Transcendentalism, the Unity of Science movement, the search for unification theories in physics, and E. O. Wilson's theory of consilience. Reviewing the history of discourse on TD, philosopher Joseph Kockelmans (1979) found it has tended to center on educational and philosophical dimensions of sciences. The search for unity today, though, does not follow automatically from a pregiven order of things. It must be continually 'brought about' through critical, philosophical, and supra-scientific reflection. It also accepts plurality and diversity, a perspective prominent in the Centre International de Recherches et Etudes Transdisciplinaire (CIRET). CIRET is a virtual meeting space where a new universality of thought and type of education is being developed, informed by the worldview of complexity in science (<http://basarab.nicolescu.perso.sfr.fr/ciret/>).

The second trendline is akin to Critical ID. Transdisciplinarity is not just 'transcendent' but 'transgressive'. In the 1990s, TD began appearing more often as a label for knowledge formations imbued with a critical imperative, fostering new theoretical paradigms. Ronald Schleifer (2002) associated the new interdisciplinarity in humanities with new theoretical approaches and transdisciplinary or cultural study of social and intellectual formations that have breached canons of wholeness and the simplicity of the Kantian architecture of knowledge and art. The transdisciplinary operation of cultural studies, Douglas Kellner specified, draws on a range of fields to theorize the complexity and contradictions of media/culture/communications. It moves from text to contexts, pushing boundaries of class, gender, race, ethnicity, and other identities (Kellner 1995, pp. 27–8). Dölling and Hark (2000, pp. 1196–7) associate transdisciplinarity in women's and gender studies with critical evaluation of terms, concepts, and methods that transgress disciplinary boundaries. And, in Canadian studies, Jill Vickers links trans- and antidisciplinarity with movements that reject disciplinarity in whole or in part, while raising questions of sociopolitical justice (Vickers 1997, p. 41).

The third trendline is an extension of the OECD connotation of overarching synthetic paradigms. Miller defined TD as 'articulated conceptual frameworks' that transcend the narrow scope of disciplinary worldviews. Leading examples include general systems, structuralism, Marxism, sociobiology, phenomenology, and policy sciences. Holistic in intent, these frameworks propose to reorganize the structure of knowledge, metaphorically encompassing the parts of material fields that disciplines handle separately (Miller 1982, p. 21). More recently, a variant of this trendline has emerged in North America in the notion of 'transdisciplinary science' in broad areas such as cancer research. TD science is a collaborative form of 'transcendent interdisciplinary research' that creates new methodological and theoretical frameworks for defining and analyzing social, economic, political, environmental, and institutional factors in health and well-being (Stokols *et al.* 2008).

The fourth trendline—trans-sector TD problem solving—is prominent in Europe and North–South partnerships. A new form of TD was evident in the late 1980s and early 1990s in Swiss and German contexts of environmental research. By the turn of the century case studies were being reported in all fields of human interaction with natural systems and technical innovations as well as the development context. The core premise of this trendline is that problems in the *Lebenswelt*—the life-world—need to frame research questions and practices, not the disciplines (Transdisciplinarity Net 2009). Not all problems are the same, however. One strand of TD problem solving centers on collaborations between academic researchers and industrial/private sectors for the purpose of product and technology development, prioritizing the design of innovative milieus and the involvement of stakeholders in product development. A different type of TD research arises when academic experts and social actors contributing local knowledge and contextual interests cooperate in the name of democratic solutions to controversial problems such as sustainability and risks of technological modernizations such as nuclear power plants (Transdisciplinarity Net 2009).

The fourth trendline also intersects with two prominent concepts—'Mode 2 knowledge production' and 'postnormal science'. In 1994, Gibbons, *et al.* proposed that a new mode of knowledge production is fostering synthetic reconfiguration and recontextualization of knowledge. In contrast to the older Mode 1—characterized by hierarchical, homogeneous, and discipline-based work—the defining traits of the new Mode 2 include complexity, non-linearity, heterogeneity, and transdisciplinarity. New configurations of research work are being generated continuously, and a new social distribution of knowledge is occurring as a wider range of organizations and stakeholders contribute their skills and expertise to problem solving. Gibbons, *et al.* (1994) initially highlighted instrumental contexts of application and use, such as aircraft design, pharmaceuticals, electronics, and other industrial and private sectors. In 2001, Nowotny *et al.* extended the Mode 2 theory to argue that contextualization of problems requires participation in the agora of public debate, incorporating the discourse of democracy that is also voiced strongly in Critical ID. When lay perspective and alternative knowledges are recognized, a shift occurs from solely 'reliable scientific knowledge' to inclusion of 'socially robust knowledge' that dismantles the expert/lay dichotomy while fostering new partnerships between the academy and society.

Postnormal science, in Funtowicz and Ravetz's (1993) classic definition, breaks free of reductionist and mechanistic assumptions about the ways in which things are related and how systems operate. 'Unstructured' problems are driven by complex cause–effect relationships, and they exhibit a high divergence of values and factual knowledge. Weingart (2000, pp. 36, 38) finds a common topos among claims for new modes of knowledge production, postnormal and postmodern science, and newer forms of inter- or transdisciplinary research. They are all oscillating between empirical and normative statements, positing more democratic and participatory modes while resounding the same theme that triggered the escalation of interdisciplinarity in the context of higher education reform during the 1960s. Now, though, claims are framed in the context of application and involvement of stakeholders in systems that are too complex for limited disciplinary modes portrayed as being too linear and narrow for 'real-world' problem solving. New TD and counterpart ID forms, though, are not without their own 'blind spots', failing to recognize the opportunistic dimensions of both presumably 'internal' academic science and strategic research for non-scientific goals.

2.6 New implications for taxonomy

The most recent authoritative typology appeared in a report issued by the National Academy of Sciences in the United States. *Facilitating interdisciplinary research* (National Academy of Sciences 2004, pp. 2, 40) identifies four primary drivers of interdisciplinarity today:

(1) the inherent complexity of nature and society,
(2) the desire to explore problems and questions that are not confined to a single discipline,
(3) the need to solve societal problems,
(4) the power of new technologies.

Drivers (1), (2), and (3) are not new. They have intensified, however, in recent decades. Driver (3), in particular, escalated with a force anticipated in 1982, when the OECD concluded that Exogenous ID had gained priority over Endogenous University ID. The Endogenous originates within science, while the Exogenous originates in 'real problems of the community' and the demand that universities perform their pragmatic social mission (OECD 1982, p. 130). Driver (4) has gained force as well. Generative technologies such as magnetic resonance imaging are enhancing research capabilities in many fields. New instrumentation and informational analysis are enhancing studies of human behavior through brain mapping and cross-fertilizations of cognitive science and neuroscience. New quantitative methods and advanced computing power are also facilitating the sharing of large quantities of data across disciplinary boundaries (Yates 2004, pp. 133, 135).

In addition, the growth of interdisciplinary fields is being recognized in traditional taxonomies. When a committee affiliated with the National Research Council (NRC) proposed an updated taxonomy of research–doctorate programs in the United States, it recommended an increase in the number of recognized fields from 41 to 57. It also recommended that 'biology' be renamed 'life sciences' and include agricultural sciences, while urging that subfields be listed to acknowledge their expansion. Mathematics and physical sciences, they added, should be merged into a single major group with engineering, and the committee called attention to the problem of naming in all fields. Despite general agreement that interdisciplinary research is widespread, doctoral programs often retain traditional names (Ostriker and Kuh 2003). The final 2009 guide to methodology is especially responsive to change in the category of life sciences and added a field of 'biology/integrated biology/integrated biomedical sciences'. Other changes in the guide's taxonomy served to expand disciplines, and programs were added in agricultural fields, public health, nursing, public administration, and communication. Appendix C also includes the 'emerging fields' of bioinformatics, biotechnology, computational engineering, criminology and criminal justice, feminist gender and sexuality studies, film studies, information science, nanoscience and nanotechnology, nuclear engineering, race ethnicity and postcolonial studies, rhetoric and composition, science and technology studies, systems biology, urban studies and planning (Ostriker *et al.* 2009).

Two other recent reports signal changes to come. In 2008, the NRC commissioned a Panel on Modernizing the Infrastructure of the National Science Foundation's Federal Funds for R&D Survey. This survey provides data on R&D spending and policy in the United States. However, the taxonomy for fields of science and engineering has not been updated since 1978. It does not capture the increasingly multi- and interdisciplinary character of science. Moreover, related activities are lumped together into a large amorphous category of 'not elsewhere classified' that includes new subfields, single-discipline projects without field designations, emergent fields, established ID fields, cross-cutting initiatives, problem-focus areas, and miscellaneous 'other'. In its final report, the Panel's report (*Data*) recommends capitalizing on the affordances of new technologies in federating, navigating, and managing data. It highlights, in particular, the National Institute of Health's Research Condition and Disease Classification (RCDC) database. The RCDC demonstrates the potential of bottom-up comprehensive systems to incorporate taxonomic elements while permitting users to construct cross-walks with agency-relevant keywords (tags) in particular projects and programs. In a review of the literature on evaluation of interdisciplinary research, a second taskforce affiliated with SRI International's Science and Technology Policy Program underscored the problem of classification systems while calling for greater use of new technologies capable of mapping underlying dynamics of relationships among disciplines and specialties (Wagner *et al.* 2009).

Changes of the kind traced in this chapter put pressure on not only conventional taxonomy but also on underlying assumptions about knowledge. In an issue of the journal *Science*, Alan Leshner contended that 'new technologies are driving scientific advances as much as the other way around', facilitating new approaches to older questions and posing new ones (Leshner 2004, p. 729). New topic-based domains outside or between disciplines are also transforming the disciplinary identities of collaborating researchers while fostering new skill sets. 'Thirty years ago', Norm Burkhard observed, 'the difference between a physicist and a chemist was obvious. Now we have chemists who are doing quantum-level, fundamental studies of material properties, just like solid-state physicists. There's almost no difference' (National Academy of Sciences 2004, p. 54). Developments in one area are stimulating new understandings in multiple fields as well, a phenomenon that occurred earlier in the theory of plate tectonics and more recently in the Human Genome Project and in nanoscience. Conventional taxonomies should not be jettisoned. Yet, they need to develop open, dynamic, and transactional approaches capable of depicting research in a network representation that is more aligned with changing configurations of knowledge and education.

References

Apostel, L., Berger, G., Briggs, A., and Michaud, G. (eds) (1972). *Interdisciplinarity: problems of teaching and research in universities.* Paris: Organization for Economic Cooperation and Development.

Bal, M. (2002). *Traveling concepts in humanities.* Toronto: University of Toronto Press.

Becher, T. (1989). *Academic tribes and territories: intellectual enquiry and the cultures of disciplines.* Buckingham and Bristol: The Society for Research into Higher Education and Open University Press.

Bennett, D.C. (1997). Innovation in the liberal arts and sciences. In: R. Orrill (ed.) *Education and democracy: re-imagining liberal learning in America*, pp. 131–49. New York: The College Board.

Boden, M.A. (1999). What is interdisciplinarity? In: R. Cunningham (ed.) *Interdisciplinarity and the organization of knowledge in Europe*, pp. 13–24. Luxembourg: Office for Official Publications of the European Communities.

Boisot, M. (1972). Discipline and interdisciplinarity. In: L. Apostel, G. Berger, A. Briggs, and G. Michaud (eds) *Interdisciplinarity: problems of teaching and research in universities*, pp. 89–97. Paris: Organization for Economic Cooperation and Development.

Bruun, H., Hukkinen, J., Huutoniemi, K., and Klein, J.T. (2005). *Promoting interdisciplinary research: the case of the academy of Finland.* Publications of the Academy of Finland. Series no. 8/05. Helsinki: Academy of Finland.

Burns, R.C. (1999). *Dissolving the boundaries: planning for curriculum integration in middle and secondary schools*, 2nd edn. Charleston, WV: Appalachia Educational Laboratory.

Cozzens, S.E. (2001). Making disciplines disappear in STS. In: S.H. Cutcliffe and C. Mitcham (eds) *Visions of STS: counterpoints in science, technology, and society studies*, pp. 51–64. Albany, NY: State University of New York Press.

Dölling, I. and Hark, S. (2000). She who speaks shadow speaks truth: transdisciplinarity in women's and gender studies. *Signs* **25**(4), 1195–8.

Dogan, M. and R. Pahre. (1990). *Creative marginality: innovation at the intersections of social sciences*. Boulder: Westview Press.

Funtowicz, S.O. and Ravetz, J.R. (1993). The emergence of post-normal science. In: R. von Schomberg (ed.) *Science, politics, and morality: scientific uncertainty and decision making*, pp. 85–123. Dordrecht: Kluwer.

Geertz, C. (1980). Blurred genres: the refiguration of social thought. *American Scholar* **42**(2), 165–79.

Gibbons, M. *et al.* (1994). *The new production of knowledge: the dynamics of science and research in contemporary societies*. London: Sage.

Gunn, G. (1992). Interdisciplinary studies. In: J. Gibaldi (ed.) *Introduction to scholarship in modern languages and literatures*, pp. 239–61. New York: Modern Language Association of America.

Heckhausen, H. (1972). Discipline and interdisciplinarity. In: L. Apostel, G. Berger, A. Briggs, and G. Michaud (eds) *Interdisciplinarity: problems of teaching and research in universities*, pp. 83–90. Paris: Organization for Economic Cooperation and Development.

Hübenthal, U. (1994). Interdisciplinary thought. *Issues in Integrative Studies* **12**, 55–75.

Jantsch, E. (1972). Towards interdisciplinarity and transdisciplinarity in education and innovation. In: L. Apostel, G. Berger, A. Briggs, and G. Michaud (eds) *Interdisciplinarity: problems of teaching and research in universities*, pp. 97–121. Paris: Organization for Economic Cooperation and Development.

Kann, M. (1979). The political culture of interdisciplinary explanation. *Humanities in Society* **2**(3), 185–300.

Kellner, D. (1995). *Media culture: cultural studies, identity, and politics between the modern and the postmodern*. London: Routledge.

Klein, J.T. (2005). *Humanities, culture, and interdisciplinarity: the changing American academy*. Albany: State University of New York Press.

Kockelmans, J.J. (1979). Science and discipline: some historical and critical reflections, and Why interdisciplinarity? In: J.J. Kockelmans (ed.) *Interdisciplinarity and higher education*, pp. 11–48, 123–60. University Park, PA: Pennsylvania University Press.

Lambert, R. (1991). Blurring the disciplinary boundaries: area studies in the United States. In: D. Easton and C. Schelling (eds) *Divided knowledge: across disciplines, across cultures*, pp. 171–94. Newbury Park, CA: Sage.

Landau. M., Proshansky, H., and Ittelson, W. (1962). The interdisciplinary approach and the concept of behavioral sciences. In: N.F. Washburne (ed.) *Decisions: values and groups, II*, pp. 7–25. New York: Pergamon.

Lattuca, L. (2001). *Creating interdisciplinarity: interdisciplinary research and teaching among college and university faculty*. Nashville, TN: Vanderbilt University Press.

Leshner, A.I. (2004). Science at the leading edge. *Science* **303**(5659), 729.

Lichnerowicz, A. (1972). Mathematic and transdisciplinarity. In: L. Apostel, G. Berger, A. Briggs, and G. Michaud (eds) *Interdisciplinarity: problems of teaching and research in universities*, pp. 121–7. Paris: Organization for Economic Cooperation and Development.

Mahan, J.L. (1970). *Toward transdisciplinary inquiry in the humane sciences.* PhD Dissertation, United States International University.

Miller, R. (1982). Varieties of interdisciplinary approaches in the social sciences. *Issues in Integrative Studies* **1**, 1–37.

National Academy of Sciences (2004). *Facilitating interdisciplinary research.* Washington, DC: National Academies Press.

Newell, W. (1998). Professionalizing interdisciplinarity: literature review and research agenda. In: W. Newell (ed.) *Interdisciplinarity: essays from the literature*, pp. 529–63. New York: The College Board.

Nowotny, H., Scott, P., and Gibbons, M. (2001). *Re-thinking science: knowledge and the public in an age of uncertainty.* Cambridge, UK: Polity Press.

OECD (1982). *The university and the community: the problems of changing relationships.* Paris: Organisation for Economic Co-operation and Development.

Ostriker, J.P. and Kuh, C.V. (eds) (2003). *Assessing research-doctorate programs: a methodology study.* Washington, DC: National Academies Press.

Ostriker, J.P., Holland, P.W., Kuh, C., and Voytuk, J.A. (eds) (2009). *A guide to the methodology of the National Research Council assessment of doctorate programs.* Washington, DC: National Academies Press.

Piaget, J. (1972). The epistemology of interdisciplinary relationships. In: L. Apostel, G. Berger, A. Briggs, and G. Michaud (eds) *Interdisciplinarity: problems of teaching and research in universities*, pp. 127–39. Paris: Organization for Economic Cooperation and Development.

Salter, L. and Hearn, A. (eds) (1996). *Outside the lines: issues in interdisciplinary research.* Social Science and Humanities Research Council of Canada. Montreal and Kingston: McGill-Queen's University Press.

Schleifer, R. (2002). A new kind of work: publishing, theory, and cultural studies. In: D. Shumway and C. Dionne (eds) *Disciplining English: alternative histories, critical perspectives*, pp. 179–94. Albany, NY: State University of New York Press.

Simon, E. and Goode, J.G. (1989). Constraints on the contribution of anthropology to interdisciplinary policy studies: lessons from a study of saving jobs in the supermarket industry. *Urban Anthropology* **18**(1), 219–39.

Smelser, N. (2004). Interdisciplinarity in theory and practice. In: C. Camic and H. Joas, (eds). *The dialogical turn: new roles for sociology in the postdisciplinary age*, pp. 43-64. Lanham, MD: Bowman and Littlefield.

Stokols, D., Hall, K.L., Taylor, B.K., and Moser, R.P. (2008). The science of team science: overview of the field and introduction to the supplement. *American Journal of Preventive Medicine* **35**(2S), S77–S89.

Tashakkori, A. and Teddlie, C. (2003). Major issues and controversies in the use of mixed methods in the social and behavioral sciences. In: C. Teddlie and A. Tashakkori (eds) *Handbook of mixed methods in social behavioral research*, pp. 3–50. Thousand Oaks, CA: Sage.

Transdisciplinarity Net (2009). *Transdisciplinarity research.* Available at: <http://www.transdisciplinarity.ch/e/Transdisciplinarity/index.php>.

Turner, B. (1990). The interdisciplinary curriculum: from social medicine to post-modernism. *Sociology of Health and Illness* **12**, 1–23.

Vickers, J. (1997). [U]framed in open, unmapped fields': teaching and the practice of interdisciplinarity. *Arachne: an Interdisciplinary Journal of the Humanities* **4**(2), 11–42.

Wagner, C. S, Roessner, J. D., and Bobb, K. (2009). Evaluating the output of interdisciplinary scientific research: a review of the literature. Prepared for the SBE/SRS, National Science Foundation, by SRI International Science and Technology Policy Program (May).

Weingart, P. (2000). Interdisciplinarity: the paradoxical discourse. In: P. Weingart and N. Stehr (eds) *Practicing interdisciplinarity*, pp. 25–41. Toronto: University of Toronto Press.

Yates, S.J. (2004). *Doing social science research.* London: SAGE Publications.

CHAPTER 3

Interdisciplinary cases and disciplinary knowledge

WOLFGANG KROHN

This chapter provides an epistemological analysis of interdisciplinary knowledge and research. It points at the peculiarities of interdisciplinarity and determines its place in the context of modern social epistemology. Interdisciplinary research can be subdivided into three kinds. At the center of the following analysis there is interdisciplinary problem solving, or better said, *interdisciplinary case work*. Of no less relevance, but of less epistemological concern, there is *interdisciplinary communication* as it is cultivated by many research centers. And finally there are a few cases of *interdisciplinary fusion* creating new disciplines. Among the suggested—but contested—examples are biochemistry, cognitive science, climate research, and public health.

In the following analysis interdisciplinary fusion is excluded as a mode of discipline formation. Even if the relevance of fusion may be underrated compared with disciplinary branching, newly fused disciplines leave observers where they started. Interdisciplinary communication will also be put aside. It can be described as the 'irritation' of disciplinary work. It provides scholars with fresh ideas and triggers them to redirect their research. If organized around themes and topics by the agendas of interdisciplinary research centers (e.g. Princeton, Berlin, Budapest) or foundations (e.g. Gordon Conferences), the effect may well go beyond the individual researcher. Most importantly—and opposite to the fusion zones—the themes can have this stimulating function even if they are extremely disparate. However, the function is to push the disciplines, not interdisciplinarity.

Interdisciplinary case work remains the most important kind. The intuitive conviction supporting this view is that most problems when they first appear are too complex for just one or two disciplines. The problem-solving power of disciplines is strong only with respect to theoretically simplified versions of problems. If complexity is added interdisciplinarity is needed. The most complex problems are so-called 'real-world problems'. The simplest way of organizing interdisciplinary research on complex problems is the multidisciplinary approach. It resembles the 'organic division' of labor in industrial production. Every component of a research problem calls for a different science. The integration of results may cause some trouble and require several attempts, but need not lead to exceptional difficulties. The efficiency of multidisciplinarity can best be observed in industrial research. Champions of 'true' interdisciplinarity tend to belittle multidisciplinary work, perhaps underestimating the quasi-industrial potential of modern knowledge work and romanticizing lost ideals of intellectual craftsmanship.

If, however, the organic division of intellectual cooperation presupposes common efforts to understand and define a problem, research requires interaction between disciplines. Each participant researcher observes the others and makes his or her decisions dependent on theirs. This is time-consuming, and without clear criteria for success. Investments without returns are frequent. Whatever drives people into highly complex interdisciplinary projects—curiosity, social responsibility, or money—the need for manageable objects and presentable results in their reference community drives them out again. If, however, public and political concerns are strong enough to exert a more permanent pressure, the difficult process of discovering and shaping the components of a complex problem can continue and generate a complex field of interactive interdisciplinary research. The problem, thereby, turns into a case.

The main propositions of this chapter are:

- Interdisciplinary research constitutes a relationship between individual cases and more general knowledge bases which is untypical for disciplinary research.
- This relationship demands a new mode of knowledge, in which learning about a case is equally as important as understanding causal structures. It calls for a combination between the 'humanistic' ideal of understanding the individual specificities of just one case, and the 'scientific' search for common features of different cases.
- Reflection on the character of interdisciplinary knowledge supports a critical reassessment of the received concept of scientific law and exemplary application.

If it is taken as a point of departure that most interdisciplinary research projects are organized around real-world cases, it is implied that these cases have to be understood with all their contingent features and circumstantial conditions. Each case is more or less different from every other case and has a certain value in itself. A paradigmatic example is global climate research. It aims at understanding the climate just exactly as it is, its origins and its future, in all its complexity and vagueness. Even if climate change is a broad topic, it is a unique one. It needs to be understood by means of a highly specific or even unique model to which many specialties contribute.

Interdisciplinary research also deals with cases which seemingly exist in several exemplars: cities and buildings in urban planning and architecture, prairies, sand dunes, or estuaries in restoration ecology and adaptive management, refugees in migration research, and prototypes in technological innovation. Here it seems possible to transfer knowledge gained in one case to similar cases. However, as will be seen later, relying on similarities without respecting differences can be misleading. In any case, reference to real-world cases is the essential cognitive and political dimension of interdisciplinary research.

This approach deviates from other approaches in not attempting to define interdisciplinarity on the basis of and as a derivative of the disciplinary structure of knowledge. Rather, it is assumed that real-world cases necessarily integrate heterogeneous knowledge bases, whether these are gathered under the institutional cover of a discipline or not. Any research field or research project that addresses real-world problems is considered to be essentially interdisciplinary.

An advantage of this conceptual approach is its independence from unsatisfactory attempts to define institutionally or cognitively what a discipline is. In consequence, research fields which are rhetorically addressed as disciplines can be considered to be epistemologically interdisciplinary. Moran (2002) has nicely made this point with respect to the humanities—English, literary criticism, cultural studies, feminism, psychoanalysis, and the like. They are all interdisciplines, or disciplines with interdisciplinary features, because they tend to accept cases in their complexity and contingency.

The same point was made earlier by Donald Campbell with respect to anthropology, sociology, psychology, geography, political science, and economics, which he called 'hodgepodges' caused and shaped by real-world problems (Campbell 1969). Later in this chapter I suggest that we expand the concept of real-world cases toward a softer definition focusing on complexity and contingency. However, to start with real-world cases helps us to better understand certain features of interdisciplinarity. The main interest is not to provide managerial and methodical solutions for cooperation between disciplines but to exploit the fruitful tension between understanding a case and searching for general knowledge. The main proposition here will be that taking cases seriously implies a kind of learning considerably different from received views of inductive or deductive methods. Doing research in the context of real-world problems demands and develops types of skills and competencies that scholars are not used to.

3.1 Idiographic and nomothetic knowledge

What are 'real-world' cases? The concept is meaningful only if contrasted with some 'ideal' state of something. Every scientific experiment makes things simpler than they are and theory imagines the world yet simpler. Historically, the paradigm is set by the invention of geometry. Since there is no *real* line, curve, or body that fits the demands of mathematical definition, they are ideally constructed. The ontological status of ideal objects has always been controversial, but this is not our point. Real things, those which we can point at, are only approximations of ideal objects. The science is still called 'earth-measuring' (geo-metry), though there is not a single place on earth that fits its definitions of objects. Sciences which do care for real-world measurement, such as surveying, alignment, and mapping, have developed methods able to determine an area of any shape. Limits to precision are not set by the methods but by changing and melting borders—as between land and water, forest and prairie, city and suburban sprawl. Open boundaries is a very important issue in the analysis of real-world objects or systems. Geometry and surveying have fruitfully interacted in history. Surveying is oriented to the real world and therefore is in itself an interdiscipline. Geometry is a classical discipline (or subdiscipline if mathematics is the discipline). Both come together in the earth sciences, in which on the one hand, sites, events, and (hi)stories are important and on the other the objects, models, and methods of the lab. Frodeman (2003) has provided an epistemological analysis of the earth sciences showing how difficult it is to integrate the interdisciplinary strands into a coherent self-understanding of the discipline.

There are numerous other examples where, in a roughly identical segment of reality, strategies for grasping peculiar cases as they are coexist with strategies to construct cases as they are wanted for theory. The general proposition to be made with respect to this distinction is simply this: interdisciplinary research focuses on the peculiarities of given cases, while disciplinary research is characterized by substituting ideal features for given ones. Basically, many modern research fields relate to both foci and, therefore, have a tendency to become more of a discipline, as well as a place of integration for potential contributors from various disciplines. How this is balanced institutionally—in terms of journals, societies, handbooks, curricula, and research sites—is of no concern here.

We can call the specific features of a problem, a system, or a case its 'idiographic component'. The more general features gained by taking problems, systems, or cases as exemplifying or inducing a more abstract or idealized object of knowledge its are called the 'nomothetic component'. This terminology was introduced by the neo-Kantian philosopher Wilhelm Windelband. Idiographic literally means describing the peculiar, singular, and specific.[1] Nomothetic literally means setting the (scientific) law. The law-like quality of scientific knowledge is associated with certain features of knowledge such as the reproducibility of experimental facts, prognosis of events, the general validity of propositions, and causal explanations of correlations. The definition and relation of these epistemic features are controversial. But undoubtedly they contribute to strengthening the difference between something one happens to know and theoretically corroborated knowledge. Windelband thought the ideographic structure of knowledge was best exemplified in historiography. A historian who specializes in the founding of the United States of America does not usually wish to become a specialist on foundings in general, but builds his or her reputation on knowing everything about just this case and giving it an original and surprising interpretation. If he or she cared to analyze another founding—say of the Roman Empire, Brazil, or the European Union—neither factual knowledge nor interpretation schemata can be transferred from one to the other.

When Windelband introduced this terminology he was not only a famous philosopher but also rector of Strasbourg University. He found himself in a position to reconcile a heated controversy between the natural/technical and the cultural sciences/humanities. The rapid ascent of the natural sciences led to claims that true knowledge would only reside in laws. Eventually all knowledge fields including the humanities were to be converted into law-seeking disciplines. The counter attack aimed at the assumed weak point that the natural sciences are completely unable to develop a coherent understanding of something as complex as a culture and its history, or even some part of it, such as a specific city, not to mention art, literature, and religion.

In his rectorial lecture in 1894 Windelband suggested equal rights for both forms of knowledge. Knowledge production is guided either by an interest in identifying laws, which implies turning things into variables, or by an interest 'to describe as complete as possible a singular event or chain of events spread over a limited time'. Examples of events worth scholarly interest are, according to Windelband, 'Actions of a person, the character and

[1] The likewise usual wording "ideographic" does not refer to the Greek *idios* = peculiar, but to *idea* = poem, Gestalt, which is no less appropriate.

life of a single man, or of an entire people, the character and development of a language, a religion, a legal order, of a product of literature, art, or science: and each of these subjects demands a treatment corresponding to its peculiarities'. (Windelband 1907, p. 363) For Windelband the distinction is not built on different classes of objects—natural events versus human affairs—but on methods. In principle, everything can become the object of a nomothetic as well as an idiographic analysis. His examples are language, physiology, geology, and astronomy. If objects in these fields are considered in their specificity, 'the historical principle is carried over to the realm of the natural sciences' (Windelband 1907, p. 365). If the objects are taken as types or exemplars the methods of the natural sciences apply.

By the traditional views of philosophy of science it seems obvious that the sciences should search for laws, principles, and other forms of generalized explanations. It is less obvious why they should care for singular or even unique cases. Windelband assumes their relevance with respect to cultural heritage, identity, and value. Admittedly, one can never know in advance whether or not a single case will turn out to be culturally relevant. But if it is considered to have no potential value at all, research would not be started. Or put in more constructive language, a scholarly effort to study a case automatically attaches some sort of value to it. Windelband's neo-Kantian disciple Rickert gave the following equation: 'There is not only a necessary connection between the *generalizing* and the *value-free* observation of objects, but also an equally necessary connection between the *individualizing* and *value-laden* perception of objects' (Rickert 1924, p. 58). Even if this general statement may be doubtful, obviously all real-world problems have a value dimension, be it economical, social, cultural, or environmental. Windelband and Rickert chose historical research as their paradigmatic field because the preservation of cultural goods and values seemed to be even more important in a society that had become exposed to dramatic industrial changes. Today we would add to historians' conversational work pressing problems caused by misguided developments. Real-world problems are problems because values are at stake. Solutions are only accepted if they address these values.

Concern for idiographic cases does not invalidate more general knowledge. Usually, interdisciplinary case studies are not only expected to solve single problems but to contribute to stocks of knowledge. However, the epistemic structure of these stocks of knowledge is different from knowledge condensed in theories or paradigms. The relationship between idiographic and nomothetic orientations of interdisciplinary research needs to be analyzed and interpreted in a new way. The first step will be to better understand the nature of cases by looking at variants of the so-called case-study method practiced in professional schools. Certainly, higher education of professionals and experts aims at goals different from doing research. However, the reasons why the case-study method seems to be successful in professional training are important for understanding how cases contribute to interdisciplinary knowledge.

3.2 Learning based on case studies

The methodology of using case studies in educational programs originated in the pioneering achievements of the Harvard University professional schools. As early as 1870, the

Harvard Law School shifted the study of law from the classical systematic approach to the analysis of case studies. In 1920, the Harvard Business School developed a new curriculum based on case studies. In 1985, the Harvard Medical School followed suit with its 'New Pathway Program', which was considered revolutionary within the field of medical training. The following presentation is concerned not with an evaluation of this educational method, but rather with the question of what can be learned from individual cases.

David Garvin—himself a faculty member of the Harvard Business School—emphasizes the three dominant goals of case-study methodology: 'learning to think like a lawyer; developing the courage to act; fostering a spirit of inquiry' (Garvin 2003). Competencies from three professional fields merge here: the logical expertise of a lawyer, the decision-making capacity of a manager, the curiosity of a researcher. Cases that have been of paradigmatic importance for the development of laws are not central to the training at the Harvard Law School. The focus is rather on those cases which are controversial within the legal profession, those which were wrongly decided or were revised. Garvin cites another member of the faculty: 'We have conflicting principles and are committed to opposing values. Students have to develop some degree of comfort with ambiguity' (Garvin 2003, p. 58). The analysis of individual cases frequently does not lead to a clear result. 'Students often leave class puzzled or irritated, uncertain of exactly what broad lessons they have learned' (Garvin 2003, p. 59). On the contrary, they learn that general legal doctrines are rarely unambiguously applicable and that the smallest distinctions can play a role in their application. Furthermore, these cases help students practice dealing with unknown and unforeseen circumstances, with varying conditions and with surprises.

The description of Stanford Law School's 'situational case studies program' is similar to Harvard's: 'Case studies and simulations immerse students in real-world problems and situations, requiring them to grapple with the vagaries and complexities of these problems in a relatively risk-free environment – the classroom' (Stanford Law School 2008). The program emphasizes cases not as legal cases but as true-to-life social configurations, for which it has yet to be decided whether or not they should be treated within the justice system. The aim is to thereby improve the students' 'lawyering skills'. Far from introducing individual cases in Kuhn's sense as paradigms, these are examined as unsculpted and uninterpreted as possible. This methodology is thus quite suited to an academic policy which places value on the grasping of complex configurations, on the identification of possible action, and on the assessment of consequences.

Education at the Harvard Business School is also guided by the principle that greater competence can be acquired through constant rehashing of case studies than through studying theoretical and methodical knowledge and the intended applications thereof. Underlying the choice of these individual cases are the following criteria: 'Typically, an HBS case is a detailed account of a real-life business situation, describing the dilemma of the "protagonist"–a real person with a real job who is confronted with a real problem. Faculty and their research assistants spend weeks at the company....The resulting case presents the story exactly as the protagonist saw it, including ambiguous evidence, shifting variables, imperfect knowledge, no obvious right answers, and a ticking clock that impatiently demands action' (Harvard Business School 2008). The students are presented with about 500 of these cases in the course of their studies, the main goal being to school their decision-making behavior. The large number of cases is not seen as an inductive basis for statistically generalizable knowledge, but rather as preparation for a maximum number of diverse situations. In addition to these case studies, the program offers courses in 'analytical tools'. The following list of academic goals is presented in Garvin (2003, p. 62):

- Training of diagnostic skills in a world where markets and technologies are constantly changing.
- Assessment of the ambiguity of constellations.
- Consideration of the incompleteness of the information at hand.
- Recognition of the existence of a multitude of possible solutions.
- Preparedness to make decisions in the face of uncertainty and time pressure.
- Development of persuasive skills. Management is a social art; it requires working with and through others.

From a critical perspective, the tendency to make a quick decision should be noted. 'The case method does little to cultivate caution.... Students can become trigger-happy' (Garvin 2003, p. 62).

Inaugurated in 1985, Harvard Medical School's 'New Pathway Program' has supplanted the classic basic training in medical fields and has with some delay affected applications at the sickbed. It also highlights the point that every single case is self-contained. Garvin quotes Tosteson, the program's founder, as saying that medicine 'is a kind of problem solving' and each medical encounter is 'unique in a personal, social and biological sense... All these aspects of uniqueness impose on both physician and patient the need to learn about the always new situation, to find the plan of action that is most likely to improve the health of that particular patient at that particular time' (Garvin 2003, p. 63). The program prompts students to identify and correct their knowledge deficits. Garvin quotes Lowenstein as saying that the program's overriding goal is to 'foster a true spirit of inquiry', and quotes Moore as stating, 'I want my students to be able to identify a gap in their knowledge, feel guilty about not filling it, and have the nimbleness to learn what they need' (Garvin 2003, p. 64). Tosteson, as quoted by Garvin, adds, 'They discover that choosing what to learn is the hard part; learning it is a lot easier' (Garvin 2003, p. 64).

Further examples of curricula which have adopted the case-study method entirely or partially include engineering, sociology, psychology, education, architecture, and economics. What constitutes the success if not the superiority of the case-study method in higher education? The most notable criterion for the choice of the cases is the insistence on the individuality of cases. They are not cases in point, not exemplars of a type—at least not in the first place. The didactic concept is not to present a general structure via a number of examples, whose special features quickly retreat behind the emerging abstraction. No case can be exchanged for another, since one learns something different from each case. Concentrating on the idiographic nature of each case means to develop a sense for its details and the seemingly incidental aspects that make it special. Every case study of this kind is unavoidably connected to deficits in information, to ambivalent interpretations, and to the risky effects of possible interventions. The pressure on making a decision blocks the option of completely assembling all the relevant knowledge.

At variance with a more traditional academic education, the focus is on grasping both the differences and the similarities between cases. Identifying case-specific gaps in knowledge is as important as applying knowledge gained from other cases. The background philosophy seems to be that professional realities are not determined by general rules or even scientific laws, but are constituted by a vast network of particular cases. The competency of the professional consists in deriving operative gain from comparing similarities as well as differences between cases.

Case-study methodology obviously distances itself from both inductive and deductive learning strategies, which is why it has been closely analyzed here. Traditionally, the two pillars of scientific methodology are inductive generalization leading to theory and deductive specification via application to cases. Here, however, neither is applied, rather both are substituted by the expansion of a network of cases, in which the mesh density of analogous relationships is continually tightened. Does this indicate a third path that avoids the alternative between generalization and specification? Does such professional training develop a learning core not contained in the traditional theories of the growth of knowledge?

3.3 Knowledge and skills: the professional perspective

The launching point for the educational programs described in the previous section is the shortcoming of academic training with respect to professional competencies. The criticism is that the academy is unable to deal with the complexities of real life, but must reduce these in accord with theoretical concepts. Academic training follows the paradigm of alternating theoretical construction and experimental research by which the object of study is subjected to the ideal conditions of the laboratory. This is precisely not the reality that the professional expert confronts.

The case-study method cultivates certain capacities that are most often termed 'skills'. Skills do encompass rational pieces of knowledge, but equally important are routines, habits, and trained intuitions. These not wholly explicable components come into play not only for professional know-how, but also in many fields of teaching and learning, like the acquisition of crafts and trades, doing sports, or mastering a musical instrument. More generally, all techniques which require the coordination of physical training with the comprehension of rules are based on skills. Here the study of introductory books and instruction manuals helps a little. The observation of the masters helps a bit more. However, is the continual exercise of physical practices until these become routine is decisive. Situational assessment, spontaneous coordination of action, and a repertoire of strategies are all conditions for success. The important point in our context is this: Even when skills have been developed, each individual case retains its particular meaning. There is no overarching level of competence comparable to theoretical knowledge, in which skillful action could be adequately reconstructed as theoretical objects. Although there are attempts in sport and music sciences to construe such levels, what ultimately counts are skills in action.

The Harvard method and teaching methods practiced in the fields mentioned have in common that they build on the accumulation of analogies between related configurations,

Prospects for a philosophy of interdisciplinarity

Jan C. Schmidt

Interdisciplinarity is one of the most popular buzzwords in scientific and public discourses. At the same time, however, the term is vague.

This vagueness poses challenges to philosophy. Does the term 'interdisciplinarity' carry any distinctive epistemic content and any *differentia specifica*? In addition to what has been achieved in the field of reflection on interdisciplinarity (ID), the aim of this box is to provide a philosophical foundation for a classification *and* criticism of the innumerable usages of ID. A plurality of meanings will be shown, without a unifying semantic core. There is not *one* type of ID, but various coexisting types. With regard to established positions in the philosophy of science, different types of ID can be distinguished: the object type ('ontology'), the theory/knowledge type (epistemology), the method/process type (methodology), and the problem/purpose type.

This box is intended to foster the debate on ID. It presents elements of a pluralist *philosophy of interdisciplinarity*.

Object interdisciplinarity—entities or objects constitute the central elements of *object ID* (the ontological dimension of ID). The historically established functional differentiation into disciplines does not seem to be contingent. Rather, it mirrors aspects of the structure of reality. Edmund Husserl, Nicolai Hartmann, and Alfred North Whitehead argue in favor of a concept of layered reality. Boundaries between the layers separate the micro-, meso-, and macrocosm. Interdisciplinary objects are thought to be located on boundaries between different cosms or within border zones between disciplines, for example the brain–mind object. In order to substantiate this position one has to presuppose a minimal ontological realism, interlaced with a concept of layered reality, and, based on this, an ontological non-reductionism: Brain–mind objects can be reduced neither to the material brain nor to the mental mind but, perhaps, to other entities (neutral ontology). Old and ongoing philosophical issues about monism, dualism, and pluralism emerge in this debate. ID here does not refer mainly to knowledge, methods, or problems, but to an external, human-independent reality.

The foregoing position is a strong one. It might be called *universal object* ID. A weaker position—which can be named *real-constructivist* or *techno-object* ID—does not assert a timeless existence of interdisciplinary objects in an unchangeable reality but rather that interdisciplinary objects are created by human action. Examples include the hole in the ozone layer, or techno-objects of nanoscience: nano-objects are placed on the boundaries between physics, chemistry, biology, and engineering sciences. This ontological position is neither a classic cognitive-oriented realist's position nor a constructivist one: it can be called *real-constructivism* and it can be traced back to Bacon. Unfortunately, however, real constructivism is not fully developed in the philosophy of science.

Theory ID focuses on knowledge, theories, and concepts, and not primarily on objects and reality. It is concerned with whether interdisciplinary theories exist and how they may be specified. Can we demarcate interdisciplinary knowledge from disciplinary knowledge and from non-scientific knowledge? Is there a unique context of justification? Do interdisciplinary models, laws, descriptions, and explanations exist? Possible candidates for theory-ID are concepts which can be applied to describe objects in different disciplinary domains. They highlight structural similarities between properties of these objects. Such theories cannot be reduced to disciplinary ones. Theory ID is, therefore, based on an epistemological non-reductionism.

(*cont.*)

Prospects for a philosophy of interdisciplinarity (cont.)

Structural sciences such as complex systems theory are prominent examples. Structurally similar process phenomena, e.g. pattern formation, self-organization, bifurcations, structure breaking, and catastrophes—can be found in different disciplinary branches. The objective is an integration of general structures regardless of the disciplinary content. Alike theories are self-organization theory, dissipative structures, synergetics, chaos theory, and fractal geometry. Hermann Haken regards synergetics as an 'interdisciplinary theory of general interactions' (Haken 1980). Most of these interdisciplinary theories were established in the 1960s and 1970s. Basic ideas—and the term 'structural sciences'—however, can be found in works from the 1940s and 1950s. Structural sciences 'study their objects regardless of disciplinary domains and in abstraction from disciplinary content' (von Weizsäcker 1974, p. 22). Classic examples are cybernetics, information theory, or game theory.

Method ID refers to knowledge production, to research processes, to rule-based actions, and to languages. The central issue of methodology is *how*, and by *which rule*, can and should we obtain knowledge? In terms of interdisciplinarity the central questions are: Do interdisciplinary methods and actions exist? Is there a specific context of discovery within interdisciplinary projects? Interdisciplinary methodologies, however, are thought to be irreducible to a disciplinary methodology.

Biomimicry, for example—sometimes used interchangeably with bionics—claims to be an interdisciplinary *transfer methodology* from biology to engineering sciences. The basic idea of biomimicry is 'learning from nature' in order to 'inspire technological innovations'. Nature seems to provide excellent inventions that can be used to develop efficient technologies. However, the transfer is not a one-way street. Biomimicry constructs models of biological nature based on the perspective of engineering sciences. A robot mimics an ant, but at the same time the ant has been described from the mechanistic perspective of technology. Besides biomimicry, there are other examples of interdisciplinary methodologies. Econophysics methodologically organizes a transfer between physics and finance/economics. In addition to these transfer methodologies, a new kind of non- or meta-disciplinary methodology of knowledge production has emerged over the past 50 years: mathematical modeling and computer-based simulations (see Lenhard, Chapter 17 this volume).

Problem-oriented ID: we have to add another type of methodology that focuses on the starting points and goals, problems, and purposes of research programs—in other words, the problem-framing and agenda-setting type. Erich Jantsch argues in favor of a 'purposive level of interdisciplinarity' and a 'purpose-oriented interdisciplinarity', today sometimes called 'transdisciplinarity'. An explicit reflection on, and revision of, purposes should be regarded as the highest level of interdisciplinarity (Jantsch 1972, p. 100). Normative premises, such as problem identification and agenda setting, the volition or intention to obtain certain knowledge, precede both the context of discovery and the context of justification, i.e. the theories and the methods.

The very first step in scientific inquiry is often judged to be a contingent factor; the teleological structure in the process of knowledge production is not always acknowledged. In fact, philosophers of science have widely ignored problem identification or agenda setting, although work has been done on 'wicked problems' (see Boradkar, Chapter 19 this volume). The lack of clarification is a disadvantage for specifying problem-oriented ID and demarcating it from disciplinarity. Obviously, interdisciplinary problems are external to disciplines and to sciences in general. They are primarily societal and are defined by society, e.g. lay people, politicians, and stakeholders. To contribute to societal problem solving and to ensure societal progress, disciplinary limitations have to be overcome. In this sense ID is seen as an instrument for meeting societal demands in order to tackle pressing problems. Examples of problem-oriented ID are sustainability research, technology assessment, and social ecology (Decker 2001).

One or other of the above-listed types of ID may raise concerns. Underlying philosophical convictions determine which type might be considered most important and which of the other types will just be viewed as mere inferences.

Regarding well-established positions in the philosophy of science, we can denote:

(1) *Realists* and *real-constructivist* refer to given *or* constructed objects of reality (they prefer the ontological dimension of ID).
(2) *Rationalists* focus on knowledge, theories, and concepts; *positivists* share the same orientation toward theories (epistemological dimension).
(3) *Methodological constructivists* and many *pragmatists* reflect on methods, actions, or cognitive rules (methodological dimension).
(4) *Critical theorists*, together with *instrumentalists*, *utilitarians*, and some *pragmatists*, refer to problems and how to handle and solve problems pragmatically. The impact, effect, and consequence of ID are of utmost relevance (problem-oriented dimension).

The different approaches to interdisciplinarity depend on underlying philosophical convictions. We cannot eliminate this plurality. 'ID' is, and will always be, a multifaceted term. Philosophy of science is effectively helpful in analyzing and classifying interdisciplinarity. A *philosophy of interdisciplinarity*, however, still remains a desideratum. By the approach presented here some elements for a philosophy of ID may have been proposed.

References

Decker, M. (ed.) (2001). *Interdisciplinarity in technology assessment*. Berlin: Springer.

Haken, H. (1980). *Dynamics of synergetic systems*. Berlin: Springer.

Jantsch, E. (1972). Towards interdisciplinarity and transdisciplinarity in education and innovation. In: L. Apostel, G. Berger, A. Briggs, and G. Michaud (eds) *Interdisciplinarity: problems of teaching and research in universities*, pp. 97–121. Paris: Organization for Economic Cooperation and Development.

von Weizsäcker, C.F. (1974). *Die Einheit der Natur*, p. 22. Munich: dtv.

whereby it is as important to attend to differences as to similarities. In this way, the learner knots together a network of configurations that is fed by individual cases and used to situate further cases. The common denominator of such networks of pragmatic know-how consists in structuring an individual case, for arriving at a decision regarding action and for evaluating its effects. This is what defines the professional expert (e.g. the lawyer, doctor, or manager), the specialized expert (e.g. the craftsman, athlete, musician), and even, if one can say so, the everyday expert (e.g. the habitual walker in uneven terrain, the parent, the driver). As applied to interdisciplinary research, one can conclude that learning from case studies is suited primarily to expanding the professional know-how of experts. In keeping with the traditional concept of professions, one could coin the term of a 'professional researcher'. Such a professional would be an expert in the investigation of open problems in contingent and complex individual cases, which occur within a certain field of knowledge. Since some interdisciplinary competency is required for this type of research, this expert would work in a team with other experts so that the professional know-how would consist in the coordination of an overall cognitive understanding of the situation distributed among diverse experts.

One of the best analyses of the design of case studies in sociology (also inspired, by the way, by the Harvard methodology) confirms this grounding of research in expertise. 'Common to all experts is that they operate in their fields of expertise on the basis of an intimate understanding of many thousands of concrete cases. Context-dependent knowledge and experience constitute the core of expert praxis.... Only through experience in dealing with cases can one develop from a beginner to an expert.' (Flyvbjerg 2006, p. 222).

3.4 Individual cases and epistemic knowledge

The idiographic aspects of interdisciplinary research have now been sufficiently explored. It was important to begin with these, since they are quite removed from standard philosophy of science and from learning theories of higher education. However, to end with the case-study method would mean to declare theory-based epistemic knowledge a needless encumbrance. The important point was that sensitivity to cases cannot be derived from theory. This does not imply that theory cannot contribute to understanding cases, nor that cases cannot advance theory. The statement that contingency in interdisciplinary research cannot be eliminated gains its epistemological value only because important resources of knowledge can be tapped into, the validity and applicability of which are accepted, even if they do not suffice to grasp all the details of a specific case.

The question to be raised is how the two paths of nomothetic and ideographic research can be commonly pursued, when they are, as Windelband argued, separated by diverse explanatory ideals. Windelband showed that nomothetic and idiographic knowledge can function as alternate resources for one another. In describing an individual case, one must unavoidably reach back to some sort of general knowledge. This can even consist in prescientific everyday convictions. Among his examples are psychological background assumptions concerning the behavior of historical figures. The reverse perspective is that every nomothetic statement—as abstract as it may be—must be exemplified in a context which unavoidably plays into everyday reality, despite the idealization of the objects referred to. Windelband gives the example of an explosion. On the one hand, an explosion follows the laws of chemistry, which allow for an explanation of the process; on the other hand, it happens in the here and now under singular circumstances, whose tiniest details might possibly interest a criminal detective. The core idea derived from Windelband is that interdisciplinary research projects are usually set up in such a manner that both ideals are pursued concurrently: the goal of dissolving a concrete case down to its smallest detail, and the goal of extracting generalized causal knowledge from this case.

The interconnection between law-like causality and the singularity of a case can be illustrated using the example of an interdisciplinary research project dealing with the rehabilitation of an atrophied lake in Switzerland (Sempacher Lake, Gross *et al.* 2005,

p. 135*ff*). The starting point was the scientifically ascertained causal knowledge that the reduction of phosphate content, either by directly decreasing the input or increasing the output of phosphate, would reduce algal growth. On the basis of this knowledge a theoretical model was developed and an aeration technology implemented which already operated successfully in a nearby lake. Surprisingly, the expected effect of decreasing phosphate content failed to appear. Careful study of the sediment layers and the highly complex seasonal dynamics at the boundary led to a quite different view of the relationship between aeration and phosphate concentration. The findings did not completely invalidate the underlying model but suggested protecting it through additional assumptions (so-called '*ceteris paribus*' clauses). If the findings can explain the failure of theoretical prognosis for an individual case this unleashes less turbulence than debunking the theoretical model. In conjunction with the steadfast core of causal knowledge they describe the individual situation all the better. In the case of this project it became clear that without measures to reduce the feed charge of phosphates the project would fail. Since the intensive agribusiness of the region made stakeholders averse to general constrictions, fine-tuned analysis led to the establishment of differential ecobalances. The case shows how an explanatory model of the lake satisfactorily capturing its manifold causal interactions was achieved step by step over a period of 20 years.

Obviously, the procedure of this rehabilitation project is opposed to the radical version of the case-study method. Interdisciplinary cooperation rests upon bringing together reliable knowledge from independent disciplines into case-specific modeling. The current status of scientific knowledge as organized within disciplines presents an enormous potential for interdisciplinary work and especially for modeling a specific case. Working with the model will probably lead to surprises. It is precisely because the individual case counts as such that its investigation leads to surprises, which cannot be ignored. Only after incorporating these will the model become sufficiently fine-tuned in terms of an idiographic understanding of the lake. In turn, surprises induce causal analysis and can expand our knowledge about atrophied alpine lakes. In the case of the rehabilitation of Sempacher Lake, the specific knowledge gained from sediment core analysis can be applied to geologically similar lakes exposed to high phosphate input.

3.5 The relevance of concreteness and the questionable concept of law

3.5.1 Individual case and unconditional laws

The relationship between the specification of causal knowledge toward individual cases and the generalization of on-site findings appears at first sight to be that between a deductive strategy of applying substantiated knowledge and an inductive strategy of developing hypotheses for new knowledge. But this distinction does not allow the methodological challenge of interdisciplinary research to come to light. The challenge is to balance the tension between understanding a case in its real-life context and contributing to a stock of theoretical knowledge. This section relates this tension to current discussions in philosophy of science.

In her influential book, *How the laws of physics lie* (1984), Nancy Cartwright presented the thesis that the fundamental laws of physics hold true only for highly idealized theoretical objects that don't exist in the real world. Strictly interpreted, these laws are false when taken as empirical descriptions of reality. The well-known example is that of Galileo's law of falling bodies. Its real-world validity is modified by friction, wind force, rain drops, and the shape of the body. Cartwright loves to illustrate the problem by an example already used by the Vienna Circle philosopher Otto Neurath (Cartwright 1999, p. 27): the calculation of the trajectory of a bill dropped from St Stephan's dome in Vienna. Even the joint forces of mechanics, fluid dynamics, and computer simulation methods wouldn't come close to a correct prediction.

From a pragmatic point of view, Cartwright's objection is of no effect. In the laboratory objects are stylized to better fit theory, and theorists acknowledge practical limitations to the absolutely perfect realization of causal assumptions. Within these limits, knowledge can be put to work. From a philosophical perspective, however, her thesis continues to provoke unrest. If under close scrutiny universal laws have no empirical content, then the project of interpreting reality through reductionism remains ungrounded. At best, it can be played through for simple cases from which one cannot extrapolate, what Cartwright (1999)called 'the dappled world'. This world can be scientifically captured only by a broad variety of laws with limited range and with no consistent logical order. In describing this world we can better speak of capacities, tendencies, and potentialities than of rigid laws.

Cartwright's strong statement regarding the presence, if not the predominance, of the idiographic in the scientific description of the world is highly controversial (Earman *et al.* 2002). It has at least shattered the privileged position of the concept of natural law as the standard and compass for scientific theorizing. Moving beyond Cartwright's proposal, Giere (1999) suggested that the concept of law should be completely struck from the language of philosophy of science. He is of the opinion that we cannot rid ourselves of the theological origin of the concept. Only God as the external legislator of the world would be in the position to command by general rules completely obedient natural things. Since the Kantian project of anchoring fundamental laws in the structure of reason failed, for Giere no further candidate remains that could guarantee the universality and necessity of the laws of nature. In Giere's reconstruction, lawful regularities become systems of equations that pertain not to reality but rather to imaginary models created for their verification—an idea for which Cartwright coined the term 'nomological machine'. Real-world constellations cannot be grasped precisely. Whether, despite these objections, it will remain meaningful to speak of general and unconditional laws of nature can be left an open question here. It suffices to ascertain that the classical notion of a law's universal validity no longer fully captures the 'cases' that fall within the law's domain.

The take-home message of this philosophical discussion concerning the relationship between the nomothetic and idiographic in science is that the tension between universal validity and exemplary cases is already contained within the unconditional laws of physics.

3.5.2 Individual case and conditional laws

Some laws of physics still possess the elevated status of being general. Laws typical for sciences such as biology, psychology, and economics are burdened from the beginning with the acknowledgement that their predictions and causal explanations are valid only under specific conditions or to a certain degree. The two central problems of such laws are:

(1) that the respective specific conditions cannot be listed completely and definitively,
(2) that exceptions to the rule can always be included in the collection of excluded conditions.

The difference with regard to the laws discussed in the above section is this: although the mutual attraction of bodies and the conservation of energy and entropy are considered unavoidably and eternally valid, intervening factors arise in the calculation of concrete cases. These factors are not part of the models and are incompletely understood. For conditional laws such as Mendel's laws of heredity in genetics, the law of diminishing return in economics, or the Gestalt laws in psychology, the lawful connections are defined for objects whose uniformity, continuity of existence in time, and independence from their environment are not guaranteed.

Following in the footsteps of the evolutionary biologist Stephen Jay Gould, Sandra Mitchell asserted the following for biological regularities: 'if we rewound the history of life and "played the tape again", the species, body plans, and phenotypes that would evolve could be entirely different. The intuition is that small changes in initial "chance" conditions can have dramatic consequences downstream.... Biological contingency denotes the historical chanciness of evolved systems, the "frozen accidents" that populate our planet, the lack of necessity about it all' (Mitchell 2002, p. 332). Conditional laws can be investigated only in tandem with the historical development of the objects and their contingent context. In this manner, the idiographic is officially granted entrance into the grasp of the law-like generalization under consideration. The conjecture of a conditional empirical law usually emerges with the reservation that intervening contingencies are to remain irrelevant (the *ceteris paribus* clause). If and when they do become relevant, the question must be confronted whether they dissolve the assumed law or alter the set of conditions.

It is possible to reinterpret the epistemological problem of the validity of contingent laws as an answer to the question of how the tense relationship between the nomothetic and the idiographic can be combined. Within the realm of biological research, it is as productive to search for conditional laws as it is to identify configurations of restricted validity. It is as interesting to reduce contingency through *ceteris paribus* clauses—thereby expanding the effective domain of a law—as it is to increase contingency—thereby pursuing the relevance of configurations not yet understood. Mitchell writes, 'In systems that depend on specific configurations of events and properties,... which include the interaction of multiple, weak causes rather than the domination of a single, determining force, what laws we can garner will have to have accompanying them much more information if we are to use that knowledge in new contexts. Thus the central problem of laws... is shifted... to how do we detect and describe the causal structure of complex, highly contingent, interactive systems and how do we export that knowledge to other similar systems.' (Mitchell 2002, p. 335). It is in this manner that the analysis of the concept of law within these specific sciences approximates learning from case studies.

3.5.3 Individual case and ideal type

The diverse efforts within the social and historical sciences to formulate diachronic and synchronic generalizations have never led to results that are in any way comparable with the status of the conditional causal laws in the natural sciences. The only exception is in modern economics, which since its origins in the eighteenth century has attempted to formulate qualitative laws (like, for example, Marx's law of falling profits) and quantitative laws of market behavior (starting with Leon Walras). All such attempts remain controversial within the economic sciences and even more as applied to political economy. In the other social sciences (such as historical sciences, cultural anthropology, and sociology), the generalizations of empirical findings have not achieved the status of recognized laws. Despite this, generalizations are being considered. The concept of 'ideal type' developed by Max Weber has gained widespread recognition. Weber formulated this concept in the context of the ongoing discussion of Windelband's and Rickert's ideas. His goal was to justify that social sciences can also search for objectively valid and controllable propositions in attempting understand highly specific and complex constellations in which elements of culture, politics, religion, and economics merge. In Weber's words, an ideal type 'is a conceptual construct which is neither historical reality nor even the "true" reality. It is even less fitted to serve as a schema under which a real situation or action is to be subsumed as one *instance*. It has the significance of a purely ideal *limiting* concept with which the real situation or action is *compared* and surveyed for the explication of certain of its significant components.... In this function especially, the ideal-type is an attempt to analyze historically unique configurations or their individual components by means of categorical concepts' (Weber 1922, p. 194).

3.6 Summary: epistemic qualities of interdisciplinary research

The preceding analyses of the relationships between law-like universality and concrete cases support the conclusion that this rapport may indeed be fraught with tension, but that it in various ways contributes to the scientifically rooted description and construction of reality.

Our starting point was the observation that interdisciplinary projects are often tied to field-specific phenomena and expectations. Whereas disciplinary research too often aims at eliminating incidental factors in order to achieve concise models and causal explanations, interdisciplinary research is forced to recognize and incorporate details. Generally, these pressures are imposed by the respective interdisciplinary programs. In some cases even the voices of local actors are influential. Even when the motivation for interdisciplinary research originates with inner-scientific concerns, it aims to master a higher degree of complexity than a disciplinary research agenda would allow.

I have tried to demonstrate that the study of cases is essential in the learning of capacities and skills. The most important claim here is that learned know-how encompasses the recognition of both similarities and differences between relevant cases. Based on the presented examples, this holds true for professional training in existing fields such as business/management, law, and medicine. Modifications result when this finding is transferred from professional training to scientific research. The safety-rails of curricular regulation fall by the wayside. Research works without corrective instruction from those who have already mastered the matter at hand. Nevertheless, an important attribute of interdisciplinary research can be extracted from the comparison with professional case-study training—namely that of professional expertise.

The next step was to investigate the role allotted to the individual case within philosophy of science. Surprisingly enough, the individual case is present everywhere, even though it has been traditionally overlooked or dismissed. The conflict between the all-encompassing and simultaneously exact grasp of an individual case and its inadequate description through general laws is demonstrable deep down to the most fundamental laws of physics. The individual case becomes more and more relevant for the conditional laws of the specialized sciences and the ideal types of the social sciences until, in the end, a symmetrical balance between the investigation of universal cognitions and localized idiosyncrasies is achieved.

The goal here has been to integrate nomothetic potential and idiographic description into a model that correlates a causal explanation of reality (nomothetics) with the situational, local specifics of a case (idiography) as far as possible. In closing, this point can be briefly illustrated using the example with which this chapter began. Modern research into the effects of climate change has taken the form of a giant worldwide project. It forces the participating researchers to comprehend the singular, extraordinarily improbable case of earth's climate in its specific state and its developmental dynamic. This is an extremely idiographic situation. Enormous constraints arise from being tied into a heterogenic configuration of political and scientific actors—the Intergovernmental Panel on Climate Change, whose ultimate goal is not cognition, but rather science-based coping with climate change. The background for this effort is the consensus that a certain state of climate constitutes a principle value for life on earth. From this idiographic value component (in Rickert's sense), it follows that research into the effects of climate change does not only deliver analysis and prognosis, but also participates in articulating local and global strategies for controlling and adapting to climate change. The research is integrated into social transformation while it is being carried out, even though its conclusive end results are still out of sight. This merger of research and innovation seems to become a decisive characteristic of the so-called knowledge society. Interdisciplinary projects play a leading role in it.

The example of research into the effects of climate change furthermore demonstrates the relevance of developing methods for integrating core disciplinary knowledge. The method of choice here is integration within simulation models. The interdisciplinary goal is fitting to the singular case of earth's climatic dynamics into the most widely accepted simulation model. The process of model development mirrors the tension between the nomothetic (core disciplinary knowledge) and the idiographic (unique features of our current climate). The optimization of the model's fit resulted from integrating research results from rather heterogeneous disciplines and from an almost arbitrary model tweaking with the help of empirically non-interpretable factors. All this happens for the purpose of keeping the individual case in focus. Because of its complexity and contingency the currently accepted model could probably no longer be programmed by a single researcher. Its prognostic power can no longer be identified as stemming from certain disciplinary explanations. The unique dynamics of the individual case has been translated into the unique dynamics of the model (Lenhard *et al.* 2007).

The example of the giant project of research into the effects of climate change is used here because it explicitly brings together the case-study aspect and the disciplinary knowledge core aspect of interdisciplinary research. More than that, given its public status, it is of paradigmatic importance for the interfacing of societal innovation with scientific research. Interdisciplinary projects of all magnitudes exhibit a similar structure: Research results are expected to provide expert knowledge for case analysis and problem solving. The know-how of the researchers consists in establishing networks of individual cases, within which their similarities and differences are worked out. (This applies within climate research to comparative cases from geological history.) Their resources consist of core disciplinary knowledge. Their instruments are models, which fit together substantiated core knowledge from various disciplines and elaborate scenarios for structuring new cases.

References

Becker, E. and Jahn, T. (eds) (2006). *Soziale Ökologie: Grundzüge einer Wissenschaft von den gesellschaftlichen Naturverhältnissen.* Frankfurt am Main/New York: Campus.

Campbell, D. (1969). Ethnocentrism of disciplines and the fish-scale model of omniscience. In: M. Sherif and C. Sherif (eds) *Interdisciplinary relationships in the social sciences*, pp. 328–48. Chicago: Aldine Publishing Company.

Cartwright, N. (1984). *How the laws of physics lie.* Oxford: Oxford University Press.

Cartwright, N. (1999). *The dappled world.* Cambridge: Cambridge University Press.

Earman, J., Glymour, C., and Mitchell, S. (eds) (2002). *Ceteris paribus laws.* Dordrecht: Kluwer.

Embodied Communication in Humans and Machines Research Group (2007). Embodied communication II: an integrated perspective. In: *ZiF-Mitteilungen*, pp. 15–20. Bielefeld: Bielefeld University Center of Interdisciplinary Research.

Eucken, W. (1965). *Die Grundlagen der Nationalökonomie.* Berlin: Springer.

Flyvbjerg, B. (2006). Five misunderstandings about case-study research. *Qualitative inquiry* **12**(2), 219–45.

Frodeman, R. (2003). *Geo-logic: breaking ground between philosophy and the earth sciences.* Albany: SUNY Press.

Garvin, D. (2003). Making the case: professional education for the world of practice. *Harvard Magazine* **106**(1), 56–65, 107.

Giere, R. (1999). *Science without laws.* Chicago: University of Chicago Press.

Gross, M., Hoffmann-Riem, H., and Krohn, W. (2005). *Realexperimente: Ökologische Gestaltungsprozesse in der Wwissensgesellschaft*. Bielefeld: Transcript.

Harvard Business School (2008). *The case method*. Available at: <http://www.hbs.edu/case/hbs-case.html> (accessed 29 February 2008).

Krohn, W. (ed.) (2006). *Ästhetik in der Wissenschaft*. Hamburg: Meiner.

Lenhard, J., Küppers, G., and Shinn, T. (eds) (2007). *Simulation: pragmatic constructions of reality*. Heidelberg: Springer.

Mitchell, S. (2002). *Ceteris paribus* – an inadequate representation of biology. In: J. Earman, C. Glymour, and S. Mitchell (eds) *Ceteris paribus laws*, pp. 53–74. Dordrecht: Kluwer.

Moran, J. (2002). *Interdisciplinarity*. London: Routledge.

Rheinberger, H. (2006). *Epistemologie des Konkreten: Studien zur Geschichte der modernen Biologie*. Frankfurt: Suhrkamp.

Rickert, H. (1924). *Die Probleme der Geschichtsphilosophie*. Heidelberg: Winter.

Saint, S., Drazen, J.M., and Solomon, C.G. (2006). *New England Journal of Medicine: clinical problem solving*. New York: McGraw-Hill Professional.

Stanford Law School. (2008). *Frequently asked questions*. Available at: <http://www.law.stanford.edu/publications/casestudies/faq/> (accessed 28 February 2008).

Weber, M. (1922). Die »Objektivität« sozialwissenschaftlicher und sozialpolitischer Erkenntnis. In M. Weber (ed.) *Gesammelte Aufsätze zur Wissenschaftslehre*, pp. 146–214. Tübingen: J.C.B. Mohr (Paul Siebeck).

Windelband, W. (1907). Geschichte und Naturwissenschaft. In: *Präludien. Aufsätze und Reden zur Einleitung in die Philosophie*. Tübingen: Mohr.

CHAPTER 4

Deviant interdisciplinarity

STEVE FULLER

'Deviant interdisciplinarity' refers to a set of interdisciplinary projects that aim to recover a lost sense of intellectual unity, typically by advancing a heterodox sense of intellectual history that questions the soundness of our normal understanding of how the disciplines have come to be as they are. After contrasting the normal 'Whig' sense of intellectual history with the deviant's 'Tory' sense, I briefly present six versions of deviant interdisciplinarity that have been prominent in the twentieth century, each of which has enjoyed a hybrid existence both inside and outside the academy.

Deviant interdisciplinarity has been historically fueled by an imaginative use of mathematics to bring together seemingly different fields of knowledge under a common set of principles. In the early modern period, physics and theodicy represented two forms of deviant interdisciplinarity, the former becoming arguably the most successful of all normal disciplines, the latter consigned to a more shadowy existence in theology, philosophy, and the social sciences. Over the past 200 years, thermodynamics has spawned several forms of deviant interdisciplinarity, especially once its conservation and entropy principles were seen as equally applicable to energy and information. Molecular biology and general systems theory were two of its twentieth-century offspring, the former becoming much more mainstream than the latter. The chapter ends with a brief reflection on postmodernism as the ideology of deviant interdisciplinarity.

4.1 The parallax view: deviant interdisciplinarity as Tory intellectual history

The Slovenian philosopher and cultural critic Slavoj Žižek (2006) recently published a book entitled *The parallax view*, a reference to the 1974 Alan J. Pakula film that opens with the assassination of a US presidential candidate, apparently by one gunman, yet a second gunman flees the crime scene unnoticed. The film then follows a reporter's attempt to resolve the situation, by analogy with how our minds reconcile the somewhat different sensory inputs registered by our two eyes, aka the parallax view.

Deviant interdisciplinarity presupposes a 'parallax view' of intellectual history, whereby the normal account by which disciplines develop and give rise to interdisciplinary inquiry is taken to be only part of the whole story. There is at least one other side, which reflects a different sense of how things came to be as they are and how they might turn out to be in the future. A subgenre of science fiction, 'alternate history', is dedicated to this prospect, though in the form of 'counterfactual history' it has become a mainstay of economic and military history and, increasingly, history of science (Hellekson 2001; Fuller 2008b). But how do 'normal' and 'deviant' accounts of events differ—that is, accounts that respectively ignore and recognize the presence of parallax? There is precedent for this question in historiography more generally.

Herbert Butterfield (1965) famously used the word 'Whig' to describe 'normal' accounts of history. The Whigs were the eighteenth-century English political party that successfully championed the cause of Parliament over the King and recounted its victory as the latest phase in the long slow progress of liberty over tyranny in the governance of human affairs. 'Whig histories' are thus told from the standpoint of history's winners. They reveal the good reasons why things have happened the way they have, in the order they did, to reach the ends they have. Whatever does not fit into this narrative structure is either ignored or treated as an error of an epistemic and/or a moral kind that in the long term is ultimately put right.

However, in the midst of Whig historiography's polite silences and expressions of righteous indignation *vis-à-vis* the alleged enemies of progress, one can glimpse alternative historical narratives. In these, what the Whig finds incidental becomes central and what would otherwise seem bad and wrong now appear better and smarter, perhaps even culminating in history's winners and losers reversing their fates. These counterfactual scenarios are the stuff of 'Tory' historiography, named for the King's defenders, who lost the struggle with the Whigs but maintain that their defeat was by no means inevitable and that in the future their party may be reinstated (Fuller 2003, Ch. 9). Behind every form of deviant interdisciplinarity is a version of Tory historiography.

But how exactly does this contrast in attitudes to history bear on how 'normals' and 'deviants' conceptualize interdisciplinarity? I shall consider each in turn.

A convenient example of the normal view of interdisciplinarity is provided by the distinction between 'Mode 1' and 'Mode 2' knowledge production that has become canonical in European science policy circles, courtesy of Gibbons *et al.* (1994). This distinction presupposes a model of intellectual history whereby increasingly specialized disciplines are seen as the natural outgrowth of the knowledge production process, which is envisaged as a kind of organism that develops functionally differentiated parts (aka disciplines and subdisciplines) over time as its investigations become more deeply embedded in their fields of inquiry. Kuhn (1970) has been the most influential theorist of this 'Mode 1' side of the process. It is an internally directed—'supply driven', if you will—account of knowledge production.

Interdisciplinarity enters as 'Mode 2', a complementary 'demand-driven' process. It attempts to bridge the epistemic gaps that have emerged between the disciplines as a result of their increasing specialization. These gaps are collective blindspots, by-products of an otherwise well-ordered Mode 1 process. They emerge because disciplinary practitioners pursue the implications of what Kuhn called their 'paradigm' for a given field of inquiry until it creates more problems than it solves. Such an intense focus, perhaps inherent in the very idea of discipline, can easily leave neglected issues that cut across disciplinary boundaries; hence, the kind of interdisciplinarity associated with Mode 2 knowledge production is often called 'transdisciplinary' and include topics that are at least initially defined in categories of broader social relevance than normally found in academia. Nevertheless, presumed in this narrative is that the complementarity between disciplinary depth and interdisciplinary breadth is appropriate and even to be expected, since interdisciplinary matters are best addressed by those who can mobilize a range of highly developed expertise. In that context, the only controversial feature of Gibbons *et al.* (1994) is its suggestion that the balance of science policy resources should now be shifted from Mode 1 to Mode 2 knowledge production.

The historical narratives associated with deviant disciplinarity and interdisciplinarity differ radically from their normal counterparts. What the normal disciplinarian sees as progress in the technical mastery of an originally unruly domain of knowledge, the deviant regards as an increasingly unreflective adherence to one from among several different paths of inquiry that could have been taken—and perhaps may be taken again in the future. From the deviant's standpoint, it is not merely that normal disciplinarians knows more and more about less and less but their very narrowness of vision distorts what they purport to see. It follows that the deviant tends to treat the very presence of different disciplines as *prima facie* pathological, rather like neuroses, which Freud treated as mere coping mechanisms for a reality we cannot fully manage in its entirety. The deviant, for whom reality is 'always already' interdisciplinary, is a bit more optimistic than Freud. But how does one write history that way?

One exemplar, now neglected but highly regarded in the first quarter of the twentieth century, is Merz (1965), who gives a comprehensive intellectual history of Europe based on the idea that in the nineteenth century a cluster of competing and overlapping metaphysical world-pictures coalesced—in different ways, places, and times—into the disciplinary cartography that by World War I had come to define the structure of the modern university. Merz listed the relevant world-pictures as: the astronomical, the atomic, the statistical, the mechanical, the physical, the morphological, the genetic, the vitalistic, and the psychophysical. He seemed to think that a version of the nebular hypothesis for the universe's origin governed intellectual history, whereby initially molten states of ideation—i.e. the metaphysical world-pictures—disperse and cool down into solid disciplines that follow predictable trajectories.

What disciplines gain in stability of intellectual focus is lost in sheer energy in terms of the overall project of making sense of reality, typically because the 'molten' political and religious features of the original world-pictures are removed or sublimated in the transition from metaphysics to 'science'. Here it is worth noting that Merz, a German engineer living in England, conceived of 'science' as *Wissenschaft*, which is now best understood as any field of study that is constituted as a systematic mode of inquiry but which originally referred to the autonomy of critical-historical theology from pastoral theology in the modern German university system. In effect, academic theologians were only accountable to their fellow researchers for the findings, however much they might challenge the intuitions of the faith community. Divested of the need to regularly and directly engage with ordinary believers, theologians became increasingly liberal, and even secular, as they acquired the mindsets of academic neighbors in history, philosophy, and sociology (Collins 1998, Ch. 12).

4.2 Six twentieth-century visions of deviant interdisciplinarity

As an engineer, Merz was an expert in thermodynamics. He understood that the logic of the nebular metaphor implies that unless some new energy is injected into the knowledge system, the disciplines will simply wind themselves down over time. Indeed, writing before the full force of the relativity and quantum revolutions had been felt in physics, Merz shared the widespread *fin de siècle* view that academic disciplines had fallen into predictable patterns of what Kuhn (1970) called 'normal science,' the future of which would be devoted to technical refinement and application. Any relevant new energy into this system would amount to a deviant interdisciplinarity that revolutionizes knowledge production by recovering the original animating political and/or religious impulses that had unified fields of inquiry before they settled into discrete domains of inquiry. The twentieth century witnessed several such deviant interdisciplinarities, each of which struggled for legitimacy within the university but commanded a considerable following outside the academy.

4.2.1 Dialectical materialism

Dialectical materialism was the most ambitious and generally successful ideological project to unify the natural and social sciences in a single knowledge-and-power package, and helped to propel the Soviet Union into the top two nations for science in the third quarter of the twentieth century. Dialectical materialism was based on three dynamic principles, what Marx and especially Engels called 'laws of dialectics', which were designed to update German idealism by taking on board 'energetics', a nineteenth-century rival to the Newtonian worldview that by the end of the century had been reduced to statistical mechanics and is now the branch of physics known as thermodynamics. For Engels, this would-be merger of philosophy and physics provided a deep structural understanding of the engine of history. It was aligned to Hermann von Helmholtz's attempt to shift the paradigm case of a mechanical system from the celestial body to the human body, which signified a larger epistemic transformation in the conduct of science from idealized bodies moving in abstract space to humanly relevant material translations between forms of energy (Rabinbach 1990). Here Engels, German heir to a Manchester textile firm, took linguistic refuge in the equivocal meaning of the English word 'work' to cover both human labor (*Arbeit*) and mechanical force (*Kraft*), which in turn reflected the roots of thermodynamics in engineering.

4.2.2 Socialist feminism

The patriarchal character of mainstream economics and politics appears in their failure to cost-account women's domestic labor as both a factor of production and a form of social security—that is, in the current jargon, a 'minimizer of transaction costs'. Mothers instill values in children that inhibit their tendencies to disrupt the circulation of commodities required by the logic of capitalism, as well as provide a relatively stable home life for the male 'bread winner' to compensate for the volatility of the workplace. Were this work subject to proper accounting, the result would provide rational economic grounds for the massive redistribution of wealth and power from men to women, and possibly a reorientation of capitalist political structures in a more socialist direction (Walby 1990). Note, however, that this defense of feminism is based exclusively on the social position that women occupy in maintaining the capitalist mode of production. It is not tied to gender *per se* as a distinctive 'way of being'. In the future, the same role may be performed by some other group, an idea also present in the related 'standpoint' feminism, whose name alludes to György Lukács' original designation for the proletariat, which he understood in conventionally male factory-based terms (Harding 1986).

4.2.3 Racial hygiene

Racial hygiene, the intellectually most adventurous field of German biomedical science in the period from Bismarck to Hitler, aimed to place medicine on a sound scientific footing based on the evolutionary and ecological theories of Darwin and Haeckel. It refused to regard the indefinite extension of human life as an unmitigated good, which was seen as a residue of pre-scientific Christian thinking. Instead humans would now be treated literally as animals—that is, as populations that must remain within certain parameters to ensure global sustainability. Racial hygienists treated matters of breeding, culling, and migration as cutting across species boundaries to such an extent that 'disease' appeared as an anthropocentric way of referring to a key *modus operandi* of natural selection in maintaining ecological balance. Recall that prior to Hitler's rise to power, racial hygiene was a left-wing anti-imperialist movement dedicated to a naturalistic determination of the optimal size of welfare states—that is, ecologically sustainable populations, which were understood to be genetically homogeneous ones (Proctor 1988). A similar sensibility remains today amongst 'deep ecologists' who differ from the original racial hygienists in being less state-oriented and more sympathetic to 'bioliberal' policies of devolving responsibility to self-organizing tendencies in individuals that are themselves based on the emergent interdisciplinary fields of sociobiology and evolutionary psychology (Fuller 2006).

4.2.4 Neo-Thomism

Philipp Frank (1949), the logical positivist exiled at Harvard, regarded neo-Thomism as the most serious modern rival to dialectical materialism in unifying scientific theory and political practice by focusing on the perennial spiritual and secular needs of humanity, as originally formulated by the official philosopher of the Roman Catholic Church, Thomas Aquinas, and filtered through modern developments in the natural sciences, including Darwin's theory of evolution. This 'unification' rests on assigning the 'how' and the 'why' of natural law to science and theology, respectively. Neo-Thomism has drawn strength from both the mutual non-interference of secular and religious educational authorities in post-1870 Europe and a pervasive lingering cultural antipathy to academic specialization (aka value-free science). Its main twentieth-century standard-bearer has been the born-again Catholic existentialist Jacques Maritain (1882–1973), father figure in the Christian Democratic movement in European politics who helped to draft the Universal Declaration of Human Rights. Neo-Thomism's ultimate selling point is its resolute adherence to natural law as a fixed normative standard in terms of which any scientific or political innovation can be evaluated.

4.2.5 Psychoanalysis

Although psychoanalysis is often portrayed as an exotic hybrid, this interdisciplinary deviant shares a foundation in both philosophy and medicine common to the original schools of scientific psychology, which led to an interest in mapping psychosomatic relations—indeed with an eye to how the 'energetics' that so fascinated Marxists made its way to neurophysiology (Sulloway 1979). The key difference is that Freud failed to be appointed to a professorship, which rendered him incapable of developing his nascent science in an academically protected market. Instead he was forced to vie in a heterogeneous market for patients and readers. This explains the prize-winning literary turn of his writing, which like the best popular psychology today appeals in equal measures to perennial humanistic themes (e.g. Greek myths and Biblical tales) and clinical evidence but without relying on specifically academic jargon. The detachment from academic culture also made it easier for Freud to debunk the ideology of scientific progress as a delusion that psychoanalysis could dispel by providing a secular therapeutic replacement for the disappearing pastoral mission of the churches (Rieff 1966).

4.2.6 General semantics

In name general semantics was the brainchild of Alfred Korzybski, (1879–1950) Polish count and US intelligence officer in the two world wars, but in substance it is positivism as a popular rather than an academic movement. The founder of positivism, Auguste Comte, had already set the precedent. An academic outcast, Comte's main source of income was what we now call 'management training seminars' that were attended by many of Europe's academic and political elites. Korzybski followed suit, his most influential follower being S. I. Hayakawa (1939), author of *Language, truth and action*, a US best-seller that bears comparison with A. J. Ayer's (1936) contemporaneous UK best-seller, *Language, truth and logic*. Although both decried the tendency to reify language, Ayer traced the problem to wayward metaphysicians who then fool an impressionable public, whereas Hayakawa saw reification as a universal human liability that propagandists especially turn to their advantage. The former diagnosis led the logical positivists to a focus on purifying academic language, while the latter led the general semanticists to range over the more broadly binding concerns of therapy and legislation. The two strands perhaps reached a happy academic medium in I. A. Richards' (1930) 'practical criticism,' which aimed to reconceptualize the humanities as the cultivation of a discriminating response to literature.

4.3 Deviant interdisciplinarity as the recovery of lost unity: the cases of physics and theodicy

Tory leanings notwithstanding, the deviant interdisciplinarian is no mere trader in nostalgia. Intellectual history is full of attempts to reset the epistemic clock at zero by proposing new foundations for knowledge. These efforts are often taken to be indicative of what it means to be 'modern', but invariably they veer into deviance because they treat disciplinary differences as barriers—perhaps even deliberately posed—to some lost unified sense of knowledge and power. While Descartes' '*cogito ergo sum*' and Hobbes' *Leviathan* stand out for philosophers, more effective both in science and in society at large has been *Principia mathematica* by Isaac Newton, someone obsessed by the need to recover the closeness to God that Adam had renounced in the Garden of Eden (Harrison 2007).

Newton is the source of the still popular notion that physicists might get into the 'mind of God' by adopting the 'view from nowhere' (Nagel 1986). He represented an especially extreme but fecund version of Protestantism's 'back to basics' approach to Christianity, especially the idea that lay Christians could make sense of the Bible for themselves because of its 'univocal' mode of expression. In particular, divine attributes like omniscience and omnipotence were to be understood as not merely analogous to human knowledge and power but as infinite extensions of them. This semantic shift opened the conceptual space for measuring the distance between human and divine knowledge, the theological basis for modern notions of scientific progress (Funkenstein 1986; Harrison 1998).

For Protestants, ideas of 'univocality', 'unity', and 'union' refer to the sense that everything reflects a common design, itself the result of an intelligence in whose image and likeness the human mind was created, and from which our minds differ only by degree not kind. Thus, it was common among the Protestant and free-thinking Catholics like Galileo who ushered in the seventeenth-century Scientific Revolution to interpret the apparent diversity of nature as masking a more fundamental unity, the fathoming of which would ultimately enable us (in some material and/or spiritual form) to reunite with God. They read the official Catholic line, which interpreted diverse knowledge domains as indicative of our invariably partial and hence merely analogical understanding of ultimate reality, as a conspiracy of priests and experts against an increasingly literate public capable of taking the Bible into their own hands and judging epistemic matters for themselves (Eisenstein 1979).

Mathematics has played a special role in the modern promotion of deviant interdisciplinarity, namely, as a vehicle for identifying and modeling isomorphic patterns suggestive of a common underlying reality that goes beyond mere analogy. Whereas followers of Aristotle, including the Catholic Church, restricted the epistemic range of mathematics to domains deemed inherently measurable or countable, those following Plato, whose fortunes were revived in the Scientific Revolution, detected ontological depth in math beyond its surface instrumental value. Here it is worth recalling that prior to Newton's *Principia mathematica*, the Church routinely dismissed the sort of 'lateral' or 'associative' mode of reasoning fostered by a Platonist approach to mathematics as the stuff of dreams and literature but *not* science. That royal astrologers like Johannes Kepler led the drive to unify the materially diverse modes of divine agency under a set of mathematical relations only served to underscore the heretical, if not psychologically deranged, character of the enterprise (Koestler 1959).

Nevertheless, all this changed with the introduction of standardized mathematical notation for discussing physical reality, an outgrowth of the Newtonian revolution (Cohen 1980). Newton's vision of a unified theory of matter and motion encompassing the entire physical universe was driven by the prospect that nature would display the same mathematical structure at all levels of reality. The apparent differences between disciplines would thus disappear. Perhaps the most convincing demonstration of this point came in the 1780s with the discovery of Coulomb's law, whereby the electrical force between two charged bodies was shown to be governed by a principle homologous to Newton's law of universal gravitation, which had been grounded in the character of the planetary orbits around the sun. Thereafter physicists began to understand their task as unifying all the forces of nature under a single set of laws, what continues to be portrayed popularly as 'GUTs' (grand unified theories; Barrow 1991). In this respect, an important source of deviant interdisciplinarity in the modern era has come to be domesticated within the standard operating procedures of a single discipline.

However, not all of the original deviance has been so neatly sublimated. At most physics addresses only half of what is entailed by a literal interpretation of the Biblical claim that we are created in the image and likeness of God. Physics may address our capacity to understand *how* the divine plan works as it does, but not *why*. This latter task was taken up in the seventeenth century by what may be the most persistently deviant form of interdisciplinarity in the modern era: *theodicy*. Nowadays theodicy is a boutique topic in philosophical theology concerned with explaining the existence of evil and suffering in a divinely sanctioned world: If God is so great and good, why is the world so miserable? However, theodicy originally stood for a much more general inquiry into nature as the expression of intelligent design, operating on the assumption that however imperfect the world may appear, it is—as Leibniz notoriously put it—'the best of all possible worlds' (Nadler 2008).

Nowadays we would say that Leibniz's robust formulation of theodicy envisages God as the ultimate 'optimizer', that is, the being who achieves the best overall outcome while pursuing multiple goals in a materially limited, if not resistant, environment (Fuller 2008a, Ch. 5). It suggests that as we shift our focus on the divine plan from 'how' to 'why', God looks more like an engineer than a physicist—i.e. one who cuts corners when necessary to get what he wants. This may explain why theodicy has proved so unpopular with theologians. From a pastoral standpoint, God would appear to act by an 'ends justifies the means' principle that provides cold comfort for the suffering believer. But even from an academic standpoint, theodicy presupposes that God struggles against a resistant matter. Although the Biblical idea that Creation took 6 days—forget about 6 billion years!—would already seem to concede the point, most theologians have been reluctant to distinguish God's will from his intellect as openly as theodicy demands, given its implication that God was *forced* to act to turn an otherwise morally indifferent nature into a wholly good work. Unlike the Platonic God, who operates much more like a physicist, the Abrahamic God could not simply think things into existence: He had to make them fit to purpose. The version of deviant interdisciplinarity that nowadays most adequately captures the original comprehensiveness of theodicy strays far from theology to what the social science polymath Herbert Simon (1977) christened 'sciences of the artificial'.

While this is not the place to explore everything that has helped to diminish theodicy's significance in the 300 years separating Leibniz and Simon, two key moments are relevant to making sense of how this especially ambitious and explosive version of deviant interdisciplinarity came to be defused:

(1) The success of Newtonian mechanics, which is founded on a principle of *inertia* that marked a clear break from its precursor concept, *conatus*. The difference was subtle but significant: Both concepts refer to the default motion of physical bodies, but conatus implies that God at least imparted, if not actively maintains, the capacity of bodies for motion. However, Newton showed that a comprehensive understanding of physical reality was possible without introducing any specific conception of the deity's causal relevance, let alone motivation (Blumenberg 1987). While conatus remains a live concept among those who believe that contemporary events always bear the marks of their origins, most notably in the social sciences (Fuller 2008c), the overall effect of the ascendancy of inertia over conatus has been to divorce the study of divine means from divine ends, which in turn has rendered physics conceptually independent of theology (Proctor 1991, Part 1). The visceral unease that philosophers and other humanists expressed towards the ascendancy of physics as the foundational science in the nineteenth and twentieth centuries arguably reflects a residual yearning for a discipline—or interdisciplinary field—that would address in secular guise the fundamental value questions posed by theodicy.

(2) Many of the defining arguments and tropes of theodicy migrated into what became the foundation of modern ethics and value theory, such as the utilitarian maxim of 'the greatest the good for the greatest number', which was originally proposed as a formula for calculating divine providence *vis-à-vis* the salvation of souls rather than the welfare of bodies (Schneewind 1997, Ch. 22). Either as a branch of theology in the eighteenth century or an adjunct to political economy in the nineteenth century, theodicy aspired to apply mathematics much more thoroughly than simply to represent and measure reality, as physics managed to do: it aimed to put math to use as a tool for constructing and calculating reality, very much in the spirit of the social engineers who from the eighteenth to the twentieth centuries have most obviously followed up its leads (Passmore 1970, Ch. 10, 11).

4.4 Thermodynamics as a source of deviant interdisciplinarity in the recent period

From the nineteenth century onward, ambitious forms of deviant interdisciplinarity have been inspired by the mathematical formulation of thermodynamic principles, especially the conservation of energy and the tendency to entropy. I have already alluded to the role of energetics in the formulation of dialectical materialism and psychoanalysis. Notwithstanding their speculative application to historical forces (Marx) and nerve impulses

(Freud), thermodynamic principles have been generally seen as equally relevant to the mechanical expenditure of physical force in a steam engine, the application of human labor in manufactures, and the transaction of messages across a telegraph line. In all these cases, energy was understood as always exacting a cost to achieve a benefit, which could be quantified in the same equations. On this basis, a host of Victorian natural philosophers such as William Whewell (who coined the word 'scientist' to name a profession), James Clerk Maxwell, and Lord Kelvin saw themselves as simultaneously contributing to physics, economics, epistemology, and even theology (Smith 1998). However, their uneven reputations in these areas today testify to the deviant nature of their interdisciplinarity. They are all suspected of having overextended the metaphor of 'work' in various respects.

Nevertheless, thermodynamic principles were extended still further into information and communication in the twentieth century, especially once Claude Shannon and Warren Weaver mathematically generalized the entropy principle to capture uncertainty in signal detection over a noise-filled medium. By the 1950s, the Shannon–Weaver formulation had come to function as a *lingua franca* for 'molecular biology', a newly emerging field largely populated by recent cross-disciplinary migrants from physics and chemistry like J. D. Bernal, Max Delbrück, Linus Pauling, and Francis Crick. It resulted in the still popular idea of DNA as constituting the 'genetic code' whose information is transmitted with minor errors (i.e. mutations interpreted as biological noise) over generations of organisms (Kay 2000).

Now, more than half a century after the discovery of DNA's role in the storage of genetic information, molecular biology is dominated by a 'bioinformatic' worldview that understands its work to be that of testing all possible combinations of 'letters in the genetic code' (i.e. the nucleotides that constitute DNA) for their biological significance. At first this rather literal sense of 'code cracking' was criticized for reducing biological research to mechanical routines (Gilbert 1991). However, among those who successfully sequenced the human genome at the close of the twentieth century, Craig Venter clearly saw the commercial potential of such 'routines' for the biomedical industries.

Nevertheless, the jury is out on the long-term viability of bioinformatics as a form of deviant interdisciplinarity, especially in terms of its ability to accommodate evolutionary biology, a topic that both information theorists and molecular biologists have generally avoided until recent times (Morange 1998). Here one finds at least three versions of interdisciplinarity, each 'deviant' in its marginalization of Darwinian natural selection, the currently dominant evolutionary mechanism. In each case, evolution is identified primarily with a process other than natural selection:

(1) The identification of evolution with entropy, which would explain the pervasiveness of species extinction and evolution's apparent irreversibility (Brooks and Wiley 1986).

(2) The identification of evolution with self-organized complexity, what Kauffman (1995) calls 'order for free,' which builds on pre-Darwinian Epicurean ideas that all species—old and new—consist of recycled genetic material.

(3) The identification of evolution with the unfolding of an intelligent design in nature that may be detected in terms of the engineering properties of genes as vehicles for the storage, transmission, and retrieval of information (Meyer 2009).

While the deviant position of bioinformatics *vis-à-vis* modern evolutionary theory is still a work in progress, one other interdisciplinary application of thermodynamics deserves mention as at once more ambitious, at one time more influential but nowadays more marginal.

General systems theory (or GST) may be the most ambitious form of deviant interdisciplinarity in recent times. It was the brainchild of Ludwig von Bertalanffy (1901–72), a Viennese-trained theoretical biologist who spent most of his academic career in North America. Bertalanffy first attracted attention in the late 1920s with his discussion of 'open systems', namely, physical systems that, strictly speaking, are not closed under ordinary conservation principles because they regularly receive new energy/information from outside their borders. Bertalanffy had been inspired by nineteenth-century *Naturphilosophie*, a version of natural philosophy associated with German idealism that treated nature as animate in its own right and not simply something defined negatively by its resistance to the knowledge and power of humans. This perspective, traceable to Hegel's Berlin rival F. W. J. Schelling, acquired scientific credibility through Schelling's student Gustav Fechner, a founder of experimental psychology and the subject of Bertalanffy's PhD dissertation (Heidelberger 2004). Bertalanffy's key insight was that Fechner's ideas about the degrees of consciousness in nature could be interpreted in terms of levels of organization in systems. In effect, Bertalanffy gave mathematical expression to concepts that had been previously treated as purely 'qualitative', if not downright subjective.

In most general terms, Bertalanffy saw GST as decisively handling the problem of entropy, the inevitable disintegration of order in a classical physical system that maintained itself only by recycling its own energy. He called such a system 'closed'. Much of the nineteenth-century scientific and literary imagination was dedicated to conquering entropy, ideally by the invention of a 'perpetual motion' machine. However, repeated theoretical and practical failures led to resigned expectations of the 'heat-death' of a universe that in the long term would simply run itself down (Smith 1998). Nevertheless, dreams of transcending our current energy/information limitations remain in the continued talk of 'open systems' in the economics of innovation and so-called critical versions of realism that promise science access to ontologies of potentially limitless depth (Bhaskar 1975).

More specifically, Bertalanffy also wanted to explain the phenomenon of 'equifinality' that had been demonstrated by the embryologist and natural philosopher Hans Driesch (Cassirer 1950, Ch. 11). Equifinality is the tendency of individual organisms to overcome any of a variety of environmental obstacles—including those that leave their bodies radically transformed—in order to reach the developmental goals of their species. Bertalanffy did not want to explain this phenomenon by postulating some physically ambiguous vital force, or 'entelechy', as Driesch had. Instead, Bertalanffy called for a more abstract characterization of the organism as a system, whose proper parts consist of functions that may be performed by a variety of replaceable physical units, the identity and efficacy of which are determined by the organism's energy/information transactions with its environment. Thus, an organism 'lives' by virtue of its principled exchanges with something outside itself, as opposed to its being driven by some mysterious internal principle (Bertalanffy 1932).

A sense of Bertalanffy's short-term influence may be found in the volume devoted to theory construction in the *International encyclopedia of unified science*, the principal collective logical positivist project in North America (Woodger 1947). GST itself consolidated as a deviant interdisciplinary movement in 1954, when Bertalanffy—along with the economist Kenneth Boulding, the game theorist Anatol Rapaport, and the neuroscientist Ralph Gerard—were among the first fellows at the newly established Stanford Center for Advanced Study in the Behavioral Sciences (Hammond 2003). Soon thereafter most of them found a common academic home at the University of Michigan, courtesy of James Grier Miller, head of the Institute of Mental Health, following a path that sometimes intersected with the general semantics movement. By the 1960s, GST had joined with Norbert Wiener's cybernetics in the promotion of a 'science of systems', in which 'organization' is proposed as an operational definition of life that can be applied across all levels of reality from the simplest to most complex system, regardless of its material composition—which is to say, it could be applied equally to carbon- and silicon-based organisms, or for that matter intermediate 'cyborg' entities (Buckley 1968; Heims 1991).

GST's Achilles' heel has been its failure to reach agreement on two fundamental aspects of its overall interdisciplinary vision: first, the exact nature of the energy/information exchanges across systems that reverse entropy, other than that they can be mathematically redescribed in terms of the laws of thermodynamics; second, the exact logic that underwrites the relationship between 'levels of organization' in a given system beyond the mere stacking of analogies. While advocates of GST (notably Ervin Laszlo) continue to tolerate these conceptual ambiguities for the purpose of pronouncing on both global and personal matters, the balance of the movement's influence has shifted from academia to the more speculative reaches of science policy, futurology, and the New Age movement. As for Bertalanffy's reframing of the 'problem of life' in terms of the emergence of open systems, it has suffered a threefold fate that is characteristic of deviant interdisciplinarity:

(1) The reframing was simply ignored by experimental researchers into the 'origin of life', who abandoned the systems-led perspective to focus instead on identifying the specific physico-chemical conditions needed for the emergence of a certain kind of self-reproducing molecule. In this bottom-up view, 'system' is treated as a by-product of the emergence of complexity (Kauffman 1995).

(2) The reframing was actively misunderstood by biologists, who in practice adopted a compatible, if not similar, position but mistook GST for its opposite, vitalism. This explains the failure of molecular biologists, especially reductionists like Jacques Monod, to see systems theorists as kindred spirits (Rosen 2000, Ch. 1). Monod treated the process by which gene expression is switched on and off as a 'mechanism' whose historic emergence he explained in terms of natural selection, whereas GST aimed to explain what made gene expression a mechanism in the first place, namely, its being part of a larger living system.

(3) The idea of an open system became scientifically domesticated by a narrowed definition that avoided controversies like those in (2) above. Open systems are now routinely presumed without being theorized. One grants that systems whose energy/information does not completely dissipate over time must be replenished from the

outside, but from where and by what means is left open. Thus, one can methodologically remain indifferent to whether the design features of systems arose by intelligent or unintelligent means.

4.5 Conclusion: postmodernism as the ideology of deviant interdisciplinarity

Jean-François Lyotard's (1983) *The postmodern condition* is reasonably read as a deconstruction of the Whig disciplinary narratives that dominate academia's self-understanding. Lyotard shows that the most innovative and influential developments in knowledge in the twentieth century have largely come from deviant interdisciplinary formations—including both molecular biology and GST—that had at least one foot outside the academy. Much of what has been subsequently described by both friends (e.g. De Landa 1997) and foes (e.g. Sokal 2008) as 'postmodern science' fits very much the pattern of deviant interdisciplinarity described in this chapter, not least the rather imaginative extension of mathematical modes of reasoning to seemingly unrelated domains of reality.

Just as the Church anathematized first heretics and later Protestants who believed that they could take the Bible into their own hands in order to reorient themselves to God and nature, something similar has repeatedly occurred in the modern period, as academic and, more specifically, scientific authorities have censured deviant interdisciplinarians for their heterodox appropriations and extensions of knowledge derived from several disciplines. In this context, postmodernism stands very clearly for the systematic reinterpretation of undisciplined errors as the realization of otherwise lost potential for thought. Only a linear view of intellectual history would immediately damn errors simply because they reflect a lingering attachment to an earlier mindset. What postmodernism promises—and the various historic forms of deviant interdisciplinarity have realized—is ample food for sustained thought, if one is willing to recover and pursue 'paths not taken'. But at the very least this would involve a commitment to a branching view of time, if not parallel possible worlds, perhaps blurring the boundary between epistemology and science-fictional accounts of 'alternate histories' (Hellekson 2001).

References

Ayer, A.J. (1936). *Language, truth and logic*. London: Victor Gollancz.

Barrow, J.D. (1991). *Theories of everything*. Oxford: Clarendon Press.

von Bertalanffy, L. (1932). *The problems of life: an evaluation of modern scientific and biological thought*. London: Watts.

Bhaskar, R. (1975). *A realist theory of science*. Brighton: Harvester.

Blumenberg, H. (1987). *The genesis of the Copernican world*. Cambridge, MA: MIT Press.

Brooks, D.R. and Wiley, E.O. (1986). *Evolution as entropy: toward a unified theory of biology*. Chicago: University of Chicago Press.

Buckley, W. (ed.) (1968). *Modern systems research for the behavioral scientist.* Chicago: Aldine.

Butterfield, H. (1965). *The Whig interpretation of history.* New York: W.W. Norton.

Cassirer, E. (1950). *The problem of knowledge: philosophy, science, and history since Hegel.* New Haven, CT: Yale University Press.

Cohen, I.B. (1980). *The Newtonian revolution.* Cambridge: Cambridge University Press.

Collins, R. (1998). *The sociology of philosophies.* Cambridge, MA: Harvard University Press.

De Landa, M. (1997). *A thousand years of non-linear history.* Cambridge, MA: Zone Books.

Eisenstein, E. (1979). *The printing press as an agent of change.* Cambridge: Cambridge University Press.

Frank, P. (1949). *Modern science and its philosophy.* Cambridge, MA: Harvard University Press.

Fuller, S. (2003). *Kuhn vs Popper: the struggle for the soul of science.* Cambridge: Icon Books.

Fuller, S. (2006). *The new sociological imagination.* London: Sage.

Fuller, S. (2007). *New frontiers in science and technology studies.* Cambridge: Polity.

Fuller, S. (2008a). *Dissent over descent: intelligent design's challenge to Darwinism.* Cambridge: Icon Books.

Fuller, S. (2008b). The normative turn: counterfactuals and the philosophical historiography of science. *Isis* **99**, 576–84.

Fuller, S. (2008c). Conatus. In: M. Grenfell (ed.) *Pierre Bourdieu: key concepts*, Ch. 10. Durham: Acumen.

Fuller, S. and Collier, J. (2004). *Philosophy, rhetoric, and the end of knowledge*, 2nd edn. Hillsdale, NJ: Lawrence Erlbaum Associates.

Funkenstein, A. (1986). *Theology and the scientific imagination from the middle ages to the seventeenth century.* Princeton, NJ: Princeton University Press.

Gibbons, M. *et al.* (1994). *The new production of knowledge.* London: Sage.

Gilbert, W. (1991). Towards a paradigm shift in biology. *Nature* **349**(10 January), 99.

Hammond, D. (2003). *The science of synthesis.* Boulder, CO: Colorado University Press.

Harding, S. (1986). *The science question in feminism.* Ithaca, NY: Cornell University Press.

Harrison, P. (1998). *The Bible, Protestantism, and the rise of natural science.* Cambridge: Cambridge University Press.

Harrison, P. (2007). *The fall of man and the foundation of science.* Cambridge: Cambridge University Press.

Hayakawa, S.I. (1939). *Language, thought and action.* New York: Harcourt, Brace, Jovanovich.

Heidelberger, M. (2004). *Nature from within: Gustav Fechner and his psychophysical worldview.* Pittsburgh, PA: University of Pittsburgh Press.

Heims, S. (1991). *Constructing a social science for postwar America: the cybernetics group 1946–1953.* Cambridge, MA: MIT Press.

Hellekson, K. (2001). *The alternate history.* Lawrence, KS: University of Kansas Press.

Kauffman, S. (1995). *At home in the universe: the search for laws of self-organization and complexity.* London: Viking.

Kay, L. (2000). *Who wrote the book of life? A history of the genetic code.* Palo Alto, CA: Stanford University Press.

Koestler, A. (1959). *The sleepwalkers*. London: Hutchinson.

Kuhn, T.S. (1970). *The structure of scientific revolutions*, 2nd edn. Chicago: University of Chicago Press.

Lyotard, J.-F. (1983). *The postmodern condition*. Minneapolis, MN: University of Minnesota Press.

Merz, J.T. (1965). *A history of European thought in the nineteenth century*, 4 vols. New York: Dover.

Meyer, S.C. (2009). *Signature in the cell: DNA and the evidence for intelligent design*. New York: Harper Collins.

Morange, M. (1998). *A history of molecular biology*. Cambridge, MA: Harvard University Press.

Nadler, S. (2008). *The best of all possible worlds*. New York: Farrar, Strauss and Giroux.

Nagel, T. (1986). *The view from nowhere*. Oxford: Oxford University Press.

Passmore, J. (1970). *The perfectibility of man*. London: Duckworth.

Proctor, R. (1988). *Racial hygiene*. Cambridge, MA: Harvard University Press.

Proctor, R. (1991). *Value-free science?* Cambridge, MA: Harvard University Press.

Rabinbach, A. (1990). *The human motor: energy, fatigue, and the origins of modernity*. New York: Harper Collins

Richards, I.A. (1930). *Practical criticism*. London: Kegan Paul.

Rieff, P. (1966). *The triumph of the therapeutic*. New York: Harper and Row.

Rosen, R. (2000). *Essays on life itself*. New York: Columbia University Press.

Schneewind, J. (1997). *The invention of autonomy*. Cambridge: Cambridge University Press.

Simon, H. (1977). *The sciences of the artificial*. Cambridge, MA: MIT Press.

Smith, C. (1998). *The science of energy*. Chicago: University of Chicago Press.

Sokal, A. (2008). *Beyond the hoax*. Oxford: Oxford University Press.

Sulloway, F. (1979). *Freud: biologist of the mind*. Cambridge, MA: Harvard University Press.

Walby, S. (1990). *Theorizing patriarchy*. Oxford: Blackwell.

Woodger, J.H. (1947). *The technique of theory construction*. Chicago: University of Chicago Press.

Žižek, S. (2006). *The parallax view*. Cambridge, MA: MIT Press.

CHAPTER 5

Against holism

DANIEL SAREWITZ

'Holism', as I will use the term, encompasses those various approaches to scientific inquiry that investigate complex systems—ecosystems, climate systems, economic systems, social systems—whose behavior cannot be understood by studying the individual system components in isolation. Holism from this perspective encompasses a wide range of methods aimed at characterizing complex system behavior, from mathematical models built up from first physical principles (such as the general circulation models used to study the climate) to neural network approaches, game theoretical and system-dynamics frameworks, stochastic, agent-based, and other modeling techniques, and so on. Holistic approaches to scientific inquiry are inherently interdisciplinary, and offer a response to the familiar limits of disciplinary science.

The need to transcend these limits seems almost too apparent to demand justification. Reality is not divided up along disciplinary lines; if we are to understand the world that we live in, we need to find ways to investigate and portray the world as it actually is, not as it is constructed from different, necessarily incomplete, and sometimes competing or even contradictory disciplinary perspectives.

Disciplinary views of the world distort as much as they reveal by artificially isolating and simplifying particular components of natural and social systems—components that happen to be amenable to precise measurement, mathematical description, or experimental replication—and then treating those edited versions as if they are discrete, puzzle- or clockwork-like pieces of a reconstructable whole. Disciplinarity is not absolutely synonymous with reductionism, but in general the disciplines support an inductive, reductionist view of understanding, where larger-scale insight is supposed to arise from the accumulation of facts and insights acquired through inquiry focused at smaller scales. Reductionist, disciplinary approaches to knowledge acquisition thus encourage mechanistic views of nature and society, views that treat the subjects of reductionist analysis as more significant than the interactions among such subjects; views that obscure, for example, the importance of emergent (non-reducible) phenomena in real-world systems, and that feed belief in the ability to exercise control on nature and society on the basis of knowledge about individual components of the whole. At the limit, however, reductionism may harbor holistic ambitions that go far beyond the scope of individual disciplines. An extreme version of this approach has been articulated by the biologist E. O. Wilson, who posited a 'consilience' of all knowledge that, building from the ground up, would construct a seamless web of causal relations to ultimately connect say, quarks to the creative acts of a painter. But as even Wilson (1998) admits, such ambitions are expressions of faith in reductionism that cannot be justified by the current state of scientific knowledge or method.

Intellectual inquiry—research—has many justifications and goals, but this chapter focuses on what is certainly the most important modern rationale for pursuing scientific knowledge about the world: the idea that such knowledge, when formalized, transmitted, and applied, can enhance the capacity of humans to act effectively in the world, to achieve the practical goals that they aim to achieve. From this perspective, the tangible value of inquiry for society is created along two main paths. The first is the embodiment of knowledge within technologies—devices and procedures that allow regularities in the behavior of natural phenomena to be exploited for the performance of particular tasks. The goal here is to increase the control that humans have over their environment and their creations, as in, for example, the use of a weakened or killed polio virus to induce immunity to polio, or the use of knowledge about aerodynamics to design a projectile or an aircraft. The second path along which knowledge is applied to yield social value is in the application of improved understanding about the world to human decisions. For example, knowledge about tomorrow's weather influences social planning at scales from the picnic to the air transport system. Of course these two paths may be related. Knowledge about the effectiveness of a polio vaccine stimulates governments to develop policies that encourage people to get vaccinated. As a general matter, reductionist and disciplinary approaches are most useful in pursuing the technological path, whereas the pursuit of holistic understanding is more often justified for its potential to support decision making.

5.1 Path number one

Technologies are reductionist; they are manifestations of knowledge-out-of-context. For example, projectiles exist in nature (meteorites, volcanic ejecta, and so on), but projectiles that go in predetermined ways to particular places at particular speeds (bullets, rockets, and so on) do so because the conditions of their production and use are precisely specified in highly controlled settings. The excruciating irony here is that the idea of natural science as the explicator of the natural is most powerfully declaimed through its capacity to do what is entirely unnatural. A more subtle, deeper irony is that for many, perhaps most, important areas of technology, the disciplinary scientific explanations come *later*: you start with a technological capacity achieved through tinkering (steam engines) or folk practices (smallpox variolation), and the promise of this new capacity stimulates the production of new, explanatory knowledge (thermodynamics and immunology, respectively) in an effort to improve performance. Indeed, to a very considerable extent the advance of disciplinary, reductionist inquiry about nature has been driven by the desire to improve technological performance outside of nature.

Disciplinary, reductionist science and its embodiment in technology are the most powerful sources of social transformation in the world today. Technologies are cause-and-effect machines that allow people to do things with often incredible degrees of control and reliability (for example, according to the US National Transportation Safety Board, there were 22 million scheduled US commercial air flights in 2007 and 2008, flying a total of about 16 billion miles—and zero fatal crashes). But we all know that the world made and continuously remade by technology is hardly a world under control. It is rather a world where the social, technological, and natural are increasingly indistinguishable, and the cause-and-effect chains that connect the use of a particular technology to a consequence are unpredictable beforehand, and extraordinarily difficult to fully comprehend, let alone manage, afterward. For example, super-reliable passenger jets enable not only rapid transport of humans and artifacts, but also rapid dissemination of disease vectors like the swine flu virus. Passenger jets also provided a convenient and potent weapon for the 9/11 terrorists whose actions provoked an inflection point in global politics. And jets' emissions contribute to the alteration of the earth's atmospheric chemistry.

Two things are happening. Humans are increasing the reliability with which they achieve their aims, by increasing their control over and manipulation of phenomena that formerly were invisible, and which often acted on society with impunity. Yet this very process of exercising control over particular natural phenomena also amplifies complexity, contingency, and surprise across virtually all dimensions of human endeavor, due to the continual introduction of novelty—of things that never existed before—into the world. Technologies are conceived and developed in the socially decontextualized setting of a laboratory but are released into a world of complex interconnectivity. This dilemma, or tension, is as reasonable a definition of modernity as one is likely to find, for it captures the connection between the reduced and whole worlds: the former, the venue of increased scientific knowledge and technological control, continually remaking the latter, the site of complexity, contingency, and surprise, where, in the realms of science, only the practitioners of holism dare to tread.

Another way to look at this tension is by revisiting the unfashionable notion of progress. Only a crank would claim that technological advance has not enabled huge advances in human material well-being, and there should be no embarrassment in describing such advance as 'progress'. Yet it takes a naiveté verging on the mystical to equate the ever-expanding domain of local control at the level of, say, individual human health, agricultural productivity, energy generation and distribution, transport, or information-processing power, with progress in managing the social and natural systems that contain these more parochial examples of technological effectiveness. The dictum 'think globally, act locally' is all well and good, but its implementation demands some capacity to understand the connections between the two levels—technological action and system transformation. Yet technological advance, catalyzed by reductionist inquiry, makes those connections continually more complex, less comprehensible. Indeed, when one considers the array of daunting challenges facing the world as I write these words—an emerging threat of a flu pandemic, the vortex of global economic decline, the rising specter of climate change, the geopolitical complexities of the Middle East, and so on—the inability of local actors to understand the cumulative and emergent consequences of their actions is revealed as a central cause of the problems to begin with. Notions of individual responsibility [if I cover my mouth when I sneeze; if I save more, spend less (or should it be the other way around?); if I drive less, or drive a more efficient vehicle, if I send a pilotless drone to attack a terrorist camp] dissolve into quaintness at best. Technology creates a logic for action at the local level that virtually no one manages to escape, a logic that cannot carry us far outside the context and consequences of our own local actions, even as it creates consequences that can reasonably be thought of as global.

5.2 Path number two

And so, we see decision makers in recent decades increasingly calling upon scientific research to improve the ability of humans to manage the techno-social complexity world that we are always, but unwittingly, creating. The idea is obvious: we will make better decisions if they are based upon factual understanding—knowledge of the world as it is, rather than as we wish it to be, or fear it might be. And when decisions are being made about difficult problems—managing the economy; dealing with immigration or drug use; fighting terrorism; countering global warming; regulating powerful new technologies—the knowledge that can help is necessarily holistic. We need knowledge about the system we are trying to manage (economic; geopolitical; socio-technical), about how all the important components interact, about how pushing on one part of the system leads to a consequence in another part of it. This type of knowledge is different in its essence from reductionist understanding. Reductionism strives for insight that is independent of its context in time or space (e.g. projectiles on earth always accelerate downward at the same rate; killed polio virus injected into a human induces an immune response regardless of where a person lives or what their standard of living might be). This context independence of science allows reductionist knowledge to be embodied in engineered artifacts that themselves behave predictably, regardless of context. Holistic scientific inquiry, in stark contrast, strives for insight that embraces and explains context and complexity, that enhances comprehension of human and natural systems—the very systems that are continually being rendered more complex and incomprehensible due to the technological fruits of reductionist inquiry.

Holism would seem to be everything that reductionism is not, and yet from a different perspective they are the same: a response to the appeal for more and better scientific understanding as a foundation for improving the effectiveness of human action in the world. And who could possibly argue with this call? But there is an empirical embarrassment here. Naïve technological optimism aside (everyone, by now, knows that the artifacts we create can often bite back in surprising ways), reductionist science and its technological embodiments do, in fact, often do what they claim (plus much more, of course, and not all of it desirable; but a vaccine protects you against polio, a jet gets you to your destination). Where, in contrast, can we find powerful evidence that holistic approaches to inquiry improve our ability to act, to make effective decisions? The body of formal knowledge on ecosystems, for example, is housed in something like 160,000 peer-reviewed scientific publications written between the years 1970 and 2005 (Neff and Corley 2009). We know so much more, and in limited cases have sometimes moved away from bad ecosystem management practices (monocropping; fire suppression)—or at least recognized that they are bad. But the return on this knowledge investment seems incredibly modest; one would be hard-pressed to demonstrate that humans are, on the whole, doing a better job stemming ecosystem decline than they were in 1970. Nor is there much reason (and perhaps there is none) to think that enormous and ever-expanding bodies of formal knowledge on other complex systems have translated into general improvement in the practice of such crucial human activities as international diplomacy, military intelligence, management of complex organizations and economies, or governance of democratic societies.

The only truly holistic representation of a system would be the system itself. Anything else, however mathematically or analytically sophisticated or clever, is incomplete and thus in some sense a lie. How good does a lie—or, more delicately put, an approximation—need to be to provide a useful guide to decision making? This might seem to be the key question here: perhaps a truly holistic view of a problem is impossible to achieve, but certainly the accumulating insights of a holistic research agenda—systems-level scientific understanding—must be a contribution not just to science but to the capacity of humans to manage their affairs? Literal holism may be impossible, but we can use successive and converging approximations, probabilistic distributions, and general insight about system dynamics to get better at making decisions about our complex world.

This sensible line of argument seems to fail on several counts. To begin with, we often cannot know if the right thing is being approximated; nor can we judge in advance the conditions under which the approximation will hold. Climate scientists have been given tens of billions of dollars of research support over a period of almost two decades to gain a holistic view of the coupled ocean–atmosphere system. The idea was supposed to be that scientific understanding of climate change would both motivate and enable the social action necessary to address the climate problem. Yet it has turned out that growing knowledge about the climate system—increasingly holistic knowledge, to be sure—is pretty much irrelevant to the problem of managing climate change. The reason is that human behavior is deeply embedded in complex, interdependent, non-deterministic socio-technical systems that no one knows how to alter in particular ways to yield particular outcomes (e.g. a reduction in greenhouse gas emissions). Similarly, for all the prowess that economists display in their complex systems models, these abilities could neither anticipate nor forestall the 2007 US mortgage meltdown and simultaneous global spike in energy and food prices that blossomed into global economic crisis. An important point here is that the tools for managing these crises are the standard incremental system tweaks—interest rate adjustments, government stimulus packages, institutional bail-outs, and so on. As another poignant example, consider that earth scientists have for decades known pretty much everything there is to know about the vulnerability of New Orleans (a complex urban–natural system) to a major hurricane (including the utter inevitability of such an event) but this knowledge had little if any bearing on the consequences of hurricane Katrina when it did occur, or on the response to the catastrophe.

A predictable rejoinder to these points is that holistic knowledge should not be held accountable for the failings of politics or bureaucracies. If political leaders lack the will to decisively confront climate change, if corporate executives behave irresponsibly, if bureaucrats fail to carry out their duties, why lay such failure at the door of holistic inquiry? Why indeed. But in that case, how whole is holistic? Why are political and social failings outside of the systems that constitute the whole? For the fact is that the complex systems of interest to humans are those that are affected by and affect humans. Humans, in turn, tend to hold diverse beliefs and values about how the world works and what the appropriate goals of action ought to be. Here the problem with approximate holism is that there is always room for disagreement, both about what the science means, and about which science is relevant. Is nuclear power a viable option for reducing carbon emissions? Well it depends (for example) on how you balance the benefits of reduced emissions with the risks from creating more nuclear waste. Are genetically modified foods too risky to be used? Well it depends on whether you are mostly worried about protecting the economic viability of small farmers, fighting the power of global corporations, preserving ecosystems, or increasing agricultural productivity.

There are no right answers to these sorts of questions. Humans bring diverse and often conflicting values to complex problems, and such value conflicts are only rarely amenable to resolution through factual arguments. Moreover, even if people agree on a value that they would like society to pursue, system complexity means that it often remains difficult to know what actions to take to actually advance toward that value. Diverse values, and uncertainty about cause–effect relations associated with decision options, allow questions to be framed in many ways, involving different systems, each of which merits a holistic approach. The resulting knowledge will in turn reflect those framings. Knowledge of gene flows in ecosystems may help assess the risks of genetically modified organism (GMOs) to ecosystem health; knowledge of agro-economic system dynamics in developing countries may help assess the potential contribution of GMOs to nourishing global population. These bodies of knowledge are not reconcilable through holistic synthesis because they describe different wholes whose boundaries reflect different human concerns, different problem definitions. There is, in other words, no whole to be 'istic' about; there are only subwholes that reflect various combinations of the (always imperfect) research tools available and the (often incommensurable) way the questions are framed. To make matters worse, complete knowledge even within a subwhole is impossible anyway, so there is always room for disagreement and conflicting interpretation on technical grounds as well.

The point here is not the oft-made one about the need to separate the science from the politics, but quite the contrary: when the system is complex—when it, basically, can be expanded to include everything—declaring that the science is separate from the politics is an arbitrary act of boundary-drawing. The decision about which approximation of the whole one seeks to construct—or about which construction one finds most compelling—itself reflects a preference that helps define both the scientific and political framing of the problem. Is the western spotted owl a symbol of the decline of old growth forests and the rapacious habits of the logging industry? Or is it a symbol of how environmental regulations stress and transform local economies? Yes. Plenty of approximately holistic knowledge can be mobilized to support either position—and plenty of uncertainty can undermine each one, as well.

5.3 Now consider air transportation

Let us come at this from a different angle. Unlike, say, managing environmental problems, or the long-term behavior of national economies, there are some apparently difficult things that society actually manages to do quite well, with clear evidence of improvement over time. Consider, for example, the rather chaotic air transport system in the United States, which on many levels is associated with the same sorts of dysfunction, inequity, and resistance to effective reform that we see in other complex socio-technical systems. Most passengers are dissatisfied with service, and most airline corporations are continually flirting with fiscal disaster. Yet the core function of the system, to move passengers safely from one location to another, is carried out with a quite incredible consistency and safety, as I've already noted. Another crucial attribute of air travel is that the number of passengers and flights has continued to rise without sacrificing overall safety. So the system somehow maintains, and even improves, its core reliability even as it becomes more complex, and even as other system functions (flight delays; lost baggage; food quality) may worsen (La Porte 1988).

To a considerable extent this reliability is embedded in a technology—the jet airplane—and is thus a beneficiary of technological reductionism. And technology plays a supporting role in many other ways, for example through successful weather monitoring and forecasting, air traffic monitoring and warning systems, and the increasing automation of the actual flying of the aircraft. Nevertheless, fallible and fractious humans remain a central component of the air transport system.

Many other 'systems' are suffused with technological capacities but do not achieve anything like the reliability of air transport, or its ability to maintain core performance with increased complexity. What, then, besides the reliability of the airplane itself, distinguishes air transport from human-technical systems that perform less well? Three other attributes are crucial. First, when it comes to flying in airplanes, people's primary values and interests are pretty unitary: the goal is to arrive at one's destination safely. (Rare exceptions to this rule, e.g. terrorists, can introduce major perturbations that the system moves quickly to counteract.) Second, the metrics of success are both obvious and easy to measure. Success means that planes don't crash; failure means they do. Negative feedback, and the learning it provokes, is thus strong and clear. Third, the consequences of failure are dire. Even 10 fatal crashes a year—a one-in-a-million safety record—would likely be deemed unacceptable by the public. This means that incentives to maintain excellent performance are high.

All four of these conditions are essential to the successful management of the air transport system. The whole is radically constricted not just in terms of the techno-scientific but also the political. Other permutations of these variables (e.g. strong technical capability but divergent values, as with the expanded use of nuclear power; or generally convergent high-level values but weak technical capability, as with the need to improve public education) expand the realm of the whole—that is, the definition of the system that is relevant to its performance—enormously. Yet it is also the case that the air transport system is a complex whole that does not fit into any disciplinary framework, and cannot be fully comprehended by any individual. The types of expertise that can design super-reliable jet engines overlaps little if at all with the expertise that designs cockpit instrumentation, provides weather forecasts, guides air traffic in a crowded urban sky, or pieces together the shards of an aircraft after a crash.

The reliability of air transport, that is, does not emerge from holistic theory that can guide long-term planning, but from a process of continual learning, leveraged by technical advance, and made possible by convergent values, all of which are embedded in both the organization of the system (as complex as it is) and the engineering of its central components.

5.4 Toward Orwellian humility

Now imagine a system that does not possess these qualities of technological reliability, convergent human values, and consensus measures of success or failure. Many, perhaps most, of the world's most pressing challenges are characterized by and associated with systems that lack these qualities. Such systems therefore also lack a strong internal capacity to improve system-wide performance through progressive innovation and learning. How, then, is improvement to occur?

If one looks at, say, the American political scene, one striking feature is that big problems may recede from attention, but they rarely go away for very long or get 'solved' in any sense that is comparable to the consistently high levels of air traffic safety. Reform of the medical system or tax code, reduction of poverty and racial inequality, improvement of citizen literacy or corporate governance, management of immigration, protection of the environment, reduction of illicit drug use: these and other difficult issues are always simmering on the back burner and every few years one or more of them makes it to the front burner due to some confluence of crisis and publicity. Often some action will be taken—new legislation or regulations or more research or the rise of new organizations dedicated to fixing the problem—that satisfies the political pressure to do something, or quells an immediate crisis, but in most cases the problem is not solved but merely put back on a simmer, awaiting its next boil-over.

Such problems are often described as if they were located within a coherent system. We talk about the medical, education, welfare, and financial systems, with some sense that we are referring to a particular array of institutions, rules, participants, and practices. But managing these systems—that is, intervening in them in particular ways to achieve desired *system-wide* results—seems damnably difficult, and is usually associated not only with surprising consequences but also widespread disagreement about levels of success. New knowledge relevant to the management of these systems does not seem to lead to more successful or lastingly satisfying results.

I've already offered an empirical and theoretical critique of the expectation that more, and more holistic, inquiry is the key to progress in such systems. Now I want to suggest that the goal itself is poorly specified. If people hold different values related to system performance then there cannot be any single way to talk about, or measure, 'improvement', and irreconcilable tensions between alternative objectives are likely. For example, if the goal is to 'reform the medical system', then maintaining attractive salaries for doctors, ensuring good health care coverage for all people, and maximizing progress in medical technology are probably internally incompatible subgoals. This is the point where we must face the politically troubling essence of holism. Any definition of the whole is in part a reflection of the preferences that define that particular subwhole. Any description of a subwhole is necessarily incomplete in ways that can never be known. We find conflict and indeterminacy everywhere we turn—in the scientific descriptions of the whole, in the boundaries we place around the whole, in the values about the proper performance of the whole, in the metrics for measuring that performance.

No one can know enough about any complex system to accurately assert the connections between system inputs and outcomes—to claim accurately, based on 'science' or 'rigorous analysis', or 'complex systems models' or some such, that a particular decision made locally will lead to a desired consequence at the system level. This means that experts wielding credentials of holistic insight in arguing for action in the human world, experts on the medical system, or the climate system, or whatever—anyone who claims to know what is to be done—are always in fact political entrepreneurs. *Claims of holistic expertise are always political claims.* They are political claims because they reflect a choice process—about how to define the system, about what system functions and outcomes are important, about what is to be done to make things better, about what 'better' means.

Is this political essence of holistic expertise a bad thing? A cultural commitment to the Enlightenment, to notions of rationality and inquiry and their contribution to progress, encourages people to see deviations from the world they would like to live in as invitations to make the world better. The promise of rationality and inquiry, that is, seduces us into interpreting various states of the world as 'problems' amenable to '"solutions' that can be approached through better understanding, more knowledge. Wielders of more knowledge—experts—are given particular privilege in discussions about how to solve problems, and the most powerful aspect of this privilege is that it shields knowledge claims from being discussed in political terms. Experts may—and usually do—argue over the apparently technical aspects of their interpretations of the world, but in political forums they are not expected to associate their expert views with their political views. We act as if scientific rigor can shield inquiry from politics. Indeed, when it seems that an expert has mixed their politics with their analysis, we call this 'bias' and discount the value of the expertise.

When it comes to holism, though, everyone is conflicted. From the Enlightenment perspective, this unavoidable politicization of science renders incoherent the notion of action based on rationality. If the system cannot be understood in a purely rational manner, and translated into prescriptions for action that can reliably lead to desired consequences, what good is rationality? Yet can we possibly give up on rationality as the key to solving those very problems that rationality allows us to recognize in the first place? All hope of a better world would then be lost—or at least all hope that the path to such a world can be uncovered by experts bearing special new holistic knowledge about complex system dynamics.

The idea that holistic knowledge can define a path towards solving problems that are embedded in complex human–natural systems is both the child and the assassin of Enlightenment ambition. As problems come to light (often as a result of scientific inquiry) we continually seek to create more systemic knowledge to help solve them, and are continually suffused with a sense of frustration about how much we seem to know, and how little progress we seem to make.

Is there a future for holism? Can it be rescued from the expectations of the Enlightenment and the indeterminacy of reality? I've recently been reading some essays by George Orwell, and besides the enviable clarity of his prose and his thinking, a couple of things are noteworthy. First, he is incredibly unscientific—his observations and thinking are fraught with generalizations and assertions that are at best supported by telling anecdote and in any case completely untestable. ('People can foresee the future only when it coincides with their own wishes, and the most grossly obvious facts can be ignored when they are unwelcome. For example, right up to May of this year the more disaffected English intellectuals refused to believe that a Second Front [the US invasion of France] would be opened. They went on refusing while, bang in front of their faces, the endless convoys of guns and landing-craft rumbled through London on their way to the coast. One could point to countless other instances of people hugging quite manifest delusions because the truth would be wounding to their pride.' [Orwell 1968, p. 297])

Second, he never shies from making his own biases and preferences totally clear. The reader always sees the connections between Orwell's ruthless logic and his guiding principles. ('Any thinking Socialist will concede to the Catholic that when economic injustice has been righted, the fundamental problem of man's place in the universe will still remain. But what the Socialist does claim is that that problem cannot be dealt with while the average human being's preoccupations are necessarily economic.' [Orwell 1968, p. 64])

Third, he nevertheless often conveys an almost mystical capacity to penetrate to the essence of what matters, to see things as they are, achieving moments of crystalline insight by connecting aspects of the world that might seem to be unrelated. ('Snobbishness, like hypocrisy, is a check upon behaviour whose value from a social point of view has been underrated.' [Orwell 1968, p. 224])

And fourth, Orwell is relentlessly, painfully humble and self-critical. ('A man who gives a good account of himself is probably lying, since any life when viewed from the inside is simply a series of defeats.' [Orwell 1968, p. 156]) This explicit tying together of analytical and moral judgment ('I believe that it is possible to be more objective than most of us are, but that it involves a moral effort.' [Orwell 1968, p. 298]) is of course totally against the rules of scientific holism, but once we recognize that 'the whole' is as much constituted of moral (and other subjective) conditions as it is of the facts on the ground, this becomes a strength not a fault. Indeed, the most important part of Orwell's analytical authority comes from his moral clarity: we know where he stands, so we understand why he sees things as he does. And what makes this clarity most compelling is the *soto voce* 'of course I may well be wrong about this' that seems to shadow his every observation.

Scientific experts wielding holistic knowledge almost never display such virtues. Their claims to authority derive from the unstated assumptions that their analysis is on the one hand unsullied by their own individual commitments and on the other beyond criticism of all except equally credentialed experts. Yet in making these assumptions, both the expert's claim to holistic insight, and our reason to have confidence in the expert's reliability, are vitiated, as the scientific and the political collapse into one. We see this phenomenon on stark display in two dominant challenges of our time, global terrorism and global climate change. In the former, the strongest holistic expert claims for action are associated with the political right; in the latter, expertise lies to the left. Each challenge has proven to be a powerful lever for suppressing dissent and driving political agendas. Opponents of the dominant expertise on terror are 'soft on defense'. Opponents of the dominant expertise on climate are 'anti-science'. Various solutions are touted, but progress toward them remains elusive.

If progress is the ideal, then perhaps we are consigned to two paths that often seem divergent. On the one hand there is the technological fix, the taming of some part of the whole that leaves most of reality in its feral state. On the other there is the slow, frustrating, uneven, unpredictable, tortuous pursuit of political change that can, maybe, bring values more into alignment and allow better decisions to evolve. In either case, holism is nowhere to be found, except in the technocrat's dreams, and in the opposing corners of the political boxing arena. In its place, however, we might recognize a different sort of expertise in those that can master the Orwellian synthesis: relentlessly pragmatic, unapologetically tendentious, excruciatingly humble, exceedingly patient. Perhaps this is the whole that we need to aim for, a whole that emerges from the acceptance of our frailty, rather than the pursuit of our hubris.

References

La Porte, T. (1988). The United States air traffic system: increasing reliability in the midst of rapid growth. In: R. Mayntz and T. Hughes (eds) *The development of large technological systems*, pp. 215–44. Boulder, CO: Westview Press.

Neff, M. and Corley, E. (2009). 35 years and 160 000 articles: a bibliometric exploration of the evolution of ecology. *Scientometrics* **80**(3), 657–82.

Orwell, G. (1968). *As I please: collected essays, journalism and letters, 1943–1945*. New York: Harcourt, Brace, Jovanovich.

US National Transportation Safety Board. *Aviation accident statisitcs*. <http://www.ntsb.gov/aviation/Table6.htm> (accessed 3 May 2009).

Wilson, E.O. (1998). *Consilience: the unity of knowledge*. New York: Knopf.

PART 2
Interdisciplinarity in the disciplines

CHAPTER 6

Physical sciences

ROBERT P. CREASE

Few more striking sites of interdisciplinary collaboration exist than the experimental hall of the National Synchrotron Light Source (NSLS) at Brookhaven National Laboratory on Long Island, New York. Synchrotron radiation is light given off by charged particles bending in a magnetic field. The light is emitted in a continuous spectrum and can be finely tuned, giving it a broad range of applications, from imaging biological tissues and determining chemical structures to etching computer chips. The NSLS contains two rings of electrons that emit such light: an ultraviolet ('U') ring with a radius of 51 meters that emits light through 17 ports, and an X-ray ('X') ring with a radius of 170 meters that spins out light from 48 ports. Each port is equipped with devices to customize properties of the light for specific techniques or purposes, and many ports are subdivided two or three times (Fig. 6.1). The researchers who interact on projects there are employed by a variety of institutions, including industry, universities, and national laboratories in the United States and abroad, and come from branches of physics, biology, chemistry, condensed matter, geology, environmental science, and medicine (Fig. 6.2).

Port X5, once the site of an experiment in nuclear physics, is sandwiched between two ports used for protein crystallography and for imaging polymers, artificial biomembranes, and other soft matter. Port X17 was converted from angiography to high-pressure geoscience. A diffraction enhanced imaging device at port X15A allows an optical physicist to work with medical researchers on producing sharp, 10-micron resolution images of soft tissue such as cartilage and ligaments (the team also works at Brookhaven's nearby MRI facility). An X-ray microfluorescence facility at port X26A for detecting trace elements is used by environmental scientists, Alzheimer's researchers, and scientists studying samples of interstellar dust collected by the Stardust spacecraft as it flew through the tail of Comet Wild 2 and returned to earth. The Stardust scientists, who include astronomers, chemists, physicists, and geologists, also used ports X1A, U10A, U10B, and ports at the other major synchrotron radiation sources in the United States. Scientists from Exxon Corporation use five ports (U1A, X2B, X10A, X10B, and X10C) to examine proteins, catalysts, and minerals. On the corridor wall adjacent to port U10B, used for infrared microspectroscopy, a bulletin board displays dozens of snapshots of the diverse international teams

Figure 6.1 The National Synchrotron Light Source (NSLS) at Brookhaven National Laboratory. Photograph courtesy of Brookhaven National Laboratory.

who have used the port, including forensic scientists, art historians, volcano researchers, medical investigators, soil scientists, food scientists, and more.

The NSLS experimental hall thus illustrates a variety of ways that interdisciplinary research occurs in the physical sciences. First, it is a single facility that supports multiple projects in different fields. Second, each project, large or small, requires integration of knowledge, techniques, and perspectives from several disciplines or specialized subfields. Third, in a few instances the projects are helping to create new disciplines. Finally, many projects require more than a single instrument, and incorporate knowledge, techniques, and perspectives from additional facilities at Brookhaven and other labs.

The NSLS, and its interdisciplinary research progeny, arose not from any desire to be interdisciplinary for its own sake, but from the developing interests and specific goals of solid state researchers. What is distinctive about interdisciplinary research in the physical sciences as compared with that in the humanities is that the physical sciences do it more and without fanfare or self-congratulation, face the problems practically, and theorize about it less. Interdisciplinary research is seductively easy to theorize about, and can give rise to high-minded glorifications of 'boundary-crossing', 'transgression', and the production of 'new objects'. Such talk can deliver the impression that boundaries are good to cross so long as they are on someone else's property—NIMBY ('not in my back yard') interdisciplinarity, one might call it. Interdisciplinarity is more treacherous than it looks. The advantage of the case of the physical sciences is that interdisciplinarity can be looked at concretely in ways that can help to weed out much posturing and ideology.

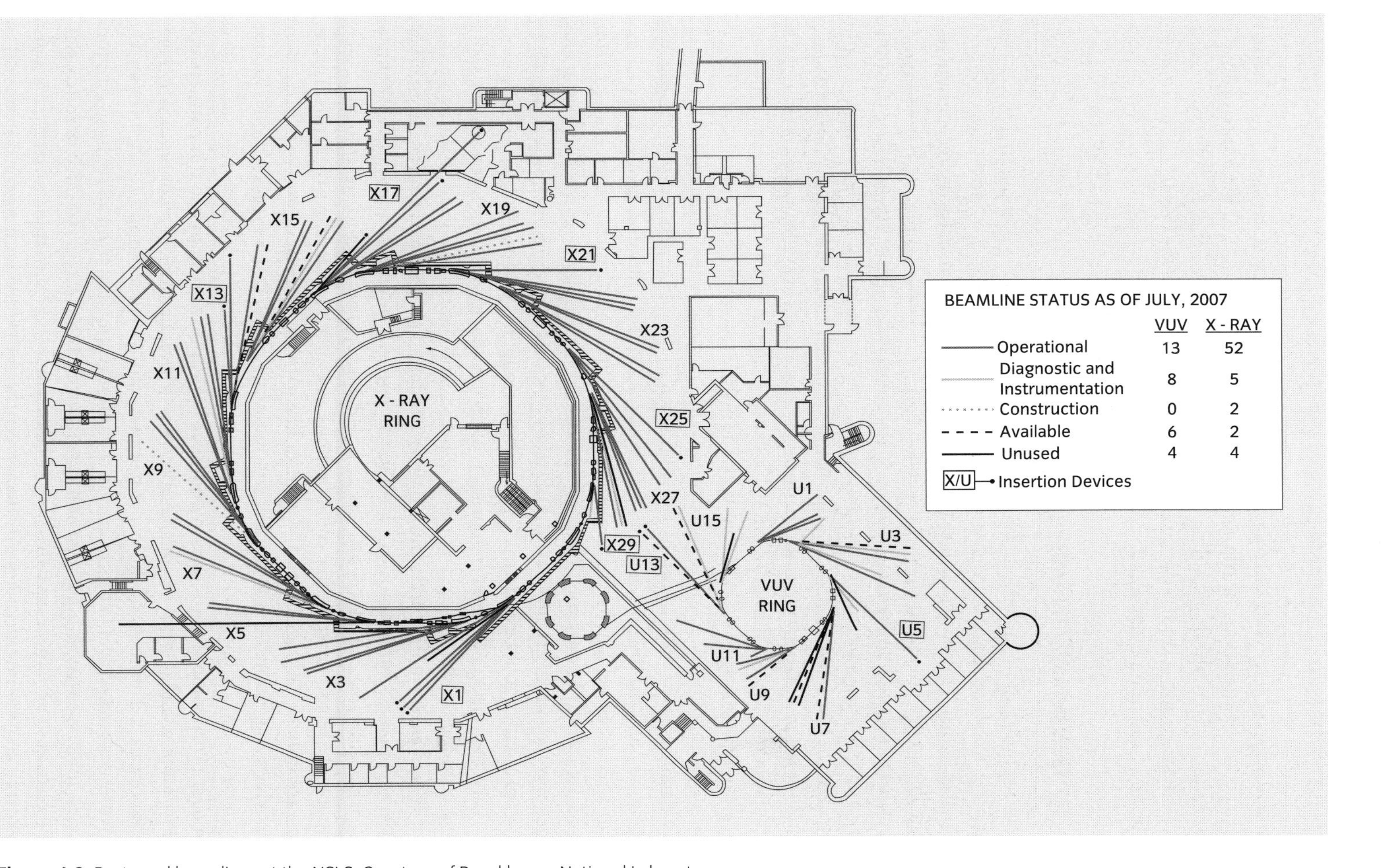

Figure 6.2 Ports and beamlines at the NSLS. Courtesy of Brookhaven National Laboratory.

6.1 History

Interdisciplinary research and collaboration is surely as old as science itself. In a three-part article in *Scientometrics*, Beaver and Rosen (1978, 1979) examined the entire history of scientific collaborations, including interdisciplinary ones, through a co-authorship study. Yet the scope and scale of what is done on the NSLS floor, and the impact on instruments, facilities, and techniques, are of recent vintage.

6.1.1 Emergence of interdisciplines

The beginning of the nineteenth century witnessed a disciplining of modern science, when it came to be conceived as consisting of relatively discrete and specific bodies of knowledge or 'logies'. However, this development was also accompanied by the recognition that the knowledge embodied in each discipline bore on others, and that understanding any particular slice of human life involved a spectrum of fields. These two poles are illustrated by Humphry Davy's famous introductory lecture on chemistry at the Royal Institution in 1802, in which he extolled the value of chemical knowledge for a multitude of sciences and throughout human life and experience, and by Michael Faraday's equally famous discourse at the same institution about half a century later, in which he showed that the complete understanding of a single, simple candle involves many different fundamental laws of nature, from capillary action to gravitation. When Auguste Comte propounded his scheme of classification of the sciences, he argued that while 'the division of intellectual labor' was necessary and the disciplines would have to be separately cultivated, he also stressed that the sciences all belonged to a 'greater whole' and that any division was 'at bottom artificial'. He warned against 'too great a specialization of individual researches' as 'pernicious', because the end of science was to understand the world around us, which is inherently complex and cannot be addressed by any single discipline (Comte 1988, pp. 16–17).

Early interdisciplinary research projects often took the form either of researchers applying techniques (whether theoretical or experimental) cultivated in one field to another, or of researchers in one field working at the frontier of another. Warren Hagstrom (1964, 1965) compared early forms of collaborative research in science to medieval forms of economic organization. Professor–student relationships, for instance, resembled master–apprentice relations, while 'free collaborations' resembled medieval partnerships. The latter are initiated informally, and Hagstrom likened their initiation process to courtships in which suggestions of interactions are cautiously initiated and explored, often accompanied by fear of rejection (Hagstrom 1965, p. 114). But Hagstrom wrote that just as modern corporations have come to dominate both apprenticeships and free partnerships, so a more complex form of collaboration was arising that would soon dominate scientific research. The roots of this more complex and corporate form of collaboration, he wrote, were threefold: (1) centralization of authority imposed from above by institutions or funding agencies and by large and expensive facilities, access to which was necessarily restricted; (2) a necessary division of labor among various kinds of technicians and experts; and (3) interdisciplinarity, which can be contrasted with *multidisciplinarity*, or mere division of labor among disciplines.

Such more complex collaborations began to emerge early in the twentieth century. As Davy had prophesied, chemistry was often a principal ingredient of interdisciplinary collaborations in fields such as biophysics, physical chemistry, and chemical engineering. Other interdisciplinary fields to emerge in the early twentieth century included radiation science, which combined elements of physics, chemistry, engineering, biology, and medicine; and cybernetics, which brought together pieces of architecture, control systems, electronics, game theory, logic, mechanical engineering, neuroscience, psychology, and philosophy. Sometimes interdisciplinary projects were a function of the goal of a specific set of researchers, such as the famous BBFH astrophysics paper, 'Synthesis of the elements in stars', that sought to explain the formation of heavy elements in stellar interiors (Burbidge *et al.* 1957). At other times, interdisciplinary research was deliberately cultivated by individuals at funding agencies, such as Warren Weaver of the Natural Sciences Division of the Rockefeller Foundation (Kohler 1991). Interdisciplinary research often forced laboratories such as the Radiation Laboratory at the University of California at Berkeley, and projects such as astronomical and space programs, to devise efficient ways to handle it (Everitt 1992; Seidel 1992).

The discovery of the molecular structure of DNA in 1953 was an important landmark, and generated a special set of problems for researchers. One was a certain amount of *disciplinary anxiety* that biology was about to be colonized by other fields, leading to A. V. Hill's rejoinder that 'Physics and chemistry will dominate biology only by becoming biology' (cited in Pantin 1968, p. 24). It also inspired some rudimentary reflection about interdisciplinary research; Carl Pantin, for instance, was moved to propose what he called a 'real' distinction between *restricted* and *unrestricted* sciences, or those (like physics, he thought) that do not require investigators 'to traverse all other sciences', and those (like biology) where the 'investigator must be prepared to follow their problems into any other science whatsoever' (Pantin 1968, p. 24).

In the 1960s, when Hagstrom wrote, applied research, especially industrial research such as DuPont's, already tended to be interdisciplinary. 'Better living through chemistry' was then a popular adaptation of an advertising slogan adopted by the DuPont chemical company in 1935 and used for almost half a century to market its research and development projects across many fields (for the past decade the company has used the more generic slogan 'The miracles of science'). But Hagstrom remarked that interdisciplinarity was much less common in basic research. When it did exist, he wrote, it experienced strains of the sort that befall 'inherently heterogenous' emerging disciplines (Hagstrom 1965, p. 215), manifested for instance by behaviors such as obsessive celebration of a field's founders. Interdisciplinary work indeed can create not just disciplinary anxiety but also an intense kind of personal anxiety. When boundaries that have been taken for granted come to appear moveable, it not only opens the question 'What is the discipline?' but concomitantly the more personal questions 'What am I doing?' and 'Who am I?'.

Today, the situation faced by Hagstrom has changed, and interdisciplinarity is common throughout basic research in fields such as addiction research, bioengineering, biological physics, biophysics, climate change, nanotechnology, and polymers. In 2000, the Nobel Prize for Chemistry was awarded to three scientists—two chemists and a physicist—for 'the discovery and development of conducting polymers'. In his acceptance speech,

Alan J. Heeger, the physicist of the trio, remarked that simply by attempting 'to understand nature with sufficient depth', he had 'evolved... into an interdisciplinary scientist', for the field was 'inherently interdisciplinary' (Heeger 2000).

6.1.2 Interdisciplinary instruments, facilities, and techniques

Interdisciplinary research has affected instruments, facilities, and techniques involved in experimental research by fostering their deliberate planning and construction. Many new devices and techniques, particularly imaging technologies, apply to more than one field. X-rays are a classic example; within 3 weeks of their discovery in January 1896, physicians had used them to help reset a child's broken arm. But the scale and expense of modern instruments makes it necessary to maximize their constituency and design and promote facilities from the outset as dedicated for interdisciplinary use. The NSLS—the first facility planned from the outset for synchrotron radiation research—is a classic example (Crease 2008a). Supercomputers are another.

Yet the impact of interdisciplinarity on research takes still more complex forms. All experimentation is a species of *performance*, for it involves bringing together well-understood pieces of equipment and material in staging an event or series of events that seek to make some phenomenon appear, and let it be examined, in a way that would not otherwise be possible (Crease 1993, 2003). Staging performances requires *production*, or an advance set of behaviors and decisions necessary to assemble elements created for other purposes. The production of research equipment thus sometimes requires a kind of improvised engineering that John Law has called *heterogeneous engineering* (Law 1987). But the equipment of modern interdisciplinary research is of such a scale that not just pieces of knowledge and apparatus, but entire fields of knowledge are sometimes transformed and whole instruments reconstructed for new purposes, resulting in what Catherine Westfall has called *recombinant science*.

Recombinant science does not occur as a natural outgrowth of previous research, but involves researchers combining 'insights and expertise from various subfields in new ways to create a brand new outlook' (Westfall 2003; Crease 2008b). In small-scale interdisciplinary collaborations, such as those commonly found at the NSLS, the end is generally a natural outgrowth of traditional interests, and the means require recruiting and coordinating researchers from different fields. Recombinant science, however, involves an untidier story, in which the ends as well as the means have arisen as the result of contingencies and convergences that require researchers to adapt their intentions and methods, sometimes awkwardly.

6.1.3 The example of the RHIC

A case study in recombinant science is the construction of the Relativistic Heavy Ion Collider (RHIC), a $486 million nuclear physics facility at Brookhaven, located not far from the NSLS but an entirely separate facility. It sprang from a high-energy physics proton collider named ISABELLE, on which construction began in 1978 (Crease 2005a,b). But various problems caused the US physics community to lose enthusiasm for the ISABELLE project (briefly renamed the Colliding Beam Accelerator or CBA), and it was terminated in 1983. In a remarkable turn of events, the facility was converted into a facility of a new sort to explore a new field, relativistic heavy ion physics. To justify this transition, scientific subfields were invoked that did not exist at the time of ISABELLE's birth, and the transition was made possible by certain key hardware components that also did not exist when ISABELLE was conceived. The new field of heavy ion physics effectively blended, initially with difficulty, nuclear and high-energy physics (Crease 2008b).

6.1.4 The age of interdisciplinarity

Why has interdisciplinarity become so routine in the physical sciences? Several theories have been advanced.

One, advanced by Hagstrom, is corporate; the scale of scientific projects and facilities now requires corporate-style organization and management in which different disciplinary components are coordinated (see Stokols, Chapter 32 this volume). Indeed, such organizations have now been around long enough that patterns have developed. In their study of multi-institutional collaborations, for instance, Shrum *et al.* (2007) identified five different patterns of collaboration formation and four organizational types of collaboration, and note several bureaucratic features that have evolved to stabilize such interdisciplinary research.

Another theory, advanced by historian of science Paul Forman (2007), is epochal; the rise of interdisciplinarity is tied to the shift from modernity to postmodernity. The assumptions of modernity—especially the priority of theory over practice, of basic over applied research, and of disinterested over interested knowledge—produced the traditional disciplinary borders, and served to reinforce them. These disciplinary structures have all but collapsed as an inevitable consequence of the reversal of the priority of science and technology characteristic of postmodernity, with its 'pragmatic-utilitarian subordination of means to ends, and of the concomitants of that predominant cultural presupposition, notably, disbelief in disinterestedness and condescension toward conceptual structures' (Forman 2007, p. 2).

A third theory is historical; that two seminal events—the development of quantum mechanics and the massive expansion of computational power—made interdisciplinarity all but inevitable. Quantum mechanics forced the reworking of the foundations of physics, chemistry, biology, materials science, electronics, thermodynamics, and other fields. It provided scientists with the confidence to claim that enough was known about the structure of matter so that, even if only in principle, large-scale substances and many real-world behaviors could be traced back to, if not entirely explained by, small-scale structures and forces. And the sciences of these large-scale substances and real-world behaviors—from proteins to superconductors—were not abstract domains like particle physics or cosmology but inherently interdisciplinary 'real-world' systems.

The expansion of computational power, meanwhile, also transformed nearly all the physical sciences not only through codes and calculations—which have often made it possible to trace back the behaviors of large-scale substances to small-scale structures and forces—but also through data analysis and fitting, search techniques, simulations, visualization methods, and other tools. This has led to what Wilson (1984) called the 'computerization of science'. It also led to the interdisciplinary field (applied mathematics, computer science, and science and engineering) of computational science and engineering (CSE) which itself is a field that participates in other interdisciplinary fields (on its impact just on physics see Landau *et al.* 2008). Computation has also profoundly affected disciplines outside of the natural sciences, including art. Recognition of the relevance of mechanics and optics to painting dates back at least to Leonardo da Vinci's *Treatise on painting* and Hermann von Helmholtz's lectures *On the relation of optics to painting*. Yet the recent expansion of computational power (plus technological developments such as the development of selective laser sintering devices) has transformed the practice of artists in striking ways, such as in the recent emergence of the field of 'mathematical sculpture' (Grossman and Hart 2008; Zalaya and Barrallo 2008), which includes representations of four-dimensional objects—the creation of a 'new object' if there ever was one (Fig. 6.3).

Yet a fourth theory offers a Comtean-style teleological explanation involving the purpose of science itself. The point of science is to allow the prediction and control of nature, and if we have divided science into disciplines it is only so that we can better cultivate them to the point where we can do this. We have had a learning curve while the disciplines were being cultivated, but at last we can bring them together again in interdisciplinary

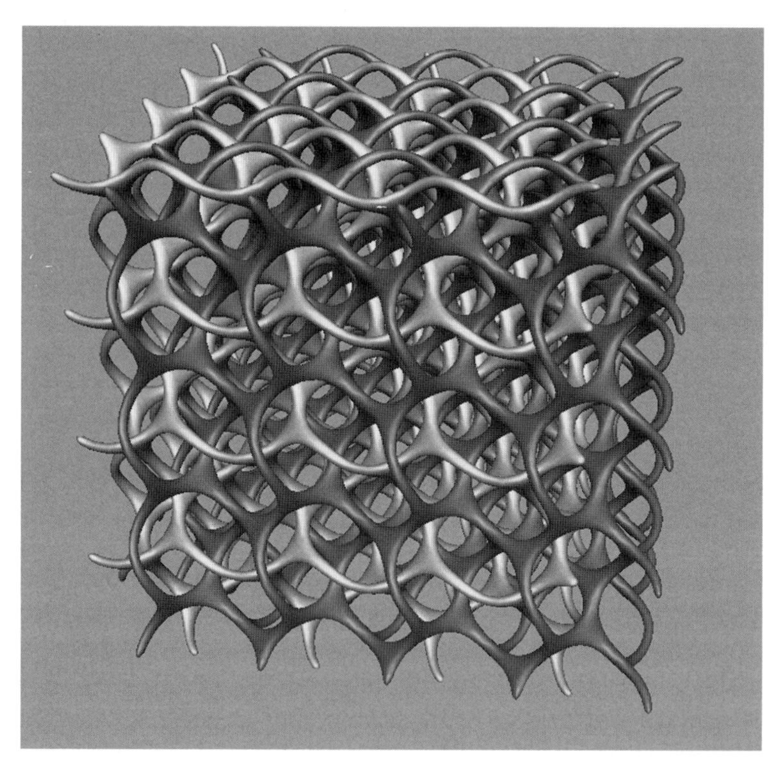

Figure 6.3 A 'mathematical sculpture'. Courtesy George W. Hart.

research. Any obstacles to so doing are the result of what Comte called the 'pernicious effects of an exaggerated specialism'.

For whatever reason—corporate, epochal, historical, or teleological—interdisciplinarity is here to stay. Many people have referred to the 'frontier of complexity', whose manifestations include biotechnology and nanotechnology, and which ensures that interdisciplinary research will dominate the natural sciences in the twenty-first century, a period sometimes referred to as the 'Age of Interdisciplinarity' (Marburger 2008).

6.2 Practical issues

Promoting the growth of interdisciplinary research is surely a fine goal. But as the Spanish proverb says, it is one thing to speak about bulls and another to be in the bullring. Fortunately, scientists and science administrators have had decades of experience trying to meet the concrete and practical challenges of interdisciplinary research. One speaker at a 2006 American Association for the Advancement of Science (AAAS) workshop on quality assessment in interdisciplinary research prefaced his remarks by recalling US President Grover Cleveland's blunt remark, in vetoing a tariff bill, that 'This is a condition we face, not a theory'. Interdisciplinarity is indeed a condition with pressing challenges that often do not respond nicely to theory. Its challenges vary throughout the phases of a project such as the construction of a big facility—from construction to operation to data analysis—and are also different for theoretical research. Practical challenges of the condition of interdisciplinary research in the physical sciences include coordination, quality assessment, communication, and culture.

6.2.1 Coordination

One set of practical issues arises in laying out the conditions in which the various disciplines can work comfortably. Again, the example of the RHIC is illustrative. Nuclear and high-energy physicists would not only have to learn and adapt techniques from each other, but also learn to work comfortably together—yet their existing practices were quite different, using different kinds of instruments and different sized teams, with nuclear physicists used to working with a handful of collaborators, and high-energy physicists used to collaborations of dozens or even hundreds. At a key meeting at the beginning of the project, Arthur Schwartzchild, the chairman of Brookhaven's physics department, outlined a plan to address the problem by initiating an interim program of heavy ion physics at existing facilities at the lab that would run while RHIC was under construction. This, Schwarzschild said, would address the looming 'manpower and sociology issues' by 'building a constituency for collider experiments, effecting collaborative efforts between nuclear and particle physicists, and providing an appropriate arena and stimulus for detector development necessary for collider experiments'. In an interesting unwitting echo of Hagstrom's relationship metaphor, Schwarzschild concluded by saying that 'The new physics calls for a marriage between nuclear and high energy experimenters, and this conference looks like an engagement party to me' (Ludlam and Wegner 1984, p. 377c).

Mathematics and root interdisciplinarity

Erik Fisher and David Beltran-del-Rio

In the typology that distinguishes between multi-, cross-, and transdisciplinarity it is also possible to think of a discipline that lies at the root of other disciplines. Mathematics enables work in many different disciplines, from the natural and physical sciences to the social sciences and fine arts. It can also support knowledge integration across disciplines. Furthermore, developments in mathematics can be correlated to cultural style periods – as in the case of Kurt Gödel's (1906–78) incompleteness theorems and postmodern theory (Thomas 1995). Mathematics can thus be thought of as a foundation for other disciplines both because of its direct applicability to a wide set of phenomena, practices, and developments in many other fields of learning and human endeavor and because it represents a fundamental form of knowledge.

'Mathematics', from the ancient Greek *mathemata* (that which can be learned), was originally broader than modern use implies. It encompassed all learned knowledge, not just that which can be characterized through number. In modern usage, however, mathematics can be defined as the study of patterns and order within structure, space, and change. It employs logical reasoning and quantitative calculation to make statements that can be shown to be true or false based on first principles or axioms and rules of inference. Mathematics is considered 'pure' and outside of the natural sciences insofar as it investigates the properties of and relationships among idealized objects. Yet insofar as the knowledge it generates approximates physical phenomena, it can aid in their conceptualization and control, and can thus be considered 'applied'.

Mathematics is generally considered to have originated prior to any clearly written historical record with practical problems, mostly involving commerce and agriculture, eventually extending into natural science and military applications. In the European tradition, pure mathematics arose much later with Pythagoras (*c.* 569–475 BCE). However, emergence of the systematic study of natural phenomena did not historically coincide with mathematics. Aristotle (384–322 BCE) developed a phenomenological science of nature based on understanding four distinct causes of natural phenomena, with mathematics being merely one mode of causality.

Mathematics played a more prominent role in ancient astronomy than in physics. In the cosmology that framed scientific thinking before the sixteenth century, the celestial sphere on which the moon was thought to travel defined a dividing line between heaven and earth, with the corruptible and imperfect beneath this line, while the heavens were a realm of perfection and perfectly circular motions. Hence early astronomers, particularly Claudius Ptolemy of Alexandria (*c.* 85–165 CE) concluded that mathematics, also being perfect, was the appropriate tool to describe heavenly motion. Ptolemy's insistence on using only perfect circles required him to employ epicycles, circles whose centers moved upon the circumferences of larger circles centered nearly (but not exactly) on the position of the earth and around which orbited each planet. Likewise, mathematics was considered largely *inappropriate* for the study of sublunar phenomena, since a perfect tool could not describe an imperfect, corruptible world. Such a view of mathematics limited its interdisciplinary potential perhaps intentionally (Bolotin 1998).

It was Galileo Galilei (1564–1642) who first clearly grounded the study of nature in mathematics. The Galilean revolution is often taken to involve his acceptance of Copernican heliocentric cosmology and insistence on experimentation and empirical 'proof' of physical theories. Of even more importance, however, was Galileo's application of mathematics to sublunar motions and his insistence that physicists should not seek causes, as Aristotle held, but generate only mathematical descriptions of natural phenomena.

The calculus, which enables a more precise description of Galileo's chief interest, motion, was independently invented 50 years later by Isaac Newton (1642–1727) and Gottfried Wilhelm von Leibniz (1646–1716). Calculus allows physicists to model and predict natural motions with previously unknown accuracy. In a similar development, by assuming an inverse square relationship of gravitational force and then by developing and applying a geometry of limits, Newton demonstrated what Johannes Kepler (1571–1630) had only posited, namely, that planets move in elliptical orbits. In one of several disclaimers, however, Newton significantly pointed out in a 1693 letter that his mathematical work did not reveal the more philosophical '*cause of gravity itself*'.

With calculus, Newton succeeded in unifying formerly disparate fields of physics. Attempts to further unify or even 'end' physics with a mathematical theory of everything (TOE) continue to the present. String theory is one such attempt, and as in Newton's day new physics and new mathematics are developed simultaneously.

Calculus is the mathematics of motion and change. Since these feature so prominently in our world, it is no surprise that calculus has found its way into so many fields. If one's best tool is calculus, everything begins to look like a differential equation. It is interesting to note that calculus requires the quantities it studies to be the result of continuous processes. In other words, it builds in the assumption that the physical world is made up of unbroken, rather than discrete operations. Whether such continuous processes exist at all in nature is still an open question; we do not know whether space and time are continuous. Hence calculus, and mathematics in general, should be thought of as an approximation, rather than an explanation of many, if not all, of the phenomena it seeks to predict. Ptolemy's epicycles can be recalled here as a classic example of a mathematical device that predicts but cannot explain natural behavior.

Whereas Newton, like Aristotle, distinguished mathematical knowledge of nature from 'causal' knowledge, Galileo, like Ptolemy, may have been more interested in a computational and predictive tool. Meanwhile, by the time of Pierre-Simon Laplace (1749–1827), scientists widely held that the universe was akin to a gigantic deterministic clockwork machine, and began to envision an end to physics with a colossal system of differential equations that predicted every natural event.

The Galilean revolution produced still another result—the attempt to quantify or otherwise find a secure mathematical foundation for as many fields as possible. Many believe a field of study becomes truly scientific only insofar as it can be made mathematical. Due to its tremendous success in describing natural phenomena, mathematics is often thought of as a root of scientific knowledge and, by extension, of knowledge in general. This largely means the application of calculus, but calculus is a deterministic tool, and in many cases cannot be applied. When calculus fails, another modern form of mathematics known as probability and statistics tends to take its place.

Probability and statistics have their own specifically modern origins in specifically modern problems, such as those being presented to bankers and investors in high-risk potentially high-gain shipping ventures. About the time of the development of the calculus, for instance, Blaise Pascal (1623–62) famously proposed that one employ a statistical wager regarding the decision of whether or not to believe in the existence of God. Shortly thereafter, the nation state also sought to develop statistical tools that could facilitate its control and manipulation of large-scale populations and the monitoring of commercial activities. Modern scholars often propose that mathematics in the form of probability and statistics be directly applied to moral contexts and

(*cont.*)

Mathematics and root interdisciplinarity (cont.)

questions. Economic and risk assessment methods such as cost–benefit analysis and probabilistic risk assessment are presented as tools for moral and public policy decision making.

The modern notion of mathematics as a root interdiscipline has nevertheless been challenged on a number of levels. Jacob Klein (1899–1978) argued that the ancient Greek understanding of *arithmos* differs importantly from the modern understanding of 'number'. According to Klein, *arithmos* always means a definite number of definite things, whereas the modern 'number' replaces 'the real determinateness of an object with a *possibility* of making it determinate' (Klein 1992, p. 123). The symbolic characteristic of 'number' is based on seemingly paradoxical assumptions about the ontological status of mathematical objects, since it identifies mind-independent 'things' with a mind-dependent 'concept', namely, *quantity*.

This brief history of mathematics suggests that the application of mathematics to other domains is partially a function of what mathematics is considered to be. Moreover, different conceptions of what counts as knowledge have at various times both limited and enabled the integration of mathematics within these other domains. The extensive modern employment of mathematics in describing and predicting phenomena can on the one hand be taken as evidence of the primacy of mathematics as a form of knowledge, and on the other comes at the cost of other forms of description and ways of knowing. In this case, at least, interdisciplinary success appears to be rooted in what constitutes the discipline.

References

Bolotin, D. (1998). *An approach to Aristotle's Physics: with particular attention to the role of his manner of writing*. Albany: State University of New York Press.

Klein, J. (1992). *Greek mathematical thought and the origin of algebra*. Mineola, NY: Dover Publications.

Thomas, D.W. (1995). Gödel's theorem and postmodern theory. *PMLA* **110**(2), 248–61.

6.2.2 Quality assessment

But such relationships still need to be monitored for their long-term health. In 2008, Boix Mansilla and Gardner wrote that 'a re-emerging awareness of interdisciplinarity as a pervasive form of knowledge production is accompanied by an increasing unease about what is often viewed as "the dubious quality" of interdisciplinary work'. One factor is that the traditional method of quality assurance—peer review—can prove difficult in practice in the absence of true 'peers'. A step in alleviating this concern, the authors continued, is to develop suitable processes, criteria, and contexts for assessing interdisciplinary work, including ways of selecting appropriate reviewers and of effectively managing their collective expertise in review sessions. One must find, as Martin Blume, the former Editor-in-Chief of the American Physical Society put it at the AAAS quality assessment meeting mentioned above, 'referees who have open minds and a deep knowledge of the fields'. Among the problems is 'a tendency of physicists to believe that another area of science is not significant until it can be understood in terms of the techniques of physicists, and for, say, economists to believe that physicists have nothing to teach them' (Blume 2006). Another problem involves metrics for evaluation, such as citation counts or publication in 'high impact factor' journals, for different fields may be of different sizes and differ, too, in the shelf-life of influential articles. Groups such as the Council of Environmental Deans and Directors of the National Council for Science and the Environment provide online resources for interdisciplinary hiring, tenure, and promotion (CEDD 2008).

Other special measures that may be required to ensure the quality of fields include making sure that the appropriate spectrum of journals turns up in citation indexes; that once articles in journals such as *Physical Review E* and *Physical Review Letters* become relevant to medical research, for instance, these journals are listed in Medline. Special awards for interdisciplinary research may be necessary to ensure that noteworthy research that may otherwise slip through the cracks is appropriately recognized. The New York Academy of Sciences, for instance, sponsors an annual award, the Blavatnik Award for Interdisciplinary Research. And interdisciplinary research poses special problems for librarians and information scientists: 'It is imperative for information scientists to understand the characteristics of interdisciplinary research and the researchers' information need(s) to better serve the scientific community' (Tanaka 2008, p. 41).

6.2.3 Communication

Thomas Kuhn famously argued that disciplines are defined by paradigms. If so, then any crossing of disciplines can only be either undisciplined, or a trade or exchange of something between disciplines: multidisciplinary rather than interdisciplinary. How is genuine cross-communication possible? Peter Galison provided a twofold answer involving the claim that paradigms are not that monolithic plus the idea of a trading zone, or special kind of place where different cultures meet and interact. What takes place in such a zone, he claims, is not 'translation' with its implication of one-step transpositions of meaning from one holistic context into another. Rather, local languages emerge—inter-languages, 'pidgins and creoles'—that 'grow and sometimes die in the interstices between subcultures'. In this way, 'trading partners can hammer out a *local* coordination, despite vast *global* differences' (Galison 1997, p. 783; Collins *et al.* 2007).

6.2.4 Culture

But interdisciplinary research involves more than language. Seligman *et al.* (2008) point out that, in genuine communal interaction, it is often more important to examine what people do rather than what they say or mean. One must beware of overtextualizing the world, of overemphasizing the efficacy of language and belief in human action. Despite the detached, third-person style of research papers, what matters is not whether the result is epistemically justified, but whether the goal has been reached. The language of science is subservient to the practical requirement of achieving its goal. This signals the importance of another set of subjects critical to interdisciplinary research—its 'immaterial culture', so to speak—including trust and expertise, to be mentioned below.

Cooperation, for instance, may require overcoming cultural differences, not just learning a new language. An example is provided by what happens to Stony Brook University computer engineer Steven Skiena each time he teaches his graduate course in computational biology. The two largest groups who take his class are biologists and computer engineers, and these have diametrically opposed backgrounds, experience, interests, and educational attitudes. From the beginning, it was difficult. 'The biology students took for granted the existence of a strict hierarchical pecking order that leads from professor to postdocs to grad students to lab assistants to undergraduates, and assumed that they must start at the bottom and work up. The computer students, by contrast, saw no such hierarchy, described themselves simply as working in the "Skiena lab", and treated everyone as peers, including Skiena himself. The biology students tended to feel violated if asked to program a computer, and computer engineers tended to feel likewise if asked to learn something about proteins' (Crease 2006, p. 226). It is two disciplines, one might say, divided by a common subject. Skiena must get the class at least to mingle intellectually. He begins by mirroring back these cultural differences in a slide (Fig. 6.4). The PhD students in this class tend to retain their disciplinary affiliations after graduation—the computer science students tend to get jobs in computer science departments, the biologists in life science departments—which no doubt is a function of teaching, tenure, and funding factors. However, they do tend to wind up publishing or co-publishing much more in the other discipline—thus engaging more in interdisciplinary work—than their disciplinary peers.

6.3 Theoretical issues

What's distinctive about interdisciplinary research in the sciences, I said above, is that they do it more and theorize about it less. Scientists are accustomed to redrawing their disciplines, and live and work with their boundaries under reconstruction. The practical, goal-oriented focus of the researchers allows them to bypass the need for reflection and intersubjective inquiry. Moreover, theorizing about scientific practice is the task of other kinds of scholars.

6.3.1 Disciplines and interdisciplines

One way to understand interdisciplinarity is through understanding disciplinarity. What constitutes a discipline: Objects? Methods? Concepts? Culture? Are the RHIC researchers, and NSLS researchers, actually being interdisciplinary, or merely retreading within what is essentially the same large discipline of physical science? And is interdisciplinarity in the physical sciences different from what happens in social sciences and the humanities? Are there different kinds of boundaries? Examining such questions using case studies from the physical sciences can help clarify what we mean by a discipline.

A *realist* conception of disciplines would picture science as seeking to describe territories of knowledge or of objects that are out there independently of how we come to know them—where nature is divided at its joints. If we make changes in what our sciences

Computer Scientists vs. Biologists

There are many different types of life scientists (biologists, ecologists, medical doctors, etc.), just as there are many different types of computational scientists (algorists, software engineers, statisticians, etc.).

There are many fundamental cultural differences between computational/life scientists:

- *Nothing* is ever completely true or false in biology, where *everything* is either true or false in computer science/mathematics.
- Biologists strive to understand the very complicated, very messy natural world; computer scientists seek to build their own clean and organized virtual worlds.
- Biologists are *data* driven; while computer scientists are *algorithm* driven.

  One consequence is CS WWW pages have fancier graphics while Biology WWW pages have more content.

- Biologists are much more obsessed with being the first to discover something; computer scientists invent more than discover.
- Research biologists have to know more than computer scientists; computer scientists know how to do more.
- Biologists are comfortable with the idea that all data has errors; computer scientists are not.
- Biologists are live in stronger hierarchies than computer scientists: PI $\rightarrow$ postdocs $\rightarrow$ graduate students $\rightarrow$ lab assistants.

  Genetics students seeking to work with me ask to join the "Skiena lab".

- The Platonic ideal of a biologist runs a big laboratories with many people. The Platonic ideal of a computer scientists is a hacker in garage.

  Biologists can get/spend infinitely more research money than computational scientists.

- Biotechnology/drug companies are largely science driven, while the computer industry is more engineering/marketing driven.
- Biologists seek to publish in prestigious journals like *Science* and *Nature*. Computer scientists seek to publish in prestigious refereed conference proceedings.

  One consequence is life science journals get refereed faster than computational science journals.

- Computer scientists can get interesting, high-paid jobs after a B.S. Biologists typically need to complete one or more postdocs..

Figure 6.4 Introductory slide from Steven Skiena's Computational Biology class. Courtesy Steven Skiena.

encompass we are correcting these boundaries to be more in accord with what is out there, rather than transforming the sciences or acting interdisciplinarily. In this view, the skeptics are right, and interdisciplinary research is arbitrary, hybrid, a disciplinary mule—sterile, not creative, and dependent for its continued existence on further seminations. But it has proven difficult to differentiate disciplines by their global object, or what the scholastics called their material object. Each discipline comes at its objects in a different way, so the disciplinary objects differ—what the discipline sees in the global object is based on the discipline's own ways of investigating. Indeed, there seem to be only nominal and historic differences between physics, chemistry, biology, and so forth, in terms of their global and formal objects.

For this and other reasons, following the appearance of Kuhn's *Structure of scientific revolutions*, we have seen the emergence of what might be called a *postmodern* conception of interdisciplinarity, exemplified by Forman. In this view, the boundaries of disciplines are essentially arbitrary, as a function of how these sciences emerged and the social forces exerted on them, susceptible to change as these forces change. *We* created nature's joints. Indeed, if the disciplines make any attempt to resist the transformation of their boundaries they become suspect as ideologies, subject to a hermeneutics of suspicion of their justifications of their interests, claims, and narratives. The postmodern conception of disciplinarity valorizes, even celebrates, interdisciplinary work and its heterogeneity.

A third possibility is a *hermeneutical* conception of disciplines, in which the sciences are about the world as it presents itself to us and with which we are creatively engaged through our laboratory experiences. The world does not present itself to us as undifferentiated, but as being landscaped, certain of its regions being nearer or farther from others. We inherit, adapt, and transform this landscape—you first have to recognize and accept boundaries in order to reorganize or transgress them—both the areas comprising it and how these are related (Ginev 1997, 2006). When X-ray instruments first appeared, they could be used in different fields without significantly affecting the boundaries. By the time of the NSLS, however, the engagement with nature to which X-ray technology belonged—the scales and energies involved—had been sharply altered. The NSLS was not simply a bigger X-ray bulb in the same landscape; human beings and nature were positioned very differently in a changed landscape.

6.3.2 Trading zones

Another path to understanding interdisciplinarity involves looking at what happens in interdisciplinary projects. Collins *et al.* (2007) sought to develop a more general form of Galison's notion of trading zones, or places where different cultures interact. Noting that in the absence of communications problems there is only trade, they defined trading zones as 'locations in which communities with a deep problem of communication manage to communicate'. How, then, can such a 'deep problem of communication' be overcome? In several possible ways, say Collins *et al.*, depending on the kind of trading zone it is. They propose a four-fold division of such zones by mapping interdisciplinary collaborations onto a graph with two axes. One involves whether the collaboration is cooperative or coerced, the other whether the end product is a heterogeneous or homogenous culture (Fig. 6.5). In this way, the creation of new scientific disciplines like astrophysics, biophysics, or relativistic heavy ion research is only one of several possibilities for interdisciplinary collaborations. But the diagram is based on the assumption of a neat distinction between cooperation and coercion—which reminds philosophers of the old Aristotelian distinction between natural and enforced motions, and inspires wonder about the grounding of this distinction. How is this distinction reflected in scientific practice? Is the interaction between nuclear and high-energy physicists at the RHIC collaborative or coerced? On the one hand, the interaction moves scientists towards a goal—further understanding particles and nuclei—that they have always sought, which might suggest collaboration; on the other hand, it was political necessity for the laboratory, stitched together because of the failure of a big science project, which might suggest coercion. When someone makes the claim that the collaboration was cooperative or coerced, who then is speaking and why? The collaboration was both cooperative and coerced at the same time, or neither; it arose from the scientists living in the midst of the scientific world, motivated by dissatisfaction, and using what tools they had to achieve what they could in pursuing their inquiries. They were making their way intelligently in an atmosphere whose elements were not separable into categories like 'cooperative' and 'coerced'. Maintaining cultural heterogeneity is not always natural, and transforming it is not always slavery. The notion that all transformation of the boundaries of science is enforced is the product of a Forman-like postmodern conception of disciplinary boundaries.

What if interdisciplinary research, instead, were looked at from the perspective of its participants themselves, rather than from the outside? For someone joining a RHIC collaboration, say, it is not a matter of contributing a block of information to the project the way that a jigsaw piece contributes to the whole. Rather, it is a matter of working *with* other participants, oriented toward the practical realization of a goal. Being in such

	Homogeneous	Heterogeneous	
Collaboration	**Inter-language** Biochemistry Nanoscience	**Fractionated** Boundary Object Cowrie shell Zoology	 Interactional Expertise Interpreters Peer Review
Coercion	**Subversive** McDonalds Relativity	**Enforced** Galley Slaves Use of AZT to treat AIDS	

Figure 6.5 A general model of trading zones. From Collins *et al.* (2007).

a project cannot be conceived of in terms of a space of disciplines or departments but is rather more like participation in a community, with the life of the community determining and altering its structures rather than the other way around.

6.3.3 The immaterial culture of interdisciplinarity: trust and expertise

Interdisciplinary collaborations thus involve a matrix of intangible elements. To collaborate, you do not have to share the culture, or the same understanding, of the project on which you are collaborating; less tangible elements may come into play (Seligman *et al.* 2008, p. 8). All that may be required for one to help build or operate an X-ray machine may be things like a desire to help out. To be sure, this matrix and these less tangible elements tend to be drowned out by the task, the topic, the goal, and it is difficult to speak about something that is so easily overwhelmed by the discourse of facts and results. But these things are part of the atmosphere that allows us to inquire and act intelligently.

One of these elements is trust. Trust is a key, if often overlooked, concept in science. Trust here does not mean a moral virtue. Rather, to put it briefly, trust means deferring with comfort to others, in ways sometimes in our control, sometimes not, about a thing or things beyond our knowledge or power, in ways that can potentially hurt us. Trust, which has both a cognitive and a non-cognitive dimension, is extremely important in different kinds of interactions within science and between science and society. Science depends on trust in the form of all those bonds of mutual cooperation that have to exist between scientific colleagues in their various roles. Shrum *et al.* (2007) note the importance of trust in interdisciplinary collaborations.

The correlate of trust is expertise; an expert is often the one to whom one defers to obtain knowledge on which one is dependent. Collins *et al.* (2007) describe 'interactional expertise', or fluency in the discourse of a field without the ability to contribute, as a particular kind of expertise necessary for at least one of their four categories of interdisciplinary collaboration. One of their key examples is Steven Epstein's (1996) description of San Francisco AIDS activists, who collaborated with researchers. Shall we call this a collaborative or coerced interaction? Here, too, the inquiry's the thing, and the activists' recognition of the need for scientific expertise is behind the interaction. When there is no common inquiry—antinuclear activists versus a research reactor, say—and the atmospheres are fundamentally different, interactional expertise cannot happen. Expertise breaks down in the absence of trust and a shared life-world. Without that shared life-world, there is the possibility of reading the meaning of that expert advice differently—that the experts are hired guns, misguided, ignorant, ideologically or politically motivated—conspiracy theories thrive, and the value of expertise vanishes.

6.3.4 Fractionation

Many studies have discussed the fractionation of fields of knowledge under various rubrics: internal differentiation, cross-stimulation, clusters of specialization, hybridization, and so forth (Tanaka 2008, p. 24). Collins *et al.* (2007) note that while many fields, such as that of gravitational wave detection and, we might say, relativistic heavy ion research, appear from the outside to be coherent, when viewed more carefully they can be seen to be divided into numerous subspecializations with no move toward homogeneity—that there is discontinuity when looked at closely. They propose that this may well be the real state of all science—that it is like a surface that seems smooth to the naked eye, but turns jagged when magnified enough. 'It may be that, when examined closely, what appear to be integrated networks of scientists are really conglomerations of small groups bound together by rich interactional expertises' (Collins *et al.* 2007). They add, 'One can always choose to "zoom in" on any area of social life and, as the scale increases and ever more detail is exposed, as with a polished metal surface, what appeared smooth turns out to be jagged.' In this event, they claim, scientific disciplines are like 'fractals' whose structure is reenacted at every scale.

This interesting observation raises many questions. Is the fractionation of the same type throughout science, or does it vary throughout the phases of a construction project like that of a giant telescope or accelerator? And is there a limit to this behavior? Isn't research 'quantized', in the sense that a basic unit of research is the researcher, who builds expertise and competence by being cultivated in a particular area in a particular kind of research context? That person's career and advancement are also determined by rewards and institutional structures, which also seek to keep that person focused on individual areas. This focus on individual areas may thus be for social reasons—prestige, advancement, coping with the administrative structure. The researcher may eventually join with others in a goal met jointly, but begins by mastering one area or set of areas. Research involves not the achievement of a collective oneness but an endless task of integrating and splitting in a communal context. Research is dominated by the practical goal at hand, whose attainment is often negotiated rather than solved like a puzzle. Solutions are always changing, giving rise to new kinds of goals with new expectations of attainment. Research takes place in a 'plain we do not totally control, one that is always also open to the other, to strange and different, beyond power of the center' (Seligman *et al.* 2008, p. 21). Adapting Whitehead's famous remark apropos of the way science treats its founders, we might say that a science that hesitates to move its boundaries is lost, but add that one that seeks to abandon them is lifeless.

6.4 Integrative systems

The interdisciplinary research described above involves regions of knowledge and interactions between researchers. A different, though related, set of issues are raised when such knowledge is considered as arising within *integrative technological systems* that have been planned and promoted for practical applications. Now not only scientists but also administrators, politicians, evaluators, lawyers, and businessmen are involved in a nexus that Klein (Chapter 2 this volume) calls transdisciplinary. A classic example is the Biopolis, established in Singapore, to promote not just medical research but also interactions with clinical applications, and to facilitate the construction of a proper legal and economic infrastructure in which these applications will thrive.

Rüdiger Wink, for instance, refers to *innovation systems* and *integrative technologies*, by which he means 'the systemic linkages between single innovation networks to enhance interaction of knowledge between the networks and their members and to increase the innovative capacity of the whole system' (Wink 2008). These systems connect abstract and theoretical scientific knowledge with 'incumbent technologies'; involve 'no clear boundaries between basic and applied science' insofar as new scientific knowledge can be plugged directly into new goods and services; and involve scientists serving as researchers, managers, and entrepreneurs. Such systems encompass the 'whole knowledge production process', or the entire 'knowledge value chain', extending from knowledge production through review and exploitation, in which the laboratory is only a part—but the rest of the system/chain affects what happens in the laboratories. Wink stresses the importance of *gatekeepers* as the connections between the elements of this process—the parallel to interdisciplinary research—and notes facilitating factors such as *cognitive*, *social*, and *organizational proximity*.

An example of integrative systems at work is human embryonic stem cell research. Here a science with a variety of direct and urgent practical applications is subject to a variety of regulations that cannot be ignored in research, and with huge effects on laboratory research, involving ethics, capital markets, intellectual property rights, and so forth. Different countries have different integrative networks for dealing with stem cell research with different kinds of legal frameworks, and different kinds of links to industries, in play that affect how research takes place. A country's integrative networks may facilitate or hinder its ability to link with networks in other countries.

Justus Lentsch, meanwhile, discusses the need to develop better *boundary institutions* that are accountable both to scientists and to policy makers (Lentsch 2006). Frequently cited examples of institutions with such *dual accountability* include the Dutch Sector Council Model, the European Food Safety Authority, and the European Environment Agency.

6.5 Interactional networks

Even more issues are raised when the public reaction to an integrative network is taken into consideration. A vast distance exists between the knowledge about a subject that circulates in a laboratory and the knowledge about the same subject among the public. A gap exists between the 'load', as it were, born by the discourse in the two cases (Crease 2000). Connecting the two requires a kind of 'impedance matching', in which the load is stepped down. This cannot be a one- or two-step process—education plus science popularization, say—but requires an entire spectrum of *interactional networks* between discourses with different loads. Without it, in public controversies with a technical dimension, positions become not argued but dramatically presented by people who think in slogans and communicate in images. Anti-biotechnology protesters dress up as Frankenstein monsters, protesters call shipments of low-level radioactive wastes 'mobile Chernobyls' while carrying placards of the skull-and-crossbones—actions which serve to displace, in public arenas, those who would argue or inform. The issues—especially highly significant ones like genetically modified organisms, nuclear power, the safety of low levels of toxins, and the ethics of stem cell research—are treated as if they were entertainment, and the public is effectively precluded from engaging them. The German philosopher Jurgen (Habermas 1989) refers to such patronizing tactics as a 'refeudalization' of the public sphere. Interactional networks attempt to overcome such refeudalization, requiring yet another kind of interdisciplinary activity, one that reaches not from one discipline to another but from one way of life to another.

6.6 Conclusion

The physical sciences present excellent case studies of interdisciplinarity, its problems, and its prospects. Interdisciplinary research in the physical sciences is a particularly interesting case because of the amount of experience, the practical challenges, and the theoretical issues it raises in connection with science and its practice. Theorizing about interdisciplinarity can involve considerable posturing and self-congratulation. The physical sciences present clear examples of the inheriting, adapting, and transforming of disciplines—which can transform not only our understanding of science but also of all research. Interdisciplinary research is not simply changing science—its disciplines and the boundaries between them—but forcing the question of what science itself is. Its boundaries are shifting, in ways that make us mindful that it could have been otherwise, and doubtless will change still more in the future. And interdisciplinary research in the physical sciences, its integrative systems and interactive networks, is becoming ever more important to the welfare of the planet, making its study essential. Sites such as the NSLS floor would be a good place to start.

Acknowledgements

Thanks to Harry Collins, Lee Miller, Adam Rosenfeld, Robert C. Scharff, and the German–American Fulbright Commission.

References

Beaver, D. de B. and Rosen, R. (1979). Studies in scientific collaboration. Part III professionalization and the natural history of modern scientific co-authorship. *Scientometrics* **3**, 231–45.

Beaver, D. de B. and Rosen, R. (1979). Studies in scientific collaboration. Part II scientific co-authorship, productivity and visibility in the French scientific elite. *Scientometrics* **1**, 113–49.

Beaver, D. de B. and Rosen, R. (1978). Studies in scientific collaboration. Part I the professional origins of scientific co-authorship. *Scientometrics* **1**, 65–84.

Blume, M. (2006). *Quality assessment of interdisciplinary research.* Remarks at AAAS Conference on 'Quality assessment in interdisciplinary research and education', personal communication.

Boix Mansilla, V., Feller, I., and Gardner, H. (2006). *Quality assessment in interdisciplinary research and education.* Meeting report. American Association for the Advancement of Science, 8 February 2006.

Boix Mansilla, V. and Gardner, H. (2008). *Assessing interdisciplinary work at the frontier: an empirical exploration of symptoms of quality*. Available at: <http://www.interdisciplines.org/interdisciplinarity/papers/6>.

Burbidge, M., Burbidge, G., Fowler, W., and Hoyle, F. (1957). Synthesis of the elements in stars. *Reviews of Modern Physics* **29**, 547–650.

CEDD (Council of Environmental Deans and Directors of the National Council for Science and the Environment) (2008). *Interdisciplinary hiring, tenure and promotion: guidance for individuals and institutions*. Available at: <http://ncseonline.org/CEDD/cms.cfm?id=2042> (accessed 2 July 2008).

Chubin, D.E. (1976). The conceptualization of scientific specialties. *Sociological Quarterly* **17**, 448–76.

Collins, H., Evans, R., and Gorman, M. (2007). Trading zones and interactional expertise. In: H. Collins (ed.) *Case studies of expertise and experience* [special issue of *Studies in History and Philosophy of Science* **38**(3)], 686–97.

Comte, A. (1988). *Introduction to positive philosophy*. Indianapolis, IN: Hackett.

Crease, R.P. (1993). *The play of nature: experimentation as performance*. Bloomington, IN: Indiana University Press.

Crease, R.P. (2000). A top ten for science and society. *Physics World* **13**(12), 17–18.

Crease, R.P. (2003). Inquiry and performance: analogies and identities between the arts and the sciences. *Interdisciplinary Science Reviews* **28**(4), 266–72.

Crease, R.P. (2005a). Quenched! The ISABELLE saga, part 1. *Physics in Perspective* **7**(Sept.), 330–76.

Crease, R.P. (2005b). Quenched! The ISABELLE saga, part 2. *Physics in Perspective* **7**(December), 404–52.

Crease, R.P. (2006). From workbench to cyberstage. In: E. Selinger (ed.) *Postphenomenology: a critical companion to Ihde*, pp. 221–9. Albany, NY: State University of New York Press.

Crease, R.P. (2008a). The national synchrotron light source, part I: bright idea. *Physics in Perspective* **10**, 438–67.

Crease, R.P. (2008b). Recombinant science: the birth of RHIC. *Historical Studies in the Natural Sciences* **38**(4), 535–68.

Crease, R.P. (2008c). Life at the frontier. *Physics World*, **21**(October), 45–48.

Dogan, M. (1996). The hybridization of social science knowledge – navigating among the disciplines: the library and interdisciplinary inquiry. *Library Trend* **45**, 297–314. Available at: <http://findarticles.com/p/articles/mi_m1387/is_n2_v44/ai_18928468>.

Epstein, S. (1996). *Impure science: AIDS, activism, and the politics of knowledge*. Berkeley, CA: University of California Press.

Evans, R. and Marvin, S. (2006). Researching the sustainable city: three modes of interdisciplinarity. *Environment and Planning A*, **38**(6), 1009–28.

Everitt, C.W.F. (1992). Background to history: the transition from little physics to big physics in the Gravity Probe B relativity gyroscope program. In: P. Galison and B. Hevly (eds) *Big science: the growth of large-scale research*, pp. 212–35. Stanford, CA: Stanford University Press.

Forman, P. (2007). The primacy of science in modernity, of technology in postmodernity, and of ideology in the history of technology. *History and Technology* **23**, 1–152.

Galison, P. (1997). *Image and logic: a material culture of microphysics.* Chicago: University of Chicago Press.

Ginev, D. (1997). *A passage to a hermeneutic philosophy of science.* Amsterdam: Rodopi.

Ginev, D. (2006). *The context of constitution: beyond the edge of justification.* Dordrecht: Springer.

Grossman, B. and Hart, G. (2008). Deux sculpteurs de mathematiques. *Pour la Science* no. 369 (July), 90–5.

Habermas, J. (1989). *The structural transformation of the public sphere: an inquiry into a category of Bourgeois society.* Cambridge: MIT Press.

Hagstrom, W. (1964). Traditional and modern forms of scientific teamwork. *Administrative Science Quarterly* **9**, 241–64.

Hagstrom, W. (1965). *The scientific community.* New York: Basic Books.

Hargens, L.L. (1986). Migration patterns of U.S. PhDs among disciplines and specialities. *Scientometrics* **9**, 145–64.

Heeger, A.J. (2000). *Nobel Lecture. Semiconducting and metallic polymers: the fourth generation of polymeric materials.* Available at: <http://nobelprize.org/nobel_prizes/chemistry/laureates/2000/heeger-lecture.html> (accessed 2 July 2008).

Kohler, R.E. (1991). *Partners in science: foundations and natural scientists, 1900–1945.* Chicago: University of Chicago Press.

Kuhn, T. (1996). *The structure of scientific revolutions.* Chicago: University of Chicago Press.

Landau, R.H., Páez, M.J. and Bordeianu, C.C. (2008). *A survey of computational physics: introductory computational science.* Princeton, NJ: Princeton University Press.

Law, J. (1987). Technology and heterogeneous engineering: the case of Portuguese expansion. In: W.E. Bijker, T.P. Hughes, and T. Pinch (eds) *The social construction of technological systems: new directions in the sociology and history of technology*, pp. 111–34. Cambridge, MA: MIT Press.

Lentsch, J. (2006). *Quality control in scientific policy advice.* Report of the Expert Symposium at the Berlin-Brandenburg Academy of Sciences and Humanities. Available at: <http://www.easst.net/review/mar2006/lentsch>.

Ludlam, T.W. and Wegner, H.E. (eds) (1984). Quark matter '83: Proceedings of the Third International Conference on Ultra-relativistic Nucleus–nucleus Collisions, Brookhaven National Laboratory, 26–30 September 1983. *Nuclear Physics* **A418**.

Nissanni, M. (1997). Ten cheers for interdisciplinarity: the case for interdisciplinary knowledge and research. *The Social Science Journal* **34**, 210–16.

Palmer, C.L. (1996). Information work at the boundaries of science: linking library services to research practices – navigating among the disciplines: the library and interdisciplinary inquiry. *Library Trends* **45**(2), 129–33.

Palmer, C.L. (1999). Structures and strategies of interdisciplinary science. *Journal of the American Society for Information Science* **50**, 242–53.

Pantin, C.F.A. (1968). *The relations between the sciences.* Cambridge: Cambridge University Press.

Seidel, R. (1992). The origins of the Lawrence Berkeley Laboratory. In: P. Galison and B. Hevly (eds) *Big science: the growth of large-scale research*, pp. 21–45. Stanford, CA: Stanford University Press.

Seligman, A.B., Weller, R.P., Puett, M.J., and Simon, B. (2008). *Ritual and its consequences: an essay on the limits of sincerity.* New York: Oxford University Press.

Shrum, W., Genuth, J., and Chompalov, I. (2007). *Structures of scientific collaboration* Cambridge, MA: MIT Press.

Swanson, D.R. (1986). Undiscovered public knowledge. *Library Quarterly* **56**(2), 103–18.

Tanaka, M. (2008). *Toward an understanding of scholarly communications: scientific computing group at Brookhaven National Laboratory*. PhD dissertation, C. W. Post Campus of Long Island University.

Westfall, C. (2003). Rethinking big science: modest, mezzo, grand science and the development of the Bevalac, 1971–1993. *ISIS* **94**, 30–56.

Wilson, K.G. (1984). Science, industry, and the new Japanese challenge. Proceedings of the IEEE. **72**, 6–18.

Wink, R. (2008). Integrative technologies and knowledge gatekeepers: bridging the gap between epistemic communities in the case of stem cell research. *International Journal of Learning and Change* **3**, 57–74.

Zalaya, R. and Barrallo, J. (2008). *Classification of mathematical sculpture*. Available at: <http://www.mi.sanu.ac.rs/vismath/barrallo1/Index.html>.

CHAPTER 7

Integrating the social sciences: theoretical knowledge, methodological tools, and practical applications

CRAIG CALHOUN AND DIANA RHOTEN

The distinctions among the social science disciplines are historically forged and largely arbitrary. Nonetheless they are reproduced not only in boundary struggles but also in the training of graduate students, the writing of textbooks, and the review processes of scholarly journals. They may be more matters of style than method—characteristic structures of attentions, values, and ways of solving problems—but they are jealously defended and also maintained simply by habit and social networks.

By contrast, it is common to speak of physical sciences in the singular. The common scientific method suggests unity despite the differences among disciplines. There is a tacit commitment to commensurability in science (Wilson 1998). There is no analogous expectation in the humanities, but instead a celebration of different perspectives. University arts and humanities faculties are always plural, a reflection of divergent methods as well as topics. Oddly, these are both responses to the decline of theological authority. On the one hand a new unifying faith, on the other the recovery of a classical idea of multiple liberal arts, each distinct as craft skills are. In this as much else, the social sciences occupy an in-between position. Since the *methodenstreit* of late nineteenth-century Germany, the social sciences have been torn by recurrent struggles over scientific universalism versus humanistic particularism.

In any case, social scientists have divided increasingly between those who understood human behavior as a natural phenomenon to be studied using 'objective' scientific methods and those who believed human behavior could not be understood outside of particular histories, cultural contexts, or subjective understandings. Universalizers were most prominent (though not universal) in economics, sociology, and political science and much less dominant in history and anthropology (Wallerstein 2003). This oversimplified historical account fits the Anglo-American model most closely, but this model of disciplinary differentiation has diffused globally so that these five disciplines have become the common core of social sciences around the world (Abbot 2001b).[1]

Be this as it may, the social sciences have been shaped by interdisciplinary projects and communication almost as long as they have been divided into disciplines. Some of these have coalesced into new disciplines; some have been at the heart of professional training programs; some have created centers for combining different disciplinary perspectives on a specific topic; most have been more temporary projects in which researchers from two or more disciplines combine their different methods, analytical frameworks, or empirical knowledge to try to advance an intellectual agenda that may change one or both disciplines or may simply result in new knowledge incorporated eventually into the separate fields. Of course the transformation can also be more personal. At least as far back as the great early twentieth-century anthropologist Ralph Linton, leading social scientists have been quoted reminding their colleagues that the most effective interdisciplinary relations were those that took place under a single skull.[2]

In this chapter, we look to the development of area studies and quantitative research methods as the two great interdisciplinary engagements of the postwar era. These movements reveal the different agendas of pursuing a comprehensive understanding of concrete patterns in social life—in this case, geopolitical regions and/or civilizations—and of pursuing tools to support innovation and greater rigor inside different disciplines. These different intellectual agendas are linked to different patterns of institutional autonomy and interrelationship. We then turn to interdisciplinary fields organized in terms of problems, issues of public concern, and/or professional practice. These have been central to the growth of interdisciplinary social science, especially since the 1960s. They also represent a third pattern of institutional organization.

7.1 Prehistory and early history

Early social science grew out of a predisciplinary context. As one of us has written elsewhere, 'Hobbes and Locke could integrate politics and psychology without need for an interdisciplinary field of political psychology. Vico and Montesquieu informed anthropology, history, sociology and political science in equal measure.... Adam Smith was not one to distinguish theoretical from applied economics, just as he saw the intimate connections of both to the "moral sentiments" and other concerns of what would later be called psychology and sociology' (Calhoun 2001). The early modern period did come to make a distinction among economy, polity, and society as spheres of life, but the development and separation of corresponding academic disciplines was minimal before the late nineteenth century.

[1] Psychology is an ambiguous case—often, as in the current period, closely linked to the biomedical sciences but through much of the middle twentieth century a 'core' social science.

[2] Versions of this quotation are in fact attributed to a variety of scholars—much in the manner of the famous remark about standing on the shoulders of giants studied by Robert Merton (1965).

All this changed with reforms of the university designed to emphasize the production and communication of new knowledge. These were anticipated by the founding of new universities in Britain, were given dramatic expression in the restructuring of German universities, and then framed transformation of American higher education after the Civil War. It was in this new context that the university became the breeding ground of new disciplines, partly because it underwrote careers of academic employment defined by research attainment. PhDs enabled people to join universities faculties; publications supported advancement in faculty ranks. Even undergraduate study was rethought to reflect this new orientation to research; majors were introduced, displacing much of the old curriculum that was both 'classical' and general with specialized knowledge based (at least in principle) on new research.

The social sciences were consolidated as distinct disciplines in the late nineteenth and early twentieth centuries. The configuration varied slightly from country to country, as did the relative importance of ties to humanities and natural science. But in both Europe and America the major disciplinary departments and associations date from this period. They were formed partly by subdividing existing academic fields. In the United States, for example, history was a sort of parent discipline. Economics divided from it first, taking sociology along as a subfield of economics, but leaving government (or political science) behind as a subfield of history. A few years later, sociologists branched off from economics and political scientists branched off from history. Even within the academy this had some political edge: history and political science were more conservative, economics and sociology more challenging, and by the time sociology split it included many of the more radical erstwhile economists (Ross 1991).

But at least as important, this era of disciplinary formation was also one of gathering fields into the university that had initially grown partly outside of the academy. The American Social Science Association, thus, was as much a parent of social science as academic history, though it was a mixture of scholars, reformers, and educated lay people (Haskell 1977). Its members were not really disciplined by academic divisions, and were commonly engaged with inquiries that cut across what would later be disciplinary lines. The divisions were not random, though, since for more than 200 years thinkers had seen the modern world as shaped by its organization into three big domains—polity, economy, and society—each to some extent autonomous of the others. Similarly, the divisions of the West from other civilizations, of European colonial powers and American conquerors of a continent from indigenous peoples encouraged a kind of expertise in the other that would inform anthropology.

It was in this context, thus, that the idea of interdisciplinarity was born. It reflected the new consolidation of disciplines. Without some special effort, many worried, academics might talk only to each other in ever narrower specialties and subspecialties. The worry was particularly acute in the social sciences. On the one hand, these lacked the strong sense of a common underlying scientific method that helped to unify the natural sciences. On the other hand, social scientists were often moved by the desire to inform public affairs and even solve public problems. If advances in scientific research and the pursuit of rigor divided fields, the effort to solve problems called them back together.

In the United States, these concerns informed the 1923 creation of the Social Science Research Council (SSRC), where the idea of interdisciplinarity received its first explicit formulation. Charles Merriam, a political science professor at the University of Chicago, helped conceive the SSRC, calling for the 'closer integration of the social sciences themselves':

> The problem of social behavior is essentially one problem, and while the angles of approach may and should be different, the scientific result will be imperfect unless these points of view are at times brought together in some effective way, so that the full benefit of the multiple analysis may be realized. (Worcester 2001, p. 16)

By September 1930, the SSRC was already re-examining and restating its existing policy when it declared:

> The Social Science Research Council is concerned with the promotion of research over the entire field of the social sciences. The Council's thinking thus far has been largely in terms of social problems which cannot be adequately analyzed through the contributions of any single discipline. It is probable that the Council's interest will continue to run strongly in the direction of these interdiscipline inquiries (Barnett *et al.* 1931, p. 286).[3]

The Depression, New Deal, and World War II gave social scientists plenty of problems to address. Interdisciplinary teams examined issues from job creation to the training of soldiers. But it was after the war that leading social scientists and a number of universities took on a significant restructuring of academic attention, giving interdisciplinary agendas deeper roots. Two broad agendas were most influential in this: the development of international knowledge and the improvement of research methods. Neither was narrowly problem-focused, though better capacity to solve future problems was a rationale for funding each.

7.2 Area studies

After the war, social scientists returned to universities both invigorated by their wartime experiences and challenged by a sense of their own previous limits. They sought widespread improvements in social science in order to make it an effective source of objective knowledge that could inform government policy. And as former soldiers swelled university enrolments to record numbers, social science departments grew. Graduate education and research programs grew with them. Former soldiers and returning professors alike brought life experiences—including service in intelligence branches of the military—that they wanted to understand. They brought a hope that scientific knowledge could help avert war and deal with domestic social problems.

[3] While the *Oxford English Dictionary* cites a 1937 article in the *Journal of Educational Sociology* as the first printing of the term 'interdisciplinary', it had in fact appeared annually since at least 1930 in the Social Science Research Council awards listings and reports printed in the journals of various professional societies including *Journal of the American Statistical Association*, *American Sociological Review*, and *American Economic Review* (Sills 1986; Prewitt 2002).

The SSRC founded a Committee on World Area Research in 1946. In 1950 the Ford Foundation began the Foreign Area Fellowship Program, which it later turned over to the joint SSRC–American Council of Learned Societies (ACLS) committees on different world areas to administer. Ford put nearly 300 million dollars into this project, and it was in due course joined by other foundations and by the US government which made major investments in foreign language teaching and foreign area research. Today many social scientists regard this as largely an investment in the humanities. Social scientists were, however, central to the area studies project. It is only in the last 30-some years that area studies programs have tilted towards humanities fields. And this is largely the result of secession by social scientists, not conquest by humanists.

Area studies programs responded to a widespread sense that Americans—and in particular American intellectuals and academics—were too ignorant of other world regions to adequately shoulder the burden of world leadership the country was assuming. But the problem focus was in the background; it explained the investment but not the organization. The organization of the area studies programs was interdisciplinary, not only because the goal of ramping up knowledge cut across disciplines, but also because grasping another region of the world (or civilization) in its fullness seemed to demand bringing together the perspectives of different disciplines.

The area studies fields differed from each other in the extent to which research and teaching focused on contemporary politics or civilizational history and different disciplines accordingly figured more or less prominently. Thus the Cold War put politics at the center of Russian and East European studies, and even contributed to the demarcation of the region itself. South Asian studies certainly confronted political issues, but focused more on civilization and culture. And there were other characteristic thematic foci in different area studies fields. Economic development was front and center for Latin American studies, and the formation of 'new nations' was a key theme for African studies.

During the postwar period, however, all the area studies fields shared a broad intellectual orientation associated with the idea of modernization. Economic development, political reform and the creation of new national institutions, transformation of social institutions, expansion of literacy and consequent cultural production, and even change in psychological attitudes were all seen as parts of a common process. And if modernization described what was shared in this process, different histories and cultures shaped distinctive patterns in each region. This connected work in the area studies fields to disciplinary agendas.

The connection, however, came unstuck. There was always a fault line. Some of this reflected the ways in which disciplines pursued generalizations, especially in an era when the notion of 'covering laws' was prominent in philosophy of science (and the vague appropriations of this philosophy filtered into social scientific understanding). The area studies fields, by contrast, seemed to be particularizing, focused on the specifics of local conjunctures of history, culture, politics, and even environment. Disciplinary knowledge was understood as ideally abstracting from such specifics to establish more universal laws.

This was always a caricature of area studies research, and perhaps a misunderstanding of what disciplines themselves achieved. It is easy to mock either side: the psychologist who thought human nature could be found in experiments involving only white, middle class, male American undergraduates; the anthropologist who responded to every assertion of a more general causal pattern with 'well, it's not exactly so on the island I studied'. But there is a point of more basic significance.

The area studies projects at their best were not so much about idiographic particulars as about the notion that there were and are different ways to be human, to be social, to be political, and even to have markets—and therefore that the pursuit of more general knowledge required working with attention to specific historical and cultural contexts and patterns. Such knowledge could be of broad application without being abstractly universal. And indeed, the area studies fields contributed to major analytic perspectives that far transcended their initial sites of development. Benedict Anderson's account of nationalism as a matter of imagined communities was informed by Southeast Asian studies, but not contained by it (Anderson 1991). So too James Scott's effort to understand states and the ways states viewed societies (Scott 1998). Dependency theory developed as an effort to understand specifically Latin American problems, as did Albert Hirschman's work on development assistance and unbalanced growth (Prebisch 1950; Hirschman 1958; Frank 1967; Cardoso and Faletto 1979). The 'world systems theory' of Immanuel Wallerstein was deeply shaped by African studies as well as by Braudelian global history and Marxist political economy and indeed the earlier Latin American dependency theories (Wallerstein 1974). And so forth.

Each of these examples became part of interdisciplinary discussions—of development and underdevelopment, class and power, power and knowledge, states and nations. Of these, only the first really became an academic field of its own—and in the United States development studies is only weakly established; it is more substantially institutionalized in Britain and some other settings, and more dominated by disciplinary economics in the United States. Marxism was for a time a vital interdisciplinary discussion, with strong social movement links, but never with strong academic institutionalization outside the communist countries. Wallerstein's Braudel Center at Binghamton was influential but not widely imitated. And if political economy remains a topic or perspective that many social scientists would claim, its base of intellectual reproduction is not well-established.

This points to a more general problem with interdisciplinary work. When it lacks institutional conditions of reproduction, it is at the mercy of disciplines which may either claim it or ignore it or, most often, incorporate some ideas from interdisciplinary projects without providing ways of sustaining the intellectual ferment that produced them. So while a few universities set up autonomous departments of Latin American or East Asian studies, many more set up interdisciplinary committees or centers and left the awarding of PhD degrees to disciplinary departments.

In the 1950s and 1960s, the area studies fields were relatively well financed and often able to offer funding to students not funded by their disciplinary departments. There was a new infusion of students—once again like the soldiers with motivating life experiences, but this time also often with language skills and local knowledge forged in the Peace Corps. More generally, while the university system expanded, there were jobs for the political scientists and sociologists with area studies emphases. This began to change in the 1970s and became a crisis by the 1990s. The system stopped expanding and suffered a shortage of faculty jobs, sharp tightening of tenure standards, and new pressures on graduate students to demonstrate disciplinary publications before entering the job market.[4] In this context, disciplinary departments exercised discipline by rewarding intradisciplinary achievement. At the same time, area studies programs saw their proportionate funding decline, not least as graduate student financial aid became widely tied to teaching assistantships administered by departments. By the 1980s and 1990s, efforts to shrink cohorts further consolidated disciplinary control.

In many disciplines, academic initiative turned away from critical theory and toward more or less formal methods. Economics effectively seceded from area studies as it relied increasingly on mathematical models and on theories (some lumped together as neoliberal) that stressed more or less universal microfoundations. Economists who retained strong area interests often wound up in interdisciplinary programs rather than economics departments—not just area studies but urban studies, policy analysis, and development studies—or working for the World Bank or other non-academic institutions. In varying degrees sociology and political science followed suit, leaving the area studies fields increasingly tilted towards the humanities.

Despite the strong and influential work rooted in area studies in the later 1960s and 1970s, these fields languished without capacity for autonomous reproduction (not just graduate training programs but capacity to hire). Then the end of the Cold War came as a further blow. This encouraged a dramatic expansion in international work conceived as directly global—that is, about what might in principle happen everywhere—rather than as context specific. Ironically, in other words, attention to globalization came to a considerable extent at the expense of attention to the specific regional and other contexts through which globalization was refracted and in which it took on different meanings.

It would be an error, nonetheless, to write the obituary of area studies programs—still probably the fields of interdisciplinary social science and humanities with the strongest record of achievement. In the first place, the new fields of international studies and globalization studies have not produced the major intellectual accomplishments many expected, and are perhaps at their best when integrated with area studies rather than imagined as alternatives. A strong example is the struggle many in international studies—and especially the quasi-discipline of international relations—have faced recently in trying to figure out how to take religion seriously after assuming it had been defined out of international relations in 1648. The difficulties have to do with reigning theories, of course, but also with the institutional and intellectual distance from those with more knowledge—often area studies specialists and researchers from the humanities and fields like history that straddle humanities and social science.

Shifting concerns drive new engagements with area studies today. Perhaps more inclusively described as a concern for situating knowledge in contexts, these new engagements take up themes like Islam (or Christendom) as they cut across traditional regions. They examine interconnections between regions (like those that long linked the Middle East to South Asia, that ran along the Silk Road or coastal trading routes, but also those that shape

[4] These pressures were in general more acute higher in the acad emic pecking order, but they were present throughout.

pan-Asian economic integration today). They take up previously neglected areas like Central Asia, which sits at the intersection of the Middle East, the Eurasian region defined by the former Soviet Union, and South Asia, and which attracts attention not just because of an idea of coverage but because of its geopolitical significance.

7.3 Research methods

The area studies fields were challenged in the late twentieth century by a reassertion of more or less 'universalizing' disciplinary agendas married to quantitative research methods. But of course neither these agendas nor the methods arose in order to 'discipline' area studies.

Statistics had long been important to social science, both in the sense of technique and in that of an accumulated knowledge base. The typical use focused on descriptive statistics: absolute numbers, percentages, and distributions. One wanted to know 'the statistics' on crime or employment. If the state was a central collector and user of statistics (as the name suggests) the non-disciplinary pioneers of social science recurrently mobilized statistics to make cases for social reform (Stigler 1986; Porter 1995).

Statistics grounded in probability theory made uneven headway in the social sciences before World War II (Hacking 1990). They mattered most in economics and psychology, though even there much work continued to focus on absolute numbers and measures of association. The problem of establishing patterns of heredity was especially influential, inspiring figures such as Galton, Edgeworth, and Pearson who developed multivariate approaches. As both Pearson and Yule analyzed families of curves this work moved out of evolutionary theory and into economics and analyses of phenomena like welfare reform. Gradually efforts to compare groups, and especially differential rates of occurrence of certain phenomena, grew more influential.

The rise of testing in psychology (and related interdisciplinary education research) was prominent, spurred on by its use in World War I efforts to classify military recruits as well as by the expansion of public schooling. Durkheim's study of suicide pursued intergroup variation in rates as part of a disciplinary project intended to establish the autonomy of social explanations from psychological ones (Durkheim 2006). Certain interdisciplinary fields and issues were especially important, including notably crime and public health, in each of which the attempt to introduce treatments and measure changes in rates became basic.

A variety of applied and problem-oriented fields were influential in pushing the collection and analyses of social statistics forward. Statistics were demanded by governments and reformers alike. The SSRC was once again influential, both launching specific statistical projects (like the work of its first president, Wesley Clair Mitchell, establishing the business cycle) and launching more general interdisciplinary efforts to promote the development and use of better quantitative research techniques. With money from the Rockefeller Foundation it paid for the creation (and even the physical buildings) of interdisciplinary institutes for social science research at several universities such as Chicago and North Carolina. The Rockefeller goal was always concrete, 'realistic' solutions to pressing social problems. Realism was identified with quantification (but also with short-term politically palatable solutions—and on the later count Rockefeller was often disappointed) (Richardson and Fisher 1999; Camic 2007). Through all of this, statistics maintained a hybrid organization—partly a field of its own (though increasingly divided between biostatistics and social statistics) and partly a subfield within different disciplinary structures.

This encouraged an interdisciplinary flow of knowledge about new techniques. The structure was different from area studies, where scholars often felt a divided loyalty. The reason may be that the pursuit of knowledge about a place (or civilization, culture, etc.) appeared as an alternative project to the pursuit of knowledge about a dimension of social organization (economy, society, politics). The modernization framework suggested a two-way trade in which area knowledge was fitted into and completed the larger framework (sometimes modifying previous generalizations) while the larger framework gave structure to area knowledge. But area knowledge appeared to most in the disciplines more as a set of positive claims—the results of research—than like the equipment for new research. Of course one might say that before attempting a generalization about, say, the nature of politics in a particular area, one ought to be well informed about all the different varieties of political systems and practices. But in fact this was just what the creation of political *science*, as distinct from older more historical approaches to the study of government, resisted. It sought the capacity to abstract from such comparative historical knowledge, to make it less important to the work of generalization. And a key to this pursuit was reduction to statistics, disembedding the study of specific phenomena—say party systems and voter turnouts—from mastery of detailed contextual knowledge. This created a place within disciplinary structures for statistics as part of a general toolkit rather than an alternative specific field of study.

The transformation of the social sciences by this statistical research and more generally by the disembedding of particular research topics from detailed contextual knowledge accelerated in the 1950s and 1960s. The New Deal and World War II once again had a major impact. Many social scientists recruited into the war effort worked with larger data sets than they had previously encountered. More basically, many found arguments in terms of broad analytic-interpretative frameworks unpersuasive in their new work situations, while claims straightforwardly about 'the facts' and perhaps their causes could matter a great deal. During the postwar era, the study of more or less disembedded research topics, aided by statistics, became a program for improvement of the social sciences. It would make them less ideological and more scientific, many thought, and it would give a clear basis to specialized training in research skills (and perhaps theory, which was in some of its formulations equally disembedded from history and culture) rather than accumulated substantive knowledge.

Indeed, many of the specific research techniques taken up by the early 'behavioralists' in political science came from sociology and social psychology. Above all, they were associated with the rise of survey research. This had early roots outside academic social science and was aided by the appeal of some of its results to journalists. Accounts of 'the average American' joined statistics on various sorts of deviation (Igo 2007). Opinion polls followed in their wake, informing not only political campaigns but also market research. Academic survey research developed as an interdisciplinary field dedicated to raising the standards of this partly extra-academic pursuit at the same time as advancing social science. While it had its most enduring associations with sociology and political science, it is worth noting that it drew much from its roots in the 'golden era' of a kind of interdisciplinary social psychology (Sewell 1989).

Survey research became central to an interdisciplinary field of specialists in data collection, closely related to but somewhat distinct from statisticians as specialists in data analysis. Survey data informed (and transformed) the study of elections, inequality, race, education, and other topics. In some cases, a large, complicated data set (say the Panel Study of Income Dynamics (PSID) begun in 1968) developed its own cadre of experts and became the focus on interdisciplinary discussion (in the PSID case mainly of economists and sociologists). In some cases a survey program was not topically specific but opened an integrated data collection effort to researchers from different fields. Thus the National Opinion Research Center's (NORC) General Social Survey has since 1972 provided researchers from different disciplines the opportunity to purchase questions or modules to gain data on their specific concern that could be related to a common background of demographic and attitudinal data.

Survey data remain central to social science, if somewhat less dominant than a generation earlier. But survey research is less prominent as an interdisciplinary field. The reason is that it has moved largely outside research centers to contract organizations that deliver data to researchers' specifications. Some of these are based at universities and some are for-profit companies. But the crucial point is the same, a disconnection of technical performance of research data collection from the intellectual centers of the social sciences. There is major work to be done on data archiving and accessibility, but it is not in itself social science. Likewise, the collection of survey data has become increasingly a facility for social science, not substantively part of social science as it was in the postwar period.

Although many of the specific techniques used in social science have much older provenance, often in other sciences, their use became widespread only in the 1960s and after (Calhoun *et al.* 2005). Both multivariate data analysis and mathematical modeling were, for one thing, greatly aided by greater computational power and easier access to it. Graduate training programs substantially increased the numbers of social scientists able to use sophisticated methods. And the proportionate time such programs give to quantitative data analysis as distinct from quantitative data collection has typically increased. Most of the training and much of the intellectual work is discipline specific, but throughout the period analytic techniques have spread in a pattern of interdisciplinary diffusion aided by a series of particularly intense nodes of exchange and innovation.

At the same time as capacity to handle large-scale data sets inspired one set of methodological projects, capacities for mathematical modeling inspired another. The two were joined in the cybernetics of the 1950s and the systems theory that grew from it but were often divergent later. Formalization took the logic of abstraction a step further than the disembedding of specific empirical concerns from comparative or historical contextual knowledge. It pursued clarification of causal relationships abstracted from data. The mathematicization of economics offers a relatively extreme disciplinary example, one imitated in some degree in other social sciences (as economics had imitated physics).

But this turn to applied mathematics and away from data peaked in the 1990s. Since then mathematical formalization has remained a crucial tool but innovations have more often been driven by empirical analysis.

The link forged between economics and psychology in order to study decision making—including decision making shaped by imperfect information and different patterns of learning and culture—is exemplary. This is perhaps more of a 'cutting edge' in economics than in psychology, where the dominant tendency is interdisciplinary connections into the brain sciences and the study of cognition. But social scientists from several disciplines have taken up the pursuit of an understanding of social phenomena rooted in 'microfoundations' including rational choice theory, agent-based modeling, and investigations of the ways in which particular empirical situations from social networks to cultural contexts shape the behavior of agents.

The spread of chaos theory and related approaches to the analysis of complexity further illustrates the pattern. In this case much of the foundational work came from the natural and especially physical sciences. Interdisciplinary centers like the Santa Fe Institute were important nodes in the networks of diffusion and also sites of collaboration and innovation. Still another example is network analysis, which spread from foundations in mathematics (especially graph theory) and anthropology to achieve a center of gravity in sociology but influenced work throughout the social sciences.

The distinctive pattern here is of interdisciplinary fields (and both looser communication and specific collaborations) organized by abstraction from concrete particulars rather than an attempt to integrate dimensions in a holistic picture. The various waves of innovation in research methods were occasionally accompanied by dramatic declarations of revolutionary leaps forward and capacity to integrate all of the social sciences (or to integrate social science fully into science generally). For the most part, though, they operated in practice as providers of tools to individual disciplines. Interdisciplinarity was most intense in a phase of innovation. Sometimes there was a topic as well as a technique—networks for example—and this helped to sustain a quasi-autonomous field. But in general, as techniques became more routinely available the involvement of leading social scientists in the interdisciplinary domains faded and the role of technical support staff and facilities increased. Beyond the provision of specific techniques, the interdisciplinary movements for quantification and 'behavioral' analysis have shaped the intellectual orientation and style of the social sciences generally.

Area studies and quantitative research methodology can be seen as opposite ends of a continuum. They represent 'humanistic' and 'scientific' approaches to social science, and for better or worse these are often opposed to each other. But by looking at the interdisciplinary character of each we can get a better sense of the issues than simply by applying those nineteenth-century labels.

Area studies mobilize different disciplinary perspectives in order to achieve an inclusive, integrated view of societies or cultures in different settings. It is in a sense holistic, trying to bring all different aspects of knowledge about its particular focus together. Work done in area studies fields may in turn inform and improve disciplinary inquiries, but this is usually not its *raison d'etre*. Quantitative research methodology—the actual study of methods, not just the use of them—does not pose any particular topic or focus for investigation. It improves tools. But with the tools comes an orientation to abstracting particular aspects of social life from their contexts—whether variables, or mechanisms, or dimensions of structure. The methods have affinities to certain ways of thinking about social reality.

One may of course use quantitative methods in area studies. The point is not radical mutual exclusion. It is the contrast between seeking context-specific knowledge and seeking the capacity to disembed findings from specific contexts. But the disciplines—especially the so-called core social science disciplines of economics, political science, and sociology—are set up in terms of the disembedding project, for each claims a domain of life distinct from the others and from any one context in which examples of that aspect of life appear. Accordingly, it is much easier for the disciplines to appropriate the results of quantitative research methodology as new tools, while the area studies fields look like competitors.

The image of a continuum is informative, then, but also problematic. As we have noted, for example, both area studies and quantitative research agendas were informed by the desire to address public problems. They each had a 'pure science' or 'purely scholarly' dimension and ambition, but they also engaged researchers and perhaps especially attracted funders because of the belief that they could help the country and the world address social issues. Many other interdisciplinary projects, arguably falling between the poles of the continuum, reflect this orientation to public problems and practical action more directly.

7.4 Conclusion

Area studies and quantitative research methods were the two most influential interdisciplinary movements in the social sciences of the postwar era, and in a sense the most 'general', but they were not the only ones. A wide variety of other interdisciplinary projects appeared then and continue to appear all the time. While some involve new research methods or reorganized foci of international attention, most follow a third pattern. They focus on an issue of public concern or professional practice. This is often framed in terms of a social problem. Of course, one needn't think business is a social problem to think that social science can contribute to management education or research on organizational behavior. Gender is of widespread public interest in ways not limited to specific problems of gender discrimination.

There are at least three different patterns in the organization of interdisciplinary social science giving attention to matters of public concern:

- There is work that brings researchers together around a specific problem to which relatively short-term solutions are sought. For example, the sense that welfare reform was urgent mobilized many researchers in the 1990s. Likewise, funders today have committed significant resources to studies of obesity and potential societal responses and seek to mobilize social and behavioral as well as biomedical scientists.
- There is work that combines different disciplinary perspectives to address a topic—like cities and urbanization, or media and communication. This may or may not

involve a problem focus. It may simply reflect the prominence of the topic, often as a result of social change, and sometimes as a result of its relative neglect in disciplinary research.

- There is work designed to underpin professional practice. This is sometimes left out of accounts of interdisciplinary social science, but in fact professional schools are major sites of both research and pedagogical collaboration. Schools of social work and public affairs, public policy, or public administration are essentially social science programs with an emphasis on practical action. Schools of business and education are overwhelmingly rooted in social science, though in varying degrees their different component programs—like marketing or testing—have gained quasi-disciplinary standing on their own. Schools of law, nursing, medicine, and journalism all employ social scientists and provide interdisciplinary bases for their work.

Despite interdisciplinary innovations, the social sciences have retained substantially the same basic disciplinary structure since their formation in the late nineteenth and early twentieth centuries. Interdisciplinary programs have been added, without great effect on the disciplines themselves. Abbott sees this as likely to continue, partly because the disciplines control much of the allocation of academic resources and partly because they occupy ecological niches that allow them to reproduce (Abbott 2001b). Others see the disciplinary system as more likely to decay or be transformed (Whitley 1967; Fuller 1991; Turner 2000). There have been several calls for the reorganization of social science, such as Wallerstein's suggestion that there be a regrouping around quantitative, ethnographic, and historical methods. Wallerstein suggests that some proportion of each of these new groups would come from each of the traditional social science disciplines but that they would more truly reflect the research engagements of academics.

Ironically, while the disciplines have considerable power over academic turf, they generally have much less capacity to command mutual intellectual engagement (let alone agreement) among their members. They are like states that still tax and police and even elect governments but have relatively weak national unity. In the United States, economics is perhaps most unified with a strong sense of a common method of economic analysis. Even here, the unity is sometimes deceptive, both because disagreements within 'mainstream' economics are deep and because there are vocal and sometimes influential communities of 'non-standard' economists. Psychology is deeply split between practitioners and researchers, but its researchers are also divided between those gravitating increasingly to the natural sciences, often focusing on the brain as such, and those sometimes referred to as applied psychologists, joined more to the social sciences, more likely to focus on development, social psychology, environmental psychology, and so forth. Political science has long been sharply divided into the four domains of political theory, international relations, American politics, and comparative politics. Political theory once had ambitions to unify the field, but has become a relatively distinct subfield where the history of theory and normative theory are strong. Anthropology also traditionally has four fields, and in several late twentieth-century instances departments came apart along those seams, particularly those dividing physical anthropology (with its strong links to the biomedical sciences) from cultural anthropology (which has partially absorbed anthropological linguistics). Archaeology, with its somewhat different material conditions of work, has long been a quasi-autonomous field and anthropological archaeologists often have strong ties to classical archaeologists.

Geography is divided between physical and cultural geographers (and new technologies like global positioning and spatial information systems have reinforced divisions as often as they have built new bridges across them). Sociologists are scattered among literally dozens of different subfields, mostly topical, from sociology of medicine to urban sociology and social stratification. Several suggest sociological perspectives on domains claimed also by other disciplines: political sociology, economic sociology, sociology of culture. History's long-standing organization by period and place remains but is increasingly cross-cut by topical research communities like women's history and indeed an encompassing to world history as a research field (rather than simply synthesis). In short, each of the disciplines has become intellectually an interdisciplinary field.

We have identified three basic patterns in interdisciplinary social science: the pursuit of a comprehensive view of social life that requires different perspectives, the pursuit of innovation based on learning skills or acquiring tools from other disciplines, and the pursuit of better understanding of a social problem, public concern, or object of professional practice. Of course these three abstract patterns lend themselves to combination in innumerable concrete projects. Urban studies reveal aspects of both the 'comprehensive view' and the 'social problem' frame; it is sometimes pursued in professional schools. Theories can circulate in circuits similar to research methods. A professional field can develop links to social science disciplines through the use of common data sets or techniques.

The most distinctive new agendas are those that integrate lines of research from social science disciplines with natural and physical sciences, engineering, and design. These appear disproportionately in connection with 'problem' foci and sometimes professional fields. From the many different versions of environmental studies (water resources, climate change, pollution reduction) to the many different public health agendas (obesity, nutrition, biosecurity) there are calls for greater social science engagement with other kinds of science in pursuing issues of public concern. Sometimes these are pursued by division of labor (economists will design schemes for trading carbon credits) and sometimes by new training programs designed to produce researchers with integrated sets of skills.

But if these are the most prominent lines of interdisciplinary development—partly because of the resources that follow from identification of public problems—it needs also to be seen that links to the natural and physical sciences are forged in the basic patterns too. Science studies itself is a field seeking a comprehensive view of science as a major dimension of modern social life, a view that combines sociology and history and philosophy of science with reflexive analyses by researchers trained in the various science disciplines (see Jasanoff, Chapter 13 this volume). Likewise, research based on neural imaging techniques is spreading from the biomedical sciences into the social and behavioral sciences (especially psychology and economics).

In any case, as we have tried to show through some major examples, interdisciplinary engagements, learning, and research come in several forms. These have different institutional supports and involve different goals, practices, and social networks. They are crucial to the continuing vitality of social science research, yet are often in tension with disciplines. The disciplines have material bases that make it unlikely that they will fade completely in the near future, which means that interdisciplinarity will remain important.

References

Abbott, A. (2001a). *Time matters: on theory and method.* Chicago: University of Chicago Press.

Abbott, A. D. (2001b). *Chaos of disciplines* Chicago: University of Chicago Press.

Anderson, B. (1991). *Imagined communities: reflections on the origin and spread of nationalism.* London: Verso.

Barnett, G.E., Secrist, H. and Stewart, W.W. (1931). Report of the representatives of the Association to the Social Science Research Council. *American Economic Review* **21**(1), 286–88.

Bourdieu, P. (1990). *The logic of practice.* Stanford: Stanford University Press.

Calhoun, C. (1998). Explanation in historical sociology: narrative, general theory, and historically specific theory. *American Journal of Sociology* **104**(3), 846–71.

Calhoun, C. (2001). Foreword. In: K. Worcestor, *History of the social science research council.* New York: SSRC.

Calhoun, C. (ed.) (2007). *Sociology in America: a history.* Chicago: University of Chicago Press.

Calhoun, C., Rojek, C., and Turner, B.S. (2005). *The Sage handbook of sociology.* London: Sage Publications.

Camic, C. (2007). On edge: sociology during the Great Depression and the New Deal. In: C. Calhoun (ed.) *Sociology in America: a history*, pp. 225–80. Chicago: University of Chicago Press.

Cardoso, F.H. and Faletto, E. (1979). *Dependency and development in Latin America.* Berkeley, CA: University of California Press.

Durkheim, E. (2006). *On suicide.* London: Penguin.

Frank, A.G. (1967). *Capitalism and underdevelopment in Latin America: historical studies of Chile and Brazil.* New York: Monthly Review Press.

Fuller, S. (1991). Disciplinary boundaries and the rhetoric of the social sciences. *Poetics Today* **12**(2), 301–25.

Habermas, J. (1988). *On the logic of the social sciences.* Cambridge, MA: MIT Press.

Hacking, I. (1990). *The taming of chance.* Cambridge: Cambridge University Press.

Haskell, T. (1977). *The emergence of professional social science.* Champaign, IL: University of Illinois Press.

Hirschman, A. O. (1958). *The strategy of economic development.* New Haven, CT: Yale University Press.

Igo, S. (2007). *The averaged American.* Cambridge, MA: Harvard University Press.

Jencks, C. and D. Riesman. (1968). *The academic revolution.* Garden City, NY: Doubleday.

Kuhn, T.S. (1970). *The structure of scientific revolutions.* Chicago: University of Chicago Press.

Merton, R.K. (1965). *On the shoulders of giants: a Shandean postscript*. New York: Free Press.

Porter, T.M. (1995). *Trust in numbers: the pursuit of objectivity in science and public life*. Princeton: Princeton University Press.

Prebisch, R. (1950). *The economic development of Latin America and its principal problems*. Lake Success: United Nations Dept. of Economic Affairs.

Prewitt, K. (2002). *SSRC: a brief history of the council*. Available at: <http://www.ssrc.org/workspace/images/crm/new_publication_3/%7Bbe08b034-f560-de11-bd80-001cc477ec70%7D.pdf>. Accessed 2 December 2009.

Richardson, T.R. and Fisher, D. (1999). *The development of the social sciences in the United States and Canada: the role of philanthropy*. Westport, CT: Greenwood Publishing Group.

Ross, D. (1991). *The origins of American social science*. Cambridge: Cambridge University Press.

Scott, J.C. (1998). *Seeing like a state: how certain schemes to improve the human condition have failed*. New Haven, CT: Yale University Press.

Sewell, W.H. (1989). Some reflections on the golden age of interdisciplinary social psychology. *Social Psychology Quarterly* **52**(2), pp. 88–97.

Sills, D. (1986). A note on the origin of 'interdisciplinary'. *Items* **40**(1), 17–18.

Stigler, S. (1986). *The history of statistics*. Cambridge, MA: Harvard University Press.

Turner, S. (2000). What are disciplines? And how is interdisciplinarity different? In: P. Weingart and N. Stehr (eds) *Practising interdisciplinarity*, pp. 46–66. Toronto: University of Toronto Press.

Wallerstein, I. (1974). *The modern world system*. New York: Academic Press.

Wallerstein, I. (2003). Anthropology, sociology, and other dubious disciplines. *Current Anthropology* **44**, 453–66.

Wallerstein, I. and Young, N.D. (1997). *Open the social sciences: report of the Gulbenkian Commission on the Restructuring of the Social Sciences*. Chicago: American Library Association.

Whitley, R. (1967). Umbrella and polytheistic scientific disciplines and their elites. *Social Studies of Science* **6**, 471–97.

Wilson, E.O. (1998). *Consilience: the unity of knowledge*. New York: Knopf.

Worcester, K. (2001). *Social Science Research Council, 1923–1998*. New York: Social Science Research Council.

CHAPTER 8

Biological sciences

WARREN BURGGREN, KENT CHAPMAN, BRADLEY KELLER, MICHAEL MONTICINO, AND JOHN TORDAY

Curiosity, an innate human characteristic, is inevitably directed to the biological world of which we are an integral part. Indeed, curiosity about our biological surroundings and its role in it pre-dates the written word, as evident in ancient cave drawings at Lascaux (*c.* 16,000 BP) depicting the living world around the artist. Likely as ancient is the interplay between biology (as our ancestors perceived it) and other human endeavors, including religion, art, and the emergence of technology.

From these origins has arisen the discipline of biological sciences—a discipline that is fundamentally shaped by its interdisciplinary activities. Moreover, interdisciplinarity in the biological sciences is constantly shifting as new technologies and theories arise, evolve, and mature and—sometimes—fade away. Thus, the biological sciences, like many scientific disciplines, are constantly subjected to an 'the interdisciplinary cycle' shown schematically in Fig. 8.1. The merger of biology and chemistry, forming the new discipline of biochemistry (discussed below), is a classic example. Emerging as a new discipline (steps 4 and 5 in Fig. 8.1), biochemistry is now a long-standing discipline that is itself going through another turn of the interdisciplinary cycle through its interactions with information science and nanotechnology.

Against this dynamic backdrop of constantly changing associations with other disciplines, we first offer a series of vignettes or case studies on how interdisciplinary studies between the biological sciences and other science and engineering fields have yielded a wealth of new insights and practical products. We then discuss the advantages and challenges of undertaking interdisciplinary activity in the biological sciences. Befitting the task, as well as reflecting the collaboration typical of the biological sciences, the 'we' represents a collaboration of authors with backgrounds in physiology, biochemistry, medicine, and mathematics who collectively have experienced and benefited from team approaches to problem solving.

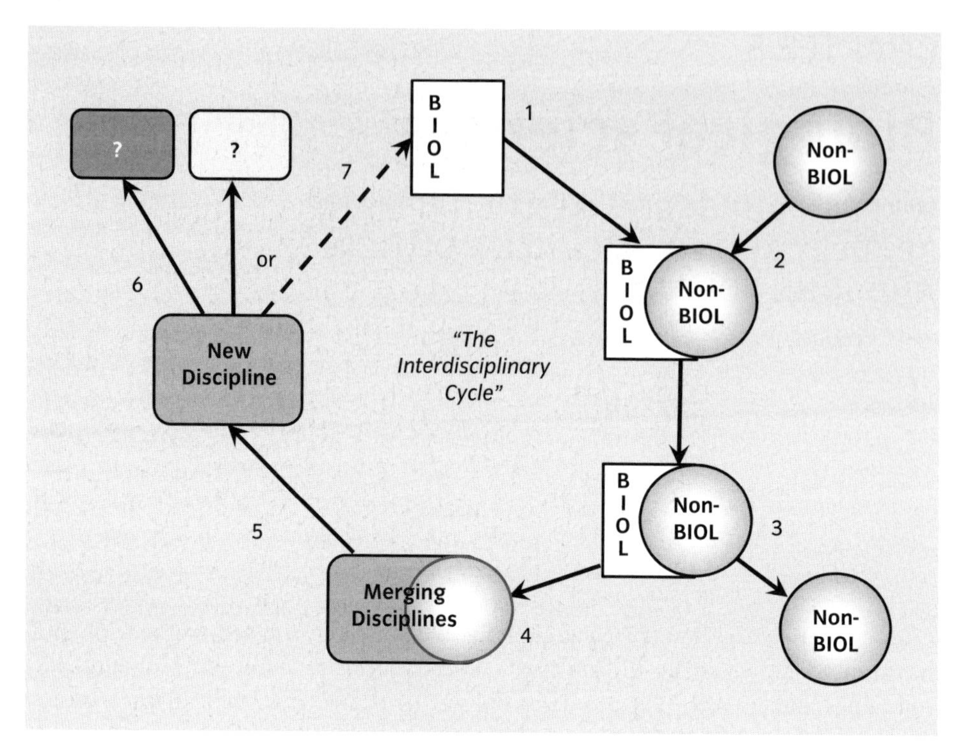

Figure 8.1 The interdisciplinarity cycle, in the biological sciences and other disciplines experiencing interdisciplinarity cross-fertilization. (1) A free-standing discipline such as biology regularly experiences the influence of non-biological disciplines. (2) This co-mingling of ideas and techniques can ultimately be only fleeting (3), or it can result in a true merger of the disciplines (4). This new discipline, formed from the merger of biology and non-biology (5), may eventually fragment into new disciplines (6), or can persist (7) to enter the interdisciplinarity cycle once again.

8.1 Case studies in biological interdisciplinarity

The interdisciplinary reach of the biological sciences is extensive—indeed, too far-reaching to cover completely in this chapter. As an alternative approach that is hopefully illustrative, we present several case studies of how the biological sciences have effectively interfaced with engineering, medicine, mathematics, and chemistry.

8.1.1 Biology and medicine

Biology and medicine—two distinct disciplines each with their own approaches, currencies, and outcomes—have nonetheless coexisted as intertwined disciplines for more than two millennia. The study of animals has been used to understand principles in medical science since Aristotle (384–322 BCE) and Erasistratus (304–258 BCE), who were among the first to experiment with living animals. Aelius Galen (129–200 CE), a second-century Roman clinician, dissected pigs and goats. In the seventeenth century, William Harvey (1578–1657) described the circulation of blood in mammals. In the eighteenth century,

Stephen Hales (1677–1761) measured blood pressure in the horse and Antoine Lavoisier (1743–94) placed a guinea pig in a calorimeter to prove that respiration was a form of combustion. Claude Bernard (1813–78) established animal experimentation as part of the standard scientific method. In the 1870s, Louis Pasteur (1822–85) demonstrated the germ theory of medicine by giving anthrax to sheep.

It should be noted that up to that point in human thought, Western society embraced the belief in a Great Chain of Being, ordering all of existence is a continuous natural hierarchy that placed humans between God and all other animals. It was Charles Darwin (1809–82) who destroyed the belief in purpose in nature with the publication of *The origin of species* in 1859. By the turn of the twentieth century Ivan Pavlov (1849–1936) used dogs to describe classical conditioning, and Sir William Osler (1849–1919) developed the field of pathophysiology, creating a systematic way of understanding disease and health and their relationship to the biological sciences.

The descriptive approaches alluded to here dominated biology and medicine until the advent of transgenic animals, and the publication of the human genome and those of other model organisms (mouse, chicken, fruit fly). Recent discoveries that patterning genes are common to all of animal life from flies to humans, and that humans have fewer genes than a carrot (25,000 versus 40,000), have emphasized the importance of comparative studies at the cell/molecular level for understanding both biology and medicine. As a result, modern biology is now expected, by analogy to physics, to generate a periodic table, formulate its own equivalent to $E = mc^2$, and develop a quantum mechanics of a predictive biology that relies less on time-honored empirical observation and much more heavily on prediction (Torday 2004).

Unfortunately, such advances cannot be achieved by a direct analysis of available data from species, because organisms have evolved through an emergent and contingent process in which new species have appeared while others have become extinct. But the fossil record of evolution is embedded in the developmental processes involved in the formation of existing organisms, and can be elicited by determining the genetic basis of phenotypes across species as they develop, i.e. ontogeny recapitulates phylogeny. Based on this knowledge, the indirect methods of developmental and comparative biology, reduced to cells and molecules, have been used to connect the dots between first principles of physiology (e.g. homeostasis, acclimation) and the scientific basis for a more prediction-based medicine in the future (Torday and Rehan 2009a).

For centuries biology has used disease to leverage our knowledge and understanding of health, and vice versa, since all biologists had available were descriptive phenotypes (the outward appearance of an organism, as opposed to its genotype, its genetic makeup). However, with the merger of genetics, molecular biology, and physiology into the subdiscipline of genomics (the study of genes and their functions), we can now address the questions of health and disease as a continuum, based on genetic mechanisms as they apply to the relevant phenotypes. Furthermore, the discovery of so-called 'patterning genes' has led to the recognition of fundamental commonalities between very different appearing phyla (e.g. fruit flies and mice, nematode worms and humans). Along with the sequencing of the genomes of fishes, amphibians, and birds, it is now possible to exploit evolutionary–developmental biology to provide a Rosetta Stone for helping to decipher the organic nature of disease, rather than describing health as the absence of disease. These advances have primarily arisen through the interdisciplinary commingling of basic life sciences and advanced medicine.

Evolutionary biology is fundamental to the biological sciences. Plugging genes and related phenotypes of interest into an evolutionarily robust model of animal development will allow us to decipher causes of human diseases. Using this approach will allow biologists to see the continuum from adaptation to maladaptation and ultimately to disease. Such a perspective would finally offer a scientific basis for monitoring health independently of disease, ushering in a new era of preventive medicine (Torday and Rehan 2009a). The key to such an approach is to identify the developmental cellular and molecular mechanisms that are fundamental to an organism's structure and function. Such studies are conventionally conducted by developmental biologists. Unfortunately, they only rarely involve biologists familiar with comparative, phylogenetic analyses across species. As a result, collaboration primarily occurs through the passive capture of data in the biological and medical literature that examines the development of phenotypes at the cell/molecular level. A number of national and international meetings have fostered more activist approaches involving developmental biologists and medical researchers, and a website (<http://evolutionarymedicine.labiomed.org/>) has been established to draw attention to this unconventional, but biologically sound and effective, interdisciplinary approach.

It has previously been suggested that life scientists should generate the biologic equivalent of the periodic table. Drilling down to the molecular pathways that have given rise to complex physiologic traits is the key to understanding the first principles of physiology as the scientific basis for predictive medicine (Torday and Rehan 2009a). For example, by identifying the genes that mediate the 'cross-talk' between the cells of the mammalian lung, gene regulatory networks responsible for lung evolution have been traced back to the swim bladder of fishes, genetically linking mechanisms for buoyancy, gas exchange, and nutrition at the cell–molecular level (Torday and Rehan 2009b). By systematically reducing complex traits to their genetic phenotypes developmentally across species, and by then sharing this vast number of data via public databases, biologists will ultimately be able to unravel complex physiologic principles relevant to both biology and medicine. There are, of course, dangers in unmitigated reductionism, such as the failure to identify emergent properties that result from the interactions across components and levels. However, the success of a reductionist approach as one of many concurrent approaches shows great promise in medical advances.

The discipline of biology is on the verge of a sea change in the interactions between biology and medicine, if only it can utilize the huge data sets being created (via exploiting yet another interdisciplinary field—bioinformatics). By abandoning the old paradigm of descriptive biology, and moving into a mechanistic paradigm based on evolutionary principles, it may be possible to progress towards an era of predictive biologic science. This will enable biologists to address counterintuitive aspects of biology such as why the lens of the eye is composed of digestive enzymes, or why the lung is a hormone-secreting endocrine organ. With the anticipated interdisciplinary activities between predictive biology and predictive medicine (Torday and Rehan 2009a) society's burden of chronic diseases may be significantly diminished.

8.1.2 Biology and chemistry

One of the oldest and most productive interdisciplinary amalgamations within the life sciences is that of biology and chemistry into 'biochemistry'. Biochemistry involves many different areas of research, but at its heart it is the study of the organic molecules (those containing carbon) and their chemical reactions within living systems. Biochemists today may not readily imagine themselves as interdisciplinary, yet their work bridges both the life and physical sciences.

That biochemistry is now less frequently thought of as at the interface of two disciplines is due in large part to its maturation as a discipline in its own right over the last 150 years. Generations of scientists are now trained with a common view of their science and are comfortable using the languages of either biology or chemistry. In fact by the mid twentieth century, entire departments of biochemistry were commonplace among many colleges and universities, where none had existed 50 years earlier.

The early history of biochemistry developed from the general concept that living materials catalyze chemical reactions. Probably most exemplary are the studies of the fermentation process by yeast. Most of the early research was carried out in the late 1800s and early 1900s by scientists trained as chemists. Indeed, Eduard Buchner received the Nobel Prize in Chemistry in 1907 for his pioneering discoveries of the biochemical fermentation of sugar by cell-free systems, a clear recognition of the emerging science of 'biological chemistry'.

The popularity and power of biochemical approaches led to widespread exploration of biological systems where chemists, familiar with the properties and analysis of organic molecules, sought to work with biologists experienced in physiology. The products of these interactions have created a remarkable knowledge base over the last 100 years, and have spawned new interdisciplinary lines of research. In this short space it is not possible to provide a complete list of the many scientific contributions that can be attributed to the field of biochemistry. However, the quest to understand the mechanisms that drive biological systems has been the major driver for the emergence and maturation of biochemistry. Indeed, numerous major discoveries have been made possible through the interdisciplinary research of biochemistry, including the identification of:

- the structural features of major classes of macromolecules such as DNA (which contains the gene sequence of an animal, RNA (involved in the replication of DNA) and proteins,
- the basis of enzymes, which facilitate metabolic reactions,
- the mechanisms of photosynthesis, for the biological conversion of light to chemical energy and reduction of inorganic carbon,
- the machinery of cellular respiration and membrane transport, for biological energy conversion and nutrient and waste movement in and out of the cell,
- the genetic code, whereby variations in the sequence of just four nucleotide bases universally explains the nature of proteins from bacteria to human,
- the basis for protein synthesis and turnover, for the production, regulation, and recycling of cellular machinery,
- the enzymes that regulate gene expression.

Since 1901, at least 35 Nobel Prizes in Chemistry and many more in Physiology and Medicine (<http://almaz.com/nobel/>) have been awarded for discoveries in biological chemistry, illustrating the tremendous rewards of working at the interface of chemistry and biology. Many of these discoveries have led to entirely new fields of interdisciplinary research. For example, out of the structural determinations of DNA, RNA, and proteins has developed the new discipline of structural biology; out of the enzymology of transcription and the genetic code has arisen the discipline of molecular biology. These two newer disciplines, much like the newly emerging area of systems biology, have been driven by the scope of biological questions, but have depended upon the contributions of scientists from many disciplines—including mathematics, computer science, chemistry, biology, and physics.

Interdisciplinary collaborations in the life sciences—or any other area, for that matter—are most successful when the overall outcome is greater than the sum of its parts, and when all collaborators have a vested commitment in and benefit from that outcome. An excellent example of this in biochemistry is in the recent area of comparative metabolomics—essentially the simultaneous profiling and quantification of all metabolites from a tissue or cell type. This has analytical biochemistry at its basis, but on a high-throughput, massive scale (many thousands of chemical components). These types of experiments have required the development of sophisticated mass spectrometry-based instrumentation, the know-how for sample preparation, expertise in separation technologies and robotics, computational capabilities for data analysis, and someone to know the relevant questions to address. Success depends upon contributions from chemistry, biology, computer science, mathematics, and instrument design and engineering, and it could not be achieved without any one of these components.

Biochemistry continues to evolve as an interdisciplinary activity. This is evident now with the era of 'omics'. In the 1970s and 1980s, a combination of biochemical and molecular genetic approaches toward biological questions resulted in an interdisciplinary area of research referred to as 'biochemical genetics'. This term has fallen out of favor, but the emphasis on understanding the biochemical functions of genes remains at the forefront of life sciences today. The areas of genomics, proteomics, metabolomics, etc. are an extension of the concept of understanding gene function, but on a genome- or system-wide scale. With the rapidly advancing tools for analyzing DNA sequences, monitoring gene expression, identifying proteins, and quantifying metabolites, information is being gathered on an enormous scale.

Instead of an individual research laboratory experimentally addressing the function of a single gene over many years, teams of scientists are attempting to understand biology from an entire 'systems-wide' approach. This requires expanded capabilities orders of magnitude greater than those of two decades ago when the first gene sequences were being collated in a database called GenBank. For example, as of February 2008, there were over 190 billion bases of nucleotide sequence information archived in the GenBank databases (<http://www.ncbi.nlm.nih.gov/Genbank/index.html>). To accommodate these increasing numbers of gene sequence, gene expression, protein structure, and metabolic data, requires new computing power, expertise in predictive programs, powerful statistical methods, and computational algorithms. Questions can now turn to the functions of thousands of genes, proteins, and metabolites at once, helping to address everything from human health to agricultural production. These grand challenges require the collaboration of scientists with expertise in many disciplines in addition to biochemistry, and will involve tools and languages yet to be developed; but it is certain that interdisciplinary activity across traditional boundaries of science and engineering is the way forward.

8.1.3 Biology and engineering

The relationship between biology and engineering is an old one. For example, Leonardo da Vinci (1452–1519) was a prototypic artist/inventor/anatomist/engineer. His studies on human form and function revealed the interdependence between biological processes and biomechanical function and physical forces. However, da Vinci's efforts extended beyond the reduction of complex processes to include the design and fabrication of structures as representations of what he observed in nature. Additionally, da Vinci showed us the great potential in the marriage of biology and mechanical functions. Indeed, modern engineering disciplines now encompass a broad matrix of biological topics, including developmental biology, bioenergetics, biomechanics, biomaterials, artificial intelligence, and bionics related to the development of artificial organs.

Yet the marriage between biology and engineering is neither easy nor automatic. Consider the comments of Fung and Tong (2001) in their classic engineering text, *Classical and computational solid mechanics*:

> Engineering is quite different from science. Scientists try to understand nature. Engineers try to make things that do not exist in nature. Engineers stress invention.... Most often, (engineers) are limited by insufficient scientific knowledge. Thus they study mathematics, physics, chemistry, biology and mechanics.

Unfortunately, the inverse is not true—biologists, who also are often limited by insufficient knowledge, are not (yet) drawn in great numbers to study engineering. Yet, many biological processes occur within biophysical environments that are dynamic and rapidly changing. Analytic engineering principles and paradigms have been developed and applied to investigate and quantify many of these dynamic interrelationships, and emerging biology–engineering interdisciplinary partnerships are now poised to take advantage of them. Here we consider a few representative vignettes that highlight the unique opportunities and insights that have gained through the interface of biology and engineering.

Our current understanding of the developmental biology of the heart and blood vessels has been substantially influenced by interdisciplinary interactions between biologists and engineers. One of the most fundamental processes during animal development is the growth and remodeling of the embryonic heart from a single cell, to a peristaltic tube and finally into a multichambered organ with functioning unidirectional valves, a specialized conduction system for electrical impulses, and optimized blood flow to correctly direct deoxygenated and oxygenated blood to the tissues. Complex processes of heart tissue formation, including how heart cells, tissues, and structures (chambers, valves) grow and change, initially quantified by developmental biologists and physiologists, have now been analyzed by bioengineers. Using computer technology, cardiovascular physiologists working with bioengineers can now actually visualize previously only imagined forces in the wall of the beating embryonic heart. This interdisciplinary partnership has provided new understanding of how sheer and strain in the heart walls actually help shape the heart and its growth. In fact, the interdisciplinary interactions of developmental biology and engineering used so effectively in cardiovascular biology have now been expanded to provide relevant insights and identify novel questions across an extremely broad landscape of developmental and comparative biology, ranging from protein configurations to whole embryo structure (Davidson *et al.* 2009).

Regenerative medicine (the creation of replacement tissues and organs) is another example of the emerging products of interdisciplinary collaborations between biologists, physicians, and engineers. By exploring developmental processes in tissue and organ generation, bioengineers have developed new technologies being applied to the design and fabrication of biomaterials (materials that can become part of or even replace original tissues). Bioengineering approaches have also led to a large potential commercial market for biotherapeutics (therapeutic substances produced by biological means, e.g. vaccines).

Such insights have led to the rapid expansion of regenerative medicine. In fact, tissues and organs generated *in vitro* ('in the test tube') have approached the critical phase of initial clinical trials. Bioengineering regeneration techniques have also been employed to investigate the biomechanics and regulation of cardiac valve formation, with the goal of generating robust replacement heart valves *in vitro* (Engelmayr *et al.* 2008). At the cellular and molecular levels, biologists and engineers are contemplating the creation of nanomachines that are injected into the bloodstream of a patient, travel to their targeted tissues, and then carry out a specific suite of activities which can include actually permanent assimilation into the tissue. Of course, ethical issues arise from regenerative medicine (see Callahan,Chapter 29 this volume), with an uncontrolled extrapolation leading to the specter of 'borg-like' creatures where the boundary between human and machine is blurred.

Successful collaborations between biologists and engineers, such as the examples of heart development and organ regeneration outlined above, are being catalyzed by a targeted expansion of funding in the United States by the National Institutes of Health, the National Science Foundation, and numerous foundations that support interdisciplinary teams. But even as collaborations that lead to advances in health are expanding, there is also a great deal of attention being paid to potential military applications resulting from interdisciplinary activities between biologists and engineers. As just one example, the exoskeleton of invertebrates such as insects and crabs is being studied with a view of providing an external 'exoskeleton' for soldiers. This external, motor-driven scaffolding would allow them not only to carry more gear, but potentially even to be remotely activated to march wounded soldiers out of danger.

The interdisciplinary collaborations between engineers and biologists often revolve around mathematical analyses. We now turn to the highly productive collaborations between biologist and mathematicians.

8.1.4 Biology and mathematics

Biology, as a quantitative science, has always depended heavily on mathematics. Collaborations between biologists and mathematicians, as a focused area of research, began in the early twentieth century with the study of disease transmission (epidemic models), population dynamics, and genetic frequency models. Building upon this foundation, the past few decades have seen tremendous growth in the advances made through mathematical methods in almost every area of biological research, especially with respect to modeling biological processes. Agent- or individual-based models, supported by increased computational capabilities, have joined classic mathematical models to enhance understanding of population dynamics, including processes of disease transmission.

Early mathematical population models have also provided the groundwork for significant advances in cellular systems modeling, with direct applications to oncology as biological knowledge of cellular responses and computational methods have enabled more realistic models for the treatment of cancer. Mathematical models of the genetics of organisms and their resulting features continue to develop, finding new applications in epidemiology. This, in turn, has motivated meta-analysis of databases of genetic sequences of different animals (and plants), which has resulted in ideas for new disease therapies. At the same time, examination of the molecular basis of the formation of new species has deepened our understanding of evolutionary relationships. Recent advances in the theory of complex systems are providing new insights into physiological systems with multiple feedbacks and interacting components (Burggren and Monticino 2005). Indeed, the list of biomathematical applications is ever-expanding, including the analysis of complex images (e.g. the three-dimensional images of cells provided by confocal microscopes), and the interpretation of the complex folding of proteins, of data from new genetic techniques (e.g. microarrays), and of complex nerve networks in the brain.

The rich diversity of progress described above, and the promise of future advances, has led to the establishment of strong interdisciplinary programs in biomathematics and bioinformatics at a wide variety of institutions. Graduates from these programs hired into traditional mathematics and biology departments at universities are influencing departmental culture (including promotion and tenure criteria; see Phirman and Martin, Chapter 27 this volume). While significant challenges remain, there is a growing realization among mathematicians that not being involved in interdisciplinary work with biologists means missing out on some of the most exciting discoveries of our time.

Biologists and mathematicians cannot just decide to work together, and then do so. Mathematicians entering into collaboration with biologists often require a crash course in the basic biology underlying the research, and must relearn (or learn for the first time) what most undergraduate biology majors know. It also requires patience from one's biology colleagues who take for granted a certain knowledge base when interacting with colleagues. The biologists will appreciate the same patience about topics a mathematician may assume that every educated person knows—when the reality is that very few people know (or care) about 'Kolmogorov–Sinai entropy' or 'isomorphism groups'.

Even simple vocabulary can be an early stumbling block in collaboration between biologists and mathematicians. Not only may terms mean different things in different disciplines, but there are different levels of precision in how terms are used. Confusion can especially arise with words that have both common English and technical definitions. For example, the term *chaotic* is a commonly used term that nonetheless has a precise and much narrowed mathematical meaning. A biologist may be perfectly comfortable characterizing a system as chaotic based on perceived disorder; while a mathematician would argue that the system does not meet the definitional requirements, and merely has a complicated response function. It is important to calibrate vocabulary early in a collaboration to reveal common core ideas and avoid misunderstandings.

Much discussion in this chapter has focused on why and how biologists engage in interdisciplinary work with non-biologists. A complementary question is, why would non-biologists—in particular, mathematicians—collaborate on problems with biologists? A compelling reason is intellectual curiosity. It is refreshing to venture out of increasingly narrow disciplinary subfields to gain a substantive understanding of research questions in other fields. Collaboration can also provide a rewarding opportunity to make significant contributions to problems that have importance outside of mathematics, especially to bioscience questions that have clear applicability. Consider that the very top mathematics journals typically have a so-called 'impact factor' (a calculation of overall impact based on the frequency with which its articles are cited) of less than 3, while some biological journals have impact factors over 20. This is not a judgment on the relative intrinsic worth of disciplines. It does, however, suggest a certain insularity of pure mathematics research and the prospect for extending reach that collaborations with biologists afford.

Effective collaboration also requires flexibility. It is often not clear going into an interdisciplinary project which mathematical tools will be needed to best address the problem. So, broad mathematical awareness is extremely valuable, as well as the willingness to learn and apply mathematics outside of one's immediate area of expertise. Interdisciplinary work thus provides mathematicians opportunities to learn new areas of science as well as occasions to apply a variety of mathematical techniques.

Of course, these very same arguments apply to biologists attempting to work with mathematicians, and each has much to offer to their colleagues across the disciplinary boundaries. Mathematicians bring not only a toolbox of modeling and analysis techniques to biological projects, but also a useful level of rigor and clarity of analytical thought. The challenge is to engage more biologists and mathematicians in interdisciplinary studies that make significant contributions to the increasingly quantitative field of biology. This can be partially achieved by training a new generation of mathematicians within undergraduate and graduate programs that have substantive interdisciplinary components. However, it is also important to encourage traditional, established mathematics programs to participate. This is an incredible time, filled with opportunities, for mathematicians to apply their distinctive training and expertise in fields outside their discipline.

8.1.5 Biology and beyond

The case studies described above show the fruits of the mergers of the biological sciences with the major disciplines of medicine, chemistry, engineering, and mathematics. But many other established interdisciplinary bioscience-based fields exist, including biogeography (Lomolino *et al.* 2006), bioinformatics (Lesk 2008), biophysics (Nölting 2003), biostatistics, (Glantz 2005), and biotechnology (Pisano 2006). Particularly exciting developments are occurring in the interdisciplinary merger of biology and nanotechnology. For example, materials scientists intent on manufacturing machines at the molecular level are using the effective molecular recognition properties of DNA to allow this molecule to act as a template, generating novel materials with useful properties at highly controllable rates (Priyadarshy and Shankar, in press).

Of course, the reach of biology extends well beyond the sciences and technology into interdisciplinary interactions within the social sciences, arts, and humanities. Space does not allow us to expand into this rich area of discussion, but a few examples can suffice as an introduction.

Environmental issues have become a very active area of collaboration between humanists and biologists. Environmental problems typically involve an intricate mix of bio- or environmental science, environmental philosophy, and policy concerns (see Callicott, Chapter 33 this volume). Bioethics, a related field founded in the 1960s, addresses ethical and philosophical questions that arise from advances at the intersection of biology and medicine (see Callahan, Chapter 29 this volume). Political science, government, and history are interwoven with biological principles. For example, studies of peace and war are often interpreted in the context of sociobiology, and evolutionary theory has been turned towards an understanding of human conflict (e.g. Vergata 1995). Indeed, human behaviors for good or ill, are often placed within a biological context, most notably using E. O. Wilson's (1975) concept of sociobiology. As computing and robotic technologies continue to evolve, the field of human–computer interactions will have relevance to social behavior as well as to the investigative sciences.

Biology has, of course, long been a topic for the arts (consider Claude Monet's *Water lily pond* or Van Gogh's *Sunflowers*). However, biology has also depended on art in the form of medical illustration. This dependence has existed for millennia, with medical illustration likely originating in Hellenic Alexandria during the fourth century BCE, and evident as mature interdisciplinary activity in the work of such famous illustrators as Leonardo da Vinci and Andreas Vesalius (1514–64).

Biology and religion have a long and sometimes uneasy history of coexistence, most notably in recent years in debates over evolution, creationism, and intelligent design. More fruitfully, perhaps, interdisciplinary studies are helping to understand the origins of social morality, cooperation, peace, and war (e.g. Bekoff 2001), as well as the evolution of religion (Dow 2006).

8.2 What are the impediments to interdisciplinarity in the sciences?

Given the richness of interdisciplinary collaborations described above, why don't more mathematicians, biologists, chemists, physicists, and others step across disciplinary lines? There are myriad potential impediments—none insurmountable, but many quite formidable. Since scientists often use jargon, or have specialized knowledge that other team members lack, frequent communication is essential for all to work productively together. Thus, it can be difficult to develop a common working knowledge, or understanding of the complementary discipline's perspective, capabilities, and limitations. Scientists are most comfortable within the confines of their narrow disciplines, but much less so when venturing into unfamiliar territory. Overcoming the obstacle of a common understanding may take many frustrating discussions, much like learning to communicate in another language. Some potential interdisciplinarians lack the patience for this process.

Interdisciplinary collaborations also take time to bear results. The tenure clock doesn't recognize the extra time it takes to absorb the key concepts in the secondary discipline, to develop a shared view of a problem, and then search for appropriate techniques. Consequently, it is not unusual for junior faculty to be advised by their mentors not to pursue interdisciplinary work until after tenure. All too often, however, by the time tenure has been achieved, research paths have developed into deep ruts for which there are few institutional incentives to climb out of (see Phirman and Martin, Chapter 27 this volume).

It is also difficult for many academic (and non-academic) evaluators to judge the value of interdisciplinary projects. Consider, for example, the challenges to mathematicians proposing to work with biologists. Mathematics departments, like all academic departments, evaluate the research productivity of their faculty by the number of articles published in disciplinary journals and the quality of the journals in which articles are placed. Mathematics journals follow an exacting theorem–proof format. A collaboration with biologists will typically not produce a fundamental advance in mathematics (of course, sometimes this does happen, enriching both mathematics and biology). Even if it does, the theorem–proof exploration of the result would rarely find its way into a biology journal article. Traditional mathematics departments are challenged to evaluate the worth of an article that does not contain a proof, no matter how innovative or useful the application. Often, faculty members are admonished to translate the application into work that can stand on its own in a conventional mathematics journal. Thereby, the work necessary to attain evaluations similar to those of departmental colleagues not collaborating outside their discipline is at least doubled.

Even when scholarly work can be easily evaluated with regard to content, there may be an attached stigma (or at least lack of appreciation) for the venues in which interdisciplinary work appears. Front-line, cutting-edge interdisciplinary journals that are the 'must-publish' targets for interdisciplinarians may nonetheless have low impact factors and very small circulations of a few thousand compared with the disciplinary 'usual suspects' such as *Science* (with a very high impact factor and a paid circulation of more than a million). Put differently, scientists and mathematicians working in interdisciplinary areas still face the significant challenge that a paper in *Science* is typically regarded as far, far more significant than a paper in, for example, the interdisciplinary journal *Science Studies*, targeting not only scientists but sociologists, philosopher, historians, and psychologists.

Although there are many impediments to interdisciplinarity, there are many ways to actively promote such approaches. Interdisciplinary scientists need to remain open to new ideas, commit to learning alternative approaches and, perhaps above all else, be patient with respect to their own advancement, that of their colleagues, and ultimately of the project. Beyond the individuals, institutional practices need to be implemented that provide clear incentives to departments and faculty to engage in interdisciplinary research projects. This can start proactively with, for example, workshops and other educational opportunities for evaluators so that they can learn of both the promise and pitfalls of interdisciplinary research.

8.3 Conclusion

The case studies described above demonstrate the power of interdisciplinary approaches in the biological sciences, drawing upon a variety of disciplines in the sciences and beyond, for generating new perspectives, approaches, hypotheses and ideas for future experiments. Also apparent is that interdisciplinarity in the biological sciences is typically not just a single person working in an interdisciplinary area, but rather 'sympathetic' disciplinarians working together to bring the best of their training and knowledge together in new and innovative ways. Environments such as think tanks, centers, and institutes have all proven to be highly useful for getting dissimilar types of people together to work on interdisciplinary issues in biological sciences.

Yet, interdisciplinary work in the biological sciences can be challenging. Communicating with collaborators in other disciplines requires (re)learning disciplinary-dependent concepts, adopting new vocabulary, and committing to new approaches. Even when successfully completed, interdisciplinary science may not be fully appreciated by conservative or more traditionally inclined evaluators.

Yet interdisciplinarity in the biological sciences is burgeoning, driven by a spectrum of motivations ranging from unbridled intellectual curiosity to demonstrated practical solutions to engineering and medical problems. Clearly, in the future the biological sciences will continue to operate within an interdisciplinary cycle (Fig. 8.1), spawning new subdisciplines and, in time, changing the fabric of biology itself. As stated by Thomas Kuhn (1962), we won't recognize the most fundamental paradigm shifts in science until after they have occurred.

References

Bekoff, M. (2001). Science, religion, cooperation, and social morality. *BioScience* **51**(3), 171.

Burggren, W.W. and Monticino, M.G. (2005). Assessing physiological complexity. *Journal of Experimental Biology* **208**, 3221–32.

Davidson, L., von Dassow, M., and Zhou, J. (2009). Multi-scale mechanics from molecules to morphogenesis. *International Journal of Biochemistry and Cell Biology* **41**(11), 2147–62.

Dow, J.W. (2006). The evolution of religion: three anthropological approaches. *Method and Theory in the Study of Religion* **18**(1), 67–91.

Engelmayr, G.C., Jr, Soletti, L., Vigmostad, S.C. *et al.* (2008). A novel flex-stretch-flow bioreactor for the study of engineered heart valve tissue mechanobiology. *Annals of Biomedical Engineering* **36**, 700–12.

Fung, Y.C. and Tong, T. (2001). *Classical and computational solid mechanics*. Singapore: World Scientific Press.

Glantz, S. (2005). *Primer of biostatistics*. New York: McGraw-Hill Medical.

Kuhn, T.S. (1962). *The structure of scientific revolutions*. Chicago: University of Chicago Press.

Lesk, A. (2008). *Introduction to bioinformatics*. New York: Oxford University Press.

Lomolino, M.V., Riddle, B.R., and Brown, J.H. (2006). *Biogeography*, 3rd edn. Sunderland, MA: Sinauer Associates.

Nölting, B. (2003). *Methods in modern biophysics*. New York: Springer.

Pisano, G.P. (2006). *Science business: the promise, the reality and the future of biotech*. Cambridge, MA: Harvard Business School Press.

Priyadarshy, S. and Shankar, L. (in press). *DNA nanotechnology*. Boca Raton, FL: CRC Press.

Torday, J.S. (2004). A periodic table for biology. *The Scientist* **12**, 32–3.

Torday, J.S. and Rehan, V.K. (2004). Deconvoluting lung evolution using functional/comparative genomics. *American Journal of Respiratory Cell and Molecular Biology* **31**, 8–12.

Torday, J.S. and Rehan, V.K. (2009a). Exploiting cellular-developmental evolution as the scientific basis for preventive medicine. *Medical Hypotheses* **72**(5), 596–602.

Torday, J.S. and Rehan, V.K. (2009b). Lung evolution as a cipher for physiology. *Physiological Genomics* **38**, 1–6.

Vergata, A.L. (1995). Darwinism, war and history: the debate over the biology of war from the 'Origin of species' to the First World War. *Medical History* **39**(3), 378–9.

Wilson, E.O. (1975). *Sociobiology: the new synthesis*. Cambridge, MA: Harvard University Press.

CHAPTER 9

Art and music research[1]

JULIE THOMPSON KLEIN AND RICHARD PARNCUTT

In taxonomies of Western knowledge, 'art' and 'music' are grouped conventionally within a cluster of disciplines labeled 'humanities'. Research on art and music, however, encompasses a wider array of disciplines and interdisciplinary fields. Music research also spans natural sciences (e.g. acoustics, physiology), formal sciences (mathematics, computing), and social sciences (psychology, sociology). Likewise, art research draws on the concepts, theories, and methodologies of social sciences as well as science and technology. Moreover, in both cases research spans not only historical studies of aesthetic forms and movements but also the nature of creativity, the physiology and cognition of perception, the reception of aesthetic works within particular groups and cultures, and the institutional formations of research and education. While much has been written about the universality of art and music across cultures and in general theories of the arts, the disciplines are constructed differently. In the Anglo-American academy, for example, 'art history' became an umbrella label for research centered initially on the history of Western art but later including other disciplines and art forms. In European practice 'musicology' (*Musikwissenschaft*) is often thought to include both humanities and sciences, whereas in North America it was more confined traditionally to humanities.

Like all collaborators who cross knowledge boundaries to work on a common project, the co-authors of this comparative reflection grappled with differences shaped by their academic and cultural backgrounds. One, a professor of humanities, was trained in literary studies in North America, then her research and teaching interests broadened to include interdisciplinary humanities, American cultural studies, and the theory and practice of interdisciplinarity. The other studied music, physics, and psychology in Australia, then conducted research at the interface of music psychology, music theory, and music performance, and as a professor in Austria developed new infrastructures to encourage musicological interdisciplinarity. Even such basic terms as 'humanities' and 'science' became boundary objects

[1] Portions of the chapter adapt material from Klein, J.T. (2005). *Humanities, culture, and interdisciplinarity: the changing American academy*. New York: SUNY Press.

in our conversation. The term 'art' was also an intersection for negotiating meaning. It has referred traditionally to a large family of visual and performing arts, including not only visual and plastic arts of painting and sculpture but also music and theater. For the sake of comparison, and heeding separations in major taxonomies and divisions of scholarship, in this chapter the singular form of the term 'art' designates visual and plastic forms, and the plural form 'arts' refers to a more general compass inclusive of both art and music.

The disciplines of art and music have several elements in common that make for a compelling comparison. They both inherited an identity vested in creativity and traditional values of the liberal arts that were transmitted through European tradition to the American colonial college, then challenged and reinvigorated in new critical approaches to scholarship and teaching. They both enjoy a presence beyond the academy in performance venues, museums, other cultural institutions, and electronic media. They are both non-verbal media whose data are more resistant to verbal explication than the data of other disciplines classified as humanities (Parker 1997). And, over the course of the twentieth century, they underwent parallel changes. Their canons expanded, new scholarly approaches fostered new understandings, and the objects of research were profoundly changed by technological developments.

This chapter explores the interdisciplinary nature of the two disciplines, with particular attention to patterns of origin, new practices, and (re)constructions of disciplinarity and interdisciplinarity. Throughout the discussion, we also consider the place of art and music in the larger relationship of humanities and science and technology.

9.1 Origins

Research on both art and music has a long history. Giorgio Vasari's evolutionary approach to style history influenced the general outlines of historical studies of art for several centuries. In *Lives of the most eminent painters, sculptors, and architects* (1550), Vasari extolled the European High Renaissance as the pinnacle of excellence. His methodology was a series of aesthetic and moral value judgments interwoven with anecdotes and references to purported facts (Kraft 1989). Barbara Stafford (1988) traces the formal origin of art history to the eighteenth century. Art history was a borrower from the start, taking attitudes and vocabularies from prior or canonical disciplines and constructing a hybrid identity from mathematics, rhetoric and poetics, and philosophy. Its founders were mindful of an intellectual deficit. Artists were inclined to offer inductive and artisanal conjectures about visual and aesthetic matters, not deductive or exact knowledge about a fixed or stable mental territory with objects of intellection. The formation of the modern disciplines during the late nineteenth and early twentieth centuries had a dramatic impact on both art and music research, shaped by the cumulative forces of professionalization, specialization, and a growing scientification of knowledge.

The attempt to orient humanities toward a new positivist paradigm began in the early nineteenth century, initially in Germany and in linguistics. The dominant model was a form of grammatical study that differed from the normative and philosophical approach of the eighteenth century. Imbued with new empirical values of scholarship, James Stone

(1969) recalled, discovery became the primary purpose of research by professionalized experts. The tendency towards painstaking investigation and minute methodology became as evident in historiography as in the sciences. For the humanities scholar, the equivalent of the laboratory was analytic abstraction, reinforced by description, classification, comparison, and compilation. Like laboratory specimens, humanities objects could be manipulated, dissected, and embalmed; measured, counted, and calibrated; and subjected to precise methodologies. A credo, a comedy, a portrait, an idea, or a hero could be subdivided by analysis and abstraction into the propositions of philosophers, the techniques of literary and art historians, and the events of historians. As scholars in growing numbers embraced ideals of objectivity, precision, and specialization, the notion of a shared culture diminished. Decentralization and fragmentation of education hastened. Older unified fields of inquiry and principles of the university eroded, and new unifying hypotheses were foreshortened. The positivist paradigm that took root in the modern discipline of art history, Kraft notes, accentuated empirically grounded facts, historicism, and style analysis defined by formal characteristics such as color, shape, line, texture, and space. Works, artists, styles, and national or ethnic groups could be compared and classified, explicated, and interpreted in a systematic fashion. In addition, art historical research could be separated from other fields, interpretations legitimized with truths parallel to scientific laws, and the contextual dynamics of production and reception placed beyond positivist science.

Pulling in the contrary direction, the most powerful early basis for interdisciplinary relationships among the arts, including music, was periodization and interart comparison. Common motifs, themes, and genres suggested synchronic relations within chronological eras (e.g. medieval, romantic) and stylistic categories (e.g. classical, mannerist). During the nineteenth century, interart comparison was typically formulated in terms of historical criteria, such as a *Zeitgeist* or a *Formwille*. The theory of Goethian or romantic organicism also treated 'arts' in the general sense, as a holistic entity and, in the twentieth century, E. H. Gombrich's concept of norms and theories of social reflexes and technical achievement provided a basis for unity. The discourse of unity centered on the spirit of movements, periods, or moments thought to convey coherence among all cultural activities and a complete parallelism of arts (Fowler 1972; Steiner 1982). Synoptic theorizing and the generalist tradition of humanities fostered different kinds of connection. Albert William Levi, for example, proposed that music involves a 'radiating theory' of overlapping value-concerns in arts and humanities, and the generalist tradition furnished a holistic model of moral, social, and religious development that aligned cultivation of taste in music with cultural literacy and moral character (Levy and Tischler 1990; Sibbald 1992).

The relationship between the sciences and the arts changed over time, though it has been dominated by two views. One, Sheldon Richmond (1984) recounted, treats them as polar opposites, the other as different expressions of a single voice. Richmond disputes both views, arguing that they are based on a false dichotomy that posits rationality (cognitivity) as the realm of science and irrationality (imagination) as the realm of arts. This dichotomy, though, minimizes correspondences over time. The period of the Renaissance, for instance, saw revolutions in both arts and sciences that turned on the same point—the discovery of linear perspective and the development of optics, allowing three-dimensional interpretation of images whether they appeared in a painting or through the lenses of a telescope. Older hierarchies were also displaced. Aristotelian cosmology and Platonic ideals were initially supplanted by a non-perspectival ordering of objects on a canvas. Later, the Einstienian revolution overturned Galilean–Newtonian space and time and the Impressionist–Cubist revolution recognized the plurality of visual fields and other kinds of spatial representation. Developments in science and technology have contributed to art research in diverse ways. On the one hand they have provided new subjects for the arts. In the case of music, the noises and rhythms of factories, production lines, and other aspects of modern life inspired and enabled new genres from *musique concrète* to punk rock. On the other hand, they have offered new means of asking research questions about all of the arts—such as how art and music are physically represented, perceived, and cognitively processed. Art history and music have increasingly interacted with scientific disciplines such as physics, physiology, psychology, and information sciences.

Defined broadly as 'the study of music', musicology includes all research about all music. Like art history, music was a borrower from the start. Before about 1600, most music research corresponded to the late nineteenth-century concept of 'systematic musicology'. Ancient and medieval musical thought was dominated by precursors of modern sciences such as mathematics, acoustics, physiology, and psychology. In medieval universities, the liberal arts were organized into a *trivium* of grammar, logic, and rhetoric, and a *quadrivium* of arithmetic, geometry, music, and astronomy; early music research involved mathematical analysis of intervals and string length ratios in the Pythagorean tradition. 'Systematic' understanding of music improved gradually with the scientific revolution in the seventeenth century, the Enlightenment in the eighteenth century, the rise of the modern university in the nineteenth century, and the rise of computer technology in the twentieth century.

Just as art history became a dominant practice in art research, in the nineteenth century historical musicology dominated music research. Historical musicology depended on art history for the paradigm of style history and on literary studies for paleographic and philological principles (Treitler 1995). Since then, musicology has been institutionally grouped in the humanities, in spite of the important role of sciences throughout the history of musical thought.

The bias towards Western cultural elites in musicology has various origins. First, despite the long chronicle of musical thought musicology did not enter the academic canon until the nineteenth century—soon after the word 'musicology' was coined—so that Western culture and identity could be documented. Second, the humanities were academically central during the nineteenth century, and history has always been central to the humanities. Third, the music of Western cultural elites (also misleadingly, tautologically, or discriminatingly referred to as *classical music*, *art music*, or *high culture*) was regarded in the nineteenth century as intrinsically superior. In the language of alterity research (Taussig 1993), music that was not Western and elitist was (and often still is) considered to be Other music and thereby classified outside the taxonomy of sanctioned forms; Real music (or just 'music') was composed by Bach, Beethoven, Brahms, and other white males, and mimetically reflected European identity and superiority. The Eurocentric worldview of nineteenth-century humanities lives on today in conservative academe (cf. Cook 1998);

most societies, conferences, and journals of 'musicology' still focus on the history of Western cultural elites, and Western music is still the main focus of other musicological subdisciplines such as music psychology, in which research on expressive performance focuses on Chopin and Mozart.

9.2 New art histories and new musicologies

By the mid twentieth century, the positivist paradigm was still prioritized in humanities scholarship. During the 1980s and 1990s, though, talk of 'new art history' and 'new musicology' signaled the growing impact of expanding canons and scholarly practices. Donald Preziosi (1989) likens modern art history to an Ames Room or a Foucauldian heterotropic space of contradictory practices and theoretical positions. New stylistic movements such as pure form, color field painting, and minimal art had little in common visually with earlier traditions. The notion of art, Selma Kraft (1989) adds, expanded to include the works of women and different cultural groups. The boundary between high and low or popular art eroded, legitimating once-excluded objects such as popular, traditional, and folk music, furniture and quilts, cartoons and graffiti, commercial illustrations and tattooing. The repertoire of works of both art and music expanded on a global scale and large exhibits on Chinese painting and excavations, African art, and the art of the Mamluks and the Mughals. New hybrid genres also emerged. Performance art, for instance, combines music, visual art, literary expression, and theatrical performance and multigenre forms emanated from cultural movements for identity and equality.

Scholarship changed in kind. Kraft (1989) identified two general directions in art research. One strand of influence—from the social sciences—accentuated production and use, focusing on the political, cultural, social, and economic conditions under which art is made and on subjects such as patronage, the art public, and workshop practices. The other strand—closer to the humanities—drew on critical, semiotic, and deconstructionist approaches, especially from literary theory and philosophy. Both strands differed from the interdisciplinarity of Erwin Panofsky, a prominent figure in the early twentieth-century formation of the modern discipline. Panofsky was interested in the inherent meaning of works regarded as exact reflectors of attitudes and values. The new art history critiqued assumptions about self-evident meaning and uniformities of interpretation that ignore differences of ethnicity, gender, and class. Scholars began treating artworks as texts and as structures of signification. They weighed the relative merits of disciplinary methods and explored processes of professionalization. They expanded history's relationships with criticism, aesthetic philosophy, markets, exhibitions, and museology. They used insights from Marxism to understand social and economic determinations. And they imported explanations of repressed instincts from psychoanalysis, power relationships from political theory, institutions from sociology, and structures from anthropology.

The relationship of art and science also changed. Science has long been a theme in art, providing narrative content and imagery. One of the fundamental properties of art—beauty—continues to be studied in terms of mathematical elements of proportion, pattern, and (a)symmetry. Domains of inquiry, though, have expanded. The science known as psychology of art, which began at the end of the nineteenth century, now has ties with mainstream psychology, all disciplines of the humanities, art therapy, art education, and practitioners such as architects and curators (Lindauer 1998). Scholarly understanding of visuality also changed with new sciences and technologies of imaging, a development especially prominent in cognitive science and neuroscience. Emerging technologies have led to new forms and conceptions of art as well. Some new media, Sian Ede (2005) points out in her recent book on *Art & science*, make demands on multiple and differing mental processes. This complexity is evident in live art, sound art, digital art, Internet art, and film and video. In his international overview of the current intersections of art, science, and technology, Steven Wilson (2002) documents the many ways in which artists today are working at the frontiers of scientific inquiry and emerging technologies and the Internet, forging new relationships with biology, physical sciences, mathematics and algorithms, kinetics, telecommunication, and digital systems.

Ede highlights, among other examples, Andrew Carnie's *Magic forest*. Carnie worked with a developmental neurobiologist to find ways of visualizing the structure and growth of neurons. The resulting artwork was a walk-through installation in which viewers moved through a floating woodland of lacy winter trees. The trees were actually images of living brain cells in the process of conducting signals, captured using a laser-scanning confocal microscope, drawn with the aid of computer-imaging techniques, stained with fluorescent dyes, and projected onto layers of fine fabric. Images produced by new scanning technologies have a beauty that has led some to call them a new form of abstract art. Yet, Ede questions whether they are instead information and data, lacking inherent aesthetic properties of expression and abstraction.

Equally striking changes and implications have appeared in music research. While traditional musicology focused on Western music, cultural elites, and history, today's scholars are moving beyond these intellectual strictures. Musicology, broadly defined, addresses the music of all cultures (any country, group, language, or religion), all subcultures (such as modern youth subcultures), and all classes (owning versus working, privileged versus suppressed, and so on). The expanse of relevant disciplines includes acoustics, aesthetics, anthropology, archaeology, art history and theory, biology, composition, computing, cultural studies, economics, education, ethnology, gender studies, history, linguistics, literary studies, mathematics, medicine, music theory and analysis, neurosciences, perception, performance, philosophy, physiology, prehistory, psychoacoustics, psychology, religious studies, semiotics, sociology, statistics, and therapy. The cultural and epistemological diversity of musicology also became increasingly evident during the twentieth century, as smaller subdisciplines grew faster than traditionally dominant historical musicology. Comparative musicology, for example, in which different musical cultures are compared, emerged in a context of nineteenth-century colonialism. In the twentieth century it was largely superseded by ethnomusicology, which aims to view each culture in its own terms. Ethnomusicology grew rapidly due to increasing interest in non-Western cultures, and their sheer number and diversity.

Music research today can be subdivided in various ways. At the highest level, all music research is either historical/ethnological or systematic. That classification is not quite the same as the humanities–sciences split, though. Ethnomusicology comprises mainly humanities but also includes social sciences, while systematic musicology is mainly oriented to sciences but also includes humanities. There is no widely accepted definition of systematic musicology, and implicit definitions shifted during the twentieth century. Modern systematic musicology is often held to include music philosophy, aesthetics, psychology, sociology, acoustics, computing, and physiology. These disciplines tend to address *general* musical questions. In contrast, historical musicology and ethnomusicology tend to focus on *specific* performances, pieces, genres, traditions, cultures, styles, and composers (Parncutt 2007).

To echo a point made in the introduction, subdivision of the main branches in music research also differs internationally. In Central Europe, musicology is thought to comprise historical musicology, ethnomusicology, and systematic musicology (cf. Adler 1885). The numbers of participants attending international conferences suggest that these three areas are now roughly equal in size. But, this structure does not clearly accommodate important and growing fields such as popular music or music performance research, or the traditionally central disciplines of music theory and analysis focused on musical scores. The North American tripartite division comprises historical musicology, ethnomusicology, and music theory, but has explicitly excluded scientific approaches such as music psychology, computing, acoustics, physiology, and empirical sociology, all of which are independently growing and thriving.

If musicology is defined as the study of music, it also includes the knowledge of musicians, regardless of their musical roles. This classification includes the history of performance practice, studies of composers' sketches, and psychological/sociological studies of performance. Recent research on group creativity in popular music is also focusing on the political, cultural, social, and economic conditions under which music is produced. This development suggests a third possible tripartite subdivision of musicology: into humanities, sciences, and practice. Musical practice includes not only performance, composition, and improvisation, but also education, medicine, and therapy (applied musicology). This alternative tripartite structure exposes the most salient epistemological differences within music research.

9.3 Rethinking (inter)disciplinarity in art history

Change, as always, provokes debate. The primary faultline in both art history and musicology was, and continues to be, the divide between 'inside' and 'outside' the discipline proper. Some scholars see themselves as custodians of tradition and the internal purity of the object of 'art' or 'music'. Others see themselves as interventionists, critiquing and expanding the construction of their domains. Interdisciplinarity has been not only a driver of heterogeneity, fostering new communities of practice; it too is heterogeneous. In art history, one scholar might be investigating the social history of a genre with the goal of understanding how aesthetic forms are shaped by connotations of taste in particular time periods and cultural groups. Another might be borrowing from other disciplines to answer a historical question about the provenance of a particular painting. Others might be using new technologies to read Paleolithic imagery or to create digital-born art. The latter examples are 'scientific' in the sense of capitalizing on affordances of new technologies and instrumentation. Yet, they are not scientific in the same sense as music psychology or empirical aesthetics, which have different methods and epistemologies and are based on scientific questions or hypotheses that can be confirmed or rejected on the basis of statistical analysis of data obtained in empirical investigations of music perception.

The shift from older forms of interart comparison to subsequent interdisciplinarities of word-and-image studies, the cultural turn in the humanities, visual culture, and images studies are further indicators of change. Interart comparison focused traditionally on similarities and differences between particular forms of arts, the influence of one art on another, common origins comparable to the ancient unity of stem languages, experiential psychological studies, and structural semiotic analyses that view arts as alternative language systems (Greene 1992). In its most ambitious forms, Mitchell (1990, 1994) explains, interart comparison argued for the existence of extended formal analogies across the arts capable of revealing structural homologies united by historical styles. At its best, it resisted compartmentalization of media into particular disciplines and the academic administration of knowledge. Yet, interart comparison had three major limitations. It was based on the presumption of a unifying, homogeneous concept (e.g. meaning or representation) and a positivist methodology of comparing and differentiating propositions. The strategy of systematic comparison and contrast also ignored other forms of relation. And, interart comparison was a ritualistic and generalized historicism that affirmed the dominant sequence of periods and a canonical masternarrative leading to the present. Alternative histories, counter-memories, or resistant practices were neglected. By and large, too, the insularity of disciplines was not challenged and superficial comparisons ignored crucial differences across art forms and genres.

The difference between the interdisciplinarities of interart comparison and word-and-image studies is illustrated by the work of Svetlana Alpers and Mieke Bal. In *The art of describing* (1985), Alpers expanded the traditional canon of privileged works while drawing on both visual and verbal documents pertaining to visual culture in the Netherlands during the seventeenth century, including ideas about vision in science and instrumentation. In *Rembrandt's enterprise* (1988), Alpers treated the artist as both the product and an instrument of change by analyzing the materiality of painting, economics of the art business, and Rembrandt's use of theatricality. In treating visual arts as sign systems, she demonstrated how paintings, photographs, sculptures, and architecture are imbued with textuality and discourse. In *Reading 'Rembrandt'*, Bal (1991) drew on methods from art history, gender analysis, and reader-response theory to explore theoretical and interpretive problems pertaining to relations between verbal and visual art. The context of works becomes a text that can be 'read' using a semiotic methodology that treats medium-bound terms such as spectatorship, storytelling, rhetoric, reading, discursivity, and visuality as aspects, not essences. Shifting attention away from the intrinsic properties of discrete visual and verbal domains opens up larger questions of representation and interpretation that facilitate systematic interrogation of the ways arts emerge, circulate, and are intertwined within a culture.

The cultural turn in humanities stimulated expansion of interdisciplinary theory and practice propelled by both strands of influence identified by Kraft in describing changes associated with the 'new art history'. The older text-bound and elite concept of 'culture' broadened, influenced by the more encompassing anthropological notion of the lifeways of a people. Investigations of the political, cultural, social, and economic conditions of artistic production and consumption were framed by newer critical, semiotic, and deconstructionist approaches. In introducing a 1999 collection of essays on *The practice of cultural analysis*, Bal took a graffito on a wall as the starting point for defining cultural analysis as the central interdisciplinary practice for humanities. This short and uncanonical text-image is a public exhibit that even in its simplicity engages the complex interdiscursivity of visual performance and verbal argument. Cultural analysis is an 'interdiscipline' with a specific object as well as a set of collaborating disciplines that includes the social sciences as well as the sciences. It is primarily analytic and is representative of much of the interdisciplinary work that goes on in humanities and cultural studies today.

In *Travelling concepts in the humanities* (2002), Bal highlighted the methodological potential of concepts as the backbone of interdisciplinary study of culture. The major exemplars are image, *mise en scène*, framing, performance, tradition, intention, and critical intimacy. Concepts such as these exhibit both specificity and intersubjectivity. They do not mean the same thing to everyone, but they foster common discussion as they travel between disciplines, between individuals, between periods, and between academic communities. In the process of travel, their meaning and use change, and that changeability becomes part of their usefulness.

The term 'image studies' signals a further horizon in the changing relationship of art, science, and technology. Mitchell (1994, 1995) introduced the concept of a 'pictorial turn' to name the challenge that visuality presents to the dominant textual model of the world in humanities. Vision is a mode of cultural expression and communication as fundamental and widespread as language. The term 'visual culture' evolved from a phrase used in several fields, including art history, film and media studies, semiotics, history of science, comparative arts, and philosophical inquiries into art and representation. Conversations about visuality also occur in cultural studies, queer theory, and African-American studies, and among psychoanalysts, anthropologists, phenomenologists, theorists, and optical technologists (Mirzoeff 1998, 1999). Widening interest in images and visual knowledge led James Elkins to suggest that the proper term is no longer 'art history' but 'image studies'. Barbara Stafford has been influential in opening up this terrain in terms of research on scientific and other non-art images (Jones and Galison 1998). In *Echo objects* (Stafford 2007), she explores the cognitive work of images with insights from neuroscientific discoveries and evolutionary biology. New data confirm some traditional assumptions about cultural objects, but also exert pressure on and turns them upside down, whether talking about historical emblems or electronic media.

9.4 Rethinking (inter)disciplinarity in musicology

The heterogeneity of practices is evident in musicology as well. One scholar might be studying the integrative dynamics of opera, a performance art that requires the collaboration of experts from multiple disciplines and professions. Another might be examining questions of race in jazz studies. One might be studying the perception of emotion by listeners from different cultures. Another might study the effect of the acoustics of a given musical instrument on the syntax (pitch, loudness, timbre) of music played with that instrument. One might study the role that music plays in determining or manipulating the psychological identities of young people, or even work on a collaborative team with an anthropological interest in the hybridization of cultural forms in contemporary genres. Yet another might study the relationship between hearing or performing music and linguistic abilities, as a possible strategy against dyslexia.

More generally, musicology is epistemologically diverse for several reasons. First, definitions of music itself are diverse and dependent on cultural and historical context. If music is defined as an acoustic signal that evokes recognizable patterns of sound, implies physical movement, is meaningful and intentional, is accepted by a cultural group, and is not lexical (i.e. is not language), each point in this list implies different disciplines or epistemologies. Second, music may be represented in different ways. Popper and Eccles (1977) divided reality into three 'worlds': in the physical world, music is signal and vibration (physics, physiology); in the subjective world, it is private experience (phenomenological psychology, cultural studies); and in the abstract world of information and knowledge, it is scores (notation) and sampled waveforms (music theory, computing). To this we might add a fourth world of *agents* (selves, egos, souls, spirits) in which music is constituted by the social interactions of performers, composers, and listeners (sociology, cultural studies). A third reason for musicology's epistemological diversity involves its contexts. Scientific subdisciplines such as acoustics, physiology, psychology, and computing tend to focus on music itself in different representations, whereas cultural subdisciplines such as history, ethnomusicology, and cultural studies focus on music's historical, geographical, and cultural contexts.

How should musicologists deal with this epistemological diversity? Clearly, no individual can claim to be an expert in all relevant disciplines—or even just two of them. Nor is any specific epistemology central to musicology. That implies that good musicology must be multi- and interdisciplinary. High-quality synergetic interactions between epistemologically distant disciplines can only be achieved by interdisciplinary teams, which should be promoted by musicological institutions. Several impediments, though, remain.

Institutionally, there is a deep divide in musicology between ingroup and outgroup subdisciplines. The ingroup is traditionally headed by music history and may also include music theory/analysis and cultural studies. The outgroup (the musicological Others) is largely scientific: acoustics, psychology, physiology, and computing. Between the two is an intermediate group of subdisciplines that, in conservative music schools and departments, are politely tolerated but relatively powerless: ethnomusicology, pop/jazz research, music sociology, music philosophy, music performance research. Internationally, approaches to research often depend on nationality. That is understandable if the research object is music of different cultures, but surprising when general questions are asked. Consider, for example, music theory, whose principle object of research is the structure of Western musical scores (melody, harmony, counterpoint, tonality, rhythm, meter, form). North American music theory is traditionally characterized by a formalist, mathematical, positivist, 'scientific' approach and inspired by Heinrich Schenker's reductional approach to analysis (for tonal music) and Milton Babbitt's mathematical pitch-class theory (for atonal music). German music theorists are traditionally more intuitive and holistic, and more likely to relate musical meaning to the social and historical context. There is a clear need for better cross-Atlantic communication.

Politically, musicological institutions also tend to assume that power is wielded by historical musicologists, and that it is legitimate to maintain that power by exploiting the ambiguity of the word 'musicology'. Musicology sounds more interesting, relevant, and important when it is broadly defined—as in leading (historically dominated) music encyclopedias such as *Grove's dictionary of music and musicians* or *Musik in Geschichte und Gegenwart*. But, when it comes to decisions that affect all musicologists, such as what aspect of musicology should dominate in programs of study or which 'musicological' grant applications should be funded, 'musicology' is tacitly assumed to be historical. The International Musicological Society aims and claims to represent all of musicology, but most of its members, officers, conference papers, and journal contributions are historical; the ethnological proportion is increasing, but the systematic (scientific) proportion remains negligible. As a result, outgroup researchers do not identify with musicology (they do not call themselves 'musicologists') and are not motivated to contribute to musicological institutions.

The challenge of diversity is amplified in interactions between humanities and sciences. They have never been easy, because of enormous differences in basic beliefs and assumptions. While humanities scholars may primarily be interested in specific examples of a given phenomenon, scientists may prefer to focus on phenomena whose frequency of occurrence is statistically significant. While humanities scholars strive to experience a research object directly, personally, and vividly, while tending to trust personal experience, intuition, and introspection as sources of evidence, scientists try to distance themselves from research objects in order to objectively compare hypotheses with data.

The tension between humanities and sciences is deeply ingrained in academic structures and traditions (Snow 1959). The conflict dates to Ancient Greece. Aristotle's philosophy was similar to that of modern empirical psychology, focusing on perception, systematic observation, and association, while Pythagoras anticipated modern humanities scholars in his opposition to empiricism. For Pythagoras, music was divine, perfect, and inseparable from pure mathematics and astronomy. His spirit, which dominated medieval music theory, lives on the formalism of North American music theory, while the tendency of British music psychology and sociology to regard music as a phenomenon of everyday life (DeNora 2000), not manuscripts and concert halls, harkens back to Aristotle. The emergence and rapid growth of a new, US-led approach to music psychology in the 1980s and 1990s also led to a general increase in interest in empirical approaches and methods, which expanded to include qualitative methods and spilled over into other musicological subdisciplines such as music sociology and music theory.

As the twentieth century drew to a close, technological developments enabled new advances in music physiology, and especially neuroscience. Music psychology has grown rapidly since the 1960s following psychology's *cognitive turn* (a reaction against behaviorism); the cognitive paradigm has been applied extensively to the perception of musical structure, and advances in computer technology have expanded the methodological possibilities. Music information sciences include computer music (computing and composition), computing in musicology (computer-based analysis of musical scores), and music information retrieval (automatic extraction of culturally and perceptually interesting information from sound files). Advances in brain imaging technology during the 1990s provoked rapid growth in music physiology and the neurosciences of music. Within the sciences, the growth in music psychology has attracted representatives of other disciplines such as information sciences, acoustics, and physiology to collaborate with psychologists and achieve a new modern realization of Adler's systematic musicology.

9.5 Conclusion

Two overriding points emerge from a comparative reflection on interdisciplinarity in art and music research. First, it is harder to talk in the singular anymore. In a pluralistic conception of discipline, Pasler (1997) suggests, the question is not so much what is new or old, or what needs to be replaced or superceded, but rather how each perspective can be enriched by the presence of the other. Research on both art and music has been likened to a web and a network of cross-secting, sometimes conflicting, and sometimes cross-fertilizing influences. However, the second point follows from the first. In the midst of an expanding repertoire, institutional challenges continue to haunt both disciplines and related fields. Talk of increased interdisciplinarity must be accompanied by strong educational programs that provide training both in the methodological and epistemological foundations of individual disciplines and in integrative methods. As the twentieth century progressed, for instance, the three main European subdisciplines of musicology became increasingly independent and isolated. Constructive interactions among historical, ethnological, and systematic subdivisions are rare, both locally (within departments) and globally (in conferences and journals). If musicology is to make progress as a unified, socially relevant, and financially buoyant discipline, musicological institutions should take decisive action. A similar challenge confronts research on art, visual culture, and image studies. Professional associations and curricular categories still segment differences, and links with interdisciplinary fields such as cultural studies and media studies are neither fully identified nor robust. Differences should not be erased, but relations and boundary questions need to be addressed proactively and reflexively in the practice of research and education. That can only be achieved, however, if experts work together—creatively, constructively, and as intellectual equals.

The same may be said of the larger relationship between humanities and sciences. Conflict resolution techniques (Deutsch *et al.* 2006) offer one means of unification. Yet, their power remains unequal. In proportion to the social relevance and pragmatic value of sciences, humanities have too little influence both within universities and in society generally, when one considers the central and undiminished importance of identity and meaning for all humans and all cultures. Competition for resources should be fairer and more transparent. And researchers in both humanities and sciences need to learn more about each other by improving communication in both scholarship and teaching, and by facilitating mutual critical reflection on the strengths and weaknesses of interactive fields and methods. Reflecting on the track record in the psychology of art and creativity, Martin Lindauer (1998) lamented that it is still framed by opposition more than cooperation.

He called for greater reciprocity and mutual illumination, a call echoed by across the expansive domains of art and image, music and sound. Superficial generalization, *ad hoc* borrowing, and reductive use of one discipline in the interests of another are not enough; deep, detailed, synergetic interactions are necessary. The history and practices in this chapter illustrate the richness of interdisciplinary research but affirm the importance of more systematic attention to both its intellectual and its institutional dynamics.

References

Adler, G. (1885). Umfang, methode und ziel der musikwissenschaft. *Vierteljahresschrift für Musikwissenschaft* **1**, 5–20.

Alpers, S. (1985). *The art of describing: Dutch art in the seventeenth century*. Chicago: University of Chicago Press.

Alpers, S. (1988). *Rembrandt's enterprise: the studio and market*. Chicago: University of Chicago Press.

Bal, M. (1991). *Reading 'Rembrandt': beyond the word-image opposition*. Cambridge: Cambridge University Press.

Bal, M. (1999). Introduction. In: M. Ball (ed.) *The practice of cultural analysis: exposing interdisciplinary interpretation*, pp. 1–14. Stanford: Stanford University Press.

Bal, M. (2002). *Travelling concepts in humanities*. Toronto: University of Toronto Press.

Cook, N. (1998). *Music: a very short introduction*. Oxford: Oxford University Press.

DeNora, T. (2000). *Music in everyday life*. Cambridge: Cambridge University Press.

Deutsch, M., Coleman, P.T., and Marcus, E.C. (2006). *The handbook of conflict resolution: theory and practice*, 2nd edn. San Francisco, CA: Wiley.

Ede, S. (2005). *Art & science*. London: I. B. Tauris.

Fowler, A. (1972). Periodization and interart analogies. *New Literary History* **3**(3), 487–509.

Greene, D.B. (1992). Music and the humanities: coming to new questions. *Interdisciplinary Humanities* **9**(4), 55–64.

Jones, C.A. and Galison, P. (1998). Introduction: picturing science, producing art. In: C.A. Jones and P. Galison (eds) *Picturing science, producing art*, pp. 1–26. New York: Routledge.

Kassabian, A. (1997). Introduction: music, disciplinarity, and interdisciplinarities. In: D. Schwarz, A. Kassabian, and L. Siegel (eds) *Keeping score: music, disciplinarity, culture*, pp. 1–12 Charlottesville, VA: University Press of Virginia.

Kraft, S. (1989). Interdisciplinarity and the canon of art history. *Issues in Integrative Studies* **7**, 57–71.

Levy, A. and Tischler, B.L. (1990). Into the cultural mainstream: the growth of American music scholarship. *American Quarterly* **42**(1), 57–73.

Lindauer, M.S. (1998). Interdisciplinarity, the psychology of art, and creativity: an introduction. *Creativity Research Journal* **11**(1), 1–10.

Mitchell, W.J.T. (1990). Against comparison: teaching literature and the visual arts. In: J.-P. Barricelli, J. Gibaldi, and E. Lauter (eds) *Teaching literature and other arts*. New York: Modern Language Association.

Mitchell, W.J.T. (1994). *Picture theory: essays on verbal and visual representation*. Chicago: University of Chicago Press.

Mitchell, W.J.T. (1995). Interdisciplinarity and visual culture. *The Art Bulletin* **77**(4), 540–4.

Mirzoeff. N. (1999). *An introduction to visual culture*. London: Routledge.

Mirzoeff, N. (1998). What is visual culture? In: N. Mirzoeff (ed) *The visual culture reader*, pp. 3–13. London and New York: Routledge.

Parker, R. (1997). Literary studies: caught up in the web of words. *Acta Musicologica* **69**(1), 10–15.

Parncutt, R. (2007). Systematic musicology and the history and future of western musical scholarship. *Journal of Interdisciplinary Music Studies* **1**, 1–32.

Pasler, J. (1997). Directions in musicology. *Acta Musicologica* **69**(1), 16–21.

Popper, K.R. and Eccles, J.C. (1977). *The self and its brain*. Berlin: Springer.

Preziosi, D. (1982). Constru(ct)ing the origins of art history. *Art Journal* **42**(4), 320–7.

Preziosi, D. (1989). *Rethinking art history: meditations on a coy science*. New Haven, CT: Yale University Press.

Richmond, S. (1984). The interaction of art and science. *Leonardo* **17**(2), 81–6.

Sibbald, M.J. (1992). The search for a humanistic approach to music education. *Interdisciplinary Humanities* **9**(4), 17–21.

Snow, C.P. (1959). The two cultures. *Science* **130**(3373), 419.

Stafford, B.M. (1988). The eighteenth century: towards an interdisciplinary model. *The Art Bulletin* **70**(l), 6–24.

Stafford, B.M. (2007). *Echo objects: the cognitive work of images*. Chicago: University of Chicago Press.

Steiner, W. (1982). *The colors of rhetoric: problems in the relations between modern literature and painting*. Chicago: University of Chicago Press.

Stone, J.H. (1969). Integration in the humanities: perspectives and prospects. *Main Currents in Modern Thought* **26**(1), 14–19.

Taussig, M.T. (1993). *Mimesis and alterity: a particular history of the senses*. London: Routledge.

Treitler, L. (1995). History and music. In: R. Cohen and M.S. Roth (eds) *History and... histories within the human sciences*, pp. 209–30. Charlottesville, VA: University Press of Virginia.

Wilson, S. (2002). *Information arts: intersections of art, science, and technology*. Cambridge, MA: MIT Press.

CHAPTER 10

Engineering

PATRICIA J. CULLIGAN AND FENIOSKY PEÑA-MORA

This chapter uses civil engineering as a focus to explore concepts of interdisciplinarity in engineering. The choice of civil engineering arises from the fact that civil engineering is the oldest formal engineering discipline, originally including all forms of engineering other than military engineering. In what follows, civil engineering is traced from an engineering branch that first produced 'master integrators' to a field that now has numerous well-established subdisciplines, including geotechnical, structural, transportation, construction, and environmental engineering—a subdiscipline often considered a discipline unto itself. An examination of the history of civil engineering in the United States reveals the evolution of a discipline-centric system that currently requires candidates for professional engineering licensure to have breadth in five different civil engineering areas and depth in one, accredited undergraduate programs to provide training in at least four civil engineering areas, and research faculty to develop narrowly focused research programs in a specific area of civil engineering in order to improve their chances of academic tenure. A re-evaluation of the role of civil engineers in the context of twenty-first century challenges is used to argue the need for civil engineers to, once again, become master integrators and leaders of specialists, rather than specialists unto themselves.

10.1 The evolution of engineering and engineering disciplines

The term engineer originated in the eleventh century from the Latin *ingeniator*, meaning one with *ingenium*, the ingenious one (Auyang 2004). Indeed, Leonardo da Vinci held the official title of Ingegnere Generale. According to de Camp (1963), civilization as we know it today owes its existence to ancient engineers of Syria and Iraq, who around 8000 BCE provided the crucial technological inventions needed to move societies forward. These ancient engineers were responsible for the irrigation, architectural, and military projects of the time, with the same person often expected to have expertise in all three areas.

The first phase of modern engineering began in the sixteenth century with the advent of the Scientific Revolution. The engineers of the Scientific Revolution sought to supplement the intuition, rules of thumb, and empirical data of their predecessors with scientific reasoning, mathematical analysis, and controlled experiments. Learned scientific societies established in the seventeenth and eighteenth centuries, such as The Royal Society of London (founded 1660), started to publish journals archiving some of the fundamental principles of engineering science. The eighteenth century also witnessed the creation of some of the first engineering schools in the world. As early as 1702 there was a school of mining and metallurgy in Freiberg, Germany. Other schools of mining were established in the latter half of the eighteenth century in Austria, Spain, Sweden, and Russia. Nonetheless, L'Ecole Nationale des Ponts et Chaussées (the National School of Bridges and Highways), founded in 1747 in Paris to provide the French Corps de Ponts ēt Chaussées with qualified engineers, is often considered to be the first formal school of engineering in the world (Grayson 1980). According to the official website of L'Ecole Nationale, its early students would gain knowledge and skills in algebra, mechanics, hydraulics, and other domains, in addition to practical experience in the field.

In contrast to the formal system of education that was becoming available to aspiring engineers on continental Europe, aspiring eighteenth-century engineers of Great Britain learned their trade via a system of apprenticeship, in which learning and training went side by side on the job (Watson 1982). Nevertheless, a British engineer, John Smeaton, who was trained as an instrument maker, is usually accredited with establishing the first formal discipline of engineering when, in 1768, he described himself as a 'Civil Engineer' in order to differentiate himself from the 'Military Engineer'. Unlike members of the French Corps de Ponts et Chaussées, who engaged in public engineering works as officers of the state, Smeaton ran his own private practice, pioneering a pattern of work still reflected in the offices of many modern professional engineers. During his career, Smeaton initiated the use of hydraulic lime (a form of concrete that will set under water) to construct the first successful Eddystone lighthouse in Devon, England. He also designed windmills, watermills, canals and bridges, steam pumps, ports, mines, and jetties.

In 1771, the non-military engineers of Britain founded the Society of Civil Engineers, which is thought to be the world's first professional engineering society. The society became the Institution of Civil Engineers (ICE) in 1818. In 1828, when the ICE petitioned for a royal charter to further legitimize itself, it defined civil engineering as 'being the art of directing the great Sources of Power in Nature for the use and convenience of man'. Soon after, the civil engineers started to mark their professional territory, and a class distinction began to form between civil engineers and other kinds of engineers.

With the increasing demarcation of professional territories, fast growth in engineering-based industries in nineteenth-century Britain, emergence of new specialized engineering groups, and the civil engineers' growing disinclination to embrace others, it became clear that the ICE could not cater for the needs of multiple disciplines. Thus, non-civil engineers began to establish their own institutions. In 1847, the Institution for Mechanical Engineering was formed, followed by the Institution of Naval Architects in 1860 and then the Institution of Gas Engineers and Managers in 1863. These four original British engineering institutions remain active in 2009. However, the total number of engineering institutions licensed by the United Kingdom (UK) Engineering Council has expanded in 180 years to over 35.

In the United States the foundation of new professional engineering societies, and thus to some extent the creation of new engineering disciplines or subdisciplines, has been even more prolific than in the UK. In the 100 plus years following the foundation of America's first national engineering society, the American Society of Civil Engineers (ASCE), the number of national-level engineering societies in the United States increased at a rate of more than one per year, rising from one in 1852 to 130 in 1963 (Auyang 2004).

Today, the world's largest professional engineering society is the Institute of Electrical and Electronics Engineers (IEEE), which was founded in the United States in 1963. The stated mission of the IEEE is to promote the engineering profession for the benefit of humanity and the profession. The IEEE encourages public involvement in engineering as well as the integration of different disciplines and technologies. Tensions still remain, however, between what are considered the more aristocratic views of discipline-centric engineering societies, like the ICE and ASCE, and the more populist views of societies like the IEEE.

10.2 The growth of disciplines within disciplines

The existence of well over a hundred national-level engineering societies within the United States can, to a large extent, be explained by the evolution of engineering disciplines within a primary discipline. Here, a primary engineering discipline will be termed a 'branch' of engineering, and engineering disciplines within the primary discipline will be referred to as 'subdisciplines'.

Today, the five so-called 'traditional' major branches of engineering are civil, mechanical, electrical, chemical, and industrial engineering. Other major branches of engineering that grew from the traditional branches include aerospace, bio/biomedical, computer, environmental, and nuclear. There are also a myriad of what are considered more specialized branches of engineering, including the older established fields of naval engineering and mining and metallurgy.

The number of subdisciplines within each branch of engineering varies from field to field. In what follows, the factors contributing to the evolution of subdisciplines within a branch of engineering are explored using civil engineering, the first formal branch of engineering, as an example. Discussion is focused on the history of civil engineering within the United States from the perspectives of professional practice and licensure, engineering education, and engineering research.

10.2.1 Professional practice and licensure

The ASCE was founded in 1852 by a group of engineers meeting in offices of the Croton Aqueduct in New York City. The group, who wished to establish a national civil engineering society that emulated the British ICE, used the constitution of the Boston Society of Civil Engineers (founded 1848) as their legal framework (Griggs 2003). They also made a list of potential society members from their own knowledge of practicing engineers of the time, which contained about 230 names. Most of the engineers on the list had received no formal engineering education and, instead, had been trained via the British apprenticeship model. These early US engineers had readily aligned themselves with business and capitalist values, because it was business people and their capital that supported their great works (Noble 1984). Nonetheless, the claim to professionalism and social responsibility that accompanied the establishment of the ASCE enabled the first ASCE members to assert some independence from business, despite the fact that the new society deliberately shied away from adopting a code of ethics. Like the ICE before them, the ASCE proved unable to serve the needs of the bulk of those who considered themselves professional engineers. According to Layton (1971), by the turn of the twentieth century the ASCE had established itself as an elitist society, believing its standards and professionalism to be higher than those of the other founder societies. Discussions on the social responsibility of engineers and the application of engineering to societal problems were discouraged, and the younger, as well as the more progressive, engineers were alienated by the conservatism of the society. As a result, those wishing to reform the ASCE led 'the revolt of the civil engineers' beginning in 1909. The result was a set of concessions by the ASCE that included the 1914 adoption of a code of ethics.

During the time period immediately preceding 'the revolt of the civil engineers', anyone in the United States could work as an engineer without proof of competency. Consequently, many self-declared 'engineers' lacked the qualifications and experience of the professional civil engineers of the time, who were designing and constructing major works such as the Brooklyn Bridge in New York, the first skyscraper in Chicago, the first water and sewage treatment works in Chicago, and the New York City Interborough Rapid Transit subway. In 1907, concern in the state of Wyoming over unqualified water speculators who were identifying themselves as engineers resulted in the passage of the first US engineering licensure law in 1907, 'in order that all the surveying and engineering pertaining to irrigation works should be properly done'. Although the ASCE supported this piece of legislation, it resisted the notion of state-controlled licensure, believing that only engineers should pass judgment on other engineers. Nonetheless, the state of Louisiana quickly followed Wyoming's example in 1908, and thus began a trend of requiring that a person needed to pass a state examination in the theory and practice of engineering before being allowed to practice civil engineering (Grayson 1980). By 1950, all 50 US states plus the District of Columbia and Puerto Rico had adopted some kind of registration laws for practicing engineers that restricted use of the designation 'professional engineer'. Almost 60 years later, professional engineering licensure in the US still remains governed by the rules and regulations of the individual states. Nonetheless, the milestones to licensure are common across many states.

10.2.2 Education

About the same time that the ASCE was being founded, US college education in engineering was becoming established. The first structured civil engineering curriculum in the United States, which was based on the French system, was introduced at the Military

Academy at West Point, NY, in 1817. Soon after, structured civil engineering curricula began to emerge at the non-military schools of technical education that were being established at the time. In 1835, the first degree in civil engineering was granted to a graduating class of four by Rensselaer Institute in Troy, NY, which later changed its name to Rensselaer Polytechnic Institute. By 1862, there were about a dozen engineering schools in the United States and a model of a 4-year engineering curriculum, which featured a parallel sequence of humanities, mathematics, science, and technical subjects, was in place. More than 150 years later, this educational model remains the backbone of most undergraduate engineering programs in the United States. Of note is the fact that this education model for civil engineering was developed without any input from the practicing engineers of the time.

In 1862, the passage of the Morrill Land Grant Act allowed for the creation of land-grant colleges to promote 'liberal and practical education of the industrial classes in the several pursuits and professions in life'. In the following three decades, the number of people attending college increased and the purpose of college education became more closely associated with training for skilled vocations and professions. Nonetheless, engineering education remained largely in the hands of the engineering educators, with little input from practicing engineers. In 1894, a group of engineering educators who were dedicated to improving the rigor of engineering education in the United States formed the Society for the Promotion of Engineering Education (SPEE), which later became today's American Society for Engineering Education (ASEE). Concern about the diverse structure of engineering education and its relationship with the professional engineering societies prompted the SPEE to initiate an evaluation of engineering education in the United States. The outcome was a report which came to be known as the Mann Report, which criticized over-specialization in engineering schools and called for more unified engineering curricula with a greater focus on fundamentals.

In the period during and immediately following World War II, engineering education in the United States was retooled to address issues of national defense, and engineering curricula and accreditation policies became more narrowly focused and technically specific. In 1955, another major evaluation of engineering education, termed the Grinder Report, found that many engineering schools were treating non-technical subjects in a dismissive or even hostile fashion. In response to the recommendations of the Grinder Report, the Engineers Council for Professional Development (ECPD) adopted additional criteria for accreditation that specified that the equivalent of a half to one full year of engineering undergraduate studies be dedicated to humanities and social science studies. Still, with increasing diversity of technical fields in engineering, the degree accreditation procedure became more precise and rule-laden.

In 1980, the ECPD undertook a major reorganization and was renamed the Accreditation Board for Engineering and Technology (ABET). Following intense dialogue with the engineering educational and professional communities, ABET revised, substantially, its accreditation criteria in its 1995 announcement of Engineering Criteria 2000 (EC 2000). The EC 2000 criteria, which became mandatory in 2001, represented a paradigm shift in focus from what is taught in an engineering program to what is learned. Under EC 2000, engineering programs were also required to demonstrate that their graduates 'had an understanding of professional and ethical responsibility, an ability to communicate effectively, the broad education necessary to understand the impact of engineering solutions in a global and societal context and knowledge of contemporary issues' (Stephan 2002). The EC 2000 criteria have undergone several, mostly minor, changes since their release. In civil engineering, the criteria for accrediting engineering programs during the 2008–9 accreditation cycle (<http://www.abet.org/>) have been amended to require that students need to be able to apply knowledge of four 'technical areas appropriate to civil engineering'. The new phrase 'technical areas appropriate to civil engineering' is intended to allow programs to accommodate 'non-traditional' emerging engineering areas like biotechnology and nanotechnology, provided that programs can justify such areas as 'appropriate to civil engineering'. Hence, an opportunity has been provided to reverse a century-long trend of discipline-centric *technical training* in civil engineering. Nonetheless, linkages between a broader technical approach to civil engineering education and the requirements for professional engineering licensure have yet to be made. In addition, a serious examination of whether the inherited 150-year-old model for undergraduate engineering education is even appropriate for training aspiring engineers has yet to take place.

10.2.3 Research

At the time that engineering education was becoming established in the United States, faculty in most US colleges were affiliated with a particular religion and there was an expectation that they would support the views of that religion and its benefactors. Thus, contracts between faculty and their institutions were rare: instead 'gentlemen's agreements' were the norm.

In 1913, concern about the vulnerability of faculty to dismissal from their institutions should they upset those in power led to the formation of the American Association of University Professors (AAUP). In 1915, the AAUP issued a declaration of principles, which laid the foundation for academic tenure (AT) in the United States. In 1940, the AAUP further recommended that the academic tenure probationary period be 7 years, and proposed that a tenured professor could not be dismissed without adequate cause, except 'under extraordinary circumstances, because of financial emergencies'. In 2009, the AT probationary period has remained close to 7 years at most US institutions. However, AT standards have varied over time.

Following World War II, the US government began significant investment of research in US universities via the awarding of competitive grants. To increase their chances of securing such grants, US engineering schools shifted faculty recruitment away from engineering educators and practitioners to people oriented toward academic research. However, by the 1960s, the Soviet–American race for technological leadership had created such shortages in the academic workforce that, in an effort to make engineering professorships attractive, AT standards began to lose their rigor. Over the past 40 years several factors, including decreasing state and federal support for academic institutions, have contributed to reversing this trend and raising AT standards in most engineering schools to new heights.

Today, AT criteria in most US PhD-granting universities have become increasingly focused on research accomplishments—as measured by publication productivity, funding success, and 'being known for something'. Indeed, in order to address the latter criterion, faculty in their AT probationary period can be specifically advised by senior colleagues to create their own research niche within a research field, in order that they can be identified as the acknowledged expert in a particular area at the time of tenure evaluation. As a result, the growth of subdisciplines within branches of engineering from the research perspective has been almost unbounded. For example, in the geotechnical engineering subdiscipline within the branch of civil engineering, the US National Science Foundation recognizes four research areas alone: These are geoenvironmental engineering, geomaterials, geohazards, and geomechanics. The total number of active research subdisciplines within civil engineering at the time of writing this article is actually unknown to the authors, but it is thought to be well over 30. Thus, the contribution of the current AT process to the parsing of the major engineering branches into multiple subdisciplines is significant. Indeed, today's tenure system is much more likely to reward faculty who adopt a reductionist research approach over faculty whose approach is more holistic (Rhoten and Parker 2004). As a result, the promotion of interdisciplinarity in engineering research within a university environment can be severely hampered by the normative AT structures.

10.3 Visions for a twenty-first century engineer

The start of the twenty-first century has prompted many engineering societies and educators to re-evaluate the role of engineers in the world (e.g. Duderstadt 2008). Societal and environmental challenges such as climate change, ecosystem vulnerability, the growth of megacities, water scarcity, the globalization of market places, clean energy needs, etc., all require new ingenuity (i.e. engineering) if they are to be tackled in a meaningful way. At the same time, the science and engineering knowledge base, in what is becoming known as the info-bio-nano technology (information technology; biotechnology; nanotechnology) domain, is growing at an exponential rate. Thus, predicting future engineering and technology tools that will be available to the twenty-first-century engineer is becoming increasingly problematic, if not impossible. Consequently, identifying the education that will produce engineers competent in the application of the new tools that will be at their disposal is also a challenge. Finally, the United Nation's (UN's) commitment at its 2000 Millennium Summit to eight goals to end global extreme poverty by the year 2015 (United Nations 2000) has forced attention on a need to shift engineering from a profession that focuses on the development of products and services for the richest 10 per cent of the world's customers to a profession that considers the requirements of the other 90 per cent (Polak 2007).

'The engineer of 2020' report by National Academy of Engineering (NAE 2004) has highlighted the skills required for the engineering profession to endure the next 20 years. The report states: 'In the next twenty years, engineers and engineering students will be required to use new tools and apply ever-increasing knowledge in expanding engineering disciplines, all while considering societal repercussions and constraints within a complex landscape of old and new ideas'. The report also highlights the various skills and attributes that engineers will need to play an effective role in the twenty-first century, including science and practical ingenuity, creativity, business and management skills, leadership, dynamism, agility, resilience, analytical skills, professionalism, and flexibility. Three years after the release of the NAE report, the ASCE published its proposal for the twenty-first-century civil engineer in the report 'The vision for civil engineering in 2025' (ASCE 2007). The ASCE report cites the following aspirations for the role of future civil engineers:

... civil engineers serve competently, collaboratively, and ethically as master:

- planners, designers, constructors, and operators of society's economic and social engine, the built environment;
- stewards of the natural environment and its resources;
- innovators and integrators of ideas and technology across the public, private, and academic sectors;
- managers of risk and uncertainty caused by natural events, accidents, and other threats; and
- leaders in discussions and decisions shaping public environmental and infrastructure policy.

To achieve the vision articulated above, 2007–8 ASCE President David Mongan stated 'civil engineers need to become master integrators and leaders so that they can orchestrate and draw upon the efforts of specialists in a broad variety of disciplines' (Mongan 2008). Both the NAE report on 'The engineer of 2020' and the ASCE report on 'The vision for civil engineering in 2025' support the need for engineers to be more broadly trained, have the ability to work in multidisciplinary collaborative teams, and have awareness of socio-politico-economic issues and cultural diversity. The ASCE vision also highlighted the need for civil engineers to be more proficient in non-technical areas, such as globalization, communication, ethics and professionalism, and leadership (Galloway 2008). Thus, in many ways the visions for the twenty-first-century engineer/civil engineer call for an engineer not unlike the first civil engineers, who used interdisciplinary approaches and integrative skill sets to meet the societal needs of the time; they do not call for an engineer whose training is focused primarily on acquiring specialized technical knowledge in a particular subdiscipline of a branch of engineering.

Given the body of knowledge and skill sets that civil engineers are projected to need in the future, there is a growing swell of opinion within the ASCE that structured engineering education needs to go beyond a bachelor's degree from an ABET-accredited program. In February 2008, the ASCE released a new 'body of knowledge' report, termed BOK2 (ASCE 2008), that addresses knowledge the society believes students and young professionals will need if they are to become licensed in the United States and 'reach their full potential as engineers in the twenty-first century'. BOK2 proposes 24 learning 'outcomes', each with six individual achievement levels based on Bloom's taxonomy of education objectives (Bloom 1956). It also proposes a master's degree, or equivalent, as a minimum requirement for professional practice.

BOK2 acknowledges that 'the civil engineering profession is continually becoming more interdisciplinary'. Indeed, specific attention is paid to interdisciplinarity in the report's discussion of a proposed new technical outcome in sustainability. Nonetheless, the means for achieving 'interdisciplinarity' in the retooling of the civil engineering profession are not addressed. Moreover, multiplication of the proposed number of BOK2 outcomes (24) by the number of required achievement levels (6) results in a figure large enough (96) to cause genuine concern that BOK2 might be too prescriptive and unwieldy to produce engineers worthy of the da Vinci title 'Ingegnere Generale'.

10.4 Real problems and community partnerships

The paradigm shift from a focus on teaching to a focus on learning outcomes that was embedded in ABET's EC 2000 criteria caused a re-examination of the way engineering was being taught in the United States. Recognition dawned that it was no longer sufficient, or even practical, to cram students full of technical knowledge and expect them to fulfill the foundational learning requirements needed for a life-long career in engineering. Instead, engineering education needed to provide students with an understanding of the context within which they would work (Beder 1999). In order to meet that need, engineering schools began to incorporate problem-based learning experiences, a pedagogical model developed in medical education in the early 1970s (Savery and Duffy 2001).

Around the same time that engineering schools were rethinking approaches to teaching, the publicity generated by the UN Millennium Development Goals (MDGs) and events such as the 2004 Asian tsunami and the landfall of hurricane Katrina in 2005 prompted engineering students, faculty, and professionals in the United States to question how they could direct their resources and skills more immediately toward the alleviation of human suffering. Organizations such as Engineers Without Borders, USA and Engineers for a Sustainable World were established and rapidly gained popularity as university and professional chapters were founded. Students started to push for curriculum-based projects that involved sustainable development and capacity building for communities in need: in other words, problem-based learning experiences with a *societal context*. Some university centers and institutions, such as the Earth Institute at Columbia University (<http://www.earth.columbia.edu>) started to direct resources and intellectual capital toward solving sustainability challenges faced by communities in the developed and developing worlds. At the Massachusetts Institute of Technology (MIT), the D-Lab, which is a series of courses and field trips that focus on international development, appropriate technologies, and sustainable solutions for communities in developing countries, was introduced into the curriculum in 2003, while at the University of Colorado, Boulder (UC-Boulder), the Engineering for Developing Communities Program was started in Spring 2004. Forward-thinking programs, such as the Engineering Projects in Community Service (EPICS) at Purdue University, which was founded in 1995, began to receive more national attention.

In what follows, two case studies are provided which illustrate how the growing impetus among engineering students, faculty, and professionals to use their skills to provide technical services and appropriate technologies to some of those who are most vulnerable in today's society is helping to engender interdisciplinary approaches and thinking in civil engineering. For convenience, the two case studies are drawn from the authors' own experiences. However, that work by others, including work by those involved in the aforementioned initiatives of MIT, UC-Boulder, and Purdue University, could equally well illustrate the points that are made. Thus, no claims are made that the following case studies represent either unique or pre-eminent work.

10.4.1 Engineering for communities in need

Pressured by increasing student demand to be involved in 'engineering that matters', the first author—Culligan—introduced a new course entitled 'Engineering for developing communities' (EDC) into the undergraduate engineering curriculum in the School of Engineering and Applied Science (SEAS) at Columbia University in Spring 2005.[1] The course is open to any undergraduate engineering student in SEAS provided that they have taken the freshman introductory course to engineering design. In this respect, the class is fostering education that integrates the different branches of engineering. Arts and sciences (A&S) undergraduate students can also petition to enroll in the class. Thus, the class is seeding collaborations between engineering, science, and liberal arts students.

The primary goals of the class are to introduce students to engineering problems faced by developing communities, and to explore design solutions to these problems in the context of a real project with a community client. Emphasis is placed on the design of sustainable solutions that take account of social, economical, and governance issues, and can be implemented now or in the near future. Because of the significance of water and sanitation to the health of developing communities, designs for delivering potable water and improving waste management, which are traditional civil engineering challenges, have been a key point of focus.

Columbia University's EWB Chapter (CU-EWB) identified the first community client for the class, which was a subsistence farming community located in the eastern region of Ghana, Africa. A neighboring community was added as a client the following year. Both communities have remained clients for the class up until the writing of this chapter. Biannual field trips to the communities have contributed to building relationships with the community members and a database of information about the settings of the communities, including their physical, cultural, social, economic, and political settings.

Unlike many classes in engineering, the EDC class is not prescriptively structured in advance, meaning that the students drive the design direction and thus, to some extent, the learning experiences. Nonetheless, the class always starts with a review of the UN MDGs and a discussion of how engineering science and technology can contribute to meeting the goals. Next, the class researches the fate of science and technology interventions in developing countries that were geared toward improving health and alleviating poverty. Of interest is the fact that the class always concludes that the failure or success of such interventions rarely correlates with a failure in the design. Instead, issues of cultural appropriateness, community education, cost and expertise needed for operation and maintenance, and the availability of spare parts, etc., influence most outcomes. Hence, the class begins to grasp the significance of non-technical context to technical design success.

Following on, the class explores the history, physical, cultural, social, economic, and political settings of the community clients and discusses the projects that the clients themselves have identified as important community priorities. Although the class is not constrained to working on designs that address these priorities, it is encouraged on the basis

[1] That the title of the course matched the one year old UC-Bolder program was a coincidence.

that a community's prioritization of its needs is often more important than an engineer's perception of the community's needs. If students decide to work on addressing a problem or issue not given community priority, they must provide justification.

The class deliverable is the design project, which is then archived in a standard design report on the class website (<http://www.civil.columbia.edu/edc/>), presented to an expert panel of faculty, scientists, and professionals who judge the final products, and put in a format that can be presented to the community during the next field trip. This stresses to the students the importance of being able to communicate technical work in a variety of formats to a range of audiences. An estimated 30% of class time is spent on non-technical topics, yet students are producing designs that have been actually constructed and, at the time of writing this article, are still functioning properly.

10.4.2 A collaboration network for preparedness, response, and recovery under extreme events

One of the most urgent and vital challenges confronting society today is the vulnerability of urban areas to extreme events (XEs), such as earthquakes, hurricanes, and floods, as well as accidental and intentional human-made disasters like fires and terrorist attacks. At the global level, a total of 608 million people were affected by these disasters in 2002, out of which 24,500 died. Furthermore, the resulting economic damages were estimated at $27 billion (IFRC 2003). These significant societal and economic costs emphasize the urgent need for the civil engineering profession to improve: (1) the preparedness of physical infrastructure to a disaster and the readiness of first response plans for a disaster, (2) the response process to reduce the impact of XEs involving critical physical infrastructure (CPI), and (3) the recovery of the affected areas to reduce post-disaster consequences. Going forward, the manner in which XEs are addressed will undoubtedly influence the future of the world's cities.

In 2003, concern about the vulnerability of present and future populations to XEs promoted the second author—Peña-Mora—to develop an initiative to improve collaboration among the key actors who should be involved in disaster response and recovery involving critical physical infrastructure. The key actors in the collaboration were envisioned to be firefighters, police officers, medical personnel, and civil engineers. Information technology (IT) components, including sensors and systems of sensors embedded in the critical physical infrastructure itself, were also considered to be important. Thus, interdisciplinary interactions combined with technology were pictured as the foundations for the collaboration. As the project has evolved, a recipe for success in this interdisciplinary environment has been developed that includes: (1) providing robust communication and collaboration support in a disaster setting, (2) improving upon the collaboration process by including IT components, (3) involving the original designers, engineers and constructors of the infrastructure in the collaboration, both locally and remotely, and (4) providing a collaboration framework to prepare against, respond to, and recover from disasters (termed CP2R).

An interdisciplinary team comprising researchers from civil engineering, computer science, psychology, sociology, speech communication, epidemiology, and entomology is developing the CP2R collaboration framework. The initiative is supported by professional experts in the field of disaster relief from the US Army Construction Research Laboratory and the Illinois Fire Service Institute (IFSI). Special facilities have been used to implement an integrated project space composed of a physical, a digital, and a mobile laboratory.

In addition to the social, organizational, and technological aspects of this interdisciplinary project, the applications of epidemiology and entomology in disaster preparedness, response, and recovery have also been explored. Many microorganisms and insects follow predefined interaction rules using only local information in order to carry out major collaborative tasks, similar to emergency personal who need to be able to perform multiple functions in a coordinated way in their management of a disaster. Here, the interaction rules in epidemiology and entomology, which have been improved during many years of evolution, have been used to define patterns of collaboration among first responders during preparedness, response, and recovery activities in order to better hone the CP2R model (Aldunate *et al.* 2005). As a result, the project has generated new fertile links between fields that are not normally connected in the research community.

This project has attracted federal funding and institutional support and is generating peer-reviewed publications. Thus, the research component of the work is no less successful in an academic context than a project narrowly focused in a specific subdiscipline of civil engineering. However, unlike a more traditional civil engineering research project, non-traditional connections are being generated between discipline fields that could potentially lead to paradigm shifts in knowledge. Hence, interdisciplinarity in engineering research promoted by an urgent societal need has opened new doors without necessarily compromising the academic contributions of the civil engineering faculty and research students who are involved. A focus on real problems of societal relevance has, once again, proven an effective means of generating the interdisciplinarity needed to shift civil engineering closer to twenty-first-century needs.

10.5 Conclusions

The engineering profession in the United States is facing a crisis in terms of its ability to attract and retain a diverse workforce. Although women represented more than 50% of undergraduate students in 2008, less than 20% identify engineering as their major: the representation of African, Hispanic, and Native Americans among undergraduate populations is even worse. Engineering for societal needs requires a workforce that reflects the society it serves. Furthermore, as the demand for engineers continues to increase, so does the requirement to draw upon talent from all available pools. A European study of women and engineering education found that interdisciplinary engineering degrees appear to be more attractive to female students than single/traditional/classical engineering degrees (Beraud 2003). Thus, the role of interdisciplinarity in engineering, and the means by which this can be achieved, might be key to multiple ways of changing the 'face' of engineering.

Finally, it is important to note differences between engineering interdisciplinarity approaches in countries and socio-cultural contexts other than the United States. For example, unlike the United States where the majority of engineering societies are associated with a specific branch of engineering, Germany has a unified engineering society; the Verein Deutscher Ingenier or VDI. Members of Germany's VDI include technicians and craft technical workers, who would not normally participate in the activities of engineering societies like the ASCE. As a result, discipline-centric engineering approaches appear to be much less entrenched in the German culture. Furthermore, a widespread German interest in environmental problems has provided a focus for interdisciplinary collaborations involving technical as well as socio-economic factors. In France, a civil engineer is someone who works in a private corporation where he or she can do civil engineering work as well as work in mechanical engineering, electronics, etc. (Didler 1999). Hence, the integration of engineering disciplines within the workplace can be a more common occurrence in France than the United States.

A review of approaches in developing countries reveals that, to a large extent, a traditional discipline-centric engineering approach has been adhered to. However, in a few of these countries there is an increasing awareness of the need for interdisciplinarity. A case in point is India, where many universities took a conscious decision in early 1990s to focus their engineering education so as to optimize the commercialization of technical know-how. As a result, Indian engineering graduates now hold senior positions in other fields—among them business, consulting, and finance (Galloway 2008). Thus, as demonstrated by German, French, and Indian approaches, interdisciplinarity can be fostered via the structure of engineering societies, within the engineering workplace, and/or within university environments, provided that there is a motivating will and an identified societal need.

Acknowledgements

The work reported in this chapter was partially supported by funds received from the Earth Institute at Columbia University, the National Science Foundation CAREER and PECASE award CMS-9875557, and the National Science Foundation awards EEC-0431946, CMS-0427089, and CNS-0709249. The authors would like to express their gratitude for this funding support. The authors would also like to acknowledge the contributions provided by Mani Golparvar-Fard, William E. O'Neil, predoctoral fellow, and Dr Zeeshan Aziz, postdoctoral research fellow, in the preparation of this work. The work was considerably strengthened by the comments provided by two anonymous reviewers. Any opinions, findings, and conclusions or recommendations expressed in this work are those of the authors and do not necessarily reflect the views of the Earth Institute nor the National Science Foundation.

References

ACSE (American Society of Civil Engineers) (2007). The vision for civil engineering in 2025. Available at: <http://www.asce.org/files/pdf/professional/summitreport12jan07.pdf> (accessed March 2009).

ACSE (American Society of Civil Engineers) (2008). *Civil engineering body of knowledge for the 21st century*, 2nd edn. Available at: <http://www.asce.org/professional/educ/> (accessed June 2008).

Aldunate, A.G., Peña-Mora, F., and Robinson, G.E. (2005). Collaborative distributed decision making for large scale disaster relief operations: drawing analogies from robust natural systems. *Complexity* **11**(2), 28–38.

Auyang, S.Y. (2004). *Engineering – an endless frontier*. Cambridge, MA: Harvard University Press.

Beder, S. (1999). Beyond technicalities: expanding engineering thinking. *Journal of Professional Issues in Engineering Education and Practice* (January), 12–18.

Beraud, A. (2003). A European research on women and engineering education (2001–2002). *European Journal of Engineering Education* **28**(4), 435–51.

Bloom, B.S. (1956). *Taxonomy of educational objectives, handbook I: the cognitive domain*. New York: David McKay Co., Inc.

de Camp, L.S. (1963). *The ancient engineers*. New York: Dorset Press.

Committee on Evaluation of Engineering Education (1955). Report of the Committee on Evaluation of Engineering Education. *Journal of Engineering Education* **45**, 25–60.

Didler, C. (1999). Engineering ethics in France: a historical perspective. *Technology in Society* **21**(4), 471–86.

Duderstadt, J.J. (2008). Engineering for a changing world: a roadmap to the future of engineering practice, research and education. *The Millennium Project*. Ann Arbor, MI: The University of Michigan.

Galloway, P.D. (2008). *The 21st century engineer: a proposal for engineering education reform*. Reston, VA: ASCE Press.

Grayson, L.P. (1980). A brief history of engineering education in the United States. *IEEE Transactions on Aerospace and Electronic Systems* **AES-16**(3), 373–92.

Griggs, F.G. (2003). 1852–2002: 150 years of civil engineering in the United States of America. In: J.S. Russell (ed.) *Perspectives in civil engineering: commemorating the 150th anniversary of the American Society of Civil Engineers*, pp. 111–22. Reston, VA: ASCE Press.

IFRC (International Federation of Red Cross and Red Crescent Societies) (2003). *World disasters report 2003: focus on ethics in aid*. Geneva: IFRC.

Layton, E. (1971). *The revolt of the engineers: social responsibility and the American engineering profession*. Cleveland, OH: The Press of Case Western Reserve University.

Mongan, D.G. (2008). Vision 2025 – making it a reality, *ASCE News* **33**(7).

NAE (National Academy of Engineering) (2004). *The engineer of 2020: visions of engineering in the new century*. Washington, DC: National Academies Press.

Noble, D. (1984). *The forces of production: a social history of industrial automation*, p. 44. New York: Knopf.

Polak, P. (2007). World designs to end poverty. In: C.E. Smith *Design for the other 90%*. Distributed by Assouline Publishing, New York for the Smithsonian Institution.

Rhoten, D. and Parker, A. (2004). Risks and rewards of an interdisciplinary research path. *Science* **306**(5704), 2046.

Savery, J.R. and Duffy, T.M. (2001). Problem based learning: an instructional model and its constructivist framework. *CRLT technical report*, no. 16-01. Bloomington, IN: Center for Research on Learning and Technology, Indiana University.

Stephan, K.D. (2002). All this and engineering too: a history of accreditation requirements. *IEEE Technology and Society Magazine* (Fall), 8–15.

United Nations (2000). *Millennium declaration*. Available at: <http://www.un.org/millennium/summit.htm> (accessed March 2008).

Watson, J.G. (1982). *A short history: the institution of civil engineers*. London: Thomas Telford Ltd.

CHAPTER 11

Religious studies

SARAH E. FREDERICKS

Religion. This term encompasses Hopi Kachina dolls, baptisms, the Hajj, puja, Scientology, the Dalai Lama, and Passover meals. 'Religion' rightfully recalls compassion and social justice as much as it recalls hatred and war. 'Studying religion' may connote educating children about their family's religious beliefs, proselytizing masked as objectivity, or describing religion in a supposedly objective manner. Because the subject is so vast, potentially involving all cultures and times, one cannot include everything in one article about the study of religion.

This chapter will focus upon religions in the Western world, particularly in the United States, because the development of religious studies as an academic discipline distinct from religious practice arose in the West and because debates about the proper study of religion in the United States illustrate the varied nature of religious studies. Preliminary remarks about disciplines will ground a discussion of interdisciplinarity in contemporary Western religious studies and in the history of religious studies. Special attention will be given to the tensions between studying religion as a practitioner and scholar and to the ways in which ecumenical and interfaith activities can be described using interdisciplinary literature. Though the language of interdisciplinarity is not sufficient to describe all interaction between religions, religious studies itself is inherently multi- and interdisciplinary.

11.1 Discipline, interdisciplinarity, and religious studies

In an academic discipline, a group of scholars uses particular methods to answer a set of questions to enhance a shared body of knowledge. Scholars in a discipline tend to share terms and epistemological and ontological assumptions, though these elements may be an implicit rather than explicit part of the discipline. Scholarly organizations, degree programs, journals, and textbooks perpetuate the work and assumptions of the discipline.

Yet, disciplines are not static entities nor are they rigidly separated from each other (Lakatos 1970; Squire 1992; Laudan 1998; Kuhn 1996). To account for the overlap between disciplines, Geoffrey Squire (1992) constructs a three-dimensional definition of a discipline:

> The first of these dimensions manifests itself in the content, topics or problems which are addressed [by a discipline]; the second in the methodologies, techniques and procedures which are used; and the third in the extent to which the discipline treats its own nature as the subject of reflexive analysis.

According to Squire, dimensions can overlap with those of other disciplines. Changes within a dimension in a discipline may spark modifications in other dimensions, in that discipline or other disciplines. Thus, Squire's account recognizes that disciplines are not completely isolated from each other and acknowledges change within disciplines (Squire 1992).

Such a multifaceted understanding of a discipline is necessary to begin to describe the academic study of religion. Religious studies—to some extent dominated by the history of religions but also known in Germany as *Religionswissenschaft* and in France and the francophone world as *sciences de la religion*—is a well-established multidisciplinary field that draws on anthropology, sociology, philosophy, and other disciplines.

The variety of topics studied and methods used by religious studies scholars stretch Squire's definition of a discipline. After all, scholars studying the history of the rise of Buddhism and those constructing a Christian ethic in the face of contemporary environmental destruction do not necessarily share a narrow subject, method, or degree of reflexive analysis yet both are considered part of religious studies. Certainly, Squire's impulse for a multifaceted description of a discipline that allows change over time and relates to other disciplines characterizes the contemporary study of religion to some degree. Yet, subdisciplines must be added to Squire's analysis to fully describe religious studies. These subdisciplines are themselves a fluid bunch, but may include the sociology of religion, religion and literature, ethics, Biblical studies, and philosophy of religion. At times, the various subdisciplines within religious studies, say the sociological study of religion, and the study of religious literature may have more to do with sociology and literature studies, respectively, than with each other or other subdisciplines within religious studies.

One may ask why these subdisciplines should constitute a discipline of 'religious studies' given that they may have more to do with other disciplines than with each other, and that scholars of the various subdisciplines may know little about the content or methods of each other's work (Gill 1994). Though scholars of religion may argue about the character of their discipline, many implicitly recognize the importance of a diverse discipline as they hold joint conferences with different subdisciplines, and as they read journal articles and do research outside their specialty area. Through these activities, scholars from different subdisciplines may borrow a significant amount from each other and develop new integrative methods. Thus, while the subdisciplines may relate in a multidisciplinary fashion, investigating their subject sequentially, the subdisciplines of religious studies may also practice 'true interdisciplinarity' in which they integrate methods and insights (Klein, Chapter 2 this volume). The continual cross-fertilization among subdisciplines in religious studies and between religious studies and other disciplines indicates that the various sorts of religious studies scholars make up a discipline in the broadest, dynamic sense of the word.

11.2 Religion and interdisciplinarity in history

The complex disciplinary nature of modern Western religious studies has its roots in the complexities of religion itself. Throughout most of human history what is now called 'religion' was coterminous with culture. All sorts of actions were ritualized to connect people to the sacred and each other in one overarching worldview. While shamans, priests, or other religious leaders may have had distinct access to religious knowledge or experience, these esoteric elements of religion were typically understood to be unified with all other knowledge and were studied to develop religious practice. Thus, the modern view of religious studies as a discipline distinct from both other disciplines and religious adherence breaks down when we look across history.

Even when distinct academic disciplines arose, it was assumed that their content was connected. For instance, traditional Islamic scholarship was based on explicit metaphysical principles which formed the foundation of all thought (Nasr 1996). This schema linked areas of study including the religious 'sciences' of Quranic exegesis, Hadith studies, and jurisprudence as well as other branches of thought including astronomy, alchemy, medicine, and mathematics. (The term 'science' is used here in its medieval meaning of a form of knowledge and learning. It does not imply the experimental, law-based potentially reductionistic vision of science popular in the modern world.)

One such principle is balance (*al-mīzān*). Use of this term across the sciences continually reminded medieval Islamic scholars of their belief that all knowledge is connected even though the term had different connotations in various sciences. In some philosophical schools balance was defined as (1) the way 'consequences of human action are weighed in the next world', or (2) 'the necessity of leading a morally balanced life in this world', or (3) 'the discernment that allows us to establish balance in all aspects of life' (Nasr 1996). It was also used as a physical term in studies of weights, mechanics, and hydrostatics. In alchemy Islamic scholars used the term to indicate that the proper proportion of qualities of nature (hot, cold, moist, dry) was reached enabling the physical properties of the substance, the human's soul, and the 'tendencies of the World Soul' to be in balance (Nasr 1987, 1996).

Thus, medieval Islamic scholars did conceive of disciplines with distinct methods and subjects; however, insofar as these disciplines shared terms, a metaphysical foundation, and were committed to a vision of the unity of knowledge established by God, they also transcended our modern bounds of disciplinarity. Multi-, inter-, or transdisciplinarity, as defined by Julie Thompson Klein and other contemporary scholars, do not adequately describe these historical concepts of disciplines because the new terms imply intentional efforts to overcome a separation between the disciplines that was not a part of the Muslim understanding of disciplines and knowledge (Klein, Chapter 2 this volume).

Moving ahead in time to the early modern period and traveling from the Middle East to Western Europe, we see that the study of religious ideas was involved in a variety of disciplinary endeavors even though the early modern period is often understood as the time when academia was secularized. As John Hedley Brooke points out, many significant advances in early modern science explicitly used theological claims (Brooke 1991). For instance, after trying many physical theories to explain the gravitational force, Isaac Newton eventually decided that God must be the source of all forces and must periodically intervene to keep planets in their orbits (Brooke 1991). He also pondered the theological implications of new scientific theories; Newton maintained that Cartesianism's limited emphasis on God (as compared with other contemporary theories) would lead to atheism (Westfall 1986).

The relationship of religion to other fields of thought was not, however, limited to terms and concepts. Peter Harrison suggests that the move from allegoric and symbolic to more literal biblical interpretations during the sixteenth-century Protestant reformation influenced scientific visions of nature. Scholars increasingly investigated the world as a series of events that had integrity in and of themselves—not just because they stood for something else (Brooke 1991; Harrison 1998). In the short term this trend contributed to a new connection between science and religion: scholars began to study Biblical stories, such as the Flood story, factually. Janet Browne argues that literalist readers of the Bible who studied the dispersal of animals from the ark after the flood helped spark interest in species development and migration in the seventeenth and eighteenth centuries (Browne 1983).

Thus, we see that religious concepts and methods permeated other fields of thought such that stark delineations between the study of religion and other academic disciplines are anachronistic until the twentieth century. Applying the term 'interdisciplinary' here is not appropriate because disciplines were not separated enough that they needed to be intentionally recombined. Instead, the above examples show that scholars recognized the reality of the fluid boundaries between fields of study, something we have ignored in modern times as disciplines became more distinct in practice and as scholars assumed that they needed to be distinct.

Diffuse boundaries between fields have not only characterized the study of religion; they have often been a hallmark of religious practice. For example, religions have long utilized art in a variety of ways. Art can enrich the worship experience by setting ritual space apart from the everyday. Art can also remind people of central religious themes or illustrate stories for the laity; stained glass, mosaics, and murals have been used for these purposes throughout much of Christian history. Works of art may also be used as symbols through which the Ultimate is worshiped (Hindu statuary) or to focus attention during worship (icons in Eastern Orthodox Christianity). Artwork can also reinforce deep theological beliefs, as when geometric patterns are used by Muslims to ensure that decorations are not taken as images to be worshipped, a form of idolatry since they believe that God is One and beyond representation.

While some subjects such as art have enriched religious experience for millennia, some such as psychology are relatively new fields, both in themselves and in relationship to religious practice. Insights from psychology and social work have been used to develop techniques of pastoral counseling. These methods, often taught to religious leaders during their training, are integrated with religious concepts of human nature, guilt, responsibility, and divine action to help religious leaders to understand individuals and discern how to assist them.

Religious practitioners also engage with, develop, and utilize knowledge from the natural sciences. Sometimes this knowledge is used instrumentally, as when astronomers created calendars in part to determine the proper dates of religious rituals in ancient India, Mesopotamia, Central America, and Egypt (L. P. Williams 2007). Visions of the Ultimate have also been significantly influenced by knowledge of the natural world. Additionally, knowledge from the natural sciences has been used by religious people in order to develop ethical positions about new health care dilemmas and environmental destruction in the last few decades.

As religious people work to bring their religion to bear on contemporary issues, they often wish to maintain continuity with their tradition. Thus, they also rely on history, studies of ancient languages, textual analysis, and archaeology to understand the belief systems, legal codes, rituals and texts of their religious ancestors.

From these brief examples, it is clear that religious practitioners have relied, and continue to rely, on a number of bodies of knowledge to practice their religions. This observation is at the root of the interdisciplinary nature of religious studies: the formal study of religion and religion itself are and have been varied, that the discipline that studies religion today must also be diverse.

11.3 Religiosity and secularity in religious studies: another mode of interdisciplinarity

So far, our discussion of the disciplinary and interdisciplinary potential of the study of religion has overlooked one implicit requirement of contemporary academic disciplines—that the discipline's knowledge is publicly available to anyone who wishes to study it (Weibe 1999). In many disciplines this assumption is so strong that it is rarely discussed. Yet, some scholars maintain that the full meaning of rituals and beliefs or the depth of religious experience can only be understood by believers, adherents, or devotees. Others claim that such privileged knowledge is not necessary to study religion and that it should have no place in the academy. This debate, sometimes framed as one between theology and the social scientific study of religion, has been the most contentious element of the development of religious studies (Cady and Brown 2002) and illustrates another way in which it is multidisciplinary.

The academic study of religion in the West grew out of faith-based endeavors. After all, for much of human history the people who studied religion were religious leaders—shamans, priests, legal experts, and monks. Most focused on their religion, with some study of the traditions from which they came or with which they interacted. Thus, Buddhists knew about Hinduism; Jewish, Christian, and Muslims scholars in medieval Spain studied together. Yet, until the Enlightenment, there was no significant study of religion as a scholarly endeavor divorced from the belief in and practice of a particular religion.

With the Enlightenment and rise of Cartesianism, some Western scholars tried to identify the 'essence' of religion. Whether the essence was identified with morality (Immanuel Kant), 'the feeling of absolute dependence' (Friedrich Schleiermacher), the '*mysterium tremendum*' (Rudolf Otto), or 'ultimate concern' (Paul Tillich), essence theories persisted well into the twentieth century. Developmental, comparative, and phenomenological approaches arose as competitors to essence theories, yet all of these claims of a general theory of religion were typically grounded in Christianity and prioritized Christian concepts (Gill 1994; Capps 1995). Indeed, this cultural context has often led scholars to ignore elements of religion not central to Christianity, whether oral traditions, sacred land, or the belief in multiple or no deities. Such bias has made it difficult for scholars to understand religious diversity.

The prioritization of particular, typically Christian, belief systems in academia was a sign of Protestant dominance in the United States as well as a means for reinforcing this domination throughout society. Well into the nineteenth century most institutions of higher learning in the United States were founded with religious goals, and children learned to read using primers infused with Protestant ideals (Gaustad and Schmidt 2002). Only in 1962 did the Supreme Court rule that prescribed prayer in public schools was unconstitutional. In 1963, the court clarified the status of religion in school by separating the practice of religion through prayer or ritualized Bible reading (unconstitutional) from the study of religion (constitutional and encouraged to help children understand history and culture) (Gaustad and Schmidt 2002).

The prioritization of Protestant belief systems and rituals in United States history has had implications well beyond education. Many Protestant habits have been adopted by non-Protestant religious groups in order to fit into the society of the United States. For example, the Native American Church was incorporated in 1918 in order to gain legal protection for their religious rituals (Thompson 2005). Jews in the United States, especially Reform Jews, adopted new practices including mixed-gender seating for worship and singing hymns accompanied by classical European organ music (Corrigan *et al.* 1998, pp. 229–32). Japanese-American Buddhists were targeted for special treatment in the World War II internment camps in the United States because of their religion. Many in the camps adopted new rituals to become more 'American', i.e. more Christian. These practices included worshipping in English, adopting Christian-style hymns in new prayer books, destroying cultural and religious objects with Japanese writing on them, and starting 'Sunday schools' to teach their children about Buddhism (D. R. Williams 2006).

While religious studies scholars are not solely responsible for such societal trends, limited knowledge about religious traditions other than one's own contributes to prejudice. Seeking to avoid these dangers and the scholarly bias that has come from religiously motivated studies of religion, scholars such as Sam Gill and Donald Weibe argue that religious studies should not require, support, or evaluate religious beliefs and practices. Anyone, they argue, should be able to arrive at the same conclusions when studying religion.

Weibe looks to the work of Max Müller to delineate the ideal shape of religious studies (Weibe 1999). According to Weibe, Müller's general goals for the study of religion included impartiality and critical scholarly methods resting on history and comparative analysis. He also emphasized the search for the truth through pre-existent facts rather than through the creative development of ideas as in philosophy and theology (Weibe 1999).

Müller's influence on the discipline of religious studies was most strongly felt by social scientists of religion. William M. Newman's 1974 study of the first 25 years of the

Society for the Scientific Study of Religion (SSSR) reveals that over time dialogue between 'religious believers' and social scientists who study religion was de-emphasized in favor of the social scientific study of religion (Newman 1974). Wiebe would applaud such trends. He writes (Weibe 1999):

> If the academic study of religion wishes to be taken seriously as a contributor to knowledge about our world, it will have to concede the boundaries set by the ideal of scientific knowledge that characterizes the university. It will have to recognize the limits of explanation and theory and be content to explain the subject-matter – and nothing more – rather than show itself a form of political or religious behavior (or an injunction to such action).

Similarly, Sam Gill sees the tendency of religious studies scholars to segregate by religion and the frequency with which they study the tradition to which they adhere a step away from *academic* study of religion. Instead, Gill advocates comparative work and the study of overarching religious questions such as what religion reveals about personhood. Thus, both Gill and Wiebe think that the discipline of religious studies should be a unified endeavor without sectarianism that focuses on explaining religious phenomena, not developing religious ideas (Weibe 1999).

Gill, Weibe, and others, academics and citizens alike, have good reasons to be wary of the biased nature of much of religious studies throughout the twentieth century. Yet Weibe draws too sharp of a line between 'objective', 'social scientific' studies of religion and the study of religion for religious reasons. Gill and Weibe's assumption that a narrow definition of a discipline is necessary and their over-reliance on social scientific guidelines for religious studies cause them to reject significant elements of the field such as the literary, theological, ethical, and philosophical and to overlook the blurring of objective, constructive, and advocacy-based approaches to religious studies that may arise out of a social scientific approach.

In recent decades, the academic study of religion in the United States has shifted toward the study of world religions and away from studying Christianity alone. Faith-based studies of religion are yielding to critical, constructive, comparative approaches involving a variety of methods from multiple disciplines, religions, and cultures. These moves encourage students to 'examine and engage religious phenomena, including issues of ethical and social responsibility, from a perspective of cultural inquiry and analysis of both the other and the self' (The Religion Major and Liberal Education Working Group 2007, pp. 3–6). Thus, they are more than an 'objective' study of religion but less than indoctrination into a particular religious tradition.

The debate over whether religious studies should be theological, social scientific, or a new critical, constructive, intercultural method of inquiry demonstrates another way in which religious studies is inherently multidisciplinary.

11.4 Ecumenical, interfaith, interdisciplinarity

The practice of religion and the scholarly development of religious beliefs and practices include other possible loci of interdisciplinarity: the ecumenical and interfaith movements.

These activities between different denominations of one tradition (ecumenical) or between different religious traditions (interfaith) aim to develop rigorous concepts of religious similarities and differences, promote peace and other social goals, and encourage proper relationships between religions. Ecumenical work may also advocate unity in the religion at large and may lead to mergers or blurred boundaries between denominations. Several types of inter- and intrareligious activity illustrate these trends: ecumenical organizations involving a wide number of religions (the Parliament of the World's Religions) or denominations (the World Council of Churches); interfaith movements arising out of conflict (post-Holocaust Jewish–Christian dialogues or Islamic–Christian dialogues after 9/11), curiosity (Buddhist–Christian study in the United States), or concern about social problems (interfaith environmental movements).

Many parallels exist between these ecumenical and interfaith activities and interdisciplinary work. First, there are parallels between the structure of a denomination or religion and a discipline. Like disciplines, religious groups have some defined subject (the Ultimate, the human condition, myths), rely on epistemological, ontological, and metaphysical presuppositions, and have favored methods. Second, these elements change over time to meet the needs of their religious communities as they interact with similar segments of other religious traditions much as Squire's disciplines mutually shape each other. Third, we see that despite differences in subject, presuppositions, and method, people involved in ecumenical or interfaith movements work across religious boundaries to address questions unsolvable by any one tradition—How do theologies, rituals, ethics, and histories relate? How should people within a tradition conceive of and relate to others? How can religions address social problems together? As ecumenical and interfaith activities rely on resources of various religious groups to address these issues, they are, in a sense, involved in interdisciplinary work. All of the challenges of interdisciplinarity arise here as well: the communication barriers between groups with different terms, methods, and presuppositions; the suspicion and distrust of groups different from one's own; the potential for one group to dominate the activity.

Yet, Mircea Eliade warns that as much as other disciplines can aid understanding of religion, they cannot fully describe religion because they do not have the terms to appreciate and understand the sacred; thus, the study of religion is a specialized field (Eliade 1995, 1996). Extending his argument, completely subsuming ecumenical and interfaith activities under the heading of interdisciplinarity will threaten to impoverish our understanding of religious activity. Certainly religious activity and the activity of academic disciplines have much in common, but the scope of a discipline is much narrower than the scope of religious worldviews. Disciplines, especially in our modern world, focus on narrow segments of or limited approaches to reality, while religion typically involves ideas about the human condition, ultimate reality, and their relationship to the world. A discipline may have a code of ethics, but its norms focus on behavior related to the discipline and are not sufficient to guide one's entire life, whereas religious norms typically aim to guide an adherent's entire life. Debates between disciplines may be vigorous and can even lead to the destruction of careers and suppression of ideas; conflicts between religions have led to persecution and war. These contrasts between disciplines and religions are just a few indications that religions are deeper, wider, and involve more commitment than disciplines.

Thus, if our understanding of ecumenical and interfaith interactions was reduced to interdisciplinarity, we would miss significant facets of these movements.

Despite the dangers of limiting our knowledge of religion by overusing the language of interdisciplinarity, using its various terms may help identify and understand the many ways in which religious groups interact since religious studies has not defined terms for all of the types of relationships and goals of inter- and intrareligious dialogue and action.

The first major modern interfaith endeavor was the 1893 Parliament of World's Religions held in Chicago. A part of the cultural counterpart to the technical-focused World Columbian Exposition, the Parliament aimed to promote cross-cultural understanding through religions. The Parliament was dominated by Christians both in sheer numbers and by the groups underrepresented (Africans, South Americans, Indigenous traditions), by the groups not invited (Mormons, African-Americans, Native Americans), and those groups not present (Muslims, Sikhs, and Tibetan Buddhists). The Parliament was successful insofar as it enabled Asian religions such as Hinduism and Buddhism to formally introduce themselves to the West and as it promoted understanding among Christians (Kuschel 1993). The Parliament of 1893 is a prime example of multidisciplinary encounter; religious people wanted to learn about each other but did not aim to collaborate or integrate their ideas.

The centennial celebration of the Parliament in 1993 had a different aim: to articulate the global ethics already found in the world's religions. It expressly did not seek to establish a universal religion or obliterate the religious ethics of individual religious traditions. Rather it sought to identify common ethics (the golden rule; do not lie, steal, kill, or commit sexual immorality) thought to be necessary for a world with increasingly global structures of economics, politics, and society. Much more diverse than the first Parliament, with 6500 representatives from nearly all world religions (evangelical and fundamentalist Christians were notably absent), it has been praised for its movement toward a global ethic even as elements of its process and content have been criticized (Küng and Kuschel 1993). This contemporary event illustrates one facet of much interreligious and ecumenical work: the discovery of existing commonalities between groups that can be the basis of future study, collaboration, and peace even as participants recognize and affirm the differences between their traditions.

The World Council of Churches (WCC), organized in 1948, has exhibited similar trends within Christianity. It is the largest ecumenical Christian organization, with 349 member churches comprising over 500 million individual members from over 120 countries. 'Church' is often equated with a local congregation but can also indicate an organizational body which unites many individual congregations, often according to theological, ritual, and regional or national ties. Churches of the WCC come from all over the world and 'include nearly all the Eastern and Oriental Orthodox churches; Anglicans; diverse Protestant churches, including Reformed, Lutheran, Methodist, and Baptist; and a broad representation of united and independent churches' (Wogaman 1993; World Council of Churches 2009a).

The WCC aims to recognize and reinforce the significant common beliefs among Christians through worship and action. It does *not* intend to be a monolithic church body where all differences are wiped away. The WCC's decisions are not binding on its members.

Rather, its activities are supposed to enable debate and prophesy through which members will be challenged to live lives of faith and service (World Council of Churches 2009b). WCC activities often involve theological, ritual, and ethical innovations as experimentation is possible within its non-binding format. When these actions transcend denominational boundaries WCC members participate in transdisciplinary activity, but since they do not seek to obliterate all differences, there is a sense of multidisciplinarity here as well. The WCC members tend to operate in an instrumental mode as they seek to resolve religious problems about ecumenical worship and to resolve social problems related to economics, war, racism, environmental degradation, and human rights.

Cooperative study and reconciliation is not, however, limited to ecumenical discussions. For example, the scholarly study of and community reflection upon Jewish–Christian relationships has grown considerably since World War II. Many factors led to this interfaith work including the horrors of the Holocaust, the establishment of the state of Israel, the ecumenical movement, the Second Vatican Council, and Enlightenment visions of human dignity and equality (Kessler and Wenborn 2005; Kessler 2006).

Insofar as these dialogues aim to articulate constructive new relationships, they engage in activities similar to the critical interdisciplinarians who study how the relationship of knowledge between fields are made (Thompson 2005). As Jews and Christians collaborate to promote peace, a goal many argue cannot be achieved by either group alone, their activities parallel instrumental interdisciplinarity. Yet interdisciplinarity, whether critical or instrumental, does not quite fit this situation because each religion intends to remain distinct even as they learn from one another. Thus, there will be barriers to the amount of integration either group is willing to entertain.

Muslims and Christians have recently begun similar dialogues. On 13 September 2006 Pope Benedict XVI gave an address which was widely regarded as implying that Islam was violent and immoral. In the wake of this address, Islamic leaders and scholars wrote an open letter to the Pope to discuss their faith and promote understanding of Islam. The next year 138 Muslim leaders from all branches of Islam and all major Islamic nations and regions released *A common word between us and you*. This document brought Muslims together in a way not experienced since the time of the Prophet. Through this document, and a series of conferences in 2008 with hundreds of Catholics, Anglicans, and Protestants in turn, Muslim leaders hope to promote understanding of what the faiths share and to promote peace. Importantly, participants in the conferences did not wish (1) to convert each other, (2) to make the other adopt ideas of their own theology, or (3) to reduce the two religions to a common denominator or new religion. Rather, the document looked to the sacred texts of the Bible and Quran to discover what Christianity and Islam have in common in order to begin to work for peace ('Final declaration of the Yale Common Word Conference' 2008).

Not all interfaith dialogue, however, is motivated primarily by violence. Buddhist–Christian dialogue in the United States has primarily been an academic affair in which scholars expert in each of these traditions have studied the major ideas of Buddhism and Christianity in a comparative fashion (Lai and von Brück 2001). (Buddhist–Christian dialogue in countries with significant Buddhist populations has spent more time on the social implications of contact between the religions and has involved religious communities and scholars.) Studies cover a wide range of topics, but issues of ultimate reality (meditation, contemplation, and prayer; suffering; and ethics) have been most popular as is demonstrated in the journal *Buddhist–Christian Studies*. 'Multidisciplinarity' describes some of these endeavors as scholars study the same phenomenon sequentially using different theories or utilize the same theory to explore different phenomena. Buddhist–Christian dialogue, however, is more often interdisciplinary or transdisciplinary as it aims to integrate insights from various religious traditions, and academic disciplines. Though most of the comparative work between Buddhism and Christianity in North America has occurred within academic circles, rather than in churches or *sangha*, this does not mean there is a clear distinction between academics and religious practice in Buddhist–Christian studies (*Buddhist–Christian Studies* 1998). Many scholars engaged in this dialogue are themselves Buddhists, Christians, or adherents of some beliefs and/or practices from each system. These scholars engage in dialogue in part to develop their own religious ideas, a form of transdisciplinary endeavor.

People of different religions also come together to resolve pressing social issues that do not directly stem from their religious differences, a type of collaboration that can be classified as instrumental interdisciplinarity as groups rely on their various methods and beliefs to reach a common goal. For example, various faith communities collaborate to promote environmental protection. The Evangelical Environmental Network and the National Religious Partnership for the Environment, a group of mainline Protestants, Catholics, and Jews, campaigned to save the Endangered Species Act in the 104th Congress (1995–6) (Kearns 1997, p. 349). Interfaith Power and Light organizations located in most states also educate religious communities about how to simultaneously save energy, money, and the planet; band together to purchase cheaper energy; and provide a support network to help achieve such changes.

Though interfaith and ecumenical activities should be distinguished from interdisciplinarity because of their connection to religious belief and practice, there are enough parallels between them that scholars engaged in interdisciplinary endeavors can learn from these activities. First, they could learn of the dangers, both to understanding and to relationships, of evaluating other disciplines with the criteria of one's own and calling it a dialogue. One does not need to look far to find prejudicial (intentional or inadvertent) descriptions of religious traditions unfamiliar to the adherent or scholar. For instance, Christians have long ignored the importance of land to Native Americans. Secondly, ecumenical and interfaith activities may teach interdisciplinarians about forging terms that resonate with multiple perspectives to avoid privileging or ignoring one viewpoint. For example, 'Ultimate Realities' or 'Ultimate Reality' are terms used to avoid the limitations of 'God' language (Neville 2001). Third, interdisciplinarians could learn something about how to link communities who not only have different methods, assumptions, and subjects, but also experience deep distrust or animosity toward the other based on centuries of prejudice, persecution, and power imbalances. Working to resolve practical problems about the environment, peace, and other social issues can often be a starting point to deeper collaboration. Academics may find that working to address community issues can build bridges between hostile disciplines. For all of these reasons, interdisciplinarians would do well to learn from the experience of the ecumenical and interfaith movements.

11.5 Conclusion

Though interdisciplinarity should not replace terms like ecumenism and interfaith, something like the integrated result of interdisciplinarity has long been a facet of religious study and practice. After all, religion is a subject that pre-dates the rise of modern disciplines and has often been seen as connected to all modes of thought and experience. We see this inter- and transdisciplinarity today as religious practitioners utilize ideas and methods from art, psychology, history, languages, and the sciences. Thus, it is not surprising that the subdisciplines of the academic study of religion are often multi- or interdisciplinary as is the field of religious studies in relationship to other fields. Of course, this diversity has and does lead to quarrels about the proper ways to study religion. In a fitting move given the history of religious studies, religious studies scholars are forging a new path between the extremes of objectivity and evangelism to encourage description and critical reflection so that religious studies becomes as an openly interdisciplinary discipline.

References

Buddhist–Christian Studies (1998). Buddhist–Christian studies and academia. Special issue [introduction and eight articles] **18**, 87–132.

Brooke, J. (1991). *Science and religion: some historical perspectives.* New York: Cambridge University Press.

Browne, J. (1983). Descent from Ararat. In: *The secular ark: studies in the history of biogeography*, pp. 1–31. New Haven, CT: Yale University Press.

Cady, L. and Brown, D. (ed.). (2002). *Religious studies, theology, and the university: conflicting maps, changing terrain.* Albany, NY: State University of New York Press.

Capps, W. (1995). *Religious studies: the making of a discipline.* Minneapolis, MN: Fortress Press.

Corrigan, J., Denney, F., Eire, C., and Jaffee, M. (1998). *Jews, Christians, Muslims: a comparative introduction to monotheistic religions.* Upper Saddle River, NJ: Prentice Hall.

Eliade, M. (1996). *Patterns in comparative religion*, transl. R. Sheed. Lincoln, NE: University of Nebraska Press.

Eliade, M. (ed.) (1995). *The encyclopedia of religion.* New York: Simon and Schuster.

Final declaration of the Yale Common Word Conference (2008). Available at: <http://www.yale.edu/faith/downloads/Yale_Common_Word_Conf_2008_Final_Decl.pdf> (accessed 13 February 2009).

Gaustad, E. and Schmidt, L. (2002). *The religious history of America: the heart of the American story from colonial times to today.* New York: HarperCollins.

Gill, S. (1994). The academic study of religion. *Journal of the American Academy of Religion* **62**(4), 965–75.

Harrison, P. (1998). *The Bible, Protestantism, and the rise of natural science.* New York: Cambridge University Press.

Kearns, L. (1997). Noah's ark goes to Washington: a profile of evangelical environmentalism. *Social Compass* **44**(3), 349–66.

Kessler, E. (2006). *Dabru Emet* and its significance. In: J.K. Aitken and E. Kessler (eds) *Challenges in Jewish–Christian relations*, pp. 195–202. Mahwah, NJ: Paulist Press.

Kessler, E. and Wenborn, N. (eds) (2005). *A dictionary of Jewish–Christian relations*. New York: Cambridge University Press.

Kuhn, T. (1996). *The structure of scientific revolutions*, 3rd edn. Chicago: University of Chicago Press.

Küng, H. and Kuschel, K.-J. (1993). *A global ethic: the declaration of the Parliament of the World's Religions*. New York: Continuum.

Kuschel, K.-J. (1993). The Parliament of the World's Religions, 1893–1993. In: H. Küng and K.-J. Kuschel (eds) *A global ethic: the declaration of the Parliament of the World's Religions*, pp. 77–97. New York: Continuum.

Lai, W. and von Brück, M. (2001). *Christianity and Buddhism: a multi-cultural history of their dialogue*. Maryknoll, NY: Orbis Books.

Lakatos, I. (1970). Methodology of scientific research programmes. In: I. Lakatos and A. Musgrave (eds) *Criticism and the growth of knowledge: Proceedings of the International Colloquium in the Philosophy of Science, London 1965*, pp. 91–196. New York: Cambridge University Press.

Laudan, L. (1998). Dissecting the holistic picture of scientific change. In: M. Curd and J. Cover (eds) *Philosophy of science: the central issues*, pp. 139–69. New York: W.W. Norton and Co.

Nasr, S.H. (1987). *Science and civilization in Islam*. Cambridge: Cambridge University Press.

Nasr, S.H. (1996). *Religion and the order of nature*. New York: Oxford University Press.

Neville, R.C. (ed.) (2001). *Ultimate realities*. Albany, NY: State University of New York Press.

Newman, W.M. (1974). The society for the scientific study of religion: the development of an academic society. *Review of Religious Research* **15**(3), 137–51.

Squire, G. (1992). Interdisciplinarity in higher education in the United Kingdom. *European Journal of Education* **27**(3), 201–10.

The Religion Major and Liberal Education Working Group (2007). *The religion major and liberal education: a white paper*. Atlanta, GA: American Academy of Religion.

Thompson, W. (2005). *Native American issues: a reference handbook*. Santa Barbara, CA: ABC-CLIO.

Weibe, D. (1999). *The politics of religious studies: the continuing conflict with theology in the academy*. New York: St Martin's Press.

Westfall, R. (1986). The rise of science and the decline of Orthodox Christianity: a study of Kepler, Descartes, and Newton. In: D. Lindberg and R. Numbers (eds) *God and nature: historical essays on the encounter between Christianity and science*, pp. 218–37. Berkeley, CA: University of California Press.

Williams, D.R. (2006). From Pearl Harbor to 9/11: lessons from the internment of Japanese American Buddhists. In: S. Prothero (ed.) *A nation of religions: the politics of pluralism in multireligious America*, pp. 63–78. Chapel Hill, NC: University of North Carolina Press.

Williams, L.P. (ed.) (2007). *Encyclopaedia Britannica*. Encyclopaedia Britannica online.

Wogaman, J.P. (1993). *Christian ethics: a historical introduction*. Louisville, KY: Westminster/John Knox Press.

World Council of Churches (2009a). *What is the World Council of Churches?* Available at: <http://www.oikoumene.org/en/who-are-we.html> (accessed 13 February 2009).

World Council of Churches (2009b). *Self-understanding and vision*. Available at:<http://www.oikoumene.org/en/who-are-we/self-understanding-vision.html> (accessed 13 February 2009).

CHAPTER 12

Information research on interdisciplinarity

CAROLE L. PALMER

In this digital age, the way people search for information is changing while the amount of information continues to accelerate unabated. An abundance of valuable information can now be found using standard Web search engines. At the same time, reliance on easily accessed digital information could deter the advancement of research by narrowing the historical base and diverse range of information used by scientists and scholars (Evans 2008). The field of library and information science (LIS) is concerned with how to collect, represent, organize, store, manage, preserve, retrieve, and disseminate society's vast stores of information for all kinds of user communities. Not surprisingly, some of the field's most formidable problems stem from the need to develop effective information systems and services for interdisciplinary researchers.

Scholars and scientists are continually influenced by 'the push of prolific fields and the pull of strong new concepts and paradigms' (Klein 1996a, p. 56). Disciplinary dynamics—how disciplines, grow, split, and merge—and their influence on how researchers work with, produce, and disseminate information have been a focus of study in LIS for decades. Statistical studies of the relationships among disciplines as represented in catalogs of literature, patent records, and other bibliographic sources date back to at least the 1920s (Hulme 1923). By the 1950s, interdisciplinarity was becoming a common theme in the professional discourse of research librarianship in relation to managing the complex interrelations among information systems and improving researchers' ability to find information from outside their field (Clapp 1954). By the 1960s a trend in user-centered research emerged, with an initial emphasis on scientific information, which began to document the special problems involved when scientists cross disciplinary lines, such as the limits of information available within a single social network of colleagues and the high amount of cross-disciplinary reading required in applied sciences (Allen 1966).

The importance of interdisciplinary research is now widely accepted, with universities and funding agencies increasingly expecting and supporting collaborative interdisciplinary approaches to solving complex societal problems, while also recognizing the need for integrative cyberinfrastructure to support the conduct of this research (National Science Board 2005; American Council of Learned Societies 2006). As research that draws on knowledge and expertise from multiple disciplines becomes more commonplace, along with the proliferation of digital information and technologies, LIS is facing growing challenges in organizing, preserving, and making accessible the range of digital information systems, resources, and tools in alignment with the practices and needs of scientific and scholarly communities.

The field of LIS is itself highly interdisciplinary, with influences historically coming from fields ranging from communication, cognitive science, computer science, linguistics, and philosophy of science. Cognate areas contributing to the study of interdisciplinarity include information systems and information management, especially for statistical approaches, human computer interaction (HCI), computer-supported cooperative work (CSCW), science and technology studies, and social informatics for studies of scholarly information use. LIS has a unique orientation to interdisciplinarity, however, since it is concerned with the organization, preservation, and mobilization of knowledge across the entire landscape of disciplines and is guided by a core mission in research librarianship to promote the transfer and exchange of knowledge among scholars.

As noted by the late economist Kenneth E. Boulding (1968), without professions of scholarly exchange like librarianship, the body of knowledge would be a 'mere pile of intellectual accumulations instead of an organic and operating whole' (p. 147). Jesse Shera, considered to be the father of LIS education, and his colleague Margaret Egan, referred to this theoretical foundation of librarianship as 'social epistemology' (Shera 1972) to reflect the field's basic interest in how knowledge is coordinated, integrated, and used by people. This conceptualization emerged prior to, and remains somewhat distinct from, contemporary academic discourse on social epistemology in philosophy (Goldman 1987; Fuller 1988). It was a direct response to the practical needs of the profession to build collections and services for library users informed by an understanding of the production, distribution, and utilization of the intellectual products held by libraries. From this perspective, information problems associated with interdisciplinarity are a natural focus and have proven to be a point of synergy for research and practice in LIS, and between LIS and other social sciences with interests in knowledge production (Klein 1996b; Palmer 1996).

The interdisciplinary integration of knowledge disrupts structures and other methods of representing and managing information that are essential to the performance of libraries and information systems (Iyer 1995). For instance, discipline-based indexing vocabularies and classification schemes tend to be inadequate for subject access to interdisciplinary intellectual content, and mapping semantic relationships remains a major research challenge. These problems greatly complicate the selection of materials for building scholarly collections, how information is cataloged or encoded with metadata for retrieval, and the organization of information services for interdisciplinary research communities. Web-based search engines have made a significant contribution to information access and retrieval, but they do not provide all the content or functionality required for sophisticated scholarly purposes.

This chapter provides an overview of research on interdisciplinarity in the field of LIS by first introducing the concept of information scatter, a theme that runs through the studies conducted over the past decades. The focus then turns to two core bodies of research that address interdisciplinarity from very different perspectives: *bibliometric research* that provides statistical analyses of patterns and flows of information within and among disciplines, and *information behavior research* that investigates the information practices and needs of interdisciplinary scholars. For the purposes of this chapter, discussion is limited to a small number of studies selected to illustrate the scope and direction of research in the two areas. The chapter concludes by pointing to important directions in information research aimed at advancing information systems, services, and infrastructure to support and foster interdisciplinary science and scholarship.

12.1 Information scatter

Interdisciplinary scholars use information from an array of sources scattered across a range of fields. This notion of information scatter is a salient concept found throughout LIS research on interdisciplinarity (Bates 1996). The distribution of information, intellectually across subject areas but also physically across sources and organizations, has generated considerable interest in information science, including principles such as Bradford's law of scattering (Bradford 1948), which has been applied primarily to disciplinary information phenomena such as the concentration of information on a subject in a core set of journals. For interdisciplinary scholars, scatter has to do with the great dispersion of potentially useful information across disciplines which results in serious problems in information seeking and use. At the same time, the scattered state of knowledge serves important functions. Innovation often comes not from the core of a discipline but from the margins where knowledge is more diffuse, and scatter outside the core promotes discovery and integration of disparate knowledge rather than isolation within a domain (Crane 1972; Chubin 1976).

In response to the problems associated with information scatter, information scientists have developed strategies for interdisciplinary searching in bibliographic databases to manage the variant vocabularies and exploit markers, such as concepts, terms, and names, for retrieval across disciplines (Smith 1974; Weisgerber 1993; White 1996). Studies have also examined levels of scatter by measuring the number and range of domains and information resources associated with a field or a research area. For example, an early study aimed at improving information services for scientific research and development analyzed the information queries made at a corporate research center, determining that subject fields of greater 'width' required more specialized information retrieval services (Mote 1962). Another study showed that chemists working in self-proclaimed high-scatter research areas were less efficient in keeping up to date and argued that professional current awareness services should concentrate on high-scatter fields (Packer and Soergel 1979). Comparative analysis of the distribution of documents across databases for different topics has been applied to demonstrate variations in scatter across research areas and the challenges of interdisciplinary information retrieval (Hood and Wilson 2001).

In these kinds of investigations, determinations of high and low scatter tend to be relative within the scope of the domains covered by the study. For example, in the corporate research center study, scatter was considered low for scientists involved in basic organic chemistry, higher for chemists in an engineering environment that required application of both physics and chemistry, and highest for scientists working on specific applied problems with no associated body of literature, such as 'thermal properties of frozen soils' (Mote 1962, p. 171). Thus, a small and very specialized domain can be more highly scattered than a larger, more general domain. In the database comparison, out of 14 test topics 'family violence' was found to be the most interdisciplinary, with literature scattered across 10 databases; 'dark matter' and 'cladistics' were the least interdisciplinary, with more than 50 per cent of their literature covered by one database. An interesting study of scatter applying broader disciplinary distinctions to library book circulation data determined that materials borrowed by social scientists were widely scattered across disciplines, while borrowers from mathematics and the physical and life sciences were focused on their own discipline (Metz 1983).

A particularly influential body of work with important implications for the provision of information for interdisciplinary researchers has been developed around Don Swanson's notion of 'undiscovered public knowledge' (1986a). The concept refers to new knowledge that can be uncovered by identifying implicit connections, or missing links, between disconnected literatures. Swanson made significant biomedical discoveries using a technique he developed for literature searching in the MEDLINE database (Swanson 1986b), and his approach has been further developed into a literature-based discovery (LBD) technology to support scientists in mining promising connections and for testing hypotheses in the medical literature (Smalheiser and Swanson 1998). Related systems and approaches have advanced the technique and its applications, primarily in biomedicine (Bruza and Weeber 2008). There has been limited development of LBD outside the sciences; however, one preliminary investigation made progress on locating hidden analogies in literature in the humanities (Cory 1997).

Unlike LBD, bibliometric approaches and information behavior research covered below do not attempt to manage scatter directly to aid searching or management of information. Bibliometric studies are applied primarily to illustrate the influences of scatter on knowledge structures, and studies of information behavior provide evidence of the actual work performed and problems encountered by scholars trying to find and use scattered information. Both kinds of studies have practical applications for the development of libraries and information systems, but they have also contributed to our basic understanding of how disciplines interact and conditions that promote and deter the conduct of interdisciplinary research.

12.2 Bibliometric research

Bibliometric analyses of interdisciplinarity are generally applied to produce statistical interpretations of relationships among disciplines rather than to define or determine what constitutes a discipline. Disciplinary relationships are represented by bibliographic indicators in documentary sources, such as the disciplines referenced or cited by authors, disciplinary affiliations of co-authors, structures of co-citation clusters, or co-word associations among titles or texts. Bibliometric data became plentiful with the introduction of citation indexing services, such as the Science Citation Index, first made available by the Institute for Scientific Information (ISI) in 1961, and became eminently easier to access and manipulate when these and other bibliographic data sources became available online in digital form. Bibliometrics, and now webometrics, which analyzes similar associations and structures on the Web, can provide unique and valuable insights on interdisciplinarity based on the cross-disciplinary connections and patterns found among information sources (Borgman and Furner 2002; Thelwall *et al.* 2005).

Citation analysis and co-citation analysis are standard bibliometric approaches, and various techniques have been developed to study interdisciplinary information transfer. As symbols used to represent the ideas of authors (Small 1978), citations can be tracked and measured to determine intellectual lending and borrowing across disciplinary boundaries (Porter and Chubin 1985; Pierce 1999). One common measure is 'citation outside category' where assignment of citations to disciplines is usually based on the subject area of journals or books listed in bibliographic entries or less frequently by academic affiliation of authors. Studies have been designed, for example, to test highly cited classic papers for interdisciplinary content and usage, track the influence of highly interdisciplinary articles in a given field, and trace the development of narrow interdisciplinary research specializations, such as developmental dyslexia and molecular motors in bionanotechnology (Perry 2003; Rafols and Meyer 2007).

Localized case studies have been used to inform the organization of information resources and services and collection development in research libraries. Citation evidence has shown high use of biology and physics journals by chemistry faculty, suggesting that general science libraries are more effective than specialized units within an academic library system (Hurd 1992). Studies have also examined citation patterns of faculty in particular academic programs or research centers to determine the base of literature that should be collected, and methods have been developed for compiling lists of core journals and evaluating the state of primary and secondary literature in interdisciplinary fields.

The cross-disciplinary dynamics illustrated by citation patterns are of interest outside the field of LIS, particularly in terms of their policy implications for research organizations and funding agencies. The scope of these studies may cover an organization, research specialization, or discipline, or provide broader cross-field analysis. Methodological insights are also often documented, as in a study of an interdisciplinary research organization that compared results from both bibliometric and social network measures and found no strong relationship between the two kinds of network ties, with citation motivated more by a researcher's disciplinary perspective than by their interpersonal associations (White *et al.* 2003).

More global data sets are used to track research fronts and growing concentrations in interdisciplinary activity, to indicate levels of knowledge transfer and the impact of interdisciplinary research, and to determine the interdisciplinarity of journals. For example, Rinia *et al.* (2002) analyzed disciplinary impact using a set of 643,000 articles derived from Science Citation Index data, applying three different measures of journal 'relative openness' to articles from other disciplines. While the measures produced different results, basic life sciences ranked highest on impact on other disciplines and computer science was among the lowest ranked. Applying social network measures of degrees, closeness, and betweenness to ISI citation data from 7379 journals, Leydesdorff (2007) illustrated the value of betweenness centrality as an indicator of interdisciplinarity for local citation environments and for identifying those journals most central to an emerging field, with the analysis showing that for biotechnology interdisciplinary influences were coming primarily from journals in the engineering sector.

Typologies and assessments of disciplines and research areas by degree of interdisciplinarity have also been derived. Morillo *et al.* (2003) measured external links, diversity, and strength of relationships among nine research areas based on the multi-assignation of journals to disciplinary categories in ISI's science, social science, and humanities citation indexes. In general, engineering and biomedicine were highest in disciplinary relationships and humanities the lowest. There was also growth of specialized journals in interdisciplinary categories, and new categories were more interdisciplinary than old categories. One study of the 'global structure of all of science' applied five intercitation and three co-citation frequency measures to 16.24 million references from ISI data representing 7121 science and social science journals (Boyack *et al.* 2005). The resulting maps highlighted biochemistry as the most interdisciplinary hub, and indicated that fields such as medicine, ecology/zoology, social psychology, clinical psychology, and organic chemistry also function as hubs but with fewer strong links to other disciplines. Collaboration patterns as represented in citations are also used to measure interdisciplinarity, again with studies ranging from local investigations within a given institution to broader tracking of collaboration patterns across fields. The results of one study suggested that the field of nanotechnology is multidisciplinary, not interdisciplinary, in that it 'consists of an artificial composition of different research fields with little to no relation to each other' (Schummer 2004, p. 448).

12.3 Information behavior research

The citations and other bibliographic indicators extracted from documents, databases, and the Web are important sources of data for analyzing how information moves and coalesces across disciplines. They are, however, abstractions of the actual practice of research. Other types of research are needed to understand information behavior in the context of day-to-day research. Studies of scholarly information-seeking and use complement the more general results obtained through statistical bibliometrics by providing a deeper understanding of how scholars work with information across disciplines and the problems they encounter in doing so. Many recent studies take an 'information practices approach' that emphasizes the social dimensions of disciplines as a primary influence on the information activities of scholars and scientists (Palmer and Cragin 2008), and, where bibliometric studies have tended to focus on the sciences, information behavior research has consistently covered the humanities and the social sciences as well as the sciences. Social science research methods, particularly surveys and interviews, have been favored, with many investigators applying multiple techniques to increase the validity of results derived from the smaller samples typical of qualitative methods.

Transcending discipline-based library classifications

Richard Szostak

As Carole Palmer has noted, the systems of document classification used in modern libraries are each many decades (usually more than a century) old, are firmly grounded in disciplines, and are immensely complex due to the fact that classes have been subdivided to make room for new areas of research. Interdisciplinarians suffer because the same subject is addressed separately, often using different terminology, in different disciplines. Less obviously, all scholars suffer because documents are only classified in terms of what a work is 'about'; scholars generally also want to know which theories or methods were applied (among other things). This is a particular burden for interdisciplinary scholars who will usually want to identify works applying different theories and methods to a particular question.

The explosion of digital databases creates a unique historical opportunity for the development of a novel classification system. This system could take advantage of the low costs of modern information storage: existing classification systems were developed during an age of card catalogues and thus naturally limit the dimensions along which works are classified. The new system could simplify document classification by replacing cumbersome discipline-based systems with a universal classification of phenomena studied and theories and methods applied.

The Leon Manifesto, issued after the 2007 conference of the Spanish chapter of the International Society for Knowledge Organization (ISKO)—the conference theme was interdisciplinarity in knowledge organization—calls for the development of such a classification. It can be found at <http://www.iskoi.org/ilc/leon.htm> The Italian chapter of ISKO also hosts the 'Integrative levels classification' project (<http://www.iskoi.org/ilc/>) which draws on an international body of collaborators to develop such a classification.

Some information scientists worry that language is too ambiguous to allow a universal classification: they argue that only within narrow groups of scholars can the words used to classify carry an agreed meaning. Interdisciplinarians must always grapple with such claims; in the end interdisciplinarity is only possible if some considerable degree of shared understanding is possible across communities of scholars. Advocates of a universal classification argue that the very act of classification can serve to provide the necessary degree of shared understanding. These positions are articulated in a recent debate in the *Journal of Documentation* between Rick Szostak and Birger Hjørland (Hjørland 2008; Szostak 2008a,b).

The alternative to a discipline-based classification lies foremost in the development of a universal classification of the phenomena that scholars study. Note that the bulk of scholarly research involves examining the influence that one set of phenomena exerts on another set of phenomena. These 'causal links' can be classified using synthetic/synthesized notation: a scholar interested in the physiological effects of certain drugs can thus find the studies they want without being sent to general works on the drugs in question or the physiological effects in question. This particular notational device reflects a larger strategy of using 'faceted notation' to identify different aspects of any phenomenon or the relationships between any pair of phenomena. Faceted notation is often advocated in information science but is rarely employed on a large scale. The ILC affords a further notational efficiency by using the same notation for works about a particular theory or method as is used (albeit in a different place in the notation for a particular work) for works applying that theory or method.

References

Hjørland, B. (2008). Core classification theory: a reply to Szostak. *Journal of Documentation* **64**(3), 333–42.

Szostak, R. (2008a). Classification, interdisciplinarity, and the study of science. *Journal of Documentation* **64**(3), 319–32.

Szostak, R. (2008b). Letter to the Editor. Interdisciplinarity and classification: a reply to Hjørland. *Journal of Documentation* **64**(4).

As with citation studies, scholarly information behavior research is often 'domain analytic' in that it is designed to examine a particular discipline or research area. Domain-based studies of scholars strive to capture information use in ways that reflect the socio-cultural context of the unit analyzed, sometimes drawing comparisons across fields or specializations. Interdisciplinary information behavior has been investigated within fields ranging from environmental science to music, women's studies, and ethnic studies, and the findings continually document significant levels of information use across disciplines and problems finding and using information scattered across unfamiliar subject areas.

Since interdisciplinary researchers 'constitute a significant and distinctive class of scholars' in their own right (Bates 1996, p. 163), numerous studies have examined interdisciplinary information behavior within larger or more diverse groups not constrained by domain parameters. For example, a survey of users of the Finnish National Electronic Library corroborates many earlier observations about scatter and interdisciplinarity, suggesting that scholars in high-scatter fields use more databases and have more difficulty keeping up with information across fields. The results also raised interesting questions about the role of browsing and the adequacy of cross-database keyword searching for interdisciplinary topics (Vakkari and Talja 2005). Interdisciplinary organizations and teams are particularly effective sites for investigating active programs of research. Studies of humanities scholars and scientists working in interdisciplinary research centers have documented how different modes of inquiry and collaboration are linked to approaches for gathering information from far afield and for building the knowledge needed to solve research problems (Palmer 2001; Palmer and Neumann 2002). Interdisciplinary collaborative research teams offer a microcosm of interaction for examining information transfer. One study showed that the knowledge exchanged in both science and social science teams is often fact based, but 'know-how' related to research processes and methods is also vital in supporting interdisciplinary projects (Haythornthwaite 2006).

12.3.1 Models of scholarly information processes

To date, studies of information behavior have not produced a complete account of the differences between disciplinary and interdisciplinary information behavior or differences in practices among scholars in various interdisciplinary fields. The models that have emerged provide a base of understanding on how scholars manage and use information in the research process, usually representing activities and workflows involved in searching for and assimilating information, which have implications for how information resources and tools can be linked together to support stages of research or groups of research practices. One well-known model of scholarly information seeking developed by David Ellis was based on a set of qualitative, comparative studies of academic faculty in the social sciences, physical sciences, and literature. It identified a series of information activities—starting, chaining (footnote and citation chasing), browsing, differentiating, monitoring, and extracting—common to researchers across the fields (Ellis 1993). The model does not represent interdisciplinary scholarship, but it does provide a scheme for contrasting how interdisciplinary and disciplinary information processes might differ. For example, a subsequent study of a specialized group of interdisciplinary social scientists determined that the framework lacked important activities, such as networking, verifying, and managing information (Meho and Tibbo 2003).

Based on analysis of information use and knowledge development strategies of interdisciplinary scientists, Palmer (2001) identified three modes of interdisciplinary inquiry. Specifically, 'collaborators' tend to work in consultation with colleagues from other domains, and their information practices focus on locating specific information and expertise needed to move research projects forward. 'Team leaders' gather information more systematically and recruit collaborators into their research groups, while 'generalists' explore widely to find information outside their field and prefer to learn for themselves rather than consult or collaborate with others. This framework served as a point of departure for a model of information seeking based on an investigation of interdisciplinary scholars from departments spanning an entire university (Foster 2004). The resulting non-linear model accounts for important social and organizational structures in interdisciplinary research within three core research processes—opening, orientation, and consolidation—that each encapsulate a range of activities important to interdisciplinary research. Opening includes information-seeking activities involved in elaborating a research topic, such as keyword searching and browsing; orientation activities relate to problem definition and identification of existing research, key themes, and disciplinary communities; consolidation consists of the refining, sifting, and verifying used to judge and integrate information during the research process.

12.3.2 Weak information work

Interdisciplinary scholars and scientists develop strategies that help them move into other domains, and to interpret and communicate information across disciplines. Two activities—*probing* to find information and *translation* of terminology, concepts, and ideas—are prominent practices that have been identified in multiple studies. These practices typify 'weak information work', very difficult and time-consuming information activities associated with specific research conditions, particularly ill-structured problem space, lack of domain knowledge, and unsystematic research steps (Palmer *et al.* 2007a). These exploratory and integrative activities are important targets for the development of information systems and services, since they are intellectually challenging and require non-routine tasks that take a high toll on researchers' time but also have great potential for stimulating and producing innovative research and new discoveries.

Probing is essentially a highly investigative form of browsing. Browsing has been widely studied and is understood to be an important general scholarly information practice (Chang and Rice 1993) that is especially essential for interdisciplinary researchers working in new and rapidly developing fields. Web surfing is one kind of browsing activity, but interdisciplinary scholars also browse issues of journals, publisher's catalogs, bibliographic indexes, library collections, and personal collections of books and papers, all of which may be available as physical documents or digitally on the Web. But, while browsing has the connotation of being somewhat *ad hoc* in nature, probing is a deliberate strategic approach for identifying scattered or remote information. Researchers probe into peripheral areas outside their expertise to increase breadth of perspective, generate new ideas, or to explore a wide range of types and sources of information. However, probing is not necessarily always broad in scope. For instance, a study of high-impact information in interdisciplinary neuroscience laboratories (Palmer *et al.* 2007a) showed that scientists probed the literature outside their domain to investigate specific concepts or diseases in the clinical literature.

Through probing researchers encounter new ideas, theories, methods, and terms that they do not fully understand, which prompt the need to gather further information. This translation work can be the most difficult and laborious part of interdisciplinary research, collaboration, and communication. Studies across interdisciplinary fields have indicated that most interdisciplinary researchers need to be familiar with the terminology of other disciplines in order to understand the literature they consult and to carry out their research projects. For instance, scholars must translate terminology to construct a successful search in a database outside their discipline. Then, as they encounter information that is intellectually distant or from unfamiliar sources, the new material needs to be assessed, interpreted, and grounded with background information from the original domain. Vocabulary differences are at the heart of translation activities, but research conventions and cultures must also be learned and navigated. Valid interdisciplinary research is necessarily based on a deep understanding of how concepts, methods, and results fit in the body of discourse and practice in which they were developed. Only then can judgments be made about how ideas can legitimately be applied in a new area. Interestingly, while the imprecise language used in the humanities and social sciences complicates the identification of analogies and applications across domains, the very technical nature of scientific vocabulary can also be a serious barrier to cross-domain searching and browsing.

Interdisciplinary research teams are continually engaged in translation activities as a basic feature of the give-and-take involved in collaborative projects. In fact, learning enough about the perspectives and problems driving the interests of collaborators appears to be a key factor in the success of interdisciplinary research groups. This is evident in applied informatics projects where the goal of domain researchers is to analyze data that contribute to the solution of a scientific or social problem, while computer science and technology collaborators are working to make progress on an algorithm or a prototype. These two groups can have very different views of what constitutes the end point in the same project. In the humanities, collaboration is less formal in nature, but those involved in interdisciplinary scholarship can have a strong dependence on local colleagues or outside experts who serve as translators of concepts and ideas (Palmer and Neumann 2002). Social networking is essential in making and maintaining the greater number of personal contacts needed to stimulate and validate ideas, open doors for sharing information, and for the exploration of interdisciplinary subject matter (Foster 2004). Not surprisingly, the joint authorship that results from collaboration adds another layer of intensive translation work. Writing a co-authored document requires negotiations and decisions on what needs to be explained for a particular audience and refinement of terminologies and formats for different fields.

Both probing and translation tend to be non-routine and time-intensive, that is, weaker than typical information activities performed by disciplinary scholars. Data curation is another example of weak information work, and is emerging as an important area in information behavior research. Data management and preservation problems are acute in modern science, where researchers are generating vast numbers of digital data, which is more fragile than any preceding data format. The case of networked sensor systems in the environmental sciences provides a telling example of the growing complications with the management of data as a resource for interdisciplinary research (Borgman *et al.* 2007). With instruments now automatically generating environmental data, the contextual information previously surrounding data acquisition is no longer systematically recorded and preserved in lab or field notebooks. Additionally, the different disciplinary cultures within a research community may employ seemingly incompatible processes and terminologies in describing and coordinating data sets. Currently, the needs and expectations for managing and mobilizing data for interdisciplinary research purposes are not well understood or supported, and progress will require the development of professional procedures and standards for data representation and archiving, as well as techniques for cross-disciplinary data curation and integration (Palmer, *et al.* 2007b).

12.4 Conclusion

There is no doubt that LIS research on interdisciplinarity will continue to intensify as the centrality interdisciplinary inquiry increases. A recent report on library service needs at the University of Minnesota (2006) is compelling, finding that among 50 participants from across the social sciences and humanities 'nearly every faculty member interviewed considered his or her work to be interdisciplinary' (p. 20). As networked information systems become more advanced, scholars will find it easier to draw from and make connections to other fields and to place their research in a broader context, compounding the amount and rate of knowledge exchange and collaboration across disciplines.

One important unexplored research area is interdisciplinary relevance. There is an extensive existing body of research on information relevance, but it has not addressed interdisciplinarity as an influential factor in relevance judgments or how relevance criteria might differ for interdisciplinary researchers. Existing studies on the differences between domain experts and novices have some applicability; however, interdisciplinary researchers are not novices in the same sense as the students usually examined in these studies. Clearly, information technologies designed for typical non-experts will not be optimal for sophisticated scientists and scholars taking interdisciplinary approaches to their research. A second area of research is the value of traditional information sources, such as discipline-based handbooks, textbooks, encyclopedias, and other reference materials, which play a vital role in the interpretation and learning needed for researchers to integrate knowledge from outside their own areas of expertise. With many of these materials now in digital form, their functionality can be enhanced for the translation work involved in interdisciplinary scholarship.

There is also need for the development of cross-disciplinary ontologies and thesauri to assist in mapping content to provide smoother digital access across fields, but not at the expense of deep access within a curated collection. Current broad-based search engines are assisting scholars in casting a wide net to gather information concurrently from an array of sources. However, as the digital information environment increases in size and complexity, the original, intended context of a scholarly work can get lost as many collections are fused together into one immense networked information space. As digital information continues to grow and be reconfigured, it will be necessary to make explicit the relations among the multitude of sources, not only for discovering resources across institutions, repositories, and disciplines, but, just as importantly, for retaining the meaning of content within its original domain context.

There has long been a vision of a comprehensive, integrated information ecology. It motivated early information scientists such as Paul Otlet and Vannevar Bush, and continues to drive current digital technology research, as evidenced by sustained work on the semantic web (Shadbolt *et al.* 2006). But, the overall digital information environment is far from complete and unified at this time. The Web has produced a powerful, distributed network of information resources that is larger and more complex, perhaps, but which is also more *ad hoc* than that once held and organized primarily in research libraries. The imagined high-functioning information infrastructure remains elusive, at least for the short term, but there is great promise in future networked cyberinfrastructure for enhancing cross-disciplinary access to information, data, and services that can foster interdisciplinary research. The ideal of a seamless web of information is misplaced, however, unless it includes the ability to systematically track back to the disciplinary heritage of a research result, idea, concept, or theory. Thus, the greatest information research challenges in the future may not be about enhancing the ability of scholars to move across disciplinary boundaries but about maintaining the increasingly long and mutable intellectual paths to our disciplinary past.

References

Allen, T. (1966). *Managing the flow of scientific and technical information.* Washington, DC: US Department of Commerce, National Bureau of Standards, Institute of Applied Technology.

American Council of Learned Societies (2006). *Our cultural commonwealth: the report of the American Council of Learned Societies Commission on cyberinfrastructure for humanities and social science.* New York: American Council of Learned Societies.

Bates, M.J. (1996). Learning about the information seeking of interdisciplinary scholars and students. *Library Trends* **45**(2), 155–64.

Borgman, C.L. and Furner, J. (2002). Scholarly communication and bibliometrics. *Annual Review of Information Science and Technology* **36**, 3–72.

Borgman, C.L., Wallis, J.C., and Enyedy, N. (2007). Little science confronts the data deluge: habitat ecology, embedded sensor networks, and digital libraries. *International Journal on Digital Libraries* **7**, 1–2.

Boulding, K.E. (1968). *Beyond economics: essays on society, religion, and ethics*. Ann Arbor, MI: University of Michigan Press.

Boyack, K.W., Klavans, R., and Börner, K. (2005). Mapping the backbone of science. *Scientometrics* **64**(3), 351–74.

Bradford, S.C. (1948). *Documentation*. London: C. Lockwood.

Bruza, P. and Weeber, M. (eds) (2008). *Literature-based discovery*. Berlin: Springer.

Chang, S.-J. and Rice, R.E. (1993). Browsing: a multidimensional framework. *Annual Review of Information Science and Technology* **28**, 231–76.

Chubin, D.E. (1976). The conceptualization of scientific specialties. *Sociological Quarterly* **17**(4), 448–76.

Clapp, V.W. (1954). The problem of specialized communication in modern society. In: M.E. Egan (ed.) *The communication of specialized information*, pp. 1–13. Chicago: American Library Association for the University of Chicago Graduate Library School.

Cory, K.A. (1997). Discovering hidden analogies in an online humanities database. *Computers and the Humanities* **31**(1), 1–12.

Crane, D. (1972). *Invisible colleges: diffusion of knowledge in scientific communities*. Chicago: University of Chicago Press.

Ellis, D. (1993). Modeling the information-seeking patterns of academic researchers: a grounded theory approach. *Library Quarterly* **63**(4), 469–86.

Evans, J.A. (2008). Electronic publication and the narrowing of science and scholarship. *Science* **321**(5887), 395–9.

Foster, A. (2004). A nonlinear model of information-seeking behavior. *Journal of the American Society for Information Science and Technology* **55**(3), 228–37.

Fuller, S. (1988). *Social epistemology*. Bloomington, IN: Indiana University Press.

Goldman, A.I. (1987). Foundations of social epistemics. *Synthese* **73**(1), 109–44.

Haythornthwaite, C. (2006). Learning and knowledge networks in interdisciplinary collaborations. *Journal of the American Society for Information Science and Technology* **57**(8), 1079–92.

Hood, W.W. and Wilson, C.S. (2001). The scatter of documents over databases in different subject domains: how many databases are needed? *Journal of the American Society for Information Science and Technology* **52**(14), 1242–54.

Hulme, E.W. (1923). *Statistical bibliography in relation to the growth of modern civilization: two lectures delivered in the University of Cambridge in May 1922*. London: Butler & Tanner; Grafton & Co.

Hurd, J. M. (1992). Interdisciplinary research in the sciences: implications for library organization. *College and Research Libraries* **53**(4), 283–97.

Iyer, H. (1995). *Classificatory structures: concepts, relations and representation*. Frankfurt am Main: Indeks Verlag.

Klein, J.T. (1996a). *Crossing boundaries: knowledge, disciplinarities, and interdisciplinarities*. Charlottesville, VA: University Press of Virginia.

Klein, J.T. (1996b). Interdisciplinary needs: the current context. *Library Trends* **45**(2), 134–54.

Leydesdorff, L. (2007). Betweenness centrality as an indicator of the interdisciplinarity of scientific journals. *Journal of the American Society for Information Science and Technology* **58**(9), 1303–19.

Meho, L.I. and Tibbo, H.R. (2003). Modeling the information-seeking behavior of social scientists: Ellis's study revisited. *Journal of the American Society for Information Science and Technology* **54**(6), 570–87.

Metz, P. (1983). *The landscape of literatures: use of subject collections in a university library*. Chicago: American Library Association.

Morillo, F., Bordons, M., and Gómez, I. (2003). Interdisciplinarity in science: a tentative topology of disciplines and research areas. *Journal of the American Society for Information Science and Technology* **54**(13), 1237–49.

Mote, L.J.B. (1962). Reasons for the variations in the information needs of scientists. *Journal of Documentation* **18**(4), 169–75.

National Science Board (2005). *Long-lived digital data collections: enabling research and education in the 21st century*, publication no. NSB-05-40. Washington, DC: National Science Foundation.

Packer, K.H. and Soergel, D. (1979). The importance of SDI for current awareness in fields with severe scatter of information. *Journal of the American Society for Information Science* **30**(3), 125–35.

Palmer, C.L. (ed.) (1996). Navigating among the disciplines: the library and interdisciplinary inquiry. *Library Trends*, **45**(2) [Special Issue].

Palmer, C.L. (2001). *Work at the boundaries of science: information and the interdisciplinary research process*. Dordrecht: Kluwer.

Palmer, C.L. and Cragin, M.H. (2008). Scholarship and disciplinary practices. *Annual Review of Information Science and Technology* **42**, 165–212.

Palmer, C.L. and Neumann, L.J. (2002). The information work of interdisciplinary humanities scholars: exploration and translation. *Library Quarterly* **72**(1), 85–117.

Palmer, C. L., Cragin, M.H., and Hogan, T.P. (2007a). Weak information work in scientific discovery. *Information Processing and Management* **43**(3), 808–20.

Palmer, C.L., Heidorn, P.B., Wright, D., and Cragin, M.H. (2007b). Graduate curriculum for biological information specialists: a key to integration of scale in biology. *International Journal of Digital Curation* **2**(2), 31–40.

Perry, C.A. (2003). Network influences on scholarly communications in developmental dyslexia: a longitudinal follow-up. *Journal of the American Society for Information Science and Technology* **54**(14), 1278–95.

Pierce, S.J. (1999). Boundary crossing in research literatures as a means of interdisciplinary information transfer. *Journal of the American Society for Information Science* **50**(3), 271–9.

Porter, A.L. and Chubin, D.E. (1985). An indicator of cross-disciplinary research. *Scientometrics* **8**(3–4), 161–76.

Rafols, I. and Meyer, M. (2007). How cross-disciplinary is bionanotechnology? Explorations in the specialty of molecular motors. *Scientometrics* **70**(3), 633–50.

Rinia, E.J., van Leeuwen, T.N., Bruins, E.E.W., van Vuren, H.G., and van Raan, A.F.J. (2002). Measuring knowledge transfer between fields of science. *Scientometrics* **54**(3), 347–62.

Schummer, J. (2004). Multidisciplinarity, interdisciplinarity, and patterns of research collaboration in nanoscience and nanotechnology. *Scientometrics* **59**(3), 425–65.

Shadbolt, N., Hall, W., and Berners-Lee, T. (2006). The semantic web revisited. *IEEE Intelligent Systems* **21**(3), 96–101.

Shera, J.H. (1972). *The foundations of education for librarianship*. New York: Becker and Hayes.

Smalheiser, N.R. and Swanson, D.R. (1998). Using ARROWSMITH: a computer-assisted approach to formulating and assessing scientific hypotheses. *Computer Methods and Programs in Biomedicine* **57**(3), 149–53.

Small, H.G. (1978). Cited documents as concept symbols. *Social Studies of Science* **8**(3), 327–40.

Smith, L.C. (1974). Systematic searching of abstracts and indexes in interdisciplinary areas. *Journal of the American Society for Information Science* **25**(6), 343–53.

Swanson, D.R. (1986a). Undiscovered public knowledge. *Library Quarterly* **56**(2), 103–18.

Swanson, D.R. (1986b). Fish oil, Raynaud's Syndrome, and undiscovered public knowledge. *Perspectives in Biology and Medicine* **30**(1), 7–18.

Thelwall, M., Vaughan, L., and Björneborn, L. (2005). Webometrics. *Annual Review of Information Science and Technology* **39**, 81–135.

University of Minnesota Libraries (2006). *A multi-dimensional framework for academic support: a final report*. Submitted to the Andrew W. Mellon Foundation. Available at: <http://purl.umn.edu/5540>.

Vakkari, P. and Talja, S. (2005). The influence of the scatter of literature on the use of electronic resources across disciplines: a case study of FinELib. In *Research and advanced technology for digital libraries: 9th European Conference, ECDL 2005*, pp. 207–17. Berlin: Springer.

Weisgerber, D.W. (1993). Interdisciplinary searching: problems and suggested remedies: a report from the ICSTI group on interdisciplinary searching. *Journal of Documentation* **49**(3), 231–54.

White, H.D. (1996). Literature retrieval for interdisciplinary syntheses. *Library Trends* **45**(2), 239–64.

White, H.D., Wellman, B., and Nazer, N. (2003). Does citation reflect social structure? Longitudinal evidence from the 'globenet' interdisciplinary research group. *Journal of the American Society for Information Science and Technology* **55**(2), 111–26.

PART 3

Knowledge interdisciplined

CHAPTER 13

A field of its own: the emergence of science and technology studies

SHEILA JASANOFF

In 2001, the field of science and technology studies (STS) was included for the first time in the *International encyclopedia of social and behavioral sciences* (IESBS; Smelser and Baltes 2001). The editors classified STS as an 'intersecting field' rather than a discipline; the latter label was reserved chiefly for fields with well-established, one-word names (e.g. anthropology, economics, history, law, and philosophy) and for various branches of psychology. STS shared the rubric 'intersecting' with a cluster of amorphous, ill-assorted, and relatively recent fields such as genetics and society, gender studies, religious studies, and behavioral and cognitive neuroscience. Unlike media studies or public policy, however, STS was not banished to the status of 'applications'. Inclusion in the IESBS was a breakthrough of sorts. This was the first time that STS was named as a card-carrying field in a comprehensive roster of the social and behavioral sciences. It validated years of effort by many more or less loosely networked scholars to establish the social studies of science and technology as a recognized, and recognizable, domain of intellectual activity. As a member of that network and as the editor of the IESBS section on STS, I was understandably elated, even though putting the section together entailed many difficult and initially unforeseen choices of what to include or exclude in defining the field for outsiders.[1]

How did STS come to take its place in an encyclopedia of the social sciences as a well-demarcated territory on the map of knowledge? What major contributions has it made to research and teaching, and what have been its principal successes and failures? How does the future look for STS? Responses to these questions help shed light on the meanings and challenges of interdisciplinarity, illuminating the potential that the spaces between disciplines offer for novel contributions to human self-awareness, understanding, and action. At the same time, the track record of STS illustrates how difficult it is to populate those

[1] I had already co-edited the second edition of the field's own handbook (Jasanoff *et al.* 1995), but inclusion in the *IESBS* meant more overtly staking out a claim for STS in relation to other disciplines.

in-between spaces with well-trained scholars, new curricular offerings, and long-term research programs. At the heart of the story, too, are questions about the capacity of STS to overcome entrenched status differentials among disciplines, especially between the more humanistic social sciences and the increasingly more commercial enterprises of university-based science, medicine, and engineering.

13.1 An interdisciplinary history

At the risk of oversimplification, and of flattening cross-cultural differences, STS can be described as a merger of two broad, mid-twentieth century streams of scholarship.[2] One is a body of research that looks at the *nature and practices* of science and technology (S&T) as social institutions possessing distinctive normative commitments, structures, practices, and discourses that nevertheless change over time and vary across cultural contexts. The other is a tradition that concerns itself mainly with the *impacts and control* of science, and even more of technology, with particular focus on the risks that S&T pose to human health and safety, as well as to peace, security, privacy, community, democracy, development, environmental sustainability, and other human values. STS's claim to recognition at the beginning of the twenty-first century is largely a consequence of these once-discrete lines of concern coming together, in increasingly more fruitful collaboration, around a shared core of theoretical orientations, texts and topics, research methods, and professional infrastructures (e.g. programs, departments, journals, societies). In short, STS is the product of decades of effort by people who perceived important gaps in academic engagements with the analysis of science and technology, and who gradually, painstakingly, and with mixed success built institutional foundations to advance their shared interest in filling those blank spaces.

The resulting field is interdisciplinary in a very particular sense. One way to capture its characteristics is through a cartographic metaphor. Underlying any definition of interdisciplinarity is an ideal-typical map of the relationship among pre-existing disciplines. Two maps come immediately to mind: in one, the disciplines are tightly lined up, one against another, as in a map of the contiguous United States, with shared boundaries and no gaps between; in the other, as in a map of the Indonesian archipelago, the disciplines are oddly and idiosyncratically bounded formations, haphazardly scattered across a sea of ignorance, with unexplored waters in between.[3] On the first map, a new 'interdiscipline' comes into being principally through exchanges among scholars already belonging to one or another established disciplinary community and trained in its forms of reasoning and research practices. On the second, an '*inter*discipline' may be literally that—an indepen-

[2] The story told in this chapter is unavoidably US-centric, given the author's experiences and knowledge limitations. STS is an international field, whose past, present, and future rest on global networks of scholarship and exchange. One way to strengthen future formations is to tell parallel stories of the emergence of STS from other national and regional vantage points, an impossible undertaking in a chapter of this scope.

[3] Note that the two alternatives captured by my cartographic metaphor correspond roughly to the categories of multidisciplinarity and transdisciplinarity presented in Klein's taxonomy in this volume. The taxonomic approach, however, does not problematize the taken-for-grantedness of disciplinary boundaries, nor emphasize their contingency or question their claims to coherence as I implicitly do in this chapter.

dent disciplinary formation situated among other disciplines. Such a field may come into being through topical exploration and theoretical or methodological innovation as well as through exchange, coalescing into an autonomous island of knowledge-making with its own native habits of production and trading. STS looks more like the latter than the former: it is less a program of interstate highway construction among existing disciplinary states than an attempt to chart unknown territories among islands of disciplined thought in the high seas of the unknown.

13.1.1 The nature and practices of science and technology

Beginning in the interwar period and continuing into the start of the Cold War, sociologists and historians, and not infrequently scientists and engineers, became interested in the relationship between scientific practice and its work products. Perhaps the best known result of these explorations was the publication in 1962 of Thomas Kuhn's hugely influential *The structure of scientific revolutions*. That book helped crystallize a new approach to the history of science, in which scientific facts were seen as the products of scientists' communal, purposive, knowledge-generating efforts, conditioned by specific contexts of discovery. In brief, Kuhn's work helped turn scholarly attention away from the theoretical content and coherence of scientific claims to the social means of their production. Among the ramifications of this shift was an effort by a group of mainly British scholars to probe how far questions about the nature of science that had once been asked mainly by philosophers could be productively addressed, and consequently reframed, by sociologists and historians (Bloor 1976). These inquiries laid the basis for a distinctive school of 'sociology of scientific knowledge' (SSK)[4] and the founding of 'science studies' centers at a number of UK universities in the 1970s, including Edinburgh and Bath. The impetus behind SSK was more imperial than interdisciplinary: it was to render social what had previously been seen as mainly epistemic (crudely, how scientists think); it was to appropriate for the qualitative and interpretive social sciences what had once securely belonged to philosophy (roughly, by asking what scientists do, why they do it, and how their work achieves authority).

While SSK was emerging from conversations, and sometimes quarrels, between philosophy and sociology of science, scholars from diverse backgrounds recognized the value of ethnographic methods for studying scientists at work. An early, influential product of this approach was the 1979 book by Bruno Latour and Steve Woolgar, *Laboratory life: the social construction of scientific facts*; the word 'social' was dropped from the title of the 1986 edition. In this and subsequent writing, Latour urged students of science to 'follow the scientist' if they wished to understand how observations in the lab or the field attain the status of facts (Latour 1987). Work in a related vein used participant-observation to explore the cultural characteristics of different scientific disciplines (physics, molecular biology, genomics,

[4] SSK contrasted, in particular, with then dominant trends in US sociology of science which concentrated more on the social organization and roles of scientists than on their specific knowledge-producing practices. American sociology of science was led by a number of distinguished practitioners, such as Robert K. Merton of Columbia, but their work increasingly diverged in aims and methods from the more epistemologically, metaphysically, and semiotically inclined European schools.

climate modeling) and organizations ('big science', university laboratories, interdisciplinary research centers). A metaphysician by training and inclination, Latour, together with his colleague Michel Callon in the Paris-based school of STS, also produced important works on the relations between the human and non-human or social and material elements of S&T. Their 'actor-network theory', which proposes that non-human agents known as 'actants' be treated symmetrically with human agents, emerged as another salient conceptual foundation for STS research. By highlighting the material elements of knowledge-making, that line of work foregrounded technology as a significant object of STS study.

A third important strain of research looked at science and technology as cultural formations and asked how they relate to other aspects of culture. Engaging anthropologists, feminists, postcolonial scholars, discourse analysts and other theorists of language and power, this body of work most explicitly crossed the line between the humanities and the social sciences, particularly in its preoccupation with the meanings people attach to the products of S&T. In works such as the historian Donna Haraway's (1989) investigations of primatology or Evelyn Fox Keller's (1986) studies on gender and science, cultural studies of science and technology questioned how social power translates into scientific authority and vice versa. A flourishing body of scholarship emerged around medical S&T, with a focus on such topics as reproductive medicine, patient activism, and hereditary disease; unlike classical studies of the physical sciences and technologies, these more human-centered investigations emphasized themes of identity and subjectivity, especially of those affected by disease classifications. More generally, the influx of research funds from the Human Genome Project spurred broad-based exploration of the ethical, legal, and social implications of genetic science and technology, contributing new dimensions to the cultural studies of science. Less well represented, but equally agenda-setting, was work on the relations between science and other powerfully institutionalized belief systems, such as law, politics, and religion; of central interest here were the effects of cultural norms of legitimacy and reasonableness on the production and reception of policy-relevant scientific facts (Jasanoff 1990, 2005).

13.1.2 The invention of technoscience

An example of the way that STS charts its own course is the field's distinctive treatment of technology in relation to science. Unlike historians of science and technology, who maintain largely distinct identities through professional training and associations, STS scholars have made a point integrating the study of scientific discovery with the analysis of the technological systems that assist in or result from advances in science. The term 'technoscience', widely used in STS research, and adopted as the name of the Society for Social Studies of Science (4S) newsletter, signals a deep commitment to the view that science and technology are inextricably interwoven. STS scholars often claim that technological innovation would not be possible without scientific problem-solving; nor could scientific discovery be imagined without technological means to enable new experimental methods and approaches. Accordingly, whether in studying high-energy physics or molecular biology, Bakelite or musical synthesizers, stem cells or Golden Rice, the internet or the human genome, STS researchers pay particular attention to the interplay of ideas and instruments in the practices of the discoverers, inventors, and users of S&T. By invoking the term technoscience, the field thus draws its own distinctive boundaries across the subject matter it investigates.

For illustration, one can turn to the third handbook of STS sponsored by the Society for Social Studies of Science, one of the field's major professional societies (Jasanoff 1995; Hackett *et al.* 2007). The handbook's final section, headed 'emergent technosciences', deals with systems that cross the lines between the cognitive and the material as well as the natural and the social. Included in this section are articles on genomics, medical biotechnologies, finance, environment, communications, and nanotechnology. All are areas in which scientific and technological breakthroughs are intimately linked, conform to no simple temporal or causal relationships, and depend on multifaceted engagement by actors ranging from individual discoverers, inventors, and entrepreneurs to expert communities, economic sponsors, policy makers, and consuming (or sometimes resisting) publics.

13.1.3 Impacts and control of science and technology

The second major thrust within STS derives from scientists'—and, with increasing intensity, citizens'—concerns about the impacts of S&T developments on health, safety, and fundamental human values. No event did more to spur these concerns than the dropping of the atomic bombs on Hiroshima and Nagasaki in 1945, and the ensuing arms race between the United States and the former Soviet Union during the Cold War. Themes of scientists' complicity in war and violence, and technology's lack of democratic accountability, gained added prominence during the Vietnam War, which also spurred linkages between earlier worries about the ungovernability of science with nascent concerns about S&T's environmental impacts. The marine biologist Rachel Carson (1962), for example, is widely credited with launching the modern US environmental movement with *Silent spring*, an attack on the indiscriminate use of chemicals, which appeared in the same year as Kuhn's book on scientific revolutions. Closer to the present, genetic, information, and nanotechnologies, and their rapid convergence in areas such as synthetic biology, have aroused new anxieties about technological risks. Observers have questioned whether the benefits of these promising developments might be offset or even undermined by negative consequences for liberty, privacy, autonomy, equality, and other cherished liberal ideals. Increasingly, too, the consequences of global imbalances in S&T innovation, and their implications for human rights and social justice, have emerged as focal points of STS scholarship.

In the late 1960s, several US universities, including Cornell, Harvard, MIT, Penn State, and Stanford, reacted to these social and political concerns by forming programs in 'science, technology and society' (also abbreviated as STS). These programs were founded, and often led, by senior scientists or engineers with years of experience in science advice and policy formation. A presupposition of these programs was that they had to be cross-disciplinary in the sense of highway-building described above, engaging natural scientists and engineers, as well as humanists, social scientists, and professionals in law, business, and public policy. Familiarity with the technical content of S&T was seen as a prerequisite for research on STS topics, which meant that early contributions to both research and teaching were made either by scientists (or ex-scientists) and engineers, or by teams whose members included technically trained researchers at their core. Humanists and social scientists were tacitly assumed not to have significant independent insight into the functioning of S&T, although their participation was considered essential for illuminating the 'soft', value-laden, societal dimensions of S&T's impacts. The genealogy of Cornell's STS program provides a small marker of these attitudes and assumptions. It was established in 1969—interestingly, also the year of the traumatic takeover of the Cornell student union by a group of African-American student activists—by a chemist (Franklin Long), a physicist (Raymond Bowers), a biologist (Richard D. O'Brien), and a philosopher of language and mathematics (Max Black).

The prominent role of scientists and engineers initially helped establish the credibility of STS research, but also introduced a number of constraints: emphasis on empirical case studies rather than social theory; reaffirmation of scientists' necessarily partial perceptions about the cultures and practices of science and technology; reliance on anecdotal practitioner narratives rather than systematic research to explain science–technology–society relationships; and acceptance of public 'scientific illiteracy' as the main explanation for popular concerns about science and technology. The topics treated by first-generation STS scholars reflected some of these limitations. Research was usually directed toward the public controversies of the day (airports, nuclear power, supersonic transport, vaccines, environmental pollution), with results that were sometimes indistinguishable from robust journalism. More seriously, such research failed on the whole to win the interest of major scholars in established humanistic or social scientific disciplines, and many STS programs in the United States, such as Harvard's and Cornell's, either died a quiet death or substantially lost momentum by the mid-1980s.

To explain that attrition and loss of energy, one should note that STS scholars in the 1970s drew on fairly conventional social theory—for example, attributing technical controversies to differences in participants' taken-for-granted interests—and hence they neither contributed to nor drew from seminal insights from other fields. At a time when many social sciences were looking to quantitative methods and rational choice theory, it was also easy to dismiss qualitative STS findings as merely anecdotal or subjective. Unlike many scholars preoccupied with the practices of scientists, however, researchers focusing on the impact and control of science and technology were attracted from the first to issues of power and governance. Their work highlighted how dominant processes of technical decision making tended to marginalize weaker social groups and their views of the world; neo-Marxist theorists tied these dynamics to class, capital, and hegemonic beliefs. In these respects, even first-generation STS research shared significant concerns with later cultural studies of science and technology. Openings thus existed for a productive synthesis, which began in the United States in the late 1980s, usually under the rubric of 'science and technology studies'.

13.1.4 Common ground

Convergences between the two major precursors of contemporary STS—work on the nature of scientific production and on the impacts of science and technology—occurred on both intellectual and institutional levels. Maturing research programs brought earlier disparate projects and practitioners into closer communion and helped define common theoretical approaches and topical interests. In brief, research on the nature of science became more concerned with how social understandings or arrangements are taken up into the production of knowledge and artifacts, while research on the impacts of science and technology recognized that the interactions of science and society begin long before the material products of technology enter the market and affect peoples' lives. As a result of the new synthesis, the power of S&T was no longer seen as wholly separable from other kinds of power. Nor were the formation and application of knowledge considered entirely distinct from their eventual uses and impacts. Thus, the ways in which scientific authority interpenetrates with other kinds of social and psychological authority emerged as a leading topic in the field's evolving agenda of inquiry.

By the end of the 1990s, a new generation of STS scholars began examining issues such as the following: the nature of expertise in various historical periods and cultural settings; the persuasive resources used to forge agreement on 'facts'; the relationship between scientific representations and wider visual culture; the disciplining effects of instruments, measuring techniques, and administrative routines; the use of non-human agents, including model lab organisms such as flies or mice, in the work of science; the methods of maintaining or challenging boundaries between scientific, technological, and other cultural practices; and the intermingling of expert and lay cultures around such issues as genetic disease. The field's long-standing concerns with fact, truth, and method did not vanish, but they 'thickened' to include a new preoccupation with how novel ideas, entities, and belief systems appear and make their way in the world (and how old ones die out). More than simply accounting for 'truth', STS became concerned with the social dimensions of the accreditation and diffusion of knowledge and its technological manifestations. There was also growing interest among STS scholars of all stripes in examining the relations between scientific and other modes of belief, expression, and power: law, literature, culture, religion, art. Science in non-Western contexts was a late-blooming topic, but was included, for instance, in the 1995 STS handbook (Jasanoff, *et al.* 1995).

With such projects on the rise, older disciplinary divisions no longer made much sense within STS, particularly in the training of young scholars. For example, since the field's research questions frequently cut across historical periods, centering on the nexus of knowledge and power, budding STS scholars saw benefit from exposure to historiography as well as social theory, ethnography as well as metaphysics, and political as well as moral philosophy. The methods used by some of the best-known senior academics in the field were increasingly difficult to localize by discipline. Equally, the work they produced was read across the field as a whole and found its way into many neighboring disciplines. STS books were reviewed in journals running the gamut from *Science* and *Nature* to the *New York Times Book Review* and the *Times Literary Supplement*, with the whole range of the field's professional journals between. The unifying feature in all cases was the *subject* of study, namely, the human investment in science and technology. While many STS researchers could still be characterized as mainly historians, sociologists, or anthropologists, it seemed increasingly appropriate to distinguish them also on grounds of their research fields and theoretical commitments. By the early years of the new century, it became less common to find mature STS scholars who defined themselves in terms of a 'pure' discipline (history, philosophy, sociology, anthropology, politics, economics) applied to a single scientific or technological area (biology, physics, chemistry, engineering, medicine, risk analysis).

To be sure, there was never a complete merger of assumptions and methods across the entire spectrum of STS, any more than there is between identifiable subfields within the most traditional disciplines. Specialties endure and thrive, as in any disciplinary context. For example, boundary-spanning subjects such as risk, scientific evidence, bioethics, or the public understanding of science figure more prominently in the work of modern STS scholars who are descendants of the tradition of concern with the impacts of science and technology; by contrast, historically or philosophically trained STS researchers have tended to look more at the evolution and practices of disciplinary scientific knowledge and technological communities. By the same token, attention to visual representation and instrumentation, widespread in historical and cultural studies of science, is not so common in the work of those with a primary interest in the politics of S&T. Ethnographic approaches have been used more often to study lab cultures and patients' groups than, say, environmental controversies or legal proceedings using scientific evidence. More generally, constructivist theoretical approaches have made greater headway in modern than in historical studies of science and technology, possibly because historical methods are poorly adapted to observations of science in the making. Comparable differences of theory, method, research styles, and topical emphasis, however, may be encountered within the most securely established disciplines.

13.2 Academic institutionalization

Despite its creativity and originality, the branch of STS concerned with the nature and practices of modern S&T was relatively slow to gain a foothold in university structures. In part, this simply reflected the field's growing pains: at the turn of the twenty-first century, there were not many senior scholars of unquestioned eminence whose careers were unambiguously identified with STS. In part, too, the field suffered from the balkanization that sets in when resources are insufficient: seeing no benefit from self-identification in STS, people reverted to better recognized disciplinary affiliations, such as anthropology, history, or sociology, or to topical subfields within STS for which there was current market demand, such as bioethics, environmental studies, or even nanotechnology and society. In turn, such moves militated against the recognition of commonalities that cut across the field, with particularly negative consequences for graduate education, which thrives in a stable environment of accredited teaching centers and job opportunities.

Non-negligibly as well, STS in the 1990s earned a reputation for relativism that evoked distrust from working scientists, other social scientists, and some university administrators. Labeled the 'science wars', a subset of the wider culture wars afflicting the universities, that turbulent period called into question whether constructivist approaches fairly portray progress in science or advances in technology. Although difficult to document, such worries about the field's intellectual soundness and descriptive accuracy, coming at a time when universities were becoming increasingly dependent on their links to science-based industries, may have inhibited the institutionalization of STS in the upper reaches of academia in several Western countries. The widely decried hostility toward science during the US presidency of George W. Bush, coupled with a growing perception that scientific progress and technological innovation are crucial for economic growth, may also have undermined institutional support for scholarship seen as questioning the authority of science.

Until the late 1980s, graduate studies of science and technology in US universities were mostly organized in one of the following ways: departments or programs in the history (and sometimes philosophy) of science and technology (HPST); programs (occasionally departments) in science, technology, and society (STS); and programs in science, technology, and public policy (STPP). These arrangements reflected a number of tacit intellectual boundaries. Historical and modern studies were thought to belong in separate compartments; even at the University of Pennsylvania, where history and sociology of science were nominally represented in the same department, the focus remained on social histories of science and medicine. The frequent pairing of history with philosophy of science reflected a union of interests in these fields around the content of scientific ideas. This alliance worked well for 'internalist' historians, but not so well for those venturing into social and cultural history in order to place scientific ideas in broader contexts. Another implicit boundary sequestered studies of science, technology, and public policy within professional schools, as a supposedly 'applied' field, away from the more 'fundamental' humanities and social sciences (as at Harvard, Michigan, and Wisconsin). A few programs and departments did not respect these divisions, but they existed for the most part at engineering colleges and technical universities, where they did not compete with traditional disciplines. Members of those programs, too, tended to define themselves as historians, sociologists, anthropologists, or political scientists, rather than as representatives of an integrated field of STS.

Two external developments in the mid-1980s helped to rewrite this configuration. First, the processes of global academic exchange brought about closer contact between European and North American scholarship, thereby narrowing the gap between research traditions on the two sides of the Atlantic. Bridges were built between what were on the whole more structuralist approaches to studying the S&T enterprise in the United States and greater concern with philosophical issues in Europe. Second, the US National Science Foundation (NSF) decided to open a nationwide competition to support interdisciplinary graduate training in science and technology studies. Three programs were founded one after another in the early 1990s as a result of this initiative, at the University of California-San Diego (UCSD), Cornell University, and the University of Minnesota. The Cornell grant proved to be unique in that it spurred the establishment in 1991 of a Department of Science and Technology Studies in the College of Arts and Sciences. Created through a merger of the earlier HPST and STS programs, the new department comprised about a dozen faculty members offering both undergraduate and graduate training in STS. By the late 1990s, all three NSF-supported programs had begun to produce doctorates and postdoctoral trainees whose entry into the academic market strengthened the field's profile and foundations.

While large-scale center awards were deemed unnecessary after the first three, the NSF continued to support research-based graduate training in STS on a more modest scale. A program of small grants for training and research (SGTR) provided support for limited numbers of graduate students and postdocs to work on thematically well-defined areas within the field. SGTR recipients in early years included Carnegie Mellon, Cornell, Harvard (JFK School), Minnesota, Oklahoma, and Rensselaer Polytechnic. In addition, the NSF supported conferences and workshops designed to promote curricular innovation and theoretical integration in particular thematic areas, such as diversity in science and engineering, or biology and the law.

Unlike the earlier STS programs, the new science and technology studies maintained strength where it put down solid institutional roots and made gradual inroads elsewhere. Thus, the NSF-funded programs at Cornell, Minnesota, and UCSD added faculty strength over time and, in some cases, branched into new areas of research, such as genomics, information technologies, and nanotechnology. MIT, where the STS program long controlled its own faculty lines, also grew during this period, partly by adopting a new doctoral program, although the faculty remained organized along mostly disciplinary lines, with greatest strengths in history and anthropology. STS departments or programs at some prominent technical universities (e.g. Georgia Tech, Rensselaer Polytechnic, Virginia Polytechnic, University of Virginia School of Engineering and Applied Science) made additional professorial appointments. In the Midwest, the University of Michigan appointed STS scholars in several departments and created an STS undergraduate certificate program. The University of Wisconsin, home to well-established history of science and history of medicine departments, appointed a cluster of STS scholars and established a graduate certificate program in STS. Rapid expansions in research and graduate training at Arizona State University included a build-up of STS scholars and the establishment in 2008 of a doctoral program in the human and social dimensions of science and technology.

During the 1970s, STS became established as a field of specialization in a number of northern European countries (the Netherlands, Scandinavian countries, Switzerland, United Kingdom). Subsequently, from the turn of the century, the European Union began supporting a widening network of universities offering a standardized master's level curriculum in STS, administered through the University of Maastricht in the Netherlands in collaboration with the European Interuniversity Association on Society, Science and Technology. In the same period, the French government added a required component of history and philosophy of science to graduate training in science and technology, while other state-funded initiatives looked to strengthen research and training in STS more broadly. Initiatives in Germany included most importantly the STS graduate programs at the University of Bielefeld, a leading center for interdisciplinary studies. Several southern (and eventually eastern) European countries also began building strength in STS during the 1990s, usually through professional societies and European research collaborations. Japan formed an STS network of its own in 1990, and by the late 1990s actively participated with China, South Korea, and Taiwan in an East Asian STS Network served by its own specialist journal and professional meetings. From the mid-1990s the Chinese Academy of Social Sciences undertook a major effort to publish STS work, often with an emphasis on the impacts and social control of technology.

13.3 Research frontiers

Interdisciplinary research is often driven by questions that call for input from more than one area of study. Policy research is a prime example: to know how best to control greenhouse emissions from automobiles, one needs to know something about the design of cars, the economics of innovation, the dynamics of the automobile market, the impact of incentives on consumer behavior, and the laws regulating air pollution at state and federal levels. No single field or person possesses all the necessary knowledge; collaboration among disciplinary frameworks and their distinctive knowledge systems—on the model of interstate highway construction—is therefore crucial. The development of STS, however, has been driven by different kinds of questions: those that one field sought to appropriate from others, and those that no field had thought to investigate before. In each case, the impetus was the same. The fundamental shift in thought that gave birth to STS was to view scientific and technological production as social domains deserving fine-grained study, and to bring the full-blown apparatus of social analysis, including interpretive methods, to elucidating those dynamics. The results, in research terms, are more consistent with the model of charting the unknown seas for new islands of insight and learning.

Published in 2007, the third edition of *The handbook of science and technology studies* ran to 1080 pages, comprising 38 chapters organized under five topical headings (Hackett *et al.* 2007, cf. Jasanoff *et al.* 1995). The 2008 joint meeting of the European and American societies for STS showcased around a thousand presented papers. Clearly, any attempt to characterize the research frontiers represented by all this activity risks simplification to the point of caricature. Nevertheless, some broad strokes may help convey the unique nature of STS's interdisciplinarity.

Some of the earliest foundations for STS were laid by sociologists and anthropologists who visited laboratories and provided minute but eye-opening accounts of the practices that lead to the creation of facts. The resulting genre of laboratory studies remains a staple of STS, but its focus has widened to include many more dimensions of scientific practice than the moments of significant discovery or revolutionary paradigm shifts that concerned early historians of S&T. One expansion has been in the conception of science itself to include wider domains of systematic knowledge production and technological uptake, from automobile engineering and weapons development to environmental and financial modeling, the creation of markets and fiscal instruments, and varied indicator systems, such as the metrics used to measure scientific productivity. A second direction has been to investigate not just the leading figures associated with breakthroughs and prizes, but also the ordinarily invisible technicians, instrument-makers, nurses, counselors, forensic practitioners, and even patent writers without whose involvement scientific knowledge could not be produced or disseminated beyond the lab or clinic. A third extension was to pay closer attention to the role and standardization of the non-human elements that play a part in the discovery process, from mice to microscopes to microarrays.

A more subtle shift occurred as researchers considered not only the production of new knowledge but also its circulation in society. A seminal history of experimental practices in Restoration England by Shapin and Schaffer (1985) called attention to the importance of credibility and witnessing in the spread of experimental science—themes that these and other authors developed in later work. While many STS researchers addressed the topics of reception and uptake within the perimeters of specific expert communities, subsequent work showed that broader social analysis is needed to understand the authority of science in the modern world. Thus, studies of the public understanding of science (Wynne 1995) and science used in public policy (Jasanoff 1990, 2005) followed science out of its contexts of production into contexts of interpretation and use where science acquired substantial power over human advancement and well-being.

Questions of reception—whether inside or outside the recognized circles of science—are intimately linked to an abiding STS concern with the relationship between science and power, especially in democratic societies. Though this research area has shifted in focus and methodology over more than 40 years, it too provides powerful justification for the existence of STS as a distinct academic field. Salient insights include the following: controversies are productive social moments, offering windows on the ambiguity of scientific observations and the possible existence of alternative interpretations; technological systems are agents of governance because, like laws and social norms, they both enable and constrain behavior; science and technology policies, in both the private and public sector, build on tacit and inarticulate imaginations of publics and what they need; public participation and engagement are essential for ensuring that the imaginations of states and industries will be held to critical scrutiny and democratic oversight. Some of these findings are now so taken for granted that they underwrite operational rules of citizen participation in most technologically advanced societies; others are inchoate and remain to be translated into political and administrative action. STS scholars have become increasingly involved not only in generating knowledge about the relations between science and politics, but also in the translation work needed to convert knowledge to action.

13.4 Outlook for STS: barriers and opportunities

Some 50 years into the life of a new field, and a decade into a new century, STS remains weakly institutionalized in the upper reaches of academia. Despite growing attention to the field's intellectual contributions, there are few fully fledged STS departments in the United States and even fewer in Europe. Departments, moreover, tend to cluster in engineering schools and, with a few exceptions, have not taken hold in high-prestige research universities, where STS has to compete with better established social sciences and humanities. Large hurdles need to be overcome to achieve a stronger institutional presence. They are built into the political economy of the disciplines in contemporary higher education, as well as the field's own contradictory self-understandings. Briefly, there are three challenges of disciplinarity and interdisciplinarity that STS will have to meet before it can take its place as a necessary, indeed indispensable, component of higher education: establishing its relations to its objects of study (science and technology); defining its relations to other analytic disciplines; and asserting its sense of its own boundaries and mission. The good news for the field is that it has the resources to meet all three; the bad news is that STS scholars have not yet chosen, as a community, systematically to tackle any.

First, STS faces the not inconsiderable problem of 'studying up': it presents a classic case of a less established, less accredited field commenting on ones that are far more securely established, generously endowed, and seen as conferring greater public benefits. It is well known that such power differentials affect the content and credibility of academic analysis. With respect to science and technology, in particular, S&T practitioners are often skeptical that anyone not trained in a technical field could have legitimate things to say about that field's workings. Indeed, many of the earliest entrants into STS held postgraduate degrees in science or engineering before becoming professional observers of those fields. Physicists routinely became historians of physics, while biologists took up the historical or sociological study of biology, and engineers became major contributors to the history of technology. Yet, the requirement that one must be formally qualified in a field in order to speak authoritatively about it not only restricts access but also narrows the analyst's imagination and capacity to ask probing questions; an insider perspective develops that does not always accommodate the outsider's questioning gaze. A consequence of this attitude in early STS work was to turn disproportionate attention to the production of scientific knowledge at the expense of understanding better how scientific claims and practices circulate through and are incorporated into society. Only with the emergence of STS as a field of its own is this imbalance between production and reception gradually being righted.

Second, STS has to confront charges of redundancy. STS claims special status as *the* field that observes and interprets the work of science and technology, but this privileged position is by no means universally accepted. Indeed, at many universities the traditional social sciences and humanities are reluctant to concede territory to an autonomous STS. Disciplinary scholars insist more or less openly that the map of existing disciplines provides a sufficient foundation for constructing any of the highways needed for traffic in STS. Thus, it is difficult to persuade a sociologist that STS is not synonymous with the sociology of science, or an anthropologist that there is anything to study about science or technology that is not subsumed in the notion of S&T as cultures. Accordingly, strict disciplinarians argue that there is little value to STS as sovereign academic currency. It unlocks no new doors to research questions or methods, let alone to permanent positions and successful careers. Would-be STS graduate students are often told that they would be better off with a degree in a recognized discipline, with a sideline or subspecialty in studying science or technology. These are, to some degree, self-serving assessments. Few of the disciplines named in the IESBS have been prepared to recognize the study of science and technology as a legitimate specialty within their own intellectual configurations. More usual is the reaction of a political scientist at a major research university who once told me, 'My department would never hire someone in the politics of science'. Regrettably, blocking appointments and degree programs in STS effectively dries up the pipeline of human resources dedicated to comprehensive studies of S&T. University administrators for their part can rarely be counted on to create new conditions of possibility. Faced with interdisciplinary boundary struggles and resource constraints, they are more likely to draw back from the hard work of adjudicating among competing claims, to the disadvantage of any new island in the academic high seas.

Third, it is worth noting that many scholars who see themselves as members of the STS community are hesitant to support disciplining in either sense of that term: either importing order and coherence into the delightfully unruly territory they came to know as STS in the 1970s; or constituting STS as what some have dismissively called a 'high-church', and therefore an elitist and exclusionary academic discipline (Fuller 1993). External funding initiatives, whether from governments or private donors and foundations, can overcome some of these hesitations, to the point of grounding new programs and reviving old ones (e.g. at Cornell, UCSD, and Wisconsin in the United States). Forging new transdisciplinary identities, however, demands an intensity of effort and engagement that may seem unnecessary or excessive to academics whose own histories are discipline based. Even the most secure STS programs in the United States and elsewhere have endured identity crises at points in their development; at such times, moreover, new fields are substantially more likely than old ones to succumb to abiding administrative pressures for efficiency and cost-cutting.

Fields demand organization for their survival and continuity, both to demarcate them from neighboring territories and to set up internal markers by which to measure such academically essential attributes as originality, quality, progress, and contributions to fundamental knowledge. Yet in a field's emergent, formative phase, attempts to develop a curriculum, create a canon, evaluate students and faculty for professional advancement, or even represent the field in an encyclopedia or handbook all arouse high tension and anxiety. Who will be brought in and celebrated; who will be left out? Many therefore prefer the quieter option, which is to retain STS as a loosely constructed society to which anyone with a passing interest can gain easy entry. This broad-church approach satisfies liberal academics' often deep-seated desire for intellectual democracy, but it also gets in the way of critical stock-taking, meaningful theorizing, and methodological innovation—in short, of *disciplining*. In this respect, STS operates as its own most effective critic, and it ratifies a status quo that militates against the field's maturation as a self-defining, self-governing area of inquiry.

13.5 Conclusion

The problem of interdisciplinarity is often posed as one of harmonization, or bringing disparate perspectives into alignment so that different discourses can speak productively with one another. Much as independent nation states have trouble marrying their divergent interests and political cultures into agreements on common problems, so the traditional disciplines encounter frictions in their efforts to focus on phenomena—from climate change to the roiling of global financial markets—that seems to demand investigation from multiple perspectives. How should number-crunchers speak to qualitative analysts, or critical theorists engage with advocates of game theory and rational choice? How should inductive, evidence-based, and practice-oriented scholarship find common ground with principled approaches that draw authority from historical texts and frameworks that seem to have little bearing on the issues of the present? Is integration possible and desirable, as in behavioral science or area studies (see Klein, Chapter 2, this volume), or are exchange and bridge-building the only realistic alternatives? And who decides when and by what criteria participants in an interdisciplinary venture have made sufficient contributions to the purposes of the academy to merit their own charter of independence?

STS has encountered all of these problems, and to some extent coped with them, but in a structural context that makes the field's challenges larger and more consequential than those of interdisciplinarity more generally. For what is at stake in the success of STS is the underlying self-understanding of the disciplines themselves as coherent and unified entities. By contesting such dominant understandings, as a field with epistemology as its primary focus *must* do, STS enters into troubled and uncertain territory. In the terms sketched here, the future of STS depends on redrawing the map of the disciplines to demonstrate that they are all islands of happenstance, with unmapped waters between; STS can then claim a space for itself as another fertile territory in these wide waters, offering resources for understanding some of humanity's most impressive accomplishments, but without threatening anything achieved, or yet to be achieved, in other quarters of the disciplinary archipelago. What is needed to make this case, first and foremost, is an abiding conviction on the part of STS-islanders that they have shared crafts and practices, and valuable goods to offer, in the ongoing enterprises of pedagogy and scholarship. There are major obstacles to achieving such agreement, both internal and external to the field. Equally, however, there are growing numbers of ambassadors abroad who confidently wear the badge of STS as their primary academic credential. Their diplomatic acumen and intellectual ambition will define the future of the field.

References

Bloor, D. (1976). *Knowledge and social imagery*. Henley: Routledge and Kegan Paul.

Carson, R. (1962). *Silent spring*. New York: Houghton Mifflin.

Fuller, S. (1993). *Philosophy, rhetoric and the end of knowledge: the coming of science and technology studies*. Madison: University of Wisconsin Press.

Hackett, E.J., Amsterdamska, O., Lynch, M., and Wajcman, J. (eds) (2007). *Handbook of science and technology studies*, 3rd edn. Cambridge, MA: MIT Press.

Haraway, D.J. (1989). *Primate visions: gender, race, and nature in the world of modern science*. New York: Routledge.

Jasanoff, S. (1990). *The fifth branch: science advisers as policymakers*. Cambridge, MA: Harvard University Press.

Jasanoff, S. (2005). *Designs on nature: science and democracy in Europe and the United States*. Princeton, NJ: Princeton University Press.

Jasanoff, S., Markle, G., Petersen, J., and Pinch, T. (eds) (1995). *Handbook of science and technology studies*. Thousand Oaks, CA: Sage Publications.

Keller, E.F. (1986). *Reflections on gender and science*. New Haven, CT: Yale University Press.

Kuhn, T.B. (1962). *The structure of scientific revolutions*. Chicago: University of Chicago Press.

Latour, B. (1987). *Science in action: how to follow scientists and engineers through society*. Cambridge, MA: Harvard University Press.

Latour, B and Woolgar, S. (1979). *Laboratory life: the social construction of scientific facts*. Los Angeles: Sage.

Shapin, S. and Schaffer, S. (1985). *Leviathan and the air-pump: Hobbes, Boyle, and the experimental life*. Princeton, NJ: Princeton University Press.

Smelser, N.J. and Baltes, P.B. (eds) (2001). *International encyclopedia of the social and behavioral sciences*, 26 vols. Amsterdam: Elsevier.

Wynne, B. (1995). Public understanding of science. In: S. Jasanoff, G. Markle, J. Petersen, and T. Pinch (eds) *Handbook of science and technology studies*, pp. 361–88. Thousand Oaks, CA: Sage Publications.

CHAPTER 14

Humanities and technology in the Information Age

CATHY N. DAVIDSON

Although the rhetoric of the humanities is often about traditionalism, historical continuity, and foundational stability, the humanities have a remarkable capacity to change and grow over time. Not just new paradigms, but whole new areas of study are constantly changing the landscape of humanistic knowledge. Today, the interdisciplinary field that may best symbolize the ability of humanists to grapple with key concerns of an era is a form of technological humanism that spans all of the discrete humanistic departments and fields as well as the interpretive social sciences while also reaching far into other divisions of the contemporary university—into engineering, technology, computational sciences, industrial design, natural sciences, business, law, and medicine. Indeed, the interdisciplinary humanistic field that is addressing the new arrangements of intellectual, social, political, and economic life in the Information Age is as complex and dynamic as the Information Age that is its object of study.

There is no consensus about what this interdisciplinary endeavor should be named. Donna Haraway calls it 'technohumanism', a useful descriptor that will be used throughout this chapter (Haraway 2003). Others, however, place the emphasis elsewhere. Richard Miller dubs it simply the 'new humanities' (Miller and Spellmayer 2006). Timothy Lenoir's term is 'critical studies in new media' while Henry Jenkins addresses 'comparative media studies' (Lenoir 2002; Jenkins 2006). A new program at Concordia University in Montreal prefers a different neologism: 'program in intermedia'. The interdisciplinary program at Duke University has a different titular emphasis: 'information science + information studies' (ISIS). And the John D. and Catherine T. MacArthur Foundation has developed a $50 million initiative to build the new interdisciplinary field that they call 'digital media and learning'.

In 2002, scholars at Duke began talking about a new kind of academic organization suitable for such a transdomain vision, and worked together to create a virtual organization, a network of networks called the Humanities, Arts, Science, and Technology Advanced Collaboratory, or HASTAC.[1] By pronouncing the acronym 'haystack', the name was meant to signify

[1] HASTAC was co-founded by Cathy M. Davidson and David Theo Goldberg, Director of the University of California's system-wide Humanities Research Institute. The first meeting was held at UCHRI in 2003. The infrastructure for HASTAC is supported largely at Duke University where the first international conference was held in May 2007.

a collective, collaborative interdisciplinary model of humanistic-based learning (Davidson and Goldberg 2004). Branding its online information commons 'Needle' (<http://www.hastac.org/needle>), HASTAC also meant to signal the centrality of pointed and purposeful critique as an essential and defining element of the contemporary humanities, especially the theorizing of social and cultural issues of race, gender, class, region, ethnicity, sexuality, nationalism, and transnationalism, all of which have an impact on technology and which, in turn, have been transformed by the World Wide Web and its manifold affordances.

No name ever encompasses a field, either at its moment of inception or in its evolution over time. However, names do serve as important historical reference points, pinpointing the converging energies of a given intellectual moment. Names of new fields provide us with 'keywords' in Raymond Williams' sense of a 'vocabulary of culture and society', a snapshot of pertinent issues not yet crystallized into an existing disciplinary formation (Williams 1976). As Williams also notes, keywords similarly define lacunae as well as demarcating that which is tangential, intersectional, or even orthogonal to the new field.

Unfortunately, none of the terms currently in play admits of a value-free relationship between the constituent parts. 'Technohumanism', the rubric selected for this chapter, still privileges 'technology' and 'humanism' over other components (such as the social sciences and arts). With time, it may well turn out that a term such as 'technology studies' (by analogy to 'science studies') or 'history of technology' (analogous to 'history of science') will gain widespread acceptance. At present, the neologism 'technohumanism' is adopted as a rubric for a series of interests, problematics, questions, and concerns of interest in this historical moment. That the term is a neologism may well serve to underscore how new and transitional this moment is, and that no term fully comprehends all of the disciplinary components or interdisciplinary cross-currents currently shaping the larger discourse.

Technohumanism's closest—if orthogonal—interdisciplinary field is 'digital humanities', previously and sometimes still referred to as 'humanities computing'. Although some scholars make distinctions between those terms, most agree that digital humanities', as a term, has evolved from the earlier 'humanities computing' and is now the more prevalent usage (and, for most purposes, interchangeable with it). Digital humanities is dedicated to the integration of the newest forms of technology into scholarly humanities disciplines and includes the wide range of practices, analyses, and methods deriving from the digitizing of textual and multimedia archives. Digital humanities involves representation, analysis, manipulation, interpretation, and investigation of humanistic knowledge while using computational media ranging from databases and digital archives in literature, visualization or sonification in art or music history, or GPS in archeology. Digital humanities has its roots in the long history of bibliographic methods going back to the great bibliographers of the nineteenth century, in philological and archival traditions, and in newer fields including library and information science (McCarty 2005).

As an example of the capaciousness of digital humanities one might consider the call for papers for the 2007 conference sponsored by the Alliance of Digital Humanities Organizations. That call lists several 'suitable subjects' including 'text analysis, corpora, corpus linguistics, language processing, language learning, delivery and management of humanities digital resources, collaboration between libraries and scholars in the creation, delivery, and management of humanities digital resources, computer-based research and computing applications in all areas of literary, linguistic, cultural, and historical studies, use of computation in such areas as the arts, architecture, music, film, theatre, new media, and other areas reflecting our cultural heritage', as well as the 'role of digital humanities in academic curricula' (<http://digitalhumanities.org/dh2007/cfp/>). A call for the 2009 Digital Humanities Conference, designed in part to celebrate the tenth anniversary of the Maryland Institute for Technology in the Humanities (MITH), one of the most distinguished digital humanities centers, emphasizes 'early adopters' doing innovative 'new work on tools, text analysis, electronic editing, virtual worlds, digital preservation, and cyberinfrastructure' (<http://www.mith2.umd.edu/dh09/?page_id=2>).

As these conference agendas make clear, digital humanities uses new computational tools to help in the analysis and interpretation of what are often considered to be traditional or field-based humanistic objects of study (Schreibman *et al.* 2008). By contrast, technohumanism uses theoretically inflected humanistic critique to explicate the role and implications of technology in all aspects of human life and, indeed, in the environment as well. While technohumanists, like digital humanists, may well develop and utilize cutting-edge computational tools as part of their mission, the critical objectives, scope, and objects of study in these fields vary significantly.

As has happened in many other interdisciplinary fields, differences among and between these different versions of the new humanities in the Information Age may well change over time. Certainly, there are signs of confluence, such as the recent appearance of Planet DHASS (<http://hass.informatics.uiuc.edu/>), a metablog that collects through RSS feeds the contents of dozens of blogs in the digital humanities, arts, and social sciences as well as blogs from HASTAC and other sites dedicated to technohumanism. Similarly a journal such as *Vectors: Journal of Culture and Technology in a Dynamic Vernacular* perfectly encapsulates the areas of overlap.[2] Most recently, in 2008 the National Endowment for the Humanities established a Division of Digital Humanities, an effort that also signals the coming together of many different energies within the many versions of technohumanism and digital humanities.

Defining technohumanism requires not only identifying its constituent features and its intellectual ambitions, but also understanding the time and place in which it was conceived. In the 1990 essay 'The emergence of cultural studies and the crisis of the humanities', a retrospective account of the origins of the interdisciplinary field of cultural studies at the University of Birmingham, UK, in the mid-1960s, Stuart Hall emphasizes that the establishment of any new field must be situated within the political, theoretical, educational, and economic circumstances from which it arises (Hall 1990). To understand the birth of a new interdisciplinary field, then, is to be cognizant of those historical and institutional exigencies and urgencies that inspired it and against which it responds. Technohumanism is emerging as a field now, in the early twenty-first century, because, as

[2] *Vectors: Journal of Culture and Technology in a Dynamic Vernacular* is edited by Tara McPherson, one of the earliest HASTAC leaders, and is part of the HASTAC consortium. It is also perfectly encapsulates the areas of overlap by using all available existing technologies to present non-linear, multimedia scholarly articles, largely in the humanities and interpretive social sciences, and that are clearly committed to cultural critique, including on issues of race, class, gender, sexuality, and so forth. Each article is co-designed by a scholar and a working media artist or multimedia designer further strengthening technohumanism's tie to the arts. See also McPherson (2008).

historian Robert Darnton notes, it is clear that the current transformations of culture and society are epistemic. Darnton (2008) argues that the world has only seen four great Information Ages. He defines the first happening with the beginning of writing systems in the Middle East in around 4000 BCE. The second is the invention of movable type in China in the tenth century CE and, with Gutenberg, in Europe in the fifteenth century. He sees the democratization of mass printing and mass literacy in the West in the late eighteenth- to mid-nineteenth centuries as the third great Information Age. The present Information Age, he argues, is by far the most influential, rapid, extensive, and global in impact and nature. The technological changes of the last decade are so vast that they are affecting social, political, cultural, scientific, and economic arrangements worldwide.

Seen within this larger frame, the Information Age is less significant for its technology than for its rearrangement of all of the aspects of human life with which the humanities concern themselves. In literary fields, this might include such crucial issues as narrative, authorship, publication, and the creation of new multimedia and interactive imaginative and virtual worlds (environments such as Second Life, for example, but also narrative games, fantasy games, and other imaginative virtual spaces). In linguistics, the social codes embedded in computer code are a ripe new area of study. So is careful analysis, from a multicultural perspective, of the cultural and scientific assumptions about mind, nature, logic, cognition, and categorization that form the basis of artificial intelligence as well as hypertext and other markup languages. These are also key issues in philosophy. Finally, virtually all of the arrangements in the arts and music are changing with new technologies. History puts all of these vast and various changes into perspective.

Pedagogically, technohumanism also takes the World Wide Web and its potentials for interactive, non-hierarchical learning as part of its mission. The 2008 HASTAC/MacArthur Foundation Digital Media and Learning Competition devised a label for this pedagogical method: participatory learning. Some might also call it Learning 2.0, playing off media prognosticator Tim O'Reilly's famous definition of social networking and other collaborative sites as 'Web 2.0' (O'Reilly 2005). Participatory learning includes the various ways that learners (of any age) use new technologies to share research, data, and ideas, in the classroom or in their scholarship. Unlike traditional humanistic pedagogies, participatory learning is a collaborative, interactive, and non-hierarchical version of authorship. This form of collaborative, interactive authorship yields what the research group at the Massachusetts Institute of Technology, led by the eminent media theorist Henry Jenkins, calls participatory culture or, more recently, 'spreadable media', that form of interactive media that 'generates active commitment from the audience'. Spreadable media empowers users to create new online and interconnected communities and to interact with others in forms (such as fandom or shared social, political, or intellectual interests) that they value (Jenkins *et al.* 2009).

As Hall reminds us, a new interdisciplinary field not only fills existing gaps or takes advantage of new possibilities but sometimes, quite overtly, seeks to remedy, redress, respond to, or in other ways compensate for lacks, problems, rigidities, blindspots, and incapacities inherent in existing or traditional disciplinary structures, omissions made evident by the unfilled needs at a specific historical moment. Technohumanism requires the concerted collaborative research, thinking, and teaching of those in quite literally every division of the university. It requires a new methodology—the 'collaboratory'—where participants combine aspects of wet-bench lab culture in the sciences with the

Vectors: an interdisciplinary digital journal

Tara McPherson

For over 5 years, the multimedia journal *Vectors* (<http://www.vectorsjournal.org/>) has served as a test bed for interdisciplinary digital scholarship. In starting the journal, we aimed to create a sustained space of experimentation with emergent modes of multimodal scholarship and to explore the value of deep interdisciplinary collaboration for humanities scholars. In particular, we focused on the potential for new visual, affective, sensory, and computational aspects of humanities research and on the possibilities gleaned from rich collaboration across diverse skill sets.

The journal grew out of conversations among faculty and staff at USC's Institute for Multimedia Literacy in the late 1990s, where experiments in digital pedagogy were well under way. In that milieu, we came to understand that there were virtually no warranted spaces (like a peer-reviewed journal) where scholars could publish work realized in the expressive and interactive languages of new media. We continue to push far beyond the 'text with pictures' format of much online scholarly publishing and continue to encourage work that takes full advantage of the multimodal and networked capacities of computing technologies. Simply put, *Vectors* doesn't publish work that can exist in print.

We recognized too that many scholars were interested in undertaking such work but hardly had the time or support structures they would need to create their own interactive scholarship. In launching the journal, we designed a fellowship model to pair scholars with first-rate designers and programmers. Since 2004, groups of 10 to 12 *Vectors* fellows have come together for a week-long summer seminar during which the potentials for multimodal humanities research is discussed and explored. The residency period is then followed by a sustained and iterative cycle of collaboration between the *Vectors*' fellow and a subset of our team (typically comprising a designer, a programmer, and an editor). This collaborative process has been especially crucial to our ability to work with humanities scholars who have limited or no background in creating digital scholarship.

The interdisciplinarity layered into the *Vectors* process is two-fold. First, the journal's content is itself diverse, bringing together scholars from various disciplines for theme-based issues. While it might have been easier to lodge the journal firmly in one field, we felt we would learn more about the potentials and challenges of digital scholarship by bringing together scholars from different departments. By working across the interpretative humanities and 'soft' social sciences, we have been able to zoom out to several large questions that cut across multiple fields and to begin to think through what types of digital platforms and tools might generalize across the humanities.

For instance, in a summer residency with 'evidence' as one of its central themes, scholars from literary studies, sociology, art, and performance together interrogated the status of evidence in each of their disciplines. These discussions directly impacted the second mode of interdisciplinary work at *Vectors*: our structuring of diverse development teams. These development teams bring together in equal collaboration interactive designers, scholars, and programmers in a sustained process of iteration and exchange. This process offers humanities scholars a supportive environment in which they can explore the affordances of the computer for their research endeavors, while also providing the designers and programmers with new insight into the central questions motivating humanities research. Through this process, our team has learned new ways to scaffold the digital humanities.

In close collaboration with humanities scholars, the *Vectors* design team began to develop a relational database tool that better supported the types of evidence our scholars were most interested in exploring. This tool was developed in a bottom-up mode, beginning from our conversations with humanists about how they might re-imagine their scholarship in a digital vernacular. Rather than build a tool and then 'sell' it to the scholars, we listened carefully and experimented with scholars in collaborative teams to generate a middleware package that could address the needs of contemporary methodologies in the humanities. We learned a great deal from these collaborations. In turn, these tools aided and also changed the relationship that scholars had to their work and to digital environments. As a team, we have seen the transformative role such tools can play for humanities scholars as they begin to imagine next-generation, interactive scholarship.

The scholars we have partnered have found that a fusion of scholarly writing with database practices has resulted not only in a deeper understanding of, or new approaches to, their own research, but also in a scholarly endeavor that proved to be one of the most intellectually satisfying of their careers. They have discovered new contours and nuances in their work through the restructuring afforded by a database environment and have also experimented with new forms of argument and expression. The ability to deploy new experiential, emotional, and even tactile aspects of argument and expression has opened up fresh avenues of inquiry and research.

At *Vectors*, successes have emerged from a careful calibration of peer-to-peer collaboration and also of 'scholar-to-machine' collaboration. The growth and evolution that our scholars' experience result both from their engagement with database forms and their deep interaction and partnership with technologists, artists, and interactive designers. These two forms of collaboration are deeply intertwined and support one another. They begin in our summer workshop and continue once our fellows have returned to their home institutions, with each project spanning 4 to 5 months.

Such a process is profoundly interdisciplinary, bringing together a team with several diverse and distinct skills. While we sometimes encounter a humanities scholar able to realize interactive work without the support of an interdisciplinary team, the work process we have experimented with might also serve as a model for modes of research in the humanities that push beyond the 'single scholar stereotype' that so characterizes the humanities. This team-based model may have much to offer the humanities.

participatory online elements of Web 2.0. Like many transformative interdisciplinary fields, technohumanism realigns research, teaching, and publishing. It also necessitates a rethinking (as Hall implies) of the disciplinary compartmentalization of knowledge that has been instantiated within and rewarded by the research university as an institution. Finally, technohumanism requires a breaching of the barriers between the divisions of the university and, as we shall see, a crossing of the 'two cultures' of the arts and the sciences that have been a hallmark of Western thought for the last 100 years (Snow 1959).

14.1 Technohumanism, academe, and education

In 2003, at the time the HASTAC academic network was formed, not a single school of education had significantly altered its curriculum to address the new forms of participatory, engaged, online collaborative learning enabled by the Internet and its digital affordances. There were, however, individual, isolated, and embryonic attempts (typically a single faculty member or an innovative interdisciplinary program) to rethink how students could be more informed, careful, and introspective about the social and intellectual implications and limitations of their digital lives. Many of these thinkers were, in fact, in the humanities. Among them were Henry Jenkins (MIT), Kathleen Woodward (University of Washington), Donna Haraway (UC Santa Cruz), Timothy Lenoir (Stanford), Jeffrey Schapp (Stanford), Katherine Hayles (UC Los Angeles), and Tara McPherson (University of Southern California). There were, of course, many others as well, but these were some of the faculty in the humanities engaged in thinking through the critical humanistic issues of technology in culture and society.

The schools most invested in the radical transformations of the Information Age were schools of library science. In fact, in the 1990s, many of those schools renamed themselves schools of information science. Trying to keep pace with the rapidity of the technological changes, many universities then took the expedient of placing offices of information technology (IT) either directly under or directly over the school of information science within the university hierarchy. This was beneficial in that it allowed libraries to promote expertise in digital media, digital archiving, and a host of related issues of data and meta-data.

However, some would argue that the rapid transition from library schools to schools of information science had an unanticipated consequence as well: it cordoned off the research and thinking about technology as a pre-professional specialization, a subject important to those wishing to go into library science as a profession but not to university education more generally. Even the development of instructional technology within schools of information had the unanticipated consequence, in many instances, of being developed apart from the faculty in arts and sciences who were teaching the majority of undergraduate students. One interesting early exception to this pattern was in the more traditional digital humanities. Here, there was a confluence of archival interests, humanistic bibliographic interests, and technological interest and ability.

At the turn into the twenty-first century, very few universities were attending in research and teaching programs to the massive changes in social, economic, scientific, artistic, cultural, and political arrangements afforded by computational innovation. It was rare to find interdisciplinary programs dedicated to studying the implications of a digital world in anything like a systemic, interdisciplinary, and sustained way. There were occasional courses in individual departments but very little crossing of departmental boundaries, and most of these courses were considered 'add ons' to other departmental offerings. They were neither requirements within departments nor by university-wide distribution.

This is starting to change. Technohumanism (by any name) is beginning to play a more prominent role as an interdisciplinary research field at least partly in response to the demands of entering students. Given the importance of the desktop computer (whose official 'birthdate' is typically said to be 1985) to this generation of students, part of the mission of technohumanism is to understand how new technologies help us to learn and, concomitantly, how learning through new technologies, from preschool on, changes how youth today understand the world. While IT is evident in every aspect of the university campus, academe (and formal education in general) has been slower than commercial industries in exploiting the new forms of learning and the new skills of young learners.

Technohumanism thus addresses the need for structural changes in educational institutions, the importance of new curricula, the possibilities of collaborative learning in virtual environments, and the need for radical interdisciplinary restructuring of the academy. Similarly, new forms of teacher training—at all levels—are crucial if there is going to be a pedagogy for a new global, participatory form of learning (Brown *et al.* 2008; Davidson and Goldberg 2009).

In *Facilitating interdisciplinary research*, a comprehensive report by the National Academies of Science, four drivers are listed for interdisciplinarity: the inherent complexity of nature and society; the desire to explore problems and questions that are not confined to a single discipline; the need to solve societal problems; the power of new technologies (Committee on Facilitating Interdisciplinary Research *et al.* pp. 2, 40). Technohumanism would also add a fifth driver: the desire for an interactive, collaborative, participatory method of research and learning that capitalizes upon the power of new technologies and the customizing skills that youth bring to the college classroom today.

One particularly notable feature of technohumanism is its overt attempt to cross the divide of the 'two cultures' that C. P. Snow famously mapped long ago (Snow 1959). For Snow, there was a virtually unbridgeable gap between the world of the arts, humanities, and interpretive social sciences on the one side and, on the other side of the divide, the world of science and technology.

Snow demarcates this shift as beginning in the late nineteenth century. Until then, divisions between the 'scientific' and 'humanistic' were by no means fixed. Isaac Newton, for example, was an astronomer, a physicist, a mathematician, a theologian, a philosopher, and an alchemist—with no contradiction across those domains. Galileo, too, was a mathematician, astronomer, and a philosopher. In the mid-nineteenth century, Charles Darwin's Cambridge degree was in theology and his motivation to study biological diversity came as much from his abolitionist leanings as his scientific ones. Even in the early twentieth century, Einstein credited philosopher David Hume, as well as contemporary physics, for his thinking about relativity.

However, since the late nineteenth century formal education has strongly reinforced a divide between science and the arts. The research university has contributed by promoting an increasingly fragmented curriculum and methods of training. Along with the schism, there has been a value judgment, with more and more weight being placed on the scientific versus the humanistic and artistic side of the disciplinary equation. There is a hierarchy, with science at the top of the intellectual heap. One manifestation of this disparate valuation of the scientific is in the ways more and more areas of the social sciences have sought to define their methods as 'scientific'. Another is seen in the application of so-called scientific (and often, pseudo- or quasi-scientific) assessment and evaluation measures for education, from K-12 (primary and secondary) to university level. Unquestionably, for several decades institutional power and cultural authority has accrued to the quantitative side of the 'two cultures' equation. Even now there is a clear divide in digital humanities between the more technology-oriented scholars and others who are interested in the social and cultural implications of technology but have no interest or expertise in developing technological skills of their own. There remain, even within the interdisciplinary field, differing expectations of technical literacy.

Technohumanism is certainly not going to rectify a balance that has tipped too much in one direction for the last 100 years. However, technohumanism does require interdisciplinary revaluing, relaying, and remixing across, between, and among opposite areas in this cultural divide. Indeed, one of its pedagogical motivations is a conviction that youth today, especially those born after 1991 (the official 'birthdate' of the commercially available Internet), do not, as a matter of everyday and informal learning, intuitively make the distinction between 'art' and 'science'. In contemporary customizing digital media culture, a young person might, for example, be writing code for a multiplayer game or for a better interface on a MySpace page one moment and, the next, designing a new, fanciful avatar for Second Life. To which side of the 'two cultures' divide does that belong?

This question was given specific intellectual heft in a recent online public forum sponsored by the HASTAC Scholars, a network of 55 undergraduate and graduate students across the country who are engaged in forward-looking conversation on various topics of interest to technohumanists. The subject of a recent public forum was 'Metaverses and scholarly collaboration'. The forum was presented on the HASTAC website in a Seesmic-enabled vlog-to-vlog format as well as in a text-based linear blog format, by the University of Texas graduate student Ana Boaventura. The topics in this online forum ran the gamut from the role of empathy in the role-playing interactions among avatars in Second Life; to the social situations necessary for the sponsorship, sustaining, creation, preservation, and display of digital art (in Second Life and also at a concurrent exhibit in Beijing, China); to the importance of 'glitches' in examining the shortcomings and benefits of technology and of face-to-face pedagogy; to issues of interoperability with other Internet applications and innovative end-user licensing agreements and other intellectual property matters; to concerns with unpaid consumer labor and corporate capitalism; to issues of race, gender, class, and sexuality played out in virtual role-playing environments. To which side of the 'two cultures' debate do such extended and intertwined intellectual categories belong?

Another example serves to illustrate the role of the arts in this interdisciplinary field. One winner of an award in the 2007 HASTAC/MacArthur Foundation Digital Media and Learning Competition is the Mobile Musical Networks project, based at Princeton University. The principal investigators of this project are Daniel Trueman, a professor of music composition, and Perry Cook, professor of computer science. Working together with students in music and computer science, they are collaboratively developing networked, portable musical laptops that allow musicians all over the world to co-compose, improvise, and perform online together simultaneously and also to customize the laptop 'instruments' together themselves. To accomplish this requires not only rethinking music composition and principles of composition but also solving the engineering problem of time lag in audio transmission (a phenomenon with which we are all familiar from newscasts between people in different parts of the world; the picture is visible before the sound is transmitted). Faculty and students working on this project collaborate on designing the expressive mobile musical laboratory, and also do research on musical acoustics, social networking, instrument design, human–computer interfacing, procedural programming, signal processing, music history, and musical aesthetics. As with the previous examples, one realizes that, with such a collaboration, the scientific/art divide is crossed; the hierarchy of science *over* art is leveled; and the hierarchy of teacher and student is also more balanced since co-development and co-critique are incorporated into the learning process.

Musical performance and writing code go hand-in-hand. Thinking and doing also go hand-in-hand. Projecting future applications of their project requires humanistic and sociological examination of principles. Who will own the compositions produced collectively? What is 'authorship' when a network creates a musical composition together? Will these compositions be archived, produced, manufactured, marketed, and distributed as albums or downloads? If one becomes profitable, how will the profits be dispersed? What if musical laptops become the next Wii, a highly profitable technology game system? Who will capitalize that venture, and under what IP rules? Will this be high-end audio equipment that symphonies around the world might use, or scalable for use on cell phones so kids can network music together? These social, economic, and intellectual issues are all *consequences* of digitality, and thus key to the interdisciplinary technohumanist's research and pedagogical concerns.

14.2 Technohumanism, interdisciplinary programs, and administrative barriers: what works, what does not

Technohumanism requires the collaborative efforts of a dizzying array of other fields that span not only departments but also the traditional administrative divisions of the university. 'Divisions' are defined differently at different institutions and, as David Scholle has reminded us, inter- and intradisciplinary challenges are not only constructed differently depending on different institutional structures but, in turn, structure the forms of intellectual work that can occur across and within departments and divisions (Scholle 1995). At most research universities, divisions serve to aggregate disciplines and departments as well as interdisciplinary programs into distinct and separate organizational units (sometimes called 'silos'): the arts and humanities; the social sciences; the natural sciences and engineering; and then, in parallel and overlapping but distinctive relationship to the divisions, the various professional schools (such as law, medicine, business, divinity, and so forth). Colleges and universities have elaborate administrative and financial structures supporting these silos and individuals (typically, deans) whose responsibility it is to maintain the excellence, the mission, and the bottom-line of their particular silo.

The divisional structure poses special obstacles to interdisciplinarity. To do its job, technohumanism requires, for example, partnerships, trades, and shared responsibility across the silos of departments and schools. It requires administrative oversight one level up, in the office of a provost (or whoever at a university serves as the chief academic officer presiding over all the educational units). Such issues as distribution of indirect costs from federal grants across school budgets, infrastructure costs, reporting lines, accreditation, evaluation procedures, and disparate requirements for tenure and promotion in different schools or departments within schools all have an impact on the organization of such radically interdisciplinary and interdivisional programs and on the faculty they are able to attract.

Another case study might be useful to illustrate the range of obstacles such cross-divisional programs face. When Duke University created ISIS, the reach was so extensive that an entirely new structure had to be developed. In 2000, ISIS became the university's first freestanding interdisciplinary program offering an undergraduate certificate, essentially an interdisciplinary minor. Previously interdisciplinary certificate programs reported to a specific department. Creating ISIS prompted the rethinking of various forms of support, curricular matters concerning cross-listed courses (and which department would get the credit for which enrollments), faculty rewards for faculty, distribution requirements for students, and so forth. Although it took a few years to work these issues through, they eventually became precedents (as was feared by some and hoped for by others) for other programs. At the same time, as with any program that crosses division of universities, the accounting of student hours and faculty full-time equivalences (FTEs) proved almost impossibly difficult. On any dean's balance sheet, the cross-divisional program is an anomaly—and, for deans, the anomalous is rarely a good thing. For ISIS, the upshot was that, once it was no longer under the protective eye of the administrator who oversaw its creation, ISIS was moved into a more conventional structure, one that required reporting to just one dean (not several), the Dean of the College of Arts and Sciences. Interestingly, however, because of the success of the program, it is currently being considered, along with a program in visual studies, for being named as its own independent, cross-divisional institute, with separate hiring lines, bylaws, reward structures, and so forth. There are currently sixth of these 'über-institutes' at Duke and this may well become a seventh.

These narratives of success, failure, compromise, change, and complication are, of course, familiar to anyone pioneering interdisciplinary academic structures. Technohumanism is challenging in the double sense that it is challenging to organize and it challenges existing administrative structures. It requires collaboration across areas of the university that rarely speak to one another and often do not even find themselves in the same college, never mind the same room.

14.3 Needs and prospects: the role of critique in an interdisciplinary field

Most fields do not define themselves by their ability to critique their own chosen object of study. Yet a crucial function of new interdisciplinary technohumanism is exactly that, to provide critique of technology and its operations historically as well as in contemporary intellectual, educational, and political life and, concomitantly, to use the new arrangements allowed by digitality to highlight areas of the humanities that seem narrow, provincial, outmoded, Eurocentric, or ill-conceived (when judged against new paradigms of access, for example). For example, as Timothy Lenoir and Henry Lowood have shown, military training simulations depend on first-person shooter games, but the military actually uses the appeal of such games to recruit young men and women into the military (Lenoir and Lowood 2000). To extend Edward Castronova's caution about gaming and other virtual realms to all digital spaces, one needs to be cautionary, as well as celebratory, when engaged in any project where the object of study is an 'expansive, world-like large group environment made by humans, for humans' but which is 'maintained, recorded and rendered by a computer' (Castronova 2005, p. 11). A range of ethical and practical issues (including privacy, security, identity, and intellectual property) arise as we consider human interaction in digital spaces that can be mined for data, commercialized, or otherwise exploited.

To return to Stuart Hall's point about interdisciplinary fields emerging in specific historical moments, it is clear that digital technology is going to continue to transform our lives, which makes it important for humanists—who are, after all, trained in cultural interpretation and critique—to apply their methods of close reading, historical perspective, social engagement, and linguistic attentiveness to new technologies. A September 2008 study by the Pew Internet and American Life Project, the first major ethnographic study of youth and games, showed that 97% of Americans between the ages of 12 and 17 play digital games. The study was controlled for gender, class, race, region, and used the most sophisticated of procedures for making statistical inferences. It found that youth who play games are more, not less, likely to be readers, to have friends, to engage in social life that involves game-playing with their friends, to have interests in science and technology as well as in literary, narrative, and artistic interests, and are also likely to be interested in social and civic engagement (Kahne *et al.* 2008). This is no longer a social problem; it is a *changed environment*. Understanding that changed environment in all of its dimensions is one goal of technohumanism.

14.4 Conclusion

The potentials for abuse of new technology, as well as the potentials for positive transformation, loom so large that it requires scholars in many fields thinking together to understand the implications of our age. It also requires humanists moving out of their comfortable disciplinary niches to assay the interdisciplinary scope of technology's impact. Technohumanism requires collaborative thinking from the development of new tools all the way through to the implementation stage, from ideas to application. We are posed to answer, collectively, one of the most complex and important questions of our time: how do we collaborate, from the beginning, across domains so we are not creating a technology whose potential for abuse is greater than our capacity for limiting potential abuses?

Anne Balsamo, a leading theorist of technohumanism, has coined the term 'epistemological humility' (Balsamo, at a SECT seminar in 2006) to address the intellectual predisposition necessary for radical interdisciplinarity to succeed. For collaboration to be fully successful, one must begin from the assumption that one's own definition of what counts as 'knowledge' may not be the right or the only definition. Such collaboration requires accepting that there are reasons (practical, historical, philosophical, or simply traditional) for the practices of another field, including those that seem most antithetical (or even annoying) to one's own practices.

Radical transdomain interdisciplinarity across the humanities, arts, social and natural sciences, engineering, and technology requires translation of the most minute and the least examined disciplinary assumptions that we all hold (sometimes without knowing it) in order to communicate with those who share almost nothing in the way of training, expertise, skills, or knowledge. Such translation is worth it because it is the only way that we are able to answer a question or face a challenge which is shared across the disciplinary divide. It is that shared commitment, in fact, that crosses the divides of practices, traditions, and deep affective relations to one's subject areas that are, on the deepest level, what binds us (in all senses) to traditional disciplines, even when we think we have migrated away from them.

Typically, when such collaboration happens, the result is, in Julie Klein's taxonomy, something closer to multidisciplinarity (see Klein, Chapter 2 this volume). Each person contributes but there is no actual transformation. Klein notes, 'When integration and interaction become proactive, the line between *multidisciplinarity* and interdisciplinarity is crossed' (Klein 1996, p. 6). In digital learning, the collaboration leads to new questions, new challenges, and (as is appropriate for the field) constant customization, repurposing, retooling, and redesigning—from the conceptual level through to implementation—of the objects of study and, indeed, of the definition of the field itself. In Klein's terms, that is not just interdisciplinary but transdisciplinary, with an emphasis on the 'transcendent' qualities that inform the most basic assumptions that the participants bring to the enterprise.

As many people have noted in regard to many fields (from music to science), new digital technologies and tremendously accelerated computational capacities are driving advances in knowledge as much as the other way around. This means that technohumanism is also a driver of monumental and even foundational conceptual changes in many disciplines. Technohumanism is not only developing new areas for the new computational tools but is also, while developing such tools, expanding our understanding of the implications and consequences of their development. Finally, because how one learns underlies every part of a university, technohumanism has the secondary consequence of pressing change in all of the component areas from which it draws and to which it contributes.

References

Benokraitis, N.V. (ed.) (1997). *Subtle sexism: current practice and prospects for change.* Thousand Oaks, CA: Sage.

Brown, J.S., Iiyoshi, V. and Kumar, M.S.V. (eds) (2008). *Opening up education: the collective advancement of education through open technology, open content, and open knowledge.* Cambridge, MA: MIT Press.

Cassell, J. and Jenkins, H. (eds) (1998). *From Barbie to Mortal Kombat: gender and computer games.* Cambridge, MA: MIT Press.

Castronova, E. (2005). *Synthetic worlds: the business and culture of online games.* Chicago: University of Chicago Press.

Committee on Facilitating Interdisciplinary Research; Committee on Science, Engineering, and Public Policy; National Academy of Sciences; National Academy of Engineering; and Institute of Medicine. (2005). *Facilitating interdisciplinary research.* Washington, DC: National Academies Press.

Darnton, R. (2008). The library in the New Age. *The New York Review of Books* **55**(10), 72.

Davidson, C.N. (1986). *Revolution and the word: the rise of the novel in America.* New York: Oxford University Press.

Davidson, C.N. (2008). Humanities 2.0: promise, perils, predictions. *PMLA* **123**(3), 707–17.

Davidson, C.N. and Goldberg, D.T. (2004). A manifesto for the humanities in a technological age. *The Chronicle of Higher Education* **13**(February), B7.

Davidson, C.N. and Goldberg, D.T. (2009). *The future of learning institutions in a digital age.* MacArthur Foundation research papers on digital media and learning. Cambridge, MA: MIT Press. Available at: <http://mitpress.mit.edu/books/chapters/Future_of_Learning.pdf>.

Epstein, S. (2007). *Inclusion: the politics of difference in medical research.* Chicago: University of Chicago Press.

Hall, S. (1990). Emergence of cultural studies and the crisis of the humanities. *October* **53**(Summer), 11–23.

Haraway, D.J. (2003). *The Haraway reader*. New York: Routledge.

Harding, S. (2005). Science and technology. In: P. Essed, D.T. Goldberg, and A. Kobayashi (eds) *A companion to gender studies*, pp. 241–55. Malden, MA: Blackwell.

Hayles, N.K. (2005). *My mother was a computer: digital subject and literary texts*. Chicago: University of Chicago Press.

Hayles, N.K. (2008). *Electronic literature: new horizons for the literary*. Notre Dame, IN: University of Notre Dame Press.

Jenkins, H. (2006). *Convergence culture: where old and new media collide*. New York: New York University Press.

Jenkins, H., Li, X., and Krauskopf, A.D. (2009). If it doesn't spread, it's dead. *Spreadable Media* parts 1–8(February), 11–29. Available at <http://www.henryjenkins.org>.

Kahne, J. *et al.* (2008). *Teens, video games, and civics*. Pew Internet and American Life Project, MacArthur Foundation research paper on digital media and learning. Cambridge, MA: MIT Press. Available at: <http://www.pewinternet.org/Reports/2008/Teens-Video-Games-and-Civics.aspx>.

Klein, J.T. (1996). *Crossing boundaries: knowledge, discplinarities, and interdisciplinarities*. Charlottesville: University Press of Virginia.

Lattuca, L. (2001). *Creating interdisciplinarity: interdisciplinary research and teaching among college and university faculty*. Nashville, TN: Vanderbilt University Press.

Lenoir, T. (2002). Makeover: writing the body into the posthuman technoscape: part two: corporeal axiomatics. *Configurations* **10**(Fall), 373–85.

Lenoir, T. and Lowood, H. (2000). All but war is simulation: the military entertainment complex. *Configurations* **8**, 238–335.

McCarty, W. (2005). *Humanities computing*. Basingstoke: Palgrave Macmillan.

McCarty, W. and Kirschenbaum, M. (2003). Institutional models for humanities computing. *Literary and Linguistic Computing* **18**(3), 465–89.

McPherson, T. (2008). *Digital youth, innovation, and the unexpected*. Cambridge, MA: MIT Press.

Miller, R. and Spellmeyer, K. (eds) (2006). *The new humanities reader*. Boston: Houghton-Mifflin.

O'Reilly, T. (2005). *What is Web 2.0: design patterns and business models for the next generation of software*. Available at: <http://oreilly.com/web2/archive/what-is-web-20.html>.

Scholle, D. (1995). Resisting disciplines: repositioning media studies in the university. *Communication Theory* **5**, 130–43.

Schreibman, S., Siemens, R., and Unsworth, J. (ed.) (2008). *A companion to digital humanities*. New York: Wiley-Blackwell.

Snow, C.P. (1959). *The two cultures and the scientific revolution*. New York: Cambridge University Press.

Thorne, B. (1993). *Gender play: girls and boys in school*. New Brunswick, NJ: Rutgers University Press.

Trueman, D. (2007). Why a laptop orchestra? *Organised Sound* **12**(2), 171–9.

Williams, R. (1976). *Keywords: a vocabulary of culture and society*. New York: Oxford University Press.

Wolf, M. (2007). *Proust and the squid: the story and science of the reading brain*. New York: HarperCollins.

CHAPTER 15

Media and communication

ADAM BRIGGLE AND CLIFFORD G. CHRISTIANS

Language is commonly singled out as the essence of humanity (Cassirer 1946). Human beings are co-creators, because they give names to the plants and animals. They invent symbols to represent things in their world, which allows them to share the contents of their minds with one another. Thus, as linguistic creatures, humans are also inherently social, because they inhabit a shared symbolic order made possible by their powers of representation and communication. And because of this pervasive character of communication in the development of the human species, media and communication studies have not been contained in an explicit discipline, with its own subject matter. Interdisciplinarity has been essential for understanding it.

As core features of humanity, communication and media clearly pre-date academic disciplines. They are in this sense non-disciplinary. Yet, they have for centuries been the subject of inquiry by those concerned to understand and improve human correspondence. Since the early twentieth century, such studies of media and communication have proliferated. In the process, they have adopted nearly all of the forms of interdisciplinarity identified in the taxonomy provided by Julie Thompson Klein (Chapter 2 this volume). The 'bridging' and 'restructuring' of knowledge communities to form new interdisciplinary domains of 'communication studies' and 'media studies' has been a particularly important development in this regard.

This chapter surveys the historical development and present form of multi-, inter-, and transdisciplinary studies of media and communication. It begins with a brief historical sketch of media and communication in order to indicate the kinds of phenomena motivating the studies. This sketch indicates that the four primary drivers of interdisciplinarity are present in this field. Media and communication are (1) inherently complex, (2) raise questions that are not confined to a single discipline, (3) pose societal problems that transcend the academy, and (4) are tightly linked to new technologies. Indeed, media and communication studies are motivated in large part by the complex questions and social changes brought about by new technologies.

15.1 A brief historical sketch of media and communication

For millennia, the oral medium was the sole form of communication, and techniques such as chanting were crafted for memorizing the essential stories of a people. Though epics could be told, the ephemeral nature of the oral medium established a natural governor on the production of knowledge. The inventions of the alphabet and of writing heralded a seismic shift in both human consciousness and social order (Ong 1982). Though writing made systematic inquiry and knowledge production possible, Socrates famously reacted to it with skepticism. Not only does the written transmission of knowledge betray a softness of mind (as one no longer has to rely solely on memory), it exposes one's most serious commitments to attack and degrading treatment while one is not there to defend them.

Subsequent innovations slowly prepared the way toward a modern world drowning in technological media and suffused with knowledge about media and communication—knowledge that is itself communicated, conveyed, and shaped by various media. These innovations include the index, punctuation, and other twelfth-century developments that lifted the 'text' from the page, transforming reading from a communal mumbling to a silent, solitary affair (Illich 1993). Gutenberg's mid-fifteenth-century printing press is often cited as the most important watershed in the development of media. Movable type revolutionized European culture (by standardizing expression), politics (by broadening access to ideas and fostering nationalism), and religion (by making the Bible widely available, thereby upsetting the Church's monopoly).

Electrification brought about the next major wave of change. This was primarily a shift toward broadcast media (waves encoded as transmission signals) as opposed to the mass production and circulation of physical artifacts (e.g., newspaper copies). But it also heralds the birth of film as it progressed beyond the daguerreotype and other early photograph technologies of the mid nineteenth century. The beginning of this era can be symbolically dated on 24 May 1844, when American inventor Samuel Morse first publicly demonstrated his electrical telegraph by sending a message from Washington, DC to Baltimore that read: 'What hath God wrought'. Wireless telegraphy, or radio, soon followed with the 1896 construction of the first radio station on the Isle of Wight, UK by the Italian inventor Guglielmo Marconi. The broadcasting of images through television first occurred in the early twentieth century, and in the years after World War II television sets became common household items.

At this time, 'the media' became an established singular collective term referring to (1) the institutions and organizations in which people work with communication media (the press, cinema, broadcasting, publishing, etc.), (2) the cultural products of those institutions (genres of news, movies, radio and television programs, etc.), and (3) the material forms of media culture (newspapers, books, broadcasting towers, radio sets, films, studios, tapes, discs, etc.).

The arrival of the digital computer in the mid twentieth century and later development of the Internet are widely credited as enabling the latest wave of change in media and communication. The shift here is from broadcast to network communication—arguably implying a shift from state control and masses to democratization and individuality. Digitality (the conversion of input data into discrete abstract symbols such as numbers) is a distinguishing characteristic of 'new media' (Lister *et al.* 2003). Other distinguishing features include interactivity (active involvement and many-to-many communication as opposed to the passive consumption of the one-to-many broadcast media), hypertext (texts that link to other texts), dispersal (the decentralization of the production and distribution of media), and virtuality (in a strong sense as immersion or in a weaker sense as the cyberspace where participants in online communication feel themselves to be, including virtual worlds such as Second Life).

New media have also developed a further stage, often signified by the term 'Web 2.0', which is characterized by enhanced social networking affordances and user-generated content delivery systems such as Facebook, Blogger, and Twitter. The latest revolution may be precipitated by the increase of mobile media such as cellular phones and iPods. Developments in the near future may include the rise of wearable (and perhaps implantable) multimedia technologies that serve as cameras, phones, entertainment systems, and even meta-information devices for accessing and displaying information about anything encountered in one's environment. And artifacts—from products at a store to home appliances—may soon be connected in a communication network or an 'Internet of things' (e.g., a refrigerator linked to a car capable of updating the driver on his or her milk supply before driving home from work). Another potentially revolutionary change is germinating with regard to user interfaces and the shift away from keypads toward natural gestures and perhaps even toward direct brain–computer interfaces. Children born on the crest of this accelerating wave of change, as 'digital natives', are thrown into a world so pervaded by media that it is now known as the 'Information Age' (Castells 1996).

Scholarly reflection on these developments is motivated by the increasingly profound implications of media in modern society. For example, media are making good on the popular image of a 'global village', by intertwining the cultural, political, and economic fates of more and more people. Life via the Internet poses questions about personal identity, as people come to develop their sense of self in cyberspace, sometimes through the use of 'avatars' or digital representations of people . The pervasive matrix of information and communication technologies now aids cognition to such an extent that it could be seen as an extension of the human mind beyond the confines of the skull. Other questions pertain to the quality of online communities, relationships, and education. Websites such as Wikipedia and WebMD muddy the categories of 'expert' and 'lay', while cultures suffused with screens and images confront questions about the meaning and relative value of reality and virtuality. Media of all sorts continue to be implicated in the fate of democracies around the world, entangled in thorny issues about censorship and legal jurisdictions. The 'old' media suffer under the influence of new technology, posing questions about the future of journalism and the academy. For these and other reasons, media and communication studies have grown into a thriving and bewildering constellation of academic study.

15.2 Studies of media and communication: an overview

Media and communication studies have drawn conceptual distinctions, formed methodologies and theories, hewn specialized discourses, coalesced communities of experts, created journals, awarded degrees, and become housed in institutions. They have at times developed within existing academic disciplines. At other times they have created their own disciplinary trappings or remained more nebulous in terms of disciplinary identity. And all of this is currently taking place in a context where new media are challenging many of these traditional academic endeavors by changing the way in which knowledge is produced, disseminated, and consumed.

Attempts at understanding, evaluating, and improving communication have roots in the study of rhetoric, the art of oratory and persuasion, in ancient Greece and Rome. They branch upward through the medieval university and its *trivium* of logic, grammar, and rhetoric—the arts of thinking, inventing and combining symbols to express thought, and communicating from one mind to another. The modern research university with its emphases on specialization and knowledge production has scattered and multiplied academic inquiry. The profusion of academic studies has also been fueled by the increasing diversity and importance of media in modern society. The resulting cornucopia of titles, programs, methods, and theories mirrors the jumbled labyrinth of the contemporary media technologies and cultures under consideration.

It is possible, however, to discern two main streams of academic study of media and communication, one social scientific and one humanistic. The first stream dates back to World War I, fueled by the problem of war propaganda and by radio technology that linked nations into mass media markets for the first time. Scholars in sociology, psychology, journalism, and political science began researching such developments, using the methodologies of their disciplines. Charles Horton Cooley, Walter Lippmann, and John Dewey were influential, because they all gave communication a central role in the attempt to understand social relations. In terms of Klein's taxonomy, this stream has adopted several identities. Especially in its early stages, it was predominantly a multidisciplinary juxtaposition where the disciplines retained their original identity. Yet it has increasingly featured versions of composite, methodological, and theoretical interdisciplinarity where integration occurs around a common problem and via conceptual frameworks, organizational principles, and methods.

The second stream comprises contributions from philosophers, historians, cultural anthropologists, cultural theorists, and scholars of art, literature, and film. Its origins are diverse and thus more difficult to pinpoint, although critical theory and post-structuralism are two major sources of much present-day humanistic study of communication and media. More concretely, in 1947, Wilbur Schramm, the 'father of communication studies', founded the Institute for Communications Research at the University of Illinois at Urbana-Champaign. Holding a PhD in literature, he argued that communication theory will emerge out of language and linguistics, and established appointments in these areas. This stream can be roughly distinguished from the social science stream by its tendency toward critical interdisciplinarity and transdisciplinarity. Broad interdisciplinarity between these streams—let alone with the natural sciences or engineering—remains rare.

Insofar as they remain elements of existing disciplines, studies of media and communication do not acquire their own disciplinary identity. Students use established methods and theories and receive traditional degrees in philosophy, sociology, economics, etc., although with a dissertation topic focused on media and communication. This situation characterized communications studies at Columbia University. Through the Bureau of Applied Social Research at Columbia, Paul F. Lazarsfeld and others produced work that was highly influential in shaping the field. Yet while this work began in 1944, Columbia did not create a degree-granting graduate program in communications until the 1990s. Prior to then, communication studies fell under the umbrella of sociology.

In a dialectic familiar to students of interdisciplinarity (see Krohn, Chapter 3 this volume), many forays into interdisciplinary media studies have been driven back into disciplinarity. Or as Klein notes, today's interdiscipline is tomorrow's discipline. This is caused by the 'need for manageable objects and presentable results' within a reference community. Indeed, it is caused by the academy's need for reference communities to define the nature and judge the quality of scholarship and to perpetuate themselves by initiating students and obtaining financial and institutional support. Furthermore, the diversity of communication and media phenomena is also partly responsible for the fracturing of inquiry. The appearance of new media and new social landscapes calls for and creates ample opportunities to fashion the new theories, concepts, and methods that become the intellectual lifeblood of institutionalized disciplinary communities (McQuail 2003).

These epistemological and institutional requirements have caused the current abundance of university degree-awarding programs operating under a variety of titles and housing scholars publishing in a growing array of specialized journals. A sampling of the dozens of journals supporting this field of inquiry includes *Journal of Communication*, *Communication Theory*, *Human Communication Research*, *Critical Studies in Media Communication*, *Media, Culture & Society*, and *Feminist Media Studies*. Of course, the boundaries of this field are shifting and porous, and could be drawn more widely to include such journals as *Ethics and Information Technology* and *Journalism Studies*.

A sampling of some common university programs shows them grouped under such terms as 'communications', 'communication studies', 'rhetorical studies', 'communication science', 'media studies', 'mass communication', and 'media ecology'. Many of these programs self-identify as multi- or interdisciplinary because they juxtapose or integrate traditional disciplines. Some programs claim to be transdisciplinary, because they frame research questions and practices around real-world problems and coalesce around conceptual frameworks that transcend disciplinary worldviews. Yet, they are also disciplines in their own respect, because they sustain and perpetuate specialized communities of discourse (via majors and advanced degrees) around a shared set of problems, theories, methods, and/or concepts. As one way to indicate the disciplinization of this field, when Schramm established the first PhD in communications in 1947 at the University of Illinois at Urbana-Champaign, all faculty members held their PhDs in the established disciplines. Today faculty members of most of these degree-granting programs are recipients of doctorates from communication programs.

15.3 History of mass communications research

Wireless broadcasting achieved technical excellence during World War I and swept rapidly through society as peace returned and military need subsided. The war symbolized the late-modern breakdown of traditional society and the emergence of not just mass media, but 'mass society', including the mass production of transportation, goods, and education. Formal studies of mass media originated in the same post-war period as a central part of attempts to understand the massification of society. Many such attempts shared the idea that the masses, as formed by the disintegration of traditional society, were in need of mechanisms of incorporation to ensure social integration.

The history of mass media research could be told through a variety of narratives, including: (1) disputes about goals, (2) incremental progress, (3) revolutionary change, and (4) disagreements about methods. This section briefly glosses each narrative. The take-home message is that these disputes, advances, changes, and disagreements create the various fault lines in the intellectual subsurface underlying the current panoply of departments and programs. That is, much of the institutional diversity in terms of (inter)disciplinary identities stems from the different positions staked within these narratives.

First, a basic divide in media and communication studies exists between the goals of serving mass media and critiquing it—or between what Klein identifies as instrumental and critical interdisciplinarity. Understanding media could be considered an independent goal, but often understanding is sought as a means to improved service or criticism. Of course, these goals are often reconciled, as both service and critique can lead to reform. Radio advertising, one of the original loci of mass media research, illustrates this overly simplistic but instructive dichotomy. In the 1920s, radio became a promoter's dream. Pepsodent sponsored *Amos and Andy*, one of the first radio comedy serials, and its sales increased by 70 per cent in the first year. A host of today's prominent products achieved their recognition initially from the newly formed networks: Bayer, Goodrich, Wheaties, Pepsi-Cola, Bulova, Texaco, and more. In fact, early radio history could be written around combinations of program and brand name: Lucky Strike Orchestra, Eveready Hour, Voice of Firestone, Ipana Troubadours, A & P Gypsies, and Sieberling Singers.

In order to secure more such advertising success, official market research received abundant commercial funding. Such instrumental interdisciplinary research has allowed media messages to be delivered with more substantial impact. This is attributable to increased understanding of the significance of audience demographics (the age, gender, etc. of those tuned into a given media outlet) for optimizing exposure to advertising and other content. It is also a result of the stipulation of differences among the media—especially their varying technological affordances. This is a clear example of research serving media. Yet commercial radio also became a site of critical interdisciplinarity. For example, Theodor Adorno and Max Horkheimer coined the term 'culture industry', arguing that popular culture is akin to the factory production of standardized goods. Like political propaganda, this culture industry manipulates the masses into docility and passive consumption of easy pleasures. It creates the false needs satisfied by capitalism and threatens the true needs of freedom, creativity, and flourishing.

Second, the history of broadcast media research could be told in terms of linear progress. In several cases, media and communication studies have advanced knowledge by progressing in the manner commonly thought typical of science—theories or models are put forwarded, tested, and either tentatively accepted or rejected. For example, in the 1940s many researchers drew from the pioneering work of Harold Lasswell on propaganda to develop the hypodermic needle or magic bullet model of communication (Lasswell and Casey 1946). This model (a variant of the then predominant stimulus–response model) holds that mass media have a direct, uniform, and immediate impact on their audience. The mass hysteria caused by the 1938 broadcast of *The war of the worlds* was cited as evidence for this model. But Lazarsfeld and others would go on to use this incident and other empirical evidence to challenge the model. Their studies demonstrated that broadcast media typically have selective and diverse impacts on people, depending on their beliefs and on contextual factors. Building from such studies, they offered the two-step flow model, with its greater emphasis on human agency, as an alternative.

Third, the history of mass media research could also be told as one of major conceptual rifts that resemble what Thomas Kuhn (1962) called 'paradigm shifts' rather than stepwise linear progress. The most important paradigm shift occurred in the 1960s and 1970s as a transition from content to form. Prior to this time, studies tended to conceptualize media as tools for the transmission of content, with an emphasis on the nature of the content or message. For example, the earliest studies of political communication conceived of media as a vehicle for either education or propaganda. Concerns were raised by the pervasiveness of propaganda in totalitarian governments and its success in undermining critical thinking by the public. In 1937, an interdisciplinary group of US scholars founded the Institute for Propaganda Analysis with the goals of studying illegitimate manipulation, fostering critical thought, and contributing to intelligent engagement with mass media. In place of propaganda, early Marxist critiques conceptualized mass media as a vehicle for the transmission and reproduction of ideology, hegemony, or class domination. Whether propaganda or ideology, the emphasis was on the content of the messages rather than the structure of the medium.

By contrast, the French Marxist Louis Althusser initiated a 'paradigm earthquake', by arguing that ideology should be understood as the structure or form of mass media, not just its content (Holmes 2005). For Althusser, ideology is not just found in the ever-shifting content of the messages absorbed by 'given' or pre-existing individuals. Rather, 'ideology-in-general' constitutes individuals as subjects—it is the very condition by which an individual comes to have a representation of self and world. This subjectivity is created by the communication process itself. Thus, the kind of selfhood that emerges and the world it takes as reality depend on the structure or form of the communication.

The profound implication is that media do not deliver a representation (either neutral or distorted) of reality. Rather, they create reality. This 'revolution', in Kuhn's terms, resonated widely. It can be seen, for example, in the thesis put forward both by the cultural critic Jean Baudrillard (1997) and the Heideggerian philosopher Albert Borgmann (1999) that simulacra have come to precede, determine, and crowd out the real. It is also apparent in the work of Marxist theorist Guy Debord (1977) and others advancing various spectacle or ritualistic theories of mass media. Debord argued that mass media create a certain field of visibility by concentrating the attention of the many on a particular event or representation. When this image is repeated, in time it begins to take on a life of its own—it becomes a spectacle—and that to which it refers becomes secondary and may even disappear from view.

This paradigm shift toward form or structure was also advanced, and even foreshadowed, by the two main 'medium theorists', Marshall McLuhan (1962) and Harold Innis (1964). McLuhan differed from the spectacle and ideology views by rejecting their homogeneous picture of media and culture in favor of an account of the distinct specificities of different media corresponding to different modes of perception. Yet he shared their emphasis on form rather than content: 'the medium is the message'. Innis similarly analyzed how power gathers around different media structures, which has influenced later work on new media.

Finally, the story of broadcast media could be told in terms of debates about methods. For example, Lazarsfeld *et al.* (1944) transferred empirical and inductive methods from the study of radio advertising to the analysis of the 1940 presidential election. Their work suffered somewhat because they assumed that promoting candidates and selling soap were methodological equivalents. And although Lazarsfeld utilized sample surveys innovatively, his inductivism could not ultimately specify causal relations. It proved impossible to move beyond the correlation of two factors to demonstrate a causal relationship—an issue that has long haunted research on the impacts of media on society, from pornography to violence in computer games and movies. As one report on obscenity and pornography noted: 'The research evidence is of the kind in which science follows in the wake of common sense' (Barnes 1971, p. xiii).

Carl Hovland, a psychologist working at Yale in the 1950s, produced some of communications' most suggestive studies (Hovland *et al.* 1953). These included the first report of the 'sleeper effect'—when a highly persuasive message paired with a discounting cue causes the individual to be more persuaded (rather than less) over time. Leon Festinger also adopted the experimental method, but with less emphasis on exact precision and verification of causal relations. Specifically, his 'dissonance theory' described communication effects in terms of desirable psychological states. Experiments statistically measure attitudes before and after some persuasive message under the basic presumption that humans need equilibrium, and beliefs change only to alleviate inconsistency.

As an alternative to laboratories, Norbert Wiener (1948) and Ross Ashby (1963) developed cybernetics as a formalist, mathematically based approach to the study of communication. Cybernetics is an instance of what Klein calls generalizing interdisciplinarity, because it applies a single theoretical perspective to a wide range of disciplines. This influenced Claude Shannon and Warren Weaver's *Mathematical theory of communication* (1949), which laid out the basic elements of communication as an information source, a transmitter, a channel or medium, a receiver, and a destination. It also developed the concept of a bit as a unit of information. This laid the foundations for information theory, becoming the basis for digital communications technology and the birth of networked or new media.

15.4 Networked communications research

The explosive growth of the Internet and network (as opposed to broadcast) communication in the 1990s has generated scholarship on a second or new media age (Hassan and Thomas 2006). Many of its foundational tropes—social disintegration, the virtual replacing the real, individuality, disembodiment, realignment of political power and economic order—were foreshadowed by science fiction works such as *Neuromancer* (Gibson 1986) and *Snow crash* (Stephenson 1992) and portrayed in films such as *The matrix* (1999). Here too both social science and humanities streams are discernable, though with considerable overlap. Another set of distinctions is helpful for indicating some important topographic contours, including: (1) the relationship between old and new media; (2) utopias and dystopias; (3) computer-mediated communication; and (4) cyberculture.

First, early scholarship on new media placed strong emphasis on its distinguishing features. The old media architecture is one of central media producers transmitting content to an undifferentiated mass. The individual looks to the central media source to acquire cultural identity, not 'sideways' at others in the crowd. By contrast, the new media architecture breaks down the walls separating individuals. They look at one another for a sense of self and belonging. This is why Mark Poster (1995) sets 'interactivity' at the core of new media. Placing such stress on the revolutionary differences of new media fostered a widely held thesis that the new would quickly displace the old (Manovich 2002). The contraction or demise of newspaper publishers has lent some support to this thesis.

Yet newspapers have also adopted online publishing, featuring new forms of articles and advertisements. This kind of development has led some to argue that the picture is far more complex. David Holmes (2005), for example, questioned the historical distinction between the first and second media ages. He argued that the way in which individuals connect with the different media forms is interdependent—network communication becomes meaningful because of broadcast and broadcast becomes meaningful in the context of network. Jay Bolter and Richard Grusin (1999) similarly coined the term 'remediation' to argue that newer forms of media have always refashioned older forms. A simple example of this is the way in which YouTube videos often remix popular television shows. The hit US comedy television show *The Colbert report* even responded to such creations, prompting yet more online videos. Some describe such phenomena as the 'convergence' of media functions and industries (Van Dijk 1999). The lesson seems to be that new media offer different possibilities for connectedness and creativity, but some of these engage and reshape old media rather than simply eclipse them.

Second, the view that digital, interactive media marked a revolution in communication was often wed to a utopian ideal. One expression of this ideal was the 1994 manifesto entitled *Cyberspace and the American dream: a Magna Carta for the knowledge age* (Dyson *et al.* 1994). Langdon Winner (1997) extracted its core tenants as deterministic but positive technological change, radical individualism, free-market capitalism, and a rebirth of the public sphere and participatory democracy. Overcoming the passivity and homogeneity of the broadcast architecture means emancipation, enfranchisement, and creativity—indeed, individuals are free to experiment with identity in radically new ways (Turkle 1995). No longer does the mass media industry determine cultural or individual consciousness. Furthermore, an interactive media renews community by strengthening the bonds connecting people to their world.

As is often the case with emerging technologies, there are dystopian visions contrasting with the utopian ones. A primary motif here is the impoverishment that results when virtual and mediated experiences displace real and direct experiences. Hubert Dreyfus (2001), for example, argued that distance learning is a poor substitute for classroom education and more generally that lives increasingly spent online lack the defining commitments that sustain meaning and community. Cass Sunstein (2001) deflated claims about cyber-democracies, by arguing that cyberspace is far more a private than a public space. It allows those online to see, hear, and read only what they like. This egocentrism is not only narcissistic, but weakens the exchange of ideas necessary for democracies. Nicholas Carr (2008) argued that the Internet diminishes cognitive capacities by fostering a staccato style of reading and thinking. The interpretive ability to make imaginative mental connections and relate new information to one's biography remains largely disengaged online. Other dystopian themes center on increased risks of identity theft, cyberstalking, an acceleration of the pace of life, and the threats to privacy posed by surveillance and data mining.

A third important story about new media is the growth of social scientific and psychological studies of computer-mediated communication (CMC) (Joinson 2003; Thurlow *et al.* 2004). CMC research examines the social and psychological dimensions of communication through two or more networked computers in formats ranging from e-mail to instant messaging to social networking sites and virtual worlds. Examples include research on identity construction online and behavioral changes under conditions of anonymity. Researchers often compare CMC to face-to-face relating. The umbrella term of CMC has created new communities of academic discourse via such outlets as *The Journal of Computer-Mediated Communication* and *Cyberpsychology*. The CMC literature can be mapped onto the utopia–dystopia landscape, especially regarding disputes about whether online communication is better or worse than offline forms. But by in large it strives for value neutrality and empiricism. Furthermore, this literature tends to adopt a narrower focus on individual interactions rather than the overall contexts by which those interactions form a meaningful whole.

Fourth, and by contrast, 'cyberculture' has become a term of art in the humanities to draw attention to the ways in which media are shaping entire value systems, basic concepts, and patterns of life (see Davidson, Chapter 14 this volume). Culture, communication, and media are tightly interlinked (Langer 1977; Carey 1988). Cultures are interconnections of symbolic forms, those fundamental units of meaning are expressed in words, gestures, and graphics. Realities called cultures are inherited and built from symbols that shape action, identity, thoughts and sentiment. Communication, therefore, is the creative process of building and reaffirming cultures through symbolic action. Although not identical to what they symbolize, symbols participate in their meaning and power; they share the significance of that to which they point.

The concept of cyberculture intrinsically links such humanistic theories of culture with technical concepts from computer science, robotics, artificial intelligence, and genetics. It has thus become the site of both interdisciplinary collaboration—including wide interdisciplinarity across the humanities, engineering, and sciences—and turf wars as various traditions and disciplines seek to make claims to a superior understanding of the unprecedented mixture of artifacts and ideas that characterize our times. Cyberculture, far more than CMC, maps onto the utopia–dystopia dialectic, because it conjures forth fundamental reflection on culture, technology, and nature—including how these basic categories are blurring through such phenomena as androids, cyborgs, and virtual ski slopes. By blurring these categories, cyberculture theorists tend to adopt a non-linear sense of causality—things do not determine ideas nor do ideas determine things, but they are co-constitutive.

This means that the traditional humanist views of agency as solely the preserve of human beings and the human agent as separable from culture and technology are called into question. Thus, semiotics (the study of signs, symbols, and the construction of meaning) is an important wellspring for cyberculture studies, but it is often modified such that non-humans become actors rather than just signs. Furthermore, semiotics traditionally maintains a narrow definition of culture as the products of the arts and language. Though this definition long dominated communication studies, cyberculture expands it by including the physical, technological media as intrinsic to culture. Culture is not just the content conveyed by media, but the structures and forms of media technologies and the other artifacts in which they are embedded and systems with which they are networked.

15.5 Conclusion

As the diversity of narratives and concepts in these histories would suggest, there is substantial disagreement and turmoil in the current study of media and communication. For example, the standard textbook on approaches to mass communication study (Severin and Tankard 2000) added critical theory and cultural studies to its overview in its fifth edition. But only, the authors note, because 'they have become popular with scholars. Nevertheless, we remain committed to the scientific approach, with its emphasis on observation, evidence, logic and hypothesis testing' (p. xv). Most research funding still supports studies which measure observable behavior, finding in such results the statistical precision desired by private and public benefactors.

The current aim of the more scientific approach is to develop more elaborate and finely tuned procedures, more complex multivariate scales, faster computer banks, and longer-range experiments, trusting that greater development of method will eliminate previous weaknesses, confusions, and uncertainties. Severin and Tankard (2000) summarize this scientific trajectory in terms of the incremental progress narrative:

> Communication researchers have not yet come up with a unified theory that will explain the effects of mass communication. Instead, we have a number of theories, each attempting to explain some particular aspect of mass communication. As communication research advances, perhaps we shall see several of these mini-theories combined into one overall theory of mass communication effects. Or, perhaps some of these theories will not survive the test of empirical research and will be winnowed out, while others survive (p. 286).

The scientific trajectory in communication studies is currently bolstered by the surge of new forms of physical, computer, and biological sciences as well as new technical capacities. Indeed, this trend is toward communication studies as a form of 'big science'. Media and communication studies find themselves in the current transdisciplinary trend line toward the unification of knowledge. Some see this unification in terms of the cognitive and natural sciences swallowing social scientific and humanistic approaches. Communication and media studies, then, would become 'scientistic', implying the importation of natural scientific methods for the study of social and cultural phenomena:

> Scientific advances, particularly in neurobiology, genetics, and neuropsychology, are encouraging researchers to consider re-theorizing 'cultural' problems to take the new knowledge generated by science into consideration. Added to this, the achievement of the technical capacity to process large and complex data fields, a feature of the computerized knowledge environment, now suggests that alternative methods and approaches for the study of cultural phenomena may be possible. In other words, some research that we previously believed could only be solved by cultural approaches may be recast as questions for science and scientific inquiry (Nightingale 2003, p. 361).

Yet this trajectory is not likely to yield that magical universal theory or homogenize the current diversity in the topography of media and communications studies. Rather, this infusion of natural, cognitive, computer, and physical sciences will most probably map onto existing landscapes and create ever more niches. This trend has occurred before—for example, in the fact that sociobiology became just another approach to human social life rather than a grand consilience marking the demise of approaches rooted in the humanities or social sciences. Indeed, this seems inevitable given that the mechanistic tropes central to the natural sciences are incapable of accounting for the spontaneities of the human lifeworld.

Joseph Klapper (1965), a proponent of scientific rigor in communication studies, regretted that after years at the 'inexhaustible fount of variables', systematic description and prediction 'becomes the more distant as it is the more vigorously pursued' (p. 316). The Enlightenment dream of mirroring nature would mean that at some point we could close the book of knowledge, having adequately transcribed reality. But the pursuit of knowledge is 'inexhaustible'—especially in an information society where everyone is a publisher. The only governors on its growth are external and relatively contingent—the availability of funding and the interests of citizens, politicians, provosts, and CEOs.

Thus, there are contrary reactions to the growing confusion about the nature of communication and media studies. Some desire multidisciplinary juxtapositions to address narrowly defined academic questions. Others want interdisciplinary integrations to unify knowledge. And still others seek a transdisciplinary transcendence of academic disciplines in order to either serve or critique society. The danger of the transdisciplinary path is that in seeking to become relevant, media and communication studies will lose the disciplinary trappings that ensure academic viability. Yet as the academy continues to evolve under the influence of new media, it may be that transdisciplinary structures of knowledge production become more stable than the traditional disciplinary forms. In many respects, the future of media and communication research depends on how the phenomena under study will impact those very studies.

References

Ashby, W.R. (1963). *Introduction to cybernetics.* London: Routledge, Kegan, Paul.

Barnes, C. (1971). Introduction. *US Commission on Obscenity and Pornography.* New York: Bantam.

Baudrillard, J. (1997). *Simulacra and simulation.* Ann Arbor, MI: University of Michigan Press.

Bolter, J. and Grusin, R. (1999). *Remediation: understanding new media.* Cambridge, MA: MIT Press.

Borgmann, A. (1999). *Holding onto reality: the nature of information at the turn of the millennium.* Chicago: University of Chicago Press.

Carey, J.W. (1988). *Communication as culture.* Boston: Unwin Hyman.

Carr, N. (2008). Is Google making us stupid? *Atlantic Monthly* July/August, 75–80.

Cassirer, E. (1946). *Language and myth,* transl. S. K. Langer. New York: Harper and Row.

Castells, M. (1996). *The rise of the network society.* London: Blackwell.

Debord, G. (1977). *The society of the spectacle,* transl. F. Perlman and J. Supak. Detroit: Black and Red Publications.

Dreyfus, H. (2001). *On the internet.* London: Routledge.

Dyson, E., Gilder, G., Keyworth, G., and Toffler, A. (1994). *Cyberspace and the American dream: a Magna Carta for the knowledge age.* Available at: <http://www.pff.org/issues-pubs/futureinsights/fi1.2magnacarta.html>.

Gibson, W. (1986). *Neuromancer.* London: Grafton.

Hassan, R. and Thomas, J. (2006). *The new media theory reader.* New York: Open University Press.

Holmes, D. (2005). *Communication theory: media, technology, and society.* London: Sage.

Hovland, C.I., Janis, I.L., and Kelley, H.H. (1953). *Communications and persuasion: psychological studies in opinion change.* New Haven, CT: Yale University Press.

Illich, I. (1993). *In the vineyard of the text: a commentary to Hugh's didascalicon.* Chicago: University of Chicago Press.

Innis, H. (1964). *The bias of communication.* Toronto: University of Toronto Press.

Joinson, A. (2003). *Understanding the psychology of internet behaviour: virtual worlds, real lives.* New York: Palgrave.

Klapper, J. (1965). *The effects of mass communication.* New York: Free Press.

Kuhn, T. (1962). *The structure of scientific revolutions.* Chicago: University of Chicago Press.

Langer, S.K. (1977). *Feeling and form.* New York: Prentice Hall.

Lasswell, H.D. and Casey, R.D. (1946). *Propaganda, communication and public opinion.* Princeton, NJ: Princeton University Press.

Lazarsfeld, P.F., Berelson, B., and Gaudet, H. (1944). *The people's choice: how the voter makes up his mind in a presidential campaign.* New York: Duell, Sloan and Pearce.

Lister, M., Dovey, J., Giddings, S., Grant, I., and Kelly, K. (2003). *New media: a critical introduction.* London: Routledge.

McLuhan, M. (1962). *The Gutenberg galaxy: the making of typographic man.* Toronto: University of Toronto Press.

McQuail, D. (2003). New horizons for communication theory in the new media age. In: A, Valdivia (ed.) *A companion to media studies,* pp. 40–9. Oxford: Blackwell.

Manovich, L. (2002). *The language of new media.* Cambridge, MA: MIT Press.

Nightingale, V. (2003). The cultural revolution in audience research. In: A. Valdivia (ed.) *A companion to media studies,* pp. 360–81. Oxford: Blackwell.

Ong, W. (1982). *Orality and literacy: the technologizing of the world.* New York: Routledge.

Poster, M. (1995). *The second media age.* Cambridge: Polity.

Severin, W.J. and Tankard, J.W. Jr (2000). *Communication theories: origins, methods, and uses in the mass media*, 5th edn. New York: Longman.

Shannon, C., and W. Weaver. (1949). *Mathematical theory of communication.* Urbana, IL: University of Illinois Press.

Stephenson, N. (1992). *Snow crash.* New York: Spectra.

Sunstein, C. (2001). *Republic.com.* Princeton, NJ: Princeton University Press.

Thurlow, C., Lengel, L., and Tomic, A. (2004). *Computer mediated communication: social interaction and the internet.* London: Sage.

Turkle, S. (1995). *Life on the screen: identity in the age of the internet.* New York: Simon and Schuster.

Van Dijk, J. (1999). *The network society: social aspects of new media.* London: Sage.

Wiener, N. (1948). *Cybernetics: or control and communication in the animal and the machine.* Cambridge, MA: MIT Press.

Winner, L. (1997). Cyberlibertarian myths and the prospects for community. *Computers and Society* **27**(3), 14–19.

CHAPTER 16

Cognitive science

PAUL THAGARD

There are many reasons why a budding academic might want to avoid interdisciplinary research. It is difficult enough to acquire expertise in one field of research, let alone two or more. The time required to read the literature in a field outside your own main area can be hard to find, and the additional time investment to learn novel methods from another field can be huge. Moreover, the hiring and reward systems in academia still run strongly along disciplinary lines, so that work that draws on or contributes to other fields may not be fully valued in one's own field. Interdisciplinary research may not be appreciated by narrow-minded colleagues. Some interdisciplinary projects have a bogus air about them, looking like they were designed more to bring in big research grants than to accomplish intellectual goals. The interdisciplinary scholar can look a bit like a dilettante, dabbling in multiple fields in order to avoid tackling the difficult problems in an established field. Grants for interdisciplinary research can be difficult to get, because most granting agencies are organized along disciplinary lines.

Despite these deterrents to interdisciplinary research, there are powerful intellectual reasons why work that oversteps the ossified boundaries of established fields can have great intellectual benefits. Such benefits are vividly apparent in the interdisciplinary field of cognitive science, which attempts to understand the mind by combining insights from the fields of psychology, philosophy, linguistics, neuroscience, anthropology, and artificial intelligence. After a brief review of the history of the field and its contributing disciplines, this chapter will examine some of the main theoretical and experimental advances that cognitive science has accomplished over the past half century, deriving lessons that might be useful for researchers in any emerging interdisciplinary area.

16.1 History

Construed broadly, cognitive science is as old as philosophical reflections about the nature of mind, and so dates back at least to Plato and Aristotle. Philosophers such as Francis Bacon, John Locke, David Hume, Immanuel Kant, and John Stuart Mill generated ideas about the contents and processes of thinking. Experimental psychology originated in the late nineteenth century with the establishment of laboratories by Wilhelm Wundt, William James, and others.

Modern cognitive science began in the 1940s when visionaries such as Alan Turing (1950), W. S. McCulloch (1965), Norbert Wiener (1961), and Donald Hebb (1949) began to apply emerging ideas about computing, engineering, and brain systems to develop new hypotheses about mental mechanisms. Previous mechanistic theories of mind, ranging from the atomism of Lucretius to the behaviorism of B. F. Skinner, were much too impoverished to explain the complexities of human thinking. But in the mid-1950s there emerged a panoply of powerful ideas about how mental processes could be understood by analogy to computational ones. The major contributors included the psychologist George Miller (1956), the linguist Noam Chomsky (1957), and researchers in the nascent field of artificial intelligence, including Herbert Simon, Allan Newell, Marvin Minsky, and John McCarthy (McCorduck 1979). The year 1956 was particularly notable, as it marked publication of Miller's famous paper on information processing, 'The magical number 7 plus or minus 2', and the Dartmouth Conference that initiated the field of artificial intelligence. The fundamental hypothesis of cognitive science, that thinking consists of computational procedures applied to mental representations, began to influence research in psychology and other fields.

The term 'cognitive science' was only coined two decades later (Bobrow and Collins 1975). Events in the late 1970s included the formation of the Cognitive Science Society, the creation of the journal *Cognitive Science*, and the establishment of cognitive science programs at many universities. Today, evidence that interdisciplinary research and teaching in cognitive science is thriving includes multiple successful journals, international societies with regular conferences, and active teaching and research programs in many universities and organizations around the world. For detailed treatments of the history of cognitive science see Gardner (1985), Thagard (1992, 2005b), and especially Boden (2006).

16.2 Patterns of collaboration

The interdisciplinary structure of cognitive science is displayed in the hexagon in Fig. 16.1, the original version of which appeared in a report for the Sloan Foundation in 1978 (Gardner 1985, p. 37). The 13 lines in the hexagon indicate the range of possible connections between the six main disciplines of cognitive science, but the links are misleading in several respects. First, the disciplines have been highly unequal participants in interdisciplinary research. For example, although anthropology has contributed some highly interesting work on mental representations and processes in non-Western cultures, most anthropologists have shown little interest in cognitive science. More significantly, some of the most widely read philosophical discussions of cognitive science have been highly critical of it, for example attacks by Herbert Dreyfus (1979) and John Searle (1980) on the computational view of minds. The field of artificial intelligence has moved away from the interest in human thinking that inspired its early decades to a more engineering-oriented concern with the building of intelligent computers. In contrast, most cognitive psychology research is naturally dedicated to understanding the operation of human intelligence.

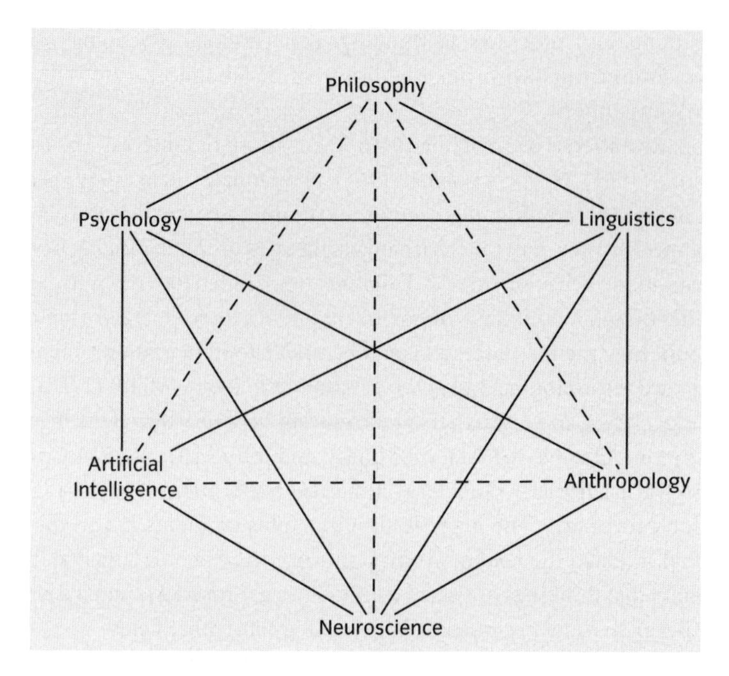

Figure 16.1 Connections among the cognitive sciences (based on Gardner 1985, p. 37). Unbroken lines indicated strong interdisciplinary ties *c*. 1978, and broken lines indicate weak ones. The ties between philosophy and both neuroscience and artificial intelligence are much stronger today.

Second, the hexagon does not convey the historical fact that some combinations of the fields have been much more active than others and that levels of activity have varied over time. When cognitive science began officially in the 1970s, by far the most prominent kind of interdisciplinary collaboration occurred at the intersection of psychology and artificial intelligence, continuing a pattern established in the 1950s by pioneers such as Herbert Simon (1991). Psycholinguistics also flourished early on. Neuroscience became much more central starting in the 1980s and 1990s, with the increased sophistication of neurally inspired computational models and the development of brain scanning technology that greatly expanded the possibilities for neuropsychological experiments. Philosophers' involvement in cognitive science has been highly variable, ranging from dismissal on the grounds that philosophy must transcend the merely empirical (Williamson 2007), to systematic reflection on controversial issues such as the extent to which knowledge is innate (Stainton 2006). Since the 1980s there has been much philosophical discussion of issues that arise in cognitive psychology and neuroscience (e.g. Thagard 2007; Bechtel 2008). Most strikingly, the application of psychology and neuroscience to traditional philosophical problems in ethics and epistemology has become an active enterprise (e.g. Appiah 2008; Knobe and Nichols 2008; Thagard 2010). For example, progress in neuroscience raises serious challenges to traditional ideas about free will and responsibility. In contrast, philosophers' interest in linguistics has waned, probably because language is no longer seen as so central to philosophy as it used to be; and work at the intersection of philosophy and anthropology has always been rare.

The third misleading feature of the hexagon is that the lines only indicate binary relations between disciplines, whereas some important developments have involved collaborations across several fields. For example, computational psycholinguistics draws on ideas from three disciplines to develop formal models of how minds use language. Current work in theoretical neuroscience combines the study of brains with psychological and computational ideas. Recent work on emotion attempts to address philosophical issues about rationality by means of computational models that are psychological, neurological, and even sometimes social (Thagard 2006). In sum, although Fig. 16.1 provides a useful diagram of possibilities for interdisciplinary connections, it does not display the shifting patterns of disciplinary involvement in such research.

There are at least three styles of interdisciplinary interconnection. The first is when an individual alone does research at the intersection of two or more disciplines. This requires the researcher to acquire mastery not only of the ideas but also of the methods of more than one field. For example, there are psychologists who have learned to do computational modeling, and a few philosophers who have learned to do experiments in psychology or neuroscience.

A second powerful kind of interdisciplinary interconnection involves collaboration, in which two or more individuals work together on a project combining their knowledge and skills in ways that require some mutual comprehension but not full duplication of abilities. This pattern of research has often been the most successful one in cognitive science, which has benefited from collaborations involving people whose original backgrounds combined, for example, psychology and artificial intelligence, psychology and neuroscience, and linguistics and anthropology.

The third style of interdisciplinary research does not require such collaboration or even individuals who have mastered more than one field. There has been much valuable work by more narrowly disciplinary researchers that draws on ideas from related fields. For example, Eleanor Rosch's influential work on concepts as prototypes was inspired in part by ideas of the philosopher Ludwig Wittgenstein (Rosch and Mervis 1975). Many articles published in the journal *Cognitive Science* are not internally interdisciplinary, as they lack a combination of methods. However, most articles that appear there are intended to be of interdisciplinary interest in that they address concerns inspired by or relevant to work in various fields concerned with the nature of mind and intelligence. For example, an experimental paper on the nature of human concepts falls squarely within cognitive psychology, but should be relevant to philosophical, computational, neurological, linguistic, and cross-cultural issues about mental representations. This third style of interdisciplinary research requires less personal investment than the individual mastery and collaborative styles, but it usually presupposes at least some acquaintance with relevant literature in other fields.

In the introduction, I mentioned some of the impediments to interdisciplinary research, but have described how cognitive science has provided a strong example of a successful effort to combine insights and methods from at least six disciplines. Now I want to depict more fully what that success has consisted in, by discussing the theoretical and experimental benefits of being interdisciplinary.

16.3 Theoretical benefits

A scholar has been defined as someone who knows more and more about less and less. Pursuing minutiae is often an effective strategy in academic research, since becoming an expert in some narrow niche is often a good way to publish and secure tenure. For the more intellectually ambitious, however, it is much more exciting to pursue theoretical ideas that are both important and novel. How can such creativity be achieved?

It helps, of course, to be a genius, with cognitive resources such as unusually powerful memory, imagery, or speed in connecting previously unrelated ideas or facts. But creativity is not only for the swift, because others of more modest intellectual capacities can still be creative by putting together ideas that have not been associated by other thinkers. Perhaps it takes a genius to work in a well-trodden area and manage to come up with something totally novel, but for the rest of us there is an easier road to creativity. Instead of focusing narrowly on one academic field, a researcher can cast a broader intellectual net and make new connections by tying together ideas from different disciplines. Cognitive science has thrived intellectually by making such creative theoretical connections.

In the mid 1950s, the dominant psychological theories, especially in the United States, were behaviorist, claiming that a scientific approach to the mind should restrict itself to considering how environmental stimuli are correlated with behavioral responses. Behaviorism was encountering difficulties in explaining the complex performance of rats, let alone humans, but theories are rarely rejected because of empirical problems alone. Rather, it is only when an alternative theory comes along with a new way of explaining recalcitrant data that a dominant theory comes strongly into question (Kuhn 1970; Thagard 1992). What happened around 1955 was that ideas from the rapidly emerging study of computers provided a new way to think about mental processes that was as rigorously mechanistic as behaviorism but possessed much more explanatory power.

A computer program consists of a set of structures, such as numbers, words, and lists, and a set of algorithms, which are mechanical procedures that operate on those structures. Those not familiar with computer programs can think of how people add up a list of numbers, where the structures are the numbers and the algorithm is the procedure for addition learned in elementary school. Or consider a recipe book, in which the recipe consists of a list of ingredients (the structures) to which people apply a set of procedures such as mixing and baking. Computer programs provide a highly suggestive analogy about how minds might work: mental representations may be like the structures used in computer programs, and mental procedures may be like the algorithms that make computers run. The strongest claim to consider is not only that thinking is like computing, but that thinking in fact is a kind of computing (Thagard 2005b).

The analogy just described has been fertile in suggesting many new ideas about how representational structures and computational procedures might be responsible for mental processes such as perception, memory, learning, problem solving, language use, and so on. Many productive specific theories have been developed about how rules, concepts, images, and analogies might operate in the mind. This theoretical productivity could never have happened if psychologists had stuck with the intellectual resources of behaviorism. Instead, by importing ideas from the study of computers, it became possible to formulate creative new theories of mental functioning. Whereas behaviorism restricted itself to stimulus–responses connections, cognitive science investigates how behavior and thought result from mental representations and computational procedures that integrate perceptual stimuli and produce responses based on complex inferences.

Another interdisciplinary source of ideas about how the mind works is the study of the brain. Some early ideas about how the mind works drew on neural mechanisms, but brain-style computing only took off in the 1980s through the development of an approach known as connectionism or parallel distributed processing. Brains operate differently from conventional computers. Neurons are slow, firing on average fewer than 100 times per second, but they perform powerful computations by virtue of the fact that that there are so many of them (around 100 billion) operating in parallel. In contrast, computer chips are very fast, with billions of cycles per second, but they usually operate serially, one step at a time. Today there is a flourishing field called theoretical neuroscience that develops new computational ideas about how brains support various kinds of thinking (Dayan and Abbott 2001).

Besides computer science and neuroscience, psychology has also been influenced by ideas from other fields, including philosophy and linguistics. Psychology is not just a recipient of theoretical ideas, but has also served as a donor. Psychology has contributed to the field of artificial intelligence that tries to build computers capable of some of the impressive feats of problem solving accomplished by people. For example, some expert systems that are engineering projects with the aim of making computers capable of tasks such as medical diagnosis have drawn on psychological ideas about mental representations like rules, analogies, and neural networks. Philosophy of mind and cognitive anthropology have also been heavily influenced by developments in cognitive psychology. Oddly, cognitive science has had little influence on fields such as literary theory and history, which could greatly benefit from richer ideas about how minds find meaning and make decisions.

Many more specific examples of the development of new theoretical ideas in cognitive science through interdisciplinary collaboration could be given, but here are two illustrations. The study of analogy has blossomed since the 1980s as the result of theoretical ideas that have combined insights from philosophy, psychology, artificial intelligence, and neuroscience. The goal of trying to understand how minds can often so productively apply ideas from one domain to another was studied by philosophers such as Mary Hesse (1966), but was greatly fostered by the development of new psychological ideas about how minds can use representations of one problem to solve another. Psychologists such as Dedre Gentner and Keith Holyoak devised new ideas about how people use analogies, partly on the basis of their own experiments but also drawing heavily on computer models, including ones that employ artificial neural networks (e.g. Gentner 1983; Holyoak and Thagard 1995; Gentner *et al.* 2001).

Recent work on emotion has also been highly interdisciplinary, drawing on philosophical ideas about norms, psychological ideas about representations, and most recently neurological ideas about how brains process emotions (Thagard 2006; Thagard and Aubie 2008). The intellectual goal holding all this together is the attempt to build computational models of how the brain produces emotions and uses them in other cognitive processes.

Like research on analogy, it is hard to imagine how theoretical progress on emotion could have proceeded without combining ideas from multiple fields of cognitive science.

16.4 Experimental benefits

Like physics and biology, cognitive science is not a purely theoretical enterprise, but also requires experimental investigations that can be used to evaluate competing theories. Interdisciplinary collaboration has contributed to experimental work in psychology in two ways: through suggesting new kinds of experiments to test interesting theoretical ideas, and through providing new measurement tools for performing experiments.

In the 1960s, the young field of cognitive psychology evolved by developing new kinds of experimental techniques. The growing availability of computers made it much easier to perform experiments that measured the reaction times of subjects performing complex tasks, and the resulting data were used to test the information-processing models of thinking suggested by the new computational theories of mind. The computational models of analogy generated new experimental work to test their predictions. Linguistics also provided new theoretical ideas through Chomsky's work on rules and representations, which inspired new kinds of experiments in psycholinguistics (Pinker 1994). Philosophical ideas have sometimes suggested psychological experiments, as in Rosch's experiments on prototypes. A huge line of experimental research in developmental psychology concerning the ability of children to understand false beliefs originated with philosophical ideas about intention (Boden 2006, p. 488).

In recent years, experiments in cognitive psychology have been most influenced by developments in neuroscience. Ideas about how the brain works have suggested valuable new experiments, but even more importantly neuroscience has provided a whole new set of tools for measuring mental activity. The 1980s saw the development of powerful machines for scanning brains using techniques such as functional magnetic resonance imaging (fMRI) and positron emission tomography (PET). It has become common for cognitive psychologists not only to measure the behavior of experimental subjects when they are performing various tasks, but also to scan their brains while the performance is taking place. Different scanning techniques provide different kinds of detail about the brain regions and temporal courses of neural operations. It is even possible to temporarily disrupt neural processing using transcranial magnetic stimulation. Information about neural processes is also sometimes obtainable by implanting electrodes deep in the brain to stimulate particular regions. Thus, the field of cognitive psychology has been transformed in recent years by the development of new experimental techniques made possible by neuroscience.

Science is most powerful when theoretical ideas mesh with experimental ones; such meshing is very apparent in current attempts to use computational models of brain operations to explain the results of many different kinds of brain scanning experiments. By combining ideas and techniques from psychology, computer science, and neuroscience, cognitive science is successfully pursuing fundamental questions about how the brain works. Answers to these questions are directly relevant to ancient philosophical questions about how minds know reality, make judgments about right and wrong, and appreciate the meaning of life. For example, Thagard (2010) uses psychological and neurological research about vital human needs to argue that the meaning of life is love, work, and play.

Other practical applications include the prospect of improving education by a deeper understanding of the neural mechanisms by which people learn (Posner and Rothbart 2007). The rapidly emerging interdisciplinary field of neuroeconomics is using new knowledge about how brains make decisions to identify the causes of good and bad decisions (Camerer *et al.* 2005). Similarly, political decisions such as voting choice can be illuminated by investigations in psychology and neuroscience (Westen 2007).

16.5 Lessons

The successes and attractive prospects of cognitive science can be attributed to five factors: ideas, methods, people, places, and organizations (Thagard 2005a). It is only useful for people from different disciplines to try to collaborate if there are theoretically powerful ideas that cross disciplinary boundaries. For cognitive science, the main integrative ideas have been representation and computation, which can illuminate the nature of thinking in ways that are useful for all fields of cognitive science—psychology, neuroscience, artificial intelligence, philosophy, linguistics, and anthropology. A representation is a mental structure that can stand for things and events in the world, and inference is a computational mental process that transforms representations. There are other more specific ideas that find valuable applications in many fields, for example particular kinds of representations such as rules and concepts. For instance, some psycholinguists hold that knowledge of language consists primarily of rules such as 'To put an English verb in the past tense, add -ed'. For cognitive scientists, a concept is not a word or an abstract entity, but a mental representation with complex internal structure (Murphy 2002).

In addition, successful interdisciplinary collaboration requires complementary methods. Cognitive science employs many different methods, including psychological experiments, neurological experiments, computer simulations, conceptual analysis, linguistic theorizing, and ethnography. Few people have the time and aptitude to master more than one or two of these methods, but cognitive science benefits from the ways in which methods can be combined to help develop and evaluate explanatory theories about how the mind works. For example, a theory about the nature of concepts can be evaluated on the basis of all of the following: psychological experiments about how people form new concepts; neurological experiments about multiple brain areas involved in the use of concepts; computer simulations of concept learning and application; philosophical reflection on how concepts attach to the world; linguistics studies of concepts in different languages; and ethnographic studies that compare concepts such as color across different cultures. The goal of cognitive science is to arrive at theories that are strongly supported by evidence acquired through all these methods.

The initiation and progress of an interdisciplinary enterprise requires the participation of extraordinary people with the energy and vision to combine the insights of multiple fields. The origins of cognitive science in the 1940s and 1950s benefited from the efforts of exceptional intellectual talents such as Alan Turing, Herbert Simon, George Miller, Noam Chomsky, and Marvin Minsky. Each of these thinkers combined powerful theoretical ability with an appreciation of the insights and methods provided by a variety of different fields. The development of cognitive science organizations in the late 1970s depended on the intellectual vision and organizational skills of another generation of interdisciplinary talents, including Allan Collins, Donald Norman, and Roger Schank. Today, cognitive science depends on a host of people who are active both intellectually and practically in organizations such as the Cognitive Science Society.

Ideas, methods, and people cannot operate in isolation from each other, and occasional conferences are not sufficient to bring about the theoretical and experimental benefits possible from interdisciplinary research. It is therefore important to have places where disciplines can come together on a much more regular basis, at universities or other research institutions. In the 1960s, the Center for Cognitive Studies at Harvard, led by George Miller and Jerome Bruner, brought together many of the early contributors to the interdisciplinary study of mind. Carnegie Mellon University also provided a lively center of activity because of the presence of Herbert Simon and Allen Newell. In the 1970s, other universities such as Yale, Pennsylvania, Berkeley, Michigan, and Edinburgh developed active cognitive science programs, and by the beginning of the twenty-first century there were many places that played the crucial role of fostering such interdisciplinary work. Some do so by explicitly having cognitive science programs, but there are many other related enterprises with different names, such as Harvard's Mind/Brain/Behavior initiative.

Finally, the successful pursuit of an interdisciplinary field is greatly helped by the development of organizations that foster the communication of ideas and methods across fields. For cognitive science, the main organization is the Cognitive Science Society, which began in 1979 and is now complemented by smaller societies operating more locally in Europe and Asia. There also are more specific organizations operating at the intersection of particular pairs of fields, such the Society for Philosophy and Psychology, the Cognitive Neuroscience Society, and the International Conference on Cognitive Modeling. The Cognitive Science Society holds annual conferences that bring together people from many institutions and fields, although psychologists are by far the most heavily represented. The Cognitive Science Society publishes the journal *Cognitive Science* and the new *Topics in Cognitive Science*, which are complemented by a host of other interdisciplinary journals as well as a huge range of periodicals in the various fields of cognitive science. Thus organizations such as societies and journals are an important part of the flourishing of an interdisciplinary field. Goldstone and Leydesdorff (2006) used citation patterns to show that *Cognitive Science* plays a unique bridging role in transferring information across psychology, computer science, neuroscience, and education. Interdisciplinarity can be measured not only by the number of articles produced by multidisciplinary teams, but also by the role that publications play in connecting fields, thereby merging perspectives, tools, and methods.

Like narrower fields, interdisciplinary ventures are far from static, but benefit from changes in ideas, methods, people, places, and organizations. Much cognitive science work has shifted dramatically in recent years toward neuroscience, as many researchers see the study of the brain as providing much of the currently most exciting work on cognition. But not all psychologists, philosophers, or other practitioners share this view, which is just as well. The last thing needed by an interdisciplinary field, or any particular discipline for that matter, is a monolithic approach that narrows down to only a small set of ideas or methods.

In contrast, the full benefits of interdisciplinarity require integration, interaction, and blending of ideas and methods, not their mere juxtaposition and sequencing as found in multidisciplinarity (Klein, Chapter 2 this volume). Cognitive science is sufficiently mature to have its own textbooks, but some are still structured sequentially, describing separately the approaches taken by philosophy, psychology, neuroscience, linguistics, and artificial intelligence (Stillings *et al.* 1995; Friedenberg and Silverman 2006). In contrast, Thagard (2005b) discusses issues about mental representation in an integrated manner that intertwines issues and contributions from different disciplines.

16.6 Conclusion

This chapter has tried to show succinctly how the intellectual benefits of interdisciplinary research can dramatically outweigh the personal and social difficulties of operating in more than one field. Cognitive science provides an excellent illustration of the theoretical and experimental advantages of leaping beyond the confines of particular disciplines. The project of trying to understand the nature of mind is inherently interdisciplinary, requiring the ideas and methods of many different fields. There is still a place for researchers who prefer to restrict themselves to a narrow set of intellectual tools, but progress, especially of the most dramatic sort, requires the mingling of concepts, hypotheses, and methodologies from multiple disciplines. The human brain is so astonishingly complex that we should expect not decades but centuries of ongoing investigations in cognitive science.

Acknowledgements

I am grateful to Robert Frodeman and an anonymous referee for comments on an earlier draft, and to the Natural Sciences and Engineering Research Council of Canada for funding.

References

Appiah, A. (2008). *Experiments in ethics.* Cambridge, MA: Harvard University Press.

Bechtel, W. (2008). *Mental mechanisms: philosophical perspectives on cognitive neuroscience.* London: Taylor and Francis.

Bobrow, D.G. and Collins, A. (eds) (1975). *Representation and understanding: studies in cognitive science.* New York: Academic Press.

Boden, M.A. (2006). *Mind as machine: a history of cognitive science.* Oxford: Clarendon.

Camerer, C., Loewenstein, G.F., and Prelec, D. (2005). Neuroeconomics: how neuroscience can inform economics. *Journal of Economic Literature* **34**, 9–64.

Chomsky, N. (1957). *Syntactic structures.* The Hague: Mouton.

Dayan, P. and Abbott, L.F. (2001). *Theoretical neuroscience: computational and mathematical modeling of neural systems.* Cambridge, MA: MIT Press.

Dreyfus, H. (1979). *What computers can't do*, 2nd edn. New York: Harper.

Friedenberg, J. and Silverman, G. (2006). *Cognitive science: an introduction to the study of mind.* Thousand Oaks, CA: Sage.

Gardner, H. (1985). *The mind's new science.* New York: Basic Books.

Gentner, D. (1983). Structure mapping: a theoretical framework for analogy. *Cognitive Science* **7**, 155–70.

Gentner, D., Holyoak, K.H., and Kokinov, B.K. (eds) (2001). *The analogical mind: perspectives from cognitive science.* Cambridge, MA: MIT Press.

Goldstone, R.L., and Leydesdorff, L. (2006). The import and export of cognitive science. *Cognitive Science* **30**, 983–93.

Hebb, D.O. (1949). *The organization of behavior: a neuropsychological theory.* New York: Wiley.

Hesse, M. (1966). *Models and analogies in science.* Notre Dame, IN: Notre Dame University Press.

Holyoak, K.J. and Thagard, P. (1995). *Mental leaps: analogy in creative thought.* Cambridge, MA: MIT Press/Bradford Books.

Knobe, J. and Nichols, S. (2008). *Experimental philosophy.* Oxford: Oxford University Press.

Kuhn, T.S. (1970). *The structure of scientific revolutions*, 2nd edn. Chicago: University of Chicago Press.

McCorduck, P. (1979). *Machines who think.* San Francisco: W. H. Freeman.

McCulloch, W.S. (1965). *Embodiments of mind.* Cambridge, MA: MIT Press.

Miller, G.A. (1956). The magical number seven, plus or minus two: some limits on our capacity for processing information. *Psychological Review* **63**, 81–97.

Murphy, G.L. (2002). *The big book of concepts.* Cambridge, MA: MIT Press.

Pinker, S. (1994). *The language instinct: how the mind creates language.* New York: William Morrow.

Posner, M.I. and Rothbart, M.K. (2007). *Educating the human brain.* Washington, DC: American Psychological Association.

Rosch, E.B. and Mervis, C.B. (1975). Family resemblances: studies in the internal structure of categories. *Cognitive Psychology* **7**, 573–605.

Searle, J. (1980). Minds, brains, and programs. *Behavioral and Brain Sciences* **3**, 417–24.

Simon, H.A. (1991). *Models of my life.* New York: Basic Books.

Stainton, R. (ed.) (2006). *Contemporary debates in cognitive science.* Malden, MA: Blackwell.

Stillings, N.A., Weisler, S.E., Chase, C.H., Feinstein, M.H., Garfield, J.L., and Rissland, E. (1995). *Cognitive science: an introduction*, 2nd edn. Cambridge, MA: MIT Press.

Thagard, P. (1992). *Conceptual revolutions.* Princeton, NJ: Princeton University Press.

Thagard, P. (2005a). Being interdisciplinary: trading zones in cognitive science. In: S.J. Derry, C.D. Schunn, and A. Gernsbacher (eds) *Interdisciplinary collaboration: an emerging cognitive science*, pp. 317–39. Mahwah, NJ: Erlbaum.

Thagard, P. (2005b). *Mind: introduction to cognitive science*, 2nd ed. Cambridge, MA: MIT Press.

Thagard, P. (2006). *Hot thought: mechanisms and applications of emotional cognition*. Cambridge, MA: MIT Press.

Thagard, P. (ed.). (2007). *Philosophy of psychology and cognitive science*. Amsterdam: Elsevier.

Thagard, P. (2010). *The brain and the meaning of life*. Princeton, NJ: Princeton University Press.

Thagard, P. and Aubie, B. (2008). Emotional consciousness: a neural model of how cognitive appraisal and somatic perception interact to produce qualitative experience. *Consciousness and Cognition* **17**, 811–34.

Turing, A.M. (1950). Computing machinery and intelligence. *Mind* **59**, 433–60.

Westen, D. (2007). *The political brain: the role of emotion in deciding the fate of the nation*. New York: Public Affairs.

Wiener, N. (1961). *Cybernetics: or control and communication in the animal and the machine*, 2nd ed. Cambridge, MA: MIT Press.

Williamson, T. (2007). *The philosophy of philosophy*. Malden, MA: Blackwell.

CHAPTER 17

Computation and simulation

JOHANNES LENHARD

This chapter treats the two concepts of computation and simulation as a pairing. Although this pairing occurs regularly in the literature, and although it makes perfect sense to link these terms when discussing interdisciplinarity within computer science, it is appropriate to start by distinguishing the terms. Clearly, they are neither identical nor is one a subset of the other. Computation *per se* is a mathematical activity. It has been speeded up tremendously by electronic computing devices, but taken as an isolated fact this is hardly relevant to interdisciplinarity. What makes it highly relevant is the way in which electronic computing machines are embedded into a scientific—technological context. Simulation modeling is perhaps the more accurate term here as it indicates that it is a special brand of scientific modeling. It stands for a way—or rather a variety of ways—to make use of computational resources in activities that are not strictly mathematical but are of a fundamentally interdisciplinary nature.

The word computer was initially used for human workers, typically women, who were employed to carry through huge numbers of elementary calculations (Grier 2005). This activity was replaced by machines—much in the vein of the replacement of human work skills by machine tools during the Industrial Revolution. At first, analogue devices, like Vannevar Bush's differential analyzer, were used to solve specific classes of mathematical problems (Mindell 2002). During the 1940s and early 1950s, digital computers were developed. Both types of machine coexisted for a while, but eventually digital computers took over and became a synonym for 'computer'. One decisive reason for this was the establishment of computers not as tailor-made instruments for specific purposes, but as general-purpose machines—machines that can be instructed to do virtually everything that can be described in a formal way (Ceruzzi 2003). This has extraordinary implications. Simulation translates everything, not just numerical algorithms, into digital information and uses computation to construct any object, be it an airplane, social trend, or cultural belief.

Computational power is at the heart of the computer, but why does it matter? One could argue that mathematization has been one of the driving forces for science and technology since the Scientific Revolution. Is there a point in calculating faster? Yes, there is.

The new instrument has widened the scope of science and technology in unforeseen ways. Every argument about these matters will have to introduce the more specific aspects of applying computers, and in the following simulation will play this role.

Simulation, too, has analogue precursors—mimicking models like flight simulators that worked with sophisticated arrangements of Bowden cables, etc., to create a model cockpit that for the novice pilot inside should feel like a real one (Rolfe and Staples 1988). Today, however, simulation is normally conceived of as digital computer simulation and this is the sense in which it is relevant for interdisciplinarity.

From now on, the present chapter will concentrate on the digital computer and consequently will consider simulation as 'digital computer simulation'. Roughly in parallel to the development and spread of the computer, simulation has become established as a new means for knowledge production. The amount of computing power that has become available over the last decades is undoubtedly one reason for this development. Another reason is that simulation is an especially generic instrument, quickly convertible to different contexts of application. However, computer simulation affects the social, cognitive, and organizational spheres of science and technology—and even significant parts of broader contemporary culture. In particular, simulation practices involve the concept of interdisciplinarity in many (and partly new) ways. This concept might provide a lens to view the particularities of computation and simulation (C&S) and thus help to get a grip on its strengths and weaknesses.

17.1 Historical development

We can distinguish three phases of C&S: a pioneering one lasting from 1940–60; a phase of disciplinary specialization, roughly from 1960 to the 1980s; and thirdly, the recent phase of ubiquitous diffusion. All phases are connected to interdisciplinarity in a different way.

17.1.1 Pioneering phase 1940–60

Digital computer simulation emerged in the scientific–military complex of World War II. This first pioneering phase saw an interdisciplinary effort to establish C&S in the context of 'big science'. This time witnessed a number of interrelated technological innovations that make it hard to single out a linear story. Perhaps the *First draft of a report on the EDVAC*, written by John von Neumann and based on the ideas of many researchers (von Neumann 1945), can be seen as a founding document of the modern mainframe computer—a computing machine that can be programmed and has a serially working central processing unit. This achievement was based on a close encounter between engineers and mathematicians, disciplines that do not meet regularly but did so in the context of war-related big science (Heims 1980; Aspray 1990; Edwards 1996; Akera 2006). Funding increased rapidly, particularly in the United States, and consequently a great number of people got involved in research and development and also organizational issues. Progress in C&S happened in relatively small interdisciplinary groups that were highly interconnected, like those at the Moore School of Electrical Engineering at the University of Pennsylvania or at MIT; sometimes these groups were embedded in bigger and more hierarchically structured endeavors in computation like those in the Aberdeen Proving Ground (USA) or Bletchley Park (UK).

Interestingly, basic simulation concepts and techniques were invented at the same time in direct correspondence to the anticipated growing resources for doing calculations. Monte Carlo methods, their sibling Markov chain Monte Carlo (MCMC), cellular automata, artificial neural networks, and finite difference methods all go back to this early phase (Metropolis 1980; Aspray 1990; Galison 1996; Johnson 2004). The so-called Cybernetics Group is an example where scientists of very different disciplines tried to spell out the new possibilities of C&S. The Macy Conferences of the 1940s and 1950s documented their highflying hopes for interdisciplinary achievements and even a new epoch of science that surrounded the computer (Heims 1991; Pias 2003). The joint interest of cognitive scientists, engineers, mathematicians, and psychologists led to new fields of research like human–machine interaction (cybernetics) or artificial intelligence.

Though the mentioned simulation techniques are all different, two basic approaches are manifest. (1) Simulation is taken as a numerical solution, i.e. a way to solve mathematical problems intractable by other means. This view recognizes the instrumental power but tends to downplay the conceptual significance of simulation. (2) Simulation is taken as a new kind of modeling that makes use of the new possibilities of computer syntax but is not derived from traditional mathematics. This second stance ascribes much more independence to simulation and tends to underline claims about (radically) new features. Finite differences, for instance, transform differential equations into a discrete system to be treated in a stepwise procedure, or MCMC uses statistical experiments to approximate theoretically defined probability distributions. Both simulation methods build on a pre-existing mathematical model and therefore fit more to the first approach. Cellular automata or artificial neural networks fit more to the second approach. However, these two approaches coexist, pure cases are rare, and the tension between these approaches is extant in most practical applications of simulation.

17.1.2 Disciplinary specialization 1960–85

In the course of a second historical phase, electronic computation and computer simulation methods ceased to be extraordinary. The often *ad hoc* mixture of interdisciplinary teams located at the frontier of research that developed prototypes of machines changed into a professional—and often industrial—configuration of research and development. The early estimation that the potential demand for electronic computing machines would be merely a handful worldwide was disproved, and the computer became a commercial product. IBM is the icon of the commercial facet of this phase. Moreover, this icon also signaled the end of this phase, as the introduction of the personal computer by IBM in 1981 heralded the next phase in which the influence of IBM became marginal.

The advance of computers still happened in interdisciplinary research teams, in the context of industry as well as, to a smaller extent, academia, and showed some continuity with the first historic phase. The military sector remained a significant source of money and interest, as steady progress was made with regard to decreasing size and increasing computational power. The so-called Moore's law observes the exponential advance in these measures. Importantly, the whole field underwent professionalization to a very high degree. One facet was the emergence of trained computer scientists that in a sense secured and stabilized the interdisciplinary mixtures of the pioneering phase. Computer science, however, despite professionalization, was—and continues to be—a vaguely defined discipline. While computer science departments split off from mathematics departments in the early 1980s, the second generation of computer scientists are more like software engineers. It is still controversial whether computer science constitutes a discipline in its own right or is of an interdisciplinary nature (Mahoney 1992). A main reason for this is that the instrument itself became so complex. Hence it turned out to be unfeasible to draw an abstract picture of how computer models work or how the computer itself works—at least there is still no unanimity about it, and different viewpoints that favor abstract mathematical approaches versus practice oriented engineering ones still vie with each other (MacKenzie 2001).

Expertise was distributed among hardware developers and software engineers; standardized components began to foster exchangeability both in terms of instruments and people. The development of higher-level computer languages, like FORTRAN, which is still in use today, provides a particularly significant example. Information and computer science were established as scientific and engineering disciplines. In sum, a corona of specialized disciplines accompanied the further development of C&S during that phase, disciplines that divided labor and ranged from developing optimal materials to the investigation of effective algorithms. That is, a prominent strain of interdisciplinarity was structured professionally with the computer at the center.

Specialization and professionalization also took place in simulation. On the one hand, tentative and often merely heuristically based methods were supplemented by more succinct conditions of convergence and rules of applicability, for example, the question of which sort of space-time grids compromise speed (a not-too-fine grid) and accuracy (a not-too-coarse grid) has become its own subdiscipline. On the other hand, abstract mathematical treatment is restricted to 'nice' cases, whereas simulation is not. Quite the contrary: the range of applications steadily increased and has been extended to 'nasty' cases where the experience of practitioners has had to guide simulation approaches. Thus, both mathematical knowledge and experience in application have contributed to simulation, which has become its own field of research.

While at this time C&S was becoming a more and more viable alternative approach in many scientific areas, it was still seen as a secondary option to more established and traditional methods of scientific research. C&S covered both parts of the scientific culture: representing and intervening (Hacking 1983). The socialization of the researchers played a key role in the assignment of an inferior status to C&S compared with traditional approaches to science: inferiority of machine- and instrument-related work compared with theoretical skills was deeply ingrained in most disciplines. Also, the association of simulation with imitation bestowed on C&S a somewhat distanced or even deceptive character.

Beginning in the 1960s, a new generation of engineers and scientists used computer simulation more and more like a 'scientific instrument', and even oriented future research plans toward the possibilities that were suited to computational methods. However, these groups remained in a minority. Steadily, C&S acquired more connectivity to various fields of application, aptly symbolized by the regular location of equipment in the basement of buildings,—not the finest address, but a base technology for many. A clear indicator of the specialized and partly autonomous status of C&S is the emergence of journals and conferences devoted to it—most prominently the Winter Simulation Conference with its own series of proceedings. Also, a number of books were published that aimed to summarize and explain the specifics of C&S, prominent among them those by McLeod (1968) or Zeigler (1976; cf. Simon 1969). The C&S community was not centrally organized, and the actual application of C&S was much wider than reflected in the journals. So it remained open and controversial whether C&S constituted a separate discipline or an interdisciplinary field.

17.1.3 Ubiquitous diffusion from 1985 on

The third phase in the evolution of C&S is linked to the wide availability of smaller machines—personal computers, workstations, and networked architectures of them—that made C&S largely independent of 'big science' and helped transform it into a virtually ubiquitous phenomenon. There is no exact starting point for this phase, but it is closely connected to the so-called 'PC Revolution'. C&S left the somewhat restricted space of computationally intensive special sciences and gained ground in many areas of modern culture, from movies ('animation') to the internet's 'virtual reality'. At the same time, C&S acquired a new status in the more restricted area of science and technology: from a professional but second-rank method during phase two it developed into an approach of equal rank, now openly accepted and widely hybridized with all sorts of traditional approaches and disciplines. Relying on C&S simply became a matter of course. From the beginning on, most of these fields, from computer animation to computational sciences of various types, perceived themselves as transforming existing disciplinary fields into interdisciplinary ones. Basically, an interdisciplinary mixture of computer science became embedded into a much wider system than in phase two. To speak in terms of the historian and sociologist of science Terry Shinn, C&S became a 'generic instrument' (Shinn 2001). Two of the main things that enabled C&S to become so widely available and widely applied were the decreasing cost and increasing transferability that have been made possible by a series of transformations.

Standardization of hardware and software came with the PC Revolution. Networked architectures of personal computers and workstations—which connect the workplace directly to local computing facilities—now replaced the big supercomputing machines. The latter still exist and are the basis for popular rankings of which countries and institutions own the most powerful computers. However, viewed statistically, the majority of scientific and technological research and applications run on relatively small machines. Applications in extrascientific culture rely entirely on these smaller architectures. They are affordable to a wide audience of research groups, commercial firms, the entertainment business, and others. Not only are hardware components widely sold as commercial goods, but the system software has also been standardized so that, for instance, trained students can do the job of a system administrator at an academic research institute. Software programming also has been standardized to an important degree: so-called higher-level languages are effective in black-boxing the immediate contact with the computing kernel, many functions can be imported from software libraries, and languages like C++ have acquired ISO certification, which turns them into an officially standardized and therefore highly exchangeable product (Shinn 2006).

A spectrum of computational sciences appeared. Every classical natural science nowadays has one or more computational siblings, like computational fluid dynamics, or computational molecular biology, not to speak of engineering disciplines—computer-aided design is going to do away with paper and pencil. Also sociology, linguistics, and other branches of social sciences and humanities have embraced this development (Bynum and Moor 1998; see the *Journal for Artificial Societies and Social Simulation*). Furthermore, medicine, cognitive, and neuroscience often rely crucially on C&S—examples are so abundant that any attempt to give an exhaustive list seems forbidding. Computational sciences emerged by the coalescence of C&S with formerly independent disciplines or without any disciplinary forerunner. Obviously these changes affect the configuration of the disciplines generally. This is true also for large parts of the engineering and design sciences. Computer-aided design, for instance, has changed from having a somewhat exotic status at the end of phase two to a now everyday process—without even an alternative in many branches. Last, but not least, the entertainment industry and large parts of the educational sector have been affected greatly by simulation and gaming. The entertainment sector based on C&S constitutes a multibillion dollar industry; in 2008, for the first time, videogame software accounted for more than half the electronic entertainment media revenues in the global market. This does not imply that the military has lost influence; some researchers hold that military and entertainment sector have converged into one complex and are the driving force of C&S (Lenoir 2000).

These developments all came together and created a new, computer-based, decentralized type of interdisciplinarity. Whereas during phase two researchers and developers with different disciplinary expertise met in the (physical) vicinity of the computer, in phase three the maturing of C&S comes with a distribution of instruments, people, and expertise. Researchers of one kind rely on many elements and modules that others have provided: for instance, molecular biologists may use C&S models and plug-in software that have been developed in completely different areas. This effect of black-boxing—of using some device without detailed knowledge of its internal set-up—is well known from all kinds of instruments that have become established and widely used.

17.2 Dynamics of interdisciplinarity

17.2.1 Complexity, experimentation, visualization

Are the dynamics of C&S a driver or rather an outcome of other recent changes in science and science-related fields? A number of much-debated diagnoses identify fundamental transformations of research (Latour 1993; Gibbons *et al.* 1994; Ziman 2000) the evaluation of which is not an issue here. However, there is basic agreement that science has increasingly entered real-life problems with all their complexities, and that the strategies of simplification, idealization, and confinement have ceased to work properly because these methods would have to reduce complexity which in turn would destroy the crucial details of phenomena under investigation.

Two specific features of C&S support and highlight the recent drift into complexity: first, an experimental and explorative mode of research, and second, the use of visualization. Complex computational models are often opaque in an epistemic sense and therefore work on them has to proceed by experimentation. Researchers experiment on simulations by changing, adding, and adapting parameterization schemes and submodules and by running the simulations repeatedly (Dowling 1999; Hughes 1999; Fox Keller 2003; Winsberg 2003; Lenhard 2007). The typical goal is to compare and adapt the behavior of these tweaked models to the phenomena they should simulate. Such a practice goes back to phase one but became fully fledged in phase three, because the exploratory nature of computational modeling is greatly enhanced by easy and cheap access to computational power. Now researchers work on computational models, whether they consider themselves experimentalists or theorists. 'Working on' models mean exploring the relationships between input data and output data in order to produce verifiable predictions or better (that is, more accurate) models. With the mature desktop computer, model exploration became a key practice in scientific research—not one limited to computational scientists, but rather one which is commonplace among most scientists and engineers.

This exploratory mode depends upon researchers' ability to quickly assess the outputs of various models. Assessment of model outputs is especially feasible in visual, as opposed to numerical, forms. Visualization is also a feature of the desktop computing revolution, although one which did not emerge from scientific computing (Lynch and Woolgar 1988; Galison 1997; Jones and Galison 1998; Ihde 2006; Johnson 2006). Desktop computers, with their many different kinds of users, have been at the center of a series of changes in the visual display of information. From computer games, to animation, to web pages, computers have become devices that are focused on visual display. These extrascientific demands have created capacities critical for scientific uses—especially in three-dimensional display. Visualization in science has changed because of the ability of computers to generate images. These images are tremendously powerful; they carry information more efficiently than do tables of numerical outputs, and as a result they yield compelling results—sometimes in misleading ways. Consequently, visualization reinforces the exploratory mode of scientific research by making possible the quick uptake of results from computational models (Johnson and Lenhard 2008).

Some examples illustrate the relevance to C&S for science, engineering, education, participation, and entertainment:

- Flight simulators are C&S-based devices for pilot education that comprise a very good imitation of many facets of the 'real' situation encountered by a pilot. Flight simulators are built on physical theory—fluid dynamics—as well as various engineering sciences and, at the practical end of the spectrum, on the experience of pilots that has helped to tune the simulator so that it 'feels real' for the person inside. Education, research, entertainment, and military applications can hardly be separated. In fact, the videogames industry is a major player in graphics development.

- Driving simulators are a related example. You can find them as educational devices (to avoid real risk and cost) and also as entertainment machines (often to seek virtual risk). A different kind of simulation (largely oriented in computational fluid dynamics) is employed in motor design where geometry and other parameters can be changed and experimented with to explore and find an optimal setting. Finite element methods divide up a complex object into small but finitely big elements. They constitute still another kind of simulation, most widely used in engineering, for instance to imitate the behavior of a car body for virtual crash tests and to avoid expensive auto body destruction. C&S accompanies much of the research and development process for many technological artifacts.
- Endoscope simulation is a newly developed device to train surgeons to practice endoscopic operations. Clearly this device needs to be an adequate imitation in several respects: the shape, color, and texture of what the surgeon sees on the screen; how the hand movements of the surgeon feel (to the surgeon); what effects these movements have on the simulated body of the patient. All these aspects have to be simulated to a high degree of accuracy for the simulation to make sense for educational purposes.

For a simulation to work properly two conditions are crucial. (1) The purpose-independent instrumental features—technical possibilities for display, feedback and interaction—need to be in place. The degree to which this first condition can be met rests itself on an interdisciplinary achievement. (2) C&S methods have to be guided adequately so that the performance of the resulting models or devices are actually sufficiently good imitations. This second condition depends heavily on the intended purpose, for instance when an endocrinological simulation counts as realistic, or when a simulated car crash can substitute for a real one. Whatever the specific content of this condition its very specification rests on an interdisciplinary accomplishment—what determines a good imitation and what characterizes a good performance are interdisciplinary questions. Hence, depending on the intended application, medical doctors, pilots, or other experienced specialists will typically take part in the development of a particular simulation. With increasing refinement of simulation technology, the interdisciplinary task grows as a more and more diverse range of aspects are simulated.

Consider the field of simulation and gaming that conceives of itself as its own cross-cutting discipline that has an interdisciplinary nature. (Producing an appropriate virtual environment is a highly interdisciplinary task for the programmer). Many simulation games only work if the imitation is good enough for the players to be immersed in the simulation. Movements, body shape, language processing, and many more aspects are crucial for that. The renaming of a pertinent scientific journal in the field serves as illustration: *Simulation & Gaming: an International Journal of Theory, Practice, and Research* was changed in 1995 to *Simulation & Gaming: an Interdisciplinary Journal of Theory, Practice, and Research.*

17.2.2 'Research technology' and the 'trading zone'

A number of views exist that draw widely different pictures of the dynamics of C&S, and especially its interdisciplinary and cultural significance. A major reason for this diversity is surely the heterogeneity of C&S itself that allows for various perspectives. The French philosopher Jean Baudrillard, for instance, aims to give a broad panorama of culture and of the way reality, symbols, and society interact. He takes 'simulacra' as a key notion to name the tendency to act in symbolically constructed worlds without a real underpinning and connects this to simulation in particular. The philosopher of science Paul Humphreys, to give another instance, analyzes C&S from a science-based viewpoint. In particular, he identifies 'computational templates' as a key for the dynamics of C&S, that is, pieces that code mathematical formulae and which can travel widely, with these formulae showing up in totally different contexts and disciplines (Humphreys 2004).

The analyses of Terry Shinn and Peter Galison take a middle ground in this spectrum and especially take into account of the social aspects of the interdisciplinary dynamics of C&S. They subsume it under a broader framework and see these dynamics as an instantiation of what they call 'research technology' (Shinn) and 'trading zone' (Galison). These frameworks are of course different, but each captures essential aspects of the C&S revolution.

Shinn claims that four elements characterize research technologies. First they are produced by interstitial communities; i.e. they do not arise from single institutional, disciplinary, or industrial problems or uses. Second, 'the devices that research technologists deal with are generic' (Shinn 2001, p. 9), meaning that they are not designed to respond to any specific industrial or academic demand. Third, research technologies 'generate novel ways of representing visually or otherwise events and empirical phenomena' (Shinn 2001, p. 9). Lastly, they are disembedded from their context of invention, a direct consequence of their general nature, becoming non-local to any one scientific community. In Shinn (2006) and Küppers *et al.* (2006), the concept of research technology is employed to interpret C&S as a 'generic instrument' that can be used in a large number of heterogeneous domains. The generic quality of C&S is key to its great economic success.

Shinn identifies a simulation-linked 'lingua franca' as an important factor to bridge different disciplinary backgrounds and refers to Peter Galison's work on the concept of a 'trading zone' (Galison 1996, 1997). In his article on the early history of the Monte Carlo method, Galison captures the interdisciplinary dynamics between the researchers as follows:

> 'Their common activity centered around the computer. More precisely, nuclear-weapons theorists transformed the nascent "calculating machine," and in the process created alternative realities to which both theory and experiment bore uneasy ties. Grounded in statistics, game theory, sampling, and computer coding, these simulations constituted what I have been calling a "trading zone," an arena in which radically different activities could be *locally*, but not globally, coordinated' (Galison 1996, p. 119, original emphasis).

Galison's concept, like that of Shinn, embraces a strong social component and links inter- or transdisciplinarity to a C&S-related language. From this viewpoint, Galison describes the passage from the historical phase one to phase two (see above) as a transformation of language: 'By the 1960's, what had been a pidgin had become a full-fledged creole: the language of a self-supporting subculture with enough structure and interest to support a research life without being an annex of another discipline…' (Galison 1996, p. 153).

17.2.3 Simulations at the edge

Simulation models of climate science present probably the most prominent simulation models. They are one of the most complex exemplars of their kind, progressing at the edge of C&S technology. Furthermore, they have become objects of debate among a wider public and are of great relevance in the policy arena; hence, C&S also inhabits the edge between science and policy. A famous forerunner in this respect is *Limits to growth* (Meadows *et al.* 1972), a study commissioned by the Club of Rome for which Jay Forrester's Whirlwind computing project at MIT was operational (Akera 2006). This C&S-based study—and in particular its mission to make predictions for complex systems—received enormous publicity and introduced simulations as objects not only in science but also in the political arena. Today, climate simulations have inherited this role. They have triggered a flurry of analyses, typically in the field of science and technology studies (STS), that investigate how these models function, how researchers and politicians argue with them, etc. (Miller and Edwards 2001; Lenhard *et al.* 2006). This case exemplifies the potential usage of C&S-based models in different disciplinary contexts and even different sectors of society.

A pertinent feature in climate science is the division between paleoclimatological data, mainly obtained from ice core drillings, and C&S-based model results. Paleodata are obtained according to the long-established empirical methodology of science. They are used to validate climate simulations: are they able, if simulating the past, to reproduce the observed data adequately? A lot of adjustments and *ad hoc* tuning, for instance sounding out which parameter values suit best, is necessary to optimize the performance of complex simulation models. Nevertheless, the C&S-based model results play a more prominent role in the public view and among the climate science and policy community. One main reason for this surely is that they can be used for forecasting. Another is that there is significant political pressure to analyze the entire climate system. Recent projects speak of 'earth system analysis'. This cannot be done using any disciplinary approach of meteorology, oceanography, biology, economy, or other scientific discipline. C&S offers a kind of technological, instrumental pathway to a network-like interdisciplinary integration.

17.2.4 Network-like interdisciplinary integration

The C&S-related dynamics of interdisciplinarity has developed over time, largely in parallel with the historical phases of C&S. Again, climate simulations can illustrate this. The historical origins of climate analysis are rooted in models of the circulation of the atmosphere—general circulation models (GCMs) that have been developed since the mid 1950s. The theoretical core of these models is built by the so-called fundamental equations, a system of partial differential equations from the physics of motion and thermodynamics. With the growing interest in climate change in the 1980s, a period of substantial growth of these models was inaugurated because more and more facets of the climate system had to be included while aiming at a comprehensive picture. The growth both included the resolution of more subprocesses, like the dynamics of aerosols in the atmosphere, and also the addition of subprocesses in parameterized form, like clouds which are included via certain parameters that express the effects of clouds but not their internal dynamics.

One aspect of the development of more comprehensive models is of particular importance. A multitude of submodels had to be included into the atmospheric GCMs that had little to do with the theoretical physical basis of the atmospheric circulation, e.g. ice cover, circulation of the oceans, or land use. Today, atmospheric GCMs—and with them physics as a discipline—have lost their central place; coupled models entertain a deliberatively modular architecture and comprise a number of highly interactive submodels. These have been developed by groups of diverse disciplinary affiliations. They constantly interchange data during the runtime, but do not share a common theoretical framework. Thus, hierarchical integration around atmospheric GCMs has been replaced by a network-like integration of exchangeable modules. Küppers and Lenhard (2006) argue that the architecture of simulation models reflects a new style of interdisciplinary modeling.

These observations are not restricted to climate simulations, but rather exemplify a typical phase three approach. The work of Merz (2006), for instance, shows quite similar developments in the organization of the particle collider at CERN which is based on a complicated and extensive phase of simulation. Various kinds of distributed computing also illustrate this: so-called grid computing is a recent issue that deals with the question of how to couple or even integrate various C&S resources at different locations.

17.3 Outlook

Simulation and computing will continue to change the face of science, engineering, and many facets of modern culture, driven by ongoing developments of all elements of C&S as well as by demands for highly complex models and applications. Hence any diagnosis of the future of C&S may be outdated quickly. A critical understanding of the technological nature of this instrument and its implications is still missing. Earlier accounts of the history of C&S still have to be adapted to the recent phase three. In particular, computer simulation often comes along with a distributed architecture, involving a multitude of working disciplinary researchers. Although some important discussions are going on already (e.g., Sismondo and Gissis 1999; Lenhard *et al.* 2006), the impact on the type and organization of interdisciplinarity is not yet fully conceptualized.

Important unresolved questions concern the issue of validation: C&S is especially attractive when direct comparison with real phenomena is difficult or impossible for reasons of risk, cost, or time. But how should these simulations be validated? How is this actually done? It is not clear what the meaning of 'validation' is or should be in this context. Again, climate science provide a good starting point as the Intergovernmental Panel on Climate Change (IPCC) regularly publishes an assessment report that summarizes current C&S-based knowledge. These assessment reports document and exemplify problems of displaying complex simulation results and of dealing with developing valid, policy-relevant simulation-based models and forecasts.

References

Akera, A. (2006). *Calculating a natural world: scientists, engineers, and computers during the rise of U.S. cold war research*. Cambridge, MA: MIT Press.

Aspray, W.F. (1990). *John von Neumann and the origins of modern computing*. Cambridge, MA: MIT Press.

Baudrillard, J. (1998). Simulacra and simulations. In: M. Poster (ed.) *Jean Baudrillard: selected writings*, pp. 166–8. Stanford, CA: Stanford University Press.

Bynum, T.W. and Moor, J. (1998). *The digital phoenix: how computers are changing philosophy*. Oxford: Blackwell.

Ceruzzi, P.E. (2003). *A history of modern computing*. Cambridge, MA: MIT Press.

Dowling, D. (1999). Experimenting on theories. *Science in Context* **12**(2), 261–73.

Edwards, P. (1996). *The closed world: computers and the politics of discourse in cold war America*. Cambridge, MA: MIT Press.

Fox Keller, E. (2003). Models, simulation, and computer experiments. In: H. Radder (ed.) *The philosophy of scientific experimentation*, pp. 198–215. Pittsburgh: University of Pittsburgh Press.

Galison, P. (1996). Computer simulations and the trading zone. In: P. Galison and D.J. Stump (eds) *The disunity of science: boundaries, contexts, and power*, pp. 118–57. Stanford, CA: Stanford University Press.

Galison, P. (1997). *Image and logic: a material culture of microphysics*. Chicago: University of Chicago Press.

Gibbons, M. *et al.* (1994). *The new production of knowledge: the dynamics of science and research in contemporary sciences*. London: Sage.

Grier, D.A. (2005). *When computers were human*. Princeton, NJ: Princeton University Press.

Hacking, I. (1983). *Representing and intervening: introductory topics in the philosophy of natural science*. Cambridge: Cambridge University Press.

Heims, S.J. (1980). *John von Neumann and Norbert Wiener: from mathematics to the technologies of life and death*. Cambridge, MA: MIT Press.

Heims, S.J. (1991). *Constructing a social science for postwar America: the cybernetics group, 1946–1953*. Cambridge, MA: MIT Press.

Hughes, R.I.G. (1999). The Ising model, computer simulation, and universal physics. In: M.S. Morgan and M. Morrison (eds) *Models as mediators*, pp. 97–145. Cambridge: Cambridge University Press.

Humphreys, P. (2004). *Extending ourselves: computational science, empiricism, and scientific method*. New York: Oxford University Press.

Ihde, D. (2006). Models, models everywhere. In: J. Lenhard, G. Küppers, and T. Shinn (eds) *Simulation: pragmatic construction of reality*, pp. 79–86. Dordrecht: Springer.

Johnson, A. (2004). From Boeing to Berkeley: civil engineers, the cold war and the origins of finite element analysis. In: M.N. Wise (ed.) *Growing explanations: historical perspectives on recent science*, pp. 133–58. Durham, NC: Duke University Press.

Johnson, A. (2006). The shape of molecules to come: algorithms, models, and visions of the nanoscale. In: J. Lenhard, G. Küppers, and T. Shinn (eds) *Simulation: pragmatic construction of reality*, pp. 25–39. Dordrecht: Springer.

Johnson, A. and Lenhard, J. (2008). *Towards a culture of prediction: computational modeling in the era of desktop computing*. Unpublished manuscript.

Jones, C.A. and Galison, P. (1998). *Picturing science, producing art*. New York: Routledge.

Küppers, G. and Lenhard, J. (2006). Simulation and a revolution in modelling style: from hierarchical to network-like integration. In: J. Lenhard, G. Küppers, and T. Shinn (ed.) *Simulation: pragmatic construction of reality*, pp. 89–106. Dordrecht: Springer.

Küppers, G., Lenhard, J., and Shinn, T. (2006). Computer simulation: practice, epistemology, and social dynamics. In: J. Lenhard, G. Küppers, and T. Shinn (eds) *Simulation: pragmatic construction of reality*, pp. 3–22. Dordrecht: Springer.

Latour, B. (1993). *We have never been modern*. Cambridge, MA: Harvard University Press.

Lenhard, J. (2007). Computer simulation: the cooperation between experimenting and modeling. *Philosophy of Science* **74**, 176–94.

Lenhard, J., Lücking, H., and Schwechheimer, H. (2006). Expert knowledge, Mode-2 and scientific disciplines: two contrasting views. *Science and Public Policy* **33**(5), pp. 341–50.

Lenoir, T. (2000). All but war is simulation: the military-entertainment complex. *Configurations* **8**(3), 289–335.

Lynch, M. and Woolgar, S. (1988). *Representation in scientific practice*. Cambridge, MA: MIT Press.

MacKenzie, D. (2001). *Mechanizing proof: computing, risk, and trust*. Cambridge, MA: MIT Press.

McLeod, J. (1968). *Simulation: the dynamic modeling of ideas and systems with computers*. New York: McGraw-Hill.

Mahoney, M.S. (1992). Computers and mathematics: the search for a discipline of computer science. In: J. Echeverria, A. Ibarra, and T. Mormann Berlin (ed.) *The space of mathematics*, pp. 347–61. New York: de Gruyter.

Meadows, D.H., Meadows, D.L., Randers, J., and Behrens, W.W. III (1972). *Limits to growth*. New York: Potomac Associates.

Merz, M. (2006). Locating the dry lab on the lab map. In: J. Lenhard, G. Küppers, and T. Shinn (eds) *Simulation: pragmatic construction of reality*, pp. 155–72. Dordrecht: Springer.

Metropolis, N., Howlett, J., and Rota, G.-C. (1980). *A history of computing in the twentieth century*. New York: Academic Press.

Miller, C.A. and Edwards, P.N. (eds) (2001). *Changing the atmosphere*. Cambridge, MA: MIT Press.

Mindell, D. (2002). *Between human and machine: feedback, control, and computing before cybernetics*. Baltimore, MD: Johns Hopkins University Press.

von Neumann, J. (1945). *First draft of a report on the EDVAC*. University of Pennsylvania, Moore School of Electrical Engineering.

Pias, C. (2003). *Cybernetics: the Macy Conferences 1946–1953*. Zürich: Diaphanes.

Rolfe, J.M. and Staples, K.J. (eds) (1988). *Flight simulation*. Cambridge: Cambridge University Press.

Shinn, T. (2006). When is simulation a research technology? Practices, markets, and lingua franca. In: J. Lenhard, G. Küppers, and T. Shinn (eds) *Simulation: pragmatic construction of reality*, pp. 187–203. Dordrecht: Springer.

Shinn, T. (2001). A fresh look at instrumentation: an introduction. In: B. Joerges and T. Shinn (eds) *Instrumentation between science, state and industry*, pp. 1–13. Dordrecht: Kluwer Academic Publishers.

Simon, H.A. (1969). *The sciences of the artificial*. Cambridge, MA: MIT Press.

Sismondo, S. and Gissis, S. (1999). Practices of modeling and simulation. *Science in Context* **12** [Special Issue].

Winsberg, E. (2003). Simulated experiments: methodology for a virtual world. *Philosophy of Science* **70**, 105–25.

Zeigler, B.P. (1976). *Theory of modelling and simulation*. Malabar: Krieger.

Ziman, J. (2000). *Real science: what it is, and what it means*, Cambridge: Cambridge University Press.

CHAPTER 18

Interdisciplinarity in ethics and the ethics of interdisciplinarity

ANNE BALSAMO AND CARL MITCHAM

This chapter explores interdisciplinarity dynamics in the context of ethics. It starts by observing the origins of ethics in the predisciplinary plentitude of philosophy followed by a historical overview of ethics in terms of significant interdisciplinary formations broadly construed. Next is an extended discussion of interdisciplinarity in contemporary fields of applied ethics. A third section reverses the approach, to consider the possibility of ethical guidelines for the practice of interdisciplinarity, through reflection on the character of knowledge production in the digital age.

Ethics itself is constituted by systematic, critical reflection on human action with the aim of both increasing knowledge about and improving culturally or personally acceptable behavior. In that *ethics* involves critical reflection, it is distinct from *morality* which exists largely independent of conscious thought. Because ethics bridges knowing and doing, it constitutes a site for multiple inter- and transdisciplinary engagements. Most commonly, however, it is identified as a major branch of the discipline of philosophy—alongside logic, epistemology, and metaphysics. Yet upon closer examination, ethics can be seen as drawing on a number of disciplines that may also be described as having emerged from it; disciplines such as psychology and anthropology, as well as politics and economics, significantly influence the domain of ethics. Especially in what are called its *applied* or *practical* versions, ethics takes form as hybridizations of disciplines and specialized concerns in the cases of biomedical ethics, environmental ethics, or computer ethics—each of which depends on multidisciplinary interactions with other domains. Finally, ethics manifests strong transdisciplinary elements, insofar as it is heavily dependent on life experience. Informed by the development of multiple disciplines, implicated in the creation of new hybrid research fields, and constituted by transdisciplinary questions, ethics is an inherently interdisciplinary endeavor.

18.1 Historical overview

In its classical form philosophy was a term that referenced all learning and was thus deeply interdisciplinary. Indeed, it may be more accurate to describe philosophy as originally *pre*disciplinary, since it preceded the demarcation of disciplines. The history of the pursuit of knowledge in the European tradition unfolds as a progressive spinning off of multiple specialized forms of learning from within a nondisciplinary matrix known as philosophy. Physics, astronomy, natural history or biology, psychology—including what are now termed the human sciences and the humanities—were all once part of philosophy as exhibited in the works of Plato, Aristotle, and their followers.

The emergence of moral theory or ethics as an explicit dimension of learning exemplifies what social scientists have termed the process of structural differentiation. Over the course of human history, there has been a tendency to disaggregate many aspects of human culture that once formed a more synthetic unity. For instance, one distinctive feature of the period between 500 BCE and 500 CE, especially among those peoples inhabiting the northeastern shores of the Mediterranean Sea, was a gradual movement to distinguish among the elements of law (in Greek, *nomos*), moral custom (*ethos*), narrative story (*mythos*), rational thinking (*logos*), and nature (*phusis*).

Since the 1500s in Europe and then the Americas, this process continued, such that a plethora of structural differentiations emerged in science (where physics, chemistry, geology, biology, and more were separated out as specialized forms of knowledge), in industry (through the division of labor), in government (separation of powers), and in religion (multiple church denominations). This systole of differentiation has in turn repeatedly given rise to the diastole of counter efforts promoting relationships or interactions among the associated socio-cultural structures, thus constituting in broad if non-standard terms multiple manifestations of interdisciplinarity: interdisciplinary research in science, team management in industry, constitutional formation and civil religion in the state, ecumenism in religion, and universal human rights in culture. One basic description of philosophy today could reasonably characterize it as the most general effort to reflect on and understand these differentiations and their countermovements. This is especially the case with that structural differentiation in philosophy known as ethics.

In a broad-brush historical overview, the development of ethics took shape in five overlapping interdisciplinary interactions or formative efforts to bridge other structural differentiations in culture. The first formation took shape in Greece in the centuries preceding the common era as an orientation toward understanding certain social norms as perfecting human nature by integrating humans into natural or cosmic orders. The macrocosm of cosmic reality was thought to be mirrored in the human microcosm, as summarized in the phrase 'as above, so below'. From this perspective, study of the natural world was itself of ethical significance. Morality served to mediate not just among human beings but also between social and natural orders. This view of the relation between human behavior and non-human orders of reality that can be found articulated in related ways in classical Hinduism and Daoism.

Insofar as the social order is itself viewed as the instantiation of a cosmic order, morality could also be understood, in a second interdisciplinary formation, as mediating between individuals and the social order. The ethics of Confucianism illustrates such an approach. Confucian ritual aims to unify not only heaven and earth but also individual persons with the present and past social fabric, and might thus be described as a practice of intergenerationality. Greek philosophy, Hinduism, Daoism, and Confucianism all emerged in what Karl Jaspers (1949) identified as the Axial Age of human history—the period between 800 and 200 BCE—which gave rise to a set of basic ethical understandings of morality as distinct from but mediating cultural formations.

A third basic formation of ethics emerged in conjunction with Judeo-Christian-Islamic notions of divine revelation, especially as articulated in Christian theology. In revealed theology, divine or supernatural infusions of knowledge from above can be variously understood as enclosing nature or opposing it. In the view of Thomas Aquinas, for instance, the natural law ethics of reason was simply confirmed by and raised to a higher level by the ethics of revelation; the natural virtues of courage, moderation, practice-wisdom, and justice were understood as complemented by the supernatural virtues of faith, hope, and love. By contrast, for Augustine, revelation functioned to relativise the importance of nature; the virtues of the ancients were no more than 'shining vices'. In both cases, however, ethics functioned as a handmaid to theology by mediating between revealed and natural knowledge. These two conceptions of ethics also manifested in the ethical theories of Judaism and Islam.

A fourth formation of ethics as interdisciplinarity has emerged opposed to the idea of ethics as handmaid of revelation. In this case, ethics functions instead as the handmaid of a new kind of science: modern natural science. During the Enlightenment, ethics became an interdisciplinary mediation not so much between nature and society as between the socio-political order and the pursuit of science in its distinctly modern form, which conceives non-human realities as devoid of moral significance except insofar as value is attributed to them by humans. In one version, ethics is cast as the protector of science. This notion of ethics argues for the autonomy of science and its support by the state because of its benefit to society. In another version, that of ethics as Romantic critique, it argues for delimitations on science in order to protect humans from dominance by science.

Finally, in a fifth formation, morality can be conceived as a practice leading to some kind of enlightenment or revelation from below. This view of morality may be interpreted as having roots in the Axial Age, through the example of Buddhism, but is also illustrated in the belief that adherence to the scientific method leads to the production of true knowledge. Additionally, modern psychology has proposed various methods, from psychoanalysis to educational techniques, as productive of knowledge. Although the enlightenments of Buddhist meditation and scientific methods are quite different, according to this formation of ethics both may be described as emerging in a natural manner from the disciplining of experience. The meaning of 'discipline' in this instance exhibits different but related meanings to those customarily associated with discussions of interdisciplinarity.

This brief historical overview of five formations of ethics as interdisciplinarity may be summarized as follows. In generalized terms, ethics as interdisciplinarity has functioned as mediation and synthesis of: (1) human and cosmic reality, (2) individual and social orders, (3) reason and revelation, (4) science and human affairs, and (5) as a pathway to insight. Only in the first half of the twentieth century were efforts made to construct an ethics purified of all forms of interdisciplinarity in what has come to be known as 'meta-ethics'. In the face of manifest needs for ethical guidance with regard to new forms of science and technology, however, the meta-ethics project gave way in the second half of the 20th century to what has come to be called applied ethics.

18.2 Interdisciplinarity in applied ethics: bioethics and nuclear ethics

The meta-ethics project aimed to set aside substantive debates about good and bad, right and wrong, and to focus instead on analyzing the meaning of moral terms and the structure of ethical discourse or argumentation, as in a seminal text titled *The language of morals* (Hare 1952). According to American philosopher Stephen Toulmin (1982), the practical problems created by scientific and technological advances in medicine 'saved the life of ethics' in a more traditional or normative sense. He might have extended his insight to note that salvation also involved resuscitation by interdisciplinarity. For Toulmin, when ethical reasoning became engaged in clinical work and considered the actual practices of physicians, hospital ethics committees' and/or institutional or governmental bodies' linguistic analyse of theoretical conflicts tended to be superceded by practical reasoning. Conflicts between deontology, utilitarianism, and virtue ethics are sidestepped in favor of *ad hoc* constructions to deal with particular problems.

Such social consensus in the area of medicine on the basis of multidisciplinary ethical practice was initially adumbrated in the post-World War II creation of the Nuremberg Code for research on human subjects. While in earlier periods medical scientists had at times exhibited a certain weakness in exercising their responsibilities for the protection of human subjects, Nazi concentration camp experiments dramatized the need to develop universally agreed-upon guidelines for the conduct of medical research. Judges from several nations collaborated with medical experts to create protocols establishing applied ethical principles for technoscientific medicine. This transdisciplinary cooperation among legal experts and medical practitioners resulted in foundational statements about the basic rights of all medical research participants to free and informed consent. Subsequent debates about human stem cell research, cloning, and the patenting of genomic sequences have continued to depend on broad cross-disciplinary dialogues regarding what factors and kinds of knowledge are relevant to policy making.

Insofar as medicine engages with and is transformed by developments in biology and the life sciences, it becomes empirically interdisciplinary. This trajectory has turned medical ethics into the interdisciplinary fields of bioethics and biomedical ethics. In a well-cited article on the constitution of the field of bioethics' Maurice de Wachter (1982) has argued that even when its interdisciplinary character is assumed, the implications require careful attention. In particular, de Wachter argues that flourishing of interdisciplinarity calls for the suspension of all disciplinary approaches, even when formulating research questions.

In light of the points made by Toulmin and de Wachter, it is important to acknowledge that the very term 'applied ethics' has been contested. Yet regardless of the different terms used to describe this domain—such as 'practical ethics' or 'professional ethics'—all assert a notion of engagement between and among multiple disciplines as the proper course for determining moral action. Consider, for instance, the case of nuclear ethics and policy, which refers to the ethics of nuclear weapons development and deployment as well as the ethics of nuclear power generation and production. Here the engagement spans the technological disciplines of nuclear science and engineering, as well as the social sciences of economics and environmental policy, along with the health sciences and medicine. If such interdisciplinary engagements are to be fruitful, the particular issues posed for ethical analysis will need to be formulated from the beginning through dialogue among the disciplines, so that the results of analyse are not predetermined by any set of disciplinary concerns.

18.3 Interdisciplinarity in applied ethics: environmental and computer ethics

Similar observations apply to the fields of environmental and computer ethics. Indeed, the development of environmental ethics was originally informed by naturalist writers (such as Henry David Thoreau and John Muir) as well as by conservation biologists (such as Aldo Leopold and Rachel Carson), all of whom undertook to advance critical ethical reflection on human—nature interactions from different disciplinary contexts. In previous formulations, ethics had been concerned primarily with the relationship of human-to-human or human-to-divine. Wildlife biologist Leopold was the first to make an explicit case for an environmental extension of ethics to include what he called a 'land ethic'. In his words:

> All ethics so far evolved rest upon a single premise: that the individual is a member of a community of interdependent parts. His instincts prompt him to compete for his place in that community, but his ethics prompt him also to co-operate (perhaps in order that there may be a place to compete for)....The land ethic simply enlarges the boundaries of the community to include soils, waters, plants, and animals, or collectively: the land. (Leopold 1949, p. 204)

Extending both the foundation of an ethical relationship to encompass both humans and the land required interdisciplinary collaboration to unpack the logic of this interaction. Carson's work (1962) in particular was instrumental in establishing a context for the creation of statutory laws and governmental agencies for protection of the natural environment. In turn, environmental protection has become a global discussion that now routinely engages politicians and economists on issues of sustainable development. In the early twenty-first century, environmental ethics developed into a broad interdisciplinary field that includes thinkers in the domains of literature, science, law, economics, public policy, education, and philosophy.

The formation of computer ethics unfolded in similar ways. In the early stages of the development of computer technologies, it was scientists and engineers such as Norbert Wiener who argued for the need to direct ethical attention to the implication of the use of the new machines of data manipulation and communication. Wiener, the founding figure of cybernetics, titled his second book *The human use of human beings: cybernetics and society* (Wiener 1950). His work was from the beginning an interdisciplinary endeavor situated in the interstices between mathematics, physics, and biology (Balsamo 1996). Through the interpretation of Terrell Ward Bynum (2008), Wiener also laid the foundation for computer or information ethics.

Following this early work, discussions in the Association for Computing Machinery (ACM), the largest non-governmental organization of computer professionals, led in 1973 to the adoption of a code of ethics (Anderson 1994). Other computer professional societies formulated similar or related codes in the following decade, e.g. the British Computer Society in 1983 and the Australian Computer Society in 1987. In 1978, interdisciplinary engagement between computer professionals and philosophers led to coining the term 'computer ethics' and in 1985 to the publication of a textbook of the same name (Johnson 1985). There followed an explosion of interdisciplinary interest and collaboration among computer professionals, philosophers, social scientists, and others who shared a general recognition that the field of computer ethics could not be pursued without extensive interdisciplinary collaboration.

As illustration of the implications of interdisciplinary ethics, consider the ACM code revision of 1992. The first code emphasized professional self-promotion and discipline (for example, being competent, acting within the limits of competence, and not misrepresenting one's abilities). Only in the last of five canons did the code specify that ACM members should be concerned to use their 'special knowledge and skills for the advancement of human welfare'. By contrast, the 1992 revision elevated the principle of contributing to 'society and human well-being' to the first commitment, while professional discipline was subordinated to education and raising consciousness. In a commentary on the revised code published in the *Communications of the ACM* in 1993, the interdisciplinary team of Ronald E. Anderson (social scientist), Deborah G. Johnson (philosopher), Donald Gotterbarn (computer scientist), and Judith Perrolle (social scientist)—a subset of the code drafting committee—noted that 'a major benefit of an educationally oriented code is its contribution to the group by clarifying the professionals' responsibility to society' (p. 98).

One substantive moral commitment that stands out as a distinctive result of the interdisciplinary nature of the ACM code is an expressed respect for individual privacy. In the 1973 version of the code, this idea took the form of obligations to minimize personal data collection, to secure such data collections, and to arrange for the disposal of data when their function had been served. In the 1992 revised code, a similar respect for individual privacy is expressed in the following terms:

> Computing and communication technology enables the collection and exchange of personal information on a scale unprecedented in the history of civilization....It is the responsibility of professionals to maintain the privacy and integrity of data describing individuals....This imperative implies that only the necessary amount of personal information be collected in a system, that retention and disposal periods for that information be clearly defined and enforced, and that personal information gathered for a specific purpose not be used for other purposes without consent of the individual(s). (ACM code, 1.7)

According to Bynum, however, computer and information ethics, especially as grounded in the foundational reflections of Wiener, raise issues that are more transdisciplinary than interdisciplinary. As Bynum describes it:

> [Wiener's] way of doing information ethics does not require the expertise of a trained philosopher....Any adult who functions successfully in a reasonably just society is likely to be [able to contribute]. As a result, those who must cope with the introduction of new information technology – whether they are public policy makers, computer professionals, business people, workers, teachers, parents, or others – can and should engage in information ethics by helping to integrate new information technology into society in an ethically acceptable way. Information ethics, understood in this broad sense, is too important to be left only to philosophers or to information professionals. (Bynum 2008, p. 30)

In effect, Bynam suggests the possibility of an interdisciplinary ethics not simply of knowledge and technological production, but also of knowledge and technological utilization. The users of technologies should be able to draw on their own broad non-disciplinary knowledge to assess them, a view that has also been termed end-user conviviality (Mitcham 2009).

Indeed, most discussions about the ethics of information focus on the production side (of information goods or solutions to problems) rather than on the use side, where consumers and citizens take up and utilize information. Information production is admittedly difficult to thematize, analyze, and practice interdisciplinarily. But interdisciplinary use of information is a quite common phenomenon, well and easily practiced if seldom theorized. Most 'disciplinary' producers engage frequently in 'interdisciplinary' consumption. When information producers leave the design shop or academic classroom they become citizens, members of families, churches, and users of all sorts of information goods and services—most of which they engage not as disciplinary experts or specialists but simply on the basis of common experience. In short, the disciplinary historian becomes an inter- and transdisciplinary human being when going to a health care provider. A physician's diagnosis and treatment recommendation is only incidentally filtered through the historian's perspective, insofar as questions might be asked about the historical development and origin of a diagnosis or therapy. The historian *qua* patient and consumer of medical services has an ability and indeed a motivation to draw from any number of disciplines in the process of making sense of a diagnosis or prescribed therapy: an old general chemistry course from high school, a required science course in college, a novel about medical care, newspapers reports and TV programs. Interdisciplinary consumption and use is a largely undertheorized aspect of interdisciplinarity that in fact functions in almost all areas of applied ethics.

18.4 Interdisciplinarity in applied ethics: professional ethics of engineering and science

The 'disciplinarity' that characterized the mode of knowledge production of the last half of the twentieth century, into which 'interdisciplinarity' is an intervention, exhibits a historically specific character. In earlier periods, producers such as farmers, tailors, and even soldiers were both producers and consumers of their own goods: farmers ate their own food; tailors wore their own clothes; vernacular architects lived in the houses they built; soldiers fought with the weapons they designed, manufactured, and maintained. The hyperstructural differentiation that resulted in the proliferation of disciplines with rigid boundaries did not arise until the modern period.

In the domain of professional ethics the typically modern separation of production and use is less pronounced than in other applied ethics fields. Consider, for example, the development of engineering ethics and the ethics of scientific research. In both cases, interdisciplinary work has become the norm. Engineering is a discipline that has evolved into several specialized subfields, from civil and mechanical to chemical, electrical, electronic, industrial, nuclear, computer, and more. Especially since the 1990s, the paramount commitment, across all fields of engineering, has been articulated as the protection of public safety, health, and welfare—a commitment that has become known as the *paramountcy clause*. In the 1970s, the US National Endowment for the Humanities (NEH) and the National Science Foundation (NSF), stimulated in part by widespread public concern about a series of engineering-related failures (most prominently involving automobiles and airplanes) and in an effort to deepen understanding and practice of the paramountcy clause, jointly awarded a number of grants to research teams that included both philosophers and engineers to study the ethics of engineering research and practice. The NSF even established a special Ethics and Values in Science and Technology (EVIST) grants program. This resulted in the creation of a number of interdisciplinary team-taught engineering ethics courses and the publication of engineering ethics textbooks (e.g. Unger 1982; Martin and Schinzinger 1983; Harris *et al.* 1995).

A similar interdisciplinary dynamic transformed critical reflection and practice with regard to the ethics of scientific research. Stimulated again in part by public concern about fraud and misconduct in science, including the misuse of public funds as revealed in US Congressional hearings in the 1980s, the NSF and the US National Institutes of Health (NIH) funded interdisciplinary research and course development on the responsible conduct of research. This trajectory of scholarship was also promoted by interdisciplinary professional scientific organizations such as the US National Academies of Science (NAS) and the American Association for the Advancement of Science (AAAS). A 1989 pamphlet, *On being a scientist*—produced by an interdisciplinary team of representatives from the physical sciences, life sciences, engineering, social sciences, and humanities under general direction of the Committee on Science, Engineering, and Public Policy of the NAS, National Academy of Engineering, and Institute of Medicine and published by the National Academies Press—became a standard teaching resource at both the graduate and undergraduate levels in universities. This and related texts review basic research protocols regarding notions of integrity and honesty in the reporting of research results, the avoidance of conflicts of interest, the fair treatment of subordinates and colleagues, and respect for animal welfare for the purposes of raising awareness and fostering ethically responsible scientific practice.

18.5 Applied ethics generalized

The permutations of 'applied ethics' discussed here display at least four common features. First, the distinctions between multi-, cross-, trans-, and interdisciplinarity are of marginal concern to those who practice interdisciplinary ethics. Interdisciplinary teams engage in their work and negotiate the parameters of their interactions on the fly without feeling much need to analyze the particulars. Second, the *practice* of interdisciplinarity is often considered a transgression of proper disciplinary order. In the discipline of philosophy, for instance, those who become involved in interdisciplinary work are often professionally marginalized. Philosophers who specialize in applied ethics are seldom accepted as equal members of their departments. Sometimes they are nudged out of the discipline and into interdisciplinary units such as programs and departments of science, technology, and society (STS) studies. Julie Thompson Klein (1990, 2001) has in this regard described the character of 'interdisciplinarity' that emerges in different institutional contexts as a consequence of the movement 'out of the disciplines' by interdisciplinary researchers.

Third, the notion of interdisciplinarity often expands beyond an initial grouping comprising engineers and scientists to eventually include social scientists and humanists (Frodeman *et al.* 2001). This results, fourth, in the formation of new questions about the professional and cultural boundaries of applied ethics work. Engaging in applied ethics in the context of contemporary globalization leads researchers to ask questions about the legitimacy of, for example, standards for the responsible conduct of research developed in Europe and the United States as distinct from China, India, or South Africa (European Commission 2009).

Along with a number of different, substantive ethical ideals such as free and informed consent in bioethics, sustainability in environmental ethics, and privacy protection in computer ethics, two other commitments are often incorporated into applied ethics fields. One is that technical experts have the obligation to promote public education regarding the most relevant aspects of their work. Another is that, when appropriate, these technical experts have obligations to involve the public in decision making about technical matters. Summarizing these two ideal commitments, it has been argued that the ethical determination of boundary conditions on research and knowledge production requires co-responsible or interdisciplinary collaboration between the scientific community and the public (Mitcham 2003). This, in turn, requires ethics (in its broadest formation) to reflect upon its own unavoidable disciplinary blind spots, to shift from one frame of reference to another, in order to appreciate the specific character of different forms of knowledge, different methods of knowledge production, and different purposes of knowledge creation. Clarifying the ethical responsibility of technical experts to engage members of the public in technical decision making enacts the political adage of 'no taxation without representation'. We live in a world where scientific research and technological invention have a more significant impact on citizens than do government tax policies. This participatory principle also moves applied ethics from the realm of personal behavior into that of politics and policy (Winner 1980; Goldman 1992).

18.6 The ethics of interdisciplinarity as an ethics of shift work

The foregoing review of applied ethics indicates that it would be difficult to do such work without interdisciplinary engagements. This persistent fact in turn suggests a value: doing interdisciplinarity in the most ethical way imaginable. Although value may not be derivable from fact, facts can stimulate reflection on relevant values. To think about the right and wrong ways of doing interdisciplinarity is to anticipate an ethics of interdisciplinarity, and in such an anticipation it would be useful to begin with a general characterization of interdisciplinarity itself. Although interdisciplinary collaboration comes in many forms—multi-, cross-, inter-, and transdisciplinarity, to name only the four most common types—uniting all forms of interdisciplinary is what may be described as the phenomenon of 'shift work', literal and metaphorical. Taking this as the starting point, a manifesto on the ethics of interdisciplinarity will itself involve a shift in rhetoric and tone. This is but one more manifestation of the dynamic movement characteristic of interdisciplinary work.

Unlike shifts that start and end with the punch clock, interdisciplinary shifts from one framework to another require the on-going crossing of boundaries. Many studies treat boundary crossing as the exception rather than the rule: 'At this historical point, however, the interactions and reorganizations that boundary crossing creates are as central to the production and organization of knowledge as boundary formation and maintenance' (Klein 1996, p. 2). Yet it can be difficult to grasp the specific dynamics of this shift work across boundaries, let alone figure out how to exercise positive influence over it. In this case, *ought* need not imply *can*, but does imply *try*.

Although interdisciplinary shift work is enabled by new technologies, it is not technologically determined. Enabled and stimulated by advancing technologies of transportation, communication, and information storage and retrieval, shift work takes form in classrooms and galleries, in virtual online worlds, in networked social spaces, and through mobile access points. The traditional spaces of cultural production and reproduction—research labs, art studios, universities, museums, libraries, galleries, theaters, and community centers—are themselves being transformed by those who inhabit them as they adopt new practices of communication, community formation, knowledge production, and technology use.

Given its dependence on, without being determined by, technological change, the shift work ethos invites cultivation of what may be called the *technological imagination*, a quality of mind that enables people to think *with* and *through* technology (Balsamo, forthcoming). This is equivalent to what Albert Borgmann calls 'real ethics': an ethics that steps beyond ideas and theories and is more expansive than that focused on personal interactions. 'Real means tangible; real ethics is taking responsibility for the tangible setting of life' (Borgmann 2006, p. 11). Real ethics rests on the recognition that even as we design the world of artifacts within which we live, those artifacts design us. Additionally, the technological imagination entails performativity and improvization, the cultivation of which rests on appreciation or understanding of: (1) the hidden character of knowledge in a digital age, (2) the multifaceted consequences of technological innovation, and (3) the development of new protocols for enacting interdisciplinary activity. The richer the technological imagination, the better the questions it will bring to these three aspects of the practice of interdisciplinarity.

(1) With regard to the hidden character of knowledge: for those now considered members of a generation 'born digital' (who came to consciousness after the emergence of the internet in the 1990s), it is obvious that data $\neq$ information $\neq$ knowledge, and that daily life is a scene of constant shifts between different networked contexts. Consequently,

they often display a transgressiveness that emerges from repeated experiences of traveling across linked information flows. Their successful navigation of media flows, distributed learning, and social environments requires the fluid mutation of interests, identities, and affiliations. Mutability becomes one of their strongest attributes. It is exactly this sensibility and fluid mutability that can serve as the foundation for a lifetime of learning.

But mutability is not easily accommodated within established institutions that govern and sanction knowledge production and which depend on specific structures, conventions, and often highly traditional rituals of production. In order to successfully navigate contemporary digital network worlds, participants must learn not only how to transform data into information, but then be able to integrate information that comes from different sources for purposes of creating knowledge. In short, they must learn how to synthesize material harvested from diverse information flows. To affect this, digital shift workers learn to produce knowledge through dialogues among disciplines, social negotiation, and collaboration with peers, experts, and multiple others. Knowledge-producing activities will thus depend on understanding how disciplinarity functions as the institutionalized practice of knowledge verification. Shift workers learn how to engage in conversations with those who hold diverse cultural values or intellectual commitments. The everyday experience of the digital generation already incorporates creative synthesis practices such as data mining, remixing, and modding. But to create knowledge, interdisciplinarians must also learn how to critique the information flows they remix. In this sense, 'critical reading' is not an outmoded text-based literacy.

(2) The multifaceted consequences of innovation can be summed up in the formula: technological innovation = social transformation. Research that produces technological innovation is socially and culturally transformative insofar as all technologies shift social and cultural arrangements. Transformation takes shape through the formation of new publics, policies, social protocols, services, cultural narratives, as well as new technological applications and devices. Innovation thus calls for multidisciplinary collaboration not just in pursuit of the creation of new technological products, but also in the formation of political alignments for interdisciplinary collaborations. In the past, such collaborations have been unnecessarily limited. As the technohumanists Cathy Davidson and David Goldberg (2004) point out in their 'Manifesto for the humanities in a technological age', those who call for interdisciplinary collaboration that focuses on applied social problems frequently disregard the participation of humanists. As an example they cite Jeffrey Sachs, as Director of the Earth Institute at Columbia University and Special Adviser to the United Nations Secretary-General on Millennium Development Goals. Although insisting 'that interdisciplinarity was the only way to solve world problems', proposed bringing together only the earth sciences, ecological sciences, engineering, public health, and the social sciences with a heavy dose of economics—leaving out the arts, culture, and philosophy. Yet complex social problems call for hybrid solutions that benefit from the incorporation of intellectually nuanced cultural analyses. The cultural aspects of technology design, use, deployment, implementation, maintenance, and disposal are fundamental to the process of forming adequate responses to variegated social problems.

(3) Having called for including a broader range of disciplinary participants, it is important to note that those who collaborate as members of interdisciplinary

shift teams must resist any facile division of labor that relegates scientists to studying conditions, engineers to designing artifacts, and social scientists and humanists left to practicing critique. While different roles are to be played by different types of participants, all must be willing—indeed, eager—to learn new skills, analytical frameworks, methods, and practices. This is the starting point for a practical ethics of interdisciplinarity. When people with different disciplinary or even interdisciplinary backgrounds come together, it is important to acknowledge that everyone has something to contribute and to learn.

The following virtues may thus be as indicative of the ethical habits appropriate to shift work interdisciplinarity:

Intellectual generosity. A genuine acknowledgment of others' work. This should be explicitly expressed to collaborators as well as mentioned via citation practices. Showing appreciation for other ideas in face-to-face dialogue and throughout a collaborative process stimulates intellectual risk-taking and creativity.

Intellectual confidence. A belief that one has something important to contribute. Confidence avoids boastfulness and includes a commitment to accountability for the quality of a collaboration. Everyone's contribution to a collaboration needs to be reliable, rejecting short cuts and guarding against intellectual laziness.

Intellectual humility. A recognition that one's knowledge is partial, incomplete, and can always be extended and revised. This is a quality that allows people to admit they do not know something without suffering loss of confidence or self-esteem.

Intellectual flexibility. The ability to change one's perspective, especially based on new insights from others. This can include a capacity for play, for suspending judgment and imagining other ways of being in the world and other worlds to be within.

Intellectual integrity. The exercise of responsible participation. Such a habit serves as a basis for the development of trust, and is a quality that compels colleagues to bring their best work and thinking to collaborative efforts.

Beyond such particular virtues, however, the practical ethics of interdisciplinarity assumes that more effective interdisciplinary production and use is a natural good. It is pragmatic in orientation, seeking only to improve interdisciplinary output—making interdisciplinarity work well rather than questioning whether interdisciplinarity should work.

At the same time, is there no need for a questioning of greater and greater productivity? Questioning is generally seen as legitimate with regard to material productivity. Why not also with interdisciplinary productivity, whether practiced in the realm of tangible or cognitive goods? Is it not possible to be overwhelmed by knowledge and innovation? Are these not the ultimate issues for any ethics of interdisciplinarity?

18.7 Conclusion

Ethics as a form of interdisciplinarity has been described in a broad-brush historical survey from the Axial Age to the twenty-first century. This survey identified five basic frameworks in which ethics has mediated interactions between human beings and other aspects of reality. This constitutes an admittedly metaphorical extension of the notion of interdisciplinarity, but one that suggests possibilities deeper than those customarily imagined—and going beyond the commonplace of interdisciplinarity as useful for problem solving.

Beginning in the second half of the twentieth century it is also possible to identify a form of ethics—applied ethics—in which interdisciplinarity in a less metaphorical sense plays an increasingly constitutive role. A selective review of work in various applied ethics fields, from bioethics to engineering, reveals how integral interdisciplinarity is to almost any critical reflection on life in a technoscientific world.

The prominence of interdisciplinarity in ethics in turn suggests the need for an ethics of interdisciplinarity, that is, for reflections on what ethos could best guide interdisciplinary practitioners in their shift work. In a sense, of course, the whole handbook of which this chapter is a part has the same aim. But as its own special contribution to such a general goal, the ethics chapter concluded by identifying shift work as a key characteristic of all interdisciplinary activity and then ventured to explore differences between right and wrong ways to enact its ever recurring shifts in perspective and method. The result was to describe five virtues for interdisciplinary practice. One way of testing the adequacy of these virtues would be to consider their relevance to the many other analyses of interdisciplinarity that inform the present handbook.

References

Anderson, R.E. (1994). The ACM code of ethics: history, process, and implications. In: C. Huff and T. Finhold (eds) *Social issues in computing: putting computing in its place*, pp. 48–71. New York: McGraw-Hill.

Anderson, R.E., Johnson, D.G., Gotterbarn, D., and Perrolle, J. (1993). Using the new ACM code of ethics in decision making. *Communications of the ACM* **36**(2), 98-1-7.

Balsamo, A. (1996). *Technologies of the gendered body: reading cyborg women*. Durham, NC: Duke University Press.

Balsamo, A. (forthcoming). *Designing culture: a work of the technological imagination at work*. Durham, NC: Duke University Press.

Borgmann, A. (2006). *Real American ethics: taking responsibility for our country*. Chicago: University of Chicago Press.

Bynum, T.W. (2008). Milestones in the history of information and computer ethics. In: K.E. Himma and H.T. Tavani (eds) *The handbook of information and computer ethics*, pp. 25–48. Hoboken, NJ: John Wiley.

Carson, R. (1962). *Silent spring*. Boston, MA: Houghton Mifflin.

Committee on Science, Engineering, and Public Policy (1989). *On being a scientist: a guide to responsible conduct in research* [2nd edn 1995, 3rd edn 2009]. Washington, DC: National Academies Press.

Davidson, C.N. and Goldberg, D.T. (2004). A manifesto for the humanities in a technological age. *Chronicle of Higher Education* **50**(23), B7.

European Commission Expert Group on Global Governance of Science (2009). *Global governance of science*. Brussels: Directorate-General for Research; Science, Economy and Society.

Frodeman, R., Mitcham, C., and Sacks, A. (2001). Questioning interdisciplinarity. *Science, Technology & Society Curriculum Newsletter* **126–127**, 1–5.

Goldman, S.L. (1992). No innovation without representation: technological action in a democratic society. In: S.H. Cutcliffe, S.L. Goldman, M. Medina, and J. Sanmartin (eds) *New worlds, new technologies, new issues*, pp. 148–60. Bethlehem, PA: Lehigh University Press.

Hare, R.M. (1952). *The language of morals*. Oxford: Clarendon Press.

Harris, C.E., Jr, Pritchard, M.S., and Rabins, M.J. (1995). *Engineering ethics: concepts and cases*. Belmont, CA: Wadsworth.

Jaspers, K. (1949). *Vom Ursprung und Ziel der Geschichte*. Zurich: Artemis Verlag.

Johnson, D. (1985). *Computer ethics*. New York: Prentice Hall.

Klein, J.T. (1990). *Interdisciplinarity: history, theory, and practice*. Detroit, MI: Wayne State University Press.

Klein, J.T. (1996). *Crossing boundaries: knowledge, disciplinarities, and interdisciplinarities*. Charlottesville, VA: University Press of Virginia.

Klein, J.T. (2001). The interdisciplinary variable. In: B.L. Smith and J. McCann (eds), *Reinventing ourselves: interdisciplinary education, collaborative learning and experimentation in higher education*, pp. 391–418. Bolton, MA: Anker Press.

Leopold, A. (1949). *A sand county almanac*. New York: Oxford University Press.

Martin, M.W. and Schinzinger, R. (1983). *Ethics in engineering*. New York: McGraw Hill.

Mitcham, C. (2003). Co-responsibility for research integrity. *Science and Engineering Ethics* **9**(2), 273–90.

Mitcham, C. (2009). Convivial software: an end-user perspective on free and open source software. *Ethics and Information Technology* **11**(3), 299–310.

Post, S.G. (ed.) (2004). *Encyclopedia of bioethics*, 3rd edn. Detroit: Macmillan Reference.

Toulmin, S. (1950). *An examination of the place of reason in ethics*. Cambridge: Cambridge University Press.

Toulmin, S. (1982). How medicine saved the life of ethics. *Perspectives in Biology and Medicine* **25**(4), 736–50.

Unger, S.H. (1982). *Controlling technology: ethics and the responsible engineer*. New York: Hold Rinehart and Winston.

de Wachter, M.A.M. (1982). Interdisciplinary bioethics: but where do we start? *Journal of Medicine and Philosophy* **7**(3), 275–88.

Weiner, N. (1950). *The human use of human beings*. Boston: Houghton Mifflin.

Winner, L. (1980). Do artifacts have politics? *Daedalus* **109**(1), 121–36.

CHAPTER 19

Design as problem solving

PRASAD BORADKAR

> The central theme of design is the conception and planning of the artificial
>
> —Buchanan (1990, p. 78)

The word design is most frequently employed to refer to the action or process of planning and making (designing something), but it is also used to describe the end result or artifact of this action (a design). The etymological root of the word design can be traced back to *designare*, Latin for 'to mark out' or 'devise'. Both marking out and devising signify an intent to create concepts that can be realized as objects. In other words, a designed object is 'reified intention' (Mitcham 1994, p. 220); the outcome of this intention can be a machine, a building, a product or a logo. Thus, design finds itself used and claimed by a variety of disciplines including, but not limited to, architecture, engineering, automotive design, industrial design, graphic design, and interior design; new additions like experience design and service design continue to appear on a regular basis. In fact, all activities directed towards the materialization of an intent—developing a business plan for a new venture, writing a public policy or creating an artwork—can be described as forms of design. In order for the designed artifact to truly valuable to all those who interact with it, designers have to consider issues of aesthetics, usability, ergonomics, safety, marketability, manufacturability, functionality and sustainability. This requires a wide range of skills and knowledge. Design has therefore been described as science and art, as communication and argumentation, and as thinking and inventing.

Designers involved in such activity often refer to their task as problem-solving, and view their work as a response to opportunities and needs in the market identified by corporations, entrepreneurs, consumers, governments, and non-profit organizations. Design practice takes on problems that can range from the creation of such small things as business cards to the planning of entire urban systems. If design's task, as Max Bill of the Hochschule für Gestaltung in Ulm once explained, is 'to participate in the making of a new culture – from a spoon to a city' (Lindinger 1991), its scope can be vast and its impact significant. And while the design of a spoon might only call for the collaboration between a designer and a metalsmith, the planning of a city certainly is not possible without the involvement of a large number of experts including urban planners, city officials, transportation engineers, citizens, and other experts representing a variety of disciplines, points of view and interests. As the problems of climate change and environmental degradation worsen, the concerns of design have extended beyond the city to include the planet.

Horst Rittel argues that the problems design handles are 'wicked' (as well as incorrigible and ill-behaved) and require new methodologies to tame them (Rittel and Webber 1973). He outlines several factors that define the nature of this wickedness. He lists some of the key characteristics of these problems: they are very difficult to formulate, they do not have right or wrong solutions, they do not have a logical end, and they are often symptoms of other problems. In such situations, the only way to devise comprehensive solutions that meet the needs of a large and diverse group of stakeholders is through an intense and integrated collaboration among disciplines. Transdisciplinarity is one of the most promising strategies for dealing with and taming the wicked, ill-behaved and incorrigible problems of design.

19.1 Design's wicked problems

Contemporary design practice is conducted in a world that is globally more connected, technologically more complex, and economically more intricate than it has ever been before. According to socio-cultural anthropologist Arjun Appadurai, 'the scale, penetration, and velocity of global capital have all grown significantly in the last few decades of this century' (Appadurai 2001, p. 18). The complexity engendered by these global capital flows is expected only to increase in the future, complicating design's task even further. In response to these impending developments, design has already started to re-imagine its scope. New conceptions of design now define its charter as the development of systems rather than individual artifacts. There is recognition that designers need to consider global needs rather than individual wants. New design thinking emphasizes concerns of social equity and environmental responsibility, pushing design's purview beyond its historical fixation on form. It is now also commonly recognized that design alone cannot solve these problems in isolation. The sheer wickedness and complexity of these issues warrants engagement with other disciplines.

The health care system in the United States, for example, presents a series of wicked problems. Take the problem of designing an effective patient transfer system for a hospital. Patients are generally transported between ambulances, emergency rooms, waiting rooms, laboratories, surgical wards, and pharmacies and using gurneys, stretchers, rolling beds, wheelchairs, lifts, hoists, and other devices. While being transferred, they are often hooked up to IV poles, oxygen tanks, or vitals monitors, and the transfers might involve, in addition to the patients themselves, nurses, nurses aides, family members, social workers, and paramedics. Health care workers moving patients from one position (reclining in a bed) to another (sitting in a wheelchair) often hurt their backs, and research shows that nurses experience more injuries on the job than any other professionals. This has led to lost work, reduced pay, and workers' compensation claims, which become financial burdens for health care workers and hospitals. In addition, the problems posed by the growing rate of obesity and the increasing average age of nurses pose additional difficulties for hospital personnel who might have to move bariatric patients (those weighing more than 152 kg).

A patient transfer system will not only have to handle the problems listed above, but it will need to be cost-effective, able to accommodate patients who represent a wide range of body types and cultural backgrounds, easy to install, effortless to use, and above all, safe for patients and health care professionals. And unless it is able to adapt to existing as well as new hospital buildings, it will not be compelling enough to hospital administrators and purchasing departments. This problem is difficult to understand thoroughly: it possesses no single right, wrong, or objectively perfect solution; and it lacks finite and reliable evaluative criteria. It is clear that developing such a system would need to involve teams of hospital staff, health care workers (nurses, nurses' aides, paramedics, ambulance drivers), engineers, product designers and marketing professionals, all working to inform and transform each other's thinking. This is merely one example of design's wicked problems that demand transdisciplinary efforts.

The process of new product development is inherently interdisciplinary, and is typically conducted in cross-functional teams of designers, engineers, and business professionals. In addition, depending upon the nature and scope of the project, other disciplines such as anthropology, medicine, psychology, nursing, etc. may be brought in to guide the design development. As design takes on the vast challenges of environmental pollution, global poverty, and lack of clean drinking water, it will need to inform and to learn from other disciplines: 'Interdisciplinary skills are also particularly important for problem-solving in areas where there are a large number of variables together with high levels of uncertainty and risk. As Nobel Laureate Gunnar Myrdal commented "problems do not come in disciplines"' (Gann and Salter 2001, p. 99).

19.2 Design praxis: a taxonomy

> No single definition of design, or branches of professionalized practice such as industrial or graphic design, adequately covers the diversity of ideas and methods gathered together under the label. Indeed, the variety of research reported in conference papers, journal articles, and books suggests that design continues to expand in its meanings and connections, revealing unexpected dimensions in practice as well as understanding (Buchanan 1992, p. 5)

The variety of domains in which designers operate and the range of outcomes they produce have made it difficult to establish a thorough taxonomy of design disciplines. In addition, as it evolves, design takes on new meanings, adopts new methodologies, addresses a broader range of problems and redefines its scope, making it challenging to keep taxonomical structures current. If one imagines the totality of the built environment (from

InnovationSpace at Arizona State University

Prasad Boradkar

In 2004 Arizona State University established a program called InnovationSpace to engage faculty, students, and industry in the process of transdisciplinary design and innovation. Faculty from industrial design, visual communication design, business and engineering co-direct the program and teach classes in which teams of senior-level undergraduate students from these four disciplines work together over a period of a year on solving socially critical problems on such topics as health care, assistive technologies, and alternative energy.

The program is built on the premise that a traditional, discipline-specific education no longer provides enough expertise or variation in thinking to handle the complex challenges of new product development. The program's goal is to teach students how to create sustainable product concepts that can anticipate the challenges presented by industry, consumers, society, and the environment: 'One strand of TD problem solving centers on collaborations between academic researchers and industrial/private sectors for the purpose of product and technology development, prioritizing the design of innovative milieus and involvement of stakeholders in product development' (see Klein, Chapter 2 this volume). In order to engage stakeholders from industry, InnovationSpace collaborates with corporate partners and has, to date, completed projects with Herman Miller, Inc., Intel Corporation and Procter and Gamble.

Students enroll for two courses over fall and spring semesters for a total of 10 academic credits, and are placed in a studio team environment where they can learn from each other, from a multidisciplinary team of faculty members, and from industry leaders. The students, faculty, and industry experts bring to class a new set of resources, theoretical approaches, specialized methodologies and unique tools from a variety of disciplines that advance the level of general and specific knowledge of the entire group. This prepares the students to be professionals who are not only trained in their disciplines but who quickly learn that other areas of expertise can in fact improve the quality, depth and impact of their own work.

Integrated innovation

One of the critical guiding principles of InnovationSpace is integrated innovation. The model of innovation asks four questions that demand a holistic approach to new product development:

- What is valuable to people?
- What is possible through engineering?
- What is desirable to business?
- What is good for society and the environment?

Using the model of integrated innovation, students aim to create products that:

- Satisfy user needs and desires.
- Apply innovative but proven engineering standards.
- Create measurable value for business.
- Benefit society while minimizing impacts on the environment.

This model serves as a framework that guides the students' research and analysis and helps them collect information about existing and potential users, the market, and emerging tech-

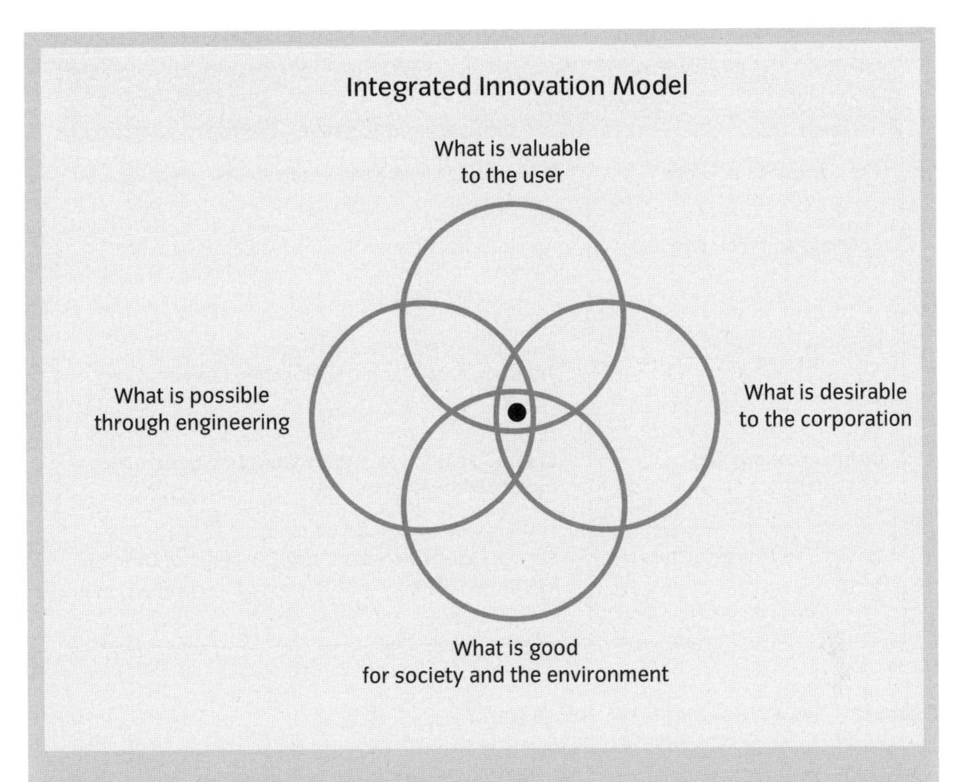

nologies, as well as critical social and environmental issues. In addition, as they start generating product ideas, this model also serves as an evaluative tool, helping them gauge the efficacy of their solutions for all stakeholders. This model serves as a very effective teaching tool because it is simple and it encourages in-depth exploration of all the constituencies that will be positively and negatively impacted by their solutions.

Students also start to recognize the dynamic tensions that exist among these factors of innovation and realize that design involves making trade-offs. For example, certain commodity plastics may possess the right material properties and meet manufacturing requirements, but also may pose higher risks due to human and environmental toxicity. Or, certain executive decisions regarding business process offshoring to poorer nations that have economic advantage for the corporation may not always translate into positive societal impacts for the workers in those countries. This model of innovation emphasizes that responsible design involves the delicate task of negotiating the tensions that exist between what's valuable to users, what's desirable to corporations, what's possible through engineering, and what's good for society and the environment.

Students also recognize that trade-offs extend beyond the scale of design—such as material capabilities versus formal decisions or appropriate ergonomics versus compactness—to larger-scale compromises involving such business decisions as outsourced labor cost versus corporate social responsibility, brand strategy versus advertising budget, etc. This also helps the industrial design students realize quickly that the birth and development of a product involves the expertise of several disciplines and compromises are central to moving products to market. The primary objective of Arizona State University's InnovationSpace program is to equip designers, business professionals, and engineers with the knowledge and skills needed to transform the world into a better place, by design.

The Built Environment	The Design Discipline
The city and its environs	Urban Design, Landscape Architecture, Planning
Buildings and their interiors	Architecture, Interior Design, Exhibit Design, Set Design
Products	Product Design, Industrial Design, Toy Design, Transportation Design, Engineering Design
Communications and new media	Graphic Design, Visual communication Design, Web Design, Interaction Design
Services and infrastructures	Service Design, Process Design, Experience Design, Systems Design

Figure 19.1 The domains and disciplines of design.

the spoon to the city) to be the domain of the designer, it can be broadly (and incompletely) classified into the domains and disciplines shown in Fig. 19.1.

Though the divisions that exist among the various forms of design practice fracture the discipline, they do serve a critical role: 'There are, of course, some good reasons why these practices were separated in the first place, and the issue is not to meld them all into a new, comprehensive profession that is at once everything and nothing' (Margolin 1989, p. 4). The design and manufacture of a hand-held device presents a set of challenges that are far different from those faced by an architect who is called on to oversee the design and construction of a hospital. Similarly, the design of a car interior demands the attention of transportation designers, ergonomists, mechanical engineers, and others, making it a vastly different challenge from the design of an archeological exhibition that tells the complex story of Egypt's rich history. The level of granularity in the division of design labor encourages the development of domain-specific knowledge and allows designers to refine their craft. However, it also presents the danger of narrow and compartmentalized thinking that can seriously limit design's impact. In order to generate holistic and comprehensive solutions to problems of the built environment, collaboration among disciplines is imperative.

Mitcham (1994) has classified design into two broad categories—engineering design and artistic design; the former driven by performance specifications and the latter by form; the first by efficiency and the second by beauty. Engineering design uses physics and mathematics in visualizing its material outcomes, and artistic design relies on the senses and intuition in creating its results. However, with growing interdisciplinarity and the emergence of new disciplines, the boundaries of such classifications become blurred. For instance, web designers create graphic user interfaces that determine the aesthetic character of a website (a form of artistic design), but many of them are also required to know some computer programming (a form of engineering design) in order to make the sites functional: 'Design is partly rational and cognitive, and partly irrational, emotive, intuitive, and noncognitive. It is rational to the extent that there is conscious understanding of the laws of nature; it is irrational to the extent that the sciences have not yet succeeded in revealing the laws of complex phenomena' (Buchanan 1995, p. 50). Most designers do not see their practice as purely artistic; although imparting beauty to everyday objects is certainly of importance, the agenda for design also includes solving problems that can improve people's lives and minimize impacts on the environment. In other words, engineering and artistic work are both central to design and not easily separable.

19.3 Design: practice and theory

Most definitions of design refer primarily to design practice as manifest in the professions of architecture, engineering, planning, etc. In addition to the professional occupations, though, it is important to recognize the emergence of *design studies*, an interdisciplinary activity established to study design itself and develop a theory of practice. Bruce Archer, one of design's leading voices and advocate for the establishment of design studies, created a taxonomy outlining ten topics within which further research would be needed to develop a theoretical body of domain knowledge. These ten topics represent the earliest formation of design studies (Fig. 19.2).

Design studies seeks to develop reflexive knowledge about design itself, especially in the areas of history, theory, and criticism. As the labels imply, these are interdisciplinary areas of inquiry that depend upon thorough engagement with such disciplines as philosophy, history, education (pedagogy), etc., for their development. Scholarship in these topics has been growing steadily through a variety of books and such journals as *Design Issues*, *Design Studies*, *The Design Journal*, and *Design and Culture*.

19.4 Wicked, incorrigible, ill-behaved problems

For Rittel, design's wicked problems are also ill-behaved because they frustrate the designers' efforts of wanting to create and follow a clear pathway to the solution from analysis to synthesis (Rittel 1971). He attributed this ill behavior to the following issues:

> 1. They are not well-defined; i.e., every formulation of the problem is already made in view of some particular solution principle [...]. 2. For design problems there is no criterion which would determine whether a solution is correct or false. Plans are judged as good, bad, reasonable, but never correct or false [...]. 3. For design problems there is no rule which would tell the designer when to stop his search for a better solution. He can always try to find a still better one. Limitations of time and other resources lead him to the decision that now it is good enough.... (Rittel 1971: 19)

Topics of study	Definition
Design Technology	The study of the phenomena to be taken into account within a given area of design application
Design Praxiology	The study of the design techniques
Design Language	The study of the vocabulary, syntax and media for recording, devising, assessing and expressing design ideas in a given area
Design Taxonomy	The study of the classification of design phenomena
Design Metrology	The study of the measurement of design phenomena, with special emphasis on the means for ordering or comparing non-quantifiable phenomena
Design Axiology	The study of goodness or value in design phenomena, with special regard to the relations between technical, economic, moral and aesthetic values
Design Philosophy	The study of the language of discourse on moral principles in design
Design Epistemology	The study of the nature and validity of ways of knowing, believing and feeling in design
Design History	The study of what is the case, and how things came to be the way they are, in the design area
Design Pedagogy	The study of the priniciples and practice of education in the design area

Figure 19.2 Areas of work and research likely to be involved in the future development of design studies (Baynes *et al.* 1977).

While not all problems that designers tackle behave so badly, there are several, especially in the health care and transportation industries, that certainly do. Buchanan argues that design problems are wicked because 'design has no special subject matter of its own apart from what a designer conceives it to be' (Buchanan 1992, p. 16). Designers tackle problems from a variety of domains, and the products of their labor range from paper clips to airplanes. For example, while a biologist may focus his or her life's scientific efforts on the narrow and highly specialized examination of butterfly coloration, an industrial designer may focus his or her efforts on the unique problems of the creation of a car, a guitar, and a chair within the span of a few months. The domain knowledge required to practice design needs to be abstract enough to be applicable in a variety of contexts, while being specific enough to appropriately address the challenges at hand. And, the difficulty in being able to develop content expertise in several domains makes it even more attractive to engage disciplines that possess deep knowledge in those topical areas.

19.5 Dealing with wickedness

If design problems pose a unique set of challenges, designers need a unique set of tools with which to tackle them. Brainstorming, mind mapping, visualization, prototyping, storyboarding, scenario development, etc., are some of the commonly used methods in design praxis. However, while these methods can help with discrete segments of the problems, they do not serve as overarching strategies for taming or coping with wickedness. Roberts (2000) classifies problems as simple, complex, and wicked, and offers three unique coping strategies that she titles *authoritative*, *competitive*, and *collaborative*. She cautions that no single approach can present itself as a panacea, and decisions about selecting the most appropriate strategy will depend upon the specificity of the problem. Authoritative strategies are recommended when a few key stakeholders are in positions of power in the problem-solving group, competitive strategies work best when power is dispersed and contested, and collaborative strategies serve well in the remaining situations (Fig. 19.3).

There is no question that the design process—whether played out in small and medium-sized design consultancies or in large corporations—does involve power hierarchies and disputes among stakeholders (as well as disciplines). While authoritative or competitive strategies might lend themselves to simple problems that involve few stakeholders or small projects that can be quickly executed, it is the collaborative strategy that can work best for design's wicked problems. Collaboration offers the benefits of shared costs, the possibility of more comprehensive solutions, better problem prediction, and so on.

19.6 Interdisciplinarity in design

In examining the inter- and transdisciplinary nature of design, it is important to assess both *design practice* and *design studies*. Design straddles craft and science, the humanities and the social sciences, as well as art and engineering in its practice and in its theory. Design is generative and analytical; it demands creative thinking and critical problem solving. If such is the task of design, its practice necessitates the practitioner and the theorist to draw upon the type of knowledge that resides in disparate disciplines, and requires a type of thinking that is flexible enough to fluctuate among them. Ken Friedman refers to design as an integrative discipline that resides at 'the intersection of several large fields' (Friedman 2000). For Friedman, the natural sciences, humanities, and liberal arts as well as the social and behavioral sciences constitute the 'Domains of Theory' while the human professions and services, creative and applied arts, and technology and engineering make up the 'Domains of Practice and Application'. However,

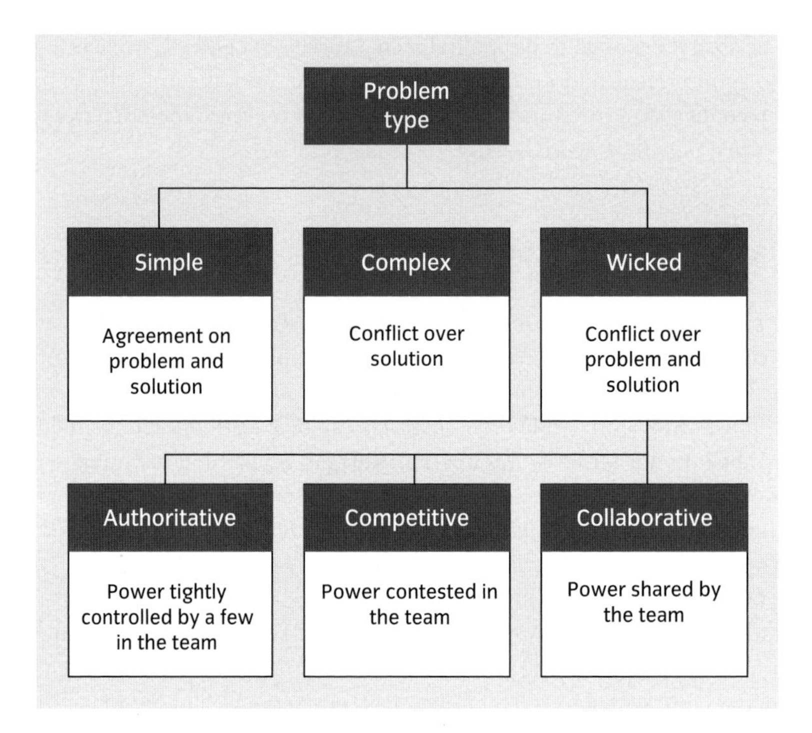

Figure 19.3 Coping strategies for wicked problems (developed from Roberts 2001).

classifying these domains on the basis of theory and practice presents problems; just as there are theories of engineering, there is application in the humanities. These disciplines should instead be conceived of as contiguous areas of study so as to demonstrate the interaction among them.

Figure 19.4 represents a model where design problems can be mapped out on the basis of their engagement with other disciplines. Locating design within this web of traditional disciplines speaks to its interdisciplinary nature. The domain map of the design project therefore takes form on the basis of the nature of the problem and the disciplines required to be involved.

While it is clear that wicked design problems require transdisciplinary approaches, scholars in design studies (Friedman 2000; Cross 2002) have sought to demonstrate that design theory and praxis possess components that are unique and distinct from other disciplines: 'The underlying axiom of this discipline [of design] is that there are forms of knowledge and ways of knowing that are special to the awareness and ability of a designer, and independent of the different professional domains of design practice' (Cross 2006, p. 100). Their goal is to set up design as an independent discipline, but one that enriches itself in transdisciplinary engagements: 'This is the challenge for design research – to construct a way of conversing about design that is at the same time both

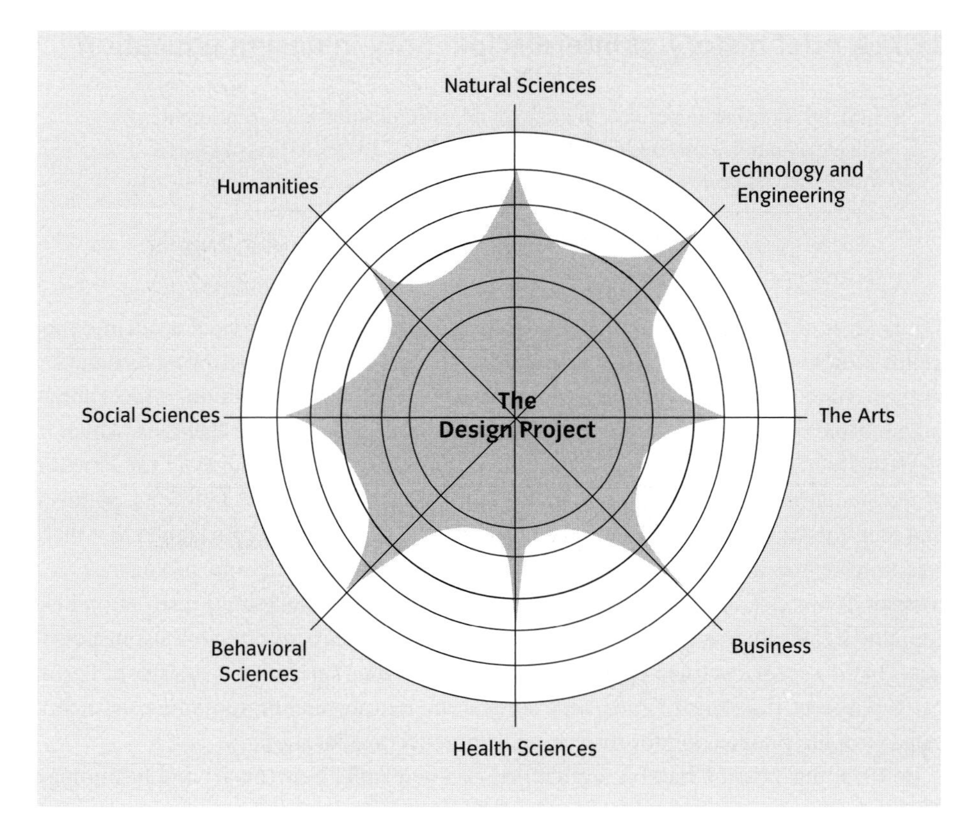

Figure 19.4 Domain map of a transdisciplinary design project.

interdisciplinary and disciplined. It is the paradoxical task of creating an interdisciplinary discipline. This discipline seeks to develop domain-independent approaches to theory and research in design' (Cross 2006, p. 100). Cross describes design's unique activities as 'designerly ways of knowing, acting and thinking' (Cross 2001). Multidisciplinarity, interdisciplinarity, and transdisciplinarity too can be described as three related yet distinct forms of knowing, acting, and thinking. As Klein explains in Chapter 2 of this volume: these three terms 'constitute a core vocabulary for understanding both the genus of *interdisciplinarity* and individual species within the general classification'. These three terms also represent varying levels of integration among disciplines; while multidisciplinarity might signify a mere juxtaposition of several disciplines aligned to tackle a specific problem, transdisciplinarity refers to the transcending of disciplines in developing transformative solutions to complex problems. Design praxis and design studies can be described as Klein's 'individual species', and they represent different degrees of interdisciplinarity. Design praxis can be described as Klein's 'trans-sector transdisciplinary problem solving'. Design studies, on the other hand, fits the model Klein labels 'critical interdisciplinarity'.

19.7 A brief history of interdisciplinarity in design education

> Students who wish to become designers in the postindustrial knowledge economy will enter an inherently multidisciplinary profession. This profession involves a wide variety of professionals, including scientists (physical, biological, and social), engineers (industrial, civil, biological, genetic, electrical, and software), and managers, as well as the many kinds of artists and artisans now called designers (Friedman 2000, p. 200)

While most of the coursework that students undertake in design-related academic programs is tightly circumscribed by the individual disciplines, there is a growing recognition of the need to create transdisciplinary opportunities. The formal tradition of incorporating multiple perspectives into design education can be traced to the Bauhaus, which in 1919 strove to create a unity between arts and the crafts. Walter Gropius, in the *Program of the Staatliche Bauhaus* published in 1919 in Weimar, summoned architects, painters, and sculptors to return to the crafts (Wingler 1969). This manifesto proclaimed that there was no essential difference between the artist and the craftsman, and proficiency in a handicraft was essential to every artist. Therefore, *Werkstatt* [workshop] instruction held supreme significance, and made up a large part of the students' quotidian learning activities. That the academic title of professor was supplanted by *Formmeister* [Master of Form] or *Werkmeister* [Master of Craft], and 'student' by journeyman or apprentice, authenticated Gropius' predilection for the artisanal approach to education.

By 1923, this mission had been redefined as a new unity between art and technology. This shift in focus was exemplified in Gropius' lecture, 'Art and technology: a new unity', when the Bauhaus embraced the ideals of mass production over craft-romanticism. They had decided to train not craftsmen but collaborators for industry, craft, and building. The workshops were renamed laboratories with the purpose of building prototypes of designs suitable for mass production. Towards the end of its life, the Bauhaus became an architectural school, and it was eventually closed in 1933.

In 1937, László Moholy-Nagy founded the New Bauhaus in Chicago to continue the initial Bauhaus mission by forming art, science, and technology as the three primary dimensions of design. Moholy-Nagy sought advice from the philosopher Charles Morris, who was then developing his theory of semiotics. Morris established coursework at the New Bauhaus in order to achieve 'intellectual integration' among these three key pillars of design: 'Morris considered the design act to be a kind of semiosis, and he drew a parallel between the syntactic, the semantic, and the pragmatic dimensions of a sign and, respectively, the artistic, the scientific, and the technological dimensions of design' (Findeli 2001, p. 7). Though these theories did not take root at that time, the attempt does demonstrate Moholy-Nagy's desire to introduce philosophical and linguistic concepts in design education. The New Bauhaus, which later merged with the Illinois Institute of Technology, continues to function today as the Institute of Design.

The Hochschule für Gestaltung (HfG) founded in 1951 in Ulm expanded the Bauhaus vision and outlined design's task as participating in the making of a new culture. Tomás Maldonado, who led the school from 1957 for a period of 10 years, suggested a more rigorous interdisciplinary education that included social psychology, sociology, anthropology, cultural history, and perception theory. The arts were no longer considered a critical foundation for design, and there was a heavier emphasis on developing a stronger scientific basis. Maldonado was interested in developing scientific design methodology and turned to several new disciplines emerging at that time: 'cybernetics, information theory, systems theory, semiotics, ergonomics' (Maldonado 1990, p. 223). Though these disciplines were not thoroughly integrated into the curriculum, engaging them allowed Maldonado and the Ulm school to investigate and develop design's own scientific base. This school, which eventually closed in 1968, has been singled out as having influenced design pedagogy all over the world.

The three schools—Bauhaus, New Bauhaus, and HfG—developed interdisciplinary curricula around three primary concepts: art, science, and technology. As design itself has evolved, design education has extended its interdisciplinarity beyond these three to include new disciplines. At the undergraduate level, most programs require students to take courses in the natural sciences, social sciences, and the humanities (mathematics, physics, psychology, etc.) as a part of their general studies requirements. In addition, design programs also encourage or require courses in marketing, economics, anthropology, etc. However, this does not qualify as transdisciplinary design education, and therefore several design programs have set up team-based learning environments where students from a variety of disciplines (frequently business, engineering, and anthropology) work together on projects. At the graduate level, design programs exhibit a higher level of transdisciplinarity, and it is not uncommon to find thesis and dissertation projects that critically engage several disciplines. Today, with varying degrees of integration, several departments and schools of design have partnered with programs in business, engineering and the social sciences across campus and at times across universities. Arizona State University, Art Center College of Design, Carnegie Mellon University, Illinois Institute of Technology, Rhode Island School of Design, Stanford University, and the University of Cincinnati are but a few examples of academic programs actively engaged in interdisciplinary design education.

19.8 Conclusion

In order to devise solutions to the significant challenges facing our world today, (such as clean drinking water, access to health care, climate change, renewable energy sources, and so on) we will need creative and innovative design thinking from transdisciplinary teams. These are wicked problems that elude quick solutions, and educational programs should prepare students by providing them with the tools they can utilize in their professional careers to be able to tackle these issues: 'But if knowledge is to be genuinely interdisciplinary, it needs to do more than simply reach across campus...Our academic research portfolio must include an account of how to effectively integrate knowledge within the decision-making context faced by governments, businesspeople, and citizens' (Frodeman and Mitcham 2005, p. 513).

It is clear that active participation from a large number and diversity of stakeholders is critical to doing transdisciplinary design in practice and teaching it in the university: 'The concept of superimposing various disciplines to address the problem or project in question could spawn a new hybrid category of design activity, which will emancipate itself from traditional disciplinary concepts' (Meurer 2001, pp. 52–3). This superimposition can be effective in design praxis and in design studies only if the boundaries among the overlapping disciplines can be made porous through truly integrated transdisciplinarity. Over the years, design's function has evolved from a craft-based practice of creating artifacts to the planning of artificial systems. The collaborative strategy which transdisciplinarity brings to problem solving can tame some of the wickedness of design problems. However, the highest possible level of integration among disciplines is necessary for this strategy to be truly effective. Only thus can society's wicked problems be tamed.

References

Appadurai, A. (2001). Grassroots globalization and the research imagination. In: A. Appadurai (ed.) *Globalization*, pp. 1–21. Durham, NC: Duke University Press.

Archer, B. (1979). Design as a discipline: the three Rs. *Design Studies* **1**(1), 17–20.

Baynes, K., Langdon, R., and Myers, B. (1977). Design in general education: a review of developments in Britain. *Art Education* **30**(8), 17–21.

Boradkar, P. (2001). Re-establishing the established: probing the foundations laid by the Bauhaus. *Proceedings of the 2001 IDSA Education Conference*. Dulles, VA: IDSA.

Buchanan, R. (1990). Myth and maturity: toward a new order in the decade of design. *Design Issues* **6**(2), 70–80.

Buchanan, R. (1992). Wicked problems in design thinking. *Design Issues* **8**(2), 5–21.

Buchanan, R. (1995). Rhetoric, humanism and design. In: R. Buchanan and V. Margolin (eds) *Discovering design: explorations in design studies*, pp. 23–66. Chicago: University of Chicago Press.

Cross, N. (1982). Designerly ways of knowing. *Design Studies* **3**(4), 221–7.

Cross, N. (2001). Designerly ways of knowing: design discipline versus design science. *Design Issues* **17**(3), 49–55.

Cross, N. (2002). *Design as discipline, designing design (research) 3: the inter-disciplinary quandary*. Available at: <http://nelly.dmu.ac.uk/4dd//DDR3-Cross.html> (accessed 5 May 2009).

Cross, N. (2006). *Designerly ways of knowing*. Berlin: Springer-Verlag.

Cross, N., Naughton, J., and Walker, D. (1981). Design method and scientific method. *Design Studies* **2**(4), 195–201.

Findeli, A. (2001). Rethinking design education for the 21st century: theoretical, methodological, and ethical discussion. *Design Issues* **17**(1), 5–17.

Friedman, K. (2000). Design knowledge: context, content and continuity. In: D. Durling and K. Friedman (eds) *Doctoral education in design: foundations for the future*, pp. 5–16. Stoke-on-Trent: Staffordshire University Press.

Friedman, K. (2002). Toward an integrative design discipline. In: S. Squires and B. Byrne (eds) *Creating breakthrough ideas: the collaboration of anthropologists and designers in the product development industry*, pp. 199–214. Westport, CT: Bergin and Garvey.

Frodeman, R. and Mitcham, C. (2005). New directions in interdisciplinarity: broad, deep, and critical. *Bulletin of Science, Technology and Society* **27**(6), 506–14.

Gann, D. and Salter, A. (2001). Interdisciplinary education for design professionals. In: R. Spence, S. Macmillan, and P. Kirby (eds) *Interdisciplinary design in practice*. Reston, VA: Thomas Telford.

Klein, J. (1990). *Interdisciplinarity: history, theory, and practice*. Detroit: Wayne State University Press.

Lindinger, H. (1991). *Hochschule für gestaltung (Ulm, Germany)*, transl. D. Britt. Cambridge, MA: MIT Press.

Maldonado, T. (1990). Looking back at Ulm. In: H. Lindinger (ed.) *Ulm design: the morality of objects*, pp. 222–4. Cambridge, MA: MIT Press.

Margolin, V. (1989). Introduction. In: V. Margolin (ed.) *Design discourse: history, theory, criticism*, pp. 4–28. Chicago: University of Chicago Press.

Meurer, B. (2001). The transformation of design. *Design Issues* **17**(1), 44–53.

Mitcham, C. (1994). *Thinking through technology: the path between engineering and philosophy*. Chicago: University of Chicago Press.

Rittel, H. (1971). Some principles for the design of an educational system for design. *Journal of Architectural Education (1947–1974)* **25**(1/2), 16–27.

Rittel, H. and Webber, M. (1973). Dilemmas in a general theory of planning. *Policy Sciences* **4**, 155–69.

Roberts, N. (2000). Wicked problems and network approaches to resolution. *International Public Management Review* **1**(1). Available at: <http://www.ipmr.net/>.

Roberts, N. (2001). Coping with wicked problems: the case of Afghanistan. In: L. Jones, J. Guthrie, and P. Steane (eds) *Learning from international public management reform*, pp. 353–75. Bingley: Emerald Group Publishing.

Simon, H. (1996). *The sciences of the artificial*. Cambridge, MA: MIT Press.

Wingler, H. (1969). *The Bauhaus*. Cambridge, MA: MIT Press.

CHAPTER 20

Learning to synthesize: the development of interdisciplinary understanding

VERONICA BOIX MANSILLA

> With the Vietnam Veterans memorial, I needed to ask myself the question 'what is the purpose for a war memorial at the close of the twentieth century?'... Perhaps it was the empathic idea about war that led me to cut open the earth, an initial violence that heals in time but leaves a memory – like a scar.
>
> —Maya Lin, *Boundaries*

> Water resource are critically influenced by human activity, including agriculture and land use... changes in population, food consumption, economic policy... In order to assess the relationship between climate change and freshwater, it is necessary to consider how freshwater has been and will by affected by changes in these non-climatic drivers.
>
> —IPCC Climate Change Synthesis Report 2007

The tension is clear. Excerpts like the ones above shed light on forms of interdisciplinary reasoning that are increasingly in demand at the dawn of the twenty-first century, and yet psychological studies of interdisciplinary learning and cognition to date have been surprisingly sparse and non-paradigmatic. Missing, in my view, is a generative epistemological foundation for the study of interdisciplinary cognition—one that can embrace a broad range of interdisciplinary intellectual agendas, while attending to the disciplinary foundations on which such insights are built and the intellectual processes required to integrate them in a coherent whole. Consider the two excerpts above. In the first, artist Maya Lin describes the Vietnam Veterans Memorial as a scar. Her metaphor frames the Vietnam War experience in terms of a country divided by the war and in need of healing. In Lin's work, detailed analysis of military records gives room to names chronologically engraved on reflective granite, where living selves and lost others meet and reconcile—where art and history intertwine to illuminate past and present human experience. Resulting from a scientific collaboration of unprecedented scope, the second vignette highlights factors affecting the availability of fresh water in times of climate change. In the report produced by the Intergovernmental Panel on Climate Change, biological, chemical, and physical laws are used to determine the quality and quantity of available water resources. Superimposed on this natural phenomenon is an analysis of the human drivers that intensify natural water cycles: population growth, climate variation, and economic development.

How do memorials and complex explanations function as interdisciplinary learning achievements? How do individuals come to integrate disciplinary traditions, and what cognitive demands does interdisciplinary learning impose? A striking array of metaphors have been deployed to describe the nature of interdisciplinary intellectual activity—from working at 'crossroads' and in 'trading zones' to engaging 'boundary objects' and 'bridges' (Klein 2005). Metaphors have served us well as evocative approximations to interdisciplinary cognition. However, they have proven less productive in their ability to structure strong research agendas or to design empirically grounded programs on interdisciplinary learning and its assessment. This chapter seeks to move beyond evocative language to examine the phenomenon of interdisciplinary learning in epistemological and cognitive terms.

Interdisciplinary learning is a process by which individuals and groups integrate insights and modes of thinking from two or more disciplines or established fields, to advance their fundamental or practical understanding of a subject that stands beyond the scope of a single discipline (National Academies 2005; Boix Mansilla 2005, 2006, 2008). Interdisciplinary learners *integrate* information, data, techniques, tools, perspectives, concepts, and/or theories from two or mores disciplines to craft products, explain phenomena, or solve problems, in ways that would have been unlikely through single-disciplinary means. Conceived as a cognitive phenomenon, understanding interdisciplinary learning demands an empirical examination of the mental processes involved, such as analogical reasoning, conceptual blending, and complex causal reasoning. However, because key to interdisciplinary learning is the integration of knowledge forms that respond to distinct epistemologies (preferred units of analysis, methods, validation criteria), a psychological study of interdisciplinary learning requires a strong epistemological foundation. It requires an articulation of the nature of disciplinary knowledge and the methods and criteria by which such knowledge is produced and deemed acceptable. It also requires an epistemological theory that enables us to make sense of—and validate—the insights that emerge at multiple interdisciplinary crossroads.

In this chapter, an epistemological foundation for interdisciplinary learning is proposed. I argue that a *pragmatic constructionist* view of interdisciplinary learning can account for the variety of enterprises considered 'interdisciplinary'. Such a view can illuminate the process of considered judgment and critique involved in advancing an understanding that integrates multiple specialties effectively with a purpose in mind. Interdisciplinary learning, it is argued, involves a series of delicate adjustments by which new insights are weighted against one another and against antecedent commitments about the subject matter under study. To advance the case for a *pragmatic constructionist* theory, I first review available literature on interdisciplinary learning, considering its strengths and limitations. I show how epistemological assumptions frame (and limit) our understanding of interdisciplinarity by revisiting two classic approaches—logical positivism and E. O. Wilson's *Consilience*. I then introduce a pragmatic constructionist framework for interdisciplinary learning and test it against the two learning examples described above: creating historical monuments and explaining water availability. The chapter concludes with recommended future avenues for research and instruction in which interdisciplinary learning is considered as a cognitive phenomenon with deep epistemological roots.

20.1 Interdisciplinary learning

Empirical studies of interdisciplinary learning today unfold without a generative epistemology or convergent lines of research that would render knowledge accumulation possible. Interdisciplinary learning has been linked to critical thinking skills; more sophisticated conceptions of knowledge, learning and inquiry, and heightened learner motivation and engagement (Hursh *et al.* 1988; Huber and Hutchings 2004; Minnis and John-Steiner 2005; Baxter Magnolia and King 2008). Occasionally, authors have advanced conceptual models of interdisciplinary learning mechanisms rooted in specific learning theories. Models vary in the degree to which they are empirically or conceptually based and the dimensions of interdisciplinary learning they seek to explain. For example, the conceptual blending theory, advanced by Gilles Fauconnier and Mark Turner, captures a key human cognitive computation: the capacity to combine two existing concepts to produce a new unit of meaning. Blended concepts such as 'problem solving' or 'handwriting' are pervasive in every day language and contribute to our capacity to make efficient sense of the world around us. Matthew Miller (2005) showed how compound concepts—for example, *empirical bioethics*—and concepts of expanded meaning—such as *innovation* in evolution, cell development, technology and organizations—enabled individuals to integrate disparate bodies of information. His study illuminates microrepresentations of interdisciplinary integrations; it does not address the process by which integrated concepts are constructed.

Researchers following a neo-Piagetian tradition put a premium on the construction and revision of knowledge structures of increasing levels of complexity and abstraction. Understanding the connection between two concepts must be preceded by a lower-level understanding of each concept in isolation. Further, understanding the connections between sets of related concepts builds on a prior low-level understanding of each participating set. Higher-order concepts such as 'systems' or 'systems of systems' organize lower-order ones rendering such abstractions a desirable mark of learning success. Applied to interdisciplinary contexts, a neo-Piagetian approach suggests that, at first, learners construct abstractions in one relevant discipline. They then acquire knowledge in two or more disciplines but do not draw connections among them. Third, they integrate knowledge from two disciplines around a central and more abstract theme. Eventually, Lana Ivanitskaya and her colleagues suggest, learners build an overarching knowledge structure of still further complexity and abstraction that can be applied to new interdisciplinary themes (Ivanitskaya *et al.* 2002).

Emphasizing the social dimension of learning and cognitive development, researchers such as Svetlana Nikitina and Rebecca Burns characterize progressive appropriation of disciplinary discourses and modes of thinking among individuals trained in different fields. These authors take the social mediation of learning as their point of departure. Their proposed learning progressions begin with an individual's sensitivity toward foreign concepts and terms from a colleague in another discipline, followed by a growing capacity to define such constructs and eventually utilize them productively in interdisciplinary contexts as part of an established personal repertoire. Similarly emphasizing communicative dimensions of interdisciplinary learning, others examine collaborative learning in the construction of a common ground—a shared definition of a problem or approach on the part of a two or more individuals (Bromme 1999; Nikitina 2005).

The emerging research on interdisciplinary learning has benefited from multiple approaches to studying human cognition and their corresponding assumptions about the nature of knowledge. Perspectives are also limited. For example, an emphasis on integrative concepts must be complemented with a sense of how concepts are learned. A neo-Piagetian commitment to complexity and abstraction as markers of learning sophistication must be complemented with an account of learning in which other cognitive goals are pursued, such as effectiveness or innovation. Models differ in their assumptions about the nature of knowledge and the process by which it is acquired. In fact, to be informative, a comprehensive framework for the study of interdisciplinary learning begins with greater clarity about the nature of what is being learned. What constitutes interdisciplinary knowledge or understanding? Can we discern key dimensions of this elusive epistemological phenomenon to inform a theory of interdisciplinary learning?

20.2 The problem of reductionism

'Interdisciplinary learning' encompasses diverse cognitive endeavors, from aesthetic interpretations of past events to comprehensive explanations of water availability. It engages concepts and modes of thinking in a broad range of specialties. An epistemological foundation for interdisciplinary learning must account for such variety, while illuminating the processes of learning involved. Generally speaking, epistemological theories seek to illuminate the nature, scope, and utility of knowledge. They differ, however, in the way they characterize the landscape of human knowledge and insight, the relative significance they attribute to particular knowledge forms, and the standards and criteria by which knowledge is deemed acceptable (Elgin 1997). As a result, epistemological frameworks also differ in their utility to shed light on interdisciplinary knowledge integration.

The search for an integrated theory of knowledge has galvanized thinkers in a number of intellectual traditions (Gould 2003). Scholars have sought to distill underlying patterns across apparently disconnected disciplinary facts or claims. While efforts to make reasonable connections across knowledge spheres are laudable, their results have all too often prioritized a single preferred mode of explanation typically stemming from logic and mathematics or, more recently, from biology. In what follows, two such reductions are considered: logical positivism as exemplified in the classic work of A. J. Ayer, and *Consilience* as introduced by E. O. Wilson (1998).

Logical positivism dominated English-speaking philosophy since its origins in the School of Vienna of the early 1920s. It regarded logic and mathematics as sources of

ZiF

Ipke Wachsmuth

The ZiF (*Zentrum für interdisziplinäre Forschung*/Center for Interdisciplinary Research) of Bielefeld University is an internationally operating institute for advanced study which supports and houses interdisciplinary research projects from all fields across the natural and social sciences, engineering, and the humanities. ZiF's aim is to promote and realize interdisciplinary basic research by offering residential fellowships, grants, and conference services to scholars and scientists from all fields and from all over the world.

Five principles frame all ZiF projects: interdisciplinarity, internationality, topic-orientation in the choice of projects and fellows, open application procedures, and scientific excellence of the projects, ensured by peer reviewing of all proposals. Any member of the international scientific community may propose a topic to ZiF, throughout all fields of science and the humanities. Applications are decided upon by the board of directors after external reviewing and—for longer-term projects—a hearing before the scientific advisory council. Support by the ZiF comprises financial assistance as well as provision of its infrastructure. Languages of communication and correspondence at the ZiF are German and English. The 'ZiF *Mitteilungen*' newsletter comes out in four German–English issues per year.

ZiF is the oldest institute for advanced study in Germany and has been a model for numerous similar centers elsewhere in Europe. Renowned scientists and scholars, as for instance the later Nobel laureate for economics Reinhard Selten, have worked at the ZiF. One of the most famous ZiF fellows certainly was Norbert Elias, who, after being rediscovered by a new generation of scholars in the 1970s and eventually becoming one of the most influential sociologists in the history of the field, worked and lived at the ZiF from 1978 to 1984.

The founding of the ZiF in 1968 preceded, but was directly related to, the creation of the University of Bielefeld in North Rhine-Westphalia, which itself was founded in 1969. In this process a central idea from the very inception was to reinsert philosophical–humanistic ideals in the setting of a modern university, with the ZiF being the seed institute and intellectual center. Its principal design was laid out by the prominent German sociologist Helmut Schelsky who became the first director of ZiF (1968–71). ZiF came to emphasize the role of thematically coherent research groups but it insisted that all fields of knowledge should be included. These features have been characteristic of ZiF until the present day.

Usually running from October through August or September of the following year, ZiF research groups are the primary format to support long-term interdisciplinary collaboration. For several months, up to a year, the fellows reside at the ZiF and work together on a broader research theme. To name but a few recent ones: '*E pluribus unum*?—ethnic identities in processes of transnational integration in the Americas' (2008–9), 'Control of violence' (2007–8), 'Science in the context of application' (2006–7), 'Embodied communication in humans and machines' (2005–6), or 'Emotions as bio-cultural processes' (2004–5). About two-thirds of the ZiF fellows come from outside Germany.

Besides the research years, ZiF workshops support interdisciplinary collaborations on a short-term basis. They range from invitational colloquia on specific topics to larger-scale conferences on a variety of themes. More than a thousand scholars come to visit the center every year this way, giving ample evidence of ZiF's great attraction as a meeting place for interdisciplinary exchange. The ZiF Author's Colloquia are a special type of workshop featuring a distinguished scholar; they create a platform for discussion and critique by selected experts. To trigger new developments in the research landscape, the ZiF network of young scientists was founded in 2002. Also started in 2002 and convening smaller-size interdisciplinary research collaborations, ZiF cooperation groups offer the opportunity to invite fellows to the ZiF for up to 6 months.

Located in the west of Bielefeld in northwest Germany, the ZiF campus is situated on the edge of the Teutoburg Forest. ZiF's main building houses a spacious plenary hall, five conference rooms, and foyers for casual meetings. A satellite of the university library system, the ZiF library offers individual support to all fellows and moreover provides a comprehensive resource of literature on interdisciplinary research. More than 20 staff employees help create comfortable conditions for working and living. Situated on the park-like campus, apartments of various sizes and an indoor swimming pool with sauna are available to the fellows and their families. The proximity of the university campus facilitates contacts among scholars beyond their collaboration at the ZiF.

By no means an ivory tower, the ZiF regularly opens its doors to the interested public. Once or twice a year scientists give public evening lectures to a wide audience, and well-known authors from Germany or abroad are invited to read from their work. Adding to the inspiring atmosphere of the place, ZiF hosts six art exhibitions per year which are, like the public lectures and authors' readings, often connected to ongoing research themes.

In October 2008, ZiF celebrated its 40th anniversary—a development highly appreciated by Bielefeld University of which the ZiF is an integral part and whose international reputation the ZiF has contributed to decisively. The center also plays an important strategic role, as an incubation center and scope for development enhancing the university's stature in the German Excellence Initiative.

analytical truths and the natural and social sciences as the only way to reveal verifiable truths about the world. It placed a premium in propositional knowledge, restricting the universe of meaningful claims to those that can be, in principle, verified or falsified by experience or logical proof (Ayer 1956; Popper 1965). Yet, as Ayer's emotive theory of ethics suggests, propositions pertaining to the moral or aesthetic realm remain outside of the logical positivist worldview. They cannot be empirically or logically confirmed or disconfirmed. Similarly non-propositional knowledge embodied in images or movements cannot be considered properly meaningful. Logical positivism in its strictest form sought to guarantee that if a claim satisfies its validation criteria it is highly credible. However, it does so at a cost. It restricts the kinds of knowledge it seeks to understand to science and logic, excluding important human cognitive achievements in the realms of art and normative moral reasoning (Goodman 1976, 1978).

When applied to knowledge representation in interdisciplinary learning, logical positivism emphasizes the acquisition of propositional knowledge in the disciplines and the development of deductive and inductive reasoning skills. Yet, confronted with Maya Lin's Vietnam Veterans Memorial, it falls short. It remains unable to make sense of Lin's aesthetic experience and is silent about her visually nuanced interpretation of the past. Too complex and uncertain to be encoded in a system of irrefutable premises and logic, too semantically dense for modeling and verification, the monument falls outside the purview of the positivist mind and explanatory framework.

Edward O Wilson's theory of *Consilience* stands out as a rather recent effort to bridge C. P. Snow's two cultures of sciences and humanities (Snow 1998; Wilson 1998). Consilience admits, at least in principle, a diversity of intellectual endeavors. It seeks, in practice, to bring specialists together to 'agree on a common body of abstract principles and evidentiary proof' (Wilson 1998, p. 10). Wilson characterizes it as 'a new 21st century enlightened unity of knowledge' (1998, p. 14). Consilience, he proposes, can unite the humanities and the sciences legitimately. It grants the humanities the right to articulate human and cultural constructs to be studied—consciousness, beauty, altruism, cooperation—and entrusts the biological sciences with the power to explain them.

Yet, to understand Maya Lin's aesthetic and evocative achievement, consilience sidesteps history, art, and architecture to look at the human biology of visual perception. Lin's aesthetic use of notations on reflective granite or her symbolic violation of the natural landscape is overlooked by consilience; as is the way in which multiple interpretations of the monument give new meaning to the experience of war. Rather, an unwavering aspiration to scientific truths leads consilience to seek a biological explanation of our aesthetic mind. Aesthetics as a form of knowing in its own right is 'black boxed' and evolutionary hypotheses about our universally wired preference for slightly symmetrical visual patterns and open prairie landscapes prevail. Consilience is useful for framing interdisciplinary endeavors when learners seek to explain the biological foundations of human life. It is limited, however, when learners seek other goals, such as understanding the emotional cost of war, or exploring how to use the arts to heal and reconcile. A more pluralistic epistemological theory is in order—one that embraces the multiple knowledge forms on their own terms and at the same time discerns between more and less trustworthy insights.

20.3 A pragmatic constructionist epistemology

Interdisciplinary pursuits are diverse: the learning demands of designing a historical monument contrast substantively with those of explaining climate change. Substantive cognitive transfer across tasks can rarely be expected. What constitutes a productive epistemological framework for interdisciplinary learning? Arguably, four criteria are required. First, an epistemological framework must be *pluralist* in its capacity to account for multiple forms of disciplinary understanding on their own terms and embrace various intellectual agendas. Second, it must be *relevant* to the phenomenon of interdisciplinary learning, illuminating the processes of interdisciplinary integration. Third, the theory must *explain* how knowledge advances from less to more accomplished instantiations; shedding light on the essential dynamics of learning. Finally, it must offer some form of *knowledge quality assurance*—an epistemic mechanism that diminishes the likelihood of error by putting forth robust and relevant standards of acceptability across interdisciplinary endeavors.

To shed light on knowledge integration in interdisciplinary learning, an epistemological theory must neither limit its reach to the realm of empirically validated propositions, nor reduce all forms of knowledge to a privileged one, such as biology. Such emphases, as we have seen, constrain the types of interdisciplinary learning that can be legitimately examined. Instead, a productive epistemology offers insight into how understanding of a subject matter can be advanced, whether such understanding entails an aesthetic interpretation of the Vietnam War or a comprehensive explanation of a shortage of fresh water. Relevant to interdisciplinary learning is an epistemology that sheds light on how humans can make increasing and better sense of the world, themselves, and others through the integration of available disciplinary insights.

A *pragmatic constructionist* epistemology rooted in the work of philosophers Nelson Goodman and Catherine Elgin meets the criteria above (Goodman and Elgin 1988). As *constructionist*, the epistemological framework proposed posits that the purpose of inquiry (and learning) is the advancement of understanding. Inquiry is not the accumulation of propositional knowledge in search for certifiable truths. Rather, inquiry seeks a broad, deep, and revisable understanding of its subject matter. Taking a *pragmatist* stance, the proposed epistemology puts a premium on the purpose of knowledge construction. It judges the worth of an emerging insight by its effectiveness in advancing the desired understanding.

Ultimately, understanding involves the construction of what Elgin defines as a *system of thought in reflective equilibrium*. Elgin argues that a system of thought is in reflective equilibrium when its components are reasonable in light of one another and the account they comprise is reasonable in light of our antecedent convictions about the subject at hand. Such a system, she notes, affords no guarantees. It is rationally acceptable not because it is certainly true but because it is reasonable in the epistemic circumstances (Elgin 1996, p. ix). Building and validating understanding involve a series of delicate adjustments by which new insights are weighed against one another and against antecedent commitments about the subject matter. A conclusion is deemed acceptable not through a linear source of argumentation but through a host of sources of evidence (much of which may not precisely 'match up', but paints a telling picture) which include findings, statements, and observations, as well as useful analogies, telling metaphors, and powerful exemplifications. The acceptability of a knowledge system is to be measured against the purposes of inquiry that guide its production. Justification is also provisional. In Elgin's view, considered judgment recognizes the unfortunate propensity for error of the human mind and adapts to it by demanding corrigibility. This epistemology demands that we be prepared to criticize, revise, reinterpret, and abandon intellectual commitments when more reasonable ones are conceived.

The implications of a *constructionist pragmatic* approach for interdisciplinary learning theory are potent. By shifting our attention from accumulation of propositional knowledge to deep and broad understanding, the proposed epistemology recognizes that prior knowledge matters in the ways in which individuals make sense of the world. Prior knowledge sets the stage for the insights to come, by informing questions, affording hypotheses, and providing an initial representation of a problem under study. By broadening the admissible sources of knowledge and inquiry beyond strictly certified propositions, this pluralistic epistemology invites the inclusion of other symbol systems (visual, musical, kinesthetic) and ways of knowing such as artistic interpretations or literary fictions. Interdisciplinary understanding can thus be viewed as a 'system of thought in reflective equilibrium'—embodying insights and tensions across disciplines, representing an improvement over prior beliefs, and remaining open for review. A cognitive process for interdisciplinary learning can be derived.

Creativity and interdisciplinarity

Thomas Kowall

Creativity has been widely discussed, and in a seminal paper David J. Sill (1996) has examined at length arguments about creativity in relation to interdisciplinarity. Underlying all forms of creativity as original production, cognitive problem solving, and subjective experience is the simple and pervasive use of language.

Speakers of natural languages exhibit a remarkable linguistic creativity as they find an unending number of new things to say (or produce) across all disciplines. This occurs equally in contexts that manifest multidisciplinarity, transdisciplinarity, and other forms of interdisciplinarity. Such creativity is present in both young and mature speakers and is widely appreciated in cognitive science, linguistics, and philosophy.

There is a parallel to cross-disciplinary linguistic creativity in perceptual experience (which includes both problem solving and subjectivity). As we assimilate experience, we understand an unending number of things perceived for the first time. No one discipline has a monopoly on creative perception any more than on linguistic expression.

Because understanding the newly created infuses our everyday linguistic and perceptual experience, any theory of creativity needs to account for its ubiquity not just in but across disciplines. This account implies, perhaps surprisingly, that creativity is intersubjective and learnable. To the extent that creativity exhibits these properties, it does not serve as a criterion for distinguishing between disciplines but bridges them in unexpected ways and illuminates the character of interdisciplinarity.

Language versus perception disciplines

Many things are said for the first time. Some things said once are never said again by anyone, anywhere, anytime. On many of the occasions when we say something for the first or only time, we communicate as successfully as we do through utterances spoken or heard many times before. Language users, speakers and hearers, have the remarkable ability to communicate as successfully by means of the rare, even unique, as by the familiar, commonplace, or banal.

A significant theory of linguistic communication must provide strategic insight into how language users achieve their remarkable success when uttering the novel and the unique. Additionally, by providing an explanatory description of the creative use of language, a theory of linguistic communication becomes an important component of a more encompassing theory of creativity. Because language and perception are immersed in the novel and unique, they contribute to the construction of a broader understanding, even a theory, of human creativity and its role in understanding interdisciplinarity.

The same holds for perceptions. The parallel between utterances and perceptions is seamless because some of the things we perceive, we perceive only once. Some things perceived once are never perceived again by anyone, anywhere, anytime. Many of the things we perceive, we perceive a number of times until they become entirely familiar to us. We successfully perceive something when we understand what we perceive. This success is achievable through perceptions of the rare and the unique as well as through perceptions of the familiar and commonplace.

One contribution of the relation between language and perception to understanding creativity is to clarify the relation between disciplines that place relatively more emphasis on natural language (linguistics, literature, philosophy) and those that emphasize perception (art, architecture, design).

For some, language-emphasis disciplines and perception-emphasis disciplines appear to be distinct. Popular and scholarly views alike propose distinguishing between these two discipline clusters. One view construes the difference as characterized by the degree of intersubjectivity required to practice in each cluster. Under this view, language-emphasis disciplines are treated as more intersubjective than disciplines that depend more heavily on perception.

Standards of intersubjectivity are required for teaching and learning natural languages. In the absence of parallel standards for perception-based disciplines, it is difficult to see how the disciplines of art, architecture, and design can be teachable. In the absence of the capacity to teach and learn about art, architecture, and design, these disciplines run the risk of being dismissed as not disciplines at all but as practices clustered by circumstance and chance. Standards drawn from language-emphasis disciplines would be, on this view, one attractive model for building standards for the teaching and learning of perception-emphasis disciplines.

In the quest to make perception-emphasis disciplines more intersubjective, efforts have been made to interpret them through explicitly language-based vocabularies. Examples include such expressions as 'the languages of art' and 'the reading of images' (Goodman 1976) and 'the vocabulary of design' and 'the syntax of architecture' (Bloomer and Moore 1977). This approach aims to extend the strengths of language-emphasis disciplines to perception-emphasis disciplines.

This approach of modeling visual education after language education is nevertheless contrary to educational assumptions adopted in many post-secondary art and design programs. In the words of architect Philip Johnson: 'Art has nothing to do with intellectual pursuit... it shouldn't be in a university at all. Art should be practiced in gutters – pardon me, in attics. You can't learn architecture anymore than you can learn a sense of music or painting. You shouldn't talk about art, you should do it'. (Johnson 1978, p. 14)

Another view, with a consequence similar to Johnson's, argues that the abilities of artists cannot be taught because what constitutes the successful practice of art is too indeterminate to be generally accepted. When we create, we cause something to come into existence, to change, or to cease to exist. The range of these causes includes ideas, objects, patterns, and practices. Sometimes the creativity most highly prized is causing something to change or come into existence when what results is rare, even unique. From this perspective, the new and the novel are at the conceptual heart of creativity. But the new and the novel are also present quite naturally in language and perception.

There are popular views that creativity is a form of rare inspiration (Plato's *Ion* and William Blake). From this perspective, creativity is thought to be caused by uncommon, even non-natural powers. Accepting this view of the creative process, some artists and authors await inspiration from the muse. Others, however, see creativity as the outcome of a quotidian process. Hans Bethe, for instance, described his creative process as follows: 'Well, I come down in the morning and I take up a pencil and I try to think, to put things together in a new way' (quoted in Miller 1987, p. 517). This view of creativity calls for no external intervention because creativity is construed as a property emerging from, perhaps supervening on, the structured, daily activity of authorship and studio practice.

The second view of creativity is appealing for two reasons. First, the naturalization and consequent demystification of the creative process implies success through effort rather than election. Second, naturalization is consistent with the pervasiveness of creativity. Rather than being rare, creativity is a common, even ordinary, part of language and perception. The naturalization of creativity takes everyday experience seriously.

(*cont.*)

Creativity and interdisciplinarity (cont.)

Interdisciplinary implications

This naturalized account of creativity suggests at least four implications for interdisciplinarity. First, because creativity is ubiquitous it offers no criterion for distinguishing alleged creative disciplines (architecture, the fine and performing arts, design) from non-creative ones (craft, engineering, history). This consequence holds for discipline clusters (art, science) as much as it does for specialized disciplines (industrial design, chemical engineering).

Second, eliminating creativity as a criterion for distinguishing disciplines points toward other criteria better suited to distinguish disciplines. History and practice rather than ontology or epistemology provide stronger identity conditions for the disciplines. The standing and prerogatives granted by universities to current disciplines are not etched in stone but are the by-products of a pervasive incrementalism analogous to that found in saying and seeing, both of which are imbued with creativity.

Third, because creativity is ubiquitous, there is every reason to expect it to be present in interdisciplinary thought and practice. There is no reason therefore to either stigmatize interdisciplinarity as lacking in creativity nor to praise it for possessing a hitherto undiscovered and unique form of creativity.

Fourth, understanding disciplines as created artifacts reflecting the historical trajectories of saying and seeing sets the stage for more than disciplinary creativity. As with all effective creative endeavors, interdisciplinary creativity requires sound practice and good management. When we abandon the model of disciplines as silos, interdisciplinarity is a strong candidate for creating new forms of method and management. This is the fundamental challenge for first understanding creativity and interdisciplinarity and then acting on that understanding.

References

Bloomer, K. and Moore, C. (1977). *Body, memory, and architecture.* New Haven, CT: Yale University Press.

Goodman, N. (1976). *The languages of art.* Indianapolis, IN: Hackett.

Johnson, P. (1978). *Philip Johnson: processes, the glass house, 1949 and the AT&T corporate headquarters, 1978.* New York: Institute for Architecture and Urban Studies.

Miller, A. (1987). *Timebends.* New York: Grove Press.

Sill, D.J. (1996). Integrative thinking, synthesis, and creativity in interdisciplinary studies. *Journal of Education* **45**(2), 129–51.

20.4 Interdisciplinary learning as the construction of systems of thought in reflective equilibrium

The epistemological framework outlined below offers a dynamic picture of interdisciplinary integration (see Fig. 20.1). Four core cognitive processes are involved: *establishing purpose*; *weighing disciplinary insights*; *building leveraging integrations*; and *maintaining a critical stance.* In interdisciplinary learning such processes interact dynamically, informing one another as learning progresses. The result is a system of thought in reflective

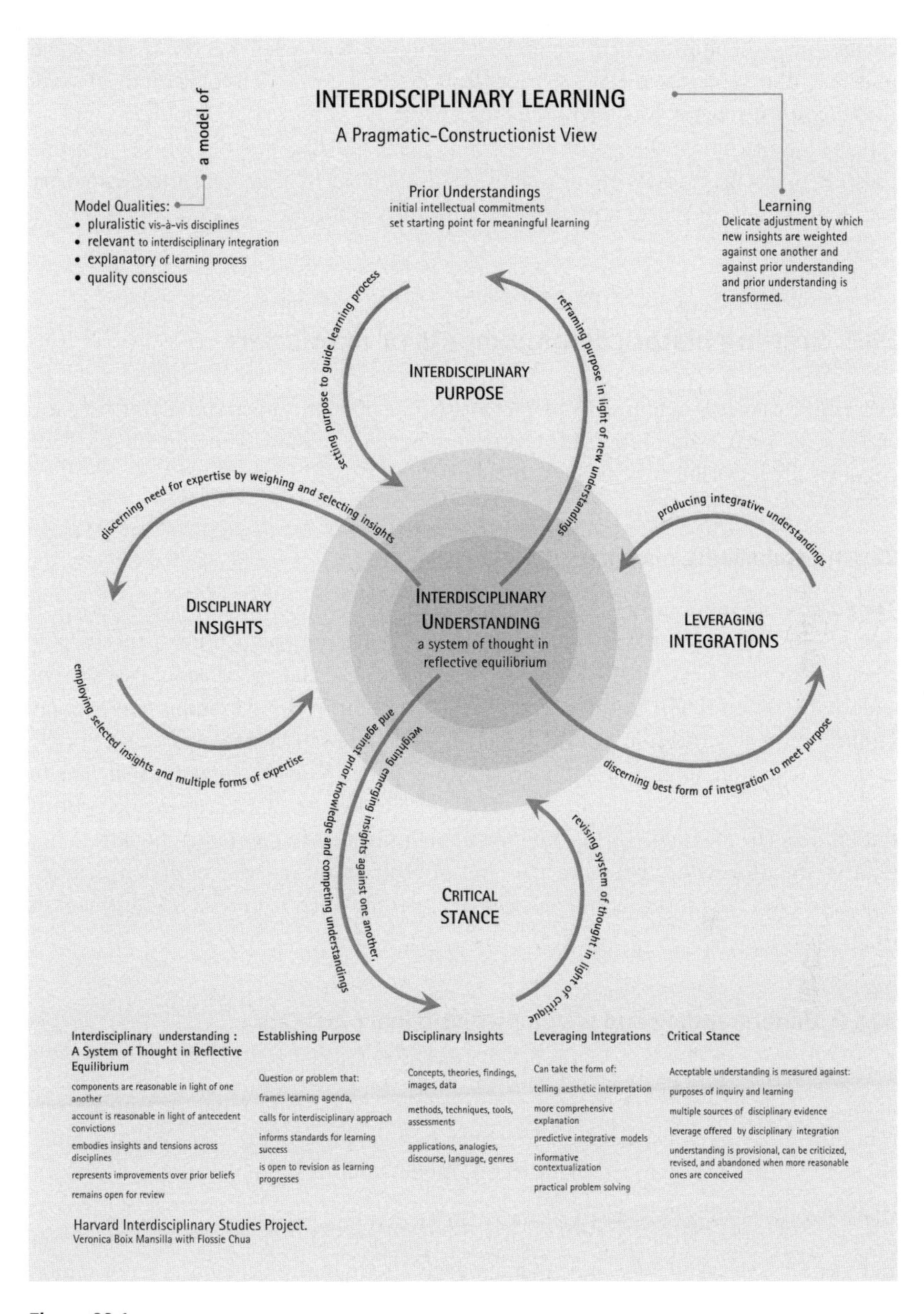
a model of
INTERDISCIPLINARY LEARNING
A Pragmatic-Constructionist View
Model Qualities:
• pluralistic vis-à-vis disciplines
• relevant to interdisciplinary integration
• explanatory of learning process
• quality conscious
Prior Understandings
initial intellectual commitments
set starting point for meaningful learning
Learning
Delicate adjustment by which new insights are weighted against one another and against prior understanding and prior understanding is transformed.
setting purpose to guide learning process
reframing purpose in light of new understandings
INTERDISCIPLINARY PURPOSE
discerning need for expertise by weighing and selecting insights
producing integrative understandings
DISCIPLINARY INSIGHTS
INTERDISCIPLINARY UNDERSTANDING
a system of thought in reflective equilibrium
LEVERAGING INTEGRATIONS
employing selected insights and multiple forms of expertise
weighting emerging insights against one another, and against prior knowledge and competing understandings
discerning best form of integration to meet purpose
revising system of thought in light of critique
CRITICAL STANCE
Interdisciplinary understanding : A System of Thought in Reflective Equilibrium
components are reasonable in light of one another
account is reasonable in light of antecedent convictions
embodies insights and tensions across disciplines
represents improvements over prior beliefs
remains open for review
Establishing Purpose
Question or problem that:
frames learning agenda,
calls for interdisciplinary approach
informs standards for learning success
is open to revision as learning progresses
Disciplinary Insights
Concepts, theories, findings, images, data
methods, techniques, tools, assessments
applications, analogies, discourse, language, genres
Leveraging Integrations
Can take the form of:
telling aesthetic interpretation
more comprehensive explanation
predictive integrative models
informative contextualization
practical problem solving
Critical Stance
Acceptable understanding is measured against:
purposes of inquiry and learning
multiple sources of disciplinary evidence
leverage offered by disciplinary integration
understanding is provisional, can be criticized, revised, and abandoned when more reasonable ones are conceived
Harvard Interdisciplinary Studies Project.
Veronica Boix Mansilla with Flossie Chua

Figure 20.1

equilibrium—an improvement in understanding *vis-à-vis* prior beliefs as well as an understanding subject to further revision. How do these dimensions of learning play out in the construction of two interdisciplinary artifacts—a historical monument and an explanation of climate change? What learning demands does interdisciplinary learning present? To test the capacity of our proposed epistemology to illuminate interdisciplinary learning and the challenges it presents, we now turn to the learning examples described earlier.

20.5 Crafting historical monuments or memorials

Successful historical monuments or memorials integrate an understanding of the past, and use of space, symbolism, and materials to advance evocative interpretations. Four cognitive processes are involved in their creation.

20.5.1 Establishing purpose

The *purpose* of a monument is to commemorate the memorable, to make past experiences part of our present. Memorials—a particular kind of monument—offer a special precinct, a segregated place where we come to honor the dead and reflect about past present and future (Danto 2005). A study of interdisciplinary learning must examine how learners set their epistemic intention. For example, Lin seeks to re-represent the past aesthetically to invite reflection about war and reconciliation. Other potential intentions such as to *explain* why the war happened are not addressed. The success of a learning enterprise of this kind will be measured by the monument's effectiveness in provoking thought rather than by the explanatory power or the level of abstraction and generalization of one's vision. In turn, the process of interdisciplinary learning often requires a readjustment of its purpose.

20.5.2 Understanding and weighing disciplinary insights

To construct such a system of thought in reflective equilibrium learners also come to *understand disciplinary contributions* and *weigh their role* in informing the whole. Contributing disciplinary insights vary. They may take the form of theories, findings, models, methods, tools, techniques, characteristic modes of thinking, applications, discourses, languages, exemplifications, powerful analogies, or explanations. The example of the Vietnam War memorial challenges the learners to distill the *significance* of the Vietnam War and identify a relevant story to be told about it. The cognitive demands are not minor. Without inquiry experience in history, even post-adolescents tend to view significance solely as an intrinsic quality of key events, not one *attributed* to them in light of their consequences or shifting interests in present societies (Seixas 2004, Danto 1985). Similarly, learners may construe historical accounts as stories unproblematically pasted together from literal interpretation of primary sources. In fact historical accounts are constrained by historians' choices of perspective (military persons, political leaders, antiwar youth) time frame (the Tet offensive versus colonialism or the Cold War), and forms of explanation (individual triggers or long-standing cultural forces). These too become options for the learner who must, through considered judgment, decide on a representation of the past that will inform his or her monument. Weighing options is not simple. Deciding, for instance, where to draw the line marking the beginning and the end of the Vietnam War is still a contested matter.

The arts and architecture too impose important challenges on monument design. They call upon the artist to envision detailed versions of the monument in his or her mind; consider competing materials and techniques as well as provocative symbolisms. The artist will need to overcome deeply rooted misconceptions such as believing that the quality of art depends merely on its decorative beauty or that an artist's intention is unequivocally the last word on a piece. Thinking aesthetically will require a commitment to multiple interpretations, some intended, some emerging.

20.5.3 Building leveraging integrations

Interdisciplinary learning yields a system of thought in reflective equilibrium typically organized around a preferred form of disciplinary integration. Learning to create a historical monument involves learning to reframe a significant past in terms of a central visual metaphor that drives the aesthetic design of a piece. In Maya Lin's work, the devastating consequence of the Vietnam War on the individual minds and social cohesion of American society is represented as a scar—a cut in the earth to be healed by time. When the purpose of learning is aesthetic synthesis, examining how the mind constructs metaphors becomes key.

Metaphors frame reality in terms of similarities between constructs pertaining to different realms. In them, a *vehicle* concept (e.g. the scar) highlights certain features of the *topic* one (e.g. the consequences of war), while obscuring others (Goodman 1976). Framing the Vietnam War as a scar sheds light on the personal emotional experience of war and its long-lasting impact. It does not illuminate, for instance, the political and military conundrums that the war presented to American administrations at different points in time. To the extent that the mind can explicate the tacit analogy presented by a metaphor, the metaphor offers parsimony and impact in our representation of reality. Visual thinking metaphors create a holistic synthesis and operate in a physical medium—in this case, the landscape, the stone, the engravings. (Arnheim 1960, 1966; Bruner 1986; Hetland *et al.* 2007).

Learning to interpret and produce metaphors of this kind imposes important challenges on the developing mind. Early in life children can make sense of metaphors based on concrete similarities 'the wrinkled apple is an old lady'. However, the sophisticated interdisciplinary synthesis of the Vietnam War as a scar requires that learners understand the content of each portion of the statement to establish adequate analogy between vehicle and topic (Vosniadu 1994) Furthermore, creating a workable metaphor about the past involves assessing tenable metaphors for their capacity to portray essential aspects of the past accurately, to lend themselves to powerful visual representation, and to maximize the likelihood that the overall purposes of commemoration, healing, and reconciliation are served. A workable metaphor stands in a delicate tension among these three forces: historical accuracy, visual generativity, and power to heal.

20.5.4 Critical stance

Understanding is an endless and cyclical task. Our informed conclusions about a topic are challenged by novel contexts, insights, or experiences. A pragmatic constructionist epistemology draws its strength not from the attainment of final infallible truths but from the recognition of the limitations of our knowledge. Understanding must stand the test of competing interpretations of the subject matter. The debate that followed the publication of Lin's design centered on a reflection about the purpose of a veterans' memorial and the aesthetic choices that were or were not fit.

Researchers studying critical thinking and the development of epistemological beliefs have documented the role of meta-cognition in student learning. The capacity to reflect about the nature of knowledge, learning, and thinking has been associated with more complex understanding of subject matter and growing preparedness for independent learning. In interdisciplinary work, navigating multiple knowledge landscapes demands a meta-cognitive—and often a meta-disciplinary—stance. Students must recognize the preferred units of analysis in different domains or their sometimes conflicting standards of validation. Lin's defense of her design involved a clarification of her view of the significance of the Vietnam War. Lin is also aware of limits in her interpretation—the many Vietnamese lives that were not engraved in her design. Such limitations often function as a pathway to further revision of understanding, new setting of purpose, novel disciplinary insights, integrations, and the construction of yet a new system of thought in reflective equilibrium.

20.6 Explaining the shortage of fresh water under climate instability

Clearly not all interdisciplinary integrations seek an aesthetic synthesis of a past or present phenomenon. As the chapter's opening excerpt suggests, understanding the availability of freshwater resources during times of climate change involves examining both natural and anthropogenic factors affecting the quality and quantity of available water. Framed in this way, the purpose of learning is primarily explanatory, e.g. to understand why water resources may be at risk in order to decide what to do about it. Advancing an interdisciplinary system of thought in reflective equilibrium in this arena demands that learners make sense of selected concepts and findings produced by fields ranging from climate science to oceanography and chemistry, from demography to economics and political science (i.e. weighing disciplinary insights). It also requires that they integrate such insights in a complex and productive explanation of water availability (i.e. leveraging integration) and that they remain aware of the limitations and provisionality of their conclusions (i.e. critical stance).

For example, understanding this topic invites learners to make sense of observed changes in temperature and precipitation and how these are measured. It demands understanding the chemical composition of usable freshwater and the variations that may indicate particular forms of pollution. Learners will need to understand observed and expected population growth, changes in land use and irrigation demands, as well as the role of pollutant emissions such as pesticides and thermal pollution. The insights and modes of thinking depicted above stem from work carried out by chemists, climate scientists, geographers, and demographers.

An interdisciplinary understanding of water availability involves more than the juxtaposition of factors outlined above. Learners must integrate these factors in a complex explanation that serves as the driving cognitive structure for integration. In it, climate-related factors are mediated by anthropogenic ones in a comprehensive explanation of water availability.

Building complex explanations is a demanding task for learners. From early in life children are prone to linear explanations, in which causes and consequences stand in temporal and spatial proximity (Perkins and Grotzer 2005). Only through careful instruction can they advance explanations rooted in multiple mechanisms and agents. For example, learners face the challenge of understanding reciprocal causality where causes and consequences intertwine in feedback loops—loss of glacial reflective surface contributes to atmospheric temperature rise which in turn augments melting and further loss of glacial reflective surface. Learners face the challenge of understanding causal variations associated with the temporal distance between cause and effect, e.g. glacial reduction increases river flow and flooding in the short term but decrease it in the long run. They face the challenges of understanding multiple non-linear causal mechanisms such as the emergent demands on water resources caused by population growth and growing food and irrigation demands.

An explanatory and interdisciplinary system of thought in reflective equilibrium integrates these general and local causes into a complex account of water availability. Yet it also demands that learners remain critical of their resulting conclusion. Important factors may have been missed, the evidence used holds varying levels of confidence, future developments may call for revisions in the account proposed. In sum, interdisciplinary learning as here conceived is clearly more than recording information about stated causes of water availability risk.

20.7 Toward a research agenda

This chapter has sought to advance an epistemological foundation for the study of interdisciplinary learning. I have argued for a pragmatic constructionist epistemology that offers a pluralistic view of knowledge forms able to account for a broad variety of interdisciplinary endeavors. Moving beyond metaphor, the proposed epistemology offers a dynamic construct to represent the phenomenon of interdisciplinary integration: a system of thought in reflective equilibrium. Its articulation and dynamics invite further empirical work. For example, future research on interdisciplinary learning may reveal additional forms of interdisciplinary integration.

As the call for interdisciplinary education expands to both primary education and the graduate years, understanding developmental progressions in interdisciplinary learning capacities will become key. We may expect to see young children able to produce aesthetic syntheses as long as metaphors refer to concrete dimensions of the problem under study—the shapes of leaf cells under a microscope or the reflection of a face on a fishbowl. The pre-adolescent mind may begin to find more abstract metaphoric representations more engaging. As least two dimensions of interdisciplinary learning will need to be addressed in a developmental study: the capacity to integrate disciplines, and the capacity to think and act in disciplinary informed ways. At the other end of the developmental spectrum, studies of interdisciplinary learning may examine the ways in which young adults manage the tensions, incompatibilities, and complementarities among insights from multiple domains. Studies may examine the role that students' beliefs about the nature of knowledge and inquiry may play in their capacity to construct systems of thought in reflective equilibrium. They may address how a given interdisciplinary understanding (a system of thought in reflective equilibrium) moves through phases of stability and contestation.

Finally, challenging interdisciplinary learning often demands collaboration. Research on group learning has addressed dimensions that range from leadership to group composition, from dilemmas of power to the nature of tasks, from the construction of trust to challenges of communication (see Stokols *et al.*, Chapter 32 this volume). A pragmatic constructionist epistemology can add systematicity to our study of interdisciplinary collaborations. It can focus our attention on the key learning demands a group experiences in the construction of systems of thought in reflective equilibrium: negotiation of intellectual purpose, the weighing of disciplinary contributions, the advancement of leveraging integrations, and the disposition toward critical review. Expanding beyond the cognitive realm, such a study of interdisciplinary collaborations could also benefit from examining how cognitive, social, emotional factors interact to advance understanding.

In sum, whether we focus on the construction of a generative taxonomy of interdisciplinary endeavors or a progression of interdisciplinary capacities over the lifespan or socio-cultural conditions for collaborative work, understanding interdisciplinary learning necessitates a clear articulation of 'the kind of knowledge being learned'. The approach promises lines of research in the area of interdisciplinary cognition that are as generative as those in historical, scientific, mathematical, or artistic cognition. It also promises to set the foundation for a new form of 'pedagogical content knowledge'—an understanding of the unique teaching and learning demands presented by particular kinds of knowledge—to ensure quality interdisciplinary assessment and instruction (Shulman 2004).

Acknowledgements

The author would like to thank the Atlantic Philanthropies and the Canadian Institute for Advanced Research for partial funding for this chapter and Flossie Chua for her contributions to our graphic model.

References

Arnheim, R. (1960). *Perceptual analysis of a symbol of interaction.* Basel: Karger.

Arnheim, R. (1966). *Toward a psychology of art: selected essays.* Berkeley: University of California Press.

Ayer. A. J. (1956). *The revolution in philosophy.* London: Macmillan.

Baxter Magolda, M. & King, P. (2008). Toward reflective conversations: an advising approach that promotes self-authorship. *Peer Review* **10**(1), 8–11.

Boix Mansilla, V. (2005). Assessing student work at disciplinary crossroads. *Change* **37**(1), 14–21.

Boix Mansilla, V. (2006). Quality assessment of interdisciplinary research: toward empirically grounded validation criteria. *Research Evaluation* **14**(4), 17–29.

Boix Mansilla, V., Duraisingh, E. D., Wolfe, C. R., and Haynes, C. (2008). Targeted assessment of interdisciplinary writing: an empirically grounded rubric. *Journal of Higher Education* **80**(3), 334–353.

Bromme, R. (1999). Beyond one's own perspective the psychology of cognitive interdisciplinarity. In: P. Weingart and N. Stehr (eds) *Practising interdisciplinarity.* Toronto: University of Toronto Press.

Bruner J. (1986). *Actual minds, possible worlds.* Cambridge, MA: Harvard University Press.

Burns, R. (1994). *Interdisciplinary teamed instruction: development and pilot.* Washington, DC: Educational Research and Improvement Office.

Danto, A.C. (2005). Mute point. *Nation* **281**(12), 40–4.

Danto, A.C. (2007). *Narration and knowledge.* New York: Columbia University Press.

Elgin, C.Z. (1996). *Considered judgment.* Princeton, NJ: Princeton University Press.

Elgin, C.Z. (1997). *Between the arbitrary and the absolute.* Ithica, NY: Cornell University Press.

Fauconnier, G. and Turner, M. (2002). *The way we think: conceptual blending and the minds' hidden complexity.* New York: Basic Books.

Fischer, K.W. and Bidell, T.R. (1997). Dynamic development of psychological structures in action and thought. In: R.M. Lerner and W. Damon (eds) *Handbook of child psychology: vol. 1. theoretical models of human development*, 5th edn, pp. 467–561. New York: Wiley.

Goodman, N. (1976). *Languages of art: an approach to a theory of symbols.* Indianapolis: Hackett.

Goodman, N. (1978). *Ways of worldmaking.* Hassocks, Sussex: Harvester Press.

Goodman, N. and Elgin, C.Z. (1988). *Reconceptions in philosophy and other arts and sciences.* Indianapolis, IN: Hackett Publishing Co.

Gould, S.J. (2003). *The hedgehog, the fox, and the magister's pox: mending the gap between science and the humanities*, 1st edn. New York: Harmony Books.

Hetland, L., Winner, E., Veenema, S., and Sheridan, K.M. (2007). *Studio thinking: the real benefits of arts education.* New York: Teachers College Press.

Huber, M. and Hutchings, P. (2004). *Integrative learning: mapping the terrain* Washington, DC: American Association of Colleges and Universities.

Hursh, B., Haas, P., and Moore, M. (1988). An interdisciplinary model to implement general education. *The Journal of Higher Education* **54**(1), 42–59.

Intergovernmental Panel on Climate Change (2007). *Climate change 2007: synthesis report*, No. 4. Geneva: IPCC Secretariat, World Meteorological Organization. Available at: <http://www.ipcc.ch/pdf/assessment-report/ar4/syr/ar4_syr.pdf>.

Ivanitskaya, L., Clark, D., Montgomery, G., and Primeau, R. (2002). Interdisciplinary learning: process and outcomes. *Innovative Higher Education* **27**(2), 95.

Klein, J.T. (2005). *Humanities, culture and interdisciplinarity*. Albany, NY: State University of New York Press.

Lattuca, L.R., Voight, L.J., and Fath, K.Q. (2004). Does interdisciplinarity promote learning? Theoretical support and researchable questions. *Review of Higher Education* **28**(1), 23–48.

Lin, M.Y. (2000). *Boundaries*. New York: Simon & Schuster.

Miller, M.L. (2005). *Integrative concepts and interdisciplinary work: a study of faculty thinking in four college and university programs*. Harvard Graduate School of Education qualifying paper. Cambridge, MA: Harvard Graduate School of Education. Available at: <http://pzweb.harvard.edu/interdisciplinary>.

Minnis, M. and John-Steiner, V. (2005). The challenge of integration in interdisciplinary education. *New Directions for Teaching and Learning* **2005**(102), 45–61.

National Academies (2005). *Facilitating interdisciplinary research*. Washington, DC: The National Academies Press.

Nikitina, S. (2005). Pathways to interdisciplinary cognition. *Cognition and Instruction* **23**(3), 389–425.

Perkins, D. and Grotzer, T. (2005). Dimensions of causal understanding: the role of complex causal models on students' understanding of science. *Studies in Science Education* **41**, 117–66.

Popper, K.R. (1965). *The logic of scientific discovery*. London: Routledge.

Seixas, P. (1997). Mapping the terrain of historical significance. *Social Education* **61**(1), 22–8.

Seixas, P. (2004). *Theorizing historical consciousness*. Toronto: University to Toronto Press.

Shulman, L. (2004). *The wisdom of practice: essays on teaching learning and learning to teach*. San Francisco: Jossey-Bass.

Snow, C.P. (1998). *Two cultures and the scientific revolution*. Cambridge: Cambridge University Press.

Vosniadu, S. (1994). Analogical reasoning in cognitive development. *Metaphor and Symbolic Activity* **10**(4), 297–308.

Wilson, E.O. (1998). *Consilience: the unity of knowledge*. New York: Knopf.

PART 4
Institutionalizing interdisciplinarity

CHAPTER 21

Evaluating interdisciplinary research

KATRI HUUTONIEMI

While interdisciplinarity has become a major topic in discussions of knowledge production and research funding, criteria for its evaluation remain poorly understood. Analyses of the topic do exist, but they have been scattered across different genres and forums (e.g. the special issue of *Research Evaluation* 2006; Klein 2008). Whenever research crosses boundaries between disciplines, the problem arises that each discipline carries specific and sometimes conflicting assumptions about quality. If research also integrates expertise outside of academia, the dilemma of multiple standards is even more challenging. Following standard usage of terms (see Klein, Chapter 2 this volume) the former is 'interdisciplinary' and the latter 'transdisciplinary' research. This chapter will use the term 'interdisciplinary' in its broad sense, as a genus of integrative research activities that combine more than one discipline, field, or body of knowledge. In contrast, the term 'transdisciplinary' is more focused: it will refer to trans-sector problem solving where various stakeholders in society are actively involved in knowledge production.

In both interdisciplinary and transdisciplinary research, the presence of various perspectives raises questions about the contents and procedures of research evaluation—How can a balance be achieved between different epistemic viewpoints, and what criteria may be used to assess them? How should we organize the evaluation of research, and who should we include in the process?

The present chapter provides neither a recipe for conducting interdisciplinary evaluations nor a normative framework for successful integrative work. Instead, it highlights some key features of interdisciplinarity as a special type of challenge for research evaluation, and the practices through which those features have been reflected. These issues are considered by including insights from conceptual and descriptive discussions of interdisciplinary research, empirical analyses of evaluation activities, and experiences and initiatives of participants.

The chapter is structured as follows. This introductory section summarizes the functions of evaluation for both interdisciplinary research itself and society at large. The next section considers the merits and criteria of interdisciplinary research from three conceptual perspectives, since competing positions on interdisciplinarity shape competing assumptions about quality and how it is best determined. We then look beyond the normative canon of appropriate standards and illustrate the contextual and situated aspects of the evaluation process, revealed by the empirical analyses of quality judgments. In the light of empirical findings and the recent literature on science studies and research policy, the chapter then concludes by considering whether and how the challenges of interdisciplinary evaluation are distinct from the more general problematic of research evaluation today.

'Research evaluation' means the systematic determination of the merit, worth, and significance of a research activity. While the bulk of the evaluation literature deals with the general mechanisms and techniques of evaluation, this chapter is framed using as a point of departure the defining characteristics of interdisciplinary research, rather than evaluation methodology. In terms of evaluation theorists, the concern here is with the *subject* of an evaluation, as well as with the process of *valuation*: that is the evaluator placing value on the subject. According to this approach, evaluation is not only about detecting evidence of quality, but 'concerns the making of value judgments about the quality of some object, situation or process' within some value framework (Eisner 1998, p. 80). The subject of evaluation, as understood in this chapter, is an *interdisciplinary research effort*, ranging from a single study to a cluster of work carried out within a network of researchers.

As we will see, it is not easy to avoid arbitrariness in determining which is a better interdisciplinary or transdisciplinary research effort. Nonetheless, evaluations are needed for organizational learning and the improvement of performance and quality of research activities. Another essential function of evaluation is to bolster the credibility of research. Whether or not evaluation has an impact on an ongoing research activity, it helps legitimate research and its results both across the academy and within society. The question of legitimacy is prominent among the obstacles confronted by people engaged in inter- and transdisciplinary research. While there is a strong political and social pressure to cross disciplinary boundaries, the quality of integrative research is often seen to be dubious because of the lack of epistemological standards and proper peer reviewers—the cornerstones of disciplinary quality control (Gibbons and Nowotny 2001).

The current momentum for the topic derives from political demands for accountability and transparency in the use of public funds. While interdisciplinarity and especially transdisciplinarity have gained an informal status of being indices of accountability in knowledge production (Strathern 2004, pp. 79–80), more explicit discussion is needed about the criteria and procedures which make these activities themselves accountable. In the current budgetary setting, scholarly excellence needs to be clearly demonstrated. Evaluations are now used not only for separating the qualified from the unqualified, but for distinguishing between competing *types* of high-quality research. It is in this context that inter- and transdisciplinary research face the most serious problems. When competing with disciplinary research, the benefits of integrative approaches confront the warranted quality of more established forms of inquiry.

21.1 Values in interdisciplinary research

Quality is not a unitary concept. It has a distinctive meaning within each discipline. 'Discipline' denotes a set of codified rules, beliefs, perceptions, and procedures with regard to producing and evaluating knowledge. These often tacit standards determine the criteria for admission into that community of researchers or scholars, the range of problems considered important, the approaches considered appropriate, and the criteria for legitimizing findings as new knowledge (Russell 1983). The existence of rules provides clear indicators for assessing performance within disciplines. Empirical studies of quality judgments in research have illustrated the differences between disciplines in this sense. Quality criteria such as originality, soundness, feasibility, and relevance have different meanings and roles within each disciplinary community. Such differences are likely to be highest between conceptually distant disciplines, like those between 'hard' versus 'soft' fields, or basic versus applied fields (Montgomery and Hemlin 1991).

Interdisciplinary research is, by definition, a hybrid compounded by more than one discipline, and is thus a form of scholarship that is not easily amenable to evaluation. It is widely acknowledged that rigorous criteria for judging interdisciplinary quality are strongly needed. However, scholars are equally unanimous in their observation that there is no single phenomenon of interdisciplinarity; instead, multiple interdisciplinarities exist. Quality standards for these activities have thus to accommodate the various types of work across boundaries. Competing formulations of interdisciplinarity also shape competing assumptions about quality (Klein 1996).

In the literature on the topic, three identifiable, albeit overlapping, conceptual approaches appear to answer the question of quality criteria for interdisciplinary research. While the first approach deals exclusively with interdisciplinary research within the academy, the second and third ones include insights relevant for the evaluation of both interdisciplinary and transdisciplinary activities. The three approaches are summarized in Table 21.1 and discussed below.

21.1.1 Mastering multiple disciplines

The most common approach to the assessment of interdisciplinary research has been to prioritize disciplinary standards, premised on the understanding that interdisciplinary quality is ultimately dependent on the excellence of the contributing specialized components. This view treats interdisciplinary research as one more form of the general division of labor in the production of knowledge. For example, the Organisation for Economic Cooperation and Development (OECD) states in its discussion of best practice in interdisciplinary that 'highly competent proficiency in a single discipline is the only acceptable basis for interdisciplinary success' (OECD 1998, p. 18). A similar conclusion, though less exclusive, was reached by a recent symposium of interdisciplinary experts, convened by researchers from Harvard University and the American Association for the Advancement of Science: 'A basic premise of quality interdisciplinary work is that it satisfies quality standards arising from the disciplines involved' (Boix Mansilla *et al.* 2006, p. 73).

Table 21.1 Three approaches to the question of quality criteria for interdisciplinary research (IDR).

	Mastering multiple disciplines	Emphasizing integration and synergy	Critiquing disciplinarity
Epistemic assumptions about IDR	Enriching disciplinary knowledge production by cross-fertilization	Alternative, integrative model of knowledge production	Diverting narrowly focused disciplinary trajectories; redefining knowledge
Scholarly standards	Standards of contributing disciplines are combined	Standards of contributing disciplines cannot be by-passed, but new criteria of ID expertise are needed	Standards of contributing disciplines are transformed in the act of interpenetration; emphasis on external criteria
Valuation context	Relevant disciplinary communities	Integrative research environment	Knowledge production–consumption system with permeating boundaries
Policy implications	More flexibility in current evaluation and funding mechanisms	Choices among modes of research support; specific mechanisms for promoting IDR	Reassessment of the governance of knowledge production
Proponents	Most funding organizations and academic institutions	Interdisciplinary organizations and practitioners	Theorists of science in society; critical interdisciplinarities

From this vantage point, *disciplinary originality or excellence* is the baseline for assessment. The specific meaning of excellence is defined according to the standards of the relevant scholarly or scientific communities. It is assumed that good interdisciplinary research must fulfill existing methodological requirements and theoretical standards. This is a common perception among funding agencies and science administrators, but also among the frontline researchers who serve as peer reviewers. As illustrated by an empirical study of interdisciplinary peer review panels in five American funding organizations in the social sciences and the humanities (Lamont 2009), panelists considered interdisciplinarity as a separate criterion of excellence. Interdisciplinary attributes of proposals and applicants were considered a 'plus', but not substitutive for disciplinary markers of quality. Good interdisciplinary proposals successfully combine breadth, parsimony, and soundness, the panelists concurred, but to meet these stringent standards researchers have to gain adequacy in several fields. According to Lamont's conclusion, 'Combining traditional standards of disciplinary excellence with interdisciplinarity presents a greater challenge and creates the potential for double jeopardy for interdisciplinary scholars, because *expert and generalist criteria... have to be met at a same time*' (Lamont 2009 p. 210, emphasis added).

Since the evaluation community of interdisciplinary research is usually a hybrid of existing disciplinary communities, detecting the scholarly networks that the particular research is built on may require extra effort. In the assessment of proposals, funding organizations often acknowledge the need to identify proposals that do not fall within discipline-based evaluation panels, and to be flexible in the allocation of projects, cross-referencing several panels and relying on joint evaluation. Some funding organizations have also developed new review procedures to ensure expertise in each discipline represented in a proposal. For example, the Australian Research Council has set guidelines for the handling of what they call 'multi-panel' proposals and navigating them through the existing discipline-based panel structure (Grigg 1999, p. 48). Apart from the challenge of identifying the multiple disciplines involved and locating the right peer communities, most of the standard means for evaluating disciplinary research are assumed to be applicable to interdisciplinary research as well (National Academy of Sciences 2005, p. 152). In general, the stance has been to enhance equality between disciplinary and interdisciplinary research in evaluation outcomes rather than actively promote interdisciplinary research.

However, basing the evaluation of interdisciplinary research on the traditional disciplinary standards of scholarly excellence often punishes boundary-crossing activities. From a specialist's point of view, an interdisciplinary endeavor may not be original enough, or may not represent cutting edge research. An obstacle called the 'reception problem' (Salter and Hearn 1996, p. 146) is likely to occur: 'Interdisciplinary work…easily falls between the cracks. It is "outside the lines". It finds no easy audience in the literature either because it appears to deal with issues that are not being debated or because it draws on methodological and paradigmatic assumptions that are unfamiliar (and thus not likely to be acceptable) to the established disciplines'. Another problem may be caused by the incompatibility of different disciplinary standards—what is valuable from the perspective of one contributing field may not be similarly prized from the other. For example, while humanists often define interpretative skills as essential for the production of high-quality scholarship, social scientists, especially those who champion empiricism, deride interpretation as a corrupting force in the production of truth (Lamont 2009). Moreover, putting a premium on the excellence of disciplinary components does not tell us much about the success of their interplay within an interdisciplinary research effort.

21.1.2 Emphasizing integration and synergy

Another approach is to argue that the evaluation of knowledge integration across multiple fields is an enormously complex effort requiring a novel conceptual framework with distinctive methodological tools. While the conventional ideology of scholarly excellence rests on the assumption of a standard body of knowledge or a fixed body of content, according to this view the basis of interdisciplinary work is different. The appropriate approach to judging quality is thus not to impose disciplinary standards but to create a *new model of excellence* for interdisciplinary research. The common bond shared by integrative activities is the need to combine knowledge resources in order to develop an integrated product, either a conceptual advance or a solution to a practical problem. This goal implies that these activities should be judged by the quality, novelty, and degree of integration of knowledge they achieve. On this view, transdisciplinary research is seen as even less amenable to evaluation by disciplinary standards, since the involvement of different non-academic stakeholders as research participants broadens the domain in which the values and merits of research are to be assessed. Transdisciplinary research is thus understood as a part of a broader process of innovation, which brings into play an interactive and iterative pattern of mutual influence between different actors (e.g. Spaapen *et al.* 2007).

The strongest proponents of this view are typically those who work within boundary-crossing organizations, or as consultants or practitioners themselves. Many of these organizations include both interdisciplinary and transdisciplinary research in their analysis. For instance, commissioned by two interdisciplinary research programs of the Swiss National Science Foundation (SNSF), Defila and DiGiulio built a heuristic profile of 'building blocks' to be operationalized for inter- and transdisciplinary evaluation, based on the special characteristics of this kind of research. According to their criteria, cooperation and cross-fertilization should be at the core of evaluation, even though 'the sub-projects or research groups of such a project group may not always be at the cutting edge of disciplinary and specialized research' (Defila and DiGiulio 1999, p. 7). Disciplinary excellence is thus regarded as less essential than a rigorous process of integration. While mastery of the participating disciplines is not required, it is still acknowledged that interdisciplinary study should build explicitly and directly upon the work of disciplines.

As a new, integrative model of competence, under this second approach inter- and transdisciplinarity call for revisions to the research evaluation system at large (see Stokols *et al.*, Chapter 32 this volume). To evaluate such research in a reliable way, Spaapen *et al.* (2007) suggest, an assessment needs to be both comprehensive—consisting of a review of the various activities of the research group—and interactive—allowing for the influence of stakeholders in the evaluation process. Both the benefits and challenges of integrative research are often found primarily in the collaboration and communication process itself, which is best to assess on an *ongoing* basis alongside the traditional *ex ante* and *ex post* evaluations. Accordingly, some scholars have suggested organizational learning or integrative skills as a criterion of inter- and transdisciplinary research, instead of whether or not it makes a contribution to an existing body of knowledge. The replacement of a 'jury model' in evaluation by an interactive coaching model or an interactive peer review can facilitate self-reflection about what members are supposed to be doing as well as mutual learning and trust between researchers and reviewers. The participation of researchers in the definition of criteria and the selection of reviewers ensures that all aspects of the work could be competently assessed. This ongoing dialogue and feedback between researchers and reviewers also supports a mutual commitment to long term goals.

A downside of emphasizing the unique features of inter- and transdisciplinary research is the unintentional creation of a new boundary within the academy, the one between disciplinary and interdisciplinary research. The existence of 'double standards' of quality may lower the credibility of boundary-crossing activities by signaling that they do not meet the approved quality standards of academic research, which can thus 'ghettoize' inter- and transdisciplinary research, distancing them from the ongoing development within disciplines (Feller 2006). Moreover, the impact of integrative endeavors may be undermined if the performance of each effort is evaluated in relation to its mission, without an attempt to evaluate the mission itself from an outsider's perspective. Another problem inherent in the creation of distinct criteria and practices for inter- and transdisciplinary evaluations is that it would require an operational definition of such research, plus a set of viable parameters to empirically distinguish it from disciplinary research—a problem that has not been solved, despite decades-long definitional debates (Huutoniemi *et al.* 2010).

21.1.3 Critiquing disciplinarity

The most radical approach to the question of evaluation criteria is that offered by numerous theorists who argue for a critical mode of interdisciplinarity which moves away from the idea that interdisciplinary pursuits draw their strength from building on the methods and findings of established fields, or on their integration. Unlike the two approaches discussed above, this standpoint views disciplinarity and interdisciplinarity as strongly opposed. Disciplinary expertise or excellence, if understood as the key measure of quality research, serves to heighten the incommensurability among disciplinary trajectories. This, in turn, leads towards the fragmentation of research, failure to communicate across disciplinary boundaries, and the separation of epistemology from politics.

The major contribution of this approach is to *undermine the prevailing status of disciplinary standards* in the pursuit of a non-disciplinary, integrated knowledge system. For example, drawing on Popper's falsificationist methodology, Fuller (1993) argues against disciplinary 'pluralism', a position which implies that issues resolved in one discipline leave untouched the fate of cognate issues in other disciplines. Similarly, Campbell (1969) has criticized the 'ethnocentrism' of disciplines, which has the unintended consequence that 'the disciplinary clusters may at their edges overlap other clusters, but as ships that pass in the night, they fail to make contact'. These and other critical voices imply that the difficulties of evaluating interdisciplinary research will not be overcome by creating new quality standards for that type of research, but by *transforming* the prevailing ethnocentrism and mutual ignorance between disciplines.

A complementary trend is the 'transdisciplinarization' of knowledge, the erosion of the distinction between academic and non-academic contexts of research. The concepts of 'Mode-2 science' (Gibbons *et al.* 1994), 'post-normal science' (Funtowicz and Ravetz 1993), and 'knowledge policy' (Fuller 1993, 2002) are examples of the overarching trend to attach interdisciplinarity to a wider set of processes that have become the hallmark of modern knowledge societies, such as democratization and accountability. These authors would replace disciplinary values by a form of integrated societal values that emphasize comprehensive knowledge responsive to political and social needs. Unlike the more instrumental or opportunistic stance towards transdisciplinarity, discussed in the previous section, this position on transdisciplinarity is aligned with principles of democratic community and addressing problems caused by technology.

As inter- or transdisciplinarity is viewed in this third approach as the force that diverts the discipline-driven direction of knowledge production, the success of those efforts cannot merely be measured by asking how well the research is done. Instead of judging quality on the basis of internal criteria, that is, answering the question above by using the criteria that are generated within the field or by the practitioners themselves, judgments should be made by using external criteria. According to Weinberg (1962), such criteria should answer the question 'Why pursue this particular research?', and can only be generated outside the field. The importance of external criteria is evident when making choices among different, often incommensurable, fields of science. As noted by Fuller (1993), internal criteria are equally worthless when evaluating cognitive transfer between disciplines, since the interaction itself leaves the constitutive disciplines and their criteria permanently transformed. The same goes for cognitive interaction in a transdisciplinary context. Unlike internal criteria, external criteria are relational and depend on the particular context in which the knowledge is produced. Thus, there is no recipe for how to do it.

By this standard, the revisions needed for the evaluation system are profound. Competitive evaluation by one's peers is the capital sin in the current disciplinary overproduction: specialist evaluators are experts in judging whether a research effort is an excellent example of the kind of research they themselves are conducting, and their evaluation criteria will thus function to support continued research efforts in directions that a broader perspective might deem unproductive. Peer judgments would be improved if peers were selected from a *larger population*, including also non-elitist members of academia, such as teachers and administrators, and representatives of neighboring fields. On another front, Funtowicz and Ravetz (1993) argue that to guarantee the quality and safety of scientific research, 'extended peer communities' should be established, consisting not only of persons with some form of institutional accreditation, but rather of all those with a desire to participate in the resolution of the issue at stake.

There are several points of debate in an approach that rejects disciplinary merits altogether. First, most theorists have taken purely ideological, programmatic, or antagonistic stances to the question of criteria, while both the alternative measures of quality assurance and the empirical puzzles of conducting evaluations are for the present left intact. Second, and partly due to the lack of empirical tests, it is not clear that simply extending the process of evaluation and the pool of judges to practitioners in other disciplines or even non-academics will result in knowledge that is truly socially responsive. Third, if traditional academic values are to be replaced by external goals, there is a risk that research will simply become pressed into the service of the innovation system, a wealth-creating technoscientific motor of the whole economy. It can be argued that the pursuit of high-quality science needs an autonomous space, where curiosity is the driving force, pursued by creative minds in a nourishing academic culture (Nowotny 2006).

21.2 Evaluators and the process of valuing

From a conceptual point of view, the lack of appropriate quality criteria introduces a remarkable degree of uncertainty in the evaluation of interdisciplinary research. However, as illustrated by empirical studies of reviewers' conceptions of scientific quality, criteria always have some fuzziness when applied in concrete evaluation settings, whether interdisciplinary or not. 'Quality' appears to be an intangible character of research, which is very difficult to articulate but which experienced evaluators nevertheless seem able to recognize (Lamont 2009). Moreover, even within a scientific community, different experts may weigh evidence differently and adhere to different standards of demonstration. Quality is thus not an objective property of research, but is located to some extent 'in the eyes of the beholder'. Accordingly, evaluation is not only about detecting evidence of quality, but also placing value on evaluation data. The role of evaluators as valuing subjects and the act of making judgments are thus crucial for evaluation.

This perspective allows one to see beyond the prescribed quality criteria, which so often are counter-intuitive for interdisciplinary assessments, and to focus on the more tangible aspects of evaluation—the procedure itself, including the selection of reviewers and the way different evaluators use their expertise during the review. From a pragmatic perspective of research administrators, 'getting the process right' may partly compensate for what remains unsettled in the characterization of the quality of interdisciplinary research efforts (Boix Mansilla *et al.* 2006). In order to understand the process of valuing and the way in which judgments are reached in the face of epistemological diversity, scholars have studied evaluation procedures from the perspective of the evaluators themselves. These studies have shown that evaluation is essentially a social and emotional undertaking, rather than a purely rational or cognitive process. Lamont and colleagues (Lamont *et al.* 2006; Lamont 2009) analyzed the social conditions for building a consensus in interdisciplinary funding panels, and found that group context may prevent disciplinary cronyism by the reviewers, since reviewers often lose their credibility with colleagues by pushing their own fields. Not everyone agrees, though. Langfeldt (2004), for example, found in her research on the decision making of various expert panels in Norway that most collective evaluation processes were characterized by a clear division of scholarly tasks, little interaction, and tacit compromises.

The role of evaluators includes not only making final judgments about evaluation data, but also making judgments about what questions to ask and on what to focus. To some extent, evaluation is based on the experience and 'connoisseurship' of those who conduct the evaluation. In regard to interdisciplinary research, however, there are few pre-existing peer communities from which to draw connoisseurs. To avoid cognitive particularism and disciplinary parochialism, evaluators should be open-minded, respectful of various traditions, and tolerant to approaches other than their own. As noted by Boix Mansilla *et al.* (2006), reviewer selection is important not only for producing valid evaluations, but also to convince researchers that their work will be appropriately evaluated by those truly able to do so. Researchers are particularly suspicious about the evaluators' expertise in broad, interdisciplinary areas, because the reviewers themselves are often believed to represent established disciplines and narrow mainstream thinking. Given the uncertainties around interdisciplinary evaluations, the need for mutual trust and credibility should not be underestimated.

21.3 Conclusion

While the evaluation of interdisciplinary and transdisciplinary research has recently raised special concern, it is more rarely asked to what extent this discussion should be extended to other research as well. Even without disciplinary differences, there is dissatisfaction with, for example, the peer review system and the insufficient operationalization of quality criteria (see Holbrook, Chapter 22 this volume). Although interdisciplinary research has by definition many characteristics that make it particularly difficult to evaluate, it is important to note that there is also much contingency and variation within disciplinary research. Quality and performance are relative not only to disciplinary standards, but also to the goals, expectations, norms, and values of stakeholders and thus vary from one evaluation context to another.

In the recent literature on science studies and research policy, there are clear indications about a shift in academic quality control, which is in line with the reforms called for in inter- and transdisciplinary evaluations. According to this literature, much of today's academic quality control is transforming into a monitoring system that has a process rather than product orientation, uses new criteria, has other foci and goals, and uses different peers and evaluation times. In a knowledge society, the contextualization of problems and social accountability of knowledge have become just as important indicators of quality as scientific reliability and disciplinary rigor (Hemlin and Rasmussen 2006). At the same time, comparative studies of research evaluations by, for example, funding organizations are not particularly alarming from the perspective of interdisciplinary research. Comparisons between the success rates of disciplinary and interdisciplinary proposals imply that the current system is relatively non-biased, and only few studies have found evidence of practices inhibiting interdisciplinary applications and their success in the evaluation process.

These observations raise the question of whether the evaluation challenges of interdisciplinary research are, in the end, so distinct from the general problematique of research evaluation in the current age, or whether they simply illustrate the profound needs of the whole evaluation system. One lesson learned from the analyses of inter- and transdisciplinary evaluations is that the recent emphasis on accountability has limits, too, as the concept is increasingly linked with quantification and other forms of standardized knowledge that constrain the local, subjective, and personal. While the core premise has been that efforts to produce a mechanical uniformity across evaluators and settings promote fairness, attention diverted from goals, context, or distributive effects can produce poor outcomes (Espeland and Vannebo 2007). Inter- and transdisciplinary evaluations reveal the invaluable importance of the latter aspects. Trust in persons and their expertise cannot be entirely replaced by trust in numbers and evaluation mechanics.

References

Boix Mansilla, V., Feller, I., and Gardner, H. (2006). Quality assessment in interdisciplinary research and education. *Research Evaluation* **15**(1), 69–74.

Campbell, D.T. (1969). Ethnocentrism of disciplines and the fish-scale model of omniscience. In: M. Sherif and C.W. Sherif (eds) *Interdisciplinary relationships in the social sciences*, pp. 328–48. Chicago: Aldine.

Defila, R. and DiGiulio, A. (1999). Evaluation criteria for inter- and transdisciplinary research. *Panorama* [newsletter of the Swiss Priority Program Environment, Swiss National Science Foundation], special issue 1/99.

Eisner, E. (1998). *The enlightened eye: qualitative inquiry and the enhancement of educational practice*. Upper Saddle River, NJ: Merrill.

Espeland, W.N. and Vannebo, B.I. (2007). Accountability, quantification, and law. *Annual Review of Law and Social Science* **3**, 21–43.

Feller, I. (2006). Multiple actors, multiple settings, multiple criteria: issues in assessing interdisciplinary research. *Research Evaluation* **15**(1), 5–15.

Fuller, S. (1993). The position: interdisciplinarity as interpenetration. In: *Philosophy, rhetorics, and the end of knowledge: the coming of science and technology studies*, pp. 33–65. Madison, WI: University of Wisconsin Press.

Fuller, S. (2002). *Knowledge management foundations*. Boston: Butterworth-Heinemann.

Funtowicz, S.O. and Ravetz, J.R. (1993). Science for the post-normal age. *Futures* **25**(7), 739–55.

Gibbons, M. and Nowotny, H. (2001). The potential of transdisciplinarity. In: J.T. Klein, W. Grossenbacher-Mansuy, R.W. Scholz, and M. Welti (eds) *Transdisciplinarity: joint problem solving among science, technology, and society*, pp. 67–80. Basel: Birkhäuser.

Gibbons, M., Limoges, C., Nowotny, H., Schwartzman, S., Scott, P., and Trow, M. (1994). *The new production of knowledge: the dynamics of science and research in contemporary societies*. London: Sage.

Grigg, L. (1999). *Cross-disciplinary research: a discussion paper*. Commissioned report no. 61. Canberra: Australian Research Council.

Hemlin, S. and Rasmussen, S.B. (2006). The shift in academic quality control. *Science, Technology and Human Values* **31**(2), 173–98.

Huutoniemi, K., Klein, J.T., Bruun, H., and Hukkinen, J. (2010). Analyzing interdisciplinarity: typology and indicators. *Research Policy* **39**, 79–88.

Klein, J.T. (1996). *Crossing boundaries: knowledge, disciplinarities, and interdisciplinarities*. Charlottesville. VA: University Press of Virginia.

Klein, J.T. (2008). Evaluation of interdisciplinary and transdisciplinary research: a literature review. *American Journal of Preventive Medicine* **35**, S116–S123.

Lamont, M. (2009). *How professors think: inside the curious world of academic judgment*. Cambridge, MA: Harvard University Press.

Lamont, M., Mallard, G., and Guetzkow, J. (2006). Beyond blind faith: overcoming the obstacles to interdisciplinary evaluation. *Research Evaluation* **15**(1), 43–55.

Langfeldt, L. (2004). Expert panels evaluating research: decision-making and sources of bias. *Research Evaluation* **13**(1), 52–62.

Montgomery, H. and Hemlin, S. (1991). Judging scientific quality. A cross-disciplinary investigation of professorial evaluation documents. *Göteborg Psychological Reports* **21**(4).

National Academy of Sciences (2005). *Facilitating interdisciplinary research*. Washington, DC: National Academies Press.

Nowotny, H. (2006). Real science is excellent science: how to interpret post-academic science, Mode 2 and the ERC. *Journal of Science Communication* **5**(4), 1–3.

OECD (1998). *Interdisciplinarity in science and technology*. Paris: OECD, Directorate for Science, Technology and Industry.

Research Evaluation (2006). Special issue on interdisciplinary research assessment. *Research Evaluation* **15**(1).

Russell, M.G. (1983). Peer review in interdisciplinary research: flexibility and responsiveness. In: S.R. Epton, R.L. Payne, and A.W. Pearson (eds) *Managing interdisciplinary research*, pp. 184–202. New York: John Wiley and Sons.

Salter, L. and Hearn, A. (1996). Interdisciplinarity. In: L. Salter and A. Hearn (eds) *Outside the lines: issues in interdisciplinary research*. Montreal: McGill-Queen's University Press.

Spaapen, J., Dijstelbloem, H., and Wamelink, F. (2007). *Evaluating research in context: a method for comprehensive assessment*, 2nd edn. The Hague: Consultative Committee of Sector Councils for Research and Development (COS).

Strathern, M. (2004). *Commons and borderlands: working papers on interdisciplinarity, accountability and the flow of knowledge*. Wantage: Sean Kingston Publishing.

Weinberg, A.M. (1962). Criteria for scientific choice. *Minerva* **1**(2), 159–71.

CHAPTER 22

Peer review

J. BRITT HOLBROOK

Peer review serves a gatekeeping function both within and outside academe. *Sub specie academicus*, academic excellence is validated by the process of peer review. Academic excellence, however, is often inversely proportional to societal relevance. Interdisciplinary research is increasingly encouraged as a way of making academic research more societally relevant. *Sub specie societatis*, academic research is also called upon to help societal decision makers craft evidence-based policies, and peer review is the preferred tool for ensuring the integrity and reliability of the research used by decision makers.

These trends toward interdisciplinarity and transdisciplinarity for research strain the process of peer review. The key issue for advocates of peer review is whether a tool that has been used mainly to determine academic excellence can be adapted to judge societal relevance without undermining the foundations of knowledge production (Sarewitz 2000).

22.1 Background: the view from inside academe

Peer review is a process by which a group of individuals renders judgment on the work of others in order to determine whether that work is meritorious enough to warrant consideration (e.g. for publication or tenure) or support (e.g. in the form of a grant or fellowship). Typically, the individuals asked to render such judgments are selected from a pool of reviewers who are considered to be 'peers' of whoever has produced the work to be judged. What constitutes a peer is more complicated than one might think; but given the uses to which the process of peer review has been commonly put, a peer has traditionally been characterized in terms of shared disciplinary expertise.

The a priori justification for using peer review as an assessment tool is relatively straightforward: no one is in a better position to assess the merit of work in a particular area than experts in that particular area. Thus, in order to judge whether work in area P is meritorious, it makes sense to ask individuals renowned for their expertise in area P rather than people who know comparatively little or nothing about P. Although individual nonconformists exist, along with several quasi-disciplines, which may or may not be evolving toward disciplinary status, areas of academic expertise are most often carved out by and within academic disciplines. Indeed, the connection between academic excellence and disciplinary expertise is so common that interdisciplinarity among academics is often perceived as amateurism (cf. Frodeman and Mitcham 2007).

Despite the fact that the standards of one academic discipline are incommensurable with those of other disciplines, relying as they do on expert (and often tacit) knowledge within the field, there is universal agreement across academe that peer review is essential for determining what counts as academic excellence. Indeed, publications that are not peer-reviewed typically do not count—either at all or as much as—peer-reviewed articles when it comes to tenure and promotion standards for higher education faculty; and the majority of grants from public funding agencies are allocated only after and on the basis of some form of peer review. For this reason, the process of peer review is usually characterized in terms of 'quality control' or as having a 'gatekeeper' function, and it is no exaggeration to say that peer review is the *sine qua non* of academic excellence.

The most common uses of peer review are in academic publishing (e.g. to determine whether a paper submitted for publication in an academic journal is worthy of being published in that journal) and in the review of proposals for grants (e.g. to determine whether the proposed activities deserve to receive funding). Both prepublication peer review and grant proposal peer review are *prospective* uses of peer review, which puts a great deal of pressure on reviewers to predict the future: will this paper (or this proposed research) ultimately be well-received by the field (see Rip 2000)? In most, though not all, cases of prospective peer review, the identity of the reviewers is withheld from the reviewee (a process known as blind peer review); and in many cases of prospective peer review, the identity of the reviewee is also withheld from the reviewers (a process known as double-blind peer review).

The process of peer review is also increasingly employed to conduct *retrospective* analyses of particular people, practices, or institutions. Thus, for instance, peer review may be employed within an academic department to rank the performance of individual members of the department relative to other members of the department. Often, 'external' reviewers are brought in to assess the strengths and weaknesses of the business practices of a particular company or to identify strengths and weaknesses on an institutional level, judging a university, a particular program within a research funding agency, or the agency as a whole. Usually, cases of retrospective peer review make fewer, if any, attempts to hide the identity of reviewers and reviewees from one another through blinding. Because of dissimilarities with the typical peer-review process, which relies heavily on the use of disciplinary peers as reviewers, many are reluctant to call retrospective institutional review peer review at all, preferring instead to refer to this practice as *expert* review. There also exist other 'extensions' of the peer-review process, i.e. atypical uses of peer review, such as the use of peer review in relation to regulatory decision making (Jasanoff 1990).

Typical criticisms of the process of peer review include the worry that it may be potentially biased against people for reasons unrelated to the merit of their work (Wennerås and Wold 1997). Blinding reviewers and reviewees to the identity of the other is an attempt to allay this criticism. Some critics suggest that peer review is inefficient and unwieldy as a tool for evaluating large volumes of research. In response, some funding agencies have taken the step of limiting grant proposal submissions, e.g. by shortening the allowable length of proposals, by previewing letters of intent and accepting only invited full proposals, limiting the number of proposals particular institutions may submit for particular calls, or limiting the number of submissions a particular researcher may make of the same proposal.

Another common criticism of peer review is that it is inherently conservative, tending to favor work conducted along traditional lines (in the sciences this concern is often expressed in terms of bias toward existing paradigms and against novel, transformative, or revolutionary ways of thinking). To counter conservatism, reviewers are sometimes instructed to value paradigm-shifting or 'transformative' ideas. Another tactic that funding agencies use to counter conservatism is to put out calls for interdisciplinary research proposals. Reviewing interdisciplinary proposals, however, presents special difficulties (Lamont 2009; Huutoniemi, Chapter 21 this volume).

One of the most notorious criticisms of peer review is that it is ineffective at determining quality and/or detecting errors (e.g. the so-called Sokal affair or the widely publicized failure of reviewers to detect the falsification of data by Hwang Woo-Suk in publications on stem cell research in 2004 and 2005 in the journal *Science*). The typical response to this criticism is to deflect it with humor: Winston Churchill's quip about the value of democracy is paraphrased, and peer review is admitted to be the worst form of research evaluation, except for all the others. In this way, advocates of peer review effectively divert the conversation back to considerations that do not threaten the very existence of peer review: how to improve its efficiency, reliability, responsiveness, and fairness (and hence its overall effectiveness).

22.2 A history of peer review

It is a commonly held belief that the process of peer review is venerable because it is ancient, as opposed to merely respectable because it is institutionally well-entrenched. Searching for 'the first documented description of a peer review process', the *2007–2008 peer review self study* published by the National Institutes of Health (NIH) cites two articles published in a 1997 issue of the *Annals of Saudi Medicine* that note a peer-review process described 'more than a thousand years ago in the book *Ethics of the physician*, authored by Syrian physician Ishaq bin Ali al-Rahwi (CE 854–931)' (NIH 2008, p. 8). *Ethics of the physician* 'outlines a process whereby a local medical council reviewed and analyzed a physician's notes on patient care, to assess adherence to required standards of medical care' (NIH 2008, p. 8). This description seems most reminiscent of medical peer review, which is a quasi-judicial, retrospective fact-finding procedure to determine whether (as with a grand jury) a hearing is necessary. Of course, according to a sufficiently broad definition of peer review, one might also cite the Athenian judicial system: Socrates' trial (as documented in Plato's *Apology*) might be seen as a kind of peer-review process, and whose practice of confronting and examining his 'peers' in the *agora* (as documented throughout Plato's early dialogues) could also count.

Most histories of peer review trace the origin of prepublication peer review to the Royal Society of London and its journal *Philosophical Transactions*, founded by The Royal

Society's first joint secretary, Henry Oldenburg, in 1665. Although no one questions whether Oldenburg deserves credit as the founder of the world's longest-running scientific journal, whether his practice of passing manuscripts around to members of the Royal Society prior to publishing them in the *Philosophical Transactions* actually constitutes the 'real' origin of the prepublication peer-review process is the matter of some debate (Kronick 1990; Spier 2002; Royal Society 2009). Regardless of its 'real' origin, Spier (2002) notes that both the practice of prepublication peer review and the time of its adoption vary from journal to journal, and that the practice did not become widespread until after the Xerox photocopier became commercially available in 1959.

Scarpa (2009) dates the very first (*ad hoc*) peer review of grant proposals to 1879, and Germany's *Notgemeinschaft der Deutschen Wissenschaft*, predecessor of the *Deutsche Forschungsgemeinschaft* (DFG), had a review system during the 1920s, which was later adopted by the DFG in 1951. But the robust institutionalization of grant proposal peer review began around the middle of the twentieth century with the passage of the Public Health Service Act of 1944 in the United States, which authorized the NIH to make grants, an extension of the power that in 1938 had been limited to the National Cancer Institute. The NIH quickly established a Division of Research Grants to oversee the NIH's peer-review process. In the late 1940s, the US Office of Naval Research (ONR) also began making grants, although no process of peer review was required. Instead, grants officers sometimes asked experts to review proposals in order to help them make their decisions. In 1950, the US National Science Foundation (NSF) was founded, and NSF adopted a process of grant allocation that not only copied the strong program manager model from the ONR, but also incorporated a process of peer review like the NIH. The NSF's peer-review process remains to this day less standardized than that of the NIH, but more standardized than that of the ONR.

Two salient features regarding peer review stand out from the foregoing historical account: (1) peer review is not as ancient a practice as many assume—it was not widely practiced in either publication or grantmaking until after the middle of the twentieth century; and (2) in both prepublication peer review and grant proposal peer review, practices vary widely. Nevertheless, despite some criticisms of the process, members of the academic community are almost unanimous in their support of the peer review as a decision-making tool, both for publication and for grantmaking purposes (Boden Report 2006). This near unanimity of support cannot stem from the fact that peer review is the way things have always been decided in academe, for that simply is not the case.

22.3 Autonomy and expertise: the disciplining of peer review

In part, the institutionalization of peer review is motivated by the growth of academic disciplines, both in terms of the *fact* of their growth (i.e. the fact that academic disciplines became, in the nineteenth century, the new model for how research was to be conducted within the German and American research universities) and in terms of the *need* for growing particular disciplines (a need generated by the invention of this new model of the university). Along with the disciplinary division of labor advocated by Kant at the end of the eighteenth century, this new model for the university incorporated a strong demand for autonomy. Wilhelm von Humboldt's 'On the spirit and organizational framework of intellectual institutions in Berlin' proclaims: 'The state must always remain conscious of the fact that it never has and in principle never can, by its own action, bring about the fruitfulness of intellectual activity. It must indeed be aware that it can only have a prejudicial influence if it intervenes. The state must understand that intellectual work will go on infinitely better if it does not intrude' (von Humboldt 1970, p. 244). According to Humboldt's vision, the state's only role should be to facilitate the conditions necessary for the greatest production of knowledge (for the sake of knowledge, rather than for the sake of the state)—to serve an instituting, but not an institutional, role *vis-à-vis* the university. Humboldt's justification for the state's playing this facilitating role is that the state will ultimately benefit from supporting the unfettered pursuit of knowledge in the university.

Incorporating both a division of labor and a strong sense of autonomy, the new universities produced both more knowledge and more specialized knowledge, thus simultaneously cultivating depth (as defined by particular disciplines) as the mark of excellent research and reinforcing the divisions between disciplines. Just as the desire to form the 'new science' led to the formation of The Royal Society of London and to Oldenburg's establishment of the *Philosophical Transactions*, the desire to form new disciplines led to the establishment of new, disciplinary journals. As disciplines grew, they produced both more and more specialized knowledge, which spawned both more and more specialized journals. Competition for resources between universities, between different disciplines within universities, and between faculty members within departments eventually led to the 'publish or perish' mentality, as well as to increasingly sophisticated ways of judging whether one journal was better than another, ranging from the relative prestige of the editors or the academic home of the journal to circulation and impact factors. The most widely used—and crudest—measure of the worth of any particular journal, however, is whether that journal is peer reviewed. This is true despite the fact that the peer-review process varies widely across journals. The case is much the same for the outputs of research, i.e. publications. Indeed, that a particular line of research does not appear in the peer-reviewed literature is taken as *prima facie* evidence of its lack of quality (e.g. the case of intelligent design theory); and publication in peer-reviewed journals is the coin of the realm of many disciplines, largely determining the outcome of many tenure and promotion cases. The close link between peer review and disciplines also presents problems for those who are seeking to explore interdisciplinarity in their own scholarship (Graybill and Shandas, Chapter 28 this volume).

There is a remarkable unity of themes between Kant's call for the division of labor in research, Humboldt's plea for facilitated autonomy for the university, and the canonical document of post-World War II science funding policy in the United States, Vannevar Bush's *Science – the endless frontier* (Bush 1945). Echoing both Kant and Humboldt, Bush argues for state support of autonomously pursued basic research, that is, research pursued for its own sake, without concern for the practical ends that are the proper province of applied research. According to the Bush conception, applied research, which yields technological, medical, and military advancements, fundamentally depends on basic research. Just as Humboldt had argued at the turn of the nineteenth century, Bush suggests that although the particular uses of basic research and the eventual benefits that will accrue are difficult to predict, societal benefits cannot occur unless scientists are allowed to pursue science without interference from the state—a notion that was later labeled as the linear model (or sometimes, the linear-reservoir model) of science.

Because Bush was asking for large outlays of public funds, and on a continuing basis, in support of the unfettered pursuit of basic scientific research, some form of accountability needed to be built into the system. Indeed, there was a great deal of debate between the strong-autonomy advocates in the Bush camp and the more pragmatic adherents of the views expressed in the Steelman Report (Steelman 1947), which advocated more limited scientific autonomy in the name of a stronger connection to public benefit. Bush's advocacy of a strong form of autonomy ultimately won the day when the NSF was created in 1950. Arguably, however, one reason why the NSF abandoned the ONR model for grants decision making, in which a program officer can make funding decisions without subjecting proposals to peer review at all, was the controversy over the demands for the autonomy of research and the demands for more closely linking research to societal benefits. Peer review of grant proposals is meant to guarantee that scientists have a large degree of autonomy when it comes to making decisions about which particular research proposals ought to receive funding, while simultaneously demonstrating their accountability for making wise use of public funds.

The success of the process of peer review in guaranteeing autonomy for the academic pursuit of knowledge, along with concomitant financial support in the form of public funding for research, are key drivers of academe's love affair with peer review. But the fact that society allows peer review to serve this dual function—providing autonomy and asking only self-regulation as accountability—perhaps needs some explanation, given society's ambivalence, or what Jasanoff (1990, p. 9) terms 'oscillation between deference and skepticism', toward experts. Even as we profess our distrust of experts, we evidence faith in expertise. In part, this faith can be attributed to what Chubin and Hackett (1990) call 'enclaves of expertise' in the face of which 'we usually delegate to experts the authority for making decisions in areas we do not understand' (Chubin and Hackett 1990, p. 4). We routinely follow the advice of doctors when it comes to our health and of mechanics when it comes to our cars. Indeed, we ignore the advice of experts at our own risk. It is also the case that what constitutes an autonomous academic discipline, at least in part, is there being something it is, some field of knowledge, which is its special task to pursue. Academic journals mark out this disciplinary territory, and prepublication peer review ensures that this territory is marked well (i.e. according to the standards of the discipline). Academics are experts, and even within academe, perhaps especially so in the context of peer review, scholars from different disciplines display a remarkable deference to the expertise of scholars from other disciplines (Lamont 2009). The experts trust the other experts; is it really any wonder, then, that non-academics should have some faith in peer review?

There is also a growing political problem for anyone who would question society's faith in peer review, as much of the current rhetoric surrounding global climate change attests: so-called climate deniers are routinely characterized as having ulterior motives (something other than truth, such as greed), and decision makers who question scientific consensus—which was gained only after a thorough trial by peer review—run the risk of being charged with the politicization of science (Mooney 2005). Although Sarewitz (2009) is correct that the Obama administration's attempt to 'restore science to its rightful place' in US policy making—in contrast to the presumably wrongful place science occupied in the Bush administration—is yet another politicization of science, the political appeal of Obama's strategy rests on a more basic faith in the value of knowledge and a philosophical presumption about what knowledge actually is.

Academics and non-academics tend to share the presumption that knowledge is something that comes along with specialization and the depth that such specialization brings—what Frodeman (2004) critiques as an epistemology of external relations and opposes to a kind of epistemological holism. An epistemology of external relations—or epistemological reductionism—tends to support analysis: knowledge is gained by examining parts of reality, which can later be pieced together to generate a view of the whole. Epistemological holism, however, holds that knowledge of the whole is always greater than the sum of knowledge of its parts. Epistemological reductionism tends to support the idea of expertise, whereas an epistemological holism tends to undermine the idea of expertise (Sarewitz, Chapter 5 this volume). Epistemological reductionists tend also to think that more knowledge is always a good thing, whereas epistemological holists tend to believe in limits to knowledge. Discipline-based peer review is essentially founded upon an epistemology of external relations, and part of the explanation for our overall acceptance of the process of peer review is that we tend—whether we realize it or not—to view knowledge in (reductionist) terms of external relations. Because we tend to view knowledge in reductionist terms, the notion of expertise seems intuitively obvious to us. (Note that although this last point is a holistic claim, there is no necessary incompatibility between holism and reductionism. The seeming opposition between the two ways of viewing knowledge simply reveals our own reductionist tendencies.)

Another factor supporting our faith in peer review is that we tend to ignore the fact that peer review has a history—and it has a far shorter one than many presume. Adhering to the process of peer review is not simply a disinterested matter of scholarly housekeeping on the part of academe or objectivity on the part of grant-making institutions or societal decision makers. Rather, the process of peer review has its roots in the institutional disciplinization of knowledge production, a process that has always been as political as it has been epistemological. Within the university setting, disciplines deserve at least as much identification with power as knowledge does: in its role as the valuator of academic and scholarly work, the process of peer review acts to wall off disciplines from each other, guaranteeing the existence of disciplinary islands where petty princes (or tyrants) rule. In its role as guarantor of autonomy from societal influence, peer review also walls off academe from the rest of society, guaranteeing autonomy at the price of isolation. Discipline-based peer review *is* the gatekeeper—not only of the little disciplinary hearths within academe, but also of the ivory tower itself.

22.4 Interdisciplinary and transdisciplinary pressures on peer review

Academic excellence is one thing; relevance to anything in the real world outside academe, however, is something altogether different. Often, academic rigor—and relevance within disciplinary scholarship—is achieved only at the price of irrelevance to anyone outside that academic discipline or subdiscipline. Put differently, academe has disciplines and the real world outside of academe has problems—none of which are 'merely academic'.

Interdisciplinarity is often touted as the way to free academics of their disciplinary blinders so that they can begin to develop real solutions to real problems. Yet interdisciplinarity creates all sorts of problems within academe, not the least of which are problems with peer review. As Huutoniemi (Chapter 21, this volume) points out, evaluating interdisciplinary research is exceedingly difficult given the lack of agreed upon standards that disciplines provide. Graybill and Shandas (Chapter 28, this volume) also point to problems for early career academics trained as interdisciplinarians, who are caught between publishing for the discipline that houses them or for a 'new academy' that is yet to materialize: promotion and tenure decisions invariably turn on a record of publication in high-quality journals, which, with a few notable exceptions, are organized (and peer reviewed) along disciplinary lines. Both of these chapters raise the fundamental question for academic interdisciplinarity: who counts as a peer?

Although this question does arise for the 'old academy'—for instance, it is typical to question whether more established investigators within a field are truly peers of early career academics or vice versa—the typical answer is that *disciplines define peers*. It is this answer that brings into relief the difficulty of evaluating interdisciplinary research (whether publications or grant proposals). Lamont (2009) provides a way of viewing the process of peer review—as an interactive social process in which the participants (all multidisciplinary panels of reviewers in her study) aim at a kind of Habermasian ideal speech situation, in which reviewers from different disciplines respect each other's differing disciplinary standards and aim to reach a consensus decision—that may prove useful in the review of interdisciplinary grant proposals. She also suggests that more intensive training of personnel at public funding institutions may be necessary in order to sensitize agencies to the exigencies of evaluating interdisciplinary research. Since many journal editors do not aim for consensus among reviewers, but treat reviews as a way to improve submissions, it may be easier for them to navigate the difficulties presented by an interdisciplinary submission, provided they are attuned to those difficulties and sympathetic to the approach the author takes. It may not be intellectually satisfying, but it may simply be a case of waiting things out until more and more of the old guard is replaced by members of the 'new academy' for which Graybill and Shandas yearn, in much the way that Kuhn suggested paradigm shifts might ultimately occur. Once the Graybills replace the graybeards, it is likely that things will be different.

Although it is tempting to think of interdisciplinarity as only the labor pains that accompany the birth of 'new disciplines' for a 'new academy'—a kind of organic-developmental timeline view—interdisciplinarity within academe could also be seen as a kind of mean between the extremes of isolated disciplinarity and engaged transdisciplinarity. Disciplines serve both to carve out territory within academe and to separate academe from the real world. Interdisciplinarity breaks down disciplinary boundaries within the halls of academe; but transdisciplinarity is needed to tear down the walls of the ivory tower. This may sound like what Huutoniemi (Chapter 21, this volume) terms a critical approach to disciplinarity, in which case it would make sense to reference 'Mode 2 science', 'post-normal science', and 'knowledge policy'—one might also add 'well-ordered science' (Kitcher 2001) and 'Pasteur's quadrant' (Stokes 1997)—and to call for some form of extension of peer review beyond academe to include not just reviewers from different academic disciplines but also other stakeholders in the decision-making process. But such an approach can always be criticized as overly theoretical (or even ideologically committed to epistemological holism).

Rather than approaching the issue of transdisciplining peer review from an ideological or theoretical standpoint, i.e. from an academic point of view, let us begin with a problem in the real world, one for which some empirical evidence already exists, and on which experiments could be conducted: the Government Performance and Results Act (GPRA) of 1993. The GPRA is designed to focus US federal agencies on measuring and improving results, which, once communicated to Congress, will provide decision makers with the necessary data to assess the 'relative effectiveness and efficiency of Federal programs and spending'. The GPRA's explicit mandate is to require three things of all federal agencies: (1) multiyear strategic plans; (2) annual performance plans; and (3) the development of metrics that would gauge adherence to the annual performance plans. The underlying message of the GPRA is that agency plans must be tied to societally relevant outcomes. This presented a particular challenge to the NSF, since it is the one federal agency devoted to supporting basic research.

Basic research, as Vannevar Bush has so clearly articulated, is conducted without consideration for the results. With the passage of the GPRA, the NSF found itself, more starkly than before, caught between politics and science. The NSF is what Guston (2000) refers to as a 'boundary organization'—as the federal agency responsible for supporting basic research, it owes allegiance both to the government and to scientists. While the government wanted to see results, basic scientists wanted still to be able to pursue basic, rather than applied, research (Kostoff 1997). How did the NSF respond to these conflicting demands?

Not surprisingly, the NSF did not respond as an academic might, by turning to the literature about post-normal, well-ordered, Mode 2, use-inspired science to create a new knowledge policy. Instead, the National Science Board (NSB), the NSF's policy branch, restructured the NSF's peer-review process (known as 'merit review') to enlist the scientific community—both as proposers and as reviewers—in the task of articulating the societal relevance of the basic research NSF funds (Holbrook 2005). In 1997, the new merit review criteria were introduced, and they asked only two questions: 'What is the intellectual merit of the proposed activity?' and 'What are the broader impacts of the proposed activity?'. Essentially, the NSF engaged in what Miller (2001) calls 'hybrid management'. Peer review has always served both academic and political purposes—the NSF simply manipulated these elements to place a greater emphasis on the political function of peer review, without stripping scientists of the academic autonomy they demand. Proposers and reviewers were still asked to articulate and evaluate the intellectual merit of proposals (for which they could still appeal, in most cases, to disciplinary standards of excellence); but they were also asked to articulate and evaluate the impact of basic research on society (for which they lacked the expertise).

In effect, the NSF was asking scientists to break free from their disciplinary bounds and to engage in activities that involve interdisciplinary and transdisciplinary interactions (e.g. communicating one's research beyond one's discipline, either to academics in different fields or in novel ways to non-academic society; communicating one's research to political decision makers in useful ways; enhancing diversity in ways that go beyond a simple head count of minorities; training graduate students and mentoring postdoctoral researchers in the ethics of research; etc.). Scientists, to put it baldly, balked (Frodeman and Holbrook 2007). In part, this is because most scientists trained along disciplinary lines to conduct basic scientific research are generally *not* trained either to articulate or evaluate the societal impacts of their work. The broader impacts criterion (BIC) was at first simply ignored, until the NSF announced that they would begin returning without review proposals that failed to address the BIC, at which point compliance began to rise. Even after more than a decade, however, the quality of responses to the BIC remains a persistent problem.

Beginning in this way with a real world problem—the NSF's response to the GPRA, scientists' response to the BIC—allows for an important point: science studies scholars need not call for a 'transdisciplinarization' of the process of peer review, for the transdisciplining of peer review has already begun. Moreover, the case of the NSF, unique as it is, is not unlike changes to peer-review processes at other public science funding agencies around the world, many of which have incorporated similar societal impacts criteria into the process of peer review (CSID 2009).

22.5 Evaluating disciplinary, interdisciplinary, and transdisciplinary relevance

Disciplinary expertise is required to assess disciplinary excellence. Hence, reviewers charged only with assessing the disciplinary merit of a grant proposal (or article submission) need only be selected from the particular discipline under consideration. A mix of disciplinary expertise(s) is required to assess academic excellence beyond a single discipline. Hence, reviewers charged with assessing the merit of multidisciplinary or interdisciplinary proposals ought ideally to be selected from all the disciplines included in the proposals. Although review of such multidisciplinary and interdisciplinary proposals is more complicated than monodisciplinary review, it nevertheless takes place within academe, where each reviewer is ideally accorded a kind of authority over his or her own disciplinary domain. What sorts of expertise are required to address and assess societal relevance?

To the extent that societal impacts criteria ask proposers and reviewers to address issues that can be addressed from within academe, experts can be drawn from the relevant disciplines to address those issues in the proposal and its review. For example, some societal impacts criteria can be addressed in terms of educational impact—in which case it would seem necessary to employ experts in education both in writing and in reviewing the proposals. This would simply present another case of interdisciplinarity with which peer review must cope. However, some societal impacts criteria take peer review beyond the disciplines to such issues as offering policy-relevant knowledge for societal decision makers. When societal impacts criteria go beyond the realm of academe to address societal relevance, if proposers are to make their research societally relevant and reviewers are to judge societal relevance, then who counts as a peer must be extended to include non-academic members of society at large.

Although these claims are normative, they are not based on an ideological imposition of theory onto reality. The claim is not that peer review should be de-disciplined, and either interdisciplined or transdisciplined in order to pursue some ideal form of knowledge. There is no ideology of epistemological holism at work here. Instead, the point can be expressed as a hypothetical imperative: If we introduce transdisciplinary criteria into the process of peer review, then we should expand the definition of who counts as a peer beyond the boundaries of the disciplines.

There is also a more comprehensive lesson to be learned: instead of thinking of peer review only in terms of its academic disciplinary use as an evaluation tool (according to which interdisciplinarity presents a special problem for peer review), peer review must also be addressed in terms of its larger social context. Doing so will allow us to see that peer review has never been only a disciplinary activity, one that ought to be jettisoned as an artifact of prepostdisciplinarity, but has always been a transdisciplinary activity, as well. Patrolling the border between academe and society, peer review can be the ultimate tool of transdisciplinary hybridization.

References

Boden Report (2006). *Peer review: a report to the advisory board for the research councils from the working group on peer review*. Available at: <http://www.mrc.ac.uk/Utilities/Documentrecord/index.htm?d=MRC003951> (accessed 27 May 2009).

Bush, V. (1945). *Science – the endless frontier*. Washington, DC: United States Government Printing Office. Available at: <http://www.nsf.gov/about/history/vbush1945.htm> (accessed 27 May 2009).

CSID (Center for the Study of Interdisciplinarity) (2009). *Comparative assessment of peer review (CAPR)*. Available at: <http://www.csid.unt.edu/research/capr.html> (accessed 29 July 2009).

Chubin, D. and Hackett, E. (1990). *Peerless science: peer review and U.S. science policy*. Albany, NY: State University of New York Press.

Frodeman, R. (2004). Environmental philosophy and the shaping of public policy. *Environmental Philosophy* **1**: 7–16.

Frodeman, R. and Holbrook, J. (2007). Science's social effects. *Issues in Science and Technology* **23**(3), 28–30. Available at: <http://www.issues.org/23.3/p_frodeman.html>.

Frodeman, R. and Mitcham, C. (2007). New directions in interdisciplinarity: broad, deep, and critical. *Bulletin of Science, Technology and Society* **27**: 506–14.

Guston, D. (2000). *Between politics and science: assuring the integrity and productivity of research*. Cambridge: Cambridge University Press.

Holbrook, J. (2005). Assessing the science – society relation: the case of the U.S. National Science Foundation's second merit review criterion. *Technology in Society* **27**(4), 437–51.

von Humboldt, W. (1970). On the spirit and the organizational framework of intellectual institutions in Berlin. University reform in Germany. *Minerva* **8**, 242–50.

Jasanoff, S. (1990). *The fifth branch: science advisers as policymakers.* Cambridge, MA: Harvard University Press.

Kitcher, P. (2001). *Science, truth, and democracy.* Oxford: Oxford University Press.

Kostoff, R. (1997). Peer review: the appropriate GPRA metric for research. *Science* **277**(5236), 651–2.

Kronick, D. (1990). Peer review in 18th-century scientific journalism. *Journal of the American Medical Association* **263**, 1321–2.

Kronick, D. (1994). Medical 'publishing societies' in eighteenth-century Britain. *Bulletin of the Medical Library Association* **82**(3), 277–82.

Lamont, M. (2009). *How professors think: inside the curious world of academic judgment.* Cambridge, MA: Harvard University Press.

Miller, C. (2001). Hybrid management: boundary organizations, science policy, and environmental governance in the climate regime. *Science, Technology, and Human Values* **26**(4), 478–500.

Mooney, C. (2005). *The Republican war on science.* New York: Basic Books.

NIH (National Institutes of Health) (2008). *2007–2008 peer review self study, final draft.* Available at: <http://enhancing-peer-review.nih.gov/meetings/NIHPeerReviewReportFINALDRAFT.pdf> (accessed 27 May 2009).

Rip, A. (2000). Higher forms of nonsense. *European Review* **8**(4), 467–86.

Royal Society (2009). <http://www2.royalsociety.org/campaign/strategic/increase.htm> (accessed 24 November 2009).

Sarewitz, D. (2000). Science and environmental policy: an excess of objectivity. In: R. Frodeman (ed.) *Earth matters: the earth sciences, philosophy, and the claims of community*, pp. 79–98. Upper Saddle River, NJ: Prentice Hall.

Sarewitz, D. (2009). The rightful place of science. *Issues in Science and Technology* **25**(4): 89–94.

Scarpa, T. (2009). *Assessing and advancing funding of biomedical research benchmarking: values and practices of different countries.* Available at: <http://www.vr.se/download/18.72e6b52e1211cd0bba880005479/Toni+Scarpa+%5BKompatibilitetsl%C3%A4ge%5D.pdf> (accessed 27 May 2009).

Spier, R. (2002). The history of the peer-review process. *Trends in Biotechnology* **20**(8), 357–8.

Steelman, J.R. (1947). *Science and public policy: the president's scientific research board*, Vol. 1. Washington, DC: United States Government Printing Office.

Stokes, D. (1997). *Pasteur's quadrant: basic science and technological research.* Washington, DC: Brookings Institution Press.

Wennerås, C. and Wold, A. (1997). Nepotism and sexism in peer-review. *Nature* **387**, 341–3.

CHAPTER 23

Policy challenges and university reform

CLARK A. MILLER

In the twenty-first century, humanity faces an array of policy challenges that are likely to demand the kind of broad, sweeping policy reforms reminiscent of the Progressive and New Deal eras of a hundred years before. Like those previous eras of policy upheaval, many of the challenges of the twenty-first century are driven by rapid changes in the scientific and technological foundations of every aspect of human society, from agriculture and health to economic production and global security. Unlike those prior transformations, however, universities seem ill-prepared to provide the necessary ideas and human resources to successfully address the policy challenges of the twenty-first century (Crow 2007). What will it take for universities to reverse course? A lot depends on the possibility of developing new and innovative approaches to interdisciplinary policy research and education.

23.1 Twenty-first-century policy challenges

The twenty-first century has brought an array of novel policy problems that have fundamentally challenged the capacity of states and societies to develop meaningful policy responses. Consider, for example, the impacts of globalization on the social fabrics of societies, the risks of long-term climate change and biodiversity loss, the rise of powerful terrorist networks, the continued threat of nuclear proliferation, and the emergence of a wide range of novel epidemics, from AIDS and SARS to avian and swine flu.

Part of the complexity of these policy challenges derives from their transnational scope. The fundamental lodging of sovereign decision-making authority in states, coupled with the general illegitimacy and ineffectiveness of international policy institutions, has made policy responses to these challenges piecemeal, *ad hoc*, and, in some cases, deeply oppressive. Just as importantly, however, the challenges posed by these policy problems arise from their fundamental interdisciplinarity. In recent years, social and political historians have illuminated just how interconnected the social science disciplines are, institutionally and in their basic conceptual frameworks, with the origins of the modern welfare state (Rueschemeyer and Skocpol 1996; Calhoun and Rhoten, Chapter 7 this volume). It is not surprising, therefore, that problems that lie beyond the capacity of any individual state also lie beyond the capacity of any one discipline.

Perhaps even more challenging to universities, however, are the interdisciplinary linkages that these policy problems demand, between on the one hand the humanities and social sciences and on the other the natural sciences and engineering. In the wake of the transformation wrought in funding for the sciences and engineering after World War II, no more fundamental divide exists in universities today than between the humanities and social and policy sciences and the 'harder' physical and life sciences that the former aspire to imitate. For all that the basic infrastructure of the university—promotion and tenure, departments and disciplines, undergraduate and graduate degrees—appears the same, fundamental differences have evolved in universities over the past half century in how graduate education, career development, and professional advancement are organized and evaluated between these two halves of the university. This transformation has been driven almost entirely by external funding, which as will be described below, has fundamentally altered the political economy of the natural sciences and engineering, reducing the internal controls exercised within the university by disciplinary units.

These differences in academic reward systems and their impacts on society will be described in greater detail below, but for the moment let me simply note that the natural sciences and engineering, today, are in many respects fundamentally organized around key policy problems of the twenty-first century. By contrast, top-ranked social science or humanities programs in the United States often have few, if any, faculty working on problems of global environmental change, international pandemic policy, regulation of science and technology, or even nuclear proliferation or global terrorism (with the exception of a few scholars in international relations). From the standpoint of the social sciences, these problems are not theoretically sexy, not likely to result in publication in the top journals, too complex to be analyzed with what is taken to stand for methodological rigor, and, therefore, neglected. Thus, while universities might be argued to be ideal places for combining the diverse knowledge and forms of expertise necessary to address twenty-first-century policy challenges, their current organization seriously limits their capacity to respond in ways that meaningfully include the social and policy sciences and humanities.

23.1.1 The food crisis

One illustrative example of a current policy challenge is the rise in food prices across much of the globe during 2007 and 2008. From an economic perspective, rising food prices seem relatively easily explained. During this period, the demand for grain in global markets consistently outpaced supply. Why this occurred is a more complicated story.

One piece of the story begins in the 1960s, with the Green Revolution. While much of the attention on the successes of the Green Revolution has focused on the use of plant breeding to increase yields, the key to plant breeding in many cases was the ability to design cereal varieties that would respond greatly to increases in agricultural inputs—particularly water, fertilizer, and pesticides—while another aspect of increases in productivity came from the introduction of mechanization to agriculture. As a consequence, global agriculture today, in many parts of the globe, consumes a great deal of energy, both in the production of fertilizer and pesticides and the use of farm equipment. With the rapid rise in prices of oil and gasoline, the costs of agricultural production have risen rapidly. Among producers working at low profit margins this could and did have a significant effect on, to some extent, profitability and production levels, but more importantly on commodity prices, which rose alongside the rise in production costs.

A second piece of the story relates to changes in consumption patterns. In recent decades, meat consumption has risen significantly on a per capita basis (while the number of people has risen as well) in all regions of the globe, and especially in East and South Asia. In relation to its nutritional content (in terms of calories), meat production requires far higher quantities of grain than if people consume the grains directly. Shifts in daily calorie consumption from grain to meat therefore require significant increases in global consumption over and above population growth rates.

A third piece of the story appears to be drought conditions in parts of the world, especially Australia, that have significantly reduced supplies of rice. Australian droughts are consistent with, but may be entirely unrelated to, climate change, which has the potential to require large-scale shifts over the next few decades in the geographic location of crop-growing zones as temperature and precipitation patterns shift, driven by the accumulation of greenhouse gases like carbon dioxide and methane in the atmosphere.

Responding to the threat of long-term climate change, the United States and several European countries have also put in place new policies that call for long-term shifts in energy production toward biofuels. Coincidentally, the amount of grain purchased for use in biofuel production is roughly equivalent for each of the past 3 years to the shortfall between global production and consumption of grains. Thus, it is perhaps not surprising that many developing countries have called on the United States and Europe to back away from their policy commitments to increasing biofuel production in order to relieve pressure on global commodities markets and help to halt further price increases, which are creating severe disruptions for poor communities around the globe.

Ironically, just at the time that food riots were beginning to appear around the globe, the US government announced its intentions to eliminate its share of funding for the Consultative Group on International Agricultural Research (CGIAR), a collection of international agricultural research centers located around the globe and widely credited with the success of enhancing yields during the Green Revolution. This policy likely originates with large agricultural seed companies in the United States, who see the CGIAR institutions as competitors in the business of providing seeds to agriculture in developing countries. Eliminating the primary international agricultural research institutions at a time of global food crisis seems counterintuitive, however.

23.2 Policy research and training in the twentieth century

The complexity of social and policy challenges at the start of the twenty-first century mirrors a similar set of social and policy challenges that faced nations at the end of the nineteenth century. Driven by the Industrial Revolution, society had changed dramatically, accommodating the rapid rise of factory labor and the increasing centralization of production in large-scale enterprises (oil, steel, railroads, etc.). The result was the emergence of several novel social problems, including worker accidents, unemployment, boom-and-bust business cycles, and extreme poverty among a handful of social groups, including the elderly, women, and veterans. Responding to these problems became a key policy challenge for states in the late nineteenth and early twentieth centuries (Skocpol 1992).

From this period emerged a 'double institutional transformation' of modern societies: the creation of the administrative or welfare state and the research university, each linked to each other. The specialization of labor within the university helped to give rise to new disciplines, especially but not entirely in the social sciences—economics, sociology, political science—each oriented around specific aspects of social problems and each involved in training a new class of public sector managers who could take social science research methods and insights and apply them to solving society's problems: managing the economy, promoting social welfare, and improving public administration (Nowotny 1991; Porter 1995; Wittrock and Wagner 1996). Through the 1940s, much of the most important social and economic legislation in the United States developed from ideas first articulated in the social sciences.

After World War II, universities took a different direction. While the social sciences had helped to secure universities' prestige in the early twentieth century, postwar prestige came to be dominated almost exclusively by prowess in the natural sciences and engineering. The rise of military research dominated postwar research funding, while the research budgets of other science and technology oriented agencies, from the National Aeronautics and Space Administation (NASA) and the National Oceanic and Atmospheric Administration (NOAA) to the National Institutes of Health (NIH) and the National Science Foundation (NSF), created in 1950, also rose dramatically. The postwar emergence of MIT and Stanford as rivals for the nation's top science and engineering schools, both driven by huge influxes of military research and development funds, helped radically restructure the nature of American science and American universities (Leslie 1993).

In the social sciences, the rise of behaviorism in the 1950s shifted attention away from social problems toward more theoretical questions with less immediate and intuitive implications for social policy. By the 1960s and 1970s, policy researchers and educators at many universities had become increasingly dissatisfied with the direction of disciplinary departments and had begun to look outside the disciplines for support for their work. From the 1960s to the 1980s a number of US universities established interdisciplinary schools of public policy that brought together political scientists, economists, and, to a lesser extent, sociologists, with an explicit interest in policy-oriented research. Training changed, too, as these new policy schools moved away from academic training in disciplinary methods toward novel methods and approaches to policy and cost–benefit analysis and the administrative skills required of government managers.

23.3 The rise and consequences of interdisciplinarity in US universities

The emergence and consolidation of policy schools at many of the largest US universities occurred as part of a broader trend toward interdisciplinarity during the last few decades of the twentieth century. From the 1970s onward, universities pursued a handful of interdisciplinary programs like policy schools and occasional research centers (e.g. centers for research on poverty, which flourished alongside policy schools) in the social sciences, and even more in the humanities (e.g. programs in American studies, gender studies, African-American studies, regional studies, religious studies, and environmental studies with a humanities focus).

By far, however, the largest number and greatest scale of interdisciplinary programs created during this period occurred within the natural sciences and engineering. To see this, walk through the halls of America's premier science and engineering research universities. At major research universities, interdisciplinary research centers in the sciences and engineering number in the multiple dozens. Across the United States, the number of environmental science and science-oriented environmental studies programs far outnumbers the number of environmental studies programs with a primary humanities and social science orientation.

The primary driver for this shift towards interdisciplinarity in the natural sciences and engineering was large-scale government funding of research oriented toward the solution of societal problems, from curing disease and ensuring national security to protecting the environment. Although the NSF has long maintained discipline-based funding programs for science and engineering research, most federal agencies never made that distinction in their research funding, and by the 1980s, even the NSF was fully committed to high-profile funding streams devoted to breaking down and transcending the disciplinary boundaries of research. Recent programs have included the $50 million Biocomplexity Initiative and the multi-agency National Nanotechnology Initiative that has been funded at approximately $1 billion per year. Indeed, by the time the concept of interdisciplinarity had emerged as a key focus of NSF funding programs in the 1990s, the NSF already had several decades of experience with large-scale, interdisciplinary programs. It is also useful to note that external funding has also been crucial to those areas of the humanities and social sciences that exhibit high levels of interdisciplinary research, such as regional studies centers and programs. For these, US Department of Education Title VI grants have proven essential to long-term stability.

The broad shift of attention of the funding agencies away from disciplinary boundaries gave rise to the rapid proliferation of new, interdisciplinary fields in the sciences and engineering, from biochemistry and atmospheric chemistry to bioengineering, environmental engineering, and nanotechnology. In almost all cases, training programs followed research, especially at the graduate level, with students in electrical engineering, for example, being trained in semiconductor physics, power engineering, plasma physics, ionospheric physics, circuit design, and more, each requiring its own unique combination of interdisciplinary activity. Disciplinary departments became little more than shells, as multiple departments hired researchers with very similar research skills and programs, whose work tended to take place as much or more in interdisciplinary networks or centers. Today, one of the most common developments in engineering colleges is, for example, the creation of centralized instrumentation facilities that hold instruments that are routinely used by faculty and graduate students from many different departments.

A significant consequence of the shift toward interdisciplinarity in the natural sciences and engineering has been a substantive differentiation of reward systems in the social and natural sciences. In the social sciences and humanities, hiring and tenure evaluation decisions have evolved to focus almost exclusively on the theoretical contributions of individual scholars to disciplinary canons. By contrast, in the natural sciences and engineering, hiring and tenure decisions focus more pragmatically on whether the individual in question has the intellectual and managerial skills necessary to develop and maintain a highly productive research facility (most frequently a laboratory). Where informal, quantitative benchmarks in the social sciences tend to focus on numbers of publications in top disciplinary journals, their counterparts in the sciences and engineering tend to focus on the scale of annual research expenditures (and the successful grantsmanship skills necessary to secure them). While natural scientists and engineers are often expected to have graduated at least one PhD student (and have several more in the pipeline) by the time they come up for tenure, social scientists are often discouraged from taking on any graduate students at all in order to reserve more time for their own work. In my own experience and among my colleagues at several major research universities, social scientists are even discouraged as junior faculty from pursuing large-scale research grants that would support research teams, this being seen most frequently as 'service' rather than 'research', which is evaluated solely in terms of the researchers' own personal work.

23.4 The disciplinary limits of policy schools

In the United States, policy schools have arguably been among the strongest and most consistent exemplars of interdisciplinarity in the social sciences. Most policy schools include significant faculties with training in economics, political science, and statistics, with smaller faculties trained more broadly in the social sciences (e.g. history and sociology) and, increasingly, in interdisciplinary public administration, public policy, or public affairs PhD programs. This emphasis on building interdisciplinary research and teaching programs as well as the focus of policy research on concrete policy problems—rather than advancing social theory—have tended to isolate policy school faculties from their disciplinary colleagues in the social sciences, especially where policy schools retain independent hiring and tenure authority.

One cause—and consequence—of the drift of policy schools away from the traditional disciplines of economics, politics, and sociology has been the emergence of distinct research trajectories and problems in policy research. Here we come to the core of the problem of interdisciplinarity, in my view, and one that policy schools illustrate nicely. Too often, the view of interdisciplinarity is that it occurs when problems have multiple disciplinary dimensions (say, political and economic), requiring disciplinary specialists to collaborate to create robust knowledge. This view presumes that disciplines cover the

Transdisciplinary efforts at public science agencies: NSF's SciSIP program

Erin Christine Moore

Publicly funded scientific research exists in a boundary space: as science, it is purportedly value-free; but being publicly funded, it is supposed to serve societal (that is, political) goals. The 60-year history of publicly funded science is marked by a continuing struggle to square this circle.

The dominant model for public research funding in the United States was established in the years after World War II. The most famous—and still prototypical—articulation of this model can be found in Vannevar Bush's 1945 report to President Roosevelt, *Science, the endless frontier*. This document, formally a recommendation for shaping the policy for science funding in the United States, is in fact an *apologia*—an explanation and defense of the scientific enterprise. Bush argues that progress in science will result in societal progress: 'New products, new industries, and more jobs require continuous additions to knowledge of the laws of nature, and the application of that knowledge to practical purposes' (Bush 1945, p. 1).

But note the means by which knowledge benefits society: Bush insists that the path from knowledge to benefit cannot be anticipated or controlled. Benefits will automatically ensue simply by adding to the 'reservoir' of scientific knowledge. Although Bush justifies the public funding of science by pointing to future societal benefit, he simultaneously claims that it is inappropriate to use societal needs or priorities to guide the use of that funding. In the Bush model one simply takes it on faith that science funding will always result in societal benefit. While this model has been called into question over the years, it has shown remarkable longevity and influence in terms of how national science policy is pursued and justified.

This is especially true for the National Science Foundation (NSF), the major funding agency for basic science in the United States. In fact, the NSF was the federal answer to Bush's recommendation. The NSF was founded on the assumptions that Bush put forth in 1945: that basic (non-applied, non-targeted) research is the primary limiting factor for societal progress, including the growth of wealth, full employment, relief from disease, and national security in both wartime and peacetime. According to Bush, 'Scientific progress on a broad front results from the free play of free intellects, working on subjects of their own choice, in the manner dictated by their curiosity for exploration of the unknown' (Bush 1945).

In the past 15 years, however, policy makers have placed new demands for accountability and social relevance on the research funding system. It is no longer sufficient to assume that more federal dollars automatically equals more societal benefit; policy makers (and tax payers) increasingly want to see evidence of the outcomes of investments in science. In 1993 Congress cancelled the Superconducting Super Collider project after having already spent $2 billion on it, largely because the massive project costs (cost projections ballooned from $4.4 billion estimated at project inception to over $12 billion by 1993) could not be justified as having sufficient societal benefit. Also in 1993 Congress passed the Government Performance and Results Act (GPRA), demanding strategic planning and proof of the return on investment for all government-funded programs including science. Some in the science policy community—especially agencies like the NSF who funded so-called basic research—reacted to the GPRA with dismay, as its demand for performance-driven management threatened the autonomy and isolation from social concerns they had grown accustomed to.

Responding to such calls for science to prove its social relevance, the NSF has implemented a suite of ways to require and demonstrate the societal impacts of the research funded through this agency. In 1997 the NSF updated its merit review criteria for evaluating research proposals. All

(*cont.*)

Transdisciplinary efforts at public science agencies (cont.)

proposals were required to include a 'broader impacts' statement, anticipating how the proposed research would impact society. In principle, the broader impacts criterion negates the assumption that basic science is best pursued without regard to societal relevance. It also calls into question time-honored disciplinary boundaries—by requiring science to prove its relevance, it can no longer be claimed that disciplinary knowledge is autonomous and self-justifying. Scientists had (and continue to have) mixed reactions to the broader impact criterion, some even claiming that it conflicts with the intellectual merit criterion: to advance basic science. But Congress continued to support a focus on broader impacts, twice calling for an investigation to ensure compliance within NSF.

Within the science policy community it became apparent that more knowledge was needed about the connection between publicly funded science and social impacts. SciSIP, a NSF research program newly funded in 2005 in the 'science of science and innovation policy', is an attempt to rethink that relationship by investing in research on the science policy process. SciSIP-funded projects are aimed toward understanding the relationship between science and society and predicting the outcomes of public investments in science and technology.

It is notable, however, that almost as soon as this transdisciplinary research effort was implemented, its inter- and transdisciplinary aspects were cast aside in pursuit of disciplinary stability. Of the 81 projects funded by SciSIP over its first 3 years, over half of the awards centered on economic analysis. Further, the vast majority of all funded projects emphasize the process and measurement of 'innovation', where innovation is almost always associated with economic profitability. Nearly all the successful proposals state economic gains as a main goal of the proposed research. Contributing to this orientation is an overt attempt by the directors of SciSIP to 'disciplinize'—to solidify its research portfolio in order to establish a recognizable field that can be associated with the title 'science of science policy'. Program officers have stated that their goal is to have PhD programs in the area within the next 10 years.

But has SciSIP correctly identified the sort of knowledge it actually needs to address the problems that gave rise to the program? Rather than challenging the reservoir model of scientific knowledge, SciSIP has put most of its resources in measuring the inputs and outputs of research and development (R&D). It can thus be seen as reinforcing the very model it was intended to critique. This is in large part because the SciSIP research portfolio is focused on economic growth instead of examining the multifaceted relationship between science and society. Continuing along this path will provide measures of the *outputs* but not the *outcomes* of investments in science and technology.

To better understand the societal impacts of investment in R&D it is necessary to question the real outcomes of science—both positive and negative—and to evaluate how science is used in society, and to what end. In order to do so we must first ask: What knowledge is needed in order to understand the relationship between science and society? How might that knowledge be integrated with the economic data SciSIP is currently gathering in order to inform decision makers about the best way to go about funding research in science and technology, in order to best benefit society?

Reference

Bush, V. (1945). *Science – the endless frontier*. Washington, DC: National Science Foundation.

terrain of potential research topics fully, so that all interdisciplinarity needs to do is to bring divergent disciplinary perspectives together in collaboration.

By contrast, one finds an alternative model of interdisciplinarity operating in policy schools (and other fields, such as science and technology studies). In these cases, what we clearly see is that, at any given moment in time, disciplines constitute a broad but not infinite set of interesting problems and potential methods that can be applied to those problems. At that same point in time, however, for whatever historical reason, disciplines leave other sets of research problems unexamined, or partially examined, as well as methods unapplied. Part of the value of interdisciplinarity lies in expanding the range of potential problems and methods that are available for use in research and teaching. Policy schools, for example, have emerged largely to fill the gap that was left when the social science disciplines moved away from important but supposedly theoretically uninteresting research oriented toward understanding and solving the kinds of problems that face governments in their day-to-day activities. But they have also, freed from disciplinary evaluation, advanced research into these problems in new ways, identifying new problems, and adopting new methods that were never taken up by prior disciplinary approaches.

Despite their leadership in interdisciplinary research and teaching, however, policy schools remain limited in key respects in their ability to address the policy challenges of the twenty-first century. The gap between reward structures in the social and natural sciences has complicated the ability of policy schools to extend their faculty to the natural sciences. Policy schools have thus also typically been isolated from science and engineering schools and faculties. Although science and technology are increasingly central to a wide array of policy domains (energy, environment, defense, health, etc.), only a handful of policy schools have significant research foci that examine science and technology policy or that include significant numbers of science and engineering faculty. Scientists and engineers with an interest in policy have typically sought out opportunities for careers in Washington, DC, or other policy capitals, and significant resources have been created by organizations such as the American Association for the Advancement of Science (AAAS) for facilitating the transition from the laboratory to policy agencies. By contrast, very few policy schools have explicitly sought to foster closer connections with science and engineering faculty or to make available career paths for them that would bring them into greater engagement with policy research and training. Similar, if not larger, gaps exist between many policy schools and other professional schools at research universities, including law, medicine, public health, and agriculture.

Policy schools are also increasingly limited by the rise of disciplinary tendencies within the policy sciences. Driven as much by the desire for standardized curricula for professional-level policy education as anything, policy schools have become increasingly narrowly focused on a suite of policy analytic methods and tools that include econometrics, microeconomic policy analysis, cost–benefit analysis, and advanced statistical methods at their core. This development has created something of a tussle between proponents of these quantitative and statistical approaches, who tend to focus narrowly on efficiency as the primary goal of policy making, and others who view policy analysis as an art, as much as a science; who are broadly interested in public value rather than a sole focus on private, economic value; and who emphasize the dimensions of ideas, meaning, and identity in policy-making processes.

23.5 The need for university-level reform and strategic policy research initiatives

Universities need to pursue high-level reform if they are going to position their research and teaching to contribute meaningfully to understanding and addressing the policy challenges facing humanity in the twenty-first century. In particular, I believe universities need to invest in new policy research initiatives or centers that explicitly cross interdisciplinary boundaries. Like the highly successful policy research centers focused on poverty at the University of Michigan and the University of Wisconsin, these centers would bring together a wide range of social science researchers to concentrate on high-profile policy problems (I do not necessarily believe that these initiatives must specifically focus on individual policy problems as their primary focus; there are lots of reasons to believe that cross-cutting initiatives that focus on similarities across a range of policy problems would also work very well).

These initiatives would need to develop a new model of interdisciplinarity that recognizes and promotes transformational research on new problems, with new methods, that significantly extends existing domains of scholarship into new territories. While this has happened in the sciences and engineering (e.g. in the development of nanotechnology and synthetic biology), there is a need to invest strategically in comparable programs in the social and policy sciences and to bridge these fields with the sciences and engineering. Twenty-first-century policy problems cannot be readily differentiated into scientific and policy problems and demand more integrated approaches to research and teaching (I describe each of these elements below in greater detail). In this fashion, universities can leverage this foundation of interdisciplinary capacity by investing strategically in coordinated, university-wide research initiatives that specifically address key policy challenges for the twenty-first century.

23.5.1 A new foundation of interdisciplinarity

Problems like the current global food crisis, climate change, or the transition to sustainable energy systems cannot be addressed by any single discipline. This fact is well recognized. What is less well recognized is that even combinations of existing disciplines will not provide the intellectual foundations necessary to solve these problems. The economics problems associated with climate change are fundamental, complex, challenging, and deserving of careful university research. Sadly, such work will not, in most economics departments, be rewarded with recognition, hiring, or tenure. Comparable examples can be found across most of the major disciplines, especially in the social sciences.

What is needed is a new foundation of interdisciplinarity in universities that starts from an archipelago model of disciplines: namely, that disciplines are like islands in the ocean, reflecting limited definitions of what counts as significant research problems and viable research methods and maintained by social practices and institutions that reward certain kinds of research but not others. Disciplines have come over the past century to be seen, especially in the social sciences and humanities, as the ultimate arbiters of what does and what does not count as legitimate research. If we are to be able to grapple with the grand policy challenges of the twenty-first century, universities need to engage meaningfully with how to allow inquiry in the broad domains of intellectual turf that lie beyond and outside of the conventional disciplines. Acknowledging that there is lots of room for valuable inquiry in intellectual domains surrounding contemporary policy problems, a new foundation of interdisciplinarity would build new social structures and reward systems that allow researchers to focus energy and initiative on new kinds of research that extend well beyond the shallow waters surrounding the existing disciplines. In many respects, this foundation—and this model of interdisciplinarity—already exists in the natural sciences and engineering. Now is the time to expand this foundation to the social and policy sciences and humanities.

23.5.2 Strategic investments in interdisciplinary social sciences

To create a new foundation for interdisciplinary research and education in the social sciences will require substantial shifts in the distribution of financial resources in universities. Disciplinary departments have maintained their control over the terrain of acceptable problems and methods largely through their control over hiring, tenure, and the financing of graduate education. All three will require direct attention, but arguably the most important piece is the last. For the most part, teaching assistant positions fixed to large undergraduate majors fund all but a tiny fraction of graduate education in the social sciences. So long as disciplinary departments control the allocation of these resources, graduate education in the social sciences will continue to make it very difficult for students to acquire training in interdisciplinary research opportunities.

Two approaches to altering the reward structure in the social sciences can be imagined. On the one hand, teaching assistant positions can be distributed to a broader and more diverse community of graduate students. There is little a priori reason, for example, that teaching assistants for a freshman-level introduction to American government class must go to disciplinary political science graduate students. Presumably, many graduate students in policy fields with substantial exposure to the structure and functioning of governments, and with adequate preparation and training, could teach this class very well. On the other hand, universities could also invest new resources into interdisciplinary social science research and teaching programs. Much as interdisciplinary funding from federal agencies has provided a financial incentive to alter the structure of disciplines in the natural sciences, universities could opt to create similar financial incentives in the social sciences.

23.5.3 Bridging the humanities, social and natural sciences, and engineering

Meeting the policy challenges of the twenty-first century will also require new and substantial efforts to bridge the humanities and social sciences, on the one hand, with the natural sciences and engineering, on the other. In recent decades, a handful of programs at the NSF have begun to emphasize the incorporation into science and engineering research programs of research on the societal dimensions of major scientific challenges. As a consequence, a growing array of experiments exist that could provide valuable lessons for how to make such collaborations work and what pitfalls to avoid. Yet, other federal funding agencies, whose budgets dwarf those of NSF, have largely failed so far to follow NSF's lead in this area. Thus, the vast majority of research funding in the United States today does not aim to foster strong linkages between the social and natural sciences.

The experience of those participating in experiments in bridging the humanities, social and natural sciences, and engineering suggests that differences in reward structures make collaborative work extremely difficult. Publication strategies vary widely between the social and natural sciences, for example, with the former largely insisting on single-authored publications while the latter emphasize collaboration and co-authorship. Social science fields that rely on books further complicate the publication picture. Graduate student research assistant positions often work differently, too. Social science research assistants often work part time on faculty projects, while doing their own, separate topic in their spare time. In the natural sciences, grants typically fund students to work on their own dissertation research (or, to put it differently, students' dissertation research topics are typically selected to match available grant-funded projects).

Bridging these gaps frequently requires even more careful attention to collaborative strategies than the challenges of learning problem framings, methods, and jargons across disciplinary bounds. Here, again, universities need to recognize the challenges of building interdisciplinary programs and be flexible in accommodating creative solutions. The value of bringing research and training from diverse facets of the university into dialogue cannot be underestimated, however, as universities seek to contribute to addressing policy challenges from climate change and sustainable energy to emerging diseases, hunger, and poverty.

References

Crow, M. (2007). None dare call it hubris: the limits to knowledge. *Issues in Science and Technology*, Winter 2007, 1–4.

Leslie, S.W. (1993). *The Cold War and American science: the military-industrial-academic complex at MIT and Stanford.* New York: Columbia University Press.

Nowotny, H. (1991). Knowledge for certainty: poverty, welfare institutions, and the institutionalization of social science. In: P. Wagner (ed.) *Discourses on society: the shaping of the social science disciplines*, pp. 23–41. New York: Springer.

Porter, T. (1995). *Trust in numbers: the pursuit of objectivity in science and public life.* Princeton, NJ: Princeton University Press.

Rueschemeyer, D. and Skocpol, T. (1996). *States, social knowledge, and the origins of modern social policies.* Princeton, NJ: Princeton University Press.

Skocpol, T. (1992). *Protecting soldiers and mothers: the political origins of social policy in the United States.* Cambridge, MA: Harvard University Press.

Wittrock, B. and Wagner, P. (1996). Social science and the building of the early welfare state: toward a comparison of statist and non-statist western societies. In: D. Rueschemeyer and T. Skocpol (eds) *States, social knowledge, and the origins of modern social policies*, pp. 90–113. Princeton, NJ: Princeton University Press.

CHAPTER 24

Administering interdisciplinary programs

BETH A. CASEY

Higher education has entered a transformational period in the twenty-first century. 'This implies both a new social contract for universities and a restructuring to support interdisciplinary collaborations. In the latter half of the nineteenth century governments and institutions of higher learning sought to make higher education more accessible and more directly connected to society with the establishment of land-grant colleges and universities. In the last half of the twentieth century another transformation occurred, as the multiversity emerged in response to increased pressure for research, to resolve complex societal problems, and to educate large and diverse student populations (Scott and Awbrey 1993).

As the multiversity responded to societal demands, interdisciplinary programs, centers, and institutes proliferated and flourished to assist problem-focused research. This twentieth-century transformation in higher education sparked controversy, especially with the seeming alignment of the multiversite with the military–industrial complex and its disciplinary hegemonies (Casey 1990). Such controversy will surely emerge again with the present transformation—one in which interdisciplinary programs, centers, and institutes will support networking both within institutions of higher education and outward to society. As Julie Thompson Klein (1996, p. 1) has noted 'Crossing boundaries is a defining characteristic of our age'.

Today, however, the university is but one element in a suffused knowledge society, and one member of the triple helix of academia, industry, and government (Scott 2000; Becher and Trowler 2001). To continue to have significance it must both increase its ability to respond to the need for the resolution of systemic problems in economies, the environment, and medical practice while meeting the demands of mass education. What is especially stressful is that restructuring research, teaching, and service must also take place in a changing economy in which it is difficult for states to support public education adequately and for private education to acquire funding, except at the most elite levels.

An examination of the newer structures for interdisciplinary research and teaching indicates that a transformation of the university is under way. We are moving, as

Klein and Newell (1996) have noted, from simple to complex systems. Burton Clark has called for each institution to develop 'a steering core, an expanded developmental periphery, a diversified funding base, a stimulated academic heartland, and an integrated entrepreneurial culture' (Clark 1998, pp. 137–45). The steering core, most critics agree, must be a devolvement—a movement of hierarchic authority downward to a flexible, flat, heterarchal level for decision making. Collegiate and school structures will seek decentralization, team focus, and the infrastructures that will enable faculty to cross departmental boundaries. Interdisciplinary centers, programs, and institutes are central to Clark's 'developmental periphery' and amplify the 'diversified funding base'. Interdisciplinary study also occurs within the departments that constitute 'the academic heartland'.

This chapter first examines general policies for the support of interdisciplinary programs, centers, institutes, schools, or colleges devoted to both teaching and research. It then turns to administrative organization, beginning with interdisciplinary programs. Each of these structures has a steering core, a developmental periphery, and an entrepreneurial culture.

24.1 Setting the context: administrative organization and policies

Administering interdisciplinary programs, centers, institutes, or schools is a challenge requiring entrepreneurial leadership, knowledge of the best processes of interdisciplinary scholarship, curricular design, pedagogy, and assessment, as well as the ability to network for collaboration both within and without the university or college. University and college missions should specifically address problem-focused interdisciplinary teaching and research that serves societal mandates within a well-connected curriculum. In turn, interdisciplinary programs, centers, and institutes need a clear mission congruent with institutional goals. A structure must exist to coordinate interdisciplinary programs, centers, and institutes and establish a working context. For programs and centers, the simplest context for support is an interdisciplinary program council chaired by an associate dean of the college who will meet monthly with directors. Directors should also sit on the dean's council with department chairs.

Contracts for directors should be equitable with respect to administrative versus research/teaching time. Since contracts are negotiated individually, disparities may be perceived as injustices. Deans should have an institution-wide policy for a dual appointment of tenure in an interdisciplinary program and department. Normally, genuinely hybrid faculty are most desirous of tenure in an interdisciplinary program as well as a disciplinary department, though institutions vary according to how frequently this practice is invoked. Deans must also have policies in place to reward departments for releasing faculty for participation in interdisciplinary programs as well as rewarding the faculty participants themselves on annual evaluations. Frequently, divisions or centers are created to support and nurture interdisciplinary programs. These may, however, result in the directors finding the interests of their programs inadequately represented or rendered invisible to higher administration.

24.1.1 **Faculty**

Program directors in need of teaching faculty must normally recruit and arrange their participation by negotiating with the chairs of departments to release faculty. Most frequently, departments loan them as part of the faculty member's desired teaching interests, but chairs may request remuneration for replacement of instruction. If faculty 'loaning' is to continue for a number of years, a contract should be arranged with the department chair and the dean for a term or continuing appointment in the program if desired, though tenure most often remains in the department. Policies must be established as well for fair and equitable evaluations of teaching, research, and service in the consideration of merit salary, tenure, and promotion. The Council of Environment Deans and Directors of the National Council for Science and the Environment's website document on guidance for hiring, tenure and promotion is an excellent guide to this topic (<http://ncseonline.org/CEDD/cms.cfm?id=2042>) as is the material in *Facilitating interdisciplinary research* (National Academy of Sciences 2004). The reports of program directors on the evaluations of teaching, research, and service for probationary faculty must be included in institutional portfolios for committees on tenure and promotion. As we will see in case histories, programs sometimes become departments in order to offer tenure tracks to all new faculty when enrollments become very large in both graduate and undergraduate programs.

24.1.2 **Committee structure**

Interdisciplinary programs, centers, and institutes most commonly have both an executive committee for central administrative matters composed of participating faculty and a sizable advisory committee of relevant chairs and faculty across the college and the university. The associate dean or associate provost in charge of programs may sit on the executive committee as well. Creative planning for college- and campus-wide initiatives, such as lectures, conferences, and symposia; for civic or business initiatives; or for external funding can take place through discussions with members of the advisory committee. Committed members of that committee can often actively assist in initiatives that help make the unit visible on campus. In larger programs a curriculum committee should be formed to discuss the assessments of core courses in the program and the design of the interdisciplinary curriculum. In small programs the executive committee in consultation with all faculty may fulfill that function.

24.1.3 **Students**

The director's duties include organized planning for recruitment; the development of brochures that explain the nature and purpose of the program as well as vocational pathways, employment opportunities, or graduate school placement; and the development of internships where relevant. Advising is of central importance and should be a strength of the program. It can be distributed among faculty, though large programs should have one or more staff advisors. Continuing contact should be made with graduates of the program. In addition, the director should plan with faculty to conduct productive learning assessments of the program and course learning outcomes on a regular basis and discuss the results of these with the curriculum and executive committees. Interdisciplinary units normally undergo external review every 5 years.

24.1.4 Collaboration

The community for any interdisciplinary unit is both external and internal. The director's leadership is entrepreneurial and developmental in this respect and of central importance. The unit must connect to relevant faculty both on campus and community, scientific, or business institutions to seek out opportunities for service, for creative changes in mission and program direction, as well as for the formation of flexible research teams or more informal networks of interdisciplinary communication and connection. Normally this also involves fundraising for unit-sponsored symposia, conferences, lectures, or film series that make the program or center visible and invite broad participation by students and faculty. A large part of the director's time may go to seeking external funding to support research, internal events, or external service initiatives.

24.2 Interdisciplinary programs

In the 1960s and 1970s, when the prevailing structures of knowledge could not supply new societal needs, programs such as environmental studies, urban studies, area studies, women's studies, or studies in technology and society emerged. In the early decades, programs sprang up without inclusion in the institutional mission or adequate administrative connections or structure. In 1978 such programs were still perceived by departments and colleges as 'marginal' and appeared to float on administrative charts (Eckhardt 1978). Often that 'floating' was an actuality and not an appearance. In the past three decades, however, growth in student enrollments in most of the interdisciplinary fields cited above as well as growth in interdisciplinary understanding both within and without disciplines have produced expansion.

Some of the expansion of programs has meant moving conservatively from program status to departmental status, as part of seeking after tenured faculty and organizational security in spite of the dangers of closing off collaboration that that might entail. However, innovative schools embracing both programs and departments for better research and instructional collaboration among faculty have also emerged, and some successful structures might well be emulated. The latter schools may foster the continued existence of interdisciplinary programs and even small departments in times of economic downturn. Three examples of these changes and structures will suffice.

24.2.1 Women's studies

Courses in women's studies began in the late 1960s; by 1990 such courses were estimated to enroll between 30,000 and 50,000 students across 630 programs (Messer-Davidow 2002). Women's studies has advanced with a constant refocusing and broadening of mission. In the 1970s urgent political needs often seemed to be more pressing than the evolution of feminist theory and cultural critique. At present many women's studies programs characterize their mission as one focused both nationally and globally on issues that disproportionately affect women in all cultures, such as poverty, violence, and political disenfranchisement. A well-focused mission congruent with institutional goals is essential for interdisciplinary programs in a time of financial duress. Moreover, interdisciplinary programs appearing to be disconnected or isolated from departments and colleges are vulnerable for elimination.

On the other hand, growth can present new requirements for restructuring even in what may be regarded as a conservative direction. Growth in enrollment has encouraged a move from program status to that of an interdisciplinary department at Minnesota and Michigan. The University of Minnesota's Department of Gender, Women and Sexuality Studies has both a large graduate and undergraduate program. The graduate program alone involves faculty from 28 different programs, departments, colleges, and institutes. The University of Michigan's program in women's studies became a department in 2008. Michigan's women's studies program had 30 budgeted faculty and 40 unbudgeted as well as four joint doctoral programs with other departments.

24.2.2 Environmental studies

Environmental programs first arose in the late 1960s with the rise of ecology and of environmental awareness. Most of these were coordinated multidisciplinary programs focused on 'environment' with ecology as a locus for integration (Klein 1996, pp. 96–8). Environmental studies and, indeed, other interdisciplinary programs as well, can be placed in structures which foster dialogue with other relevant disciplines, giving rise to interdisciplinary graduate programs and an infrastructure of community networks, meetings, and opportunities for collaborative research.

Stanford University's School of Earth Sciences is composed of four departments: geophysics, geological and environmental sciences, energy resources engineering, and environmental earth system science; and three interdisciplinary programs: earth systems, graduate program in earth, energy, and environmental sciences, and the Emmett interdisciplinary program in environment and resources (<http://pangea.stanford.edu/about/>). The primary goal of the School of Earth Sciences is to integrate, synthesize, and apply scientific and engineering knowledge to societal problems. For example the graduate program in earth, energy, and environmental sciences has the goal of complementing the disciplinary departments of earth science and engineering by training graduate students to integrate knowledge from these disciplines through tools and methods needed to evaluate the linkages among physical, chemical, and biological systems of the earth. The responsibility to share understandings of the earth with the greater community is recognized by offering programs for students aged 5–17, their teachers, and the general public. The Earth Science Advisory Board provides perspectives for industry, government, and other academic institutions, and faculty work collaboratively with corporations in a formal program of industrial affiliates.

Many colleges and universities do not have the resources to emulate Stanford's example, but the environmental studies program in the College of Arts and Sciences at Bowling Green State University in Ohio provides a more typical case. In 2007 the environmental studies program joined with the program in environmental health and the departments of geology and geography to form a School of Earth, Environment, and Society. Though more modest in scale, the school has increased opportunities for external funding by this collaboration. The department and programs in the school have worked together for many years, with the department faculty often on split time appointments in environmental studies. The mission of the school is to strengthen interdisciplinary and disciplinary approaches to issues dominant in the twenty-first century such as sustainability, human health issues, non-renewable resources, population growth, or global climate change which require cross-disciplinary collaboration. Environmental studies and environmental health merged into one large department and, hence, may now tenure faculty. Significantly, at least one tenure track line has been created which will be in the school itself across the departments, and it is assumed that other lines will follow. Two interdisciplinary programs have emerged thus far in the school to address interdisciplinary issues: geospatial science and environmental quality. At least one common course has emerged for all undergraduate majors as well.

This structure has been of sufficient success at Bowling Green to encourage discussions on the formation of a similarly constructed School for Cultural Studies that would include the American culture studies and women's studies programs and the ethnic studies and popular culture departments. These interdisciplinary units have long worked closely together in an informal network and are strong candidates for integration. Such schools can also protect faculty teaching in small interdisciplinary programs and departments that may not survive in times of economic downturn. Such schools can catalyze opportunities for more entrepreneurial collaborations and a better external funding base than might emerge from solitary 'silo' status.

24.3 **Centers and institutes**

The growth of interdisciplinary research and teaching has caused centers and institutes to increase on campuses at a more rapid rate than programs. Over the past four decades they have fostered new areas of research; enabled institutions to recruit and retain important research faculty; increased and strengthened funding possibilities from a range of sources; served business, community, state, and national needs for resolving complex problems; and above all fostered and supported collaboration among faculty in departments and programs across the college and university (Ikenberry and Friedman 1972; National Academy of Sciences 2004). Centers abound for social philosophy and policy, medical ethics, neuroscience, material science, biotechnology, public policy, and hundreds of other subjects. Significantly, institutes are playing new roles in helping centers collaborate with one another, fostering cross-disciplinary conversations, assisting with grant preparation, and fulfilling the institutional mandates to serve societal needs. Collaboration among centers is more productive than the simple proliferation of centers.

Centers and institutes are also under pressure to fulfill needs for what has been termed Mode 2 research (Gibbons 1995) to resolve society's most complex issues when appropriate. Mode 2 research involves practitioners, is most often interdisciplinary and transdisciplinary in nature, is transient in character, and is carried out in a context of application. An extensive amount of such research takes place in university centers with close ties to industry. Over a thousand of these existed as early as 1990 (Brooks 1994), and as many as 12,000 faculty participated in them in that year (Mode 1 research, according to this paradigm, is knowledge production based on disciplines). Mode 2 research is continuously negotiated with the resolutions to problems beyond any single discipline.

What is needed to support Mode 2 research in centers and institutes is contextual planning rather than strategic or long-range planning. For example, policies need to be developed by the Office of Sponsored Research for agreements between corporate universities and industrial and commercial organizations. Such agreements should be consistent with the university's mission and with the need to maintain a balance between the pursuit of research as an integral part of education and industry's need for applied knowledge. Policies must exist for procedures such as the center's authority to make contracts, to develop publications or patent policies, for liability and risk, and the shared general administration of projects.

The examples which follow present the administrative structure of one dynamic, collaborative and entrepreneurial center in the physical sciences; one institute in the social sciences; and one university center for the humanities, providing new kinds of assistance, experiences, and infrastructure for the support of interdisciplinary research and teaching based in several centers.

24.3.1 The Center for the Study of Complex Systems (CSCS)

The study of complex systems, one of the newest areas of interdisciplinary research, is based on the recognition that different kinds of complex adaptive systems have a common underlying structure despite apparent differences. Methods of analysis can therefore be transferred from one field to another. Ecology, economics, immunology, physics, mathematics and public policy, cognitive science, political science, biology, and sociology are among the disciplines where the study of complex systems is important.

The CSCS at the University of Michigan (UM), Ann Arbor is representative of the way in which centers in general can encourage and catalyze research, expand educational opportunities, and form a university-wide community of researchers and students. Fifty faculty members currently participate in the center, representing nearly every college in the university. More than half are considered primary faculty. These take an active role in grant proposals, research groups, administration, and teaching. They teach a complex systems course, do research on complex systems, and help administer the center, creating a democratic or 'devolved' administrative group. A smaller group called the Bach Group represents a kind of 'steering core'. (The original planning group for the center called themselves the Bach Group based on the first initials of their names, hence a historical connection is maintained to provide foundation for a new interdisciplinary area.) Tenured split and dual appointments are utilized for the Primary Group. Associated faculty whose research includes complex systems attend weekly CSCS seminars and events as desired and constitute a 'developed periphery'.

The goal of the CSCS is to catalyze research on complex systems at the university and in the regional area through weekly seminars, conferences co-hosted with other research groups on campus, regular workshops, an annual Nobel symposium, connections to units such as the Ford Research Laboratories and Argonne National Laboratory, and the acquisition of external funds through grants and corporate and private donations. A community of complex systems researchers and students at UM and involved practitioners throughout southeast Michigan is thus formed. The center supports numerous research groups and projects, publications, and technical reports and has received strong external support.

24.3.2 Social Science Research Institute (SSRI)

The mission statement of Duke University includes two significant goals: to increase the capacity of faculty to develop and communicate disciplinary and interdisciplinary knowledge through a faculty enhancement initiative, and to strengthen the engagement of the university in real-world issues by continuing institutional commitment to flagship interdisciplinary programs and advancing new initiatives that build on distinctive university strengths. This focus supports the work of the university-wide SSRI, which fosters the creation and dissemination of new interdisciplinary knowledge in the social sciences by serving as a gathering point for collaboration.

Specific academic partners include numerous centers such as: child and family policy; globalization, governance, and competitiveness; social demography and ethnography, population research; race, ethnicity and gender. Programs and labs support faculty and graduate student research projects from both centers and departments. The mission is to contribute to knowledge in the service of society by connecting research and theory to policy and practice. A highly interactive scholarly community is fostered among centers and programs where basic researchers can work collaboratively with more problem-oriented researchers. SSRI has an educational core offering workshops and seminars as well as a data and statistics core to support research. It also provides administrative support for externally funded research. The faculty fellows program identifies specific topics, selects faculty participants, and provides release time for collaborative investigation. Fellowships exist for graduate students as well.

24.3.3 Center for Social Justice Research, Teaching, and Service

Georgetown University in Washington, DC has a founding mission for education in the service of justice and the common good. The Center for Social Justice Research was created as a concrete manifestation of that university-wide commitment. It was initiated in 2001 to promote and integrate community-based research, teaching, and service by collaborating with diverse partners and communities in the District of Columbia. The center involves students in several large community service programs from local schools and helps faculty to develop both interdisciplinary and disciplinary courses across centers and departments in the university to incorporate community-based work and service to justice.

The center advances this work through faculty workshops, course development grants, and continued support of conferences that enable faculty to learn the pedagogy of service learning, design courses to incorporate it, and link theory to practice. The center trains college students to mentor and tutor in schools throughout the city, supports a large service learning credit program, provides job development training, and serves as a base for urban research combined with service learning. The center also supports the program on justice and peace, an interdisciplinary unit offering an undergraduate minor in the emerging area of peace studies with special emphasis on developing practical solutions to problems of social inequality and injustice. The office of research in the center supports the collaboration of teachers, students, and community members and validates multiple sources of knowledge and multiple methods of discovery and dissemination of the knowledge produced.

24.4 Interdisciplinary schools and colleges

Interdisciplinary schools and colleges, many of which began in the 1960s and 1970s, have been described as 'telic reforms' (Grant and Reisman 1978), meaning that the founding faculty and administrators attempted to change undergraduate education to embody a distinctively different set of ends or purposes. Many of these did not survive unless they had very experienced leadership. As has been noted above, congruence and fit with the host university or college is important for connection, support, and collaboration. It is also essential, as Trow (1998) also noted, that the founding administrators and faculty of interdisciplinary colleges or schools see themselves as adding significant innovations to academic culture rather than producing 'a counterculture'.

24.4.1 The Evergreen State College

Located in Olympia, WA, Evergreen State College is a public college enrolling 4300 students. Founded in 1971 with a mission to serve as a non-traditional institution, evergreen maintains a special relationship with state government, and provides service to southwestern Washington State (Smith 2001). Early development was fortunate in that a core of 18 planning faculty and administrators, most of whom had experience with earlier interdisciplinary alternative institutions, had a fully funded planning year to design the college. The faculty brought with them progressive ideas such as narrative evaluations, team teaching, and collaborative and community-based learning and internships. The college's first president was committed to placing responsibility in the hands of faculty and students to a degree almost unknown in the annals of administration. Departments were avoided to ensure faculty would join together to create coordinated studies. The first central committee on governance was termed 'the Disappearing Task Force', a name that continued to be used for primary policy recommending committees that are dissolved after they complete their tasks. More radically, tenure and ranking were rejected to reduce hierarchy. A uniform salary scale was created based on years of experience, which has never proved a problem in recruiting good faculty. Retention of faculty is based on evaluations by students and by the faculty with whom they have worked.

The founding deans at Evergreen agreed to a team-taught interdisciplinary theme-based curriculum in which students and faculty would work in year-long programs rather than discrete courses creating a practice. Students still enroll in a single comprehensive 'program' rather than a series of separate courses with the curriculum being renewed each year. Programs vary from 'animal behavior' to 'Greece and Italy' or 'the extraordinary science of everyday life'. Faculty work in teams of two to four in each program and plan labs, seminars, and field trips. In addition, the college acted dynamically on the mission to serve the community by establishing four public service centers to support innovation and collaboration with key communities, including labor, Native American tribes, the K-12 system, and the rest of higher education. With support from Exxon Foundation and the Ford Foundation, Evergreen also established the Washington Center for Improving the Quality of Undergraduate Education, a statewide public service initiative to share Evergreen's experience. The latter has been extraordinarily successful and played a major role in creating the Learning Community movement. As Barbara Leigh Smith has noted, the Washington Center has shown that many aspects of the Evergreen experience are transferable, and that learning communities and collaborative learning are successful in diverse institutional settings (Smith 2001, p. 79).

24.4.2 Hutchins School of Liberal Studies

Interdisciplinary colleges are formed in times when the need to experiment is powerfully felt and deemed essential and when funding for such institutional developments is available. Historically and economically this is rare, though a college of 'sustainability' is currently being developed as part of the State University of New York at Stonybrook. However, interdisciplinary schools that offer alternative programs in general education or a Bachelor of Arts in liberal studies or integrative studies often endure even in economic downturns. These bachelor's degrees may often involve individualized majors, much like the standard liberal studies degrees that exist in most universities and are often mandated by the state either to serve non-traditional students or simply encourage college enrollment.

The Hutchins School of Liberal Studies at Sonoma State University, California was founded in 1969 by a group of faculty and administrators with experience in innovative programs much like the faculty at Evergreen. Hutchins offers a lower-division general education program consisting of four semester-long thematically organized, integrative courses. As at Evergreen, faculty working collaboratively design and discuss the curriculum to be offered each year. The administrative structure is heterarchal, with multiple and overlapping patterns of relationship. Seminars are small and experiential learning is common. As at most schools of this design, students may take elective courses and prerequisites for future majors in departments while enrolled in Hutchins. Upper-division students, however, may choose to remain in the school and take a Bachelor of Arts in liberal studies offering three tracks to the degree. Community outreach is deemed important, and four public service programs have been created. Recently the school initiated the Hutchins Institute for Public Policy Studies and Community Action, which offers a masters program and promotes discussions about environmental and socio-economic issues on and off campus.

24.5 Interdisciplinary general education

At present, interdisciplinary, integrative, or multidisciplinary education plays an extensive role in the general education curriculum of most colleges and universities. Directors of interdisciplinary programs have realized that general education courses can become a part of their offerings if they can recruit the faculty to teach them. Such general education courses bolster enrollment in a program and may gain increased funding for faculty participation. Furthermore, requirements for multicultural education and international education, often interdisciplinary in nature, are now common to most institutions. Environmental studies and women's studies have experienced significant increases in enrollments through general education.

In acknowledgment of the benefits to be derived from interdisciplinary problem-focused study, some universities have created an upper level interdisciplinary required curriculum or even implemented it as an option if faculty resources were not available for a requirement. Liberal arts colleges often create one or more required interdisciplinary courses for the freshman curriculum and a capstone for seniors, sometimes thus providing a 'signature' curriculum or one unique to the mission of the college. Interdisciplinary general education can exist in both large and small institutions, but all upper and lower implementations require an administrative structure still lacking in most institutions.

In large universities the development and coordination of any general education program—interdisciplinary or disciplinary—requires the establishment of a university-wide office for general education with a director or coordinator. In larger institutions an undergraduate studies dean might be appointed to coordinate advising and supervise other academic needs of undergraduate students. The goal of the office is to ensure that interdisciplinary general education development is continuous and not an activity that takes place periodically after a 10-year review.

A general education committee in large universities should involve representation from all colleges and participating departments. It may be elected and be a part of the faculty senate committee structure, or it may be appointed. It is often good to begin with an appointed committee of faculty experienced in general education. Small development committees can also be organized to focus on interdisciplinary development or specific curricular areas of the program, and thus distribute program leadership. Informal luncheon groups can be organized as open forums at which pedagogical strategies can be shared by faculty and graduate students. Plans to develop pedagogy for an interdisciplinary program and to work collaboratively with faculty on curriculum development as well as long-term assessments of it must also be in place. Not only will such plans lead to the enhancement of student learning, but they can also create a community among faculty who are otherwise without cross-departmental academic engagement.

Let us look administratively at two universities whose institutional size might be thought to preclude such possibilities, at a representative liberal arts college, and at the learning communities now being developed on many campuses.

24.5.1 Portland State University

In administration, ironically, adversity often leads to innovation and good fortune. When Portland State University, Oregon, suffered budgetary shortfalls and declining retention rates, administrators and faculty decided to design a distinctive interdisciplinary 4-year general education program that would benefit not only students and faculty but the community as well. Numerous obstacles had to be surmounted. An urban university, Portland had to establish a clear identity and a sense of involvement for its many part-time students. It also had to deal with large numbers of transfer students, often a formidable barrier to general education development in large institutions. Portland's administrators found funding by reducing administrative support staff by 13 per cent and middle management by a third to generate savings of more than $35 million. In addition, tenure and promotion guidelines for faculty were changed to support the new program, and faculty replacements in departments were linked to program participation—a necessity for success.

A director and a Center for Academic Excellence were created to support faculty teamwork in the development of interdisciplinary curriculum design, pedagogies for implementation, and a portfolio-based program assessment that would create a culture of evidence for learning. The three-tiered curriculum concludes with a required senior capstone designed to build cooperative learning communities, taking the students out of the classroom to apply learning from both the major and general education to issues and problems in the city itself. Retention at Portland rose to 80 per cent for first-year students after implementation and is expected to continue to grow. Applications increased by 40 per cent. In 2008 the Portland administration committed to 25 new tenure track lines to continue to maintain and develop the interdisciplinary curriculum.

24.5.2 Michigan State University

Michigan State University formerly had three cluster colleges in the social sciences, natural sciences, and the humanities and arts, which eventually closed. However, physical spaces existed as well as a significant cultural memory. This inspired faculty and administrators to create a distinctive integrative general education program organized around three centers for integrative studies. In each of the centers a director aided by an advisory committee of faculty and students appointed by the appropriate dean is charged with the responsibility of soliciting new courses from college faculty members. The faculty from two or more departments collaborate in the construction and delivery of courses, some of which have large enrollments. Arts and humanities, for example, offer a single multimedia course called 'America and the world' which is required of all students. Many courses strive to introduce students to such cross-disciplinary subjects as global diversity, world urban systems, or social differentiation and equality. More complex plans for upper-division courses were abandoned, which often happens eventually in the interest of easier and more effective management.

24.5.3 St Lawrence University

In 1986, St Lawrence University, a small college of about 1200 in Canton, NY, piloted an interdisciplinary first-year program in general education consisting of a residentially based team-taught course in the fall and a single instructor, research-skills-oriented, first-year seminar in the spring. The founders also explored the possibility of providing more of an international base to the rest of the general education program as whole or even an alternative general education curriculum. Planning began with an interdisciplinary faculty reading group entitled 'cultural encounters'. Describing their experience, the founders have stated that the first-year program seemed to spawn an alternative faculty culture, which then became a resistant one, and finally an 'emergent one' (Cornwell and Stoddard 2001). The structural outcome, they report, has been a substantially transformed faculty culture engaged in sustained and critical thinking about teaching (Cornwell and Stoddard 2001, p. 164). St Lawrence now offers a number of in-service workshops, seminars on teaching, and training sessions for faculty entering the program. To assure progress, language about the program is included in all faculty job advertisements, and some departmental positions are tied to participation. Participation also has a stronger role in mid-probationary and tenure reviews.

24.5.4 **Learning communities**

Residence halls have become the site of learning communities for entering freshmen focused along broad lines of interest such as the humanities, science, international studies, or integrated arts. Spurred by a seminal book entitled *Learning communities: creating connections among students, faculty, and disciplines* (Gabelnick *et al.* 1990), the movement has met with success. Residence-based learning communities for one or two interdisciplinary general education courses are now the place where most faculty encounter the pleasures and intellectual rewards of team teaching and cross-disciplinary communication. The movement itself has fostered an overarching principle of cooperation, enhanced collegiality and intellectual community, and creating a new reciprocity between academic and student life. Experiential and service learning has also become an expected part of these endeavors. Learning communities have had an impact on retention in most cases and a positive impact on budgets as well.

However, learning communities are encountering the same problems that interdisciplinary programs once had on most campuses. Faculty are given released time to head a community, but it is increasingly difficult to recruit faculty, and there usually is no central coordinating office. This was not dealt with early in the movement nor thought to be a problem. A number of possible resolutions would assist with this, including placing the programs under the Office of General Education, creating a university college to coordinate advising and the community-based programs, or assigning the units to a vice provost for academic affairs or an associate dean with time to resolve the numerous administrative problems that the communities encounter.

24.6 **Conclusion**

This examination of interdisciplinary programs, centers, institutes, and schools or general education suggests some useful administrative principles. First, the goals, structure, and praxis of an interdisciplinary curriculum must be developed as democratically as possible by administrators and faculty working in collaborative groups and continually renewed. Second, a pedagogical philosophy and strategies for its implementation must provide a foundation for the delivery of instruction, and new faculty must be carefully socialized to the educational practices of the institution. Third, innovation in any interdisciplinary institution, large or small, requires decentralized decision making in order to increase the desired commitment and ownership. The latter is essential and requires entrepreneurial thinking. Leaders of interdisciplinary colleges and schools need the ability to create structures for work in collaborative groups to manage the problems associated with the very team work by which instruction is delivered. Hence the institutions model the 'devolvement' universities at large are seeking. Fourth, problem-focused, experientially based interdisciplinary education produces graduates who have a sense of social responsibility and the ability to participate in a democratic society. These simple values in a market society assist universities and colleges to move toward placing all education—disciplinary and interdisciplinary—in a meaningful context, and one that should fulfill pressing societal needs. Fifth and finally, interdisciplinary research and teaching require networking both within the university or college and outward to society for the resolution of societal issues and systemic economic, social, political, or environmental problems where needed.

The discussions above also make manifest some of the structural transformations occurring within interdisciplinary programs, centers, institutes, schools, colleges, and general education and demonstrate a movement from simple to complex systems. Significantly, many of the administrative arrangements for interdisciplinary learning suggest some of the more holistic transformations that universities and colleges must create to foster faculty empowerment and better opportunities for collaboration with industry and government. It is possible that departments and programs might be organized around research issues of common interest to them in cooperative boundary-crossing schools such as the school of earth, environment, and sustainability at Bowling Green noted above. The administrative lessons of such institutions as the Evergreen State College should be studied to develop ideas about 'devolvement' or the movement of decision making downward to include faculty. Center and institutes in particular demonstrate the power to create 'a diversified funding base' and extend to a developed periphery. The extensive outreach for research and service that interdisciplinary endeavors can foster will contribute abundantly to new societal expectations. Lastly, the restoration of an ethically based community for liberal education can be seen in the movements toward interdisciplinary general education. Examples exist in sufficient number to suggest that it is indeed possible to construct an entrepreneurial culture in the university with a developed periphery and a differential funding base ready to fulfill a new social contract and better serve students.

References

Becher,T. and Trowler, P.R. (2001). *Academic tribes and territories. Intellectual enquiry and the culture of disciplines*, 2nd edn. Buckingham: Open University Press.

Brooks, H. (1994). Current criticisms of research universities. In: J.R. Cole, E.G. Barber, and S.R. Graubard (eds) *The research university in a time of discontent.* Baltimore, MD: Johns Hopkins University Press.

Casey, B.A. (1990). The administration of interdisciplinary programs: creating climates for change. *Issues in Interdisciplinary Studies* **8**, 87–110.

Casey, B.A. (1994). The administration and governance of interdisciplinary programs. In: J.T. Klein and W.G. Doty (eds) *Interdisciplinary studies today*, pp. 53–68. San Francisco: Jossey-Bass.

Casey, B.A. (2003). Developing and administering interdisciplinary programs. In: J.T. Klein (ed.) *Interdisciplinary education in K-12 and college*, pp. 159–75. New York: The College Board.

Clark, B.R. (1998). *Creating entrepreneurial universities: organizational pathways of transformation.* London: Pergamon.

Cornwell, G.H. and Stoddard, E.W. (2001). Toward an interdisciplinary epistemology: faculty culture and institutional change. In: B.L. Smith and J. McCann (eds) *Reinventing ourselves: interdisciplinary education, collaborative learning, and experimentation in higher education,* pp. 160–78. Bolton, MA: Anker.

Eckhardt, C.D. (1978). *Interdisciplinary programs and administrative structures: problems and prospects for the1980s.* University Park, PA: Pennsylvania State University Center for the Study of Higher Education.

Gabelnick, E. *et al.* (1990). *Learning communities: creating connections among students, faculty, and disciplines.* San Francisco: Jossey-Bass.

Gibbons, M. (1995). The university as an instrument for the development of science and basic research: the implications of Mode 2 science. In: D.D. Dill and B. Spoon (eds) *Emerging patterns of social demand and university reform: through a glass darkly*, pp. 90–104. London: Pergamon.

Grant, G. and Riesman, D. (1978). *The perpetual dream: reform and experiment in the American college.* Chicago: University of Chicago Press.

Ikenberry, S.O. and Friedman, R.C. (1972). *Beyond academic departments: the story of institutes and centers.* London: Jossey-Bass.

Klein, J.T. (1996). *Crossing boundaries: knowledge, disciplinarities, and interdisciplinarities.* Charlottesville, VA: University Press of Virginia.

Klein, J.T. (1999). *Mapping interdisciplinary studies: the academy in transition.* Washington, DC: Association of American Colleges and Universities.

Klein, J.T. and Newell, W.H. (1996). Advancing interdisciplinary studies. In: *Handbook of the undergraduate curriculum*, pp. 393–415. San Francisco: Jossey-Bass.

Messer-Davidow, E. (2002). *Disciplining feminism: from social activism to academic discourse.* Durham, NC: Duke University Press.

National Academy of Sciences (2004). *Facilitating interdisciplinary research.* Washington, DC: National Academy Press.

National Council for Science and the Environment (NCSE), Council of Environment Deans and Directors. *Interdisciplinary hiring, tenure and promotion: guidance for individuals and institutions.* On-line resource, available at: <http://ncseonline.org/CEDD/cms.cfm?id=2042>.

Scott, D.K. and Awbrey, S.M. (1993). Transforming scholarship. *Change Magazine* **25**(4), 38–43.

Scott, P. (2000). A tale of three revolutions? Science society and the university. In: P. Scott (ed.) *Higher education reformed*, pp. 190–206. London: Falmer Press.

Smith, B.L. (2001). Evergreen at twenty five: sustaining long-term innovation. In: B.L. Smith and J. McCann (eds) *Reinventing ourselves: interdisciplinary education, collaborative learning, and experimentation in higher education.* Bolton, MA: Anker.

Trow, M. (1998). interdisciplinary studies as a counterculture: problems of birth, growth and survival. In: W.H. Newell (ed.) *Interdisciplinarity: essays from the literature*, pp. 181–94. New York: College Entrance Examination Board.

Wilson, A. (2000). Strategy and management for university development. In: P. Scott (ed.) *Higher education reformed*, pp. 29–44. London: Falmer Press.

CHAPTER 25

Undergraduate general education

WILLIAM H. NEWELL

Across the twentieth century, undergraduate general education in the United States became increasingly identified with the development of core curricula general education and interdisciplinarity. This chapter provides a broad overview of the evolving role of interdisciplinary studies in undergraduate education in the United States, especially its relationship with disciplinarity and with various pedagogical innovations, and concludes with a brief survey of interdisciplinary general education in other countries.

25.1 Historical background

Seen in the broad sweep of Western civilization, interdisciplinarity is the latest response to the dominant intellectual tradition of rationality and reductionism that is ultimately grounded in dichotomous thinking. Unlike earlier responses, such as romanticism, that sought to replace reason with affect and reductionism with holism, interdisciplinarity takes a both/and approach, embracing reductionism as well as holism, and the dichotomies and systemic thinking lying behind them, respectively, as it draws on disciplinary perspectives and integrates their insights. Instead of rejecting the increasingly narrow specialization bred by reductionism, interdisciplinarity embraces salient specialties while transcending them as it constructs a more comprehensive understanding from their insights. As such, interdisciplinarity can be understood as an attempt to right the balance of Western thought.

The choice of interdisciplinarity as the response to the continuing disciplinary hegemony of the twentieth century reflects the distinctive nature of the intellectual and pragmatic challenges currently confronting humans as individuals, as societies, and as a species. It is now commonplace to observe that contemporary societal problems have become increasingly complex, and to the extent that those problems reflect globalization of transportation, communication, and markets the claim is undoubtedly accurate. But to some extent complexity has always been with us; it was merely obscured by more obvious simple or complicated problems (e.g., the eradication of some pervasive infectious diseases or the dampening of business cycles) whose solution drew largely on the insights of a single discipline (e.g., biology or chemistry, and economics). Thus, the complexity facing modern human societies reflects to some extent the inherent limitations of the disciplines. As the most effective means available for addressing complex problems, interdisciplinarity was the obvious response to the burgeoning of academic disciplines taking place at the end of the nineteenth and the beginning of the twentieth centuries.

From the perspective of higher education, interdisciplinarity offers desirable general education outcomes that extend well beyond preparing students to cope with complexity, as important as that is. Without a general education, human beings tend to be somewhat parochial. We are disinclined to think beyond the scope of direct human experience—to factors or forces that operate on different scales of time or space, that function systemically rather than individually, or that have multiple causes; nor are we inclined to see a problem from other perspectives (be they grounded in cultures, religions, or disciplines). Even well educated humans have some difficulty moving back and forth between the general and the specific, theory and application, the abstract and the concrete. Interdisciplinary studies provide an approach in which such skills become habits of mind; they fall naturally out of the interdisciplinary process. Indeed, a host of intellectual skills, sensitivities, and sensibilities valued by educators are developed as by-products of interdisciplinarity. Thus, it is not surprising that early experiments with interdisciplinary general education began in the 1930s, only a few decades after the modern ascendancy of disciplinarity.

25.2 Interdisciplinary general education in the United States

Interdisciplinary general education started with pioneering efforts at a few prominent universities in the first half of the twentieth century, became a hallmark of experimental colleges on the radical fringe of higher education in the 1960s, gained legitimacy as part of the liberal mainstream through its embrace by national movements in honors, women's studies, and environmental studies in the 1980s, emerged as a small but normal part of a university education in the 1990s, and achieved the (somewhat dubious) distinction of being the latest academic fad in the first decade of new millennium. As the location and standing of interdisciplinarity evolved within the academy, so did its conception and application. The story of interdisciplinary studies over the last century, especially in American general education, is bound up in the joint evolution of how interdisciplinarity was understood and where and how it was applied.

The first significant effort at interdisciplinary general education was probably the contemporary civilization program at Columbia University in 1919. Its approach to interdisciplinarity 'was not a survey of the subject matter in history, economics, government, and philosophy; instead it applied the inherent perspectives and methods of these disciplines to [help] the student understand his present-day world, so he could more effectively develop values, make judgments, and participate in the world' (Miller 1988, p. 41). Like the program at Columbia, founded by President Nicholas Murray Butler, many of the pioneering experiments in at least quasi-interdisciplinary general education were established by influential education leaders such as John Maynard Hutchins at the University of Chicago ('great books'), Alexander Meiklejohn at the University of Wisconsin ('experimental college'), and John Dewey (the inspiration for the 'general college' at the University of Minnesota, among others). Their educational philosophies differed considerably, though: Hutchins took an elite view of education as a decontextualized conversation among prominent mainstream intellectuals; Meiklejohn experimented on the fringe of higher education to create an academic community that used mental problem-solving skills to extract lessons of living adaptable from one context to another; while Dewey sought a balance between passing on the cultural heritage and critiquing it. Dewey advocated starting with student interests and then drawing them out into larger contexts, where insight into pressing social and political issues could be achieved through discussions among people with different perspectives. Nonetheless these thinkers shared a concern for creating an integrated educational experience that prepared students for modern life.

In the 1960s and very early 1970s, interdisciplinarity became identified with the experimental college movement and radical curricular experiments within more traditional institutions. Many of these experiments were indirectly influenced by Dewey, while others recognized the legacy of Meiklejohn or Hutchins, but they were typified by a 'thin veil' of 'Apollonian consciousness' (Apollo being the god of light) covering 'the whole Dionysiac realm' (Dionysus being identified with rapture). The Apollonian–Dionysian tension played itself out between those who wanted to embrace disciplines and then transcend them, and those who rejected the legitimacy of disciplines; those who sought rigor in interdisciplinarity, and those who saw interdisciplinarity as freedom; and those who strove for intentionality in integration, and those who embraced serendipity (Newell *et al.* 2003). Most interdisciplinary programs dealt with these tensions by presuming but not discussing their interdisciplinarity.

In the early 1980s the National Collegiate Honors Society declared that 'honors' was 'synonymous' with interdisciplinarity, thus linking it with quality and rigor. Women's studies programs asserted that they were interdisciplinary by their very nature, linking interdisciplinarity with fundamental critiques of the academy in general and the disciplines in particular. And environmental studies, seeking to pull together insights from a variety of disciplines into holistic conceptions such as ecosystems, likewise embraced the interdisciplinarity impulse. By the middle of the decade, an examination of interdisciplinary general education programs revealed that 'when faculty wish to revitalize the core of the liberal arts, promote excellence, or fundamentally reexamine orthodoxy, they are turning increasingly to interdisciplinary studies' (Newell 1986, p. vi). While such movements did much to legitimize interdisciplinarity within the liberal mainstream of the academy, they encouraged divergent views about the relationship between the disciplines and interdisciplinarity (are they complementary or antagonistic?) and perpetuated the impression that the nature of interdisciplinarity is self-evident. Professional associations such as INTERSTUDY (in the sciences and research) and the Association for Integrative Studies (see Box, p. 364) did much to clarify the nature and practice of interdisciplinarity, but they were working against a major national trend.

In the 1990s, interdisciplinarity was widely recognized as part of a package of curricular and pedagogical innovations including collaborative learning, multicultural education, learning communities, inquiry- and problem-based learning, writing-across-the-curriculum, civic education, service learning, and study abroad. Thanks in part to the leadership of the Association for American Colleges and Universities, those innovations were widely adopted within mainstream higher education. By late in the decade, several studies revealed that a majority of all colleges and universities in the United States included at least one interdisciplinary experience in their institution-wide general education requirements. These interdisciplinary experiences were typically only at the introductory level, though interdisciplinarity also began to seep into upper-division general education and capstone courses. While those innovations could be argued to implicitly share a kind of generalized interdisciplinary process, what I called 'integrative learning' (Newell 1999), most saw them as loosely interconnected at best; indeed, most faculty members were inclined to focus on one or two of these innovations and leave the rest to others. (Most new doctoral programs established after 1990 were interdisciplinary as well, but that development is outside the mandate of this chapter). Inevitably, as interdisciplinarity became accepted by a wider range of discipline-based faculty members unfamiliar with its origins, the historic roots of interdisciplinarity were lost and the range of conceptions of interdisciplinarity grew wider and even more fuzzy, at the same time that the sometime antagonism between interdisciplinarity and the disciplines was being greatly reduced.

In the first decade of the new millennium, interdisciplinarity became the new 'in thing' in higher education, not just in general education (and disproportionately in the humanities and social sciences) but also in scientific research (a major development extending into governmental funding mandates, but again beyond the scope of this chapter). Any faculty, program, or university wishing to appear cutting edge was likely to claim they were doing interdisciplinary work. One ironic casualty of this popularity were prominent long-standing interdisciplinary programs remaining from the 1960s and 1970s (notably at Appalachian State University, Miami University, and Wayne State University), which came to be seen as obsolete experiments that had out-lived their usefulness.

On the other hand, thanks to the burgeoning of scholarship on interdisciplinarity and the tireless work of prominent and highly networked individuals such as Julie Thompson Klein, considerable clarification of the nature of interdisciplinarity was finally achieved during the first decade of the twenty-first century. A spate of highly visible national reports by prestigious groups as well as path-breaking books or articles by key scholars revealed the details of an emerging consensus first tentatively identified in the mid 1990s (Klein and Newell 1996): an interdisciplinary study has a specific substantive focus that is so broad or complex that it exceeds the scope of a single perspective; interdisciplinarity is characterized by an identifiable process that draws explicitly on disciplines for insights into that substantive focus; those insights must be integrated; and the objective of integration is instrumental and pragmatic—to 'solve a problem, resolve an issue, address a topic, answer a question, explain a phenomenon, or create a new product' (Newell 2007).

While the extent of agreement on interdisciplinarity is refreshing after decades of ignoring or papering over differences in conception, important issues remained:

- Is there a single interdisciplinarity that manifests itself differently depending on the task at hand, or are there different interdisciplinarities, e.g. instrumental and critical (and, if so, what do they have in common)?

- Should interdisciplinarity draw from the perspectives of disciplines, professions, areas of instruction, fields of research practice, approaches from outside the academy, or any or all of the above (and how should the choice be determined)?
- Just how broad or complex must a specific substantive focus be before an interdisciplinary approach is required (and must it be broad at all, or merely complex)?
- What is the interdisciplinary process (or are there many legitimate processes)?
- From what, exactly, do interdisciplinary studies draw their disciplinary insights (information, data, techniques, tools, concepts, theories). How do those insights get integrated, and are there prerequisites for successful integration?

The motivations for the new millennium appeal of interdisciplinarity extend beyond the confluence of factors that led to its mainstream acceptance a decade earlier. Those factors include general education reform, professional training in fields that were becoming increasingly interdisciplinary (e.g., medicine, public administration, social work), real-world problem solving, fundamental epistemological and structural critique of knowledge production in the academy, faculty development, down-sizing by administrators, and the production of new knowledge (Klein and Newell 1996). The twenty-first-century appeal of interdisciplinarity also includes the recognition that a globalizing and thus interdependent world is increasingly characterized by complexity, which, in the absence of other viable alternatives for dealing with complexity, turns public intellectuals (e.g. Thomas Friedman) as well as academicians (e.g., Jared Diamond) towards interdisciplinarity. Societal, and especially global, problems are increasingly systemic, produced by multiple causes and influenced by factors studied separately by a variety of disciplines. Individual disciplines, indeed individual perspectives whatever their source, can illuminate some single aspect of those complex problems, and multiple perspectives can offer alternative partial solutions, but only interdisciplinarity holds out the hope of moving towards full or comprehensive solutions.

The Association for Integrative Studies

William H. Newell

The Association for Integrative Studies (AIS) was founded in 1979 at a conference on the teaching of interdisciplinary social science to create the profession of interdisciplinary studies and serve as its national association. The decision to use 'integrative' rather than 'interdisciplinary' in the name of the organization reflected a belief that the term 'interdisciplinary' was debased by association with too many non-rigorous courses and programs, while 'integrative' highlighted what the founders felt is the key distinguishing characteristic of good interdisciplinarity. Membership today totals 2000. The AIS journal, *Issues in Integrative Studies*, publishes annually.

Membership has been a roughly equal mix of undergraduate faculty and administrators, typically from second-tier institutions, e.g. from the California State University system rather than the University of California system, associated with interdisciplinary studies programs. With the recent provision of student travel grants to conferences and the publication of Repko's (2008) *Interdisciplinary research: process and theory*, graduate students more interested in research than teaching are being attracted as well. While AIS was initially conceived as a national organization, it has become increasingly international in leadership and in those attending conferences; there have been Canadians on the Board of Directors for most of the last decade, and a number of conference participants come from Europe and Australia/New Zealand.

In the first decade, papers at annual conferences and articles in *Issues in Integrative Studies* focused on the nature of interdisciplinarity or on undergraduate interdisciplinary curriculum and pedagogy. Since then, scholarship has broadened to include interdisciplinary research, institutional politics, complex systems, and most recently the details of interdisciplinary theory and process. The AIS has been prominent in sponsoring the publication of a substantial number of books on interdisciplinarity in higher education, publishing directories of undergraduate interdisciplinary programs and now an on-line directory of interdisciplinary doctoral programs, sponsoring the development of assessment tools and strategies for interdisciplinary programs, managing a Listserv, and making available a cadre of consultants and external evaluators on interdisciplinary higher education who have collectively served hundreds of American colleges and universities. Its website also features an electronic job market, a collection of exemplary syllabi and supporting documents, and a North American teleconference on interdisciplinary studies today.

The AIS has worked collaboratively over the years with the Association of American Colleges and Universities, the Association for General and Liberal Studies, the Society for Values in Higher Education, and a variety of other national professional associations, and is an affiliate of the American Association for the Advancement of Science.

25.3 Interdisciplinary general education and pedagogical innovations

It has already been remarked that interdisciplinary general education was embraced by mainstream higher education at about the same time (roughly, the 1990s) that a variety of other pedagogical and curricular innovations were gaining prominence. That synchronicity deserves closer examination than it usually gets.

Multicultural education and educational practices based on multiple intelligences and varied learning styles have in common with interdisciplinary studies the belief that there are legitimate and useful alternative ways of perceiving the world. Teachers must seek out the intellectual strengths, learning styles, and cultural perspectives of their students and find diverse ways of communicating their material that are responsive to the diverse ways students perceive, understand, and learn. They also have in common a predisposition towards a more constructivist epistemology than has characterized much of traditional education.

While these pedagogies share with interdisciplinary studies a belief in the value of diversity, interdisciplinary study moves beyond the mere celebration of diversity to a more critical stance, one that requires attention to the weaknesses or limitations of each perspective as well as its strengths. In their classic article, Cornwell and Stoddard (1994) make the case that we must move beyond multicultural to intercultural education, in which cultural diversity is not just celebrated but confronted. Their essentially interdisciplinary argument applies with equal force to multiple intelligences and learning styles: teachers need to move beyond mere validation, celebration, and utilization of diverse ways of thinking and learning to confront their limitations and conflicts as well as their strengths and complementarities.

Where this interdisciplinary approach becomes especially important is in other pedagogical innovations that came to the forefront in the 1990s, such as collaborative learning, learning communities, and service learning, as well as in study abroad that took on new urgency in response to globalization. These pedagogies all require students to confront differences in perspective not just from disciplines but also from people who think, learn, and perceive differently from them in order to complete a group task or interact productively in a community. Teachers and administrators who recognize and present such 'integrative learning' (Newell 1999) as a generalization of the interdisciplinary process can make these pedagogies mutually reinforcing and thus enhance their educational impact.

Other innovations such as problem-based learning, team teaching, and writing across the curriculum focus on some single aspect of interdisciplinary education. By the beginning of the twenty-first century, there was consensus among interdisciplinarians that interdisciplinary study needs to focus on a problem, question, or issue that is broad or complex. Thus one could argue that interdisciplinary study is (complex) problem based, though the inherent connection between problem-based learning and interdisciplinary studies was largely overlooked. Team teaching, on the other hand, had long been presumed to be desirable if not necessary for interdisciplinary education. Thanks in large part to literature sponsored by the Association for Integrative Studies, interdisciplinary studies was uncoupled from team teaching. While team teaching can serve useful functions in the classroom (e.g., ensuring that contrasting perspectives are ably presented), it has potential disadvantages as well, such as permitting faculty members to function as advocates of their discipline instead of serving as models of the solo interdisciplinarian, not to mention its exorbitant cost (for the relations among a variety of innovative pedagogies, including team teaching, and interdisciplinary studies, see Haynes 2002). Writing across the curriculum, with its focus on the distinctive writing conventions of different disciplines, has an obvious affinity with interdisciplinary studies. Both seek to make disciplinary perspectives visible, but writing across the curriculum has essentially taken a multidisciplinary 'when in Rome...' approach and largely ignored the unique writing challenges of interdisciplinary studies.

Perhaps because they were introduced into the mainstream of higher education around the same time, it has been tempting to think of interdisciplinary studies as just another curricular or even pedagogical innovation. But while interdisciplinarity is complementary to these other innovations, making their joint entrance into the mainstream of higher education quite understandable, it is much more than just another curricular innovation. Indeed, as discussed above, interdisciplinarity represents a fundamental rethinking of the reductionist, dichotomous strategy of knowledge production that characterized mainstream Western thought for nearly four millennia. It represents not the abandonment of that strategy, but its expansion to include holistic, inclusive (both/and) thinking. By embracing disciplines while transcending them, and by pulling together disciplinary insights and applying them to real-world problems, interdisciplinary studies makes general education relevant; more generally, it forms a much-needed bridge between the ivory tower and the rest of society. Looking ahead, interdisciplinary studies offers our best collective hope for harnessing knowledge from the disciplines to confront the looming challenges of a complex, interconnected world.

25.4 The status of interdisciplinary general education

The strengths of interdisciplinary general education at the end of the first decade of the twenty-first century are practical, institutional, and intellectual. It is widely recognized that modern societies require leaders, experts, and citizens who can function effectively if not thrive in a world characterized by globalization, interdependence, and diversity—in short, by complexity—and that academic disciplines alone are insufficient to meet that need. While complex systems thinking is useful in developing general intellectual tools (see Strijbos, Chapter 31 this volume), only interdisciplinary studies can integrate what insights the various disciplines have to offer into the most comprehensive understanding currently possible of any particular complex problem. Major governmental funding agencies in the United States, such as the National Science Foundation and the National Institutes of Health, are earmarking ever-larger programs for interdisciplinary research, while prominent national research organizations such as the Social Science Research Council and the National Academies of Science, Engineering, and Medicine issue reports promoting interdisciplinary research. University administrators are eager to respond to these new opportunities by inventing procedures and institutional structures. While the funding initiatives and national reports are usually aimed at interdisciplinary research, they eventually create a derived demand for general education programs connecting the results of the new interdisciplinary research to the future careers, civic activities, and personal lives of undergraduate students. Finally, interdisciplinary theory has blossomed at the start of the new millennium, as captured in Allen Repko's *Interdisciplinary research: process and theory* (2008), and early formulations, while quite controversial, have made possible the first tentative steps of a complete step-by-step interdisciplinary process.

The weaknesses of interdisciplinary general education are professional and institutional. The professional literature on higher education has never been held in high regard by faculty, and the professional literature on interdisciplinarity suffers from being, at best, the third field for most faculty teaching interdisciplinary general education courses—after their discipline, and then the substantive topic on which they are teaching. Consequently, few faculty members involved in general education will voluntarily seek out the professional literature on interdisciplinary higher education, putting the onus on administrators to provide incentives and lower costs for faculty to encounter that literature. Even when faculty members become aware of the new generation of professional literature on interdisciplinarity, many of them may find the intellectual demands of implementing the full interdisciplinary process excessive, since it requires significant time and intellectual energy devoted to learning about other disciplines as well as developing a whole set of interdisciplinary skills. Most faculty teaching general education courses see their discipline as the primary source of professional advancement and general education as a luxury they permit themselves or as an institutional obligation. Again, administrators will need to find ways to increase the payoff for the faculty teaching university-mandated general education courses.

Finally, higher education institutions are notoriously slow to change. Over a century ago, a president of Harvard University famously quipped that the pace of institutional change is inversely proportionate to the lifetime of the faculty. Intellectually, interdisciplinary integration in particular runs against the grain of Western thought, requiring holism not reductionism, both/and not either/or thinking, and multiplicities and continuous gradations, not dualities. Institutionally, interdisciplinarity similarly subverts the hierarchical, decentralized (if not semi-autonomous) departmental structure by requiring horizontal links and preferably a degree of integration, affecting everything from staffing to budgets to promotion/tenure policies.

A number of professional organizations have a stake in the success of interdisciplinary general education: the Association for Integrative Studies (see Box), the Association for General and Liberal Studies, the American Association of Colleges and Universities (AAC&U), and The Carnegie Foundation for the Advancement of Teaching, to name just a few. These organizations may make the difference in determining if the strengths of interdisciplinary general education programs outweigh their weaknesses. Some, like the AAC&U, have the visibility to focus national attention and the initiatives, conferences, and institutes (many of which are already focused on general education) to facilitate coherent institutional change in member institutions.

25.5 Interdisciplinary general education outside the United States

'General education' is primarily an American term, one that is not in use in the UK, Australia, or New Zealand, for example. Similarly, US-based distinctions among general education (the breadth component of a college or university undergraduate education), liberal education (non-professional, intellectually liberating undergraduate education that can include depth as well as breadth), and interdisciplinary studies (a process of drawing on disciplines and integrating their insights that is equally applicable in general, liberal, and professional education at the graduate as well as undergraduate levels) do not always translate well to the educational systems of other countries. Having said that, some useful connections can still be made between the preceding US-focused discussion of interdisciplinary undergraduate education and emerging educational developments in developed Western countries (especially Canada, Europe, and Australia) as well as in developing nations in the East and Middle East (including portions of the former Soviet Union such as Tajikistan, Kyrgyz Republic, Kazakhstan, and Russia itself; Pakistan; and the United Arab Emirates).

Because general education requirements in Canadian universities are traditionally determined by each faculty, not institution wide, interdisciplinary efforts have been more spotty than in the United States. There are a few long-standing interdisciplinary graduate programs (particularly Green College at the University of British Columbia) and recent institution-wide efforts towards interdisciplinarity in graduate education and research at York University, Toronto, but much less interdisciplinarity at the undergraduate level (perhaps reflecting in part a national resistance to American cultural hegemony). Notable exceptions are the culture division in the Faculty of Communication and Culture at the University of Calgary (adapted from the former general studies), a series of interdisciplinary majors and minors at the University of Alberta, and efforts at York University to move in the direction of interdisciplinarity at the undergraduate level through enhanced team teaching and double major interdisciplinary BA programs.

In European universities, interdisciplinary efforts became focused towards the end of the twentieth century primarily on transdisciplinary research and graduate education, notably at the University of Bielefeld (Germany) and Linköping University (Sweden). Globalization and the knowledge society helped shift the role of higher education from 'an emphasis on social mobility to wealth creation' (Nowotny 1995), contributing to the 'professionalization of general education' (Sporn 2001, p. 15976). With rare exceptions, such as the liberal arts and sciences program at Utrecht University or the Institute for Interdisciplinary Studies at the University of Amsterdam, the call of the Council of Europe for education for democratic citizenship that encourages 'multidisciplinary approaches and actions combining civic and political education with the teaching of history, philosophy, religions, languages, social sciences and all disciplines having a bearing on ethical, political, social, cultural or philosophical aspects' (Council of Europe 2002) has led to few efforts at interdisciplinary liberal, much less general, education.

In Australia, Melbourne University's 'new generation' degrees represents a limited revival of interest in interdisciplinary undergraduate education. The Melbourne model incorporates US-inspired liberal education objectives grounded in the belief that 'the ability to use multiple knowledges, methods, skills and ways of knowing is critical in the global knowledge era and for lifelong learning', and includes interdisciplinary units (university breadth subjects) such as 'Introduction to climate change', which draws on science, history, economics, public policy, and law (Devlin 2008). At Murdoch University, Perth, founded in the 1970s on models such as the Open University and Sussex University in the UK and The Evergreen State College in the United States, the commitment to interdisciplinary is sustained in the curriculum through team-designed cross-university foundation units that aim 'to expose students to different viewpoints and modes of inquiry centered on an integrated and coherent theme'. Examples of arts interdisciplinary foundation subjects at Griffith University, Queensland, include Democracy, From Homer to Hollywood, Knowing Nature, Australian Indigenous Studies, Globalization, Understanding Asia, and Self and Other. This burgeoning interest in broad interdisciplinary undergraduate courses, however, is largely confined to the 'new wave' universities founded in the 1960s and 1970s (Franks *et al.* 2007).

In sharp contrast, a number of nations seeking to democratize or modernize have become acutely aware of the importance of undergraduate liberal arts education, and some of their newest universities make extensive use of US-inspired interdisciplinarity, albeit applied to subject matter reflecting their own cultural legacy. The University of Central Asia, a private international university with campuses in Tajikistan, the Kyrgyz Republic, and Kazakhstan, now offers an interdisciplinary core humanities curriculum and is setting up a School of Undergraduate Studies featuring required foundation courses emphasizing critical thinking and interdisciplinary approaches. Aga Khan University, the only private university in Pakistan, follows a liberal arts model that develops problem-solving skills and prepares students to integrate information from a spectrum of academic fields. Zayed University for women in the United Arab Emirates requires an interdisciplinary core general education curriculum featuring a colloquy on integrated learning. Smolny College, part of St Petersburg University in Russia and in collaboration with Bard College, offers interdivisional programs including interdisciplinary liberal arts courses. These scattered examples demonstrate that interdisciplinary general education can take root in diverse cultures, even those without a history of liberal education.

25.6 The future of interdisciplinary undergraduate education

All indications are that interdisciplinary studies will become increasingly prevalent in coming decades. The economic and technological forces that continue to drive globalization create conditions of increasing complexity and scale that require an interdisciplinary approach to understanding and coping with twenty-first-century life and the myriad of problems it presents. Diverse perspectives grounded not just in academic disciplines but in ethnicity, race, religion, culture, gender, geography, education, class, and so on are brought into ever-closer proximity and are becoming increasingly volatile, creating the need for citizens as well as leaders who know how to create common ground, bridge differences, and construct understandings that are more inclusive as well as more comprehensive. As the effects of globalization spread across the planet, even nations new to general or liberal education seem to be drawn to interdisciplinarity as an organizing approach for reform in higher education.

A second factor that shows promise of undermining disciplinary hegemony and promoting interdisciplinarity is the continuing intellectual shift from modernism to postmodernism. Disciplinary meta-narratives are increasingly forced to compete with a cacophony of contesting perspectives, and many new voices in the academy are finding common causes that cut across disciplinary lines. Even if this trend follows a Hegelian dialectic process, as I expect it will, the synthesis of modern thesis and postmodern antithesis seems likely to embrace interdisciplinarity at least as much as the disciplines.

A third factor driving the academy toward a more prominent role for interdisciplinarity is the rising chorus of voices throughout the rest of society demanding greater accountability from the academy. As the relative cost of higher education continues to increase, state legislatures and public intellectuals alike are insisting that higher education demonstrate its real-world relevance. Since interdisciplinary studies provide the most promising vehicle for the disciplines to demonstrate the applicability of their research and teaching to the problems facing society, it is hard to see how higher education can meet those demands without greater and greater use of interdisciplinarity.

What is now a pipeline effect, with undergraduate students exposed to interdisciplinary ways of thinking becoming graduate students who chafe under disciplinary constraints, seems likely to become a generational effect as newly minted PhDs with more interdisciplinary doctoral training enter the academy and slowly gain positions of more influence within the academy as their careers advance. Thus, these shifts towards interdisciplinarity have a positive feedback loop that will intensify them in the coming decades.

The predictable cumulative effect of these separate trends is the continued burgeoning of interdisciplinarity well into the twenty-first century.

References

Cornwell, G.H. and Stoddard, E.W. (1994). Things fall together: a critique of multicultural curricular reform. *Liberal Education* (Fall), 40–51.

Council of Europe (2002). *Recommendation Rec (2002) 12 of the Committee of Ministers to member states on education for democratic citizenship*. Available at: <http://www.coe.int/t/dg4/education/edc/Documents_Publications/Adopted_texts/092_Rec_2002_12_EDC_en.asp>.

Devlin, M. (2008). *An international and interdisciplinary approach to curriculum: the Melbourne model*. Keynote address at the U21 conference in Glasgow, Scotland, 21–22 February. Available at: <http://www.universitas21.com/TandL/Papers/DevlinKeynote.pdf>.

Franks, D. *et al.* (2007). Interdisciplinary foundations: reflecting on interdisciplinarity and three decades of teaching and research at Griffith University, Australia. *Studies in Higher Education* **32**(2), 167–85.

Haynes, C. (ed.) (2002). *Innovations in interdisciplinary teaching*. Westport, CT: Oryx Press.

Klein, J.T. and Newell, W.H. (1996). Advancing interdisciplinary studies. In: J. Gaff and J. Ratcliff (eds) *Handbook of the undergraduate curriculum*, pp. 393–415. San Francisco: Jossey-Bass.

Miller, G.E. (1988). *The Meaning of general education: the emergence of a curriculum paradigm*. New York: Teachers College Press.

Newell, W.H. (1986). *Interdisciplinary undergraduate programs: a directory*. Oxford, OH: Association for Integrative Studies.

Newell, W.H. (1999). The promise of integrative learning. *About Campus* **4**(2), 17–23.

Newell, W.H. (2007). Six arguments for agreeing on a definition of interdisciplinary studies. *AIS Newsletter* **29**(4), 1–4.

Newell, W.H. *et al.* (2003). Apollo meets Dionysius: interdisciplinarity in long-standing interdisciplinary programs. *Issues in Integrative Studies* **21**, 9–42.

Nowotny, H. (1995). Mass higher education and social mobility: a tenuous link. In: D.D. Dill and B. Sporn (eds) *Emerging patterns of social demand and university reform: through a glass darkly*, pp. 72–89. Oxford: Pergamon.

Repko, A.F. (2008). *Interdisciplinary research: process and theory*. Thousand Oaks, CA: Sage Publications.

Sporn, B. (2001). Universities and science and technology: Europe. *International encyclopedia of the social and behavioral sciences*, pp. 15974–8. Amsterdam: Elsevier Science.

CHAPTER 26

Interdisciplinary pedagogies in higher education

DEBORAH DeZURE

Interdisciplinary teaching and learning do not claim any unique set of pedagogies. Instead, interdisciplinary teachers employ an array of instructional methods to promote and support interdisciplinary learning outcomes. This chapter identifies an array of productive pedagogies, providing the background and context in which they emerged and their relevance to interdisciplinary teaching and learning. These include:

- advances in cognitive science and the scholarship of teaching and learning that support active and experiential approaches to teaching and learning;
- efforts to promote diversity in higher education through multicultural curricula and inclusive pedagogies designed to ensure that all students can succeed;
- accreditation, external calls for accountability, and the Assessment Movement that focuses attention on what students know and can do upon graduation;
- the shift from mastery of content to competencies, and the importance of student learning outcomes;
- the emergence and development of pedagogies that support the skills needed to engage in interdisciplinary problem-solving;
- the emergence of the World Wide Web, the Internet, and instructional technologies; and
- the proliferation of faculty development and teaching centers to disseminate pedagogical innovations.

26.1 Defining interdisciplinary and integrative learning

Interdisciplinary outcomes for student learning include learning to solve complex problems that are too broad to be addressed through a single disciplinary lens (Klein 1990) as well as the related abilities to analyze problems from several perspectives, including disciplinary ones, to compare and contrast, to critically analyze resources, to place problems and solutions within a larger context, to develop critical arguments, to empathize with multiple perspectives and stakeholders, and to tolerate ambiguity and complexity (Haynes 2002). More recently, Boix Mansilla (2005) defined the goal of interdisciplinary understanding as:

> The capacity to integrate knowledge and modes of thinking in two or more disciplines to produce a cognitive advancement – e.g., explaining a phenomenon, solving a problem, creating a product, raising a new question – in ways that would have been unlikely through single disciplinary means. (Boix Mansilla 2005, p. 4)

One additional term, *integration*, requires clarification because it is used frequently in defining interdisciplinary student learning outcomes. Interdisciplinary learning is a special case of integrative learning, requiring several of the same skills and habits of mind:

> Integrative learning comes in many varieties: connecting skills and knowledge from multiple sources and experiences; applying theory to practice in various settings; utilizing diverse and even contradictory points of view; and, understanding issues and positions contextually. Significant knowledge within individual disciplines serves as the foundation, but integrative learning goes beyond academic boundaries. Indeed, integrative experiences often occur as learners address real world problems, unscripted and sufficiently broad to require multiple areas of knowledge and multiple modes of inquiry, offering multiple solutions and benefitting from multiple perspectives. (Association of American Colleges and Universities 2004)

These are all useful working definitions for instructors. They can help faculty and students differentiate interdisciplinary learning outcomes from other forms of discipline-based problem solving and integrative learning. However, these definitions are not widely understood or employed by instructors across the disciplines and not everyone concurs that interdisciplinary learning should be deeply rooted in or dependent on disciplinary knowledge. For a comprehensive discussion of interdisciplinary definitions, see Klein (Chapter 2, this volume).

Many faculty use the term *interdisciplinary learning* variably and loosely to mean: (1) multidisciplinary learning outcomes that engage students in the study of two or more disciplinary perspectives on a problem or phenomenon without producing an integrated analysis or solution, (2) cross-disciplinary learning in which one discipline is used in the service of another, or (3) proto-disciplinary outcomes that enable students to draw on resources without knowledge of the disciplinary modes they represent. Other faculty describe their courses as interdisciplinary when they present their own interdisciplinary syntheses of disciplinary materials without formal explication or instruction on how to employ disciplines to arrive at integrated interdisciplinary solutions, while some assign interdisciplinary tasks to students without instruction on how to proceed or what interdisciplinary work entails.

The definitions by Klein (1990), Haynes (2002), and Boix Mansilla (2005) are deeply rooted in knowledge of the disciplines and disciplinarity. While many faculty embrace these definitions, there are others who challenge the centrality of disciplinary knowledge inherent in these definitions, preferring interdisciplinary approaches to teaching and learning that reduce the hegemonic influence of the disciplines in higher education, focusing instead on general skills in critical and analytical thinking and integrative problem solving.

Nonetheless, the definition by Klein is widely used and cited by those engaged in interdisciplinary studies and those who have read her prodigious body of work in this field. Many more faculty are engaging in interdisciplinary work today, entering through disciplinary pathways, often prompted by federal funding sources for research, such as the National Science Foundation (NSF) or National Institutes of Health (NIH), and are now looking for definitions to guide and describe their research and teaching. Klein's work is foundational for many who have worked with intentionality on interdisciplinary teaching and learning during the last few decades. With the current exponential expansion of interest in interdisciplinary teaching and learning, these definitions continue to guide thought and practice while inviting elaborations by new waves of scholars in interdisciplinary teaching and learning, including Augsburg (2005) and Repko (2008), among others.

26.2 The context in higher education

Significant changes have occurred in teaching and learning in higher education during the past 30 years. These changes were propelled not by a single engine, but by many different developments acting as levers—shaping attitudes, creating opportunities, and promoting shifts in policies, practices, and programs. Together they provided the critical mass to enable higher education to make unprecedented strides in the development of teaching and learning generally and interdisciplinary teaching and learning specifically (DeZure 2000, p. 423).

These factors can be seen as both causes and effects of the changes that occurred in teaching and learning during this period. None of them is discrete and the interactive effects are profound and ongoing. The factors cannot be offered in strict chronological order because many of these developments emerged concurrently, albeit in different sectors of higher education, gaining momentum and significance at different rates, e.g. open admissions and affirmative action introduced more diverse learners to higher education, the Assessment Movement, proliferation of teaching centers, and the dissemination of research on teaching and learning, among others. Others have had an ongoing impact that is periodically energized by innovations in the field, as in the case of new technologies. Collectively the significant changes in pedagogy in higher education were made possible by other shifts; some antecedent, some concurrent; some proactive, some reactive. They help to explain why, for example, many of the student-centered active learning methods particularly relevant for interdisciplinary teaching and learning that were advocated four decades ago are now taking root and flourishing as never before. These methods are gaining acceptance because a sufficient number of preconditions now exist, enabling innovations to be adopted, assessed, rewarded, and sustained. Taken together, these developments represent a cultural shift, one that increasingly promotes and supports an active culture of interdisciplinary teaching and learning.

As noted by Newell (Chapter 25, this volume), interdisciplinary curricula and programs are proliferating in disciplinary departments (with the ironic reduction in interdisciplinary studies programs), offering compelling evidence that interdisciplinary curricula are increasingly mainstream in higher education. With the proliferation of interdisciplinary programs and courses, there has been an increased interest in how to design, teach and assess them—enabling instructors to document that students have attained competence in interdisciplinary problem solving and integration.

While interest is high, institutions and faculty are struggling with how to meet this challenge. In a study of 139 institutional applications to participate in a national project entitled 'Integrative Learning: Opportunities to Connect', designed to promote integrative and interdisciplinary learning, DeZure *et al.* (2005) found that campuses that already employed numerous integrative and interdisciplinary curricular and pedagogical practices nonetheless had fundamental questions about what are integrative and interdisciplinary learning, what teaching methods are most effective, and what methods can be used to assess and document student mastery. In sum, interdisciplinary teaching and learning are alive and well in higher education, and there are models to inform instructional decision making; but there is much more work to fulfill their promise and potential to enable graduates to solve the magnitude and diversity of the challenges we face as a global society.

26.3 Levers for change

What were the levers that supported the growth of interdisciplinary teaching and learning in higher education? Advances in cognitive science and neuroscience have affirmed the efficacy of teaching methods that actively engage students, requiring students to be active agents in the learning process rather than passive recipients of information. Students should interact with the materials to be learned and reflect on their work to reinforce their learning and to promote meta-cognitive skills. Cognitive science also spawned research into students' preferred learning styles, underscoring the need to diversify approaches to teaching and learning and to include opportunities for experiential learning and peer interaction.

These insights from brain research reinforced the proliferation of active and experiential approaches to teaching and learning, including collaborative, cooperative, and team-based learning; case studies; role-playing, simulations, and serious gaming problem-based learning; discovery-based learning; and field experiences, including internships, service learning, and study abroad (Klein 2006). While these methods can be enacted through a disciplinary lens, they also invite multidisciplinary perspectives and opportunities for interdisciplinary integration. The more the pedagogy engages students in experiences based in the complexities of the real world, the more there is a need to employ interdisciplinary approaches to problem solving and authentic assessment.

In the mid 1990s, new conceptions of the social construction of knowledge began to take hold in higher education, particularly in the humanities and social sciences, leading to the emergence of constructivist teaching and assessment methods. In constructivist methods, students actively construct knowledge with their peers, often in the context of collaborative and cooperative learning groups. These methods differ, particularly with regard to the level of structure and guidance provided by the instructor. But both approaches involve working with peers to construct knowledge, invite multiple perspectives as part of the critical examination of solutions, and require analysis and synthesis skills, often leading to an integrative solution. This is true whether the task is discipline-based, multidisciplinary, or interdisciplinary. These methods are all learner-centered, although the degree to which power and authority shifts from teacher to students can vary considerably. Collaborative learning often has a looser structure with low levels of instructor intervention, and cooperative learning often has a tighter structure with high levels of instructor design and oversight.

Collaborative learning (Bruffee 1995) is a philosophy and process of interaction with roots in the humanities in which individuals are responsible for their actions, including learning and respect for the abilities and contributions of their peers. It is not primarily concerned with converging on a correct or predetermined answer; rather, it is concerned with the nature of reasoning, questioning, and informed conversation, or what Bruffee has called the 'conversation of mankind'.

Cooperative learning (Johnson *et al.* 1991) is a structure of interaction that involves students working in teams to accomplish a common end product or goal and includes all the following elements:

(1) Positive interdependence. Team members are obliged to rely on one another to achieve the goal. If any team members fail to do their part, there are consequences for all team members.
(2) Individual accountability. All students in a group are held accountable for doing their share of the work and for mastery of all of the material to be learned.
(3) Face-to-face promotive interaction. Although some of the group work may be divided and done individually, some must be done interactively, with group members providing one another with feedback, challenging one another's conclusions and reasoning, and teaching and encouraging one another.
(4) Appropriate use of collaborative skills. Students are encouraged and helped to develop and practice trust-building, leadership, decision-making, communication, and conflict management skills.
(5) Group processing. Team members set group goals, periodically assess what they are doing well as a team, and identify changes they will make to function more effectively in the future.

Cooperative learning emerged from the social sciences, particularly work with K-12 (primary and secondary school) students who benefited from the high levels of structure and teacher guidance.

Although Bruffee emphasizes that originally collaborative learning was designed for adults engaged in the higher-order critical thinking skills of reasoning and questioning, and cooperative learning was designed for younger students mastering foundational knowledge, both of these approaches have proven to be highly effective with students in higher education across the disciplines for a range of learning outcomes. Both have a place in promoting the skills inherent in interdisciplinary analysis and problem solving, and methodological borrowing is common across these two models based on the nature of the students, the disciplinary paradigms, and the goals for student learning. Instructors may also take a developmental approach to these models, providing beginning students with the tighter structure of cooperative learning for introductory group work and mastery of material and more advanced students with the looser structure and more open-ended critical thinking goals of collaborative learning.

A 'jigsaw' is a highly structured model of cooperative learning that is particularly well suited to interdisciplinary problem solving because it breaks down complex problems into more manageable pieces and presents them in a sequence that students new to interdisciplinary studies can more readily handle. The jigsaw method has two stages. In the first stage, students are assigned to primary groups in which they address one dimension of a large, complex problem. Each primary group focuses on a different dimension of the same problem. In the second stage, students are dispersed into secondary groups comprising one member from each of the primary groups. In these secondary groups, students share the insights from their primary groups and then work collaboratively to integrate their insights into a holistic solution to the complex problem (Smith 2000, pp. 32–3).

This model is particularly useful for interdisciplinary problem solving. In phase one, students are assigned to primary groups in which each group studies a different disciplinary perspective on a complex problem (with materials often selected by the instructor). In phase two, students are reorganized into secondary groups comprising at least one student from each of the primary groups. The secondary groups then endeavor to bring together their disciplinary insights into an integrated interdisciplinary solution to the larger problem. The model is complex and time-consuming, but it models the systematic and challenging steps in interdisciplinary problem solving and demonstrates the relationship between discipline-based and interdisciplinary solutions.

While collaborative and cooperative learning were beginning to take hold in the humanities and social sciences, the sciences, technology, engineering, math, and medicine have long been developing and pursuing problem-based learning (PBL), discovery-based, and inquiry-based learning which resonated as a variant of the scientific method. These approaches also reinforced the integration of multiple sources of information and perspectives, higher-order critical thinking skills, and student-centered learning in groups and teams—all required for authentic work in these fields.

Beginning with open admissions and affirmative action in the 1970s, efforts to promote diversity and multiculturalism in higher education led to the identification of 'inclusive pedagogies', characterized by approaches that invite multiple perspectives and discussion to ensure that the voices of all members of the learning community are heard. As noted by Patricia Cross (1973), the new clientele for higher education in the 1970s consisted of everyone who wasn't there in the 1940s, 1950s, and 1960s, including low academic achievers, adults and part-time students, ethnic minorities, and women in disciplines traditionally dominated by men.

Diverse learners brought more variation in learning styles, requiring diverse approaches to teaching. Increasingly, dialogue, panel discussions, reflective journals and narratives, and more relational and feminist pedagogies preferred by minorities and women were introduced to structure the expression of different positions toward an issue—promoting empathy and understanding for other viewpoints, critical analysis, and synthesis.

The proliferation of academic service learning (i.e. tying academic course goals to required service experiences in the community) and study abroad further broadened the exposure of students to diverse populations and perspectives—requiring students to integrate learning from in-class and out-of-class with exposure to diverse people in real world contexts. One of the hallmarks of good practice in service learning is critical reflection (often in the form of student journaling and a culminating reflective essay) to integrate multiple sources of insight and experience. All of these approaches support the skills required for interdisciplinary integration and solutions that are relevant for real world contexts.

Beginning in the late 1960s, writing across the curriculum (WAC) underscored that effective writing differs according to its purpose, audience, and disciplinary context. Advocates for WAC proposed that writing be taught and assessed by faculty across the disciplines and not relegated entirely to a single freshman composition course, often located within an English department. In response, institutions established upper-division intensive writing courses in the disciplines to assist students to learn the paradigm of writing in at least one discipline beyond English, usually in their major. This movement also reflected the move away from the traditional liberal arts curriculum and general education toward greater emphasis on the major and electives, a movement often fueled by the wishes of faculty to teach their areas of narrow specialization and the needs of departments to increase required courses in the major, often for financial reasons.

By the mid-1980s, the Assessment Movement was emerging in American higher education, focusing on the articulation and assessment of student learning outcomes along with institutional accountability for their attainment. Initially driven by accrediting bodies, these efforts were reinforced by employers, who found that college graduates were weak in fundamental skills such as critical thinking, written and oral communications, quantitative literacy, civic responsibility, ethics, and teamwork—all core skills in the workplace. As institutions sought to comply with assessment mandates, they began by conducting college-wide evaluations to identify the outcomes they most valued for their college graduates: they too identified these same cross-cutting skills and attributes for their outcomes of liberal learning and general education. These campus processes accelerated the gradual shift in higher education from mastery of content to competencies, from passive to active pedagogies, from traditional testing methods to assess student learning to more authentic methods that modeled and mirrored the complexities of the real world (DeZure 2000).

In their efforts to define critical thinking, campuses relied heavily on Benjamin Bloom's *Taxonomy of educational objectives*, focusing primarily on goals in the cognitive domain, that identified levels of critical thinking—knowledge, comprehension, application, analysis, synthesis, and evaluation—with the latter four levels as higher-order and inclusive of all the lower-level skills. All of these shifts reinforced the relevance of interdisciplinary problem solving and its capacity to engage students actively in the complexities of analysis, synthesis, and evaluation in addressing real-world problems.

The Assessment Movement also reasserted the importance and centrality of writing to attainment of student learning outcomes, both in terms of 'writing to assess' student learning and 'writing to learn' in which writing is used to promote learning and to provide formative feedback to instructors and students about whether students are understanding the material and are able to express concepts in their own terms. Classroom assessment techniques (CATs) were popularized by the publication of Tom Angelo and Patricia Cross's (1993) *Classroom assessment techniques*, offering instructors dozens of informal, quick, and easy methods to determine whether students are learning and to integrate writing across the disciplines. Many of these techniques were tied to integrative skills, particularly the most famous of the CATs, the minute paper, which asks students at the end of class to take a couple of minutes to summarize the key thesis or theme of the class session—requiring them to focus on the big picture, the forest, not just the trees. For many students, listening to lectures is an exercise in verbatim note-taking, often without thought, documenting everything that is said. The minute paper challenged that pattern by requiring students to listen for the big theme and weave the disparate threads together into an integrated whole at the end of each class session. CATs are generally ungraded and anonymous, offering a low-risk way to learn this important integrative skill.

Writing continues to be the primary means by which to teach and to assess written communications and critical thinking generally and interdisciplinary critical thinking more specifically. This trend was reinforced by the emergence of more rigorous and multidimensional approaches to the evaluation and documentation of learning required by external accrediting agencies and the designation of assessment units on campuses to promote, support, and evaluate those efforts. By 1999, a study of American colleges and universities by the National Center for Post-Secondary Improvement indicated that everyone was doing assessment of student learning outcomes, but few institutions were using the data to inform practices or guide curriculum. Accrediting agencies then revised their expectations to require institutions to document student learning outcomes and close the feedback loop by indicating how they used or intended to use their assessment data to improve teaching and learning.

To address the challenge to document student learning outcomes for interdisciplinary learning, interdisciplinary studies scholars developed a writing curriculum to teach and assess interdisciplinary learning and related critical thinking skills. Haynes (2004), for example, identified a developmental sequence of writing assignments to assist students to support disciplinary analysis and interdisciplinary synthesis. The preliminary step was to help students identify and make explicit the thinking of the disciplines under study: their assumptions, frameworks, foci, methods, and key questions. This is not to suggest that disciplines speak with one voice; there are deep divides within disciplines, and they too should be identified. Subsequent assignments ask students to analyze a problem using different disciplinary frameworks, later moving to assignments that require comparison and contrast among the analyses. This is followed by integrative assignments in which students must draw on the methods and insights of several disciplines, reconciling them in an integrated approach. Repko (2008) elaborates on reconciliation in a step he calls 'finding common ground' among disciplinary perspectives on a problem, which adds a distinctive and important phase to this process.

It is important to note that these methods are grounded in the disciplines. Further, they require meta-cognitive skills to critique the limits, biases, and unique opportunities offered by both disciplines and interdisciplinary solutions; and they require all the higher-order critical thinking skills—from analysis through evaluation—to arrive at viable interdisciplinary solutions. These processes are designed to increase awareness through which we seek and perceive information and by which we construct knowledge.

One model proposed by Wolfe and Haynes (2003) for assessing interdisciplinary writing identifies four key dimensions: (1) drawing on disciplinary sources; (2) critical argumentation; (3) multidisciplinary perspectives; and (4) interdisciplinary integration. The evaluation of writing and interdisciplinary critical thinking continue to pose significant challenges to teachers of interdisciplinary studies, but scholars like Christopher Wolfe and Carolyn Haynes (2003), Michael Field and Donald Stowe (2002) and more recently Veronica Boix Mansilla (2005) are providing promising approaches to clarify this terrain.

In the twenty-first century, in addition to the liberal learning outcomes identified earlier, e.g. written and oral communications, teamwork, etc., integrative learning has been identified as an additional core competence in undergraduate education. In 2003, the Association of American Colleges and Universities released its 'Statement on integrative learning', which defines *integrative learning* as the ability to integrate from multiple sources of knowledge and experience, that is, to integrate theory and practice, in-class and out-of-class learning, learning in general education courses and in the major, and learning across the entire collegiate experience.

This competence is a response to concerns that the collegiate experience is atomized with too many fragmented experiences and too few connections. To promote integration, institutions are turning to student portfolios, particularly e-portfolios, keystone and capstone courses, learning communities, living-learning communities, and interdisciplinary courses. All of these provide opportunities for reflective writing and structured integrative assignments to enable students to bring together their disparate collegiate experiences and to make meaning of them. Interdisciplinary pedagogies offer models of how to foster connected learning that leads to integration. These approaches honor disciplinary ways of knowing while enabling students to reach across the disciplines and their disciplinary coursework to create robust integrated interdisciplinary solutions to real-world problems.

Advances in instructional technology that make online resources readily accessible have also enabled innovations in interdisciplinary teaching and learning. This generation of traditional-aged college students has grown up with the Internet. For them, traversing its interdisciplinary universe is comfortable and normative and each new form of technology is welcomed and easily adopted. The Internet is not discipline-based. Most online searches take you to resources that represent multidisciplinary and interdisciplinary perspectives, both challenging and inviting users to think outside of disciplinary boxes.

The use and development of interdisciplinary materials have never been easier nor more affordable for both instructors and students. The Internet also enables access to disciplinary and interdisciplinary experts from around the world to engage with classes and, more importantly, the ability to bring news and real-world challenges into classes as they occur, giving an immediacy and urgency to the complexity of local, national, and global events as they occur. Whether it was news of 9/11 or hurricane Katrina or the 2004 tsunami, students were afire with questions about what caused these events. While single-disciplinary perspectives were helpful in answering student questions about important dimensions of these events, campuses also responded with multidisciplinary and interdisciplinary discussions of the complex interaction of causes and effects. When students ask, as they did after 9/11 and Katrina and the tsunami, 'What caused this?' they are asking questions that require interdisciplinary responses.

After hurricane Katrina, for example, Michigan State University sponsored a series of campus-wide forums entitled, 'Katrina and its aftermath', including three interdisciplinary events:

- the cost of marginalization: place, race, class and media in the Katrina catastrophe;
- the city at the end of the river: the unique setting of geology, engineering, history, and culture of New Orleans; and
- Katrina's impact on human and animal health: now and for generations.

While clips from the Internet were shown during these presentations, providing artifacts about the event and deserving critical analysis themselves, the Internet had been the key factor motivating students to attend and to learn more about this complicated national tragedy.

Technology also provides the means to store student work for future reference, be it for integrative student reflection in capstone courses or student advising. In their study of integrative learning nation-wide, DeZure *et al.* (2005) found that e-portfolios were identified most frequently as the method campuses were using (or planned to use) to promote, document, and assess integrative and interdisciplinary learning. E-portfolios have all the advantages of traditional portfolios that contain artifacts selected by the student to represent his or her learning and progress over time.

E-portfolios take traditional hard-copy portfolios to the next level, eliminating the problems of physical storage, retrieval, and transport, while expanding their multifunctionality and access by current and future instructors and advisors as well as students themselves. E-portfolios can provide an organic document that can be used throughout the collegiate experience. It has huge capacity to store and archive many types of artifacts, including CDs and other audiovisual materials, photos of projects, and traditional written documents written by the student. It can be used by instructors across the disciplines as well as advisors and student affairs staff who may interact with students on co-curricular and other experiential learning activities, such as service learning, study abroad, engagement in leadership activities, and student government.

As with all instructional technologies, issues of security, privacy, and expense for ongoing technology infrastructure are non-trivial. Emerging e-portfolio technologies will provide students and administrators with options to control levels and types of access by advisors, instructors, peers, future employers, and others selectively, greatly enhancing the desirability and flexibility of this new and promising technology. And last but not least, e-portfolios can provide compelling data about student learning for the purposes of interdisciplinary program review and institutional assessment.

26.4 New approaches to supporting and evaluating interdisciplinary teaching

Faculty development programs and teaching centers have proliferated across the United States in the last 25 years, with recent increases in the number of programs that explicitly support interdisciplinary teaching and learning. Coupled with the dramatic increase in the availability of research on teaching and learning (both discipline-based and interdisciplinary), faculty efforts to expand their repertoire of instructional strategies are supported at most institutions.

For example, for several years the Center for Research on Learning and Teaching at the University of Michigan sponsored a year-long facilitated cohort program entitled Interdisciplinary Faculty Associates to enable teams of faculty to study the research on interdisciplinary teaching and learning, design interdisciplinary team-taught courses, and produce course portfolios documenting their experiences with interdisciplinary teaching, learning, and assessment.

Although many individual faculty employ interdisciplinary approaches in their own teaching, many interdisciplinary courses are team taught by faculty with expertise in different disciplines. Team teaching has the potential to be a powerful source of interdisciplinary learning *if* students are able to witness how experts from different disciplines approach issues and how they negotiate integrative solutions between them. Interdisciplinary faculty teams have to determine the degree of integration they wish to use and how it will impact their planning, curricular choices, instructional methods and delivery, assignments, and assessment practices (Davis 1995). But having faculty present their disciplinary perspectives in serial fashion is not sufficient. If students are to engage in complex intellectual tasks to integrate the insights of different disciplines, then faculty should join in this endeavor, modeling it and sharing the difficulties and the richness of doing so.

It is often challenging for faculty members to describe succinctly the basic epistemology within their discipline, and harder still to describe the range of beliefs and contested topics within their field. Nonetheless, it is such unpacking that helps students to understand disciplinary perspectives and the contributions that different disciplines can make toward interdisciplinary solutions. In team-taught classroom settings (or in distance-learning contexts) in which faculty panelists or presenters are asked to represent a disciplinary perspective on an issue, it may be appropriate to begin with a brief introduction to their discipline, identifying the key questions or topics that concern practitioners in the field, what counts as evidence, and what methods they commonly use to search for answers. Depending on the context, it can be useful to identify key points of contention or difference in the field to avoid over-simplifications that suggest the field speaks with one voice. This type of introduction grounds the discussion, enabling students to better appreciate how the views of faculty on an issue are shaped by the differences in their disciplinary ways of thinking and why interdisciplinary syntheses are challenging to create but often provide more robust solutions to complex problems.

New approaches to documenting and evaluating teaching go beyond the use of student ratings to enable faculty to capture the complexity of their instructional efforts and their impact on student learning. These newer forms of peer review are appropriately multi-dimensional and include teaching portfolios, course portfolios, and the scholarship of teaching and learning (i.e. research on the instructor's own teaching and its impact on student learning that is critiqued by peers and made public). The proliferation of these approaches is a particularly important development for instructors who teach interdisciplinary courses because it enables them to clarify the complexity of interdisciplinary teaching and learning to their disciplinary colleagues and administrators.

Course portfolios are particularly well suited for documenting interdisciplinary courses by individuals or teams. In addition to describing the design of the course, course portfolios provide opportunities for instructors to convey their assessment of the actual enactment of the course as well as recommendations for ways to improve the course and to enhance the climate to promote interdisciplinary teaching and learning in the future.

The movement to promote the scholarship of teaching and learning (SoTL) endeavors to foster research on teaching and learning, to open it to critical review, and to make it public so that others can build on it. Course portfolios provide an important model for the scholarship of interdisciplinary teaching and learning because they meet all the criteria for SoTL work and are able to document the complex processes by which faculty and students engage in interdisciplinary problem solving (Bernstein *et al.* 2006). Course portfolios were originally developed as part of the Peer Review of Teaching Project, sponsored by the Carnegie Foundation for the Advancement of Teaching. Course portfolios have proliferated so that there are now online repositories of disciplinary and interdisciplinary examples to make them more visible to the public.

Course portfolios are generally 12 to 15 pages in length plus appendices and are increasingly electronic documents. They begin by clarifying the audience (1 page) for the document, most commonly other faculty who will teach the course in the future; evaluation committees or department chairs for purposes of reappointment, tenure or promotion; and/or faculty at their institution or across higher education who are interested in interdisciplinary teaching and learning and ways to document its complexity.

The genesis and context section (2–3 pages) includes the motivation for the development of the course, the course background and history, and administrative logistics. Given the challenges of administrative logistics in planning interdisciplinary courses, this type of information provides a useful resource about emerging practices. The anatomy section (4–5 pages) describes the course design, including intended goals for student learning, content, teaching methods, assessment methods, and approaches to interdisciplinarity and collaboration throughout the course and its planning. The course enactment and implementation then describe what really occurred. The outcomes section (2–4 pages) reflects on student learning (with support from evidence of student learning in the appendices) as well as faculty learning. The recommendations section (1–3 pages) provides thoughts for the future of the course, advice to colleagues who may teach similar courses, and ways to promote and support interdisciplinary teaching and learning at the institutional level. The appendices provide the syllabus, the assignments, activities, and other course materials, including websites, and evidence of student work with instructor feedback. If the course is team taught, the instructors can create a collaborative document, with points of agreement and difference identified by each instructor, offering a form of public dialogue and mutual feedback about their teaching.

Beyond faculty development on individual campuses, there are national associations that provide excellent faculty development seminars, resources, and access to networks of peers committed to interdisciplinary teaching and learning. The Association for Integrative Studies (AIS) provides annual meetings, research, and pedagogical resources on interdisciplinary teaching and learning (<http://www.units.muohio.edu/aisorg/>). Likewise, the Association of American Colleges and Universities (AAC&U) sponsors national and regional meetings, peer-reviewed publications. and campus projects on liberal learning, general education, and integrative and interdisciplinary learning (<http://www.aacu.org/>).

Faculty development focused on interdisciplinary endeavors is increasingly being supported by external funding agencies, e.g. the NSF and NIH, to promote large-scale interdisciplinary research projects. These projects often engage faculty, graduate students, and undergraduate research assistants on interdisciplinary cross-generational research teams. The inclusion of faculty is to increase their ability to work on interdisciplinary research teams, and to productively engage graduate and undergraduate students in those efforts. The inclusion of graduate students is to foster a pipeline of early career faculty who wish to pursue interdisciplinary work. The inclusion of undergraduates is to promote undergraduate research as a form of active, inquiry-based learning that is fast becoming a hallmark of undergraduate education at research universities. In the process, undergraduates are gaining experience with interdisciplinary teamwork. In this time of limited resources, significant funding opportunities to pursue interdisciplinary work are likely to generate great faculty interest, even among those who have not previously done interdisciplinary work.

Interdisciplinary teaching and learning is a burgeoning phenomenon, moving from the margins to the mainstream of higher education. Discipline-based faculty are eager to learn more about what interdisciplinary teaching entails, how to design and deliver instruction, and what impact it is having on student learning and on the instructors themselves. These new approaches to documenting teaching offer productive vehicles to open the classroom door on interdisciplinary teaching and learning—and to inspire others to explore its possibilities.

26.5 Conclusion

Echoing the thesis of this chapter, Haynes (2002) concludes that there is no single, unique method associated with interdisciplinary teaching and learning:

> Interdisciplinary pedagogy [...] is not synonymous with a single process, set of skills, method or technique. Instead, it is concerned primarily with fostering in students a sense of self-authorship and a situated, partial and perspectival notion of knowledge that they can use to respond to complex questions, issues and problems. While it necessarily entails the cultivation of the many cognitive skills such as differentiating, reconciling, and synthesizing [...] it also involves much more, including the promotion of student's interpersonal and intrapersonal learning. Because interdisciplinarity is a complicated psychological and cognitive process, it cannot be taught with one approach. (Haynes 2002, p. xvi)

Indeed, interdisciplinary teaching and learning requires a host of powerful pedagogies to inspire and enable teachers and students to grapple effectively with the complexity of problems we face in the twenty-first century. This work is challenging for students, but in many ways it is even more challenging for faculty who will be crossing borders and charting new terrain in higher education, leaving the relative safety of disciplinary expertise behind for what Donald Schön (1995) has called the 'swampy lowlands' of important and real-world problem solving. Just as students benefit from peers to support them in this endeavor, it is equally important for faculty to have colleagues to share their wisdom of practice, their triumphs, and their challenges—recognizing that this is, as Haynes notes, a 'complicated psychological and cognitive process', requiring creativity, commitment and courage. In an era in which interdisciplinary teaching and learning is coming into its own, interdisciplinary instructors no longer have to go it alone.

References

Angelo, T. and Cross, K.P. (1993). *Classroom assessment techniques: a handbook for college teachers*, 2nd ed. San Francisco: Jossey-Bass.

Association of American Colleges and Universities (2004). A statement on integrative learning. Available at: <http://www.aacu.org/integrative_learning/pdfs/ILP_Statement.pdf>.

Augsburg, T. (2005). *Becoming interdisciplinary: an introduction to interdisciplinary studies.* Dubuque, IA: Kendall/Hunt.

Bernstein, D., Nelson Burnett, A., Goodburn, A., and Savory, P. (2006). *Making teaching and learning visible: course portfolios and the peer review of teaching.* Bolton, MA: Anker Publishing.

Boix Mansilla, V. (2005). Assessing student learning at disciplinary crossroads. *Change* **37**(1), 14–21.

Brookhart, S.M. (2004). Assessment theory for college classrooms. In: M.V. Achacoso and M.D. Svinicki (eds) *Strategies for evaluating student learning. New Directions for Teaching and Learning* (special issue) **100**, 5–14.

Bruffee, K.A. (1995). Sharing our toys: cooperative vs. collaborative learning. *Change* **27**(1), 12–18.

Cross, K.P. (1973). The new learners: changing perspectives on quality education. *Change* **5**(1), 31–4.

Davis, J.R. (1995). *Interdisciplinary courses and team teaching: new arrangements for learning.* ACE Series on Higher Education. Phoenix: Oryx Press (now an imprint of Greenwood Press).

DeZure, D. (ed.) (2000). *Learning from change: landmarks on teaching and learning in higher education from* Change *magazine (1969–1999).* Sterling: Stylus Press/American Association for Higher Education.

DeZure, D. (2002). Innovations in the undergraduate curriculum. *Encyclopedia of education*, 2nd ed, Vol. 2, pp. 509–14. New York: Macmillan Reference USA.

DeZure, D., Babb, M. and Waldmann, S. (2005). Integrative learning nationwide: emerging themes and practices. *Peer Review* **7**(4), 24–8.

Field, M. and Stowe, D. (2002). Transforming interdisciplinary teaching and learning through assessment. In: C. Haynes (ed.) *Innovations in interdisciplinary teaching*, pp. 256–74. Washington: American Council on Education/Oryx Press.

Haynes, C. (ed.) (2002). *Innovations in interdisciplinary teaching.* ACE Series on Higher Education. Westport, CT: Oryx Press/Greenwood Press.

Haynes, C. (2004). Promoting self-authorship through an interdisciplinary writing curriculum. In: M.B. Baxter Magolda and P.M. King (eds) *Learning partnerships: theory and models of practice to educate for self-authorship*, pp. 63–90. Sterling, VA: Stylus Publications.

Johnson, D.W., Johnson, R.T., and Smith, K.A. (1991). *Cooperation in the classroom.* Edina, MN: Interaction Book Co.

Johnson, D.W., Johnson, R.T., and Smith, K.A. (1998). Cooperative learning returns to college: what evidence is there that it works? *Change* **30**(4), 27–35.

Klein, J.T. (2006). *A platform for shared discourse of interdisciplinary education.* Available at <http://www.jsse.org/2006-2/klein_platform.htm>.

Klein, J.T. (1990). *Interdisciplinarity: history, theory, practice.* Detroit, MI: Wayne State University Press.

Klein, J.T. and Newell, W.H. (1996). Advancing interdisciplinary studies. In: J. Gaff and J. Ratcliff (eds) *Handbook of the undergraduate curriculum: a comprehensive guide to purposes, structures, practices, and change*, pp. 393–415. San Francisco: Jossey-Bass.

National Center for Post-Secondary Improvement (1999). Gauging the impact of institutional student assessment strategies: revolution or evolution? *Change* **31**(5), 53–6.

Repko, A.F. (2008). *Interdisciplinary research: process and theory.* Thousand Oaks, CA: Sage Publications.

Schön, D. (1995). The new scholarship requires a new epistemology: knowing-in-action. *Change* **27**(6), 26–34.

Smith, K.A. (2000). Going deeper: formal small group learning in a large classes. In: J. MacGregor, J.L. Cooper, K.A. Smith, and P. Robinson (eds) *Strategies for energizing large classes. New Directions for Teaching and Learning* (special issue) **81**, 25–46.

Wolfe, C. and Haynes, C. (2003). Interdisciplinary writing assessment profiles. *Issues in Integrative Studies* **21**, 126–69.

CHAPTER 27

Facilitating interdisciplinary scholars

STEPHANIE PFIRMAN AND PAULA J. S. MARTIN

> Through the control over faculty appointment, promotion, and tenure decisions, departments shape the intellectual capital of the campus in tune with the cultures, norms and intellectual orientations of the disciplines. Universities also derive recognition and prestige through the collective achievement of their departments as probed by discipline-based mechanisms (e.g. the National Research Council's assessment of doctoral programs)
>
> Sá (2008)

As innovation increasingly occurs at the boundaries of disciplines, scholarship is now breaking out of the lines that make the 'discipline within a department' structure troublesome for the interdisciplinary scholar. Traditionally, power, money, hiring, and promotion are allocated by departments, posing barriers to interdisciplinarity at every turn—in the universities' organizational design, lack of motivation within the institutional power structure, and lack of institutional incentives. These lead to difficulties in managing the complexity of interdisciplinary relations—the transaction costs—both within and outside of the institution. Even students feel the strain when they undertake interdisciplinary programs of study and find themselves being taught and advised by faculty on loan from departments, unstable course offerings leading to difficulty in completing requirements, and a lack of facilities, community, and information on potential career trajectories. This chapter reviews the dynamics of interdisciplinary scholars functioning within a disciplinary tradition and provides guidance for better support mechanisms to facilitate interdisciplinary scholarship.

27.1 Personal approaches to interdisciplinarity

In many cases, a variety of issues are conflated with the pursuit of interdisciplinary scholarship and teaching (Table 27.1). This adds a level of complexity to the life of interdisciplinary scholars as they work within the traditional disciplinary framework. Many interdisciplinary endeavors are in new fields, requiring the establishment of new scholarly communities, with new resource needs (e.g. the need for shared space or additional travel) and new relationship demands (e.g. the need to learn a shared language). Interdisciplinary research and education is often collaborative, using informal, *ad hoc* teams (Evaluation Associates Ltd 1999; Lattuca 2001). Informal arrangements result in questions about credit for leadership, and challenges in negotiating group interactions. Assessment of an individual's contribution (crucial for promotion and tenure) is also problematic in that interdisciplinary scholars tend not to specialize (Porter *et al.* 2007). And researchers who don't specialize pay a productivity penalty (Leahey *et al.* 2008), as coming up to speed in new fields and setting up new collaborations slow down publication rates. An additional complication is that members of groups underrepresented in the academic elite (women and perhaps minorities) appear to be disproportionately drawn to interdisciplinary research and education (Evaluation Associates 1999; Beraud 2003; Rhoten and Pfirman 2007a,b).

However, not all interdisciplinary research is collaborative, nor is it all applied—as Rhoten and Pfirman (2007a,b) point out, there are many ways to be interdisciplinary (Fig. 27.1). One can approach interdisciplinarity at a variety of scales, ranging from intrapersonal—where an individual decides to tackle research from multiple perspectives; to interpersonal—working with others; to interfield—working with non-academic stakeholders, for example business and policy makers. Interdisciplinary teaching ranges through the same categories, each with their own set of administrative issues: the need for course release to develop the intrapersonal expertise, co-teaching credit for the interpersonal approach, departmental buy in for the interfield class, and adjunct support for practitioners when external stakeholders are involved.

Table 27.1 Characteristics often associated—and conflated—with disciplinary and interdisciplinary research and education.

Disciplinary	⇔	Interdisciplinary
Departmental	⇔	Interdepartmental
Mainstream	⇔	Non-mainstream (WISELI 2003)
Specialized	⇔	Diverse (Leahey 2006)
Discovery	⇔	Integration, application (Boyer 1990)
Specialization	⇔	Integration (Porter *et al.* 2007)
Laser	⇔	Searchlight (Gardner 2007)
Disciplinary	⇔	Synthesis
Basic	⇔	Applied
Hierarchical	⇔	Collaborative, democratic
Formal	⇔	Informal (Lattuca 2001)
Established	⇔	New ('fringe') (Spanner 2001; Choucri *et al.* 2006)
Established order	⇔	Dissolution and amalgamation (Weingart and Stehr 2000)
Majority	⇔	Minority (Rhoten and Pfirman 2007a,b)

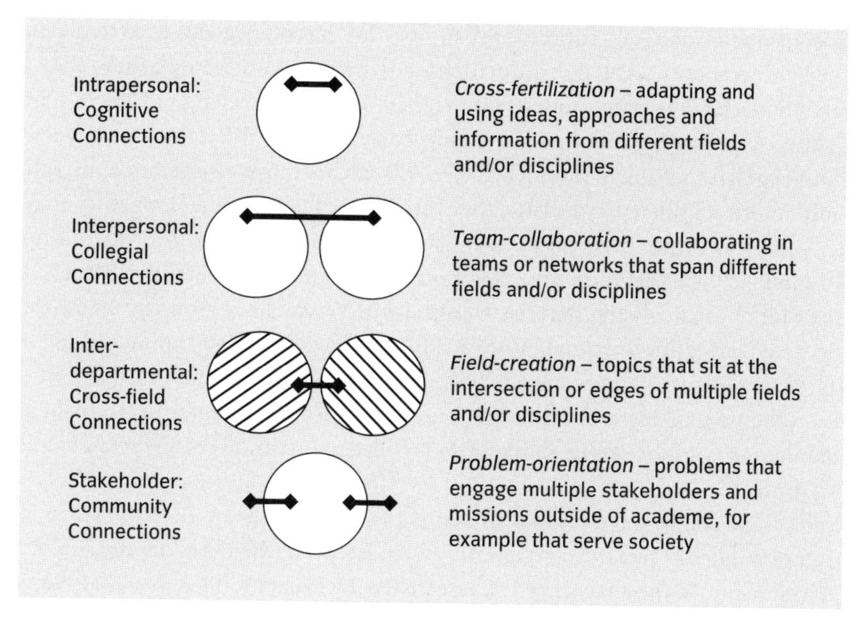

Figure 27.1 Interdisciplinary ways of conducting research and of teaching (Pfirman *et al.* 2007; based on Rhoten and Pfirman 2007a, b).

What this means in practice, for the establishment and fostering of interdisciplinary scholars, is that they are dealing with many issues at the same time: not only the gathering of their own, individual expertise and gaining recognition for their creative contributions, but also the need to justify both their field and their approach (Langfeldt 2006), even within small interdisciplinary research teams:

> Each of us has had the experience of feeling as though we do not 'really' belong to the research team, or that, upon returning to our scholarly 'homes' after a research meeting, we do not really belong there either. Working at the boundaries of communities of practice, team members can feel uprooted, alien, frustrated. [...] When data from one's discipline is under scrutiny in an analysis session, the insider may perceive a need to defend her turf, provoking a sense of resentment and conflict with the rest of the team [...] (Lingard *et al.* 2007)

Spanner (2001) found that most interdisciplinary scholars believed that they operated in a more complex environment than disciplinary scholars, and many thought that they needed to know more information—with significant problems in locating useful information scattered across diverse fields (see Palmer, Chapter 12 this volume):

> When one considers the unique problems and barriers ID [interdisciplinary] scholars face, particularly in tandem with their expressed discomfort levels in non-affiliate fields and dissatisfaction with available resources, the lot of the average ID scholar might be more stressful and pressure ridden than those conducting traditional single-discipline research. (Spanner 2001)

Co-teaching an interdisciplinary course raises similar issues in the classroom as faculty feel compelled to justify their teaching methods and content selection (Jang 2006).

These continued self-examinations and appeals for acceptance can lead to a sense of personal vulnerability, tension, insecurity, and demoralization. Many believe they must continually declare, and be modest about, their limited knowledge of other fields in which they are working, or they risk being considered as 'dilettantes who knew too little and claimed too much' (Lattuca 2001). As scholars move away from a disciplinary base into interdisciplinary endeavors, they often report that they no longer fit in as well as they once did: while their peers establish identity and status within the discipline, interdisciplinary scholars have to live without the comfort of expertise' (Lattuca 2001). It is therefore not surprising that the University of Wisconsin's Women in Science and Engineering Leadership Institute (WISELI) found the critical determining factor in the quality of workplace interactions (including informal departmental interactions, colleagues' valuation of research, isolation and 'fit', and departmental decision-making) was whether or not individuals thought their colleagues considered their research to be 'mainstream'.

Despite these challenges to personal identity, many scholars are determined to follow their interdisciplinary research and teaching agendas, even when they are not in supportive environments. In the analysis of UK researchers by Evaluation Associates (1999), 30% of highly interdisciplinary scholars were based at institutions where they rated the 'overall environment in your institution for interdisciplinary research worse than that for single disciplinary research'. For others it is a choice between interdisciplinarity or something else completely: as Kinzig commented in Haag (2006) 'I think we have an increasing number of students who aren't that interested in being disciplinary. I think if I had had to focus narrowly within a particular discipline, I would not have finished graduate school. I just would have gotten bored'.

Junior interdisciplinary scholars are especially affected by issues of academic community, evaluation, and administrative responsibility. When they first embark on interdisciplinary research and education, they are buoyed by excitement and see mainly the positive aspects of venturing into new territory (Table 27.2). By breaking new ground, they are able to set themselves apart from others, have a lot of autonomy in their research agenda, and can work with colleagues from a variety of disciplines and communities. But then, as they continue their research, often moving toward tenure consideration, the negatives become more and more problematic (Choucri *et al.* 2006), and many of the conflated issues (Table 27.1) raise difficult challenges. It is harder to publish interdisciplinary research in traditional journals well known by the disciplines. Collaborative projects take a long time to get up and running due their high transaction costs. Additionally, if scholars have affiliations with more than one department, they may be getting conflicting advice (or none at all) on how best to demonstrate their research contributions.

Similar issues arise from the perspectives of education, community participation, and service (Table 27.2). Given these challenges, junior scholars are often wary—or warned off—of embarking on interdisciplinarity, while senior scholars tend to be more open and willing. Mentors, champions, and role models are often helpful in easing the personal anxieties of junior scholars at the same time that they provide professional guidance and support.

Table 27.2 Positive and negative aspects of conducting interdisciplinary research (1) and education (2) are disproportionately skewed towards positive in the early stages, followed by negative at later stages (adapted from Pfirman *et al*. 2007).

(1) Interdisciplinary research

	Often early attraction...	**But later difficulties...**
New area	Can break new ground	Less recognition by established scholars
	Less competition Less urgency	Fewer sustained funding opportunities Fewer journals Fewer peer reviewers Career trajectory not known Long start-up time
Social/applied connections	Appeals to social conscience Connect with public good	Less prestigious research area Considered less rigorous
Complex questions	Holistic approach required	Considered less rigorous
Collaborative	Build on strengths of others	Time to cultivate and maintain
	Use people skills	Critical literature in other field Dependent on collaborator Idea origin not clear
Between depts/centers	Freedom because outside of established hierarchy	Less administrative support
Interinstitutional	Broadens network for letter writers	Requires travel
		Less visibility on home campus

(2) Interdisciplinary education and community

	Often early attraction...	**But later difficulties...**
Teaching	Exciting subject Student interest Co-teaching Field experiences Service learning No textbooks, resources	Fewer textbooks, resources Less infrastructure and fewer rewards to sustain 'extra' activities (field, service) Co-teaching Heavier student advising load
Campus life	Campus programming Community connections Bridge between disciplines: search committees, presentations Become known on campus	More service and outreach expectations
Scholarly participation	Field more open, can initiate programs	Fewer high level, prestigious committees Fewer honors than in disciplinary fields
Promotion and tenure		Criteria often disadvantage interdisciplinary scholars

27.2 Institutional support for interdisciplinary scholars

Institutions have recognized that departmental structures create barriers for scholars working between departments and are adjusting to the needs of interdisciplinary scholars (Table 27.3).

While most institutions have now made at least modest efforts to include interdisciplinary educational programs through establishment of minor courses of study, many others have established interdisciplinary centers and programs, created interdisciplinary departments, and hired senior interdisciplinary scholars: some have gone as far as breaking down the disciplinary departmental structure altogether (Collins 2002; Feller 2002). The Consortium on Fostering Interdisciplinary Inquiry (2008) includes 10 large research universities who are in partnership to build upon their interdisciplinary policy and process experiences, aiding each other to improve institutional structures affecting interdisciplinary activities.

The greatest stress seems to occur at intermediate levels of investment as institutions and individuals attempt to adjust to the needs of interdisciplinary scholars. Because their needs are novel, the scholars often fall between the cracks of administrative responsibility (Fig. 27.2). As Monroe, *et al.* (2008) commented (with regard to women in academia) the '[...] lack of established procedures reflect an institution in flux, not one that is biased so much as unfamiliar with the needs [...] and struggling hard to catch up to a new institutional reality and culture'.

Being intentional about supporting interdisciplinary scholars requires thinking through the potential challenges in advance. The individual should not be put in the position of having to create their own process at the same time as they are attempting to navigate it. Creating an awareness of differences between interdisciplinary and disciplinary experiences—as we discuss below—can be helpful, from structuring a new hire, to understanding issues related to productivity, teaching, recognition, and evaluation. Awareness, however, is not enough. Funding and administrative support must also be provided. It is critical that

Table 27.3 Spectrum of institutional interdisciplinary commitment, investment, and therefore also responsibility (adapted from Pfirman *et al.* 2007).

Commitment and investment	Modest	Intermediate	Significant
Students and curriculum	Minor. General education elective	Concentration. Special major	Major. General education requirement
Administration	Committee	Center, program	Interdisciplinary department. Dissolution of departments
Faculty	Affiliated hire in disciplinary department. Adjunct hire	Off-ladder. Joint hire	Tenure-track interdisciplinary appointment
Research scientists	Soft-money support for single or short-term project	Multiyear support	Institution-committed career interdisciplinary research scientist line

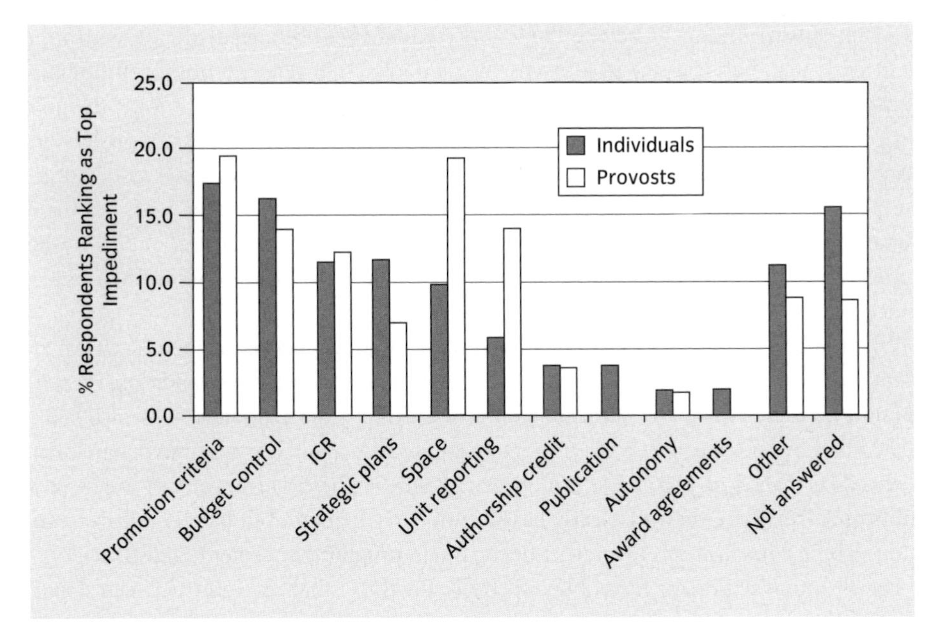

Figure 27.2 Impediments to interdisciplinary research identified by individuals and provosts in response to a request to rank the top five impediments to interdisciplinary research at their institutions (Committee on Facilitating Interdisciplinary Research 2004). Note the high ranking of promotional criteria as well as structural/administrative concerns: budget, indirect costs, space, and unit reporting. ICR = indirect cost recovery. Reprinted with permission from the National Academies Press. Copyright 2004, National Academy of Sciences.

institutions make commitments at the level of provost, vice president for research, or dean to the *implementation*—not just to the *initiation*—of interdisciplinarity (Feller 2002).

27.3 Structuring an interdisciplinary hire

The process of creating a new interdisciplinary position and the negotiation of the hire often determines the administrative framework of a position, and it is this framework that needs special attention for interdisciplinary scholars. Decisions about new interdisciplinary positions require more extensive cross-institutional preparation than for traditional disciplinary hires. At the start of position creation, roles and expectations must be clarified and agreed upon, by all the departments and academic administrators involved, ideally including representatives of promotion and tenure committees, and those responsible for allocating facilities and resources (Fig. 27.2; Pfirman *et al.* 2007).

While joint appointments (department–department or department–center) appear to make sense for interdisciplinary scholars, such appointments often lead to mismatches in their professional life. One is the expectation for service, an expectation that is often double for the joint appointment, serving the needs of two entities (or being penalized for appearing not to serve—for example when teaching is bought out by a research center).

Joint appointments may be held to the tenure standards of both departments, which may be at odds (e.g. publications in journals versus books, sole versus multiple authorship). Because responsibility for joint hires is divided, the junior scholar may not get the guidance that they would within a disciplinary department or even through a professional association (Table 27.2). The annual meeting of a discipline's professional association is the place to give presentations, test ideas, and meet the leaders in the field. Interdisciplinary scholars either contribute at the fringe of disciplinary meetings, or risk limited mainstream visibility when they participate in smaller workshops closer to their field of endeavor.

When interdisciplinary faculty are joint hires, it becomes imperative that each department manages their expectations, so that the time and activity demands on the joint appointment are reasonable and not doubled. Having a departmental split of 60:40 or 70:30 may be preferable over a 50:50 split to provide immediate clarity about departmental service (Pfirman *et al.* 2007). For junior faculty, an even better arrangement might be an 'affiliated hire' where they are clearly based in one department, but have specific research and teaching contributions to another department, program, or center (Table 27.2).

For all interdisciplinary hires, but especially for those that are joint between departments, the scope of the position should be articulated in a memorandum of understanding (MOU) that spells out scholarship expectations, promotion criteria, teaching responsibilities, departmental and community service, budget, indirect costs, graduate student/technician support, and space (Table 27.2, Fig. 27.2). These expectations can then be shared with potential candidates, and later adjusted as part of the negotiation package for the new hire.

Interdisciplinary teaching expectations need particular attention. Co-teaching classes with scholars from other departments can result in difficult negotiations with the administration and each department about course load, credit, responsibility, content, and classroom management methods. Also, many interdisciplinary educational goals would be best served by student-centered pedagogy—taking students out into the field, interacting with stakeholders, getting involved in civic engagement, or conducting student-led research (e.g. van Hecke *et al.* 2003). While these types of programs are often cited by students as transformative educational experiences, they are generally considered by the administration to be optional for faculty where the academic program has traditionally been delivered through in-class lectures and structured laboratories. Faculty who choose to incorporate these aspects in their teaching therefore do so at the expense of time they could spend on research, and may even risk having their teaching considered 'soft' or 'not rigorous' in comparison with colleagues who use more traditional approaches.

An institutional structure that can work well for interdisciplinary hires is a cluster hire (usually two to about eight new positions) (Sá 2008) to support a general theme or initiative, such as environmental sustainability. University of California campuses have been using cluster hires to quickly build interdisciplinary research teams. The administration, relevant departments, and centers, work to create the cluster, setting the stage for broad acceptance of the theme. Departments can compete to be the home department of the new hires, thereby creating greater departmental acceptance of the interdisciplinary scholar.

27.4 Productivity and the interdisciplinary scholar

One of the most critical aspects to the success of any scholar is that they are productive: the number of publications is the factor first reviewed for faculty hires and candidates for tenure (Steinpreis *et al.* 1999). Interdisciplinary scholars face hurdles in being productive beyond those of other researchers for a variety of reasons, the field may be new, the scholarly community not yet established; collaborative research requires high overhead/transaction costs in terms of communication, administration (Tables 27.1 and 27.2, Collins 2002; Shanken 2005; Sá 2008), and additional training requirements. Moreover, each discipline has its own convention for writing grants and publications, and disciplinary-based reviewers often raise issues and request revisions inappropriate for the scope of the interdisciplinary project or difficult to reconcile because they are at odds.

An interdisciplinary scholar could deal with this situation by building expertise in their particular interdisciplinary area—effectively specializing in that area—and then branching into related research topics and publishing in related journals. Leahey *et al.* (2008) showed that in sociology and linguistics, researchers who specialized (had a more limited set of key words associated with their publications) were twice as productive as researchers who pursued a research agenda that changed fields substantially over the course of their career trajectory.

Although junior researchers in any field are often admonished not to 'spread themselves too thin' this advice might be especially important for interdisciplinary scholars. Research by Porter *et al.* (2007) indicates that scholars who are highly integrative tend not to specialize (Table 27.1). It may be that people with a 'synthesizing mind' (Gardner 2007) use integration as part of their methodology, just as a lab scientist may enjoy addressing lab research problems through experiments in their lab throughout their career. Spanner (2001) also found that interdisciplinary researchers—especially those at the junior level—reported that they often deviated from their research agenda as they received input from another field.

Börner has tracked intersections among the disciplines by mapping knowledge domains—in the process creating a communication tool (e.g. Shiffrin and Börner 2004; Börner, Chapter 31 Box this volume). Interdisciplinary scholars can use this approach to work through related communities in linked networks, expanding their connections (e.g. Ginsparg *et al.* 2004), and therefore spreading their professional recognition. Mapped knowledge domains not only connect scholarly communities but can act as another measure of interdisciplinary productivity (Börner 2006; Palmer, Chapter 12 this volume).

27.5 Recognition of the interdisciplinary scholar

Along with productivity, assessment of research performance relies on reputation, especially recognition for creativity and achievement (Avital and Collopy 2001). Recognition arises from scholars reading and discussing each other's work, often within disciplinary boundaries. As noted by Csikszentmihalyi (1996), individuals who act as gatekeepers for a field have the responsibility to decide whether a new idea or product should be included in the domain. It is much easier for gatekeepers to recognize innovation when the advance is a direct extension of their own work or that of known colleagues. This issue is compounded by publication in new, interdisciplinary journals with nascent reputations (Campbell 2005). Without a process and a community for achieving recognition for creativity, the interdisciplinary scholar is faced with significant hurdles in promotion and tenure as well as in funding.

One way to create an interdisciplinary culture on campus, as well as to raise the profile of specific interdisciplinary scholars, is for interdisciplinary scholars to invite leading researchers to give presentations locally. This allows the local scholar to be the host: they get to know the external speaker better, they have the opportunity to talk about their own research, and issues of common interest become something known and talked about on campus. Such interactions are useful for any junior scholar, but are particularly important for those who are interdisciplinary or are in emerging fields (Tables 27.1 and 27.2). Lee and Bozeman (2005) showed that scholars typically spend about 50% of their time working with members of their own department, 15% working alone, 10% with others in the same institution, and 25% with outside collaborators. This means that, within a discipline-based department, there is a tremendous amount of shared knowledge. Interdisciplinary scholars often work with a broader community outside of the department, with collaborators who may be unknown to their departmental colleagues; it is helpful in gaining trust if departmental members get a chance to meet prominent interdisciplinary experts first hand.

While our focus thus far has been mainly on junior interdisciplinary scholars, senior scholars also experience recognition challenges (Pfirman *et al.* 2007). Most disciplinary societies have something along the lines of a 'lifetime achievement award' that identifies major accomplishments and gives credit for accumulated success. In emerging interdisciplinary areas, the scholarly community structures, and therefore the opportunities for recognition, are not well formed (Tables 27.1 and 27.2). Also, if the interdisciplinary scholar has not specialized, their contributions may be spread over a number of different communities and therefore may not rise to the level of an award in any one of them. Less likely to be the targets of recruitment from other institutions, interdisciplinary scholars may not get the offers that stars do within the disciplines. It is essential that institutions recognize these fundamental differences, and that they support their interdisciplinary scholars—perhaps through the establishment of institutional awards and medals that recognize their overall impact.

27.6 Interdisciplinary evaluation and promotion

Conventional, disciplinary-based procedures and standards to assess the work of interdisciplinary scholars ignore the real asymmetries between disciplinary and interdisciplinary research and teaching. In the 2004 Committee on Facilitating Interdisciplinary Research study, concern about 'promotion criteria' was the most frequent issue raised by both individuals and provosts in response to a request to rank the top five impediments to interdisciplinary research at their institutions (Fig. 27.2). Mismatched metrics include: the number of publications (as noted above, interdisciplinary, multi-authored work often has a slower production rate), focus on single- or first-authored papers (interdisciplinary publication often involves multiple authors), prioritizing well-known, disciplinary journals (not always an outlet for interdisciplinary scholarship), and citation indexes (interdisciplinary research is often new and must build its own constituency) (Table 27.2).

While tacit knowledge including unwritten guidelines for tenure within a department are passed along through informal collegial interactions and following the outcomes of individual cases, the interdisciplinary scholar is commonly the test case that establishes the criteria through their own performance. But it is not their responsibility to do so—institutions that hire interdisciplinary scholars should create appropriate procedures and metrics, and then be clear about expectations. A compelling way to address this situation is to change how scholarship is evaluated. Boix Mansilla (2006) noted that interdisciplinary work can be viewed through the lens of 'consistency with multiple disciplinary antecedents, balance of disciplinary perspectives in relation to research goals, and effectiveness in advancing knowledge through disciplinary interventions'. Lattuca (2001) recommends judging all scholarship simply 'on the basis of its contribution to the advancement of knowledge'. Another option is to shift from using only 'discovery' as the critical component, to the use of Boyer's (1990) expanded set of criteria: 'discovery', 'integration', 'application', and 'teaching' (e.g. Porter *et al.* 2006). Individuals can be asked to provide information on their contributions in each of these areas in their annual performance reports and then the same categories can be used in tenure review. The University of Southern California, Duke University, the University of Michigan Medical School, along with some small liberal arts colleges and some large US land grant universities do this now, because of their historic mission. However, a word of caution, one study of applied health researchers found that even when interdisciplinarity is at the core of an institution's mission, the chairs of promotion committees, and to a lesser extent the deans, tend to accord significantly more value to traditional scholarly outputs, ranking the importance of non-traditional research output at or below the level of teaching (Phaneuf *et al.* 2007).

Reviews of interdisciplinary scholars and proposals can also be facilitated by providing institutional clarity in terms of overall staffing/budget priorities and helping evaluators understand their mission. Letter writers, reviewers, and evaluation committees, can be alerted that the scholar or request for proposals is interdisciplinary, and then providing them with the original position or program description. Other options are to collect input from more areas of expertise, permit proposers to provide input on reviewer selection, and allow for proposer response to initial reviews (Langfeldt 2006).

In the case of a tenure review, the make up of the review committee itself can be critical: it is frequently helpful to include an external expert in the field of the candidate on a tenure review. A problem that can arise, particularly with new areas of interdisciplinary endeavor, is that the outside expert may not be a senior scholar, and therefore may not carry the same professional capital that the external member typically wields in this situation. In order that the review does not depend on this one scholar, individuals can also provide an annotated *curriculum vitae*, detailing their specific contributions to co-authored publications and grants, co-taught classes, informal advising, and standing of journals/publications; venues which may not be known to members of the committee (Pfirman *et al.* 2005; Pfirman *et al.* 2007).

27.7 Funding for interdisciplinary research and education

Traditionally, funding sources, whether internal or external to the university, have been channeled through disciplines. Therefore, support for interdisciplinary research is less stable than that for disciplinary research. When interdisciplinary calls for proposals are issued, they often have incredible proposal pressure, resulting in low funding rates: for example 4% for the National Science Foundation (NSF)-Environmental Protection Agency (EPA) Water and Watersheds (NSF AC-BIO 1996). Then, the funding area is often either discontinued, or moved to another administrative structure. The NSF, after a period of attention toward an interdisciplinary area, frequently migrates support back into the disciplinary directorates, with the goal of changing the culture in the directorates, as well as allowing for new areas of focused attention at the cross-directorate level. However, because the established, disciplinary communities are strongly manifested in the directorates, the emerging interdisciplinary areas may not fare well (especially under conditions of budgetary stress). As a result, interdisciplinary scholars lack continuity in programs and program managers to go to for support. When responsibility for the program shifts, interdisciplinary researchers must establish new contacts, spending considerable effort in rebuilding professional capital.

Funding agencies and institutions can help support interdisciplinary scholars by first recognizing that as they initiate interdisciplinary activities, the individual will move 'out on the limb' with their infrastructure lagging behind their needs (Collins 2002). They can be provided with release time, co-funding, matching funds, and other support for crafting and implementing complex or major research proposals, as well as new interdisciplinary or co-taught classes. Investing and promoting a small number of high-profile projects likely to have success can help institutions develop models that will then reduce resistance to tackling more risky endeavors.

The second major need in terms of funding is to explicitly support all four approaches to interdisciplinarity: intrapersonal, interpersonal/collaborative, interfield/departmental, and working with external stakeholders. The scholar wishing to develop intrapersonal expertise will need seed funding, sabbatical time, course release as well as perhaps travel support to learn from other institutions, along the lines of the Andrew W. Mellon Foundation 'New Directions' grants.

Scholars who are pursuing collaborative, interfield, and stakeholder approaches can be facilitated by funding agencies allowing more than one principal investigator (PI) to be listed on the grant, permitting PIs to be rotated or even added as the grant evolves, and crediting non-academic products.

Support is also required to develop opportunities for collegial contact, both professional and social: time and space is needed for collaboration to occur. Co-funding of research centers is one way many institutions are supporting interdisciplinarity. But funding for informal interactions is also helpful. As noted above, most interdisciplinary research is conducted in *ad hoc*, rather than formal, research teams (Evaluation Associates 1999). Similarly, 91% of the interdisciplinary scholars in the 2001 Spanner study rated collegial contact as being very important for their work. Trust, in addition to serendipitous connections, can be built through shared experiences such as social occasions and field trips, as well as through the more usual academic paths such as seminar series and workshops. Managing teams is difficult, but managing *ad hoc* interdisciplinary teams is even more challenging, due to issues conflated with interdisciplinarity (Table 27.1). Explicitly training interdisciplinary scholars in team management could lessen stress and increase effectiveness.

Interdepartmental and interinstitutional initiatives also face major hurdles in negotiating terms of budgets, indirect cost recovery, and space. In fact, the Committee on Facilitating Interdisciplinary Research found that, after promotion criteria, these are the most critical issues faced by interdisciplinary scholars (Fig. 27.2). Having a particular person within the institution's administrative structure whose job it is to sort out these issues greatly reduces the transaction costs of initiating new projects.

27.8 Concluding Remarks

Clearly, institutions serious about interdisciplinarity need to invest in support for individuals conducting interdisciplinary scholarship and teaching. Discretionary resources, incentives, and administrative support such as seed funding, incubation grants, co-funding and matching grants, cross-disciplinary workshops and seminars, leaves, travel, and joint appointments can go a long way toward helping people overcome personal and professional challenges. Funding agencies and donors can also help improve the working environment for interdisciplinary scholars by supporting research on the reform of faculty reward systems and investing in research on ways to evaluate and facilitate interdisciplinarity.

Institutions interested in fostering interdisciplinarity should review their administrative processes to determine whether there are impediments to fair and objective review and support of scholars working across disciplines. Administrative structures must be flexible to address the needs of interdisciplinary scholars, as one size does not fit all. Attention to the particulars of the interdisciplinary scholar's position is crucial, starting from the point of position creation to those of a senior faculty member. The lifecycle analysis by the Council of Environmental Deans and Directors (Pfirman *et al.* 2007) provides guidance for overcoming typical questions and challenges at each stage of career development.

High-level administrative leadership—through a committee or an individual with strong support from the provost-level—should oversee the implementation of interdisciplinary activities and fostering of interdisciplinary scholars. Leadership actions include: crafting MOUs for interdisciplinary hires, hosting an interdisciplinary speaker series, running interdisciplinary faculty research and pedagogy events, providing training on managing and evaluating interdisciplinary scholarship, and recognizing junior and senior scholars who have contributed significantly to interdisciplinary scholarship or teaching. An interdisciplinary faculty pedagogy forum, joint with schools or departments of education, can be designed to foster sharing of best practices, as well as an increased awareness of new educational approaches and challenges faced within different disciplines. It can also open up education as an area of common ground, building ties between disciplinary and interdisciplinary academic professionals.

Interdisciplinary faculty can thrive when institutions make the investments necessary to create the support structures commensurate with those provided—and taken for granted—by departments and professional societies for those within the established disciplines.

References and further reading

Alves, J., Amorim, C., Saur, I., and Marques, M.J. (2004). How to promote interdisciplinary R&D in the Academia: the case of the 'House of the Future'. *The R&D Management Conference 2004: Managing People and Managing R&D, Sesimbra, Portugal.* Available at: <http://www.casadofuturo.ua.pt/En/Papers/AlvesAmorimSaurMarques%20-%20RDMgmtConf2004.pdf>.

Austin, J. (2003). *Career advice: interdisciplinarity and tenure.* Science Careers from the journal *Science*. 10 January. Available at: <http://sciencecareers.sciencemag.org/career_development/previous_issues/articles/2100/interdisciplinarity_and_tenure/>.

Avital, M. and Collopy, F. (2001). Assessing research performance: implications for selection and motivation. *Sprouts: working papers on information environments, systems and organizations* **1**(1), 40–61. Available at: <http://sprouts.aisnet.org/1-14/>.

Beraud, A. (2003). A European research on women and engineering education (2001–2002): potentials of interdisciplinary courses in engineering, information technology, natural and socio-economic sciences in a changing society. *European Journal of Engineering Education* **28**(4), 435–51.

Boix Mansilla, V. (2006). Symptoms of quality: assessing expert interdisciplinary work at the frontier: an empirical exploration. *Research Evaluation* **15**(1), 17–19.

Börner, K. (2006). Mapping the structure and evolution of science. *Symposium on knowledge discovery and knowledge management tools at NIH Natcher Conference Center, Bethesda, 6 February 2006.* Available at: <http://grants.nih.gov/grants/KM/OERRM/OER_KM_events/Borner.pdf>.

Boyer, E.L. (1990). *Scholarship reconsidered: priorities of the professoriate.* Princeton, NJ: Princeton University Press and The Carnegie Foundation for the Advancement of Teaching.

Bronstein, L.R. (2003). A model for interdisciplinary collaboration. *Social work* **48**(3), 297–306.

Campbell, L.M. (2005). Overcoming obstacles to interdisciplinary research. *Conservation Biology* **19**(2), 574–77.

Campbell, W.H., Anderson, W.K., Burckart, G.J. *et al.* (2002). Institutional and faculty roles and responsibilities in the emerging environment of university-wide interdisciplinary research structures: report of the 2001–2002 research and graduate affairs committee. *American Journal of Pharmaceutical Education* **66**(4), Supplement.

Carayol, N. and Thi, T.U.N. (2005). Economics of science: why do academic scientists engage in interdisciplinary research? *Research Evaluation* **14**(1), 70–9.

Chamberlin, J. (2006). Uncommon ground: interdisciplinary scholars share the benefits – and challenges – of teaching and conducting research alongside economists, historians and designers. *APA Monitor on Psychology* **37**(5), 26. Available at: <http://www.apa.org/monitor/may06/uncommon.html>.

Choucri, N., de Weck, O. and Moavenzadeh, F. (2006). Editorial: promotion and tenure for interdisciplinary junior faculty. *MIT Faculty Newsletter* (January/February). Available at: <http://web.mit.edu/fnl/volume/183/editorial.html>.

Collins, J.P. (2002). May you live in interesting times: using multidisciplinary and interdisciplinary programs to cope with change in the life sciences. *Bioscience* **52**(1), 75–83.

Committee on Facilitating Interdisciplinary Research (2004). *Facilitating interdisciplinary research.* Washington, DC: National Academies Press. Available at <http://www.nap.edu/catalog.php?record_id=11153#toc>.

Consortium on Fostering Interdisciplinary Inquiry (2008). Fostering interdisciplinary inquiry: a fall 2008 invitational conference and creation of a new multi-institutional consortium. Minneapolis, MN: University of Minnesota. Available at: <https://www.myu.umn.edu/metadot/index.pl?id=1562406>.

Csikszentmihalyi, M. (1996). *Creativity: flow and the psychology of discovery and invention.* New York: HarperCollins.

Domino, S.E., Smith, Y.R., and Johnson, T.R.B. (2007). Opportunities and challenges of interdisciplinary research career development: implementation of a women's health research training program. *Journal of Women's Health* **16**(2), 256–61.

Evaluation Associates Ltd (1999). *Interdisciplinary research and the research assessment exercise.* UK Higher Education Funding Bodies report RAE 1. Available at: <http://www.rae.ac.uk/2001/Pubs/1_99/>.

Feller, I. (2002). New organizations, old cultures: strategy and implementation of interdisciplinary programs. *Research Evaluation* **11**(2), 109–16.

Feller, I. (2006). Assessing quality: multiple actors, multiple settings, multiple criteria: issues in assessing interdisciplinary research. *Research Evaluation* **15**(1), 5–15.

Gardner, H. (2007). *Five minds for the future.* Cambridge, MA: Harvard Business School Press.

Ginsparg, P., Houle, P., Joachims, T., and Sul, J.-H. (2004). Colloquium paper: mapping knowledge domains: mapping subsets of scholarly information. *Proceedings of the National Academy of Sciences USA* **101**, 5236–40.

Haag, A. (2006). A testing experience. *Nature* **443**(21), 265–7.

van Hecke, G.R., Karukstis, K.K., Wettack, F.S., McFadden, C.S., and Haskell, R.C. (2003). The interdisciplinary laboratory: an integrative chemistry, biology and physics. In: L.R. Kauffman and J.E. Stocks (eds) *Reinvigorating the undergraduate experience.* Washington, DC: Council on Undergraduate Research.

Jang, S-J. (2006). Research on the effects of team teaching upon two secondary school teachers. *Educational Research* **48**(2), 177–94.

Kenna, T., Pfirman, S., Selleck, B., Son, L., Land, M., and Cronin, J. (2006). *An integrated learning experience: 'river summer' on the Hudson.* White paper for the Teagle Foundation. Available at: <http://environmentalconsortium.org/taskforces/fieldstudies/riversummer/teagle.htm>.

Klein, J.T. (1990). *Interdisciplinarity: history, theory and practice.* Detroit, MI: Wayne State University Press.

Langfeldt, L. (2006). Risk avoidance: the policy challenges of peer review: managing bias, conflicts of interest and interdisciplinary assessments. *Research Evaluation* **15**(1), 31–41.

Lattuca, L.R. (2001). *Creating interdisciplinarity: interdisciplinary research and teaching among college and university faculty*. Nashville, TN: Vanderbilt University Press.

Lattuca, L.R. (2002). Learning interdisciplinarity: sociocultural perspectives on academic work. *Journal of Higher Education* **73**(6), 711–39.

Lau, L. and Pasquini, M. (2008). 'Jack of all trades': the negotiation of interdisciplinarity within geography. *Geoforum* **39**, 552–60.

Leahey, E. (2006). Gender differences in productivity: research specialization as a missing link. *Gender and Society* **20**(6), 754–80.

Leahey, E. (2007). Not by productivity alone: how visibility and specialization contribution to academic earnings. *American Sociological Review* **72**(4), 533–61.

Leahey, E., Crockett, J.L., and Hunter, L.A. (2008). Gendered academic careers: specializing for success? *Social Forces* **85**(3), 1273–309.

Lee, S. and Bozeman, B. (2005). The impact of research collaboration on scientific productivity. *Social Studies of Science* **35**, 673–702.

Lingard, L., Schryer, C.F., Spafford, M.M., and Campbell, S.L. (2007). Negotiating the politics of identity in an interdisciplinary research team. *Qualitative Research* **7**, 501–19.

Marzano, M., Carss, D.N., and Bell, S. (2006). Working to make interdisciplinarity work: investing in communication and interpersonal relationships. *Journal of Agricultural Economics* **57**(2), 185–98.

Monroe, K., Ozyurt, S., Wrigley, T., and Alexander, A. (2008). Gender equality in academia: bad news from the trenches, and some possible solutions. *Perspectives on Politics* **6**(2), 215–33.

National Institutes of Health (2006). *NIH roadmap for medical research. Interdisciplinary research.* Available at: <http://nihroadmap.nih.gov/interdisciplinary/>.

NSF AC-BIO (1996). *Meeting of the BIO advisory committee. Summary minutes 7–8 November 1996.* Available at: <http://www.nsf.gov/bio/bioac/meetings/minutes/9611.jsp>.

Pfirman, S.L., Collins, J.P., Lowes, S., and Michaels, A.F. (2005). Collaborative efforts: promoting interdisciplinary scholars. *Chronicle of Higher Education* **51**(23), B15.

Pfirman, S.L., Collins, J.P., Lowes, S., and Michaels, A.F. (2005b). *To thrive and prosper: hiring, fostering and tenuring interdisciplinary scholars.* Project Kaleidoscope resource. Available at: <http://www.pkal.org/documents/Pfirman_et-al_To-thrive-and-prosper.pdf>.

Pfirman, S.L, Hall, S., and Tietenberg, T. (2005c). The frontlines of interdisciplinary education: environmental programs at liberal arts colleges. *Environment, Science and Technology* **39**(10), 221A–224A. Available at: <http://pubs.acs.org/doi/pdf/10.1021/es053270w>.

Pfirman, S., Martin, P.J.S., Berry, L. *et al.* (2007). *Interdisciplinary hiring, tenure and promotion: guidance for individuals and institutions.* Washington, DC: Council of Environmental Deans and Directors. Available at: <http://ncseonline.org/CEDD/cms.cfm?id=2042>.

Phaneuf, M.-R., Lomas, J., McCutcheon, C., Church, J., and Wilson, D. (2007). Square pegs in round holes: the relative importance of traditional and nontraditional scholarship in Canadian universities. *Science Communication* **28**, 501–18.

Porter, A.L., Roessner, J.D., Cohen, A.S., and Perreault, M. (2006). Interdisciplinary research: meaning, metrics, evaluation. *Research Evaluation* **15**(3), 187–96.

Porter, A.L., Cohen, A.S., Roessner, J.D., and Perreault, M. (2007). Measuring researcher interdisciplinarity. *Scientometrics* **72**(1), 117–47.

Preston, L. (2000). *Mentoring young faculty for success: rewarding and encouraging involvement in cross-disciplinary research.* ASEE Engineering Research Council Summit. Available at: <http://www.erc-assoc.org/topics/Interd%20Resch%20&%20Tenure-Preston.ppt>.

Rhoten, D. and Pfirman, S. (2007a). Women in interdisciplinary science: exploring preferences and consequences. *Research Policy* **36**(1), 56–75.

Rhoten, D. and Pfirman, S. (2007b). Women, science, and interdisciplinary ways of working. *Inside Higher Education* 22 October 2007. Available at: <http://www.insidehighered.com/views/2007/10/22/rhoten>.

Sá, C.M. (2008). 'Interdisciplinary strategies' in US research universities. *Higher Education* **55**, 537–52.

Shanken, E.A. (2005). Artists in industry and the academy: collaborative research, interdisciplinary scholarship and the creation and interpretation of hybrid forms. *Leonardo* **38**(5), 415–18. Available at: <http://muse.jhu.edu/journals/leonardo/v038/38.5shanken.pdf>.

Shiffrin, R.S. and Börner, K. (2004). Mapping knowledge domains: mapping knowledge domains. *Proceedings of the National Academy of Sciences USA* **101**, 5183–5.

Spanner, D. (2001). Border crossings: understanding the cultural and informational dilemmas of interdisciplinary scholars. *Journal of Academic Librarianship* **27**(5), 352–60.

Stahler, G.J. and Tash, W.R. (1994). Centers and institutes in the research university issues, problems, and prospects. *Journal of Higher Education* **65**(5), 540–54.

Steinpreis, R.E., Anders, K.A., and Ritzke, D. (1999). The impact of gender on the review of the *curricula vitae* of job applicants and tenure candidates: a national empirical study. *Sex Roles* **41**(7–8), 509–28.

Teodorescu, D. and Kushner, C. (2003). Trailblazing or losing course? Views of interdisciplinary scholarship at Emory. *The Academic Exchange.* Atlanta, GA: Office of Institutional Research, Emory University. Available at: <http://www.emory.edu/ACAD_EXCHANGE/2003/octnov/teodorescukushner.html>.

Weingart, P. and Stehr, N. (eds) (2000). *Practising interdisciplinarity.* Toronto: University of Toronto Press.

WISELI (2003). *Study of faculty work life at the University of Wisconsin-Madison.* WISELI (Women in Science and Engineering Leadership Institute), University of Wisconsin-Madison. Available at: <http://wiseli.engr.wisc.edu/docs/Report_Wave1_2003.pdf>.

Woodward, K. (2005). *Encouraging recruitment, promotion and tenure, and awarding merit to interdisciplinary faculty.* Interdisciplinary Initiatives Working Group, The Graduate School, The University of Washington. Available at: <http://www.grad.washington.edu/Acad/interdisc_network/Meeting%201%20June%2005/Issue%20Brief-Tenure%20Merit%20Promotion.pdf>

CHAPTER 28

Doctoral student and early career academic perspectives

JESSICA K. GRAYBILL AND VIVEK SHANDAS

Interdisciplinarity is heralded as a new educational and research paradigm that can effectively address complex problems at disciplinary boundaries. Proponents of cross-disciplinarity, particularly inter- and transdisciplinarity, proclaim that it creates new kinds of researchers and educators 'at the leading edge' by promoting new forms of communication and collaboration among disciplines. Developing interdisciplinary research and education has become a goal for many universities, within both programs and individual courses, as noted in the mission statements of universities and funding institutions. Often, interdisciplinary research and training (IDRT) programs are conceptualized and implemented by faculty interested in conducting cross-disciplinary research, but they are also often managed by those most entrenched in the structure and experience of traditional universities, colleges, and departments.

While collaborative initiatives have typically occurred in the realm of established senior scientists (Dubrow and Harris 2006), they are now appearing earlier in academic careers, including undergraduate and graduate education. Those who actually experience and potentially reap the longest-term benefit from these programs are graduate students and early career academics. Moreover, as increasing numbers of doctoral students trained in interdisciplinary approaches enter the academic workforce, they carry with them their experiences, influencing how or whether they decide to continue such work in their future careers. The challenges and opportunities they face during doctoral training and in the transition to academic careers provoke questions about interdisciplinary research, training, and pedagogy within academia.

Extensive theoretical discourse exists regarding interdisciplinarity (e.g. Tress *et al.* 2003; Max-Neff 2005), but understanding the experience from the viewpoints of graduate students and early career academics who are explicitly trained within an IDRT frame provides a unique perspective on the benefits and challenges of pursuing interdisciplinarity. In fact, few publications address the graduate student and early career experiences in IDRT programs in traditional university settings. Published perspectives are largely those of well-established faculty or researchers, for whom it is difficult to 'understand and empathize with the ways students experience the institution. Faculty and staff tend to see the institution from their own perspective' (Hunt *et al.* 1992, p. 103). Additionally, guidance is virtually non-existent for students in interdisciplinary programs and early career faculty considering the pursuit of interdisciplinary academic trajectories in traditional university settings. Thus, the experiences of doctoral students and junior faculty—perhaps the most useful source of information on conducting interdisciplinarity—remain largely unheard, despite the proven utility of investigating such perspectives to understand innovative pedagogy (Anderson *et al.* 2000).

When heeded by faculty, departments, and universities, this perspective provides useful information for the creation and management of successful, long-term interdisciplinary programs. Whether universities are able to adequately support newly minted interdisciplinary scholars may determine the future success of these efforts.

This chapter contributes to an emerging body of literature about 'doing' interdisciplinarity (Graybill *et al.* 2006; Morse *et al.* 2007). By 'doing' interdisciplinarity, we refer to the formal and informal mechanisms that enable scholars from multiple disciplinary backgrounds (and epistemological persuasions) to engage in mutually beneficial and effective collaborations that may address a pressing societal challenge. We draw on existing literature and our own experiences to ground the pursuit of interdisciplinary research and pedagogy (IDRP)[1] in practical questions in how to guide multiple participants (individuals, departments, institutions, and disciplines) in cross-disciplinary endeavors. We address the benefits and challenges of participating in IDRP from the perspectives of doctoral students and early career academics, the latter defined as recently graduated PhDs in professorial academic positions.

Specifically, we address three major topics. First, we identify the transitional stages of the interdisciplinary career, from doctoral student to early career academic. Second, we provide an account of the overarching concerns arising for doctoral students and early career academics simultaneously pursuing disciplinary and interdisciplinary research and pedagogy in traditional university settings. Finally, we visualize potentially ideal IDRP institutions for newly minted interdisciplinarians. When addressing each topic, we draw on the existing literature and our experiences to pose questions and suggest pragmatic answers for promoting successful interdisciplinarity experiences for doctoral students and early career academics. Each topic contains opportunities and challenges for students, faculty, and institutions alike. Each of these must be addressed if interdisciplinary endeavors are to be successful.

28.1 Transitional stages

Four major transitional stages for doctoral students and early career academics trained in interdisciplinarity can be identified. In what follows we define and describe each stage, concluding each stage with questions aimed to promote reflection of IDRP across multiple institutional settings. The focus here is on critical questions that students and early career academics face concerning academe and the purpose and intent of interdisciplinarity as they progress through doctoral programs into academic positions.

28.1.1 Initiation

Doctoral students being initiated into interdisciplinary research and training must understand and meet a number of expectations that are not necessarily integrated across intellectual communities, subject matter, or mode of conducting research (e.g. qualitative versus quantitative approaches, empirical versus theoretical foci, etc.). Interdisciplinary doctoral students are often asked to develop dual intellectual communities—disciplinary and interdisciplinary—simultaneously, while also balancing disciplinary and interdisciplinary expectations and becoming familiar with disparate knowledge bases, all of which compete for time as students contemplate their advancement through their doctoral programs. In this early stage, students must develop identities in disciplinary departments and interdisciplinary programs. Combined with establishing solid theoretical footing and fulfilling expectations in both disciplinary and interdisciplinary realms, these tasks can be liberating, yet also exhausting and potentially disorienting if an individual's grounding in disciplinary research is not solid before interdisciplinary links are attempted. In this stage, students grapple with the process of doing interdisciplinary research and are most fully immersed in the training aspect of becoming interdisciplinarians.

Questions confronted during the initiation stage include: Where do I situate my scholarship? What is my identity in my disciplinary department and interdisciplinary program, and are they (should they be) the same or different? How can I strategize to obtain maximum benefits from the daunting task of creating one or more research projects that must be undertaken simultaneously? How do I craft research project(s) so that (a) I may complete them in a timely manner, and (b) they are rigorous and acceptable in both disciplinary and interdisciplinary realms?

28.1.2 Familiarization

Once initiated into interdisciplinary research and training, familiarization occurs as students become comfortable with creating potentially dual research paths and disciplinary and interdisciplinary identities, and have learned to anticipate some of the competing intellectual inputs, expectations, and reward mechanisms coming from disciplinary and interdisciplinary directions. Familiarization implies that students have met with some degree of success in initiating interdisciplinary research, are fully immersed in the training aspects of it, and have grappled with many of the 'nuts and bolts' issues of doing interdisciplinarity discussed at length elsewhere (e.g. Rhoten and Parker 2004; Lélé and Norgaard 2005), such as learning/creating a common language, developing the professionalism needed for cross-disciplinary interaction, accommodating the extra time needed for team work, and practicing appreciative inquiry (learning to understand and value different kinds of scholarship).

The familiarization stage also requires that doctoral students have oriented themselves enough to know how to navigate the dual loyalties to disciplinary and interdisciplinary intellectual communities and research projects, and have learned to juggle the demands of completing all requirements in each intellectual realm (including interdisciplinary team research, individual disciplinary coursework, general exams, dissertation proposals and defenses, fieldwork, fellowships, and publications). Navigating through these multiple requirements and expectations requires students to learn and practice sophisticated negotiation techniques in order to define the limits of individual and team research possibilities. Navigation is a skill developed and used by individual students with peers, faculty advisors, and potentially with departments, as the boundaries of feasible research must be carved out to maintain timely progress through the degree.

Finally, in the last phases of familiarization, doctoral students have become comfortable with their dual identities in home departments and in interdisciplinary working groups, and may be nearing completion of all degree requirements. At this point, publication outlets are identified, and academic career searches begin. Doctoral students who are now already familiar with the process of doing interdisciplinarity on a daily basis must demonstrate in publications and to potential employers their breadth and depth acquired by dual accomplishment of disciplinary and interdisciplinary research and training agendas. Further strategizing and negotiation may be involved, as students emphasize either disciplinary or interdisciplinary training, research, and potentially also pedagogy, to potential employers based on the job descriptions and the characters of institutions, disciplines, and individual departments.

Questions confronted during the familiarization stage include: How do I maintain rigor and depth in my disciplinary and interdisciplinary research, yet still complete the PhD in a timely manner? What qualifies as legitimate amounts and types of research in my disciplinary and interdisciplinary fields? Where will my interdisciplinary research be published, and how will choice of publication venues impact my hire-ability in a university setting (e.g. within or external to a discipline)? In a job interview, how do I describe the benefits and value of my interdisciplinary training to scholars entrenched in disciplinary knowledge bases?

28.1.3 **Adaptation**

Once placed in academic posts, newly minted interdisciplinary PhDs face new challenges and potential benefits. Most challenges derive from adapting to what interdisciplinarity means across different institutions, departments, and among new sets of colleagues. While there is certainly excitement at being in a new institutional setting and facing the opportunity for new collaborations and funding arrangements for conducting IDRP, there is also a caution that may develop as recent graduates from interdisciplinary programs and early career academics find that new colleagues and institutions may have differing interpretations of—and value for—interdisciplinary research, training, and pedagogy. As Palmer (2001) points out, interdisciplinarity is now essential to dialogues about knowledge production, yet 'because the notion of interdisciplinary research has not solidified, debate about what it really means goes on' (p. ix). Recent attempts to delineate and describe the range of cross-disciplinary endeavors (e.g. Klein 2005) aid in solidifying understandings of multi-, inter-, and transdisciplinarity, but as increasing numbers of practitioners across disciplines and institutions continue to pursue cross-disciplinarity, the range of interpretations and practices also increases.

It is into this milieu that early career academics trained in interdisciplinarity today arrive in their first job postings. From the point of view of an early career academic, it is exciting to enter a new institution with the possibility of conducting new collaborations as a doctorate holder. However, one formidable challenge is the reality that academics from different disciplines often only come together (1) when research proposals are formulated, or (2) when cross-disciplinary courses are taught. In either case, these collaborations often remain among established faculty who are already familiar with each other's research or pedagogical *modus operandi*. Although the reasons may vary by department or by institution (e.g. collaboration is seen as risky to obtaining promotion, the quality/record of a junior faculty's work is unknown by established faculty, collegial relations are already established among certain departments/colleagues, etc.), early career academics are often held at bay from such collaborations.

In the case of faculty coming together only when research proposals are the site of collaboration, deeper ties may not form among faculty, as concern rests largely with obtaining funding to support student research, attend conferences, purchase data/equipment, etc. In the case of faculty joining forces for pedagogical reasons, the focus may only be on providing students with multiple viewpoints on a particular topic rather than on faculty creating deeper, collaborative ties related to research and teaching interests. Without these deeper ties developing in either case, the interaction between researchers may remain limited—or may even become strained—as space for trust and social bonding has not been created.

Early career academics trained in IDRP can become particularly frustrated by these issues, because they already know the challenges of conducting IDRP (see stages 1 and 2 above), but must now weigh the benefits and challenges of pursuing it on top of institutional and departmental requirements. For some, this adds to the exhilaration of continuing to pursue dual career interests (disciplinary and interdisciplinary), but for others it is overly taxing or unrealistic in their individual institutional settings, which may lead to reduced capacity (personal or institutional) to continue with IDRP.

Questions confronted during the translation stage include: How much should I introduce and promote my vision for interdisciplinarity in my new institution and/or department as an early career academic? Does my institution consider interdisciplinary research or pedagogy risky in the tenure process, and how much should I risk as a non-tenured faculty? Should new pedagogical techniques that incorporate interdisciplinarity be introduced while I am in a pre-tenure phase? How will I manage the time commitments to building new IDRP collaborations on my new campus, when my time is already apportioned by the university's pre-existing expected research, teaching, and service commitments? Am I conducting IDRP alone, or are there other faculty who share my commitment and interests on whom I can call?

28.1.4 **Protected enthusiasm**

Within the first few years of placement in traditional academic institutions, early career academics trained in IDRP reach a point at which they know what is expected for advancement within their individual institutions. While enthusiasm for and commitment to the ideals of IDRP may remain, they may become protected or even diminished as the individual must consider self-preservation within the institution in order to remain a viable candidate for tenure. Palmer (2001) attributes this to the fact that there are 'few concrete incentives' for academics to prove themselves outside their disciplines, largely due to entrenched systems of rewards and promotion based on *individual* achievements and awards. Without incentives, academics 'engender more potential risks than rewards' for pursuing cross-disciplinary, collaborative endeavors (p. 71). After all, '[t]here is a serious disaccord between what leads to a successful scientific project and what leads to advancement at a university' (p. 80). The tenure and promotion system remains primarily based upon the construct of division and competition among disciplines, and evaluation is about whether a pre-tenure academic has contributed to the scholarship *within a discipline*. In other words, the luxury afforded to doctoral students in today's IDRT programs—the time and space to think creatively and collaboratively—is removed for early career academics who must concentrate their energy on becoming experts in their disciplinary fields in time-limited, promotional academic positions.

Additionally, early career interdisciplinarians may arrive at their new institutions unaware of real and pre-existing disciplinary divisiveness and of the protectionist mechanisms that may exist for disciplinary preservation within any given institution. A hypothetical example of this is the following. Prior to the arrival of an early career academic at a new institution, cross-campus ties between academics could have become strained due to an event involving two departments or the larger institution, perhaps what one individual perceived to be an unfair or unethical decision regarding their department or the institution, or perhaps disciplinary 'dissing' had become the norm (e.g. 'social sciences are not "rigorous"' or 'natural sciences are only concerned with narrow, laboratory-based learning and don't address the "larger picture"'). As a result, an early career academic's hopes to bring together specific researchers/departments on an IDRP research project may be difficult or impossible.

During this stage, these issues could halt the best intentions of any individual seeking interdisciplinary research or pedagogical collaboration across disciplinary divides, but may be particularly devastating to early career academics who are interested in jump-starting new collaborations in the early years of a tenure-track career. The reality sets in that some of these divides are best not battled in pre-tenure years, which may change the nature of how collaboration or the possibilities for interdisciplinary research or pedagogy are considered. Early career academics may find that instead of seeking new collaborators within their new institutions, they rely on existing, trusted connections to continue researching and publishing. While the phenomenon of continuing to conduct research and publish with prior collaborators is not unique to interdisciplinarians, the trick for those trained in IDRP becomes to maintain meaningful and truly interdisciplinary research and pedagogical collaboration between not only disciplines, but also perhaps across different institutions scattered worldwide, which logistically may be very difficult.

Questions confronted during the protected enthusiasm stage include: How should I best represent my dual identity as a disciplinarian *and* interdisciplinarian and my dual research agendas to internal and external reviewers for promotion purposes? What does my discipline/institution consider 'risky' in the pre-tenure years (e.g. co-teaching across disciplines, conducting more interdisciplinary research than disciplinary research, publishing more co-authored interdisciplinary research than single-authored research)? How do I maintain enthusiasm for interdisciplinarity in my new institution when I may not be able to pursue it as I was trained as a doctoral student? Should I seek to build interdisciplinary bridges within my institution, or maintain them with collaborators trained as I was and who are likely external to my institution? How can I seek to challenge or change my institution's views and practices of interdisciplinarity as an early career academic?

28.2 Overarching concerns

While this handbook attests to the growing body of literature on 'doing' interdisciplinarity, many colleges and universities claim that they have been conducting IDRP for several decades, and the 'doing' is already being done. What is different for doctoral students and early career academics now, however, is the fact that many institutions are only just acquiring newly minted PhDs who are formally trained in interdisciplinarity (coincident with the rise of large-scale, funded interdisciplinary programs nationwide, such as the National Science Foundation's Integrative Graduate Education and Research Traineeship [IGERT] program). Formal training of interdisciplinary doctoral students still occurs largely by disciplinarians, although this will change over time. Currently, one primary challenge for institutions is to engage these newly minted interdisciplinary PhDs in ways that support their scholarship while also ensuring that traditional disciplinary-focused research continues.

The simultaneous pursuit of disciplinary and interdisciplinary research and pedagogy raises one major question (which in turn spawns several other questions) for doctoral students and early career academics: *are they performing for their disciplines, or for a new academy*? Posing this provocative question pushes past the rhetoric of interdisciplinarity to address today's pressing scientific problems that lie at the boundaries of disciplinary interests and capabilities. It is the one overarching question that serious interdisciplinarians must address explicitly—and as early as possible—in their research and pedagogical agendas.

For example, imagine a scenario in which an interdisciplinary doctoral student seeking a tenure-track post composes her list of publications and notes that there are many interdisciplinary contributions, but only one disciplinary contribution. Her publications reflect the proportion of interdisciplinarity and disciplinarity in her degree, which has been well received by her doctoral committee. However, as she applies for jobs, she notes that most remain disciplinary based, which she knows will most likely require performance as a disciplinarian first, and as an interdisciplinarian second. She begins to doubt if her publication record speaks to this ability, and wonders how potential employers will evaluate her for a tenure-track job, or if she will be free to perform in the future in cadence with her doctoral training *if* she is hired. Will she need to adapt her scholarly trajectory to be more disciplinary? If so, then why did she go through the exciting, cutting edge interdisciplinary program? She questions if existing, largely traditional disciplinary programs and universities can (or will) accept this new type of academic in their junior faculty positions. Thus, should she continue to perform for a new academy that values interdisciplinarity, when that focus may not be present in the universities where job postings are open, or should she revert to performing for a discipline, which requires her to become more traditional in approach to research and pedagogy?

If research and pedagogy are truly aimed at advancing interdisciplinarity, then some serious questions must be raised about the structuring of knowledge accumulation, learning processes, and promotional structures entrenched in institutions of higher education. Three major points related to these overarching concerns are discussed below.

28.2.1 Meaning of the interdisciplinary PhD

One concern is what interdisciplinary training means in the transition between academic career phases (from doctoral students to early career academics and beyond). Recognizing that a major challenge for IDRP is that it is still being formulated—and variably—across different institutions and disciplines, early career academics must evaluate the level of risk associated with remaining involved in IDRP at different institutions. While important breakthroughs or scientific advancement are more likely in settings that allow for early risk-taking and failure, this is often not an opportunity for many tenure-track academics (regardless of career stage). In fact, it is rare for any department to include risk-taking as an explicit agenda in research or pedagogy (one exception is the promotion and tenure standards in the Department of Philosophy and Religion Studies at the University of North Texas). Typically, if risk is encouraged at all, it may reside only in individual classes or advising relationships; but activities involving actions that do not result directly in typical scholarly outputs (e.g., grant awards, publications)—such as consulting with public or private agencies or developing new educational programs—are usually not rewarded. This is not to argue that IDRP academics should not conduct these activities, but rather to point out that early career interdisciplinarians are particularly vulnerable because of their formative training in problem-based, innovative research approaches and educational curricula.

The challenges of 'doing' interdisciplinarity are magnified by the shifting requirements, expectations, and understandings (theoretical and practical) of interdisciplinarity across disciplines and institutions. Ironically, perhaps the one thing that remains constant in the academic setting is an individual's disciplinary home base. As doctoral students and early career academics explicitly trained in IDRP continue to move into the professoriate, defining the notions and expectations of IDRP for the individual and the academic community will become increasingly important.

This raises the issue of *clarity* related to interdisciplinarity. As increasing numbers of academics formally trained in IDRP join the academic workforce, questions about how interdisciplinarity is defined and practiced (theoretically and practically), and who has 'ownership' of it (e.g., those who have formal training in it versus those who have tried to design it without formal training) will increasingly emerge as central to 'doing' interdisciplinarity. Related to clarity are the following questions: if academics in multiple stages of their careers are responding to calls for increased interdisciplinarity, how do we know

(1) what it is, (2) if it has arrived, and (3) how it will be sustained? These questions speak to the need to continuously reflect on the use of the term, and to evaluate our individual and collaborative practices of IDRP.

28.2.2 Evaluation

Interdisciplinary doctoral students and early career academics are trained to evaluate and consider knowledge, to conduct research, and potentially to teach in new ways. How these academics are evaluated, however, has not changed either in graduate school or in the early career stages. This raises the issue of how the existing systems of internal institutional and external evaluation could also be changed to make the pursuit of interdisciplinarity within largely traditional institutional settings holistic instead of dualistic (i.e., on top of existing performance requirements). Addressing transdisciplinarity, Klein *et al.* (2001) write that such 'projects should be evaluated in a different mode than disciplinary projects' (p. 16). The same can be argued for interdisciplinary projects, which also have what Klein and co-authors call a 'special context of application, team process and participation [and] outcome and problem solving' (p. 17). Part of that special context is that interdisciplinary grants and projects are often undertaken to advance innovative research in new directions. As such, expected findings may be unknown ahead of time, and consensus among multiple reviewers on the purpose or merit of interdisciplinary grants and projects is difficult to obtain (Porter and Rossini 1985; Holbrook, Chapter 22 this volume).

External peer evaluation of interdisciplinarity in grant writing and publishing is also an overarching concern for graduate students and early career academics. Of particular concern are (1) what aspects of interdisciplinary research should be evaluated, and (2) by whom? Are the criteria commonly considered in the peer review of disciplinary research sufficient for interdisciplinary research? Are the same peer evaluators qualified to judge disciplinary and interdisciplinary research? Russell (1983) argues that 'criteria which acknowledge the unique qualities of interdisciplinary research' are necessary, including consideration of a research team's need to 'maintaining cooperation and communication towards a common goal' (p. 190). This may suggest that peer reviewers attuned to competition-driven grants and publications may not appreciate the structuring of interdisciplinary endeavors, and thus may not be suitable reviewers for such research. With increasing numbers of interdisciplinarians seeking grants and new publication venues, this is a critical issue that deserves new attention today.

28.2.3 Institutional adaptation

Interdisciplinary doctoral students and early career academics have been asked to become 'agents of change' and to accommodate a 'new academy' in their doctoral training, yet the institutions in which they function have largely not been asked to change. How this mismatch will be addressed in the future is important for the success of interdisciplinarity. For example, many innovative doctoral training programs bring together students from multiple disciplines to address a regional problem with global significance (e.g., urban sprawl, infectious disease migration, climate destabilization). These programs produce PhDs and early career academics who are already familiar with the substantive, linguistic and syntactical, conjunctive, and value domains of multiple disciplines (see Phenix 1986). As a result, such early career academics instinctively engage with those outside their own disciplines, but often institutional barriers preclude a deeper integration of disciplines.

Previously mentioned, of course, is the primary barrier of the existing awards and penalties system of tenure. Additionally, accommodating interdisciplinarity could be done by finding ways for faculty to 'share' student course credit hours across divisions or colleges within one institution, establishing interdisciplinary centers/programs that assist in creating cross-disciplinary dialogue, recognizing and accommodating the time requirements for teaching interdisciplinary courses in addition to 'required' courses, and assisting doctoral students and early career researchers in finding other new IDRT researchers on campus. In other words, good intentions to conduct (by doctoral students and early career academics) or encourage (by institutions) interdisciplinarity may not be enough for 'doing' IDRT in higher education, and specific institutional strategies for encouragement may be needed.

28.3 Interdisciplinary futures: strategizing for and visualizing success

A commitment to interdisciplinary training and research taken on by doctoral students and continued into further research and pedagogy by early career academics is admirable, as it is often conducted on top of existing requirements during the PhD or the pre-tenure years. Some authors have suggested that the commitment interdisciplinarians make to go 'above and beyond' in research or pedagogy is done by a self-selecting kind of person or group (e.g., Stone 1969; McCorcle 1982; Cassell 1986). Regardless of the type of person who may pursue IDRP in their career, successful encounters with it require strategizing for success in it and, at the next level, visualizing what individuals, departments and institutions need in order for IDRP to gain in stature and to be recognized as a force both internally and externally to any discipline. Below we briefly discuss some individual strategies for success and propose some scenarios for visualizing success for individuals, departments, and institutions. This discussion and the scenarios are not exhaustive or conclusive; rather they are meant to provoke responses and further thought from those most interested in the overall, long-term success of interdisciplinary research and pedagogical endeavors.

For interdisciplinary doctoral students and early career academics, part of the process of bridging disciplines is recognizing the need to negotiate a 'best' career path with oneself, one's department, and one's institution. This requires identifying strategies for *individual* success, which is counter to much collaborative, interdisciplinary research training. Particularly for doctoral students and early career academics trained as interdisciplinarians, strategizing for success means examining one's ethics and making choices about the pursuit of disciplinary or interdisciplinary endeavors and individual or team projects based on individual needs. For example, academics at these stages must decide what the balance of disciplinary and interdisciplinary research and pedagogy will be and how to maintain integrity in research and pedagogy in their individual disciplines while remaining 'true' to interdisciplinary interests. Indeed, Palmer (2001) writes that 'even the most interdisciplinary scientists have to make "cold-blooded" decisions about how to concentrate their energy' (p. 82). Decision making involves strategizing the routes for survival through internal (departments, institutions) and external (greater disciplinary) career pathways. Every individual trained in IDRP navigates differently based on these three sites of knowledge and career production, largely in response to the tenure system of promotion fostered by their institutions.

A common ethical concern that arises from strategizing is maintaining and building on IDRP momentum when also asked to operate in the current tenure-track system of promotion and penalties that is so strongly tied to individual disciplines. Individual responses to this task become 'ethical strategies' for survival, as interdisciplinarians seek to remain ethical to themselves, to their colleagues, and to the project of interdisciplinarity. For many, this requires strategizing about a 'best path' to carve through the dissertation and early career academic phases. That best path may come across as disingenuous to one's identity as an interdisciplinarian or to the project of interdisciplinarity when individuals are forced to negotiate their research and pedagogical agendas to accommodate increased disciplinarity. This is particularly poignant for pre-tenure faculty.

Table 28.1 attempts to visualize what ideal institutions could look like for interdisciplinary doctoral students and early career academics. The scenarios in the table directly address the substantive issues that arise for individuals, departments, and institutions as interdisciplinarity continues to be sought as a new paradigm for research and pedagogy within multiple institutions. Specifically, we target solutions to the needs of interdisciplinary doctoral students and early career academics so that they may thrive in the sustained rise of this new paradigm.

Conceptualizing and exploring the five dimensions of support (for research and pedagogy, by organizational structures, and through incentives and evaluation) needed to do interdisciplinarity provides this handbook with tangible recommendations for supporting doctoral students and early career academics through the early years of an interdisciplinary career in academia. What is key is to recognize that younger generations of academics committed to interdisciplinarity are being asked to perform for a new academy, not necessarily only for their disciplines. As such, these individuals champion interdisciplinarity, but wonder how and when institutions will also commit to this new *modus operandi*, while also promoting and protecting individuals and disciplines. If institutions continue to respond positively to interdisciplinary endeavors, then promoting it *and* protecting those who practice it will become increasingly necessary at institutional levels (departmental, administrative). As more interdisciplinarians enter academia, it will become more urgent and imperative that universities increase their support. Fostering it now will pave the way for long-term and successful engagement with and growth of interdisciplinarity.

Table 28.1 Framework for encouraging institutional support of interdisciplinary training, research, and pedagogy for doctoral students and early career academics

Dimension of support	Need	Selected ideal scenarios	Career stage
Research	Opportunities for formal and informal interaction with scholars from multiple disciplines	Create multiple cross-disciplinary forums for interaction: Research symposia Methodological workshops Social events	Doctoral student (with other IDRT students and faculty). Early career academic (with other early career and interdisciplinary faculty)
Pedagogy	Encouragement for innovative and interdisciplinary approaches to engage students in cross-disciplinary coursework	Give faculty time to develop service- or enquiry-based courses (e.g. course releases for development of IDRP courses) Provide institutional flexibility for cross-disciplinary classroom interactions (e.g. sharing course credit hours, listing courses across divisions) Promote campus-wide initiatives to practice appreciative inquiry of other disciplines in courses Create topical seminars and workshops on interdisciplinary pedagogy	Early career academic
University/institutional structures	Opportunities to meet outside departments. Support for collaboration	Fund IDRP centers that include research and training facilities Create on-line networks to aid learning about and participating in cross-disciplinary research/pedagogy clusters Provide workshops on 'doing' interdisciplinarity and team research with external facilitators	Early career academic
Incentives	Explicit and effective rewards system for pursuing IDRT/P	Develop language and guidelines in PhD programs to support IDRT students	Doctoral student. Early career academic

(*cont.*)

Table 28.1 Continued

Dimension of support	Need	Selected ideal scenarios	Career stage
		Develop language and guidelines in tenure and promotion cases to support early career academics	
		Create institutional or departmental mission statements and goals regarding disciplinary and interdisciplinary research	
		Fund internal grants for promoting team-based, cross disciplinary efforts	
Evaluation	Recognition of contributions made outside disciplines (e.g. interdisciplinary journals, professional reports).	Laud disciplinary and interdisciplinary contributions to scholarship	Doctoral student. Early career academic
	Recognition of interdisciplinary teaching that draws together students with diverse disciplinary backgrounds.	Promote enrollment of students from multiple disciplines in disciplinary courses	
		Evaluate the process ***and*** product related to efforts to conduct research with members of other disciplines (e.g. recognize the time and 'leeway' for IDRP efforts to succeed)	
	Space for students and faculty to express needs to administrations	Instill adaptive management of IDRP in the administration as a mode for faculty to actively engage with institutional structures	

References

Anderson S.K., MacPhee, D., and Govan, D. (2000). Infusion of multicultural issues in curricula: a student perspective. *Innovative Higher Education* **25**: 37–57.

Cassell, E.J. (1986). How does interdisciplinary work get done? In: D.E. Chubin (ed.) *Interdisciplinary analysis and research: theory and practice of problem-focused research and development*, pp. 339–45. Mt Airy, MD: Lomond.

Chubin, D.E. (ed.) (1986). *Interdisciplinary analysis and research: theory and practice of problem-focused research and development.* Mt Airy, MD: Lomond.

Dubrow, G. and Harris, J. (2006). *Seeding, supporting, and sustaining interdisciplinary initiatives at the University of Washington: findings, recommendations and strategies.* Seattle, WA: Network of Interdisciplinary Initiatives, University of Washington. Available at: <http://www.grad.washington.edu/acad/interdisc_network/ID_Docs/Dubrow_Harris_Report.pdf> (accessed 20 April 2008).

Graybill, J.K., Dooling, S., Shandas, V., Withey, J., Greve, A., and Simon, G. (2006). A rough guide to interdisciplinarity: graduate student perspectives. *Bioscience* **56**(9), 757–63.

Hunt, J.A., Bell, L.A., Wei, W., and Ingle, G. (1992). *Monoculturalism to multiculturalism: lessons from three public universities. In: Adams (ed.) Promoting diversity in college classrooms: Innovative responses for the curriculum, faculty, and institutions.* San Francisco: Jossey-Bass.

Klein, J.T. (1996). *Crossing boundaries: knowledge, disciplinarities, and interdisciplinarities.* Charlottesville: University of Virginia Press.

Klein, J.T. (2005). Interdisciplinary teamwork: the dynamics of collaboration and integration. In: S.J. Derry, C.D. Schunn, and M.A. Gernsbacher (eds) *Interdisciplinary collaboration: an emerging cognitive science*, pp. 23–50. Mahwah, NJ: Erlbaum.

Klein, J.T., Grossenbacher-Mansuy, W., Scholz, R.W., and Welti, M. (eds) (2001). *Transdisciplinarity: joint problem solving among science, technology, and society.* Basel: Birkhäuser.

Lélé, S., and Norgaard, R.B. (2005). Practicing interdisciplinarity. *Bioscience* **55**(11), 967–75.

McCorcle, M.D. (1982). Critical issues in the functioning of interdisciplinary groups. *Small Group Behavior* **13**, 291–310.

Max-Neff, A.M. (2005). Foundations of transdisciplinarity. *Ecological Economics* **53**, 5–16.

Miller, N. and Brimicombe, A. (2004). Mapping research journeys across complex terrain with heavy baggage. *Studies in Continuing Education* **26**, 405–17.

Morse, W.C., Nielsen-Pincus, M., Force, J.E., and Wulfhorst, D. (2007). Bridges and barriers to developing and conducting interdisciplinary graduate-student team research. *Ecology and Society* **12**(2), 8.

Pallas, A.M. (2001). Preparing education doctoral students for epistemological diversity. *Educational Researcher* **30**, 6–11.

Palmer, C. (2001). *Work at the boundaries of science.* Dordrecht: Kluwer.

Phenix, P. (1986). *Realms of meaning: a philosophy of the curriculum for general education.* Los Angeles, CA: National/State Leadership Training Institute.

Porter, A.L. and Rossini, F.A. (1985). Peer review of interdisciplinary research proposals. *Science, Technology, and Human Values* **10**(3), 33–8.

Rhoten, D. and Parker, A. (2004). Risks and rewards of an interdisciplinary research path. *Science* **306**(5704), 2046.

Russell, M.G. (1983). Peer review in interdisciplinary research: flexibility and responsiveness. In: S.R. Epton, R.L. Payne, and A.W. Pearson (eds) *Managing interdisciplinary research*, pp. 184–202. Chichester: John Wiley and Sons.

Stone, A.R. (1969). The interdisciplinary research team. *Journal of Applied Behavioral Sciences* **5**(3), 351–65.

Tress, B., Tress, G., and Fry, G. (2003). Potentials and limitations of interdisciplinary and transdisciplinary landscape studies. In: B. Tress, G. Tress, A. van der Walk, and G. Fry (eds) *Interdisciplinary and transdisciplinary landscape studies: potential and limitations*, pp. 182–91. Wageningen: Delta Program.

Weingart, P. and Stehr, N. (eds) (2001). *Practicing interdisciplinarity*. Toronto: University of Toronto Press.

Young, L.J. (2001). Border crossing and other journeys: re-envisioning the doctoral preparation of education researchers. *Educational Researcher* **30**, 3–5.

CHAPTER 29

A memoir of an interdisciplinary career

DANIEL CALLAHAN

It was the late 1960s, searching about for a professional niche in life, that I became interested in the emerging ethical and policy problems of what was then called 'the biological revolution', but could no less have been called the 'medical revolution'.

Where was biology taking us? In the air were utopian speculations about remaking human nature, allowing us to choose the genetic characteristics of our children, and radically extending human life expectancy. How should we ethically assess those possibilities? At the same time, complaints were rising about medicine's power to keep us alive too long and too miserably, anxieties were emerging about the rising costs of health care, and revelations were surfacing about wrongful exploitations of human beings for research purposes. How should those issues be evaluated?

29.1 Creation of the Hastings Center

With such questions in mind, in 1969 I helped create a research center devoted to the ethical and policy problems of medicine and biology. It was eventually called the Hastings Center, named after the town in which we started, Hastings-on-Hudson, NY (Callahan 1999). I recruited a neighbor to work with me, Willard Gaylin, a psychiatrist who had been an English major at Harvard before going into medicine, and the author of interesting books on various social problems having little to do with the technical problems of psychoanalysis, his specialty. So far as I know, he never wrote a single article for professional journals in psychiatry, but managed to make a solid name for himself in the field, and far beyond it. For my part, I had a PhD in philosophy, focused on moral philosophy. We made a fine interdisciplinary pair.

Not only because of our own proclivities, but also because of the breadth of the issues, we quickly agreed that our Center, and the field (still to be created), should be interdisciplinary. I was at that time working with demographers at The Population Council in

New York on the ethical problems of trying to lower birth rates in developing countries; that was my introduction to demography and reproductive biology. Gaylin, meanwhile, had simultaneous appointments at the Columbia University College of Physicians and Surgeons, Union Theological Seminary, and the Columbia University School of Law, an uncommon mix.

How did I become interested in interdisciplinary work? I have to go back in my personal history to explain that. As an undergraduate at Yale in the late 1940s I was unsure what field interested me the most. Then Yale created what it called an 'interdisciplinary major', allowing students to take a variety of courses but without any single disciplinary focus or any effort to coordinate the courses one took. My combination was literature, history, and psychology. It was not much of a program, and was subsequently improved with seminars and course coordination. But it was quite enough to whet my interdisciplinary appetite.

I finally decided on graduate work in philosophy, though I had only one philosophy course as an undergraduate. I chose the Harvard philosophy department, then in the throes of Oxford-dominated analytic philosophy. Along the way to getting my degree I taught a freshman writing course and also served as a research assistant to a cultural historian. By the time I received my degree I decided I was not enthralled with analytic philosophy, and particularly its narrow approach to ethics; nor did I see myself settling down in an academic philosophy department. My work with historians made me acutely aware of the cultural setting of ideas, and how even analytic philosophy was the child of a special Anglo-American outlook on philosophy, narrowly indifferent to other ways of thinking and insensitive to the unexamined culture that had shaped it.

Instead of teaching, I then took a job in New York as a magazine editor, a position I kept for 7 years. I wanted to make some use of my philosophy education, but outside of the university; that desire—combined with my interest in medicine and health care—led to the idea of starting the Hastings Center. The combination of my own checkered academic and journalistic life, and the nature of the field we were shaping—which came to be known as bioethics—made the need for interdisciplinary work obvious.

Why was interdisciplinarity necessary for bioethics? After all, its focus is on ethics, a standard academic discipline in philosophy and religious studies. It seemed to us, however, that the ethical problems emerging from medicine and biology required that they be set in a wide context, not only to understand them well but no less to see the way they played out in the wider world (Callahan 1973).

Our first project was on the changing definition of death, which was then moving from death traditionally defined as the cessation of heart and lung activity to the cessation of whole brain electrical activity. The change was being driven by two forces. One of them was that new technologies were able to keep hearts and lungs going for indefinite periods of time, making it unclear when to stop treatment. The other was that the fast-developing technology of organ transplantation made it necessary to have a more precise definition of death, one that decisively allowed treatment to be stopped, thus making transplantation morally acceptable.

Or did it? Our research project aimed to look at this change to determine how morally valid the move from one definition to another was, and particularly to examine whether the desperate desire for transplantable organs was seductively shaping the debate in a dangerous way. We put together an interdisciplinary team of philosophers and theologians (to include their modes of dealing with ethics), neurologists (to help us understand the concept of brain death), lawyers (to determine how laws and regulations needed to be changed to deal with the new definition), physicians (to understand the implications of the change for the care of dying patients), sociologists and psychologists (to get a sense of how the public would react to the new definition, and whether it would help or hinder the procurement of organs), and medical historians (to grasp the history of definitions of death and the force of the symbolic move from the heart to the brain as the locus of death) (Pernick 1999).

It was a fine and interesting exercise, with each of the participants learning from the others. It led to articles in medical, law, and social science journals as well as contributions to public policy in effecting the changed definition. Yet it had already become clear that perhaps our success with that first project was a fluke. Other efforts were not so easy. We were asked to initiate a course in medical ethics at a major New York medical school. That sounded interesting and we agreed to do so. Problems quickly surfaced. What kinds of credentials were necessary to teach such a course? Physicians said that obviously they should be the teachers since one could hardly teach such a course without actual medical experience in caring for patients, and that, anyway, to teach ethics was in the end to shape character and that no formal education in ethics was necessary to do that; good role models would do the job. Those with a training in ethics said that ethics was itself a discipline, one that required far more analytical skills than simply serving as a role model.

The compromise solution was to have team teaching, a doctor and a philosopher, sometimes supplemented by a social scientist or nurse. As time went on, however, it gradually became acceptable to have either a physician only, or a philosopher only to teach such courses. It came to be understood that the teacher should have a good disciplinary background in one field but be an educated amateur in the other: a physician teacher should have some education in ethics, and a moral philosopher should understand reasonably well what it was like to practice medicine and understand its culture.

But that was not the end of the problems. Some people held that while the field should be interdisciplinary in the end one field should have a privileged position. Not surprisingly, philosophers believed that their discipline deserved that position and doctors that theirs should. That argument has more or less faded away now, but that is perhaps because the earlier interest on the part of philosophers in the field has cooled considerably. During the 1970s moral philosophy took a more applied form, with the journal *Philosophy and Public Affairs* leading the way. Of late that interest appears to have faded. A larger number of physicians and lawyers are entering the field as the number of philosophers seems to have declined. Bioethics remains a fringe branch of philosophy, widely taught but with no special prestige, and in some departments not taught at all (at Yale, for instance, where I have a 'senior scholar' appointment but do not teach).

Another problem was that, in medical schools, there were no departments ready made for those hired to teach ethics. In what have been called 'convenience appointments', various medical departments were willing to give them a home. But when the time came for promotions and tenure decisions, matters often got complicated. If they were housed in, say, the department of surgery, the standards of judgment in surgery would not make sense. Yet if they were philosophers, but not in the philosophy department (much less doing standard philosophy research), that department was not in a position to evaluate them either. Who then were their peers, able to judge their research and teaching (see Holbrook, Chapter 22 this volume)? A common answer was to make use of ethicists at other institutions, doing the same kind of academic work. Moreover, since their work was interdisciplinary, they had to be judged by interdisciplinary, not disciplinary, standards. To ask a philosophy department to use its usual standards of judgment was thought to be unfair. Yet it is probably also fair to say that there are to this day no clear standards about what counts as good interdisciplinary work (see Huutoniemi, Chapter 21 this volume).

While the work of the Hastings Center is carried out by research groups drawn from national talent, that talent has always been interdisciplinary as well: mainly philosophers, lawyers, and social scientists. Staff at Hastings have had to learn how to work with people from other disciplines, understanding their modes of reasoning, their criteria for good work, and the folkways of different disciplines. Three traditions of the staff developed over the years. I told those newly hired that I hoped they would, after a time, learn how to talk in ways that did not reveal their own disciplines; what they said should just sound like ordinary language common sense. That meant, in effect, picking up enough law, philosophy, or social sciences over the years to be able to converse comfortably with experts from those fields—yet without quite sounding like them.

There was also an unwritten rule that no one, under any circumstances, should try to pull disciplinary rank; that is, to claim some special deference in a discussion because of one's discipline. Physicians at our conferences were often the worst offenders, opening their interventions with a sentence that typically began 'well, as a physician…', implying that no one but a physician could have the necessary experience and knowledge to say anything of value.

In order to keep interdisciplinarity alive and well at the Center, our hiring practices focused on making certain no one discipline had a disproportionate number. My rule when I was president was that there should be at least one person from every major discipline, but no more than two. Two from a major discipline was the ideal number, so that every staff member would have at least one colleague from the same discipline to talk and work with—but more than two would make it too easy for them to fall into shop talk and not be forced to talk with colleagues in different disciplines. It is a rule that has worked well over the years.

Moreover, as is true with many other free-standing research centers, the fact that we are not part of the university culture—which is heavily organized around disciplinary departments, and with few rewards for serious interdisciplinary work that truly cuts across departments and schools—is a great advantage. My colleagues do not as a rule much care what people in their own disciplines think about them, and they don't feel that they are held hostage by disciplinary peer review. Of course there is a price to be paid for that independence: their work may not be highly thought of by those in their disciplines, in great part because they feel no compunction about working outside of the traditional boundary lines. At the same time it can be said that, though not conforming to disciplinary traditions or worrying much about getting published in peer-reviewed journals, many staff members over the years have been lured away from us to take university jobs.

I have been drawn to interdisciplinary work over the years for some personal and some professional reasons. I like interdisciplinary work because I enjoy reading the literature of different fields, learning things I don't know, and taking on large topics that spill over disciplinary boundaries. Most disciplinary work I think of as painting carefully crafted miniatures: a very careful working through of a small and manageable problem. But I prefer to paint murals and panoramas. And the kinds of panoramas I most enjoy professionally are those that can't be fully assessed without crossing many disciplinary boundaries. The last point can be illustrated by showing how I have approached three major problems in health care in recent years: the clash between government-oriented and market-oriented ideologies in health care reform (Callahan and Wasunna 2006); the control of technology costs in health care (Callahan 2009); and the relationship between birthrates and health care for the elderly (Callahan, manuscript).

29.2 Medicine and the market in health care

A key issue in the American debate on health care reform—driven in great part by some 47 million uninsured—is that of the comparative roles that government or the private market should play. If there is to be universal care, should it be put in the hands of the government, which most liberals would like, or should it be managed by private insurers, as most conservatives would prefer? The European health care systems cover all citizens and are either financed directly by government (tax-based systems) or by mandated contributions from employers and employees (social health insurance). Both systems are heavily regulated by government. The American system, *sui generis* among developed countries, is a 50% mix of government financed and managed care and employer-provided private care.

When I first became interested in that issue, I turned to the work of health care economists, who have done research on the performance of both government and private care (Rice 2002). What I wanted to know was the comparative empirical evidence on the market versus government in light of health outcomes, access, quality, and costs. By that standard, European health care systems are easy winners. But as I was reading the economists, I was also following the debate in the pages of *The Wall Street Journal* and other conservative publications and the simultaneous push by President George W. Bush for more private market-driven care. Their test of a good health care system is an ideological one, based on a profound distrust of government and a full embrace of the market. They are far less worried about access and far more worried about consumer choice and private competitive insurers and providers.

The economist heroes of conservatives are Friedrich von Hayek and Milton Friedman, who believed that democracy requires a strong private market, and that a strong private market requires democracy. To understand that ideology requires a grasp of the history of market thinking, going back to Adam Smith. To understand the specific force of market thinking in the United States I read political scientists, ever returning to Thomas Jefferson who said that 'the best government is the least government'. To understand the historical resistance of American physicians to a government-run system, it was necessary to understand their objection to any outside force that would interfere with the practice of medicine as they saw fit or threaten their incomes (Starr 1982). To understand the problem of inequity in health care I turned to philosophers, who have given problems of justice a high place. And to understand why health reform has been so difficult to achieve in the United States—with sporadic efforts going back over 60 years—I studied public opinion surveys. They show that a majority of Americans have for years said they favor universal care (75–80%), but that the surveys also showed profound ideological differences on how it might acceptably be achieved.

Now it may seem self-evident to the reader that if one wants to understand the government–market conflict about health care one should be doing some reading in all those fields. Remarkably enough (and maybe a bit depressingly so) hardly anyone in each of those disciplines cites or makes use of the knowledge or insights from the others. There were exceptions, but those who write on the problem tended to stay within their own disciplines. Health economists do not as a rule cite historians or political scientists and the latter did not cite them. Astonishingly, I was berated by a health care economist for even taking on the topic of medicine and the market, which she believes belongs entirely to her field, and which I as a philosopher have no competence to discuss.

29.3 Medical technology and health care costs

The United States now spends $2.2 trillion a year on health care. Fueled by a projected 7% cost increase per year for the foreseeable future, the national care costs in a decade will be $4 trillion, some 20% of the gross domestic product. The Medicare program for the elderly, now costing $421 billion a year, will be bankrupt in a decade without radical reform. The rising costs are a major reason for the steady rise of the uninsured, now 47 million, for a decline in employer-provided health care, and for rising out-of-pocket expenses even for those with good health insurance.

While costs rise because of overall inflation pressure and a variety of other causes, the main driver of costs is medical technology, taken to account for 50% of the annual increase (Congressional Budget Office 2008). New technologies and the intensified use of old ones—primarily drugs and medical devices—are the leading accelerants. While many efforts at controlling costs in general and of technology costs in particular have been pursued for over 30 years, none of these efforts have made much difference.

At the heart of the technology problem, I believe, is American culture, one that is uncommonly dedicated to medical progress and technological innovation. American patients seem enamored with progress in general and technology in particular, doctors are trained to use it, and industry makes billions of dollars a year selling it. There are dozens of incentives to use technology in our health care systems, and few disincentives. How is this phenomenon to be understood and dealt with?

While the discipline of health care economics has provided much of the data on rising health care costs as well as offered a solid analysis of its proximate causes, that is simply not enough to get the full picture. Why has Congress forbidden the Medicare program to take cost into account in determining the benefits it will provide for the elderly? That is a question for political scientists and legal scholars. Why is it that, compared with other countries, Americans have a much higher respect for medical technology and greater expectations for its benefits? That is a question for medical and cultural historians. The control of costs will of necessity require rationing, not giving patients all they will want and may need. That is a question for philosophers since it must deal with matters of justice, and for legal scholars who will have to cope with the almost certain recourse to the courts by patients denied care. No one of those disciplines can by itself offer a road map to control costs, and at least one reason for a failure to make much progress is the lack of an integrated plan that blends economics, political science, medicine, cultural studies, policy analysis, and ethics.

29.4 Birthrates and elder care

Every developed country faces serious problems with a rising number and proportion of the elderly. Health care costs for the elderly under Medicare are shortly projected to rise sharply as the baby boom generation begins to retire, and not too long into the future the Social Security program will come under pressure as well. Yet American problems with an aging society are mild compared with those projected in most other developed countries, where the ratio between the young (whose taxes pay for the old), and the old themselves (who need the financial and social support of the young) is changing to the disadvantage of both the young and the old (United Nations 2004).

At first glance, it might seem that it is medical progress that has greatly lengthened average life expectancy and thus the number of the aged that lies at the root of the cost problem. That is true to some extent (average life expectancy beyond 65 has increased by about 6–7 years since 1970), but a more important influence has been a decline in birthrates. The American baby boom generation (those born between 1947 and 1964) was a very large one, with three or four children for most women. But those baby boomers, now on the verge of retirement, did not themselves have comparably large families, and have been averaging slightly fewer than two children per woman. What had been a ratio of about four younger working people for every retired person is now declining to a ratio of 2.5.

In short, there is a declining base of young taxpayers to support a rising base of elderly retirees. And the situation is far worse in countries like Spain, Italy, Japan, and Poland, where every woman now bears only 1.3–1.4 children. As if the social and medical and economic problems generated by the increased proportion of the elderly is not enough to generate some anxiety, there is considerable agreement among economists that a steady stream of young people is necessary for economic vitality and stability. Many European countries are threatened with declining populations, which will create some unprecedented economic problems (Grant 2004).

To get a good sense of those trends, and to devise public policies to manage them, requires research in a variety of disciplines. I have turned to demographers to get a grasp of the history of procreation and birth rates as well as explanations of various historical trends. Economists have taught me a great deal about the place of young workers in an economy. I make use of historical and sociological knowledge to understand the changing place of family and childbearing in modern societies. Feminists have a good deal to say about governmental calls to increase birth rates and that literature must be consulted. Gerontologists provide important information and studies of the changing place of the elderly in society, as do geriatricians on the medical situation of the elderly. And of course there is a considerable literature—novels, plays, poems—that touch on those topics. To take on the combined topics of bearing children and getting old is to enter a complex, rich, and challenging arena, with hardly any limit to the range of disciplines that can help make sense of it all.

29.5 Passing it on

Looking back on many years of interdisciplinary work, this approach still has an insecure standing in universities and American intellectual life. Interdisciplinary fields and programs in universities are marginal in clout and prestige in comparison with traditional disciplines. I have in mind urban studies, black studies, feminist programs, American studies, and my own field, bioethics. Bioethics now has a place in every medical school, but it hardly has the academic standing of the department of surgery or internal medicine, and it never will. Universities and professional schools are still organized along traditional disciplinary lines, and young faculty members are asking for trouble if they do not publish standard disciplinary articles and research in the mainline disciplinary journals. In what I have come to think of as the tyranny of peer review, CVs frequently now focus on peer-reviewed articles and books, consigning everything else (including op-ed articles in the *New York Times*) to the second-string miscellaneous category, as if it is a kind of waste basket in comparison with the real stuff. It is, however, far harder to get a short article published in the op-ed section of the *New York Times* than a long article in a peer-reviewed journal.

In the early days of the Hastings Center we recruited a number of distinguished scientists—Rene Dubos, Ernst Mayr, Theodosius Dobzhansky, for example—to work with us. For our purposes, they had in common one trait: they had all been educated in Europe, going through the gymnasium system. This meant they had a fine education in the humanities. They were well read, enjoyed mixing it up with philosophers and historians, and were quick to see the important social problems and inclinations of their own research. They had no problem in talking about ethical problems that could well impede scientific research; they were often the first to raise them. They did not treat those from the humanities with skepticism about various scientific developments as enemies of science and research. They thought, that is what we are supposed to do.

By the 1990s those distinguished scientists were either dead or retired. The next generation of scientists have been more narrow. Their education in the humanities is often scant. Even though it is not necessarily true, applicants to medical schools or PhD programs in biology believe that anything but a straight scientific background in science is a hazard to one's chances of acceptance. Meanwhile, medical and biological research had become more professionally competitive, more expensive, and more grant-driven. A track record of getting grants was not quite as important as publishing in (of course) peer-reviewed journals, but it became a required section on a scientist's CV. Getting grants proves one is a winner, a sign of the right stuff, and a good way to win points in medical schools and science departments always looking for money.

A parallel development began appearing in young applicants for jobs at the Hastings Center. They mainly come out of universities, drawn to us because of a strong reputation in the field. In principle they like our interdisciplinarity, but are nervous about it as well. They have been trained to do rigorous disciplinary work, not to adventurously explore other disciplines. They cannot understand how those of us who were in bioethics in its early years could blithely move from genetics to end-of-life care to health policy to reproductive biology. Why not, I respond: you know how to read, don't you? That casual response rarely persuades them, and the example of those of us who have done just that is often brushed aside. It was easier in the beginning, they respond, but each one of the issues the Center works on is now a large research area with a huge literature; it is unrealistic to expect them to do what we did.

Nonetheless, I continue to chip away at them, noting that there are some old and cross-cutting problems that can be found in the subareas of bioethics. The classical tension between, say, individual good and common good, or medical progress and unforeseen social pitfalls, can be found in almost every discrete topic of bioethics. Over the years, in fact, two distinct streams have emerged in bioethics, going back to its very beginnings, and they have different implications for the field.

One of them, the earliest, came from worries and speculations on the part of leading scientists in the 1960s about where the new biology and medicine were taking us. Would they lead us to think differently about human nature? Would they bring changes in our common life that would lead us to live radically different kinds of lives than our parents and grandparents? How much power should be granted to science to remake our lives? The other stream focused on some more immediate medical, policy, and legal issues. How can we give patients more autonomy in determining their care at the end of life? Is it possible to establish a better balance of power between doctors and patients? Can we do a better job of protecting research subjects from harm?

Over the years, practical moral problems of that latter kind came to dominate the field. The larger initial questions came to be overshadowed. Foundations and the media, among others, were far less interested in the future of humanity under the impact of science than they were of ethics at the bedside, laws on care at the end of life, and specific policy recommendations on stem cell research. The larger speculative questions do not fare well in competition with the more immediate issues, nor do they admit of the kind of specificity that can be achieving in devising, say, guidelines for genetic counseling. Interdisciplinary work can and does take place at both levels, but the larger questions are more likely to attract the attention of theologians, a few philosophers, and some stray social scientists. The more policy oriented ones draw on the wider range of disciplines I have noted earlier.

The future of interdisciplinarity in bioethics is not clear. It was easier to pursue it in the 1970s than at present, mainly because at the beginning there was no formal field, just a group of people from the biological and medical sciences, the social sciences, philosophy, law, and theology who were interested in the emerging issues, and who saw the value of working together. As time went on and bioethics flourished it took on the traits of a subdiscipline and one with increasingly many sub-subdisciplines. At the same time, so it seemed to me, the traditional disciplines became stronger, with pressure on students to stick with the straight and the narrow.

Interdisciplinary work is often lauded as a fine thing, but the underlying message in most universities is to be careful: don't get carried away and don't stray far from established rigor of established disciplines. And if you go into bioethics, pick one area and work at it diligently. Don't flit around from topic to topic. For me, I have to confess the fun and adventure of the field comes in flitting about. I have published only one article in my own field of philosophy, but am far more proud of the fact that I have been published in medical and health policy journals, law reviews, social science journals, science journals, and have had op-ed pieces in every major American newspaper. I have had a good time.

References

Callahan, D. (1973). Bioethics as a discipline. *Hastings Center Studies* **1**(1), 66–73.

Callahan, D. (1999). The Hastings Center and the early years of bioethics. *Kennedy Institute of Ethics Journal* **9**(1), 53–70.

Callahan, D. (2009). *Taming the beloved beast: how medical technology costs are destroying our health care system*. Princeton, NJ: Princeton University Press.

Callahan, D. (manuscript). *Two few babies? The clash of economics, culture, and religion.*

Callahan, D. and Wasunna, A. (2006). *Medicine and the market: equity v. choice*. Baltimore, MD: Johns Hopkins University Press.

Congressional Budget Office (2008). *Technological change and the growth of health care.* Washington, DC: Congressional Budget Office.

Fox. J. (2005). Medicare should, but cannot, consider cost: legal impediments to a sound policy. *Buffalo Law Review* **53**(2), 577–633.

Grant, J. *et al.* (2004). *Low fertility and population aging: causes, consequences, and policy options.* Santa Monica: Rand Corporation.

Pernick, M. (1999). Brain death in a cultural context: the reconstruction of death, 1967–1981. In: S. Youngner, R. Arnold, and R. Schapiro (eds) *The definition of death: contemporary controversies*, pp. 3–33. Baltimore, MD: Johns Hopkins University Press.

Rice, T. (2002). *The economics of health care reconsidered*, 2nd edn. Chicago: Health Administration Press.

Starr, P. (1982). *The social transformation of American medicine.* New York: Basic Books.

United Nations (2004). *Policy responses to population decline and aging.* Population bulletin of the United Nations, special issues 44/45 2002. New York: United Nations.

PART 5

Knowledge transdisciplined

CHAPTER 30

Solving problems through transdisciplinary research

GERTRUDE HIRSCH HADORN, CHRISTIAN POHL, AND GABRIELE BAMMER

Problem solving in the real world is an important driver for integrative and collaborative research. Problem-directed research transgresses academic cultures and engages in mutual learning with societal actors in order to account for barriers in real life and possible unintended effects of problem solving. This chapter explores two approaches to integrative and collaborative research: transdisciplinary research from Europe, and integration and implementation sciences from Australia.

Transdisciplinary research (TR) aims at better fitting academic knowledge production to societal needs for solving, mitigating, or preventing problems such as violence, disease, or environmental pollution. TR strives to grasp the relevant complexity of a problem, taking into account the diversity of both everyday and academic perceptions of problems, linking abstract and case-specific knowledge, and developing descriptive, normative, and practical knowledge for the common interest. Integration is a core feature and major challenge of TR. Practitioners of TR call for a recursive approach to problem solving, focusing on problem identification and structuring, investigation, and bringing results to fruition as the three phases of the TR process.

The following account of the integration and implementation sciences lays out five core concepts, a set of five methods, and an overall framework for describing and planning integration. The concepts are: a systems approach, attention to problem framing and boundary setting, attention to values, a sophisticated understanding of ignorance and uncertainty, and understanding the nature of collaborations. The methods are dialogue-based, model-based, product-based, vision-based, and common metric-based. The overall framework involves six questions that focus on aims, processes, actors, context, and outcomes.

30.1 Background

Academic research is an integral component of the knowledge society as it has developed across the twentieth century. Stokes identifies the linear model—'the belief that scientific advances are converted to practical use by a dynamic flow from science to technology' (Stokes 1997, p. 10)—to be the leading paradigm for the science-practice interrelation after World War II. The linear model is based on the idea of a one-way transfer of allegedly reliable knowledge from experts to 'ignorant' users (Wynne 1993; Maranta *et al.* 2003). Scientific theories and models (for instance molecular and microbiological mechanisms, patterns of epidemiological spread underlying infectious diseases, the *homo oeconomicus* model) idealize and simplify complex correlations and, by that, make them scientifically treatable. This means that such theories and models have a restricted potential for dealing with concrete societal problems and that measures and technologies developed according to the linear model may lead to unexpected side-effects. Towards the end of the twentieth century the risks of modern science and technology have triggered a debate about the need for a more reflexive relationship between science and the knowledge society (Beck 1992).

The debate surrounding transdisciplinarity emerged in the 1970s, as questions were raised concerning the orientation of knowledge production in research, education, and public and private institutions. Concerned systems scientists such as Erich Jantsch and intellectuals from the humanities such as Joseph Kockelmans initiated a debate about how to deal with complexity and value issues related to human activities. Three developments in the history of science, the humanities, and the social sciences have nurtured the debate on transdisciplinarity over the years: (1) systems theory and analysis, (2) the interpretative paradigm in social research, and (3) theories of social action and learning (see also Hirsch Hadorn *et al.* 2003, 2008):

(1) A systems theory approach for conceiving of complexity in research had been developed as early as the eighteenth century by Johann Heinrich Lambert, who proposed to structure complexity as a set of interrelated elements and applied his approach to systems of scientific knowledge and to belief systems of cultures, religions, and narratives, including systems that are constructed to realize desired states. But it was not until the twentieth century, when systematic theoretical approaches had been developed independently in various fields, that systems theory became a blueprint for structuring complexity against the background of the progressive fragmentation of science into more and more specialized disciplines and thematic fields (Cash *et al.* 2003; Midgley 2003; Strijbos, Chapter 31 this volume).

(2) Debates concerning how to best understand the social sphere date back to the dissociation of the humanities and the social sciences from philosophy starting in the nineteenth century. The German philosopher Wilhelm Dilthey advocated a hermeneutic approach to understanding the cultural ideas and historical configurations that constitute the identity of an epoch, for instance the age of enlightenment or of historicism. Wilhelm Windelband stressed the difference between natural sciences and history. He argued that history must investigate the individuality of empirical phenomena by understanding their historical meaning and importance, which stems from the role they play in societal ideals. Max Weber based his theory of social action and his methodology of ideal types on the hermeneutic relation between empirical phenomena and societal values. He became one of the founders of the interpretative paradigm in social sciences. In TR, the interpretative paradigm is particularly relevant for bridging idealized theories or models and concrete problem situations

(Rayner and Malone 1998; Krohn 2008), and also for investigating and interrelating normative, practical, and empirical knowledge.

(3) Theories of social action and learning changed toward the end of the nineteenth century. The social effects of industrialization and migration gave rise to sociology in Europe and America and stimulated innovative developments such as the Chicago School of Sociology in the United States. Since then, the changing circumstances of individuals and institutions have been a prominent subject of investigation in the social sciences. Action research aims at mutually benefiting theory and practice in understanding and dealing with societal problems. Action research starts with people's interpretation of reality, basing research on action in the field to learn about the consequences of social action. People studied should both be researched and research themselves. Conceptions of how to link action and learning, e.g. in double-loop or second-order learning, have been further developed by Chris Argyris, Donald Schön, Ernest T. Stringer, and others in their work on experiential and organizational learning (Argyris and Schön 1996; Stringer 2007).

Being shaped by these and further lines of thinking, many terms and a large body of approaches and concrete projects emerged in TR. A synthesis of the literature in this field has identified core defining elements of the concept, key characteristics of knowledge for problem solving, and the basic structure of a transdisciplinary research process (Pohl and Hirsch Hadorn 2007). They are providing the conceptual basis for the activities of the Swiss Academies network for TR (see box on td-net—the Swiss Academies of Arts and Sciences' forum for transdisciplinary research) and are described in the following section.

30.2 Transdisciplinary research

30.2.1 Subject and definition

The central position of research within society today has recently intensified efforts to integrate separate bodies of knowledge. The US National Academies defined interdisciplinarity thus: 'Interdisciplinary research (IDR) is a mode of research by teams of individuals that integrates information, data, techniques, tools, perspectives, concepts, and/or theories from two or more disciplines or bodies of specialized knowledge to advance fundamental understanding or to solve problems whose solutions are beyond the scope of a single discipline or area of research practice' (National Academies 2005, p. 188). This definition gives two reasons for knowledge integration: (1) to advance fundamental understanding and (2) to solve problems. The terms 'problem' and 'problem solving' emphasize the usefulness of knowledge for addressing real-world issues as opposed to the search for fundamental scientific understanding. In a similar way, Ann Bruce *et al.* (2004) refers to (1) as 'Mode 1 interdisciplinarity' and to (2) as 'Mode 2 interdisciplinarity'. In the European context—and specifically in the German-speaking part of Europe—'transdisciplinary research' has become the familiar term for research that is driven by solving real-world problems. Moreover, some scholars conceive of transdisciplinarity as a unifying principle for knowledge integration, which is determined by universal formal structures or patterns

td-net—the Swiss Academies of Arts and Sciences' forum for transdisciplinary research

Christian Pohl and Gertrude Hirsch Hadorn

Transdisciplinary research (TR) has developed under various names, for instance 'interdisciplinary problem solving' or 'implementation and integration sciences' as well as around diverse subjects: public health, technology assessment, environmental research or migration, and peace and gender studies. In addition, cooperative and mutual learning of researchers from various institutional contexts and with heterogeneous backgrounds often takes place on a project-related basis that does not outlast the temporary context of the research project or program. Strong institutional structures, which force scientific communities to evolve in the art of TR, are lacking. Special efforts and initiatives are needed to systematize experiences, enable cross-issue learning and capacity building, and to advance concepts, methods, and practices of TR within a college of peers.

Therefore in 2000 the Swiss Academic Society for Environmental Research and Ecology launched the precursor of td-net. This occurred at the Swiss Priority Program Environment's international conference 'Transdisciplinarity: joint problem-solving among science, technology and society' at ETH Zurich. In 2003 the network was more broadly anchored in the Swiss Academies of Arts and Sciences as td-net (<http://www.transdisciplinarity.ch/>).

In the Swiss context the Academies are the appropriate institution for facilitating TR: they comprise the broad range of problem fields addressed by means of TR (in contrast to a specific research institute); they stand for national research in general (as opposed to universities in competition); they are more closely related to research than public agencies, NGOs, or the private sector; and they are experienced in hosting long-term projects for national and international collaborations.

td-net is a small initiative. It consists of an office at the Swiss Academy of Sciences, a president, and a strategic advisory board. The board members are researchers who act as gatekeepers for transdisciplinary projects in their specific fields as well as representatives of public agencies and foundations supporting TR. The aim of td-net is to establish TR as a form of research of its own to address complex, socially relevant, and contested problems. As has also been pointed out by the US National Academies (National Academies 2005), such a strategy requires a number of elements (see the figure).

We started td-net as a facilitator of exchange and collective learning on methods, concepts, and success stories between the diverse fields of TR, thus 'giving a face to TR'. We addressed this challenge mainly with two publications, *Principles for designing transdisciplinary research* (Pohl and Hirsch Hadorn 2007) and the *Handbook of transdisciplinary research* (Hirsch Hadorn et al. 2008). The Principles analyze the current literature on TR for concepts and methods and depict them as challenges from the perspective of the researcher who runs a TR project. Researchers cannot learn simply by a collection of methods, but also need concrete examples to understand problems and how to structure them. Therefore the *Principles* are complemented by the *Handbook* presenting practical research experiences: researchers from various thematic fields involving 19 projects present their research and reflect on how they met the challenges of TR.

A first aim of the *Handbook* is to make successful projects known. A second aim is to prevent researchers from overburdening their projects by promising too many results. The 19 projects presented in the *Handbook* address selected challenges of TR in an exemplary

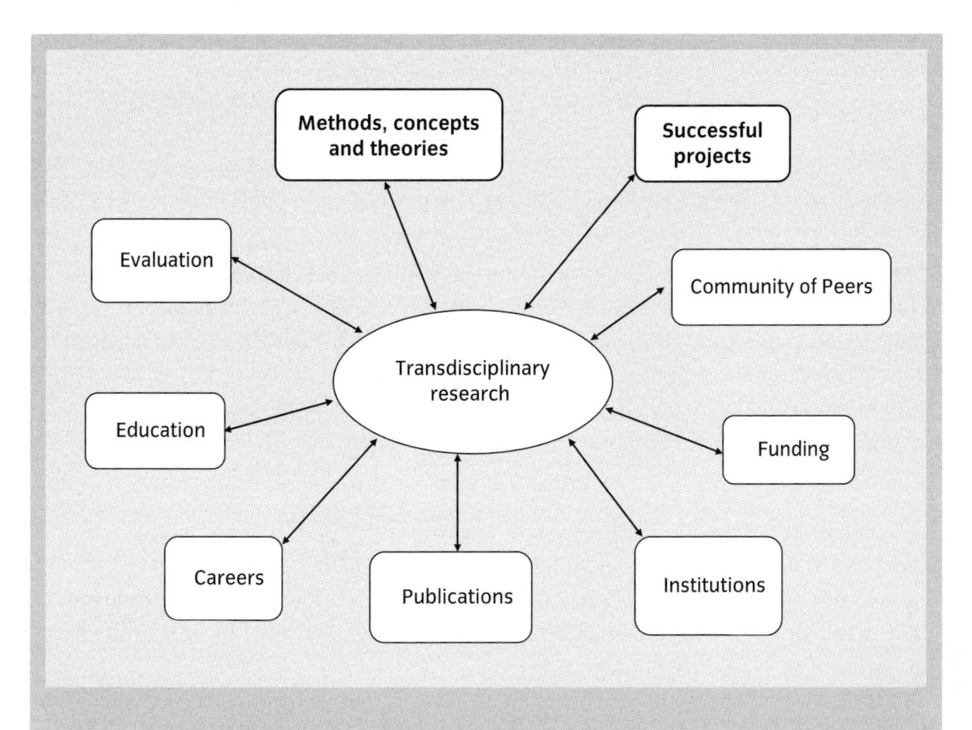

Elements that are needed to establish transdisciplinary research. td-net has zeroed in on the bold elements during recent years.

way, but none of them is a perfect transdisciplinary project in every respect. A third aim is cross-project learning. For this reason in the second part of the *Handbook* authors discuss cross-cutting themes of TR—such as participation, education, management, integration, values, uncertainties, and how to learn from case studies—whenever possible by relating their thoughts to the projects presented in the first part. The *Handbook* is not a readymade cookbook of TR, but an organized and structured sample of good projects and reflections on them.

As the figure shows, with td-net we zeroed in only on some of the elements needed to establish TR: the state of methods and concepts and successful projects, the latter also by biannually awarding outstanding projects with the Swiss Academies award for transdisciplinary research (about $67,000), which is made possible by the Gebert Rüf Foundation and since 2008 by the Foundation Mercator Schweiz. In addition, TR projects struggling with problem structuring and integration are provided with advice on demand.

Regarding the other elements we also followed several steps: We facilitated the locating of literature and journals for publication through our online bibliography and by proposing strategies for publication (Kueffer et al. 2007), and we prompted community building with our monthly news mail, our yearly colloquium in spring, and our yearly td-conference in the fall. However, to successfully establish TR internationally much more has to be done. For the future we see td-net ideally as one initiative among others, which closely collaborate in order to work on all the elements needed to establish TR in a concerted way.

(cont.)

td-net—the Swiss Academies of Arts and Sciences' forum for transdisciplinary research (cont.)

References

Hirsch Hadorn, G., Hoffmann-Riem, H., Biber-Klemm, S. et al. (eds) (2008). *Handbook of transdisciplinary research*. Dordrecht: Springer.

Kueffer, C., Hirsch Hadorn, G., Bammer, G., van Kerkhoff, L., and Pohl, C. (2007). Towards a publication culture in transdisciplinary research. *GAIA* 16, 22–6.

National Academies (2005). *Facilitating interdisciplinary research*. Washington, DC: National Academies Press.

Pohl, C. and Hirsch Hadorn, G. (2007). *Principles for designing transdisciplinary research*. Munich: Oekom.

at the basis of pluralistic processes and their dynamics (Nicolescu 1996). Klein (Chapter 2, this volume), quoting the OECD classification (OECD 1972), uses the term 'endogenous interdisciplinarity' in contradistinction to exogenous (i.e. motivated by 'real problems of the community') knowledge integration.

TR, as a form of research, provides knowledge to solve, mitigate, or prevent issues in dispute across society such as violence, hunger, poverty, disease, and environmental pollution. Those involved—academic researchers as well as societal actors—may not agree on either the relevance of the problem or on its causes and consequences, or on the type of strategy required. TR defines knowledge production in terms of four fundamental requirements: 'TR deals with problem fields in such a way that it can (a) grasp the complexity of problems, (b) take into account the diversity of scientific and life-world perceptions of problems, (c) link abstract and case-specific knowledge, and (d) develop knowledge and practices that promote what is perceived to be the common good' (Pohl and Hirsch Hadorn 2007, p. 20). Figure 30.1 describes a transdisciplinary research project as a system. The elements of the system are: the problem field, academic researchers, and societal actors. The term 'system' refers to the interaction of these elements during the research process, i.e. by discussing what the problem is about, by investigating the problem, by deliberating about values and goals, or by developing measures. The reason why the societal actors and the academic researchers interact is the shared aim to improve a particular situation in a problem field, hunger in this case.

The starting point of transdisciplinary research exemplifies Funtowicz and Ravetz's description of post-normal science. The 'facts are uncertain, values in dispute, stakes high and decisions urgent' (Funtowicz and Ravetz 2008, p. 365). The requirement to grasp the relevant complexity of a problem is included in Erich Jantsch's definition of transdisciplinarity, which is inspired by systems theory thinking (Jantsch 1972, pp. 105–6). Taking into account the diversity of perceptions is a requirement that calls for the collaboration of researchers from different disciplines and the participation of actors from the public and private sector as well as civil society—the latter being a mandatory requirement for some scholars (Lawrence 2004). Interrelating abstract and case-specific knowledge is derived

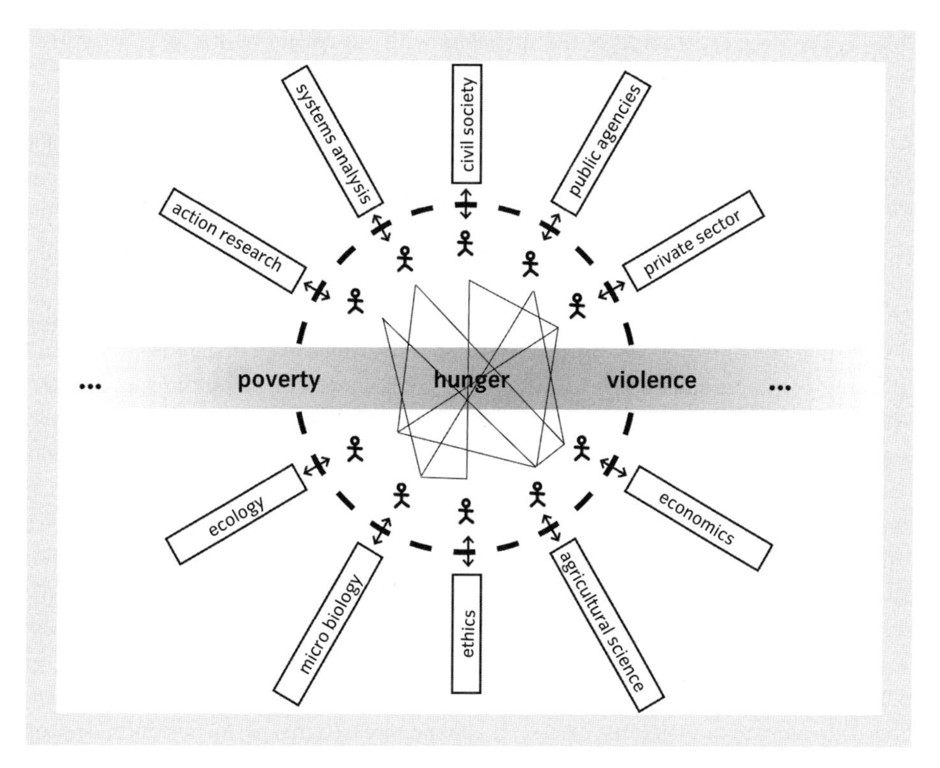

Figure 30.1 In transdisciplinary research scientific disciplines (represented by individual researchers) and sectors of the real world (represented by individual actors) become interrelated and transformed through a problem field. A transdisciplinary research project is the system built by the collaborative research process (adapted from Pohl and Hirsch Hadorn 2008, p. 13).

from the field of intervention research. Orienting problem solving towards the common interest is implicit to fitting knowledge production to problems of real life rather than disciplinary boundaries. Some authors in the policy sciences (Clark 2002, p. 13) and in technology assessment call it an explicit goal of research.

Including actors from the public and the private sectors as well as civil society within the design and implementation of research, and emphasizing collaboration among researchers and other actors, serves to transcend and integrate different perspectives. The integration of knowledge from different academic disciplines and practical contexts in public or private agencies and civil society poses major challenges for working effectively within the project team during the research process. The first step for such integration is to acknowledge, respect, and explore the diversity of perspectives. As Loibl (2006) states, this diversity is not a handicap to be overcome, but an invitation to creative interaction. To engage effectively in creative interaction participants have to learn each other's language. Baccini and Oswald developed a transdisciplinary method in urban design, integrating architecture, urban planning, environmental sciences, and engineering. They started with 'daylong field trips, where each described to the other the people, things and movements he observed, how and why he perceived them in the specific way he did at that time.

This was a guiding experience for the later work with our collaborators, too' (Baccini and Oswald 2008, p. 80).

Integration is guided by overarching values as described in point (2) of Section 30.1. In TR the overarching value or purpose is the idea of the promotion of the common good. In the history of ideas, promoting the common good has been understood as the goal of the state or community as opposed to striving for private interests. Today, promoting the common good can be conceived as sustainable development, which means aiming at meeting the needs of present and future generations according to principles of equity. The common good and sustainable development, respectively, serve as regulative ideas for reflecting on controversial attitudes towards issues. As socio-political ideals, both concepts are open to various interpretations. As a consequence of this pluralism, neither a particular theory such as utilitarianism nor a particular position in society such as being a pastor or politician can lend reasonable authority to a certain concept of the common good or sustainable development, or to how such concepts should be applied to a specific situation. So, how to analyze and specify the concepts in view of the particular problem field is one of the research questions to be addressed in deliberating on and providing normative, descriptive, and practice-oriented knowledge. In Klein's typology TR embodies critical rather than instrumental interdisciplinarity. Instrumental interdisciplinarity suits the framework of applied research, which is client-serving (Funtowicz and Ravetz 1993) and therefore does not enter into value conflicts and uncertainties in the wider societal context. Dealing with infectious disease in applied research means, for instance, developing pharmaceutical therapies for profit in the market.

30.2.2 Knowledge for problem solving

Problem solving in TR essentially consists of bringing about change in the private and public sectors as well as civil society by developing and experimentally implementing better practices, products, and policies (van den Daele and Krohn 1998). The knowledge requirements encompass ends, means, and contexts of human agency. The overarching goal for integration is a vision or societal ideal, such as sustainable development. Sustainable development integrates ecological, economic, and social aspects of real-world issues according to principles of inter- and intragenerational justice (World Commission on Environment and Development 1987). For instance, dealing with health services for nomadic pastoralists and their animals in the Republic of Chad (central Africa) includes: studying the epidemiological pathways of disease and disease patterns of the affected communities; studying the microbiological processes in the system of nomads and their animals; getting acquainted with traditional health knowledge and practices; jointly developing and implementing innovative ideas such as mobile joint vaccination services for nomadic pastoralists and their animals; and promoting services through movies by local artists for information and trust building (Schelling *et al.* 2008).

In TR the three types of knowledge requests have been termed 'target knowledge', 'transformation knowledge', and 'systems knowledge' by Swiss researchers in their visions of research on sustainability and global change (ProClim 1997). Similar distinctions of knowledge forms, sometimes using different terminology, are widespread among scholars in environmental and sustainability research (Costanza *et al.* 1997; Grunwald 2004). To produce

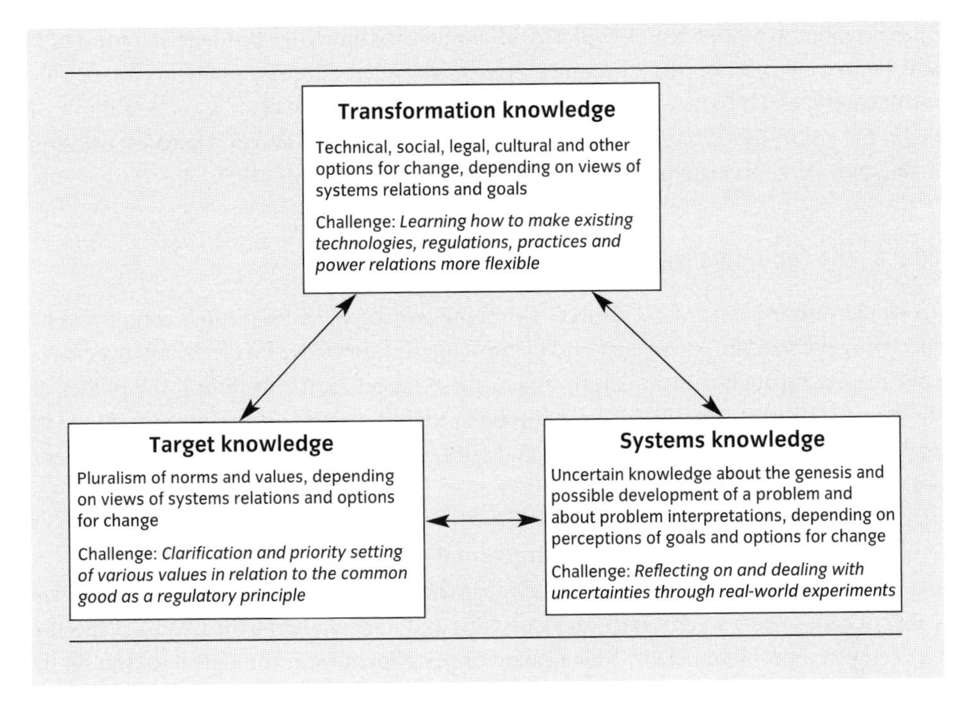

Figure 30.2 Interdependences between systems, target, and transformation knowledge and their particular challenges (from Pohl and Hirsch Hadorn 2007, p. 38).

valid results in research, the three forms of knowledge must be treated according to their particular challenges by taking into account their interdependences (see Fig. 30.2).

For example, to learn about obstacles and possible unintended negative effects in advance, not only must the systemic processes be taken into account when developing transformative knowledge, but the needs, interests, and motives of the practitioners and the stakeholders involved must be reflected upon as well. This means that the investigation of systemic processes has to be related to the societal purposes and practices upon which the systemic processes depend and which they influence (see box on Sustainability Foresight).

TR frames decisions that eventually have to be made in the private and public sector or civil society. Such problem solving through TR includes dealing with uncertainties and conflicting values and options. Borders between research and politics become fuzzy. One role of researchers in such contexts consists of making the pros and cons of alternatives transparent to decision makers. These pros and cons encompass instrumental values such as effectiveness and efficiency, as well as ethical issues of fairness and justice among and between generations. Dealing with instrumental values has long been familiar in the technological development of applied research. The additional challenge in TR consists of developing concepts and methods for reflecting on the ethical issues of fairness and justice in developing, deliberating, and deciding upon technical and institutional devices and in using the common good or sustainable development as regulative ideas. By including such reflections in problem solving, TR bridges instrumental and critical forms of interdisciplinarity (see Klein, Chapter 2 this volume). Knowledge production of the pros and cons of alternatives in TR becomes a collective endeavor, instead of being a task only for science (Pohl 2008; see also the box on Sustainability Foresight). A good example of this is the vision-based integration work by the World Commission on Dams for decision making on water and energy management as described in Section 30.3.2.4.

30.2.3 The recursive research process

To avoid overburdening of TR projects, two things are important: to reduce complexity by specifying the need for knowledge and identifying the knowledge involved, and to achieve effectiveness through contextualization. For these purposes, it is helpful if the phases of the research process, namely (phase 1) problem identification and structuring, (phase 2) problem investigation, and (phase 3) bringing results to fruition, do not follow a sequential order but a recursive one (see Fig. 30.3).

30.2.3.1 Problem identification and structuring

In problem identification and structuring, researchers and actors from the public or private sector or civil society work jointly on identifying and understanding the nature of specific problems in a problem field. A broad range of participants and competencies should be involved in order to frame and structure the unclear issues jointly, properly identify the

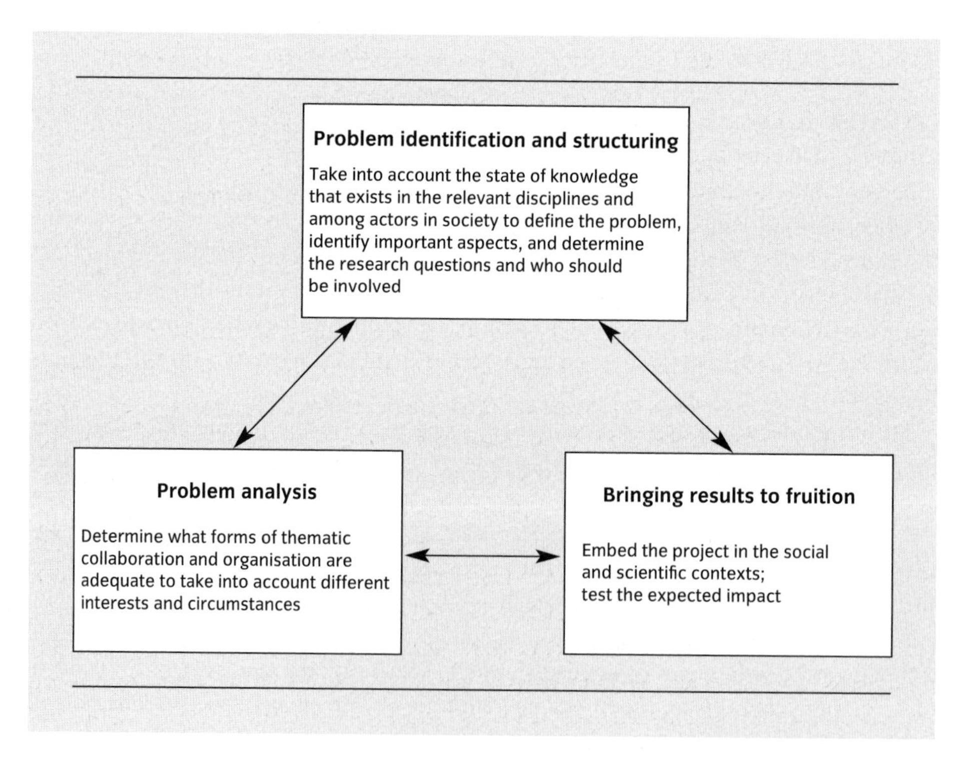

Figure 30.3 The three phases of transdisciplinary research (Pohl and Hirsch Hadorn 2007, p. 42).

Sustainability foresight: participative approaches to sustainable utility sectors

Bernhard Truffer

The sustainability foresight project was an initiative (2002–6) mandated by the German Ministry for Education and Research within the transdisciplinary socio-ecological research program (<http://www.sozial-oekologische-forschung.org/>). The project aimed to developing sustainable development alternatives for German utility sectors in a timeframe of 25 to 30 years (Truffer *et al.* 2008). The project focused on the electricity, gas, water, sanitation, and telecom sectors. It asked whether the historical paradigm of utilities as public services could give way to a new overarching sustainability paradigm, or whether each sector was more likely to develop according to its own specific logic. Finally, the project was intended to provide strategies for promoting system transformations towards new more sustainable forms of utility services.

Foresight methods have often been used to develop strategies in science, technology, and innovation policy. Sustainability foresight puts emphasis on sectoral production and consumption systems as coherent configurations of institutions, technologies, and cultural and environmental structures. Moreover, it encompasses an elaborate assessment and strategy phase after the scenario construction. Although sustainability foresight was developed specifically for the German utility sector, it is kindred to a number of recently formulated foresight and system management approaches.

The sustainability foresight method was based upon a broad stakeholder process in which utility firms, technology developers, environmental and consumer NGOs, government officials, and researchers elaborated their expectations. This broad setting was chosen because each actor group may contribute to the overall analysis based on his or her own rationality and in accordance with specific knowledge. This knowledge may relate to the structure and potential future dynamics of the sector (system knowledge), to goals which a sector should try to fulfill, as well as trade-offs that might exist when trying to reach specific goals (goal knowledge), and finally knowledge about potential actions that might support a transformation (transformation knowledge). As a consequence, expectations are likely to differ not only on their substance but also with regard to the access to supporting evidence and even with regard to wording and framing.

The sustainability foresight procedure was structured in three phases:

(1) Exploration of expected transformation dynamics: this consisted of an analysis of implicit visions about the future of the selected utility sectors. Each analysis was concluded by an expert workshop with representatives of different stakeholders. Based on perceived development trends, four overarching scenarios were constructed in three participatory scenario workshops.

(2) Sustainability assessment: The four scenarios were characterized with regard to their challenges and opportunities relating to sustainability criteria. The evaluation criteria were determined by scientific experts from different fields. Determination of preferences was carried out in a stakeholder workshop. Stakeholders were carefully selected in order to represent the whole range of different value positions.

(3) Identification of transformation strategies: Based on the four sector scenarios and the risks and opportunities derived from the sustainability assessment, potential development trajectories for three critical innovation fields were worked out by the project team. Roadmaps for development were then presented in a final workshop to representatives of the utility sectors

(*cont.*)

Sustainability foresight: participative approaches to sustainable utility sectors (cont.)

and experts in the selected technologies in order to derive potential coordinated strategies that could lead to more sustainable utility structures in the long run.

Each of the three project phases ran over a year and involved a broad range of stakeholders, who were invited in order to guarantee a broad and balanced spectrum of knowledge types and perspectives. The core analytical steps were carried out conjointly between project team and participants. Overall about 150 experts participated in the project in one way or another. Among these, about 120 stakeholders participated in the nine workshops. Each workshop ran over 2 days and encompassed roughly 20 participants.

The process allowed stakeholders to move from implicitly held visions on sustainable future utility sectors into a potentially more widely shared agenda. Carrying out this procedure as a participatory process was necessary because, in general, no shared understanding of the system dynamics existed and a high number of potential and actual areas of conflict could intervene in any attempt to challenge sustainability strategies for these sectors. The process yielded an elaborate set of arguments for coordinating the different individual strategies, which could lead to formulating conjoint innovation projects. In that sense, the sustainability foresight method has the potential to contribute to an actor-spanning research and innovation program oriented at sustainable transformations of entire sectors that explicitly relies on transdisciplinary forms of knowledge production.

References

Truffer, B., Voss, J.-P., and Konrad, K. (2008). Mapping expectations for system transformations. Lessons for sustainability foresight in German utility sectors. *Technological Forecasting and Social Change* **75**, 1360–72.

relevant scientific disciplines and actors in the real world, evaluate the existing knowledge about problems in academia and in practical situations, and learn about the needs and interests at stake. This information provides the knowledge base for problem solving, the questions that need to be addressed in research, and the competencies required for the investigation of and deliberation about results. The Swiss Man and Biosphere program, for instance, followed a recursive approach to problem identification and structuring. The researchers combined an analysis of the systems dynamics of an alpine valley with target knowledge about land use, aesthetic and recreational qualities, the analysis of management options, and the development of long-term strategies in a participatory process with regional stakeholders (Messerli and Messerli 2008).

One outcome of problem identification and structuring could be that all the knowledge required for designing and experimentally implementing measures is already there and that phase three (Section 30.2.3.3) should be launched. Another possible outcome is that different competencies and participants are required from those initially expected, so that problem identification and structuring has to be repeated. Furthermore, problem identification and structuring, on the one hand, and problem analysis, on the other, can overlap. All this makes a recursive treatment of phases a more rational approach for achieving valid results than a sequential treatment.

30.2.3.2 Problem investigation

In order to grasp the relevant complexity of relations in detailed problem analysis, an adequate understanding is needed of the way in which diverse aspects and perspectives are integrated. In addition, quality assurance of knowledge has to take into account mutual influences between systems knowledge, target knowledge, and transformation knowledge, giving rise to conceptual, epistemological, and methodological uncertainties. Instead of defining standard conditions for idealization, generalization of knowledge has to be achieved by transferring models and methods from the context in which they have been developed to other contexts, while carefully validating the conceptions of each setting (Krohn 2008). Therefore, problem analysis and bringing results to fruition are best conceived of as recursive and integrated steps. Developing integrated assessment methods for climate change mitigation, Held and Edenhofer (2008) combined approaches from natural science, economic growth theory, engineering, and ethics to identify climate policies that integrate both knowledge of the climate and economic systems as well as competing values of interest groups.

30.2.3.3 Bringing results to fruition

Bringing the results to fruition, as a phase of the transdisciplinary process, also relies on the integration of knowledge. It is important that practitioners and researchers jointly learn about the strengths and weaknesses of problem-solving strategies and develop competencies for implementing and monitoring progress in order to be able to adapt strategies and purposes. Constructive technology assessment of nanotechnologies (Rip 2008) is an instructive example highlighting the reflexive learning of societal actors and researchers. Learning is based on the mapping and analysis of ongoing dynamics, on the articulation of socio-technical scenarios about further developments and impacts, and on real-time experiments.

The recursive transdisciplinary research process changes the nature of problem solving from the implementation of definitive scientific solutions to social learning, experimental implementation, and adaptation of problem-solving measures. Thus, the core of problem solving through transdisciplinary research is integration (Pohl *et al.* 2008).

30.3 Integration

There is no shortage of people with experience in the integration of disciplinary and practice knowledge in research. But, as yet, the publication culture on integrative issues is poorly developed (Kueffer *et al.* 2007) and there is no systematic approach to integration. This section lays out five core concepts, five groups of methods, and a framework for describing and planning interdisciplinary integration. These concepts, methods, and framework are taken from the newly evolving cross-discipline of integration and implementation

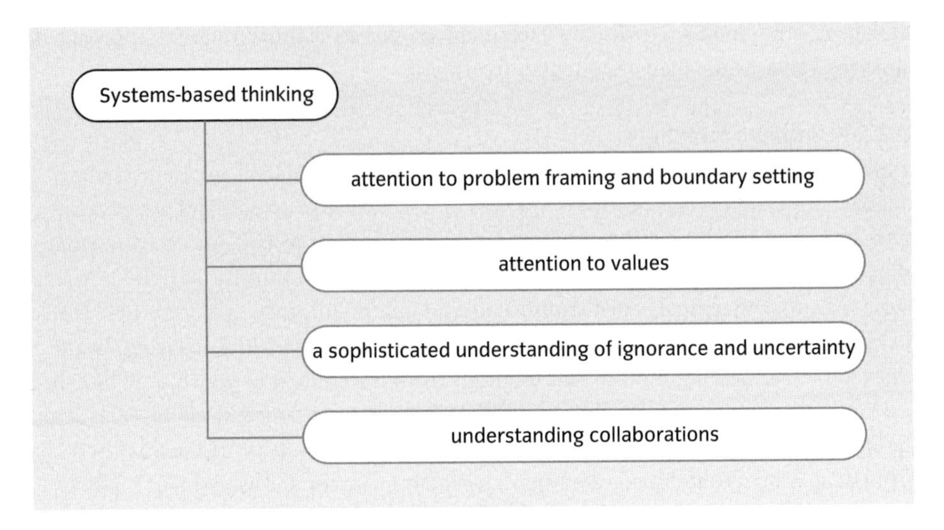

Figure 30.4 Five core concepts for integration (Bammer 2007).

sciences (Bammer 2005, 2006, 2007). Others have also recognized the importance of integration for inter- and transdisciplinary research (Klein 2008).

30.3.1 Core concepts for integration

The five concepts are: (1) a systems approach, (2) attention to problem framing and boundary setting, (3) attention to values, (4) a sophisticated understanding of ignorance and uncertainty, and (5) understanding collaborations. A systems approach is a core underpinning concept, from which the others naturally flow (Fig. 30.4).

30.3.1.1 Systems-based thinking

Systems-based thinking is at the heart of integration. This helps one to look at the whole problem and its relationship to its parts. To put it simply, everything is interconnected. Systems-based thinking emphasizes that problems have many dimensions. There are many factors involved in any problem and there can be various types of connections between them. For example, the connections may be linear, so that two factors increase in direct proportion, or they may be more complex. Because of the extensive interconnections, changes made in one area often have consequences elsewhere, and these may occur in unexpected ways. Systems-based thinking also recognizes that the political, cultural, disciplinary, and sectoral contexts of a problem are important. There is now an extensive body of knowledge encompassed by systems thinking and complexity science (see Strijbos, Chapter 31 this volume).

30.3.1.2 Problem framing and boundary setting

Although a systems view is important, no research project or program can cover everything, so the way in which any particular issue or problem is tackled has to be delimited.

This is done both through the way the problem is defined or framed and where the boundaries around the problem are set. Frames and boundaries will determine what is included, excluded, and marginalized in the research. In terms of problem framing, the way problems are defined and the language used to describe them can play a powerful role in setting the basis for research integration.

The way a problem is framed already implicitly sets some boundaries around the problem. The boundaries specify what will be attended to, what will be ignored, and what will be marginalized (Midgley 2000). An important aspect of this for research integration is determining which disciplines and which non-academic or practice perspectives will be included in the project and which dimensions of uncertainty will be incorporated.

30.3.1.3 Values

The way the problem is framed and which boundaries are set are closely aligned with underpinning values. Even though all research is located in a values framework, this is often implicit and researchers may be unaware of how values shape their work. Furthermore, research that brings together the perspectives of different disciplines and practice groups often has to find ways of managing different values.

One way in which differences in values are highlighted is through epistemology. For example, positivism sees research as value-free, with values having no place except when choosing a topic; interpretive social science considers values to be an integral part of social life, with no group's values being seen as wrong, only different; and critical social science maintains that all research has a value position and that some positions are right while others are wrong (Neuman 2003).

30.3.1.4 A sophisticated understanding of ignorance and uncertainty

A systems approach also helps us realize that there are vast areas which may be relevant to the problem of interest where nothing is known or where available knowledge is uncertain. Such appreciation orients research integration to give more emphasis to a sophisticated understanding of ignorance and uncertainty and to more refined ways of dealing with them.

In dealing with any complex problem, there will always be many unknowns, including about facts, causal and associative relationships, and effective interventions. Some unknowns result from resource limitations on research, some result from methodological limitations, and some things are simply unknowable. There are epistemological, ethical, organizational, and functional aspects to dealing with ignorance and uncertainty. A more sophisticated understanding of and approach to ignorance and uncertainty involves better appreciation of the nature of ignorance and uncertainty, of the underpinning motivations and moral orientations to ignorance and uncertainty, as well as strategies for coping and managing under ignorance and uncertainty (Smithson 1989; Bammer and Smithson 2008).

30.3.1.5 Principles of collaboration

A systems-based approach involves bringing a range of perspectives and skills to bear on the issue of interest and therefore involves collaboration with relevant people from both

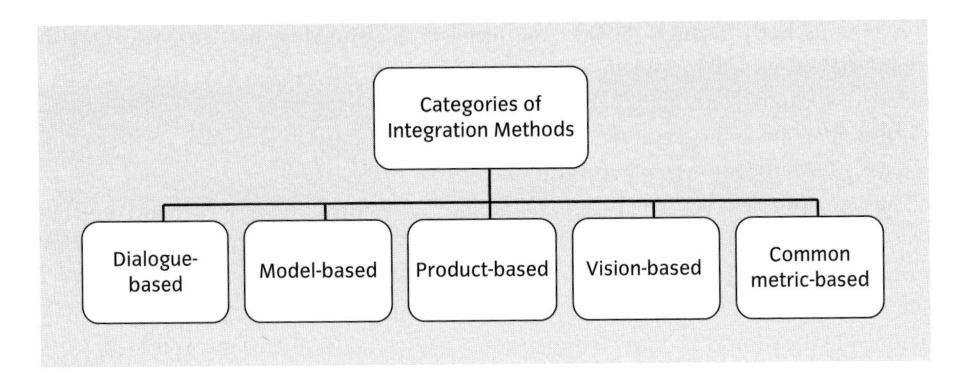

Figure 30.5 Five classes of integration methods (Bammer 2006).

disciplines and practice. Collaborations are all about harnessing difference. The whole point of working with someone else is that he or she has an alternative perspective, skills, or some other attribute that contributes something relevant to addressing the issue either in improving understanding about it or in implementing that understanding in decisions and action.

However, the differences between research partners cannot be limited to those which advance understanding of, or effective action on, the problem. Differences in ideas, interests, and personality will also provide potential sources of unproductive conflict. Harnessing differences involves integrating those differences that provide the rationale for the collaboration and managing those which will get in the way of partnership (Bammer 2008).

30.3.2 Core methods for integration

In this subsection, five strategies for carrying out integration are outlined, each of which encompasses a range of methods. They are (1) dialogue-based, (2) model-based, (3) product-based, (4) vision-based, and (5) common metric-based (Figure 30.5).

30.3.2.1 Dialogue-based methods

Dialogue is the most common strategy for achieving integration of discipline and practice perspectives and is an essential component of the other strategies, as well as being an approach in its own right. Franco (2006, p. 814) draws on key references in the field to provide a useful definition: 'The goal of dialogue is to jointly create meaning and shared understanding between participants…'. From an integration perspective, a key question is 'jointly create meaning and shared understanding' about what? The 'about what' question is answered by the particular aspects of research integration under consideration. Thus, some dialogue methods are well suited to creating meaning and shared understanding about the judgments people have about how best to move forward on a problem. Others can provide mutual insights into different interests involved in the problem, and still others into different visions for how the problem might ideally be solved. To date, five categories of dialogue methods have been identified, namely for integrating judgments, visions, worldviews, interests, and values, as well as methods that are useful for integrating more than one of these elements (McDonald *et al.* 2009).

30.3.2.2 Model-based methods

The second primary group of methods for integration are model based. Models are a key way to represent systems and to provide aids to thinking about complex issues. Different modeling types focus on different aspects of systems. For example, a system dynamics model concentrates on feedback and demonstrates how vicious and virtuous cycles are, sometimes unwittingly, established. An agent-based model focuses on the different actors involved (the agents) and the key determinants of their behaviors ('rules' for their actions). Among other things, an agent-based model seeks to understand if there are simple rules or behavioural determinants which can explain even quite complex behaviors. When modeling is used as an integrative tool, the emphasis is on the process of developing the model and its utility in helping decision makers. The model is therefore a device which provides a focal point for discussion and action between people representing different disciplinary perspectives and different types of practical experience relevant to the issue under consideration.

30.3.2.3 Product-based methods

Like model-based strategies for integration and implementation, product-based strategies use the artifact as a device around which to build interaction between people representing different disciplinary perspectives and different types of practical experience relevant to the problem under consideration.

An example of large-scale product-based integration comes from building the atomic bomb in the 1940s. The atomic bomb project brought together basic science (such as achievement of controlled fission), solutions to a vast range of technical problems (such as developing an implosion trigger device), engineering and manufacturing prowess (as in generating adequate amounts of fissionable material), and military and political judgment in terms of its use (Rhodes 1986).

30.3.2.4 Vision-based methods

The World Commission on Dams, which was active between 1998 and 2000, provides an example of vision-based integration. The vision was to achieve 'development effectiveness', where 'decision-making on water and energy management will align itself with the emerging global commitment to sustainable human development and on the equitable distribution of costs and benefits' (World Commission on Dams 2000, p. xxxiii). The Commission integrated wide-ranging considerations in terms of issues, evidence, countries, and participants; including diverse technical, social, environmental, financial, and economic evidence from case studies, country studies, a survey, technical reports, submissions, and fora. It eschewed a 'balance sheet' approach to assessing costs and benefits in favor of multicriteria analysis. A guiding set of values based on United Nations declarations and principles about human rights, social development and environment, and economic cooperation underpinned the approach.

30.3.2.5 Common metric-based methods

The idea behind the common metric is to convert all the discipline and practice perspectives of a problem to the same measure, which allows integration through simple arithmetic. The most widely used common metric is monetary value, such as the dollar. Much can be learnt from the discipline of economics about the conversion of a range of aspects of a complex problem to a dollar value, such as putting a value on life or fresh water. Apart from the dollar, other common metrics include global hectares of land, metric tons of carbon dioxide equivalent, and disability-adjusted life-years.

30.3.3 **A framework for integration**

One reason why our understanding of integration is not further advanced is that there is no unified way of thinking and writing about it. This volume and other literature (Rossini and Porter 1979; Klein 1990, 2004) provide ways forward. In addition, a new level of specificity could be introduced by addressing the following six questions:

(1) Integration for what and for whom? In other words, what are the aims of the integration and who is it intended to benefit?
(2) Integration of what? This addresses the diverse perspectives being synthesized and applied and the actors involved.
(3) What is the context in which the integration is occurring? This involves the political or other action context, which influences priorities in terms of the framing of the issue and the people seen to be key actors, as well as the focus of action resulting from the integration.
(4) Integration by whom? Even though integration often requires partnerships, the process of synthesis and application does not need to be collaborative. It can be undertaken by an individual (often the leader), a subgroup, or the whole group.
(5) How is the integration undertaken? This takes us back to the methods outlined in the previous section. Different methods will be suitable for different integration purposes and the choice of methods will also depend on the expertise of those undertaking the integration.
(6) What are the measures of success? Success is often not reported in integrative studies and there are no standard procedures to evaluate success. The questions described above provide the substrate for evaluating success. First, how well were the integration aims met? Were influential new insights produced? Did effective action result? Second, some process issues can also be evaluated. Were all the necessary elements included in the integration? Were effective integrative methods used? (Bammer 2006).

30.4 **Conclusion**

This chapter has described the principles, conceptions, methods, and tools of integrative and collaborative research that are developed and used in TR and in integration and implementation sciences. Although a considerable stock of competences and knowledge has grown in the last decades, major scientific, institutional, and societal challenges are lying ahead (Wiesmann *et al.* 2008). Focusing on the scientific challenges, good examples of transdisciplinary practice have to go hand in hand with systematization of this practice and methodological innovations. Methodological innovations are needed, for instance in how to design and interpret real-world experiments in validating the complexity of knowledge, or how to better account for the diversity of values in evaluation methods. Facing the scientific challenges calls for a college of peers gathering researchers who are specialized in how to integrate and make research socially effective. On the other hand, to take problem orientation seriously, integrative and collaborative research should not develop as a detached specialization but bring its potential to fruition in close interaction with disciplinary specialists.

References

Argyris, C. and Schön, D. (1996). *Organizational learning II: theory, method, and practice.* Reading: Addison-Wesley.

Baccini, P. and Oswald, F. (2008). Designing the urban: linking physiology and morphology. In: G. Hirsch Hadorn, H. Hoffmann-Riem, S. Biber-Klemm *et al.* (eds) *Handbook of transdisciplinary research*, pp. 79–88. Dordrecht: Springer.

Bammer, G. (2005). Integration and implementation sciences: building a new specialization. *Ecology and Society* **10**(2), article 6. Available at: <http://www.ecologyandsociety.org/vol10/iss2/art6/>.

Bammer, G. (2006). A systematic approach to integration in research. *Integration Insights* no. 1. Available at:<http://www.anu.edu.au/iisn/activities/integration_insights/integration-insight_1.pdf> (accessed 25 July 2007).

Bammer, G. (2007). Key concepts underpinning research integration. *Integration Insights* no. 5. Available at: <http://www.anu.edu.au/iisn/activities/integration_insights/integration-insight_5.pdf> (accessed 25 July 2007).

Bammer, G. (2008). Enhancing research collaboration: three key management challenges. *Research Policy* **37**, 875–87.

Bammer, G. and Smithson, M. (eds) (2008). *Uncertainty and risk: multi-disciplinary perspectives.* London: Earthscan.

Beck, U. (1992). *Risk society: towards a new modernity*, transl. M. Ritter. London: Sage.

Bruce, A., Lyall, C., Tait, J., and Williams, R. (2004). Interdisciplinary integration in Europe: the case of the fifth framework programme. *Futures* **36**, 457–70.

Cash, D.W., Clark, W.C., Alcock, F. *et. al.* (2003). Knowledge systems for sustainable development. *Proceedings of the National Academy of Sciences USA* **100**(14), 8086–91.

Clark, T.W. (2002). *The policy process: a practical guide for natural resource professionals.* New Haven, CT: Yale University Press.

Costanza, R., Cumberland, J., Daly, H., Goodland, R., and Norgaard, R. (1997). *An introduction to ecological economics.* Boca Raton, FL: St Lucie Press.

van den Daele, W. and Krohn, W. (1998). Experimental implementation as linking mechanism in the process of innovation. *Research Policy* **27**, 853–68.

Franco, L.A. (2006). Forms of conversation and problem structuring methods: a conceptual development. *Journal of the Operational Research Society* **57**, 813–21.

Funtowicz, S.O. and Ravetz, J.R. (1993). Science for the post-normal age. *Futures* September, 739–55.

Funtowicz, S. and Ravetz, J.R. (2008). Values and uncertainties. In: G. Hirsch Hadorn, H. Hoffmann-Riem, S. Biber-Klemm *et al.* (eds) *Handbook of transdisciplinary research*, pp. 361–8. Dordrecht: Springer.

Grunwald, A. (2004). Strategic knowledge for sustainable development: the need for reflexivity and learning at the interface between science and society. *International Journal of Foresight and Innovation Policy* **1**, 150–67.

Held, H., and Edenhofer, O. (2008). Climate protection vs. economic growth as a false trade off: restructuring global warming mitigation. In: G. Hirsch Hadorn, H. Hoffmann-Riem, S. Biber-Klemm *et al.* (eds) *Handbook of transdisciplinary research*, pp. 191–204. Dordrecht: Springer.

Hirsch Hadorn, G., Pohl, C., and Scheringer, M. (2003). Methodology of transdisciplinary research. In: G. Hirsch (ed.) *Unity of knowledge (in transdisciplinary research for sustainability)*. Oxford: Eolss Publishers.

Hirsch Hadorn, G., Hoffmann-Riem, H., Biber-Klemm, S. *et al.* (2008). The emergence of transdisciplinarity as a form of research. In: G. Hirsch Hadorn, H. Hoffmann-Riem, S. Biber-Klemm *et al.* (eds) *Handbook of transdisciplinary research*, pp. 19–39. Dordrecht: Springer.

Jantsch, E. (1972). Towards interdisciplinarity and transdisciplinarity in education and innovation. In: L. Apostel *et al.* (edS) *Problems of teaching and research in universities*, pp. 97–121. Paris: Organisation for Economic Cooperation and Development (OECD) and Center for Educational Research and Innovation (CERI).

Klein, J.T. (1990). *Interdisciplinarity: history, theory and practice*. Detroit, MI: Wayne State University Press.

Klein, J.T. (2004). Disciplinary origins and differences. *Australian Academy of Science Fenner conference on the environment: understanding the population-environment debate: bridging disciplinary divides*. Canberra, ACT: Australian Academy of Science. Available at: <http://www.science.org.au/events/fenner/fenner2004/klein.htm>.

Klein, J.T. (2008). Integration in der inter- und transdisziplinären forschung. In: M. Bergmann and E. Schramm (eds) *Transdisziplinäre forschung: integrative forschungsprozesse verstehen und bewerten*, pp. 93–116. Frankfurt: Campus.

Krohn, W. (2008). Learning from case-studies. In: G. Hirsch Hadorn, H. Hoffmann-Riem, S. Biber-Klemm *et al.* (eds) *Handbook of transdisciplinary research*, pp. 369–83. Dordrecht: Springer.

Kueffer, C., Hirsch Hadorn, G., Bammer, G., van Kerkhoff, L., and Pohl, C. (2007). Towards a publication culture in transdisciplinary research. *GAIA* **16**, 22–6.

Lawrence, R.J. (2004). Housing and health: from interdisciplinary principles to transdisciplinary research and practice. *Futures* **36**, 487–502.

Loibl, M.C. (2006). Integrating perspectives in the practice of transdisciplinary research. In: J.P. Voss, D. Bauknecht, and R. Kemp (eds) *Reflexive governance for sustainable development*, pp. 294–309. Cheltenham: Edward Elgar.

McDonald, D., Bammer, G., and Deane, P. (2009). *Research integration using dialogue methods.* Canberra: ANU E-Press. Available at: <http://epress.anu.edu.au/dialogue_methods_citation.html>.

Maranta, A., Guggenheim, M., Gisler, P., and Pohl, C. (2003). The reality of experts and the imagined lay person. *Acta Sociologica* **46**(2), 150–65.

Messerli, B. and Messerli, P. (2008). From local projects in the Alps to global change programmes in the mountains of the world: milestones in transdisciplinary research. In: G. Hirsch Hadorn, H. Hoffmann-Riem, S. Biber-Klemm *et al.* (eds) *Handbook of transdisciplinary research*, pp. 43–62. Dordrecht: Springer.

Midgley, G. (2000). *Systemic intervention: philosophy, methodology, and practice.* New York: Kluwer Academic/Plenum Publishers.

Midgley, G. (2003). *Systems thinking.* London: Sage.

National Academies (2005). *Facilitating interdisciplinary research.* Washington, DC: National Academies Press.

Neuman, W.L. (2003). *Social research methods: qualitative and quantitative approaches*, 5th edn. Boston, MA: Allyn and Bacon.

Nicolescu, B. (1996). *La transdisciplinarité: manifeste.* Monaco: Éditions du Rocher.

OECD (1972). *Interdisciplinarity: problems of teaching and research in universities.* Paris: Organization for Economic Cooperation and Development.

Pohl, C. (2008). From science to policy through transdisciplinary research. *Environmental Science and Policy* **11**, 46–53.

Pohl, C. and Hirsch Hadorn, G. (2007). *Principles for designing transdisciplinary research. Proposed by the Swiss Academies of Arts and Sciences.* Munich: Oekom.

Pohl, C. and Hirsch Hadorn, G. (2008). Methodological challenges of transdisciplinary research. *Natures Sciences Sociétés* **16**, 111–21.

Pohl, C., van Kerkhoff, L., Hirsch Hadorn, G., and Bammer, G. (2008). Integration. In: G. Hirsch Hadorn, H. Hoffmann-Riem, S. Biber-Klemm *et al.* (eds) *Handbook of transdisciplinary research*, pp. 411–24. Dordrecht: Springer.

ProClim (1997). *Research on sustainability and global change: visions in science policy by Swiss researchers.* Bern: Conference of the Swiss Scientific Academies and Swiss Academy of Sciences. Available at: <http://www.proclim.ch/Reports/Visions97/Visions_E.html> (accessed 3 December 2006).

Rayner, S. and Malone, E.L. (1998). The challenge of climate change to the social sciences. In: S. Rayner and E.L. Malone (eds) *Human choice and climate change*, pp. 33–69. Columbus, OH: Battelle Press.

Rossini, F.A. and Porter, A.L. (1979). Frameworks for integrating interdisciplinary research. *Research Policy* **8**, 70–9.

Rhodes, R. (1986). *The making of the atomic bomb.* London: Simon and Schuster.

Rip, A. (2008). Nanoscience and nanotechnologies: bridging gaps through constructive technology assessment. In: G. Hirsch Hadorn, H. Hoffmann-Riem, S. Biber-Klemm *et al.* (eds) *Handbook of transdisciplinary research*, pp. 145–57. Dordrecht: Springer.

Schelling, E., Wyss, K., Diguimbaye, C. *et al.* (2008). Towards integrated and adapted health services for nomadic pastoralists and their animals: a north-south partnership. In: G. Hirsch

Hadorn, H. Hoffmann-Riem, S. Biber-Klemm *et al.* (eds) *Handbook of transdisciplinary research*, pp. 277–91. Dordrecht: Springer.

Smithson, M. (1989). *Ignorance and uncertainty: emerging paradigms.* New York: Springer.

Stokes, D.E. (1997). *Pasteur's quadrant: basic science and technological innovation.* Washington, DC: The Brookings Institution.

Stringer, E.T. (2007). *Action research.* Los Angeles: Sage.

Wiesmann, U., Hirsch Hadorn, G., Hoffmann-Riem, H. *et al.* (2008). Enhancing transdisciplinary research: a synthesis in fifteen propositions. In: G. Hirsch Hadorn, H. Hoffmann-Riem, S. Biber-Klemm *et al.* (eds) *Handbook of transdisciplinary research*, pp. 433–41. Dordrecht: Springer.

World Commission on Dams (2000). *Dams and development: a new framework for decision-making.* London: Earthscan. Available at: <http://www.unep.org/dams/WCD/report.asp>.

World Commission on Environment and Development. (1987). *Our common future.* Oxford: Oxford University Press.

Wynne, B. (1993). Public uptake of science: a case for institutional reflexivity. *Public Understanding of Science* **2**, 321–37.

CHAPTER 31

Systems thinking

SYTSE STRIJBOS

Systems thinking is one form that interdisciplinarity has adopted since the middle of the twentieth century. It is a catchall term for different postwar developments in a variety of fields, such as cybernetics, information theory, game and decision theory, automaton theory, systems engineering, and operations research. These developments concur, however, inasmuch as, in one way or another, they relate to a basic reorientation in scientific thinking attempting to overcome ever-increasing specialization, and trying to make a shift from reductionist to holistic thinking, while acknowledging the unity of reality and the interconnections between its different parts and aspects.

There have been a number of attempts to define interdisciplinarity and identify its different types. Of particular interest in the present case is Margaret Boden (1999), who distinguishes six forms ranging from weak to strong: encyclopedic, contextualizing, sharing, cooperative, generalizing, and integrative types of interdisciplinarity. Encyclopedic interdisciplinarity requires no exchange or sharing between any disciplines involved, whereas integrative interdisciplinarity demands rigorous interaction. The latter is thus, according to Boden, the most genuine kind of interdisciplinarity as 'an enterprise in which some of the concepts and insights of one discipline contribute to the problems and theories of another—preferably in both directions'. Artificial intelligence (AI), a field in which Boden has a scholarly reputation, is in her view an excellent example of integrated interdisciplinarity. Each of the main types of AI, traditional or symbolic AI, connectionism, and 'nouvelle AI', has borrowed concepts from other disciplines such as philosophy, logic, psychology, and neurophysiology.

How does systems thinking fit into this typology? Boden labels the proposal for a 'general systems theory' that was launched by Ludwig von Bertalanffy and others in the middle of the twentieth century and Norbert Wiener's closely related idea of cybernetics as examples of 'generalizing interdisciplinarity', defined as 'an enterprise in which a single theoretical perspective is applied to a wide range of previously distinct disciplines'. Also the more recent developments in the area of complexity studies can be regarded as an example of this type. Boden (1999, p. 20) correctly notes that it is no accident that these examples are all heavily mathematical: 'The abstractness of mathematics enables it to be applied, in principle, to all other disciplines'.

Boden nevertheless fails to note some of the ways systems thinking has developed. In his later work von Bertalanffy for instance has distinguished between general system theory *in a broader sense* and *in a narrower sense*. Although von Bertalanffy's own theoretical work focuses on the latter, he stresses in his *General system theory: foundations, developments, applications* (1968), a collection of articles published over a period of more then 20 years, that he had both in mind from the outset. His concern is not just with a certain theory but the breakthrough of a new paradigm in science. Different postwar developments, such as cybernetics, information theory, network theory, game theory, systems engineering and related fields culminated in the birth of the systems movement when von Bertalanffy joined with Boulding, Rapoport, and Gerard to establish in 1954 the Society for General Systems Research, an association that still exists under the name of the International Society for the Systems Sciences. Stimulated by this new scientific association, a dynamic, broad-based field has developed and a multiplicity of approaches and trends has arisen.

With the increasing expansion of systems thinking, von Bertalanffy felt the need to distinguish different domains. Following his distinctions, the wide range of studies in the systems field—general system theory in a broader sense—can be divided into three realms or basic types. The first is *systems science*, which can be defined as the scientific exploration and theory of 'systems' in the various sciences, such as biology, sociology, economics, etc., while general system theory concerns the principles that apply to all. The second realm is *systems approach in technology and management* that concerns problems arising in modern technology and society. While philosophy is present in the areas of systems science and systems technology, *systems philosophy* can be distinguished in the systems field as a third domain in its own right. In the view of leading systems thinkers such as von Bertalanffy the introduction of 'system' as a key concept entails not only a total reorientation in science and technology, but also in philosophical thought.

To explore the implications of systems thinking for interdisciplinarity it is appropriate to consider each of the domains more in detail. In what follows some main lines will be sketched, rather than pursuing an encyclopedic overview of the developments in each domain. A broad and rather up-to-date documentation of the systems field can be found in *Systems thinking* (2002), a four-volume collection edited by Gerald Midgley that includes more than 70 classic and contemporary texts, including some critical evaluating studies.

31.1 Systems science

The well-known stock phrase that 'a whole is more than the sum of its parts' stems from a tradition in Greek philosophy, older than the conceptual use of the term 'system', that speaks of wholes that are composed of parts (Harte 2002). This whole–part relationship attracted renewed scientific interest in wholeness and the whole arising in the early twentieth century. Exploring the genealogy of contemporary systems thinking, reference has been made to Jan C. Smuts (1870–1950), a South African statesman and philosopher who is often depicted as a white supremacist supporting a racially segregated society (cf. Shula Marks 2000). In his book *Holism and evolution* (1926) he created the concept and word 'holism' (derived from the Greek *ὅλος*, *holos*, meaning whole, and entirety), expressing the idea that all the properties of a given system (biological, chemical, social, economic, mental, linguistic, etc.) cannot be determined or explained by its component parts alone. Instead, the system as a whole determines in an important way how the parts behave. It has also been claimed by Mattessich (1978) and others that the Russian philosopher and scientist Alexander A. Bogdanov (1873–1928) worked out the first version of a general systems conception in his book *Tektologiya: vseobschaya organizatsionnaya nauka* [*The universal science of organization: essays in tektology*] (1922). Both Smuts and Bogdanov had thus anticipated systems ideas at the beginning of the twentieth century. However, the conceptual use of 'system' as a technical term in science and technology arose some decades later and has become ubiquitous since the 1950s.

The philosopher–biologist Ludwig von Bertalanffy (1901–72) became one of the leading figures in the rise of systems thinking by coining the concept of a 'system', or more precisely the concept of an 'open system', as a key concept in the quest for a unified science incorporating all the disciplines, each corresponding to a certain segment of the empirical world. Just like Smuts, von Bertalanffy was also inspired by the debate in the biological sciences in the first decades of the twentieth century. Struggling with the controversy between two competing views, the dominant mechanistic-causal approach and vitalist-teleological conception, he did not take one or other side but proposed what he called an 'organismic' view. At issue was the possibility of an explanation for the phenomena of life that would have the status of an exact science, not through a *reduction* of biology to physics but through the *expansion* of classical physics into a broader, exact natural science. Von Bertalanffy considered this idea of expansion of scientific concepts as a key that opens the door to very far-reaching scientific developments. The extension of the domain of exact science from physics to biology must be carried further. Organismic biology, he argued, which focuses on the study of the organism as an open system (in contrast to the study of closed systems in classical physics) becomes in its turn a borderline case of the so-called 'general system theory'. The concept of the 'open system' was for him the truly 'general system' concept enabling the integration of all the sciences into a general system theory.

Like von Bertalanffy, the economist Kenneth E. Boulding (1910–95) was one of the early pioneers and founders of the systems movement. Being aware of the increasing difficulty of profitable exchange among the disciplines the more science breaks into subgroups, Boulding started pursuing the unity of sciences as an economist within the social sciences. Early in his scientific career he became convinced that all the social sciences were fundamentally studying the same thing, which is the social system. In his book *The image: knowledge in life and society* (1956a) Boulding introduces the 'image' concept, apparently inspired by Shannon and Weaver's concept of information, serving as a basis for the desired integration of the social sciences. And in a classic article 'General systems theory: the skeleton of science', published in the same year (Boulding 1956b), he pointed out the next step towards a general systems theory, incorporating all the sciences. Boulding sketched two possible approaches in the interdisciplinary quest for a general systems theory. A first approach is to identify general phenomena which are found in many disciplines, such as the phenomenon of growth. A second, more systematic, approach is to arrange the empirical fields in a certain hierarchic order, a hierarchy of systems in which each higher systems level has a higher degree of complexity. This issue of hierarchy has subsequently been widely discussed in the systems literature, e.g. by Herbert Simon in an often reprinted paper about 'The architecture of complexity: hierarchic systems' originally published in 1962.

Looking back over a period of more than 40 years Peter Checkland (1999, p. 49) made the observation that the original interdisciplinary project of the founders cannot be declared a success. A meta-level kind of approach leading to a greater unification of the sciences as envisaged has not occurred. However, one can admit that systems ideas and concepts have been incorporated in many disciplines. And sometimes new systems concepts and insights born in one discipline have contributed to the problems and theories of another. An impressive example of such an exchange between disciplines—or integrative interdisciplinarity, speaking in Boden's typology—is the work of the social scientist Niklas Luhmann (1927–98).

Aiming for a unified social theory, a general theory of social systems, Luhmann argues in his *Social systems* (1995) that two subsequent paradigm changes have taken place on the level of general systems theory, showing a shift from an ontological to a more functionalistic systems concept, i.e. from thinking in terms of wholes as unchangeable substances to systems that maintains themselves in a dynamic exchange with their environment. The first move in this direction was due to von Bertalanffy in the mid 1950s. By proposing the concept of the 'open system' a transformation of thinking took place in which the traditional difference between *whole and part* was replaced by *system and environment*. Like any paradigm change, Luhmann notes, this implies a conceptual broadening. What has been conceived of previously as the difference between whole and part, the old paradigm, was reformulated by this new schema as system differentiation and thereby built into the new paradigm. Systems differentiation can be understood as the repetition within systems of the difference between system and environment.

The second paradigm change and move towards a more radical functionalistic way of thinking is due to developments in systems science leading to a theory of *self-referential systems*. Initial efforts in the 1960s, in which Heinz von Foerster (1911–2002) played a leading role, employed the concept of self-organization. Self-organization is the phenomenon of self-reference with regard to the structure of a system, that is to say that structural changes are produced by the system itself. Self-reference in a more encompassing way, however, also include the elements composing a system. For this purpose the biologists Humberto Maturana and Francisco Varela (1946–2001) created the term *autopoiesis* (self-creation). Autopoiesis thus means that a system has the ability to reproduce itself at the level of its own elements.

According to Luhmann, a theory of self-referential systems as the most recent general system theory opened up important avenues for a general theory of social systems. This broadening of the general system concept from 'open system' to 'self-referential system' enabled Luhmann to avoid criticisms of the views of Talcot Parsons, his great predecessor in sociology, whose social systems theory was the dominant paradigm in sociology during the 1950s and 1960s. While very influential for a few decades, Parsons' systems theory was also widely criticized as a legitimization of the status quo. It was charged that Parsons' systems approach was inherently conservative in its focus on the maintenance of social order and in emphasizing consensus at the expense of acknowledging social change and conflict. Profiting from newer developments in systems science, Luhmann succeeded in the 1980s to propose a new social systems theory, turning around Parsons' structural-functionalism into a functional-structural systems approach.

Mapping interdisciplinary research

Katy Börner and Kevin W. Boyack

This box reviews existing approaches to visualizing interdisciplinary research from a science mapping perspective (Börner *et al.* 2003). Although visualization is often the way in which the results of analysis are communicated, visualization (or, in our case, mapping) is not analysis. The purpose of mapping is simply to visually display the results of analysis to enhance communication of those results. Maps often play the role of templates upon which the results of previous analyses are overlaid.

Conceptualizing science for mapping

Measurement of the degrees of interdisciplinarity and generation of maps on which those measurements can be displayed can only be done within a recognized framework or conceptualization. Although such conceptualizations can be very detailed, containing (1) units of analysis, (2) their interactions, (3) basic mechanisms of growth and change, and (4) system boundaries (Börner and Scharnhorst 2009), the conceptualization can be highly simplified for the specification of interdisciplinarity. Although the units for analyzing and mapping interdisciplinarity could be authors, journals, disciplines, or even countries, the key facet is to be able to define their disciplinary inputs and outputs. Thus, a map showing the results of an analysis of interdisciplinarity would need to have units and show the relationships between those units. In addition, it might include a time dimension to see changes in structure and/or dynamics. Last, but not least, it would need to describe the boundaries of the analysis (e.g. neuroscience by itself; neuroscience in the context of all of science; or all of science). By way of example, we discuss co-author collaboration, journal citation, and paper citation flow analyses and maps here.

Collaboration maps

Co-authorship networks can be generated for authors from a specific institution, country, or a specific field of science. They are often visualized in a node-link diagram (authors as nodes; co-authorships as lines linking authors) that places linked authors in close spatial proximity and unconnected authors further apart, while minimizing the number of link crossings. If the author nodes are colored by discipline, authors with interdisciplinary co-authorships are easily identified in the author map.

Journal network maps

The measure of *betweenness centrality* has recently been promoted as a measure of journal interdisciplinarity, and to good effect. Given that betweenness is a network measure, based on the links that would be shown in a network map if it were drawn, such maps are a natural way to display the results of these analyses. For example, Leydesdorff and Schank (2008) show maps of the local citation networks for several different journals, each showing the key 'central' position of a particular journal. In one case, for the journal *Nanotechnology*, they animate a sequence of annual maps generated from citation statistics. The visual maps correlate well with the betweenness measure for *Nanotechnology*; a dramatic rise in betweenness correlates with *Nanotechnology* taking the central linking position in the field away from the journal *Science* in the early 2000s.

(*cont.*)

Mapping interdisciplinary research (cont.)

Knowledge flow maps

While collaboration studies focus on the assembly and impact of (inter)disciplinary teams, knowledge flow studies try to answer questions related to the diffusion of expertise and/or knowledge over time, geospatial space, and topic space.

Knowledge diffusion is often measured using citation relations. The fact that paper A cited paper B is taken as an indication that knowledge diffused from B to A, although it is difficult to quantify just how much and what type of information was truly transferred. Papers can be aggregated to journals resulting in journal citation networks. Journals can be aggregated into disciplines or scientific fields. *Historiographs* are visualizations of small, localized, paper citation networks over time. Other network visualizations help to understand the 'super highways' of information diffusion as well as knowledge hubs and authorities.

Typically, the topical composition of nodes, as well as the type and strength of their interlinkages, changes over time. The figure illustrates this effect by showing the topical composition and changes in citation flows for 14 major subdisciplines of chemistry, biology, biochemistry, and bioengineering (see Boyack *et al.* 2009 for details).

Each node in the figure represents a cluster of journals, where the disciplinary composition of journals is denoted by pie charts and where the disciplinary assignments for each journal are based on their Thomson Reuters categories. Seven categories are used in the figure (those denoted by the six colors, and 'Other'). The areas of the pie charts scale with the number of papers, thus accurately representing the relative sizes of the different subdisciplines. Knowledge flows among these 14 subdisciplines in terms of number of direct citations are represented by arrows. Arrows denote the flow of information from the cited subdiscipline to the citing subdiscipline. Arrows inherit the color of the knowledge source, and are proportional in thickness to the square root of the number of citations. Changes in topical composition and knowledge flows can be animated over time. Maps at 5-year intervals along with specific observations drawn from the maps are available in the original work.

Outlook

Current research on mapping science includes the creation of standards for sharing scientific and technical data, including means to connect different types of data organized by different taxonomies and classifications across fields and languages. For example, it is desirable to interlink publications with patents and funding as well as with the impact on education and training, and economic activity.

Clarification on what actually constitutes 'interdisciplinarity' for the purpose of measurement and mapping is also needed. The pie charts in the figure are more aptly described as multidisciplinarity than as interdisciplinarity. Such definitions may also be discipline- or even team-specific. Better understanding of the real-life types, mechanisms, and amounts of knowledge generation and transfer and how they could be approximated using data from bibliographic databases is needed.

The communication of results, via network drawing or other map types, for example, has to meet the needs of the intended user group and their tasks. Today, it is not clear what metaphors work best to depict something as abstract as science—charts, networks, geospatial maps. How many dimensions does it take to render science?—do one-dimensional time lines suffice, are topic maps best, which map types are best to communicate interdisciplinarity, an so on.

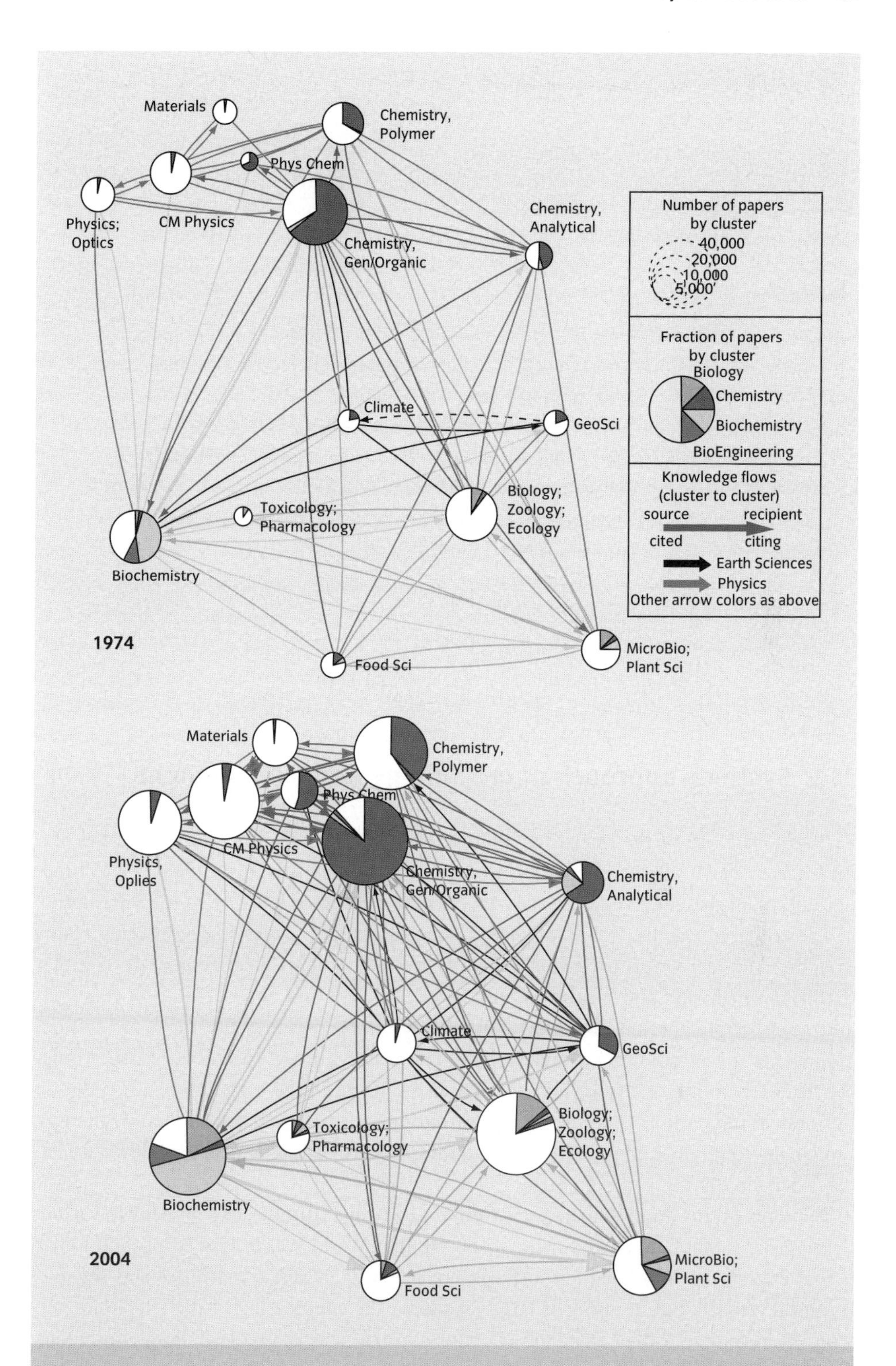

Map of 14 major subdisciplines, fractions of papers by field for each subdiscipline, and knowledge flows between subdisciplines for 1974 (top) and 2004 (bottom).

(cont.)

Mapping interdisciplinary research (cont.)

These are just few of the many questions asked by the Mapping Science exhibit (<http://scimaps.org/>). Maps featured in this exhibit provide first answers for specific user groups such as science policy makers, researchers, or commercial decision makers. Ultimately, the inner workings and impact of interdisciplinary research should be communicated and understood by scholars and the general public alike.

References

Börner, K. and Scharnhorst, A. (2009). Visual conceptualizations and models of science. [Special issue on the science of science: conceptualizations and models of science.] *Journal of Informetrics* **3**(3), 161–72.

Börner, K., Chen, C., and Boyack, K. (2003). Visualizing knowledge domains. In: B. Cronin (ed.) *Annual review of information science and technology* **37**, 179–255. Medford, NJ: American Society for Information Science and Technology.

Boyack, K.W., Börner, K., and Klavans., R. (2009). Mapping the structure and evolution of chemistry research. *Scientometrics* **79**(1), 45–60.

Leydesdorff, L. and Schank, T. (2008). Dynamic animations of journal maps: indicators of structural changes and interdisciplinary developments. *Journal of the American Society for Information Science and Technology* **59**(11), 1810–18.

31.2 Systems approach in technology and management

Parallel to the rise of the interdisciplinary movement in the sciences, the need has also increasingly been felt for integration and general frameworks in the fields of technology and management. While science concerns the pursuit of knowledge for the solution of theoretical problems, technology and management aim at shaping or altering reality in addressing real-world problems. However, these problems have become so complex that traditional ways and means are no longer sufficient and approaches of a generalist and interdisciplinary nature have become necessary. Nowadays the realm of systems approaches in technology and management comprises a broad spectrum of issues ranging from environmental modeling and world modeling in the early 1970s, to studies in business strategy and management of organizations, medical practice and family therapy, human development and poverty issues, to the quickly developing field of industrial ecology since the 1990s.

The roots of this domain in systems thinking are quite complex and go back to various developments that happened during or shortly after World War II. One important aspect is that engineering has been led to think not in terms of single machines and separate technical artifacts but in terms of larger 'systems': the engineering of the telephone network, for example, rather than the telephone instrument or the switching equipment. Traditionally, engineers are used to tackling practical problems by analyzing their parts and finding a solution for the different parts. As the name *systems engineering* suggests, the idea took hold that the traditional approach of engineering separate components needed to be extended to approach systems made up out of many components that are interacting. Engineers speak about electric systems, power systems, transportation systems, computer systems, etc. The initial use of the term 'systems engineering' with roughly its present meaning probably began in the early 1940s at the Bell Telephone Laboratories (Schlager 1956). A leading pioneer was the electrical engineer Arthur D. Hall (1925–2006) who worked for many years at Bell Labs and in 1962 published the first significant book on systems engineering entitled *A methodology for systems engineering*.

A development closely related to systems engineering is *operations research* or 'operational research' as it is known in the United Kingdom. Briefly discussing the difference between both fields, Hall (1962, p. 18) noted that operations research is usually concerned with the operation and the optimization of an existing system, including both humans and machines, while in contrast systems engineering focuses on the planning and design of new systems to better perform existing operations or to implement new ones never performed before. In the aftermath of the war C. West Churchman (1914–2004) and Russell L. Ackoff, who were inspired by American pragmatism and aimed to apply this philosophy to societal issues, became leading scholars in North America in the incipient fields of operations research and systems thinking. Together with E. L. Arnoff they published one of the field's first textbooks *Introduction to operations research* (Churchman *et al.* 1957), which became internationally recognized. The book emphasized an interdisciplinary team-based approach, characterizing operations research as 'the application of scientific methods, techniques and tools to problems involving the operations of a system so as to provide those in control of the system with the optimum solution to the problem'.

Simultaneously with the development of systems engineering and operations research, an approach emerged in the 1950s that was known as *systems analysis*; at that time it was closely associated with the RAND Corporation (RAND being an acronym for 'Research ANd Development'), a not-for-profit organization in the advice-giving business established in 1948. From the 1960s, RAND-style systems analysis began to find broader industrial and governmental uses, leading to a 1972 initiative by 12 nations to set up a non-governmental interdisciplinary research institute in Austria—the International Institute for Applied Systems Analysis (IIASA). Systems analysis was defined by Quade (1973, p. 121) as 'analysis to suggest a course of action by systematically examining the costs, effectiveness and risks of alternative policies and strategies—and designing additional ones if those examined are found wanting'. A case described by Miser and Quade (1985) is a policy analysis clarifying the issues for a governmental decision in the Netherlands after the North Sea flood of 1953 about the protection of the Oosterschelde estuary from flooding.

Acknowledging the differences that are present in their background and concerning particular features of systems engineering, systems analysis, and operations research, these systems approaches show important commonalities. They all rely heavily on the methods of the natural and technical sciences. Consequently they aspire to describe phenomena by mathematical-statistical models, while holding the assumption that an optimal solution exists for a problem situation which may be uncovered in this way. Another member of this family of approaches is *systems dynamics* which gained a certain reputation in the

1970s in the work of Forrester (1971) and Meadows (1972) on world modeling for the Club of Rome.

Examining the origins and nature of systems engineering and systems analysis, Checkland (1978, p. 107) concluded that a single view underlies these approaches: 'there is a desired state, S(1), and a present state S(0), and alternative ways of getting from S(0) to S(1). "Problem solving," according to this view, consists of defining S(1) and S(0) and selecting the best means of reducing the difference between them'. This constitutes what Checkland called *'hard' systems thinking*, defined as any kind of systems thinking which adopts the means–end schema. Although this model may be useful for engineering-type problems, it has a very limited applicability. Hard systems thinking demands that objectives can be clearly defined; however, an important aspect of many 'soft' problem situations is that the involved parties are likely to see the problem situation differently and define objectives accordingly. Checkland was thus faced with the challenge of rethinking the failing concept of a systems approach rooted in the engineering tradition. This led to his conceptualization of a soft systems approach in the 1970s that admits the human dimension, dealing with multiple perceptions of reality, values, and interests of the people involved (Checkland and Haynes 1994).

The later work of Churchman and Ackoff in North America is similar to the scientific program started in the 1970s by Peter B. Checkland and his colleagues at Lancaster University in the UK. Dissatisfied or even disillusioned with the course of operations research, Ackoff (1973, p. 670) argued that mainstream operations research as it had developed since 1950 was only useful in dealing with problem areas that can be decomposed into problems that are independent of each other. However, major societal problems such as discrimination, inequality within and between nations, increasing criminality, and so on, must be attacked holistically, with a comprehensive systems approach. Ackoff's dispute with the operations research community culminated in two papers (Ackoff 1979a,b) in which he called for a new paradigm breaking away from the ever-increasing 'mathematization' of operations research and for a return to true interdisciplinarity, involving in the research of all those affected by it.

In their plea for a systems approach Ackoff and Churchman not only triggered debate in the operations research community about the nature and characteristics of the field but also delivered a fresh input to the debate in the systems movement on interdisciplinarity. In 1963 Ackoff published an article in the Yearbook of the Society for General Systems Research in which he argued for a new vision of an integrating systems science and the difference between the conception of general systems theory. According to Ackoff the conception of a general system theory endeavors to achieve integration using the results that are available in the mono-disciplines, that is to say it attempts a unity *afterwards*. However, in his view 'the integral' precedes the disciplinary splitting of a problem into disjoint chunks—'Therefore, posing the problem of unifying science by interrelating disciplinary output either in the forms of facts or concepts (i.e. logical positivism), or laws or theories (i.e. general system theory), is to try to lock the barn door after the horse has gone' (Ackoff 1963, p. 120).

Ackoff's idea that integration has to take place a priori, i.e. in the phase of knowledge production, implies that he put emphasis on science as an activity and the scientific method employed in that activity. Integral knowledge requires an integration of the disciplines involved within an interdisciplinary framework. The integration must come during, not after, the performance of the research. In his conception of systems science, systems research is on sounder ground than von Bertalanffy's general systems theory because it takes systems as it finds them, in all their multidisciplinary glory. For the realization of interdisciplinary research Ackoff formulates three important conditions. First, it is necessary to unify the variables and concepts of the different disciplines to a common denominator. This enables the construction of interdisciplinary systems models. Second, for a healthy development of systems research an appropriate methodology is required. There is a need, for example, to develop scientific methods to evaluate and compare the performance of systems such as cars, planes, production systems, or health care systems. Third, the realization of programs of interdisciplinary 'systems research' involves special educational requirements.

31.3 Systems philosophy

The worlds of science, technology, and philosophy do not exist in isolation from each other. Because philosophy raises questions that are fundamental for science and technology one could argue that philosophy is by nature an interdisciplinary endeavor. For the sake of clarity it is therefore useful to distinguish some of the various meanings in which the term systems philosophy can be used, each standing for different themes and a different role of philosophy in the systems field.

First, systems philosophy deals with the fundamental philosophical issues involved in the realm of systems science. Such a fundamental issue in biology is the question 'what is life?' or 'how do we understand the phenomena of life?'. As we discussed, von Bertalanffy advocated a so-called organismic conception—the view that the organism is a whole or system, transcending its parts when these are considered in isolation. Searching for a satisfying understanding of the Aristotelian dictum of the whole that is more than its parts, von Bertalanffy at the same time takes a stand on another fundamental problem of Greek philosophy. There is the famous statement of Heraclitus: '*panta rhei*', everything is in flux, arguing against Parmenides who taught that only the static being was real, the fixed, and that change is an illusion. In this controversy, which has persisted in one form or another across the whole of Western philosophy and science, systems science adopts the Heraclitean point of view. The model of the organism as an open system implies that life has to be understood as primarily a stream of life. Forms and structures that manifest themselves in living nature are in von Bertalanffy's view secondary, just like social structures are secondary in Luhmann's understanding of social phenomena. Systems science thus manifests a totally dynamic view of reality in which enduring structures seem to evaporate and become volatile and dynamic.

Second, systems philosophy concerns the philosophical foundations of the systems approach in technology and management. Comparing Ackoff with von Bertalanffy, one notices that they agree that society is going through an important intellectual revolution that will usher us into a new era of science and society—in Ackoff's wording, going from a Machine Age to a Systems Age. One of the important characteristics of systems science, as we have seen above, is the priority given to the dynamic and flowing character of reality. The same characteristic seems to hold for systems research when Ackoff (1981, p. 16) points out that there is a turn from analysis to synthesis, which implies a turn to a functional understanding of the thing to be explained in terms of its role or function within its containing whole or environment. The synthetic approach does not exclude analysis, but in the Systems Age synthesis has priority over analysis, and function over structure. The turn from the Machine Age to the Systems Age even implies a different understanding of reality. Characteristic of the Machine Age is the deistic view in which God is regarded as the creator of the world as a machine which runs according to fixed laws. While the Machine Age and deism personify God as the Creator God, who is independent from his handiwork, God loses this personal and independent character in the Systems Age. Like Smuts' holism, Ackoff's (1981, p. 19) systems thinking is also infused with a rationalist pantheistic view in which the world coincides with God as the largest, all-embracing whole.

In a more elaborate way this is also the case in Churchman's work. In his view, the most fundamental and serious issues of the systems approach concern the problem of improvement. If we assume that we have the capability to improve systems, then what exactly do we mean by 'improvement' in designing interventions for our social systems? Churchman (1968, p. 2) concisely describes the fundamental problem right at the start of his book *Challenge to reason* as follows: 'How can we design improvement in large systems without understanding the whole system, and if the answer is that we cannot, how is it possible to understand the whole system?'. In a line of reasoning similar to Ackoff's, Churchman points to the tradition of analysis in Western thought that presumes that parts of the whole system can be studied and improved more or less in isolation from the rest of the system. And comparable to Ackoff, Churchman also discerns two differing views of the whole system and its relationship to God. If we assume that a Supreme Being exists, Churchman (1979, p. 41, italics added) says, 'then we have the conceptual problem of describing [modelling] His relationship to the rest of reality'. And he continues: 'Two plausible hypotheses come to mind. The Augustinian hypothesis [...] is that *God is the designer* of the real system, as well as its decision maker. [...] The other hypothesis, the one chosen by Spinoza, is to say that *God* **is** the whole system: He is the most general system'.

Third, there is the aspiration to formulate a systems philosophy as a new philosophy, of which Archie Bahm, Mario Bunge, and Ervin Laszlo are the chief proponents. As a prolific author of many books Laszlo became the most influential. Building on von Bertalanffy's ideas for a new scientific world view he developed in the 1970s the framework for a systems philosophy in tune with the latest developments in science and technology, representing a total reorientation of thought which aims to overthrow and replace the dominating mechanistic worldview and its incarnation in the industrialized and commercialized society of today. The dynamic view of reality that, as we noticed, underlies von Bertalanffy's and Luhmann's theoretical ideas and concepts, is a typical feature of the systems view of the world that has been summarized by Laszlo (1972, pp. 80–1) as follows: 'Imagine a universe made up not of things in space and time, but of patterned flows extending throughout its reaches. [...] Some of the flows tie themselves into knots and twist into a relatively stable pattern. Now there is something there – something enduring [. . .] "Things" are emerging from the background of flows like knots tied on a fishing net'.

Laszlo's philosophical conceptions culminate in his view on the future of humankind in our globalizing world. The general thrust of the many books that he published over a period of nearly 40 years is that contemporary society is in a critical stage of development. World society can get out of the danger zone if there is a complete turnabout at the immaterial-spiritual level. In Laszlo's view there is thus not only the need to bridge the gap between the sciences, gaining an integral scientific view of the world—more important even is the integrating role of systems thinking in bridging the divide between science and religion, between science and spirituality. The interdisciplinary challenge for systems thinking is thus extended in Laszlo's view to the search for a new uniting spirituality for humankind. From his *Introduction to systems philosophy* and *The systems view of the world* originally published more then 30 years ago up to his more recent books such as *Science and the reenchantment of the cosmos* (Laszlo 2006), such a spirituality is linked to an evolutionary dynamic view of the universe, arguing that there exists an interconnecting cosmic field that conserves and conveys information, a subtle sea of fluctuating energies from which all things arise. Similar to the pantheism of Churchman and Ackoff, Laszlo also thus rejects a personal God who is separated as creator from the universe. In his systems view of the universe, God is the all-embracing cosmic consciousness, and we are part of that.

31.4 Subsequent developments

Although systems science is perpetuated in newer developments such as systems biology, chaos theory, and the study of complex systems (Santa Fe Institute, NM, United States), the original interdisciplinary program of the founders of the systems movement has largely failed in its early aspirations to create a greater unification of the sciences, setting out general laws and principles governing the behavior of any type of system. On the contrary the systems movement was more successful in creating interdisciplinary approaches for tackling practical real-world problems. Jackson (2001, p. 234) offers two reasons why systems approaches in technology and management should have proven so successful. First, practical problems are by nature interdisciplinary and do not correspond to a single mono-discipline. Second, the systems idea provides a useful antidote to reductionism and enshrines a commitment to looking at real-world problems in terms of wholes and interconnected elements. With the work of Ackoff and Churchman in North America and that of Checkland in the UK this domain has not come to a standstill. Moving from 'hard systems thinking' to 'soft systems thinking' they in principle opened the way to further debates and advances. Ideas that have inspired subsequent developments derive from social theory, philosophy, and theology. The account I shall give here is necessarily biased by the role played by myself and the programmatic research efforts in which I am involved.

In the 1980s a program entered the stage that has been called 'critical systems thinking', a program that involved many people and gained a strong basis at the University of Hull in the UK since the appointment of Michael C. Jackson in 1979 (Jackson is also the editor-in-chief of a central journal in the systems community, *Systems Research and Behavioral Science*). An important source that supplies information about the broader context of critical systems thinking is a collection of articles *Critical systems thinking* (1991) edited by two of its main proponents Robert L. Flood and Michael C. Jackson.

Inspired by the social theorist and philosopher Jürgen Habermas, critical systems thinking tried to overcome shortcomings in soft systems thinking. Similar to Checkland's critical analysis of the origins and nature of hard systems thinking in 1978, Jackson embarked upon a similar critique of the ambitions of soft systems thinking in an early article published in 1982 on the nature of soft systems thinking. He arrives at the conclusion that although soft systems thinking has attacked the technical rationality embodied in hard systems thinking, one crucial element was never targeted—it still proceeds from existing power relationships. In Jackson's own words: 'Soft systems thinking is most suitable for the kind of social engineering that ensures the continued survival, by adaptation, of existing social elites. It is not authoritarian like systems analysis or systems engineering, but it is conservative-reformist' (Jackson 1982, p. 28). In an overview article about 20 years later Jackson (2001, p. 233) pointed out how critical systems thinking gradually made progress towards realizing its goal. After it became obvious that all systems approaches have their limitations, it was critical systems thinking which supplied the bigger picture at a meta-methodological level and 'has set out how the variety of methodologies now available can be used together in a coherent manner to promote successful intervention in complex societal problem situations'.

Independently of the group at Hull University, an important contribution to the strand of critical systems thinking was made in the 1980s by Werner Ulrich from the University of Fribourg in Switzerland. As a student of Churchman, and inspired by Kant's critical philosophy and Habermas' critical social theory, Ulrich launched a program that led to the conception of 'critical systems heuristics', exposed in his main publication *Critical heuristics of social planning: a new approach to practical philosophy* (Ulrich 1994). A distinguishing feature of this dialect of critical systems thinking is its methodological core principle, known as 'boundary critique'.

The latest development is a program that emerged in the late 1990s. This program involves a variety of disciplines, ranging from engineering to philosophy, executed by an international group of cooperating scholars affiliated with universities in different countries. In view of the need for an independent organizational basis, the Centre for Philosophy, Technology and Social systems (CPTS) was established in 1996 and is linked with the philosophy faculty of the Vrije Universiteit in Amsterdam. Inspired by the legacy of philosophers from this university, Herman Dooyeweerd (1894–1977) and his student Hendrik van Riessen (1911–2000), this program attempts to break with the Western idea of an autonomous human rationality and the absolutization of a scientific view of the world as the final horizon for human understanding. It aims to break with deism and a mechanistic-technical worldview in which God and reality are separated, but also with pantheism and a dynamic worldview blurring the boundary between God and the world. Dooyeweerdian thinking, that often has provided common ground in the CPTS program, is based on a theistic worldview that distinguishes a personal God from created reality and relates God and reality in a living, continuous, and sustaining creator–creation relationship. Churchman (1987, p. 139) once formulated as the most important question for systems thinking 'Does God exist?'. Of equal importance, however, is the next question 'If God exists, how does he relate to reality?'. Both questions are also fundamental in Christian theology and were rephrased by John Calvin (1509–74) in terms of the two connected questions about our knowledge of God and that of ourselves (Calvin 2008).

With the appearance of *In search of an integrative vision for technology*, edited by Strijbos and Basden (2006), the results of the CPTS program during its first decade have been documented. There are at least three important features that distinguish the interdisciplinary scope and character of this program. In the first place, interdisciplinarity concerns the shaping of a philosophical integrative framework that depicts the relationship between 'technology' and 'society', aiming for a normative-ethical basis to guide the development of science and technology for the benefit of society. For that purpose a systems view on 'technology and society' has been conceived in which different systems levels are distinguished (Strijbos and Basden 2006). With the help of this model it is possible to connect research—in engineering, management methodology, philosophy—on a specific systems level with research on other systems levels.

Second, an important part of the research program to which a number of people have contributed deals with the second realm of systems thinking, the study of practice-oriented systems methodologies for the fields of engineering and management. While making use of key notions of Dooyeweerdian philosophy, and in a critical conversation with hard, soft, and critical systems thinking, a new strand of systems thinking has been explored, labeled 'multi-modal systems thinking' by de Raadt (1997) or 'disclosive systems thinking' by Strijbos (2000).

Third, the CPTS program involves a wide spectrum of disciplines and thus seems to fit nicely with what Boden has classified as integrated interdisciplinarity. It even takes this type of interdisciplinarity further, aiming to bridge the gap between the natural sciences and the humanities, and between theory and practice. Borrowing distinctions from Frodeman *et al.* (2001) and Frodeman and Mitcham (2007), the CPTS research can also be characterized as a 'wide' and 'deep' interdisciplinarity, a type of interdisciplinary research that aims to be 'wide' rather than 'narrow' and 'deep' rather than 'shallow.' The narrow–wide distinction refers to whether only the natural and engineering sciences are involved or whether these are integrated with the human and social sciences. The shallow–deep distinction refers to whether interdisciplinarity is limited to scientific experts or whether people are also involved who are not academic researchers, but are experts with practical experience concerning real-world problems.

31.5 Final remarks

The discussion in this chapter focuses on the ambitions of systems thinking to attain general integrative frameworks that will enable relevant communication and exchange between the disciplines. Reviewing its now more then 50-year history, one can conclude that this interdisciplinary movement has stimulated fruitful theory formation in a broad variety of fields in the natural and social sciences but has not succeeded in achieving its original far-reaching goals. Furthermore, one can conclude that integrative, interdisciplinary systems approaches in technology and management have become well-accepted and have put normative considerations and ethical issues firmly on the agenda. With respect to this there still remains much to be done. An important challenge for the future is to foster an open and critical debate between the different systems approaches about their normative sources and underlying worldview (Strijbos 1988; Eriksson 2003). Another vital element is the establishment of links with other interdisciplinary fields, such as development studies and science, technology, and society (STS) studies, which also struggle for a better understanding of the forces shaping our times and search for strategies to address the big societal problems facing us.

References

Ackoff, R.L. (1963). General system theory and systems research: contrasting conceptions of systems science. *General Systems* **8**, 117–24.

Ackoff, R.L. (1973). Science in the systems age: beyond I.E., O.R. and M.S. *Operations Research* **21**, 661–71.

Ackoff, R.L. (1979a). The future of operational research is past. *Journal of the Operational Research Society* **30**, 93–104.

Ackoff, R.L. (1979b). Resurrecting the future of operational research. *Journal of the Operational Research Society* **30**, 189–99.

Ackoff, R.L. (1981). *Creating the corporate future: plan or be planned for.* New York: John Wiley and Sons.

von Bertalanffy, L. (1968). *General system theory: foundations, developments, applications.* New York: Penguin Books.

Boulding, K.E. (1956). General systems theory: the skeleton of science. *Management Science* **2**(3), 197–208.

Boulding, K.E. (1977). *The image: knowledge in life and society*, 11th printing (originally published 1956). Ann Arbor, MI: University of Michigan Press.

Boden, M.A. (1999). 'What is interdisciplinarity?'. In: R. Cunningham (ed.) *Interdisciplinarity and the organisation of knowledge in Europe*, pp. 13–24. Luxembourg: Office for Official Publications of the European Communities.

Bogdanov, A.A. (1922). *Tektologiya: vseobschaya organizatsionnaya nauka* [*The universal science of organization: essays in tektology*]. Berlin and Petrograd-Moscow (Original Russian printing). [Translated and reprinted as *Bogdanov's tektology*, ed. P. Dudley. Hull: Centre for Systems Studies, University of Hull, 1996.]

Calvin, J. (2008). *Institutes of the Christian religion.* Peabody, MA: Hendrickson Publishers.

Checkland, P.B. (1978). The origins and nature of 'hard' systems thinking. *Journal of Applied Systems Analysis* **5**(2), 99–110.

Checkland, P.B. (1999). Systems thinking. In: W.L. Currie and B. Galliers (eds) *Rethinking management information systems*, pp. 45–56. Oxford: Oxford University Press.

Checkland, P.B. and Haynes, M.G. (1994). Varieties of systems thinking: the case of soft systems methodology. *Systems Dynamics Review* **10**(2–3), 189–97.

Churchman, C.W. (1987). Systems profile: discoveries in an exploration into systems thinking. *Systems Research* **4**, 139–46.

Churchman, C.W. (1979). *The systems approach and its enemies: basic concepts of systems and organization*. London: Basic Books.

Churchman, C.W. (1968). *Challenge to reason*. New York: McGraw Hill.

Churchman, C.W., Ackoff, R.L., and Arnoff, E.L. (1957). *Introduction to operations research*. New York: Wiley.

Eriksson, D.M. (2003). An identification of normative sources for systems thinking: an inquiry into religious ground motives for systems thinking paradigms. *Systems Research and Behavioral Science* **20**(6), 475–87 [an extended version is reprinted in Strijbos and Basden 2006].

Flood, R.L. and Jackson, M.C. (1991). *Critical systems thinking: directed readings*. Chichester: John Wiley and Sons.

Forrester, J.W. (1971). *World dynamics*. Cambridge, MA: Wright Allen Press.

Frodeman, R. and Mitcham, C. (2007). New directions in interdisciplinarity: broad, deep, and critical. *Bulletin of Science, Technology and Society* **27**(2), 506–14.

Frodeman, R., Mitcham, C., and Sacks, A.B. (2001). Questioning interdisciplinarity. *Science, Technology, and Society Newsletter* **127**(Winter/Spring), 13–24.

Hall, A.D. (1962). *A methodology for systems engineering*. Princeton, NJ: D. Van Nostrand Co.

Harte, V. (2002). *Plato on parts and wholes: the metaphysics of structure*. Oxford: Oxford University Press.

Jackson, M.C. (2001). Critical systems thinking and practice. *European Journal of Operational Research* **128**, 233–44.

Jackson, M.C. (1982). The nature of 'soft' systems thinking: the work of Churchman, Ackoff and Checkland. *Journal of Applied Systems Analysis* **9**, 17–29 [followed by a response from Ackoff, Churchman, and Checkland].

Laszlo, E. (1972a). *Introduction to systems philosophy: toward a new paradigm of contemporary thought*. New York: Harper and Row.

Laszlo, E. (1972b). *The systems view of the world: the natural philosophy of the new developments in the sciences*. New York: George Braziller. [An update of this book is Laszlo, E. (1996). *The systems view of the world: a holistic vision for our time*. Cresskill, NJ: Hampton Press].

Laszlo, E. (2006). *Science and the reenchantment of the cosmos: the rise of the integral vision of reality*. Rochester, VT: Inner Traditions, Bear and Company.

Luhmann, N. (1995). *Social systems*. Stanford, CT: Stanford University Press. [First published in German as *Soziale Systeme*. Frankfurt am Main: Suhrkamp Verlag, 1984.]

Marks, S. (2000). *Before 'the white man was master and all white men's values prevailed'?: Jan Smuts, race and the South African war*. Lecture given on the invitation of the Institute of Economic and Social History and the Institute of Africanistic Studies, both at University of Vienna, and of the Southern Africa Documentation and Co-operation Centre (SADOCC), 24 October 2000, Vienna. Available at: <http://www.sadocc.at/publ/marks.pdf>.

Mattessich, R. (1978). *Instrumental reasoning and systems methodology: an epistemology of the applied and social sciences*. Dordrecht: D. Reidel.

Meadows, D.L. (1972). *The limits to growth: a report for the Club of Rome project on the predicament of mankind.* New York: Universe Books.

Midgley, G. (ed.) (2002). *Systems thinking*, Vols 1–4. London: Sage Publications.

Miser, H.J. and Quade, E.S. (eds) (1985). *Handbook of systems analysis: craft issues and procedural choices.* New York: Wiley.

Quade, E.S. (1963). *Military systems analysis.* Santa Monica, CA: The RAND Corporation. [Reprinted in Optner, S.L. (ed.) (1973). *Systems analysis*, pp. 121–40. Harmondsworth: Penguin Books.]

de Raadt, J.D.R. (1997). A sketch for humane operational research in a technological society. *Systems Practice* **10**(4), 21–41.

Schlager, K.J. (1956). Systems engineering: key to modern development. *IRE Transactions* **EM-3**, 64–6.

Simon, H.A. (1962). The architecture of complexity: hierarchic systems. *Proceedings of the American Philosophical Society* **106**, 467–82. [Revised and reprinted in *The sciences of the artificial*, 3rd edn. Cambridge, MA: MIT Press, 1996.]

Smuts, J.C. (1926). *Holism and evolution.* London: Macmillan. [Reprinted 1999, Sherman Oaks, CA: Sierra Sunrise Publishing.]

Strijbos, S. (1988). *Het technische wereldbeeld: een wijsgerig onderzoek van het systeemdenken* [*The technical worldview: a philosophical investigation of systems thinking*]. Amsterdam: Buijten and Schipperheijn. Available at: <http://hdl.handle.net/1871/15599>.

Strijbos, S. (2000). Systems methodologies for managing our technological society: towards a 'disclosive systems thinking'. *Journal of Applied Systems Studies* **1**(2), 159–81.

Strijbos, S. and Basden, A. (eds) (2006). *In search of an integrative vision for technology: interdisciplinary studies in information systems.* New York: Springer.

Ulrich, W. (1994). *Critical heuristics of social planning: a new approach to practical philosophy.* New York/London: John Wiley and Sons. [Original edition 1983, Bern, Switzerland/Stuttgart, Germany: Paul Haupt.]

CHAPTER 32

Cross-disciplinary team science initiatives: research, training, and translation[1]

DANIEL STOKOLS, KARA L. HALL, RICHARD P. MOSER, ANNIE FENG, SHALINI MISRA, AND BRANDIE K. TAYLOR

Over the past two decades, the US government has devoted significant financial resources to the creation of large-scale team research projects, many of which involve hundreds of scientists working together from a wide range of different fields. In the health arena, some initiatives such as the Transdisciplinary Tobacco Use Research Centers (TTURCs) have been operational for nearly 10 years and span multiple university-based centers. Such projects pose considerable logistic, infrastructural, and conceptual challenges that must be overcome to ensure their success. The surge of interest and investment in team science (TS) and training initiatives (Nass and Stillman 2003; National Academy of Sciences 2005) has spawned a rapidly emerging field that focuses on understanding and managing the circumstances that facilitate or hinder the effectiveness of large-scale research, training, and translational initiatives[2] (Kessel *et al.* 2008; Klein 2008; Stokols *et al.* 2008).

This chapter provides a broad overview of the *science of team science* (SOTS) in terms of its historical and conceptual foundations, methodological approaches, training, and translational concerns. It draws on the authors' experiences in designing and implementing evaluative studies for assessing the collaborative processes, scientific training, public policy implications, and potential health outcomes associated with cross-disciplinary TS initiatives funded by the National Cancer Institute (NCI) of the National Institutes of Health (NIH).

[1] Portions of this chapter are based on: Stokols, D., Hall, K.L., Taylor, B.K., and Moser, R.P. (2008). The science of team science: overview of the field and introduction to the supplement. *American Journal of Preventive Medicine* **35**(2S), S77–S89. The authors thank the editors for their valuable comments on an earlier version of the manuscript.

[2] Throughout this chapter, the term *translational* refers to the use of scientific knowledge to create evidence-based health promotion policies, community interventions, and clinical practices.

We summarize recent developments in the evaluation of TS initiatives, using the NCI large-center initiatives as exemplars, and propose future directions for this burgeoning field.

The development of large-scale, institutionally based (and, sometimes, multi-institutional) TS initiatives exemplifies one possible arrangement for promoting cross-disciplinary collaboration in health science, training, practice, and policy. Other approaches include scientists working independently to bridge the disciplinary perspectives of multiple fields, and relatively small cross-disciplinary networks and teams whose members are dispersed across locations and who meet regularly through electronic means (e.g. via telephone and video conferencing, e-mail, and intranet) and occasional face-to-face meetings. The SOTS, utilizing evaluation methodologies, network techniques, and organizational and management theories and frameworks, represents a subarea within the field of science studies (Hess 1997) concerned with understanding and enhancing the processes and outcomes associated with team-based initiatives that are undertaken to promote cross-disciplinary research, training, and translations of science into improved practices and policies.

The remainder of this chapter consists of four sections. Sections 32.1 and 32.2 summarize important *conceptual* and *methodological* developments in the SOTS, including the definition of key terms and the development of logic models or program theories to guide the study of TS initiatives; the creation of new methods and metrics for examining the processes and outcomes (including both scientific and translational products) of TS; and the establishment and evaluation of TS training programs. Section 32.3 examines *emerging concerns and research directions* within the SOTS field. These new avenues of research include the development of more rigorous quasi-experimental research designs for assessing the scientific and translational contributions of TS initiatives in comparison with smaller scale, non-team-oriented projects; more comprehensive models of cross-disciplinary training and strategies for evaluating training outcomes; a more nuanced typology of cross-disciplinary TS initiatives and funding models as, well as strategies for evaluating their relative efficacy in achieving the intended goals of TS; and a broader understanding of the contextual factors that determine the 'collaboration readiness' of a particular team and scientific field. Section 32.4 is an epilogue.

32.1 Conceptual developments in the science of team science

The SOTS integrates and builds on many of the efforts highlighted in this book. For instance, Klein (Chapter 2 this volume) offers a rich discussion explicating the complexities of defining disciplines, *per se*, and explores different kinds of disciplinary interactions by developing a taxonomy of interdisciplinarity. Such complexities among key SOTS concepts drive the field forward as well as create challenges for defining the scope of the field and evaluating the work therein. In this section we provide succinct definitions that have served as the basis for our earlier research within the SOTS field. Furthermore, we define our units of analysis with respect to process and outcome goals of TS initiatives and then describe how they can be integrated into theoretical models and conceptual frameworks.

32.1.1 Key terms and conceptual models

We distinguish between TS initiatives themselves and the SOTS field whose principal units of analysis are the large research and training initiatives implemented by public agencies and non-public organizations. TS initiatives are designed to promote collaboration and facilitate cross-disciplinary approaches to conceptualizing and analyzing research questions about particular phenomena. In contrast, the SOTS field is construed more broadly as a branch of *science studies* concerned especially with understanding and managing circumstances that facilitate or hinder the effectiveness of TS initiatives (Stokols *et al.* 2008a).

32.1.2 Characteristics of scientific initiatives and teams

Research teams may comprise investigators drawn from either the same or different academic fields (i.e. unidisciplinary versus cross-disciplinary teams). These teams vary not only by their disciplinary composition, but also by their size, organizational complexity, and geographic scope, ranging from a few participants working at the same site to many investigators dispersed across different geographic and organizational venues. Furthermore, the goals of TS initiatives are diverse (e.g. spanning scientific discovery, training, clinical translational, public health, and policy-related goals), and both the quality and level of intellectual integration between disciplines that is intended and achieved varies from one program to the next, for example along a continuum ranging from unidisciplinary to multidisciplinary, interdisciplinary, and transdisciplinary integration.

Because TS initiatives differ along so many dimensions, it is important to differentiate between various types of research and training initiatives (Klein, Chapter 2 this volume). Team-based projects can include a handful of scientists working together at a single site, all the way to larger and more complex initiatives comprising many (e.g. 50–200) investigators who work collaboratively on multiple research projects and are dispersed across different departments, institutions, and geographic sites. Expenditure is another dimension; Trochim *et al.* (2008), for example, define large research initiatives as grant-funded projects solicited through specific requests for applications (RFAs) with an average annual expenditure of at least $5 million. The usual duration of these initiatives, e.g. National Institutes of Health (NIH) P50 and U54 centers, National Cancer Institute (NCI) specialized programs of research excellence (SPOREs), is 5 years and may be renewed, in some cases, extending over one or more decades. Some broad-gauged initiatives such as the NIH roadmap programs provide an over-arching framework and funding source for scores of interrelated research and training initiatives, all of which are designed to promote cross-disciplinary scientific collaboration (National Institutes of Health 2003). Often, large research initiatives incorporate career development and training components as well as clinical translation, health promotion, and policy-related functions.

Team science initiatives also vary with respect to the collaborative orientations and disciplinary perspectives of team members. This chapter focuses primarily on initiatives intended to promote *cross-disciplinary* (CD) rather than *unidisciplinary* (UD) collaboration. CD teams strive for leverage and, in some cases, integrate concepts, methods, and theories drawn from two or more fields. Three different approaches to CD collaboration, which have become relatively standard in this nascent field, have been described by Rosenfield (1992). *Multidisciplinarity* (MD) is a process in which scholars from disparate fields work independently or sequentially, periodically coming together to share their individual perspectives to achieve broader-gauged analyses of common research problems. Participants in MD teams remain firmly anchored in the concepts and methods of their respective fields. *Interdisciplinarity* (ID) is a more robust approach to scientific integration in the sense that team members not only combine or juxtapose concepts and methods drawn from their different fields, but also work more intensively to integrate their divergent perspectives, while remaining anchored in their own respective fields. *Transdisciplinarity* (TD) is a process whereby team members representing different fields work together over extended periods to develop shared conceptual and methodological frameworks that not only integrate but also transcend their respective disciplinary perspectives. Our definition for TD is distinct from others such as Huutoniemi (Chapter 21, this volume), who suggest that TD is different from ID based on the inclusion of contributors from outside academia rather than the distinction being the degree of synergy or integration as Rosenfield (1992) proposes. Considering both Rosenfield's and Huutoniemi's definitions, TD collaborations perhaps have the greatest potential to produce highly novel and generative scientific outcomes, but they are more difficult to achieve and sustain (compared with UD, MD, and ID projects) due to their greater complexity and loftier aspirations for achieving transcendent, supradisciplinary integrations.

The ensuing discussion focuses on ID and TD (rather than UD and MD) science initiatives in which an explicit goal of collaboration is to integrate theories, methods, and training strategies drawn from two or more fields. Examples of large-scale ID and TD team initiatives are the NCI TTURCs, the Centers for Excellence in Cancer Communications Research (CECCR), the Centers for Population Health and Health Disparities (CPHHD), and the Transdisciplinary Research on Energetics and Cancer (TREC) Centers; and the National Center for Research Resources (NCRR) Clinical and Translational Science Centers (CTSC).

The distinctions between UD, MD, ID, and TD forms of scientific collaboration are directly relevant to the development of criteria for gauging the success of TS initiatives. In particular, measures of scientific collaboration and its outcomes should be appropriately matched to the research, training, and translational goals of a particular initiative (Huutoniemi, Chapter 21 this volume). A major goal of ID and TD initiatives, for example, is to bridge the perspectives of different fields through collaborative development of integrative conceptualizations, methodological approaches, and training strategies. Thus, an important criterion for gauging the success of these initiatives is the extent to which cross-disciplinary integrations are actually achieved by research teams.

32.1.3 Units of analysis, theoretical models, and conceptual frameworks

It is important to define the major units of analysis and core subject matter of the SOTS field. A major challenge in this regard is to specify the dimensions of *program effectiveness* or *success* as they pertain to TS initiatives. For example, the quality of scientific work may be defined differently in the context of ID and TD team initiatives as compared to UD projects. Traditional criteria of scientific quality include conceptual originality, methodological rigor (e.g. validity and reliability of empirical findings), and the amount of research output produced, such as peer-reviewed publications. In the context of TS initiatives, however, the *quality and scope of ID and TD integration* (e.g. development of integrative conceptualizations, methodological approaches, training programs bridging two or more fields, and/or the emergence of new hybrid fields of inquiry, organizational structures, and management strategies for supporting TS) are important facets of TS initiatives that must be considered in view of their explicit mission to promote scientific integration (Stokols *et al.* 2003). Finally, it is important to identify the defining features of successful *ID and TD training* (e.g. multi-mentor training models, trainees' subsequent career trajectories, and the intellectual contributions of current and former trainees).

To date, a number of conceptual models have been proposed by SOTS scholars to identify key *antecedent conditions, intervening processes, and outcomes* (including near-term, mid-term, and long-term outcomes) associated with TS initiatives and to explain the interrelationships among them (e.g. the presence of institutional supports or constraints at the beginning of an initiative and their impact on subsequent collaborative processes and outcomes). For instance, Trochim *et al.* (2008) used concept mapping techniques to derive an empirically based logic model, which guided the NCI TTURC Initiative-wide Evaluation Study. The TTURC logic model, shown in Fig. 32.1, posits a series of temporal links between early processes of intellectual collaboration and TD integration and subsequent outcomes, including scholarly publications, TD training programs, community health interventions, and public policy initiatives. Using the TTURC logic model, both the constructs of interest for evaluation (e.g. the degree of collaboration achieved; the emergence of integrative conceptual frameworks) as well as the temporal sequence in which one would expect to see changes are clearly delineated.

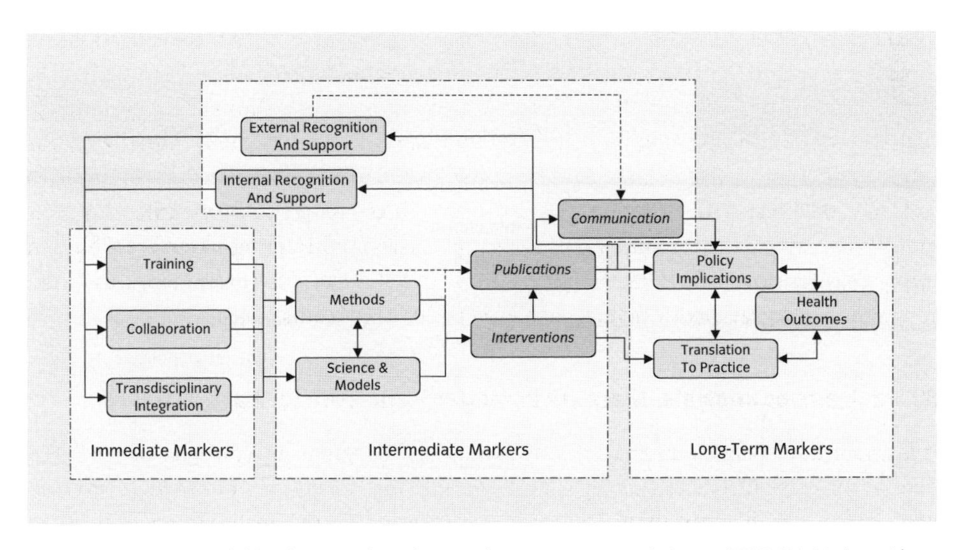

Figure 32.1 Logic model for the Transdisciplinary Tobacco Use Research Center(TTURC) initiative-wide evaluation study. Reprinted from Trochim *et al.* (2008).

In their work, Stokols and colleagues have proposed a different antecedent–process–outcome model of TD science in which several interpersonal, environmental, and organizational antecedents of collaboration are considered, such as the leadership styles of center directors, scientists' commitment to team research, the availability of shared research and meeting space, electronic connectivity among team members, and the extent to which they share a history of working together on prior projects (Stokols *et al.* 2003). Intervening processes examined in this model include intellectual, interpersonal, and affective experiences as well as observed and/or self-reported collaborative behaviors. Examples of these processes are: brainstorming strategies to create and integrate new ideas; cross-disciplinary biases and tensions that often arise in collaborative situations; and strategies for negotiating and resolving conflicts. The antecedent and process variables specified in the model, in turn, influence several near-term, mid-term, and long-term outcomes of scientific collaboration including the development of new conceptual frameworks, research publications, training programs, and translational innovations over the course of the initiative (see Fig. 32.2). Empirical support for the hypothesized links among antecedent, process, and outcome variables was derived from a longitudinal (5-year) comparative study of the TTURC centers.

More recently, Hall, Stokols *et al.* (2008), Holmes *et al.* (2008), and Warnecke *et al.* (2008) developed multistage conceptual frameworks that have guided TD research, training, and community intervention efforts within the NCI TREC and CPHHD initiatives, respectively. From its inception, the CPHHD initiative has placed greater emphasis on community-based participatory research strategies (as compared with the TTURC and TREC initiatives) for the purposes of translating scientific knowledge about the causes of health disparities in the United States into university–community partnerships and collaborative interventions to mitigate these disparities. Thus, the CPHHD evaluation model incorporates a 'community stakeholder–investigator incubator' component (see Fig. 32.3) not

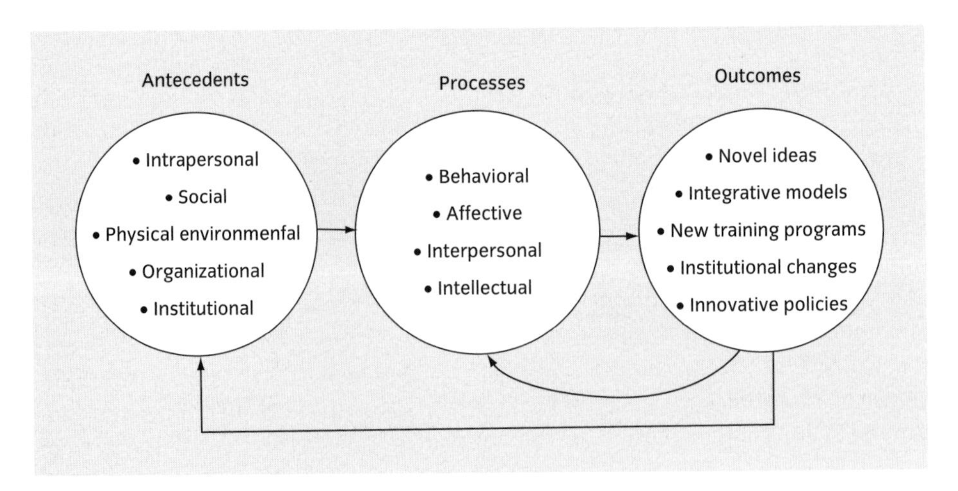

Figure 32.2 Antecedents, processes, and outcomes of cross-disciplinary scientific collaboration. Reprinted from Stokols *et al.* (2005).

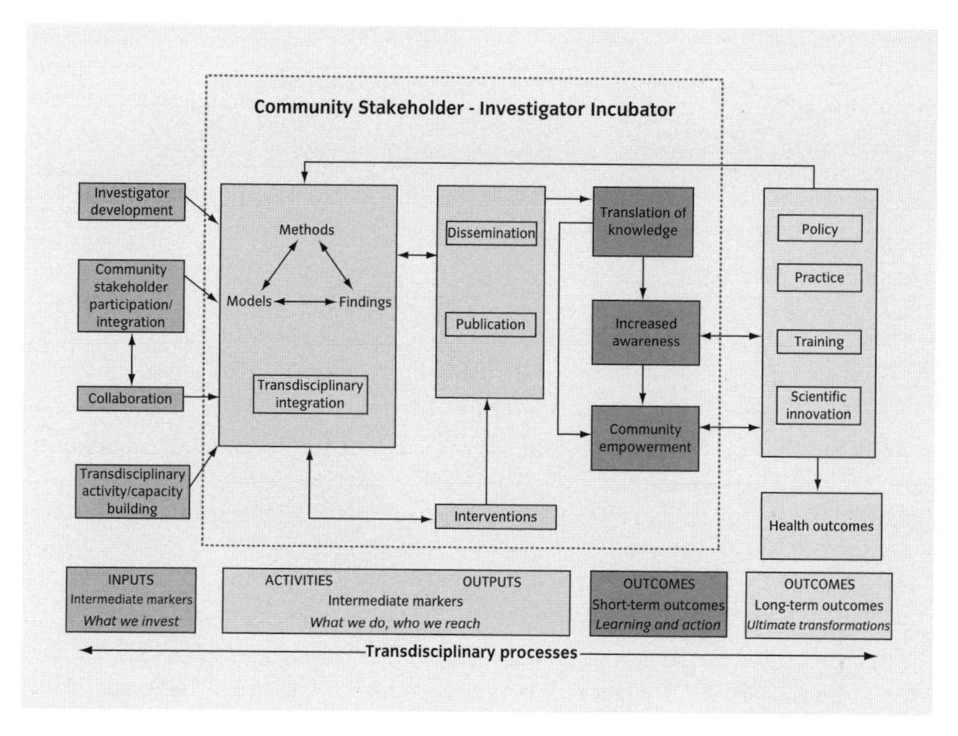

Figure 32.3 Centers for Population Health and Health Disparities (CPHHD) evaluation logic model. Reprinted from Holmes *et al.* (2008).

found in the previously described TTURC logic model (Fig. 32.1) or in the TREC evaluation logic model (see Fig. 32.4) outlined by Hall and colleagues.

Existing models of ID and TD collaboration raise several questions for future research. For example, antecedent conditions present at the outset of a TS initiative can be conceptualized as *collaboration readiness* (CR) factors that jointly influence a team's prospects for success over the course of an initiative (Hall *et al.* 2008a). However, the relative contributions of individual CR factors (e.g. leadership skills of center directors, availability of shared office and laboratory space, team members' experiences of working together on earlier projects) to specific dimensions of collaborative effectiveness (e.g. number of team publications produced as well as their integrative quality and scope; development of sustainable partnerships with community organizations) are not well understood and warrant further study.

Also, earlier conceptual models and the field studies on which they are based suggest that the scientific outcomes of TS initiatives are strongly influenced by social and interpersonal processes, including team members' collaborative styles and behaviors, interpersonal skills, and negotiation strategies. Yet the precise ways in which these social processes—such as team members' disagreements about scientific issues, interpersonal trust, 'group-think' among scientists who have worked together over extended periods—influence scientific productivity and TD integration are not known. The empirical links between interpersonal and intellectual dimensions of scientific collaboration remain unclear.

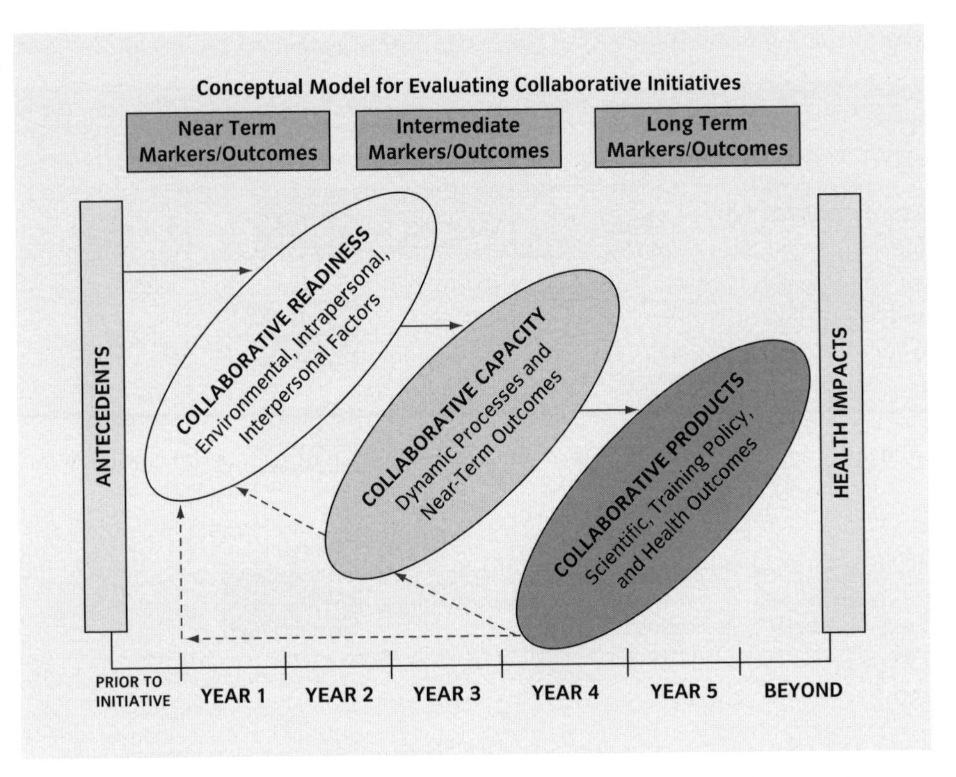

Figure 32.4 Transdisciplinary Research on Energetics and Cancer (TREC) evaluation logic model. Reprinted from Hall *et al.* (2008a).

32.2 Methodological developments in the science of team science

A variety of *qualitative* and *quantitative* methods and metrics are available to assess the antecedents, processes, and outcomes of TS initiatives. Examples of these multiple methods and measurement strategies include:

- The use of unstructured and semi-structured interviews with team members to assess key goals and concerns surrounding their collaborative activities.
- Content analyses of interviews with team members to reveal important themes in narrative accounts of collaborative experiences (e.g. perceptions of progress toward cross-disciplinary scientific integration).
- Administration of standardized survey scales to assess participants' collaborative readiness, research orientations and values, and collaborative experiences.
- Protocols to evaluate cross-disciplinary qualities of near-term written products (e.g. co-authored manuscripts and developmental project proposals).
- Social network analyses of cross-disciplinary collaboration.
- Bibliometric and peer-review assessments of scientific impact and productivity.

These methodological approaches and measurement strategies for evaluating TS initiatives are discussed below.

32.2.1 Qualitative methodologies

32.2.1.1 Appreciative inquiry to identify team goals and aspirations

Team-wide discussions among all relevant stakeholders (e.g. participating scientists and trainees, representatives of funding organizations, other partners) can establish a valuable foundation for a smooth and comprehensive evaluation. Appreciative inquiry techniques (Cooperrider and Witney 2005) can be used to identify team members' goals and collaborative aspirations, as well as their major strengths and assets for achieving team goals. These early planning activities offer opportunities, in a supportive environment, to identify and address challenges of particular relevance to ID and TD collaborations (e.g. issues stemming from divergent disciplinary cultures including language barriers, product development priorities related to tenure, scientific discovery) and may enhance 'buy-in' from all relevant stakeholders to the evaluation process.

32.2.1.2 Investigator interviews

Unstructured and semi-structured interviews with individuals or groups of team members can yield valuable insights into the divergent as well as shared perspectives of participating scientists, trainees, research staff, and community partners. Such interviews can be valuable for generating anecdotal evidence and narrative accounts of collaborative 'success stories'. Additionally, interviews can identify areas that need improvement, which can be shared with various stakeholder groups for the purposes of providing feedback about collaborative processes, and offer suggestions for enhancing team projects and outcomes over the course of an initiative.

32.2.1.3 Self-directed qualitative discussions

Alternatively, individuals within a TS initiative can be encouraged to lead self-directed or investigator-led discussions in an attempt to gather qualitative information about the collaborative processes and outcomes of an initiative. Such discussions may be conducted in meetings among investigators, trainees, and/or research staff within a single project, working group, or entire center. Another option is to organize a large scientific retreat that is self-directed and involves the participation of all members within a particular research project, center, or entire TS initiative. Scientific retreats have been found to be highly effective strategies for encouraging informal social communication as well as scientific dialogue and sparking ideas for integration among the members of TS centers (Stokols *et al.* 2003; Fuqua *et al.* 2004).

32.2.1.4 Document review of narrative accounts of team experiences

A valuable qualitative methodology for revealing major themes and patterns reflected in team members' collaborative experiences is to conduct content analyses of transcripts of the proceedings from scientific retreats and brain-storming meetings. Similarly, content analyses of team progress reports and other written products can provide valuable information about the processes and outcomes of TS initiatives. The data from content analyses can be supplemented by quantitative sources of evaluative information as described below.

32.2.1.5 External peer reviews

Analogous to the peer-review process for grants applications and publications, external peer reviews can be conducted to evaluate TS initiatives. The external panel can examine written products of initiative members (e.g. scientific progress reports, pilot research project proposals, peer-reviewed manuscripts, reports of trainees' research accomplishments and career development), and through periodic site visits assess the day-to-day operations and relative progress being made at particular centers over the course of an initiative.

32.2.2 **Quantitative methodologies**

In addition to the qualitative methods summarized above, several quantitative strategies can be incorporated into evaluations of TS initiatives, as summarized below.

32.2.2.1 Standardized surveys

Structured self-report surveys are used to assess facets of cross-disciplinary collaboration (e.g. focusing on constructs gleaned from the logic model guiding an evaluation), taking into account the temporal course of the initiative. For the TTURC initiative, scales and indices were created to measure each of the 13 constructs delineated in its logic model (see also Fig. 32.1). The results from the year-three TTURC survey revealed greater progress toward achieving proximal outcomes (e.g. transdisciplinary integration) as compared to more distal ones (e.g. community interventions and their impacts on health outcomes; cf. Mâsse *et al.* (2008) and Trochim *et al.* (2008)).

32.2.2.2. Ratings of written products

As part of the TREC evaluation study, a written products protocol was developed to evaluate the cross-disciplinary qualities of these types of collaborative scientific products. An example of written products includes the research proposals submitted by TREC investigators (during the first and second years of the initiative) to obtain TREC center developmental project funds. The written products protocol was designed to evaluate the conceptual breadth and integrative scope of the developmental projects, as well as the number and type of scientific disciplines and levels of analysis represented by the project staff. Reviews of the proposals were conducted by trained reviewers in consultation with a moderator and expert consultant to reach consensus on the identification and rating of each construct. The written products protocol is intended to be implemented several times over the course of the initiative to examine hypothesized changes over time (Hall *et al.* 2008a), such as progress toward TD integration.

32.2.2.3 Financial analyses

Because of the large monetary investment in cross-disciplinary scientific initiatives and the corresponding need to ensure that the management of these funds is handled effectively, it is important to assess project-specific and center-wide financial expenditures as the initiative progresses from its initial to later phases. Mandatory annual grant reports were used in the TTURC evaluation to compare actual versus proposed yearly spending and to identify any resulting carry-over funds from one year to the next. In addition, data for any financial carry-over were obtained from budget justifications included in investigators' annual reports. Using pre-existing, mandated budget and expenditure reports as a source of evaluative data reduced the participant burden of the overall evaluation. Results from these analyses can be used to identify centers that are having difficulty allocating their financial resources as planned. Large discrepancies between proposed and actual expenditure levels may indicate, in some instances, that team members are encountering collaborative difficulties in their efforts to implement proposed research projects. Often, these discrepancies are larger at the outset of a complex TS initiative when administrative structures and procedures are still in the early stages of development.

32.2.2.4 Social network analyses

Social network analysis (SNA) is another useful tool for evaluating TS programs. TS, by definition, involves people working collaboratively and a SNA can help reveal how these scientific networks develop over time, the density of the networks (i.e. the number of relationships within any particular network), and who the 'brokers' are (i.e. those who facilitate linking others together) within and between participating centers. SNA techniques also can be used to assess constructs such as the quantity and quality of TD work accomplished by the members of a particular network. The CPHHD and TTURC evaluations (see Stokols *et al.* 2005; National Cancer Institute 2007) utilized this methodology to obtain quantitative and visual evidence of collaborative relationships among scientists over time. Within the TTURC and CPHHD initiatives, for example, SNA data were used to assess network densities and the prevalence of collaboration among participating scientists within and across centers. It is also possible to correlate these networking outcomes with other measures such as publication history and scientific impact using bibliometric data.

32.2.2.5 Bibliometric analysis

Bibliometric analysis is another method for assessing the impact of scientific initiatives. This technique provides an assessment of the quantitative and qualitative scholarly impact of publications—those produced either by an individual or a group of individuals—and can be conducted with minimal or no participant burden. Within the TTURC initiative, several bibliometric indices were derived, including the number of times a published work is cited in subsequent publications, the impact factor of the journals in which an article is published, and a measure of how multidisciplinary a journal is. For scientific teams that remain intact over several years, bibliometric analyses can be used to identify temporally lagged changes in the scientific productivity of team members.

As the SOTS field matures, however, there is a need to develop more sophisticated methods and research designs for evaluating processes and outcomes of TS—for example, prospective quasi-experimental (and perhaps truly experimental) research designs, as compared to the retrospective case studies that have been predominant in the SOTS

Managing consensus in interdisciplinary teams

Rico Defila and Antonietta Di Giulio

Collaborative problem framing and consensus

One of the most important requirements of interdisciplinary research consists of what we call 'consensus': By means of suitable procedures and methods, the participants have to arrive at a shared view of the problem under investigation. Consensus here does not mean 'agreement' in an everyday sense. Rather, it means the development of models and theories that integrate the various disciplinary viewpoints in such a way that the result, for example the description of the research objective, is shared by all. Problem framing that incorporates diverse perspectives therefore lies at the heart of successful inter- and transdisciplinary work; success being defined as achieving a synthesis (the integrated result) that is more than just the addition of different points of view and individual results.

Problem framing consists of the following elements:

- Defining the problem (what exactly the problem is and for whom, what is the factual background of the problem, which assumptions and which type of knowledge help us to understand the problem, what is the context in which the problem is meaningful?).
- Figuring out possible solutions to the problem (what is causing the problem, how can and should the problem be approached, where could the most promising solutions be found?).
- Identifying the resources needed to solve the problem (in terms of money, time, perspectives and methods, types of knowledge, stakeholders).

Problem framing in interdisciplinary teams is collaborative problem framing. In the case of wildland fire, Brooks *et al.* (2006, p. 3) state that 'Problem framing involves the different ways that stakeholders define the problem and the terminology and concepts related to it [...]. Different frames allow stakeholders to see what they want to see, or what they are guided to see [...]. The existence of many different frames, or definitions of the problem, suggests a need to develop common goals and a common language'.

The importance of common goals, a common language, and a common theoretical basis is confirmed by the results of the survey conducted by the international cooperation DACH. In 1999 researchers from four inter- and transdisciplinary environmental research programs in Switzerland, Austria, and Germany were sent a written questionnaire about their experiences of and recommendations on research management, leadership and personal skills, methods of knowledge integration, and development of theories etc. Out of 649 questionnaires sent out, 294 completed questionnaires were returned, which corresponds to a response rate of 45%. The results quite clearly show significant differences concerning common goals, a common language, and a common theoretical basis between those who had achieved a synthesis and those who had not (see Figs. 1–3).

Of course, research in an interdisciplinary project will not be carried out collectively with the whole team acting all the time as a group. Rather, there is a division of labor. The common goals, the common questions, and the shared description of the research object are the starting points for individual research work and the point to return to after this work has been done (see Fig. 4).

The role of management in achieving consensus

Many projects fail in their efforts at collaborative problem framing, and, consequently, in developing integrated results. This is often due to a deficit concerning theory and methodology

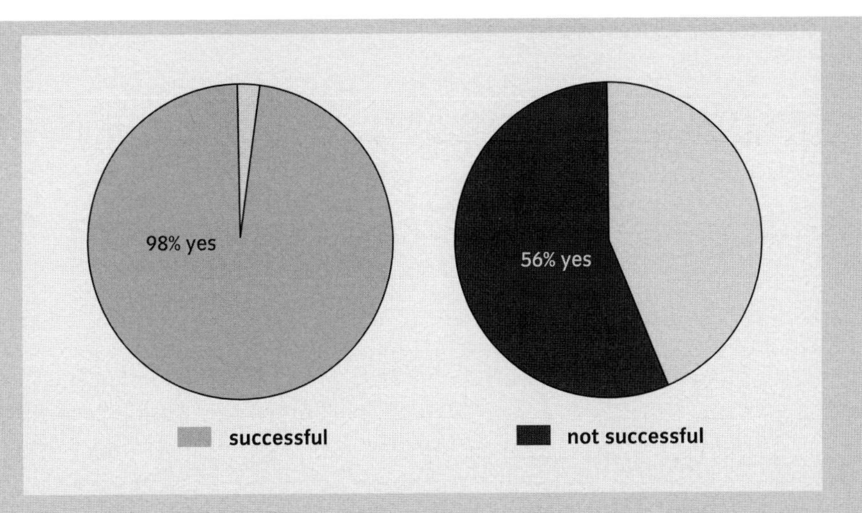

Figure 1 Common goals: of those who achieved a synthesis, 98% said that they had had common goals, whereas only 56% of those who had not achieved a synthesis had common goals (Defila *et al*. 2006, p. 72).

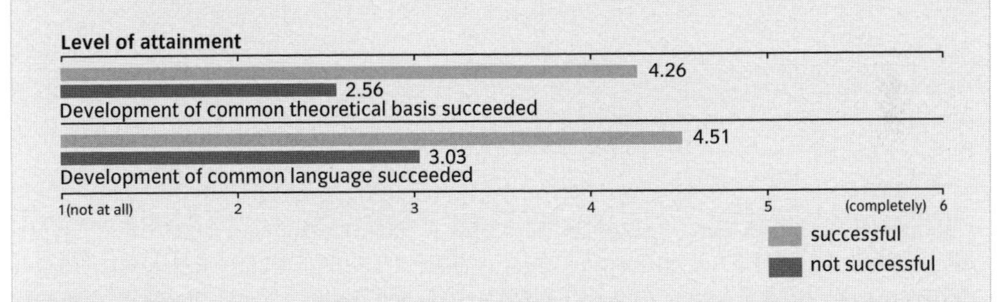

Figure 2 Common language, common theoretical basis: those having achieved a synthesis (the successful ones) were also successful concerning the development of a common language and a common theoretical basis (Defila *et al*. 2006, p. 118).

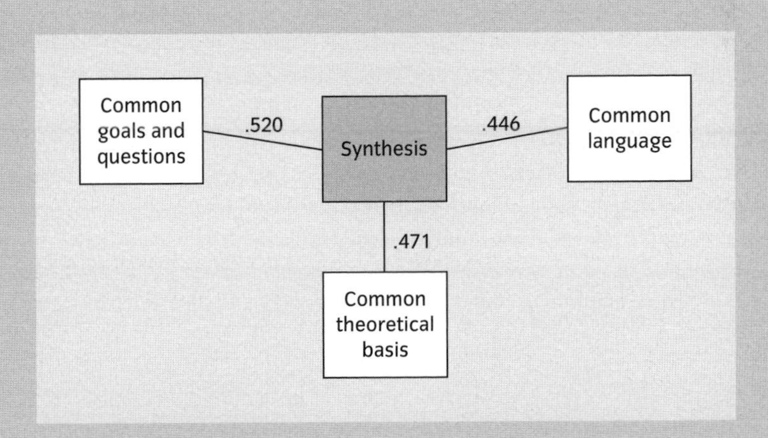

Figure 3 Success factors: the correlation between common goals and achieved synthesis is rather high, as is the correlation between common language and synthesis and between a common theoretical basis and synthesis (Defila *et al*. 2006, p. 49).

(*cont.*)

Managing consensus in interdisciplinary teams (cont.)

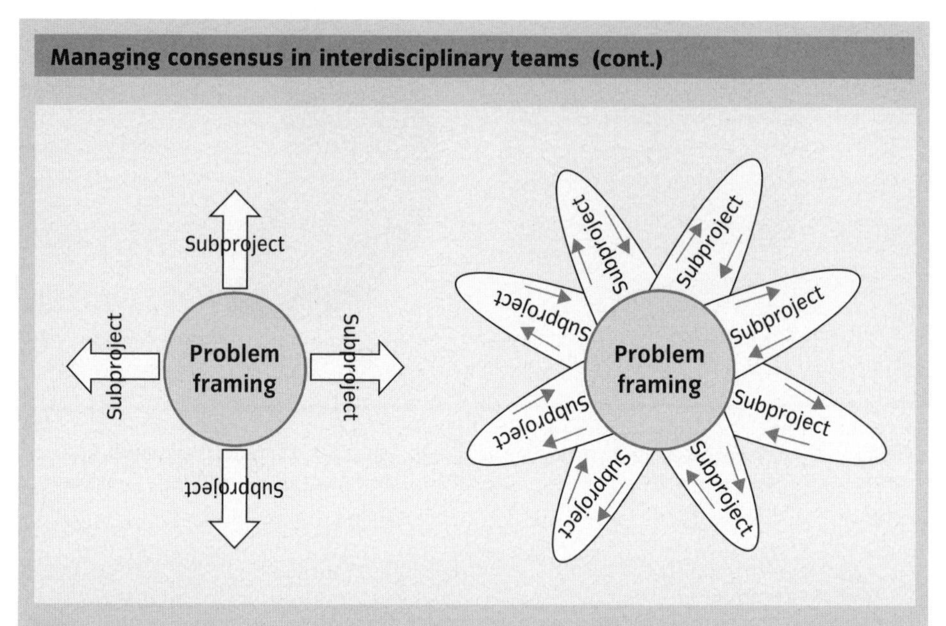

Figure 4 Balancing teamwork and individual work: put metaphorically, the division of labor and the collaborative problem framing together form a flower, the petals standing for the individual research work, respectively the research work in the subprojects.

with regard to interdisciplinary processes. Second, it is due to disciplinary socialization. Interdisciplinarity does not mean the disposal of disciplinarity, ignoring or covering up disciplinary differences. On the contrary, to be successful, the members of a team have to make substantial contributions from their disciplinary ways of thinking and of investigation. To this end they need a strong disciplinary identity and a deep understanding of their disciplinary way of thinking and tackling scientific problems. The specific disciplinary perspectives have to be made productive for interdisciplinary research. This just won't happen if the team members aren't able to relate their disciplinary way of problem framing to the ways of others. If all are convinced that their way of seeing the world is the only possible and right way, that only the questions asked by their discipline are relevant, that their methods of investigation are the only ones possible and the only ones leading to success, collaborative problem framing is out of reach.

So, in interdisciplinary research a strong disciplinary identity is both the *conditio sine qua non* of success and a serious obstacle to success. Collaborative problem framing in interdisciplinary teams therefore has to be balanced with disciplinary problem framing.

Collaborative problem framing in interdisciplinary teams as well as the balancing of teamwork and individual (disciplinary) work will not just happen, even if all team members are strongly committed and willing. Rather, it has to be properly managed. But which processes have to be managed in order to establish successful interdisciplinary work right from the beginning?

Most teams concentrate only on the process of integration. That, as should be clear by now, would be dangerous. Great attention has to be paid to the process of collaborative problem framing, including the process of defining common goals and questions. To ensure integrative problem framing, managers of interdisciplinary research projects have to make sure that this is accompanied by careful reflection on the disciplinary way of structuring the world and on the disciplinary contributions to the solution of the problem by the members of their teams. Then, during the research work, managers of interdisciplinary research have to make sure that the research is actually informed by the common goals and the common questions and always refers to the common description of the research object.

Other tasks in the management of inter- and transdisciplinary projects are (for details see Defila *et al.* 2006):

- to coordinate the research of the members and subprojects, and support joint surveys;
- to ensure the development of common results (synthesis) and ascertain that the research ends up with common products;
- to support team development by discussing the expectations of the team members towards each other as well as concerning the interdisciplinary project, by monitoring the team's working with an eye to possible conflicts due to the disciplinary socialization of the individuals involved;
- to support the participation of practitioners and cooperation between researchers and the stakeholders involved by negotiating the goals and forms of the cooperation, by reaching an agreement concerning the contribution of the stakeholders involved in terms of time and effort as well as products, and by ensuring that the stakeholders really benefit from the cooperation;
- to design and monitor internal and external communications by defining the different disciplinary and non-scientific target audiences to be addressed, by defining the different media and languages needed for addressing the target audiences, and by discussing specific assignments concerning communication within the project team;
- finally, to organize the work within the project group by negotiating rights and duties, by discussing the criteria to be used in the evaluation of the processes and the achieved results, and by tuning the different disciplinary work schedules.

In conclusion, the management of interdisciplinary research can by no means be reduced to simple technical management. Rather, it is a complex task that researchers have to be specially trained for.

References

Brooks, J.J., Bujak, A.N., Champ, J.G., Williams, D.R. (2006). Collaborative capacity, problem framing, and mutual trust in addressing the wildland fire social problem: an annotated reading list. *General Technical Report RMRS-GTR-182*. Fort Collins, CO: US Department of Agriculture, Forest Service, Rocky Mountain Research Station.

Defila R., Di Giulio A., Scheuermann M. (2006). Forschungsverbundmanagement. *Handbuch für die Gestaltung inter- und transdisziplinärer Projekte*. Zürich: vdf Hochschulverlag an der ETH Zürich.

literature to date. A major challenge facing the development of quasi-experimental designs for evaluating TS initiatives is the identification of appropriate comparison or control groups. One effort to address this challenge consists of a novel bibliometric analysis of the TTURC initiative recently begun by our team that incorporates a quasi-experimental interrupted time-series design (Cook and Campbell 1979), supplemented by science visualization maps for examining the 'footsteps' of an initiative in its broader field of research (e.g. tobacco use) over an extended period of time. This new direction of the SOTS field is examined in Section 32.3.

32.3 Emerging concerns and new developments in the science of team science

Four major emerging directions of the SOTS field are considered in this section:

(1) implementing more rigorous quasi-experimental research designs for evaluating the contributions of TS research;
(2) developing and evaluating TD training programs;
(3) creating a more nuanced typology of cross-disciplinary TS initiatives and funding models, and evaluating their relative efficacy in achieving the intended goals of TS; and
(4) achieving a broader understanding of the contextual factors that determine the *collaboration readiness* of particular teams and scientific fields.

32.3.1 Implementing quasi-experimental research designs

An important direction of research in the SOTS field is the implementation of multimethod, quasi-experimental research designs to evaluate the processes and outcomes of cross-disciplinary scientific projects and initiatives. Both quantitative and qualitative methods are now being used to assess the scholarly productivity and impact of scientists working within particular research areas (e.g. tobacco use research). A key component of these evaluation studies is the inclusion of multiple comparison groups—for example, scientists participating directly in a large-scale initiative such as the TTURCs compared with other scientists who are working in the same field of research (e.g. tobacco use and control) individually on single investigator grants.

32.3.2 Development and evaluation of TD training programs

Another important frontier of the SOTS field is the development and evaluation of comprehensive strategies for training the next generation of TD scholars and professionals. The availability of a critical mass of scientists ready for cross-disciplinary inquiry is crucial for the success of TS initiatives. TD training is increasingly being encouraged and required by both public and private funding agencies (e.g. NIH, NSF, RWJF, McArthur Foundation). Although TD training shares many features with more traditional, unidisciplinary training programs at university and other institutional settings, it is unique in that it focuses on fostering a set of attitudes, knowledge, and skills among trainees that will equip them to transcend disciplinary frameworks and methodologies through creative syntheses and to produce significant advances in scientific research. Although still in an early phase of development, evaluative studies of TD training models and programs have been undertaken over the last decade.

A number of different TD training models have been examined in earlier studies. Nash (2008) summarized some of these approaches to TD training, one of which includes university-based programs offering structured TD curricula. In this approach, students receive their formal undergraduate and/or graduate training within an interdisciplinary academic unit (e.g. the University of California Irvine's School of Social Ecology; Cornell University's College of Human Ecology, Ithaca, NY). Another TD training model incorporates multi-mentor apprenticeship whereby a trainee is assigned, through mutual agreement, to more than one faculty advisor and is designated as an apprentice of two or more established mentors in their domains (e.g. as occurs within the TTURC at Brown University, Providence, RI). In yet a third TD training prototype, sometimes referred to as 'residential scholars', junior investigators are encouraged to expand their training experiences by conducting research projects at a center other than their own over a specified period (e.g. 2 to 3 months). In this model, which is being pursued by the TREC centers, trainees have the opportunity to collaborate with colleagues in other institutions by working on a joint project with them and overcoming geographic barriers to TD collaboration. With the growing interest and investment in cross-disciplinary TS, TD training models are likely to continue to evolve and expand in future years.

32.3.2.1 Challenges in TD training

There are several challenges inherent in developing and refining TD training models to respond to the rapidly changing landscape of TS. In addition to the need to overcome language and cultural barriers associated with exposure to multiple disciplines, first-generation TD trainees face many uncertainties in their career development owing to traditional academic reward systems that encourage first or sole authorships on scholarly publications, principal investigator status on grant-funded projects, and the publication of peer-reviewed articles in prestigious unidisciplinary journals (see Graybill and Shandas, chapter 28 this volume).

Moreover, prior TD training models have been dominated by programs designed for advanced graduate and postdoctoral training. A more comprehensive training model is needed for senior investigators who are charged with greater management responsibilities within large research initiatives. Broader models of TD training that encompass the needs of all stakeholders including senior investigators, junior investigators, postdoctoral scholars, graduate students, and research support staff, should be incorporated into the overall infrastructure of collaborative TS. Also, future TD training programs should offer innovative mentoring practices, and expose trainees to collaborative leadership styles and communication strategies, interpersonal and managerial skills, and technological expertise (Gray 2008).

32.3.2.2 Evaluation of TD training processes and outcomes

In earlier evaluations of TD training programs and outcomes, a variety of metrics have been used to assess the quality, novelty, and scope of disciplinary integration reflected in the work completed by trainees. Further refinement and validation of these metrics are sorely needed in the SOTS field. Quantitative and qualitative measures of trainees' career trajectories as they evolve within various TS initiatives can provide a deeper understanding of the near-term, mid-term, and long-term outcomes of these programs, and the ways in which TD trainees gain entry to various academic, government, and private sector jobs; as well as whether their collaborative (e.g. multi-mentor) training leads to sustained TD research efforts as they move forward in their careers. For example, the assessment of changes in trainees' research orientations over time may be used to model and subsequently predict the long-term career outcomes of these individuals. Systematically tracking the career development of TD trainees over time and examining the influence of collaborative training programs on their subsequent productivity will ultimately help to gauge the returns on TS investments at both individual and societal levels.

32.3.3 A typology of team science initiatives and funding models

A substantial increase in funding initiatives for ID and TD science and training has occurred in recent years. Critics of TS, in addition to being concerned about the volume of funds directed towards cross-disciplinary TS initiatives and away from unidisciplinary research, contend that once TD-specific funding is withdrawn from a research group, center, or institution, the collaborative efforts will cease (Weissmann 2005; Marks 2006). To date, this contention has not been tested directly by evaluating whether TD teams remain productive and cohesive once their original sources of funding are gone. These critiques of TS initiatives raise important concerns about the continuity of and strategies for funding collaborative research.

Currently, a variety of different funding model is used to support and sustain TD science and training initiatives. Funding for TD research varies along several dimensions that may affect the continuity of funding and the sustainability of TD science. For instance, an initiative may have several sources of funds, including private foundations and federal agencies as well as combinations thereof (e.g. an initiative funded by both a foundation and a federal agency). Funding mechanisms also vary with regard to the scope of the science addressed: medical research (broadly) versus tobacco use (specifically); the breadth of disciplines spanned (molecular to policy versus social science to policy); the amount of funding (e.g. thousands versus millions of dollars per project/center); and whether or not a strategic plan is developed at the outset to guide the research agenda over the course of the project.

Future evaluations of TS initiatives, therefore, should address the following kinds of questions. To what extent will the collaborative research supported by various funding models produce integrative conceptual models, methodological approaches, and empirical advances spanning multiple fields and extended periods? What happens when the funding for a TS initiative is withdrawn—will the TD science stagnate? Will the lack of long-term funding commitment lead researchers to revert to more traditional small, incremental scientific development processes? What situational factors facilitate the sustainability of cross-disciplinary inquiry communities? Can substantial gains in cross-disciplinary integration and translations to health practice be achieved through small-scale TD science teams? Is small-scale TD science more sustainable with respect to funding streams, or do we ultimately need large-scale TD science to create a critical mass of researchers and infrastructure for the sustainability of TD science? What are the necessary conditions to ensure the continuity of funding for TD science projects of different scopes and sizes? Are large initiative-based TD science centers (as compared with smaller-scale projects) required to foster sufficient levels of cross-disciplinary expertise in order to propel collaborations as well as theoretical and methodological advances in resolving the most urgent societal health problems?

32.3.4 **Understanding multiple determinants of collaboration readiness**

'Collaboration readiness' (CR) in cross-disciplinary TS can be conceptualized and measured in a variety of ways—for instance, in terms of individual and group research orientations; organizational and technological resources that enhance capacity for collaboration; and the scientific readiness of different fields for collaborative integration. Stokols, Misra *et al.* (2008) offered a social ecological framework for identifying multiple factors that enhance collaboration readiness, including the availability of specific communication tools and cyberinfrastructural resources, shifts in individuals' research orientations and their attitudes toward collaboration, and funding agencies' willingness to invest in center-based, multiple-PI (Principal Investigator) grants. A diagrammatic representation of the typology of contextual influences on the effectiveness of TD team science initiatives is shown in Fig. 32.5. Key categories of contextual influences on CR are organized according to intrapersonal, interpersonal, organizational, institutional, physical environmental, technological, and political and societal levels of analysis.

Future research in the SOTS field should explicitly consider multiple levels and dimensions of CR for TD team science and undertake in-depth case studies to identify which types of readiness factors (e.g. psychological, interpersonal, organizational, societal, technological, scientific) exert the greatest influence on the effectiveness of TS projects and initiatives. A readiness framework can help generate appropriate multilevel interventions to increase the success of TD team science. For instance, at the interpersonal level, understanding a team's readiness to engage in group processes to create common ground, common language, and shared goals can lead to the development of workshop modules to foster improved communications skills and team cohesiveness (Stokols *et al.* 2008b). To date, evaluations of TD initiatives have not examined the joint influence of these diverse readiness factors on the effectiveness of TS and training. This is a potentially fruitful direction for future research in the SOTS field.

Finally, the demands for cross-disciplinary and cross-national collaborations in health science, engineering, and technology will continue to grow in the coming decades. TD team science at a global scale requires an understanding of and sensitivity to cultural differences and their impact on teamwork to ensure success. Also, as funding streams ebb and flow, the need to coordinate and integrate health research efforts among academic institutions, government agencies, and private corporations and foundations will become increasingly important.

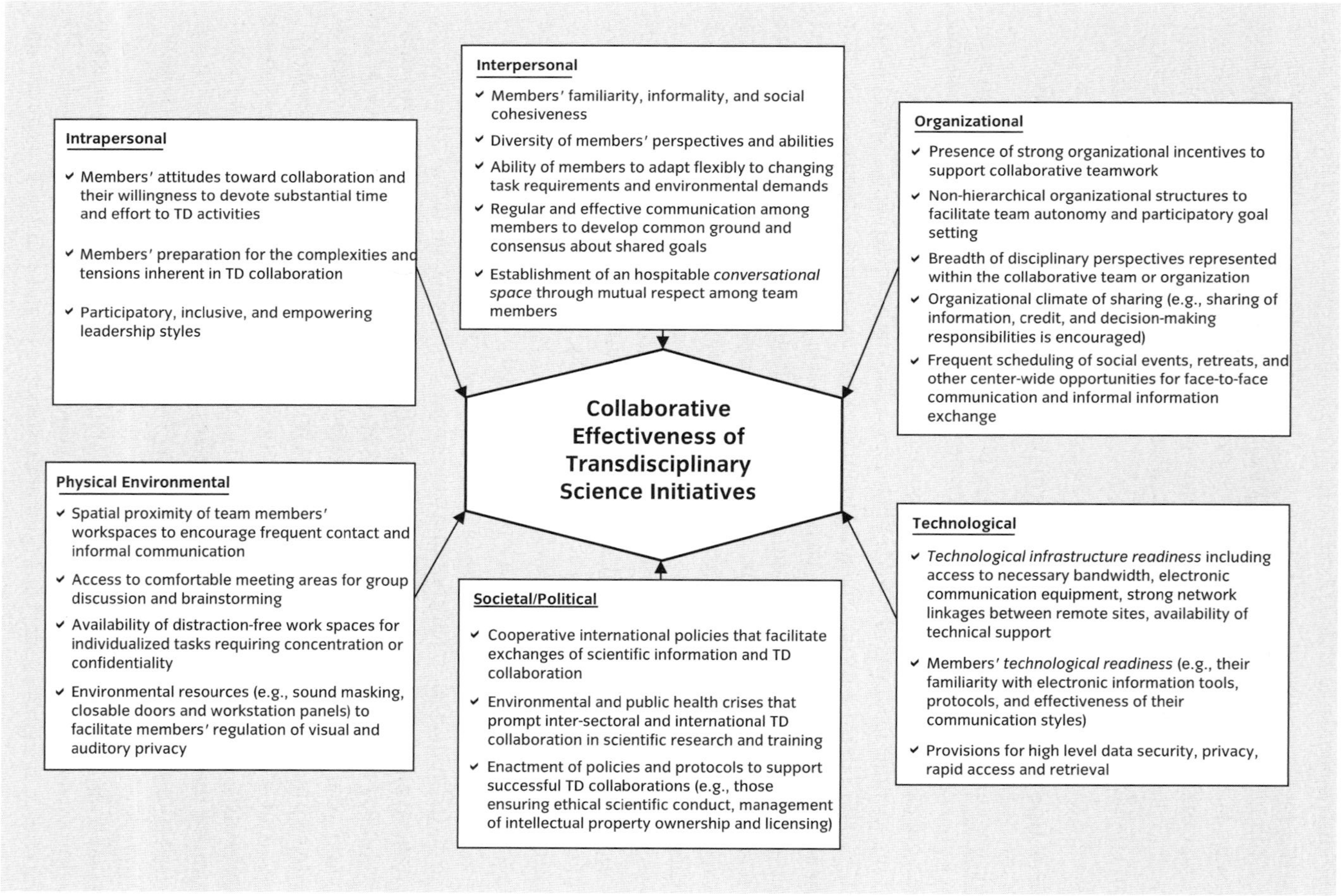

Intrapersonal

- Members' attitudes toward collaboration and their willingness to devote substantial time and effort to TD activities
- Members' preparation for the complexities and tensions inherent in TD collaboration
- Participatory, inclusive, and empowering leadership styles

Interpersonal

- Members' familiarity, informality, and social cohesiveness
- Diversity of members' perspectives and abilities
- Ability of members to adapt flexibly to changing task requirements and environmental demands
- Regular and effective communication among members to develop common ground and consensus about shared goals
- Establishment of an hospitable *conversational space* through mutual respect among team members

Organizational

- Presence of strong organizational incentives to support collaborative teamwork
- Non-hierarchical organizational structures to facilitate team autonomy and participatory goal setting
- Breadth of disciplinary perspectives represented within the collaborative team or organization
- Organizational climate of sharing (e.g., sharing of information, credit, and decision-making responsibilities is encouraged)
- Frequent scheduling of social events, retreats, and other center-wide opportunities for face-to-face communication and informal information exchange

Collaborative Effectiveness of Transdisciplinary Science Initiatives

Physical Environmental

- Spatial proximity of team members' workspaces to encourage frequent contact and informal communication
- Access to comfortable meeting areas for group discussion and brainstorming
- Availability of distraction-free work spaces for individualized tasks requiring concentration or confidentiality
- Environmental resources (e.g., sound masking, closable doors and workstation panels) to facilitate members' regulation of visual and auditory privacy

Societal/Political

- Cooperative international policies that facilitate exchanges of scientific information and TD collaboration
- Environmental and public health crises that prompt inter-sectoral and international TD collaboration in scientific research and training
- Enactment of policies and protocols to support successful TD collaborations (e.g., those ensuring ethical scientific conduct, management of intellectual property ownership and licensing)

Technological

- *Technological infrastructure readiness* including access to necessary bandwidth, electronic communication equipment, strong network linkages between remote sites, availability of technical support
- Members' *technological readiness* (e.g., their familiarity with electronic information tools, protocols, and effectiveness of their communication styles)
- Provisions for high level data security, privacy, rapid access and retrieval

Figure 32.5 Typology of contextual factors that influence the effectiveness of TD scientific collaboration. Reprinted from Stokols et al. (2008b).

32.4 Epilogue

Whereas the SOTS field is at a relatively early stage in its development, it is likely to expand substantially in several new directions owing to the increasing investment in cross-disciplinary TS initiatives by both public and private funding organizations, and the corresponding need to evaluate the scientific, training, and societal outcomes of these programs. Several promising directions for future SOTS research were identified in earlier sections of the chapter, including: the development of more rigorous research designs for evaluating the outcomes of TS initiatives; new and innovative models for funding TS research and training programs as well as systematic evaluations of them; and a broader understanding of the multiple factors that influence the collaboration readiness and capacity of particular teams, institutions, and scientific fields.

In addition to the emerging concerns and research directions of the SOTS field identified above, several other topics are likely to arise as a focus for future evaluations of TS initiatives. Although not explicitly addressed in this chapter, the challenges of creating and sustaining new collaborative partnerships among government research agencies, non-government organizations, and private corporations and foundations for the purposes of promoting and supporting cross-disciplinary science and training programs is likely to become more salient in future SOTS research. How can the disparate and sometimes seemingly incommensurate goals of these diverse (e.g. for profit versus non-profit) entities be aligned in ways that encourage and sustain cross-disciplinary collaboration in science, training, and the translation of research knowledge into effective health promotion policies and interventions? What new theoretical frameworks, metrics, and research designs will be required to evaluate the success of these multi-sectoral partnerships?

Yet another promising direction for SOTS research that has received little attention to date is the set of challenges associated with managing information overload in TD team science initiatives. Whereas many contextual factors contribute to collaborative readiness, the capacity of a TD team to sustain effective collaboration over extended periods depends to a large extent on how well team members are able to manage the enormous amounts of new information they are exposed to as they work with colleagues trained in multiple fields, who are often dispersed across several geographic locations. Future studies on TD science, training, and research initiatives should investigate the impact of information overload on the processes and outcomes of TD work. Further, new organizational, infrastructural, and intra- and interpersonal models and strategies that can better help manage information overload need to be developed to enhance the effectiveness of TD collaboration. These are among the exciting and important questions that are likely to command the attention of participating scholars as SOTS field moves forward.

References

Cook, T.D. and Campbell, D.T. (1979). *Quasi-experimentation: design and analysis issues for field settings*. Chicago, IL: Rand McNally College Publishing Company.

Cooperrider, D.L. and Witney, D. (2005). *Appreciative inquiry: a positive revolution in change.* San Francisco: Berrett-Kohler Publishers.

Fuqua, J., Stokols, D., Gress, J., Phillips, K. and Harvey, R. (2004). Transdisciplinary scientific collaboration as a basis for enhancing the science and prevention of substance use and abuse. *Substance Use and Misuse* **39**(10–12), 1457–514.

Gray, B. (2008). Enhancing transdisciplinarity research through collaborative leadership. *American Journal of Preventive Medicine* **35**(2S), S124–S132.

Hall, K., Stokols, D., Moser R. *et al.* (2008a). The collaboration readiness of transdisciplinary research teams and centers: findings from the National Cancer Institute TREC baseline evaluation study. *American Journal of Preventive Medicine* **35**(2S), 161–72.

Hall, K., Feng, A., Moser, R., Stokols, D., and Taylor, B. (2008b). Moving the science of team science forward: collaboration and creativity. *American Journal of Preventive Medicine* **35**(2S), 243–9.

Hess, D.J. (1997). *Science studies: an advanced introduction.* New York: New York University Press.

Holmes, J.H., Lehman, A., Hade, E. *et al.* (2008). Challenges for multi-level health disparities research in a transdisciplinary environment. *American Journal of Preventive Medicine* **35**(2S), S182–S192.

Kessel, F.S., Rosenfield, P.L., and Anderson, N.B. (ed.) (2008). *Interdisciplinary research: case studies from health and social science.* New York: Oxford University Press.

Klein, J.T. (2008). Evaluating interdisciplinary and transdisciplinary collaborative research: a review of the state of the art. *American Journal of Preventive Medicine* **35**(2S), S116–S123.

Marks, A.R. (2006). Rescuing the NIH before it is too late. *Journal of Clinical Investigation* **116**(4), 844.

Mâsse, L., Moser, R., Stokols, D., Taylor, B.K., Marcus, S., Morgan, G. *et al.* (2008). Measuring collaboration and transdisciplinary integration in team science. *American Journal of Preventive Medicine* **35**(2S), S151–S160.

Nash, J.M. (2008). Transdisciplinary training programs: key components and prerequisites for success. *American Journal of Preventive Medicine* **35**(2S), S133–S140.

Nass, S.J. and Stillman, B. (2003). *Large-scale biomedical science: exploring strategies for future research.* Washington, DC: National Academies Press.

National Academy of Sciences (2005). *Facilitating interdisciplinary research.* Washington, DC: National Academies Press.

National Cancer Institute (2007). *Cells to society: overcoming health disparities: report of the Centers for Population Health and Health Disparities to the NIH Board of Scientific Advisors.* Bethesda, MD: NCI.

National Institutes of Health (2003). *NIH roadmap–accelerating medical discovery to improve health: interdisciplinary research.* Available at:, <http://nihroadmap.nih.gov/interdisciplinary/index.asp> (accessed 26 April 2004).

Rosenfield, P.L. (1992). The potential of transdisciplinary research for sustaining and extending linkages between the health and social sciences. *Social Science and Medicine* **35**, 1343–57.

Stokols, D., Fuqua, J., Gress, J., Harvey, R., Phillips, K., Baezconde-Garbanati, L. *et al.* (2003). Evaluating transdisciplinary science. *Nicotine and Tobacco Research* **5**(1), S21–S39.

Stokols, D., R. Harvey, J. Gress, J. Fuqua, and K. Phillips. (2005). In vivo studies of transdisciplinary scientific collaboration: lessons learned and implications for active living research. *American Journal of Preventive Medicine* **28**(2S2), 202–13.

Stokols, D., Hall, K.L., Taylor, B., and Moser, R.P. (2008a). The science of team science: overview of the field and introduction to the supplement. *American Journal of Preventive Medicine* **35**(2S), S77–S89.

Stokols, D., Misra, S., Hall, K., Taylor, B. and R. Moser, B. (2008b). The ecology of team science: understanding contextual influences on transdisciplinary collaboration. *American Journal of Preventive Medicine* **35**(2S), S96–115.

Trochim, W.M., Marcus, S., Masse, L.C., Moser, R., and Weld, P. (2008). The evaluation of large research initiatives: a participatory integrated mixed-methods approach. *American Journal of Evaluation* **29**(March), 8–28.

Warnecke, R.B., Oh, A., Breen, N. *et al.* (2008). Approaching health disparities from a population perspective: the NIH Centers for Population Health and Health Disparities. *American Journal of Public Health* **98**(9), 1608–15.

Weissmann, G. (2005). Roadmaps, translational research, and childish curiosity. *FASEB Journal* **19**, 1761–2.

CHAPTER 33

The environment

J. BAIRD CALLICOTT

In a 1942 speech, 'The role of wildlife in a liberal education', published that same year in the *Transactions of the North American Wildlife Conference*, the great American conservationist, Aldo Leopold, powerfully expressed the inherent interdisciplinarity of environmental concerns and problems:

> All the sciences and arts are taught as if they were separate. They are separate only in the classroom. Step out on the campus and they are immediately fused. Land ecology is putting the sciences and arts together for the purpose of understanding our environment (Leopold 1999b, pp. 302–3).

Leopold seems to have thought of ecology less as a discipline among disciplines, than as a transdiscipline (cf. Klein, Chapter 2 this volume), much in the way that modern conservation biology is defined by its practitioners as a transdiscipline (Meffe *et al.* 1994, 1997). To understand the energy flows and nutrient cycles of an ecosystem, an ecologist must fuse thermodynamics and biogeochemistry with organismal biology. To understand predator–prey relationships, a community ecologist must fuse genetics and population biology with a knowledge of the life histories of a biotic community's interacting organisms.

Certainly, as Leopold claims, 'understanding our environment' is an inherently interdisciplinary task requiring a fusion of all the sciences and arts; but the environment has come to represent much more than what Leopold comprised under his own preferred simpler rubric of 'land'—which in his view consisted of 'soils, waters, plants, and animals' (Leopold 1949, p. 204). Conspicuously absent from these components of the land or the environment, as Leopold conceived it, is 'airs', the atmosphere, the central object of what is perhaps the gravest of our twenty-first-century environmental concerns.

Understanding and dealing with global climate change is perhaps the greatest call to interdisciplinary arms, requiring that such sciences as astronomy and astrophysics (e.g. the study of solar cycles and variations in the earth's orbital eccentricity) be added to thermodynamics and biogeochemistry on the list of the sciences needing to be fused—not to mention new sciences undreamed of by Leopold's generation, such as computer modeling and plate tectonics. And, in addition to the other 'arts' that Leopold may have had in mind (history, cultural geography, and anthropology prominent among them), environmental philosophy and ethics, including environmental justice, are among the disciplines that must be leavened into the mix of fused natural and social sciences to address global climate change.

To keep this chapter within reasonable limits—and within the limits of the author's experience and expertise—the discussion of interdisciplinarity and the environment will be confined to an examination of one self-consciously interdisciplinary trans- (or better, meta-) discipline, conservation biology. I then relate my own experience, as a humanist, participating in interdisciplinary research on environmental issues.

In a plenary address to the joint meeting of the Ecological Society of America and the Society of American Foresters in 1939, Leopold articulated the idea that ecology is itself an interdisciplinary science:

> Ecology is a fusion point for all the natural sciences. It has been built up partly by ecologists, but partly by the collective efforts of the men charged with the economic evaluation of species. The emergence of ecology has placed the economic biologist in a peculiar dilemma: with one hand he points out the accumulated findings of his search for utility, or lack of utility, in this or that species; with the other he lifts the veil from a biota so complex, so conditioned by interwoven cooperations and competitions, that no man can say where utility begins or ends (Leopold 1999a, pp. 266–7).

Leopold regards ecology as inherently interdisciplinary, 'a fusion point for all the natural sciences'. And, to all the natural sciences, Leopold himself is most famous for adding the humanities—with his sketch of a 'land ethic'—into the crucible from which ecology was forged. But ecology is also commonly regarded as a discipline among disciplines—and certainly it was even in Leopold's day. Indeed, Leopold acknowledges as much by referring to practitioners of the discipline, namely to 'ecologists'. According to the ecologist and historian, Robert P. McIntosh (1985), ecology became a 'self-conscious science' in the last decade of the nineteenth century. Successful fusing of disciplines, successful interdisciplinarity, thus sometimes results in a *meta*discipline—that is, an emergent new discipline born of the interdisciplinary fusion of previously existing separate disciplines. Such a discipline is ecology, Leopold thinks, and so, certainly, is conservation biology. Indeed, conservation biology may be a meta-metadiscipline, as one of the 'disciplines' it comprises is ecology, itself a metadiscipline, if Leopold's characterization of it is correct.

33.1 What is conservation biology?

'What is conservation biology?' was the title of an article by Michael Soulé (1985), published in *BioScience*, that announced the emergence of a new metadiscipline. The Society for Conservation Biology was also formed in 1985. And *Conservation Biology* (the journal) began publication in 1987. The emergence of conservation biology was one response to what has been called the second wave of the environmental crisis.

The first wave of the environmental crisis began to swell in the 1960s, as local and regional pollution from heavy industry; municipal sewage; petroleum extraction, transport, and refining; the exponential growth of the automobile culture; and broadcast agricultural pesticides became evident to the senses and a palpable public nuisance. The flames of popular outrage were fanned by Rachel Carson (1962)—whose book, *Silent spring*, galvanized public concern about pesticides—and by Stewart Udall (1963), whose book, *The quiet crisis*, may have given the phenomenon its 'crisis' cachet. By the end of the decade Congress passed the National Environmental Policy Act (1969). The first wave of the environmental crisis crested in 1970 with the Congressional declaration of Earth Day and the creation of the United States Environmental Protection Agency. There followed a spate of palliative national legislation enacted in the 1970s including, most notably, the Clean Air Act (1970), the Clean Water Act (1972), and the Endangered Species Act (1973).

The second wave of the environmental crisis arose in the 1980s. A seasonally fluctuating 'hole' in the layer of stratospheric ozone, which protects the earth's surface from dangerous levels of ultraviolet solar radiation, was discovered over the Antarctic in 1985. The principal culprit was Freon, a gaseous chlorofluorocarbon refrigerant, leaked into the atmosphere. An unusually hot North American summer in 1988, accompanied by a massive out-of-control wildfire in Yellowstone National Park, seared 'global warming' into the American mind. And, after bulldozers and chainsaws had made their way deep into the planet's moist tropical forests, scientists—followed shortly by public concern—began to realize that the rate of species extinction on earth was catastrophic and accelerating at an alarming pace. To more generally characterize this loss, the term 'biodiversity' (a contraction formed from *biological* and *diversity*) was coined in 1985 in preparation for the National Forum on Biological Diversity—a multidisciplinary conference, incidentally—organized by the US National Research Council and convened in 1986, the (multidisciplinary) proceedings of which were edited by Edward O. Wilson (1988) and entitled *Biodiversity*.

The second wave of the environmental crisis differed from the first in both spatial and temporal scale. The spatial scale of the concerns dominating the first wave was, as noted, local and regional—fouled beaches in and around Santa Barbara, CA; smog over Houston, TX; a fire on the Cuyahoga River in Cleveland, OH. Such phenomena are reversible and the temporal scale of environmental recovery is calibrated in years and decades. Four decades later, the beaches of Santa Barbara are relatively clean again; the air quality in Houston is better in 2008 than it was in 1958; and the Cuyahoga River is no longer combustible. Spatially, the ozone hole, the loss of biodiversity, and global climate change are planetary. Temporally, the recovery times vary, but they are all calibrated in decades, centuries, and millennia, not years. The earth's ozone shield is already showing signs of recovery and the Antarctic ozone hole is annually shrinking. Barring additional ozone-destroying gases leaked into the atmosphere, in about a hundred years earth's ozone shield should return to its preindustrial condition (Son *et al.* 2008). Lost biodiversity can only be restored by Darwinian evolutionary processes, over millions of years, and the species currently lost will be *replaced* by others—the nature of which no one can predict—not re-created (Eldridge 1992).

The ozone hole was the easiest to address, as well as the quickest to fix. The 1987 Montreal Protocol mandated that the production of chlorofluorocarbons be phased out—with the full cooperation of the chemical industry, incidentally, mainly because patents on them had run out and profits from ozone-safe substitutes were promising. The solution to this problem involved transdisciplinarity. The geosciences and chemistry, diplomacy and economics were the disciplines principally engaged, but bureaucrats and chief executive officers also participated as vital players.

The problem of global climate change is much harder to address, although the principle is the same as that in the case of the ozone hole: discontinue the use of its causal agents, most importantly discontinue the production of carbon dioxide from the burning of fossil fuels. Still harder to address is the loss of biodiversity—which is exacerbated by global climate change—now so ominous a prospect that, if uncontrolled, a sixth mass extinction across the entire earth's biography may occur before the current loss of biodiversity bottoms out.

Addressing the problem of biodiversity loss and avoiding a sixth great extinction is the remit of conservation biology. Because conservation biology is, according to Soulé (1985) in his field-defining paper, a 'crisis discipline' it is not a 'pure' natural science like physics or chemistry, but is more aptly compared to a science like medicine. Similar to medicine, conservation biology is openly value driven. Medicine is driven by the value of human health; conservation biology by the value of biodiversity. Indeed, like medicine, which is based on the *intrinsic value* of human health, conservation biology is based on the *intrinsic value* of biodiversity—although biodiversity also has many instrumental uses as well, such as providing a pool of resources for new medicines, foods, and fuels and the maintenance of vital ecological services, such as pollination and nitrogen fixation. Like physicians, practitioners of biodiversity preservation must first triage species at risk of extinction and work to save only those that (first) can be preserved and (second) will be preserved only if efforts to do so are expended. And, like physicians, conservation biologists must practice their art in a climate of great scientific uncertainty, because the crisis they hope to avert advances more rapidly than basic research. Finally, like medicine, conservation biology is metadisciplinary. Medicine fuses such scientific disciplines as anatomy, physiology, biochemistry, and neurology—to that extent it is interdisciplinary—but it is itself a well-established scientific discipline in its own right, a metadiscipline. Similarly, conservation biology fuses such scientific disciplines as evolutionary biology, ecology, biogeography, and geology into a metadiscipline.

So self-aware a metadiscipline is conservation biology that most textbooks in the field provide a table or chart listing the disciplines that the metadiscipline comprises. Figure 33.1 is taken from the 2006 edition of the leading graduate textbook in the field, *Principles of conservation biology* (Groom *et al.* 2006).

Inadvertently (one assumes), the figure looks back to Leopold's dichotomous distinction between the sciences and arts as it collapses the humanities and social sciences into one category. The greatest part of the list of natural sciences fused into conservation biology—biology, ecology, evolution, genetics, biogeography, geology, chemistry, and statistics—is predictable. Among the humanities and social sciences, one also finds the usual suspects—sociology, economics, policy, and law. But one also finds anthropology. The reason for the express inclusion of that social science is because conservation biology is becoming ever more a transdiscipline as well as a metadiscipline, acknowledging the extradisciplinary input from various stakeholders in various cultures in the formation of successful conservation policy. Indeed, one of the latest buzz words in conservation policy is 'community-based conservation', in which conservation initiatives are planned in close consultation with the affected stakeholders. Going further still, ecologist and

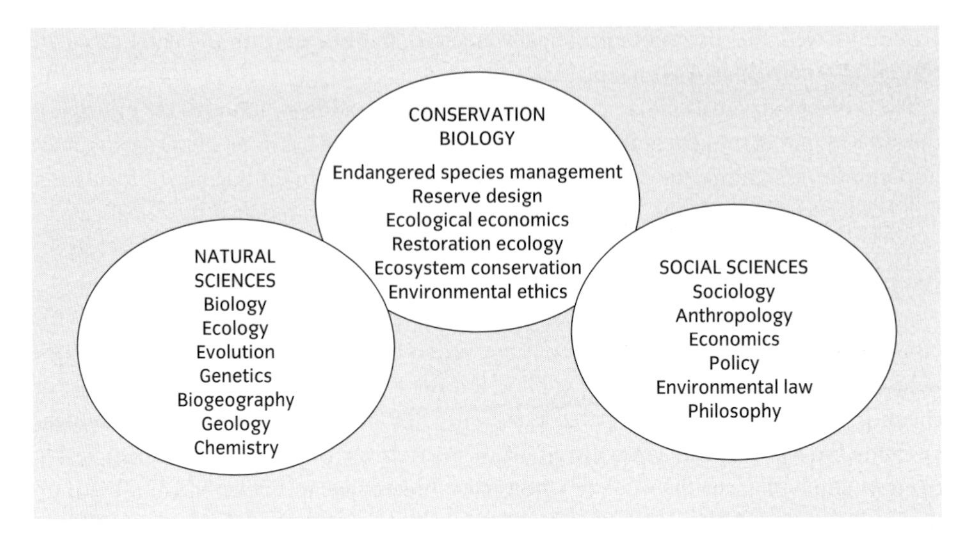

Figure 33.1 The interdisciplinary nature of conservation biology merges many traditional fields of natural and social sciences. The list of relevant subdisciplines and interactions is not meant to be exhaustive.

environmental philosopher Ricardo Rozzi now identifies the end served by conservation biology as the preservation of bio*cultural* diversity (Rozzi *et al.* 2004).

Among the humanities listed in Fig. 33.1 one finds philosophy (but not the critical study of literature, an unfortunate oversight, as much might be garnered relevant to conservation biology from a critical study of, for example, *Moby Dick*). The part of philosophy that is expressly fused into conservation biology, as the figure indicates, is environmental ethics. From its earliest expression in the mid-1980s, conservation biology has been informed by environmental ethics, and especially by the distinction, central to much discussion in environmental ethics, between the intrinsic and instrumental value of species severally and biodiversity collectively. The authors of the introductory chapter of third edition of *Principles of conservation biology* note, citing Soulé (1985), that:

> Michael Soulé, a cofounder of the Society for Conservation Biology, lists four postulates and their corollaries that characterize value statements relevant to conservation biology.[...] The final postulate is that biotic diversity has intrinsic value, regardless of its utilitarian value. This postulate recognizes inherent value in non-human life, regardless of its utility to humans, and carries the corollary that destruction of diversity by humans is bad. This is perhaps the most fundamental motivation for conservation biology (Groom *et al.* 2006, pp. 17–18).

And a chapter in each of the three editions of that textbook has been devoted to 'Conservation values and ethics' written by a professional philosopher specializing in environmental ethics (Meffe *et al.*1994, 1997; Groom *et al.* 2006). Although many of the chapters of this textbook are written by contributors other than the principal authors, as these quotations from the introductory chapter by the principal authors indicate, environmental ethics is deeply integrated into the very foundations of the metadiscipline of conservation biology.

Biocultural conservation in Cape Horn: the Magellanic woodpecker as a charismatic species

Ximena Arango, Ricardo Rozzi, Francisca Massardo, and J. Tomás Ibarra

Interdisciplinarity is often looked to for producing insights of practical use to policy makers in the public and private sectors. Yet there are many challenges facing interdisciplinary approaches: difficulties in integrating different disciplinary perspectives, scales, and methods; differences in vocabularies; and timelines for research that often do not match well with the exigencies of practical affairs. Transdisciplinary approaches—in which academic research is integrated with non-academic concerns—are often fruitful in linking interdisciplinary research with societal (or what we prefer to call biocultural) needs.

At the southernmost tip of the Americas, the Cape Horn Biosphere Reserve (CHBR) encompasses one of the world's most pristine remaining wilderness areas and is home to the indigenous Yaghan (or Yamana) community, which featured so prominently in Charles Darwin's *Voyage of the Beagle*. Its remoteness and uniqueness, however, are threatened by the introduction of exotic species such as the North American beaver and American mink, increasing development pressures from new connectivity, resource exploitation, and the development of tourism. To implement the biosphere reserve and conserve its natural and cultural richness requires the active participation of the community, as well as linkages and integration between various disciplines and institutions. In an effort to achieve the goal of transdisciplinary integration, we used the strategy of identifying a charismatic species, since doing so serves to motivate people towards biodiversity conservation, to communicate ecological concepts, and to integrate both the ecological and social dimensions of sustainability. This study was developed together with the population of Puerto Williams, a town with 2200 inhabitants located on Navarino Island, and the largest human settlement within the CHBR.

Based on structured interviews, we found that the largest woodpecker in South America, *Campephilus magellanicus* or the Magellanic woodpecker, was the favorite bird of people who inhabit the reserve, especially for members of ethnic Yamana group and long-term residents (Arango *et al.* 2007). Through a process of socialization in which the local population was involved, together with governmental and military authorities and people from different disciplines and skills, from 2005–07 we developed a program for the implementation of the Magellanic woodpecker as a charismatic species of the Cape Horn Biosphere Reserve.

A strategic plan was then designed covering both the dissemination of the program and the natural history of the woodpecker and including educational and recreational activities with the community at local and regional levels. The objective of this campaign was to consolidate the previous empathy felt toward this species and increase its charismatic appeal in social–cultural groups less familiar with the avifauna of the biosphere reserve. This program of research, environmental education, and biocultural conservation was run by the Masters of Science Program of the University of Magallanes Campus in Puerto Williams in conjunction with the Institute of Ecology and Biodiversity (IEB) and the Omora Foundation. The program was based on the Dietz *et al.* (1994) model and followed a systematic process in seven stages: Stage 1, identification of priority problems; Stage 2, identification and evaluation of the target population, available resources and the scenery; Stage 3, development of positive interaction between the participants; Step 4, selection and methods analysis; Stage 5, activities implementation; Stage 6, evaluation, and Stage 7, reiteration.

(*cont.*)

Biocultural conservation in Cape Horn: the Magellanic woodpecker as a charismatic species (cont.)

The woodpecker program was focused on disseminating knowledge about the natural history of *C. magellanicus* to all parts of the local population. From the beginning a group of researchers from the CHBR Scientific Advisory Council and Omora Park Program worked with several local leaders (including the provincial governor, commanders of Beagle Naval District, the mayor of Puerto Williams, members of local tourism groups, and the director of the only school in the city) and public entities (the Municipality of Cape Horn County, the Chilean Antarctic Province, the Chilean navy, the National Forestry Corporation, the Puerto Williams school, and other community groups) to encourage their interest in this species both as a matter of local pride and in terms of potential economic benefits from the development of tourism. Together, this group selected, planned, and executed linked activities (e.g. a permanent environmental working group of school children to participate in the Scientific Regional School Congress Meeting, environmental workshops, lectures, field trips, celebrations, drawing competitions, environmental campaigns, exhibitions, and distribution of souvenirs) focused mainly on schoolchildren (about 20% of the population are preschool and school-age students). This strategy made it possible to reach an important segment of adults within the city—about 250 families have their children studying at the local school.

The media also played an important role at local and regional levels by disseminating the program and the importance of conserving this species. The Magellanic woodpecker became the symbol of the campaign. As with the case of the municipality, members of the Chilean navy—a group that had originally chosen the Andean condor as its favorite bird in their responses—expressed interest in using the woodpecker as a symbol of their own environmental campaign to contribute to the reduction of plastic bags in the biosphere reserve.

Implementing the Magellanic woodpecker as a charismatic species has been a useful tool for biocultural diversity conservation in the CHBR. The woodpecker program has disseminated accounts of the natural history of *C. magellanicus* and increased its valuation at local, regional, national, and international levels. The construction of public awareness of and support for this bird is a contribution to the conservation of the species and sub-Antarctic Magellanic forests that simultaneously opens up opportunities for local economic development for the community of this biosphere reserve—a success that could only have been achieved through interdisciplinary and transdisciplinary collaborations.

References

Arango, X. (2007). *El pájaro carpintero gigante* (Campephilus magellanicus, *King 1828): una especie carismática al fin del mundo y su implementación para la conservación de los bosques antiguos de la Reserva de Biosfera Cabo de Hornos*. Tesis Programa de Magíster en Ciencias, Universidad de Magallanes, Punta Arenas.

Arango, X., Rozzi, R., Massardo, F., Anderson, C. B., and Ibarra, J. (2007). Descubrimients implementación del carpintero gigante (Campephilus Magellanicus) como especie carismática: una aproximación biocultural para la conservación en la Reserva de Biosfera Cabo de Hornos, *Magallania*, **35**, 71–88.

Dietz, J., Dietz, L., and Nagagata, E. (1994). The effective use of flagship species for conservation of biodiversity: the example of lion tamarins in Brazil. In: P. Olney, G. Mace, and A. Feistner (eds) *Creative conservation: interactive management of wild and captive animals*, pp. 32–49. London: Chapman and Hall.

33.2 A philosopher's odyssey through multidisciplinarity and interdisciplinarity

The integration of environmental philosophy into wide-ranging environmental transdisciplines, such as conservation biology, may perhaps be best traced via the permutations of an individual career. Having grown up with the field of environmental philosophy and ethics, and having been witness to its progressive integration with other disciplines, I offer below my own (albeit limited as well as personal) account of the evolution of interdisciplinary thinking about the environment.

The first Earth Day in 1970 was a cathartic event on the campus of the University of Wisconsin–Stevens Point (UWSP). The gymnasium was packed from 8:00 a.m. until nearly midnight with a rapt audience listening to speaker after speaker analyzing, lamenting, and suggesting how to address, one way or another, the environmental crisis. I was among those speakers, an untenured instructor, in my second year at UWSP.

With regard to Earth Day, Wisconsin was special. Wisconsin's senior senator, Gaylord Nelson, a Democrat, was the Earth Day bill's Senate sponsor. (Representative Pete McCloskey, a California Republican, sponsored it in the House.) Wisconsin was also the boyhood home of John Muir and the setting for Aldo Leopold's *A Sand County almanac*. And in Wisconsin, our proudly 'teaching-oriented' campus was special, as it was given a personality by one of its few research-oriented units with a graduate program, the UWSP College of Natural Resources (CNR). The UWSP CNR resource-management programs were then definitely old school—board-feet forestry and hook-and-bullet wildlife and fishery management—but a number of nascent environmentalists were salted into its faculty and students.

After the emotional climax of that first Earth Day had come and gone, a few of us organized ourselves into an *ad hoc* committee to introduce environment-oriented courses across the curriculum and to organize them into a multidisciplinary environmental studies program. Rashly, I offered to teach a course titled environmental ethics to complement a core curriculum of various offerings in the CNR along with other satellite courses in environmental economics, environmental policy and law, and other disciplines in the social sciences. Environmental ethics at UWSP was approved, albeit skeptically, by my colleagues in the philosophy department, endorsed by the College of Letters and Science curriculum committee, and offered for the first time in 1971, as part of a new multidisciplinary minor in environmental studies.

The analysis of values is at the core of environmental ethics. In the beginning, my colleagues in the natural and applied sciences—biology, ecology, chemistry, forestry—were barely tolerant of me and I occasionally caught them smirking behind my back about 'environmental ethics'. Their reasoning was this: science is objective and value-free; our approach to resource management is scientific; thus our approach to research management is objective and value-free; yes, people other than scientists have values, but such values should play no part in scientific resource management. Scientifically informed environmental policies were thus insolated from challenge from outsiders like me. As their own fields became rent by post-Earth-Day disagreements, however, centering on values—hook-and-bullet wildlife management, for example, shifting toward management of non-game and endangered species, such as sandhill cranes—my CNR colleagues came to appreciate the importance of environmental philosophy and ethics in an environmental studies curriculum and, in the end, treated me as a valued colleague and my course as foundational to the UWSP environmental studies minor.

My first adventure in true interdisciplinarity, as opposed to multidisciplinarity, came when I was invited to join a research team sponsored by the bi-national Great Lakes Fishery Commission in the early 1990s, tasked to re-envision fishery management in its jurisdiction.

The Great Lakes had been wracked by a series of ecological convulsions (Bogue 1993). First, they were overfished. Then, with the opening of the Erie and Welland canals, the lakes were invaded by sea lamprey, an eel-like bloodsucking parasite of finfish, and alewife, a small prolifically reproducing fish of little commercial or recreational value. By the 1950s several cisco and whitefish species endemic to the lakes were extinct, and others were threatened with extinction; the lamprey had decimated the native lake trout; and the alewife, without a top carnivore to control their numbers, died whole and washed up on shore fouling beaches. To take advantage of the forage that the alewife could provide a large predator and to control their numbers, managers introduced Pacific salmon into the lakes in the 1960s, and, not incidentally, created a sport fishery that proved to be enormously popular. But the salmon-capped community was artificial and unstable. It could be maintained only with chemical lampreycides to depress the parasite population and by annually introducing hatchery-raised Pacific salmon, which were unable to reproduce naturally because the streams around the lakes were unsuitable for that purpose.

The principal investigators on the project were Ed Crossman, based in the Royal Ontario Museum, a systematist, Larry Crowder, based at Duke University's Marine Biology lab, a community ecologist, and I. We each employed a graduate student research assistant. My part of the task was to examine the values driving the actions of and conflicts among stakeholders and to identify resource-management paradigms for a better fishery-management regime in the lakes. I protested that, especially for the values analysis, the team would be better off with a sociologist, but the others insisted that I was the person for the job.

Having no idea about how to design and conduct a values survey, I settled on something an environmental ethicist can readily do—recognize an expression of value when he sees one and organize the values thus identified into a typology. This, I discovered, is called 'content analysis' in sociology. Fortunately, my research assistant, Karen Mumford, had the complementary skills—collecting content to analyze being an important one of them—necessary for the values analysis part of the research to be successful. We identified stakeholder groups—sports fishers, commercial fishers, environmental groups, and so on—collected the publications produced by their organizations and mined them for expressions of value (Mumford and Callicott 2000; Callicott and Mumford 2003). Identifying paradigms for resource management—such as those associated with the names of Gifford Pinchot, John Muir, and Aldo Leopold—was more in line with my own research agenda, so that involved no steep learning curve.

The fruits of this 3-year research project—funded at a total of $300,000—were for the most part genuinely interdisciplinary, not merely multidisciplinary (Callicott *et al.* 1998). The key to fusing disciplines, I discovered, was not for a philosopher to try to become a systematist or a community ecologist (or for that matter an *ersatz* sociologist) or for Crossman and Crowder to become philosophers; rather, the key was to become conversant, each with the others' disciplines. This distinctive level of understanding of another's discipline—what can be called conversant knowledge—is a prerequisite for truly interdisciplinary interactions. Conversant knowledge of a discipline does not mean to be able to offer new insights into the discipline's own internal workings. Rather, it means being able to follow a disciplinary argument enough to ask thoughtful questions, to offer relevant criticism, and to integrate that knowledge with one's own areas of expertise.

In this study, Crossman and Becky Cudmore, his research assistant, identified the fish fauna of the Great Lakes; Crowder and Lisa Eby, his research assistant, identified their various interactions; and Mumford and I identified the values that drove the human interactions with the Great Lakes fish fauna and the aquatic communities they composed. I also recognized various paradigms (focused more on ends than means) in accordance with which all such interactions could be managed. I was aware of and could understand the work that other members of the team were doing and they were aware of and could understand the work that Mumford and I were doing. By this means we could pull all the results of our various and disparate research activities together into a seamless and coherent whole (Callicott *et al.* 1999).

A third sortie into interdisciplinarity came at the turn of the century and is on-going. In 1995 I left the UWSP to join the Department of Philosophy and Religion Studies at the University of North Texas (UNT), then one of only two philosophy departments in the world employing a critical mass of environmental philosophers (and now the only one, since the other, at Lancaster University in the United Kingdom, no longer employs an active group of environmental philosophers). The UNT philosophy department is housed in the Environmental Education Science and Technology building with the geography department and half the biology department. Select members of these departments constitute the UNT Institute of Applied Sciences. The office layout of the building mixes faculty offices such that one's immediate neighbors are members of the other departments—a layout that was expressly so arranged to encourage interdisciplinary engagement.

It has worked. Colleagues find themselves interacting daily on an informal basis with researchers in other fields. Moreover, science graduate students are required to enroll in environmental philosophy courses, and philosophy graduate students enroll in environmental science courses. All of us gradually have become conversant in one another's fields. This institutional setting has led to a joint project involving nine UNT colleagues to apply for the US National Science Foundation (NSF) Biocomplexity-in-the-Environment Phase II request for proposals, which expressly required that more than one discipline be represented by the senior personnel.

Beginning in 1999, the NSF announced a new competition called 'Biocomplexity in the environment'. The first competition (Phase I) was limited to 'integrated research on the function of microorganisms' (National Science Foundation 2007). The second (Phase II), was opened in 2000 to support 'integrated research to better understand and model complexity that arises from the interaction of biological, physical, and social systems' (National Science Foundation 2007). We proposed to compare, by means of computer models, human land-use changes resulting in land-cover and other ecological changes—feeding back into further human land-use changes and consequently further land-cover and other ecological changes—for four study sites: Green Valley in Denton County, TX, close to the UNT; the storied Big Thicket National Preserve in southeast Texas; and the Caparo and Imataca forest reserves in Venezuela, all subject to various development pressures. Miguel Acevedo, the project Principal Investigator (PI), is a native of Venezuela and maintains working relationships with former Venezuelan colleagues, who formed a research team, sponsored in that country, complementing ours. The disciplines involved in the Texas team were biology, geography, mathematics, computer science, and philosophy.

Our first proposal was not funded, largely because no social science discipline was represented. Steve Kellert, a celebrated Yale University environmental sociologist, and Paul Harcombe, a Yale-trained Rice University ecologist, agreed to join our research team and we resubmitted our proposal in 2001. That proposal was not funded, but we persisted and received funding on our third try (2002)—for about a half million dollars spread over 3 years. Supplemental funding, bringing the total to about $700,000, and no-cost extensions enabled us to continue the project through to August 2008.

Integrating the discipline of philosophy with the others in this project presented distinctive challenges. Given philosophy's traditional role of offering a 'big picture' view of things, I found myself taking a leading role early in the proposal process in terms of conceptualizing the project as a whole. And while such skills are not uniquely philosophical, but rather characterize all of the humanities, I also found myself serving as editor-in-chief during the process of proposal writing, simplifying and integrating the sometimes jargon-ridden prose of my disciplinary co-authors. Science often deals with simple elements in complex and dynamic relationships. In our case, we were dealing with such prosaic elements as real-estate developers in north Texas, lumber companies in the Big Thicket, extra-legal swidden (slash and burn) farmers in the Caparo Reserve, and hunter–fisher–gatherers in the Imataca Reserve, on the human systems side of the dynamic, and with such pedestrian elements as various tree species and water movements on the natural systems side. Academics of all stripes have a tendency to use technical terms and convoluted constructions; as the stylistic interdisciplinarian of the group, I tried to figure out exactly what was being referred to and what was being said about those referents and then to state the matter clearly and plainly. Especially in an interdisciplinary context, such 'translation' into a common language (ordinary English) is essential if participants in different disciplines are to become conversant with each other's contributions to the whole. It is a role that philosophy has traditionally played, and was only lost in the twentieth century when professional philosophers became specialists like other academicians.

As the project progressed, I was responsible for working with the social scientists (a sociologist and statistician) to identify and codify the stakeholder values driving land-use decisions. I found that some value concepts in environmental philosophy, especially the concepts of intrinsic and inherent value, were foreign to my social science colleagues. So I sought more palatable alternatives. For example, for the Texas landowners who valued their properties in ways that could not, without violence, be reduced to dollars and cents (one indicator that they valued them intrinsically), we labeled such values not 'intrinsic value', but 'tradition value'. And in order to avoid an economistic reduction of all values to dollars and cents, we compared the relative strength of stakeholder values, not in a monetary metric but by means of utility functions (Monticino *et al.* 2005).

Further, I sometimes took the lead in clarifying and refining nascent new concepts. For example, one of our Venezuelan colleagues came up with the concept of *connectivity*—how people are connected with the landscapes they inhabit. It was immediately obvious that this was an important concept, but it was, upon first hearing, still inchoate and unformed. We developed a matrix of three general modalities of connectivity that fell on a weak-to-strong gradient: satisfaction of basic human material needs, psycho-cultural relationships, and the regulation of the use of natural resources. In the Venezuelan sites, for example, humans satisfied more of their basic human material needs directly from the natural systems they inhabit than in the Texas sites, and in the Big Thicket this modality of connectivity is somewhat stronger than in north Texas. In the Venezuelan sites, human resource use is regulated informally, by custom—and such regulation is stronger in Imataca than in Caparo—while in the Texas sites it is regulated formally, by laws—more strongly in north Texas than in the Big Thicket, which was historically a refuge for fugitives from the law and many of its current denizens still retain something of a proud outlaw spirit (Acevedo *et al.* 2008).

In retrospect, the steepest learning curve was becoming conversant with our project's approach to modeling—which was very important because the models were at the heart of the project. The natural systems models were of a form called 'cellular automata', in which cells, representing a surface quadrant on a landscape, change in type depending, in part, on changes in neighboring cells. The social systems models were of a form called 'multi-agent based' in which types of agents are constructed, such as 'land owner', 'developer', 'government regulator' who react independently to the actions of other such agents. The equations governing both the behavior of the cells and the agents in the models were non-linear (Monticino *et al.* 2007).

Perhaps the most exciting philosophical contribution to the project—once I had gotten a basic grasp of the modeling approach—was epistemological in nature. Science, classically, is all about prediction, but it would be misleading to say that our models predicted the state of any of our real-world, coupled natural and human systems at any time in the future. Given the non-linear state-change rules for each autonomous cell and agent, every run of the models produced a different resulting pattern. So, if not prediction, what could the models reveal that would be useful to policy makers?

First, they could reveal 'sensitivities'—what factors a policy maker might tweak to get the biggest change in the system dynamic. For example, our north Texas model suggested that the most important factor affecting a landowner agent's decision to sell or not to sell a property was 'neighboring land use'. Thus, to slow development sprawl, a policy maker could recommend that the local government expend its budget allocation for purchasing widely distributed small-scale green spaces rather than on a highly concentrated large-scale green space. Second, they could reveal 'thresholds' or what is now more commonly referred to as 'tipping points', states of the system in which the direction of pattern change is irreversible. In combination, these two capacities of the models offer powerful tools to policy makers: by manipulating a sensitive factor, a policy maker might avoid exceeding a threshold beyond which the coupled natural and human systems he or she was trying manage would run out of control (Acevedo *et al.* 2008).

One philosophical issue was discussed but remained unresolved. Given the non-linear equations governing the cells and agents in the coupled natural and human systems models, are patterns produced by the models fundamentally unpredictable and 'emergent' or are they fully determined but so complex that they appear to be unpredictable to our finite minds? That of course is an epistemological issue that is not unique to our set of models; it is the subject of debate in climate models; and indeed it is the subject of debate in theoretical physics.

33.3 Conclusion

Any adequate understanding of, and especially any adequate approach to, problem solving in the general domain that is called 'the environment' is inherently interdisciplinary. Both the task of understanding global climate change and meeting the challenges it poses is the clearest and most unimpeachable example of the need for interdisciplinarity and transdisciplinarity in the environmental domain.

The great conservationist, Aldo Leopold, with his knack for forceful and simple expression, clearly articulated the interdisciplinarity of environmental understanding and problem solving back in the 1930s and 1940s. And Leopold contributed perhaps the best simple characterization of interdisciplinarity—a *fusion* of disciplines.

The humanities, including philosophy, are often left out of interdisciplinary approaches to environmental issues. The humanities and humanists, in general, and philosophy and philosophers, more especially, can and should play a vital and indispensable role in the interdisciplinary study of the environment and approaches to the solution of environmental problems. One prominent environmental transdiscipline, conservation biology, has, from its inception recognized the fundamental contribution of philosophy and philosophers to that emergent metadiscipline. And the personal experiences recounted here indicates that philosophy and philosophers have a crucial role to play in addressing new environmental challenges: to construct the conceptual architecture of interdisciplinarity; to translate multiple disciplinary argots into a common language; to identify and codify stakeholder values; to assist in selecting appropriate means for quantifying and comparing stakeholder values; to refine and clarify nascent, otherwise inchoate, concepts; and to make the epistemological aspects of interdisciplinarity thinking (e.g. modeling) transparent.

References

Acevedo, M.F., Callicott, J.B., Monticino, M. *et al.* (2008). Models of natural and human dynamics in forest landscapes: Cross-site and cross-cultural synthesis. *Geoforum* **39**, 846–66, doi:10.1016/j.geoforum.2006.10.008.

Bogue, M.B. (1993). *Fishing the Great Lakes: an environmental history, 1783–1933*. Madison, WI: University of Wisconsin Press.

Callicott, J.B. and Mumford, K. (2003). A hierarchical theory of value applied to the Great Lakes and their fishes. In: G. Dorinda and G. Dallmeyer (eds) *Values at sea: ethics and the marine environment*, pp. 50–74. Athens, GA: University of Georgia Press.

Callicott, J.B., Crowder, L.B., Crossman, E.J., Cudmore, B., Eby, L.A. and Mumford, K. (1998). *Proceedings of the Great Lakes Fishery Commission Biodiversity Workshop for Citizens: biodiversity task presentations and discussion summaries.* Ann Arbor, MI: Great Lakes Fishery Commission.

Callicott, J.B., Crowder, L.B., and Mumford, K. (1999). Current normative concepts in conservation. *Conservation Biology* **13**, 22–35.

Carson, R. (1962). *Silent spring.* Boston, MA: Houghton Mifflin.

Eldridge, N. (ed.) (1992). *Systematics, ecology, and the biodiversity crisis.* New York: Columbia University Press.

Groom, M.J., Meffe, G.K., Carroll, C.R., and Contributors (2006). *Principles of conservation biology*, 3rd edn. Sunderland, MA: Sinauer Associates.

Leopold, A. (1949). *A Sand County almanac and sketches here and there.* New York: Oxford University Press.

Leopold, A. (1999a). A biotic view of land. In: S.L. Flader and J.B. Callicott (eds) *The river of the mother of God and other essays by Aldo Leopold*, pp. 266–73. Madison, WI: University of Wisconsin Press.

Leopold, A. (1999b). The role of wildlife in a liberal education. In: S.L. Flader and J.B. Callicott (eds) *The river of the mother of God and other essays by Aldo Leopold*, pp. 301–5. Madison, WI: University of Wisconsin Press.

McIntosh, R.P. (1985). *Background of ecology: concept and theory.* New York: Cambridge University Press.

Meffe, G.K., Carroll, C.R., and Contributors (1994). *Principles of conservation biology*, Sunderland, MA: Sinauer Associates.

Meffe, G.K., Carroll, C.R., and Contributors (1997). *Principles of conservation biology*, 2nd edn. Sunderland, MA: Sinauer Associates.

Monticino, M., Acevedo, M.F., Callicott, J.B., Cogdill, T., and Lindquist, C. (2005). Coupled human and natural systems: a multi-agent based approach. *Environmental Modeling and Software* **22**, 656–63.

Mumford, K.G. and Callicott, J.B. (2000). Computer-aided qualitative content analysis: a useful approach for the study of values. In: D.N. Bengston (ed.) *Applications of computer-aided text analysis in natural resources*, pp. 43–7. St Paul, MN: North Central Research Station Forest Service–US Department of Agriculture.

National Science Foundation (2007). *Biocomplexity in the environment.* Available at: <http://www.nsf.gov/geo/ere/ereweb/fund-biocomplex.cfm>.

Rozzi, R., Massardo, F., Anderson, C., and Silander, J. (2004). Ten dimensions of a biocultural conservation approach at the southern tip of the Americas. *Sustainable Community Review* **7**, 76–83.

Son, S.-W., Polvani, L.M., Waugh, D.W. *et al.* (2008). The impact of stratospheric ozone recovery on the Southern Hemisphere westerly jet. *Science* **320**, 1486–9.

Soulé, M.E. (1985). What is conservation biology? *Bioscience* **53**, 727–34.

Udall, S. (1963). *The quiet crisis.* New York: Hold, Rinehart, and Winston.

Wilson, E.O. (1988). *Biodiversity.* Washington, DC: National Academy Press.

CHAPTER 34

Health sciences and health services

JENNIFER L. TERPSTRA, ALLAN BEST, DAVID B. ABRAMS, AND GREGG MOOR

The health services and health sciences landscape has changed significantly over the last century. Advances in science and technology, as well as dramatic lifestyle changes, have resulted in a shift in the nature of, and approach to, improving societal health and well-being. Many of the primary causes of mortality and morbidity in the early twentieth century, such as smallpox and poliomyelitis, have been eliminated or significantly reduced, and replaced by new priority challenges such as HIV (Human Immune Deficiency Virus), and type II diabetes. Over the last century there have been many lessons learned in the health field. A key lesson learned is that health is a complex phenomenon and the underlying causal pathways for disease and illness are more than just biological.

Identification of the biological mechanisms underlying disease and illness is only one part of solving health problems. Health is a phenomenon deeply rooted within a social system, and health outcomes result from a dynamic interplay between factors across the lifetime, originating from the cellular level, to the socio-political level (Glass and McAtee 2006). As such, efforts to improve health must consider the multifactorial nature of the problem and integrate appropriate knowledge across disciplines and levels of analysis. For example, HIV was discovered as the cause of Acquired Immune Deficiency Syndrome (AIDS) in 1983 (Montagnier 2002). Despite this discovery, and the subsequent understanding of routes of transmission, HIV/AIDS continues to be a formidable threat to health globally, and far from a 'problem solved'. Health research has implicated a myriad of factors involved in HIV prevention and produced a large body of knowledge to inform prevention efforts. Unfortunately, incidence rates continue to rise because the knowledge is not being applied in the unified manner necessary to address the complexity of the problem.

The purpose of health sciences and health services research is to provide knowledge that can be used to inform decision making in health services in order to better address these complex health problems. Unfortunately, the majority of health research is conducted for the sake of science, and not for the sake of dissemination and implementation. Knowledge created for science's sake tends to be discipline specific and reductionist, producing results that are not easily applied to inform practice and policy decisions. The reality is that health and health service challenges cannot be handled well by any single discipline or social sector, and the traditional reductionist approach to science does not work well for the majority of health service problems. Disciplinary knowledge and levels of analysis are intertwined in health service problems, and, as such, application requires integrative theoretical models and knowledge. As stated by Rosenfield (1992, p. 1344), 'to achieve the level of conceptual and practical progress needed to improve human health, collaborative research must transcend individual disciplinary perspectives and develop a new process of collaboration'.

Klein (Chapter 2 this volume) introduces the concept of *transcendent interdisciplinary research* and transdisciplinary science—a science 'that creates new methodological and theoretical frameworks for defining and analyzing social, economic, political, environmental, and institutional factors in health and well-being'. This vision is clearly the ideal driving many current efforts to improve health services, but the reality is that true transdisciplinarity is usually beyond our grasp. On a continuum from unidisciplinary to transdisciplinary, interdisciplinary rests between multidisciplinary and transdisciplinary. Moving to a transdisciplinary, or even an interdisciplinary, research and practice model is difficult when the infrastructure, culture, practice, and policies are deeply rooted in the traditional single-disciplinary approach. Nonetheless, this vision increasingly frames the research endeavor along what Klein calls 'the fourth trendline'—trans-sectoral problem solving. Under this vision, complex health problems, not the disciplines, frame the research questions and practices (see Hirsch Hadorn *et al.*, Chapter 30 this volume). Problem-based research, or Mode II science (Denis *et al.* 2004), represents a fundamental shift in health sciences and health services in which the problems frame the research and practice models. The traditional reductionist research approach does not allow for transdisciplinarity or problem-based research. In contrast, systems thinking (Strijbos, Chapter 31 this volume) is an alternative paradigm that provides a framework and methods for integration of knowledge across disciplines and traditionally disparate perspectives.

This chapter relies on the understanding of interdisciplinarity and transdisciplinarity provided by other chapters to explore the role of these integrative models in health sciences and health services. In particular, the chapter examines in more depth three driving factors for the recent impetus for interdisciplinarity: (1) increased recognition of the complexity of health, (2) the demand for solutions to health service questions, and (3) the need for an approach that is more aptly suited to the problems of societal interest. To better understand these factors, the chapter will explore the complexity of health and health services, and consider the role of interdisciplinarity and transdisciplinarity, systems thinking, and problem-based research as alternative and complementary approaches to solving complex problems.

34.1 The complexity of health and health systems

The lesson learned from the HIV/AIDS epidemic, as well as other health challenges such as diabetes, cancer, and asthma, is that health is a complex phenomenon, and solving health problems will require an integration of diverse knowledge. Complex problems are often described as being unpredictable, having uncertain outcomes and competing evidence or information, and being influenced by a network of interdependent factors. Health service questions and problems generally fall into the realm of complexity, as there is rarely one single causative factor for disease, and equally no one successful mechanism for treatment or prevention.

From the etiology of disease to its prevention and management, health is the result of an interaction of factors ranging from the micro- to the macro-level. For example, there is an interdependent relationship between environmental factors and genetics for schizophrenic effect, in that environmental factors have 'an impact on developing children and adolescents to increase the later expression of psychosis-like at-risk mental states and overt psychotic disorders' (Krabbendam and van Os 2005). In this example, social factors (e.g. urbanicity) interact with biological factors (e.g. genetics) to determine the health outcome (e.g. mental illness). Health services must consider such linkages between biological and social factors for successful prevention and treatment of illness.

Health sciences and health services have been applying the reductionist approach and attempting to solve health problems from silos for too long. Such efforts are not working for the majority of health challenges today, thus creating the impetus for the merging of knowledge from traditionally disparate domains. Efforts to elucidate the causal pathways of disease and illness require a systems approach and the application of a new paradigm with a more complex view for causality of health (Glass and McAtee 2006). Type II diabetes, for example, is a complex multifactorial health care challenge that cannot be successfully addressed from silos. Despite significant research conducted to elucidate the causal pathway of the disease, incidence rates have continued to rise at an alarming rate. Type II diabetes, once referred to as 'adult onset diabetes', is now at epidemic proportions in youth and children (Pinhas-Hamiel and Zeitler 2005). A major challenge is that evidence from single-discipline reductionist studies does not translate easily to practice and policy settings. Although scientific research is producing a wealth of knowledge, this knowledge is not being applied to health systems as quickly and successfully as it should. There are several reasons for the difficulty in applying evidence created in the traditional research approach: (1) the world and real-world problems are not divided into silos; (2) context is an integral and active component; and (3) research fails to answer questions originating from the health system.

In order to begin to solve health challenges, it is necessary that solutions be developed specifically to answer questions originating in the health system, while also addressing the elements of complexity inherent in the application of knowledge to the real world. As such, health research requires the collaboration of professionals from different disciplines and sectors, as well as consideration of the multiple determinants influencing health outcomes.

34.2 The need for interdisciplinarity in health sciences and health services

Recognition of the complexity of health and the need to integrate knowledge from multiple perspectives has pushed health services past traditional boundaries in search of more appropriate models for care and problem solving. According to Boon *et al.* (2004) there is a continuum of models for team care ranging from parallel practice to integrative, in

Telehospice: a case study in healthcare intervention research

Elaine M. Wittenberg-Lyles, Debra Parker Oliver, and George Demiris

In 2007, an estimated 1.4 million people received services from a hospice, approximately 38.8% of all deaths in the United States (National Hospice and Palliative Care Organization, 2008). Hospice care is provided to people facing a life-limiting illness or injury and emphasizes quality of life rather than focusing on curing the disease/illness. Currently there is a lack of interventions that include caregivers (e.g. family members, friends) in the ongoing decision-making process for hospice services. Telemedicine, defined as the use of telecommunications and information technology with the goal of bridging geographical gaps and enhancing the care delivery process, is one tool that can enhance the role of hospice caregivers. Researchers in the Telehospice project tested the feasibility and effectiveness of video-based interventions to determine improvement in caregiver pain management and problem-solving skills. This case study highlighted the successful interdisciplinary research of the Telehospice project by illustrating the interplay of interdisciplinary roles, team problem solving, the use of collaborative resources, and important lessons learned from field testing.

The Telehospice project is grounded in the complementary roles of interdisciplinary perspectives in a single research program. Academicians in three fields, social work, health informatics, and communication studies, comprise the core of the research team, all united by the single goal of using video-based technology to enable hospice caregivers to participate in routine hospice interdisciplinary team meetings. The social work perspective gives voice to the needs of hospice caregivers, and identifies a concrete role for social workers in the hospice team meeting. Health informatics, the application of information technology in the field of health care, provides a channel for enabling caregivers to virtually participate in care plan meetings. Finally, communication studies focuses on the content of the communication between the team and the caregiver as enabled by technology and the role of the social worker. Although geographically dispersed in Missouri, Texas, and Washington, the Telehospice team extensively communicates via email and utilizes technologies such as voice over internet protocol (VOIP) to facilitate team meetings and schedules time at conferences for team work. Over the last 5 years, the team has produced more than 30 peer-reviewed published articles, obtained funding from the National Institutes of Health, and presented numerous conference papers. Collaboratively, each team member enriches the feasibility, utility, and testing of the intervention as it is refined through pilot studies, incrementally working towards a formal role for the intervention in the standard practice of care.

Several transitions occur as the collaborative work of academics shifts toward use in practical settings. To capture these transitions, members of the Telehospice project communicate regularly to generate new ideas and problem-solve issues that emerge from working in the field. New ideas come easily, as different discipline-specific perspectives and methodologies are introduced and influenced by each other. With the generation of new ideas, one team member is designated as the lead on the project, and varying levels of contribution by other team members are identified. This process informally creates expectations for team productivity and defines the role for each team member. As the project develops, team members pool resources to ensure the project is carried out successfully.

Resourceful problem solving emerges from team collaboration as the pool of resources and experiences are larger than any one person. Team resources include shared information networks such as departmental colleagues, access to graduate student research assistants, and equipment. Resources are shared across universities as our graduate students are brought into the collaborative team model and get to experience working with academics from other disciplines.

(*cont.*)

Telehospice: a case study in healthcare intervention research (cont.)

For example, the lead on a project might be in the field of communication studies, but due to the availability of resources the ground work is completed by a graduate student in social work. In this manner, the Telehospice team is able to move forward with projects in a timely fashion. Likewise, when one team member experiences problems initiating a study then another team member with more appropriate resources will take the lead on the project. Finally, given that all team members contribute to the development of ideas, manuscripts are edited within the team by all team members (including students) prior to submission. The ability to edit each other's work has refined the manuscript writing process and contributed greatly to the ability to procure the external funding needed to conduct field testing.

In the Telehospice team complementary academic perspectives aid in the production of translational work that moves from theory to practice through field testing. This field testing has allowed the team to refine implementation of the intervention as well as refine aspects of the intervention. Following numerous pilot studies, the team recently undertook a 2-year exploratory study to examine the impact of a videophone intervention for hospice caregivers. Two very important lessons were learned from field testing. First, the team encountered unexpected recruitment challenges. Working in conjunction with student research assistants, the Telehospice team engaged in problem solving that resulted in new strategies, including mailing participants a letter from the hospice medical director and expanding the research criteria. Second, the team learned that patients often have more than one caregiver, an unexpected finding as much caregiver research has focused on the patient–caregiver dyad rather than patient and multiple caregivers.

The tools necessary to develop translational work were developed at the intersection between an interdisciplinary team and actual field testing of the proposed intervention. The integration of interdisciplinary perspectives allowed for quick problem solving, more resources, a larger understanding of the contextual challenges, and synergy created through diversity in analysis and interpretation. The inclusion of graduate research assistants in the team process led to quick and successful changes in the field. Ultimately, findings from research produced by the Telehospice team are shared with multiple audiences as study results are presented in publication outlets and conferences across disciplines. Still, some of the limitations of the collaborative nature of the Telehospice team include meeting the demands of institutional review boards at multiple universities, management of subcontracts, and conflicting demands of institutional policies and procedures. As members of the Telehospice team address these challenges they develop trust, mutual respect, honesty, and an appreciation of the collaborative process. Their ability to problem-solve and pool collaborative resources yields a productive interdisciplinary-based research agenda.

Reference

National Hospice and Palliative Care Organization (2008). *NHPCO facts and figures: hospice care in America. 2008* [updated December 2008]. Available from: <http://www.nhpco.org/>.

the following order: parallel practice, consultative, collaborative, coordinated, multidisciplinary, interdisciplinary, and integrative. As the models move toward the integrative end of the continuum, there is a greater emphasis on the whole person, an increase in the number of determinants of health considered, less reliance on hierarchy and defined roles, a greater need for consensus, and a decrease in practitioner autonomy. In addition, complexity of both the structure and the outcomes increases as the models move towards integrative care on the continuum, with integrative care deemed the most complex.

Complexity has been suggested as a prerequisite for interdisciplinarity, which allows for the integration of multiple perspectives for developing and applying solutions appropriate for complex phenomena. According to Newell (2001), 'in order to justify the interdisciplinary approach, its object of study must be multifaceted, yet its facets must cohere'. Newell also suggests that if the phenomenon is not multifaceted then a reductionist, single-disciplinary approach is appropriate. Although some might argue that an interdisciplinary approach may be equally appropriate for studying non-complex phenomena, few would claim that a single-disciplinary approach alone is appropriate for studying complex phenomena.

Interdisciplinarity is quickly becoming a goal in public health, in part because of evidence to support the efficacy of integrative models for health. For example, smoking cessation rates are significantly improved if behavioral therapy and pharmacotherapy are used in conjunction, as opposed to either applied independently (Hughes 1995). The Office of Behavioral and Social Science Research (OBSSR) at the US National Institutes for Health presents interdisciplinarity as a pillar for future health research efforts. Vertical integration in particular is at the core of the OBSSR's strategic prospectus and is defined as the 'integration across rather than within the three broad domains (i.e. the biomedical; the individual behavioral (intra-individual variation); and the population (inter-individual variation) levels) of systems structure' (Mabry *et al.* 2008, p. S219). Integration across disciplines and domains of knowledge is clearly becoming the vision for public health efforts.

It is important to note, however, that acknowledging the need for interdisciplinarity in no way undermines the value of single-disciplinary research. In the same way that reductionism and holism are different yet necessary approaches to studying a problem, so single-disciplinary and interdisciplinary study are both vital to achieving a comprehensive understanding of health. Knowledge from single-disciplinary perspectives is important for interdisciplinary and transdisciplinary science, because it provides a solid foundation on which to build. However, single-disciplinary research cannot provide comprehensive solutions to questions originating in the health services.

Unfortunately, moving towards an integrative model for both research and practice is difficult due to the constraints of the current paradigm rooted in reductionism and a single-disciplinary system. In reductionism, the goal is to eliminate contextual influences and drill down to the micro-level to identify single causative relationships. To move towards a transdisciplinary, or even an interdisciplinary, approach to health services, a new paradigm for both research and practice is needed that considers context and facilitates the integration of multiple perspectives.

34.3 A paradigm shift

A review of the knowledge translation (i.e. linking research to practice and policy) literature over the last 50 years suggests a paradigm shift in the way that health sciences and health services research are conducted and conceptualized (Best *et al.* 2008, 2009). Conceptual models for knowledge translation have evolved through three generations:

from linear, to relationships, and, finally, to systems. The first generation of thinking about knowledge translation applied a reductionist, linear approach to conducting and translating evidence from research to practice and policy settings. In this generation, knowledge/research evidence was seen as a product that could be packaged up and passed from evidence producers (i.e. researchers) to evidence consumers (i.e. decision makers). Unfortunately, this approach to knowledge translation does not work well for the majority of health service problems. The primary language used in generation two (relationships) is knowledge exchange, which is defined by the Canadian Institute for Health Research as 'the interactive and iterative process of imparting meaningful knowledge between research users and producers, such that research users receive information that they perceive as relevant to them and in easily usable formats, and producers receive information about the research needs of users' (Best *et al.* 2008, p. 322). The shift from linear to systems thinking has occurred in response to the inability of the linear and relationship approaches to adequately address health service questions.

Stakeholders in health services are demanding answers from researchers for problems facing the creation, delivery, accessibility, and effectiveness of health services, such as disease prevention and management programs. The questions being posed by stakeholders are often context-specific, and the traditional investigator-driven research approach does not produce knowledge that sufficiently addresses issues of context. Context is critical in understanding health outcomes as well as in the application of evidence in health practice settings. Consideration of context across the health research process requires changes in how research questions are framed, how science is conducted, how interventions and programs are developed, and how findings are disseminated and implemented.

34.3.1 Systems thinking for integration of health knowledge

A systems paradigm (see Strijbos, Chapter 31 this volume) for solving complex health challenges facilitates the merging of knowledge across disciplines and sectors, as well as the collaboration of relevant disciplinary scientists and stakeholders. The integration of multiple perspectives is critical to systems theory, which provides a framework for integrating disparate knowledge and perspectives (Williams and Imam 2007).

A systems approach requires a different type of thinking from that of researchers trained solely in reductionist approaches. The reductionist lens applies a mechanist view, in which individual parts can be studied independently to understand the whole. In contrast, a systems thinking lens assumes that the whole is greater than the sum of the parts (Meadows 2007). In systems thinking, the world is seen as composed of interdependent parts or nested subsystems with permeable boundaries, where change in one subsystem creates change in other subsystems. The result is a dynamic whole with emergent patterns that can only be seen at the whole system level and are not evident if the individual parts of the system are studied independently.

In systems thinking, it is assumed that there are different types of systems and that the methods of study should be chosen appropriately for the system of interest. For example, reductionism may be appropriate for studying systems that are linear and simple, but it is not appropriate for studying complex systems. The majority of health service systems and problems are complex, which is why the reductionist approach for both research and management has not been effective (Plesk and Greenhalgh 2001). It is important for researchers and health systems professionals to recognize how their epistemological and ontological lenses determine the type and value of the outcomes.

The health system viewed through a complex systems lens is seen as a living organism, similar to a brain in that it learns and adapts in response to a network of signals. The health system is complex because it is composed of numerous interdependent, nested systems, with permeable boundaries, such as: patient system, public health system, belief system, incentive system, funding system, physician culture system, and so on. Due to the connectivity and the dynamic, non-linear relationships, the reductionist approach of predicting specific outcomes does not work well in complex systems. To understand a complex system, it is necessary to look at the whole, and identify patterns and feedback loops in the system, which requires a merging of multiple perspectives and sources of information.

A systems approach to solving health problems also requires a new box of tools, including different types of data, methods, theories, and statistical analyses. No single discipline can account for the complexities or provide the tools necessary to address health from a systems perspective. Therefore, it is necessary to approach health research with a collaborative interdisciplinary team of investigators who bring knowledge and expertise from a variety of disciplines and sectors (see Stokols *et al.*, Chapter 32 this volume). The convergence of disciplines can produce theoretical frameworks and methodologies that traverse disciplinary boundaries to form new conceptual syntheses, spawning new measurement techniques (e.g. social network analysis) and interdisciplinary fields (e.g. behavioral genomics) that have the capacity to tackle the complex problems of population health. Systems thinking offers a robust approach to the complexity of priority health challenges, employing an integrative perspective, and incorporating multilevel interventions.

34.3.2 **Interdisciplinary health science: an example**

An example of the application of systems thinking for health challenges is the Initiative on the Study and Implementation of Systems (ISIS), a pilot project designed to explore the paradigm shift from traditional science to systems thinking. The US National Cancer Institute funded ISIS to: (1) explore how systems thinking approaches might improve understanding of the factors contributing to tobacco use; (2) inform strategic decision making on which efforts might be most effective for reducing tobacco use; and (3) serve as an exemplar for addressing other public health problems. Tobacco control provided an ideal opportunity for applying a systems perspective to a perpetually thorny health challenge. Contextually, it can easily be conceived of as a system comprising smaller systems, and existing within the broader systems of public health, economies, and society at local, regional, and global scales. Most research in tobacco control has been reductionist, and while significant gains have been made by understanding the 'parts' of tobacco use, such as the biological basis for nicotine addiction, the structure and function of cigarettes, the advertising and marketing of tobacco products, the economics of tobacco use, and the effectiveness of tobacco control programs and initiatives, tobacco use remains a leading cause of preventable death, and few strides have been made in understanding the whole, or in reducing tobacco use through system-wide change (Best *et al.* 2007, Chs 1 and 2).

ISIS was intended to become a long-term, multi-agency collaboration to create and implement transdisciplinary systems principles and methods for the discovery, development, and delivery of program and policy interventions within a research-to-practice paradigm. The ISIS team considered several core questions:

(1) How can the flow in *both* directions between research and practice be optimized.
(2) How can systems structure and function be best characterized to be useful to the public health community.
(3) In what ways can networks be better understood and optimized?
(4) In what ways do information and knowledge become the currency by which change occurs?

Based on these questions, the ISIS team concluded that systems thinking in public health cannot be encompassed by a single discipline or even a single 'systems thinking' approach (e.g. system dynamics), but instead represents a transdisciplinary integration of approaches to public health which strive to understand and reconcile linear and non-linear, qualitative and quantitative, and reductionist and holistic thinking and methods into a federation of systems thinking and modeling approaches. The resulting publications provide conceptual overviews and case studies for the following four core systems tools that ISIS explored in depth; these both illustrate systems thinking and offer practical ways to implement the approach: *system dynamics modeling*, *social network analysis*, *concept mapping*, and *knowledge management and transfer* (Best et al. 2007).

34.3.3 Problem-based research and collaboration

The demand for more relevant results, in addition to the recognition of the challenges to knowledge transfer from research to practice settings, has resulted in the emergence of a new type of science referred to as Mode II science, or problem-based research (Denis *et al.* 2004). Hirsch Hadorn *et al.* (Chapter 30 this volume) provide an elegant and comprehensive explanation of problem-based research ('problem-solving for the life world'), including a definition and description of key elements necessary to achieve problem-based research. In brief, problem-based research differs from traditional research in that the research is conceptualized, designed, and carried out in the service of generating a solution to a specific health problem, rather than in the service of generating new knowledge. Problem-based research stems from the belief that the public health imperative should be driving every research effort, and that those efforts should be solution-oriented. Therefore, by its very nature, problem-based research demands the linking of research to practice and policy, since it assumes that efforts are not complete until the problem ceases to exist. In contrast to the traditional scientific approach, which has fundamental barriers to achieving linkage between research, practice, and policy, there are several core assumptions of problem-based research that facilitate this linkage. Problem-based research is driven by stakeholders (non-researchers); it is interdisciplinary, collaborative, context specific, has an emphasis on external validity, and applies a systems lens.

Successful linking of research evidence to real-world problems will require research that places a greater emphasis on *external validity* (Green and Glasgow 2006). External validity refers to the extent to which it is possible to generalize or apply the results of a study to other populations, contexts, settings, and/or situations outside of the specific situations studied in a given investigation. External validity also concerns the representativeness of the settings, intervention agents, and participants in a study. In contrast, *internal validity* refers to the extent to which a research study has high rigor or control. For example, a randomized controlled trial is considered to have high internal validity but may have low external validity, due to the tight control, homogeneity of the population, and lack of extraneous factors influencing the outcomes. External validity is important for integration of research to practice and policy, because those responsible for applying the research findings to real-world problems must ascertain whether the study findings are relevant to their specific context.

Another key element of problem-based research is the collaboration between researchers and stakeholders who are affected by the research. Collaboration has emerged as a methodological and theoretical pillar supporting the linkage between research and practice/policy (Bammer 2005). There has been a fundamental shift in how the process of knowledge development and implementation is viewed. Not only is knowledge to be problem-based but it also needs to be co-created across disciplines and sectors, or between health service decision makers and researchers. Van de Ven and Johnson (2006) refer to engaged scholarship, a process of co-creation, and Lomas (2000) refers to a process of linkage and exchange. Both of these concepts are explicit in the need for ongoing dialogue between the researchers and decision makers throughout the research process.

The logic behind the push for co-creation of knowledge is that integration (of research and practice) is more likely to occur if knowledge is created through a process that engages both researchers and critical stakeholders. One rationale for this assumption is that the new knowledge from the research has been generated to answer a specific question that is driven by stakeholders, not researchers. Consequently, there is an investment in the utilization of the findings and a predetermined need for the specific research knowledge being generated. Secondly, because the research was conducted with the outcome and application of the findings in mind, the findings are going to be directly applicable and relevant to the context, and therefore more easily applied to the setting (Allen *et al.* 2007).

34.4 Conclusion

Increased recognition of health as a complex phenomenon, as well as the demand for solutions to health system questions, is creating an impetus for collaboration across different disciplines and health service stakeholder groups. In conjunction, there is also a move away from the mechanistic perspective and traditional managerial styles towards a focus on holism. Terms such as interprofessional, integrated, and comprehensive care are becoming standard nomenclature in health systems, as it becomes apparent that solutions will require overcoming traditional boundaries imposed by archaic policies, norms, and incentive systems. These traditional boundaries exist between disciplines, as well as organizational and social sectors, such as health care organizations, academia, funding agencies, and government (Mitton and Bate 2007). The crossing of boundaries between disciplines and organizations will allow for co-production of knowledge in the context of building bridges through collaboration between researchers and stakeholders, and should therefore be a priority.

Interdisciplinarity has the potential to impact upon both individuals and institutions, as well as health sciences and health services, in positive ways that traditional single-discipline research cannot. Interdisciplinary research is meant to add value to individual scholarship by providing a setting in which the talents of individuals can be mobilized in a collective manner to address scientific issues of common concern, and to conduct a kind of research that would not otherwise be possible. Often, individual scientists are drawn to collaborating with researchers from other disciplines when they perceive that they have 'hit the wall' in terms of their ability to make significant progress in their research endeavors using only the approaches and tools available within their own disciplines. For example, behavioral scientists have traditionally focused on socio-cognitive factors that may account for only a small portion of the variance in health outcomes; however, more recently, these scientists have begun to collaborate with geneticists and ecologists to explore additional factors and improve the explanatory value of their models. By working with, and incorporating the perspectives and methods of, scientists from other disciplines, these scientists have the opportunity to conduct a kind of research that is richer than would otherwise be possible, and that is potentially more useful in addressing major health challenges.

While, on the whole, interdisciplinary collaboration adds value to health research, it is not without drawbacks. In considering whether or not to establish or participate in interdisciplinary collaborations, individuals and institutions should take into account the challenges and drawbacks of this approach. For example, as a result of the necessary exchange process, interdisciplinary research often requires more time, which can be problematic for funding reasons, but also a distinct disadvantage for the scientists involved—especially more junior ones—particularly because of the 'publish or perish phenomenon'. Lastly, while interdisciplinary research is becoming more common, it still faces many institutional barriers and lacks acceptance as a legitimate approach in some academic institutions. In general, if interdisciplinary collaboration is to advance, changes must be made to the infrastructure that supports the research process.

Our current infrastructures were not designed for connecting the silos of disciplines, health professions, or social sectors (i.e. research and health services), so this different science in turn demands different structures and processes. The institutional barriers to interdisciplinary and intersectoral collaboration are considerable, and it will take a significant investment of research to develop effective strategies to strengthen these processes, as well as the willingness and desire to change from institutional leaders. There will also need to be a realignment of the incentive systems in both universities and health services to put a greater emphasis on ongoing communication between researchers and decision makers, as well as the dissemination/implementation of science (Kerner *et al.* 2005). The change has already been initiated, however. A shift is evident from the traditional academic value placed on knowledge for its own sake, to the view that knowledge is a social good and a potential means to solve societal problems, such as the current crisis in health services.

Despite the challenges and drawbacks of interdisciplinary research, there are many scientists and decision makers who consider interdisciplinarity to be key to developing solutions to the vexing problems of the twenty-first century. Interdisciplinary collaborative research is therefore quickly gaining acceptance and support, with the power to go beyond traditional research. Interdisciplinarity provides a setting in which the talents of individuals can be mobilized in a collective manner to address issues of common concern. In this way, interdisciplinarity will enable a kind of research that would not otherwise be possible. It is now necessary to further develop methods for monitoring and evaluating interdisciplinary efforts (Stokols *et al.*, Chapter 32 this volume), as well as develop the policies, culture, ways of working, and infrastructure necessary to support this type of research. For the health services, the promise is great—true collaboration between those who produce new knowledge and those who use it offers transformative innovations to address the needs of health system restructuring.

References

Abrams, D.B. (2006). Applying transdisciplinary research strategies to understanding and eliminating health disparities. *Health Education and Behavior* **33**, 515–31.

Allen, P., Peckham, S., Anderson, S., and Goodwin, N. (2007). Commissioning research that is used: the experience of the NHS service delivery and organization research and development programme. *Evidence and Policy* **3**(1), 119–34.

Antoni, M.H., Lutgendorf, S.K., Cole, S.W. *et al.* (2006). The influence of bio-behavioural factors on tumour biology: pathways and mechanisms. *Nature* **6**, 240–8.

Bammer, G. (2005). Integration and implementation sciences: building a new specialization. *Ecology and Society* **10**(2), 6. Available at: <http://www.ecologyandsociety.org/vol10/iss2/art6/>.

Best A., Clark, P., Leischow, S., and Trochim, W. (eds) (2007). Transforming tobacco control through systems thinking: integrating research and practice to improve outcomes. *Tobacco control monographs. Monograph 18: greater than the sum: systems thinking in tobacco control.* Bethesda, MD: US Department of Health and Human Services, Public Health Service, National Institutes of Health, National Cancer Institute. Available at: <http://cancercontrol.cancer.gov/tcrb/monographs/18/index.html>.

Best, A., Hiatt, R.A., and Norman, C.D. (2008). Knowledge integration: conceptualizing communications in cancer control systems. *Patient Education and Counseling* **71**, 319–27.

Best, A., Terpstra, J.L., Moor, G., Riley, B., Norman, C.D., and Glasgow, R. (2009). Building knowledge integration systems for evidence-informed decisions. *Journal of Health Organization and Management*, **23**(6), 627–41.

Boon, H., Verhoef, M., O'Hara, D., and Findlay, B. (2004). From parallel practice to integrative health care: a conceptual framework. *BMC Health Services Research* **4**(15), doi:10.1186/1472-6963-4-15.

Denis, J.L., LeHoux, P., and Champagne, F. (2004). A knowledge utilization perspective on fine-tuning dissemination and contextualizing knowledge. In: L. Lemieux-Charles and F. Champagne (eds) *Using knowledge and evidence in health care*, pp. 18–40. Toronto: University of Toronto Press.

Eddy, D.M., Schlessinger, L., and Kahn, R. (2005). Clinical outcomes and cost-effectiveness of strategies for managing people at high risk for diabetes. *Annals of Internal Medicine* **143**(4), 251–64, W53–W68.

Gibbons, M., Limoges, C., Nowotny, N., Schwartzman, S., Scott, S., and Trow, M. (1994). *The new production of knowledge: the dynamics of science and research in contemporary societies.* London: Sage Publications.

Glass, T.A., and McAtee. M.J. (2006). Behavioral science at the crossroads in public health: extending horizons, envisioning the future. *Social Science and Medicine* **62**(7), 1650–71.

Green, L.W. and Glasgow, R.E. (2006). Evaluating the relevance, generalization, and applicability of research: issues in external validation and translation methodology. *Evaluation and Health Professions* **29**(1), 126–53.

Greenhalgh, T., Robert, G., Macfarlane, F., Bate, P., and Kyriakidou, O. (2007). Diffusion of innovations in service organizations: systematic review and recommendations. *Milbank Quarterly* **82**(4), 581–629.

Hughes, J.R. (1995). Combining behavioral therapy and pharmacotherapy for smoking cessation: an update. *NIDA Research Monographs* **150**, 92–109.

Kerner, J., Rimer, B., and Emmons, K. (2005). Introduction to the special section on dissemination: dissemination research and research dissemination: how can we close the gap? *Health Psychology* **24**(5), 443–6.

Klein, J.T. (2004). Interdisciplinarity and complexity: an evolving relationship. *Emergence: Complexity & Organization* **6**(1/2), 2–10.

Krabbendam, L and van Os, J. (2005). Schizophrenia and urbanicity: a major environmental influence-conditional on genetic risk. *Schizophrenia Bulletin* **31**(4), 795–9, doi:10.1093/schbul/sbi060.

Lomas, J. (2000). Using 'linkage and exchange' to move research into policy at a Canadian foundation. *Health Affairs* **19**(3), 236.

Mabry, P.L., Abrams, D., Olster, D.H., and Morgan, G.D. (2008). Interdisciplinary science and the NIH Office of Behavioral and Social Sciences Research. *American Journal of Preventive Medicine* **35**(2S), S211–S224.

McKinlay, J.B. (1995). The new public health approach to improving physical activity and autonomy in older populations. In: E. Heikkinen, J. Kuusinen, and I. Ruoppila (eds) *Preparation for aging*, pp. 87–103. New York: Plenum Press.

Meadows, D.H. (2007). *Thinking in systems: a primer.* White River Junction, VT: Chelsea Green Publishing.

Mitton, C. and Bate, A. (2007). *Où sont les chercheurs*? Speaking at cross-purposes or across boundaries. *Healthcare Policy*, **3**(1), 32.

Montagnier, L. (2002). Historical essay: a history of HIV discovery. *Science* **298**(5599), 1727–8.

Newell, W.H. (2001). A theory of interdisciplinary studies. *Issues in Integrative Studies* **19**, 1–25.

Pinhas-Hamiel, O. and Zeitler, P. (2005). The global spread of type 2 diabetes mellitus in children and adolescents. *Journal of Pediatrics* **146**(5), 693–700.

Piot, P., Bartos, M., Larson, H., Zewdie, D., and Mane, P. (2008). Coming to terms with complexity: a call to action for HIV prevention. *The Lancet* **372**, 845–59.

Plesk, E.P. and Greenhalgh, T. (2001). Complexity science: the challenge of complexity in health care. *British Medical Journal* **323**, 625–8.

Resnicow, K and Page, S.E. (2008). Embracing chaos and complexity: a quantum change for public health. *American Journal of Public Health* **98**(8), 1382–9.

Rosenfield, P.L. (1992). The potential of transdisciplinary research for sustaining and extending linkages between the health and social sciences. *Social Science and Medicine* **35**, 1343–57. (Quote found on p. 1344)

Van de Ven, A.H. and Johnson, P.E. (2006). Knowledge for theory and practice. *Academy of Management Review* **31**(4), 802–21.

Williams, B. and Imam, I. (eds) (2007). *Systems concepts in evaluation: an expert anthology.* Point Reyes Station, CA: Edge Press.

Wright, R.J., Suglia, S.F., Levy, J. *et al.* (2008). Transdisciplinary research strategies for understanding socially patterned disease: the Asthma Coalition on Community, Environment, and Social Stress (ACCESS) project as a case study. *Ciência & Saúde Coletiva* **13**(6), 1413-8123, doi: 10.1590/S1413-81232008000600008.

CHAPTER 35

Law

MARILYN AVERILL

Black's law dictionary describes 'law' as 'The regime that orders human activities and relations through systematic application of the force of politically organized society, or through social pressure, backed by force, in such a society; the legal system.' (2001: 400). Law thus constitutes an early and unique form of transdisciplinarity grounded in the creation and exercise of political and social power.

Law comprises principles and rules that guide human behavior. The development of law has been inherently interdisciplinary, transdisciplinary, and problem based, reflecting and giving substance to societal norms, and in turn shaping individual and group behavior and the distribution of social costs and benefits. The legal process employs inputs from all aspects of society to define rights, responsibilities, authority, and relationships; establish rules; and resolve disputes. Law also communicates stories and educates the public about social issues.

The type and degree of transdisciplinarity varies across legal agents and institutions. Statutory law, enacted by elected officials, is openly political and most likely to reflect different community values and points of view. Administrative law, the rules enacted by governmental agencies, is constrained by enacted law but may be less directly influenced by politics. Litigation allows lawyers wide discretion in asserting creative claims and presenting different types of evidence, but judicial decisions may be decidedly less inter- and transdisciplinary. While law often interacts with and implements thinking from other academic disciplines, law also directly addresses subjects such as politics, culture, society, and the economy. Both legal education and legal scholarship have sometimes responded to and sometimes ignored the profound links between law and society.

Law is intended to solve problems, and so is heavily instrumental. Lawyers are trained to be problem solvers in a variety of contexts. Law-makers consider how best to use their authority to address societal problems. Lawyers give legal advice to protect a wide variety of client interests. Lawsuits are filed to advance the interests of particular parties. Good lawyers think beyond the law itself to try to understand how legal issues interact with economic, social, environmental, and other domains. Lawyers also must be able to recognize and manage constant uncertainties in what the facts are, how law applies to facts, what arguments will be useful, and how judges, juries, and the public will respond to decisions made and arguments presented. In a sense, all of law is transdisciplinary, as lawyers and law-makers take opportunistic advantage of any available tool to address the problem of concern. Some are more successful at this integrative task than others.

Legal systems differ across cultures. This chapter employs the legal system of the United States (US) as a case study, including the US Constitution, statutes passed by all levels of government, administrative regulations, initiatives or referenda passed by the public, and judicial decisions. Law includes the process by which law is established and enforced, as well as the legal guidelines themselves.

This chapter begins with a historical overview of law and legal thinking in the United States. It then considers the education of both lawyers and non-lawyers about the law. The next section covers various aspects of research and scholarship within and about the law. The final section considers how law evolves, and how it is likely to change in the future.

35.1 Historical development

35.1.1 Western law

Law, as a set of social rules, is as old as human civilization. These rules govern human behavior and protect rights, resolve disputes, maintain public order, regulate commerce, and give predictability to social interactions. Early law was grounded in a set of customs followed within a particular community. According to Cantor (1997), small groups have little need for formalized law. As societies become larger and more complex, they develop more formalized law to control behavior and resolve disputes.

Throughout history, law has crossed and integrated what are now regarded as disciplinary boundaries to treat all aspects of human societies. The *Code of Hammurabi*, *c.*1750 BCE, demonstrates an early form of law and of legal pre-, inter-, and transdisciplinarity. The code addressed citizen's rights and responsibilities, commercial issues, economic damages relating to different crimes, methods for ensuring accurate evidence, elements of intent, how people relate to property, cultural norms, and proper social relations among people living in Babylonian society.

While early civilizations certainly had rules and legal systems, J. M. Kelly (1992) attributes the origins of Western legal thinking to the ancient Greek city-states, maintaining that Greece was the first civilization that engaged in self-reflection about law and its place in society. The Greeks had no system of legal theory, but law was embedded in and provided structure for life in the various city-states. Our sources of Greek law themselves cross disciplines, and include philosophical and historical works, as well as literature, drama, and fragments of Greek laws. Kelly notes that the Greeks tended to view law as a social construction, and that the writings of various Greek philosophers presaged the social contract theory put forth two millennia later by Hobbes.

Western legal systems generally have grown out of either Roman civil law or English common law. Most of Europe has civil law systems, derived from Roman law, relying on laws passed by governments, with little weight given to precedents set by earlier cases. The

United States has a common law legal system, similar to that of Great Britain, from which the United States received much of its early law. A common law system relies heavily on judicial opinions, as well as on statutory law passed by elected officials.

The laws of the United States today may be grounded in history, particularly in English common law, but they also reflect modern concerns and conditions. Friedman maintains that 'Despite a strong dash of history and idiosyncrasy, the strongest ingredient in American law, at any given time, is the present; current emotions, real economic interests, and concrete political groups' (Friedman 2005, p. xi).

35.1.2 Legal thinking in the United States

Politics have always been entwined with the law, but Morton Horwitz maintains that 'The separation between law and politics has always been a central aspiration of American legal thinkers', a concern grounded in 'the fear of tyranny of the majority', (Horwitz 1994, p. 9). Beginning early in the nineteenth century, efforts were made to demonstrate that 'law is a science and that legal reasoning is inherently different from political reasoning' (Horwitz 1994, p. 10). This developed into Classical Legal Thought, a formalistic system of neutral, abstract categories that emphasized the autonomy of the law from other fields of thought, most particularly from politics. This view predominated in the United States before 1900, and its proponents included Christopher Columbus Langdell, the Dean of Harvard Law School from 1870 to1895.

Justice Oliver Wendell Holmes Jr (1841–1935) signaled a major shift in legal thinking. He maintained that 'The life of the law has not been logic: it has been experience. The felt necessities of the time, the prevalent moral and political theories, intuitions of public policy, avowed or unconscious, even the prejudices which judges share with their fellow-men, have had a good deal more to do than the syllogism in determining the rules by which men should be governed' (Holmes 1881, p. 1). Holmes saw law as a social instrument, and maintained that 'Public policy sacrifices the individual to the general good' (Holmes 1881, p. 48).

Horwitz describes how dramatic social and economic changes in the early twentieth-century continued to raise questions about whether law was separate from politics. Progressive Legal Thought, characterized by Roscoe Pound, Dean of Harvard Law School from 1916–36, rejected formalism and began to focus on the relationship between law and social issues. Pound is associated with the concept of sociological jurisprudence, which attempted to bring law more in touch with reality. Justice Benjamin Cardozo (1921), in a self-reflective series of observations about the judicial process, considered how judicial reasoning is tied to philosophy, history, tradition, and sociology. Horwitz describes how the distinct categories and bright-line rules of the nineteenth century gave way to a sense that most phenomena should be described on a continuum rather than as discrete categories.

Progressivism gave way to Legal Realism, which rejected the autonomy of law, and maintained that formalized categories themselves reflected moral values and political interests. The Legal Realists saw law as embedded in other aspects of society, and incorporated ideas from other evolving disciplines: 'The discovery of "frames of reference" in the sociology of knowledge or in the newly emerging field of anthropology marched hand in hand with an insistence that all schemes of categorization and classification embody debatable political and moral premises' (Horwitz 1994, p. 6). Nevertheless, the degree to which the Realists actually incorporated social science into law is unclear. Friedman sees Realism as more an attitude than a philosophy, notable for its skepticism about the validity of rules, and its emphasis on law as an 'instrument of social policy' (Friedman 2005, p. 546).

According to Horwitz (1992), the atrocities of World War II forced a rethinking of law as a social construct, making it more difficult to think of the way things are as the way things should be. The post-war period produced an explosion of new statutory law passed by legislatures and administrative law created by governmental agencies to regulate increasingly complex social and commercial relations. Administrative agencies assumed many regulatory and adjudicative functions. The growing administrative structure enabled specialization, which in turn allowed for more attention to other disciplines. Government professionals had the time and expertise to learn about research in areas such as environmental protection, effects of welfare funding, and other specific fields, which they incorporated into regulatory policy. This led to the Legal Process movement, which focused on 'institutional competence' rather than substance.

Horwitz identifies three interdisciplinary movements in legal thought in recent years, which vary in the disciplines emphasized. The rights-based approach is grounded in civil and human rights. Critical Legal Studies builds on leftist philosophy, rejects formalism, and sees 'law as politics' (Bix 2003, p. 82). The Law and Economics school advocates a utilitarian perspective that weighs costs against benefits.

35.2 Education

Law performs numerous educational roles in society, both deliberate and incidental. Law schools prepare students to become practicing attorneys, lobbyists, politicians, judges, law professors, bureaucrats, or for many other professional roles. Laws are designed to inform the public about acceptable norms for behavior and the consequences for violating those norms. Law sets standards for social interactions and determines rights and responsibilities. The production of law, both statutory and through court decisions, educates the public about issues relating to the laws themselves. Media reports about legal issues expose the public to underlying disputes over science and policy issues, including risk analyses, causal arguments, and economic valuations; values conflicts; possible impacts of policy alternatives; and countless other issues that cross and integrate academic disciplines.

Becoming a lawyer in the United States requires legal education at the graduate level in a professional school of law. Law schools are typically separated organizationally and physically from the rest of the university. This system binds legal education more closely to the profession, and protects it from other university influences.

Law graduates serve in virtually every sector of our society. Many politicians began their careers with a law degree. Friedman discusses the predominance of lawyers in Congress from 1790 on, and notes 'It was not that public office required legal skills; rather, the lawyers were skillful at getting and holding these offices' (Friedman 2005, p. 495).

Corporations either have a legal department or work with external lawyers to handle legal issues from incorporation to litigation to compliance. School districts require legal counsel for advice on legal requirements and litigation risks. Non-governmental organizations from all sides of the political spectrum employ lawyers as general advisors, advocates, or litigators on issues such as civil rights, the environment, or health care. Scientists and writers turn to lawyers for help with intellectual property protection for inventions, software, written products, artistic creations, and other products of human ingenuity. Many lawyers move outside the law to work in other disciplines and professions.

Legal education has not always reflected the intimate connections between law and society. According to Friedman (2005), legal education began with self-education or apprenticeships with practitioners. Law schools began to appear after the Civil War. Cantor attributes the rise of US law schools to the improvements at Harvard Law School brought about by deans Langdell (1870–95) and Pound (1916–36), including a commitment to the case method, and increasing connections to other university departments such as history and philosophy. Early instruction relied heavily on textbooks and memorization, and 'Law schools never conveyed a sense of connection between law and life, or even of the evolution of the common law', although a few schools followed the 'Blackstone Model. . . . Peppering legal education with some notions of government, politics, and ethics' (Friedman 2005, p. 467).

Langdell revolutionized legal education by introducing the case method, in which leading cases, selected to represent particular principles, were presented through the Socratic method, using questions that encourage students to analyze the material that they have read. While a focus on law should have contextualized cases, this was not Langdell's intent. Langdell 'believed that law was a "science"; it had to be studied scientifically, that is, inductively, through primary sources' (Friedman 2005, p. 468). Friedman notes that the focus on law as a science helped to establish that the legal profession requires special training and skills. The case method helped to justify a monopoly of practice for lawyers. Langdell's system was grounded in logic rather than experience. While heavily criticized in its early years, the case method became the predominant way of teaching law in America, and remains so today.

Law itself is a discipline, and lawyers constitute a community with a shared background and experience. Nevertheless, law school curricula illustrate the interdisciplinarity of legal education. Most law schools require basic courses in core subdisciplines of law that typically cross disciplinary lines. Contracts teaches how agreements are made and enforced, how promises are made, and when they will be enforced by law. Contracts may touch on almost any type of agreement produced in any type of social interaction. Constitutional law introduces the core documents of the United States, and addresses how power is distributed among institutions, the rights of citizens, and how the federal government is structured. Criminal law covers the codification of societal norms and values, and the power of law to enforce those rules. Torts describes methods and doctrines that our culture has developed to resolve disputes arising when one party causes harm to another. It includes discussions of rights and responsibilities, of how we draw lines determining liability for our actions, and typically requires translating physical, emotional, and other injuries into economic terms to calculate what liable parties owe to injured parties or to society. Property considers how wealth is distributed across society, including how ownership is established and transferred. It also covers the rights and responsibilities attached to the ownership of property.

Elective classes often are interdisciplinary. Courses such as 'Law and...' may address how the law interacts with almost any field, such as economics, existentialism, literature, religion, sociology, or psychology. Courses often follow the interests of individual professors, so one professor offering an interdisciplinary class can help others to view issues through a similar lens. Other courses seem to integrate several disciplines in order to address complex contemporary problems, as in 'Bioethics, law, and literature' or 'Law and the biology of human nature'. Course descriptions may indicate an inter- or transdisciplinary focus that is not apparent in the title. 'Foundations of natural resources law and policy', for example, 'Examines the legal, historical, political, and intellectual influences that shape natural resources development and conservation'. (These examples come from the University of Colorado Law School.) The degree to which these courses are actually integrative and interdisciplinary, as opposed to multidisciplinary, depends primarily on the professor teaching the course.

Philosophy pervades legal education. Classes explore the epistemological, moral, and other philosophical underpinnings of various legal theories. Ethical issues are constantly discussed, both in the context of professional responsibilities of lawyers and relating to procedural, distributive, and other ethical implications of legal issues.

Law school curricula track changes in society itself. Some courses such as property and criminal law have been offered since the beginnings of legal education but have evolved in response to societal changes. Other topics, such as law and gender, developed as law addressed emerging social issues. Still other classes deal with technological innovations such as biotechnology that introduced new social challenges for regulation and resolution of disputes.

As early as 1913, William Draper Lewis proposed bringing the social sciences into legal education (Tamanaha 2008). A few law schools, notably Columbia and Yale, focused on social science in the 1930s and 1940s. In 1993, George Priest, a Yale law professor, observed that viewing law as an element of social policy made it necessary to expand the study of law into other disciplines: 'Interdisciplinary work—especially in the social sciences and philosophy—consists in the study of the effects of law on the citizenry, the values embedded in the law, and how the public interest may best be achieved' (Priest 1993, p. 1943). Priest notes that law schools usually focus on the most difficult cases, which require more than the application of simple rules and where other disciplines may be useful to reach a sound decision.

Law schools train students in skills as well as knowledge. Legal analysis and persuasion are fundamental, and must be handled effectively both orally and in writing. Negotiating skills are essential to most lawyers, as lawyers must negotiate with clients, opposing counsel, judges, constituencies, experts, and others. Most law schools now provide courses in negotiations, but many lawyers must still develop these skills on the job. In recent years many law schools have also developed clinical programs in areas such as criminal law, natural resources law, or family law to encourage students to move outside of the classroom to apply legal skills and knowledge to practical problems.

Traditionally, most law schools in the United States have not encouraged students to take courses in other departments. Students are taught to 'think like a lawyer', which may require skills and concepts from many fields, but law schools generally control the way other disciplines are taught. Exceptions include joint degree programs that allow students to pursue both a law degree and an advanced degree in another field, such as business, policy, or the social sciences. The result may be more multidisciplinary than interdisciplinary, as students generally study the different fields in separate schools, leaving integration to the individual student.

Interdisciplinary legal education also has its critics. Priest (1993) describes Judge Harry Edwards' objections to what he sees as the growing separation between legal education and practice. In a blog entry, Tamanaha (2008) suggests that other disciplines have already been incorporated into legal opinions, and additional preparation in other disciplines would not be productive for students who intend to practice law rather than enter academia.

Questions remain about the advisable degree, timing, and manner of the interdisciplinary education of lawyers in law school, but educating lawyers does not end with graduation. Law firms begin to re-educate new associates as soon as they walk in the door. As attorneys specialize, they continue to learn aspects of other disciplines that will inform their practice. Attorneys must be able to understand the testimony of experts hired to explain scientific evidence. Litigators employ psychological studies on human behavior and decision making to predict how prospective jurors will respond to arguments at trial. Corporate counsel must be able to use economic information, and to understand how political pressure works at all levels, from the office to the national or even international levels. All lawyers can benefit from knowing something about disciplines outside of the law, but the exact disciplines of use to individual lawyers may not become apparent until after law school.

35.2.1 Educating non-lawyers

Laws must be enforced, and police and other law enforcers operate on the line between law and society. These officials need training in the requirements of and restrictions on their profession. They benefit from an understanding of social dynamics, individual and group psychology, and concepts from many other disciplines in addition to law.

Education about laws and the legal process is essential for an informed citizenry. Citizens need to understand how and why laws are made and how they are enforced in order to appreciate and evaluate the relevance, credibility, and legitimacy of the legal system. Citizens must also be familiar with the content of the law in order to comply with its provisions. Civic education, including teaching about the law, is part of the curriculum of virtually all secondary schools in the United States.

Many colleges offer both undergraduate and graduate courses about various aspects of the law. Political science provides courses about general legal systems, the politics of judging, or on specific areas of law such as constitutional law or international law. Sociology teaches about the links between law and society, and investigates topics such as criminology, criminal justice, and capital punishment. Anthropology addresses the cultural aspects of law, including how law develops within and across cultures, the role of law in social change, and the incorporation of local knowledge. Business schools teach about legal constraints on commercial activities and the general legal culture within which businesses operate. Philosophy addresses the nature of law and concepts of justice, jurisprudence, and theories of punishment.

Media reports about lawsuits, changes in law, police activities, or other issues provide a kind of civic education about the law. They enable and stimulate public debate on important and complex policy issues such as climate change, tobacco regulation, gay rights, police responsibilities, and the right to die. The resulting debate may stimulate political action and/or activism to change laws.

Media reports about law educate the public about many relevant issues that may extend far beyond law itself. Jasanoff argues that 'the legal system [...] has been instrumental in creating and sustaining public understandings of science and technologies in the very processes of "using science" to resolve technical controversies' (Jasanoff 1995, p. xvi). Media reports may influence opinions about the salience, credibility, and legitimacy of expert testimony from various fields. Legal decisions can affect perceptions about what is fair, and about how society should protect and reward individuals and groups in a variety of contexts.

35.3 Research

Research plays a central role in legal practice and scholarship. Students begin learning about legal research methods in their first days at law school. Practicing lawyers and judges must constantly research the law that applies to particular cases. Lawyers in areas other than litigation must research updates in law and other fields that will inform their practice. Law professors and other academics conduct detailed research to produce publishable articles about legal issues. While the primary focus is usually on research about actual law, including constitutions, statutes, administrative regulations, and judicial opinions, practitioners and scholars are increasingly looking to other fields to illuminate legal issues.

Scholarly journals called law reviews publish articles about legal cases and issues. In 1887, Harvard Law School published the first volume of the *Harvard Law Review* (HLR), a student-managed academic journal, a concept that remains an important outlet for legal scholarship. Griswold (1987), in describing the history of the HLR, notes its value to both students who work on the journal and to the legal community. During Griswold's tenure as dean, Harvard developed several additional journals, and Harvard Law School now lists 13 student-run publications, including journals that focus on such issues as human rights, law and gender, law and policy, environmental law, and law and technology. Other schools have followed a similar expansion of law review content. Lexis Academic now provides searches through more than 500 law review publications.

The adversarial method encourages lawyers to consider alternative ways of framing issues and to bring as many areas of expertise as possible to tell the story of a case. Complicated cases involve teams of experts brought in from many fields to testify about particular aspects of the case. Disciplinary experts may become part of the core team building the case or simply provide expert advice or testimony about specific issues. Lawyers must learn enough about the language, concepts, and methods of each expert's discipline in order to integrate these separate pieces into a coherent story that can be understood by the judge and jury, and to be able to defend against cross-examination.

Consider, for example, a case involving pollution from an old mining site in the American West. Who should be held responsible for injuries to people, natural resources, and property? Lawyers on both sides will put together a team of experts to provide evidence to support each side of the case, and to explain away evidence produced by the other side. Historians can tell the story of mining in the area, including the companies involved, procedures and materials employed, how the area developed and how it was affected by mining. Hydrologists, fluvial geomorphologists, water chemists, and toxicologists can talk about water quality, how materials have migrated, and how pollution becomes available to humans and the environment. Wildlife and fish biologists and botanists can testify as to effects on living things. Doctors and medical researchers evaluate possible impacts on human health and safety. Engineers evaluate the safety of existing structures and impoundments. Anthropologists study how local indigenous people live, and how their cultures have been affected by mining. Economists monetize damages, both based on actual injuries and non-use values, such as aesthetics.

Judges must be able to understand complex science presented as evidence, but judges are divided about consulting other disciplines. Some welcome information about relevant physical or social science. Others, such as Justice Antonin Scalia, generally restrict their use of other disciplines to arguments made by the parties: '[T]he Court looks to scientific and sociological studies, picking and choosing those that support its position. It never explains why those particular studies are methodologically sound; none was ever entered into evidence or tested in an adversarial proceeding [...] Given the nuances of scientific methodology and conflicting views, courts—which can only consider the limited evidence on the record before them—are ill-equipped to determine which view of science is the right one' (Ancheta 2006, p. 157 (quoting *Roper v. Simmons*, 125 S. Ct. 1183, 1195 (2005) (Scalia, J., dissenting))).

Advances in science and technology continue to shape the law. New information and techniques relating to fingerprinting, DNA analysis, and other identification techniques are routinely used by courts in linking individuals to their actions. These and other advances have required courts to consider complex new problems, such as the ownership of elements of the human genome, the risks that new products and processes present to human health and the environment, the nature and extent to threats to species, the factors affecting global climate, and other issues. Science and technology also can create legal dilemmas. Ancheta warns about advances in areas such as genetics and asks 'as scientific advancements and revisions to scientific theories portend new types of classifications and potential forms of discrimination, what are the appropriate judicial responses?' (Ancheta 2006, p. 152).

Legislators and judges are rarely trained to understand complex scientific issues. Congress typically passes general laws, and then leaves the tasks of regulating and enforcing to administrative agencies. The agencies typically employ scientific experts to propose and assess policy alternatives. Judges sometimes employ special masters to explain complex scientific material to the court. Reliance on such experts can cause problems when their advice is used to the exclusion of others.

Social sciences have been important throughout the history of law, long before they developed into formalized disciplines. Psychological issues such as intent; cultural norms in the community; the economic costs, benefits, and just compensation of various crimes are ancient concerns. In *Brown v. Board of Education*, the 1954 groundbreaking case striking down desegregation in schools, the Supreme Court relied heavily on social science evidence.

John Monahan and Laurens Walker (2002) describe four ways in which social science is incorporated into modern law. Social science can be used to determine case facts, to study some particular aspect of the case in dispute. This might include social surveys of recreational use of a natural resource at issue, or economic analyses of damages. Social science also can contribute to the law-making process, as when the Supreme Court used research on the general impacts of school segregation and desegregation as a reason for changing the law. Social science can provide context for legal analysis and shed light on the facts in the particular case. Information about characteristics likely to be associated with violent behavior might help in evaluations about individual defendants. Such studies may, however, violate constitutional rights, as when racial profiling is used to identify suspects. Social science can also assist in planning aspects of case strategy, such as jury selection.

Law also incorporates the humanities. Philosophy and religion have close ties to law. Beliefs about rights and responsibilities have been codified into laws regarding criminal conduct and civil liabilities. Logic guides thinking in the law. Epistemology forms part of legal standards for causation and the recognition of expertise. Stories about law have been told in literature and theater at least as far back as the ancient Greeks, and stories from literature are often quoted by judges, lawyers, and law-makers.

Judge Richard Posner (1993) ascribes what he sees as a decline in doctrinal legal scholarship since 1965 to the increase in interdisciplinarity and the erosion of political consensus. He also believes the shift in scholarship has moved academia away from legal practice. According to Posner, doctrinal scholarship, which maintained that legal reasoning was a unique field of study, and that legal analysis was objective, could only survive as long as it remained self-contained and apart from the rest of society, and as long as there was consensus among legal scholars. Consensus was social, a product of a tight disciplinary community, rather than epistemic, and doctrinalism could not survive when examined through the lenses of other disciplines. In addition, diverging political views have produced more contested norms, which make objectivity of the law less credible. According to Posner, the US judiciary has moved to the right while law faculties have moved to the left, increasing the separation between academia and practice.

Writing again in 2002, Judge Posner maintains that legal scholarship has become less accessible to practicing attorneys as it has become more interdisciplinary. In becoming less isolated, and more a part of the larger university community, he believes legal scholarship has also shifted its primary audiences from other lawyers to other academics, and that the introduction of less familiar discourses, theories, and methods has made legal scholarship less accessible to those in legal practice. Posner (1993) also notes that interdisciplinarity can lead to shoddy legal scholarship when law professors try to incorporate new disciplines without training in the field. Neil Duxbury, while recognizing the important creative contributions of some interdisciplinary scholars, also refers to the 'astonishing amount of cross-disciplinary chutzpah in the American law reviews' (Duxbury 2003, p. 457).

35.4 Law of the future

Law constantly evolves and provides a kind of self-executing way of responding to new ideas from all disciplines. As the needs and values of society shift over time, old rules are discarded and new ones are adopted. Emerging societal problems, innovative technologies, and other novelties constantly challenge legal theory and practice. The political process determines how, when, and why enacted laws should be changed. Processes of change within the common law are less clear, but over the years courts have managed to stretch and modify decisions to accommodate factual situations that could not have been contemplated when precedents were adopted: '[T]he genius of the common law in the United States has been its capacity to evolve over time—case by case and issue by issue—as the courts apply basic legal principles developed over the past to resolve the challenges posed by new situations' (O'Connor 2004, p. 272).

Law shapes society and continues to be shaped by it. Jasanoff employs the concept of *co-production* as 'shorthand for the proposition that the ways in which we know and represent the world (both nature and society) are inseparable from the ways in which we choose to live in it' (Jasanoff 2004a, p. 2). She describes the importance of knowledge-making, and how it 'is incorporated into practices of state-making, or of governance more broadly, and, in reverse, how practices of governance influence the making and use of knowledge' (Jasanoff 2004a, p. 3). Cantor would add that 'The legal profession and the culture of the common law binds the United States together and provides for the framework of civil society that allows for economic and technological progress' (Cantor 1997, p. 364).

Globalization is changing the way the United States thinks about both domestic and international law. Increasingly frequent and complex transboundary transactions and problems are forcing legislators, judges, academics, and practitioners to think about factors beyond the United States' national borders that affect the making, interpretation, and enforcement of laws within the United States. At the same time, US legal policy affects the way the rest of the world views issues, and the way the world reacts to the United States. Complex international problems such as climate change and poverty reduction will require holistic thinking beyond legal remedies to how law fits in with ethics, economics, religion, cultural values, social organization, and notions of national sovereignty.

Justice O'Connor describes the Supreme Court's shift from an early emphasis on property rights to a concern for individual rights. She predicts that 'Having protected, indeed exalted, the rights of the individual, we will be challenged by a world that is increasingly interdependent and that demands that we take part in a global community' (O'Connor 2004, p. 269). Participating in a global legal community will require additional emphasis on disciplines such as international relations, geography, anthropology, and sociology, as the United States and its legal system interact more intimately with other communities and legal cultures around the world.

Globalization also will force the United States to reconcile its views of the law with those of other nations, which often will require understanding unfamiliar customs and ways of viewing the world. Cantor discusses the increasing need for the United States to interact with other legal systems grounded in different traditions. He focuses primarily on increasing interactions with civil law in countries with legal systems derived from Roman law. Events of the early twenty-first century indicate that even less familiar legal traditions, such as Islamic Sharia law, will provide additional challenges for the United States and its legal system.

Patterns of agents and their activities are shifting, with new domestic and international players taking on unconventional roles in law. Intergovernmental organizations, non-governmental organizations, states, regional coalitions, industry, and others are seeking increasing influence over legal issues. Each one brings new ways of thinking about the law to light, forcing greater consideration of transdisciplinary issues such as ethical obligations among nations, human rights, sustainable development, and the interconnectedness of all elements of human societies. States and even cities are often taking on difficult legal issues, sometimes forcing action at the national level, and are themselves becoming players in international negotiations.

As science and technology continue to accelerate, the law of the future will certainly need to deal with issues that were unthinkable until recently, such as human cloning. Other innovations will be so new that they have not yet been considered even in science fiction. Laws will be needed to control undesirable research, to restrict who has access to and may use dangerous technologies, and to avoid unintended consequences. Law will also need to adapt to new technologies that change our views of what constitutes property (including human tissues and genetic information), how human relationships are defined, and how the costs and benefits of new technologies should be distributed across the United States and the world.

New political, economic, social, and environmental challenges will force law to adapt to new situations. Law will continue to use other disciplines opportunistically, to inform whatever problem is at hand or to pull scholarship in new and more creative directions. Advocates will constantly push law in new directions. Scholars will find new ways to study and teach the law. Research from other disciplines will continue to illuminate the way law operates in society, and will provide support for legislative, administrative, and judicial decisions, and for legal arguments in support of a multitude of interests. The only sure thing that can be said about the future of law in the United States is that it will be different from the law of today, but will be firmly grounded both in US legal history and in contemporary societal norms, and will be responsive to contemporary challenges.

References

Ancheta, A.N. (2006). *Scientific evidence and equal protection of the law*. New Brunswick, NJ: Rutgers University Press.

Averill, M. (2007). Climate litigation: shaping public policy and stimulating debate. In: S.C. Moser and L. Dilling (eds) *Creating a climate for change: communicating climate change and facilitating social change*. Cambridge: Cambridge University Press.

Bellow, G. and Minow, M. (1996). *Law stories.* Ann Arbor, MI: University of Michigan Press.

Bix, B.H. (2003). Law as an autonomous discipline. In: P. Cane and M. Tushnet (eds) *The Oxford handbook of legal studies.* Oxford: Oxford University Press.

Cantor, N.F. (1997). *Imagining the law.* New York: Harper Collins.

Cardozo, B.N. (1921). *The nature of the judicial process.* New Haven, CT: Yale University Press.

Code of Hammurabi, transl. L.W. King (1910), ed. by R. Hooker. Available at: <http://www.wsu.edu/~dee/MESO/CODE.HTM> (accessed 12 May 2008).

Duxbury, N. (2003). A century of legal studies. In: P. Cane and M. Tushnet (eds) *The Oxford handbook of legal studies.* Oxford: Oxford University Press.

Environmental Protection Agency (2005). Guidelines for carcinogen risk assessment, EPA/630/P-03/001F. Available at: <http://cfpub.epa.gov/ncea/raf/recordisplay.cfm?deid=116283> (accessed 3 April 2009).

Fiss, O. (2003). *The law as it could be.* New York: New York University Press.

Friedman, L.M. (2005). *A history of American law*, 3rd edn. New York: Simon and Shuster.

Garner, B.A. (ed.). (2001). *Black's law dictionary*, 2nd pocket edn. St Paul, MN: West Group.

Griswold, I.M. (1987). The Harvard law review—glimpses of its history as seen by an aficionado. *Harvard Law Review Centennial Album* (1), 11–12. Available at: <http://www.harvardlawreview.org/Centennial.shtml> (accessed 21 August 2009).

Harris, P. (2007). *An introduction to law.* Cambridge: Cambridge University Press.

Holmes, O.W. Jr (1881). *The common law.* New York: Dover Publications.

Horwitz, M. (1992). *The transformation of American law, 1780–1860.* Cambridge, MA: Haward University Press.

Horwitz, M. (1994). *The transformation of American law, 1870–1960.* Oxford: Oxford University Press.

Levi, E.H. (1949). *An introduction to legal reasoning.* Chicago: Chicago University Press.

Jasanoff, S. (1995). *Science at the bar: law, science and technology in America.* Cambridge, MA: Harvard University Press.

Jasanoff, S. (2004a). *States of knowledge: the co-production of science and social order.* London: Routledge.

Jasanoff, S. (2004b). Law's knowledge: science for justice in legal settings. *American Journal of Public Health* **95**(S1), S49–S58.

Kelly, J.M. (1992). *A short history of Western legal thought.* Oxford: Oxford University Press.

Monahan, J. and Walker, L. (2002). *Social science in law: cases and materials*, 6th edn. New York: Foundation Press.

O'Connor, S.D. (2004). *The majesty of the law.* New York: Random House.

Office of Management and Budget (2007). *Final bulletin for agency good guidance practices*, no. 07-02. Available at: <http://www.whitehouse.gov/omb/memoranda/fy2007/m07-07.pdf> (accessed 11 May 2008).

Posner, R.A. (1993). Legal scholarship today. *Stanford Law Review* **45**, 1647–58.

Posner, R.A. (2002). Legal scholarship today. *Harvard Law Review* **115**, 1314–26.

Priest, G.L. (1993). The growth of interdisciplinary research and the industrial structure of the production of legal ideas: a reply to judge Edwards. *Michigan Law Review* **91**(8), 1929–43.

Tamanaha, B. (2008). *Why the interdisciplinary movement in legal academia might be a bad idea (for most law schools)* [blog]. Available at: <http://balkin.blogspot.com/2008/01/why-interdisciplinary-movement-in-legal.html>.

Turver, S.P. and Factor, R.A. (1994). *Max Weber: the lawyer as social thinker*. London: Routledge.

Tushnet, M. (1996). Interdisciplinary legal scholarship: the case of history-in-law. *Chicago-Kent Law Review* **71**, 909–35.

Twining, W., Farnsworth, W., Vogenauer, S., and Teson, F. (2003). The role of academics in the legal system. In: P. Cane and M. Tushnet (eds) *The Oxford handbook of legal studies*. Oxford: Oxford University Press.

CHAPTER 36

Risk

SVEN OVE HANSSON

The study of risk is multidisciplinary in at least two ways. First, when talking about risks we refer to undesirable events that may or may not occur. A large variety of events answer that description, and therefore a wide array of disciplines is involved in their study. The modern discipline of risk analysis has its origin in work in the 1960s and 1970s that had a strong focus on chemicals and nuclear energy. From its beginnings, risk analysis drew on competence in areas such as toxicology, epidemiology, radiation biology, and nuclear engineering. Today, risk analysis has many other applications, including air pollution (Pandey and Nathwani 2003), radioactive waste repositories (Cohen 2003), airbag regulation (Thompson *et al.* 2002), road construction (Usher 1985), and efforts to detect asteroids or comets that could strike the earth (Gerrard 2000), to mention just a few examples. Many if not most scientific disciplines provide risk analysts with specialized the competence needed in the study of one or other type of risk—medical specialties are needed in the study of risks from diseases, engineering specialties in studies of technological failures, etc.

Secondly, several disciplines have supplied overarching approaches to risk. Statistics, epidemiology, economics, psychology, anthropology, and sociology are among the disciplines that have developed general approaches to risk, intended to be applicable to risks of different kinds. Some of these approaches are interdisciplinary, or give rise to cooperation between disciplines. Others may be thought of as transdisciplinary.

In addition to being interdisciplinary, risk studies have another feature that will be of interdisciplinary concern: they are strongly connected with normative issues. Studies of risk are often motivated and strongly influenced by debates on how societies should deal with risks. Issues of *acceptability* have an important role in many scholarly and scientific studies of risk; these issues are clearly normative. Even more importantly, the *measurement* of risk is a pervasive issue in modern risk studies. Although there is no necessary connection between measurement and normativity, the measurement of risk is in practice strongly associated with normative issues. The reason is that measures of risk refer to its severity, and severity is clearly a normatively laden concept. Once a risk analyst has concluded that a certain risk is larger or more severe than another, then this is easily taken to imply that the 'larger' risk is also the one that it is more important to avoid.

The normative nature of risk measurement has often been overlooked, and it is surprisingly common for risk analysts to express themselves as if the severity of risks could be determined in an objective and value-free way. But because of the (often neglected) normativity of central concepts in risk analysis, this discipline provides an interesting test case of how normative issues can be treated in interdisciplinary research.

36.1 A brief history

Risks have been studied by probability theorists since the beginnings of that subject in the seventeenth century. The growing insurance industry was quick to adopt probabilistic analysis of risk in the form of actuarial science. Probabilistic models of risk were also included in modern mathematical economics at an early stage of its development (Domar and Musgrave 1944). However, modern interdisciplinary risk analysis did not grow out of insurance mathematics or economic theory. Instead, its origins can be found in attempts made in the 1960s and 1970s to provide policy guidance—or policy justification—regarding complex issues involving on the one hand risks to human health and the environment and on the other hand major economic values. Much of the early discussion centered around energy production and environmental pollution, two topics that were from the beginning highly interdisciplinary.

The short history of modern risk analysis can be summarized in terms of how a series of overarching, interdisciplinary approaches to risk were introduced and put to use in order to deal with such policy issues. From the beginning, *statistics* has had a strong influence on risk analysis. Statistical expectation values were calculated for different risks, and these values used as measures of the severity of risk. Many of the earliest studies in the field aimed at determining a level of 'acceptable risk', usually in the form of an upper limit on the statistically expected number of fatalities. A common procedure was to compare new technological risks with risks accepted in everyday life. It was often assumed that a technological risk should be accepted if it was smaller than some natural risk that was already accepted.

It was soon realized that this approach is severely oversimplified (Hansson 2003b). The reason why we accept a risk is in most cases that by doing so we can obtain some benefit. As one example, we are prepared to take much larger risks in a medical treatment that has some chance of curing a serious disease than in a treatment aiming at relief from a minor headache. Risks cannot be judged alone; they must always be considered in relation to associated potential benefits.

This led to the introduction of *economics* into the core of risk analysis. In *risk–benefit analysis*, both the risks and the benefits of a technology are quantified to make them comparable. Often, a monetary value is assigned to human life in order to make fatal risks computationally comparable with economic benefits. It is important to distinguish this methodology, using hypothetical valuations of lives and other valuables as a guide to policy decisions, from standard economic analysis in which the monetary sums referred to represent actual payments on markets.

However, neither statisticians' estimates of risks nor the economists' more developed risk–benefit analyses have gained public acceptance. Members of the public are often much more worried by risks that are small, according to the experts, than by risks that experts consider large. This has often been interpreted as evidence that the public has irrational attitudes toward risk.

It was against this background that the *behavioral sciences* gained a prominent role in risk analysis, initially in the form of studies of *risk perception* that were much in vogue in the early 1980s. The ordering of risks obtained from questionnaires was said to measure 'subjective risk' and was compared to the expected number of deaths, which was called 'objective risk'. The difference was conceived as a sign of irrationality or misperception, and various explanations for such irrationalities were sought.

The next behavioral approach was *risk communication*, usually in the form of practically oriented studies aimed at providing lay people with information that would make them see risks in a certain way. Typically, risk communication is considered successful if it has made people adjust their 'subjective risk' to fit with the 'objective risk'. However, the most important conclusion from risk communication studies was that it is difficult to change public opinion on risk through communication. The public did not seem to trust the sources of information through which risk communicators operated. This gave rise to studies of *trust* that developed into a third major behavioral approach in risk studies. However, the focus on trust did not provide any easy means to influence public views on risk. There does not seem to be any recipe that corporations or public agencies can follow to gain the confidence of the public. The major conclusion from these studies is that trust is difficult to gain but easily lost.

Statistics, economics, and behavioral sciences are still the three dominant overarching approaches to risk. Other disciplines have contributed competing perspectives, including anthropology (Douglas and Wildavsky 1982), several branches of sociology (Beck 1992, Pidgeon *et al.* 2003), political science (Franklin 1998), and philosophy, including ethics (Hansson 2004b, 2007b), but none of them has been as influential as these three.

36.2 Statistics

The most common measure of risk is the statistical *expectation value* of undesired events. In other words, risk is identified with a measure that is obtained by multiplying the probability of an unwanted event with a measure of its disvalue (negative value). If only risks of death are considered (which is a surprisingly common restriction), this means that risk is identified with the statistically expected number of deaths caused by a possible event or class of possible events. Hence, if 200 deep-sea divers perform an operation in which the individual risk of death is 0.1% for each individual, then the expected number of fatalities from this activity is $0.001 \times 200 = 0.2$. Expectation values have the important property of being additive. Suppose that a certain activity is associated with a 1% probability of an accident that will kill five people, and also with a 2% probability of another type of accident that will kill one person. Then the total expectation value is $(0.01 \times 5) + (0.02 \times 1) = 0.07$ deaths.

Although expectation values have been calculated since the seventeenth century, the use of the term 'risk' to denote them is relatively new. It was introduced into risk analysis in the influential Reactor Safety Study (WASH-1400, the Rasmussen report) from 1975 (Rechard 1999, p. 776). Today it is the most widely used technical meaning of 'risk'. Some authors even claim that it is the only rational definition of risk (Cohen 2003, p. 909).

There is a fairly strong argument in favor of maximizing expected utility: when the same type of event is repeated many times, this decision method maximizes the outcome in the long run. Suppose, for instance, that the expected number of deaths in traffic accidents in a region will be 300 per year if seat belts are compulsory and 400 per year if they are optional. If these calculations are correct, then about 100 more people per year will actually be killed in the latter case than in the former. We know, when choosing one of these options, whether it will lead to fewer or more deaths than the other. If we aim at reducing the number of traffic casualties, then this can, due to the law of large numbers, safely be achieved by maximizing the expected utility (i.e. minimizing the expected number of deaths).

However, this argument is not valid for case-by-case decisions on unique or very rare events. Suppose, for instance, that we have a choice between a probability of 0.001 of an event that will kill 50 people and the probability of 0.1 of an event that will kill one person. Here, random effects will not be leveled out as in the seat belt case. In other words, we do not know, when choosing an option, whether or not it will lead to fewer deaths than the other. In such a case, taken in isolation, there is no compelling reason to maximize expected utility. In particular, extreme negative effects, such as a nuclear war or a major ecological threat to human life, cannot be leveled out in the way required for this justification of the maximization of expected utility. In spite of this, the US military has used secret utility assignments to accidental nuclear strike, and to failure to respond to a nuclear attack, as a basis for the construction of command and control devices (Paté-Cornell and Neu 1985).

Even when the leveling-out argument is valid, the maximization of expected utility is a problematic decision rule that can give rise to counterintuitive conclusions. In particular, it does not take distributional concerns into account, and it therefore requires that we expose some individuals to high probabilities of damage or death whenever this is the best way to minimize the statistically expected total damage. This can be illustrated with the following example (Hansson 1993). In an acute situation we have to choose between two ways to repair a serious gas leakage in the machine-room of a chemical factory. One option is to send in the repairman immediately (there is only one person at hand who is competent to do the job). He will then run a risk of 0.9 of dying due to a gas explosion immediately after he has performed the necessary technical operations. The other option is to immediately let out gas into the environment. In that case, the repairman will run no particular risk, but each of 10,000 persons in the immediate vicinity of the plant runs a risk of 0.001 of being killed by the toxic effects of the gas. The rule of maximizing expected utility requires that we send in the repairman to die. This is also a fairly safe way to minimize the number of actual deaths. However, it is not clear that this is the only rational response. Rational decision-makers may refrain from maximizing expected utility (minimizing expected damage) in order to avoid what would be unfair to single individuals and infringe their rights.

36.3 Economics

Risk–benefit analysis (RBA) is the major contribution of economics to risk analysis. It is a collection of decision-aiding techniques that have in common the numerical weighing of advantages against disadvantages. In a typical RBA, two or more options in a public decision are compared with each other by careful calculation of the values of their respective consequences. These consequences can be different in nature, e.g. economic costs or gains, risks of disease and death, environmental damage, etc. In the final analysis, all such consequences are assigned a monetary value, and the option with the highest value of benefits minus costs is recommended or chosen.

RBA is built on a very sound fundamental principle: advantages should be weighed against disadvantages, costs against benefits. However, the precise way in which this is done may be more controversial, and it has indeed been repeatedly criticized, not least by philosophers (Shrader-Frechette 1985a,b; Anderson 1988; Sagoff 1988; Hansson 2007a). Two problems in this methodology are particularly important in the present context.

The first and most commonly discussed of these problems is that negative outcomes can be of many different kinds, not all of which can easily be measured in monetary terms. Human death, human disease, and environmental damage are not easily made commensurate. There is no obvious answer to the question of how many cases of juvenile diabetes correspond to one death, or what amount of human suffering or death corresponds to the extinction of an antelope species. Risk analysts have often avoided this problem by restricting their calculations to human mortality. Economists performing RBA go beyond that oversimplification. By putting a 'price' on all negative outcomes, they can take them all into account and calculate the overall utilities of various options.

When price tags are put on risks and negative outcomes, they can be weighed against costs and economic benefits. It then turns out that there are large differences between different policy areas in how much we pay for risk abatement, typically expressed as differences in costs per expected life saved. Risk analysts, and in particular risk–benefit analysts, tend to be dissatisfied with these differences. They want us to make our decisions according to a uniform 'price' that covers all policy areas. The idea is to calculate the risks in all these different sectors, and then allocate resources for abatement in a way that minimizes the total risks. This is often claimed to be the only rational way to decide on risks (Viscusi 2000, p. 855).

However, risks to human health and the environment are very different phenomena from those commonly treated in economics. They do not have a market price, and the values assigned to them in a RBA should not be taken as prices in the ordinary sense of the word. Hence, if the value of a human life is set at $30,000,000, this does not mean that someone has a right to buy a human person, or the right to kill, at that price. The value is for calculation purposes only. It can be interpreted either as a report of how much we are currently prepared to pay to save a human life (in the relevant context) or as a normative statement about what amount of money we ought to be willing to pay for that purpose. The limited applicability of such values has not always been sufficiently emphasized by practitioners of RBA. Not surprisingly, then, critics have argued that RBA should not be carried out since negative outcomes such as the loss of human lives cannot be measured in terms of money (Ashby 1980; Baram 1981; Kelman 1981). Stuart Hampshire (1972, p. 9) has warned that the habits of mind engendered by this type of impersonal calculations may lead to 'a coarseness and grossness of moral feeling, a blunting of sensibility, and a suppression of individual discrimination and gentleness'. This controversy can be interpreted as a conflict about the extent to which economics can be used as an overarching discipline in studies of risk.

It should be emphasized that this is not essentially an argument about the use of monetary units to measure value. The basic problem is that in RBA, multidimensional decision problems are reduced to unidimensional ones. The common way to do this, technically, is to assign monetary values to all types of consequences, even those that are incommensurable with money. Therefore the problem of incommensurability appears as a problem of assigning sums of money to units in the analysis that do not have a monetary price. However, if we removed money from the analysis we would still have to deal with comparisons between deaths, diseases, and environmental damage. The basic, underlying, problem is not limited to valuation in monetary terms. The fundamental problem is that we need to comparatively evaluate phenomena that we conceive as incomparable. Such comparisons are unavoidable components of many of the decisions made in different social sectors. The problem does not come with RBA, but it is more clearly exhibited when a RBA is performed in order to guide the decision. It may well be seen as a virtue of RBA that it brings the problem to light.

The second major problem with RBA is its disregard for persons. In the calculations made for the analysis, the risks to which different persons are exposed are all added up. It makes no difference for the analysis who is exposed to the risk or how it is distributed. Benefits are added up in the same way, i.e. with no consideration of who receives the benefits or how they are distributed. Finally, the sum of benefits is compared with the sum of risks in order to determine whether the total effect is positive or negative.

This methodology is based on the assumption that a disadvantage to one person can always be compensated by an equally sized advantage to another person. This is a feature that RBA shares with classical utilitarianism, in which individuals have no other role than as carriers of utilities and disutilities, the values of which are independent of whom they are carried by. Such a framework excludes many types of moral considerations that we may apply in contexts of risk, such as justice and individual rights. The fact that a certain loss for Ms Black is smaller than a certain gain for Mr White does not suffice to make it allowable for Mr White, or anyone else, to perform an action that leads to this particular combination of a loss for Ms Black and a gain for Mr White. Ms Black may have a right that forbids such enforced exchanges (Hansson 2004a).

Risk analysts have often been surprisingly insensitive to issues of justice and rights. The general assumption seems to be that if a risk–benefit calculation shows that the total benefits outweigh some risk, then that risk should be accepted even if the persons exposed to the risk are not the same as those who receive the benefit. This has been particularly clear in criticisms of so-called NIMBY (not in my backyard) reactions. For example, potential neighbors of a polluting facility may refuse to sacrifice their own interests by consenting to a siting that is beneficial to a wider community but potentially harmful to themselves. NIMBY reactions have often been described as irrational. This accusation only holds under the rather implausible assumption that individual rationality requires subservience to collective interests (Luloff *et al.* 1998; Hermansson 2007).

36.4 Behavioral science

As was mentioned, risk perception is a key concept in psychological studies of risk. Unfortunately, the very notion of risk perception is in itself problematic. The notion of perception refers to how our nervous system, using signals from our sense organs, reacts to objective physical phenomena. Hence, in studies of the perception of colors, we compare our experiences of colors to physical characteristics of light such as wavelength and intensity. Similarly, in studies of auditory perception we compare auditory experiences with measurable physical properties of sound, and in studies of taste and olfaction chemical properties of substances have the corresponding role as an objective basis of the analysis.

In studies of risk perception, statistically expected numbers of deaths are treated as objective entities to which subjective 'risk perceptions' could be compared. If a person considers risk A to be worse than risk B, whereas risk B has a higher expected disutility than risk A, then this is often seen as a case of an irrational deviation of subjective risk from objective risk. This approach exemplifies the problem with measurement of risk mentioned before. The expected (dis)utility measure is not an objective or uniquely rational measure of the severity of risk, for several reasons already referred to. Different people may legitimately choose to value risks differently because they value possible outcomes differently, because they differ in their degrees of cautiousness, or because they have different views on distributive issues and individual rights. Therefore, there is no single standard against which we can measure the risk assessments of different people to determine who is more or less rational. Although the term 'risk perception' is well established, it is an unfortunate terminological choice that continues to cause confusion. The term 'risk attitude' would have been preferable.

This problem for risk perception theory illustrates a more general problem for normative decision theory. In principle, a normative decision theory is an account of how decisions *should* be made. This 'should' can be interpreted in different ways, but in practice there has been almost complete agreement among decision theorists that it refers to the prerequisites of rational decision making. In other words, normative decision theory is a theory about how decisions should be made in order to be rational (and not, for instance, about how they should be made in order to satisfy moral criteria). As a consequence of this, important normative issues, such as those concerning justice and individual rights, have often been neglected in mainstream decision theory and decision analysis.

36.5 Failures of multidisciplinarity

In summary, three major attempts have been made to provide an overarching or transdisciplinary 'umbrella' for the many disciplines that deal with risk in different ways: probabilistic risk analysis (from statistics), RBA (from economics), and risk perception (from behavioral science). Although each of these has provided valuable insights, none has given the guidance to decisions about risk that they were intended to provide. It seems difficult to avoid the conclusion that they have all failed as unified decision-guiding approaches to risk. There are at least two important reasons for this failure.

The first reason is that in all these approaches, ethical issues have been excluded. The only norms that have an explicit role are norms of rationality. In contrast, when the general public is confronted with issues of risk, ethical norms have a large role. Members of the public who have protested against siting of polluting or otherwise dangerous plants in their vicinity have seldom been impressed by total calculations of risks and benefits. Instead they have talked in terms of moral rights. They consider themselves to have a right not to be exposed to risks that they had no chance to influence. Similarly, workers protesting against dangers in their workplace have referred to individual rights and to social justice. Generally speaking, potentially risk-imposed people tend to reject the idea that they should accept risks for the sake of benefits befalling others. Proponents of probabilistic risk analysis and RBA have sometimes dismissed such moral considerations as irrational, and proponents of risk perception theory have treated them as errors in risk perception. These are of course untenable interpretations. A workable theory of risk cannot dismiss ethical considerations. Instead it has to recognize them and provide means to deal with them (Hansson 2003a).

The second reason why the three overarching approaches have failed is that they attempt to give experts too large a role in risk assessment. They are all based on the assumption that risk evaluations can in their entirety be performed as expert assessments, divorced from the political decision-making process. Some reflection will show that it is in practice impossible to separate risk issues from general political issues. Almost all social decisions include endeavors to avoid undesirable events. Some of these events can be quantified in a meaningful way, for example health risks and economic risks. Others are virtually impossible to quantify, like risks of cultural impoverishment, social isolation, and increased tensions between social strata. There is a tendency to leave the 'quantifiable' risks to experts and treat the others as more 'political'. However, even if a risk issue is accessible to quantification it may be influenced by various value issues that cannot, in a democratic society, be left to experts. Therefore the role of experts should be limited in issues of risk just as in other social and political issues. Experts should provide decision makers (in particular the public as the ultimate decision makers) with facts and information, but it is not the role of experts to dictate decisions or to be judges in normative matters. Risk decisions do not only require transdisciplinarity in the traditional sense; they also require political participation.

It should not take much reflection to realize that the notion of 'risk', as we commonly use it, is both value-laden and fact-laden. It is value-laden since risk means the possibility of something undesirable, and undesirability is a value concept. At the same time it is fact-laden since it concerns actual tendencies for certain types of events to occur. The statement that you risk losing your leg if you tread on a landmine has both a factual component (landmines tend to dismember people who tread on them) and a value component (it is undesirable to lose your leg). The propensity of these devices to mutilate is no more a subjective construct than these devices themselves.

There are discussants who deny this double nature of risk. Some maintain that risk is 'objective', devoid of any subjective component. Others claim that risk is plainly a 'subjective' phenomenon or a mere 'social construction', not concerned with matters of fact. These are both attempts to rid a complex concept of its complexity. Neither is appropriate. Any notion of risk that connects in a reasonable way to the conditions of human life will have to admit the double nature or risk—thus exhibiting interdisciplinarity in another sense—and not try to make risk either value-free or fact-free. Unfortunately, influential theories of risk often fail in this respect.

Although the interdisciplinarity of risk studies has been essential in solving many scientific problems, hopes that interdisciplinarity would solve the normative issues connected with risk have been in vain. In order to solve the normative issues of risk they must be dealt with more explicitly than has usually been done. Furthermore, we need to realize that interdisciplinary science cannot solve social and political issues. However, something important can be learnt from interdisciplinary and transdisciplinary science. The types of communication and cooperation between different academic disciplines that have been developed in inter- and transdisciplinary contexts have also to take place between these academic disciplines on one hand and more practically oriented, value-based discourses on the other. In this sense, transdisciplinarity needs to be transcended.

For that purpose we also need a social and political discourse that goes beyond science.

References

Anderson, E. (1988). Values, risks and market norms. *Philosophy and Public Affairs* **17**, 54–65.

Ashby, E. (1980). What price the Furbish lousewort? *Environmental Science and Technology* **14**, 1176–81.

Baram, M.S. (1981). The use of cost–benefit analysis in regulatory decision-making is proving harmful to public health. *Annals of the New York Academy of Sciences* **363**, 123–8.

Beck, U. (1992). *Risk society: towards a new modernity.* London: Sage.

Cohen, B.L. (2003). Probabilistic risk analysis for a high-level radioactive waste repository. *Risk Analysis* **23**, 909–15.

Domar, E.D. and Musgrave, R.A. (1944). Proportional income taxation and risk-taking. *Quarterly Journal of Economics* **68**, 388–422.

Douglas, M. and Wildavsky, A. (1982). *Risk and culture: an essay on the selection of technological and environmental dangers.* Berkeley: University of California Press.

Franklin, J. (1998). *The politics of risk society.* Cambridge: Polity.

Gerrard, M.B. (2000). Risks of hazardous waste sites versus asteroids and comet impacts: accounting for the discrepancies in US resource allocation. *Risk Analysis* **20**, 895–904.

Hampshire, S. (1972). *Morality and pessimism.* Cambridge: Cambridge University Press.

Hansson, S.O. (1993). The false promises of risk analysis. *Ratio* **6**, 16–26.

Hansson, S.O. (2003a). Ethical criteria of risk acceptance. *Erkenntnis* **59**, 291–309.

Hansson, S.O. (2003b). Are natural risks less dangerous than technological risks? *Philosophia Naturalis* **40**, 43–54.

Hansson, S.O. (2004a). Weighing risks and benefits. *Topoi* **23**, 145–52.

Hansson, S.O. (2004b). Philosophical perspectives on risk. *Techne* **8**(1), 10–35.

Hansson, S.O. (2007a). Philosophical problems in cost–benefit analysis. *Economics and Philosophy* **23**, 163–83.

Hansson, S.O. (2007b). Risk and ethics: three approaches. In: T. Lewens (ed.) *Risk: philosophical perspectives* pp. 21–35. London: Routledge.

Hermansson, H. (2007). The ethics of NIMBY conflicts. *Ethical Theory and Moral Practice* **10**, 23–34.

Kelman, S. (1981). Cost–benefit analysis: an ethical critique. *Regulation* **5**, 33–40.

Luloff, A.E., Albrecht, S.L., and Bourke, L. (1998). NIMBY and the hazardous and toxic waste siting dilemma: the need for concept clarification. *Society and Natural Resources* **11**, 81–9.

Otway, H. (1987). Experts, risk communication, and democracy. *Risk Analysis* **7**, 125–9.

Pandey, M.D. and Nathwani, J.S. (2003). Canada wide standard for particulate matter and ozone: cost–benefit analysis using a life quality index. *Risk Analysis* **23**, 55–67.

Paté-Cornell, M.E. and Neu, J.E. (1985). Warning systems and defense policy: a reliability model for the command and control of U.S. nuclear forces. *Risk Analysis* **5**, 121–38.

Pidgeon, N., Kasperson, R.E., and Slovic, P. (eds) (2003). *The social amplification of risk.* Cambridge: Cambridge University Press.

Rechard, R.P. (1999). Historical relationship between performance assessment for radioactive waste disposal and other types of risk assessment. *Risk Analysis* **19**, 763–807.

Sagoff, M. (1988). Some problems with environmental economics. *Environmental Ethics* **10**, 55–74.

Shrader-Frechette, K.S. (1985a). *Risk analysis and scientific method: methodological and ethical problems with evaluating societal hazards.* Dordrecht: Reidel.

Shrader-Frechette, K.S. (1985b). *Science policy, ethics, and economic methodology: some problems of technology assessment and environmental-impact analysis.* Dordrecht: Reidel.

Thompson, K.M., Segui-Gomez, M., and Graham, J.D. (2002). Validating benefit and cost estimates: the case of airbag regulation. *Risk Analysis* **22**, 803–11.

Usher, D. (1985). The value of life for decision-making in the private sector. In: E.F. Paul, F.D. Miller, Jr, and J. Paul (eds) *Ethics and economics*, pp. 168–91. Oxford: Basil Blackwell.

Viscusi, W.K. (2000). Risk equity. *Journal of Legal Studies* **29**, 843–71.

CHAPTER 37

Corporate innovation

BRUCE A. VOJAK, RAYMOND L. PRICE, AND ABBIE GRIFFIN

A common, if somewhat contentious, distinction is often made between invention, as the isolated creation of something new that can be patented, and innovation, as the process of converting something new into a market success, which is less often subject to patent protection (McKeown 2008). It has also on occasion been remarked that the era of the independent inventor is past; according to one biographer, Philo Farnsworth, the engineer of television, was 'the last lone inventor' (Schwartz 2002). We have seen a shift from independent invention and innovation to basic scientific discovery taking place predominantly in the academic world with follow-on technological creation taking place as firm-based invention and innovation. Most innovation in today's complex world requires inputs from multiple disciplines, and thus is accomplished by firms using interdisciplinary and transdisciplinary teams. However, there are individuals in most technology-based corporations who first act as inventors, and then go on and implement the invention as a successful innovation into the marketplace. Furthermore, these individuals are likely to do this over and over again.

Those who are repeatedly successful in the practice of truly breakthrough corporate innovations, who we will call serial innovators (SIs), exhibit a combination of broad and deep technical skills, unique insight into business issues (including customer needs, marketing, finance and manufacturing), the creativity to see connections between the two, a political savvy that gets their projects accepted for commercialization, and the facilitative capability to shepherd an innovation through an organization and into the marketplace. This is a blending of different types of expertise that creates individual interdisciplinarity that is almost never appreciated in academic discussions of interdisciplinarity. Complementing other approaches, then, we describe SIs and the processes they use. By way of interpretation, we also observe that SI skills of integration are consistent with Michael Polanyi's epistemology and offer an illustration of corporate innovation interdisciplinarity as tacit integration. We close by suggesting that a key feature of interdisciplinarity is the seeking of coherence and regularities across, and independent of, a wide range of disciplinary boundaries or restrictions, as is well exhibited by SIs.

37.1 The practice of corporate innovation

Corporations arose as distinct legal entities that could undertake larger-scale technological and commercial actions than were possible for individuals. Although corporations have existed in various forms on the margins of society since antiquity, it was not until the Industrial Revolution that they emerged as major social institutions. The modern corporation was itself a social innovation—especially when in the early 1800s it began to be granted the status of a legal person—but the focus of its activity was not initially the promotion of innovation. Instead, during the initial period of the development of the modern corporation, the conscious goal of most corporations was more simply manufacturing scale and continuity. Only in the late 1800s and early 1900s did corporations begin to become innovation-oriented by establishing research and development laboratories and other innovation-initiating activities within the firm.

In the twenty-first century, however, we live in an age of knowledge workers, with innovation one of their characteristic activities and responsibilities (Drucker 1959, 1969). One of the most important roles of knowledge workers in industry is to develop new products or processes to either increase firm revenue or decrease firm cost. Without the creative destruction of a sustained innovative output, companies gradually are reduced to irrelevance, as others repeatedly redefine the basis of competition (Schumpeter 1934). This stress on creativity and innovation has led to many studies on creativity and a plethora of literature on innovation (Freeman and Soete 1997; McKeown 2008). Yet little of this literature explicitly stresses the importance of interdisciplinarity in corporate innovation.

Innovation occurs across a broad spectrum, ranging from that which can be described as incremental (minor improvements of existing offerings, such as the addition of a new whitener to a detergent) to that which can be described as true breakthroughs (entirely new categories of products, such as the first disposable diaper—attributed to Victor Mills of Procter and Gamble, after whom the company's top honor to technologists is named) which materially change the consumer's way of life and have a significant financial impact on the innovating company. Both incremental and breakthrough innovations are today critical contributors to the continuing existence of a firm. Incremental innovation represents a lower-uncertainty, lower-return endeavor, more likely contributing to the continuance of the ongoing business. Breakthrough innovation is characterized by both high uncertainty and, if successful, high reward, and is more likely to lead the firm into new product or market arenas.

Incremental innovation typically involves insights from a single discipline or, at most, a set of minor insights from a small number of disciplines. Thus, it might be characterized by the multidisciplinary juxtaposition (rather than true integration) of different types of disciplinary knowledge. In contrast, to be successful, knowledge workers who contribute to breakthrough innovation must simultaneously navigate and, more importantly, integrate and connect a highly complex set of multidisciplinary expertise, including, but not limited to:

- marketing (customer insight, quantitative research and trends knowledge),
- manufacturing (statistical process control and cost estimation),
- technology (materials and information technology),

- product design (human factors and aesthetics),
- organizational behavior (project management and leadership), and
- legal (patents and contracts).

Creative integration across these widely ranging specialized areas of corporate practice is difficult and challenging. If it were easy to operate at this level of interdisciplinarity, someone else would have already done it.

While variations exist, the innovation literature describes the most commonly practiced innovation process as comprising two general phases, as illustrated schematically in Fig. 37.1. At the outset is the fuzzy front end (FFE) phase of innovation (Koen *et al.* 2002), a chaotic stage, where new product ideas are generated by those who, in large, technology-based corporations are typically from engineering or scientific disciplines. Next, still in the FFE phase, individuals from marketing search for opportunities in the marketplace for these new product ideas. Finally, once a connection between idea and opportunity is made, those with disciplinary expertise in project management usher the project through what is called the Stage-Gate® process (SGP) (Cooper 1990). The SGP is a relatively recipe-driven process that sees the development phase as one in which the proposed product concept sequentially undergoes refinement, with promising concepts garnering additional investment resulting in physical development, manufacturing development, and ultimately commercialization. The Stage-Gate® name, then, derives from the fact that, after each stage of development, a product concept reaches a gate, with the promising concepts passing through the gate and those without promise being stopped at that gate from further investment toward commercialization.

Of particular interest is the fact that the two basic phases of the innovation process presuppose two different epistemologies, that is, underlying philosophical assumptions about what knowledge is and how it is obtained. The FFE can be construed as a more skeptical phase, one that doubts whether we can grasp in any significant detail how innovation actually occurs. In contrast, the SGP is a more certain, methodical phase of the process that assumes knowledge comes in the form of explicit information. This explicit

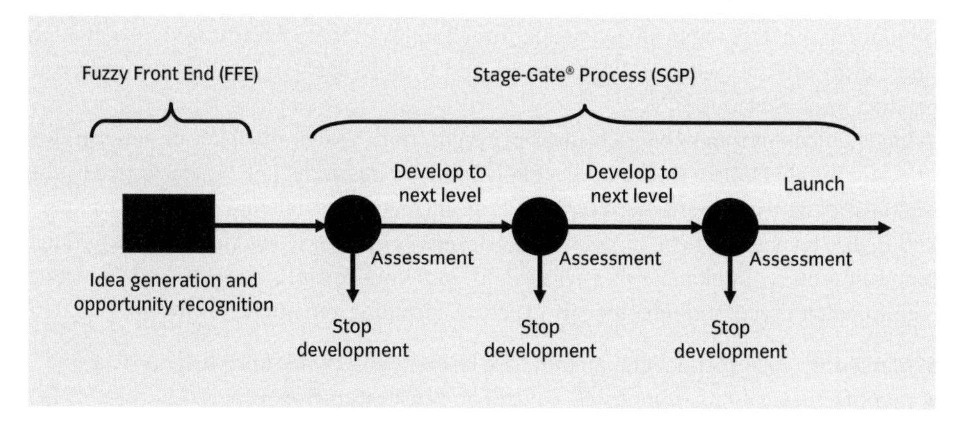

Figure 37.1 The general innovation process.

information is then systematically accessed and incorporated into the project by following a rather strict, linear methodology.

In sum, disciplinary content expertise (such as marketing, engineering design, or manufacturing technology) represents the 'know what' of innovation. The bringing of such specialized forms of expertise together (as found in navigating the FFE and SGP) represents the 'know how' of innovation (Polanyi 1983, p. 7). Both are critical to the creation of new, innovative, breakthrough products.

In spite of such representations of the innovation process, what we find in practice is that SGP- and FFE-like perspectives independently and collectively are, in fact, insufficient to describe how the integration of disparate disciplinary or specialized knowledge takes place in successful breakthrough corporate innovation activities. Further, those who are expert and accomplished in such matters in industry, especially successful individual innovation practitioners and the executives to whom they report, find both perspectives somewhat troubling. Thus, there is a need to seek another, more accurate means of understanding and describing the 'know how' of the interdisciplinary, or perhaps even transdisciplinary, integration of knowledge (i.e. the 'know what') in the act of individual corporate innovation.

37.2 Serial innovators

Research by Griffin, Price and Vojak (Vojak *et al.* 2006; Griffin *et al.* 2007), based on over 125 in-depth interviews as well as a large sample survey, has led to a clearer understanding of how at least some breakthrough corporate innovation occurs in practice. This research has investigated SIs, individuals who have operated with a high degree of interdisciplinarity, repeatedly conceiving and commercializing new breakthrough products in large, mature, technology-intensive firms. SIs are well-described by six characteristics (Sim *et al.* 2007; Griffin *et al.* 2009): how they are motivated, how they prepare to innovate, their perspective regarding innovation, their personality, how they successfully navigate the politics of corporations, and the innovation processes they follow.

37.2.1 Motivation

SIs are motivated by the basic urge to solve other people's problems. They are energized by the challenge of solving complex and difficult interdisciplinary problems of the type exemplified by breakthrough corporate innovation. SIs thrive under the guidance of managers who exhibit an understanding of how corporate innovation really occurs. They operate most effectively when neither micromanaged nor ignored, but when their manager supports them and then gives them the discretion to move forward as they see fit (Hebda *et al.* 2007).

37.2.2 Preparation

Both life experience and intentional effort prepare SIs to innovate. They have clear recollections of key learning experiences imparted from others, typically those who are their seniors, often as anecdotes or as sage advice. As they prepare professionally, these individuals exhibit multidisciplinarity by acquiring multiple and deep technical and business insights. Additionally, while immersed in the process of innovation, they add to this multidisciplinary skill set as they continually seek out new information in other disciplines or areas of life, manifesting skill as independent, lifelong learners.

37.2.3 Perspective

SIs accept responsibility for doing the right thing for the greater good: that is, what is simultaneously good for customers *and* for their companies. Many appear to have developed this perspective as a result of overcoming difficult early life experiences, such as losing a sibling or moving across cultures. Additionally, SIs recognize that technology is not to be pursued for self-aggrandizement or personal amusement, but, rather, as a means to an end—in this context, to make money for the company.

37.2.4 Personality

SIs are skilled in their ability to navigate the ambiguity typical of highly complex, interdisciplinary problems. Interestingly, however, while still in the midst of ambiguity, they exhibit a confidence about the outcome—an expectation that they will discover new, innovative insights if only they continue to work at it.

37.2.5 Politics

SIs 'cross the bridge' of accepting responsibility for a project in its entirety by taking ownership for the political process of gaining organizational acceptance for the new idea. Unlike those who assume that the burden of organizational project acceptance falls to others, SIs apply their skills of discovery to studying interpersonal interactions and organizational behavior with the same zeal that they apply to the creation of the breakthrough innovative concepts themselves. SIs seek to understand how an organization functions and what they can do to advance an innovation (Price *et al.* 2009).

37.2.6 Process

In contrast to the linear, two-phase process depicted in Fig. 37.1, SIs follow a significantly different innovation process. As depicted graphically in Fig. 37.2, SIs innovate using the 'hourglass model of innovation' (Griffin *et al.* 2007).

They first spend considerable upfront time discerning and defining the best problems to engage—those that address real customer needs, have the potential to be embraced within the organization, and provide significant financial return to the company. SIs recognize that it often takes as much time to address an unimportant problem as it does an important problem and, thus, invest their expertise only on problems worthy of their skill.

Next, SIs invest themselves in deeply understanding the problem, not from a detached, academic perspective, but in the manner and at the level required for commercial success.

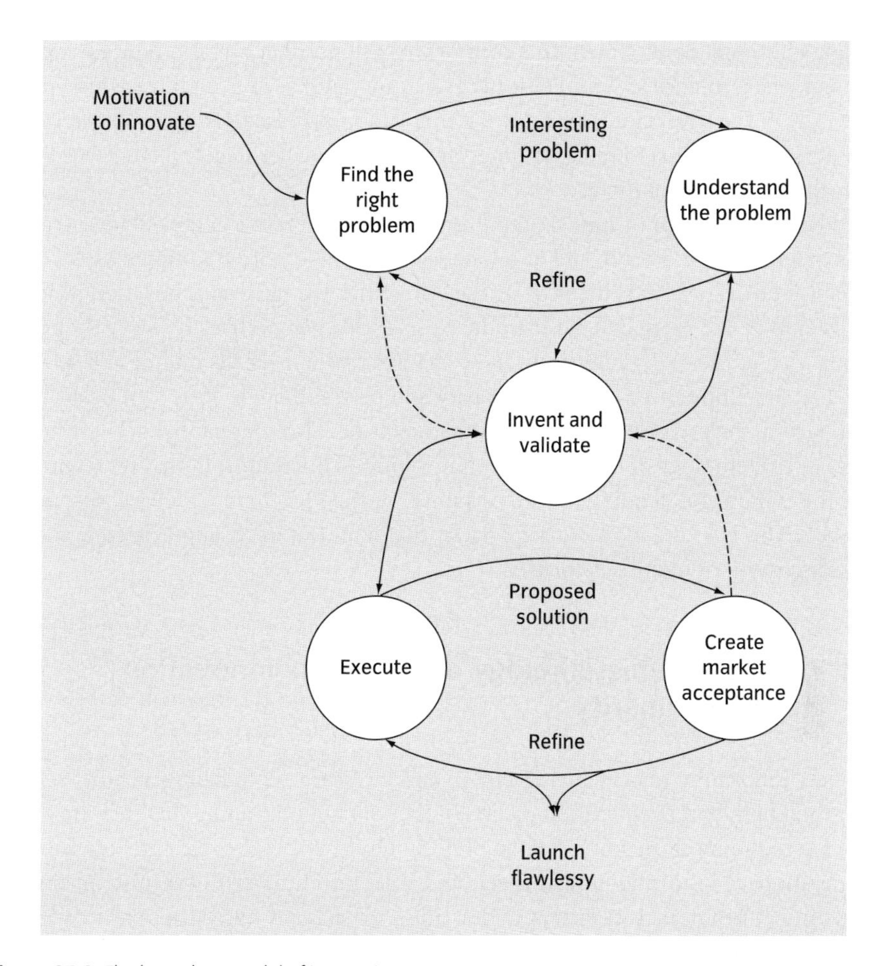

Figure 37.2 The hourglass model of innovation.

Often during the process of understanding the problem, they recognize that the problem definition they had been working within is flawed. Thus, at times, SIs return to the beginning of the process to redefine the problem.

As their knowledge of the problem matures, SIs increasingly apply their skill in interdisciplinary processes, as integrative systems thinkers, inventing and validating—simultaneously 'connecting the dots', as many SIs put it, from a vast amount of disciplinary information and background preparation. These individuals flexibly apply a wide range of techniques to 'see' the emerging innovative concept, including reframing the problem, looking beyond the obvious, and starting with a vision of what the ultimate innovative output might be and working backward. Again, the process can reverse itself here, as well. A number of SIs not only report going back to better understand the problem, but also to redefine it, even while in the midst of attempting to connect the dots.

As noted, the SIs do not consider their work complete simply with the definition of an innovative concept. They also seek to execute it. They not only shepherd the concept through additional development to a commercialized product but also typically work to drive market acceptance as well. These behaviors are rarely seen from those who typically reside only in the upstream regions of the innovation process. Not surprisingly, even at these late stages of innovation, SIs at times circle back to earlier stages when newly discovered realities direct them there.

This hourglass model of innovation illustrates in many respects how SIs transcend—that is, operate more flexibly and at a higher level than—more traditional views of the innovation process, such as the SGP. At the same time, the upfront innovation processes that SIs employ are more defined than the open-ended perspectives typically thought of as being used in the FFE. Collectively, these characteristics make SIs unique and powerful individual practitioners of interdisciplinarity and permit them to have significant financial impact on their organizations. These characteristics also suggest that a 'know how'—that is, an epistemology of breakthrough innovation—different from that represented by either the SGP (again, tending toward certainty) or the FFE (tending toward skepticism) is at work. This is a kind of interdisciplinarity that is almost never appreciated in the academic discussion of interdisciplinarity.

37.3 Toward an epistemology of corporate innovation interdisciplinarity

To review, innovation management practice and literature are touching on what we cannot or can know, that is, epistemology, even if they are not fully aware that they are engaging in a philosophical pursuit. The two philosophical perspectives that emerge from the considerations of innovation practitioners and academics are those of skepticism and certainty, but neither alone is sufficient to describe what actually happens. In order to move beyond these two perspectives, we appeal to the insight of Michael Polanyi (1958, 1983), a twentieth-century physical-chemist-turned-epistemologist. Polanyi's epistemology, which rejects the false dichotomy between skepticism and certainty, is a better way to see and understand what actually occurs in the practice of industrial innovation. Although Polanyi does not use the word interdisciplinarity, his epistemology powerfully illustrates it, as will be developed in the rest of this chapter.

A respected scientist in his own right, Polanyi developed a philosophical theory to understand the epistemological ramifications of his success in the laboratory. He argued that: (1) if knowledge is restricted to explicit information communicated to others impersonally and passively, no scientific discovery could ever occur and (2) the dominant philosophical interpretation of science as detached observation is inconsistent with practice. Polanyi connected his analysis with a problem that can be traced back to Plato's *Meno*. As Polanyi summarized Plato's statement of a fundamental epistemological dilemma involved with learning, 'the search for the solution of a problem is an absurdity; for either you know what you are looking for, and then there is no problem; or you do not know what you are looking for, and then you cannot expect to find anything' (Polanyi 1983, p. 22). Polanyi felt that the problem thus posed was significant enough for him to step away from a successful career in science in order to try to develop a fresh approach to the epistemology of search and discovery.

Many correctly associate Polanyi's work with development of the concept of tacit knowledge and the idea that 'we can know more than we can tell' (Polanyi 1983, p. 4), but misunderstand, or miss entirely, the sophisticated and helpful structure that Polanyi identified as characteristic of all efforts to know. With few exceptions (Dias 2008), most who consider the tacit aspects of innovation or systems thinking (Senker 1995; Leonard and Sensiper 1998; Cook and Brown 1999)—while insightful at many points—miss much of this richness.

For Polanyi, all achievements of knowing involve creative and active integration where the individual relies on inarticulable subsidiary clues to focus on an eventually identifiable pattern. The SIs we have observed do precisely this. They simultaneously hold fast to multiple technical domains, as well as to customer, market, finance, and manufacturing insights, while having the vision to 'see' the innovative concepts that 'connect the dots' within and between these several domains. In the language of Polanyi, SIs exhibit 'from–to' tacit integration (Polanyi 1958, p. 55, 1983, p. 10): 'from' an immersive ('indwelling,' per Polanyi) subsidiary awareness of the multiple disciplinary elements 'to' a focal awareness of the innovative product or process concept that takes into consideration all opportunities and the constraints across the subsidiary disciplinary elements. Put another way, SIs do not look 'at' the disciplinary elements of technology, customer, market, finance, and manufacturing; instead they look 'through' them, enabling SIs to see the innovation in a high-level act of inter- and transdisciplinarity. Polanyi illustrates 'from–to' tacit integration variously, such as by considering how one recognizes one person's face among a thousand others (Polanyi 1983, p. 4) or how a stereoscope functions (Polanyi 1965). We next offer an illustration of from–to tacit integration that is central to understanding innovative interdisciplinarity in the corporate context.

37.4 An illustration of corporate innovation interdisciplinarity

In a careful discussion of Magic Eye® images (Magic Eye, Inc. 1993), Esther Meek (2003, pp. 46–51) provides an illustration of Polanyi's interpretation of the act of discovering something new that is strikingly similar to the type of knowing we have observed among SIs. We thus employ it here to illustrate our research results and our understanding of how interdisciplinarity works in the corporate world. The value of the Magic Eye® illustration lies in the fact that this perspective goes beyond both the skepticism of the FFE and the certainty of the SGP, and provides a fresh illustration of corporate innovation as it actually is practiced.

Magic Eye® images are a type of two-dimensional (2D) pattern known as a random dot stereogram (RDS). A RDS is constructed in such a way so as to permit, when viewed with proper perspective, the 'seeing' of a three-dimensional (3D) image. The RDS of Fig. 37.3(a), for example, when viewed properly, permits the viewer to 'see' a 3D version of the depth map pattern presented in 2D form in Fig. 37.3(b). The key element of this illustration is the 'from–to' viewing, through the 2D surface pattern, that is required to see the embedded 3D image (analogous to the 'connecting the dots' systems thinking observed in the act of innovation as described by the SIs). Such 'from–to' viewing is depicted schematically in Fig. 37.4, as the viewer looks *through* the surface of the RDS of Fig. 37.3(a), not *at* it (step 1). The two images from the stereo-optic perspective necessary for proper viewing are then combined in the viewer's mind (step 2), enabling them to see the 3D depth map image of Fig. 37.3(b) (step 3).

The connection between viewing a RDS and the practice of seeing an innovative concept in breakthrough corporate innovation of the type discussed above is illustrated in Fig. 37.5.

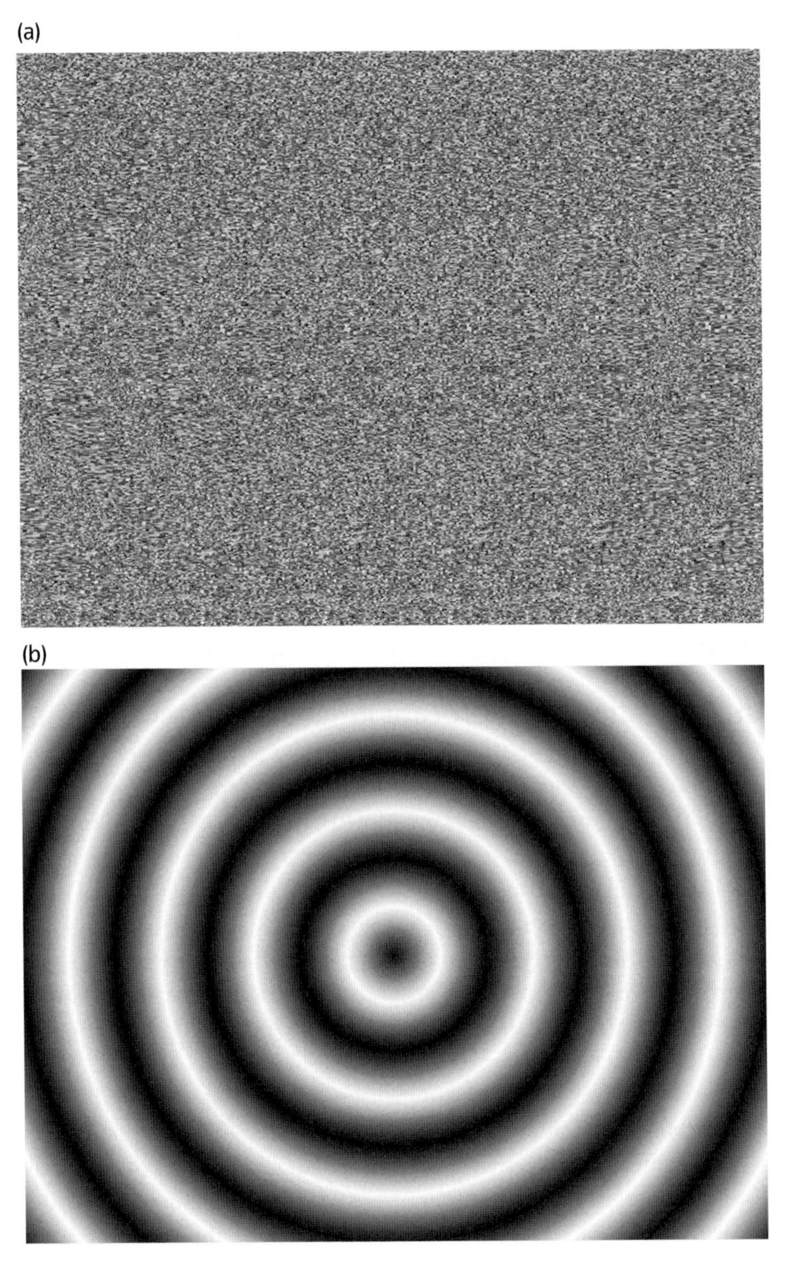

Figure 37.3 (a) Random dot stereogram (RDS). For proper viewing, photocopy at 200%. (b) Depth map associated with the RDS in (a).

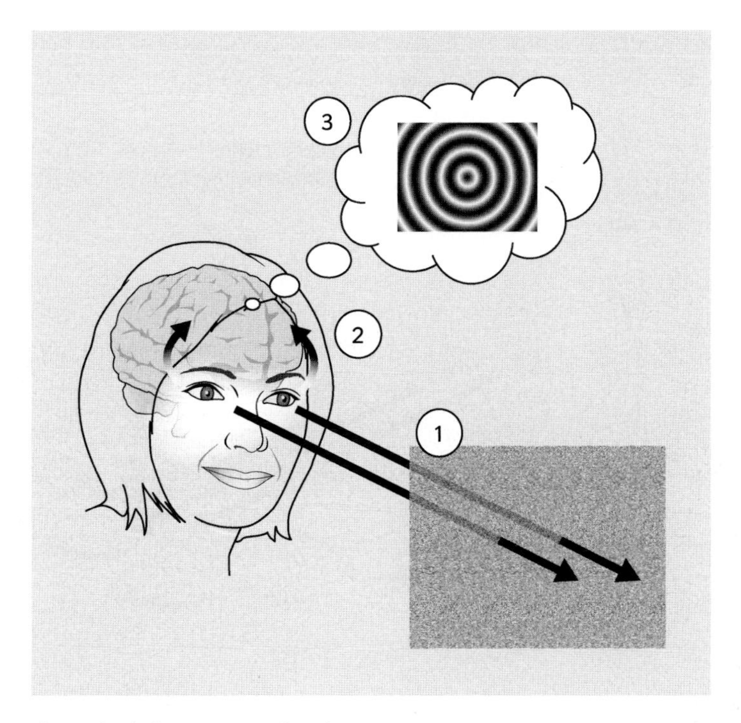

Figure 37.4 The method of viewing a random dot stereogram.

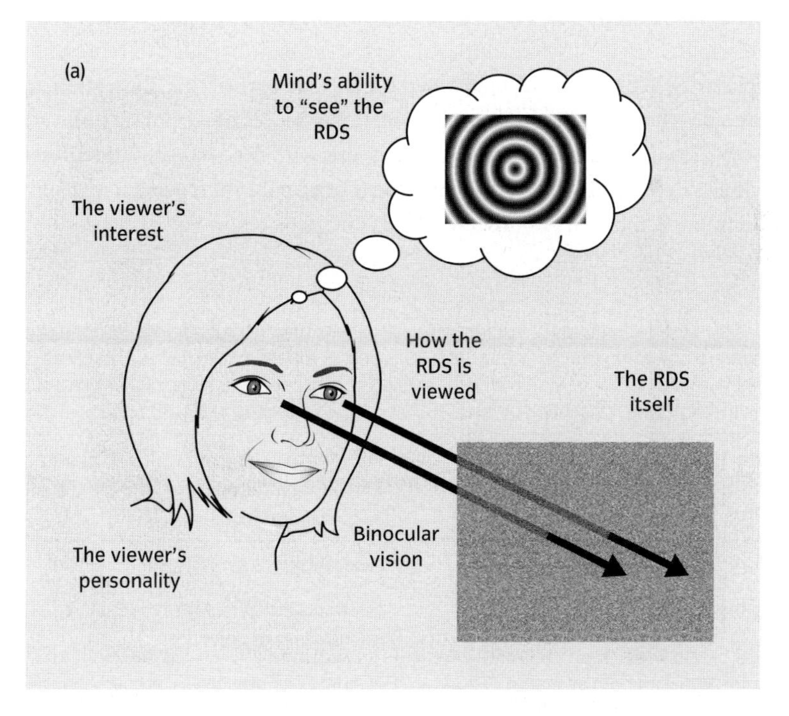

Figure 37.5 (a) Elements of viewing a random dot stereogram. (b) Elements of breakthrough innovation (see over page).

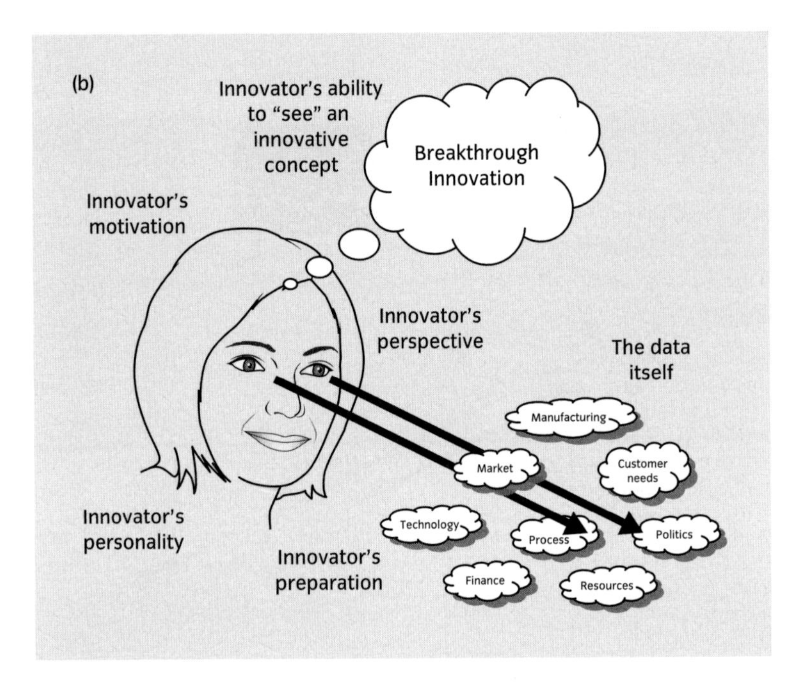

Figure 37.5 (Cont'd)

Figure 37.5(a) depicts the viewing of a RDS, while Fig. 37.5(b) depicts the innovative discovery that occurs in breakthrough corporate innovation. While this illustration presents two-fold interdisciplinarity with stereo-optic viewing, it is easily extended to represent manifold interdisciplinarity. With 'from–to' tacit integration at the core of this illustration, one is able to illustrate a number of both pitfalls and successful practices of corporate innovation interdisciplinarity. Listed in Table 37.1 are several of these, with the first column representative of RDS viewing as depicted in Fig. 37.5(a) and the second column representative of corporate innovation interdisciplinarity as observed in SIs and depicted in Fig. 37.5(b).

Table 37.1 Side-by-side summary of how variously viewing a random dot stereogram (RDS) can be used to illustrate numerous corporate innovation interdisciplinarity pitfalls and successful practices. The first column represents RDS viewing, as depicted in Fig. 37.5(a), while the second column represents what is illustrated in breakthrough innovation by such RDS viewing, as depicted in Fig. 37.5(b)

Try this with RDS viewing…	In order to illustrate various corporate. innovation interdisciplinarity pitfalls and successful practice…
Try focusing on the surface of the RDS	Illustrates not being a 'systems thinker'
Try viewing the RDS with one eye	Illustrates lack of being prepared in a multidisciplinary manner
Try viewing the RDS with both eyes shut	Illustrates no disciplinary preparation
Try viewing the RDS from too far	Illustrates poor perspective

Try forcing someone to view a RDS	Illustrates lack of self-motivation
Try without a RDS	Illustrates not having relevant data
Try hurrying someone to view a RDS	Illustrates that time is required to innovate
Try seeing how long people take to see the embedded image in a RDS	Illustrates that aptitude to 'see' an innovation varies from person to person
Try to determine if an improvement of seeing the embedded image comes with practice	Illustrates the improvement that occurs with repeated practice—serial innovation
Try with someone experienced with viewing a RDS—they just start looking for the embedded image	Illustrates that serial innovators have the confidence to just start looking for good problems and solutions
Try with more complex embedded image—the embedded image emerges gradually, at times with some mis-starts of interpretation	Illustrates that serial innovators do not see everything at first
Try starting with a mis-rotated RDS—need to rotate the RDS to see the embedded image	Illustrates that serial innovators often need to reframe the problem to gain new perspective
Try with an unusual or an obvious surface pattern, not just a set of random dots	Illustrates that individuals often come to the innovation process with preconceptions serial innovators see beyond such preconceptions

37.5 A general theory of corporate innovation interdisciplinarity

A significant implication of our use of Polanyi's epistemology to understand corporate innovation interdisciplinarity is that we see an innovator's output as a skilled creative achievement (not a random association or technical method) emerging from the tacit integration of a multidisciplinary range of subsidiary clues that can fully be expected to vary over time. For example, both technical expertise and consumer preferences are highly time variant. Viewing innovation in this manner suggests that Polanyi's epistemology provides a general theory of innovation, rather than a description of the contingent trends of some particular corporate innovation practices, such as disruptive innovation (Christensen 1997), radical innovation (Leifer *et al.* 2000), or open innovation (Chesbrough 2003).

This view is the same as that expressed by one of Polanyi's most famous doctoral students, Eugene P. Wigner. In his Nobel prize banquet speech of 1963, physicist Wigner paid homage to Polanyi in the following words: 'He taught me, among other things, that science begins when a body of phenomena is available which shows some coherence and regularities, that science consists in assimilating these regularities and in creating concepts which permit expressing these regularities in a natural way' (Wigner 1967, pp. 262–3). Wigner also suggested that this was a method that could be applied to understanding fields of learning other than those of science.

For our present purposes, then, we contend that this insight is critical to understanding corporate innovation interdisciplinarity. Indeed, we even suggest that a key feature of interdisciplinarity in general is the seeking of coherence and regularities across, and independent of, a wide range of disciplinary boundaries or restrictions. This may well be the crucial 'know how' of successful interdisciplinarity in all areas, even though we argue here merely for its value in providing a theory of corporate innovation interdisciplinarity.

Acknowledgements

The authors are indebted to their former and current graduate students who have contributed to the Serial Innovator and Technical Visionary project. They also are grateful to Professor Esther Meek (Geneva College) for several helpful discussions regarding Polanyi's epistemology and to Mr James Tucek for permission to use his random dot stereogram and depth map images found in Fig. 37.3.

References

Chesbrough, H.W. (2003). *Open innovation.* Boston, MA: Harvard Business School Press.

Christensen, C.M. (1997). *The innovator's dilemma.* Boston, MA: Harvard Business School Press.

Cook, S.D.N. and Brown, J.S. (1999). Bridging epistemologies: the generative dance between organizational knowledge and organizational knowing. *Organization Science* **10**, 381–400.

Cooper, R.G. (1990). Stage-gate system: a new tool for managing new products. *Business Horizons* **33**, 44–54.

Dias, W.P.S. (2008). Philosophical underpinning for systems thinking. *Interdisciplinary Science Reviews* **33**, 202–13.

Drucker, P.F. (1959). *Landmarks of tomorrow.* London: Heinemann.

Drucker, P.F. (1969). *Management challenges for the 21st century.* New York: Collins.

Freeman, C. and Soete, L. (1997). *The economics of industrial innovation.* New York: Routledge.

Griffin, A., Hoffmann, N., Price, R.L., and Vojak, B.A. (2007). The processes by which serial innovators innovate. *Proceedings of the 14th EIASM International Product Development Management Conference.* Brussels: European Institute for Advanced Studies in Management.

Griffin, A., Price, R.L., Maloney, M.M., Vojak, B.A., and Sim, E.W. (2009). Voices from the field: how exceptional electronic industrial innovators innovate. *Journal of Product Innovation Management* **26**, 223–41.

Hebda, J.M., Vojak, B.A., Griffin, A., and Price, R.L. (2007). The motivation of technical visionaries in large American companies. *IEEE Transactions on Engineering Management* **54**, 433–44.

Koen, P.A., Ajamian, G.M., Boyce, S. *et al.* (2002). Fuzzy front end: effective methods, tools, and techniques. In: P. Belliveau, A. Griffin and S. Somermeyer (eds) *The PDMA toolbook 1 for new product development*, pp. 5–36. New York: Wiley.

Leifer, R., McDermott, C.M., Colarelli O'Connor, G., Peters, L.S., Rice, M., and Veryzer, R.W. (2000). *Radical innovation.* Boston, MA: Harvard Business School Press.

Leonard, D. and Sensiper, S. (1998). The role of tacit knowledge in group innovation. *California Management Review* **40**, 112–32.

McKeown, M. (2008). *The truth about innovation*. Harlow, UK: Pearson Education.

Magic Eye, Inc. (1993). *Magic eye: a new way of looking at the world*. Kansas City, MO: Andrews and McMeel Publishing.

Meek, E.L. (2003). *Longing to know: the philosophy of knowledge for ordinary people*. Grand Rapids, MI: Brazos Press.

Polanyi, M. (1958). *Personal knowledge: toward a post-critical philosophy*. Chicago: University of Chicago Press.

Polanyi, M. (1965). The structure of consciousness. *Brain* **88**, 799–810.

Polanyi, M. (1983; original 1966). *The tacit dimension*. Gloucester, MA: Peter Smith.

Price, R. L., Griffin, A., Vojak, B.A., Hoffmann, N., and Burgon, H. (2009). Innovation politics: how serial innovators gain organizational acceptance for breakthrough new products. *International Journal of Technology Marketing* **4**, 165–84.

Senker, J. (1995). Tacit knowledge and models of innovation. *Industrial and Corporate Change* **4**, 425–47.

Schumpeter, J.A. (1934). *The theory of economic development*. Cambridge, MA: Harvard University Press.

Schwartz, E.I. (2002). *The last lone inventor*. New York: HarperCollins.

Sim, E.W., Griffin, A., Price, R.L., and Vojak, B.A. (2007). Exploring differences between inventors, champions, implementers and innovators in creating and developing new products in large, mature firms. *Creativity and Innovation Management* **16**, 422–36.

Vojak, B.A., Griffin, A., Price, R.L., and Perlov, K. (2006). Characteristics of technical visionaries as perceived by American and British industrial physicists. *R&D Management* **36**, 17–24.

Wigner, E.P. (1967). *Symmetries and reflections*. Bloomington, IN: Indiana University Press.

Index

abstraction 5
academic tenure (AT) 152–3
academies 7
accountability 310, 318, 326, 370
  individual 376
Accreditation Board for Engineering and Technology (ABET) 151
Acevedo, Miguel 504
Ackoff, Russell L. 461, 462–4
Acquired Immune Deficiency Syndrome (AIDS) 508, 509
action research 433
actor-network theory 194
adaptation 407–8
  institutional 412–13
administrative law 522, 525
administrative organization 346–8, 399
advertising 225
Aelius Galen 120
aesthetics 294
affiliated hire 394
affirmative action 377
Aga Khan University 370
age of interdisciplinarity 85–7
agency 230
agenda setting 40
agent-based model 447
agriculture, global 334–5
AIDS 508, 509
air transportation 71–2
Alberta, University of 369
alchemy 163
alewith 502
Alliance of Digital Humanities organizations 207
Alpers, Svetlana 140
Althusser, Louis 226
American Association for the Advancement of Science (AAAS) 341
American Association of University Professors (AAUP) 152
American Social Science Association 105
American Society of Civil Engineers (ASCE) 149–50, 154
American Society for Engineering Education (ASEE) 151
Amsterdam, University of, Institute for Interdisciplinary Studies 369
analogy, study of 239
analysis 464
analytic philosophy 420
Ancheta, A. N. 530
Andrew W. Mellon Foundation, 'New Directions' grants 398
anonymity 229
anthropology
  as cognitive science discipline 236, 237
  fields 115–16
  law courses 528–9
antidisciplines xxxi
Appadurai, Arjun 274
applied ethics
  generalized 266–7
  interdisciplinarity in 262–7
applied informatics projects 183
appreciative inquiry 406, 479
apprenticeship 148
  multi-mentor 487
Aquinas, Thomas 261
archaeology 116
Archer, Bruce 279
area studies 22, 106–10
  quantitative research methodology vs. 113–14
Aristotle 3, 56, 88, 120, 143, 260
arithmos 90
Arizona State University, Innovation Space 276–7
arms race 195
Arnoff, E. L. 461
art 86
  biology and 129
  religion and 164
art history 133, 134–5
  new 137–8
  rethinking (inter)disciplinarity in 139–41

art research 133–45
  origins 134–7
artifacts 222
artificial intelligence (AI) 235, 248, 453
  as cognitive science discipline 235–6, 239
artificial neural networks 248
artistic design 278–9
arts, sciences and 135–6, 137–8
ASCE 149–50, 154
Ashby, Ross 227
Assessment Movement 378
Association of American Colleges and Universities (AAC&U) 363, 368, 383–4
Association for Computing Machinery (ACM) 264
Association for General and Liberal Studies 368
Association for Integrative Studies (AIS) 362, 364–5, 366, 368, 383
associations
  disciplinary 8–9
  scholarly 13
  scientific 8, 11
atomic bomb 195, 447
atomism 235
attorneys 528
Augustine 261, 464
Australia
  droughts 335
  general education 369
authoritative strategies 281
autonomy, academic 324–6
autopoiesis 456
avatars 222
Axial Age 261
Ayer, A. J. 55, 291, 293

Babbitt, Milton 142
Baccini, P. 437–8
Bacon, Francis 4
Bahm, Archie 464
Bal, Mieke 140–1
balance 163
Balsamo, Anne 217
basic research 325, 329, 339
Baudrillard, Jean 226, 253–4
Bauhaus 284, 285
beauty, study of 137
behavioral sciences 21, 518
  in risk analysis 538, 542
behaviorism 235, 238–9, 336
Bell Telephone Laboratories 461
Bernard, Claude 121
von Bertalanffy, Ludwig 60–1, 453–4, 455, 456, 463, 464
Bethe, Hans 297
betweenness centrality 179, 457
bibliometric analysis 481, 485
bibliometric research 177–9
BIC 330
Bielefeld, University of 292–3, 369
Big Thicket National Preserve 504–5
Bill, Max 273
biochemical genetics 124
biochemistry 123–5
Biocomplexity in the environment competition 503–4
Biocomplexity Initiative 337
biocultural conservation, in Cape Horn 499–500
biocultural diversity 498
biodiversity 496, 497
bioethics 129, 262, 420, 426–8
biofuels 335
bioinformatics 59–60, 122, 127
biological contingency 45
biological factors, social factors and 510
biological sciences 119–31
  case studies in interdisciplinarity 120–9
biologists, computer scientists vs. 93
biology
  art and 129
  and beyond 128–9
  chemistry and 123–5
  engineering and 125–6
  mathematics and 126–8
  medicine and 120–2
  religion and 129
biomedical ethics 262
biomimicry 40
Biopolis 97
biotechnology 128–9
biotherapeutics 126
birthrates 425–6
bit 227
black-boxing 251
Blackstone Model 526
Blavatnik Award for Interdisciplinary Research 91
blending 18
Bloom, Benjamin 378
Boaventura, Ana 214
Boden, Margaret 17–18, 19, 20, 453–4, 467
Bogdanov, Alexander A. 45
Boix Mansilla, Veronica 373, 397
BOK2 154–5
Borgmann, Albert 226, 268
Börner, K. 395
Boulding, Kenneth E. 455
boundary critique 466
boundary institutions 98
boundary setting 445
Bowling Green State University 349–50
Bradford's law of scattering 176
brain scanning 240
Braudel Center 108
breakthrough innovation 547–8, 556
bridge building, restructuring vs. 21

British Columbia, University of, Green College 368
Broad ID 18
broadcasting 221, 224–5
broader impacts criterion (BIC) 330
browsing, role 181, 183
Bruffee, K. A. 376
Bruner, Jerome 242
Buchner, Edward 123
Buddhism 261
Buddhist–Christian dialogue 170–1
Bunge, Mario 464
Burns, Rebecca Crawford 17, 18, 290–1
Bush, George W. 423
Bush, Vannevar 325–6, 339
business schools, law courses 529
Butler, Nicholas Murray 361
Bynym, T. W. 264–5

calculus 89
Calgary, University of, Faculty of Communication and Culture 369
California, University of, cluster hires 394
Callon, Michel 194
Calvin, John 467
Canada, general education 368–9
Canadian Institute for Health Research 514
Cantor, N. F. 532, 533
Caparo forest reserve 504–5
Cape Horn Biosphere Reserve (CHBR) 499–500
cardiovascular biology 125–6
Cardozo, Benjamin 524
Carnegie Foundation for the Advancement of Teaching 368, 383
Carnie, Andrew 138
Carson, Rachel 195, 263, 496
Cartesianism 164, 165
Cartwright, N. 44
case method 526
case studies
  epistemic knowledge and 42–3
  learning based on 35–8
cellular automata 248, 505
cellular respiration 123
cellular systems modeling 127
Center for Research on Learning and Teaching 382
Center for the Study of Complex Systems (CSCS) 351–2
Center for Social Justice Research, Teaching, and Service 352–3
Centers for Population Health and Health Disparities (CPHHD) 474
  evaluation model 476–7
  social network analyses 481
Central Asia, University of 369–70
Centre for Philosophy, Technology and Social Systems (CPTS) 466–7
certainty 552
CGIAR 335
Chad, Republic of, health services 438
changed environment 217
chaos theory 113
Checkland, Peter 456, 462
chemistry, biology and 123–5
Chicago, University of
  general education 362
  school of social science 21
Chilean navy 500
chlorofluorocarbons 496
Chomsky, Noam 240
Christianity, ethics in 261
Churchman, C. West 461, 462, 464, 467
CIRET 24
citation analysis 178
citation indexes 397
citation indexing services 178
citation networks 457
civic education 528–9
civil engineering
  definition 148
  education 150–2, 154, 155
  future 153–5
  licensure 150
  professional practice 149–50
  research 152–3
civil law 523, 533
clarity, related to interdisciplinarity 411–12
Clark, Burton 346
Classical Legal Thought 524
classification systems 180
classroom assessment techniques (CATs) 378–9
climate change
  addressing 494–5, 497
  holistic expertise 75
  mitigation 443
  research 32, 47–8
climate simulations 254–6
Club of Rome 462
cluster hire 394
co-authorship networks 457
co-citation analysis 178
co-creation 517
co-production 532
Code of Hammurabi 523
coercion, collaboration vs. 95
cognitive proximity 98
cognitive science 234–43
  experimental benefits 240–1
  history 234–5
  journals 242
  lessons 241–3
  patterns of collaboration 235–7
  theoretical benefits 238–40
Cognitive Science Society 235, 242

cognitive turn 143
collaboration 19
coercion vs. 95
cross-disciplinary (CD) 473–4, 491
interdisciplinary 304, 518
principles of 445–6
researchers and stakeholders 517
unidisciplinary (UD) 473
collaboration maps 457
collaboration patterns 179
collaboration readiness (CR) 477, 489–90
collaborative learning 366, 375–7
collaborative strategies 281
collaborators 182
Columbia University
Bureau of Applied Social Research 224
contemporary civilization program 361
Committee on Facilitating Interdisciplinary Research 396, 399
committee structure, in interdisciplinary programs 347
common goals 482, 483
common good, promotion 438
common language 482, 483
common law 523–4, 532
common metric-based methods 448
common theoretical basis 482, 483
communication 91
see also media and communication
communication studies 511–12
community-based conservation 497
comparative metabolomics 124
comparative musicology 138
competitive strategies 281
complex systems, study of 351–2
complexity 252, 360–1, 364, 367, 432
health and health systems 510–11
as prerequisite for interdisciplinarity 513
complexity studies 453
Composite ID 18
computation and simulation (C&S) 246–56
dynamics of interdisciplinarity 251–6
historical development 247–51
outlook 256
validation 256
computational biology 92
computational power, expansion 85–6
computational procedures 238–9
computational psycholinguistics 237
computational science and engineering (CSE) 86
computational templates 254
computer-aided design 251
computer ethics 263–4
computer-mediated communication (CMC) 229
computer science, as discipline 249
computer scientists, life scientists vs. 93
computers
desktop 252
history 246
Comte, Auguste 55, 82, 87
conatus 58
concept mapping 516
concepts 241
as prototypes 237
conceptual blending theory 290
Conceptual ID 20
conditional laws, individual case and 45–6
confidence, intellectual 270
Confucianism 261
connectionism 239
connectivity 505
consensus, managing in interdisciplinary teams 482–5
conservation biology 494, 495–8
as crisis discipline 497
as metadiscipline 497
consilience 291, 294
consolidation 182
Consortium on Fostering Interdisciplinary Inquiry 392
constitutional law 526
constructivist methods 375
constructivist theoretical approaches 198
Consultative Group on International Agricultural Research (CGIAR) 335
content analyses 479–80, 502
content delivery systems 222
context, in health research 514
Contextualizing ID 18
contextualizing interdisciplinarity 453
contracts 526
'convenience appointments' 421
conversant knowledge 503
Cook, Perry 214
cooperative interdisciplinarity 19, 453
cooperative learning 375–7
coordination 87
Cornell University, STS program 196, 199

corporate counsel 528
corporate innovation 546–58
  epistemology of interdisciplinarity 552–3
  fuzzy front end (FFE) phase 548–9, 552
  general theory of interdisciplinarity 557–8
  illustration of interdisciplinarity 553–7
  practice 547–9
  Stage-Gate process (SGP) 548–9, 552
  see also innovation
corporations 547
cosmic consciousness 465
cosmic order 260–1
Coulomb's law 57
Council of Environmental Deans and Directors 399
course portfolios 382–3
'covering laws' 107
CPHHD *see* Centers for Population Health and Health Disparities
creativity 547
  and interdisciplinarity 296–8
criminal law 526, 527
Critical ID 22–4, 26
critical interdisciplinarity 225, 283, 363, 438
Critical Legal Studies 525
critical reflection 378
critical stance 298–9, 302
critical systems heuristics 466
critical systems thinking 465–6
critical theory 41, 223
critical thinking 378–80
critique, role in interdisciplinary field 216–17
critiquing disciplinarity 312, 315–16
Cross-Cutting Organizing Principles 19
cross-disciplinary (CD) collaboration 473–4, 491
cross-disciplinary learning 373
cross-fertilization 389
Crossman, Ed 502–3
Crowder, Larry 502–3
Cudmore, Becky 503
cultural analysis 141
cultural divide 213–14, 294
cultural studies 350
culture 91–2
  and media and communication 229
culture industry 225
curriculum vitae, annotated 397
Cuyahoga River 496
cyberculture 229–30
cybernetics 112, 227, 248, 263, 453
Cybernetics Group 248

D-Lab (MIT) 155
Daoism 260–1
Darnton, Robert 209
Darwin, Charles 121, 213, 499
data collection 5
data curation 184
Davy, Humphrey 82
death, definition 420–1
Debord, Guy 226
decision making 113
  public involvement in 267
deductive specification 38
Defila, A. 314
deism 464, 466
democratic community 315
dependency theory 108
desegregation 531
design 273–86
  categories 278
  disciplines 278
  domains 278
  interdisciplinarity in 281–3
  methods 281
  practice 279, 281
  problems 274–5, 279–81
  project domain map 282–3
  theory 279
design axiology 280
design education, interdisciplinarity in 284–5
design epistemology 280
design history 280
design language 280
design metrology 280
design pedagogy 280
design philosophy 280
design praxiology 280
design praxis 275, 278–9, 281, 283
design studies 279, 280, 281, 283
design taxonomy 275, 278–9, 280
design technology 280
Deutsche Forschungsgemeinschaft (DFG) 324
development studies 108, 468
developmental psychology 240
deviant interdisciplinarity 50–62
  as recovery of lost unity 56–8
  thermodynamics as source of 58–9
  twentieth-century visions of 53–5
devolvement 358
Dewey, John 362
diabetes, Type II, 510
dialectical materialism 53
dialogue-based methods 446–7
digital humanities 207–8
digital scholarship 210–11
digitality 221–2

DiGiulio, A. 314
Dilthey, Wilhelm 432
disaster relief 157–8
disciplinary anxiety 83
disciplinary associations 8–9
disciplinary originality 312–13
disciplines
  definition 162
  emergence 5–8
  fusion of 506
  growth 10
  hermeneutical conception 94
  nature 8–10
  postmodern conception 94
  realist conception 92, 94
  role in defining peers 324–31
disclosive systems thinking 467
discovery-based learning 377
disease 121–2
  transmission 126–7
dispersal 222
dissonance theory 227
distance learning 228–9
diversity, promotion in higher education 377
divisional structure 215
DNA 59, 83, 123–4
doctoral students
  interdisciplinary futures 413–16
  overarching concerns 410–13
  transitional stages 405–7
Dooyeweerd, Herman 466
Driesch, Hans 60
drivers of interdisciplinarity 26
driving simulators 253
drought conditions 335
dual accountability 98
Duke University
  ISIS 215–16
  Social Science Research Institute (SSRI) 352
DuPont 83
Duxbury, Neil 532
dystopias 228–9

e-portfolios 380, 381
early career academics
  interdisciplinary futures 413–16
  overarching concerns 410–13
  transitional stages 407–10
Earth Day 496, 501
Earth Institute (Columbia University) 155
earth sciences 33–4
'earth system analysis' 255
Eby, Lisa 503
ecology 494, 495
  see also environment
economics
  divisions 115
  laws of 46
  psychology and 113
  in risk analysis 537, 540–2
ecumenical movements 167–71
Edwards, Harry 528
Einstein, Albert 213
elder care 425–6
electrification 221
Elgin, Catherine 295
Eliade, Mircea 168
Elias, Norbert 292
Ellis, David 182
emotion 237, 239–40
encyclopedic interdisciplinarity 17, 453
end-user conviviality 265
endogenous interdisciplinarity 436
Endogenous University ID 27
endoscope simulation 253
'energetics' 53, 55
Engels, Friedrich 53
engineering 147–59, 460–1
  biology and 125–6
  branches 149
  for communities in need 155–7
  in developing countries 159
  in Europe 158–9
  evolution 147–9
  future 153–5
  heterogeneous 84
  institutions 148–9
  interdisciplinarity in 337–8
  subdisciplines 149
  women in 158
  see also civil engineering
Engineering Criteria 2000 (EC 2000) 151–2
engineering design 278–9
Engineering for developing communities (EDC) (Columbia University) 156–7
Engineering for Developing Communities Program (UC-Boulder) 155
engineering ethics 266
Engineering Projects in Community Service (EPICS) 155
Engineers Council for Professional Development (ECPD) 151
entertainment industry 251
enthusiasm, protected 408–10
entropy 59, 60
entropy principle 59
environment 494–506
  system and 456
environmental crisis
  first wave 495–6
  second wave 495, 496
environmental ethics 263, 498, 501–2
environmental issues 129
environmental philosophy 501–6
environmental protection 171, 263

environmental studies 349–50, 362
enzymes 123
epicycles 88, 89
epistemic knowledge, individual cases and 42–3
epistemological humility 217
equifinality 60
Erasistratus 120
essence theories 165–6
ethics
  as branch of philosophy 259
  history 260–2
  of interdisciplinarity 267–70
  morality vs. 259
  science and 261
  see also applied ethics
Ethics and Values in Science and Technology (EVIST) 266
ethnocentrism, of disciplines 315
ethnographic approaches 198
ethnomusicology 138, 139
Europe, general education 369
evaluation, interdisciplinarity research *see* research evaluation
Evangelical Environmental Network 171
Evergreen State College 353–4
evolutionary biology 122
excellence
  academic 321–2
  disciplinary 312–13
  disciplinary standards of 328–9
Exogenous ID 27
exoskeletons 126
expectation value 538–9
expected utility 539
experiential learning 375, 433
experimental psychology 235
experimentation 252
expert review 322
expert systems 239
expertise 96, 324–27
  interactional 96
  nature xxxiii–xxxiv
experts, teams of 529–30
external relations, epistemology of 327
external validity 517
extreme events (XEs) 157–8

faceted notation 180
faculties, hierarchy of 4–5
faculty, in interdisciplinary programs 347
familiarization 406–7
Faraday, Michael 82
Farnsworth, Philo 546
Fauconnier, Gilles 290
Fechner, Gustav 60
feedback 447
feminism 53–4
Festinger, Leon 227
field-creation 182
field testing 512
financial analyses 480–1
finite difference methods 248
finite element methods 253
Finnish National Electronic Library 181
fishery management 502
flexibility, intellectual 270
flight simulators 247, 252
Flood, Robert L. 466
flood protection 461
focusing 18
Foerster, Heinz von 456
food crisis 334–5
foresight methods 441
formalization 112–13
Forrester, Jay 255
foundation of interdisciplinarity 342–3
Foundation Mercator Schweiz 435
'fourth trendline' 509
fractionation 96–7
Franco, L. A. 446
Frank, Philipp 54
Freon 496
fresh water, and climate change 288–9, 302–3
Freud, Sigmund 55
Fribourg, University of 466
Friedman, Ken 281
Friedman, L. M. 526
Friedman, Milton 423
fundamental equations 255
funding models 488–9
funding programs 9
fuzzy front end (FFE) phase 548–9, 552

Galileo Galilei 88, 89, 213
Galison, Peter 91, 254
gaming 251, 253
Garvin, David 36, 37
gatekeepers 98
Gaylin, Willard 419–20
Gebert Rüf Foundation 435
Geertz, Clifford 20
GenBank 124
general circulation models (GCMs) 255–6
general education, interdisciplinary 354–7, 360–71
  future 370–1
  historical background 360–1
  outside United States 368–70
  pedagogical innovations and 365–7
  status 367–8
  in United States 361–4
general semantics 55
General Social Survey 112
general systems theory (GST) 60–2, 454–5, 462–3
generalists 182
generalizing interdisciplinarity 20, 227, 453

generosity, intellectual 270
genetic code 123
genomes 121
genomics 121, 124
geography, divisions 116
geometry 33
Georgetown University, Center for Social Justice Research, Teaching, and Service 352–3
German Ministry for Education and Research 441
Giere, R. 44
Gill, Sam 166, 167
global poverty 153
'global village' 222
global warming 496
globalization, effects 370, 532–3
globalization studies 109
Gödel, Kurt 88
Goodman, Nelson 295
Government Performance and Results Act (GPRA) of 1993 329, 339
grammar 223
grand unified theories (GUTs) 57
Graybill, Jessica K. 328
Great Chain of Being 121
Great Lakes Fishery Commission 502–3
Greek city-states 523
Green Revolution 334–5
Green Valley (Denton County, TX) 504–5
grid computing 256
Griffith University (Queensland) 369
Grinder Report 151
Griswold, I. M. 529
Gropius, Walter 284
group processing 376
Gunn, Gile 23

Habermas, Jürgen 99, 466
Hagstrom, Warren 82, 83, 85
Hales, Stephen 121
Hall, Arthur D. 461
Hall, Stuart 208
Hampshire, Stuart 541
Harcombe, Paul 504
Harvard Business School 36–7
Harvard Law Review (HLR) 529
Harvard Law School 36, 524, 526, 529
Harvard Medical School 36, 37
Harvard University, Center for Cognitive Studies 242
Harvey, William 120
HASTAC 206–7, 208, 211
HASTAC Scholars 214
Hastings Center 419–22, 426–7
Hayakawa, S. I. 55
Hayek, Friedrich von 423
Haynes, C. 384–5
health, complexity 510–11
health care
- costs 424–5
- market 423–4
- reform 423–4

health informatics 511
health sciences 508–19
- interdisciplinarity need 510–13
- paradigm shift 513–17

health services 508–19
- incentive systems 518
- interdisciplinarity need 510–13
- interdisciplinary example 515–16
- paradigm shift 513–17

health systems, complexity 510–11
heavy ion physics 85
Heckhausen, H. 18
Heeger, Alan J. 84
Helmholtz, Hermann von 86
Heraclitus 463
heredity, patterns of 110
heterogeneous engineering 84
hierarchy 455
Hinduism 260–1
historical monuments, crafting 300–2
historical musicology 136, 138, 139, 143
historiography 34, 135, 458
history, organization 116
HIV 508, 509
Hjørland, Birger 180
Hobbes, Thomas 523
Hochschule für Gestaltung (HfG) 284–5
holism 65, 68, 454–5
- critique 69–70
- epistemological 327
- expertise claims as political 73
- future 74–5
- in health services 517–18
- reductionism vs. 68, 327

Holmes, Oliver Wendell, Jr 524
Horwitz, Morton 524–5
hospice care 511–12
hourglass model of innovation 550–2
Houston, TX 496
Hovland, Carl 227
Hull, University of 465–6
human suffering, alleviation 155
humanities
- in environmental studies 506
- as interdisciplines 33
- in law 531
- reward systems 338
- sciences and 144–5
- technology and *see* technohumanism

humanities computing 207
Humboldt, Wilhelm von 325
humility 74–5
- intellectual 270

Humphreys, Paul 254
Hutchins, John Maynard 362

Hutchins School of Liberal Studies 354
Huutoniemi, Katri 474
hybrids 21, 22
hypertext 222
hypodermic needle model 226

IBM 248
ICE 148, 149
ideal type 46
identification techniques 530
identity construction 229
ideology 226
idiographic knowledge 34–5, 42
IEEE 149
ignorance 445
image studies 141
Imataca forest reserve 504–5
implementation sciences 431
  see also integration
improvement 464
improvization 268
incentives, for pursuing interdisciplinarity 415, 518
inclusive pedagogies 377
incremental innovation 547
Indiscriminate ID 17
inductive generalization 38
inductivism 227
Industrial Revolution 336
inertia 58
inference 241
info-bio-nano technology 153
Information Ages 209, 222
information behavior research 177, 179–84
information ecology, integrated 185
information ethics *see* computer ethics
information processes, models of 181–2
information research 174–85
  bibliometric research 177–9
  information behavior research 177, 179–84
information scatter 176–7
information sources, traditional, value of 185
information theory 227
initiation 406
Initiative on the Study and Implementation of Systems (ISIS) 515–16
Innis, Harold 227
innovation 269
  breakthrough 547–8, 556
  disruptive 557
  hourglass model 550–2
  incremental 547
  integrated 276
  invention vs. 546
  open 557
  radical 557
  see also corporate innovation
Innovation Space 276–7
innovation systems 98
inquiry, contribution to progress 73
inquiry-based learning 377
insights, disciplinary 298–9, 300–1
Institute for Communications Research 223
Institute of Design 284
Institute of Ecology and Biodiversity (IEB) 499
Institute of Electrical and Electronics Engineers (IEEE) 149
Institute for Propaganda Analysis 226
Institute for Science Information (ISI) 178, 179
Institution of Civil Engineers (ICE) 148, 149
institutional adaptation 412–13
institutional support, framework for encouraging 415–16
instrumental interdisciplinarity 22–4, 171, 225, 363, 438
instrumental values 439
instrumentalists 41
instrumentation 198
Integrated ID 20
integrated innovation 276
integration 431, 443–8
  core concepts 444–6
  core methods 446–8
  emphasizing 312, 313–15
  framework 448
  'from–to' tacit 553
  vertical 513
integrative education 17
Integrative Graduate Education and Research Traineeship (IGERT) program 410
integrative interdisciplinarity 453, 456, 467
integrative learning 363, 366
  definition 373, 380
'Integrative levels classification' project 180
integrative technological systems 97–8
integrative technologies 98
integrity, intellectual 270
intelligent design 59
intention 240
interactional networks 98–9
interactive coaching model 314
interactivity 222, 228
interart comparison 135, 140
intercultural education 365–6
interdisciplinary case work 31
  see also case studies
interdisciplinary centers/institutes 350–3

interdisciplinary communication 31
interdisciplinary cycle 119, 120
interdisciplinary education, positive and negative aspects 391
interdisciplinary facilities 84
interdisciplinary fusion 31
interdisciplinary hire, structuring 393–4
interdisciplinary inquiry, modes 182
interdisciplinary learning 288–304
  defining 372–4
  examples 300–3
  future research 303–4
  in higher education context 374–5
    levers for change 375–80
  literature review 290–1
  pragmatic constructionist view 289, 294–5, 298–300
  as systems of thought in reflective equilibrium 298–300
interdisciplinary PhD, meaning 411–12
interdisciplinary programs 348–50
interdisciplinary relevance 184
interdisciplinary research (IDR)
  definition 433
  funding 398–9
  impediments to 393
  positive and negative aspects 391
  ways of conducting 388–9
interdisciplinary research and pedagogy (IDRP)
  institutional support 415–16
  pursuit 405
interdisciplinary research and training (IDRT) programs 404
interdisciplinary scholars 387–400
  evaluation 396–7
  funding 398–9
  institutional support 392–3
  personal approaches to interdisciplinarity 387–91
  productivity and 395
  recognition 395–6
  structuring interdisciplinary hire 393–4
interdisciplinary schools/colleges 353–4
interdisciplinary teaching
  new approaches 381–4
  positive and negative aspects 391
interdisciplines, emergence 82–4
interfaith movements 167–71
Intergovernmental Panel on Climate Change (IPCC) 47, 256, 288–9
internal validity 517
International Institute for Applied Systems Analysis (IIASA) 461
International Musicological Society 143
International Society for Knowledge Organization (ISKO) 180
International Society for the Systems Sciences 454
Internet
  cognitive capacities and 229
  in interdisciplinary learning 380–1
interpretative paradigm 432–3
INTERSTUDY 362
intervention research 437
invention, innovation vs. 546
investigator interviews 479
ISABELLE 84–5
ISIS 215–16
ISI 178, 179
ISKO 180
Islam, ethics in 261
Islamic–Christian dialogue 170
Islamic scholarship 163
Islamic Sharia law 533
Ivanitskaya, Lana 290

Jackson, Michael C. 465–6
Jantsch, Erich 24, 40, 432, 436
Jasanoff, S. 529, 532
Jaspers, Karl 261
Jefferson, Thomas 423
Jewish–Christian relationships 170
jigsaw method 377
Johnson, Philip 297
joint appointments 393–4
journal network maps 457
journals, interdisciplinarity measures 178–9
Judaism, ethics in 261
jury selection 531

Kant, Immanuel 324, 325, 466
Katrina, hurricane 380–1
Kazakhstan, general education 369
Kellert, Steve 504
Kelly, J. M. 523
Kepler, Johannes 89
keyword searching, cross-database 181
Klapper, Joseph 231
Klein, Jacob 90
Klein, Julie Thompson 217–18, 267, 283, 373–4, 509
knowledge
  categorization 3–4
  co-creation 517

systematic integration 24
transdisciplinarization of 315
unification, of 230
knowledge domains, mapping 395
knowledge exchange 514
knowledge flow maps 458, 459
knowledge integration, reasons for 433
knowledge management and transfer 516
knowledge market 12
knowledge policy 315, 328
knowledge production
in digital age 269
modes 12, 25–6, 51–2
requirements 436
17th to 18th centuries 4–5
knowledge quality assurance 294
knowledge society 12, 47
knowledge translation 513–14
Korzybski, Alfred 55
Kraft, Selma 137, 139
Kuhn, Thomas 91, 193, 226
Kyrgyz Republic, general education 369

labor, division of 324, 325
laboratory studies 201
lake rehabilitation project 43
Lambert, Johann Heinrich 432
Lamont, M. 328
Lancaster University 462, 503
land ethic 263, 495
land-use changes 503–4
Langdell, Christopher Columbus 524, 526
language, use of 296
language disciplines, perception disciplines vs. 296–7
Laplace, Pierre-Simon 89
Lasswell, Harold 225–6
Laszlo, Ervin 464–5
Latour, Bruno 193–4
Lattuca, Lisa 17, 18, 20
Lavoisier, Antoine 121
law 522–33
definition 522
education 525–9
of future 532–3
historical development 523–5
media reports 529
politics and 524, 525
questionable concept of 44
research 529–32
science and technology and 530–1, 533
Law and Economics school 525
law reviews 529
law schools 525–8
layered reality 39
Lazarsfeld, Paul F. 224, 227
learning
neural mechanisms in 241
theories of 433
learning communities 354, 357, 366, 380
learning styles, preferred 375
Legal Process movement 525
Legal Realism 524–5
legal scholarship 531–2
legitimacy 194, 310
Leibniz, Gottfried von 57, 89
Lentsch, Justus 98
Leon Manifesto 180
Leonardo da Vinci 86, 125, 147
Leopold, Aldo 263, 494–5, 497, 501, 502, 506
leveraging integrations 298–9, 301–2
Levi, Albert William 135
Lewis, William Draper 527
library and information science (LIS) 174–5, 212
classification systems 180
see also information research
Lichnerowicz, Andre 24
life, meaning of 241
life experience 21
life scientists, computer scientists vs. 93
lifetime achievement awards 396
Limits to growth 255
Lin, Maya 288, 293, 294, 300–2
linear model 325-6, 431–2
linguistics, as cognitive science discipline 236, 237
linking issues 18
Linköping University 369
literature-based discovery (LBD) 177
litigators 528
living-learning communities 380
logic 223
logical positivism 291, 293
Lucretius 235
Luhmann, Niklas 456, 464
Lyotard, Jean-Francois 62

McCloskey, Pete 501
Machine Age 464
McIntosh, Ronald P. 495
McLuhan, Marshall 227
Macy Conferences 248
Magallanes, University of 499
Magellanic woodpecker 499–500
magic bullet model 226
Magic Eye images 553–7
Maldonado, Tomás 284–5
Man and Biosphere program 442
management, of interdisciplinary research 482–5
Mann Report 151
mapping 23, 457–60
conceptualizing science for 457

Mapping Science exhibit 460
Marconi, Guglielmo 221
Maritain, Jacques 55
Markov chain Monte Carlo (MCMC) 248
Marxism 108
Maryland Institute for Technology in the Humanities (MITH) 208
mass communications research 224–7
'mass society' 225
mastering multiple disciplines 311–13
mathematical modeling 112
'mathematical sculpture' 86
mathematics 56–7
  biology and 126–8
  definition 88
  history 88–90
  as root interdiscipline 90
Maturana, Humberto 456
meat consumption 335
media and communication 220–31
  content vs. form 226–7
  history 221–2
  journals 224
  mass communications research history 224–7
  networked communications research 227–30
  overview of studies 222–4
  scientific trajectory 230–1
medical ethics 421
medical S&T 194
medical technology 424–5
Medicare program 424, 425
medicine 497
  biology and 120–2
  ethics in 262
  health care market and 423–4
Meek, Esther 553
Meiklejohn, Alexander 362
Melbourne University 369
membrane transport 123
memorandum of understanding (MOU) 394
Merriam, Charles 106
Merz, J. T. 52–3
meta-cognition 302
meta-ethics 262
metabolomics 124
metadisciplnes 495
metaphors 301–2, 304
method ID 40
methodological constructivists 41
Methodological ID 19–20, 23
Michigan State University 356
Michigan, University of
  Center for Research on Learning and Teaching 382
  Center for the Study of Complex Systems (CSCS) 351–2
  women's studies 349
Midgley, Gerald 454
military research 336
military training simulations 216
Miller, George 235, 242
Miller, Matthew 290
Miller, Raymond 19, 21
Mills, Victor 547
mining, schools of 148
Minnesota, University of
  general college 362
  library service report 184
  women's studies 349
minute paper 379
MIT, postwar emergence 336
Mitchell, Sandra 45
Mobile Musical Networks 214–15
Mode 1 interdisciplinarity 433
Mode 2 interdisciplinarity 433
Mode 2 research 350–1
Mode 2 science 315, 328, 509, 516
model-based methods 447
modernity
  assumptions 85
  definition 67
modernization 107
Moholy-Nagy, László 284
molecular biology 59, 124
Monod, Jacques 61
Monte Carlo methods 248, 254
Montreal Protocol 496
Moore's law 249
moral philosophy 421
morality 260–1
  ethics vs. 259
Morrill Land Grant Act 151
Morris, Charles 284
Morse, Samuel 221
motivations
  classification 24
  serial innovators 549
Muir, John 501, 502
Müller, Max 166
multi-agent based models 505
multi-modal systems thinking 467
multicultural education 365
multiculturalism, promotion in higher education 377
multidisciplinarity (MD) 82, 217–18, 283, 474
  definition 17
multidisciplinary learning 373
multivariate data analysis 112
multiversity 345
Mumford, Karen 502–3
Murdoch University (Perth) 369
music, definitions 142
music physiology 143, 144
music psychology 143–4
music research 133–45
  origins 134–7
  subdivision 138–9
music theory 142

musicology 133, 136–7
new 137–9
rethinking (inter)disciplinarity in 139–41
mutability 269
Myrdal, Gunnar 275

nanotechnology
constructive technology assessment 443
as multidisciplinary 179
Narrow ID 18
National Academy of Engineering (NAE) 153–4
National Cancer Institute (NCI) 471–2, 515
National Center for Post-Secondary Improvement 379
National Collegiate Honors Society 362
National Endowment for the Humanities (NEH) 208, 266
National Environmental Policy Act (1969) 496
National Forum on Biological Diversity 496
National Institutes of Health (NIH) 266, 324
Office of Behavioral and Social Science Research (OBSSR) 513
roadmap programs 473
National Nanotechnology Initiative 337
National Religious Partnership 171
National Science Foundation (NSF)
Biocomplexity-in-the-Environment Phase II, 503
as boundary organization 329
and engineering ethics 266
foundation 324, 326
funding programs 337, 398
and GPRA 329–30, 339
and graduate training in STS 199–200
Integrative Graduate Education and Research Traineeship (IGERT) program 410
merit review criteria 329–30, 339–40
SciSIP program 339–40
National Synchrotron Light Source (NSLS) 79–81, 84, 94
nationalism 108
Native American Church 166
natural law, as standard 44, 55
natural sciences
bridging with social sciences 343–4
interdisciplinarity in 337–8
reward systems 338
see also physical sciences
natural selection 59
navigation 407
negotiations 527
Nelson, Gaylord 501
neo-Piagetian approach 290
neo-Thomism 54–5
network analysis 113
networked architectures 250
networked communication 221
research 227–30
networked sensing systems 184
neural imaging 116
neuroeconomics 241
neuroscience, as cognitive science discipline 236–7, 239, 240, 243
New Bauhaus 284, 285
new media
distinguishing features 222
old media and 228
'New Pathway Program' 36, 37
Newell, Alan 242
Newell, William H. 18, 513
Newman, William M. 166–7
newspapers 228
Newton, Isaac 56–7, 58, 89, 164, 213
NIH *see* National Institutes of Health
Nikitina, Svetlana 290–1
NIMBY reactions 541–2
nomadic pastoralists 438
'nomological machine' 44
nomothetic knowledge 34–5, 42
normative decision theory 542
normativity 536–7
norms 239
NSF *see* National Science Foundation
NSLS 79–81, 84, 94
nuclear ethics 263
nuclear strike 539
Nuffield Foundation 21
Nuremberg Code 262

Obama, Barack 326–7
object interdisciplinarity 39
O'Connor, S. D. 532
Office of Behavioral and Social Science Research (OBSSR) 513
Office of Naval Research (ONR) 324
Office of Sponsored Research 351
Oldenburg, Henry 324
Omora Foundation 499
oncology 127
online publishing 228
open admissions 377
open systems 60–1, 455, 456
opening 182
operations research 461, 462
opinion polls 111

Opportunistic ID 23
oratory 223
organ transplantation 420
Organisation for Economic Cooperation and Development (OECD) 311
organismic biology 455, 463
organizational learning 433
organizational proximity 98
orientation 182
originality 7
Orwell, George 74–5
Osler, Sir William 121
Oswald, F. 437–8
ozone hole 496

Pacific salmon 502
Pakistan, general education 370
paleoclimatological data 255
Palmer, Carole 182
Panel on Modernizing the Infrastructure of the National Science Foundation's Federal Funds for R&D Survey 27
paradigm earthquake 226
paradigm shifts 226
parallax view 50–1
parallel distributed processing 239
paramountcy clause 266
Parliament of World's Religions 168, 169
Parmenides 463
Parsons, Talcot 456
participatory learning 209
particle collider 256
Pascal, Blaise 89
Pasteur, Louis 121
pathophysiology 121
patient transfer systems 274–5
patterning genes 121
Pavlov, Ivan 121
PC Revolution 250
pedagogical content knowledge 304
pedagogical innovations, interdisciplinary general education and 365–7
pedagogies 372–85
  design 280
  inclusive 377
  institutional support for interdisciplinary 415
  student-centered 394
  see also interdisciplinary learning
peer communities, extended 316
peer review 8, 90, 321–31
  blind 322
  criticisms 322–3
  disciplining of 324–7
  external 480
  history 323–4
  interactive 314
  interdisciplinary pressures on 327–30
  newer forms 382
  prospective 322
  retrospective 322
  societal relevance assessment 330–1
Peer Review of Teaching Project 383
perception 542
perception disciplines, language disciplines vs. 296–7
perceptual experience 296
performance, experimentation as 84
performance art 137
performativity 268
periodization 135
personality, serial innovators 550
perspective, serial innovators 550
persuasion 223
pesticides 495–6
Pew Internet and American Life Project 217
phenotypes 45, 121, 122
philosophy xxxi
  branches 259
  as cognitive science discipline 236–7, 239
  in conservation biology 498
  history 260–2
  as interdisciplinarity xxxiii
  of interdisciplinarity xxxiii, 39–41
  law courses 529
  in legal education 527
  of science xxxii, 35, 41, 199
photosynthesis 123
physical sciences
  interdisciplinarity in 79–99
    history 82–7
    integrative technological systems 97–8
    interactional networks 98–9
    practical issues 87–92
    theoretical issues 92–7
physics 56–7
Piaget, Jean 24
pictorial turn 141
Pinchot, Gifford 502
Planet DHASS 208
Plato 56, 260, 323, 552
pluralism, disciplinary 315
Polanyi, Michael 546, 552–3, 557
policy challenges
  twentieth century 335–6
  twenty-first century 333–5
policy research 200–1, 335, 338, 341, 342–4
policy schools 336–8
  disciplinary limits 338, 341
political economy 108
political science
  creation 111
  divisions 115
  law courses 528

politics
law and 524
serial innovators 550
population dynamics 127
Portland State University 355–6
positivism 55
positivist paradigm 134–5
positivists 41
Posner, Richard 531
post-normal science 315, 328, 436
postmodernism 62, 370
postnormal science 25, 26
post-structuralism 223
Pound, Roscoe 524, 526
power, science and 202
'practical criticism' 55
pragmatic constructionist epistemology 294–5
pragmatism 41, 461
prediction 505
predictive biology 122
predictive medicine 122
preparation, serial innovators 549–50
preventive medicine 122
Priest, George 527, 528
principal investigators (PIs) 398
printing press 221
privacy, individual 264
probability 89–90
probing 182, 183, 184
problem-based learning (PBL) 366, 377
problem-based research 509, 516–17
see also problem solving
problem framing 444–5
collaborative 482–5
disciplinary 484
problem identification 40, 440, 442–3
problem investigation 443
problem-oriented interdisciplinarity 40–1, 389
problem solving
knowledge for 438–40
through transdisciplinary research 431–52
trans-sector transdisciplinary 25, 283, 509
see also problem-based research
problem structuring 440, 442–3
process, serial innovators 550–2
Procter and Gamble 547
product-based methods 447
productivity, interdisciplinary scholar and 395
professional ethics 266
professional preparation 21
professional researcher 42
progress 67
Progressive Legal Thought 524
promotion criteria 396–7
promotion system 409
propaganda 226
property 527
protected enthusiasm 408–10
proteins 123–4
proteomics 124
Protestant belief systems, prioritization 166
Protestant reformation 164
proto-disciplinary outcomes 373
prototypes 240
concepts as 237
Pseudo ID 17
psychoanalysis 55
psycholinguistics 236, 240
psychology
of art 138
as cognitive science discipline 236–7, 239
divisions 115
economics and 113
testing in 110
Ptolemy, Claudius 88
public concern, pursuing issues of 114–15, 116
Public Health Service Act of 1944 324
Puerto Williams 499–500
purpose, interdisciplinary 298–9, 300
purpose-oriented interdisciplinarity 40
Pythagoras 88, 143

quality assessment 90–1, 318
quality criteria, for interdisciplinary research 311–16
quantitative research methodology, area studies vs. 113–14
quantity 90
quantum mechanics, development 85
quasi-experimental research designs 481, 485, 486

racial hygiene 54
radio advertising 225, 227
RAND Corporation 461
random dot stereograms (RDSs) 553–7
rationalists 41
rationality
contribution to progress 73
norms of 543
RDSs 553–7
Reactor Safety Study 539
real-constructivism 39
real-constructivist ID 39
real-constructivists 41
real ethics 268
realists 41
reasonableness 194
reception 201–2
reception problem 313
recombinant science 84
reconciliation 379

reductionism 65–6, 509, 513, 514–15
epistemological 327
holism vs. 68, 327
problem of 291–2
refeudalization 99
reflective equilibrium, system of thought in 295
regenerative medicine 126
Relativistic Heavy ION Collider (RHIC) 84–5, 87
relevance 327–31
interdisciplinary 184
religion
art and 164
biology and 129
'essence' of 165–6
history 163–5
in school 166
science and 164, 165
religious studies 161–72
as discipline 162–3
ecumenical and interfaith movements 167–71
interdisciplinarity in history 163–5
religiosity vs. secularity in 165–7
as specialized field 168
subdisciplines 162
remediation 228
Repko, Allan 367
representations 239, 240, 241
research centers
co-funding 398
interdisciplinary 337, 342
Research Condition and Disease Classification (RCDC) database 27
research evaluation 309–18, 396–7, 412, 416
definition 310
role of evaluators 316–17, 412
values in 311–16
research technology 254
residential scholars 487
restricted sciences 83
restructuring, bridge building vs. 21
revelation, divine 261
reward systems 338, 343, 487
rhetoric 223
RHIC 84–5, 87
Richmond, Sheldon 135
Rickert, H. 35
van Riessen, Hendrik 466
rights-based approach 525
risk 536–44
acceptability 536, 537
behavioral sciences approach 538, 542
communication 538
ethical issues 543
expert role 543
as fact-laden 543
history 537–8
measurement 536–7
multidisciplinarity failures 542–4
objective 538
as objective 544
perception 538, 542
statistics approach 537, 538–9
subjective 538
as subjective 544
as value-laden 543
risk–benefit analysis (RBA) 537, 540–2
risk-taking 411
Rittel, Horst 274, 279
RNA 123–4
Rosch, Eleanor 237, 240
Rosenfield, P. L. 474
Royal Society of London 323–4
Rozzi, Ricardo 498
Russia, general education 370

Sachs, Jeffrey 269
St Lawrence University 356–7
St Petersburg University 370
Santa Barbara, CA 496
Scalia, Antonin 530
Schelsky, Helmut 292
Schenker, Heinrich 142
schizophrenic effect 510
scholarly associations 13
scholarship of teaching and learning (SoTL) 383
Schramm, Wilbur 223, 224
Schwartzchild, Arthur 87
science(s)
arts and 135–6, 137–8
ethics and 261
growth 5
humanities and 144–5
impediments to interdisciplinarity in 129–30
law and 530–1, 533
linear model 325–6, 431–2
politicization 73, 326–7
power and 202
purpose 86
recombinant 84
religion and 164, 165
society and 340
unity 6, 11–12
Science Citation Index 178
'science studies' centers 193
'science of systems' 61
science of team science (SOTS) 471–91
conceptual developments 472–8
conceptual models 475
emerging directions 486–90
future research 491
methodologies 478–81, 485
qualitative 480–1, 485
quantitative 479–80
units of analysis 474
see also team science (TS) initiatives
science and technology (S&T)
impacts and control 192, 195–6
nature and practices 192, 193–4
Science and Technology Policy Program 27

science, technology, and public policy (STPP) 199
science, technology and society programs 195–6
science and technology studies (STS) 191–205, 468
  academic institutionalization 198–200
  disciplining 203–4
  interdisciplinary history 192–8
  outlook 202–4
  redundancy charge 203
  research frontiers 200–2
'science wars' 198
scientific associations 8, 11
scientific research, ethics 266
scientific retreats 479
Scientific Revolution 56, 148
SciSIP program 339–40
sea lamprey 502
search engines, broad-based 185
selective attention 10
self-directed qualitative discussions 479
self-organization 456
self-organized complexity 59
self-referential systems 456
self-referentiality 8
Selten, Reinhard 292
semantic web 185
semiotics 230, 284
Sempacher Lake 43
sensitivities 505
serial innovators (SIs) 546, 549–52
  characteristics 549–52
  'from–to' tacit integration 553
  innovation interdisciplinarity in 556–7
service learning 366, 377–8
Shandas, Vivek 328
Shannon, Claude 59, 227
Shared Components 19
Shared ID 19
sharing interdisciplinarity 453
'shift work', ethics of 267–70
Shinn, Terry 250, 254
'signature' curriculum 355
Sill, David J. 296
silos 215
Simon, Herbert 242, 455
simulacra 254
simulation 246–7
  see also computation and simulation
skepticism 552
Skiena, Steven 92
skills 38, 41
Skinner, B. F. 235
sleeper effect 227
Smeaton, John 148
Smelser, Neil 21
Smith, Adam 423
smoking cessation 513
Smuts, Ian C. 454–5
Snow, C. P. 213, 294
social action, theories of 432–3
social contract theory 523
social epistemology 175
social factors, biological factors and 510
social network analyses (SNA) 481, 516
social networking 184, 222
social proximity 98
Social Science Research Council (SSRC) 21, 110
  creation 105–6
Social Science Research Institute (SSRI) 352
social sciences 103–15
  bridging with natural sciences 343–4
  disciplines 103–4
  history 104–6
  integration 455
  interdisciplinary organization
    patterns 114–15
    strategic investments 343
  in law 531
  in legal education 527
  research methods 110–14
  reward structure 338, 343
  universalizers 103
  see also area studies
social system(s) 455
social systems theory 456
social work, in hospice care 511–12
socialist feminism 53–4
Society of Civil Engineers 148
Society for Conservation Biology 495
Society for General Systems Research 454, 462
Society for the Promotion of Engineering Education (SPEE) 151
Society for the Scientific Study of Religion (SSSR) 167
Society for Social Studies of Science (4S) 194, 195
sociobiology 129, 231
sociological jurisprudence 524
sociology
  divisions 116
  law courses 528
sociology of scientific knowledge (SSK) 193
Socrates 221, 323
Socratic method 526
Sonoma State University, Hutchins School of Liberal Studies 354
Soulé, Michael 495, 497, 498
Spanner, D. 389
specialization 6–7, 10–11
specialized programs of research excellence (SPOREs) 473

spectacle theories 226
Spinoza, Baruch de 464
spirituality, uniting 465
spreadable media 209
Squire, Geoffrey 162
SSRC *see* Social Science Research Council
Stafford, Barbara 141
Stage-Gate process (SGP) 548–9, 552
Stanford Law School 36
Stanford University
  postwar emergence 336
  School of Earth Sciences 349
statistics 89–90
  in risk analysis 537, 538–9
  in social sciences 110–11
statutory law 522, 525
Steelman report 326
stem cell research 98
stimulus–response model 226
Strategic ID 23
Stoa 3
string theory 89
structural biology 124
structural differentiation 260
structural sciences 40
STS *see* science and technology studies
student-centered pedagogy 394
student portfolios 380
students, in interdisciplinary programs 347
success, strategizing for 413–16
Superconducting Super Collider project 339
Supreme Court 532
survey research 111–12
surveying 33
surveys, standardized 480
sustainability foresight project 441–2
sustainable development 155, 263, 438
Swanson, Don 177
Swiss Academic Society for Environmental Research and Ecology 434
Swiss Academies of Arts and Sciences 434
Swiss National Science Foundation (SNSF) 314
symbols 229, 254
synchrotron radiation 79, 84
synergetics 40
synergy, emphasizing 312, 313–15
synthesis 464, 482–3, 485
system dynamics modeling 516
system of thought, in reflective equilibrium 295
systematic musicology 138–9, 144
systemization 5
Systems Age 464
systems analysis 461
systems approach in technology and management 454, 460–3
systems biology 124
systems differentiation 456
systems dynamics 447, 461–2
systems engineering 460–1
systems knowledge 438–9
systems philosophy 454, 463–5
systems science 454–6
systems theory 112, 432, 514
systems thinking 444, 453–68, 509
  critical 465–6
  hard 462, 465, 466
  for health knowledge integration 514–15
  multi-modal/disclosive 467
  soft 462, 465, 466
  tacit aspects 553
Szostak, Richard 180
tacit integration, ‘from–to’ 553
tacit knowledge 553
Tajikistan, general education 369
Tamanaha, B. 528
target knowledge 438–9
taxonomy of interdisciplinarity 15–28
  new implications for 26–8
td-net 434–5
team care, models 510, 512
team-collaboration 389
team leaders 182
team management 399
team science (TS) initiatives 471–91
  antecedents 475, 476
  characteristics 473–4
  contextual influences on effectiveness 489, 490
  funding models 488–9
  information overload management 491
  outcomes 475, 476
  processes 475, 476
  program effectiveness assessment 474–5
  typology 488–9
  see also science of team science
team teaching 366, 382, 421
techno-object ID 39
technohumanism 206–18
  academe and education and 212–15
  interdisciplinary programs and 215–16
  prospects 216–17
technological imagination 268
technology/ies
  humanities and *see* technohumanism
  law and 530, 533
  as reductionist 67

see also science and technology
technoscience, invention 194–5
Telehospice project 511–12
telemedicine 511
tenure
 guidelines for 397
 review 397
tenure system 409, 413, 414
terrorism, holistic expertise 75
Thagard, Paul 243
theism 466–7
theodicy 57–8
Theoretical ID 19, 20
theoretical neuroscience 239
theory of everything (TOE) 89
theory ID 39–40
thermodynamics 53, 58–9
thresholds 505
tipping points 505
tobacco control 515–16
topics 21
torts 526–7
'Tory' historiography 51
Toulmin, Stephen 262, 263
trading zones 91, 94–6, 254
transcendent interdisciplinary research 25, 509
transdisciplinarity (TD) 12, 24, 40, 474
 debate on 432–3
 definitions 24, 283, 474
 and ivory tower 328
 trendlines 24–6
transdisciplinarization, of knowledge 315
transdisciplinary research (TR) 431, 433–43
 bringing results to fruition 443
 definition 433
 elements needed 435
 knowledge for problem solving 438–40
 project as system 436, 437
 recursive research process 440, 442–3
 subject 433, 436–8
 see also integration
Transdisciplinary Research on Energetics and Cancer (TREC) 474
 evaluation model 477, 478
 training 487
transdisciplinary science 25, 509
Transdisciplinary Tobacco Use Research Centers (TTURCs) 471
 bibliometric analysis 481, 485
 financial analyses 481
 Initiative-wide Evaluation Study 475
 logic model 475
 social network analyses 481
 survey 480
transdisciplinary training programs 486–8
transfer methodology 40
transformation knowledge 438–9
translation 182, 183–4
trans-sector transdisciplinary problem solving 25, 283, 509
TREC *see* Transdisciplinary Research on Energetics and Cancer
Trueman, Daniel 214
trust 96, 538
TTURC *see* Transdisciplinary Tobacco Use Research Centers
Turner, Mark 290
'two cultures' divide 213–14, 294
two-step flow model 226

Udall, Stewart 496
Ulrich, Werner 466
'Ultimate Reality/ies' 171
UN Millennium Development Goals (MDGs) 155, 156
uncertainty 445
unconditional laws, individual case and 43–4
undiscovered public knowledge 177
unidisciplinary (UD) collaboration 473
Unifying ID 22
United Arab Emirates, general education 370
United States Environmental Protection Agency 496
unity, recovery 56–8
Universal Declaration of Human Rights 55
universal object ID 39
university
 as institution of knowledge production 12, 13
 reforms of 105, 336–8, 342–4
University of North Texas (UNT) 503
University of Wisconsin–Stevens Point (UWSP) 501–2
 College of Natural Resources (CNR) 501
univocality 56
unrestricted sciences 83
urban design, transdisciplinary method 437–8
urban studies 116
utilitarianism 41, 541
utopias 228
Utrecht University 369

validation 256
validity
 external 517
 internal 517
value theory 58

values 35, 445
  conflicting 70, 439
  inherent 504
  instrumental 439
  intrinsic 504
values analysis 502
valuing, process of 317
Varela, Francisco 456
Vasari, Giorgio 134
Vectors (journal) 210–11
Verein Deutscher Ingenier (VDI) 158–9
vertical integration 513
Vietnam Veterans Memorial 288, 293, 300–2
Vietnam War 195
virtuality 222
vision-based methods 447
visual culture 141
visual representation 198
visuality 138, 141
visualization 252, 457
vitalism 61
von Neumann, John 247
voting choice 241
Vrije Universiteit (Amsterdam) 466

de Wachter, Maurice 262–3
Wallerstein, Immanuel 108, 115
Washington Center for Improving the Quality of Undergraduate Education 354
water availability 288–9, 302–3
weak information work 182–4
wearable technologies 222
Weaver, Warren 59, 227
Web 2.0 222
web surfing 183
Weber, Max 46, 432
webometrics 178
Weibe, Donald 166–7
welfare reform 114
Western law 523–4
'Whig' historiography 51
Whirlwind computing project 255
Wide ID 18
Wiener, Norbert 227, 263–4, 453
Wigner, Eugene P. 557
Williams, Raymond 207
Wilson, E. O. 24, 68, 291, 294, 496
Windelband, Wilhelm 34–5, 42, 432
Wink, Rüdiger 98
Winter Simulation Conference 250
Wisconsin, University of
  experimental college 362
  Women in Science and Engineering Leadership Institute (WISELI) 390
  see also University of Wisconsin–Stevens Point
Wissenschaft 3, 52
Wittgenstein, Ludwig 237
Women in Science and Engineering Leadership Institute (WISELI) 390
women's studies 348–9, 362
Woolgar, Steve 193
word-and-image studies 140
World Commission on Dams 440, 447
World Council of Churches (WCC) 168, 169–70
world modeling 462
'world systems theory' 108
writing
  across the curriculum (WAC) 366, 378–80
  invention 221
written products protocol 480

X-rays 84, 94

Yaghan (Yamana) community 499
Yale University, interdisciplinary major 420
Yellowstone National Park 496
York University (Toronto) 369
YouTube videos 228

Zayed University 370
ZiF 292–3

T3-BOC-888

JOURNAL FOR THE STUDY OF THE OLD TESTAMENT
SUPPLEMENT SERIES
327

Sheffield Academic Press

Vain Rhetoric

Private Insight and Public Debate in Ecclesiastes

Gary D. Salyer

Journal for the Study of the Old Testament
Supplement Series 327

Published by
Sheffield Academic Press Ltd
Mansion House
19 Kingfield Road
Sheffield S11 9AS
England

www.SheffieldAcademicPress.com

Typeset by Sheffield Academic Press
and
Printed on acid-free paper in Great Britain
by Biddles Ltd
Guildford, Surrey

British Library Cataloguing-in-Publication Data

A catalogue record for this book is available
from the British Library

ISBN 1 84127 181 0

To My 'Three-Cord' Strand

To Kenny,
My son, who has always been a joy
and inspiration to my heart.

And to my two life-long best friends,
Steve and Ken, strong cords
whose friendship is beyond family.

Two are better than one,
Because they have a good return for their work.
If one falls down,
His friend can help him up.
But pity the man who falls
and has no one to help him up!...
Though one may be overpowered,
A cord of three strands is not quickly broken (Eccl. 4.9-12).

CONTENTS

Preface	11
Acknowledgements	20
Abbreviations	23

Chapter 1
PROLEGOMENA: TOWARD A THEORY OF READING SCRIPTURAL TEXTS 29

1. Reading the Book of Ecclesiastes as a Text 29
2. Reading Scripture as Sacred Text Requires Different Assumptions and Interests 30
3. The Difference Textuality Makes for a Theory of Reading 41
4. 'Woven' to the Reader: How Textuality Affects the Reading Process 51
5. Sharing the Loom with the Author: Readers as Co-Authors of Meaning 54

Chapter 2
READING ECCLESIASTES AS A FIRST-PERSON SCRIPTURAL TEXT 62

1. Seeing Through Textual 'I's: Narrative Theory and First-Person Texts 62
2. Posts of Observation and Point of View in First-Person Argumentative Texts 83
3. Wolfgang Iser's Theory of Reading 90
4. Reading Theories and the Poststructuralist Perspective 108
5. Taking Stock in the Speaker—How Readers Respond to First-Person Texts 116

Chapter 3
AMBIGUITIES, RIDDLES AND PUZZLES: AN OVERVIEW OF THE LINGUISTIC AND STRUCTURAL READER PROBLEMS IN THE BOOK OF ECCLESIASTES 126

1. Ecclesiastes as a Rhetoric of Ambiguity 126

2. An Overview of Reader Problems in Ecclesiastes 132
3. Major Reading Problems in the Book of Ecclesiastes: Opacity Generated by Idiosyncratic Grammatical Ambiguities 137
4. Literary Rubik's Cubes and the Structural Ambiguities in the Book of Ecclesiastes: An Overview of Reading Strategies 143
5. Summary: A Textuality Characterized by Ambiguity 164

Chapter 4
THE EPISTEMOLOGICAL SPIRAL: THE IRONIC USE OF PUBLIC AND PRIVATE KNOWLEDGE IN THE NARRATIVE PRESENTATION OF QOHELETH 167
1. Overview of Persona Problems in the Book of Ecclesiastes 167
2. The Death of Ecclesiastes: Qoheleth as Fictional Persona 167
3. Qoheleth's Use of Emphatic 'I' and the Monologue 172
4. Qoheleth as Fictive Autobiography: Defamiliarizing the Reader's Life 177
5. The King's Fiction as a Theatrical Prop 185
6. Attractiveness, Credibility and Trustworthiness: The Rhetorical Effect of Saying 'I' 194
7. Qoheleth's Reminiscences on the Wisdom Tradition: A Dialogic Monologue That Fictively Recontextualizes the Wisdom Tradition 196
8. Endorsed Monologue: Narration Issues in the Book of Ecclesiastes 211
9. Irony and the Implied Author's Use of Public Knowledge 221
10. The Epistemological Spiral: The Ironic Presentation of Knowledge in the Book of Ecclesiastes 225
11. Summary of Reading Issues in the Book of Ecclesiastes 235

Chapter 5
ROBUST RETICENCE AND THE RHETORIC OF THE SELF: READER RELATIONSHIPS AND THE USE OF FIRST-PERSON DISCOURSE IN ECCLESIASTES 1.1–6.9 239
1. Introduction 239
2. I, Qoheleth: The Use of First-Person Discourse in Ecclesiastes 1.1–2.26 240
3. Ecclesiastes 1.1–1.11: Prologue and Preparation for Qoheleth's 'I' 241

4. Ecclesiastes 1.12–2.26: 'I', Qoheleth—The Search for Self and Knowledge 270
5. Ecclesiastes 3.1-15: Time, Darkness and the Limits of Public Knowledge 295
6. Ecclesiastes 3.16–4.6: Immorality, Mortality and the Limits of Public Knowledge 302
7. Ecclesiastes 4.7-16: Knowledge and Communal Living 307
8. Ecclesiastes 4.17–5.8: The Knowledge of Divine Duties 313
9. Ecclesiastes 5.9–6.9: Possessions and the Possession of Joyful Knowledge 318
10. Ecclesiastes 1.1-6.9 Summarized: A Rhetoric of Robust Reticence 324

Chapter 6
A RHETORIC OF SUBVERSIVE SUBTLETY: THE EFFECT OF QOHELETH'S FIRST-PERSON DISCOURSE ON READER RELATIONSHIPS IN ECCLESIASTES 6.10–12.14 326
1. Introduction 326
2. The Difference a Sceptic Makes 327
3. The Emergence of the Model Reader 329
4. Ecclesiastes 6.10-12: Epistemological Nihilism—Who Knows What is Good? 332
5. Ecclesiastes 7.1–8.17: The Ethically Blind Public 334
6. Ecclesiastes 9.1-6: The Depths of Scepticism—Who Knows about God? 353
7. Ecclesiastes 9.7-10: Reclaiming the Value of Life—Knowing How to Enjoy Life 355
8. Ecclesiastes 9.11-12: The Unpredictable and Public Knowledge 356
9. Ecclesiastes 9.13–12.7: Asking the Narratee to Fill in the Blanks 357
10. Inferring the Model Reader's Competence 359
11. Ecclesiastes 9.13–11.6: Inferring the Wisdom of Wisdom 360
12. Ecclesiastes 11.7–12.7: Youth, Mortality and the Enjoyment of Life 367
13. Ecclesiastes 12.8-14: A Public Perspective on a Private Figure 372
14. Summary of Reader Relationships in the Book of Ecclesiastes 376

15. Summary of the Effects of Reading Relationships in the Book of Ecclesiastes 378

Chapter 7
VAIN RHETORIC: SOME CONCLUSIONS 380
1. The Need for a New Loom 380
2. Vain Rhetoric and the *Sitz im Leser*: Summary of Conclusions Reached 381
3. Vain Rhetoric: The Rhetorical Backlash of Unabated Subjectivity 387
4. What Do We Mean by a Vain Rhetoric? 389
5. Three Levels of Vain Rhetoric in the Book of Ecclesiastes 397
6. Qoheleth's Ethos as Mediator Between the Logos and Pathos Dimensions of the Text 398
7. The Rhetorical Mirror: Qoheleth and the Postmodern Experience 398

Appendix
WISDOM REFLECTIONS (PUBLIC KNOWLEDGE) IN THE BOOK OF ECCLESIASTES 400

Bibliography 403
Index of References 432
Index of Modern Authors 439

PREFACE

> It is not fit the public trusts should be lodged in the hands of any till they are first proved, and found fit for the business they are to be intrusted with.[1]

This book began life as a 1997 dissertation which I submitted to the faculty of the Graduate Theological Union and the University of California at Berkeley. I have revised it with a dual purpose in mind. Obviously, I wish it to be a contribution to the field in which it was submitted as a respectable monograph. However, more than a few readers of the dissertation suggested that much of the work could function as a literary methods 'primer' for a college level or graduate student. With that perspective in mind, the manuscript was revised with the hopes that it could in some meaningful fashion function as such. All foreign languages have been translated. In the case of the Hebrew and Greek text being cited, words are fully transliterated with translations in parentheses so that a beginning student could read the book and still learn the methods being discussed. Chapter 2 is a very comprehensive introduction to literary methods, in particular, narrative and reader-response perspectives. The rest of the book could serve as a means of showing the student how such methods and perspectives can be utilized within a literary hermeneutic. Chapters 3 and 4 'recalibrate' historical scholarship for use within a reader-oriented perspective, serving as a paradigm for how to read historical scholarship from the perspective of a literary hermeneutic. Chapters 5 and 6 offer a linear reading of the entire book of Ecclesiastes. These chapters present a close reading of the book from a reader-oriented perspective. Much in this section functions like a commentary on the book, though limited to the topic at hand. As a result, one could use this work as a textbook in a hermeneutics or exegetical methods class. At the least, one of my secondary aims

1. Matthew Henry, *Commentary on the Whole Bible, Vol. 6, Acts to Revelation.* (New York: Fleming H. Revell Co., rev edn, 1925), p. 816.

when I edited the manuscript was to produce a book that could teach students literary hermeneutics as well as contribute to the scholarly guild at large.

After years of experiencing Qoheleth over and over, I have come to characterize the narrator's singular propensity to pendulate between good and bad ethos, and based on that, the book's basic literary and overall rhetorical strategy, as a 'vain rhetoric'. Most of these problems revolve around the book's literary strategy of placing all its rhetorical eggs in the strengths and weaknesses of first-person narration. In that regard, it is unique in the Canon. Though other books might extensively utilize first-person narration as a rhetorical ploy, none do so with the completeness by which the book of Ecclesiastes operates as a rhetorical unit. Specifically, I argue that by almost exclusively anchoring the book's persuasive abilities in the powers and deficiencies of a first-person narrator, Qoheleth, the implied author has made a rhetorical gamble that backfires as much as it hits the mark. This telltale effect of first-person discourse is endemic to the narrational strategy and any genres which are based upon that discourse technique. While some of this effect is surely based on the peculiar characterization which the implied author has given his literary creation, Qoheleth, it must above all be noted that such an effect is typical of many first-person discourses. In that regard, the problem readers have with the book is not entirely dependent upon the specific character of Qoheleth per se.

Privately, I compare the rhetorical strategy of first-person discourse to the baseball home run hitter who strikes out more than he hits the ball, except that when he does connect it goes a country mile. Such hitters either have spectacular results or strikeout in pitifully enemic demonstrations of futility. First-person discourse is very much like such a baseball player. For instance, think back to the latest 'confessional' sermon one might have heard from a local pastor. When such a person says, 'I believe X with all my heart', that statement has about a 50 per cent chance of failing or succeeding depending on the experiences of his or her audience. If the audience shares the experience upon which the confessional statement is made, the testimony can have startling and immediately persuasive results. But let that experience be contested, or left untried, then the speaker's use of 'I' can be woefully uncompelling. Nothing is worse than a sermon based on 'I experienced' when the audience tacitly does not agree with the experience in question. Yet such is the gamble that anyone takes when he or she places all their rhetorical eggs in the singular basket of 'I' discourse. The history of scholarship in narratological circles regarding first-person discourse is resplendent with numerous examples of just this rhetorical dynamic.

Based on this insight, my study will argue that the suasion problem which nearly every reader has experienced in Ecclesiastes is not primarily due to an underlying historical crisis such as the conflict with Hellenism or some Freudian psycho-personal dynamic as a few have argued. Rather, it is a literary problem that is endemic and inherent to all first-person discourses regardless of their historical setting. As soon as a literary work extensively utilizes the 'I' of a first-person narrrator, the discourse begins to communicate an unavoidable sense of subjectivity to the reader. When utilized too assiduously, this strategy has a history of backfiring on both authors and speakers. As a literary problem attached to the intrinsic possibilities and liabilities which has surrounded every first-person discourse throughout the ages, Qoheleth's rhetorical difficulties are first and foremost a synchronic problem, with diachronic issues supplying various complications of a problem that is not essentially anchored in any specific historical, cultural, or personal matrix. This inherent aura of subjectivity which clings to all first-person discourses creates in Qoheleth's instance what may be termed a vain rhetoric. Obviously, I am making a play on Qoheleth's use of 'vain' throughout his discourse. Still, the term more than adequately describes the reading experience that most have with the book. By choosing to base the rhetoric of the book essentially on the strengths and weaknesses of Qoheleth's 'I', the implied author spurned the aura of 'omnisciency' which surrounds so many of the Canon's third-person narrators. As a result, the book of Ecclesiastes explores the latent powers and prospects of private insight in terms of the general quest for wisdom, that is, public knowledge.

However, the weakness of an empirically-based epistemology voiced solely through first-person narration, with its built-in predilection for subjectivism, cried out for the balancing perspective of public knowledge. As such there also comes to play the subtle dynamics of third-person narration found on the outer edges of the book, with its power to produce the effect, or perhaps the illusion, of omnisciency. Once one intercalates the subtle rhetorical effects of the implied author's use of a frame-narrative, as well as his use of satiric and ironic characterizations, what one begins to see in the book is a very delicate dialogue, or perhaps better, a debate on the promises, prospects, and perils of private insight versus public knowledge as general modes of human knowing. Behind the scenes, spun throughout the narratological tapestry of the book of Ecclesiastes is an epistological debate on the role and validity of both these commodities within the tradition of Israel's sages.

At issue in the book of Ecclesiastes is the rhetorical question of how does one validate the 'truth', or perhaps, wisdom of the individual. As Matthew Henry's comment at the beginning of this preface so ably illuminates, all private voices which would seek to become public knowledge must submit to public testing in order to validate whether the insights of the individual are indeed 'fit' for public consumption. This is the case whether the individual voice is that of the philosopher, scientist, or literary scholar. It is also very much the process by which Wisdom seeks to authenticate itself within the Canon. All wisdom starts out as the 'wit of the individual' before it becomes the 'wisdom of all'. In the book of Ecclesiastes, the implied author seems to be reflecting on this process in the most subtle of fashions, calling attention to the latent perils and prospects of gaining wisdom by allowing the reader to listen into a sly debate between the narrator, Qoheleth, and that of his presenter, the frame-narrator.

In a book such as Ecclesiastes, where the protagonist speaks almost exclusively from personal experience, the necessity of public validation is immediately given prominence. As I will argue, Qoheleth threw down the gauntlet to his reading public via the radical conclusions he reached on the basis of private insight. It should therefore come as no surprise that the general public reciprocated in the voice of the Epilogist. This creates not a little literary tension between two narrative voices that essentially are coming from very different epistemological stations. The differences between these two figures creates an atmosphere that at best is characterized by literary debate, and at worst by rhetorical dissension. However, it is my thesis that this dynamic is generated in the foremost instance by the inherent and unavoidable aura of subjectivism which surrounds every and any first-person discourse. In that sense the rhetorical strategy of the book of Ecclesiastes may aptly be termed a vain rhetoric. By describing it as a vain rhetoric, I wish simply to call attention to the strong, but potentially divisive effects of first-person discourse as a discourse strategy. It is the nature of all 'I' discourses not only to convince, but also, potentially to leave a fair amount of doubt in the reader's mind. In terms of its final suasive effects, a vain rhetoric is a double-edged sword. It can be suasive, but often lacks persuasive force in any totally satisfying way. Given the fact that the book of Ecclesiastes resides in a canon wherein only authorized truth inspired by God is supposed to exist, the radical atmosphere of subjectivity that we meet in it only serves to exacerbate and amplify the 'vanity effect' which first-person discourse has on its readers.

As a result, this study reaches several conclusions regarding Qoheleth's use of first-person discourse. First, it is the nature of all I-discourses to imply their own limitations and, therefore, to invite dialogic dissension with their major premises and conclusions. They are a vain rhetoric in that the one prevalent effect of the use of 'I' is to generate an argumentative stance in the reader. A first-person discourse literally begs to be debated with, and only rarely creates unconditional rhetorical consensus between speaker and audience. As a result, the following literary analysis and reading concludes that it is the book's radical dependency upon I-discourse that has generated the problems which have created its mixed reception. To put it succinctly, the book's foundational problem is a literary problem first and foremost. What readers react to most strongly in Ecclesiastes is the over use of the subjectively-oriented properties of first-person discourse within a scriptural tradition which typically relies upon the omniciency of a third-person narrator. This extreme difference jars the scriptural reader in some very specific ways.

Once that problem is coupled with the lampooning of private insight via the satiric characterization of Qoheleth by the implied author, as happens extensively in chs. 2 and 7, the book takes on its telltale rhetorical shape as we have come to know it. Qoheleth remains an extreme character, and is so for a reason. There exists a level of ironization in the discourse that goes well beyond the intense subversive rhetoric of its narrator. The implied author of the book of Ecclesiastes utilizes Qoheleth's vain rhetoric to enact a lively debate on the adequacy of private experience as a means of achieving public knowledge worthy of scriptural or religious imagination. An important insight afforded by modern literary theory's distinction between an implied author and narrator is that there exists in this text a completely ironic interaction between private insight and public knowledge throughout the discourse. As one reads between the lines of the various narrational levels in the book, there is found to exist a vigorous epistemological debate between the implied author/Epilogist and the character Qoheleth on what constitutes valid public knowledge, that is, wisdom. Given the numerous times that the keyword 'to know' occurs in the book (especially the latter half) as well as the predominance of rhetorical questions which pepper Qoheleth's discourse, such a conclusion should not be overly surprising to anyone familiar with this text. Whether this was intended, or is simply due to the surplus of meaning which is inherent in all literary texts cannot be gainsaid. But what can be said with certainty is that if one pays attention to the interaction of the narrational levels in the text, there immediately appears to the competent reader a horizon of ironic effect generated by the relationship between the two primary textual agents at these levels.

As a result of these ironizing effects, the book educates the reader regarding the broader epistemological issues involved in the pursuit of wisdom. It pushes to the furthest limits within the constraints of the Israelite wisdom tradition the quintessential question; 'What constitutes valid religious knowledge?' In posing that question, the issues which lie just beneath the surface of Qoheleth's monologue begin to take on a very contemporary, perennial and postmodern tenor. The overall interactions between the two levels of narration in the book strongly imply a questioning of the location of true knowledge. One level suggests that it is located in the experiencing self, as postmodernism would have it. In that, postmodernism seems to lie on a trajectory with the epistemology of the monologist, Qoheleth. On the other hand, the Epilogist/implied author begs to differ with this position. That level of the book suggests that true knowledge must be found in the broad-based collective experiences of the human/religious community. This level of the book seems to lie along a trajectory with modernism. However, when one looks at the totality of the text which contains both of these positions, a compromising position seems taken up by the text as we have it. The book taken as a whole appears to suggest that true knowledge is generated in the interaction between private insight and public knowledge, that is, that both are needed and exist only as a necessary epistemological dyad. The reading offered in Chapters 5 and 6 of this study will suggest that the latter reading is ultimately the meaning of the book. Furthermore, that this mediating position is the book's implied answer to the questions raised by Qoheleth's radical centering of knowledge in the private experiences of the individual at the surface level of the book. In other words, the meaning of the book cannot be found at its surface level, but only at the deep level of its narrative structure, or more precisely, in the interactional dialogue that exists between the different narrational levels in the book. The real message of the book cannot be found solely in the monologue of Qoheleth, but also in the implied debate between the narrative levels of the discourse which can only be seen by paying strict attention to the literary sophistication of its architecture.

Qoheleth's 'I' therefore serves to sum up not only a literary character, but also functions as an indice to a much larger human problem—the problem of how to integrate individual experience into the broader experiences of the human religious community. The 'I' of Qoheleth and the Epilogist function as symbols for this broader rhetorical problem which plagues all human attempts to speak for God. Qoheleth symbolizes private knowledge while the Epilogist serves as an indice for public knowledge. By experiencing Qoheleth's monologue, the reader is drawn into the trap of solitary existence, and all knowledge that would stake its claims based solely on the knowledge of the individual self as an epistemological agent. By being drawn into Qoheleth's trap, we experience the fundamental rhetorical vanity of the human religious situation. Each of us struggles with the broad-based claims of our own unique experiences and those of the scriptural, or perhaps, human community. The interaction of these creates a never-ending, often confusing, yet absolutely necessary rhetorical and epistemological spiral which all knowledge must navigate in order to become the sort of public knowledge which is reliable and valid. Neither private insight nor public knowledge constitute true knowledge/wisdom in and by themselves at this level of reading. Rather, both depend upon each other for inspiration, renewal, mutual confirmation and existential validation. This insight is ultimately the deep-level message of Qoheleth's vain rhetoric.

Finally, by describing the rhetorical strategy of the book's literary characteristics as a vain rhetoric, this study will also call attention to the subtle effects of the text's use of ambiguity throughout the discourse. As is well known, the implied author has constructed a discourse which constantly frustrates the reader and, ultimately, allows the reader no sure answers. The narrator's choice of words often leaves the reader in a state of perplexity, confusion or indecision. By doing so, the implied author has consciously constructed a text which would recreate the same sense of *hebel* at a literary level that one often experiences in real life. Vain rhetoric therefore describes the abiding literary experience of reading the book of Ecclesiastes in a performative sense. The rhetorical effect of the text's various gapping techniques and strategies of indirection is to recreate in the reader life's penchant for absurdity and ambiguity. The use of a vain rhetoric in the performative sense allows the implied author to recreate in the reader a narrative encounter with the absurdist's experience of life. As Wittgenstein noted, language often goes on vacation when it attempts to describe the absurd dimension of life. Given this situation, absurdist writers are left to express their experience with life by means of indirect, or perhaps, non-cognitive narrative techniques in an attempt to convey to the reader what manner of absurdity fills his or her heart. The implied author of the book of Ecclesiastes seems to have intuited this and, therefore, compensated for the inability of language to say what he meant by finding ways to communicate that primal experience through literary gapping, blanking and opacity. In that regard, the type of vain rhetoric we encounter in the book of Ecclesiastes is a performative concept as well. It's chief effect is to provide the reader with a narrative experience of life's absurdity.

To sum up, vain rhetoric implies three levels of operation. First, on the surface level, it describes the peculiar characterization of the narrator and his subsequent ethos-related problems. Second, at the text's deep level, it describes how the interaction of first and third-person discourses enable the reader to become aware of the general problem of their own rhetorical existence as it relates to communally-based rhetorical systems such as those found in the Scriptures. All knowledge, both individual and communal, has specific limitations. Neither form of knowledge can be utterly relied upon in any simple manner of thinking. By paying strict attention to the narrative sophistication of the discourse, we can discern a debate between Qoheleth and the Epilogist/implied author which hints at that greater issue. Third, at the level of the text's use of ambiguity, it describes the general effects of the implied author's use of a literary gapping to generate a narrative experience which partially escapes language's inability to describe the absurd dimension of life in any completely meaningful and satisfactory manner. As such, a vain rhetoric accomplishes at the performative level what language can only vaguely hint at the descriptive level. In dealing with such important issues about human knowing, the book raises some very important issues for the postmodern reader, who also struggles with how we know truth in any reliable sense. By raising such issues, I believe that Ecclesiastes may be the most postmodern book in the Canon, and certainly, one that deserves a hearing in our age.

ACKNOWLEDGMENTS

> Apollodorus says, 'If any one were to take away from the books of Chrysippus all the passages which he quotes from other authors, his paper would be left empty'.[1]

Like Apollodorus, if one were to take away from this work all that I have gained from others, truly, very little would be left. First, I would like to thank Professor Donn Morgan, and his ever-timely advice that I consider Qoheleth as a dissertation topic. His wise advice that I look into Ricoeurian hermeneutics to complement my literary studies had more to do with the eventual slant I would take than either he or I ever imagined. Much appreciation also goes to Professor Michael Guinan who chaired the dissertation part of my program. Further thanks are due to Professor Robert Alter of the University of California at Berkeley. The seminar on Judges I took with him opened my eyes to the joy of literary studies and, in particular, Hebrew narrative techniques. Professor Alter also graciously served on my committee for the methods part of this project. In addition, I would like to pay tribute to the late Arthur Quinn, also of the University of California at Berkeley. Art showed me how to keep the rhetorical issues on the front burner when I was approaching Qoheleth.

I would also like to thank Professor Seymour Chatman and Professor Wilhelm Wuellner. Professor Chatman was an excellent guide to the field of narratology, specifically, how rhetoric dovetails with New Critical and Structuralist concerns. From Professor Wuellner I further learned the value of rhetoric for biblical studies—a value that will always be a part of my thinking. It was Professor Wuellner who introduced me to the 'power' of the rhetorical attributes of the text and therewith, to the

1. Diogenes Laertius, Chrysippus iii. Cited from *Lives of Eminent Philosophers* (2 vols.; trans. R.D. Hicks, Loeb Classical Library, ed. T.E. Page, Harvard University Press, Cambridge, MA: 1970), II, pp. 289, 291.

importance of pragmatic rhetorical theory. Not a little of his thinking lies behind the scenes of this work.

In addition, much gratitude goes to Eric Christianson. Eric and I exchanged manuscripts while he was working on the final proofs of his *A Time to Tell: Narrative Strategies in Ecclesiastes* (JSOTSup, 280; Sheffield: Sheffield Academic Press, 1998). His comments made to me privately while dining at Planet Hollywood with Dominic Rudman during the AAR/SBL Convention in Orlando have made for a better stylized text.

I would also like to thank a few mentors who were instrumental in shaping the early period of my academic career. Especially, I would like to thank Professor James Earl Massey for his initial shaping of my academic career. Thanks are also due to Professor Fred Shively. From him, I learned just how contagious enthusiasm can be and not a little about the dynamics of grace in one's faith.

But academics are not the only people who stand to be thanked at this time. If one needs guidance, one also needs the inspiration and support of friends. I owe a great deal to the tremendous faith and support of my 'three cord strand'—my two best friends, Reverend Steve Chiles and Reverend Ken Fairbanks, and my son who kept me afloat and balanced. Steve and I have always been in a covenant of friendship, and it is for his great faith in me that I dedicate this book. So too I dedicate it also to Ken, whose friendship is likewise 'foundational' to my life.

Finally, I wish to dedicate this book in the first instance to my son, Kenny. From his prayerful gift of understatement, 'Dear God, help Dad finish his *paper*' when he was younger, to the singular joy I heard from the adolescent 'war hoop' he voiced from the back of the chapel when I was hooded for my degree, Kenny has always been a joy to my heart. It has been my greatest pleasure to watch this sensitive young soul grow up to become what I know to be a quite promising young man. Most of that growth happened either while I was behind the keyboard writing this manuscript, or in the dugout as his little league baseball coach. Being a father is the greatest experience I have ever known, and not infrequently, a constant motivation to finish this book. From all those bed time stories, to watching him become my little lefty 'ace' pitcher, to those stunning 85 mph sliders and the ensuing strikeouts as a Varsity pitcher at El Cerrito High School, and of course, all those Saturday morning conversations over breakfast at Nation's, whether about life or baseball, I have experienced the one great joy that seems to have evaded Qoheleth—the love of a father and son. My thanks to God for the wonderful grace he has afforded my life with his presence has always been in the forefront of my consciousness throughout these years.

I should also like to give appreciation to Lysha Albright, one of those 'mid-life' gifts from life and God. Lysha kept me honest with myself and those self-imposed deadlines during the final years of the writing of this project. Her interdisciplinary mindset has taught me a great deal about how to intertwine spirituality and intuition in the process. She too has been a friend to whom I cannot count my indebtedness. Finally, I would also wish to thank the people who have made this possible, the staff of Sheffield Academic Press, who chose the manuscript and did all the practical work involved in publishing.

ABBREVIATIONS

AB	Anchor Bible
ABD	David Noel Freedman (ed.), *The Anchor Bible Dictionary* (New York: Doubleday, 1992)
ADP	Advances in Discourse Processes
AJBI	*Asian Journal of Biblical Interpretation*
ANETS	Ancient Near Eastern Texts and Studies
AOAT	Alter Orient und Altes Testament
ASTI	*Annual of the Swedish Theological Institute*
AUSS	*Andrews University Seminary Studies*
AUUSSU	Acta Universitatis Upsaliensis, Studia Semitica Upsaliensia
BBB	Bonner Biblische Beiträge
BDB	Francis Brown, S.R. Driver, Charles A Briggs, *A Hebrew and English Lexicon of the Old Testament*, (Oxford: Clarendon Press, 1907)
BEATAK	*Beiträge zur Erforschung des Alten Testaments und des Antiken Judentums*
BethM	*Beth Mikra*
BETL	Bibliotheca ephemeridum theologicarum lovaniensium
BH	*Buried History*
Bib	*Biblica*
BibBh	*Bible Bhashyam*
BibInt	*Biblical Interpretation: A Journal of Contemporary Approaches*
BJRL	*Bulletin of the John Rylands University Library of Manchester*
BK	*Bibel und Kirche*
BKAT	Biblischer Kommentar: Altes Testament
BN	*Biblische Notizen*
BR	*Bible Review*
BSac	*Bibliotheca Sacra*
BT	*The Bible Translator*
BTB	*Biblical Theology Bulletin*
BTF	*Bangalore Theological Forum*
BZ	*Biblische Zeitschrift*
BZAW	Beihefte zur ZAW
CB	The Cambridge Bible
CBQ	*Catholic Biblical Quarterly*

CBQMS	*Catholic Biblical Quarterly*, Monograph Series
CCC	*College Composition and Communication*
CCent	*Christian Century*
CM	*Communication Monographs*
CompCrit	*Comparative Criticism: A Yearbook*
CRGLECS	Comptes Rendus du Groupe Linguistique d'Etudes Chamito-Sémitiques
CritInq	*Critical Inquiry*
CSR	*Christian Scholar's Review*
CSSJ	*Central States Speech Journal*
CTM	*Concordia Theological Monthly*
CTR	*Criswell Theological Review*
CurTM	*Currents in Theology and Missions*
DBSup	*Dictionnaire de la Bible, Supplément*
DNEB	Die Neue Echter Bibel
EAJT	*East Asia Journal of Theology*
EC	*Essays in Criticism*
EF	Erträge der Forschung
ErI	*Eretz Israel*
ERT	*Evangelical Review of Theology*
EstBíb	*Estudios bíblicos*
ETL	*Ephemerides theologicae lovanienses*
ETR	*Etudes théologiques et religieuses*
EvQ	*Evangelical Quarterly*
EvT	*Evangelische Theologie*
ExpTim	*Expository Times*
FemTh	*Feminist Theology*
FO	*Folia Orientalia*
FOTL	The Forms of the Old Testament Literature
GKC	*Gesenius' Hebrew Grammar* (ed. E. Kautzsch, revised and trans. A.E. Cowley; Oxford: Clarendon Press, 1910)
GTJ	*Grace Theological Journal*
GTS	Gettysburg Theological Studies
GUOST	Glasgow University Oriental Society Transactions
HAR	*Hebrew Annual Review*
HAT	Handbuch zum Alten Testament
HS	*Hebrew Studies*
HTR	*Harvard Theological Review*
HUAR	*Hebrew Union Annual Review*
HUCA	*Hebrew Union College Annual*
IB	*Interpreter's Bible*
IBS	*Irish Biblical Studies*
ICC	International Critical Commentary
IDB	George Arthur Buttrick (ed.), *The Interpreter's Dictionary of the Bible* (4 vols.; Nashville: Abingdon Press, 1962)

IEJ	*Israel Exploration Journal*
Int	*Interpretation*
ISBL	Indiana Studies in Biblical Literature
ITC	International Theological Commentary
ITQ	*Irish Theological Quarterly*
ITS	*Indian Theological Studies*
JAAC	*Journal of Aesthetics and Art Criticism*
JAAR	*Journal of the American Academy of Religion*
JBL	*Journal of Biblical Literature*
JETS	*Journal of the Evangelical Theological Society*
JHStud	*Journal of Hebrew Studies (http.//www.arts.uablberta.ca/JHS)*
JJS	*Journal of Jewish Studies*
JNES	*Journal of Near Eastern Studies*
JPSV	*Jewish Publication Society Version*
JQR	*Jewish Quarterly Review*
JSOT	*Journal for the Study of the Old Testament*
JSOTSup	*Journal for the Study of the Old Testament*, Supplement Series
JSQ	*Jewish Studies Quarterly*
JSS	*Journal of Semitic Studies*
JTS	*Journal of Theological Studies*
KAT	Kommentar zum Alten Testament
L'AnThéo	*L'Année Théologique*
L&T	*Literature and Theology*
LavTP	*Laval Théologique et Philosophique*
LJLSA	*Language Journal of the Linguistic Society of America*
MR	*Methodist Review*
NAC	The New American Commentary
NCBC	New Century Bible Commentary
Neot	*Neotestamentica*
NICOT	New International Commentary on the Old Testament
NJB	*New Jerusalem Bible*
NLH	*New Literary History*
NRT	*La nouvelle revue theólogique*
OBT	Overtures to Biblical Theology
OLA	Orientalia lovaniensia analecta
OLP	Orientalia lovaniensia periodica
OLZ	*Orientalistische Literaturzeitung*
OTE	Old Testament Essays
OTL	Old Testament Library
OTM	Old Testament Message
OTS	*Oudtestamentische Studiën*
OTWSA	De Ou Testamentliese Werkgemeenskap in Suid-Afrika
PAAJR	*Proceedings of the American Academy of Jewish Research*
PEQ	*Palestine Excavation Quarterly*
PIBA	*Proceedings of the Irish Biblical Association*

PMLA	*Publications of the Modern Language Association*
POS	Pretoria Oriental Series
PR	*Philosophy and Rhetoric*
PRS	*Perspectives in Religious Studies*
PSac	*Philippinianea Sacra*
PSB	*Princeton Seminary Bulletin*
PT	*Poetics Today*
PTL	*Poetics and Theory of Literature*
PtS	*Point Series*
QJS	*Quarterly Journal of Speech*
RB	*Revue biblique*
RHPR	*Revue d'histoire et de philosophie religieuses*
RL	*Religion in Life*
RS	*Religious Studies*
RSV	Revised Standard Version
RTR	*Reformed Theological Review*
SBLDS	SBL Dissertation Series
SBLMS	SBL Monograph Series
SBLSP	SBL Seminar Papers
SBLSS	SBL Semeia Studies
SBT	Studies in Biblical Theology
SBTh	*Studia Biblica et Theologica*
SBTheo	Studia Biblica et Theologica
ScEs	*Science et esprit*
SCS	Speech Communication Series
SFEG	Suomen Eksegeettisen Seuran julkaisuja
SJOT	*Scandinavian Journal of the Old Testament*
SJT	*Scottish Journal of Theology*
SLI	*Studies in the Literary Imagination*
SLR	*Stanford Literature Review*
SM	*Speech Monographs*
SPIB	Scripta Pontificii Instituti Biblici
SpTod	*Spirituality Today*
SR	*Studies in Religion/Sciences religieuse*
SSCJ	*The Southern Speech Communication Journal*
StZ	*Stimmen der Zeit*
SubBib	Subsidia Biblica Rome
TBT	*The Bible Today*
TD	*Theology Digest*
TDOT	G.J. Botterweck and H. Ringgren (eds.), *Theological Dictionary of the Old Testament*
ThVia	*Theologia Viatorum*
TJ	*Trinity Journal*
TJT	*Taiwan Journal of Theology*
TK	Texte und Kontexte

TQ	*Theologische Quartalschrift*
TRu	*Theologische Rundschau*
TSFB	*Theological Students Fellowship Bulletin*
TZ	*Theologische Zeitschrift*
UF	*Ugarit-Forschungen*
USQR	*Union Seminary Quarterly Review*
VS	Verbum salutis
VT	*Vetus Testamentum*
VTSup	*Vetus Testamentum*, Supplements
WBC	Word Biblical Commentary
WTJ	*Westminster Theological Journal*
WW	*Word and World*
ZAW	*Zeitschrift für die alttestamentliche Wissenschaft*
ZDMG	*Zeitscrift der deutschen morgenländischen Gesellschaft*
ZDPV	*Zeitschrift für deutschen Palästina-Vereins*
ZEE	*Zeitscrift für Evangelische Ethik*
ZRGG	*Zeitschrift für Religions- und Geistesgeschichte*
ZTK	*Zeitschrift für Theologie und Kirche*

Chapter 1

PROLEGOMENA: TOWARD A THEORY OF READING SCRIPTURAL TEXTS

A sacred book...is closely involved with the conditions of its language.[1]

1. *Reading the Book of Ecclesiastes as a Text*

This study is about reading the book of Ecclesiastes. In particular, it asks the question: 'What happens to the reader when he or she attempts to assimilate the textual strategies of a first-person discourse such as the book of Ecclesiastes?' Before the advent of reader-response criticism, treatments of the reader were sparce at best. However, they could be found under a variety of 'traditional' headings such as the text, the author's intention, the design of the book, and other similar terms. Often in these older studies, critical scholarship confused the historically reconstructed 'intention' of the author with the meaning of the text. However, a postmodern perspective cannot reduce meaning to the hypothetical and often nefarious concept of 'authorial intention'. In its place, a reader-oriented approach posits that asking the question, 'How does this text function?', is a more productive place to begin the task of interpretation than beginning with the query: 'What did this author intend to mean?' Meaning is swallowed up in functionality. Textuality assumes hegemony over intentionality. The history in front of the text comes into prominence. The reader takes his or her rightful place in hermeneutic analysis.

1. N. Frye, *The Great Code: The Bible and Literature* (New York: Harvest/ Harcourt Brace Janovich Publishers, 1982), p. 3.

2. *Reading Scripture as Sacred Text Requires Different Assumptions and Interests*

The interpretation of a text necessarily follows upon the assumptions one makes about texts in general. Previous generations of biblical scholars worked with referential assumptions about texts. Texts were thereby treated as windows to another age. If given enough coaxing and historicist 'scrubbing', any text was expected to become transparent, giving a full view of both the author and his or her historical situation. Given the assumption that texts are referential windows to another age, the historical-critical method developed strategies and methods to accomplish this goal. For the last 200 years, this assumption has dominated biblical scholarship. The results have been impressive, even if they have not always been conclusive. A great deal has been learned from the referential approach. Undeniably, postmodern biblical scholarship owes a large debt to this legacy. Without the insights gained from historical research, much in the text would remain unexplainable or hard to understand. No one who has ever worked with a biblical text denies this fact.

Still, there are difficulties with a strictly historical approach to reading Scripture. Ironically, much of the problem with this approach is the very *certainty* it seeks to generate. In the last decade or two, poststructuralist, or perhaps better, postmodern philosophers and literary scholars have situated the historical-critical approach in its *own* historical matrix as a form of modernism. Though postmodernism has many contours due the diversity which is intrinsic to its ethos, its sparring partner is not so difficult to understand. Modernism is essentially the entire 'Enlightenment project'. Beginning with Descartes, most Western intellectual inquiry developed from a rational approach which accented a strict subject/object dichotomy, that is, the famed Cartesian duality. However, postmodernism does not buy into the subject/object dichotomy. Rather, it views objects as epistemologically embedded in subjects in a manner which makes them dyadic at best, and indistinguishable from their subjects at worst. This axiomatic insight therefore turns all scientific and historical approaches into acts of ideation and, ultimately, moments of interpretation which have their own biases and interests. With Philip Rice and Patricia Waugh, it is possible therefore to differentiate the two philosophical movements as follows:

> the modern and the postmodern critic hold two different views regarding what information is and how human beings process information. The modern critic assumes that information is an extrinsic phenomenon, independent of human perception, which an individual must react to and use. The postmodern critic holds that the individual is an information processing system that integrates selectively, ultimately creating whatever is perceived as information... The modern critic holds that science, morality, and art are distinct forms of logic. The postmodern critic maintains that all views are ideological, for a description can only reflect the perceptual perspective and biases of a particular symbol user in a given place at a specific time.[2]

Such philosophical advances have made the biblical interpreter painfully aware of the blinders that the Enlightenment placed over countless generations of scientists, historians, artists and interpreters. Thereby we now understand that this type of intellectual interogation is endemic to Western culture of the last 300 years and is not specific just to biblical interpretation. Stephen Toulmin has convincingly argued that the presuppositions of Enlightenment projects such as the historical-critical method are fully grounded in the 'Quest for Certainty' which began with Descartes.[3] Seen within its own historical context, Toulmin advocates that the Cartesian Quest for Certainty, which provides the philosophical moorings of all Western scientific and historical methods, is not a timeless truth, but a contextually limited and culturally bound reaction to the social anomie which surrounded the Thirty Years War in the early seventeenth Century. In its context, the Cartesian program of 'pure rationalism' satisfied Europe's craving for certainty after it was ravaged by religious wars which were themselves precipitated by the

2. P. Rice and P. Waugh, 'The Postmodern Perspective', in P. Rice and P. Waugh (eds.), *Modern Literary Theory: A Reader* (New York: Edward Arnold, 1989), pp. 428-40 (435). These two positions distinguish themselves in other strategic ways as well. Rice and Waugh also discuss how modernism tends to compartmentalize and distinguish most symbolic forms from criticism while postmodernism dissolves the two, thereby seeing both as 'power texts'. In addition, they note that the modern era 'is a period of ordering, structuring, and finding transcendent universals' (p. 435), whereas the postmodern era sees the uniqueness of all mental acts as they are found in their own contextual matrix. In that sense, postmodernism turns the historical-critical method's concern for context on its head by noting that context, as Derrida has observed, is boundless.

3. S. Toulmin, *Cosmopolis: The Hidden Agenda of Modernity* (Chicago: University of Chicago Press, 1990).

'relativism' of the early Renaissance. To accomplish this, Descartes moved the locus of knowledge from the oral to the written, from the particular to the universal, from the local to the general, and from the timely to the timeless.[4] This was the birth of modernity which assumed intellectual hegemony for Western intelligentsia until about 35 years ago.

Seen from this perspective, the pricetag of achieving some manner of certitude following the Thirty Years War was that the European academic community turned its back on the more eclectic, inductive and humane tradition of Renaissance thinkers like Montaigne and Erasmus. This, as Toulmin shows, was not only tragic but very limiting to all of Western philosophy, science and culture for about 300 years. Toulmin states:

> If uncertainty, ambiguity, and the acceptance of pluralism led, in practice, only to an intensification of the religious war, the time had come to discover some rational method for demonstrating correctness or incorrectness of philosophical, scientific, or theological doctrines... If Europeans were to avoid falling into a skeptical morass, they had, it seemed, to find *something* to be 'certain' about. The longer fighting continued, the less plausible it was that Protestants would admit the 'certainty' of Catholic doctrines, let alone that the devout Catholics would concede the 'certainty' of Protestant heresies. The only other place to look for 'certain foundations of belief' lay in the epistemological proofs that Montaigne had ruled out'.[5]

This was the natural context in which the Quest for Certainty took shape. Following Descartes, there developed a generalized cultural 'flight' from the particular, concrete, transitory and practical aspects of human experience which extended itself into all levels of intellectual inquiry, and above all, of philosophy.[6] One could say in this regard that the Cartesian paradigm functioned as a sort of 'intellectual hangover' for the Western intellectual community for the next three centuries.[7]

4. Toulmin, *Cosmopolis*, pp. 30-35.
5. Toulmin, *Cosmopolis*, pp. 55-56. Emphasis original.
6. Toulmin, *Cosmopolis*, p. 76.
7. For a fuller analysis of the relevance of Toulmin's historical analysis regarding the problem which 'modernity' poses for the study of Scripture, the reader is referred to W. Brueggemann who, relying upon Toulmin and S. Bordo, characterizes modernist models of biblical interpretation as a 'flight to objectivity' (W. Breuggemann, *Texts under Negotiation: The Bible and Postmodern Imagination* [Minnea-

However, Toulmin alertly observes that this was not the only path which was open to the European intellectual community in the seventeenth century. A more balanced perspective could have been retained. However, we should not be too quick to judge them. He also goes on to demonstrate that a similar thing re-occurred in our own culture during the early part of the twentieth century. His acute historical analysis shows how the Cartesian agenda developed over the next several hundred years and eventually had parallels in the previous century. He astutely observes how the dogmatism which grew out of the aftermath of the First World War led to the advent of Logical Positivism in the 1930s. Following the Second World War, there again arose a stultifying conservatism during the 1950s. This too was a reaction to social chaos. In a word, Toulmin shows us just how far academic and social communities will sacrifice truth for 'certain' knowledge when social climates dictate it. Understanding this all too human dynamic allows the critic to realize that times of crisis need not be resolved by an escape to certainty which operates on principles of timeless truths or single domain methods. His perceptive historical analysis serves as a

polis: Fortress Press, 1993], pp. 2-6, relying upon S. Bordo, *The Flight to Objectivity: Essays on Cartesianism and Culture* [New York: State University of New York Press, 1987]). For a more comprehensive overview of how European history and philosophy laid the foundations and context for modernist methods of textual investigation, see also W. Brueggemann, *Theology of the Old Testament: Testimony, Dispute, Advocacy* (Minneapolis: Fortress Press, 1997), pp. 1-60 (7-15). In a related vein, Levenson has also called attention to the effect that the aftermath of the Thirty Years War played on the early pioneers of biblical criticism, in particular, Hobbes, Spinoza and Richard Simon. See J. Levenson, 'Historical Criticism and the Fate of the Enlightenment Project' in J. Levenson, *The Hebrew Bible, the Old Testament, and Historical Criticism: Jews and Christians in Biblical Studies* (Louisville, KY: Westminster/John Knox Press, 1993), pp. 106-26 (117). Keegan has also written a lucid exposition on this subject. See T. Keegan, 'Biblical Criticism and the Challenge of Postmodernism', *BibInt* 3 (1995), pp. 1-14. In a similar vein, I. Spangenberg sees four paradigm changes in the last 400 years. These were located in the Reformation (sixteenth century), the Copernican and Cartesian revolutions (seventeenth century), the nineteenth-century revolution in the understanding of history, and the modern literary-critical revolution which began in the 1960s (I.J.J. Spangenberg, 'A Century of Wrestling with Qohelet: The Research History of the Book Illustrated with a Discussion of Qoh 4,17–5,6', in A. Schoors [ed.], *Qoheleth in the Context of Wisdom* [BETL, 136; Leuven: Leuven University Press, 1998], pp. 61-91 [62-67]).

reminder that there are no timeless methods which do not have a restrictive, if not oppressive underbelly.

However, beginning in the late-1960s there developed a return to earlier Renaissance models of intellectual inquiry such as those advocated by Montaigne and Erasmus.[8] These models do not function as a Quest for Certainty in the manner of the historical-critical method. Rather, they accent the rightful place of skepticism and in particular, the concrete, transitory and practical aspects of human experience. In this we see that the type of qualitative (as opposed to quantitative methodology) proposed by a reader-oriented approach is entirely in step with recent developments in both the scientific and philosophical communities. The fact that few in the modern academic world operate with a subject/object Cartesian dualism also suggests that an emphasis on the reader is both timely and necessary. Toulin aptly summarizes the postmodern perspective on the type of approach advocated here: 'Historically speaking, of course, the exclusion of practical issues from philosophy is quite recent. Those who are reviving them today find that such issues were actively debated by philosophers just 400 years ago.'[9] He then goes on to advocate that we must 'humanize modernity' by returning to the oral, that is, rhetoric, the particular, the local and the timely.[10] Thus the return to analyzing rhetorical considerations, as is being advocated by this study, can be seen as a necessary and helpful counterbalance to those methodologies such as the historical-critical method which are rooted in the Cartesian Quest for Certainty. As

8. Toulmin, *Cosmopolis*, pp. 160-67, refers to the period of 1965–75 as 'Humanism reinvented'. This is not to say, however, that postmodernism is a child of the late 1960s. I. Makarushka has shown that the origins of postmodernism can be effectively traced to Nietzsche's critique of modernity in his essay, 'History in the Service and Disservice of Life'. In that essay, Nietzche redirects the interpreter's attention away from historical matters to a consciousness of the all-pervasiveness of interpretation, and in particular, the importance of ambiguity. This legacy has been taken up and recovered by postmodern biblical interpretation, especially the post-structuralist school of Deconstruction. See I. Makarushka, 'Nietzche's Critique of Modernity: The Emergence of Hermeneutical Consciousness', *Semeia* 51 (1990), pp. 193-214. Nietzche's essay can be found in F. Nietzche, *Unmodern Observations* (ed. W. Arrowsmith; trans. G. Brown; New Haven: Yale University Press, 1990), pp. 75-145.

9. Toulmin, *Cosmopolis*, p. 191.

10. Toulmin, *Cosmopolis*, pp. 180-92.

Toulmin concludes: 'We are not compelled to choose between 16th century humanism and 17th century exact science; rather, we need to hang on to the positive achievements of them both'.[11]

Recent hermeneutical theory has therefore concluded that inspite of the admittedly monumental accomplishments of the historical-critical method, its philosphical origins in a specific historical crisis has severely limited its scope and usefulness within our postmodern cultural setting. For all of its great benefits, this approach has lacked a great deal when asked different sets of questions about texts, especially when one wanted to read these texts as scripture. One only need read the numerous articles and books given to 'canonical criticism' in the late-1970s and 1980s to see that the historical-critical method could not adequately answer all the questions that scriptural readers brought to the table in a postmodern context, especially if those questions were not particularly referential and historical. In fact, some of the problems raised in canonical-critical circles came about precisely because the questions being raised there demanded new methods and assumptions about texts. Still, because the first practitioners of canonical criticism, such as Brevard Childs and James Sanders, were historical critics first and foremost, they did not attempt to address the theoretical issues from a literary perspective which might give fresh answers to those problems. The old modernistic wineskins had burst, a new wineskin was needed. Historical-critical patches, offered under the guise of canonical criticism, could not stop the leaks brought about by 200 years of wear and tear. Questions were being raised that demanded new methods and assumptions about texts.

The dictum raised by John R. Donahue summarizes this issue: 'Any methodology is only as strong as its ability to answer questions which have been impervious to previous methodologies'.[12] While this point was raised regarding the inability of form criticism to answer those residual questions that lead to the rise of redaction criticism in New Testament Gospel scholarship, the same proposition may be tendered regarding the role of the historical-critical enterprise in general. Once a reader begins to ask non-historical questions about the text, such as its

11. Toulmin, *Cosmopolis*, p. 180.

12. J. Donahue, *Are You the Christ? The Trial Narrative in the Gospel of Mark* (SBLDS, 10; Missoula, MT: Scholars Press, 1973), p. 31.

persuasiveness for a modern audience, or perhaps to explore structuralist interests, such as how the parts of a text cohere as a discourse structure, historical methodologies lose their claim to absolute hegemony for the task at hand.

The interests which underlie this study are rhetorical and literary in nature. Analyzing how the literary use of first-person discourse induces suasion or dissuasion for the contemporary reader of the book of Ecclesiastes is the concern of this study. As one who still admires many of the goals of canonical criticism, I am interested in how this ancient Hebrew document functions as a scriptural text today.[13] In this study I hope to explain how a contemporary reader experiences both suasion and dissuasion due to the use of first-person discourse structures in the book. The historically-minded critic should note that this is not an historical question, and as such is not well-suited for historical methods. Interests, rather than methodological prejudice, dictate the synchronic bent of this study.[14]

13. This is not to say that canonical criticism is without its limitations, which have been well rehearsed in the past 15 years. Still, its religious aims are good ones for readers of Scripture, as Robert Culley noted in his Preface to *Semeia* 62. In his methodological survey of the guild, he notes how a decade ago there were two choices for most scholars—either an historical or a textual approach. He then observes how Childs' was advocating a third approach: 'While all this was going on, Brevard Childs was developing another approach, or to be more precise, trying to restate the oldest approach to the Bible, a reading of it as the text of a religious community. He proposed that the relevant starting point for a critical study of the biblical text should be a perception of the Bible as a religious text—an approach he described as canonical—rather than from historical or literary models' (R. Culley, 'Preface', *Semeia* 62 [1993], pp. vii-xiii [x]. For an excellent summary of the weaknesses of Childs' position see D. Breuggemann, 'Brevard Childs' Canon Criticism: An Example of Post-Critical Naiveté', *JETS* 32 (1989), pp. 311-26. The reader is referred to his bibliography for the usual critics of the method. Curiously, Breuggemann renounces Childs for his radical textual orientation, ultimately claiming that 'the confessing community itself is the authority' (p. 326). Such a conclusion shows how pervasive the postmodern spirit is, as he concludes in a fashion that is very much attuned to the insights of literary scholars like Stanley Fish who also locates meaning and significance within the interests of the 'interpretative community'.

14. This too is typical of the move from modernist to postmodern perspectives. Toulmin notes that as 'scientists progressively extended their scope, between 1720 and 1920, one thing working scientists did was to rediscover the wisdom of Aristotle's warning about "matching methods to problems": as a result, they edged

My specific aim is to describe the interpretative reflexes and literary competence needed by a postmodern scriptural reader in order to productively comprehend this ancient document as a scriptural text. The goal is not to attempt to make the book of Ecclesiastes a modern text in any sense, nor to anachronistically identify contemporary reading habits with those of its original audience.[15] Quite the contrary, the interests espoused in this study do not concern how the original recipients of this text were influenced per se. Instead, this study will focus on how the text as discourse possesses generic properties that will generally produce predictable responses for the reader who is skilled with the literary competence mandated by the text. I am more interested in how our generation reads Ecclesiastes as a scriptural text than in any hypothetical and historically reconstructed original audience, some of whose interests and beliefs I cannot share. We are not Iron Age or Hellenistic readers, and it cannot be supposed that an ancient reading is the only way one actualizes the document as a scriptural text. In fact, the best way to demote a scriptural text from its position as Scripture to a status as 'document' is precisely to read it as an address to another generation, thereby delimiting its meaning and significance to that time. If canonical criticism has taught biblical readers anything, it is that the nature of a Scripture is to address future generations[16] and to be able to be resignified.[17]

There is an implied 'canonical pact', or perhaps better, an implied 'synchronic reading contract' with the reader which motivates the scriptural consumer to untie the historical moorings of the text. The concept of a reading contract comes from structuralist literary scholars such as

away from the Platonist demand for a single, universal "method"…' (Toulmin, *Cosmopolis*, p. 154).

15. J. Barton, 'Reading the Bible as Literature: Two Questions for Biblical Critics', *L&T* 1 (1987), pp. 135-53 (151).

16. B. Childs, *Introduction to the Old Testament as Scripture* (Philadelphia: Fortress Press, 1979), p. 60.

17. J. Sanders, *From Sacred Story to Sacred Text* (Philadelphia: Fortress Press, 1987), p. 172. He also notes how the same type of resignification took place even after the Enlightenment: 'One need only do a diachronic study of a given passage in the successive scholarly commentaries in modern times to see the same kinds of resignification… The "original meaning" of such a passage kept changing as time marched on' (p. 62). Such observations are consistent with the presuppositions of reader-response criticism and textuality approaches which argue that it is the nature of writing to generate context-malleable texts.

Philippe Lejeune.[18] Such reading contracts utilize various codes which inform the reader about how to interpret the features of a discourse. According to Roland Barthes, a reading code is a fundamental convention for any act of verbal communication. Codes are structuralist shorthand for the system of norms, rules and constaints which determine how a reader signifies a message. For it to be effective, a code must be at least partially common to both the addresser and addressee of a message. Barthes lists several codes, such as the proairetic, hermeneutic, referential, semic, symbolic, verisimiltude and cultural codes as active ingredients in the reading process.[19] It must be noted, however, that reading codes can vary from context to context, and age to age, which creates problems when it comes to reading ancient texts.

However, as I understand the nature of Scripture, there are two competing 'macro' codes at work in these texts as religious literature. These macro codes operate at a more expansive and generic level than do the codes which Barthes discusses. Originally, the various biblical writings obviously had what may be termed for lack of a better one, an historical, or diachronic code which told the reader/hearer to signify the message as relevant for their specific context. However, the subsequent canonical process whereby a localized text was elevated to the status of Scripture entailed the overlay of yet a second reading contract, which I have termed the 'scriptural/synchronic' code or perhaps better, the 'scriptural reading contract'. In its original historical context there was a diachronic reading contract that was presumed to be operative between the work and its recipients. However, after a text was elevated to the

18. For a general discussion of a reading contract, see P. Lejeune, 'The Autobiographical Contract', in T. Todorov (ed.), *French Literary Theory Today* (Cambridge: Cambridge University Press, 1982), pp. 192-222.

19. R. Barthes, *S/Z: An Essay* (trans. R. Miller; New York: Hill & Wang, 1974), pp. 18-20. Briefly, the proairetic code tells the reader to look for a series of actions, the hermeneutic code instructs the reader to search for questions or enigmas, the referential code expects the reader to surmise given cultural background clues, the semic code tells the reader to look for characterization, the symbolic code informs the reader to ascertain meanings above and beyond the literal sense, the cultural code instructs the reader to read a text as prescribed by a specific culture's expectations (such as Greek historiography), and the verisimiltude codes tells the reader to construct the meaning of a text according to a set of truth norms that are external to it. As formulated by Barthes, all of these exist as codes within Scripture. However, we obviously have different understandings of them than would the original readers.

level of Scripture by a religious reading community, that text was given a second reading contract which operates with a synchronic set of expectations. As the ancient canonizing community read these local texts, some were seen as having additional, and perhaps global relevance beyond their original referential pact. These texts were then reclassified from a status as document to one of Scripture. As such, we see that there are two conflicting reading contracts/codes which are both intrinsic to the nature of Scripture. The first is referential and diachronic in nature. The second, which ultimately controls and determines its textual status and function, is poetic and synchronic in nature. Thus biblical texts have been run through two filters. Every text has two reading contracts which sometimes complement, but often conflict with each other.

Here again is a major reason why the historical-critical method must be utilized within the framework of postmodern perspectives. Given the dual, Janus-like nature of the canonical reading contract, it will be the task of future hermeneutical theory to more accurately define the nature of these codes/contracts, and to map out some rules which would allow them to co-exist in such a way that their diverse and sometimes contradictory claims and natures can fascilitate a meaningful encounter with the biblical text.[20] In a nutshell, the diachronic code would keep the reader from making spurious or too localized, that is, personalized readings from the text. Meanwhile, the synchronic reading contract ensures that the surplus of meaning which is generated by the specific poetics of the discourse would keep the text functionally relevant in the current context, thereby enabling the text to maintain a more global significance. Scriptural texts are therefore anchored in both yesterday and today, that is, in both local and global contexts. A theory of reading which desires to read these texts as Scripture must get to grips with both the diachronic and synchronic reading contracts which were bestowed upon these texts by the religious community. The diachronic contract expects facts, while the synchronic contract demands an emphasis on truth. The former demands an attention to meaning, while the

20. W. Randolph Tate also observes that this indeed is one of the major tasks that awaits postmodern hermeneutical theory. He concludes that modern interpretation theory should 'study the process of resignification throughout the canonical process in order to produce guidelines for the interpretative process today' (*Biblical Interpretation: An Integrated Approach* [Peabody, MA: Hendrickson, rev. edn, 1997], p. 207). Like so many others today, he too sees a legitimate level of synchronic meaning in canonical texts as they function as Scripture (p. 159).

second is satisfied only with significance and relevance. However, these are often in conflict. In my opinion, only a reading model with accents both modernist and postmodern perspectives can deal with the bi-functionality of Scripture's unique reading contract for our age.

Thus it can be seen that resignification, reinterpretation and recontextualization are necessary and, indeed, intrinsic properties of a text that has become canonical. As Jon Levenson summarizes the methodological issues involved in reading canonical texts:

> The fact of canon also challenges the most basic presuppostion of historical criticism, that a book must be understood only within the context in which it was produced. The very existence of a canon testifies to the reality of *recontextualization*: an artifact may survive the circumstances that brought it into being, including social and political circumstances
> to which so much attention is currently devoted... Because the Bible can never be altogether disengaged from the culture of its authors, historical criticism is necessary (though not necessarily in accordance with Troelsch's principles). But unless one holds that the Bible does not *deserve* to have survived its matrix—that the history of interpretation is only a history of misinterpretation—historical criticism cannot suffice. For were the meaning of the text *only* a function of the particular historical circumstances of its composition, recontextualization would never have occurred, and no Bible would have come into existence. If this be so, the tradition of historical criticism should not be abandoned within pluralistic settings, but only reconceived so as to recognize the challenge of pluralism. What must be abandoned are its totalistic claims. Room must be made for other senses of the text...[21]

That is why to the best of my ability, I will attempt to pay due regards to both reading models in an attempt to forge a new way of reading the book of Ecclesiastes. However, I admit to a dominance of the postmodern perspective in my own configuration of methods. It will be for others to decide whether my endeavours are successful.

Ultimately, one of the greatest ironies for the historical critic is the tacit acknowledgment that the canonical process itself deliberately

21. Levenson, 'Historical Criticism', pp. 122-23. Similar views have been voiced as well by John Goldingay who argues that it 'is the application of the Bible in the contemporary world that counts; there is not enough time for the luxury of the distancing, critical approach' (*Models for Interpretation of Scripture* [Grand Rapids: Eerdmans, 1995], p. 264). Like Levenson and myself, he argues that 'one of the key implications of scripture's identity...is, that this text speaks beyond its original context' (p. 156). Emphasis original.

untied the historical moorings so that these texts could have a life of their own after its original reception by the text's authorial audience, and thereby function in their new literary classification as Scripture.[22] In a complementary fashion, postmodern literary and textuality studies have done us the great service of analyzing, at least in a preliminary fashion, exactly how such classic texts accomplish this result. This is what I am terming a 'post-canonical' perspective on reading. Such a perspective demands new perspectives and methods. At the heart of this revised approach to reading the text as a scriptural address lie post-modern concerns with textuality, discourse structure, the role of the reader and how these factors interact in a suasive manner.

3. *The Difference Textuality Makes for a Theory of Reading*

The theory of reading presupposed by this study begins with understanding the difference between orality and textuality, or between reading texts and hearing speech-acts. Specifically, Paul Ricoeur's views on the nature of textuality will be foundational.[23] Those differences

22. B. Childs lists six ways by which the First Testament shapes canonical literature to lessen its historical particularity ('The Exegetical Significance of Canon for the Study of the Old Testament', *Congress Volume, Göttingen 1977*, ed. J.A. Emerton (VTSup, 29; Leiden: E.J. Brill, 1977), pp. 66-80 (70-80). See also Childs' *Introduction* in which he systematically applied these insights to the various First Testament writings. My only caveat to his axiomatic observations is that I would argue that the various reading strategies generated by the canonical process are not due to an effect *on* the text (see Childs, *Introduction*, pp. 75-76), but are rather due to the effect *of* the text on the community's reading conventions. The canonical process is a prime example of textual distanciation and how the semantic autonomy of the text, as an effect of textuality, has given us the gift of a scripture which has addressed future generations for millenia.

23. This is not to say that the model described herein is the only way to conceptualize the interface between textuality and the reading process. Recently, other post-structuralist models have been offerred which differ from the Ricoeurian perspective maintained throughout this study. For example, R. Cooper works from a Deconstructionist perspective whereby textuality is subsumed into the subject/object synthesis which follows from the programs of Foucault and Derrida. According to the Deconstructionist model, 'a text and *a fortiori* textuality are only the effects of relatively bounded systems of coherencies' ('Textualizing Determinacy/Determining Textuality', *Semeia* 62 [1993], pp. 3-18 [16]). Although my approach differs from that of Cooper, his article certainly presents some stellar insights into the contextual nature of all conceptualizations of textuality.

have been exhaustively treated by Ricoeur. The fundamental axiom in Ricoeur's theory of textual communication is the difference between oral and written discourse. Speech functions within a context of speaker and interlocutor being present to each other and to the situation they share.[24] Texts isolate and distance the author from the reader. Since the reader is no longer present with the author, sense and reference lose their grounding in authorial intentionality. Referentiality is thereby shifted from the world of the author to the world of the text. Texts become autonomous with regard to their original context and reference, creating a surplus of meaning. Dialogue is replaced by reading. Intentionality is swallowed up in textuality. The act of writing thus creates a 'double eclipse of the reader and the writer'.[25]

The essential dynamic involved in textuality is distanciation. Distanciation is 'not the product of methodology and hence something superfluous and parasitical; rather it is constitutive of the phenomenon of the text as writing'.[26] When discourse passes from speaking to writing, writing renders the text autonomous with regard to the intentions of the author. The dialogical situation inherent in oral discourse is eclipsed by the act of writing. The text no longer signifies only what the author meant. Without the presence of a physical author, our responses as readers replace, or at the very least supplement, the intention of the author. Writing creates a chasm which the reader can scarcely cross. Rather than trying to determine authorial intentionality, the critic is asked to guess at the meaning of the text as a whole, much as one resolves the meaning of a metaphor.[27] The validity of an interpretation is determined by asking if the interpretation generates the *sort* of meaning consistent with a given genre.[28] In the absence of the rules

24. P. Ricoeur, *Hermeneutics and the Human Sciences: Essays on Language, Action and Interpretation* (ed. and trans. J. Thompson; Cambridge: Cambridge University Press, 1981), p. 148.

25. Ricoeur, *Hermeneutics,* p. 147.

26. Ricoeur, *Hermeneutics,* p. 139.

27. Ricoeur, *Hermeneutics*, p. 14. He discusses the same topic elsewhere in his *Interpretation Theory: Discourse and the Surplus of Meaning* (Fort Worth: Texas Christian University Press, 1976), p. 76, where he notes: 'First, to construe the verbal meaning of a text is to construe it as a whole... A work of discourse is more than a linear sequence of sentences. It is a cumulative, holistic process'.

28. E. McKnight offers this as a criterion for validating postmodern interpretations in *The Bible and the Reader: An Introduction to Literary Criticism* (Philadelphia: Fortress Press, 1985), p. 133.

which govern social interactions, it is genre, text-types and literary competence that provide the ground rules of interactions for readers and texts.

However, I prefer the word *eclipse* to describe this aspect of distanciation when it comes to understanding the role of authorial intention while reading a text.[29] The *American Heritage Dictionary* (ed. Margery Berube *et al.*; Boston: Houghton Mufflin Co., Second College Edition, 1982) defines an eclipse as 'the partial or complete obscuring, relative to a designated observer, of one celestial body by another'. It can also mean 'to diminish in importance'. Behind this model is the concept of one body or object being cut off from sight by another, either in part or in full. Obviously, the original body remains, but it is no longer perceptible to the viewer. Ostensive reference works in the same way. There is no doubt that there was an original ostensive reference for the author of any given text. However, according to Ricoeur, that reference is now eclipsed by the reference of the world of the text. Furthermore, just as there are varieties of eclipses in the real world, with varying degrees of intensity and completeness, so it is with literary texts. How much of the background/historical reference is obscured by the foreground/poetic reference varies from text to text. For some texts, the original ostensive reference will be largely, or maybe even wholly obscurred. Such is the case for the book of Ecclesiastes. For others, the original reference will remain in sight for the reader with varying degrees of clarity and importance. Obviously, the historical situation of Ecclesiastes is more completely eclipsed than, for example, would be the case in 1 and 2 Kings. The Chronicler's work, however, would entail yet another degree of eclipsing. Part of the literary competence required

29. The problematic nature of history for reading the text can be seen in Leo Perdue's book, *The Collapse of History: Reconstructing Old Testament Theology* (OBT: Minneapolis, MN: Fortress Press, 1994). His book is a powerful presentation for the vitality of newer literary and postmodern methods. However, I do not prefer to characterize the paradigm change we are seeing as a *collapse*, but rather, as an *eclipse* of history. It is not as though biblical texts, which are fully situated in the ancient Near Eastern context, somehow fall under the weight of their own historicity. Rather, the historicity which forms the background for the text's repertoire is often hidden from the view of the reader in various degrees by the social, personal, cultural and historical foreground which constitutes the reader's history. Dynamics such as textuality also play a part. 'Eclipse' allows us to blend the best insights from the historical-critical method with postmodern perspectives without losing sight of either.

of a biblical critic is the ability to discern the degree of eclipse of ostensive reference for the various biblical texts. Rüdiger Lux has argued that because of the influence of historical methodologies, many scholars have lacked the necessary literary competence to observe the fictive signals which reside in many biblical texts. The result has been that the fictional referentiality of many texts has been confused with historical referentiality.[30] For Lux, the King's Fiction in the book of Ecclesiastes provides the reader with the Canon's 'paradigmatic' text for such literary competence and provides a model for reading the Bible's characteristic twinning of fiction and reality.

Furthermore, I would argue that the surplus of meaning in any given text will likewise be contingent upon the completeness of this fictional eclipse of historical referentiality. The more completely the original ostensive reference is obscured, the more likely it is that there will be a surplus of meaning for the modern reader, that is, given that the reader still understands enough of the text's meaning to grasp the generic type of human situation it is addressing. Hopefully, the world being poetically projected by the text will more than suffice in making up for this loss of original ostensive/historical reference and will generate a meaningful encounter with the text. For most biblical texts, this will likely be the case, though some texts will always remain enigmatic for the interpreter. Paradoxically, if Ricoeur is correct, historical obscurity or, perhaps better, fictional/poetic projection, becomes the vehicle of theological vision. Sometimes, we understand biblical texts better as readers when our sight is blocked from beholding the mundane, so that we can see the sublime reality that is being projected in front of the text. John Goldingay also notes the paradoxical relationship between meaning and contextuality. He states that to:

> talk in terms of the authors' intention may sound as if it limits the meaning of the story to what the authors were consciously seeking to achieve. In practice, authors may well have been unconscious of some of the implications inherent in what they said. Sacred texts are usually anonymous, and this is linked to the fact that they have their meaning by virtue of what they say rather than because of who says it. *There may be more depth of meaning the less we know of the author...*[31]

30. R. Lux, '"Ich, Kohelet, bin König..." Die Fiktion als Schlüssel zur Wirklichkeit in Kohelet 1.12–2.26'. *EvT* 50 (1990), pp. 331-42.

31. Goldingay, *Models for Interpretation*, p. 35 (my emphasis). He also notes that such dynamics are not limited to just scriptural works, but can be true for many

Distanciation therefore creates a surplus of meaning according to Ricoeur.[32] However, the fact that a text has a surplus of meaning does not mean that all interpretations are equal.[33] Ricoeur maintains that every text has restraints and carves out a specific audience for itself.[34] That does not, however, preclude the opportunity for multiple readings. Textuality preserves the 'right of the reader and the right of the text in an important struggle that generates the whole dynamic of interpretation. Hermeneutics begins where dialogue ends'.[35] The surplus of meaning is not inherent in the text alone, but is located in both the text and the reader. Reading becomes a 'remedy' by which distanciation is rescued from cultural estrangement, becoming an example of productive distanciation. Ultimately, reading is the act of making the text's otherness one's own.[36] However, it is also the function of texts to screen the polysemy of discourse, and so to place limits on the surplus of meaning.[37] The eclipse of reference does not mean interpretative libertinism. Rather, reference is taken to a higher level, the world of the text.

The true reference of a text becomes the world implied in that text. This creates a level of poetic autonomy for the discourse whereby the poetic function of the discourse gains immediate prominence.[38] If there is an implied author and an implied reader in narratology, Ricoeur asserts there is also an implied world, the world of the text. The goal of reading is to ascertain the 'sort of world intended beyond the text as its reference'.[39] The world of a literary text is not to be identified with any

types of literature. Goldingay cites the common interpretation of the phrase in the Declaration of Independence, 'all men are created equal', to include both black and white (as well as male and female we would add) as examples of the audience finding legitimate meaning in a text that goes beyond the strict sense of the text as meant by the original author.

32. Ricoeur, *Interpretation Theory*, pp. 29-30.

33. Ricoeur, *Interpretation Theory*, p. 79.

34. Ricoeur, *Interpretation Theory*, p. 31.

35. Ricoeur, *Interpretation Theory*, p. 32.

36. Ricoeur, *Interpretation Theory*, p. 43.

37. Ricoeur, *Interpretation Theory*, p. 17.

38. I am indebted to D. Breuggemann for this insight.; see his 'Brevard Childs' Canon Criticism', p. 326. However, unlike Breuggemann, I do not see this as a problem, but rather, a strength of Ricoeur's program.

39. P. Ricoeur, *Essays on Biblical Interpretation* (with an introduction by L. Mudge; Philadelphia: Fortress Press, 1980), p. 100.

historical situation. It is a world that is 'poetically distanced from every-day reality'.[40] This world is a 'proposed world', or perhaps a 'defamiliarized world' which projects the possibilities of human existence.[41] Each text therefore projects its own unique conception of human possibilities. It cannot be found in the world behind the text, that is, the world of the author. Rather, it is to be found in front of the text. The true task of reading becomes the ascertainment of 'the type of being-in-the-world unfolded in front of the text'.[42] Sean Freyne suggests on the basis of this model that texts cease being windows to ancient worlds and become 'mirrors of a possible world that confronts me as I grapple with the text and try to decode its meaning'.[43]

Due to the effects of distanciation, it is therefore perceived that 'the text must no longer be seen as an *imitatio* of the real world'.[44] Although there is a relationship to reality here, it is a poetically mediated one which relates to historical reality in a paradigmatic rather than in a mimetic manner. Bernard Lategan has observed that the reality preserved in biblical texts often contains a certain fictive sense (he cites the historical problem of the description of the Pharisees in Mt. 23 as an example).[45] Based on this, he concludes that biblical texts relate to historical reality in an indirect manner, preserving more of the 'essential relationships' when it comes to understanding their ostensive reference. Seen in this light, the text's poetically distanced reference relates to historical reality in a 'proportional' manner as: a:b = c:d.[46] For instance, take the example of the King's Fiction in ch. 2 of Ecclesiastes. In historical reading models, the author mimetically portrays an historical

40. P. Ricoeur, 'Philosophical Hermeneutics and Theological Hermeneutics', *SR* 5 (1975), pp. 14-33 (27).

41. Ricoeur, *Hermeneutics*, p. 142.

42. Ricoeur, *Interpretation Theory*, p. 141.

43. Sean Freyne, 'Our Preoccupation with History: Problems and Prospects', *PIBA* 9 (1985), pp. 1-18 (17).

44. B. Lategan, 'Some Unresolved Methodological Issues in New Testament Hermeneutics', in B. Lategan and W. Vorster, *Text and Reality: Aspects of Reference in Biblical Texts* (SBLSS; Philadelphia: Fortress Press, 1985), pp. 3-25 (23).

45. B. Lategan, 'Reference: Reception, Redescription, and Realty', in Lategan and Vorster, *Text and Reality: Aspects of Reference in Biblical Texts* (SBLSS; Philadelphia: Fortress Press, 1985), pp. 67-93 (87-91). For his treatment of Mt. 23 as it relates to the problem of fiction in biblical historiography, see 'Some Unresolved Methodological Issues', pp. 17-25.

46. Lategan, 'Reference', p. 92.

figure whose features portray the actual historical person (a:b). Such a model would expect the textual presentation ('b') to reflect the actual person Qoheleth ('a'). One would therefore read the King's youthful exploits as an exact account of his adolescent forays. However, as in Matthew 23, there is a very pronounced fictive representation in this text. As a result, what we get is not a person, but a poetic persona. Here, the implied author projects a fictional character to the reader which is modeled on dynamics that have their rootage in an historical situation/figure, but nevertheless, is not an actual or exact mimesis of any historical figure (c:d). Thus, when both distanciation and Scripture's penchant for fictive poetics are taken into account, we find that we do not relate directly to an historical figure. Rather, we experience whichever historical situation/figure lies beneath the text as mediated in a proportional manner through the poetic figure of 'Qoheleth' (a:b = c:d). Given this Ricoeurian view of how poetics and distantion affect the meaning and reference of a text, we perceive that at no time can the critic claim that a text's only relationship to reality or verisimilar historical reference is a strictly linear one. At the least, we see that the reference contained in biblical texts is anything but a simple matter. Frequently, there is not a direct one-to-one correspondence with the historical events which underlie the text. Textuality, distanciation, as well as the specific poetics of the text have created a very 'free' relationship to historical reality in many texts, especially the book of Ecclesiastes.

Due to the gift of textuality, scriptural texts therefore function as context-malleable acts of communication. By definition, a scripture is an ancient document which a community has decided possesses contemporary, existential, personal, social and, ultimately, theological meaning for its contemporary readers. When a community canonizes a document, it 'sees value beyond the original intentionality'.[47] In fact, much of the intentionality canonical critics do see in the process of creating scriptures is precisely to limit the historical moorings and, therefore, the original intentionality of the text. The irony of historical scholarship is that it posited authorial intention as the keystone to biblical interpretation only to learn at the end of the journey that the canonical tradents' intention was to limit authorial intention.

47. J. Ellis, *Theory of Literary Criticism: A Logical Analysis* (Berkeley: University of California Press, 1974), p. 238.

In fact, the effect of the historical approach on reading Scripture was to effectively limit its ability to function as a scripture. John Ellis has mounted a sustained and scathing criticism of the historical approach to reading literary texts. Like Goldingay, he too argues that whereas the loss of context is the beginning of literariness/scripturality, the gain of context is the destruction of the text's status as literature/scripture. In this case, more is less. Ellis contends that

> to refer them back to that original context in order to treat them as functioning primarily in that context is to make them no longer literary texts... Concentration on such factors makes our understanding more localized, and hence more superficial; in taking this path we have again reversed the process of a text becoming literature.[48]

This position argues that all that needs to be done to convert a literary text or a scriptural writing into an historical document is simply to delimit its use and meaning to that of the original context. If one bypasses textual distanciation and the decision of the community to override authorial intention, the reader effectively denies the literariness/scripturality of a text. The more localized the meaning of a text, the less likely will be its general application to the human situation. In fact, its original performance context is likely to be at odds with its function in a scriptural or canonical context. What Ellis has noted about literary texts in general also applies to biblical texts: 'Literature (*read scripture*), then, is the loss of the original performance context in order to be literature (*read scripture*)'.[49] The significance of this conclusion cannot be underestimated. This position recognizes that, in essence, a biblical text can be two textual classifications in one.[50] Read through historical methods, it can be categorized as an historical document. However, once the decision of the canonizing community is honored to change the categorization to that of a scripture, it ceases being a document. Of course, such a decision demands that the reader balance the diachronic reading contract which is inherent for any text located in an ancient setting with a synchronic reading contract. By honoring the synchronic reading code, the text becomes a scripture through the loss, or

48. Ellis, *Theory*, p. 134.

49. Ellis, *Theory*, p. 43.

50. By class I mean broad-based categories of writings that exist above the genre level. Used in this sense, class is to genre what family is to genus in biological taxonomies.

perhaps better, eclipse of its original performance context. As Edgar McKnight concludes:

> To read a text as history is to read it as a specific event, as what happened to particular individuals in geographically and temporally limited contexts. To read a text as literature is to read it as a universal truth.[51]

The preceding discussion is built upon the premise that the categorization of a text as scripture constitutes a quality that is not intrinsic to the text. Scripturality is to a large extent an extrinsic quality of the text which depends upon socially-based reading conventions. Communities of faith and their readers decide that a text is to be read as a scripture. These same communities also decide what rules and conventions are to be used during the reading process in order to consume the text as scripture. The same text can be treated as historical document (artifact) or as scripture (address) depending upon whether the reader decides to utilize a diachronic or synchronic reading contract. The dividing difference is dependent upon the set of rules/conventions the reading the community decides to apply to the text. If the community sees value in the text besides its original meaning, it decides to read the text as a trans-historical, context-malleable document. If the community decides to reverse that decision, the same text would revert to a historical document. Thus it can be seen that many of the problems which have been discussed in the diachronic versus synchronic debate really have more to do with a proper conceptualization of the nature of a sacred text than it does with methods per se.

This should come as no surprise, especially for those who are well acquainted with the canon history of the biblical text. Ellis compares the decision to read texts as historical documents or as literature (scriptures) with botanical decisions to treat some flora as flowers and others as weeds. The major difference is that the community has decided to treat some differently than others. Ellis states:

> The category of literary texts is not distinguished by defining characteristics but by the characteristic use to which those texts are put by the community...the definition of literature must, like the definition of weed, bring into a definition in a very central way the notion of value: the category is that of the texts that are considered worth treating in the way that literary texts are treated, just as weeds are the members of a category

51. McKnight, *The Bible and the Reader*, p. 10.

> of things that are thought worthy of the treatment accorded to weeds. In both cases the definition states an element of the system of values of the community. The membership of the category is based on the agreement to use the texts in the way required and not on the intent of the writer that the text shall be so used.[52]

The same process applies to the reading of texts as scriptures. A reader, or a community of readers, makes a choice to read a text as scripture or an historically delimited document. By definition, we have seen that reading a text as literature or as a scripture is a decision to read the text apart from its immediate performance context via the synchronic reading contract. Eclipse of historical context, at least to some extent, therefore seems to be the requirement for reading a text as a sacred text, as many have surmised of late. From this it can be perceived that a synchronic reading contract is not the perspective of postmodernism, but is as ancient as the canonizing community itself. All classic literature, both religious and secular, depend upon this reading convention, or perhaps better, reading code/contract in order to survive. This much seems clear to critics these days. One could therefore say that such matters have always been at issue in the reading of Scripture. The critic only has to pay attention to the debate between the Alexandrian and Antioch schools in early Christianity to see how the synchronic and diachronic codes evinced themselves for ancient readers as well. But this is another matter, and goes well beyond the scope of this study.[53] Regarding the specifics involved in reading a text will be the next topic for consideration.

52. Ellis, *Theory*, p. 51.

53. Regardig how the synchronic reading contract manifested itself in the ancient hermeneutical debate between Antioch and Alexandria, the reader is referred to D. Dawson, *Allegorical Readers and Cultural Revision in Ancient Alexandria* (Berkeley: University of California Press, 1992). For an excellent appropriation of Dawson's insights for reading theory in regards to the ancient reception history of the Abraham saga see S. Fowl, 'Texts Don't Have Ideologies', *BibInt* 3 (1995), pp. 15-34 (19-28). Ultimately, he argues that 'over its interpretive life a text can be pressed into the service of so many varied and potentially conflicting ideologies that talk about a text having an ideology will become increasingly strained' (p. 18). However, I believe this to be an overstatement. What it does show the critic, however, is that the synchronic reading contract must be balanced by diachronic concerns for an interpretation to have some measure of validity. The diachronic reading contract provides necessary limits to ward off runaway solipcistic readings. That is why this study attempts to honor both codes as they are reflected in Scripture. The

4. *'Woven' to the Reader: How Textuality Affects the Reading Process*

The importance of Ricoeur's views on textuality for biblical exegesis has been explored by numerous critics such as Edgar McKnight,[54] Robert Fowler,[55] Edgar Conrad,[56] Robert Detweiler,[57] Willem Vorster and Bernard Lategan,[58] J. Severino Croatto[59] and Sandra Schneiders.[60] In 1987, *Semeia* 40 ('Text and Textuality') was devoted to the subject to introduce its significance for biblical scholars. In that ground-breaking volume, the authors proposed that the issues of textuality and scripturality were intricately tied together. Charles Winquist introduced the volume by suggesting that the questions, 'what is a text?', 'what is a book?' and 'what is a scripture?' are in fact interrelated queries which presuppose each other.[61] These authors argued that whatever else a scripture is, it is first and foremost a text, and as a text it reads.

Perhaps the best contribution to the question of textuality in that volume is David Miller's essay, '*The Question of the Book: Religion as*

works of both Fowl and Dawson works testify to this need for balance in very powerful ways. In that respect, we can learn a lot about the pitfalls of overstressing the synchronic reading code at the expense of its partner, the diachronic code, from these ancient Hellenistic readers of Scripture.

54. E. McKnight, *Postmodern Use of the Bible: The Emergence of Reader-Oriented Criticism* (Nashville: Abingdon Press, 1988).

55. R. Fowler, *Let the Reader Understand* (Philadelphia: Fortress Press, 1992).

56. E. Conrad, *Reading Isaiah* (Minneapolis: Fortress Press, 1991).

57. R. Detweiler, 'What is a Sacred Text?', *Semeia* 31 (1985), pp. 213-30. Detweiler posits that sacred texts imply a 'faithful reader' who wills to believe such texts. For an interesting reinterpretation of Detweiler's understanding of 'sacred text' for a postmodern setting, see D. Routledge, 'Faithful Reading: Poststructuralism and the Sacred', *BibInt* 4 (1996), pp. 271-87. He argues that faithful reading of a sacred text must consider the role of both textual ambiguity and 'difference' rather view the text merely as a place where uncontestable logocentric meanings reside. As will be seen later, such an attitude is especially necessary for the reader who wishes to approach Ecclesiastes as a sacred text.

58. Lategan and Vorster, *Text and Reality: Aspects of Reference in Biblical Texts* (SBLSS; Philadelphia: Fortress Press, 1985).

59. J. Croatto, *Biblical Hermeneutics: Toward a Theory of Reading as the Production of Meaning* (Maryknoll, NY: Orbis Books, 1987).

60. S. Schneiders, *The Revelatory Text: Interpreting the New Testament as Scripture* (San Francisco: HarperSanFrancisco, 1991).

61. C. Winquist, 'Preface', *Semeia* 40 (1987), pp. i-iii.

Texture'. Building on the etymology of the word 'text', he notes that the basic meaning of its root, tex-êre, means 'to weave', concluding that 'text' means 'that which is woven'. Based on etymology, texts can be understood as weavings rather than the commonplace idea that texts are like pots which hold meanings. He asserts that:

> texts are like weavings and are not like pottings. The power of a text, including the Biblical one, would be in the pattern of its fabric, as in a tapestry, a design which can shape a life meaningfully and one in which a person or a people can be trapped, as in a web or net.[62]

In the classic 'pottery' model, textual power resides in what they contain. But in the 'tapestry' model, the vitality of a text resides in its texture and fabric. Miller concludes:

> When 'text' is taken seriously, fundamentally, in its deepest and highest literal sense, it is, not potting, but weaving, not vessel or container, but texture and fabric... The vessel-perspective is cracked. The Bible contains nothing; it opens out... It is all one rich fabric, with multifaceted patterns, shades, colorings, all weaving meanings endlessly through the life of text and through the texture of life, a thousand threads of significance, each important to the tapestry, none insignificant, all crucial to the whole picture. It is a powerful picture, this picture of the book.[63]

According to this perspective, the force of a text lies not only in its power to contain meaning or referentiality, but in the warp and woof of its pattern and the ability of that texture to elicit a *response*. Furthermore, Miller suggests that 'reading is an unweaving of the weaving that constitutes a text, that reading is at the same time a new weaving of meanings, a texturing of the world'.[64] For all readers, texts are 'woven' discourses.

A similar view is advanced by Robert Fowler. Relying on Henry James's short story, *The Figure in the Carpet*, Fowler also proposes that texts are like the patterns in a carpet which the viewer must put together in a meaningful way. Meaning is not 'there', but is constructed by the reader out of the fabric, texture and patterns which are only loosely connected. Commenting on the Gospel of Mark, he states:

62. D. Miller, 'The Question of the Book: Religion as Texture', *Semeia* 40 (1987), pp. 53-64 (57).

63. Miller, 'The Question of the Book', p. 58.

64. Miller, 'The Question of the Book', p. 61.

> What the guild needs to recognize is that the reader deals with the seemingly fragmented, pearls-on-a-string narrative by processes of gap filling, of prospection and retrospection, and of continuous encounters with many kinds of repetition or duality. In episodic narrative the discrete episodes are connected, held together in fluid and ever-changing association, and thus receive their coherence only in the act of reading. The narrative invites us to tie together its disparate pieces ourselves. At the end of the reading experience, the critic within us may return to the text in search of an innate outline or structure, but the structure, the unity, the general intention, the figure in the Markan carpet, is something we ourselves have already created in the temporal experience of reading the text.[65]

Drawing on these revisions of current textual theory, this study will argue that the reader of the book of Ecclesiastes confronts the text as a pearls-on-a-string argument. The reader does not find a container which only needs the right angle to pour out its contents. Rather, the reader confronts a weaving, the patterns and textures of which are his or her responsibility to tie together into an intelligent *Gestalt*. The threads of Qoheleth's carpet are the various discourse strategies and narrative devices used by the implied author. These threads form a fabric whose patterns the reader constructs. Ultimately, the meaning of Qoheleth is not to be found in some static concept held captive in the text, but is to be understood as the temporal experience of making sense of the different literary devices found in the text's discourse structures.

Three points stand out in this discussion regarding how textuality affects the reading of scriptural texts. First, texts, as examples of writing, stand in contrast to oral communication. How a text communicates to a reader is a vastly different enterprise from how a speaker relates to an audience. Textual communication is not dialogic. There is no author to ask questions. Textuality breaks the pipeline mentality of oral communication.[66] Yet, because both oral and written communication use words and the powers of language, it is tempting to confuse the two. To use an analogy, orality and textuality can be compared to the card games poker and solitaire. Both use cards, but they are vastly different games. In poker, one is constantly dealing with cards in the context of what one's opponent is doing, whether they are bluffing, how much they

65. Fowler, *Let the Reader Understand*, p. 150.

66. W. Ong, *Orality and Literacy: The Technologizing of the Word* (New York: Methuen, 1982), p. 166.

have to gamble, whether one knows their characteristic patterns, and so on. It is a tremendously interpersonal game. The cards actually find their significance only in the context of the otherness of one's opponent. In solitaire, on the other hand, there is no opponent. One only deals with cards and their absolute values. In the absence of another person, the cards have a different use and quality of interaction. The same goes with the use of words and language in textual and oral communication. Orality is poker, whereas textuality, and thus reading, is in many respects, solitaire.[67] Second, partly due to the self-surpassing quality of language in general, and partly due to the fact that authors are not present in a text, textual meaning does not have the same limitations placed upon it as speech-act theory has noted for oral performances. A text can mean more than its author intended. Indeed, it may have no other choice. Finally, the container theory of texts ironically holds the least amount of water. Texts are weavings consisting of various discourse techniques. These constitute a pattern which elicits a response, rather than a container which pours out some ostensive reference or univocal meaning.

5. *Sharing the Loom with the Author: Readers as Co-Authors of Meaning*

Both history and textuality have separated the author's intention from the modern reader, at least in any empirically verifiable manner.[68] The semantic autonomy of the text vis-à-vis the author's intention is grounded in the dynamics of writing and the effect of history on future readers' consciousnesses. We cannot recover their intention, pure and uncolored, based on the evidence at hand because we possess a different perceptual grid through which we read the text. Alessandro Duranti argues that this fact makes the reader a 'co-author' as well as a recipient. He reminds the critic that:

> interpretation is a form of re-contextualization and as such can never fully recover the original content of a given act... The hermeneutic circle

67. This is not to say, however, that texts lack all traits of dialogue. Some texts are intensively dialogic, though dialogic within the constraints allowed by a textual medium. For an excellent study of how texts have their own peculiar dialogic properties, see W. Reed, *Dialogues of the Word: The Bible as Literature According to Bakhtin* (New York: Oxford University Press, 1993).

68. H. Gadamer, *Truth and Method* (trans. G. Barden and J. Cumming; New York: Seabury, 1975).

> is never completed because it must be drawn while space and time change... The interpretation produced during analysis cannot provide the 'meaning'—in a causal sense, that is, the intentions, whether conscious or not...[69]

If it is true that we are 'co-authors' with the historical author(s) of a text, then we must realize that as readers, we are indeed sharing the 'loom' with those author(s), and that we participate in the weaving that makes a text.

Or, to use another analogy, we can also consider the text to be a palimpsest. The reader writes over the strictly historical meaning of the text in favor of a more contemporary and fuller understanding of the text as Scripture.[70] Juxtaposed against this postmodern perspective, the historical-critical model posits that we must understand a text in terms of the original author's historical experience and perceptual grid. John Ellis vehemently denies the validity of this model. He states:

69. A. Duranti, 'The Audience as Co-Author: An Introduction', *Text* 6 (1986), pp. 239-47 (244).

70. That this is the case in the reception history of the Bible is renowned. Take for instance the reading history of Ps. 82.1 as it is presented in the analysis of Lowell Handy. The phrase, 'God stands up in the assembly of God... In the midst of the *elōhîm* he judges', has taken some startling meanings which range from gods (its most likely original meaning in the ancient Near East as also the LXX takes it), angels (Persian/Hellenistic), the judges of the people (Rabbinic), demons (Origen), Jews (Eusebius), Christian community (Luther), and worldly magistrates (Calvin). See L. Handy 'One Problem Involved in Translating to Meaning: An Example of Acknowledging Time and Tradition', *SJOT* 10 (1996), pp. 16-27. His study demonstrates clearly how the meaning of a text does indeed march through time, and that the text has different meanings according to the specific conventions of the particular reading community to which subsequent interpreters belonged. Although we have accesss to the original text (in most instances), that does not give us access to the original intent. It is also assured that most biblical interpreters would not give validity or credence to all of these readings as 'legitimate', especially those that are too local in character or admit to uncalled for bias (as in Eusebius). Such localized readings must be submitted to a more global perspective and criteria from a rhetorically-based postmodern perspective. Delineating how the global and local perspectives on any given text may coexist is indeed the hermeneutical dilemma which confronts postmodern hermeneutical theory. The issue is thorny at best and holds no promise for consensus in the near future. Such readings suggest that the issues and concerns of modernist perspectives still lie in wait for the postmodern perspective at some level.

> criticism that proceeds by means of explanation of historical circumstances has the apparent advantage that it makes the work seem more accessible and more easily graspable to the reader from another age. But this advantage is indeed only apparent. Not only is this kind of assistance to the reader deceptive and dangerous in its substituting an acquaintance with some simple facts for the need to respond to texts, but the degree of success of this procedure often has nothing to do with the historical localization at all. For the historical situations invoked most frequently are only graspable in terms of notions with which the reader is quite familiar from his own age, and which do not need the local historical situation to exemplify them.... Indeed, if he understands the historical situation at all, it is precisely in terms of his own experience. And so the historical critic's way of looking at this situation is the reverse of what is really happening; the reader is not understanding the work through knowledge of history, but understanding history by a knowledge from his own experience of the issues raised in the work.[71]

The preceeding discussion should not, however, mislead the interpreter. As Toulmin admonishes, there is a need to balance both the modern and postmodern perspectives. For all its well-documented weaknesses, the historical-critical method is still a very necessary part of reading ancient texts. Reading theory by itself is not hermeneutical enough to unlock the meaning of the text. Historical data helps us understand the meaning or the sense of the text though, admittedly, not always its significance. In that regard, the historical-critical method is indispensable as a precursor to reading biblical texts, especially when it comes to grasping the text's *repertoire*, that is, those culturally dependent codes inscribed into the text as a matter of historical contingency. However, as Ricoeur points out above, reading is a hermeneutic activity that acts as a remedy for historical and textual distanciation. Good reading is not content to simply decipher the basic sense and reference of the text. Its goal is to ascertain the significance of the vision contained in the discourse structure. Once these latter interests are recognized, the supremacy of the historical-critical method no longer holds sway. Those interests and perspectives must give way, as Nietzche pointed out, to more hermeneutic interests and methods.[72] Yet, as co-authors, we must realize that it is our responsibility as readers to continue 'weaving' the tapestry of the text with the author in a way that

71. Ellis, *Theory*, p. 142.
72. Nietzche, 'History in the Service'.

does not do injustice to the original weaving. As such, we weave as readers in order to actualize the meaning of the text in a legitimate manner for our age. No theory of reading can advocate the absolute supremacy of the reader without compromising the historical nature of all texts. Rather, a sensible theory of reading advocates the reality that we do share the loom with authors, but that we must carry out that responsibility in a manner which enhances the text's original weaving.

At the center of the reading theory utilized in this study is the all-important realization that what we are investigating are *texts* rather than *authors* per se. Ricoeur offers the reminder that texts are distanciated from the historical dimension by the very fact of writing. In saying this, I do not mean to argue that original authors did not have intentions. Authors do have intentions even as readers have presuppositions when they come to texts. Still, the textual medium cannot carry all the information needed to reconstruct authorial intention. In the case of biblical texts, most do not have enough indirect evidence either. The text is like a two-dimensional replication of a three-dimensional object. Unfortunately, most of the author's intention is resident in the historical 'depth' dimension of the situation which can only be partially carried by any textual medium. This means that the historical author's intention is only partially available in the text. What remains is the 'textual intent' of the implied author, significance and textuality, replacing the intentionality and historicality of the real author.

From this perspective, what we recoup in texts are the intentions of a 'Dickens' or a 'Qoheleth', not those of the man Charles Dickens or the ancient sage-philosopher metonymically associated with the character 'Qoheleth'.[73] Textual intention supplants authorial intention, replacing authorial intention as the means of validating various interpretations. Working from these insights, Edgar McKnight argues that the critic should understand

> validation...not in terms of some narrow original intention of an author, but in terms of the genre, type, or langue. When an interpretation is faithful to the *sort* of meaning intended, the interpretation is valid even if it is a meaning not in the mind of the original author.[74]

73. S. Chatman, *Coming to Terms: The Rhetoric of Verbal and Cinematic Narrative* (Ithaca, NY: Cornell University Press), p. 84, makes this point which I have adapted for the book of Ecclesiastes.

74. McKnight, *The Bible and the Reader*, p. 133 (my emphasis).

A textuality-oriented approach such as the one being advocated here seeks to illuminate the text's intention as a linear-inscribed system of meaning, rather than some hypothetical reconstruction of an author's psychological mindset or intent whose logical foundations and evidential support rarely merit consensus. In this new approach, intention no longer refers to the psychological aims of an original author or redactor(s), but refers to 'a shorthand for the structure of meaning and effect supported by the conventions that the text appeals to or devises: for the sense that the language makes in terms of the communicative context as a whole'.[75] Once distanciation, which is created by the fact of writing, swallows up historical referentiality, the footing upon which authorial intention can rest is shaken. In its place stands a more reliable means to validate an interpretation—the intention of the text as an artefact of *langue* or genre when actualized by a fully competent reader.

With Robert Alter, I assert that 'what we most dependably possess is the text framed by tradition as the object of our reading'.[76] If one's interest is in the text, and not its surrogate partner, 'history', paying attention to the effects of textuality is an absolute necessity. Such a position substitutes the powers of language for the perspective of history. Language performs in the text, not some incarcerated author.[77]

75. M. Sternberg, *The Poetics of Biblical Narrative: Ideological Literature and the Drama of Reading* (ISBL; Bloomington: Indiana University Press, 1985), p. 9.

76. R. Alter, *The World of Biblical Literature* (San Francisco: Harper and Row, 1992), p. 8.

77. R. Barthes makes the point that it is language which speaks in a text, not an author. He speaks of 'the necessity to substitute language itself for the person who until then has been supposed to be its owner...to reach that point, where only language acts, "performs" and not "me"' ('Death of the Author', in R. Young [ed.], *Untying the Text: A Post Structuralist Reader* (London: Routledge & Kegan Paul, 1981), pp. 114-18 (115). For a discussion of Barthes' contribution to biblical studies, see A. Brenner, 'Introduction', in A. Brenner and F. Van Dijk-Hemmes (eds.), *On Gendering Texts: Female and Male Voices in the Hebrew Bible* (New York: E.J. Brill, 1993), pp. 1-13 (5-8). In attempting to discover female voices in the Hebrew Bible, Brenner argues that there are gendered voices which are 'textualized as well as fictionalized... Textualized voices are echoes only, disembodied and removed from their extra-verbal situation. Nevertheless, and paradoxically so, they remain grounded in "the world"' (p. 7). Of course, they discuss Ecclesiastes as an example of an 'M text', that is, as a male voice in the Hebrew Bible. See: pp. 133-57. The point of Brenner's discussion is that textuality models indebted to the views of Barthes would rather view literature as texts with 'voices' rather than surrogate manifestations of 'authors'.

The emphasis now shifts to understanding how the discourse affects the reception of the story or the message by the reader. Meaning resides in how the discourse impacts the world of the reader. In the process, the text is freed to attain the significance that it should possess as a scripture.

Obviously, one cause of concern is that an emphasis on the reader may lead to hermeneutic libertinism. However, that is an unjustified fear regarding educated readers. As Ricoeur and others such as Stanley Fish have noted, texts cannot mean anything the reader wishes. Each text 'presents a limited field of possible constructions'.[78] Reader-response criticism is not a lapse into solipsism. It is an exercise in how texts and communities collaborate in the production of viable meanings. Its only danger is that its honesty uncovers the inherently personal dimension that is latent in any reading of a text, both historical and literary.

Finally, a textuality approach to reading will help the critic resist what I have termed the anthropological trap. With the death of the author comes the death of biographism and historicism. When we read a text like the book of Ecclesiastes, we must be aware that what we are listening to is not a person, but a discourse strategy called a narrator. We respond not to an historical author, but to a textual agent called the implied author. What we are doing is neither history nor biography in the strict sense of the word. We are simply explaining the effects of various discourse devices on the implied reader in light of the text's use of its own textuality. For this reason throughout this study, I will use two terms to designate the two major sources of knowledge in the book. Whenever I wish to refer to the discourse strategies of the book, the term 'Ecclesiastes' will be utilized. Usually this will refer to the implied author. However, to distinguish this from the use of the narrator by the implied author, Ecclesiastes' use of a first-person narrator will be referred to as 'Qoheleth'. Qoheleth is simply a textual device the reader responds to while reading. It is no more a person than a hammer is a carpenter, or a canvas is an artist. It is simply another tool used by the text to communicate a message or to achieve an effect upon its encoded recipient, the reader. In this model, all aspects of person are subsumed under the aegis of textuality. What we are responding to in

78. Ricoeur, *Interpretation Theory*, p. 79.

'Qoheleth' is ultimately a life-like discourse device whom we perceive as a character. Fred Burnett summarizes a textualized view of character:

> To say that 'character' is a construct that is developed during the reading process means, on the one hand, that character can be reduced to textuality. It can be dissolved into the segments of a closed text and/or the motifs from which it was constructed... On the other hand, character as an *effect* of the reading process can 'transcend' the text. 'Character' as a paradigm of attributive propositions can give the illusion of individuality or even personality to the reader. Whether or not transcendence of the text occurs will depend both on the indicators that the text provides and the reading conventions that the reader assumes for the narrative in question.[79]

Hopefully, if such textuality issues are kept firmly in mind, it may be possible to resolve some of the problems normally associated with the book in a more persuasive and confident manner. Besides, the book of Ecclesiastes is probably the most well-suited book in all of the First Testament[80] for a radically synchronic method precisely because it is a wisdom book addressing gnomic situations. Furthermore, since its historical background is so poorly attested, it seems to cry out for such an approach. Given that so many historical studies have fallen prostrate before the 'sphinx of Hebrew literature', surely a new starting point is

79. F. Burnett, 'Characterization and Reader Construction of Characters in the Gospels', *Semeia* 63 (1993), pp. 3-28 (5).

80. The terms 'First Testament' and 'Second testament' will be utilized throughout this study. I have chosen to follow the ecumenical suggestion of James Sanders that the term First Testament replace 'Old Testament' and 'Hebrew Bible' as terms of reference for Tanak. Both of these terms contain possible offensive connotations for Jewish and Christian readers respectively. For Jewish readers, anak is simply their Bible, and is hardly an 'old' testament. On the other hand, Hebrew Bible suggests that Tanak is ether racially defined or somehow complete, which is hardly the case for Christian readers. Sanders summarizes: 'Using the expression First Testament where we have used OT, or Hebrew Bible, or Tanak, not only avoids the problems those intrinsically have, but also does what some of them do not do, and that is avoid the supersessionism of old Christendom implicit in the terms Old and New Testament, and one of the major reasons some of us want to avoid using them. It also avoids the possible implication in use of the term Hebrew Bible that it is a Bible complete in itself, which I assume Christians are not quite willing to do! The term FT can also expunge the implicit Marcionism in the use of the terms Old and New. See: 'First Testament and Second', *BTB* 37 (1988), pp. 47-49 (48).

in order. At the least, a textuality oriented study will not replicate what Santiago Bréton rued two decades ago when he complained that most studies limit themselves to problems and methods discussed by their predecessors.[81]

81. S. Bréton, 'Qohelet: Recent Studies', *TD* 28 (1980), pp. 147-51 (149).

Chapter 2

READING ECCLESIASTES AS A FIRST-PERSON SCRIPTURAL TEXT

> The sagacious reader who is capable of reading between these lines what does not stand written in them, but is nevertheless implied, will be able to form some conception.[1]

1. *Seeing Through Textual 'I's: Narrative Theory and First-Person Texts*

At the heart of this study lies a communication model which understands literature as an address between a text-immanent sender, the implied author, and a textually-encoded recipient, the implied reader. The theoretical stance argued by Seymour Chatman and Gerald Prince is presupposed by all reader-response approaches and forms the theoretical framework espoused here. Chatman's paradigm of textual communication makes a hard and fast distinction between extra-textual entities and intra-textual entities. The boundary between text and external world is uncrossable. His model stresses the differences in communication between real persons and that which involves a textual medium. The stream of communication does not flow between author and reader in an unmitigated fashion. Instead, it proceeds through the textual medium which acts both as a conduit and a barrier between those standing on either side of the text. Inside of the text, whatever privileged knowledge is necessary for understanding the story or message is transmitted from an implied author to a narrator, who conveys that information to a narratee. The narratee, who is simply the one listening to the narrator's

1. Johann Wolfgang von Goethe, *Autobiography*. Book XVIII. *Truth and Poetry*. Cited from: *The Autobiography of Goethe. Truth and Poetry: From My Own Life* (2 vols; trans. A.J.W. Morrison, London: Henry G. Bohn, 1949), II, p. 115.

voice within the story, then acts as a relay to an implied reader. As a result, real readers can respond only to implied readers, narrators, narratees and implied authors. We never respond to actual or historical persons as readers of texts, but instead, respond to textual patterns and devices which mimetically simulate real authors and persons. The act of narrative communication is conceived as follows.[2]

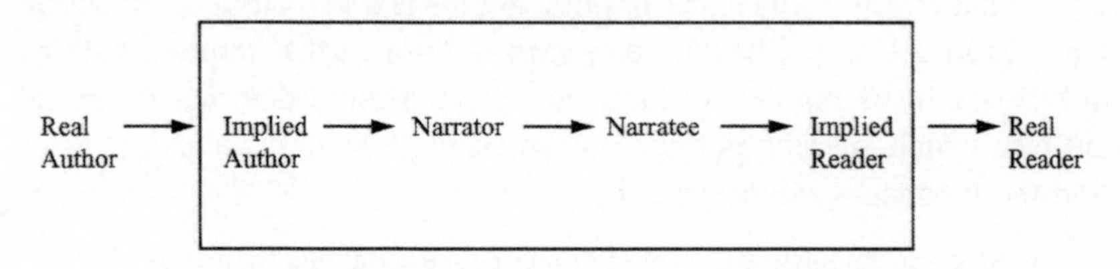

Chatman's Theory of Narrative Communication

a. *What, Not Who, is the Implied Author?*

Textual communication starts with the source of all knowledge contained in the text—the implied author. The concept of the implied author was first introduced by Walker Gibson in 1953,[3] but it was Wayne Booth who coined the term in *The Rhetoric of Fiction*. For Booth, every literary work implies a concept of the author who wrote it. Each writer imposes on his or her work an 'image' of him/herself that is different from the images we meet of other authors.[4] In this sense, the implied author is the 'second self' or persona or mask which the writer implies in his or her work.[5] Most importantly, Booth stresses that a key element for the concept of the implied author is the chief value to which he or she is committed. The emotional and moral content of each bit of action, plot, or characterization is the raw material out of which the reader infers the implied author.[6] Shlomith Rimmon-Kenan has defined the implied author as the 'governing consciousness of the work as a whole,

2. S. Chatman, *Story and Discourse: An Essay in Method* (Ithaca, NY: Cornell University Press, 1978), p. 151.

3. W. Gibson, 'Authors, Speakers, Readers, and Mock Readers', in J. Tomkins (ed.), *Reader-Response Criticism: From Formalism to Post-Structuralism* (Baltimore: The Johns Hopkins University Press, 1980), pp. 1-6.

4. W. Booth, *The Rhetoric of Fiction* (Chicago: University of Chicago Press, 2nd edn, 1983 [1961]), p. 70.

5. Booth, *The Rhetoric of Fiction*, p. 71.

6. Booth, *The Rhetoric of Fiction*, p. 74.

the source of the norms embodied in the work'.[7] For the reader of the book of Ecclesiastes, the choice of metaphors, analogies, the types of arguments, the values and judgments expressed, moral and ethical conclusions, life experiences and other related issues exist as the basic elements out of which we draw an impression of the implied author.

Restricting the implied author to a textual object means that it is not a personal entity. Rather, the implied author is a principle of invention that lies in the text. Chatman emphasizes the fact that implied authors only *seem* to be human. In fact, they are narrative devices or textual entities which merely portray or represent human personages. Meir Sternberg concurs, advocating the

> need to distinguish the person from the persona: the writer as the historical man...behind the writing from the writer as the authorial figure reflected in the writing. The person (the object of genetics) may be lost beyond recovery, but the persona (the object of poetics) is very much there, pervading and governing the narrative by virtue of qualifications denied to the historical, quotidian, flesh-and-blood self anyway.[8]

In order to escape the biographical trap inherent in the term 'implied author', Chatman stresses the textuality of the device. He states:

> He is 'implied', that is, reconstructed by the reader from the narrative. He is not the narrator, but the principle that invented the narrator, along with everything else in the narrative, that stacked the cards in this particular way, had these things happen to these characters, in these words or images. Unlike the narrator, the implied author can tell us nothing. He, or better, it has no voice, no direct means of communicating. It instructs us silently, through the design of the whole, with all the voices, by all the means it has chosen to let us learn.[9]

Chatman admits he would gladly substitute other phrases for this term such as 'text implication', 'text instance', 'text design' or 'text intent'.[10] The priority given to the textuality of the implied author is both theoretical and practical. The term keeps us focused on texts per se, rather than real authors.[11] What we get from positing such a theoretical entity is 'a way of naming and analyzing the textual intent of narrative

7. S. Rimmon-Kenan, *Narrative Fiction: Contemporary Poetics* (New York: Methuen, 1983), p. 86.

8. Sternberg, *The Poetics of Biblical Narrative*, p. 69.

9. Chatman, *Story and Discourse*, p. 148.

10. Chatman, *Coming to Terms*, p. 146.

11. Chatman, *Coming to Terms*, p. 89.

fictions under a single term but without recourse to biographism'.[12] This is especially useful since upon publication, the implied author supersedes the real author as a matter of course.[13]

The implied author must be carefully distinguished from the narrator. The implied author is the inventor of the discourse, including *all* positions or values contained therein, while the narrator is the 'utterer' of a given position whose purpose is to guide the story. Though the narrator is given words to articulate, he or she is not the source of these words. Chatman states:

> The narrator, and she or he alone, is the only subject, the only 'voice' of narrative discourse. The inventor of that speech, as of the speech of the characters, is the implied author. That inventor is no person, no substance, no object: it is, rather, *the patterns in the text which the reader negotiates.*[14]

The narrator must not be confused with the implied author, even in first-person texts such as the book of Ecclesiastes. Especially in the book of Ecclesiastes this must be taken into consideration because there are other ideological positions marked out in the text which juxtapose that of the narrator Qoheleth, such as that of the Epilogist. Since the implied author of the book of Ecclesiastes has designed all of these voices, it stands to reason that Qoheleth as narrator cannot be the implied author. Instead, narrative 'voice' belongs uniquely to the narrator, Qoheleth. The one who 'sees' in the text is the implied author. Even when the narrator Qoheleth says, as he does numerous times, 'Again, I saw…', it is really not the device of narration that literally saw that event. Rather, it was the implied author Ecclesiastes who saw that, or perhaps has reported what another has seen, but who now has chosen to speak through the textual apparatus of first-person narration known to us as Qoheleth. That perceptual grid is presented to the implied reader by the narrator who speaks. Yet, as Chatman points out, it is 'naive…to argue that this…narrator "got" this information by witnessing it. He is a component of the discourse: that is, [one] of the mechanism[s] by which the story is rendered'.[15] The perceptual grid and the guiding intelligence of

12. Chatman, *Coming to Terms*, p. 75.
13. Chatman, *Coming to Terms*, p. 81.
14. Chatman, *Coming to Terms*, p. 87 (my emphasis).
15. Chatman, *Coming to Terms*, p. 142.

the entire discourse is 'Ecclesiastes', the name I have chosen to designate the implied author. Chatman notes that especially in argumentative text-types, the one who argues is simply a tool of the one who has designed the argument and the text in general.[16] Paul Ricouer has argued a similar position. He posits that both the narrator and the implied author are simply categories of interpretation.[17]

When we interrogate a text we do not interact with persons, however textually bound they may appear. Instead, readers interact with abstract ideological positions and textual patterns manifested in the overall design of the work. In a work of literature, these patterns and positions are artfully expressed, with the effect that they mimetically depict human beings. The design of the discourse replaces the traditional emphasis upon the author. Reading focuses on the patterns, structures and devices in the text, not the persons traditionally associated with the text.

b. *The Role of the Narrator in Textual Communication*

It has already been noted that the voice of the implied author is the narrator. It is a tool used by the implied author through which events and information are expressed.[18] Chatman defines the narrator as

> the someone or something in the text who or which is conceived as presenting (or transmitting) the set of signs that constitute it. 'Presentation' is the most neutral word that I can find for the narrator's activity. As part of the invention of the text, the implied author assigns to a narrative agent the task of articulating it, or actually offering it to some projected or inscribed audience (the narratee).[19]

As a set of narrative or textual patterns which metonymically simulates a human consciousness, the narrator is best seen as 'a linguistic subject, really a metaphor for the narrative possibilities of the text as a whole'.[20]

From the perspective of a reader-oriented approach to texts, the importance of the narrator cannot be understated. It is arguably the single most demanding aspect of the text on the reader's attention. Robert Fowler has contended that what distinguishes historical from literary approaches to texts is 'the experience of reading the narrative, which

16. Chatman, *Coming to Terms*, p. 76.
17. Ricoeur, 'Philosophical Hermeneutics', p. 21.
18. Chatman, *Coming to Terms*, p. 84.
19. Chatman, *Coming to Terms*, p. 116.
20. Freyne, 'Our Preoccupation with History', p. 9.

has to do principally with the reader's encounter with the narrator's discourse'.[21] A literary reading of a text begins only when our focus settles upon the address of a narrator to its textually immanent receiver.[22] Such narrators come in a variety of sizes and shapes. Some are physically evoked, while others occupy only intellectual or conceptual space in the discourse.[23] Narrators also vary in regard to how much distance they place between themselves and the characters, the narratee(s) and the implied reader. A narrator may enjoy a close relationship with the characters or narratee(s) in a story, while maintaining an aloof position vis-à-vis the implied reader. On the other hand, the reverse may be true. Furthermore, this distance can take several different forms. Gerald Prince notes that 'one narrator may be at a greater or lesser distance from another one, that this distance may be physical, or intellectual, or emotional, or moral, and it may vary within a given narrative'.[24] The rhetorical impact of such distance is a major consequence of the narration on the reading process. While it is imperative to note that all narrations constitute a dialogue between the narrator(s), narratee(s) and the character(s), the influence of the distance evoked by the narrator's moral, intellectual, or emotional stance constantly affects the nature and rhetorical impact of that dialogue. In the case of the book of Ecclesiastes, this is especially true.

Arguably, every narrator addresses the reader as an 'I' in some sense.[25] The degree of self-effacement, intrusiveness, self-consciousness, reliability, distance and explicitness varies from text to text. But in every instance, at some level, to some degree, the reader experiences the narratorial voice as a distinct person who addresses them. Some are fleshed out while others remain mere voices, but still it must be noted that every act of narration is an address by a person, an 'I'. This is so

21. Fowler, *Let the Reader Understand*, p. 20.

22. N. Petersen, 'Literary Criticism in Biblical Studies', in R. Spencer (ed.), *Orientation by Disorientation: Studies in Literary Criticism and Biblical Literary Criticism* (Pittsburgh: Pickwick Press, 1980), pp. 25-52 (38).

23. Chatman, *Coming to Terms*, p. 123.

24. G. Prince, *Narratology: The Form and Functioning of Narrative* (Berlin: Mouton de Gruyter, 1982), p. 24.

25. Tamir has argued this point quite forcefully. She summarizes the linguistic debate by noting that every declarative statement presumes an 'I say that' in its deep structure. Such statements therefore either have an 'I' in their surface or deep structure. See N. Tamir,'Personal Narration and its Linguistic Foundation', *PTL* 1 (1976), pp. 403-29 (420-21).

much the case that Gerald Prince discusses the narrator under the rubric, 'Signs of the I'.[26] Prince lists five such signs which indicate to the reader the presence of a person who is addressing them. They are:

1. Any second-person pronoun which does not exclusively refer to a character and is not uttered or thought by him or her must refer to someone whom a narrator is addressing and therefore constitutes a trace of the narrator's presence in the narrative.
2. Any first-person plural pronoun which does not exclusively designate characters or narratees refers to a narrating self.
3. The presence of deictic terms ('now', 'here', 'yesterday', 'tomorrow', etc.) which relate to the situation of their utterance and, more particularly, to the spatio-temporal situation of the utterer which are not related to the part of a character's utterance must be related to the narrator.
4. The presence of modal terms ('perhaps', 'unfortunately', 'clearly', and so on) which indicate a speaker's attitude about what he or she says and which is not a part of the character's utterance, describing the narrator's position, is a clear sign of the narrator's presence.
5. Any sign in a narration which represents a narrator's persona, his attitude, his knowledge of worlds other than that of the narrated, or his interpretation of the events recounted and the evaluation of their importance constitutes a sign of the 'I', the narrator.[27]

c. *The Role of the Narratee in Textual Communication*

Someone at the story level must listen to the oratory of the narrator. That listener is the narratee. In Ecclesiastes, the narratee who listens to Qoheleth is explicitly referred to as 'young man' in 11.9 and 'my son' in 12.12. The narratee and the narrator are so intricately related that a study of the narrator has reciprocal significance for the study of the narratee, and vice versa. Robert Fowler notes that narrators and narratees

> represent mirror images...the diction of the narrator is reflected like a sonar wave off of the outline of the posited narratee and returns to the sender to be emitted again—each reflects the presence of the other.[28]

26. Prince, *Narratology*, pp. 8-10.
27. Prince, *Narratology*, pp. 8-10.
28. R. Fowler, 'Who is the Reader in Reader Response Criticism?', *Semeia* 31

So reciprocal is this relationship that the choice of a narratee can in turn characterize the narrator who has chosen to address such a person, thereby relating to the reader something of the cluster of values which is represented by the narrator.[29] Any study of the narrator would be incomplete without a study of the narratee and its influence in guiding the decisions of the implied reader. Since narrators and narratees are intricately related, signals pertaining to the narrator also must be considered in arriving at a portrait of the narratee. The method espoused in this study has been summarized by Gerald Prince. He states:

> By interpreting all signals of the narration as a function of the narratee, we can obtain a partial reading of the text, but a well-defined and reproducible reading. By regrouping and studying the signals of the second category, we can reconstruct the portrait of the narratee, a portrait more or less distinct, original, and complete depending upon the text considered.[30]

If there are 'signs of the I' in a discourse, there must also be 'signs of the you' who listens to that address. Prince lists seven specific signals that indicate the presence of a listening 'you' in a narrative: (1) direct references, such as 'dear reader', 'you', 'my son', and so on; (2) inclusive and indefinite pronouns, such as 'we', 'us', and 'one'; (3) questions and pseudo-questions which do not emanate from the narrator or a character; (4) negations which contradict the beliefs of the narratee; (5) demonstratives, comparisons and analogies which presuppose some prior knowledge for their comprehension; and (6) over-justifications, that is, explanations and information provided to the narratee. The latter prove very useful to the literary critic. Over-justifications are often situated at the level of meta-commentary or meta-narration, that is, narration regarding the narration. Their purpose is to 'provide us with interesting details about the narratee's personality, even though they often do so in an indirect way; in overcoming the narratee's

(1985), pp. 5-23 (13). Chatman makes a similar observation: 'In general, a given type of narrator tends to evoke a parallel type of narratee' (*Story and Discourse*, p. 255).

29. P. Rideout, 'Narrator/Narratee/Reader Relationships in First Person Narrative: John Barth's *The Floating Opera*, Albert Camus' *The Fall*, and Gunter Grass' *Cat and Mouse*' (doctoral dissertation; Tallahassee, FL: Florida State University, 1981), p. 147.

30. G. Prince, 'Introduction to the Study of the Narratee', in Tomkins (ed.), *Reader-Response Criticism*, pp. 7-25 (12).

defenses, in prevailing over his prejudices, in allaying his apprehensions, they reveal them'.[31] An example would be Eccl. 7.21: 'for well you remember the many times that you yourself have reviled others'. Such a justification enables us to see something of the humanity of the narratee presupposed by the narrator, Qoheleth. The narratee is implicitly characterized as a person who possesses the necessary honesty and humility to recognize his own dark side. Commentary, explanations, motivations, generalizations, evaluations, and other reading interludes also define the narratee and his or her role. In addition, when the narratee has been explicitly named or characterized, this information lends immediate coloring to the narratee. Thus, if a narratee is a lawyer, all information concerning lawyers in general is pertinent.[32] Phyllis Rideout therefore concludes that the narratee is 'evoked by any portion of narrative text that is not strict dialogue or a bare account of actions'.[33]

Mary Piwowarczyk summarizes the signals which designate the narratee under four broad headings: the identity of the interlocutors, their spatial-temporal location, their relative status, and their roles.[34] Under 'identity' one looks for any deviations of knowledge or personality. These include the types of experiences familiar to the narratee, the use of proper nouns or a marked common noun, use of other languages, and reference to other texts, knowledge of social customs or conventions which are assumed. Also included would be references to previously narrated elements of the story, since zero-degree narratees are by definition without knowledge and are obliged to follow the linear and temporal progression of the text. Because narratees are assumed to have perfect recall of that narration, any repetition aimed at refreshing the narratee's memory is a deviation from the zero-degree narratee, and is potentially useful in characterizing the specific narratee of the text.

Spatial and temporal location also mark deviations which further define the narratee. The critic must look for direct and explicit geographic and temporal indications, especially deictics and adverbs which cannot be attributed to characters. Words like 'here' and 'now' constitute a sign of the narratee 'whenever it situates the narratee as either

31. Prince, 'Introduction', p. 15.

32. Prince, 'Introduction', p. 13.

33. Rideout, 'Narrator/Narratee/Reader Relationships', p. 49.

34. M. Piwowarczyk, 'The Narratee and the Situation of Enunciation: A Reconsideration of Prince's Theory', *Genre* 9 (1976), pp. 161-77.

interior or exterior to the spatial and temporal situation of enunciation which is otherwise undefined'.[35] Status markers, that is, those linguistic devices which 'mark the relationship between the narrator and the narratee or their intersubjective distance',[36] are a further marker of deviation from the zero-degree narratee. These include pronouns that indicate status, like the French *tu* and *vous*, appellatives and other explicit characterizations, epithets, honorific titles, as well as inclusive and indefinite pronouns that are aimed at the narratee. Devices indicating illocutionary force, like assertions, questions, threats, orders, promises, requests, advice, warnings, greetings, congratulations and thanks that can be attributed to someone other than the characters and the narrator(s) can often mark status as well.

Finally, roles marked in the text by personal pronouns like *tu* and *vous*, as well as the use of direct or indirect speech by the narratee are signs of deviation from the zero-degree narratee. The narratee is by definition a 'listener in the text', so that any sign of speech by the narratee is a deviation marking a role assigned to the narratee by the implied author. Indirect speech includes comments, objections, questions, and so on, that can be attributed to the narratee. These take three forms: anticipations of the narratee's response by the narrator, repetitions for the sake of the narratee, and presuppositions.[37] For instance, a question on the part of the narratee may be implied if a narrator offers a rebuttal which does not answer the explicit question of a character. The narrator has anticipated the narratee's question in this instance. The most forceful are those cases where the narratee interrupts the narrator to express his or her own opinions.[38] As Prince states:

35. Piwowarczyk, 'The Narratee', p. 171.

36. Piwowarczyk, 'The Narratee', p. 171.

37. Piwowarczyk offers an excellent graph summarizing the signals which indicate the presence of a narratee ('The Narratee', p. 176).

38. Such a reading strategy has been proposed for Ecclesiastes. Some readers have viewed the conservative interpolations as neither pious additions nor as the contrary opinions voiced by Qoheleth, but instead, as the musings of a second 'voice' in the text, which narrative theory would view as the narratee whom the reader is given the privilege of overhearing. An example of such an interpretation is the reading of T.A. Perry, *Dialogues with Koheleth: The Book of Ecclesiastes, Translation and Commentary* (University Park: The Pennsylvania State University Press, 1993).

> certain parts of the narrative may be presented in the form of questions or pseudo-questions. Sometimes these questions originate neither with a character nor with the narrator who merely repeats them. These questions must then be attributed to the narratee and we should note what excites his curiosity, the kinds of problems he would like to resolve.[39]

Rhetorical questions (cf. Eccl. 1.3; 2.2, 12, 15, 19, 22, 25; 3.9, 21, 22; 4.8, 11, and so on) function in a similar fashion, acknowledging the questions of the listener or the reader. Often, such questions 'reveal a great deal about what kind of response the narrator wishes from or projects onto his narratee'.[40] Fundamentally, the literary critic looks for a deviation from the 'zero-degree narratee'.[41] Any signal which presupposes a deviation from this colorless baseline marks a characterization of the narratee by the narrator.

Summarily, the narratee is a set of attitudes brought to bear on the text by the implied author which interact with the attitudes of the narrator. Often, this produces a polyvalent reading experience. Like the narrator, the narratee expresses a point of view, except from the listener's post of observation.[42] This second point of view interacts with the point of view of the narrator. The differences must be negotiated by the implied reader if he or she is to have a productive and valid encounter with the text. As a discourse device provided by the implied author for the implied reader, narratees perform several important functions as a discourse structure. Prince lists six basic functions:

> The narratee can, thus, exercise an entire series of functions in a narrative: he constitutes a relay between the narrator and the reader, he helps establish the narrative framework, he serves to characterize the narrator, he emphasizes certain themes, he contributes to the development of the plot, he becomes the spokesman for the moral of the work.[43]

As such, the primary function of the narratee is to provide clues for reading the text. The narratee complements the implied reader as a discourse structure that governs the reading of a text. Both the narratee and the implied reader provide models for the consumption of a text.

39. Prince, 'Introduction', p. 14.
40. Rideout, 'Narrator/Narratee/Reader Relationships', p. 48.
41. Prince, 'Introduction', p. 10.
42. B. Uspensky, *A Poetics of Composition: The Structure of the Artistic Text and Typology of a Compositional Form* (trans. V. Zavarin and S. Wittig; Berkeley: University of California Press, 1973), pp. 37-41.
43. Prince, 'Introduction', p. 23.

However, the narratee's actions are something that cannot always be trusted as being reliable. Often, the act of listening by the narratee is a 'justification device' that persuades the reader to act in a similar manner, inducing belief in the reader.[44] However, the implied reader can be given clues which suggest that a given narratee is unreliable, or perhaps naive. In that case, as Robert Fowler points out, the implied reader 'may relate to the narratee, in turn, in any number of ways, ranging from a close and intimate association to an ironic distancing, if the narratee appears to the implied reader to be gullible or otherwise deficient'.[45] Basically, the implied reader looks over the narratee's shoulder and views the surrounding narrative landscape. However, in different texts, we may find ourselves induced to see through his or her eyes, sometimes over him, or sometimes around her. We may even be induced to abandon totally the narratee's reaction, viewing his or her role in an ironic light.

More importantly, the narratee may function as a relay to the reader. Seymour Chatman notes that 'direct communication of values and opinions between narrator and narratee is the most economical and clearest way of communication to the implied reader attitudes required by the text'.[46] A narratee with skeptical values would surely relay that skepticism to the reader, provided other discourse clues do not portray the narratee in an ironic light. A believing narratee would relay the opposite values. The role of the implied reader is to deduce which interpretation of the narratee is the valid one, given the norms of the work. The latter criterion protects literary analysis from overtly subjective readings. Phyllis Rideout states:

> we must ask not only whether the narrator speaks in accordance with the norms of the work, but whether the narratee 'responds' to the narrator's tale in accordance with those norms as well. If the narratee responds in what we consider an appropriate manner to the narrator we have judged unreliable, he, rather than the narrator, may be the 'spokesman' for the fundamental values of the work.[47]

The role of the implied reader is therefore to navigate the various perspectives of the narrator and the narratee, and with the help of the

44. G. Prince, 'On Readers and Listeners in Narrative', *Neophilologus* 55 (1971), pp. 117-22 (117).

45. Fowler, 'Who is the Reader?', p. 12.

46. Chatman, *Story and Discourse*, p. 261.

47. Rideout, 'Narrator/Narratee/Reader Relationships', p. 52.

norms provided by the implied author, come to a reasonable interpretation of the work. The narratee aids in this process by suggesting a possible attitude to adopt regarding the interpretation of the narrator's speech.[48] Obviously, the role of the narratee and the implied reader can become confused, especially when the narratee is an extradiegetic one.[49]

Narratees have three broad functions. They function as relays to the implied reader and as role models for those consuming the text. In addition, the relation between a narrator and a narratee can become the focus of attention itself, functioning to thematize the work. The relation between these two may 'underscore one theme, illustrate another, or contradict yet another'.[50] In a work such as the Book of Ecclesiastes, where the entire narrative focus is on the monologue between a narrator and his narratee(s), such a function gains immediate prominence.

It should be noted that there can be several different narratees in a work, each at a different diegetic level with different discourse functions in the work. In the case of the book of Ecclesiastes, there are at least two, and possibly three narratees: the 'young man' and 'my son' addressed in 11.9 and 12.12 respectively, the Epilogist who seems to listen over the student's shoulder, so to speak, and perhaps even a 'conservative' narratee in 12.12-14 should the critic not regard these verses as original. Each of these performs as an audience for the narrator, Qoheleth. Each has its own unique role to perform. Role reversals are not uncommon in this regard. In fact, narrators can turn into narratees, and narratees can turn into narrators,[51] as is the case of the Epilogist in our text.

Also important for the book of Ecclesiastes is the fact that the narratee is anonymous, going simply by the name, 'my son'. Although anonymity can sometimes function to signal the relative unimportance of a particular character, it can also serve to increase the reader's identification with a character. Commenting on the function of anonymity in the characterization process, David Beck observes:

> When names are absent, the reader has an option for the freedom of subjectivity... Anonymity erases the identity distinction of the name and

48. P. Rabinowitz, 'Truth in Fiction: A Re-Examination of Audiences', *CritInq* 4 (1977), pp. 121-41 (127).

49. Rideout, 'Narrator/Narratee/Reader Relationships', p. 152.

50. Prince, 'Introduction', p. 22.

51. Prince, 'On Readers and Listeners', p. 118.

> instead creates a gap that the reader is invited to fill with her/his own identity, entering into the narrative and confronting the circumstances and situation of the character in the text.[52]

As a result of this unconscious effect, readers who confront Qoheleth's nameless listener are invited by the most subtle of means to become one of his pupils. The desired effect would be the 'alteration and re-formation of the reader's self'.[53] Of course, the rhetorical properties of the narrator will also have a powerful influence on whether or not this ultimately succeeds.

To summarize, this study will contend that one of the central problems the implied reader of the book of Ecclesiastes faces is the problem of multiple narrators (Qoheleth and the Epilogist) and multiple narratees ('my son' and at least the Epilogist). The implied reader must navigate the responses suggested by both narratees as he or she attempts to utilize them as clues on how to read the book. The result is an ambiguous and conflicting set of reader clues, a vain rhetoric if you will, which creates contradictions at the deep level as well as on the surface level of the discourse. Not only does the content of Qoheleth's orations contain contradictions or polar structures, but also the discourse structure of the text contains the same proclivity for contradiction. This frustrates the implied reader, creating in the reader a sense of disequilibrium and ultimately, a sense of the very '*hebel*' that Qoheleth was attempting to articulate.

d. *What, Not Who, is the Implied Reader?*

The mirror image of the implied author is the implied reader. Walker Gibson first introduced the concept, referring to the feigned role of the reader as a 'mock reader'.[54] Again, it was Wayne Booth who systematically explored the significance of the concept for narrative studies.[55]

52. D. Beck, 'The Narrative Function of Anonymity in Fourth Gospel Characterization', *Semeia* 63 1993), pp. 143-58 (147). I am aware that the narratee is characterized as a male. This does limit its appeal in a postmodern setting. However, translating the references to 'my son' and 'young man' as 'my child' and 'young person' respectively would go a long way to overcome this hindrance to female readers. At the least, such a hermeneutical move seems to be in the best interests of the synchronic reading contract which Scripture relies upon to maintain its contemporary appeal.

53. Beck, 'Narrative Function', p. 148.

54. Gibson, 'Authors, Speakers, Readers', *passim*.

55. Booth, *The Rhetoric of Fiction*, pp. 119-47 (138-40).

Like its narrative twin, the implied author, the implied reader is not a real flesh-and-blood reader, but rather, a narrative pattern that functions as an interpretative construct. Its utility lies in its heuristic value and its ability to draw attention to the text itself. The seminal point in this discussion is the considerable difference between those who will actually read a text and those the author had in mind while composing the text. Peter Rabinowitz differentiates the 'Authorial Audience' from the 'Ideal Authorial Audience'.[56] Rabinowitz postulates an implied reader along the lines proposed by Chatman, but splits it into two levels: one historical and the other textual. The authorial audience consists of the basic reader competencies and skills required to minimally process the text. For the reader of the book of Ecclesiastes, this set of competencies would include a basic knowledge of Hebrew grammar and language plus a knowledge of ancient reading conventions. Cultural knowledge is also assumed at this level. It refers to all the data a reader needs to *make sense* of the text at a basic level. This aspect of the implied reader revolves around the 'axis of fact'. The ideal authorial audience refers to the basic values needed to *appreciate* what is being read. This facet of the implied reader revolves around what Rabinowitz calls the 'axis of ethics or interpretation'.[57] Obviously, a modern reader cannot function as the latter without some knowledge of the former. Both competency in First Testament reading conventions and compatible values are necessary for a modern reader to function as the implied reader of a biblical text in this sense.

This raises the issue of the relation of a text's implied reader to its various readers. First, the implied reader does contain a hint of the historical reader due to the fact that it assumes basic skills and knowledge that were present in the original audience in order to process the text. Second, an implied reader is modeled along the lines of real readers, and as such cannot be simplistically differentiated from actual recipients of texts. Implied readers are normally expected to respond to texts as real readers would. James Marra has conducted empirical studies on actual readers' responses to texts and has concluded that 'whatever cognitive or affective responses we may have...are derived from our own real life experiences and codes as they are projected onto the

56. Rabinowitz, 'Truth in Fiction', pp. 121-41.
57. Rabinowitz, 'Truth in Fiction', p. 135.

realities of the fictional illusion'.[58] Implied readers are expected to react to textual stimuli in exactly the same fashion as real readers would if they held such values and knew such facts.

Again, the problem of historical distance and textual distanciation raises its head. A modern reader of Scripture cannot be expected to respond in the same exact fashion as would the original audience because we no longer possess the same culturally inherited values. However, unless one assumes that human nature has changed vastly in these intervening millennia, it seems reasonable to assume that a level of common human response still exists between our age and primeval humanity. Given this premise, the difference between real readers and implied readers exists not only in terms of their textuality versus their historicality, but also, in terms of which aspects of our common humanity the author chooses to address or to rely upon when constructing a given text. In its most hermeneutic sense, implied readers are sets of human characteristics, values and traits which an author hopes to play upon as he or she builds a text with an anticipated response in mind. As such, it is the human dimension, rather than the historical dimension which is most useful when attempting to understand the reader implied by a text. Beyond the surface structure of historical and sociological audience characteristics lies the deep structure of human response patterns based on genetically inherited cognitive apparatuses and above all, our common species characteristics.[59] It is these response structures

58. J. Marra, 'The Lifelike "I": A Theory of Response to First-Person Narrator/Protagonist Fiction' (doctoral dissertation; Lubboc, TX: Texas Technical University, 1986), p. 193.

59. While it is beyond the scope of this book to discuss the full implications of my views here, the specifics of what these 'species traits' entail have been rather exhaustively treated by the fledgling discipline of evolutionary psychology. For a lucid and brilliant overview of the findings of this exciting new discipline which posits a genetic component to the various characteristics of the human race as they are anchored in evolutionary necessity, the reader is referred to R. Wright, *The Moral Animal: Why We Are the Way We Are; The New Science of Evolutionary Psychology* (New York: Vintage Books, 1994); *idem*, *The Moral Animal: Evolutionary Psychology and Everyday Life* (New York: Peter Smith, 1997); and P. Wilson, *Man, the Promising Primate: The Conditions of Human Evolution* (New Haven: Yale University Press, 1980). For a good overview of this discipline on the World Wide Web see 'Introduction: Darwin and Us' (http://www.clark.net/pub/wright/introduc.htm) and W. Spriggs, 'Evolutionary Psychology for the Common Person' (http://www.evoyage.com/index.html).

that are presupposed by the text which ultimately define the implied reader for a scriptural audience. However, in spite of the connection between real readers and implied readers at the level of basic competencies and generic human characteristics assumed by the text, the scholar must resist historicizing the implied reader in every way.[60]

The discussion regarding the implied reader can be summarized as follows. First and foremost, it designates a set of inferred values. Each literary work carves out for itself an audience of readers for which its designs were devised. A reader must agree on the whole with the values and norms implied by the author if he or she is to become an implied reader. Without this basic agreement between implied author and implied reader, the success of the reading is in jeopardy. Booth therefore defined the implied reader as 'the set of beliefs the story/texts presupposes for a good reading'.[61] Jeffrey Staley expands this to include not only values, but the entire affective quality of a text.[62] Second, the implied author suggests a role necessary for consuming the text by the reader. For Chatman, the implied reader is a textual device which informs the reader how to read the text. It instructs the reader regarding which choices and stances a reader must take if they are to fully consume the text. In a similar fashion, Gerald Prince argues that in many instances, the text metonymically acts like a reader. He observes how 'many a narrative text…functions as a text reading itself by commenting explicitly and directly on these constituent parts'.[63] Through the use of such reading interludes, the text 'acts frequently like a reader organizing his reading in terms of nonlinguistic codes'.[64]

60. W. Worster, 'The Reader in the Text: Narrative Material', *Semeia* 48 (1990), pp. 21-40 (36); N. Petersen, 'The Reader in the Gospel', *Neot* 18 (1984), pp. 38-51 (39-40); and W. Iser, *The Act of Reading: A Theory of Aesthetic Response* (Baltimore: The Johns Hopkins University Press, 1978), p. 28.

61. Booth, *The Rhetoric of Fiction*, p. 423.

62. J. Staley, *The Print's First Kiss: A Rhetorical Investigation of the Implied Author in the Fourth Gospel* (SBLDS, 82; Missoula, MT: Scholars Press, 1988), p. 33.

63. G. Prince, 'Notes on the Text as Reader', in S. Suleiman and I. Crosman (eds.), *The Reader in the Text: Essays on Audience and Interpretation* (Princeton, NJ: Princeton University Press, 1980), pp. 225-40 (230).

64. Prince, 'Notes on the Text as Reader', p. 232. T. Todorov argues for a similar position in 'Reading as Construction', in Suleiman and Crosman (eds.), *The Reader in the Text*, pp. 67-82.

It is important to realize that many aspects of what I am calling the implied reader are often discussed under treatments of the various textual strategies that are available to an author. Just as the implied author should not be personalized, it is important for the implied reader to be given the same abstract consideration. It is simply the *role* given to the real reader that can be inferred from the textual ordering, strategies, designs and intention of the text.[65] So textually oriented is this concept that Rimmon-Kenan can define the implied reader as 'an image of a certain competence brought to the text and a structuring of such a competence within the text'.[66]

However, there is some debate in reader-response circles over whether the implied reader is *in* the text or is to be situated somewhere *between* reader and text. While Chatman and Prince locate the implied reader strictly in the text, a reader-response critic such as Wolfgang Iser is careful to situate the implied reader in the interaction between reader and text. Rather than a textual entity, what we find in Iser is a phenomenological entity. The implied reader is part text and part human perception—an entity hovering between both worlds.[67] For Iser, the implied reader cannot be reduced to textual patterns precisely because such patterns are ultimately produced by the ideational activity of the reader. He defines the implied reader as a 'textual structure anticipating the presence of a recipient without necessarily defining him...the implied reader designates a network of response-inviting structures which impel the reader to grasp the text'.[68] The implied reader has two parts according to this conceptualization: the reader's role as a textual structure and the reader's role as a structured act of ideation.[69] Because patterns are not strictly in the text, but are something that are dependent upon the mutual interaction of the text and the ideational activity of perceiving the text by the reader, the implied reader is best understood as a patterned *Gestalt* which readers form from prestructured material in the text.

65. Iser, *The Act of Reading*, pp. 36-39.

66. Rimmon-Kenan, *Narrative Fiction*, p. 115.

67. W. Iser, 'Indeterminacy and the Reader's Response in Prose Fiction', in J. Hillis Miller (ed.), *Aspects of Narrative: Selected Papers from the English Institute* (New York: Columbia University Press, 1971), pp. 1-45 (31).

68. Iser, *The Act of Reading*, p. 34. R. Fowler discusses the significance of this for biblical studies in his essay 'Who is the Reader?', p. 16.

69. Iser, *The Act of Reading*, p. 35.

Iser makes a further contribution to the discussion of the implied reader that merits discussion. For Iser, the implied reader, sometimes termed the intended reader or the fictitious reader, is one of several textual standpoints which provide perspectives on the meaning of the text. This implied reader is no longer an addressee like the one found in Chatman's model but, rather, another textual perspective alongside those of the implied author, the characters and the plot which the reader navigates in the creation of meaning.[70] Iser states:

> The intended reader, then, marks certain positions and attitudes in the text, but these are not yet identical to the reader's role, for many of these positions are conceived ironically...so that the reader is not expected to accept the attitude offered him, but rather to react to it. We must, then, differentiate between the fictitious reader [i.e., implied reader] and the reader's role, for although the former is present in the text by way of a large variety of different signals, he is not independent of the other textual perspectives, such as narrator, characters, and plot-line, as far as his function is concerned. *The fictitious reader is, in fact, just one of several perspectives, all which interlink and interact*. The role of the reader emerges from this interplay of perspectives, for he finds himself called upon to mediate between them, and so it would be fair to say that the intended reader, as supplier of one perspective, can never represent more than one aspect of the reader's role.[71]

From this theoretical perspective, real readers must utilize the role of the implied reader as one of the tools supplied by the text as an aid to its own consumption. The other tools are those of the implied author, the narrator, the narratee and those of the plot, or, in the present case, the line of argumentation.

Iser's concept of the implied author therefore functions along two lines. It is at once an independent perspective at the discourse level and a role which facilitates the assemblage of meaning involving all textual perspectives. As a role, it is a set of *competencies* that is presupposed for assimilating the different textual perspectives into one coherent *Gestalt* or interpretation. The implied reader, understood from the model of literary competency, refers to the requisite skills necessary to join

70. W. Martin, *Recent Theories of Narrative* (Ithaca, NY: Cornell University Press, 1986), p. 161.

71. Iser, *The Act of Reading*, p. 33 (my emphasis).

these different standpoints into a meaningful *Gestalt*. However, understood from the vantage point of the communication model, it is an *addressee* to whom the implied author communicates values and information. These values constitute an independent perspective from which to evaluate other perspectives in the text. This is especially apparent in ironic texts in which the implied author communicates to the implied reader that certain positions are not reliable (e.g. the speech of a character or a narrator). An example of this phenomenon is found in the book of Ecclesiastes, in which Qoheleth's systematic reliance upon private insight is given an ironic treatment by the implied author. This communicates a certain sense of unreliability to the implied reader regarding the sufficiency of Qoheleth's method of argumentation.

The implied reader is therefore both an addressee and an assembler of viewpoints. As an addressee, it is a set of values communicated from the implied author that is deemed necessary to evaluate other textual positions and ultimately, to appreciate the text in its entirety. These values in turn may become one of the competencies the implied author can depend upon when the implied reader is asked to assume the role of producing a meaningful *Gestalt* out of the whole text. Both of these are essential roles that the implied reader must navigate when consuming texts. Obviously, then, one of the primary functions of any text is to generate the competency it takes to process the text in a productive and meaningful fashion. That competency-building function is what Umberto Eco calls the 'model reader'. Every text builds up the specific competency it takes to read it. So central is this to a text's function that Eco defines a text as 'a device conceived in order to produce its Model Reader'.[72] This is similar to Tzvetan Todorov's dictum that a 'text always contains within itself directions for its own consumption'.[73] The role of the empirical reader is to make conjectures about the kind of model reader that is postulated by the text.[74] Eco states:

> A text can foresee a Model Reader entitled to try infinite conjectures. The empirical reader is only an actor who makes conjectures about the kind of Model Reader postulated by the text. Since the intention of the text is basically to produce a Model Reader able to make conjectures

72. U. Eco, 'Intentio Lectoris: The State of the Art', in U. Eco (ed.), *The Limits of Interpretation* (Bloomington: Indiana University Press, 1990), pp. 44-63 (58).

73. Todorov, 'Reading as Construction', p. 77.

74. Eco, '*Intentio Lectoris*', p. 59.

> about it, the initiative of the Model Reader consists in figuring out a Model Author that is not the empirical one and that, at the end, coincides with the intention of the text.[75]

Eco's model reader, like the readers discussed by Fish and Iser, can be seen to have two levels. The first level is a naive one in which the model reader is supposed to understand semantically what the text says and means. At a more profound level, or a critical level, the model reader is 'supposed to appreciate the way in which the text says so'.[76]

A central part of this study will therefore be to ascertain the specific ways in which the book of Ecclesiastes builds up the competencies it needs for the reader to consume it in a skillful way. Specifically, I will note how the discourse instructs the reader to consume a first-person text and the effects that instruction has on the reader as well as the reading of the book. To briefly state what will be argued later, the model reader of the book is extensively instructed to understand the entire discourse as a first-person speech, with all the strengths and liabilities inherent in such speech. The model reader of Ecclesiastes understands nearly every word as an example of first-person speech. Not only the narrator, but the Epilogist and narratee(s) as well are understood from the limited perspective of first-person narration. By the time the model reader encounters the later chapters (which are not strictly first-person speeches, such as the proverb collection in ch. 7), he or she has already been instructed to read these chapters as examples of first-person speech as well. To put it succinctly, for the model reader of Ecclesiastes there are few third-person aspects to the book. After the opening superscription and the poetic prologue in 1.2-11, there are no 'objective' third-person perspectives within the book from which to evaluate the first-person narration of Qoheleth. Instead, what the competent reader understands is a chorus of limited first-person speeches, each with its own problems and biases, that provide various and sometimes conflicting perspectives from which to view the problems of life raised by the narrator, Qoheleth.

What then is the role of the reader? The reader's role, regardless of whether he or she is a naive consumer or critically-trained scholar, is to facilitate the convergence of the different textual perspectives offered by the discourse structure of a text. The role of reader-response criticism

75. Eco, '*Intentio Lectoris*', p. 59.

76. Eco, '*Intentio Lectoris*', p. 55.

is to give the critical reader the necessary theoretical stance from which to analyze and to appreciate how texts instruct them to become this implied or model reader.

2. *Posts of Observation and Point of View in First-Person Argumentative Texts*

Each of the preceding devices presents the reader with its own point of view. In the book of Ecclesiastes, there are four main perspectives which the reader utilizes to guide his or her reading: the values and perspectives of the implied author, the narrator(s), the narratee(s), and the implied reader. By 'perspective' I mean 'the particular angle from which we are invited by the nature of the narration to imagine the narrated personages, places, and events'.[77] In narrative texts, characters and plot also provide additional guidelines for assembling the meaning of the work as a whole. However, in spite of its many narrative-like qualities, Ecclesiastes is not a narrative text. It is an argumentative text which utilizes expository, descriptive and narrative text-types to serve its argumentative purposes.[78] However, it is possible to view Ecclesiastes as a narrative text-type. Eric Christianson has argued extensively that Ecclesiastes is a narrative text which has a plot. Like myself, he too relies heavily upon the narratological theory of Seymour Chatman. Christianson builds upon Chatman's distinction between kernels and satellites. A kernel is an event that initializes narrative motion while

77. R. Alter, *The Pleasures of Reading in an Ideological Age* (New York: Simon & Schuster, 1989), p. 172.

78. So, according to text-type theory, which attempts to go beyond genre issues to broader, more inclusive types of textual analysis. Text-types are underlying textual structures which can be actualized by different surface forms or genres. For a discussion of text-type theory, see C. Brooks and R. Warren, *Modern Rhetoric* (New York: Harcourt, Brace & World, 4th edn, 1979 [1949]) and Chatman, *Coming to Terms*, pp. 6-21. Text-type theory posits that there are only four basic text-types which all readers generally recognize. These constitute the various parts of all genres: narrative, exposition, description and argument. See L. Faigley and P.Meyer, 'Rhetorical Theory and Reader's Classifications of Text Types', *Text* 3 (1983), pp. 305-25 (320-25). Narratives basically tell the sequence of what happened. Expository texts tell us why something happened. Exposition is designed to convey information or to explain. Descriptive texts tell us what an event or object looks like. Finally, argumentative texts rely on logic and urge specific actions or beliefs based on a clear presentation of reasons for such actions or beliefs.

satellites are logically expendible, that is, the action they initialize is tangential and can be removed without damaging the major plotline of the story.[79] Essentially, Christianson argues that 'everytime Qoheleth makes his opinion known, or relates what he has done in the order to come to a certain conclusion, there is a process of change' which meets the criterion of a kernel.[80] On the basis of this insight, he concludes that 'a story-line, however small, has been created and the criterion of functionality met'.[81] Eventually, Christianson compares Qoheleth's plot to that of a character novel, where the reader does not encounter an action story per se, but rather, a plot which 'may have as the centre of its narrative logic the revelation of character'.[82]

However, for many narratologists, something more than a simple event is needed to say that a story presents a plot. That something is causality. As E.M. Forster argues in his classic description of plot: ' "The king died and then the queen died" is a story. "The king died, and then the queen died of grief" is a plot.'[83] Forster further observes that in a simple story, the reader asks 'what then?'. However, in a plot the reader asks 'why?'.[84] Although events are recounted or sometimes implied by Qoheleth, it is difficult for me to see an overarching causality which connects the various 'kernels' in Qoheleth's monologue—whether implied or stated—with each other in the way that plot usually connotes. According to R.S. Crane, plot refers to the 'material continuity' of the story.[85] What counts for Crane is 'the amount of suspense and surprise it evokes, and the ingenuity with which all the happenings in the beginning and middle are made to contribute to the resolution at the end'.[86] To my mind, Qoheleth's various reflections, if read as a storyline, possess a disjointedness which precludes such a definition of causality. I see no material continuity which would tie them together

79. E.S. Christianson, *A Time to Tell: Narrative Strategies in Ecclesiastes* (JSOTSup, 280; Sheffield: Sheffield Academic Press, 1998), p. 26, citing S. Chatman, *Story and Discourse,* pp. 53-56.

80. Christianson, *A Time to Tell*, p. 26.

81. Christianson, *A Time to Tell*, p. 27.

82. Christianson, *A Time to Tell*, p. 78.

83. E.M. Forster, 'The Plot', in R. Scholes (ed.), *Approaches to the Novel: Materials for Poetics* (San Francisco: Chandler, rev. edn, 1966), pp. 219-32 (221).

84. Forster, 'The Plot', p. 221.

85. R.S. Crane, 'The Concept of Plot', in Scholes (ed.), *Approaches to the Novel*, pp. 232-43 (237).

86. Crane, 'The Concept of Plot', p. 237.

into a coherent plot that leads from kernel A to B to C and so on. At the least, there is no sense of suspense or surprise in terms of the events to which Christianson refers. To be sure, there are events here, but I would hesitate to say that they imply a plot due to their lack of connecting causality. For instance, what causality leads the reader to proceed from the kernel which constitutes the King's Fiction in ch. 2 to the Time Poem in ch. 3? Or better, what causality connects the example story in 9.13-16 to the string of proverbs cited in ch. 10 in terms of why these events follow one another? If indeed there was a plot here, it would surely be easier to answer such questions. To me, Qoheleth's discourse appears to be better conceived as the random thoughts of an interior monologue set within the context of an argumentative text. For that reason, I still prefer to view Qoheleth's discourse as narrative-*like*. However, the distinction is slight, and may reflect my Western ideas of plot more than the ideas of plot that were current at the time of the composition of this discourse. Certainly, there is ample room for Christianson's views. My objections are not great in this regard. Furthermore, I agree with him that reading Ecclesiastes as a narrative is a reader's decision. Like Christianson, I would endorse reading the text with an 'awareness of its narrative quality'. However, I would be more hesitant in regard to the supposed 'features of its story-line'.[87] Nevertheless, his working thesis functions quite well on several levels.

Therefore I still conclude that the book of Ecclesiastes is better conceived as an argumentative text which utilizes narrative features. In an argumentative text, the flow of the argument replaces the movement of the plot in a narrative text. However, since an argument always expresses the viewpoint of the one who is arguing the point, it cannot replace plot as an independent perspective in the text. Iser's four basic perspectives must therefore be adapted for use with argumentative texts. In argumentative texts, only those whose values are expressed in the text can serve as guides to the reader. Those who express values or perspectives are the implied author, the narrator(s), the narratee(s) and the implied reader. Of course, there are argumentative texts which do use

87. Christianson, *A Time to Tell*, p. 256. However, there is more agreement between Christianson and myself than there are real differences except in regard to this subtle nuance. I whole-heartedly agree with Christianson's views that the non-narrative material does not function independently of their setting within a narrative-like monological setting (p. 257).

characters to argue points (e.g. fables). However, the book of Ecclesiastes does not utilize characters in this way. Because it is a dramatic monologue, all of the 'characters' are narrators or narratees.

Again, textuality issues must be kept clearly in mind. It is imperative that point of view, like other matters, be seen in the light of the textuality of a text. Above all, texts are a weave of various devices which express different perspectives. These devices mimetically simulate the type of consciousness we normally associate with another person's presence. At its most fundamental level, a text is a series of narrative devices and grammatical/linguistic structures which must be navigated by the reader in a strictly linear and temporal fashion. As a textual perspective, point of view refers to the expression of an ideological stance by a given narrative device, such as a narrator or implied author. It is simply a value or perspective, or a set of values or perspectives communicated to the reader through the literary magic of human representation. As such, our understanding should not be hindered by the anthropomorphic nuances implied by the term point of view. Point of view is simply a value or perspective presented to the reader in the guise of a feigned human consciousness.

To understand the rhetorical power of expressing values through the point of view of a character or narrator, we must remind ourselves that the author could have presented these values in an overt and explicit fashion, such as can be observed in a philosophical or perhaps a dogmatic textbook. However, once values are presented through a human voice, the rhetorical dynamics of presentation become more complicated. Aristotle said persuasion could be of three types: appeals to the character (ethos), appeals to the subject (logos), and appeals to the audience (pathos).[88] Although any given post of observation is the mere expression of a value or perspective at its most fundamental level, that value or perspective is given various hues when it is refracted through a lens like that of a first-person narrator. In order to understand how a first-person narrator has a persuasive effect on the reader, all three of these dimensions must be clearly kept in mind, especially the ethos dimension. The ethos of a first-person narrator is a confounding rhetorical variable in relation to the logos of the narrator's unadorned statements. A confounding situation is a circumstance in which the

88. Aristotle, 'The Rhetoric', in *The Rhetoric and the Poetics of Aristotle* (trans. W. Roberts; New York: The Modern Library, 1954), Book 1, Chapter 1.40 (1356a).

effects of two processes are not separate.[89] As such, analyzing the effect of narration in a first-person text must be extremely attentive to how that person is characterized and in what way the audience is set up to respond to that characterization. The posts of observation expressed in a first-person text operate by confounding or enhancing the logos dimension of the text with the ethos and pathos variables inherent in the characterization of the narrator(s) and narratee(s). The value or perspective expressed by any given post of observation is always colored by the ethos and pathos dimensions which accompany that post of observation. Analyzing the interplay of these levels will be the major focus of this study. When the critic understands how the ethos and pathos levels of the text confounds or enhances the logos aspects, he or she comes away with a better apprehension of the persuasive properties of the discourse. As the various posts of observation in the book of Ecclesiastes are analyzed, one must pay close attention not only to what is being said (logos), but by whom (ethos) and to whom (pathos).

In the book of Ecclesiastes, for each post of observation it must be ascertained who is speaking, from what ideological position or angle, through which means (by words, thoughts, perceptions or feelings), and with what distance from the reader, both moral and intellectual.[90] When there are multiple narrators, as is the case in the book of Ecclesiastes, these distances and their differences can generate a range of conflicting responses in the reader. A major task of the reader is to navigate both the ideological differences and the subtle distances evoked by the different posts of observation and, also, to distinguish the differing levels at which this occurs. Specifically, the reader reacts rhetorically when a narrator withdraws morally or psychologically. When this distance shifts from an intellectual to a moral or emotional level, this too has an effect on the reader.

Qoheleth's narration, like most first-person narration, sometimes draws the reader into his circle of trust, and at other times alienates the reader along varying lines. Sometimes these lines are moral, sometimes intellectual, and at other times, even emotional. In Ecclesiastes, the posts of observation provided by the implied author, Qoheleth, the

89. D. McCroskey, 'Ethos: A Confounding Element in Communication Research', *SM* 33 (1966), pp. 456-63 (463).

90. For this taxonomy, see: N. Friedman, 'Point of View in Fiction: The Development of a Critical Concept', *PMLA* 70 (1955), pp. 1160-84 (1169).

Epilogist, the narratee(s), and the implied reader offer differing perspectives, producing a chorus of voices, each with different perspectives that evoke varying distances. It is my thesis that the consonance and dissonance generated by these posts of observation is what creates the distinctive and peculiar rhetorical impact of the book. Ultimately, the reading experience centers around the reader reacting to the dialogue between these different textual agents. When one adds to this the additional influence of ethos, pathos and logos, what the reader experiences is a very rich and multi-faceted rhetorical effect.

Obviously, the analysis of psychological factors pertaining to point of view will be a major concern for my study. Boris Uspensky has analyzed point of view from the vantage point of compositional options. He posits that there are four compositional planes of expression which pertain to point of view: the plane of ideology, the plane of phraseology, the plane of the spatial and the temporal, and the plane of psychology. Particularly useful for the analysis of first-person texts is his treatment of the plane of psychology, which he defines as 'those cases where the authorial point of view relies on an individual consciousness (or perception)'.[91] Also of relevance is the plane of ideology which he defines as occurring when 'several independent points of view are present within the work'.[92] Uspensky likens the different points of view portrayed in a text with the roles an actor plays. He states:

> The author assumes the form of some of the characters, embodying himself in them for a period of time. We might compare the author to an actor who plays different roles, transfiguring himself alternately into several characters.[93]

Sometimes the perspectives that are expressed in the different levels of a composition 'concur' on a given point of view. However, the compositional aim of some texts is to set up different levels which express nonconcurring points of view. For instance, the point of view expressed on the psychological level may be at odds with the one expressed on the phraseological level. This takes place when

> the narration in a work is conducted from the phraseological point of view of a particular character, while the compositional aim of this work is to evaluate the character from some other point of view. Thus, on the

91. Uspensky, *A Poetics of Composition*, p. 81.
92. Uspensky, *A Poetics of Composition*, p. 10.
93. Uspensky, *A Poetics of Composition*, p. 91.

> level of phraseology a particular character emerges as the vehicle of the authorial point of view, while on the level of ideology he serves as its object.[94]

In the book of Ecclesiastes, Qoheleth is the one who expresses the dominant point of view on both the psychological and phraseological levels of the work. But at the ideological level, Qoheleth serves as the object of the point of view of the Epilogist. On the psychological level, the reader is given a position *inside* of the narrator. In contrast, on the ideological level, the reader is given a position *outside* of the narrator. The book deftly manipulates these two levels to give the reader a perspective that is both subjective and objective, personal and public. The tensions between such perspectives is one of the reasons why the book possesses a characteristic and overwhelming 'polar' quality.

The contrast and tension between these internal and external points of view is what gives the artistic text its basic 'deep structure'. For Uspensky, art thrives on the necessary isomorphisms built into the structure of point of view. Every point of view necessarily needs its structural counterpart. Public perspectives need private perspectives. Interior point of view needs exterior point of view. For an artistic text to succeed, it must successfully rely on these structural isomophisms.[95] Texts can therefore be viewed as 'an aggregate of smaller and smaller microtexts, each framed by the alternation of the external and internal authorial positions'.[96] This suggests that in addition to analyzing the book from the point of view expressed at the phraseological level as Addison Wright and others do,[97] perhaps another fruitful place to begin is to look at the book of Ecclesiastes from the perspective of its use of alternating posts of observation and levels of point of view, especially the internal and external points of view. In fact, the very frame of a literary text exists precisely to set up the transition from an external point of view to an internal point of view. Uspensky argues that the framing devices of a text facilitate the transition from 'the real world to the world of the representation'.[98] Since the book of Ecclesiastes is one

94. Uspensky, *A Poetics of Composition*, p. 102.

95. Uspensky, *A Poetics of Composition*, pp. 130-72 (132-37).

96. Uspensky, *A Poetics of Composition*, pp. 153-54.

97. A. Wright, 'The Riddle of the Sphinx: The Structure of the Book of Qoheleth', *CBQ* 30 (1968), pp. 313-34. I mention his study only because it enjoys wide acceptance and is typical of previous approaches.

98. Uspensky, *A Poetics of Composition*, pp. 137, 141.

of the most famous 'frame narratives' in all the Canon, such insights gain immediate relevancy for understanding the reading of this text. More precisely, the function of the initial prologue in 1.2-11 functions exactly as Uspensky described (see above, pp. 93-94), informing the reader that he or she has moved from the external world to Qoheleth's perception of the world. This prepares the reader for the brazen I-narration which begins in 1.12. It softens the shock that a competent reader of biblical literature might have with Qoheleth's unusual narration. Analyzing how a text presents the internal and external points of view is critical to understanding the total effect of the work. This is especially true for Ecclesiastes, which conspicuously plays off the subjective perspective of the narrator Qoheleth against the more public perspective of the Epilogist. Thus, in addition to the interplay between the logos, ethos, pathos and distance dimensions of the text, the literary critic must also pay attention to the structural isomorphism that exists between internal and external points of view if he or she is to comprehend the total effect of the text on the reader.

To sum up, the central rhetorical activity involved in reading a text is the negotiation of the various posts of observation found therein.[99] This is especially true for a first-person text whose natural internal orientation only exacerbates and complicates the problems associated with negotiating the various posts of observation. All texts offer the reader four or five posts of observation which they must navigate and ultimately synthesize as a part of their response to the text. However, the nature of a first-person text is to introduce the elements of ethos and pathos into those perspectives to a degree which is not found in most third-person texts. That is their unique quality which affects our response as readers.

3. *Wolfgang Iser's Theory of Reading*

Reader-response criticism analyzes how readers respond to texts in the course of their linear and temporal progression.[100] The objective is to

99. Obviously, there are other activities, such as text-type recognition, genre-recognition, grammatical and lexical competence, and a whole host of mental operations which occur when one reads. However, for the purpose of understanding the rhetorical aspects of a text, analyzing the issues involved in processing the different posts of observation is definitely the major activity engaged in by a reader.

100. There have been several excellent surveys of reader response criticism. Two

understand the text as a whole by paying strict attention to how readers assimilate the various textual patterns which constitute the structure of the text. The method's critical gaze focuses upon the reader's encounter with the text *qua* text. However, it is not so much the text, but rather, the reader's experience of the text that preoccupies the methodological interests of most reader-oriented critics.[101] The distinction between the text and the experience of reading the text is critical. The major axiom of reader-response criticism is that texts cannot be naively or simplistically equated with their physical expression. On the contrary, texts become reality for the reader only during the act of reading. Reading therefore becomes the subjective mediator between the two objective poles of reader and text. While texts and readers are objective entities that enjoy a physical reality, the process that unites them into a hermeneutical dyad is the subjective process of reading. This distinction is the principal axiom of the phenomenological theory which underlies reader-response criticism.[102] As such, there are no objective texts from a phenomenological perspective. Whenever a reader consumes a text, that text loses its objective status. During the act of reading, the physical 'text' becomes an ideated 'work' which depends on the reading process for its very existence.[103]

of the better, more recent overviews which also render due criticism to the method are: M. Brett, 'The Future of Reader Criticisms?', in F. Watson (ed.), *The Open Text: New Directions in Biblical Studies?* (London: SCM Press, 1993), pp. 13-31 and G. Aichele *et al.*, 'Reader-Response Criticism' in *The Postmodern Bible* (New Haven: Yale University Press, 1995), pp. 20-69. For an excellent contribution to the synchronic versus diachronic debate which surrounds reader response criticism from the Continental perspective, see H. Utzchneider, 'Text—Reader—Author: Towards a Theory of Exegesis; Some European Views', *JHStud* 1 (1996), pp. 1-22 (http://www.arts.ualberta.ca/JHS/). For the more standard introductions see E. Freund, *The Return of the Reader: Reader-Response Criticism* (New York: Methuen, 1987); J. Tompkins, 'An Introduction to Reader-Response Criticism', in Tomkins (ed.), *Reader-Response Criticism*, pp. ix-xxvi; S. Suleiman, 'Introduction: Varieties of Audience-Oriented Criticism', in Suleiman and Crosman (eds.), *The Reader in the Text*, pp. 3-45; S. Mailloux, *Interpretive Conventions: The Reader in the Study of American Fiction* (Ithaca, NY: Cornell University Press, 1982).

101. Fowler, *Let the Reader Understand*, p. 1.

102. W. Iser, 'The Reading Process: A Phenomenological Approach', in Tomkins (ed.), *Reader-Response Criticism*, pp. 50-69 (50).

103. Iser, 'The Reading Process', p. 50. It should be noted that the ideational nature of what we call 'text' has extreme ramifications for both diachronic and synchronic methods. Relying upon Eco's distinction between *intentio operis*, *intentio*

a. *Reader-Response Criticism as a Pragmatic Approach to Biblical Literature*

Rhetoricians such as Wayne Booth[104] have emphasized the rhetorical function of literary texts. Building upon this tradition, Seymour Chatman argues that the 'rhetoric of fiction' has two components. First, there is aesthetic rhetoric that 'suades us to something interior to the text, particularly the appropriateness of the chosen means to evoke a response appropriate to the work's intention'.[105] Aesthetic rhetoric suades the reader that there is a 'person' Qoheleth who is addressing us as an 'I' with a story or message that is appropriate to the world the work is attempting to create. It suades the reader that metaphors such as 'striving after wind' and 'vanity' are feasible given the world being created by the implied author. However, texts also function ideologically. Ideological rhetoric 'suades us to something outside the text,

lectoris, and *intentio auctoris* (intention of the text, reader, and author respectively), Utzcheider argues that the three are intrinsically related and not easily separated in spite of the preference of interpreters to do so. He cogently argues that what often goes under the name of the intention of the author is often itself a reconstruction and an act of ideation the same as any other reception of the text. Therefore, all reconstructions of the author, whether implied or historical, are seen to partake of the intention of the reader whether the historical critic desires it or not. Utzcheider states: 'we have to ask whether exegetes who are interested in the *intentio auctoris* are sufficiently aware that the author they elicit (the author of the source or the redactor) is initially a product of reception, an "implied author" or "model author", a design created by the reader—an author who cannot necessarily be equated with a real, historical author, but who is nevertheless continually, by preference, so equated... The problem about this hermeneutical circle (if one likes to call it that) is not that it exists, but that there is too little awareness of it' (Utzcheider, 'Text—Reader—Author', p. 12). Thus we see that synchronic approaches such as reader-response have a valid role to play in diachronic methodology in that they enable the reader/critic who is functioning as an historian to be more honest with what they are doing as historical readers. As Utzcheider so elequently concludes: 'But now interpretation is by no means a purely authorial activity; it is a highly crafted interweaving of reading and authorship, of "lecture" and "relecture"...*intentio lectoris* and *intentio auctoris* are bonded together—and in this order' (p. 13). Because of the ubiquitous effects of ideation, the reader-oriented perspective should indeed have a legitimate role to play in diachronic methods as well.

104. Booth, *The Rhetoric of Fiction.*

105. S. Chatman, 'The "Rhetoric" of "Fiction"', in J. Phelan (ed.), *Reading Narrative: Form, Ethics, Ideology* (Columbus, OH: Ohio State University Press, 1989), pp. 40-56 (52).

something about the world at large'.[106] In the book of Ecclesiastes, ideological rhetoric attempts to persuade the reader that the real world is absurd, that the point of living is to enjoy life, that injustice and death render wisdom a tenuous matter, and a host of other conclusions Qoheleth makes during the course of his monologue. When this study speaks of the rhetoric of first-person discourse, this latter aspect will be the foremost concern. Rhetoric will therefore be understood along the lines delineated by Douglas Ehninger, as

> an organized, consistent, coherent way of talking about practical discourse in any of its forms or modes. By practical discourse I mean discourse, written or oral, that seeks to inform, evaluate, or persuade, and therefore is to be distinguished from discourse that seeks to please, elevate, or depict.[107]

Suasion will be defined for the purposes of this study as a 'symbolic activity whose purpose is to effect the internalization or voluntary acceptance of new cognitive states or patterns of overt behavior through the exchange of messages'.[108] From the perspective of a reader-oriented approach, suasion is a textually inscribed pattern of response.[109] Ultimately, this is what is meant by a pragmatic approach to the text.

b. *The Interaction of Text and Reader Replaces the Emphasis on the Text Itself*

John Barton described reader-response criticism as an exercise in 'watching our own eyes moving'.[110] The metaphor aptly describes what reader critics do—they analyze the succession of reading activities that are required by a text during the course of its linear and temporal

106. Chatman, 'The "Rhetoric" of "Fiction"', pp. 52-55. Also see his further treatment of the subject in *Coming to Terms,* pp. 184-203.

107. D. Ehninger, 'On Systems of Rhetoric', in R. Johannesen (ed.), *Contemporary Theories of Rhetoric: Selected Readings* (San Francisco: Harper & Row, 1971), pp. 327-39 (327).

108. K. Grant-Davie, 'Between Fact and Opinion: Readers' Representations of Writers' Aims in Expository, Persuasive, and Ironic Discourse' (doctoral dissertation; San Diego, CA: University of California, San Diego, 1985), p. 6, quoting M. Smith, *Persuasion and Human Action: A Review and Critique of Social Influence Theories* (Belmont, CA: Wadsworth, 1982), p. 7.

109. G. Hauser, *Introduction to Rhetorical Theory* (SCS; San Francisco: Harper & Row, 1986), p. 109.

110. J. Barton, *Reading the Old Testament* (Philadelphia: Westminster Press, 1984), p. 132.

progression. Reader-response criticism focuses on 'the mind in the act of making sense, rather than on the sense it finally (and often reductively!) makes'.[111] While critics such as Stanley Fish do this on the sentence level, others such as Wolfgang Iser track these activities at the level of larger discourse units.[112] The common denominator between all reader critics is the emphasis that reading is not a static affair, but a temporal experience. Stanley Fish describes the method as

> an analysis of the developing responses of the reader to the words as they succeed one another on the page... In my method of analysis, the temporal flow is monitored and structured by everything the reader brings with him, by his competencies; and it is by taking these into account as they interact with the temporal left to right reception of the verbal string, that I am able to chart and project the developing response.[113]

Because the distinguishing mark of reader-oriented criticism is its emphasis on the reading experience as it develops through time, the critic must resist the tendency to concentrate on the end product, that is, the meaning that a reader makes of the text. Instead, the critic must concentrate on the entire reading experience, that is, each experience-by-experience moment as it unfolds in the course of navigating a text and its devices.[114] Reader-response criticism 'slows down' the reading experience so that the maneuvers of the reader, which occur without our conscious observation, are made explicit and become themselves a means to getting at the meaning of the text.[115] A principal axiom of reader-response criticism is that readers respond to texts not in their entirety, but in terms of minute sections of the text. Reading is intensely linear, so much so that a 'reader's response to the fifth word in a line or a sentence is to a large extent the product of his response to

111. S. Fish, *Self-Consuming Artifacts: The Experience of Seventeenth Century Literature* (Berkeley: University of California Press, 1972), p. viii, quoted by S. Mailloux, 'Learning to Read: Interpretation and Reader-Response Criticism', *SLI* 12 (1979), pp. 93-108 (100).

112. Mailloux, 'Learning to Read', p. 100.

113. S. Fish, 'Literature in the Reader: Affextive Stylistics', in S. Fish, *Is There a Text in This Class? The Authority of Interpretive Communities* (Cambridge, MA: Harvard University Press, 1980), pp. 21-67 (46).

114. Fowler, 'Who is the Reader?', p. 19.

115. Fish, 'Literature in the Reader', p. 28. Fowler makes a similar observation in *Let the Reader Understand*, p. 43.

words one, two, three and four'.[116] The same can be said of sentences, paragraphs, and sub-sections of the text. The meaning derived from a given passage is largely the result of the response to what has preceded in the text. As a result, Fish reminds the critic that one cannot go directly from the formal features of a text to its meaning, but must go 'through the mediating functions of reading'.[117]

Reader critics thus monitor the temporal flow of a text as it pertains to the *potential and probable* response of the ideal reader who has the necessary literary competence in areas of genre, conventions and intellectual background that they can make sober and relevant judgments regarding the text.[118] As a result, the critic essentially becomes a reader who 'observes his own reactions during the process of actualization, in order to control them'.[119] In so doing, the reader/critic asks

> what a reader, as he comes upon that word or pattern, is doing, what assumptions he is making, what conclusion he is reaching, what expectations he is forming, what attitudes he is entertaining, what acts he is being moved to perform... In each case, a statement about the shape of the data is reformulated as a statement about the (necessary) shape of response.[120]

To understand a text, the reader critic looks for patterns of expectation and disappointment, reversals of direction, traps, invitations to premature conclusions, textual gaps, delayed revelations, temptations, strategies designed to educate or confound the reader and any other mental operation which is induced by the structure of the text.[121]

What this method describes is, in the terms of Menakhem Perry, a 'maximal' reading of the text. It does not attempt to predict the subjective reactions of any individual reader. Instead, what reader-response criticism attempts to analyze is the probable and potential response of

116. Fish, 'Literature in the Reader', p. 27.

117. S. Fish, 'Introduction, or How I Stopped Worrying and Learned to Love Interpretation', in S. Fish, *Is There a Text in This Class? The Authority of Interpretative Communities* (Cambridge, MA: Harvard University Press, 1980), pp. 1-17 (8).

118. Fish, 'Literature in the Reader', p. 48.

119. Iser, *The Act of Reading*, p. 31.

120. Fish, 'What is Stylistics and Why Are They Saying Such Terrible Things About It', in S. Fish, *Is There a Text in This Class? The Authority of Interpretative Communities* (Cambridge, MA: Harvard University Press, 1980), pp. 68-96 (92).

121. S. Fish, 'What Makes an Interpreter Acceptable', in S. Fish, *Is There a Text in This Class? The Authority of Interpretative Communities* (Cambridge, MA: Harvard University Press, 1980), pp. 338-55 (345).

the reader implied by the text.[122] This probable and potential response is very much contingent on the reader taking into account the norms, both social and literary, that are relevant for the period of the text's composition.[123] For the biblical critic, the former prerequisite mandates the use of historical information in order to properly read the text. The historical-critical method helps the critic to understand the '*repertoire*', that is, the historical and cultural knowledge that is presupposed by a text for a maximal reading.

To sum up, in order to understand a text the reader critic must become adept at asking the quintessential question: 'what does this passage/ sentence/word do?'[124] This question replaces the former emphasis on the question: 'what does this mean?' Reader-response criticism asks, after each succeeding passage, what does this passage or word do to the reader in terms of probable responses based on the specific competency required by the text itself. It yields an analysis of the developing responses of the reader in relation to the words as they succeed one another in the course of the temporal and linear progression of the text. The result is not an analysis of the formal features of the text per se, but of the structure of response implied by those formal features.[125]

c. *Reader Critics Validate by Reading Along with Other Critics*

No critic can read a text without the subtle influence of past readings. The trap for the critic is that the multitude of readings he or she has absorbed can be a great hindrance to reading the text with vitality and freshness. Robert Fowler has noted the significance of this for biblical scholars, where the tradition of commentaries is long and extensive. He states:

> To think that we can read Mark as it was first read is a delusion. We never read the text itself, only the history of the reading of the text. The choice is either to read the history of reading with sensitivity and imagination, which is the vocation of Steiner's 'critic', or to be read by the history of reading, which is the fate of the 'reader'.[126]

122. M. Perry, 'Literary Dynamics: How the Order of a Text Creates Its Meanings (Part One)', *PT* 1 (1979), pp. 35-64 (56).

123. M. Perry, 'Literary Dynamics (Part One)', p. 43.

124. Fish, 'Literature in the Reader', p. 66.

125. Fish, 'Literature in the Reader', p. 42.

126. Fowler, *Let the Reader Understand*, p. 263, citing G. Steiner, '"Critic"/ "Reader"', *NLH* 10 (1979), pp. 423-52.

Fowler's point is well taken and has been argued by many other reader critics, as well as by literary philosophers such as Hans Georg Gadamer and Paul Ricoeur. What the biblical reader critic must realize is that for any text, and especially a traditional classic like the Bible, there is no going back to the text in any pristine fashion. There are no 'virginal' readers when it comes to the Bible—the reader critic must be aware of this fact, and must endeavor to creatively use his or her reading tradition to enlighten the reading of a text. This means that reader critics are always in conversation with previous readings of the text, rather than just the text itself. Every reading has its own contextual background. The purpose of reader-response criticism is to expose this and to creatively harness its latent powers to unleash new and vital readings of the text, as well as to explain old ones.

How does the reader critic appropriate the vast reading history of any text? According to Stanley Fish and Robert Fowler, who builds upon Fish's views for biblical critics, the answer lies in utilizing the reading history of a text to demonstrate what types of problems a text typically presents to its reader. Fish addresses this problem in his essay, 'Interpreting the Variorum'. He states:

> Typically, I will pay less attention to the interpretations critics propose than to the problems or controversies that provoke them, on the reasoning that while the interpretations vary, the problems and controversies do not and therefore point to something that all readers share. If, for example, there is a continuing debate over whether Marlow should or should not have lied at the end of the *Heart of Darkness*, I will interpret the debate as evidence of the difficulty readers experience when the novel asks them to render judgment. And similarly, if there is an argument over who is the hero of *Paradise Lost*, I will take the argument as an indication that, in the course of reading the poem, the identity of its hero is continually put into question. There will always be two levels, a surface level on which there seem to be nothing but disagreements, and a deeper level on which those same disagreements are seen as constituting the shared content whose existence they had seemed to deny. In short, critical controversies become disguised reports of what readers uniformly do, and I perform the service of revealing to the participants what it is they were really telling us.[127]

127. S. Fish, 'Interpreting "Interpreting the *Valiorum*"', *Is There a Text in This Class? The Authority of Interpretative Communities* (Cambridge, MA: Harvard University Press, 1980), pp. 174-80 (177-78). Fowler addresses the same problem for biblical scholars in 'Who is the Reader?', p. 18.

The reader critic of biblical text attempts to translate the traditional philological-historical comments about the text at hand into comments about the experience of reading that text. Hopefully, by utilizing such comments, the reader critic better understands not only the problems engaged in by informed readers, but also how readers who are trained with the necessary competence to read a text respond to its formal features. The reader critic looks for what texts have done to earlier readers in an attempt to understand what the text does to us as we respond to its linear and temporal progression. If the magic question in reader-response criticism is 'what does this X do?', then the magic question for scholarship review will be, in concomitant fashion, 'what did this X do to readers in the past?'. By so doing, we are in effect translating the 'legacy of biblical criticism into the language of readers and reading...the history of biblical interpretation is transformed into a history of reading, that is, a reception history'.[128] As such, reader-oriented methods do not valorize the traditional textual object per se, but rather, the 'experience of reading *within a tradition of criticism*'.[129]

A second reason why reader critics review the reading history of a text is to document that a suggested reading does possess sufficient intersubjective validation to be viable for the critical community. Steven Mailloux states:

> Reader-response critics make the description of reading identical to the act of criticism and claim that they accurately represent the temporal reading process in their analyses. To convince others that this descriptive claim is valid, the reader-response critic often resorts to the device of citing other reader's reactions.[130]

By citing the reading problems addressed by earlier informed readers, the reader critic circumvents the charge of solipsism so often leveled at reader-response criticism.[131] In its stead, the reader critic offers the

128. Fowler, *Let the Reader Understand*, p. 1.

129. Fowler, 'Who is the Reader?', p. 8 (my emphasis).

130. Mailloux, 'Learning to Read', p. 101.

131. There is also the matter of the inextricably social dimension of reading which could be adduced regarding this fear. All readers are held in restraint by norms of their reading communities. With A.K.M. Adam, I would argue that reading is an 'ineluctibly social matter'. Although there are no transcendent laws which would determine the meaning of a text, there are what he calls 'local constraints' which would provide a hedge against unrestrained interpretation for most good readers. What keeps reading from becoming solipsistic are the 'criteria that we share with

entire reading history of a text as an example of the reading problems addressed by informed readers. As the reader critic attempts to analyze the successive reading activities required by a text, he or she validates these descriptions with the 'evidence from other critic's reactions'.[132] Whereas historical critics excavated archeological tells for the evidence they needed to validate their readings, reader critics quarry the 'tells' of past and present readings to substantiate their analyses of texts.

d. *Reading Both Requires and Builds Literary Competency*

As Eco has already noted, the issue of how each text constructs its own model reader is of paramount inportance for understanding the reader's response to a text.[133] Analyzing how the text produces the specific sort of competency that it requires in order to understand it's meaning is a major aim of reader-response criticism. The early chapters in a work prepare the way for later ones, not simply by supplying necessary information, but principally, by 'arming the reader with interpretative habits, specific ways of reading'.[134] For instance, in the book of Ecclesiastes, the initial prologue on nature (1.4-11) defamiliarizes the reader's understanding of the world, giving the reader the necessary hermeneutic reflexes he or she will need to understand the narrator's radical worldview. It sets the tone for the book by arming the reader with the necessary values it takes to appreciate Qoheleth. The task of the reader critic is therefore to describe how the initial passages prepare the reader

particular groups of readers to whom we are accountable'; see A.K.M. Adam, 'Twisting to Deconstruction: A Memorandum on the Ethics of Interpretation', *PRS* 23 (1996), pp. 215-22 (216). On the basis of this he astutely observes: 'One can no more say that a red, octagonal road sign means whatever one likes; there is no transcendent law that obliges one to stop at such a sign, but there are effective local constraints that will enforce a particular interpretation of such a sign' (p. 217). As such, the fear that an emphasis on the reader would result in unrestrained interpretation does not adequately account for the social constraints that accompany every individual reading. By tracking the readings of competent readers, a further hedge against solipsistic readings is set in place by reader-oriented scholars. As such, the fear is not justified.

132. Mailloux, 'Learning to Read', p. 102.

133. U. Eco, *The Role of the Reader: Explorations in the Semiotics of Texts* (Advances in Semiotics; Bloomington: Indiana University Press, 1979), p. 7. See also B. Lategan, 'Coming to Grips with the Reader in Biblical Literature', *Semeia* 48 (1990), pp. 3-20 (7).

134. Mailloux, 'Learning to Read', p. 97.

to judge, interpret and understand the later passages in a text. In that sense, reader-response criticism not only describes the reading process, but teaches the reader what paradigmatic moves are required by the text.[135] As an example, Steven Mailloux describes how the early passages in *Moby Dick* prepare the reader for the 'disappearing narrator' in the later chapters. By the time the reader confronts the later passages, the narrator has already taught the reader how to 'make puzzles out of everything'.[136] As a result, the reader no longer needs the narrator's services.

A similar thing happens in the book of Ecclesiastes. Commentators have noticed that the pronoun 'I' is 'front-loaded' in the book, so to speak. Most of the occurrences of the first-person pronoun occur in the first third of the book. Furthermore, the devices of pseudonymity and the Royal Fiction are dropped after ch. 2. Roland Murphy calls attention to this reader problem by noting that 'one is left with the question ...why did Qoheleth adopt this royal identity when he uses it so sparingly and almost without need, since the experiment with riches in chapter 2 does not demand a king as the actor. Perhaps it lent some authenticity'.[137] However, by merely noting its limited appearance in the book, this judgment does not do justice to the tremendous *effect* this chapter has on the reader. The second chapter of the book of Ecclesiastes instructs the reader to respond to the entire discourse of the book as the first-person speech of a royal narrator. In the course of a few short verses, the discourse has given the reader the requisite interpretive reflexes, thereby negating the further need for utilization of this device. Here is a classic example of being so historically focused that the scholar cannot see the forest for the trees. When scholars note its limited appearance in the book, or their inability to precisely detail its historical function/origin, they fail to account for the monumental literary effect the King's Fiction has on the reader. In so doing, they have inadvertently missed the entire point being made by those formal qualities at the discourse level of the text. This is merely one insight a reader-oriented approach offers the biblical reader. By focusing on the specific competency that a text creates, reader critics make more skillful readers of both critics and readers.

135. Mailloux, 'Learning to Read', p. 107.

136. Mailloux, 'Learning to Read', p. 98.

137. R. Murphy, *Ecclesiastes* (WBC, 23A; Waco, TX: Word Books, 1992), p. 12.

e. *Gaps, Blanks, Wandering Viewpoints and Other Reader Problems*

From the vantage point of the reader critic, the reader 'is always a person with a problem'.[138] If previous generations of scholars were prone to see texts as windows to another age, reader critics tend to view texts as labrynths and puzzles which need to be solved by the reader. Reader critics attempt to describe and account for the mental processes that occur as a reader confronts these problems during the linear progression of a text. By far the fullest treatments to date of how this process unfolds in the reader is the work of Wolfgang Iser, *The Act of Reading*. Another seminal study is Menakhem Perry's essay, 'Literary Dynamics: How the Order of a Text Creates its Meanings'.[139] Both Iser and Perry, along with other reader critics such as Stanley Fish and Steven Mailloux emphasize the interaction between the ideal or implied reader and the structures of the text. Iser states:

> The model of text/reader interaction forms the basis of the communication concept. The reader 'receives' the text, and guided by its structural organization, he fulfills its function by assembling its meaning. From a communications point of view, structures are in the nature of pointers, or instructions, which arrange the way in which a text is transferred to the reader's mind to form the intended pattern.[140]

The elemental materials of a text are the repertoire and the strategies. The repertoire is composed of the 'material selected from social systems and literary traditions'.[141] It consists of references to earlier works, social and historical norms, or to aspects of the culture from which the text emerged.[142] Elsewhere Iser defines the repertoire as 'existing norms in a state of suspended validity'.[143] The selection of norms and allusions enable the background of the text to be built up, allowing for the reader to grasp the significance of the selected elements. For many texts,

138. R. Rogers, 'Amazing Reader in the Labyrinth of Literature', *PT* 3 (1982), pp. 31-46 (35).

139. M. Perry, 'Literary Dynamics (Part One)', pp. 35-64, along with M. Perry, 'Literary Dynamics How the Order of a Text Creates Its Meaning (Part Two)', *PT* 1 (1979), pp. 311-61.

140. W. Iser, 'The Current Situation of Literary Theory: Key Concepts and the Imaginary', *NLH* 11 (1979), pp. 1-20 (14).

141. Iser, *The Act of Reading*, p. 86.

142. Iser, *The Act of Reading*, p. 69.

143. Iser, *The Act of Reading*, p.70.

the repertoire represented in the text reproduces the familiar, but strips it of its current validity.[144] Iser calls this defamiliarization. He states:

> the literary recodification of social and historical norms has a double function: it enables the participants—or contemporary readers—to see what they cannot normally see in the ordinary process of day-to-day living; and it enables the observers—the subsequent generations of readers—to grasp a reality that was never their own.[145]

During the act of reading, a text causes its readers to reassess the norms it has selected. This reassessment constitutes the heart of the aesthetic response for many texts. In addition, texts are composed of strategies which organize 'both the material of the text and the conditions under which that material is to be communicated'.[146] The main function of the strategies is to offer the reader possibilities for organizing the internal network of references in the text.[147] Strategies consist of the various narration techniques utilized by the text.

Furthermore, literary texts are characterized by their indeterminacy, which includes gaps, blanks, vacancies and negations. By 'indeterminacy', Iser refers 'to the potential connectability of textual schemata which initiates ideational activity'.[148] The nature of the literary text is to be indeterminate, meaning that the reader must make assumptions, deductions, connections and other imaginative leaps to arrive at a *Gestalt* or conclusion regarding the meaning of the text. The indeterminate nature of the literary text is due to 'the fundamental asymmetry between text and reader...the lack of a common situation and a common frame of reference'.[149] Because there is no face-to-face situation between texts and readers, the reading process is asymmetrical, meaning that a reader does not possess all the facts needed to fully understand the text.

144. Iser, *The Act of Reading*, p. 74. For an excellent discussion of how Iser's theory of defamiliarization can be applied to First Testament wisdom literature, see A. McKenzie, 'Subversive Sages: Preaching on Proverbial Wisdom in Proverbs, Qohelet and the Synoptic Jesus through the Reader Response Theory of Wolfgang Iser' (doctoral dissertation; Princeton, NJ: Princeton Theological Seminary, 1994), pp. 94-120.

145. Iser, *The Act of Reading*, p. 74.

146. Iser, *The Act of Reading*, p. 86.

147. Iser, *The Act of Reading*, p. 86.

148. Iser, 'Indeterminacy and the Reader's Response', p. 37.

149. Iser, *The Act of Reading*, p. 167.

Obviously, argumentative and expository texts do not typically require the sort of ideational gap-filling that is characteristic of narrative texts (however, the Book of Ecclesiastes is an exception).[150] As such, in an argumentative text like the Ecclesiastes, one would not normally expect the sort of gap filling which characterizes narrative texts. However, the implied author's fondness for ambiguity and literary puzzles has created a text which cuts against the grain of most argumentative texts. On the basis of this characteristic, the model proposed by reader-response critics is serviceable for an argumentative text like that of Ecclesiastes. First, there are definite fictive aspects to the text, such as the Royal Fiction in ch. 2, which creates gaps for the reader regardless of the argumentative nature of its literary surroundings. Futhermore, first-person texts, because they require large amounts of characterization, are inherently 'fictive'. Due to its use of first-person narration, which profoundly affects the reading of the entire book, the book of Ecclesiastes is very much a narrative-like text, inspite of the fact that the overall book is predominately an argumentative text in which the greater amount of textual space is given to expository and argumentative text-types. Second, and more importantly, Ecclesiastes is renowned for the difficulties readers have in making sense of the text. This is due to the implied author's utilization of a rhetoric of ambiguity. As a result of the text's rhetorical strategy to utilize ambiguity to make its effect on the reader, the history of its interpretation is rife with readers who could not unravel the 'riddle of the sphinx', and a host of other such ideational conundrums. Third, Qoheleth is a text which produces a pearls-on-a-string argument, thereby inviting the sort of gap-filling that is characteristic of narrative texts. The book of Ecclesiastes therefore seems to be an argumentative text that has more in common with most narrative texts than is often the case and, as such, requires a method that can deal with the sort of gaps which are frequently encounted in the book.

As readers proceed through the text, they must connect the various segments into a coherent whole. What fills in these gaps and blanks is the ideational activity of the reader. Decisions, deductions, conclusions, connections—these are 'facts' made up by the reader in the process of diciphering the text. Menakhem Perry summarizes the process:

150. Iser, *The Act of Reading*, p. 184.

> The selection of any particular frame leads *ipso facto* to supplying information (filling gaps) which has no direct verbal basis in the text. Most of the information a reader derives from a text is not explicitly written in it; rather, it is the reader himself who supplies it by the mere fact of choosing frames... Most of what the reader infers from the text, it will be discovered, is the reader's own gap-filling.[151]

Sometimes information is deliberately withheld from the reader, resulting in a gap. Meir Sternberg defines this type of indeterminacy as 'a lack of information about the world—an event, motive, causal link, character trait, plot structure, law of probability—contrived by a temporal displacement'.[152] Such concealment of information functions to prod 'the reader into action, but this action is also controlled by what is revealed; the explicit in its turn is transformed when the implicit has been brought to light'.[153]

In addition to gaps, there are also blanks in a text. The blank is not the same as a gap. The blank signals 'a clash between adjacent textual schemata whose potential links are not made explicit in the text. We should not fill in the blanks with our own experiences, but fill it in from the system that is laid down in the text.'[154] Whereas the gap refers to missing information, blanks refer to missing connectors in the text. If the blank occurs in a marginal or nonthematic aspect of the text, it is a vacancy.[155] In the book of Ecclesiastes, Qoheleth's contradictions or his pairing of 'dueling proverbs' (4.4-6) would be examples of the textual blank, whereas the missing meaning of *'ōlām* in 3.11 would be an illustration of a gap in the text. The filling in of such gaps and the connection of blanks becomes the fundamental activity engaged in by the reader. The trick, of course, is to fill in these gaps with information garnered from the norms provided by the implied author. Failure to do

151. M. Perry, 'Literary Dynamics (Part One)', p. 45.

152. Sternberg, *The Poetics of Biblical Narrative*, p. 235.

153. Iser, *The Act of Reading*, p. 169.

154. W. Iser, 'The Indeterminacy of the Text', *CompCrit* 2 (1980), pp. 27-47 (28) (trans. R. Foster). Schemata refer to the mental 'filters' which enable us to group data together and to classify and register our experiences with the world. When narratives add to or fundamentally change the way something is perceived, these are called corrections to the schemata of the text. See Iser, *The Act of Reading*, pp. 90-92.

155. W. Iser, 'Interaction Between Text and Reader', in Suleiman and Crosman (eds.), *The Reader in the Text*, pp. 106-119 (115).

so means filling the gap with one's own projections.[156] However, Iser cautions the reader critic that gaps and blanks may be filled by textual material in different ways by different readers and, as such, 'no reading can ever exhaust the full potential' of the text.[157]

During the course of the linear progression of a text, readers make connections between various viewpoints and textual data and must reverse, change, or alter the perception of this relationship. The order of a text radically affects this relationship. Narrative ordering produces what Meir Sternberg and Menakhem Perry call the 'primacy effect' and 'recency effect'. The primacy effect refers to the influence of narrative information on the reading process at the beginning of a text. The recency effect refers to the influence of later narrative information which has recently been the object of the reader's attention during the reading process. The interaction of these two effects is a major dynamic during the reading of a text. Perry states:

> What happens in a literary text is that the reader retains the meanings constructed initially *to what ever extent possible*, but the text causes them to be modified or replaced. The literary text, then *exploits* the 'powers' of the primacy effect, but ordinarily it sets up a mechanism to oppose them, giving rise, to a recency effect.[158]

As a result, readers constantly have a double horizon set before them. In order to fill in gaps and blanks, the reader looks forward and backward, constantly attempting to create a *Gestalt* out of the two horizons.

The magnitude of the primacy effect cannot be underestimated for a first-person text. Iser notes that if a reader is concerned with the conduct of a 'hero', the reader's attitude will be conditioned by the horizon of past attitudes towards the hero, such as that of the narrator or other posts of observation.[159] In a first-person text, whatever characterizes the narrator during the initial stages of the text (such as the King's Fiction in 2.1-11) will play a substantial role in influencing the response

156. Iser, *The Act of Reading*, p. 167.

157. Iser, 'The Reading Process', p. 55.

158. M. Perry, 'Literary Dynamics (Part One)', p. 57. The terms 'recency effect' and 'primacy effect' were originally coined by M. Sternberg in *Expositional Modes and Temporal Ordering in Fiction* (Baltimore: The Johns Hopkins University Press, 1978), pp. 93-98.

159. W. Iser, 'Narrative Strategies as a Means of Communication', in M. Valdés and O. Miller (eds.), *Interpretation of Narrative* (Toronto: University of Toronto Press, 1978), pp. 100-17 (112).

of the reader. In fact, 'the most intensive closing of options occurs at the early stages. The reading tempo of actual readers is far slower at the beginning of novels than at the middle or the end.'[160] Empirical studies have shown that readers give more weight to what comes first in a text and pay more attention to the earlier material in a text.[161] Furthermore, the meaning of later words or passages often changes as a result of what precedes them in the text. Readers may actively discount later words because they are inconsistent with what precedes them. On the other hand, during the course of the temporal progression of a text, the reader must take into account the most recent data and sometimes, must engage in retrospective repatterning. Because of this dynamic, the literary text depends on the tensions between the primacy effect and the material at the present point of reading to produce a response by the reader.[162]

Because of the discrepancies between earlier and later material, readers are constantly engaged in making sense of the text by dealing with these tensions. To do so, they infer (paraphrase), query, observe, predict, evaluate and compare the various segments of the text.[163] Empirical studies have shown that of these activities, inferring and evaluating are the most common activities that readers perform, followed by comparisons of prior textual elements with current data.[164] During these moments, the reader cancels his or her previous conclusion, replacing it with another.[165] Because of the primacy and recency effects, the reader acts like Janus, 'always looking backward as well as forward, actively restructuring the past in light of each new bit of information'.[166] As a result of this phenomenon, a reader-response approach to the problem of first-person narration in the book of Ecclesiastes must pay close attention to how the primacy effect sets up a theme with which the progression of discourse ideas must interact. Because of this dynamic, the impact of the autobiography-like material used in the King's Fiction (2.1-11) cannot be underestimated.

160. M. Perry, 'Literary Dynamics (Part One)', p. 53.
161. M. Perry, 'Literary Dynamics (Part One)', p. 56.
162. M. Perry, 'Literary Dynamics (Part One)', p. 57.
163. Grant-Davie, 'Between Fact and Opinion', p. 85.
164. Grant-Davie, 'Between Fact and Opinion', pp. 100, 104.
165. M. Perry, 'Literary Dynamics (Part One)', p. 60.
166. Duranti, 'The Audience as Co-Author', p. 127.

The primacy and recency effects set up a background–foreground dynamic that is constantly shifting due to the progression of the text's presentation of the various posts of observation. This structure of theme and horizon continually guides and directs the reader.[167] The theme is what the reader is involved with at any particular moment. The horizon includes all those other perspectives in the text. The structure of theme and horizon allows all the posts of observation to be observed, expanded and changed as the reader attempts to put them together. Gaps and blanks may therefore be bridged by the reader through the shifting back and forth between theme and horizon.

Of course, the reader cannot embrace all of these perspectives at once. Since the whole text can never be grasped by the reader at any given time, a wandering viewpoint develops. Iser compares this aspect of the reading process to a stagecoach rider surveying the scenery and who must finally put it all together at journey's end. This means that 'at no time…can [the reader] have a total view of that journey'.[168] The wandering viewpoint and the lack of availability of the whole work during the act of comprehension demands that the reader builds up the text into a consistent whole bit-by-bit. This consistency building on the part of the reader constitutes another major aspect of the reading process. The reader must group together the different aspects of the text, the various strategies and the different posts of observation in order to grasp the final meaning of the text.[169]

To sum up, the individual segments of the text are usually not explicitly joined together for the reader by the text. Instead, the text consists of a series of gaps and blanks which cause the reader to expect certain things without necessarily defining them. These breaks induce the reader to reformulate the aesthetic object of the text. The reader's attention wanders between expectations and retentions, creating the continual process of fulfilled, modified or frustrated expectations. As such, the literary text consists of a series of illusion-making and illusion-breaking strategies.[170] During the reading process, every text offers the reader various challenges and problems, presenting the reader with gaps and blanks, shifting viewpoints, differing posts of observation and other indeterminacies which require the manufacturing and subsequent

167. Iser, 'The Indeterminacy of the Text', p. 29.
168. Iser, *The Act of Reading*, p. 16.
169. Iser, *The Act of Reading*, p. 119.
170. Iser, *The Act of Reading*, p. 129.

updating of preliminary *Gestalten*. The reader critic seeks to isolate the various maneuvers that are required of the reader during the course of a text's linear and temporal progression, so that the experience of the text may be more fully understood. As a result of these inferences, queries, conclusions and reversals, the text has a specific suasive effect on the reader, creating a patterned response.

4. *Reading Theories and the Poststructuralist Perspective*

Of course, poststructuralist interpreters would most assuredly take issue with the idea that the text has such a controlling effect on the reader. For instance, George Aichele and the Bible and Culture Collective criticize the approach outlined above, stating:

> To date most reader-response criticism can be characterized as the search for the implied reader or narratee of biblical texts. The blindspot of this endeavor is the neglect of the flesh-and-blood reader who claims to be able to find the implied reader or narratee suspended in the amber of the text. Most biblical reader-response criticism remains resolutely formalist—what counts is supposed to be already there in the text—and neither the psychological/subjective nor the social/structural dimensions of the reader-response critic's own agenda is given consideration. That is, much literary criticism of the Bible is comfortable with formalist-structuralist criticism but has yet to fact up to the challenges posed by poststructuralism and the broad postmodern debate.[171]

However, most 'formalist'-oriented critics are far more aware of their presuppositions than the above criticism would seem to suggest. Marianne Thompson's own confession along these lines is rather exemplary in this regard:

> And who is this reader? In the end, every reader is a mirror of the person who construes the reader... So perhaps the reader is not merely a critic's construct, but the critic. I am the reader. Not all readers bring to the text what I bring to it. But we are all reading the same 'text', if by text we mean the actual words on the page. Thus I do not suppose that the reader brings 'all the meaning' to the text nor that all the meaning is in the text. Rather, meaning is produced by the interaction of the reader and the text, both of which are shaped by their cultural location.[172]

171. Aichele *et al.*, 'Reader-Response Criticism', p. 13 n. 13.

172. M. Thompson, '"God's voice you have never heard, God's form you have never seen"': The Characterization of God in the Gospel of John', *Semeia* 63 (1993), pp. 177-204 (184).

Upon further reflection, it would appear that second-generation reader-response critics are quite fluent with the issues which Deconstruction has brought to the hermeneutic table.

However, in the light of the criticism offered by Aichele, it seems appropriate to sketch out my own background and presuppositions as a fiesh-and-blood reader. In the first instance, I would describe myself as a phenomenological formalist with definite sociological leanings (particularly in classroom settings) and an instinct for a hermeneutic of suspicion whenever it is appropriate. This is, perhaps, due in part to my training under Robert Alter and Seymour Chatman who both have unabashed formalist leanings. I should also say that I love playing 'Rubik's Cube' with texts, which is why I was drawn to narrative approaches in the first place. In addition, my readings are based on the constraints of expediency, methodology, and, for want of a better word, what I would term common sense.

In response to the criticism of Aichele and others, I should state my belief that an accent on the text and its textual constraints is important because what attracts most people to Scripture is not its readership, but the time-honored ability of the text to speak to continuing generations. To my knowledge, no readers have been canonized nor, might I add, any specific reading conventions, at least, explicitly. I will wholeheartedly admit that poststructuralist perspectives have a legitimate argument that all interpretations reflect the biases and interests of the critic advocating a given reading. Even the concept of 'textual restraints' can be shown to have a historical context.[173] As Derrida and Foucault have rightly adjudicated, texts do have social contexts which play a part in the determination of meaning. Furthermore, these contexts change with every reader and every generation. In that respect, context is boundless. I can therefore rightly agree with this insight—how could it be otherwise? Aichele *et al.* are correct to chide reader-oriented critics for

173. For an interesting viewpoint on textual constraints and the historicality of reading conventions, see B. Long, 'Textual Determinacy: A Response', *Semeia* 62 (1993), pp. 157-63. He observes in regard to midrashic texts that the Rabbis 'did not always accept consonantal order, what we might think of as a most basic, natural limit to the possibilities of meaning, as a constraint on their multivalenced readings of the Bible as a divine address. It may be that when *we* speak about the possibilities that a text offers, its constraints on allowable readings, that we mask in objectivist language our situational choices about what counts as constraint, or allowable possibility of meaning, in the first place' (p. 158).

paying too much attention to implied readers and narratees at the expense of 'the reception of biblical texts by flesh-and-blood readers'.[174] However, as will soon become apparent, this study has utilized studies of the quotidient flesh-and-blood reader in its attempt to define the implications and restraints found in the textual medium of Qoheleth's discourse. Further, the criticism is a bit slanted, as all reader-oriented critics discuss the reading history of the text in order to clarify the problems contained therein. Scholars are actual readers we are discussing even if they do often go under the general rubric of 'critics'. Given that this is so, it seems that reader-oriented critics do pay attention to real readers when they consider the reading history or reception history of the text. In regard to the formalist tendencies of reader response method, it should also be observed that most reader-oriented critics such as Fish and Fowler, to name just, freely admit that the method demands them to pay attention to whatever they as critics are doing. In this respect, the method does not resist attention to context but, rather, promotes it in a very dynamic and honest manner. No one can practice the method without becoming more aware of how one's own ideational tendencies affect what they view as implied in the text. In those ways, I see reader-response criticism as being very postmodern in its potential and ethos. Like so many methods, it is how it is applied. There is no such thing as a single reader-response method in actual practice. In this respect, the method is as potentially variegated as any deconstructionist perspective.

However, some will still object to an emphasis on the text as a determinant of meaning. Aichele *et al.* chide most reader-oriented contributions as 'remaining within the theoretical boundaries of a philologically oriented historical criticism'.[175] The major complaint here goes back to two influences which seem to bother the Bible and Culture Collective who formulated the objections outlined above. One is the continued influence of historical critics who view the text as an object which controls the reading process. The other is the influence of Iser whose theory of reading gives an 'objective status' to the text.[176] Aichele *et al.* summarize this situation:

174. Aichele *et al.*, 'Reader-Response Criticism', p. 36.

175. Aichele *et al.*, 'Reader-Response Criticism', p. 39. It should be further noted, that I do not advocate such an alignment, and so, their criticism does not seem to be true for a method that is endebted to a Ricoeurian perspective as this one is.

176. Aichele *et al.*, 'Reader-Response Criticism', pp. 40-41.

> Biblical critics, however, have traditionally engaged in a kind of close reading that has presupposed the efficacy of the biblical text to guide them to historically verifiable knowledge... Blinded by this presupposition, biblical reader-response critics continue to believe that *somehow* there must be a connection between the reader-in-the-text, the original audience, and the biblical critic...[177]

The Collective then go on to state that what is not so clearly evident in such presuppositions is the 'theological agenda' lurking behind the scenes. Citing Norman Holland, they note that what biblical scholars have *not* done is to 'study, not a text, but readers reading a text'.[178] From this flows their ultimate criticism of work done to date, charging that

> the step that biblical critics have not yet taken is to admit that the implied reader for whom they are reading is themselves, and that the implied readers whom they construct are reading strategies by which to verify their own readings... What they learn from the text is usually what they already know, and the hypostatized ideal reader is actually none other than the super-biblical critic him- or herself... Perhaps it is time for biblical critics to speak of the 'implied interpreter' instead of the implied reader...the implied author and the implied reader are interpretive constructs and, as such, participate in the circularity of all interpretation... To confess this, however, would be to admit that one's relationship to the knowledge which has been gained from reading would not be that of a subject to an objective text, but a *hermeneutical* relationship to the discursive practices of one's own discipline...[179]

From this perspective, meaning is still seen as an event of reading, but one which, more importantly, is situated in a 'sociopolitical location' as well. Furthermore, reading conventions are given priority over the data contained within the text, with the concomitant result that criticism must pay attention to the location and formation of those conventions, and especially, the politics involved in the formation of such conventions.[180] In voicing these criticisms, they rightly point out what deconstructionist readers 'have shown to be the case in a wider context'.[181]

177. Aichele *et al.*, 'Reader-Response Criticism', p. 42.

178. Aichele *et al.*, 'Reader-Response Criticism', p. 53, citing N. Holland, *Holland's Guide to Psychoanalytic Psychology and Literature-and-Psychology* (New York: Oxford University Press, 1990), p. 55.

179. Aichele *et al.*, 'Reader-Response Criticism', pp. 54-55.

180. Aichele *et al.*, 'Reader-Response Criticism', pp. 57-58.

181. Aichele *et al.*, 'Reader-Response Criticism', p. 62.

However, as has already been seen, most second-generation reader critics are quite aware of these issues and freely admit to them.

It would be difficult to deny the inherent truth in *these* observations. As a metacriticism of general hermeneutical theory, deconstructionist insights have a valid perspective that needs to be heard by all interpreters. Their great benefit to the interpreter is the degree of introspection and honesty they bring to the hermeneutical table. On the other hand, much less is gained from these insights once a critic selects a method and begins to deal with a specific text. Allow me to wax experiential on this issue since all reading is personal. At some point, as a reader/critic, I begin the actual process of looking at an individual text. In doing so, I notice that certain words are used, with a specific grammar, selecting certain aspects from its historical background for discussion, all within the constraints of genre and a host of other influences. True enough, what my guild has taught me about all of those things (or I should say guilds since my training was both in a theological setting as well as a secular university) has influenced me to understand these constraints and influences in a specific way. However, it is my belief that while texts can be read legitimately in many ways, there is a core of data that functions to restrain and to guide the meaning which we as readers both take and make from the text. Even if what I see as a textual restraint is different from other critics—this would, for example, be the case with ancient midrashic interpreters—there will still be some level of agreement that is based on what appears in the text. This seems to me to be common sense. After all, how many studies do not at some point argue on the basis of data found in the text? Texts possess an essential core of information, a quantifiable component of data which enables an editor to decide the fate of a submitted manuscript, or a professor to assign a grade to a student's exegesis. This all seems self-evident, if not in theory, then certainly *in practice*.[182] At the very least,

182. For instance, we can observe when it comes to dealing with actual texts, that even the most diehard poststructuralist reader must resort to things in the text. In summarizing how readers deal with characterization in the Gospels, F. Burnett, who is also a member of the Bible and Culture Collective, observes that readers encounter a transcendent character based on both '*the indicators that the text provides* and the reading conventions that the reader assumes for the narrative in question' ('Characterization and Reader Construction', pp. 5-6 [my emphasis]). However, if one were to read texts based solely on the metacriticism one encounters in *The Postmodern Bible*, one would walk away thinking that there were only 'reading

I have met few academics of any ideological persuasion, postmodern or otherwise, who do not exhibit objectivist tendencies when there is need for a pragmatic evaluation of critical abilities.

Regardless of the stereotyping by some poststructuralist theoreticians, readers with an appreciation of formalism do not spend hours attempting to objectivize themselves in the 'mirror of the text'. As for myself, I simply cannot reduce the idea of the text to a mere collection of social or literary conventions, as often seems popular in deconstructionist circles. There is a hyper-critical spirit in such a position which seems to reflect a bias on the part of deconstructionist critics themselves. Take, for instance, the problematic passage Eccl. 7.25-27. There the reader will find a specific text which, as poststructuralist critics will point out, has been taken to mean a variety of things due to the personal and socio-political contexts of its various interpreters. No one can deny this. In fact, such dynamics are the very grist that reader-oriented critics love to analyze. Nevertheless, regardless of the various agendas brought to bear on this text by readers (and there are several), those readings still reflect a core of issues and data found in the text which have far more in common with each other than is often implied by deconstructionist criticism. Ecclesiastes 7.25-27 contains data which many readers interpret as misogynist ideology. Even when an interpreter 'defends' Qoheleth in this regard, the reading still must deal with the data that has given rise to these various interpretations. Are there agendas and conventions that influence one's view of the implied author of this text, conventions and biases which reside in the interpreter rather than the text? The answer must be in the affirmative. However, in spite of all the diverse readings given to this text, readers must ultimately deal with the specificity of a passage which has not changed in over 2000 years. Even in their diversity, one sees a constancy of issues with which readers must deal. As such, I agree with Adele Berlin who also responds to such poststructuralist criticisms:

> The multiplication of legitimate interpretations that we find before us is not due so much to textual indeterminacy as to proliferating hermeneutical systems. And the very fact that we can make meaning at all, on

conventions' in these texts. In actual practice, even the most committed deconstructionist readers rely upon formalist, or perhaps better, objectivist concepts and methods to validate their readings.

> whatever level, is due to our having learned a hermeneutics of reading, which helps us assign meaning to various textual structures and configurations.[183]

For this reason I consider myself a *phenomenological* formalist. In my opinion texts offer constraints that help to determine meaning, even if those textual constraints are informed (but not necessarily generated) by the socially approved theory conventions that I have accepted. How else can one explain the fact that there are more similarities to what readers see in the texts to which they respond (in admittedly different fashions) than there are differences? Nevertheless, I have also benefited from the insights of poststructuralist theoreticians. In my experience, analyzing how readers gain meaning from a text adds a level of meaning and, sometimes, even creates meanings that would have escaped me had I not been paying attention to the process. Therefore, I remain alert to how the phenomenological nature of the reading process (which includes social and contextual aspects) affects the meaning gathered from those constraints. If someone wants to see this as naive, my response would be, perhaps. In response to the possible charge that my outlook is naïve, my reply would be: 'perhaps'. All I would ask is that my critics consider the question: 'what political influences have dictated such an extreme position?' For, it seems to me, the common sense approach is better served by a both-and position rather than the seemingly reductionist model as proposed by deconstructionist critics. Let me admit it now; my interests are in the text as a sacred text (that is, a classic text with synchronic significance for our generation) and how readers react to its stimuli or perhaps better, textual devices. As such, my focus is both on the text (with Formalism) and the implied reader as I abstract it from the various contributions of the text's readership (with phenomenology and to a certain extent, poststructuralism). Can there be other versions of this implied reader? Of course! I do not see a problem here. Does what I argue herein reflect my own agendas and bias? Yes. *Mea culpa*, if that suffices. But one would hope that not all that I argue is such. At least, I have certainly attempted to be self-critical even as I admit that I can only see things as I do. Again, how else could it be?

183. A. Berlin, 'The Role of the Text in the Reading Process', *Semeia* 62 (1993), pp. 143-47 (146).

Finally, I would like to think that what soon evolves in this study is different from the other few contributions that have arisen from this methodology. In that respect, some of the criticisms brought to bear on the methodology suffer from the fact that there really have been far too few reader-oriented analyses brought to bear on the biblical text for there to be much of a sustained judgment against the method per se. As the Bible and Culture Collective themselves point out, other than a few dissertations and a handful of monographs, reader response criticism has not been applied or tested often enough to allow for their conclusion that it is intrinsically flawed as they infer. Tragically, the method died in its infancy, or so it seems. Given their list and my own research, I know of only a limited number of dissertations, monographs or books that have actually used a reader-oriented method as opposed to a narrative approach which merely 'enlists' the concept of 'the reader'.[184] More importantly, many of these came out during the late 1980s, a time when the shadow of philologically oriented historical criticism was still to be seen. I can speak from experience that academic politics had a great deal to do with the fact that so few ventured to risk their careers publishing a method that was so maligned at that time. It was certainly not the method which dictated the biases of the studies published around this time.

Nevertheless, the method as I have come to view it is quite compatible with many reading agendas, both historical and deconstructionist. It therefore seems a little premature to label the method as 'objectivist' or similar until more critics using different perspectives have actually applied the method to texts. After all, the purpose of reader-response criticism is simply to enable the critic to monitor the reading process that they use, and therefore, helps them to be honest with themselves and the text. When a reader-oriented perspective is applied to other interpretive agendas, more can be legitimately garnered about the method. In the following pages I will attempt to demonstrate that the method is capable of generating new insights if given the opportunity. Unlike other contributions, this study does not stand under the long shadow of academic politics, that is, historical agenda. Rather, the Ricoeurian perspective described above has attempted to utilize the method in a

184. For a partial list of these see Aichele *et al.*, 'Reader-Response Criticism', p. 39. The other works I refer to can be found in the bibliography of this work. See in particular the works of G. De Bruin, E. Christianson, R. Johnson and A. McKenzie.

manner which is quite distinct from earlier reader-oriented contributions. Although I am influenced by Iser, I will not attempt to align the implied reader I see in this text with any historical reconstructions. If there is a criticism of this work by deconstructionist scholars, one can only hope that they recognize it for what it is—an attempt to excavate both the text and its reading history (which also includes my own history and training) with a view to understanding the book's literary problems. In addition, I also seek to generate a distinctly rhetorical reading of the book of Ecclesiastes. It is hoped that this study will be seen as a worthwhile application of the reader-oriented method and not simply as a repository for my individual biases.

5. *Taking Stock in the Speaker—How Readers Respond to First-Person Texts*

First-person texts have their own characteristic and specific built-in suasive effects. The use of 'I' forces the narrator's humanity and personality to become the center of the reader's attention. The character of the narrator—his or her ethos—utterly dominates the landscape of the text.[185] As a result, readers respond most strongly 'to the human aspect

185. It should be noted that an emphasis on character is characteristic of wisdom literature in general, and is not due solely to the effects of first-person discourse. For an insightful study which emphasizes the important role that character formation plays in the canon's wisdom corpus, see William Brown's excellent study *Character in Crisis: A Fresh Approach to the Wisdom Literature of the Old Testament* (Grand Rapids: Eerdmans, 1996). Brown demonstrates how scriptural wisdom literature functions to aid in the formation of character—both for individuals and the community. His study also shows how the individual aspect of character formation must be balanced by the role which the community plays in this process. He observes in that regard: 'the notion of character with the elements of perception, intention, and virtue provides a model of coherence to the moral life of the individual and community...character is formed in and through "socially-embodied traditions", that is, through traditions carried and passed on by the community from one generation to the next...principles and rules are part and parcel of the dynamics of character formation in that they contribute to the community's task of providing particular conceptions of the good through which character is formed' (p. 14). In reaching this conclusion, Brown has correctly perceived that the development of private insight has a role to play in the formation of character, but also, that the public has a vested interest in having a say as well. As such, Brown's study enables us to perceive that the subtle debate between private insight and public knowledge that we see in the book of Ecclesiastes is a constituent dynamic in all wisdom

of the matrix—to the "person who speaks" '.[186] While this is not to say that readers do not respond to the ethos of a third-person narrator, especially in case of unreliable narration, it is important to recognize the special character of first-person narration. An I-narrator is more than just a guiding voice in the discourse; he or she is a fully characterized person who is addressing us, instructing us, and informing us of what we as readers need to know given the aims of the text.

Readers experience a first-person text in a more direct fashion, as a direct one-on-one interchange between themselves and another person.[187] Because of this, the distinction between world and self experienced in third-person narrated texts is not present.[188] David Goldknopf calls this the 'confessional increment'.[189] According to Goldknopf, this confessional increment means that 'everything an I-narrator tells us has a certain characterizing significance over and above its data value, by virtue of the fact that he is telling it to us'.[190] I-narration forces the reader to acknowledge the role of the interpretive consciousness in the text.[191] The narrator intervenes between the reader and the discourse situation, causing the reader to see things through the narrator's eyes. As a result, the operation of the I-narrator's mind is the true subject of the discourse.[192] As a result, it becomes necessary for readers to engage themselves in a process of characterizing the narrator.

The characterization of the narrator over and above that of a textual voice gives first-person narration a specific set of suasive strengths and liabilities which are both unavoidable and pervasive. This is partially due to the specific kind of reading contract presupposed by a first-person text.[193] Philippe Lejeune suggests that for first-person texts,

literature. However, as will also be seen, the effect of first-person discourse is to exacerbate this debate in a manner that is not present in other biblical literature.

186. Rideout, 'Narrator/Narratee/Reader Relationships', p. 31.

187. Marra, 'The Lifelike "I" ', p. 59.

188. Marra, 'The Lifelike "I" ', p. 51.

189. D. Goldknopf, 'The Confessional Increment: A New Look at the I-Narrator', *JAAC* 28 (1970), pp. 13-21 (21).

190. Goldknopf, 'The Confessional Increment', p. 20.

191. Goldknopf, 'The Confessional Increment', p. 16.

192. Goldknopf, 'The Confessional Increment', p. 21. W. Booth makes a similar point in 'Distance and Point-of-View: An Essay in Classification', *EC* 11 (1961), pp. 60-79 (65).

193. Lejeune, 'The Autobiographical Contract', p. 220.

defining the reading contract means making explicit 'the inherent credibility it reveals'.[194] This suggests that it is the ethos dimension of the speaker which lies at the heart of the first-person reading contract, and as such, should occupy the greater part of the critic's attention.

While the value of the distinction between first-person and third-person discourse has been debated,[195] most literary critics have argued that there is a basic difference between the two narration techniques. The modern debate begins with the linguistic work of Emile Benveniste, who argued that 'I' is an empty linguistic sign which is both limited to and filled out by the discourse structure of a text, while the third-person pronoun is referential in nature and is limited by the reality to which it refers. Benveniste states:

> Language has...an ensemble of 'empty' signs that are nonreferential with respect to 'reality'. These signs are always available and become 'full' as soon as a speaker introduces them into each instance of his discourse.[196]

Because of this characteristic, he concludes that first-person discourse is intersubjective while third-person discourse is interobjective.[197] Subjectivity thus constitutes the realm of first-person discourse.[198] The content of this subjectivity is a linguistic blank that is filled in by each instance of discourse by a speaker. This is so much the case that 'I' never means anything other than the 'instance of discourse'.[199]

Further distinctions can be observed by using speech-act theory. Benveniste notes that 'I' is a performative while third-person pronouns are constatives. The chief difference between the two is the lack of illocutionary force in constative statements. Illocutionary force occurs when 'an act is performed *in* saying something'.[200] The illocutionary

194. Lejeune, 'The Autobiographical Contract', p. 220.

195. Booth, *The Rhetoric of Fiction*, p. 150.

196. E. Benveniste, *Problems in General Linguistics* (Miami: University of Miami Press, 1971), p. 220. For an excellent overview of the linguistic problems affiliated with first-person narration see O. Avni, *The Resistance of Reference: Linguistics, Philosophy, and the Literary Text* (Baltimore: The John Hopkins University Press, 1990).

197. Benveniste, *Problems in General Linguistics*, pp. 220-21.

198. Benveniste, *Problems in General Linguistics*, pp. 220-21. See also his chapter on 'Subjectivity in Language', pp. 223-30.

199. Benveniste, *Problems in General Linguistics*, pp. 220.

200. G. Prince, *A Dictionary of Narratology* (Lincoln, NB: University of Nebraska Press, 1987), p. 41. H. White summarizes the distinction between performatives and

force of a first-person verb operates principally by applying the force of the verb to the speaking subject. Hugh White explains the differences:

> the content of the predicate [is applied] not to the object of the subject expression, but to the speaking subject itself which exists in the form of that very instance of discourse. This creates a direct bond between the content of the predication and the speaking subject which is not created by predication in the third-person statements where the speaking subject is not verbally present.[201]

Because of these linguistic differences, White concludes that the distinction between first-person/second-person and third-person pronouns creates an invisible barrier between the framework and the direct discourse sections of a text.[202] According to Nomi Tamir, in Biblical Hebrew, the differences are even more pronounced because of the morphological distinctions between first- and third-person verbs. She concludes that first-person speech in Biblical Hebrew is linguistically double-marked as both personal and subjective.[203]

Because 'I' is both personal and empty, it is utterly dependent upon the characterization process to fill it out. Character is a paradigm of traits which persists over the whole of the discourse.[204] A reader-response approach to characterization therefore focuses on 'those constitutive activities of the reader which involve the ascription of mental properties (traits, features) or complexes of such properties (personality models or types) to human or human-like…agents'.[205] Like all reading activities, this process involves the usual series of cognitive activities, involving gaps, traps, anticipations, reversals and the like, which are induced by the text during the course of its linear development. Uri Margolin argues that the characterization process consists of two steps. The first step involves responding to local problems and textual data by 'characterizing' the discourse agent. Later, as the reader traverses more

constatives in language: 'While I judge is an engagement, he judges is only a description on the same level as he runs, he smokes' ('A Theory of the Surface Structure of the Biblical Narrative', *USQR* 34 [1979], pp. 159-73 [164]).

201. H. White, 'A Theory of the Surface Structure', p. 165.

202. H. White, 'A Theory of the Surface Structure', p. 161.

203. Tamir, 'Personal Narration', p. 404.

204. Chatman, *Story and Discourse*, p. 125.

205. U. Margolin, 'Characterization in Narrative: Some Theoretical Prolegomena', *Neophilologus* 67 (1983), pp. 1-14 (4).

of the text, he or she engages in a more comprehensive 'character building' process.[206]

In the first step, readers characterize a discourse agent by responding to local passages which contain characterizing textual signals. There are three types of characterizing textual signals: (1) dynamic mimetic elements such as verbal, mental and physical acts; (2) static mimetic elements/statements like name, appearance, customs, habits and environment; and (3) textual patterns such as analogies, parallels, contrasts, repetitions, metaphors, metonyms and other cognitive data.[207] In addition, there are 'characterization statements' made by a discourse agent, 'consisting of the ascription of mental properties or of a personality model to himself or to any other [discourse] agent'.[208] An example of characterization statements in the book of Ecclesiastes are the words of the Epilogist in 12.9-10. To sum up, characterizing signals may be found in textual data which imply character traits, acts and deeds by the character in question, explicit characterization by the discourse agent itself, direct statements by the person in question and mental and inner speech.[209]

In a dramatic monologue like the book of Ecclesiastes, the patterns of thought expressed throughout the text will provide the bulk of characterizing material for the reader. Since the reader will encounter very little static mimetic statements, he or she will have to characterize Qoheleth by the disposition and texture of his ideas. Margolin clarifies how speech characterizes a narrator:

> Topics by themselves are not significant for characterizing the act in which they occur, but the pair topic-propositional content is. For every topic discussed by a narrative agent, one can ask about the particular selection of items effected by the speaker, the relative weight and detail given to each, the proportion between the details and their organization (additive, hierarchical). From these, one may draw conclusions about the cognitive qualities of the speaker... One may also enquire into the basic categories and polarities according to which the speaker/thinker organizes the universe of his experience, and infer from them about the subject's being rational or superstitious, a believer or skeptic, etc.[210]

206. Margolin, 'Characterization in Narrative', p. 4.

207. U. Margolin, 'The Doer and the Deed: Action as a Basis for Characterization in Narrative', *PT* 7 (1986), pp. 205-25 (206).

208. Margolin, 'The Doer and the Deed', p. 222.

209. Margolin, 'Characterization in Narrative', pp. 8-9.

210. Margolin, 'The Doer and the Deed', p. 212.

The various propositions, conclusions, admonitions, metaphors, analogies, polar structures, contradictions and repetitive patterns in Qoheleth's speech supply the basic data to which a reader responds. When a reader comes across a repetitive phrase like 'striving after wind' or 'vanity of vanities', he or she begins to ask: 'what sort of person or character would argue or think such things?', and 'can this type of person be trusted?' Analyzing the temporal process through which these questions arise and the probable responses and conclusions a competent reader might arrive at regarding the character/ethos of the narrator and the implied author will be the primary task of this study.

Eventually, the reader begins to gather those tentative conclusions into a consistent *Gestalt*. When this occurs, the reader has reached the character-building stage. The step which marks this transition for the reader is 'the determination whether a given trait occurs at one/several/all times and in one/several/all situations for this narrative agent'.[211] Character-building involves

> the accumulation of a number of traits from several successive acts of the narrative agent, setting, or formal patterns; a generalization concerning their extent in terms of narrative time; the classification or categorization of these traits; their interrelation in terms of a network or hierarchy of traits; a confrontation of traits belonging to successive acts in order to infer second order traits such as 'inconsistent'; and finally, an attempt to interrelate the traits or trait-clusters into a unified stable constellation (configuration, pattern, *Gestalt*, personality model) of narrative time.[212]

The more a reader encounters a given mental property, the more likely it is that he or she will begin to engage in character-building. By continually inferring traits and revising those inferences, the reader forms a 'coherent constellation or trait paradigm' of the discourse agent.[213] However, all character inferences and conclusions are tentative in nature and will be revised if they conflict with later data. Due to the complexity involved in characterizing, Margolin cautions that successive readings will always 'actualize different subsets of the total range of possible inferences', and therefore, will result in a different image of the discourse agent.[214] As such, reader-oriented critics emphasize the

211. Margolin, 'Characterization in Narrative', p. 13.
212. Margolin, 'The Doer and the Deed', p. 205.
213. Margolin, 'The Doer and the Deed', p. 206.
214. Margolin, 'The Doer and the Deed', p. 224.

dynamic elements of characterization rather than 'static' concepts such as traits. Marianne Thompson summarizes:

> literary critics of biblical narrative prefer to speak of character not as though it were a fixed commodity simply to be unearthed from the raw materials of the text, but rather as the result of the reading of the text. Rather than mining the text for the specific virtues and traits possessed by a particular character, they mine the text for its rhetorical and literary strategies in presenting characters. Thus the emphasis falls not so much on *what* a character is (e.g. honest, virtuous, brave, pious, etc.), but on *how* that character is constructed by the reader (i.e., through actions, speech, description, etc.) and *how* these elements of characterization are progressively coordinated by the reader...[215]

Everything spreads out from the first-person pronoun and serves to contextualize it.[216] The 'I' of a first-person discourse therefore serves as the 'gravitational center' for the reader's response.[217] In addition, the reading contract required by a first-person text requires that the reader begin 'fleshing out' the linguistically blank 'I'. When a narrator is embodied or fleshed out, the narratorial role is humanized, thereby restricting his or her role to the limits of human consciousness. This transfigures the narrator. I-narration transforms the narrator 'from an abstract functional role into a figure of flesh and blood, a person with an individual history'.[218] Once a narrator becomes embodied or fleshed out, the incarnation of the narrator's function into a human personality results in predictable strengths and weaknesses. Because the narrator has become one of us, readers tend to identify more with the first-person narrator, giving the speaking 'I' a huge initial rhetorical advantage. However, the cost of this initial advantage is that the narrator must lose the aura of omniscience that is the prize of many third-person narrators. The embodiment of the narrator 'results in a restriction of his horizon of knowledge and perception'.[219] His or her knowledge becomes characterized by subjectivity. It now possesses a conditional validity for the reader.[220] Once given flesh, readers respond according to the

215. Thompson, ' "God's voice you have never heard" ', pp. 179-80.

216. Marra, 'The Lifelike "I" ', p. 43.

217. Marra, 'The Lifelike "I" ', p. 49.

218. F. Stanzel, *A Theory of Narrative* (trans. C. Goedsche; with a preface by P. Hernandi; Cambridge: Cambridge University Press, 1984 [1979]), p. 205.

219. Stanzel, *A Theory of Narrative*, p. 201.

220. Marra, 'The Lifelike "I" ', p. 89.

narrators credibility, trustworthiness and attractiveness.[221] The discourse becomes an extension of the self of the narrator, making the presentation's suasive powers dependent upon personal as well as logical considerations. Due to the above considerations, Gerald Prince describes the first-person narrator as a 'restricted post of observation'. In contrast, many third-person narrators are viewed as possessing an unrestricted or unsituated point of view.[222] Because of the restriction of the narrator's post of observation to an internal perspective, the I-narrator is 'expected to conform to the limits of observation, perception, and comprehension attendant on any individual in his relationship to the events and existents about which he speaks'.[223] As a result, 'the question of reliability is inherent to the form'.[224]

However, one must not dwell on the limits of first-person suasion without also taking into account its vast rhetorical assets. First-person narration also has the potential to establish the credibility of a narrator in ways that are only partially approached by third-person narration.[225] Chief among the suasive powers of I-narration is precisely the fact that it is more real to the reader because it is the address of a person.[226] Empirical studies with readers suggest that a lifelike narrator is more credible than that one that is not. First-person discourse has an inherent advantage at this point.[227] In fact, the unreliability of a first-person narrator can sometimes serve to 'flesh out' the narrator in a way that creates a sympathetic response on the part of the reader.[228] In addition,

221. Marra, 'The Lifelike "I"', p. 311.

222. Prince, *Narratology*, p. 51. Fowler also picks up on this distinction and discusses its merits for the biblical text; see *Let the Reader Understand*, pp. 64-65.

223. Rideout, 'Narrator/Narratee/Reader Relationships', p. 36.

224. Rideout, 'Narrator/Narratee/Reader Relationships', p. 7.

225. L. Martens, *The Diary Novel* (Cambridge: Cambridge University Press, 1985), p. 41.

226. Marra, 'The Lifelike "I"', p. 75.

227. Marra, 'The Lifelike "I"', p. 313. Empirical studies with real readers suggest that readers do respond to first-person characters in a very lifelike fashion. Marra further concludes that readers use everyday conventions to respond to fictive character (pp. 185-95). In fact, readers 'tend to move through a text as we would move through an interpersonal relationship', so personal is the effect of first-person narration (p. 213).

228. Marra, 'The Lifelike "I"', p. 86.

the self-disclosure of an I-narrator communicates trust, which is generally reciprocated by the reader, at least initially.[229] James Marra states:

> the personal nature of first-person narration has the inherent advantage of producing in the reader the perception of a trusting narrator/protagonist. As research in self-disclosure has tended to unanimously support, the receiver's perception of trust on the part of the sender leads to a reciprocation of that trust from receiver to sender. Thus, immediately, we can argue that the reader is very quickly predisposed to trusting the narrator/protagonist.[230]

Finally, the credibility of a first-person narrator can surpass that of a third-person narrator because the self-disclosure of the narrator has revealed a solid basis of expertise upon which the reader can depend.[231] The descriptions of backgrounds and occupations, such as can be found in the use of the King's Fiction in 2.1-11, frequently aids this process.[232]

As such, we see that one of the central elements in first-person narration is its ability to provide a life-like model to the reader. As Marianne Thompson points out, 'our sense that fictional characters are uncannily similar to people is therefore not something to be dismissed or ridiculed but a crucial feature of narration that requires explanation'.[233] To be sure, the formalist properties of the text have a role in shaping our understanding of any given character. On the other hand, it is our response to such traits that enables some characters to become 'transcendent' figures who capture our imagination and tell us something about our nature as homo sapiens.[234] Character is therefore a 'construct that is developed during the reading process…that is, is an *effect* of reading'.[235] Surely for a character like Qoheleth, whose legacy in the Canon is that of being the pre-eminent pessimist, thoroughly renowned for his 'melacholy', his affect on readers has been enormous.[236]

229. Marra, 'The Lifelike "I"', p. 313. However, this initial bonus can induce a severe and critical backlash if the reader's sense of trust is somehow disappointed (p. 283).

230. Marra, 'The Lifelike "I"', p. 344.

231. Marra, 'The Lifelike "I"', p. 343.

232. Marra, 'The Lifelike "I"', p. 338.

233. Thompson, '"God's voice you have never heard"', p. 184.

234. Burnett, 'Characterization and Reader Construction', p. 4.

235. Burnett, 'Characterization and Reader Construction', p. 5.

236. I. Rashkow, '"In our image we create him, male and female we create

To sum up, reading any first-person narration is an exercise in determining its inherent liabilities and assets. The reading contract that is initiated by the use of 'I' signals to the reader to begin a process of characterization, humanization, subjectivization and embodiment that essentially limits the credentials of the narrator. On the other hand, the very act of embodiment has abundant powers of suasion that act to build the credibility of the narrator. The suasive powers of any first-person discourse thus resides between these two poles. The dictum of Norman Friedman was never more true than in the case of first-person narration: 'when an author surrenders in fiction, he does so in order to conquer; he gives up certain privileges and imposes certain limits in order the more effectively to render his story'.[237] Sometimes these limits will suade, and at other times, will hinder the rhetorical power of a text. The purpose of the next chapters will be to analyze how these two effects are generated by the textual design of the book of Ecclesiastes and to suggest ways that their interaction affects both the suasive powers and the meaning of the text as a whole.

them": The A/Effect of Biblical Characterization', *Semeia* 63 (1993), pp. 105-13 (105), mentions Qoheleth as one of the outstanding characters in the Bible whose very name connotes certain qualities due to the characterization they receive in the text. In this we see that transcendent characters often become tensive symbols for certain traits. Just as Ruth stands for tenacity, Pharoah for pride, Saul for irrationality, and Joseph for virtue, Qoheleth stands for biblical realism as a transcendent character. Rashkow also notes that characterization is never entirely stable for readers, because 'each time we read a biblical narrative we see something new in its characters, not because biblical interpretation is inexhaustible but because each time we read a text we are at least slightly different people having experienced more of life's vicissitudes' (p. 109). This dynamic is what Rashkow calls the 'paradox of literary characterization' (p. 107). As she concludes: 'In other words, readers *e*ffect characters who, in turn, *a*ffect readers' (p. 112).

237. Friedman, 'Point of View in Fiction', p. 1184.

Chapter 3

AMBIGUITIES, RIDDLES AND PUZZLES: AN OVERVIEW OF THE LINGUISTIC AND STRUCTURAL READER PROBLEMS IN THE BOOK OF ECCLESIASTES

> They were not all of them blind to poetry as such. They did care to a certain extent for form, but primarily they were interested in the great problems of life, they were interested in great and noble thoughts. Doubtless many of them rather enjoyed having to dig out the thought from involved language. But probably a greater number felt a larger enjoyment in finding lofty thought expressed in language which was even more lofty than obscure.[1]

1. *Ecclesiastes as a Rhetoric of Ambiguity*

The book of Ecclesiastes confronts the critic with intricate reading problems that constantly generate a sense of ambiguity in the reader. Their cumulative effect is a very distinctive 'rhetoric of ambiguity'. By rhetoric of ambiguity I do not mean the same thing as Meir Sternberg who characterizes all First Testament poetics as a 'poetics of ambiguity'. Sternberg alludes to the fact that most biblical texts play on a system of gapping which leaves the reader, at least temporarily, caught between 'the truth and the whole truth'.[2] This, however, is characteristic of all great literary texts. What I mean by 'rhetoric of ambiguity' is a literary design which frustrates the reader in such a way that the 'whole truth' is never disclosed in any satisfactory way. The reader is left suspended in a state of literary limbo regarding the text's final meaning. An ambiguous text is characterized by the enduring and resolute pres-

1. Theodore Roosevelt, commenting on the difference between students reading Browning and Tennyson, from *History as Literature and Other Essays* (New York: Charles Scribners and Sons, 1913), pp. 211-12

2. Sternberg, *The Poetics of Biblical Narrative*, p. 166.

ence of multiple interpretations which seem equally justified.[3] Ellen Spolsky in her essay, 'The Uses of Adversity: The Literary Text and the Audience that Doesn't Understand', speaks of the 'sacred discontent' that arises when a text leads the reader to make wrong or uncertain interpretive guesses. The pivotal operation involved in reading ambigous texts is the process of rival hypothesis testing. The reader learns how to read ambigous text through the process of hypothesis testing—rejection of initial *Gestalt*—and reformulation of initial hypothesis. In the case of Ecclesiastes, this process must be repeated several times by the reader. For ambigous texts, misreading is a precondition for reading, and failure is often a prerequisite for success. The creation of an adversarial relationship between the text and the reader who is seeking clarity is the chief effect of a rhetoric of ambiguity.[4]

The meaning of an ambiguous text is often enigmatic and elusive. Sometimes it can even border on the mysterious, such as when the reader attempts to understand the cryptic meaning of *'ōlām* in 3.11. Ambiguous texts beg for a both–and, rather than an either–or paradigm when dealing with their meaning and interpretation. Like a kaleidoscope, it is their very nature to invite diverse interpretations, possessing a fertile power to inspire both the sublime and the bizarre. They have the unique ability to inspire by stultifying the reader. Ambiguous texts dangle answers in front of their readers only to pull them away at the last second. One can only stand before their complexity with a frustrated sense of awe, wonder and puzzlement. Closure is not a part of their reading experience. Partial *Gestalten* and pilgrim conclusions are the treasures these texts give, however reluctantly, to their patrons. Yet they possess a bounty that inexplicably nourishes the human spirit with a lingering sense of incompleteness. In a paradoxical way, it is their very ambiguity for which we hunger. The rhetorical effect of ambiguity

3. That ambiguity is to be seen as the presence of multi-valenced meanings in a text has also been argued by Byargeon. He too views ambiguity as implying multiple meanings, especially 'if the context supports more than one meaning'. See R.W. Byargeon, 'The Significance of Ambiguity in Ecclesiastes 2, 24-26', in Schoors (ed.), *Qohelet in the Context of Wisdom*, pp. 367-72 (368).

4. E. Spolsky, 'The Uses of Adversity: The Literary Text and the Audience that Doesn't Understand', in E. Spolsky, *The Uses of Adversity: Failure and Accommodation in Reader Response* (London: Associated University Press, 1990), pp. 17-35, see esp. pp. 30-31.

is to create a love–hate relationship in the reader. Not for nothing has Ecclesiastes been called the 'black sheep of the Bible'.[5]

Robert Fowler discusses the rhetorical tactics of ambiguity under the rubric, 'strategies of indirection'. Specifically, Fowler lists incongruity, opacity, metaphor, irony, paradox, metonymy and synecdoche as strategies of indirection.[6] Incongruity refers to the discrepancies between story and discourse in a narrative text. Since Qoheleth is not a narrative and contains no story per se (though it contains references to life's vignettes, it has no plot), this concept must be modified for an argumentative text. In an argumentative text, the flow of the argument in its logical progression replaces plot. Incongruity will therefore refer to the manner of consistent presentation or disputation between argumentative divisions within a writing. Inconsistency of argument or contradictory lines of reasoning between passages within a text would indicate the presence of incongruity. Obviously, Qoheleth's fondness for contradictions would be the major way by which the implied author utilizes a rhetoric of incongruity in his work. Opacity is a term that Fowler uses to describe 'those moments in the reading experience when the narratee "sees" something in the narrative that characters cannot "see", or vice versa'.[7] These are moments when the characters are placed in the dark concerning what is happening in the story. Again, since Ecclesiastes does not have a plot per se, opacity would not refer to the level of knowledge given to the narratee regarding the events along the axis of plot. However, as an argumentative text in which the development of the basic argument replaces the plot lines of a story, it would refer to the level of knowledge granted to the narratee/reader regarding other necessary details, such as the meaning of *hebel*, *'ôlām* or other key terms that are basic to the development of the argument, yet which are shrouded in lexical uncertainty.

Metaphor refers to the 'invitation to consider a previously unexplored similarity between acknowledged dissimilars; irony offers a challenge

5. J.S. Wright, 'The Interpretation of Ecclesiastes', *EvQ* 18 (1946), pp. 18-34 (18).

6. Fowler, *Let the Reader Understand*, pp. 156-94. Fowler acknowledges his debt to L. Thompson who first coined this term to deal with irony in Mark's Gospel. See *Introducing Biblical Literature: A More Fantastic Country* (Englewood Cliffs, NJ: Prentice–Hall, 1978).

7. Fowler, *Let the Reader Understand*, p. 209.

to see and to see through an incongruity'.[8] Metaphor not only possesses a referential function, but also serves to 'un-arrange' the mind regarding that object's nature by drawing similarities between two things that did not formerly exist in the reader's mind.[9] A metaphor functions to reconnotate the reader's understanding beyond simple reference.[10]

8. Fowler, *Let the Reader Understand*, p. 221. It should be noted that the problem of irony in the book of Ecclesiastes still presents a very thorny problem, especially, in terms of how one may define it. In regard to this problem, Spangenberg notes that there is still no consensus on how to define irony in Qoheleth's discourse. However, he suggests five prerequisites for identifying irony in Qohelet: (1) that a person needs to be 'sound in mind', that is, open to scepticism and sophistication; (2) we need to keep in mind that the ironist wants to mislead; (3) ironic statements have a double meaning and its power resides in its subtlety; (4) irony is context dependent, that is, the same statement could be ironic in one context and not ironic in another; and (5) that it is important to perceive that the book does not contain merely ironic statements, but 'entirely reflects an ironic tone'. See I.J.J. Spangenberg 'Irony in the Book of Qohelet', *JSOT* 72 (1996), pp. 57-69 (60-62).

9. McKenzie, 'Subversive Sages', p. 44.

10. J. Soskice, *Metaphor and Religious Language* (Oxford: Clarendon Press, 1985), p. 57. For an excellent discussion of the theoretical issues underlying the thorny problem of how to conceptualize metaphor from a Feminist perspective, the interested reader is referred to the collection of articles in C. Camp and C. Fontaine (eds.), *Women, War, and Metaphor: Language and Society in the Study of the Hebrew Bible* (Semeia, 61; Atlanta: Scholars Press, 1993); the articles by Camp, 'Metaphor in Feminist Biblical Interpretation: Theoretical Perspectives', pp. 3-38, and S. Elgin, 'Response from the Perspective of a Linguist', pp. 209-18, are especially insightful for their overview of the general issues involved in the processing of metaphor. Both draw on the work of G. Lakoff and M. Johnson, *Metaphors We Live By* (Chicago: University of Chicago Press, 1980), following their thesis that metaphors are linguistic means to socially structure human cognition. Elgin astutely observes the power of metaphor: 'My personal concern with them is their role as perceptual filters that have an almost holographic ability to evoke the whole from minor parts. It seems to me that this should be the primary concern to theology and religious studies' (p. 211). The insights gleaned from their discussion of the ramifications of metaphor for hermeneutics are very insightful, especially for a text like Ecclesiastes whose metaphors have powerful social ramifications (both for the good and bad) if taken to heart. Unfortunately, such broad considerations remain outside the scope of this study, though lamentably so. However, I do agree with Elgin who notes that the power of metaphor to change reality is far greater than many perceive it to be, and in some cases, is the only linguistic means available to effect changes at some levels. As she anecdotally observes regarding the effect of metaphor on her own consiousness: 'No listing of logical arguments and facts could have brought

Understanding the tension between the dissimilarities that are posited by a metaphor poses a problem for the reader, causing the metaphor to be processed like a riddle.[11] A prime example in Qoheleth would be the riddle which lies latent in the phrase *r*^{e}*'ût rûaḥ* 'shepherding the wind'. Paradox is a concealed invitation to the reader to perform a 'dance step' with the text. A paradox 'has a way of getting the reader to ask and try to answer: "How can X and Y both be? How can X be, if Y? How can Y be, if X?"'.[12]

As any reader familiar with Qoheleth is aware, Ecclesiastes teems with the use of these strategies. Metaphors, incongruities, ironies, paradoxes and opacities abound throughout the discourse. The implied author was quite fluent in the language of ambiguity, and utilized it in an assiduously shameless manner. The primary effect of an ambiguous text is to lead the reader through the process of dealing with these problems. Fowler concludes that for an ambiguous text:

> The process of working through ambiguity is more important, in its own right—and may be more lasting in its impact—than any clarity or resolution that may or may not be achieved along the way. Precisely when clarity or resolution is not achieved, we realize that the process of wrestling with the ambiguity rather than the final resolution itself is what matters in such an indirect rhetorical strategy. The experience of living in and working through...ambiguity, in the course of reading...is what... ambiguity is 'about'.[13]

In Iserian terms, metonymy, synecdoche, metaphor and other strategies of ambiguity are ways to defamiliarize reality for the reader.[14] Fowler concludes that there are four uses for literary ambiguity: (1) to promote enlightenment through parables, paradox and enigmas; (2) to

about in me this sort of instantaneous transformation of my attitudes; only a metaphor can do that. That is true power, waiting to be used, ready to hand, extremely inexpensive, and belonging to anyone who chooses to use it' (p. 212). Thus we see that metaphors are latently rhetorical in nature, and can have very powerful, though subtle effects on their readers. As a text which so resolutely depends on the effects of various metaphors to carry its meaning, Ecclesiates has utilized an extremely powerful rhetorical technique in the constant interfacing of the logos of its argumentation with the pathos of its chosen metaphors.

11. R. Bontekoe, 'The Function of Metaphor', *PR* 20 (1987), pp. 209-26 (225).
12. Fowler, *Let the Reader Understand*, p. 185.
13. Fowler, *Let the Reader Understand*, p. 209.
14. For a fuller account of how these tropes work to effect defamiliarization, see McKenzie, 'Subversive Sages', pp. 94-120.

achieve an affective response in the reader; (3) to provide a shield of obscurity behind which the implied author can hide; and (4) to provide a way to avoid rigidity and maintain flexibility in social relationships.[15]

The lasting impact of Qoheleth's rhetoric of ambiguity is precisely the unsolved problems it leaves for the reader. It is the process of being confused, and eventually becoming defamiliarized to reality that ultimately sticks with the reader. Whatever final *Gestalt* one makes of them is secondary to this effect.[16] In fact, Fowler argues that the language of ambiguity or indirection 'works predominantly along the rhetorical axis of language to affect the reader than predominantly along the referential axis to convey information'.[17] If this is so, then we have gone about solving Qoheleth's riddles with the wrong mindset. Qoheleth's text is not about giving answers that can be precisely stated. It is about recreating in the reader the same sense of profound ambiguity that Qoheleth, or perhaps Ecclesiastes experienced in the world.[18] Such an understanding of Qoheleth's language results in a Copernican revolution for the reader, who is no longer bound to solve Qoheleth's conundrums.[19] Instead, the reader is set free to enjoy and experience the life of ambiguity as narrated by the master of ambiguous language. Further-

15. Fowler, *Let the Reader Understand*, p. 196.

16. D. Levine, *The Flight from Ambiguity: Essays in Social and Cultural Theory* (Chicago: University of Chicago Press, 1985). Levine argues that Western culture, with the rise of scientific culture, has eschewed ambiguity as a mode of literary expression. However, premodern cultures were more comfortable with the ambiguous, the figurative, and the allusive. In this regard, the modern reader must make an attitude adjustment toward ambiguity in order to appreciate its rhetorical strategies, effects, goals and purposes if he or she is to become the implied reader of an ambiguous text.

17. Fowler, *Let the Reader Understand*, p. 222.

18. Caneday also has seen that this is the ultimate effect of the text on the reader. See A. Caneday, 'Qoheleth: Enigmatic Pessimist or Godly Sage?', *GTJ* 7 (1986), pp. 21-56.

19. I therefore agree with Fox who argues that Qoheleth's contradictions should be left in a state of tension. See M. Fox, *Qoheleth and His Contradictions* (JSOTSup, 71; Sheffield: Almond Press, 1989), p. 11. The result of this observation is that such solutions as the 'Zwar/Aber' or 'Yes, But' interpretation of H.W. Hertzberg (cf. 1.16-18; 2.3-11, 13-15; 3.11, 17-18; 4.13-16; 7.7, 11-12; 8.12b-15; 9.4-5, 16; 9.17–10.1; 10.2-3, 5-7) and other various harmonizations which attempt to deal with these texts (especially the idea of editorial additions or the theory of two 'voices') can, if allowed, circumvent the chief objective of the use of ambiguous language.

more, ambiguity forces the reader to deal with a more intricate set of problems, thereby requiring more mental operations on the part of the reader. Penultimately, this increased level of interaction between reader and text takes the reader to a deeper level of participation. Eventually, it recreates in the reader the same sense of disequilibrium on the affective level that Qoheleth argues for so cogently on the intellectual level. With that in mind, it is now time to turn to those seminal problems which bear most directly on the study of Ecclesiastes' use of first-person discourse.

2. *An Overview of Reader Problems in Ecclesiastes*

The result of this rhetorical strategy is a text that, for good reason, many scholars consider the single most difficult book to interpret in the entire Canon. The reading history of the book is replete with dissentious debates regarding its grammatical, lexical, historical, theological and literary riddles. For many of these discussions, there is no consensus among the interpretative community. These unsolved ambiguities challenge any critical reading of the book. Still, a reader must make some decisions regarding the basic problems in the book. How a critic tentatively solves them will have a substantial impact upon the final *Gestalt* he or she arrives at regarding the book's overall meaning. Such problems and their solutions make every reading an intensely subjective process for the critically-trained reader in a way that nearly deconstructs the entire process. This state of affairs contributes very much, albeit in an indirect fashion, to the theme of vanity or absurdity which permeates the fabric of this book. Such a 'vain rhetoric' means that no reading will ever enjoy the acceptance of the entire interpretative community. There are simply too many unsolved ambiguities for that. In this regard, the text has achieved a powerful effect.

The response of the reading community to Ecclesiastes' literary strategy of ambiguity has been surveyed by Kurt Galling, H.H. Rowley, Santiago Bréton, James Crenshaw and Roland Murphy. In 1932 and 1934, Galling isolated four main problems which had vexed the reading community: (1) the theme of the book; (2) the autobiographical form; (3) the relationship between Qoheleth and ancient Near Eastern wisdom; and (4) the influence of Greek philosophy upon the book.[20]

20. K. Galling, 'Koheleth-Studien', *ZAW* 50 (1932), pp. 276-99, and *idem*, 'Stand und Aufgabe der Kohelet-Forschung', *TRu* NS 6 (1934), pp. 355-73.

H.H. Rowley summarized the reading community's interests during the 1940s.[21] His analysis centers almost entirely on the language problems of the book and the various proposals offered by H.L. Ginsberg and Robert Gordis. During the 1970s, the reading community's interests seemed to broaden. Santiago Bréton summarized the problems in the book under the broad categories of contributions made by commentaries and those offered by special studies. Bréton described the pertinent issues as: (1) the book's peculiar language; (2) unity of the book; (3) first-person style; (4) the author's pessimism; (5) the meaning of *hebel*; (6) literary structure; and (7) the book's relation to Wisdom.[22] Nearly ten years later, the issues and interests had not changed much from his perspective.[23] James Crenshaw returned to the lines laid down by Galling by accenting the 'interpretive history of research' over the last half century. His survey stressed what he considered to be the one quintessential issue for readers that had persisted for 50 years, that being 'the search for an adequate means of explaining the inconsistencies within the book'.[24] However, in the Introduction to his commentary on Ecclesiastes, and his article in the *Anchor Bible Dictionary*, Crenshaw also deals extensively with the problems of literary structure, the book's integrity, the use of first-person observation and reflection as a means of literary expression, and the book's general historical setting.[25] Roland Murphy continues in a line similar to Crenshaw, viewing the book's chief problems as those pertaining to the book's peculiar language, its use of first-person style, form-critical issues and the genre of Qoheleth, literary integrity and structure, and the book's ancient Near Eastern background.[26]

Although these historically trained interpreters clearly perceived the ambiguous nature of the text, it has only been relatively recently that scholars have begun to look at the problem from a literary perspective. In the latter half of the 1990s, several articles have appeared which tackle the problem either as a general problem of interpretation for the book, or specifically in relation to certain texts. Both Michael Fox and

21. H. Rowley, 'The Problems of Ecclesiastes', *JQR* 42 (1951–52), pp. 87-90.

22. S. Bréton, 'Qoheleth Studies', *BTB* 3 (1973), pp. 22-50.

23. Bréton, 'Qohelet: Recent Studies'.

24. J. Crenshaw, 'Qoheleth in Current Research', *HAR* 7 (1983), pp. 41-56 (43).

25. J. Crenshaw, *Ecclesiastes: A Commentary* (OTL; Philadelphia: Fortress Press, 1987), II, pp. 23-54, and *idem*, 'Ecclesiastes, Book of', in *ABD*, pp. 271-80.

26. Murphy, *Ecclesiastes*, pp. xix-lxix.

Addison Wright have examined the artful use of ambiguity in Eccl. 4.13-16.[27] Wright's article is very lucid in its treatment of the 'internal ambiguities' which occur in the choice of verbs, the number of character's in Qoheleth's example story, and certain specific grammatical ambiguities. Although Fox and Wright note the extensive problems this creates for the reader, both attempt to 'resolve' the problems rather than attempt to understand the rhetorical effect this ambiguity has for the reader. This is not true, however, of some of the more thoroughly reader-oriented treatments offered of late. In a recent overview of the book of Ecclesiastes, Carol Newsom has duly observed that

> Since one of Qoheleth's themes is the inability of human enterprise to seize and hold, to take possession of a thing, it is perhaps no accident that the book eludes the attempts of interpretive activity to fix its meaning determinately. I think that scholars have underestimated the significance of interpretive ambiguity in Ecclesiastes by seeing it merely as a problem to be solved. Perhaps it should be seen instead as another means of communicating the book's message... Ecclesiastes is a book that makes people profoundly uncomfortable, a fact that renders its reception history particularly fascinating.[28]

From this Newson concludes that future treatments of the book's sundry literary problems will

> become less inclined to seek a simple but comprehensive resolution to the cluster of questions having to do with structure, composition, and message; instead, the contradictiveness and elusiveness of the book will be taken more into account as a part of its message, rather than an obstacle to be overcome.[29]

Indeed, her admonitions have proven to be prophetic for the *Colloquium Biblicum Lovaniense* which recently devoted a volume to Qoheleth's book. That such literary use of ambiguity has a meaningful effect on the reader seems to be a theme for several recent articles

27. See M. Fox, 'What Happens in Qohelet 4.13-16', *JHStud* 1 (1997), pp. 1-9, (http://www.arts.ualberta.ca/JHS/), and A. Wright, 'The Poor But Wise Youth and the Old But Foolish King (Qoh 4.13-16)', in M. Barré (ed.), *Wisdom, You Are My Sister* (Festschrift R. Murphy; CBQMS, 29; Washington: The Catholic Biblical Association of America, 1997), pp. 142-54.

28. C. Newsom, 'Job and Ecclesiastes', in J. Mays, D. Petersen and K. Richards (eds.), *Old Testament Interpretation: Past, Present, and Future* (Festschrift G. Tucker; Nashville: Abingdon Press, 1995), pp. 177-94 (190).

29. Newsom, 'Job and Ecclesiastes', p. 192.

offered in *Qoheleth in the Context of Wisdom*.[30] In discussing the 'artful use of ambiguity' in the prologue to Ecclesiastes (1.1-11), Lindsay Wilson observes that both the positive and negative readings of the passage have a firm basis in the text. By extensively utilizing a reader-oriented perspective, Wilson's study concludes that the reader should accept both interpretations as 'deliberate, purposeful, artful ambiguity'.[31] It is further observed that the passage is full of words with a broad semantic range, and as thus, is an 'ideal seedbed for ambiguity'. Wilson astutely observes that 'the clustering of so many words with wide ranges of meaning is surely not an accidental feature of the text'.[32] The wide-ranging meanings offered for *hebel* is therefore seen as 'purposely, deliberately, even artfully, enigmatic'.[33] Such enigmatism is not problematic, but rather, indicative of literary artistry at its best. Moreover, such deliberate ambiguity can be found in other wisdom writers as well, such as Prov. 26.4-5. Lindsay summarizes the performative nature of Qoheleth's use of ambiguity:

> What, then, can we say about the reason for this use of ambiguity in the wisdom writers?... Without denying there is order in the world, Qohelet's use of ambiguity can affirm that there is also confusion and pointlessness in this order, or at the very least in our perception of it...the purposeful use of ambiguity is a way of reminding the reader that wisdom observations usually reflect part, not all, of the truth. In other words, what is being asserted from one viewpoint might need to be qualified by other perspectives. The effect of this ambiguous opening section is that the reader is warned to tread carefully... The use of ambiguity thus does not mean that the text fails to communicate its message, but rather implies that the message is more complex than it appears at first.[34]

Douglas Miller has also looked at Qoheleth's ambiguous use of *hebel*. He too concludes that the meaning of the term cannot be restricted to

30. A. Schoors, *Qohelet in the Context of Wisdom*. See specifically the contributions by R. Byargeon, 'The Significance of Ambiguity in Ecclesiastes 2,24-26', pp. 367-72, and L. Wilson, 'Artful Ambiguity in Ecclesiastes 1, 1-11', pp. 357-65. Another contribution to Qoheleth's use of ambiguity is offered by J.M. Carrière, 'Tout est Vanité: L'un des Concepts de Qohélet', *EstBib* 55 (1997), pp. 463-77 (470-77).

31. L. Wilson, 'Artful Ambiguity', p. 358.

32. L. Wilson, 'Artful Ambiguity', p. 359.

33. L. Wilson, 'Artful Ambiguity', p. 361.

34. L. Wilson, 'Artful Ambiguity', p. 364.

mere lexical analyses of the word, but must pay attention to the performative aspects of the word within the context of the narrator's discourse. He wryly observes that there is 'nothing which requires Qohelet to be consistent with his use of terms. In fact, we may do well to consider whether such inconsistency is a part of his purpose.'[35] In fact, Miller has persuasively argued that by the time the reader has traversed Qoheleth's discourse to reach the latter passages found in the book (specifically 7.15-18 and 9.7-10), that the model reader is 'meant to recognize that any or all dimensions of *hebel* are being alluded to, and that *hebel* symbolizes all the experiences of life'.[36] In this respect, the performative function of ambiguity for the reader is to present a 'puzzle' which the reader must figure out.[37] In this respect, it can be seen that a reader-oriented perspective on the book's multiple layers of ambiguity gives the critic a new lens with which to view, and ultimately, appreciate what Qoheleth may have been saying to his readership. Literary problems, once viewed with respect to their performative function render rather than the logistic difficulties they present to the Western mindset, create a very different perspective from which to understand Qoheleth's monologue.

Of the various issues surveyed by past scholarship, several have a direct bearing on how a reader approaches a first-person text. Those issues are the peculiar language of the book, its literary structure, the issue of voice and narration in the book, the problem of the use of quotations by the narrator, the genre of Ecclesiastes as it pertains to first-person discourse, and the nature of the Solomonic/Royal Fiction. For the purposes of analyzing the reader's response to this book, this study will analyze the book of Ecclesiastes at two levels: at the level of Ecclesiastes-as-text, and at the level of Qoheleth-as-persona. The rest of this chapter will discuss the textual issues that are raised in the debates over language and structure. Chapter Four of my study will discuss the persona issues that are involved in the discussions regarding narration, quotations, genre and the King's Fiction. The purpose of these chapters will be to provide a literature review that recalibrates past scholarly contributions for utilization by a Ricoeurian/reader-oriented perspective. I must stress, however, that the decisions regarding the

35. D. Miller, 'Qohelet's Symbolic Use of *Hebel*', *JBL* 117 (1998), pp. 437-54 (443).

36. Miller, 'Qohelet's Symbolic Use of *Hebel*', p. 452.

37. Miller, 'Qohelet's Symbolic Use of *Hebel*', p. 454.

most plausible resolutions for these issues will be provisional. For each and every reader problem in the book, I have been almost equally impressed with the other side of the debate. In the meantime, I hope to make a reasonable decision regarding those persistent problems that have plagued the interpretative community and to show where past readers have confirmed or perhaps provided a reading grid for my own analysis.

3. *Major Reading Problems in the Book of Ecclesiastes: Opacity Generated by Idiosyncratic Grammatical Ambiguities*

Obviously, the first issues a linear reading of the book must deal with are the various grammatical and lexical problems involved in translating the text.[38] Whether these were intentional or not (they were probably quite unintentional), the effect of these problems is to create a sense of the text's opacity in the reader. The reader feels 'left in the dark' as to the precise meaning of many passages. Necessary information is lacking, creating gaps in the text, with the result being a sense of uncertainty and anxiety in the reader. After a while, when he or she cannot make a closed *Gestalt* of certain strategic passages, frustration and/or confusion is generated in the reader. Readers continue reading the text only by guessing, making tentative conjectures and generally going 'by the seat of their pants'. The meaning of the text is glimpsed 'as though through a darkened glass'.

Ambiguity often begins at the grammatical and lexical level for the reader of Ecclesiastes. More often than not, even at the basic level of deciding the meaning of a word or phrase, the context supports more than one meaning.[39] Simply translating the book will create a subtle feeling of uncertainty and indecision toward the book for the average critically-trained reader. The level of grammatical competence required for the modern implied reader is quite high, often proving elusive or getting lost in the history of the Hebrew language. Perhaps this was also the case even for the book's authorial audience. The extent of the text's opacity can be seen in the linguistic debates between W.F. Albright, Mitchell Dahood, W.C. Delsman, Robert Gordis, Cyrus

38. For an example, the reader is referred to Byargeon's insightful analysis of how lexical and grammatical ambiguity may radically affect the reading of a text; see 'The Significance of Ambiguity', pp. 368-72.

39. Byargeon, 'The Significance of Ambiguity', p. 368.

Gordon, Charles Torrey, Charles Whitley, Frank Zimmermann, Anton Schoors, Bo Isaksson, Daniel Fredericks and C.L. Seow. Delsman noted that there are 27 *hapax legomena* in the book and 26 words or combinations of words that occur only here in the First Testament.[40] There are also 42 grammatical *hapax legomena* and 42 Aramaicisms.[41] Except for the Song of Songs, no other book in the Canon has such a high proportion of grammatical and linguistic *hapax legomena* to tax the reader's competence.[42] The older commentaries of Franz Delitzsch, C.H.H. Wright, and C.G. Siegfried also dealt extensively with the problem of grammatical and stylistic oddities.[43] Such ambiguous language presents quite a challenge for any reader of Ecclesiastes. In many passages, there remains a high degree of opacity and uncertainty as to the precise meaning of the text.[44] For instance, one example of Qoheleth's grammatical opacity is the use of the ever-elusive *kî* ('indeed', 'because', 'when'). Anton Schoors has dealt with Qoheleth's use of *kî* in 5.6, 6.8, 7.7, 7.20 and 8.6, concluding that while emphatic *kî* occurs, the causal-explicative meaning is also quite possible in some cases.[45] Diethelm Michel has also subjected Qohelet's use of *kî* to a rigorous analysis, and found the same propensity for semantic ambiguity.[46] Roland Murphy summarizes the problems associated with translating *kî*, observing:

> When, for example, it is used four times in two verses (8.6-7; 9.4-5) or thrice in three verses (7.4-5; 2.24b-26), one almost despairs of catching the nuances, and it is difficult to find any agreement among translators... *kî*

40. W.C. Delsman, 'Zur Sprache des Buches Koheleth', in W. Delsman *et al.* (eds.), *Von Kanaan bis Kerala* (Neukirchen–Vluyn: Neukirchener Verlag, 1982), pp. 341-65.

41. Crenshaw, *Ecclesiastes*, p. 31.

42. S. Holm-Nielsen, 'The Book of Ecclesiastes and the Interpretation of It in Jewish and Christian Theology', *ASTI* 10 (1975–76), pp. 38-96 (45).

43. F. Delitzsch, *Commentary on the Song of Songs and Qoheleth* (trans. M. Easton; Grand Rapids: Eerdmans, repr. edn, 1950 [1875]), pp. 190-96; C.H. Wright, *The Book of Koheleth* (London: Hodder & Stoughton, 1883), pp. 488-500; C.G. Siegfried, *Prediger und Hoheslied* (Göttingen: Vandehoeck & Ruprecht, 1898), pp. 13-23.

44. R. Murphy, 'On Translating Ecclesiastes', *CBQ* 53 (1991), pp. 571-97 (571).

45. A. Schoors, 'Emphatic and Asseverative *kî* in Qoheleth', in H. Vanstiphout *et. al.* (eds.), *Scripta Signa Vocis* (Groningen: Egbert Forster, 1986), pp. 209-15.

46. D. Michel, *Untersuchungen zur Eigenart des Buches Qohelet* (BZAW, 183; New York: W. de Gruyter, 1989), pp. 200-12.

> is also a deictic or strengthening particle. In this function it is a signal of some subordination. Thus, 8.6-7 can be translated: 'Now (deictic *kî*) for every deed there is a time and judgment: to be sure (deictic *kî*) an evil thing weighs on humans, for (causal *kî*) they know not what will be, because (causal *kî*) who can tell them how things will turn out?'[47]

Other gramatical opacities, such as the use of *'ašer* ('which', 'so that', 'because', 'when', 'through', 'in that', cf. 8.11-12) could be adduced as well.[48]

Basically, four explanations have been advanced regarding the grammatical and linguistic difficulties encountered by the reader. All find recourse to the historical author behind the text. The grammatical idiosyncracies of the author are explained by positing either an alleged Canaanite-Phoenician (Dahood, Albright, Whitley), Aramaic (Zimmermann, Torrey), proto-Mishnaic (Schoors, Gordis) or Northern Hebrew (Gordon, Isaksson) background for the author. While no consensus has been reached in this debate, in recent times there does seem to be a trend towards the theory that the linguistic difficulties in the book are due to the influence of Aramaic and that the language of Qoheleth is a kind of 'intermediate between Biblical and Mishnaic Hebrew'.[49] However, because a Ricoeurian perspective does not consider such genetic explanations to be intrinsically valuable for understanding the textuality of a literary text, especially given the fact that every text is distanciated from its original context and author, this debate will be left for the historical grammarians to ponder until some consensus is reached (however unlikely that may be). As a result, the method will be to consult the various conjectures, weigh them on their own merits, and to set the various proposals against the broader background of the norms established by the text. If there are no compelling solutions for a passage, the confusion brought about by the text will simply be noted and the effect that opacity has on the reader will be analyzed. On the other hand, if a proposal clarifies a passage, has adequate grounding in an appropriate cognate Semitic language, makes good grammatical sense, and fits in well with the broad values of the text, then such a reading may

47. Murphy, *Ecclesiastes*, p. xxx.

48. For a more comprehensive overview of these issues, the reader is referred to the excellent article by F. Bianchi, 'The Language of Qoheleth: A Bibliographical Survey', *ZAW* 105 (1993), pp. 210-23.

49. R. Gordis, 'Koheleth: Hebrew or Aramaic?', *JBL* 71 (1952), pp. 93-109 (107).

be utilized to heighten the competence of the readers. To validate such clarifications, a second criterion would be consideration of the later reactions of the reading community. A high degree of consensus on this level would lend a commensurate degree of intersubjective validity to this unusually subjective process.

A few brief examples will clarify how a reader-oriented approach deals with such historical-critical issues. Based on cognates found in Ugaritic, Mitchell Dahood argued that Qoheleth's language was influenced by the commercial milieu of Phoenicia. Dahood observed that in Ecclesiastes there was 'the repeated use of words denoting profit and loss, abundance and deficiency, shares and wages, ownership and wealth, patrimony and poverty'.[50] Dahood located 29 terms which he thought were influenced by Qoheleth's 'northern' exposure. He concludes from the numerous Ugaritic parallels in Ecclesiastes, including such notable words such as *'āmāl* ('work'), *yitrôn* ('profit'), and *'inyān* ('occupation') that the 'distinctly commercial character of so many of the keywords and phrases is thoroughly consonant with what is known about the commercializing Phoenician culture...[it] betrays a milieu very harmonizing with the mercantile character of Phoenicia and her colonies'.[51] Referring to 12.12, Dahood observes that the roots *spr* and *hg* occur in parallelism in some Ugaritic/Phoenician contexts (e.g. Keret 90-91). Anson Rainey builds on this observation, and considers translating 12.12 as: 'Of making many accounts there is no end, and much reckoning (checking ledgers?) is weariness to the flesh'.[52] Rainey further argues that the LXX also accords well with this interpretation, in that the Greek word which translates *s^epārîm* is βιβλία here, a word which means 'accounts' in some Hellenistic papyri.

This is a very attractive reading for this verse. Many interpreters have noted and constructed readings out of the strong commercial tenor of the book. Frank Crüsemann, James Kugel, Anthony Ceresko and many others have built convincing cases for a commercialized reading of the book and its implied author based on Dahood's original insights.[53]

50. M. Dahood, 'Canaanite-Phoenician Influence in Qoheleth', *Bib* 33 (1952), pp. 30-52; 191-221 (51-52) (reprinted in *idem*, *Canaanite-Phoenician Influence in Qoheleth* [Rome: Pontifical Biblical Institute, 1952]).

51. Dahood, 'Canaanite-Phoenician Influence', pp. 51-52.

52. A. Rainey, 'A Study of Ecclesiastes', *Concordia* 35 (1964), pp. 148-57 (149).

53. F. Crüsemann, 'The Unchangeable World: The "Crisis of Wisdom" in Koheleth', in W. Schottroff and W. Stegemann (eds.), *God of the Lowly* (trans.

Given the fact that the book abounds with so many economic terms, and that the programmatic question for the book is the 'What Profit?' question (1.3) which forms a constant refrain in the book, I would hold that this reading accords well with the norms established by the text.[54] Furthermore, Dahood's observations have been taken up by several critics, or are wholly consonant with interpretations on a similar vein, such as Robert Johnston's oft-quoted classic.[55] This gives the Dahood/ Rainey interpretation of 12.12 the support of the reading community's intersubjective validation. I find no reason to rule out this as a likely or at least a possible reading of this verse. In this instance, Dahood has contributed to the reader's understanding of the text's repertoire, and enhanced our competency as readers.

On the other hand, not all suggestions have fared so well in this debate. A proposal which directly affects the characterization of the narrator is that offered by H.L. Ginsberg in 1950. Based on cognates in Arabic, Ginsberg proposed that the noun *melek* in 1.12, usually

M. O'Connell; Maryknoll, NY: Orbis Books, 1984), pp. 57-77 (first published as 'Die Unveranderbare Welt', in W. Schotroff and W. Stegemann [eds.], *Der Gott der Kleinen Leute* [Munich: Chr. Kaiser Verlag, 1979], pp. 80-104). Crüsemann speaks of the 'materialization' of Qoheleth's thought. J. Kugel, 'Qoheleth and Money', *CBQ* 51 (1989), pp. 32-49. Kugel builds directly on Dahood's list of 29 terms for his interpretation of the book (p. 32). A. Ceresko, 'Commerce and Calculation: The Strategy of the Book of Qoheleth (Ecclesiastes)', *ITS* 30 (1993), pp. 205-19. Like Kugel and Rainey, Ceresko also builds on Dahood's foundational work.

54. In addition to those already mentioned, the major scholars who hold to the centrality of the 'What Profit?' question for establishing the norms of the text are: Crenshaw, *Ecclesiastes*; G. Ogden, *Qoheleth* (Readings: A New Biblical Commentary; Sheffield: JSOT Press, 1987), esp. pp. 11-13; J.A. Loader, *Ecclesiastes: A Practical Commentary* (trans. J. Vriend; Grand Rapids: Eerdmans, 1986); *idem*, *Polar Structures in the Book of Qohelet* (BZAW, 152; Berlin: W. de Gruyter, 1979); Fox, *Qoheleth and His Contradictions*; R. Johnson, 'The Rhetorical Question as a Literary Device in Ecclesiastes' (PhD dissertation, Lexington, KY: Southern Baptist Theological Seminary, 1986); J. Williams, '"What does it profit a man?": The Wisdom of Qoheleth', in J. Crenshaw (ed.), *Studies in Ancient Israelite Wisdom* (New York: Ktav, 1976), pp. 375-89; T. Polk, 'The Wisdom of Irony: A Study of *Hebel* and its Relation to Joy and the Fear of God in Ecclesiastes', *SBTh* 6 (1976), pp. 3-17; D. Bergant, *Job, Ecclesiastes* (Wilmington, DE: Michael Glazier, 1982). This impressive list of readers suggests that a commercial characterization of the narrator enjoys the intersubjective validation of the reading community.

55. R. Johnston, '"Confessions of a Workaholic": A Reappraisal of Qoheleth', *CBQ* 38 (1976), pp. 14-28.

translated as 'king', might be better re-pointed to produce the noun *mōlek*, 'property-owner'. He calls attention to the fact that 'the difference between King (*malik*) and possessor (*mâlik*) in Arabic is exactly one morae'.[56] As a result, he proposes to read 1.12 as 'I, the Convoker, was a man of property in Jerusalem'. The result of this one change is that it effectively excises the entire King's Fiction from the text by positing that there is no real portrayal of kingship by the word *melek* ('king'). For Ginsberg, this resolves a very thorny problem that has haunted interpreters for centuries.

Nevertheless, this proposal violates the broader literary strategy of its surrounding context and has found scant support from the later reading community. David Meade has correctly adjudicated with regard to this conjecture that the 'fiction and portrayal of kingship is broader that the etymology of one word'.[57] The norms of the text, especially those found in 1.12–2.26 effectively rule out such a reading because it would be inconsistent with those norms. Furthermore, it lacks intersubjective verification. As J.A. Loader has pointed out, the parallels between Solomon and Qoheleth in 2.1-11 do presume some sort of royal characterization of the narrator and effectively weigh against Ginsberg's theory.[58] Other readers such as H.H. Rowley have concluded that this proposal is 'more ingenious than probable'.[59] My own reading of the various treatments of this passage confirms the fact that very few critics have read the text in this fashion. As such, though knowledge of cognate Near Eastern languages does hold some promise for certain passages, and 12.12 is among them, not all conjectures will withstand the rigors of the twin criteria described here. In this instance, Ginsberg's proposal violates the methodological controls established by a reader-oriented method, that is, the criteria of the norms of the text and the need for intersubjective verification. In spite of certain objections that a reader-response methodology invites solipsism, I belive that the opposite is true. If these two criteria are exercised, reader-response methods can act as a deterrent to the uncontrolled subjectivism that has

56. H. Ginsberg, 'The Designation *Melek* as Applied to the Author [Qoheleth]', in H. Ginsberg (ed.), *Studies in Koheleth* (New York: Jewish Theological Seminary of America, 1950), p. 14.

57. D. Meade, *Pseudonymity and Canon* (Grand Rapids: Eerdmans, 1987), p. 57.

58. Loader, *Polar Structures*, p. 19.

59. Rowley, 'The Problems of Ecclesiastes', pp. 87-90 (90).

plagued the interpretation of this book, such as is witnessed in this particular historical-critical solution offered by Ginsberg.

4. *Literary Rubik's Cubes and the Structural Ambiguities of the Book of Ecclesiastes: An Overview of Reading Strategies*

Ecclesiastes has been described as 'the sphinx of Hebrew literature', and for good reason.[60] The problems Western readers experience when attempting to discern a definite structure are so great that Franz Delitzsch predicted in 1875: 'All attempts to demonstrate in the overall book, not only the unity of its spirit, but also the genetic origin, overall plan, and organic arrangement, must fail hitherto and in the future'.[61] The impasses have been aptly summarized by Addison Wright:

> There is agreement that 1.4-11 and 11.7–12.8 are units and that 3.1-15 is a sub unit (or two units) of some larger piece, but there is really no agreement on anything else. A repetition which one interpreter sees as an ending formula another sees as part of a chiasm leading in a different direction and another sees as a Leitmotif. While one critic is impressed by repetitions as indicators of structure in this particular book and allows for irregularities in other stylistic features, another gives primary value to introductory formulae or to discontinuities (change of person, topic, genre) and allows for irregularities in other areas. One interpreter is quite at ease with the idea that an author may have introduced digressions into a structured composition while another finds an appeal to digressions to be a serious flaw in any structural proposal. One critic would say that if an idea occurs in two adjacent paragraphs of a book, such an occurrence precludes any division between those paragraphs, and another critic would say that such an air-tight style of composition is an extraordinary requirement to place upon any author. One commentator becomes exceedingly wary if a supposed ending formula is recessive, occurring one or two lines before the end of a section, or if a formula contains a one-word variation in two out of nine occurrences, while another commentator becomes most expansive and urges that ending formulae (if they are being used) must be conceived of far less rigidly and that one should even be prepared to allow an author to end a section without a formula now and again. One interpreter warns that structural analysis is a far

60. E. Plumptre, *Ecclesiastes* (CB, 23; Cambridge: Cambridge University Press, 1898), p. 7, was the first to coin this term for Ecclesiastes, which was subsequently taken up by B. Pick, 'Ecclesiastes or the Sphinx of Hebrew Literature', *Open Court* 17 (1903), pp. 361-71, and A. Wright, 'The Riddle of the Sphinx'.

61. Delitzsch, *Commentary*, p. 195.

> more sophisticated process than the mere effort to fit ideas into spheres of verbal patterns, and another urges that in this particular book structural analysis is precisely that simple, because of the simple technique of ending formulae which the author chose to employ. Each interpreter delineates units that make very good sense to him, and each rapidly professes to be able to make no sense out of most of the units proposed by others on the basis of alternate criteria. And advocates of a lack of structure in the book are still eager to find comfort in the disarray. In other words, there is a large element of the subjective still at work in the 'objective' attempts at structural analysis on Ecclesiastes which have been characteristic of the last decade.[62]

While I do not pretend to have solved these reading issues, it will be the purpose of this section to summarize the reading problems involved in understanding the structure of this very difficult book, to delineate some common solutions readers have offered during the course of the book's reception-history, and to describe why some solutions seem more appropriate than others. Finally, with Graham Ogden, Michael Fox, James Crenshaw, and a host of modern scholars, I accept that Ecclesiastes, with the possible exception of 1.1 and maybe 12.12-14, is the work of one sage.[63] Throughout the book, one meets the brooding

62. A. Wright, 'The Riddle of the Sphinx Revisited: Numerical Patterns in the Book of Qoheleth', *CBQ* 42 (1980), pp. 35-51 (42).

63. Many scholars would also place 12.9-10 as an addition to the core of 1.2–12.8. However, I agree with Lavoie that these verses are part of the implied author's reflections on his own work. See J. Lavoie, 'Un eloge à Qohelet (étude de Qo 12,9-10)', *LavTP* 50 (1994), pp. 145-70. Lavoie argues that these verses are 'an anonymous and postscriptive allograph which attributes the authorship of the book to its hero and narrator: Qohelet' (p. 169). The implied author takes the stage in these verses, though he hides himself from the total view of the reader through the use of third-person narration. What the reader encounters here is the 'signature' of the implied author (p. 170). Indeed, one could extend these insights, taking the whole of 12.9-14 as the concluding statement of the implied author. 12.9-14 is not experienced as an editorial addition in terms of the temporal flow of reading, but rather, as yet another narrative voice in the text. Except for the arguments adduced by G. Sheppard and G. Wilson on the 'late' sound of verses 12.12-13 and the close contact this characterization of wisdom has with Sir. 16.24–17.14, 24.3-29 and *Bar.* 3.9–4.4, one could very easily extend Lavoie's insights to the entire epilogue without much of a second thought. Still, I tend to concur with Michael Fox who takes a more cautious approach to the canon-conscious interpretation of the Epilogue. Fox argues that the Epilogue is canon-conscious only in the most vague of senses (*contra* Sheppard) and that the reference to 'words of the wise' in 12.11 refers to wisdom in general with no strict corpus such as Proverbs–Qoheleth in mind (*contra*

reflections of a single consciousness, with the result being an undeniable impression that the writing comes from a single author.[64]

Two basic solutions have been offered to explain the book's structure. Some, including C.D. Ginsberg, Georg Fohrer, Friedrich Ellermeier and Kurt Galling, view the book as a collection of aphorisms like the book of Proverbs.[65] Others, such as A. Bea, Addison Wright, George Castellino, Stephen Brown and Stephan de Jong, see a definite progression of thought in the work. The center ground is occupied by scholars such as H.W. Hertzberg who observes some development of thought within units, but not between the different chapters. The extremes of this debate has been aptly summarized by Walther Zimmerli:

> The Book of Qoheleth is not a treatise with a clearly recognizable structure and one solitary, determinable theme. It is, however, at the same time more than a loose collection of sentences, although the character of the collection in certain places is not to be overlooked.[66]

Wilson). See Fox, *Qoheleth and His Contradictions*, p. 321. Brevard Childs has also failed to see a reference to a strict corpus in these verses. See Childs, *Introduction*, p. 586. In spite of the way these verses may hypothetically function in a canonical context, I would still hold to the seminal point raised above, namely, that the reader who consumes the text in a linear fashion simply encounters another narrative voice in these verses. This voice adds a distinctly external point of view to the presiding internal point of view which dominates the bulk of the book. The external interest expressed by this voice simply discloses a point of view whose breadth includes even canonical issues such as the relation of Wisdom (which includes the book at hand) to Torah, or perhaps better, general religious duties.

64. Scholars who argue for the literary unity of the book include H.W. Hertzberg, *Der Prediger* (KAT, 17.4; Gütersloh: Gerd Mohn, 1963), p. 41; R. Gordis, *Koheleth: The Man and His World; A Study of Ecclesiastes* (New York: Schocken, 3rd edn, 1968 [1951]), p. 73; B. Lang, *Ist der Mensch Hilflos? Zum Buch Kohelet Qoh 5.9-6.6, 2.1-3.15, 7.7-16, 7.15-22, 8.10-15, 9.13-10.1* (Zürich: Benzinger Verlag, 1979), col. 195; A. Wright, 'The Riddle of the Sphinx', pp. 313-34; *idem*, 'The Riddle of the Sphinx Revisited', pp. 35-51; B. Isaksson, 'The Autobiographical Thread: The Trait of Autobiography in Qoheleth', in B. Isaksson, *Studies in the Language of Qoheleth: With Special Emphasis on the Verbal System* (AUSSU, 10; Uppsala: Almqvist & Wiksell, 1987), p. 42; and M. Fox, 'Frame-Narrative and Composition in the Book of Qohelet', *HUCA* 48 (1977), pp. 83-106. Other scholars could be adduced. Whereas past scholarship considered the idea that there were 'pious additions' throughout the book, this is no longer considered a strong possibility.

65. G. Fohrer, *Introduction to the Old Testament* (trans. D. Green; Nashville: Abingdon Press, 1965), p. 337.

66. W. Zimmerli, 'Das Buch Kohelet: Traktat oder Sentenzensammlung?', *VT* 24 (1974), pp. 221-30 (230).

Zimmerli argues that one cannot make a strong, unassailable argument for either extreme. At times, there are blanks or gaps between the textual schemata. These blanks create an abruptness between the individual units that makes the book seem like a loose collection of sayings or, to be more fair, the haphazard reflections or musings of an aging sage.[67] Some sections, like the proverb collections in 7.1-13 and 10.1-20 do remind the reader of Proverbs. Yet even here, as Robert Johnson has demonstrated, there is indisputable evidence of logical arrangement.[68] On the other hand, there are clearly recognizable sections in the work that have a definite structure to their development (1.3–3.9; 6.10–7.14; 11.7–12.7).[69] Zimmerli thus concludes that the book is more than a mere collection, but less than a treatise of some sort.

However, such blanks have a silver lining to them according to Iser. He notes that

> such breaks act as hindrances to comprehension, and so force us to reject our habitual orientations as inadequate. If one tries to ignore such breaks, or to condemn them as faults in accordance with classical norms, one is in fact attempting to rob them of their function.[70]

Positively, such breaks act as 'barbs' (cf. 12.11) to stimulate the reader's comprehension of the text. Like a Rubik's Cube, such problems are there to be solved by engaging the reader's mind. The effect of these blanks is to involve the reader at a deeper level of participation. In the end, we see that such problems are actually not a problem at all: per se, but are part of the overall effect or design of the text to involve the reader in life's ambiguities. In this regard, Iser warns the critic regarding texts like Qoheleth:

67. Eichhorn compares Qoheleth's oration to the 'musings' of an old professor. See D. Eichorn, *Musings of the Old Professor: The Meaning of Kohelet; A New Translation of a Commentary on the Book of Ecclesiastes* (New York: J. David, 1963).

68. R.F. Johnson, 'A Form Critical Analysis of the Sayings in the Book of Ecclesiastes' (PhD dissertation, Atlanta, GA: Emory University, 1973), pp. 140, 199.

69. For instance, Fischer regards Eccl. 1.3–3.15 as a 'sandwich structure' (my term). At the center of this unit is the fiction of the king in 1.12–2.26 which is sandwiched between the two poems in 1.4-11 and 3.1-8 and the two thematically motivated parenthetic remarks in 1.3 and 3.9. More discussion of this structuring will be given in the following chapter. See A. Fischer, 'Beobachtung zur Komposition von Kohelet 1,3–3,15', *ZAW* 103 (1991), pp. 72-86.

70. Iser, *The Act of Reading*, p. 18.

> His [the critic's] object should therefore be, not to explain a work, but to reveal the conditions that bring about its various possible effects. If he clarifies the *potential* of a text, he will no longer fall into the fatal trap of trying to impose one meaning on his reader, as if that were the right, or at least the best, interpretation... Far more instructive will be an analysis of what actually happens when one is reading a text, for that is when the text begins to unfold its potential; it is in the reader that the text comes to life, and this is true even when the 'meaning' has become so historical that it no longer relevant to us.[71]

What the implied reader encounters is a series of blanks and gaps which prove upon further reflection to be a succession of participatory prods. Inasmuch as it would be absurd to criticize a Rubik's Cube for the problems it presents to its user, so it is with the text of Ecclesiastes. Their effect is to draw the reader into the text, creating a sense of participation with the narrator regarding the observation of life's conundrums. In that regard, their effect is their meaning. Or, at the very least, that effect possesses a meaningful tenor for the text's model reader.

However, it is at precisely this point that the radical effect of first-person narration has not been fully appreciated by many critics. In addition, we can also see that the problem of a Western reading grid has also hampered the modern reader from becoming the text's implied reader. Aarre Lauha has hinted at how these twin problems have interfered with readers' understanding of the text's structure. He states:

> According to Western logic, the whole book must be more or less arbitrary, however from the standpoint of the Epilogist—whether it is Qoheleth himself or his student—the structure of the book is in no case accidental. Ecclesiastes is no conglomeration of loose sayings such as the book of Proverbs, but rather, the sayings construct passages in which certain topoi bestow an internal connectedness... Second, every reader notices how the entire book is held tightly together by a stylistic and mental coherence. Such a formal and above all inner connectedness can only be the product of a single personality.[72]

Lauha points to what should be the obvious, namely, that the use of first-person discourse unifies the book, giving it not only the appearance of a single work but, I believe, a very reliable means to fully interpret the book. Qoheleth's 'I' gives the work a certain structural stability

71. Iser, *The Act of Reading*, pp. 18-19.

72. A. Lauha, *Kohelet* (BKAT, 19; Neukirchen–Vluyn: Neukirchener Verlag, 1978), pp. 5-6.

even in the absence of a discernible logical progression of thought. The extensive use of 'I searched', 'I observed', 'I tested' and other such phrases, especially in the first third of the book, serve to identify the book as an expression of the narrator's worldview, thereby giving the book a unity of presentation that overrides any possibility of multiple authorship.[73] More importantly, Lauha alludes to the problem of our Western mindset that expects some sort of logical or Aristotelian progression. This often prohibits the critically-trained reader from seeing what, I believe, would have been obvious for the book's authorial audience and, by extension, its implied audience.[74] The fact that most interpreters have looked for this type of structure has greatly hindered the book's reception.[75]

This situation has been addressed by various critics. Santiago Bréton, relying upon Oswald Loretz, alerted readers to the Procrustean bed of the Western mind in the early 1970s. He warns the reader of

> the danger of attempting to discover in Qoheleth the projections of our own rational categories, of seeking there intentions and structures compatible with our present-day mentality. It is methodologically mistaken to approach Qoheleth with logical standards, be it to find the rational outline of the whole book, or to isolate the small unit. The key to the solution is of a topical, not of a rational nature.[76]

Recently, Ardel Canadey and Pauline Viviano have also noted the misleading influence of the Western mindset which expects some sort of logical progression as a means to detect the book's structure.[77]

As a result, some readers have resorted to what I call the colliding or interacting topics approach. According to this reading strategy, the book should neither be read like a string of pearls that somehow lost its

73. Fox, 'Frame-Narrative', p. 90.

74. The Epilogist spoke of Qoheleth's careful ordering of proverbs (12.9). One can only surmise that the implied author was fictively characterizing the narrator one last time, thereby providing yet another clue as to the proper response he sought from the implied reader. I suspect that the problems reside in our own reading reflexes, rather than the text's manner of presentation.

75. Murphy, *Ecclesiastes*, p. xxxv.

76. Bréton, 'Qoheleth Studies', p. 25, relying upon O. Loretz, *Qohelet und der alte Orient: Untersuchungen zu Stil und Theologisher Thematik des Buches Qohelet* (Freiberg: Herder, 1964), p. 209-12. See also T.A. Perry, *Dialogues with Kohelet*, p. 43.

77. Caneday, 'Qoheleth', p. 33. P. Viviano, 'The Book of Ecclesiastes: A Literary Approach', *TBT* 22 (1984), pp. 79-84 (80).

'string', as Galling nearly does, nor as a logical treatise with tightly interlocking gems. Instead, what we possess are structured, topical segments that bang against each other like a windchime dangling in the breeze. Kathleen Farmer has likened Qoheleth to 'the structure of a mobile or a windchime (one of those decorative constructions in which various pieces are suspended on threads or wires and balanced in such a way that each part is able to move independently in a breeze, and yet each part depends on another for its equilibrium)'.[78] André Barucq, James Crenshaw and J.A. Loader have also advocated a similar reading strategy. Like Loretz before him, Loader concludes that we 'have no logical development of thought reflected in the composition of the book, but there are various separate pericopes. These are structured carefully ...separate pericopes are compositionally related to each other.'[79]

On the other hand, there are some critics who have argued that the clues for the book's structure function along thematic or non-logical lines. These seem to be the more fruitful places to begin a reader-oriented approach. In the mid-1940s J.S. Wright pointed out the necessity of looking for the obvious reading clues that are available in most texts. He advises:

> If you pick up a book and want to find the author's viewpoint, where do you turn? The preface is usually helpful—sometimes it saves you reading the book! The conclusion also in a well-written book generally sums up the point that the author has been trying to put over. When you look through the book, you may also be struck by something in the nature of a refrain, that by its continual recurrence to drive some point home.[80]

Wright's natural yet cogent insights should be taken to heart by more readers. Ecclesiastes is characterized by refrains that provide a certain structure for the book. Qoheleth's refrains substitute for the linear

78. K. Farmer, *Who Knows What is Good? A Commentary on the Books of Proverbs and Ecclesiastes* (ITC; Grand Rapids: Eerdmans, 1991), p. 151.

79. Loader, *Polar Structures*, p. 9.

80. J.S. Wright, 'The Interpretation of Ecclesiastes', p. 22. Shank also calls attention to the use of refrains in Qoheleth. See H. Shank, 'Qoheleth's World and Lifeview as Seen in His Recurring Phrases', *WTJ* 37 (1974), pp. 57-73. In addition, Crenshaw notes that Papyrus Insinger, which is roughly contemporary with Qoheleth, also uses refrains to mark off larger literary units. The use of refrains was probably a widely utilized literary procedure for his day. See Crenshaw, 'Ecclesiastes, Book of', p. 273.

development of the text, giving the book an inner unity of thought.[81] They bequeath to the book a very specific and unifying tone. Their frequency of occurrence produces 'almost a hypnotic effect in the listener or reader'.[82] The use of specific keywords is *the* trademark of the book. Oswald Loretz tabulates that the implied author's 28 favorite words constitute about 21.2 per cent of the text.[83] The fact that Qoheleth likes to 'change channels' rather than present his case in a strictly linear fashion is simply a matter of style which serves to characterize the narrator as something of an eccentric or perhaps a 'rambler'. In fact, Michael Fox even argues that

> the structure of the text shows little structuration not because the author was incapable of creating it, but because the book is a report of a journey of consciousness over the landscape of experience (1.13), a landscape generally lacking highways and signposts, order and progression.[84]

Again, it can be seen that readers who look for a logical progression or structure have asked for something that is not in the nature of many, perhaps most, first-person discourses. Francis Hart has observed that 'the nature of an extended autobiographical act makes it self-defeating for the interpreter to expect some predictable integrity or unity. Form is too experimental, too "accidental", and at the same time too inherent in perspective still to be recovered or imposed by memory.'[85] Georg Misch, in his mammoth overview of autobiography in antiquity, also

81. G. von Rad, *Wisdom in Israel* (trans. J. Martin; Nashville: Abingdon Press, 1971), p. 227.

82. Crenshaw, 'Ecclesiastes, Book of', p. 274.

83. Loretz, *Qohelet und der alte Orient*, p. 179. The favorite terms utilized by Qoheleth are: do, wise, good, see, time, sun, trouble, evil, vanity, fool, joy, eat, there is, profit, fool, wind, die, wrongdoing, just, trouble, chase, power, remember, portion, vexation, affair, folly and succeed. Qoheleth utilizes a rhetoric of redundancy where repetition is the trademark of his discourse strategy. However, Schoors has correctly seen that four of these keywords gain the most press from Qoheleth: human being/man (49 times), to be (49 times), to see (47 times), and good (52 times). See A. Schoors, 'Words Typical of Qoheleth', in Schoors (ed.), *Qoheleth in the Context of Wisdom*, pp. 17-39. This 'typical' vocabulary serves to further characterize Qoheleth as a reflective and highly philosophical sage. The term 'god' is the fifth most frequent word, which shows that the sage's 'philosophical preoccupation has a strong component of theodicy' (p. 39).

84. Fox, *Qoheleth and His Contradictions*, p. 158.

85. F. Hart, 'Notes on the Anatomy of Autobiography', *NLH* 1 (1970), pp. 485-511 (502).

sees the same phenomena in autobiographical discourse. He observes that first-person discourse is, as a rule, 'committed to no definite form. It abounds in fresh initiatives, drawn from actual life.'[86] First-person discourse and a lack of specific, or explicit form are complementary and related features that come with the territory, so to speak. I therefore conclude that the implied author has amply alerted the implied reader to his preferred means to guide the reader, which is the use of refrains and keywords. Western readers need to allow themselves to become the text's model reader and to follow its lead when attempting to read the book. That lead is found in the luxurious use of refrains and keywords throughout the book. As a result, the approaches that follow Ecclesiastes' lead by paying strict attention to the refrains are the most reliable guides to reading the book.[87] Of the many analyses in circulation, those presented by George Castellino, Stephan de Jong, Addison Wright, Stephen Brown, Francois Rousseau and R.N. Whybray offer the most natural way to read the book. This is particularly true of Wright's analysis, which has become something of an accepted standard in the field due to the fact that many scholars have intersubjectively agreed with his analysis.[88] Both Castellino and Wright proceed from an understanding of the use of keywords and refrains in the book. The result is an analysis that follows a 'logic' that is quite different from the Western mindset.

a. *Reading with George Castellino: Reading through Literary 'I's*

The New Critical analysis of George Castellino utilizes the full-fledged use of refrains and keywords to understand the structure of Qoheleth. His work is especially insightful for an analysis that focuses on Qoheleth's use of first-person discourse. In fact, his is the first study, to

86. G. Misch, *A History of Autobiography in Antiquity* (trans. E.W. Dickes; London: Routledge & Kegan Paul, 1950 [1907]), p. 4.

87. This does not, however, diminish the fact that many passages do function in a windchime-like manner as Loretz and Loader have pointed out. I am simply giving one reading grid a dominant preference, while holding out for a both–and paradigm which allows both theories to enlighten the reader. Both Wright and Loretz have valid critical insights to offer the reader.

88. For a list of those who have accepted and rejected his proposal, the reader is referred to Wright's own very honest appraisal of the reception of his analysis by the critical reading community. See A Wright, 'The Riddle of the Sphinx Revisited', pp. 35-51. More recent scholars who have accepted his analysis are R. Johnston, R. Murphy, A. Schoors, J.S. Mulder, R. Rendtorff and S. Brown.

my knowledge, that understands how Qoheleth's 'I' directly influences the reader's perception of the structure of the book.[89] The sudden change of the text's presentation from a predominantly first-person observational style in 1.1–4.16 to an imperative form in 4.17 ('watch your steps!'), is the key to unlocking the structure of the text for Castellino. As a result, he divides the book in two: 1.1–4.16 (Part I) and 4.17–12.8 (Part II). Using a close reading of the text, Castellino observes that 4.17 marks the first imperative in the book that directly addresses the reader or the narratee.[90] The use of negative imperatives such as *'al-tᵉbahēl 'al-pîkā* ('do not be rash with your mouth'), continues in 5.1, 3, 4, 5, 7, and so on. Although Qoheleth does continue to use first-person narration after 4.17, Castellino concludes that this 'does not obscure the fact that from 4.17 on we observe a different kind of discourse'.[91] He also observes how such words as *'ᵃnî* ('I'), *hebel* ('vanity'), and *'āmāl* ('work') are more statistically prevalent in the first part than the second part of the book. On the other hand, *rā'â* ('evil') occupies more of the reader's attention in Part II.[92] Both Parts I and II begin with a prologue which clearly indicates a change in trend of thought—1.3-11 deals with the nature of the world; 4.17–5.6 deals with the nature of dealing with God, particularly the 'fear of God' which sets the tone for the second half).[93] Ecclesiastes 5.7–6.12 takes up those facts of experience touched upon in Part I. The final chapters (7.1–12.8) describe wisdom and the wise at work. In these verses, the 'problems that had been presented in Part I (especially 3.16–4.3 about injustice and oppressions in the world) are answered here'.[94] Finally, the book ends with an epilogue (12.9-14) containing biographical

89. Although Loretz's classic studies dealt extensively with the problem and nature of I-narration in Ecclesiastes, it does not directly influence his analysis of the *specific* structure of the book, as it does Castellino. See O. Loretz, 'Zur Darbietungsform der "Ich-Erzählung" des Buch Qohelet', *CBQ* 25 (1963), pp. 46-59, and *idem*, *Qohelet and der alte Orient*, pp. 161-66.

90. G. Castellino, 'Qohelet and His Wisdom', *CBQ* 30 (1968), pp. 15-28 (16). It must be pointed out however, that although this is the first place in the book where Qoheleth *directly* addresses the reader, the reader is *indirectly* addressed by the rhetorical questions which abound in the first third of this book (cf. 1.3; 2.2, 12, 15, 19, 22, 25; 3.9, 21, 22; 4.8, 11).

91. Castellino, 'Qoheleth and His Wisdom', p. 16.

92. Castellino, 'Qoheleth and His Wisdom', p. 17.

93. Castellino, 'Qoheleth and His Wisdom', p. 19.

94. Castellino, 'Qoheleth and His Wisdom', p. 20.

indications and finishes with a return to the theme of the fear of God that had been inaugurated in the second prologue (4.17–5.6).

Castellino summarizes how these two parts interact in an indirect fashion. He states:

> Summing up the impression one gains from Part I, we must say that Qohelet is consistent in his critical and negative appraisal of man and his activities in life. Having laid down his thesis at the opening of his discourse he proceeds to prove it forcibly and ruthlessly. It is no wonder that practically all difficulties for the interpretation of the book stem from Part I...Part II, the negative impression is soon relieved by more positive and orthodox language that sounds more in tune with the other wisdom books. Are we therefore entitled simply to discard the 'unorthodox' Part I and rely on Part II in order to get Qohelet's doctrine, or should we try to harmonize the two parts by reading into the first part the spirit of the second? Both ways would be faulty in method and unsound in the conclusions. Therefore, given the differences between Part I and Part II, and given...the unity of the composition, a way to account for both these facts could be to view Part I, with its characteristics, in function of Part II. That is, the true meaning of Part I can only be discovered when we consider Part I as finding its explanation and evaluation in Part II. The two parts must be looked as being complementary to each other.[95]

Castellino describes what Menakhem Perry would call a primacy and a recency effect produced by the differences in the two halves. According to this reading, although the reader is lead toward a critical and negative evaluation of humanity and its activities in Part I, that evaluation is revised toward a more orthodox appraisal in Part II. While I do not share this simplistic characterization of the spirit of the two halves, I do think that Castellino has correctly observed that the narratee/reader is given a pronounced role in the book from 4.17 onwards. In Part I, the narratee is implied, whereas in Part II, the narratee is addressed. What starts out as a soliloquy turns into a monologue. Qoheleth now begins to explicitly include the narratee/reader in his circle of intimacy. The narratee is kept at bay until 4.17, functioning as a distant confidant when, suddenly, Qoheleth turns to gaze directly into his eyes. From this time on however, the narratee is no longer an external eavesdropper on Qoheleth's internal monologue, but is drawn into the debate, becoming an intimate companion who is invited to strongly consider the ramifications of Qoheleth's argument. Narrative distance

95. Castellino, 'Qoheleth and His Wisdom', pp. 21-22.

thus characterizes narrator–narratee relations in Part I, while narrative intimacy characterizes those relations after 4.17. This functions to turn the text from a treatise into an entreaty. From the implied reader's post of observation, Qoheleth turns from being a philosopher to being a mentor who would instruct him or her on the practical ramifications of life's absurdities. The structure of the book produces a primacy/recency effect which alters the implied reader's relationship to the narrator. In the second half of the book, Qoheleth subtly reaches out his hand to the reader, offering him or her the benefits of his life experiences. The journey from distance to intimacy is an act of trust on Qoheleth's part which engenders a similar response on the part of the implied reader.[96]

In spite of the narrator's idiosyncracies, scepticism and apparent jadedness the narratee and implied reader do experience Qohelet's caring disposition toward them both. This goes a long way toward persuading the implied reader to consider Qoheleth's 'goads' (cf. 12.11). Furthermore, the text's journey from narrative distance to narrative intimacy characterizes Qoheleth as the intimate sceptic.[97] This casts a positive light across Qoheleth's dark visage. Rhetorically, the logos-level of the text is buttressed by the pathos-level of the text which rests squarely upon the narrator–narratee relationship. As most of us intuitively know, we tend to argue more with a friend than an acquaintance. But, when we disagree with them, we are less likely to dismiss their ideas in a wholesale manner because of the relationship that exists. When disagreements exist between friends, intimacy provides the dissenting partner an opportunity to have an audience that would not exist in a less intimate relationship. This is the rhetorical *coup de grâce* that Qoheleth pulls off by drawing the reader into his circle of friendship in the latter part of the book. In view of what Qoheleth will advocate to

96. Marra notes how empirical studies of readers have demonstrated that in disclosing, the narrator hints at his trust of the listener, which is reciprocated by most readers. See Marra, 'The Lifelike "I"', p. 344. By means of Qoheleth's intimate disclosures, the narratee is thereby characterized as a trusted confidant, favorite student or friend. The Epilogist, however, who is also a narratee, seems to be more of a peer or sponsor.

97. I am not the first to call attention to the level of intimacy that Qoheleth creates in the reader. Paterson observed how Qoheleth's use of 'I' turned the book into an 'intimate journal' that had sympathies with modern humanistic thinking. See J. Paterson, 'The Intimate Journal of an Old-Time Humanist', *RL* 19 (1950), pp. 245-54.

the narratee/implied reader, he will need all the rhetorical buttressing he can muster.

b. *Reading with Stephan de Jong: Observing Qoheleth's Observations*

A refinement of Castellino's analysis has been offered by Stephan de Jong. He argues that a principal structuring principle of the book is the alternation between observational complexes written in the first-person ('I saw', 'with my heart I turned to learn', 'I said to myself', 'I examined') and instruction complexes which address a 'you'. De Jong begins with Castellino's observation that there is a change of style between 4.16 and 4.17. He then notes how this pattern recurs throughout the book:

> In 4.17 the style suddenly changes. There follows a text in which various instructions are addressed to the reader. In the course of ch. 5, a complex again appears in which observations predominate (5.9–6.9). This complex is succeeded in turn by another complex composed mainly of instructions (6.10–7.22). The alternation of observation and instruction complexes is found throughout the whole book. This regularity indicates the outlines of a structure.[98]

The book begins and ends with an introduction (1.1) and epilogue (12.9-14) and a motto at 1.2 and 12.8. In between this envelope structure, the book alternates in the following manner: 1.3–4.16 (observation); 4.17–5.8 (instruction); 5.9–6.9 (observation); 6.10–7.22 (instruction); 7.23-29 (observation); 8.1-8 (instruction); 8.9–9.12 (observation); 9.13–12.7 (instruction). Although de Jong admits that instructions and observations frequently cohabit the same complexes, he stresses that 'what matters, however, is the density of these types of texts…this characteristic is also responsible for the fact that the borders between the complexes are not always as clear as one would wish'.[99] The utility of this reading strategy seems apparent. There is a very large difference between a text that centers on a self or a narrative 'I' and a text which functions as an address to a 'you'. One is inward looking while the other is outwardly focused. The caution not to expect total consistency is also appropriate. Anyone familiar with first-person discourse types knows that there is a tendency to ramble and muse a little.

98. S. de Jong, 'A Book of Labour: The Structuring Principles and the Main Theme of the Book of Qohelet', *JSOT* 54 (1992), pp. 107-16 (108).

99. De Jong, 'A Book of Labour', p. 109.

De Jong also notes that the tenor of the observational complexes is generally pessimistic while the tone of the instructional complexes tends to be more positive.[100] In addition, it is observed how the word *hebel* receives a different treatment in the two complexes. The term *hebel* occurs 38 times in the book: eight times in the frame-texts of 1.2 and 12.8 and 30 times in the body between 1.3–12.7, with 23 occurring in the observation complexes. In the observation complexes, *hebel* is used almost exclusively as a concluding remark, while in the instruction complexes, it usually marks the beginning premise of an argument or piece of advice (only 7.6 marks a deviation from this pattern).[101] This pattern suggests that the two types of complexes do different things to the reader insofar as they structure the reader's response. Generally speaking, the observation complexes establish the premises upon which the instructions will be based.

Qoheleth's positive advice is then an outcome of his predominantly pessimistic outlook. The fact that Qoheleth's call to joy and other positive admonitions are based on the premises laid down by his *hebel*-dominated observations should lay to rest the debate about whether Qoheleth was an optimist or a pessimist. He was a sceptic who simply knew how to make the best of an otherwise bleak situation.[102] Rhetorically, the reader wrestles with the ethos of a man who sees everything as one big 'zero', yet who ironically is able to find a plus, a *ḥēleq* or 'portion', out of that existential morass. Narratively, what is at stake here is the final *Gestalt* the reader forms as he or she characterizes the narrator, Qoheleth.

100. De Jong, 'A Book of Labour', p. 109.

101. De Jong, 'A Book of Labour', p. 110.

102. Johnston, however, argues that Qoheleth's intentionality, that is, his basic mind set, is optimistic while his literary intent is sceptical. See Johnston, ' "Confessions of a Workaholic" ', p. 14. Given the observations brought to bear upon the reading of the text by de Jong, I would argue that the opposite is in fact the case. The dominance of *hebel* in Qoheleth's observations and the fact that this forms the basic premise for all of his instructions proves that the narrator's worldview is sceptical, while his literary intent endeavors to put a positive spin on that negativity. The characterization of Qoheleth as a sceptic will therefore be taken as the most reasonable final *Gestalt* by this study. I therefore agree with the estimation of scholars like Murphy and Crenshaw who argue against the more optimistic characterizations offered by readers like Johnston, N. Lohfink, R.N. Whybray and A. Caneday. See R. Murphy, 'Qoheleth and Theology?', *BTB* 21 (1991), pp. 30-33 (32).

Of more importance for this study is how this alternation of complexes affects narrator–narratee relations for the reader. These two types of complexes have an effect on the reader's characterization of both the narrator and narratee. Predominantly, the observation complexes will configure the narrator for the reader, while the instruction complexes will be the major guide for understanding the narratee. The reader critic must ponder what it does to a reader when the narratee-configuring sections (instructions) tend to present the narrator in a more positive light, while in the observational complexes we meet a narrator whose musings tend toward the sceptical.

The structuring of positive and negative character-building complexes takes an ironic shape in the text's structure. Strangely, six of the seven enjoyment texts (2.24-26; 3.12-13, 22; 5.17-19; 8.15; 9.7-10) in the observation complexes are found to abound with pessimism. Only the final call to enjoyment in 11.7-10 is found in an instruction complex, probably due to the fact that it is a reaction to the earlier enjoyment texts.[103] The narrative presentation of Qoheleth's dark worldview is interspersed with texts that portray a man who desperately struggled to find the good in God's flawed creation. This has a balancing function in the narrative. The observation complexes lead the reader to form negative characterizations only to have them revised by these intermittent calls to enjoyment. There is a constant interplay between the primacy and the recency effects in these complexes. Again, the reader notes the trademark polarizing structure of the narrative at hand. The net result of this alternation between blatant pessimism and muted optimism at so many intertwined levels is a narrator who is not easily given a final characterization by the reader. Qoheleth is an ambiguous figure whose personality defies closure into a nice, neat *Gestalt*. As a dark character who ironically retains an aura of light about his psyche, Qoheleth remains an enigma to the reader. Perhaps it is this lingering sense of the enigmatic which, over and above anything else, even scepticism or optimism, characterizes the narrator of the book of Ecclesiastes.

c. *Reading with Addison Wright: A Text Riddled with Refrains*

The New Critical approach first brought to bear upon the text by Castellino is resumed and refined by Addison Wright's now classic study 'The Riddle of the Sphinx'. However, Wright breaks sharply with Castellino

103. De Jong, 'A Book on Labour', p. 110.

in a number of strategic ways. He argues that there is no major division at 4.17 because the positive advice offered by Qoheleth in 5.17-19 has already been given in 2.24 and 3.12-13, while the negative appraisal of life in 1.1–4.16 is continued in 5.12–6.9.[104] He therefore concludes that in this analysis, the 'plan does not match the thought'.[105] Nevertheless, Castellino's article alerted Wright to the importance of the refrains and repetitions in the book for understanding its structure.

Wright's methodology looks 'for repetitions of vocabulary and of grammatical forms and seeks to recover whatever literary devices involving repetition the author used, such as inclusions, mots crochets, anaphora, chiasm, symmetry, refrains, announcement of topic, resumptions, recapitulations, etc.'[106] He argues for a bifid structure which breaks the book into two equal parts separated by a median cleft. Part I, 'Qoheleth's Investigation of Life', extends from the beginning of the book to verse 6.9, while Part II, 'Qoheleth's Conclusions', extends from 6.10 to the end of the book. Wright's analysis actually begins with 1.12, which he understands as the actual point of departure for the book. The initial title (1.1) and poem on toil (1.2-11) stand outside the basic structure of the book, as do the poem on youth and old age in 11.7–12.8 and the epilogue in 12.9-14. The book's primary structure begins with a double introduction in 1.12-15 and 1.16-18. It is then observed how each of the first four sections (1.12-15, 16-18; 2.1-11, 12-17) ends with the refrain 'all is vanity and a chase after wind'. Wright follows the lead of this refrain, allowing the phrase to mark off four subsequent units (2.18-26; 3.1-15; 4.7-9; 5.12–6.9). In these four sections Qoheleth evaluates the results of one's toil, with 24 of the book's 38 occurrences of the root *'ml* occurring here. These stylistic and thematic considerations alert the reader that the subject of 2.18–6.9 concerns human effort.

Part II begins with the introduction in 6.10-12, 'who knows what is good for man'. Refrains of scepticism begin to dominate the argument, with the phrase in 8.7, 'he does not know what is to be, for who can tell him how it will be?', becoming the critical clue for reading these chapters. Chapters 7 and 8 develop the phrase, 'not find/who can find?' Chapters 9 and 10 emphasize the phrases 'do not know' and 'no knowledge'. If the reader allows these refrains to structure the text, the following sections are identified: 7.1-14, 15-24, 25-29; 8.1-17; 9.1-12; 9.13–10.15;

104. A. Wright, 'The Riddle of the Sphinx', p. 320.
105. A. Wright, 'The Riddle of the Sphinx', p. 320.
106. A. Wright, 'The Riddle of the Sphinx', p. 318.

10.16–11.2; 11.3-6.[107] The triple repetition of the phrase 'not find out' in 8.17 serves as a major division marker. As a result, the second half of the book also exhibits a bifid structure. 7.1–8.17 focuses on the theme 'humanity can not find out what is good to do', while 9.1–11.6 centers on the theme 'humanity does not know what will come after them'. Wright summarizes the overall structure of the book:

> There is the eight-fold repetition in 1.12–6.9 of 'vanity and a chase after wind', marking off eight meaningful units which contain eight major observations from Qoheleth's investigation of life, plus digressionary material. A secondary motif runs through the sections on toil (the only thing that he can find that is good for man to do is enjoy the fruit of his toil), and at the end even this is shown to have limitations. Where this pattern ceases in 6.9 there follows immediately the introduction of two new ideas: man does not know what is good to do nor what comes after him; and another verbal pattern begins. The first idea is developed in four sections in 7.1–8.17. The end of each unit is marked by the verb 'find out' and the final section ends with a triple 'cannot find out' (8.17) in an *a b a* arrangement... The second idea is developed in six sections in 9.1–11.6. The end of each unit is marked with 'do not know' or 'no knowledge' and the final section again ends with a triple 'you do not know' (11.5-6) and again in an *a b a* arrangement... When this pattern ends we are right at the beginning of the generally recognized unit on youth and old age at the end of the book.[108]

This analysis has been widely accepted by the book's critical readership. The fact that so many readers have seen the validity of his analysis gives this very insightful reading at least a claim to being intersubjectively verified.[109] Furthermore, Wright's subsequent articles have strongly augmented the force of his initial argument.[110] Though sometimes his analysis seems contrived, perhaps even bordering on the Procrustean, the cumulative effect of his 'trilogy' and the simplicity of applying his overall analysis of the book does convince me that Wright must be

107. Originally, Wright posited an analysis that kept vv. 1-6, 7-10 and 11-12 of ch. 9 as separate units. Based on critiques of his analysis, he revised this analysis. See A. Wright, 'The Riddle of the Sphinx Revisited', pp. 38-51.

108. A. Wright, 'The Riddle of the Sphinx', p. 323.

109. A. Wright, 'The Riddle of the Sphinx Revisited', pp. 35-51. Others who follow his lead are R. Johnson, R. Murphy, A. Schoors, J.S. Mulder, R. Rendtorff and S. Brown.

110. See A. Wright, 'The Riddle of the Sphinx Revisited', pp. 38-51, and A. Wright, 'Additional Numerical Patterns in Qoheleth', *CBQ* 45 (1983), pp. 32-43.

substantially correct. While that does not preclude other analyses from offering helpful insights into the kaleidoscopic structure of Ecclesiastes, it does mean that Wright's analysis should at least inform the foundational level of our understanding of the book's structure. As a result, this study will accept Wright's analysis as the foundational structure of the book, and will supplement it with insights from other compatible studies as the relevance arises.

d. *Reading with Stephen Brown: Noting How Reading Grids Affect Interpretation*

The work of Stephen Brown builds directly upon Wright's foundation and is evidence of how his analysis forms a reading grid for many readers. In addition to the use of refrains, Brown's study collaborates Wright's proposed structure by 'focusing on clusters of words and ideas at parallel positions in adjoining and complementary passages'.[111] Brown accents the importance of the seven exhortations to joy and that the book contains four chiastic quarter-sections centered around two cores at 3.1-22 and 9.1-12. With Wright, he sees a definite bifid structure to the book. The 'highly structured parallels between halves and quarters of the book…add further confirmation to the strict delimitation of paragraphs following the scheme of A.G. Wright'.[112] In each half, the central teachings can be found in the middle verse (3.12; 9.7). Furthermore, the middle verse of each quarter serves as a thematic center for those passages (2.10-11; 5.2-3; 7.25-26; 10.17-18). He concludes: 'what is true of each chiasmus or quarter section is applicable to the structure of the whole book. The centre of each half represents the central message of each half and is not fully applied until the end of a half or the end of the book.'[113] Those central messages are the futility of humanity's labor in the first half and the inscrutability of God's work in the last six chapters. The significance of Brown's analysis lies not only in its own insights, and how it functions to intersubjectively validate Wright's proposal, but also in its testimony to the pervasive influence of 'reading grids'. Brown's very insightful study is noted in order to emphasize that all texts are read within a framework of critical tradition, and that no interpretation, including this one, operates without them. One of the critical functions of a reader-oriented approach is to

111. S. Brown, 'The Structure of Ecclesiastes', *ERT* 14 (1990), pp. 195-208 (196).
112. S. Brown, 'The Structure of Ecclesiastes', p. 207.
113. S. Brown, 'The Structure of Ecclesiastes', p. 208.

make us as readers aware of how reading grids influence our experience with the texts. If a reader-oriented approach did nothing else but introduce this level of honesty and awareness into our reading of texts, it would still have a salient contribution to make. Perhaps some of the ambiguity that we as readers experience in the book of Ecclesiastes may be attributable to the conflict of reading grids we all share. In some instances, I imagine that the confusion lies not so much within the text, but within the reader as well.

e. *Reading with R.N. Whybray and François Rousseau: Listening for the Cascade of the Narratee*

Although Wright's analysis forms the foundational reading grid for my analysis of first-person discourse in Ecclesiastes, given the kaleidoscopic nature of Qoheleth's discourse, I must also heed the admonition of Stephan de Jong to use more than one reading strategy.[114] Several ancillary studies have influenced my reading of Ecclesiastes. Most notably, the studies by R.N. Whybray and François Rousseau have offered cogent insights into how readers respond to the linear progression of the text. Both of these authors advocate a final *Gestalt* for the text which differs from my own in that they posit an optimistic reading strategy for the book. However, their insights offer excellent studies of how the use of refrains influences the reading of the text. In addition, Whybray's study suggests some very cogent insights into how the text structures narrator-narratee relations.

Rousseau analyzes the prologue of Ecclesiastes in order comprehend the plan of the entire book. He finds in 1.4-11 a 'jumelage' or twinning of stichoi within the cycles of the prologue. In the prologue, various levels of parallelism are detected: 'parallelism within a stich, parallelism between stichs two by two, and parallelism between subgroups of stichs, that occur on both a primary (αα′, ββ′, γγ′) and a secondary plane (ABCB′A′)'.[115] Rousseau then argues that 'this compositional technique will aid us in better understanding the general structure of the book of Qohelet'.[116] This principle combined with the observance of the sevenfold refrain to enjoy life serves to structure the rest of the book for the implied reader. According to this reading, the call to

114. De Jong, 'A Book of Labour', p. 108.

115. F. Rousseau, 'Structure de Qohelet I 4-11 et Plan du Livre', *VT* 31 (1981), pp. 200-17 (209).

116. Rousseau, 'Structure de Qohelet', p. 209.

enjoyment provides the major structuring signal for the implied reader, dividing the text into seven parts, aside from the prologue and epilogue. He divides the book accordingly:

A.	I.	Solomon's 'confession' (1.12–2.26)
B.	II.	The sage is ignorant of God's plan in general (3.1-13)
	III.	The sage is ignorant of what will come after death (3.14-22)
C.	IV.	Various deceptions and exhortations (4.1–5.19)
	V.	Various deceptions and exhortations (6.1–8.15)
B′.	VI.	The weakness of the sage (8.15–9.10)
C′.	VII.	Deceptions and exhortations (9.11–11.10)[117]

Rousseau's analysis demonstrates how emphatically this refrain functions for many readers. The call to enjoyment halts the narrative progression of Qoheleth's presentation at key junctures in his argument, effectively functioning as a reading interlude for the implied reader. Undoubtedly, it softens the pessimistic blows which pummel the reader's consciousness. More strategically, the redundancy of the refrain trains the model reader to modify the final *Gestalt* he or she makes of each sub-section. One might therefore describe its function as an 'iterative recency effect' that modifies the implied reader's estimation of Qoheleth's advice. However, one should not take this recency effect too far, as many readers such as Whybray, Rousseau, Lohfink and others have done. It is true that the phrase ends each major sub-section in the book. Nevertheless, the fact that Qoheleth resumes his pessimistic tirade after each occurrence also trains the reader to cancel out some of the effect of the refrain. Even the last call to enjoyment in 11.9 is modified by the rather depressing poem on old age and death in ch. 12. Given this pattern, it would be wiser to say that the refrain functions more as a caveat to than a cancellation of Qoheleth's overall worldview. Nevertheless, it does break up the logical progression of the text for the reader, training the reader to stop and modify the *Gestalt* that is forming in his or her mind. In that regard, the call to enjoyment has a definite structuring function for the implied reader.

These verses also have a specific function vis-à-vis the narratee as well. Whybray has observed that these seven passages (2.24; 3.12, 22; 5.17; 8.15; 9.7-9; 11.7–12.1), in which Qoheleth recommends the whole-hearted pursuit of enjoyment, 'are arranged in such a way as to

117. Rousseau, 'Structure de Qohelet', p. 213.

state their theme with steadily increasing in emphasis'.[118] The first occurrence of this leitmotif (2.24a) is a plain statement, unadorned and without a direct relationship to the narratee. The second and third occurrences (3.12, 22a) have an attached asseverative phrase ('So I realized that', *yāda'tî kî* and *wᵉrā'îtî kî* respectively) which intensifies the first-person, confessional nature of these injunctions. The fourth occurrence in 5.17 has a more solemn introduction ('take note of what I have discovered', *hinnēh 'ᵃšer-rā'îtî*), which further intensifies the confessional nature of this refrain. The use of *hinnēh* in this verse most certainly marks this passage as one which addresses the narratee. It changes the focalization of the text to include the purveyance of its recipient, functioning as a sort of implied command addressed to the narratee to see things through the eyes of the observer.[119] Indeed, it almost functions as an imperative. The fifth occurrence (8.15a) continues this crescendo effect, expressing his advice to the narratee in more 'decided terms' ('So I praise joy'). The address to the narratee becomes explicit in the sixth occurrence (9.7-9) where the imperative mood is used. The cascade of the narratee reaches its zenith in the last occurrence of this refrain (11.9a, 10a; 12.1a) where the imperative mood is again utilized. At the end of this series, the narratee/implied reader is addressed in the most explicit of terms. The young man mentioned in 11.9 and 12.1 is clearly the narratee who is listening to the entire discourse. Certainly, this refrain lies at the very core of narrator–narratee relations in the book, providing the implied reader with a textualized role-model for

118. R. Whybray, 'Qoheleth, Preacher of Joy', *JSOT* 23 (1982), pp. 87-98.

119. T.A. Perry, *Dialogues with Kohelet*, p. 63. He notes that *rᵉ'ēh* ('see!') and *hinnēh* ('behold') are point of view shifters. Perry calls particles like *hinnēh*, *gām*, *waw*, *'ᵃnî*, *kî*, and *rᵉ'ēh* 'dialogic markers' (pp. 190-97). The particle *hinnēh* clearly functions in a similar manner to *rᵉ'ēh*. In addition, B. Isaksson calls attention to the focalization properties of this particle. He states: 'Thus *'āmartî* is realized on the *nunc* level "then I say", not in the past...in the [autobiographical] thread the author does not simply relate thoughts had in the past, but speaks out of his present condition of mind, even though as a true sage he refers to observations he has made... Thus *rā'â*, *yēš* and *hinnēh* are markers for present focalization.' See Isaksson, 'The Autobiographical Thread', p. 45. The term *hinnêh* changes the narration from the level of the younger, experience-seeking Qoheleth (*tunc* level) to that of the older, reminiscing Sage (*nunc* level). Such oscillation between the 'then' and the 'now' of the narrator is characteristic of first-person narration. Most first-person discourses alternate between the experiencing 'I' and the narrating 'I' of the person who looks back upon the experiences of the earlier self. Qoheleth is typical in that regard.

responding to the book's overall thrust. In this regard, the call to enjoyment structures not only the text, but also the reader's response. Its significance follows not only from what it specifically says or advises, but also from the various ways that it shapes or advises the reader's overall response to Qoheleth-the-narrator by providing an addressee in the text to emulate.

5. *Summary: A Textuality Characterized by Ambiguity*

This chapter has summarized the reader problems encountered at the textual level. Qoheleth's narration is presented by the implied author via a very definite rhetoric of ambiguity which is evident not only at the linguistic level, but also characterizes the structure of the text as well. At every turn, the reader must learn to cope with strategies of indirection and reading options that in the end, render tenuous *Gestalten*. He or she must read—revise—read—revise in a constantly spiraling fashion. As a text, what strikes the reader most about the book's use of textuality is the pervasive utilization of irony, paradox and, above all, ambiguity. This use of indirection has a performative function in the discourse. The illocutionary force of the implied author's use of various gapping techniques creates in the reader a sense of life's penchant for ambiguity and absurdity. As a result, Qoheleth's discourse not only has meaning, or locutionary force, but through the use of a rhetoric of ambiguity it possesses illocutionary force in that it recreates through literary indirection the implied author's own experience of *hebel* by denying the reader any sure *Gestalten* regarding the book's various literary features.

Nevertheless, the text's penchant for ambiguity does not preclude its structuring as a literary text. Given the difficulty that readers have had discerning its structure, one might very aptly describe it as a literary Rubik's Cube. Nevertheless, the book of Ecclesiastes shows evidence of a certain structuring by the implied author, though it tests the literary competency of most Western readers. With Castellino, Brown, Wright, de Jong, Rousseau, Whybray and others, this study will proceed by paying close attention to the refrains and keywords which naturally structure the text. While some readers, such as Michael Fox, still refuse to see any type of overall structure here,[120] my survey of readers suggests

120. Fox, *Qoheleth and his Contradictions*, p. 162.

that most recognize an overall design. The book of Ecclesiastes is very much like a mosaic, or perhaps even a lithographic picture that one finds in most newspapers and magazines. If one stands too close to a mosaic or blows up a picture on the front page of a newspaper, one sees that they consist of a series of unconnected dots. Only when the mosaic or newspaper picture is viewed from a distance, by taking a step back, does the picture emerge. In a similar fashion, if seen up close, the refrains utilized by Qoheleth appear as a mass of dots or tiles that stand unconnected to each other. However, when one stands back, as Wright and Castellino have done, the big picture emerges, and one can see the structure for what it is—a loose series of dots that are masterfully positioned in such a way that the longer one looks, and the more one reflects on them, a definite image or structure emerges.

In fact, if one gazes long enough, what emerges is the rhetorical face of the narrator and also the implied author who created this persona. Qoheleth's refrains and phrases have two additional functions: they characterize the 'I' of the narrator and address the narratee as well.[121] The doorway to understanding the persuasive properties of Ecclesiastes' use of first-person discourse is found in the narrator's repetitive and almost hypnotic use of refrains and keywords. All the major issues that concern first-person discourse are found here: characterization issues for the narrator, narrator relations with the narratee and by extension, the implied reader and also the characterization of the narratee. These refrains and keywords not only give the book a certain structure, but also, help shape and influence the reader's rhetorical response to Qoheleth by characterizing the narrator and addressing the narratee/implied reader through the overarching refrain to enjoy life. Through the narratee, the implied reader instinctively senses that the refrain to enjoyment is a direct address to his or her own existential situation.

Again, we see that the implied author has very subtly created a relationship with the implied reader which has a definite quality of intimacy about it, going to great lengths to present Qoheleth to the reader as a mentor and trusted guide. The importance of the enjoyment theme lies not just in its role as a balancing corrective to the negativity that permeates the book of Ecclesiastes. It has an even greater role as an intimate address to the reader. The refrain to enjoyment engenders a feeling of caring and openness between the narrator and the implied

121. Shank, 'Qoheleth's World and Lifeview', pp. 66-72.

reader. This effect builds a sense of trust, creating the sort of relationship that will bolster Qoheleth's rhetorical position. Qoheleth may be a rambling, musing and jaded sceptic who speaks with an Aramaic accent and a profound love for the ambiguous, but in the end, he is an honest and empathetic soul. I sense in the narrative presentation of Qoheleth a rhetorical persona who came to understand something of Ricoeur's 'second naivété'. On the other side of his own desert, Qoheleth may not have found faith in the classic sense, but he did find value in living, which he wanted to pass on to the next generation.

However, it is my thesis that the major rhetorical strengths and weaknesses of the book are not to be found at the textual, structural, or linguistic levels of the text, but at the persona level, in the book's audacious use of first-person discourse by the implied author. This does not diminish the effect that the structural and linguistic problems have on the reader. The structural and linguistic problems of the text have a powerful influence on the reading of the book simply because they are the first thing that a reader must deal with during the linear progression of the text as a text. In fact, such problems do tend to characterize the narrator in an indirect fashion. One only has to remember Malraux's dictum, that 'men are distinguishable as much by the forms their memories take as by their characters',[122] to perceive the tremendous effect that the form of a first-person discourse has on the reader's characterization of the main protagonist. In Ecclesiastes' case, the ambiguous nature of the book's structure certainly increases the sense of mystery that accompanies the narrator. It is my contention, however, that throughout the reading history of the book, readers have typically reacted more to Qoheleth-the-persona than to Ecclesiastes-the-text. The effects of the ambiguous structure of the text and its linguistic properties pale in comparison to the significance which the specific ethos-related qualities of the narrator as a rhetorical persona hold for the reader. To those issues I must now turn.

122. Quoted by Hart, 'Notes on the Anatomy', p. 498.

Chapter 4

The Epistemological Spiral: The Ironic Use of Public and Private Knowledge in the Narrative Presentation of Qoheleth

> When a man assumes a public trust, he should consider himself as public property.[1]

1. *Overview of Persona Problems in the Book of Ecclesiastes*

This chapter will provide an overview of those problems which relate to how the implied author utilizes the effects of first-person discourse to form an impression of Qoheleth's persona in the reader. The major issues pertain to: (1) Qoheleth's use of first-person discourse and its relationship to autobiography; (2) the nature and effects of the Solomonic/Royal Fiction; (3) the specific ways that readers construct a sense of ethos from a literary character; (4) how first-person discourse affects the model reader's understanding of Qoheleth's use of quotations; (5) narrational techniques utilized in the book; and (6) the implied author's rhetorical use of private and public knowledge. As in the previous chapter, it will recalibrate traditional historical-critical issues for use in a reader-oriented perspective.

2. *The Death of Ecclesiastes: Qoheleth as Fictional Persona*

If a literary scholar wants to understand how the past interpretative community understood Qoheleth-the-narrator or Ecclesiastes-the-implied author, he or she must search under historically-minded headings

1. Thomas Jefferson, ('Winter in Washington, 1807'), in a conversation with Baron Humbold, from B.L. Rayner, *Life of Jefferson with Selections from the Most Valuable Portions of His Voluminous and Unrivalled Private Correspondence* (Boston, MA: Lilly, Wait, Colman & Holden, 1834), p. 356.

such as 'Understanding Qoheleth the "Person"' or similar. The textually inscribed values which constitute the narrator and the implied author of the book of Ecclesiastes have usually been explained without recourse to either reader-oriented terminology or conceptualizations. Given the historical interests of the past 200 years, scholars have typically sought to locate the literary problems experienced by readers in either the author's personality (Robert Gordis, J.A. Loader and Frank Zimmermann[2]), his historical location, usually being the supposed Hellenistic 'crisis' (Martin Hengel,[3] and many others), or its twin sister, the religious/intellectual 'crisis' among the sages (Hartmut Gese and Otto Kaiser,[4] among others), his sociological location (Frank Crüsemann[5]), or his intellectual location (R.N. Whybray[6]). Recent sociological efforts to understand the values implied in the book have sought to explain them by grounding the author's worldview and consequent values in the social anomie brought about by Ptolemaic 'depoliticization' of Jerusalem (Mark Sneed[7]), or his middle-class standing vis-à-vis Ptolemaic economics (C. Robert Harrison, Jr, Stephan de Jong and A. Schoors[8]).

2. Gordis, *Koheleth: The Man and His World*; J.A. Loader, 'Different Reactions of Job and Qoheleth to the Doctrine of Retribution (Eccl 7.15-20; Prov 10-22)', in Wyk (ed.), *Studies in Wisdom Literature*, pp. 43-48; F. Zimmermann, *The Inner World of Qoheleth* (New York: Ktav, 1973).

3. M. Hengel, *Judaism and Hellenism* (trans. J. Bowden; 2 vols.; Philadelphia: Fortress Press, 1974 [1973]), esp. pp. 115-28.

4. H. Gese, 'The Crisis of Wisdom in Koheleth', in J. Crenshaw (ed.), *Theodicy in the Old Testament* (trans. L. Grabbe; Philadelphia: Fortress Press, 1983 [1963]), pp. 141-53; reprinted from *Les Sagesses du Proche-Orient ancien: Colloque de Strasborg, 17-19 mai 1962* (Paris: Presses Universitaires de France, 1963), pp. 139-51; O. Kaiser, 'Fate, Suffering and God: The Crisis of a Belief in a Moral World Order in the Book of Ecclesiastes', *OTE* 4 (1986), pp. 1-13.

5. Crüsemann, 'The Unchangeable World'.

6. R.N. Whybray, *The Intellectual Tradition in the Old Testament* (BZAW, 135; Berlin: W. de Gruyter, 1974).

7. M. Sneed, 'The Social Location of Qoheleth's Thought: Anomie and Alienation in Ptolemaic Jerusalem (Israel)' (PhD dissertation; Madison, NJ: Drew University, 1990).

8. C.R. Harrison, Jr, 'Qoheleth in Social-Historical Perspective' (PhD dissertation; Durham, NC: Duke University, 1991). See also his excellent summary of the various sociological interpretations for the book: 'Qoheleth Among the Sociologists', *BibInt* 5 (1997), pp. 160-80. He classifies Qoheleth's sociology of knowledge as a 'sociology of uncertainty' (179); S. de Jong, 'Qohelet and the Ambitious Spirit

This, however, is not the same as investigating Qoheleth's persona as a literary creation. As a persona, Qoheleth is a mask that the implied author slips over the narrator as a technique of presentation in order to bring that role to life for the reader. Rüdiger Lux observes how:

> When the author of the book of Kohelet presents himself in the sentence: 'I, Koheleth, am King over Israel in Jerusalem', then this sounds so superficial, as he makes it known to his readers. In reality, however, the self-presentation is a mask, which he holds before his face.[9]

Lux traces his concept of a fictional mask to Hans Müller's intuitive reading of 1.12, who translates the verse as 'Ich, Kohelet, bin (*hiermit*) König über Israel'—'I, Qoheleth am (*herewith*) King over Israel'.[10] Furthermore, it should be stressed that this is neither a recent nor a novel reading. Readers before Lux have also seen a mask here; indeed, Franz Delitzsch once observed: 'In the book, Koheleth-Solomon speaks, whose mask the author puts on: here, he speaks, letting the mask fall off, of Qoheleth'.[11] Milton Terry claimed that Qoheleth 'impersonated' Solomon.[12] More recently, Alexander Fischer has referred to the 'mantel' in which Qoheleth clothes himself in order to address the problem of human striving for profit:

> [in] the existing Kings Fiction...Koheleth wraps himself in the mantel of respect and surpassing Wisdom above all his predecessor kings of Israel, in order to debate in the role of an exemplary wiseman the question of the profit of human striving.[13]

Likewise, Peter Höffken is another critical reader who adopts a fictive reading for Qoheleth:

> a fictive 'I' speaks now in interesting ways, who presents himself especially in the role of Solomon and takes over this role: 1.11–2.11. It appears consequently, that the author (or redactor) felt compelled to clothe his identity, so as to conceal his 'I' under that of Solomon...[14]

of the Ptolemaic Period', *JSOT* 61 (1994), pp. 85-96; A. Schoors, 'Qoheleth: A Book in a Changing Society', *OTE* 9 (1996), pp. 68-87.

9. Lux, '"Ich, Kohelet, bin König..."', p. 335.

10. H. Müller, 'Theonome, Skepsis und Lebensfreude: Zu Koh 1,12–3,15', *BZ* 30 (1986), pp. 1-19 (3).

11. Delitzsch, *Commentary*, p. 430.

12. M. Terry, 'Studies in Koheleth', *MR* 70 (1988), pp. 365-75 (365).

13. Fischer, 'Beobachtungen zur Komposition', pp. 72-86 (72).

14. P. Höffken, 'Das Ego des Weisen', *TZ* 4 (1985), pp. 121-35 (126).

Other literarily trained scholars like Oswald Loretz, Diethelm Michel and Eric Christianson also take a similar position.[15] Kathleen Farmer argues that Qoheleth is 'playing a role in order to argue a point'.[16] Given this list of impressive readers who see a poetic persona in Qoheleth, I observe that the fictive interpretation of Qoheleth has a large following among those trained both in literary and historical methods. This provides the fictional reading of Qoheleth with a fair amount of inter-subjective verification. The use of a persona or mask allows the implied author to fully exploit the rhetorical strengths of first-person discourse. A fictional mask provides two major benefits for both the narrator and the implied author. Rüdiger Lux states:

> Every mask provides two things. It offers their bearer the possibility to hide behind it. Simultaneously, it offers the chance to appear in the figure of another. It effaces identity and complicates identification. Above all, in the disclosure of fictional texts, signals meet us which have the function of a mask.[17]

In Ecclesiastes, the use of a mask hides the narrator behind the persona of a King/Solomon and gives him the ethos of that chosen character. At that point, intertextuality dynamics begin to influence the reader's response.

Obviously, there must be clues in the text which communicate that a character is performing a fictional role. The major reading clue in the book of Ecclesiastes is the name of the protagonist. Noting that 'Qoheleth' is not a proper name at all, Lux interprets this as a fictive signal to the reader, concluding that the ensuing narrative is not oriented toward reality. He states:

> This signal of the fictional illuminates so powerfully, if we consider, that the noun Qohelet which stands here as a proper name is really not a proper name at all, but rather, it could be a designation of function. The Qal feminine participle of *qhlt* can better be accounted for as a designation of office along with 'director of collection'. What meets us in Koh 1.12 is a kind of role-play, in which the director of collection (*Qohelet*) takes over the role of the King (*melek*).[18]

15. Loretz, *Qohelet und der Alte Orient*, p. 15; Michel, *Untersuchungen zur Eigenart*, p. 81; Christianson, *A Time to Tell*, pp. 128-72.

16. Farmer, *Who Knows What is Good?*, p. 154.

17. Lux, ' "Ich, Kohelet, bin König..." ', p. 335.

18. Lux, ' "Ich, Kohelet, bin König..." ', pp. 335-36.

Again, it can be seen that the implied author's propensity for linguistic ambiguity and literary puzzles. Such clues force the reader to make interpretative guesses. This compels the reader to actively participate in the creation of meaning. Rather than offer the reader a clear and precise identity for the narrator, the implied author offers only vague indicators. This forces upon the reader yet another level of ambiguity which also has no sure final answer. Ecclesiastes is such an ambiguous text that even the identity of the narrator is enshrouded in a cloud of fictional and linguistic obscurity.

As a fictional mask, it must be emphasized that Qoheleth is first and foremost a narrative function which has been fully enfleshed into human form. The fact that Qoheleth-as-narrator has a function which has been camouflaged by its literary characterization does not diminish its role as a narrative function/entity. While I do not doubt for a moment that this character is very much related to some historical person, it still remains that Qoheleth would be Qoheleth even if it could be proven that no such person ever existed.[19] Furthermore, to treat Qoheleth as a real person does not in any way account for the radical effect of textuality on the meaning of this character for the present-day reader of Scripture. Whoever lies behind the text which explains the protagonist's peculiar outlook has been forever distanciated from the literary

19. As Fox has argued so elegantly, this is similar to the situation of Uncle Remus. He states: 'Qohelet may be recognized as a persona even if one regards him as based on a historical character, even as Uncle Remus was based on four Negroes by the author' (Fox, 'Frame-Narrative', p. 48). While I surmise that the implied author had in mind his mentor, there is nothing to disprove that, perhaps, Qoheleth is a composite personality who summed up and represented the class-consciousness of a skeptical group, much the same as 'Christian' represented Renaissance Calvinists for John Bunyon in *Pilgrim's Progress*. However, the constant refrain, 'I searched', does suggest rather strongly that one individual probably lies behind this persona. Still, it should also be noted that from a phenomenological point of view, characterization, whether it pertains to oneself, another person, or a fictive persona, always includes a perceptive grid. Even if the implied author is presenting another person to his audience, that presentation has been filtered through a mental process which characterizes him in the same manner that a fictive character is portrayed. In that respect, there are few differences between a fictive rendering of an individual, and a true autobiographical rendition. As Renza has so poignantly argued, there is always a fair amount of 'fiction' or imaginative enhancement in most autobiographical sketches. See L. Renza, 'The Veto of the Imagination: A Theory of Autobiography', *NLH* 9 (1977), pp. 1-26.

character via the effect of textuality. Qoheleth's narrative role is to present the values to the implied reader which the implied author wished to communicate. In this case, the implied author has chosen a more personal mode of presentation than, say the 'objective', 'impersonal' or 'gnomic' approach of Proverbs 10–29 in order to more fully accomplish his rhetorical purposes.

3. *Qoheleth's Use of Emphatic 'I' and the Monologue*

Qoheleth's use of 'I' is unparalleled in the First Testament. At the beginning of this century, Morris Jastrow observed that the book of Ecclesiastes 'is the only one in which an author speaks of himself by name'.[20] Although Nehemiah and some prophets come close, like Ezek. 1.1 where the prophet conspicuously begins his book with an I-discourse, the other canonical writers never emphasize themselves so blatantly. As a matter of 'style', Qoheleth's discourse is 'more individualized than that of other ancient first-person narratives'.[21] In Ecclesiastes, the speaker quickly identifies himself in an autobiographical-like manner (1.12). As Harold Fisch proclaims: 'Qoheleth could have said with Montaigne, "It is my portrait I draw...I am myself the subject of my book" '.[22] Unlike the 'I' of the Psalms, Qoheleth's 'I' is that of an autonomous subject speaking out of the depths of his soul. He is especially fond of the pleonastic use of *'ᵃnî*. This is the equivalent of saying, 'C'est moi, It's me...'[23] Grammatically, the added use of personal pronouns in classical Hebrew serves to emphasize the subject,[24] and 'gives the sentence an added weight, which may emphasize an emotional expression, an important conclusion, or the introduction of a new

20. M. Jastrow, Jr, *A Gentle Cynic: Being a Translation of the Book of Koheleth Commonly Known as Ecclesiastes Stripped of Later Additions; Also its Origin, Growth and Interpretation* (Philadelphia: Lippincott, 1919), p. 63.

21. Christianson, *A Time to Tell*, p. 35. For an alternative overview of the historical and literary studies pertaining to Qoheleth's use of the autobiographical form, the reader is referred to Christianson's survey (pp. 33-42). Like myself, he too sees Qoheleth's monologue as a fictional autobiography (p. 34 n. 57).

22. H. Fisch, 'Qoheleth: A Hebrew Ironist', in H. Fisch, *Poetry with a Purpose* (ISBL; Bloomington: Indiana University Press, 1988), p. 158.

23. B. Isaksson, 'The Pronouns in Qoheleth', in Isaksson, *Studies in the Language of Qoheleth*, pp. 142-71 (164).

24. See GKC, §135a.

line of thought'.[25] Even when the author does not use *'ᵃnî*, such as when he introduces a situation with *yēš* ('there is'), the phrase often is nothing more than a circumlocution for the first-person pronoun.[26]

During the 1950s, several early studies began to look at the First Testament's use of first-person discourse as a general literary phenomenon. Ernst Dietrich's work marks the first study to attempt a comprehensive understanding of how 'I' functions in a religious discourse, especially those with Wisdom influences.[27] His essay is an historical one, dealing extensively with the emphatic use of 'I' as it relates to the hypostatization of Wisdom. He observes that beginning with Jeremiah, the Wisdom tradition began a process of individualized thinking. This can be seen in the increased use of various first-person genres by certain major writers. Dietrich delimits this however, noting that after Jeremiah, certain Psalms, and the book of Job, 'postexilic Judaism constituted itself as a community, in which the individual was subordinated'.[28] Unfortunately, Dietrich overlooked Qoheleth, as he skipped directly to Sirach in order to track the Wisdom tradition's use of 'I'. However, Qoheleth's use of 'I' also fully utilizes first-person discourse in an emphatic or dramatic manner. By emphasizing Qoheleth's 'I' in such an emphatic way, the implied author has chosen a presentation style for his narrator which places the full weight of the reader's response on that 'I'. The chief effect of the dramatic use of 'I' is to

25. Isaksson, 'The Pronouns in Qoheleth', p. 166. He also notes that for the suffix conjugations, the emphatic function of *'ᵃnî* is likely. In addition, 'the pronoun is added in instances of greater importance where the narrative halts for a moment to make a conclusion or to introduce a new thought' (p. 171). As a result, the use of *'ᵃnî* in Ecclesiastes often serves to communicate a major transition in the discourse, to mark out either a new unit of thought or indicates the conclusion of the unit at hand.

26. Isaksson, 'The Pronouns in Qoheleth', p. 173. The term *yēš* is a way of saying 'I' in a manner that circumvents the subjectivity which goes along with the first-person pronoun. It injects a degree of externality within an internally focalized statement. Since *yēš* frequently introduces examples of what Qoheleth had observed, it really is not a true external focalization from the reader's post of observation, but is merely a way to bring some quasi-objectivity to the narrator's post of observation. In that regard, the use of *yēš* is something of a rhetorical sleight of hand.

27. E. Dietrich, 'Das Religiös-emphatische Ich-Wort bei den Jüdischen Apokalytiken, Weisheitslehren und Rabbinen', *ZRGG* 4 (1952), pp. 289-311.

28. Dietrich, 'Das Religiös-emphatische Ich-Wort', p. 289.

center the reader's response, making the implied reader focus his or her attention solely on the ethos of the narrator, Qoheleth.

During the 1960s the importance of the use of 'I' caught the eye of several scholars. Sigmund Mowinckel analyzed the use of 'I' and 'he' in Ezra. He called attention to the fact that the change of perspective between first-person and third-person narration is quite common in the ancient Near East.[29] Mowinckel's study serves as a reminder that the interplay between first-person discourse and third-person discourse is an important dynamic in any literary reading of a book. Both types of discourse have specific strengths and liabilities. How an implied author chooses to manipulate their powers and weaknesses will have an enormous impact on the reader's response. In the book of Ecclesiastes, the relationship between Qoheleth and the Epilogist is one of the chief textual devices by which the implied author controls the reader's response. Mowinckel's study reminds the critic of this feature of 'I' and 'he'.

The first truly comprehensive study of the *literary* dynamics of saying 'I' is the article by Nikolaus Pan Bratsiotis, 'Der Monolog im Alten Testament'.[30] He tracks the use of the form from the smallest one-word monologues in Genesis to the longer examples, such as Ecclesiastes. Bratsiotis concludes that 'the monologue of the Old Testament knows no particular work…however…a few use the concept, "Monologue" '.[31] In the case of Qoheleth, Bratsiotis concludes that the book is, for the most part, a classic example of the monologue form. He states:

29. S. Mowinckel, ' "Ich" und "Er" in der Ezrageschichte', in A. Kuschke (ed.), *Verbannung und Heimkehr* (Festschrift W. Rudolph; Tubingen: J.C.B. Mohr, 1961), pp. 211-33 (222-33). Typically, 'I' and 'he' function as markers which designate a change of perspective.

30. N. Bratsiotis, 'Der Monolog im Alten Testament', *ZAW* 73 (1961), pp. 30-70. A more recent contribution which discusses the use of soliloquy and free indirect discourse in the narrative sections of the canon is offered by M. Niehoff, 'Do Biblical Characters Talk to Themselves?', *JBL* 111 (1992), pp. 577-95. Niehoff classifies Qoheleth as 'one long soliloquy in which one individual attempts to make sense of life' (p. 579). He then goes on to argue that such contemplative inclinations can also be detected in the characters of earlier biblical narratives. As such, Niehoff continues the line of interpretation first brought forth by Bratsiotis that the book of Ecclesiastes is the *locus classicus* for the monologue/soliloquy genre in the First Testament. Crenshaw too classifies the book as a monologue (*Ecclesiastes*, p. 29). Fisch also has asserted that Qoheleth 'is indeed the nearest the Hebrew Bible gets to pure monologue' (*Poetry with a Purpose*, p. 158).

31. Bratsiotis, 'Der Monolog im Alten Testament', p. 32.

> Another book, in which the monological form is substantially used, is Qoheleth. One could indeed designate the entire book as one large monologue, as interspersed places prove, which again and again introduce a monologue, like, for example: (to my soul I therefore said) 'See, I have made myself great and have acquired Wisdom more than all who reigned before me in Jerusalem...' (1.16) or; (I said to myself) 'Indeed, I wanted to try it with joy...(2.1ff)', where we find the characteristic use of the address to one's own soul.[32]

Bratsiotis divides the First Testament monologue into three distinct classes: exterior, interior and mixed. An exterior monologue exists when:

> the speech is directed outside of one's inner self (to an abstract or even concrete) lifeless object, to a dumb, absent, dead or even to a person available only in one's imagination. Naturally, the object addressed during the monologue should give altogether no answer, thereby the monologue retains its literary form.[33]

The interior monologue exists when the 'the speaking person thereby turns to himself and expresses his thoughts, considerations, or feelings, with or without a self-address'.[34] Obviously, a mixed form contains both to some measure. The most frequent interior monologue is the thought-monologue (Eccl. 1.16; 2.1, 15a, d; 3.17, 18; 7.23). In Ecclesiastes, both the internal and external varieties can be seen. In fact, there is a movement from the internal monologue to the external monologue in this book. The internal monologue dominates the text from 1.12–4.16. In these verses, Qoheleth's narration is presented as the self-ruminations of an aging scholar. However, beginning in 4.17 with Qoheleth's first specific address to his narratee, the discourse shifts to the external monologue. The use of the external monologue enables the implied author to address the implied reader via the narratee a little more directly.

Other types of monologues discussed by Bratsiotis are the narrating monologue, in which the speaker presents his own thoughts, and the motto-monologue whereby a speaker reflects upon a well-known motto or theme from the Wisdom tradition.[35] Further analysis reveals that

32. Bratsiotis, 'Der Monolog im Alten Testament', p. 35.
33. Bratsiotis, 'Der Monolog im Alten Testament', p. 38.
34. Bratsiotis, 'Der Monolog im Alten Testament', p. 38.
35. Bratsiotis, 'Der Monolog im Alten Testament', p. 41.

each of these types may fall into two broad classes: the reporting monologue and the reported monologue.[36] Most monologues have an introduction such as 'I said in my heart' (1.16), though sometimes this is missing or assumed.[37] He concludes, based on the occurrence of the form in prose, poetry and proverbs, that it is a distinct '*genus litterarium*' in its own right.[38]

Bratsiotis has done a major service by locating, in an almost exhaustive manner, the corpus of First Testament monologues. His study is a classic form-critical analysis of the First Testament monologue. Its limitation, like so many form-critical studies of its era, is that there is little emphasis on the rhetorical properties of the monologue.[39] However, he does underscore some of its literary properties, such as the monologue's ability 'to characterize a person inwardly and to emphasize certain characteristics'.[40] In that, the monologue is a form distinguished by its radical individualizing of the person. He sums up the major effect of the monologue as 'perhaps the most excellent literary means by which individualism steps forward'.[41] Again, it should be noted that individualism and subjectivity are the major characteristics of the monologue as an example of first-person discourse. As the longest sustained monologue in the First Testament, Ecclesiastes manifests these properties to a quite remarkable degree. As such, one of the chief effects of Qoheleth's monologue is therefore to create a definite sense of intimacy between the narrator and the reader. For, as Baruch Hochman observes, 'we know much more about people in life... But our knowl-

36. Bratsiotis, 'Der Monolog im Alten Testament', p. 44. Conceptually, Bratsiotis's reporting monologue is similar to what narratologists today would call reporting speech. It is the speech of the narrator speaking his own words. Reported monologue, on the other hand, is like reported speech. Here the discourse of the interlocutor is given utterance under the influence of the narrator or the author. Usually there is some sort of tag clause introducing such a monologue, such as 'he thought' or the like. Uspensky defines reported discourse as 'the author's voice to some degree imitating someone else's voice' (*A Poetics of Composition*, p. 41).

37. Bratsiotis, 'Der Monolog im Alten Testament', p. 46. See also Loader, *Polar Structures*, p. 19.

38. Bratsiotis, 'Der Monolog im Alten Testament', p. 55-56.

39. For a critique of form criticism at precisely this point, see W. Wuellner, 'Where is Rhetorical Criticism Taking Us?', *CBQ* 49 (1987), pp. 448-63.

40. Bratsiotis, 'Der Monolog im Alten Testament', p. 63.

41. Bratsiotis, 'Der Monolog im Alten Testament', p. 70.

edge rarely has the definitiveness that fiction sometimes affords.'[42] Qoheleth's frank and honest monologue characterizes him as the intimate sceptic. After a while, the reader feels at a deep level as if he or she 'knows' the man. There is a depth to Qoheleth that is rarely matched in the Canon. In that respect, Qoheleth thoroughly 'adheres' as an autobiographical figure or character.[43] This seems to be the chief effect of Qoheleth's use of the monologue form.

4. *Qoheleth as Fictive Autobiography: Defamiliarizing the Reader's Life*

So beguiling is the lifelikeness of Qoheleth's character, that many scholars have mistaken the book for an autobiographical tract. The extreme of this reading grid can be seen in the life-synopsis of Qoheleth by E.H. Plumptre:

> By and by the young man travelled, and finally settled at Alexandria. Here he became acquainted with one whom he could call a true friend, 'one among a thousand', but also with a woman for whom he imbibed a passionate affection. Discovering her baseness, he barely had time to escape her net; hence his strong denunciation of the female sex in the passages of his work. At Alexandria Koheleth became also acquainted with the philosophical systems of the Epicureans and Stoics, and the natural science of physiology of the former especially attracted our student.[44]

However, as seen above, this older paradigm is now giving way to a literary model of reading. Recent scholarship has moved toward a fictional model for understanding Qoheleth. There has been a definite trend in the recent literature moving away from the older historical-critical or autobiographical paradigm which approached the book with biographical and historical interests.[45] This has been especially true of Continental scholarship. In contrast, most American studies and commentaries

42. B. Hochman, *Character in Literature* (Ithaca, NY: Cornell University Press, 1985), p. 63.

43. Christianson, *A Time to Tell*, pp. 33-36.

44. Plumptre, as quoted by Pick, 'Ecclesiastes or the Sphinx', p. 363. Christianson has also commented on Plumptre's confusion of historicality with fictionality as an example of the power of Qoheleth's individuality to confuse readers (*A Time to Tell*, pp. 33-34).

45. Michel, *Untersuchungen zur Eigenart*, pp. 78-81. Michel notes that the major issue for readers of Qoheleth's I-Reports are whether these are reflections of a

have merely mentioned the problem in passing, or have opted to summarize Continental scholarship on the subject. A few studies, such as Robert Johnson's study on Qoheleth's use of sayings,[46] have dealt more extensively with the subject, but only in a tangential way as the problem of first-person discourse was not the primary focus of his work. As a result, there are only a handful of studies that have dealt with the rhetorical and literary problem in any comprehensive way. Given the rhetorical exposure that Qoheleth's 'I' is given in the book this situation is lamentable. However, the move toward a more fictive reading model has had a tremendous effect on how readers esteem the book's autobiographical or historical value.[47]

life that has actually been lived or are simply a literary fiction. After reviewing the debate among scholars, he makes his own interesting contribution to the dispute. According to Michel, the verb *rā'â* ('to see') often means 'to consider' in the First Testament. As a result, a verse like 2.24 does not mean that the author actually 'saw' the event being described, but merely that he considered it as an assertion or claim (p. 80). This results in viewing Qoheleth's observations as considerations or reflections on the collective experience of the sages. Consequently, Michel concludes: 'The I-Report in 1.12–2.11 should likewise not "report" real experiences, but rather, show Qoheleth in the assumed role of the wise King Solomon' (p. 81, my translation). This position is an intermediary one. While he acknowledges the fictional characteristics of the narration, he argues that the fiction is based on the real experiences of the sages who comprised the author's social group. A similar position is argued by M. Schubert, 'Die Selbstbetrachtungen Kohelets: Ein Beitrag zur Gattungsforschung', in *Theologische Versuche* 24 (1989), pp. 23-34 (28-29). However, even Michel points out that this position is historically unverifiable (*Untersuchungen zur Eigenart*, p. 20).

46. R.F. Johnson, 'A Form Critical Analysis of the Sayings in the Book of Ecclesiastes' (PhD dissertation; Atlanta, GA: Emory University, 1973). Johnson's dissertation form-critically deals with the various sayings in the book. He observes that the sayings appear exclusively in: (1) relation to first-person reports, paraenesis and commentary and (2) in a series, such as ch. 7. The sayings or quotations found in Qoheleth are generally subordinated to the first-person context in which they are found by Qoheleth's use of comments. Unlike those found in Proverbs, the sayings in Qoheleth serve the purposes of Qoheleth's monologue and have no independent status, even for those which are found in a series, which are often interrupted by Qohelet's 'I' or his subtle comments. Johnson concludes that the primary function of the I-style found in Qoheleth and Proverbs is 'to authorize the sage's right to speak' (p. 254). The first-person report has the rhetorical function of legitimating Qoheleth's right to be heard or read.

47. For autobiographical interpretation of Qoheleth's discourse, see the thorough discussion of the person behind the work found in H. Duesberg and I. Fransen, *Les*

The modern literary debate regarding the autobiographical value of Qoheleth's monologue begins with Oswald Loretz' classic essay, 'Zur Darbietungsform der 'Ich-Erzählung' des Buche Qohelet'.[48] Relying heavily upon the literary theory of W. Kayser, he concludes that Qoheleth is a 'poetic persona':

> Is the I-narrative of Qoheleth autobiography or poetry? We must therefore determine, whether the I-narrative here is a presentation form without direct connection to the personal life of the Poet or even a report of his life. If it should come to light, that an autobiography exists, then it must be asked yet again, to what extent 'Poetry and Truth' are woven into one [fabric]. The generally accepted view, that the Book of Qoheleth reports the personal feelings, viewpoints, and experiences of the man Qoheleth, should be viewed hereby as questionable. It is necessary to examine, whether the usual identification of the 'I' of the book with the personal 'I' of the author can be the starting point of the interpretation of the book. It must also be explored, whether Qoheleth speaks to us as an historical person or as a 'poetic personality'.[49]

Partly because of the insights gained from literary theory, and partly due to the dearth of information regarding the historical author of the book of Ecclesiastes, Loretz takes a stance that is functionally similar to Roland Barthes' concept of the 'death of the author'.[50] Loretz advances the position that:

> As literary studies have pointed out, however, it is dangerous to explain the work of an author by his life here. The presupposition, that the author is identifiable as man and author without further ado, has been proven to be untenable. So it is also necessary in the case of Qoheleth, to strongly

Scribes inspirés: Introduction aux livres sapientaux de la Bible; Proverbs, Job, Ecclesiaste, Sagesse, Ecclesiastique (Paris: Maredsous, 2nd edn, 1966), pp. 537-93. An excellent overview of this line of interpretation is also found in Isaksson, 'The Autobigraphical Thread', pp. 39-68.

48. Loretz, 'Zur Darbietungsform', pp. 46-59.

49. O. Loretz, 'Die Darbietungsform der 'Ich-Erzählung', in Loretz, *Qohelet und der Alte Orient*, pp. xx-xx (48). He refers to the works of W. Kayser, *Das sprachliche Kunstwerk* (Bern: A. Francke, 1961), p. 276 and *idem*, *Die Wahrheit der Dicter: Wandlung eines Begriffes in der deutschen Literatur* (Hamburg: 1961), p. 7.

50. R. Barthes, 'The Death of the Author', in Rice and Waugh (eds.), *Modern Literary Theory*, pp. 114-18. Building upon the insights of narratology, Barthes argues that once a fact is narrated, 'the voice loses its origin, the author enters into his own death, writing begins' (p. 114). Written language swallows up the author upon the publication of a work, creating the 'death of the author'.

> differentiate between the work and its author. The thought of the Book of Qoheleth and the form of its fixation, must be understood as such. First of all, if it could be successful, to bring these into combination with the precisely known life-facts of the author, then the life-history of the artist could serve the explanation of his work. However, because we have no reports concerning Qoheleth outside of his writing, the possibility of a conclusion in the sense of an autobiography is to be refused as a subjective guess. The argumentation with the personality of Qoheleth is a game with a stranger and contributes nothing to the knowledge of the book.[51]

For the most part, Loretz adjudicates this judgment because of the scant historical evidence available for understanding the author. In addition, Loretz notes that for such an ostensibly autobiographical presentation, Qoheleth's presentation style is remarkably full of traditional, stereotypical phrases which seem to belie its origin in a single person.[52] As a result, the critic is offered an alternative approach which uses a literary model to offset these deficiencies. However, the critic does not actually need a paucity of historical information in order to choose such a model. As Paul Ricoeur points out, the primary effect of textuality is to distanciate a text from its author. Even if we did know more about the author, the text would still be distanciated from its original historical matrix. As readers, we can only respond to the literary presentation of Qoheleth. The narrator, much like every scriptural character or persona, exists *only* via the medium of the text. He resides in the reader's mind like every other great literary figure—as a poetic personality. In order to understand how readers respond to Qoheleth, we must first learn to read him as a character and not as a person. This is especially true since no reader has ever responded to Qoheleth-the-person. As a result, the various autobiographical/historical approaches, in particular Frank Zimmermann,[53] must be rejected outright. In that regard, Loretz'

51. Loretz, *Qohelet und der alte Orient*, p. 164.

52. Loretz, 'Zur Darbietungsform', p. 54. This, as Misch has pointed out in his mammoth overview of autobiography in the ancient world, is actually quite characteristic of the autobiographies from the ancient Near East. He classifies the autobiographical works from Egypt and Babylonian-Assyrian cultures as a 'collective kind of autobiography' (*A History of Autobiography*, p. 19). In this, the book of Ecclesiastes undoubtedly partakes of its origin in ancient Near Eastern culture. However, it is also clear that Ecclesiastes is unlike these other tracts in that there is a definite sense of individual character in the presentation of Qoheleth.

53. Zimmermann, *The Inner World of Qoheleth*, represents the extreme of the various autobiographical readings of Qoheleth. Zimmerman's reading, which uses a

literary instincts will serve the modern reader-response critic quite admirably. The modern debate witnesses the death of Qoheleth as a 'person'.[54] In the process, Qoheleth-as-character is born.

Qoheleth thereby becomes a fictional character, and a very powerful one at that. Recent readers, such as Michael Fox, have found the value of reading Qoheleth through the type of literary lens proposed by Loretz.[55] Fox argues that the implied author presents the fictional reality of Qoheleth to the reader through the vehicle of the Epilogist, or frame-narrator. This frame-narrator testifies to

Freudian reading grid, attempts to put Qoheleth on the therapist's couch, seeing a variety of Oedipal complexes and other latent psychological maladjustments. This reading goes quite beyond the requirements of literary characterization, and has not been well received. Polk admonishes readers that any such theory which rests upon the tensions and conflicts of the author 'requires more information about the author than the text or any outside source provides...we should be wary of any approach that makes our knowledge of a given author the key to understanding his work...we must concentrate on the literary work itself' ('The Wisdom of Irony', p. 4).

54. This is not to say that all readers have strictly followed Loretz's lead in this matter. Recently, Schubert has analyzed the book from a form-critical perspective, finding 23 'selbstbetrachtungen' or reflections in the book (cf. 1.13-15, 16-18; 2.1-2, 3-11, 12-14, 15-17, 20-23; 3.1-15 [mixed form], 16-22; 4.1-3 [mixed form], 4-6, 7-12 [mixed form], 13-16; 5.12-16 [= 5.13-17 [Eng = 18-20 Eng.], 17-19], 6.1-9; 7.15-22, 25-29; 8.9-15 [mixed form]; 8.16–9.10 [mixed form]; 9.11-12; 9.13–10.3; 10.5-7). See Schubert, 'Die Selbstbetrachtungen Kohelets'. However, he still seeks to maintain the book's connection to history. The approach he brings to Qoheleth seems to be close to what has been termed the 'new historicist' approach to literature. Although he admits to the fictive reality of Qoheleth, he still maintains that the various situations which are addressed in the text can be grounded in the overall living conditions in the surrounding culture (p. 28). However, in the end, Schubert must relinquish to the fact that: 'In each case real historic problems are probably addressed, which however do not allow historical verification' (p. 28). According to Schubert, the unity of style in the book is evidence that the author is present in his work. Even if we cannot find an individual here, Schubert argues that the tell-tale influence of the author's social group/location can still be felt (p. 25). However, I find this line of argument a bit pedantic. The fact that there is a unity of characterization only proves that the implied author was adept at the artful use of characterization. More importantly, grounding the author in a social group does not establish a connection between the implied author and the historical author. Both the author and his social group are superseded by the implied author upon the publication of a book. An author, whether conceived individually or as an artist-in-social-matrix, is still distanciated from his work by textuality.

55. Fox, 'Frame Narrative', p. 105.

> the reality of Qohelet, simply by talking about him as having lived, speaking about him in a matter-of-fact, reliable voice, the voice of a wise-man. Qohelet is not an entirely plausible character—with his puzzling name, with his claims of royalty and vast wealth. The epilogist indicates that we are to react to Qohelet as having lived. The reader's acceptance of the reality of literary figures is important to certain authors even when writing the most outlandish tales. Swift, for instance, created a fictitious editor for Gulliver's Travels who does not say that Gulliver existed, but simply talks about his own relationship to that character, where exactly he lived, how his memoirs came to the editor, how he edited them. What the author seeks is not necessarily genuine belief in his character's existence (though that may be the intention in the case of Qohelet) but suspension of disbelief for the purposes of the fiction... The epilogist of Qohelet succeeded in convincing many readers that he had an intimate familiarity with Qohelet, and it is clear that this is one of the epilogue's purposes. The reader is to look upon Qohelet as a real individual in order to feel the full force of the crisis he is undergoing.[56]

As a result, what we now possess is a fully-characterized narrative function called 'Qoheleth'. While the person who formed the model for this character is irretrievably lost, the persona lives on. Through the artful use of language, Qoheleth-as-character is presented in a very life-like manner. With Baruch Hochman, I note that Qoheleth, while related to Homo Sapiens, is very much a Homo Fictus. However, the two must not be confused, for; 'although we necessarily read Homo Fictus in terms of Homo Sapiens, they are not identical, and Homo Fictus must be confronted in terms appropriate to him'.[57] By appropriate, he means

> the sense of the wholeness of a person in a story or a play rests on the extent to which a writer meets the challenge of rendering character coherent from the perspective of the text's ending, which may of course be very different from the character's 'ending', or death. In this sense, character creation is always teleological, always serving the needs of the whole imaginative context but itself being generated along the way as an isolable element.[58]

In that regard, unlike the person Qoheleth who may stand behind this character, the narrative Qoheleth exists for the purpose of fulfilling the implied author's designs. Since character in fiction serves ideological aims, we must analyze Qoheleth with that preeminently in mind. Unlike

56. Fox, 'Frame Narrative', p. 100.
57. Hochman, *Character in Literature*, p. 86.
58. Hochman, *Character in Literature*, p. 105.

characterization in real life, the literary characterization of Qoheleth's persona serves to fully realize the rhetorical aims of the implied author. Qoheleth as Homo Fictus exists only to implement the teleological aims of the discourse, which in this case, is to exploit the prospects and deficiencies of private insight.

The fact that readers have seen a fictional presentation in a Wisdom tract should not surprise us. The use of fiction is widely prevalent in the various Wisdom books, especially the later ones. Proverbs 1–9, Job and Tobit, among others, are heavily indebted to the use of narrative fiction to accomplish their rhetorical goals.[59] Fictional accounts were also widely utilized in the various Pseudepigraphal books, such as the *Testament of the Twelve Patriarchs*. The use of fiction as a heuristic tool increases the rhetorical effect of a book. R.N. Whybray speaks of the 'dramatization' of gnomic stories. He states:

> A vivid account of the life of specific persons, embellished with circumstantial detail, is a hundred times more effective as a means of persuasion than a brief, bare statement of fact or principle, whether to sell a commercial product or to teach a moral or religious lesson.[60]

The use of fictional language adds liveliness and urgency to a Wisdom book. Such increased usage of fiction by later Wisdom writers is what I am calling the 'fictimization of the reader'. By that I am referring to the increased expectancy among Wisdom writers that their readers would be skilled in the specific set of competencies it takes to read fiction effectively. The model reader of the book of Ecclesiastes is a reader with finely tuned literary skills. The ability to rightly distinguish between fiction and reality is an absolute must for anyone who comes to hear Qoheleth's narration.

As a result, the model reader of Ecclesiastes requires competency in understanding the language of fiction. Relying upon Goethe's dictum that 'whoever wants to understand poetry must go into the land of poetry, while whoever wants to understand the Poet must go into the land of the Poet', Loretz has called attention to the problem of referentiality in

59. Interestingly, Misch has seen a close relationship between Tobit, which is widely regarded as fictional, and Ecclesiastes. Such comparisons lend further intersubjective support to the fictional reading model being proposed here. See Misch, *A History of Autobiography*, p. 548.

60. R. Whybray, *The Succession Narrative: A Study of II Samuel 9–20; I Kings 1 and 2* (SBTheo [Second Series], 9; Naperville, IL: Alec R. Allenson, 1968), p. 72.

autobiography. Loretz summarizes Qoheleth's use of fictional and referential language:

> The research into human autobiography allows us now to recognize, that in this area of literary presentation the 'land of Poetry' and 'the land of the Poet' are without sharp boundaries to a large extent, [becoming] inseparable wholes flowing together. In our hands we never have, therefore, an objective biographical presentation, but rather, a literarily and artistically formed report.[61]

Because literary/biblical texts are, as a rule, a mixture of fiction and reality,[62] it behoves the reader to attain the necessary competence to discern when a text is reality-oriented, and when it is fictively dealing with life. After that is accomplished the debate as to whether Ecclesiastes is autobiographical or not is solved rather easily. Once a reader realizes that Qoheleth is a writer playing a role,[63] he or she no longer asks whether the story is real or not. The competent reader is only concerned with its lifelikeness and how the use of fiction defamiliarizes reality so that the reader comes away with a better understanding of his or her own existential situation. In that regard, Qoheleth may be the most autobiographical book in the Canon, in the sense that it addresses the implied reader's life in an amazingly frank and piercing manner.

The concept of the author playing a role is especially insightful for understanding the literary use of both Qoheleth-as-narrator and the Epilogist by the implied author. Each of these ideological posts of observation are roles or masks that the implied author utilizes to fully explore the nature of Wisdom's quest for knowledge. Qoheleth represents the post of observation which views the quest for Wisdom from

61. Loretz, *Qohelet und der alte Orient*, p. 47.

62. Lux, '"Ich, Kohelet, bin König..."', pp. 333-35. In a vein similar to Renza ('The Veto of the Imagination'), Lux argues that fiction and reality are often a mix in literary texts, even for those whose ostensive intention is to deal with reality. He argues: 'The exclusively fictional text is likewise, like the text which only utilizes reality, itself a fiction. As a rule, literary texts exist out of a mixture of fictional and reality-designating parts. The task of interpretation is to recognize the signals of these parts, to examine how the mixed-relationships are procured, and thereby how the intended reception-ways of the text itself can be realized' (p. 334).

63. Uspensky likens the different points of view portrayed in a text with the roles an actor plays. He states: 'The author assumes the form of some of the characters, embodying himself in them for a period of time. We might compare the author to an actor who plays different roles, transfiguring himself alternately into several characters' (*A Poetics of Composition*, p. 91).

the point of view of the individual's experience. The Epilogist represents that post of observation which values the role of the community's corporate experiences as the source of true Wisdom. By donning one mask, and then the other, the implied author explores the role of private insight and public knowledge as the twin epistemological poles which constitute the quest for human Wisdom. In so doing, he attempts to show the reader the various strengths and weaknesses of each position, as well as their synergetic, or perhaps, symbiotic relationship to each other. As a result, it can be seen that Qoheleth fully expresses the views of the implied author, though that position does receive a muted criticism by the implied author in ch. 12. One should not therefore view Qoheleth as a 'foil' for the Epilogist or the implied author. By expressing the thought of the implied author, both the Epilogist and the narrator represent yet another 'polar structure', except at a higher compositional level than the level Loader has explored.[64] In fact, it could be argued that Qoheleth's fondness for contradictions and polar structures as a literary character is not strictly dependent upon some long-lost sage whose sayings were taken up by the implied author, but instead are wholly dependent upon the implied author's own mentality (though a previous mentor may, admittedly, have had an influence here). The only difference I can see is that the implied author has taken such polar thinking to a higher level of reflection than did his teacher.

5. *The King's Fiction as a Theatrical Prop*

A related problem which confronts the reader of Ecclesiastes revolves around the nature of the King's Fiction. Because the model reader of Ecclesiastes is a thoroughly 'fictimized' reader, those who have approached 1.12–2.26 with a referential set of competencies have been baffled by the text's use of fiction. Or worse still, lacking the competencies it takes to recognize it for what it is, they have attempted to read it with a referential reading grid. The insights originally argued by Loretz regarding the book's autobiographical value have been thoroughly imported into the discussion regarding the King's Fiction by Rüdiger Lux. For Lux, the key to reading 1.12–2.26 lies in the reader's ability to recognize the text's use of fictive signals. It becomes a reception problem, whereby a reader who lacks the competencies required of

64. Loader, *Polar Structures*, *passim*.

the text's model reader inevitably confuses the fictive world of the text with a referential world. Lux compares the historian who would read the book of Ecclesiastes without the requisite literary competencies required of the text's model reader to a medieval peasant who stumbles into a play. Having already met the play's chief actor, the peasant hears the actor playing the role of King Alexander. Not realizing the nature of a play, the peasant exclaims; 'If you are Alexander, then I am Friedrich Wilhelm!' The peasant then continues with reasons why the actor cannot be King Alexander, noting that his poverty hardly befits one who is a 'King'. The peasant concludes that the actor is a swindler and a thief! Of course, we recognize wherein the problem lies. It lies in the peasant's confusing the real world with the world of fiction. Lacking literary competence, the peasant mistakenly treats the play as if it were reality, instead as if it were *about* reality.[65]

The major task confronting the reader in 1.12–2.26 is to recognize the various textual clues which signal to the reader that a fiction is in progress.[66] In my opinion, the King's Fiction is the equivalent to a

65. Lux, ' "Ich, Kohelet, bin König..." ', p. 332.

66. Lux, ' "Ich, Kohelet, bin König..." ', p. 334. Among the recent readers who have argued that this is a consciously designed literary fiction, see N. Lohfink, '*Melek*, *Sallit*, und *Mosel* bei Kohelet und die Aufassungzzeit des Buchs', *Bib* 62 (1981), pp. 535-43 (537); T. Longman, III, 'Comparative Methods in Old Testament Studies: Ecclesiastes Reconsidered', *TSFB* 7 (1983), pp. 5-9 (7-9); *idem, The Book of Ecclesiastes* (NICOT; Grand Rapids: Eerdmans, 1998), pp. 15-20; D. Merkin, 'Ecclesiastes', in D. Rosenberg (ed.), *Congregation: Contemporary Writers Read the Jewish Bible* (London: Harcourt Brace & Jovanovich, 1987), pp. 393-405 (394); L. Perdue, ' "I will make a test of pleasure": The Tyranny of God in Qoheleth's Quest for the Good', in L. Perdue, *Wisdom and Creation* (Nashville: Abingdon Press, 1994), pp. 193-242 (238-42); Meade, *Pseudonymity and Canon*; Fox, 'Frame Narrative', pp. 83-106; R.N. Whybray, 'Qoheleth as a Theologian', in Schoors (ed.), *Qoheleth in the Context of Wisdom*, pp. 239-65 (257); M. Görg, 'Zu einer bekannten Paronomasie in Koh 2,8', *BN* 90 (1997), pp. 5-12; H. Müller, 'Kohelet und Amminadab', in O. Kaiser (ed.), *'Jedes Ding Hat Seine Zeit': Studien zur israelitischen und altorientalischen Weishet* (Festschrift D. Michel; BZAW 241; Berlin: W. de Gruyter, 1996), pp. 149-65; *idem*, 'Travestien und geistige Landschaften zum Hintergrund einiger Motive bei Kohelet und im Hohenlied', *ZAW* 109 (1997), pp. 557-74 (where he situates the fictive backdrop found in ch. 2 in a common ancient Near Eastern 'gardener fiction'); C.L. Seow, 'Qohelet's Autobiography', in A. Beck *et al.* (eds.), *Fortunate the Eyes that See* (Grand Rapids: Eerdmans, 1995), pp. 275-87.

theatrical backdrop.[67] Qoheleth's monologue reminds me of any number of monologues by the great characters in literary history. Of course, as any stage director knows, a great speech by a magnificent character must be given the proper setting to make it optimally effective. To give the character a backdrop, the stage must be given a number of props to bring the character to life. In Ecclesiastes, the King's Fiction is the fictive prop by which the implied author sets the stage for his protagonist.[68] Giving Qoheleth a royal stage setting is not much different than a theatrical production I once observed, *An Evening with Mark Twain*. The play was a simple monologue by the nineteenth century's quintessential wiseman, in which the setting of a Mississippi riverboat was given to the character to enhance the fictive reality of the monologue being given. The Royal Fiction is similar to Twain's riverboat. It gives the monologue an artistic richness by setting the narrator's speech in the midst of royal opulence. By placing Qoheleth's discourse in this context, the implied author sets in motion a powerful motif, whereby the 'call to enjoyment' is given the perfect setting. The King's Fiction artistically and thematically implies to the reader what will be made an explicit admonition shortly thereafter. The fact that the book's first call to enjoyment occurs at the end of the King's Fiction is no accident (2.24-26). The use of the King's Fiction as a literary prop therefore serves to bolster the rhetorical purposes of the book from a thematic point of view. The bold references to wealth, parks and pleasures hints at the wise counsel that will shortly be made the focal point of Qoheleth's discourse by means of the seven-fold call to enjoyment (cf. 2.24-26; 3.12-13, 22; 5.17-19; 8.15; 9.7-10; 11.7-10).

The implied author signals this fictional reality by offering several clues. Loretz has argued that the text's reluctance to provide specific

67. Quite independently of each other, both Christianson and myself have visualized Qoheleth's monologue as a one-man play. See Christianson, *A Time to Tell*, p. 257.

68. Relying upon the insights of Loretz, S. Bréton has also picked up something of this trait regarding the King's Fiction. He describes Qoheleth's use of the Ich-Erzählung as a 'theatrical' type which 'is not fictitious autobiography, but merely stems from the traditional linkage of kingship and Wisdom. The idea is that if the King can lay title to the 'wise', then the wise can lay claim to 'king' (cf. Prov. 8.15). See Bréton, 'Qoheleth Studies', p. 27. Loretz also concludes: 'Since the Kings themselves had proudly adorned themselves now and again with the Wisdom of the wise, therefore a wiseman could fictively adorn himself with the title of the King' (*Qohelet und der Alte Orient*, p. 153).

information regarding the King and his activities is *prima facie* evidence that the narrative does not have a specific regent in mind. First, Qoheleth refers to himself as king three times (1.12, 16; 2.4-9), yet no such king is attested in Israelite history. Since both modern and ancient readers know this fact, he concludes that anyone familiair with Israelite history would 'know with certainty that a man with this name never held the throne of David and the statement of Qoheleth concerning his kingdom must therefore be a fiction'.[69] Furthermore, the narrative emphasizes the wealth and Wisdom of the king. Nothing is said about his power, fame, or even his armies. Only the barest of information is given concerning his building accomplishments, which focuses almost exclusively on houses and parks—things which jump out as most essential to the enjoyment of the *individual* rather than the well-being of the state. Loretz observes:

> Qoheleth's statements concerning his royal court are similarly constituted, but distinguished in several points. So Qoheleth enters into his hymn of praise to his great wealth in no detail. In contrast to this, the report concerning Solomon's wealth, [reports] his pleasure in a detailed description of the individually valuable possessions of the king. From Qoheleth's generally held words nothing is to be taken, where his buildings stood or from where, for instance, the gold came. While Qoheleth in the framework persists with completely general reports, the report itself pleases Solomon therein, to make the most detailed statements as possible.[70]

In contrast to the account in 1 Kings 3–11, this account reads almost like a fairy-tale: 'Once upon a time, there was a very wise and rich king, Qoheleth, who had many concubines, parks, etc'. When Ecclesiastes is compared with the account in 1 Kings 3–11, which mentions the various historical specifics with rigorous detail, the fictive nature of Ecclesiastes jumps out at any reader who is skilled in the use of ornamental fiction. The secularity of the account also impresses the reader, as there is no mention of Solomon's greatest achievement—the Temple. The lack of specific detail informs the competent reader that reality was not the referent of these verses. The use of hyperbole would also impress upon the reader the fictive nature of this account. This lack of detail expressly fits the needs of a fictional account, in that it creates gaps for the reader, whereby the imagination of the reader is creatively called

69. Loretz, *Qohelet und der alte Orient*, p. 148.
70. Loretz, *Qohelet und der alte Orient*, p. 156.

into service by the text. The implied author went to great lengths to place at the center of the reader's attention the theme of parks and other pleasurable belongings. There is more than just a hint of self-indulgence in the portrait of Qoheleth's kingdom, or more appropriately, his Disneyland-like estate.[71] As Christianson summarizes: 'Whether figurative or literal, the textual ambiguity here does not seem to diminish the effect of the guise, for the real effect is not so much to fasten Qoheleth's persona immovably to that of the historical Solomon as to create a unique interpretive freedom'.[72]

Lux also argues that the text provides obvious fictional clues to the reader. He observes four major signals which provide the reader with clues that the royal experiment in 1.12–2.26 is an obvious fiction. First, he too notes the curious meaning of the name 'Qoheleth'. Since it has the definite article in 12.8, Lux argues that the competent reader would

71. In this regard, the motifs that gain prominence in the King's Fiction are very much unlike true historical autobiographies in the ancient Near East. Tadmor observes that the royal autobiographical apologies found in Assyrian literature all narrate events which have 'immanent political aims in the present or some particular design for the future', that is, events which stress the king as a military hero or as a pious master-builder. See H. Tadmor, 'Autobiographical Apology in the Royal Assyrian Literature', in H. Tadmor and M. Weinfeld (eds.), *History, Historiography and Interpretation: Studies in Biblical and Cuneiform Literatures* (Jerusalem: Magnes Press, 1984), pp. 36-57 (37). Once this background is clearly seen, the impracticality and hence, the fictiveness of the narrative is easily perceived. However, the emphasis on 'deeds', however impractical, still fully participates in the royal autobiographical genre. Seow draws attention to the Ammonite royal inscription of Amminadab which focuses on the 'deeds' of the king (cf. Eccl. 2.4, 11). Thus the King's Fiction imitates some aspects of the general style of royal autobiographies by highlighting his personal achievements as king. However, there is a key rhetorical difference between the historical autobiographies found in the ancient Near East and Qoheleth's fictive autobiography. As Seow summarizes: 'Qohelet's imitation of the genre is poignant in its irony. In the end the text makes the point that none of the deeds—even the royal deeds that are assiduously preserved in memorials—really matters… The genre of a royal inscription is utilized to make the point about the ephemerality of wisodm and human accomplishments. Qohelet itemizes the king's many deeds and surpluses only to show that kings are no better off than ordinary people' ('Qohelet's Autobiography', p. 284). As such, we see that the fictive contours of the narrative precisely fit the satiric nuances of the King's Fiction. As will be shown later, the satiric purposes of the passage include epistemological as well as royal components. See also Christianson, *A Time to Tell*, pp. 136, 156-58.

72. Christianson, *A Time to Tell*, p. 131.

have 'recognized, that it is no proper name, but rather, an appelative, a generic term, plainly presenting a function designation, as we have termed it'.[73] As a term designating an office or function, 'Qoheleth' would have raised questions in the reader's mind regarding its function as a proper name. Second, he observes that in 12.9, Qoheleth is plainly designated as a wiseman and not as a king, a disclosure which would have consciously 'blown the cover' of the King's Fiction were it meant referentially.[74] Third, he draws attention to the text's reticence to mention specifically the name of Solomon. Lux argues:

> The superscription itself in 1.1, which probably goes back to the identification with Solomon and, on the other hand, is independent of 1.12, honors in the final instance the mask of the narrator. The apposition, *ben-dawid* (son of David) belonging to *qhlt* (Qohelet) brings an apparent information surplus. But also, it does not completely disclose the deliberate secret, which hides behind the mask. It avoids, inspite of its great concretization, the simple identification, while it leaves out the name 'Solomon'.[75]

Finally, Lux notes that later Jewish exegetical tradition knew that Solomon was not the author, but still persisted in assigning the book to him.[76] This he attributes to their ability to see through the Solomonic mask utilized by the implied author.

However, there remains some debate as to whether the reader would interpret this as a Solomonic Fiction or as a more general King's Fiction. Brevard Childs maintains that one of the unresolved issues in the book is why 'the author is identified with Koheleth, and yet immediately described in a way which is only approximate to Solomon'.[77] But,

73. Lux, '"Ich, Kohelet, bin König..."', p. 336.

74. Lux, '"Ich, Kohelet, bin König..."', p. 336.

75. Lux, '"Ich, Kohelet, bin König..."', p. 336.

76. Relying upon Rabbinic sources, Lux argues: 'It was precisely this paradox, which inspired the narrative fantasy of the Haggidists. So it was told in *jSan* 2.7, that Solomon was pushed from the throne because of his sins and an Angel of equal appearance took his place. Solomon went begging through the academy. And everywhere, where he presented himself as King of Jerusalem, he was covered by a costume with shepherd stick or in the best case, reproaches and a portion of groats. How could he have maintained that he was Solomon while this one still sat in the form of an angel on the throne?' ('"Ich, Kohelet, bin König..."', p. 336).

77. Childs, *Introduction*, p. 384. While Childs does not explain the text's use of reticence here, he does argue that the function of the Solomonic Fiction is to assure 'the reader that the attack on Wisdom which Ecclesiastes contains is not to be

as David Meade points out, the entire Solomonic corpus had a 'profoundly ambiguous indifference toward a rigid identification with Solomon'.[78] Loretz has argued that no specific king is intended by this account, given its lack of specificity. He argues:

> In the description of his royal success Qoheleth intends no identification with any certain King from the history of Israel, not even with Solomon. He attributes to himself a great success in all things, what was significant for a king of the old Orient. Because the Israelite monarchy followed anyway, in many respects, the model of its Semitic neighbors, this was looked upon favorably as a model, and exists only in expectation, when in Qoheleth's imaginative picturing of his royal glory all the motives are repeated, which we know from biblical as well as extra-biblical sources.[79]

Loretz concludes that Qoheleth was content to be simply identified as a royal figure without the added burden of being associated with any specific king. His reading suggests a mask that portrays the narrator as simply the greatest and most successful King of Jerusalem. Wisdom and wealth are the prime characteristics bestowed on the narrator by the royal mask,[80] both of which would have aided tremendously the

regarded as the personal idiosyncracy of a nameless teacher…his words serve as an official corrective from within the Wisdom tradition itself' (p. 384). As will be seen, the use of a public figure is one of the ways that the implied author utilized public knowledge and traditions to balance out the subjective limitations of first-person narration.

78. Meade, *Pseudonymity and Canon*, p. 72. See the equally reticent associations found in Prov. 1.1; 22.17; 24.23; 30.1; Cant. 1.1; 8.11-12; Wis. 6–9; *Pss. Sol.* 17–18. His study suggests that the entire Solomonic tradition wanted to draw on the qualities of Solomon without invoking his name. Although Meade argues that Solomon had an increasingly positive reputation in the post-exilic period as the preeminent sage, it would rather seem that such a reluctance to explicitly mention the king suggests a less than positive acceptance of Wisdom's greatest patron. Such a situation is tacitly implied in the assessment of Armstrong, who argues that Ecclesiastes' use of the Solomonic mask is actually an attempt to raise the reader's estimation of Solomon. According to this reading, the use of a mask by Qoheleth is an effort to improve *Solomon's* ethos standing within the community. See J. Armstrong, 'Ecclesiastes in Old Testament Theology', *PSB* 94 (1983), pp. 16-25 (17). While I do not agree wholeheartedly that Qoheleth's self-absorption in the pursuit of pleasure was a great boon to Solomon's ethos, such readings do testify to the problems that readers have had with Qoheleth's appropriation of Solomon's reputation as an attempt at rhetorical accreditation and validation.

79. Loretz, *Qohelet und der alte Orient*, p. 160.

80. Loretz, *Qohelet und der alte Orient*, p. 148.

narrator's ability to counsel the reader with the categorical imperative to enjoy life. As a generic reference to royalty and its Wisdom and privileges, the mask has few, if any, negative connotations.

On the other hand, Lux and many others, have argued that a competent reader would naturally call to mind the specific image of Solomon. If the latter is the case, then the ethos of Solomon is certainly a contributing factor in how the reader characterizes and responds to the narrator. According to Lux, the parallels with Solomon would have been unmistakable for the reader who is familiar with the intertextuality issues which surround Israel's patron sage. However, the use of fictive signals clearly gestures to the reader that this is actually not Solomon, but the fictive character Qoheleth who is temporarily donning Solomonic garb. Lux states:

> The fictive signals placed in 1.12 were recognized and effective, because this Solomon...made no secret about it, that he is not Solomon. The role-playing was accepted, the text was received as fiction. For there indeed, where one identified Koheleth with Solomon, there it was not the historical Solomon, but rather, a Solomon *redivivus*, who took the word.[81]

As a reader it is often difficult to dispense with the traditional Solomonic 'baggage'. The parallels between Solomon and the characterization of royalty in these verses, even in spite of its use of reticence and its total lack of specificity, undoubtedly surrounds the narrator with some vestiges of the Solomonic ethos.[82] I believe (with Loretz) that while the narrative purposely avoided the identification of Qoheleth with any specific king, it would be difficult for any reader who was familiar with the Solomonic tradition not to think of Solomon in some sense. This is particularly likely given the fact that wealth, Wisdom and women, all salient traits of Solomon, are also prominent in Ecclesiastes. However, the reticence of the characterization protects the narrator from an overly specific, and therefore overly negative response by the reader, who can only surmise which king is intended by the author. By playing the Solomonic role in such a reticent manner, the implied author is posing a type of 'riddle' to the reader: 'Given these clues—I

81. Lux, '"Ich, Kohelet, bin König..."', p. 337.

82. For instance, both Solomon and Qoheleth are great wise men (1 Kgs 4.29-34; Qoh. 1.13-17); extremely rich (1 Kgs 4.21-28; Qoh. 2.7-10); owners of cattle and gardens (1 Chron. 27.27-31; Qoh. 2.4-7); master of many slaves and concubines (1 Kgs 10.5; Qoh. 2.7); and possessors of many musicians (1 Chron. 5.12-13; Qoh. 2.8). For a discussion of these parallels see Meade, *Pseudonymity and Canon*, p. 57.

was king over Israel in Jerusalem and known for my Wisdom—guess who I am'.[83] In effect, the implied author plays a game of charades with the reader, hoping to increase the reader's participation and to pique his or her interest in much the same way that a mystery novel sustains the reader's interest.

To sum up, the King's Fiction has several effects on the reader. The overall effect of creating a fictional mask for the narrator and the use of the King's Fiction as a literary prop increases the ideational activity of the reader. Relying upon the reader-response theory of Wolfgang Iser, Lux summarizes the effect of the text's fictionality on the reader:

> The recognition of disclosure signals of fiction had consequences for the reader/hearer. They did not settle the text on the plane of historic reality. The unspecified, barely concrete statements of the text regarding the identity of the King (his name, his time of ruling) demanded of the reader to fill these gaps with his imaginative fantasy. Through the 'holding back of information' the reader is active in the constitution of the meaning of the text. The meaning of the text opens up and does not exhaust itself... Rather, its referentiality is raised. So the text does not owe itself very much to the power of historical facts, but rather, sooner to the power of the imagination.[84]

As a literary text, Qoheleth owes much of its effectiveness to the ability of fiction to engage the reader's interest. As fiction, it enabled the narrator to defamiliarize reality, so that the reader could see human existence in a new light. In that respect, a primary effect of Qoheleth's fictional appropriation of the autobiographical style is to rewrite the reader's understanding of his or her life. Through the imaginative recasting of the world, the implied reader begins to see life in a new way, enabling him or her to defamiliarize their own existence and to come to a deeper understanding of the pains and pleasures involved in human existence. Perhaps that is the reason why the implied author used verbs which would blend the reader's 'now' with the 'now' of Qoheleth's narration?[85] Qoheleth's life serves as a defamiliarization of the reader's life. Such is the power of Qoheleth's use of fictional autobiography provided the reader has the requisite values to becomes the text's implied reader.

83. Farmer, *Who Knows What is Good?*, p. 154.

84. Lux, ' "Ich, Kohelet, bin König..." ', p. 337.

85. Isaksson, *Studies in the Language of Qoheleth*, p. 72.

6. *Attractiveness, Credibility and Trustworthiness: The Rhetorical Effect of Saying 'I'*

Qoheleth's ethos influences the reader's response at nearly every level. The effect of the narrator's character as a prism for the presentation of narrative values is immense. For example, Qoheleth's first-person oration might very well have been presented in a form similar to the sayings in the book of Proverbs. The highly personal confession in 2.17-19 (RSV) will serve as an example:

> So I hated life, because what is done under the sun was grievous to me; for all is vanity and a striving after wind. I hated all my toil in which I had toiled under the sun, seeing that I must leave it to the person who will come after me; and who knows whether he or she will be a wise man or a fool. Yet that person will be master of all for which I toiled and used my Wisdom under the sun. This also is vanity.

However, these same values could have been presented as typical proverbs written in the traditional 'gnomic' third-person style.[86] One could very well imagine an implied author who presented Qoheleth's maxims and aphorisms along the lines of typical proverbial Wisdom:

> Many a person hates life because what is done under the sun is grievous to them, and says, 'All is vanity and striving after wind'.
>
> There is a person who hates their life because they must toil under the sun and must leave the fruits of their efforts to another.
>
> Who knows if the person who inherits your wealth will be wise or foolish? Yet they will vainly be the master of the estate they have inherited.

The use of first-person discourse turns these abstract values into the worldview of a specific and limited human person. In the process, a good deal of subjective point of view replaces the objective aura of third-person narration. In the book of Ecclesiastes, the reader reacts not only to the abstract values presented therein, but also to a person's ethos as well. For traditional sayings, a value couched in a third-person form takes on a gnomic quality whereas in a first-person form it exudes a subjective quality.[87] The difference between the two forms must

86. Christianson has also noted the rhetorical effects of couching Wisdom in first-person modes. See Christianson, *A Time to Tell*, p. 37.

87. However, it must be noted that this gnomic quality is only apparent.

be kept in mind. The effect of placing a character's persona as a filter between the implied author and the reader is significant. It gives the values being presented an entirely different ring, with a unique set of rhetorical strengths and weaknesses.

The critic must ask in this regard: What is it that affects the reader's sense of Qoheleth's ethos?[88] What characteristics predispose the reader to respond to a character in a negative or positive fashion? As I track the role of Qoheleth's 'I', I will pay close attention to Qoheleth's attractiveness, trustworthiness and credibility. Aristotle argued that there are three things which 'inspire confidence' in the speaker's character; good sense, good moral character and goodwill.[89] Gerard Hauser elaborates on this tradition, delimiting seven human characteristics which provide a rhetor with the necessary ethos to be trusted by his readers: justice, courage, temperance, generosity, magnanimity (nobility of thought, willingness to forgive), magnificence (a vision that elevates the human spirit) and prudence.[90] Obviously, Qoheleth's spirit

McKenzie, like most paraemiologists, observes that 'proverbs, while they sound like general truths, due to their impersonal, generalized syntax, are actually partial generalizations, appropriate for some situations and not for others... Their impersonal, generalized syntax makes them appear to be universal truths, rather than limited generalizations made for some situations and not for others' ('Subversive Sages', p. 22). In contrast, 'An aphorism, since it is more closely associated with a particular person than a proverb and often uses the first-person, destroys the illusion of collective Wisdom and traditionality' (p. 52). In spite of these differences, the proverb which uses a third-person approach and the aphorism (such as Qoheleth's) which utilizes a first-person mode of expression actually cohabit a continuum for McKenzie. She astutely observes: 'The proverb and the aphorism, the Wisdom of many and the wit of one, exist in dialectical relationship: the proverb is an aphorism whose author has been forgotten. An aphorism, if taken to heart by the prevailing social group, can become a proverb' (p. 53). As a result, this dichotomy or distinction should not be taken too far. Qoheleth's 'doctrine of the proper time' (3.1-8) takes this fact into full consideration, showing that the ancients fully understood the limited value of the proverb's 'gnomic' qualities.

88. For the purposes of this study, I will define ethos along the lines proposed by Andersen and Clevenger: '*ethos* is defined as the image of a communicator at a given time by a receiver—either one person or a group' (K. Andersen and T. Clevenger, 'A Summary of Experimental Research in Ethos', *SM* 30 [1963], pp. 59-78 [59]).

89. Aristotle, *Rhetoric*, Book 2, Chapter 1, 1378a:5 in *The Rhetoric and the Poetics of Aristotle* (trans. W. Rhys Roberts, New York: The Modern Library, 1990).

90. Hauser, *Introduction to Rhetorical Theory*, p. 98.

possesses some of these characteristics, such as courage, temperance, prudence, and perhaps even justice. But when it comes to generosity, magnanimity and magnificence, Qoheleth is profoundly lacking. This situation leads the reader to respond both positively and negatively toward Qoheleth.

7. *Qoheleth's Reminiscences on the Wisdom Tradition: A Dialogic Monologue that Fictively Recontextualizes the Wisdom Tradition*

Qoheleth's discourse has been described by many readers as argumentative and, specifically, as a prime example of protest literature. A number of readers have noted the many verses that suggest an opposing viewpoint, another 'voice' or interlocutor in the text,[91] or perhaps the quotation of some proverbial text which Qoheleth is either reflecting upon or arguing against. While I have embraced the first explanation as a salient aspect of Qoheleth's thought, and the second as a misreading of the role of the narratee by readers influenced by historical agendas, the third option which sees Qoheleth as quoting traditional proverbial lore brings the reader face-to-face with the disputative quality of Qoheleth's reflections. Numerous readers have called attention to the dialogic aspect of Qoheleth's thought, both present and past.[92] Kathleen

91. For instance, Herder and Eichhorn both regarded the book as a dialogue between a refined sensualist and a sensual worlding. E. Podechard and D. Buzy see four literary voices in the text; Qoheleth, the Epilogist, the *hakam* scribe, and the pious glossator. See D. Buzy, 'Les auteurs de l'Ecclesiaste', *L'AnThéo* 11 (1950), pp. 317-36 (317, 322) and E. Podechard, 'La composition du livre de l'Ecclésiaste', RB 21 (1912), pp. 161-91 (186-91). Others have seen a dialogue here between an Epicurean and a Stoic, or perhaps a teacher and his pupil. See J.S. Wright, 'The Interpretation of Ecclesiastes', p. 19. The Rabbis believed that Solomon had simply changed his mind on various matters over the course of the book. See Holm-Nielsen, 'The Book of Ecclesiastes', p. 81. In these reading strategies, the critic witnesses the problems involved in misidentifying the literary role of the narratee with historical persons or types. Recently, this reading grid has been revitalized by T.A. Perry (*Dialogues with Kohelet*, *passim*). Perry views the book as one long dialogue between the pessimistic Presenter (Kohelet) and a more optimistic Antagonist, the Arguer (p. 10). However, the literary atomism it takes to achieve this reading renders his analysis suspect.

92. Murphy, *Ecclesiastes*, p. lxiii. For an overview of mediaeval readers who underscored the use of dialogue in the book, see Holm-Nielsen, 'The Book of Ecclesiastes', pp. 82-88. Arguing from a rhetorical perspective, Messner asserts that the primary effect of a citation is to 'invite the reader/listener to participate, to test

Farmer subsumes the problems associated with the various contradictions or tensions in the book under the rubric of dialogically shaped discourse. She states:

> If one assumes that the book is the result of (or is shaped in the form of) either a dialogue between a pupil and a teacher or a forum in which various individuals' opinions are aired, then variations in viewpoint are easily explained. There are, however, no indications in the text itself that this is the case.[93]

With Farmer and Roland Murphy,[94] I would argue that the recognition of dialogical thought is basic to the construal of the meaning of the work. As has already been seen, in the call to enjoyment the implied author constantly has his narratee/implied reader in mind. It makes good sense to suppose that he would address various problems by citing the traditional and proverbial lore/public knowledge upon which the debate between Qoheleth and his narratee rested. In any protest, one must at least delimit the source of one's differences. At the very least, the 'convincing effect is increased by Qohelet supporting his argument with words that are known and recognized by his listeners or readers'.[95]

his/her assumptions with those of the dialogue partners' (D. Messner, 'The Rhetoric of Citations: Paul's Use of Scripture in Romans 9') (PhD dissertation; Evanston, IL Northwestern University, 1991), p. 132.

93. Farmer, *Who Knows What is Good?*, p. 19. However, the problem involved in identifying quotations in Ecclesiastes remains a thorny critical issue. Obviously, the text demands a type of competency that modern readers can only partially comprehend. The best summaries of this debate can be found in the synopses provided by Crenshaw, 'Qoheleth in Current Research', and J. Burden, 'Decision by Debate: Examples of Popular Proverb Performance in the Book of Job', *OTE* 4 (1991), pp. 37-65. Burden's synopis appears in Appendix A of this study, though I have taken the liberty to supplement his study with the analysis of R.F. Johnson, 'A Form Critical Analysis', which was overlooked, and the recent contribution by McKenzie, 'Subversive Sages'. The basic positions in this debate have been offered by R. Gordis, 'Quotations in Wisdom Literature', *JQR* 30 (1939–40), pp. 123-47; *idem*, 'Quotations in Biblical, Oriental, and Rabbinic Literature', in R. Gordis (ed.), *Poets, Prophets, and Sages* (Bloomington: Indiana University Press, 1971), pp. 104-59; *idem*, 'Virtual Quotations in Job, Sumer, and Qumran [Eccl. 4.8]', *VT* 31 (1981), pp. 410-27; M. Fox, 'The Identification of Quotations in Biblical Literature', *ZAW* 92 (1980), pp. 416-31; R.Whybray, 'The Identification and Use of Quotations in Ecclesiastes', in Congress Volume, Vienna 1980, ed. J.A. Emerton (VTSup, 32; Leiden: E.J. Brill, 1980), pp. 435-51.

94. Murphy, *Ecclesiates*, p. lxiii.

95. B. Rosendal, 'Popular Wisdom in Qohelet', in K. Jeppesen, K. Nielsen and

Qoheleth's use of quotations fulfills this argumentative need. Rather than directly invoking the voice of one's opponent, Qoheleth goes to the core of the problem—the basic proverbial lore upon which his opponent based their beliefs. As a result, the voices some readers have heard in the text are *inferred* voices. Nowhere does Qoheleth actually give his opponent or the narratee a direct voice as Herder, Podechard or Perry have supposed. One only hears the opponents' voices in the background of Qoheleth's reflections. Again, the critic should note the implied author's fondness for reticence. Subtleness seems to always surround this author's tactics.

The fact that readers must actively infer these voices/quoted texts is important to grasp. Recently, Ellen Van Wolde has looked at the problem of intertextuality in texts such as our own in light of the insights of the French linguists Julia Kristeva and Mikhail Bahktin. She argues that it is not the author who is determinative for the reading of dialogical texts such Ecclesiastes but, rather, the reader. Van Wolde also calls attention to the role that culture plays in forming the 'collective' text which constitutes all literary texts. Relying upon Kristeva's dictum that 'every text is absorption and transformation of other texts', she insists that dialogical texts which rely on other cultural texts are both self-contained and differential at the same time. In other words, texts which comment on others exhibit the two primary characteristics of intertextuality: repetition and transformation.[96] The difficulty for the reader is to determine a method for determining when texts are in dialogue with other texts. In older models of reading, the critic had to understand which ancient Near Eastern text formed the 'genotext' which the author utilized. Then, the reader needed to recover the intention by discovering the historical relationships between author and texts being quoted. However, in the newer paradigm, she suggests that it is readers who make these connections based on perceived analogies. Rather than seeking to explain intertextuality by reference to causal connections between

B. Rosendal (eds.), *In the Last Days: On Jewish and Christian Apocalypic and its Period* (Aarhus: Aarhus University Press, 1994), pp. 121-27 (122). Rosendal sums up one of the rhetorical functions of Qoheleth's use of proverbial matter: 'By virtue of its character of generally acknowledged experiential truth the proverb functions as a common starting-point for the "conversation" in the discourse between the author and the reader' (p. 127).

96. E. Van Wolde, 'Texts in Dialogue with Texts: Intertextuality in the Ruth and Tamar Narratives', *BibInt* 5 (1997), pp. 1-28 (4).

texts, Van Wolde argues that dialogical readings actually function on the basis of analogy. That is, the reader creates the supposedly 'causal' relationships between texts based on the analogies, similarities and transformations between texts. Once it is seen that all examples of intertextuality such as quotations are actually connections made by readers, this opens up the possibility that all dialogic readings, even those which posit diachronic causality, are in reality examples of 'synchronic reading'. She summarizes:

> A reader who does not know any other texts cannot identify any intertextual relationships. The reader is the one who, through his or her own reading and life experience, lends significance to a great number of possibilities that a text offers. Consequently, the presumed historical process by which the text came into being is no longer important, but rather the final text product, which is compared with other texts in synchronic relationships. The principle of causality is rejected too; its place is taken by the principle of analogy. Words are not viewed as indexical signs but as iconic signs. Iconicity denotes the principle that phenomena are analogous or isomorphic. Similar and different texts are not explained as being directly influenced by each other, causally or diachronically, but as being indirectly related to each other and having a similar or iconic quality or image in common. Whereas indexicality works on the basis of a succession of cause and effect, iconicity works on the basis of simultaneousness and analogy. Reading intertextuality in this way is a synchronic reading. By putting two texts side by side, the reader becomes aware of the analogies, or repetitions and transformations, between texts.[97]

Seen from this perspective, all suggestions that have been offered to date on how to determine quotations in Qoheleth's discourse become reading confessions couched in diachronic garb. However, it should also be noted that this does not mean that the reader is free to create connections that are not present in the text. The responsibility of the reader is to make an inventory of all the repetitions in the texts being compared so as to prove that the intertextuality perceived in a text is not a fabrication of the reader.[98] Finally, she argues that

> productive intertextual reading must be concerned not only with the meaning of one text (T1) in its encounter with another text (T2), but also with the new text created by the interaction of both texts. This is the third

97. Van Wolde, 'Texts in Dialogue with Texts', p. 6.
98. Van Wolde, 'Texts in Dialogue with Texts', p. 7.

> stage of the analysis, which concentrates on the new network of meaning originating form the meeting of the two texts'.[99]

The fundamental insight gained from this preliminary discussion is that the dialogicity and intertextuality found in texts which quote other texts will necessarily contain both a diachronic and synchronic component to their reading process. As much as the critic may try, the diachronic relationships perceived in Qoheleth's text are firmly rooted in analogies created by the reader, making them examples of synchronic reading as well. For the book of Ecclesiastes, this situation is further exacerbated by the fact that Qoheleth's quotations seem to stem more from the general 'Wisdom culture' of his time rather than any specific Wisdom text. That there is a profound relationship between this text and some 'cultural Wisdom text', for lack of a better term, is certain. However, the ideational and synchronic nature of the reading process for such texts may never allow a consensus to be achieved here. That much can be said upfront.

As many readers have noted, traditional sayings are scattered throughout the book, and many verses have a 'proverbial' sound to them.[100]

99. Van Wolde, 'Texts in Dialogue with Texts', p. 8.

100. Most readers have an intuitive sense of what a proverb sounds like. While such a notion cannot function by itself as a criterion for the identification of proverbs in a literary text, it is a place to begin for most readers who usually have a native sense for such matters. For a discussion on the 'incommunicable quality' of proverbs, see C. Fontaine, *Traditional Sayings in the Old Testament* (Bible & Literature Studies, 5; Sheffield: Almond Press, 1982), pp. 65-71. She defines the traditional saying as: 'a statement, current among the folk, which is concise, syntactically complete, consisting of at least one topic and comment which may or may not be metaphorical, but which exhibits a logical relationship between its terms. Further, the saying may be marked by stylistic features (mneumonis, rhythm, alliteration, assonance, etc.) or be constructed along recognizable frames ("Better A than B..." etc.) which distinguish it from other genres (or folk idioms). The referents which form the image are most likely to be drawn from the experience of common, "everyday" life, but the meaning (message) of the saying may vary from context to context, and any "truth claim" for that message must be considered "relative" rather than "absolute"' (p. 64). Besides the traditional saying, Qoheleth also quotes proverbs and maxims and utilizes the aphorism quite frequently as well. McKenzie defines these three forms as: 'A proverb is a short saying that expresses a complete thought, which, while expressing traditional values, is useful in certain new situations, and offers an ethical directive that is most often implied rather than directly stated... A maxim is defined as a non-metaphorical proverb, and an aphorism as a proverb whose author we know and that does not necessarily inculcate

However, the designation of these as quotations is something of a misnomer. Because of their setting in the midst of an interior monologue, it is difficult to describe these as simple quotations. A quotation normally invokes the repetition or citation of the words of another usually for the sake of lending that person's authority to one's own point being made,[101] or to disagree with the position taken by the text being cited. However, Qoheleth's use of quotations is not quite this simple. They are presented in the midst of Qoheleth's monologue, much of which is interior monologue. Given their literary context in a monologue, it is better to view the quotations as the reminiscences of the narrator. Whatever their origin, they are now a part of Qoheleth's thought. I therefore concur with Alyce McKenzie that the quotations are in actuality the 'sometimes contradictory inner reflections of one sage'.[102] As George Savran points out, 'quotations themselves are unique in that they mark a particular intersection of repetition and direct speech'.[103]

Because Qoheleth's quotations are now part of the narrator's interior monologue, the model reader hears only the voice of Qoheleth in those verses which cite proverbial lore, even if they do stem from the views of the larger society. Reader-response theory advocates that every text arms its implied reader with the appropriate interpretative reflexes to properly consume it. A good literary text always instructs the reader in the text-specific competencies it takes to competently interpret its features. In a work which is dominated by the use of 'I', sometimes the work teaches its reader to respond to all of the discourse as if it had an 'I' in it. Naomi Tamir has observed how

> the speech of a personal narrator is an act—of arguing, confessing, telling or thinking—which is part of the fictive world. In other words, his speech is not merely referential, but performative, because it functions as an act in the structure of the text... This means that in passages in which

traditional values' ('Subversive Sages', p. 17). While the debate as to the precise definition of these forms still rages, these should be adequate for the present analysis. Furthermore, the strong possibility that Qoheleth composed his own aphorisms exacerbates this issue beyond any hope of resolution (cf. 12.9).

101. R.F. Johnson, 'A Form Critical Analysis', p. 244.

102. McKenzie, 'Subversive Sages', p. 180.

103. G. Savran, *Telling and Retelling: Quotation in Biblical Narrative* (ISBL; Indianapolis: Indiana University Press, 1988), p. 12.

> the 'I' does not appear on the surface are perceived by the reader as having a higher performative noun-phrase such as 'I say that...' in its deep structure.[104]

This raises the issue of how the model reader of Ecclesiastes consumes those texts which are not couched in a first-person form. The characterization of Ecclesiastes' model reader has been indirectly broached by Timothy Polk, who argues that exactly such a situation occurs in the latter half of the book. He observes that even when Qohelet is not explicitly speaking in the first-person, but is speaking in the third-person, the reader still understands that a first-person speech is in progress. He states:

> Where explicit first-person references are absent, one finds admonitions couched in direct address, or descriptive data, often in proverbial form, which bear directly on a personal stance toward life and in which one clearly recognizes the voice of Qohelet, *just as if the frequent 'I have observed' were present.*[105]

Polk's analysis of how Qoheleth's 'I' permeates *all* aspects of the discourse confirms my thesis that even when 'I' is not present, as in the second-person addresses and the third-person comments on the various proverbial reminiscences, the model reader has already been instructed to understand these as an 'I'-discourse and so no longer needs an explicit reminder. The implied author instead depends on the reader's competency which he has built by frontloading the book so heavily with first-person address forms, especially the pleonastic use of *'ănî*. This dynamic in the latter part of the book is a prime example of how the text educates the reader to become a model reader who consumes the entire discourse as an extended 'I'-discourse, even in those places where 'I' is absent.

Furthermore, because 'I' has an inherent performative function which is lacking in third-person discourse,[106] it can be argued that the quotations in Ecclesiastes are no longer independent 'collections' or cita-

104. Tamir, 'Personal Narration', p. 424.

105. Polk 'The Wisdom of Irony', p. 5 (my emphasis). Christianson also comes to a similar conclusion in this regard. He astutely observes regarding the the large blocks of Wisdom saying in the latter half of the book: 'the "non-narrative" material is in a narrative setting and there are no "markers" to suggest that they should be considered to be outside the story proper. The narrative integrity (of voice, person, stance and so forth) throughout the whole is firmly intact...' (*A Time to Tell*, p. 257).

106. According to speech-act theory, an I-statement is quite different from a

tions as earlier scholars presumed. Rather, they have been thoroughly subsumed into the total I-narration of Qoheleth, becoming a part of the grand scheme of the sage's reflections. For the text's model reader, they do not have an independent status in the discourse, but operate only as the thoughts of the narrator. The fact that Qoheleth so often comments on these quotations or juxtaposes contradictory proverbs[107] should suggest to the competent reader that the various citations, including those in series in chs. 4–5 and 7–10, are merely the rambling interior reflections of a man at odds with his theological heritage. Qoheleth's comments turn these proverbs into a vehicle for his own personal address to the narratee/implied reader. He borrows from the public its corporate treasures and then, through his comments, injects a good deal of 'subjective energy into proverbs whose point had been dulled, whose metaphors had been domesticated into a didacticism that confirmed the status quo of the prevailing social group'.[108]

third-person statement. 'I' always invokes a sense of illocutionary force. Unlike third-person speech, it *does* something. This is what Tamir means by first-person speech having a higher function. White explains: 'While I judge is an engagement, he judges is only a description on the same level as he runs, he smokes. "I" endows a statement with illocutionary force, while "he" does not' ('A Theory of the Surface Structure', p. 164).

107. R.F. Johnson has carefully analyzed the series, concluding that within each the sayings are 'juxtaposed for a specific reason' ('A Form Critical Analysis', p. 77).

108. McKenzie, 'Subversive Sages', p. 23. Because of this tendency, McKenzie characterizes Qoheleth as a 'subversive sage'. She states: 'In Qohelet, more so than in Proverbs, the placement of proverbs and aphorisms contributes to their subversive impact. Qohelet employs two primary strategies in his sayings placement. One is the use of a proverbial saying as a text with ironic comment. This is what some scholars have referred to as Qohelet's "yes, but" strategy in which he gives a statement of traditional Wisdom and then modifies it. A second is the juxtaposition of two proverbial sayings that offer varying interpretations' (p. 198). The latter she calls 'dueling proverbs' (pp. 198, 255). Examples of proverbs with ironic comments are: 4.9-12; 5.9-12; 7.1-10, 11-14; 8.2-4, 5-6, 11-14; 9.4-6; 10.12-15. Examples of dueling proverbs are: 4.5-6; 7.12-13; 9.16-17; 10.1-3, 12-13. Dueling proverbs are not unique to Qoheleth, as the technique was a commonplace in the Wisdom tradition (cf. Prov. 26.4-5). In that respect, we see where the ancient sages were quite aware that proverbs were situation-specific, and that 'the aphorism need not agree with its neighbor in order to be perceived as "true"'. See J. Thompson, *The Form and Function of Proverbs in Ancient Israel* (The Hague: Mouton, 1974), p. 87. By juxtaposing such contradictory proverbs in a series, Qoheleth seems to be ruminating

By including these proverbs in his monologue, Qoheleth has embedded the Wisdom tradition back into the performance context of the individual's solitary existence. In order to train the reader to have this interpretative reflex, both the prologue and the initial speech of Qoheleth's monologue cite proverbs and traditional sayings (cf. 1.4[?], 8b[?], 15, 18).[109] Verses 15 and 18 of ch. 1 end a section with the quotation of a proverbial text. These quotations prepare the reader early on to recognize the various proverbial citations as a part of the narrator's inner thoughts. Furthermore, while one could argue that these proverbs are confirming Qoheleth's thought as if he were citing a higher authority, given the context of the narrator's empirical method wherein he confirms truth solely upon his own personal observations,[110] it is just as

on the inherent weakness of the proverbial form. See R.F. Johnson, 'A Form Critical Analysis', pp. 198-200. However, most readers expect consistency in a text, and often perceive such instances as gaps or blanks. (See Fox, *Qoheleth and His Contradictions*, pp. 23-28 for a thorough discussion of this reading problem and its various solutions.) The chief effect of dueling proverbs is to increase the reader's participation in the dialogue by forcing the reader to solve the riddle of their juxtaposition. Readers have typically solved such riddles by resorting to the 'yes-but' reading strategy (Hertzberg and Galling), attributed this to the contradictory nature of the thinker (Fox), or perhaps his fondness for 'polar structures' (Loader). Other solutions such as the use of different voices (Herder) or redactive glosses (Crenshaw, Barton, Podechard, and Jastrow who actually excised them and put them into an appendix!) have been summarily dismissed in this study. The Epilogist was surely right when he said that the 'sayings of the wise are like goads' in that the chief effect of such puzzles is to stimulate the reader's thinking process.

109. Fischer has analyzed the compositional structure of 1.3–3.15 and found that there is a conscious pattern for utilizing the proverb citations in the Prologue and King's Fiction. In 1.13-15 and 16-18 the text is arranged chiastically with a pattern of project/result/proverb. The proverb citations act as conclusion markers for each of these sections. Initially, Qoheleth relies upon the confirmation of a proverb in order to authenticate the defamiliarized worldview he is presenting to the reader. Later, he will utilize proverb citations in a much more seductive, seditious, and subversive manner. See Fischer, 'Beobachtungen zur Komposition', pp. 78-83.

110. Fox has cogently argued that Qoheleth's methodology is quite unique among the sages, and is a forerunner of the modern empirical method. He summarizes Qoheleth's epistemology: 'Nevertheless, the "empirical" label is justified, first, by Qoheleth's conception of his investigative procedure, which looks to experience as the source of knowledge and the means of validation, and second, by his concept of knowledge, according to which knowledge is created by thought and dependent upon perception' (*Qoheleth and his Contradictions*, p. 86; cf. *idem*, 'Qoheleth's Epistemology', *HUCA* 58 [1987], pp. 137-55). Murphy has also commented on

likely that it is Qoheleth who is confirming the value of these texts. Given Qoheleth's subversion of the proverb cited in 1.15 at 7.13, this seems to be a reasonable deduction. Qoheleth is merely giving a soon-to-be qualified 'amen' to the Wisdom tradition at this point in the narrative.

To sum up, Qoheleth's use of quotations accomplishes three things with the reader. First, by confirming the Wisdom tradition in the early part of the discourse, Qoheleth pulls his punches, so to speak. This disarms the reader, preparing him or her for the coming onslaught. Second, it characterizes Qoheleth as a wise man who is in good standing with the sages. He begins by speaking as a bona fide member of the Wisdom tradition. Finally, and more importantly, each of these proverbs become a part of Qoheleth's reflections. They no longer possess an independent status as gnomic texts, but become a part of the 'words of Qoheleth' (1.1). As a result, the quotations retain only a vestige of their original independent status. For the model reader, they are pulled into the gravitational field of Qoheleth's 'I'. As Loretz has observed, Qoheleth's 'I' is often interrupted by proverbs, warnings and beatitudes (cf. 4.17; 5.1, 5; 7.9-10; 8.25; 9.7-9; 10.10-17).[111] In the first half of the book, Qoheleth's 'I' occurs much more frequently than the quotation of proverbs. However, as Qoheleth's monologue continues, this relationship is reversed, with proverbial material gaining more exposure. After 7.1, this is particularly noticeable. In the latter half of the book, it is Qoheleth who interrupts the quotations with his 'I'. By periodically interspersing his 'I' in the midst of these 'series', the narrator makes sure that the reader understands the proverb 'collections' as examples of his own speech or reflections. Through these 'interruptions', the implied author controls the reader's response, making sure that he or she has the specific clues to read the entire discourse as the reflections

Qoheleth's 'different' epistemology: see R. Murphy, 'Qoheleth's "Quarrel" with the Fathers', in D. Hadidian (ed.), *From Faith to Faith* (Pittsburgh: Pickwick Press, 1979), pp. 234-44 (235-37). In a very insightful article, Höffken has tracked the role of the 'I' of the sages, and concluded that increasingly, the sage's *ego* became a criterion for assessing transmitted teachings. The wisemen began to gauge the Wisdom tradition more and more by referencing their own experiences. In this regard, Qoheleth is a participant in this general development. See Höffken, 'Das Ego des Weisen', pp. 121-35. In this view, Qoheleth would simply be a symbolic example of a far more pervasive and radical development among the wise.

111. Loretz, 'Zur Darbietungsform', p. 57.

of the narrator.[112] The use of 'I' in the latter half of the book serves to remind the competent reader who is speaking in the text. As a result, the competent reader continues to read these proverbs as examples of Qoheleth's thought, and not merely as the long established Wisdom of the larger community. With Bratiotis, I would see these as examples of the 'motto-monologue' in which a character reflects on well-known Wisdom motifs.[113]

Qoheleth's 'I' thereby strips the proverb of its gnomic, absolute and transcendental qualities. There is, however, a two-way influence here. Proverbial material not only serves to characterize the narrator,[114] but the narrator begins to characterize the proverbs and traditional sayings as well. As a part of Qoheleth's monologue, the narrator's own ethos begins to effect the gnomic quality of these sayings. Having been pulled into the gravity well of Qoheleth's ethos, they become in essence rhetorical satellites whose orbit is dictated by the weight of Qoheleth's personality. Like the moon which orbits our planet, these citations enjoy only a partial autonomy, unlike their siblings in the book of Proverbs. This literary dynamic fully subjectivizes these proverbs. Eventually the differences between the proverb (the Wisdom of many) and the aphorism (the Wisdom of one) becomes almost negligible due to the monumental effect of Qoheleth's monologue. Even when he disagrees with them, they serve as ways by which the narrator's peculiar consciousness may be characterized. While it may be form-critically valid to distinguish between the aphorism and the proverb in this book, from the point of view of their literary value and rhetorical effect, this distinction holds less validity. As a part of Qoheleth's thought, the quoted proverb is only

112. Schubert, 'Die Selbstbetrachtungen Kohelets', p. 31.

113. Bratsiotis, 'Der Monolog im Alten Testament', p. 41.

114. Aristotle observed that maxims have a powerful rhetorical effect on the reader's estimation of a person's character and ethos. He states: 'Maxims…invest a speech with moral character. There is moral character in every speech in which the moral purpose is conspicuous; and maxims always produce this effect, because the utterance of them amounts to a general declaration of moral principles: so that, *if the maxims are sound*, they display the speaker as a man of sound moral character' Aristotle, *The Rhetoric*, Book 2, Chapter 21, 1395a:10-15 [my emphasis]. Obviously, in the case of aphorisms or the spurious use of maxims, a negative characterization is possible, as it surely happens in Ecclesiastes. This is especially the case when Qoheleth comments on a cited proverb. Qoheleth's evaluation of a given proverbial value will influence the reader's evaluation of his character in fundamental ways.

slightly different from Qoheleth's own aphorisms. Both become the Wisdom of one person.[115] The only difference is that the proverb has the confirmation of the public's general assent while the aphorism originates in the narrator's personal observation and, as yet, lacks this confirmation. In Ecclesiastes both eventually become intricately tied to Qoheleth's empirical method.

In order to train the model reader to have this interpretative reflex, the implied author has deliberately front-loaded a few proverbs in the prologue and at the very beginning of Qoheleth's interior dialogue (1.15, 18). Then, beginning at the end of ch. 3, and fully in ch. 4, Qoheleth's thought begins to turn more and more toward proverbial, or perhaps, dialogical matters. As a result, in the second half of the book, 'there is more of a tendency to quote, and comment upon, sayings and proverbs'.[116] It is probably therefore no accident that most of the quotations occur after 4.17 where Qoheleth begins to directly address his narratee. Qoheleth wants to engage the narratee/reader in a debate. Given the intensely dialogic nature of the proverbial genres,[117] the use of quotations enables Qoheleth to dialogue with his narratee. This draws the reader into the circle of Qoheleth's confidence, effecting yet another level of narrative intimacy between Qoheleth and his narratee/implied reader.

Finally, it should be noted that the implied author has placed these quotations into the *fictive* life of his protagonist. As I argued earlier, the

115. This in effect returns the proverb to the source from which it came. T.A. Perry argues that 'in its origins...Wisdom is the Wisdom of one' (T.A. Perry, *Wisdom Literature and the Structure of Proverbs* [University Park: Pennsylvania State University Press, 1993], p. 90). He argues that someone had to first observe and coin a given phrase. As a result, there is a dialogic tension which remains unsolved between the aphorism and the proverb. Perry concludes therefore that proverbs 'are the Wisdom of one and the wit of many' (p. 84).

116. Viviano, 'The Book of Ecclesiastes', p. 81.

117. Perry has fully discussed the dialogic nature of the Wisdom tradition and Qoheleth's discourse in particular. He notes the use of questions (cf. 1.3; 2.2, 12, 15, 19, 22; 3.9, 21; 4.8, 11; 6.6, 8, 12, and so on), the direct address form 'you' (cf. 4.17; 5.1, 4-5, 7; 7.16-17, 21-22; 9.7, 9-10; 11.2, 5-6; 12.1, also 'my son' in 12.12), the imperative (cf. 4.17; 5.6; 7.14, 17, 21; 8.2; 9.7, 9-10; 10.4, 20; 11.1, 6, 10), and the use of formulaic 'don't say' (cf. 5.5; 7.10; 12.1) as direct examples of Qoheleth's appropriation of the Wisdom tradition's dialogic nature. See T.A. Perry, *Dialogues with Kohelet*, p. 188.

sages learned that fiction can be a powerful rhetorical tool. Life situations have more potential persuasive power than the abstract maxim set in a literary text without a specific performance context. Lacking the original context for critical aspects of its proverbial lore, Wisdom writers such as Ecclesiastes attempted to introduce a fictional context for the proverb. In that new fictional context, a sage could project any number of contexts as he or she might deem appropriate and, therefore, address or perhaps debate any age regarding the hard fought insights of past generations.

In this we see the value of Ricoeur's concept of textual distanciation. Because of writing, a proverb is set free from its original context and gains a surplus of meaning. This enables it to be applied to new and unforeseen situations. However, it is also true that it is the nature of a proverb to be context flexible. William McKane contends that

> The paradox of the 'proverb' is that it acquires immortality because of its particularity; that because of its lack of explicitness, its allusiveness or even opaqueness, it does not become an antique, but awaits continually the situation to illumine which it was coined.[118]

Because the proverb is a 'portable paradigm whose fundamental role is to map one field of experience onto another',[119] the possibility of its use in an imaginative context was always present. The implied author of Ecclesiastes was a real pioneer. Having realized that fictional lives are more persuasive than abstract maxims in isolation, he has not only fictimized the reader but, by so doing, has fictimized the entire Wisdom tradition. Qoheleth's reflections upon the Wisdom tradition effectively resubmerges the Canon's proverbial lore back into the life of the individual, giving it a perennial context from which to function. Whereas the book of Proverbs had separated this tradition from its various performance contexts, the implied author of Ecclesiastes has valiantly attempted to recontextualize the Wisdom tradition by placing it in the fictive life-setting of his protagonist.

This goes a long way to explaining how the book of Ecclesiastes functions as protest literature from a Ricoeurian perspective. Claudia Camp has argued that when Wisdom's tenets become rigid and unattached to

118. W. McKane, *Proverbs: A New Approach* (OTL; Philadelphia: Westminster Press, 1970), p. 414. For a discussion of this see Fontaine, *Traditional Sayings in the Old Testament*, p. 15.

119. Fontaine, *Traditional Sayings*, p. 38.

life, the culprit is usually the complex sociological and literary dynamics involved in the process of removing proverbs from their original cultural context. By placing proverbs in a literary context, they lose their ability to function effectively in new social settings. The result of this shift is the loss of the performance context that originally clarified the purpose and meaning of a given proverbial expression. Camp states:

> It is only when the proverbs are removed from their context of real life and placed in a literary collection that the theoretical question about the relationship of common sense to the religious perspective arises in a problematic way. In the collection, context-less and hence changeless, that 'air of simple realism' of their expressed morality…begins to appear in conflict rather than support of the 'reality' of certain Yahwistic beliefs, especially those that stress the freedom and grace of the Creator and the personal care of the covenant Lord. Without the variation of performance context, the proverbial statement becomes an absolute, creating the appearance of lack of subordination to Yahweh's Wisdom. Removed from the real life situation in which it can actualize what it reveals, the proverb not only dies a slow cultural death but also, out of touch with the covenant context, a religious one as well.[120]

It is at this precise point that a Ricoeurian approach can help the reader understand what is happening in the book of Ecclesiastes. Rigidification is an unfortunate effect of the textualization of language. While the loss of intentionality often frees many literary texts in a creative way, such a loss of original context functions in a negative manner in the case of proverbial literature. They lose their ability to function as cultural models as well as their capacity to evaluate and affect real-life events. The result of the loss of the original performance context removes the flexibility built into the sociological dynamics of the original performance context, creating an aura of dogmatism around the proverb.[121] The de-contextualization of a proverb into a literary corpus entails a 'descent into platitudism'.[122] Camp's view suggests that the problem tackled by Qoheleth has hardly anything to do with historical or social crises as is so often assumed. Instead, the culprit is what Ricoeur would call the 'inscription of language'. The literary dynamics involved in the collection of proverbs causes the growth of

120. C. Camp, *Wisdom and the Feminine in the Book of Proverbs* (Bible & Literature Series, 11; Sheffield: Almond Press, 1985), p. 17.

121. Camp, *Wisdom and the Feminine*, pp. 177-78.

122. Camp, *Wisdom and the Feminine*, p. 182.

dogmatism. However, the seeds of renewal are also included in these literary dynamics. While the process of literary re-contextualization furthers the process of removing proverbs from their common sense context, it also leads to a more reflective milieu for these same proverbs, such as we see here.[123] The literary re-contextualization of proverbs plants the seeds for both dogmatism and its corrective, pessimism. As a result, one can easily argue that it is textuality dynamics which create, or at least, contribute to the debate between Proverbs and Qoheleth. According to Camp, dissent is the result of literature (or textuality) providing for its own self-correction. The implied author's solution was to imaginatively re-embed the tradition back into the life of an individual in order to discuss its potential and problems.

As a result of this discussion, I would argue that Qoheleth's use of quotations functions along four broad lines. First, they tend to characterize the narrator as a wise man. Second, they engage the narratee/implied reader in a dialogue concerning the value of collective insight for the life of the individual. Third, due to their new context in the midst of Qoheleth's monologue, the proverbs and quotations are subsumed into the narrator's consciousness, becoming a part of his thought and speech. This effectively reduces them to another instance of saying 'I'. Finally, and most importantly, by baptizing proverbial lore into the fictive life of a solitary individual, the implied author has masterfully re-embedded the Wisdom tradition back into the context of *life* and its problems. For Qoheleth, the performance context of the Wisdom tradition must be the framework of the individual's life. He is the ultimate pragmatist, who contends that a gnomic statement can only be evaluated on the basis of its 'end' or result (cf. 3.11; 7.8). A proverbial statement has validity only if it works, that is, if there is a 'profit' (*yitrôn*) for the individual life into which it is being imported. This is perhaps the most radical and far-reaching effect of Qoheleth's use of first-person discourse. The protagonist of the book of Ecclesiastes has attempted to subsume public knowledge under the aegis of individual experience by his unique appropriation of traditional material. In that regard, Qoheleth is a true example of the postmodern mentality. However, as will be seen shortly, the implied author and the larger reading community would have a few things to say about this radical thesis.

123. Camp, *Wisdom and the Feminine*, p. 224.

8. *Endorsed Monologue: Narration Issues in the Book of Ecclesiastes*

In spite of the preponderance of 'I' in the book of Ecclesiastes, the powerful effect of third-person discourse on the reader still remains to be discussed. The article by Michael Fox, 'Frame-Narrative and Composition in the Book of Qohelet', marks a quantum leap forward for the understanding of how the use of third-person discourse affects the book's implied reader.[124] Fox argues that 'the Book of Qohelet is to be taken as a whole, as a single, well-integrated composition, the product not of editorship but of authorship, which uses interplay of voice as a deliberate literary device for rhetorical and artistic purposes'.[125] His thesis begins with the observation that the voice heard in the phrase 'says Qoheleth' (1.2; 7.27; 12.8) is the voice of the frame-narrator/ Epilogist. Particularly important in this respect is the abrupt insertion of this third-person phrase into the first-person statements in 7.27 and 12.8. Fox observes:

> We have here a third-person quoting-phrase in the middle of a first-person sentence, separating the verb and its modifier. While one can speak of himself in the third-person, it is unlikely he would do so in the middle of a first-person sentence, whereas a writer quoting someone else may put a verbum dicendi wherever he wishes within the quotation. *'āmar haqqōhelet* are not Qohelet's words in 7.27 and therefore probably not in 1.2 and 12.8 either.[126]

Fox argues that such a compositional move suggests more than mere editorship. He doubts whether an editor would insert a *verbum dicendi* into the middle of a first-person sentence. This suggests that 'whoever is responsible for *'amar haqqōhelet* ("says Qoheleth") in 7.27 is far more active than a mere phrase-inserter. He is active on the level of the composition of individual sentences'.[127]

Who is this voice? It is the voice of the frame-narrator, commonly called the Epilogist. The rhetorical function of this voice is to control and shape the reader's attitude toward the main character and to set a

124. Fox, 'Frame-Narrative', pp. 83-106.
125. Fox, 'Frame-Narrative', p. 83.
126. Fox, 'Frame-Narrative', p. 84.
127. Fox, 'Frame-Narrative', p. 86.

certain distance between him or her and the implied author.[128] Fox points out that the modern reader is predisposed to expect a frame-narrator to be more prominent in the beginning of a work. For instance, one usually expects to hear some sort of 'voice over' narration by the teller of the story, introducing the reader to the whys and whereabouts of the story about to be told. However, in Ecclesiastes this is reversed. Except for 1.2, the only 'voice over' we hear from this frame-narrator comes very late in the narrative, briefly in 7.27, and then quite pointedly at the end of the discourse (12.8-14). From the beginning, the frame-narrator works in very subtle ways, allowing 'the first-person speaker to introduce himself in order to establish him immediately as the focal point'.[129] By so doing, the implied author chooses to give a place of prominence to the ethos of the narrator as the controlling impetus of the discourse above that of the staid and traditional voice of the frame-narrator. This keeps the reader's attention centered on the 'I' of the narrator, Qoheleth. Eric Christianson summarizes the effect of this narrative strategy on the reader:

> By means of this [the narrative speech-act of 1.2] and the superscription it becomes clear that Qoheleth's character (i.e. the evolution and manifestation of it in his 'own words) is to be the principal concern of what follows. This is a thrust behind much modern fiction—to break away from the traditional notions of the beginning-middle-end procedure of the novel, not relying on the 'primitive' desire to know 'what happens next'. Instead, a plot may have as the centre of its narrative logic the revelation of character. Hence the expectancy aroused concerns a character's development through what it says and/or does and not necessarily how it interacts and develops in relation to others.[130]

The use of a frame-narrator creates several layers of narration in the book. At Level 1, there is the Epilogist who reports about Qoheleth-the-reporter (Level 2a). This is the level that creates an external frame around the intensely personal presentation of the speaking 'I'. Qoheleth-the-reporter is the elderly speaker who looks back on his youthful self, Qoheleth-the-seeker (Level 2b). The levels of narration are as follows:

128. Fox, 'Frame-Narrative', p. 91.
129. Christianson, *A Time to Tell*, p. 78.
130. Christianson, *A Time to Tell*, p. 78.

1. Level 1 = The frame-narrator/Epilogist (1.1-11; 12.8-14)
2. Level 2a = Qoheleth-the-reporter (1.12–12.7)
3. Level 2b = Qoheleth-the-seeker (2.1–2.17)

However, the fact that the discourse centers the reader's attention on the 'I' of the second level narrator, Qoheleth-the-reporter, does not mean that the role played by the frame-narrator is peripheral to understanding the book's literary dynamics. By creating an external frame around the narrating 'I' of Qoheleth, the Epilogist plays an extremely important role as a frame-narrator. The use of an external point of view is a strategic part of 'the structural isomorphism' of art.[131] In art, one point of view necessarily demands its counter opposite. For literary texts, public point of view seeks out the private point of view, while interior focalization demands exterior focalization. Both need each other in order to succeed in their effects. In the case of art, Uspensky refers to G.K. Chesterton's remark that a landscape without a frame is 'almost nothing'. He then notes that

> it only requires the addition of some border (a frame, a window, arch) to be perceived as a representation. In order to perceive the world of the work of art as a sign system, it is necessary (although not always sufficient) to designate its borders; it is precisely these borders which create the representation. In many languages, the meaning of the word 'represent' is etymologically related to the meaning of the word 'limit'.[132]

In any work of art it is psychologically necessary to mark out the boundaries of the depicted world for the reader.[133] In the book of Ecclesiastes this border is established by the *hebel*-refrain in 1.2 and 12.8, the poetic prologue in 1.2-11, and the discourse of the Epilogist in 12.9-14.

The importance of the 'right' frame for any artistic work goes without saying. In the reading history of Ecclesiastes, many interpreters have noted how Qoheleth's frame cuts against the grain of the monologue it borders. In some very strategic ways, the frame has a somewhat jarring effect on the reader due to its obvious ideological differences vis-à-vis the viewpoint of the protagonist. Recently, Eric Christianson

131. Uspensky, *A Poetics of Composition*, pp. 132-37. Christianson likewise observes: 'Frames with a symmetry provide the reader with a sense of origin and ending' (*A Time to Tell*, p. 121).

132. Uspensky, *A Poetics of Composition*, p. 137.

133. Uspensky, *A Poetics of Composition*, p. 138.

has fully explored this effect from the vantage point of art history as well as narrative theory. His study is both generative and highly insightful. After noting several key ideological differences between text and frame, he argues that the frame is a poorly matched border for the picture inside the frame. On the basis of this he concludes that there were two authors: Qoheleth and his framer. Comparing the framing of Qoheleth's monologue to the inappropriate frames several great paintings have received during the course of art history (especially Picasso's 'Pipe and Sheet Music' and its frame from 1864), Christianson states:

> Let us assume that the frame of Ecclesiastes is comparable to the mismatched frame of 1864. We therefore *assume that there is no hidden agenda, no subversive strategy of presentation at hand*. Those responsible for Qoheleth's frame simply misunderstood Qoheleth's story. The book, then, does not come from one hand but has at least two authors for Qoheleth's words and for the frame, each driven by wholly different visions of Wisdom and ways of knowing.[134]

Again, the hint of a possible subversive strategy has raised it head. However, it can be maintained that a subversive rhetorical strategy does indeed go to the core of the implied author's purposes for both the frame and his literary creation, Qoheleth. If one posits two authors for this work, then an interpretation along the lines suggested by Christianson is the conclusion one is forced to make. However, as Michael Fox has demonstrated, there seems to be more than enough evidence to show that whoever framed the book also created the character to fit inside the frame. Given this thesis, the reader must ask what rhetorical agendas are matched by such a 'mismatched' frame and picture. In the final analysis, it may prove worthy to see that the mismatching was quite intentional and perhaps, even, not as mismatched as some have perceived. Christianson himself admits that 'I cannot refute anyone's belief that Qoheleth (i.e. an author) achieved a subversive literary sophistication by creating the whole, but I suggest that someone else chose the frame for him, so to speak'.[135] Likewise, I would admit the same regarding the possibility of an independent framer for Qoheleth's monologue. But for me, the evidence points more strongly to the probability that the 'bad frame' is the result of a satirized strategy being played out on the text by an implied author. Too often critics have noted

134. Christianson, *A Time to Tell*, p. 123 (my emphasis).
135. Christianson, *A Time to Tell*, p. 125.

how something jars a reader, and have therefore, must be due to some genetic consideration. However, as Robert Alter has admonished, such geneticism seems to be based on the hangover suffered due to the influence of the historical-critical method. Sensitivity to the literary artistry of the text often leads to quite different verdicts.[136]

I will therefore agree with Christianson that this frame possesses properties that jar the reader. However, I tend to view these as purposeful within a rhetorical perspective which wishes to raise questions at an epistemological level. Indeed, the very subversiveness he senses is, from a rhetorical point of view, the essence of the matter. In spite of our differences, both of our studies agree on the fundamental point:

> The frame narrator is in the paradoxical position in that he validates Qoheleth's radicalism by appearing to find his words worth relating…it is clear that the frame narrator did not agree with Qoheleth's approach to Wisdom, God and tradition, bound as they were to his wholly different epistemology.[137]

The difference between Christianson's view of the frame and the one I advocate arises from a minor divergence of opinion regarding genetic origin. Of greater importance for both of our studies are the epistemological ramifications staked out by the protagonist and his narrative companion, the frame-narrator. However, in this study I take the position that these differences are purposeful within the narrative presentation of the *whole*.[138] That is, *both* Qoheleth and the frame narrator are

136. R. Alter, *The Art of Biblical Narrative* (New York: Basic Books, 1981), pp. 3-22.

137. Christianson, *A Time to Tell*, p. 11.

138. Christianson also attempts to view the meaning gained from a study of the implied author to the 'totality of meanings that can be inferred from a text'. However, due to his conclusions regarding authorship, he restricts this totality to 1.3–12.7. See Christianson, *A Time to Tell*, pp. 119-20. According to this stance, the implied author of the book of Ecclesiates is the protagonist of the monologue, Qoheleth. However, if Qoheleth is the fictional creation of an implied author, a different view of the implied author must be taken. If this is the case, both Qoheleth and the frame-narrator are products of the implied author. For this and other reasons, I have therefore chosen to reconstruct the implied author as taken from the totality of the book, 1.1–12.14. Auwers has advocated a similar position, arguing that the editor/ Epilogist may have created the fictitious character Qoheleth and is therefore the 'real author of the entire book and Qoheleth' (J. Auwers, 'Problèmes d'interprétation de l'epilogue de *Qohèlèt*', in Schoors (ed.), *Qohelet in the Context of Wisdom*, pp. 267-82 (282). However, as a close reading of Christianson's work will show, the

literary creations whose roles dissent because they represent two epistemological poles which were perceived as conflicting by the implied author. Indeed, the implied author of Ecclesiastes created their adversarial and mutually subversive relationship for the purpose of exploiting those well-perceived differences in order to say something about the prospects and limitations of all human knowing. Therefore, in this study I will attempt to understand not just their differences, but the total effect this adversarial relationship has on the reader and how those dynamics affect the meaning of the book as a whole. Seen from this perspective, the question is not whether this is a 'bad frame'. Rather, and more importantly, the question asked here is: What does all that say about the nature and limitations of human knowing in which ideological contestants routinely 'frame' each other in ways that limit dialogue and, therefore, the very quest for knowledge that both parties seek? This, it seems to me, is the deeper significance and meaning of Qoheleth's 'bad frame'. As will be argued shortly, here is an effect which seems to lie at the very center of the Ecclesiastes' rhetorical *raison d'être*.

In a book which is basically oriented from the internal point of view (1.12–12.7), the use of an external frame takes on a very great level of importance for the reader. The reader needs a transition from his or her external point of reference to the brazenly internal orientation of Qoheleth's monologue. Uspensky argues:

> If a painting is structured from the point of view of an outside observer, as though it were a 'view from a window', then the frame functions essentially to designate the boundaries of the representation. In this instance the artist's position concurs with that of the spectator. *However, if the painting is structured from the point of view of an observer located within the represented space, then the function of the frame is to designate the transition from an external point of view to an internal point of view, and vice versa.* In this instance the position of the artist does not correlate with that of the viewer; it is, rather, opposed to it.[139]

In Ecclesiastes, the reader is invited to look at the world through the eyes of Qoheleth, but always through the window provided by the

difference this makes are not overwhelming in terms of what a reader gains from the book.

139. Uspensky, *A Poetics of Composition*, p. 141 (my emphasis). Stanzel also discusses the importance of balancing internal perspective with an external perspective in literary works (*A Theory of Narrative*, pp. 11-12).

implied author whose voice we hear in the Epilogist. In order to prepare the reader for Qoheleth's peculiar worldview, the implied author furnishes the reader with a frame in order to soften the shock of Qoheleth's narration. The *hebel*-refrain and profit-motto in 1.2-3 alert the reader to the dominant themes that will be forthcoming. In addition, the poetic prologue on nature in 1.4-11 reorients the reader to the sort of worldview that is necessary to understand Qoheleth. The implied author thereby defamiliarizes the world as it is typically presented in the First Testament, providing the reader with an artistic bridge into the protagonist's consciousness. At the end of Qoheleth's discourse, the implied author re-transitions the reader back to his or her external reality.[140] The *hebel*-refrain in 12.8 refers the reader back to the initial introduction or 'doorway' in 1.2-3, functioning as a signal to the reader that he or she is being returned to the door through which they entered into Qoheleth's consciousness. Immediately after this, the voice of the Epilogist greets the reader, much like a guide might address a group of tourists at the end of a guided tour, signaling that the tour is over. This bequeaths a sense of closure to the work and provides the reader with specific clues on how to respond to Qoheleth's narration from the perspective of a reflective reading community.

As Fox and Uspensky have pointed out, such a technique is quite common in first-person works, especially folk tales. Typically, an 'I' appears at the end of a folk tale to give the narration a fitting ending.[141] In Ecclesiastes, this dynamic is reversed. Rather than the 'I' of a first-person frame-narrator, one is introduced to the third-person voice of the Epilogist. In this utterance the reader hears the voice of Qoheleth's reading public. An address to a 'you' is also typical of such endings. This occurs with the address to 'my son' in 12.12. The second-person address to the reader also signals a return to reality for the implied reader. Uspensky notes:

140. W. Anderson has also observed that the function of the poems in 1.4-11 and 12.1-7 is to provide an 'entrance' and 'exit' for the reader, although into and out of 'their working environment or life in the world' ('The Poetic Inclusio of Qoheleth in Relation to 1,2 and 12,8', *SJOT* 12 [1998], pp. 202-13 [209]). I would only note that this entrance and exit refers to the reader gaining initiation to Qoheleth's *poetic* world given the fictional poetics of his discourse. What the reader enters into is the private sphere of Qoheleth's mental world as a character rather than life itself.

141. Uspensky, *A Poetics of Composition*, p. 146.

> Addressing a second person at the end of the narrative is compositionally justified, in particular, in those cases when the narration itself is in the first person. Thus, both the intrusion of the first person (the narrator) and the intrusion of the second person (the reader) may have the same function: the representation of a point of view external to the narrative, which is presented for the most part from some other point of view.[142]

The function of the Epilogist is therefore to offer the reader an external or reality-oriented post of observation from which to evaluate the narrator. Furthermore, the address to the reader finishes the task of returning the reader to his or her own immediate reality. From the final ending of the work (12.13-14), one surmises that this reality is one in which divine obligations have supreme importance.

Finally, the intrusion of the Epilogist further signals a return to reality by giving voice to the thoughts of the implied author regarding his literary creation. The use of a frame-narrator creates a 'signature' for the implied author. Uspensky states:

> The first person narrator who appears at the end of a narrative has the same function as a self-portrait of the artist at the periphery of a painting and as the on-stage narrator in the drama, who in some instances may represent the author. The function of the second person, representing the audience of a viewer, may be compared in some cases to that of the chorus in ancient drama, which represented the spectator for whom the action was performed. The author often finds it necessary to establish the position of a perceiver—to create an abstract subject from whose point of view the described events acquire a specific meaning (and become significative and, correspondingly, semiotic).[143]

The use of a frame-narrator gives a public signature to an 'I'-discourse. This may be compared to the voice of the publisher who sometimes appears at the end of a first-person novel.[144] In addition, the summons

142. Uspensky, *A Poetics of Composition*, p. 147.

143. Uspensky, *A Poetics of Composition*, p. 147.

144. Romberg has also noted the use of this technique in his classic analysis of first-person texts. See B. Romberg, *Studies in the Narrative Technique of the First-person Novel* (Lund: Almqvist & Wilksell, 1962), pp. 34 and 65ff. Fox has called attention to the use of an 'anonymous third-person retrospective frame-narrator encompassing a first-person narrative or monologue technique' in ancient Egyptian literary sources ('Frame-Narrative', p. 92). In the Second Testament, the Epilogue to John's Gospel would be another example. However, as noted in Chapter 2, Lavoie has argued that 12.9-10 is unlike the typical ancient Near Eastern scribal colophon. He concludes that it is, rather, an anonymous allographic postscript written by the

to the narratee/implied reader likewise encourages the reader to make a stand with the implied author regarding Qoheleth.

By 'signing off' on the discourse of the narrator in this fashion, the implied author has given an endorsement to the narrator. Qoheleth is a radical figure, whose lavish scepticism is something of a rarity in the Canon. As such, Qoheleth 'needs a plausible, normal voice to mediate him to us and show us how to relate to him'.[145] Qoheleth receives this kind of endorsement from his narrative companion, the Epilogist.[146] The role of the Epilogist is to certify or to endorse this radical narrator as possessing valid rhetorical credibility for the reader, testifying that Qoheleth was indeed a wise man (cf. 12.9). This guides the reader's response by lending the credentials of orthodoxy and normalcy to this radical voice from the depths of scepticism. It protects the discourse from an overly negative response by the implied reader.

implied author. As such, 'one must see the simple signature of the implied author, that is to say, the manner by which an author is intrinsically present in his texts by his compositional choices and by the orientations which he has given in his book' (Lavoie, 'Un éloge à Qohelet', p. 170). In light of the insights gained from Uspensky, I would argue that in these verses we return to the reader's reality, beholding the signature and voice of the implied author. In that respect, I disagree with Fox who argues that the Epilogist is merely another 'type-character' not to be confused with the implied author. The fact that the implied author has used a stock compositional technique does not mean that the Epilogist is just another 'literary creation' (Fox, 'Frame-Narrative', p. 104). It would be better to argue that the implied author has slipped behind the mask of a traditional stock character in order to present himself, just as he utilized the fictional mask of Solomon to present his mentor. The fact that the Epilogist belongs to the external, reality-oriented frame of the book suggests to the competent reader that we have left the fictional core of the book in these verses (12.9-14), and have returned to the real world. Having passed over this fictional barrier, the implied reader responds to the third-person perspective of the Epilogist as a real voice aligned with his or her own external perspective. (Cf. Uspensky, *A Poetics of Composition*, pp. 141-51). As Uspensky points out, because the framework designates an *external* perspective, the implied reader is more apt to align this perspective with a level that concurs with his or her own. That level can only be the level of the implied author, since neither the implied author nor the implied reader exist on the level of the characters of a work. Therefore, in terms of its overall effect on the implied reader of the book of Ecclesiastes, the use of a third-person *external* frame-narrator undeniably expresses the voice of the implied author who presents Qoheleth to his reading audience. With Lavoie, I see the discreet signature of the implied author in these verses.

145. Fox, 'Frame-Narrative', p. 96.

146. Fox, 'Frame-Narrative', p. 100.

Empirical research regarding the interactive effects of sponsorship on the image of a speaker lends support to the point being made. Cary Mills has analyzed the relationship between three sources of credibility in any public rhetorical situation: the speaker, message and his or her sponsor.[147] While many studies have assumed that the primary source of rhetorical viability lies in the ethos of the speaker, Mills's study with real readers suggests that much of a low-credibility speaker's rhetorical viability lies in the type of sponsor that he or she enjoys as a public speaker. This creates yet another confounding influence for the reader who would respond to Qoheleth's 'I'. Speakers and sponsors may enjoy either a high or low-credibility. Mills observes that so long as a speaker's credibility remains high, then other confounding influences such as the credibility of their sponsor remain negligible.[148] However, in a situation where the speaker's credibility is low, such as is the case with a sceptic like Qoheleth, the influence of a sponsor is quite strong. As a result, Mills study predicted that when a speaker's credibility was low, 'high-credibility conditions in evidentiary sources and sponsorship would improve the speaker's image… Conversely, when his credibility was low and the credibility of evidentiary sources and the perceived sponsor were low, the speaker's image would be lowered'.[149]

In the light of these findings the reader of the book of Ecclesiastes may well question whether any reader would trust a narrator who stealthily and reticently utilizes Solomon's ethos and reputation as a mask. Given the tremendous criticism that Solomon has enjoyed in so many canonical and extra-canonical works, that would seem very unlikely.[150]

147. C. Mills, 'Relationships Among Three Sources of Credibility in the Communication Configuration: Speaker, Message and Experimenter', *SSCJ* 42 (1977), pp. 334-51.

148. Mills, 'Relationships', p. 338.

149. Mills, 'Relationships', p. 346.

150. The Solomonic Wisdom tradition is found in 1 Kgs 3.2-15 and 2 Chron. 1.1-13. However, later tradents and readers were not so enamored with the king. Josephus, the Wisdom of Solomon and Sirach all responded to this tradition with different interests and evaluations, not all of which were affirmative. For instance, Sirach was not positively predisposed towards Solomon in his 'Praise to the Famous' (Sir. 44.1–50.24). Solomon is mentioned in 47.12-22. He addresses him by an apostrophe form in the second-person, with the final verses being an accusation. Sirach, like so many later readers, perceived Solomon in an unfavorable light, especially because of his nefarious sexual proclivity with women (cf. 47.19, 'You laid your loins beside women and let them bear sway over your body'). For the various

Furthermore, much in Qoheleth's own speech works against him. As a result, the importance of the Epilogist as a public sponsor is likely to have a tremendous effect on the implied reader. In Mills's study, the effects of a low-credibility speaker with both a high and low-credibility sponsor were empiricially measured. His study concluded that 'subjects who were exposed to the high-credibility sponsor attributed significantly greater expertise to the speaker than did subjects who were exposed to a low-credibility sponsor'.[151] This means that whoever stands behind the speaker institutionally is very important for evaluating the speaker's credibility. Mills further determined that trustworthiness, and not just expertise, was the dominant element in a speaker's ethos. In this context, sponsorship 'emerged as the most influential variable'.[152] This means that for a text like Ecclesiastes, a strictly Aristotelian or Perelmanian analysis of Qoheleth's rational argumentation would only be partially effective in terms of analyzing the reader's response to the narrator. It is not just what Qoheleth says, nor the arguments or enthymemes he utilized to make his points which influence the reader's estimation of his character. The type of sponsor he enjoys has an equal influence on the reader's final estimation of Qoheleth's character. By lending Qoheleth the rhetorical sponsorship of a larger, more orthodox reading public, the Epilogist raised the level of trustworthiness for Qoheleth. In the process, the rhetorical liabilities that accompanied the Solomonic/royal mask and Qoheleth's own ethos-related qualities were offset by the trustworthiness engendered by the affirmation of a conservative public.

9. *Irony and the Implied Author's Use of Public Knowledge*

On the rhetorical level, what is at stake in the book of Ecclesiastes is the narrative use of public knowledge to certify the claims of subjective insights. One might even call it an early attempt at intersubjective

receptions of the Solomonic Wisdom tradition, see the study by Carr which provides a very comprehensive overview of the tradition-historical issues. See D. Carr, *From D to Q: A Study of Early Jewish Interpretations of Solomon's Dream at Gibeon* (SBLMS 44; Atlanta, GA: Scholars Press, 1991); P. Beentjes, '"The countries marvelled at you": King Solomon in Ben Sira 47.12-22', *Bijdragen* 45 (1984), pp. 6-14; E. Newing, 'Rhetorical Art of the Deuteronimist: Lampooning Solomon in First Kings', *OTE* 7 (1994), pp. 247-60.

151. Mills, 'Relationships', p. 346.

152. Mills, 'Relationships', p. 350.

verification by the larger reading community. As a scripture, or at the least, a document intended for public consumption by an ancient religious community, the epilogue certifies the private insight of Qoheleth with the broader, more public knowledge of the religious community. Lloyd Bitzer has given substantial thought to this dynamic in his work on rhetoric. He defines a public as

> a community of persons who share conceptions, principles, interests, and values, and who are significantly interdependent. This community may be further characterized by institutions such as offices, schools, laws, tribunals; by a duration sufficient to the development of these institutions...[153]

Public knowledge is defined as

> a kind of knowledge needful to public life and actually present to all who dwell in community... It may be regarded as a fund of truths, principles, and values which could only characterize a public. *A public in possession of such knowledge is made competent to accredit new truth and value and to authorize decision and action.*[154]

The strategic use of public knowledge is especially important for writings such as the Scriptures. Bitzer argues that the concept of public knowledge is quite essential 'to any theory of rhetoric that regards collective human experience as the legitimate source of some truths, and, thus, the authoritative ground of a class of decisions and actions'.[155] Whether one is talking about Perelman's universal audience, the endorsements of a larger reading community or even the critical insights offered by individual reader-response critics, subjective insights have always needed intersubjective certification by a competent and knowledgeable group before they can be accepted as 'truthful'. For a Wisdom tract which must dialogue with the tradition of the 'fathers', validation by the public becomes absolutely indispensable. As Walter Breuggemann has aptly observed, 'knowledge is notoriously parochial'.[156]

Chaim Perelman's concept of the universal audience is particularly insightful for understanding the role played by the Epilogist. The

153. L. Bitzer, 'Rhetoric and Public Knowledge', in D. Burks (ed.), *Rhetoric, Philosophy, and Literature* (Lafayette, IN: Purdue University Press, 1978), pp. 67-93 (68).

154. Bitzer, 'Rhetoric and Public Knowledge', p. 68 (my emphasis).

155. Bitzer, 'Rhetoric and Public Knowledge', p. 69.

156. W. Breuggemann, *Texts Under Negotiation*, p. 9.

universal audience consists of all rational members of the human community. An argument can be validated, according to Perelman and Olbrechts-Tyteca, only if the universal audience can be persuaded. However, he cautions the rhetorician, stressing that even though a local audience may be persuaded, this does not provide the sort of rational verification which the larger human community can provide. Group verification must be truly universal for it to play the part of the universal audience. As such, he defines a convincing argument as 'one whose premises are universizable, that is, acceptable in principle to all the members of the universal audience'.[157] By employing the use of a frame-narrator who expressed the certifying presence of a larger community, the implied author has attempted to validate the argument of this text through the use of a figure who attempts to stand in for the universal audience (as anachronistically understood by the implied author, of course). However, there remains the fact that no group or institution can provide such an august service. Perelman and Olbrechts-Tyteca warn the critic that no localized community, which surely includes that of the Epilogist, is

> capable of validating a concept of the universal audience which characterizes them...On the other hand, it is the undefined universal audience that is invoked to pass judgment on what is the concept of the universal audience appropriate to such a concrete audience... It can be said that audiences pass judgment on one another.[158]

Any group that is located in space and time has a limited ability to actualize the universal audience. Because of this caveat the inter-subjective verification that is offered by any group is merely the first step towards rational validation. The danger involved in appealing to the universal audience is that a group may confuse itself with the universal audience in an unjustifiable manner. Perelman points out that for many groups who attempt to become a universal audience, 'the universal consensus invoked is often merely the unwarranted generalization of an individual institution'.[159]

157. C. Perelman, and L. Olbrechts-Tyteca, *The New Rhetoric: A Treatise on Argumentation* (trans. J. Wilkinson and P. Weaver; Notre Dame: University of Notre Dame Press, 1969), p. 18.

158. Perelman and Olbrechts-Tyteca, *The New Rhetoric*, p. 35.

159. Perelman and Olbrechts-Tyteca, *The New Rhetoric*, p. 33. For an insightful analysis of how nwarranted generalizations have worked themselves out in our own context, see Toulmin, *Cosmopolis*, pp. 84-87. Such 'unwarranted generalizations' are

The appeal to a universal audience moves the critic into issues directly involving the implied author. Michael Leff has noted that in literary works such as Ecclesiastes, the use of a speaker who represents the voice of a larger group, society, or the universal audience is 'a construct created by the speaker's notion of reason, and the source of this construct is largely dependent upon presence'.[160] This means that the universal audience as invoked in our text by the use of an Epilogist is in every way a construct of the implied author. The implied author's use of such a tactic will characterize him as much as Qoheleth's aphorisms serve to characterize the narrator. As a result, the reader also responds to the ethos of the Epilogist/implied author as well. In essence, this means that functionally, the third-person narration of the frame-narrator becomes another 'I' for the reader. Again, we see where Qoheleth's use of first-person discourse has some very subtle and far-ranging effects. It affects even the stellar properties of the book's use of third-person discourse. This is one of the reasons why the Epilogist can only approximate or simulate the aura of omnisciency that often accompanies third-person narration.

Of course, the greater irony remains that nobody can ever conceptualize the universal audience, and so the text's rhetorical strategy must remain open to the validation of successive readers. Perelman argues that the concept of a universal audience entails three stages of actualization. Stage one involves the subject himself as he or she deliberates for reasons. Stage two occurs when the speaker addresses an interlocutor in a dialogue. The third stage transcends the specific interlocutors and groups involved in the argument, extending the text's rhetorical tribunal to the 'whole of humanity' as the final arbiter of its truthfulness. Of course, this means that in reality, the universal audience always

due to the influence of what has been termed in anthropological circles as 'local knowledge'. The term comes from the anthropological work of Clifford Geertz, *Local Knowledge: Further Essays in Interpretive Anthropology* (New York: Basic Books, 1983). Geertz suggests that all knowledge is limited to various 'local spheres'. These local spheres must interact with each other in a genuine spirit of dialogue, diversity, and pluralism to achieve a more universal standing. In a manner similar to Perelman, he concludes: 'The problem of integration of cultural life becomes one of making it possible for people inhabiting different worlds to have a genuine, and reciprocal, impact upon each other' (p. 161).

160. M. Leff, 'In Search of Ariadne's Thread: A Review of the Recent Literature on Rhetorical Theory', *CSSJ* 29 (1978), pp. 73-91 (82).

remains something of a virtual entity.[161] In the case of our text, the first stage is actualized in the self-address of Qoheleth ('I said in my heart', cf. 1.16; 2.1; and so on) as he reports his findings. Stage two is only partially actualized in the discourse. Its presence is felt by the reader in the implied dialogue between the Epilogist and Qoheleth and in the unexpressed or implied dialogue that seems to exist between Qoheleth and the narratee. The third stage is played out in the course of the reading history of the book, by the actual readers of Ecclesiates, provided they have the necessary literary competence to fairly judge the work.

10. *The Epistemological Spiral: The Ironic Presentation of Knowledge in the Book of Ecclesiastes*

As a result of these dynamics, readers have a very real validating role to play in verifying the book's truthfulness. Each of us as critics/readers are invited to play the role of the Epilogist for the book of Ecclesiastes. Because each reader can only postulate an abstract entity like the universal audience, the ultimate hope of validating Qoheleth's argument will remain just as open for each reader and every generation of readers as it did for the original implied author and the group that stood behind him. In this respect, the Epilogist functions not only as the voice of the implied author, but also as a role model for each successive generation of readers. We must complete the rhetorical role of validating Qoheleth's radical insights. But even here the rhetorical circle is never closed. Just as the rivers that constantly return to the sea (cf. 1.7), the reader needs his or her readings to be validated by the larger reading community in a never-ending epistemological spiral. This is just one more of the many ways that the book of Ecclesiastes engages in a vain rhetoric. While the Epilogist attempts to play a role comparable to the universal audience, the irony of this rhetorical strategy is that successive readers must validate both Qoheleth and the Epilogist. One should therefore not overly idealize the Epilogist's role. As the reading history of this book demonstrates, the Epilogist, much like his narrative interlocutor, has enjoyed a somewhat mixed reception by the reading community. Again, we see a predilection in this book for employing rhetorical strategies that have strong effects on the reader, but in the end offer only partially convincing results.

161. Perelman and Olbrechts-Tyteca, *The New Rhetoric*, p. 30.

If philosophical and literary truths need such intersubjective validation, how much more so in the case of scriptural truths which are often based in folklore traditions and multi-generational attempts to deal with life in an ancient setting? In a book such as Ecclesiastes, where the protagonist speaks almost exclusively from personal experience, who by selecting a variety of experiences and personal deductions for public consumption offered a corrective to the tradition of the 'fathers', the necessity of public affirmation is placed at a premium. Qoheleth threw down the gauntlet to his reading public. It should therefore come as no surprise that the reading public reciprocated in the voice of the Epilogist. Since then, readers have been contributing to the epistemological, or perhaps rhetorical spiral initiated by the implied author's use of public knowledge to validate subjective insights. This dynamic lies at the heart of what I am calling a vain rhetoric. By inviting the reader to play such a role, the implied author begs his audience to argue with Qoheleth. This creates an atmosphere that, at best, is characterized by literary debate, and, at worst, by rhetorical dissension. This dynamic, which is generated by the inherent and unavoidable aura of subjectivism that surrounds first-person discourse, lies at the very heart of a vain rhetoric. By that term, I am describing the strong but divisive effects of first-person discourse. It is the nature of such a rhetorical strategy not only to convince, but also to leave a good deal of doubt in the reader's mind. In terms of its final suasive effects, a vain rhetoric is a double-edged sword. It is suasive, but in the end, lacks persuasive force in any totally satisfying way.[162] Given the fact that this is a scriptural book which by definition is supposed to speak 'truth', its status in the Canon further exacerbates and even amplifies the vanity factor in its effect on the reader.

If ever the effects of first-person discourse were felt by a reading public, this was it. By choosing to base the rhetoric of the book essentially on the strengths and weaknesses of Qoheleth's 'I', the implied author spurned the aura of 'omnisciency' that surrounds so many canonical narrators,[163] and dared to construct a book that exuded

162. As I use these terms, suasion is the 'urge' in a text, while persuasion is the 'doing' of what a text urges. This is similar to the distinction made in speech-act theory when it differentiates between the illocutionary act and the perlocutionary act.

163. Sternberg has extensively discussed this aspect of the Bible's use of third-person narration (*The Poetics of Biblical Narrative*, esp. pp. 83-88). He describes

subjectivism. However, the weakness of a rhetoric based on first-person narration, with its built in predilection for subjectivism, cried out for the buoyant powers of third-person narration, with its power to produce the effect, or perhaps the illusion, of omnisciency. Even if at some level the implied reader responds to the frame-narrator as another 'I', in the book of Ecclesiastes the Epilogist is as close as we come to this general feature of biblical narration. While the Epilogist lacks the pure 'omnisciency' of, for example, the narrator of the book of Genesis, he does lay claim to the authority of the broader community as a validating corrective. As a result of this dynamic the Epilogist was undoubtedly a factor in the book achieving its final canonical status. However, I would argue this for different reasons. Many interpreters have argued that its canonical status is due to the Epilogist pulling the book back from the frontiers of hereticism and aligning it with the Torah-piety that was prevalent at that time. Historically, there is much truth to this position.[164] However, literarily, another dynamic is at work here, one which probably had as much to do with the book's final reception as Scripture, though in a very subtle way. I would argue that it was the use of public knowledge, which loosely simulated the usual canonical propensity for omniscient narration,[165] that probably had as

the typical biblical narrator as a being who 'stands to the world of his tale as God...' (p. 83). In most texts, the narrator is a privileged narrator who has perfect knowledge of his narrative world. See also R. Alter, *The Art of Biblical Narrative*, pp. 158-61.

164. Salters has explored this issue in several of his writings. See R. Salters, 'Qoheleth and the Canon', *ExpTim* 86 (1975), pp. 339-42. For an excellent overview of the specific problems that the Rabbis had with reconciling Qoheleth to the Torah (esp. Eccl. 11.9 with Num. 15.39), see *idem*, 'Exegetical Problems in Qoheleth', *IBS* 10 (1988), pp. 44-59.

165. L. Eslinger has also drawn attention to the dynamics of third-person omniscient narration as a key to understanding how canonical literature works as scripture. He argues that the doctrine of scriptural inspiration may in fact be an attempt to 'dogmatize and prolong' the experience readers have with such narrators. Eslinger sums up the matter: 'it is the genius of biblical authors to have developed a narratorial vehicle—the external, unconditioned narrator—to explore what would otherwise be a no-man's land of misconception and ignorance. The key to understanding biblical narrative, it seems to me, is neither history nor literary history, but an appreciative acceptance of the revelations of these extraordinary narrators' (L. Eslinger, 'Narratorial Situations in the Bible', in V.L. Tollers and J. Maier [eds.], *Mappings of the Biblical Terrain: The Bible as Text* [London: Bucknell University Press, 1990], pp. 72-91 [87]). Like myself, he too juxtaposes this type of narration with the

much to do with its eventual canonization. Without this effect on the reading community one could easily surmise that Ecclesiastes would have been read as just another tract from Hellenistic Judaism, much like the Qumran or Pseudepigraphal texts are read today. Qoheleth owes a lot to his narrative companion. Never did a student do more service for his mentor than did the Epilogist. In the end, the implied author's decision to bolster Qoheleth's private insights with the public affirmation of a frame-narrator was more than a stroke of genius—it was the rhetorical *coup de grâce* that enabled a sceptic to take his rightful place alongside the other notable personas of faith who typically cohabit the Canon.

In this the perceptive reader notices yet another level of irony. In no instance in the book of Ecclesiastes, whether in the narration of Qoheleth proper or the added perspective of the Epilogist, is anything other than natural insight given as a means to understanding the ways of God and the world.[166] Both lay hold of natural, human experiences or per-

'conditioned' status of first-person narrators. According to Eslinger, only 11 per cent of biblical narrative is mediated by conditional narration found in first-person discourse (p. 81). If this is subtantially correct, it shows just how dominant third-person narration is vis-à-vis first-person discourse within the canon. It also demonstrates how relatively rare it was that authors relied upon the powers of first-person discourse to communicate divine revelation.

166. It should be further noted that this type of reliance upon private insight is unique in the ancient Near East. Fox argues that the 'idea of using one's independent intellect to discover new knowledge and interpret data drawn from individual experience is radical and, I think, unparalleled in extant Wisdom literature' (M. Fox, 'Wisdom in Qoheleth', in L. Perdue, B. Scott and W. Wiseman [eds.], *In Search of Wisdom* (Festschrift J. Gammie; Louisville, KY: Westminster/John Knox Press, 1993), pp. 115-31 (121). Relying upon R. Braun, Fox does admit, however, that similar reliance upon private insight can be found in the Hellenistic environment. See Fox, 'Wisdom in Qoheleth', p. 122, relying upon R. Braun, *Kohelet und die frühhellenistische Popularphilosophie* (BZAW, 130; Berlin: W. de Gruyter, 1973), p. 178. Loretz also smells Hellenistic influence at this point. He suggests that Qoheleth's unique style of prose may very well 'reflect an attempt, prompted by Greek philosophy, to create a Hebrew form of elevated prose to express Wisdom philosophy for the Jewish world as well' (O. Loretz, 'Poetry and Prose in the Book of Qoheleth (1.1–3.22; 7.23–8.1; 9.6-10; 12.8-14)', in J.C. de Moor and W.G.E. Watson (eds.), *Verse in Ancient Near Eastern Prose* (AOAT, 42; Neukirchen–Vluyn: Neukirchener Verlag, 1993), pp. 155-89 (157). However, Loretz's thesis that the prose and poetry represent different authors cannot be maintained. Although Hellenistic influence *may* be the soil from which Qoheleth's radical epistemology

ceptions as a means to understanding God and life. Undoubtedly, many have portrayed the Epilogist as a Torah-bound enthusiast.[167] However, a word of caution is in order here. As the voice of the implied author who has created Qoheleth as a fictional character, it should be noted that 'Qoheleth' also expresses the values of the implied author, and that the majority of textual and rhetorical prominence is given to the presentation of these values. One must assume some sort of agreement with these values since Qoheleth looms so large in the discourse. Except for the passages dealing with fulfilling one's vows (4.4-5; 5.3), a specific reference to the Law occurs only on one other occasion (12.13-14, which may not even be original), and that reads more like a concession than an epistemological statement. If this statement is read in the light of the entire discourse, a better conclusion would be that the Epilogist merely wants to retain the validity of one's covenant obligation to the Torah given the uncertainty of empirically-based human knowledge. In that respect, the admonition to Torah allegiance is actually built upon the sceptical foundation which is established by Qoheleth-the-narrator. With so much that remains hidden to humanity, at least the Torah offers some concrete advice to guide one's path. In this respect, the Torah is the 'sum of the matter' (12.13). As a concession to Qoheleth's relativistic worldview and his insistence upon empirical confirmation, this statement supports, rather than denies the skepticism argued for by Qoheleth. At least, as a reader, I do not see a radical difference between Qoheleth and his narrative presenter. The only distinction is that the Epilogist is willing to fall back on the tradition of the Torah, whereas

sprang, a precise delineation of its genetic origins is not necessary to achieve a valid reading from a Ricoeurian point of view.

167. For an excellent review of this reading strategy, the reader is referred to the following studies: G. Sheppard, 'The Epilogue to Qoheleth as Theological Commentary', *CBQ* 39 (1977), pp. 182-89; *idem, Wisdom as a Hermeneutical Construct: A Study in the Sapientializing of the Old Testament* (BZAW, 151; Berlin: W. de Gruyter, 1980). However, the influence between Torah and Wisdom went both ways. Sheppard argues that Wisdom served as a hermeneutical construct, or a reading grid for the Torah traditions. Likewise, G. Wilson, ' "The Words of the Wise": The Intent and Significance of Qoheleth 12.9-14', *JBL* 103 (1984), pp. 175-92, also rehearses the connection between the Torah and the Wisdom school. Wilson argues that the point of the epilogue was to bring out the implicit connections between Prov. 1–9 and Deuteronomy (a position which I feel reads a bit too much into the phase, 'words of the wise' in 12.9). TSee also the highly insightful article by Dell, 'Ecclesiastes as Wisdom: Consulting Early Interpreters', *VT* 44, pp. 301-29.

Qoheleth was wholly bound to his empirical methodology for guidance. In this respect, the Epilogist truly was the disciple of his master. The only difference seems to lie in the emphasis that each gives to the problem. Qoheleth was daringly willing to lay hold of personal insight as the only valid means for understanding life's problems, while the Epilogist wisely tempered this notion with the necessity of paying attention to the larger experiences of the religious community. The disciple understood what escaped the master, namely, that no single person or generation can possibly lay claim to understanding the great circle of life or God. The discourse of the Epilogist, by referring the reader to the insights of a later generation and its public knowledge, provided a 'balancing corrective' to the primary effects and weaknesses of first-person rhetoric.[168] As a result, the implied author is extending the insight of earlier sages at the micro-level to the epistemological macro-level. In this regard, Ecclesiastes could very well be considered a *metacriticism* of Wisdom's epistemological foundations, or perhaps a trend about whose veracity the implied author had some reservations. In that we see where Qoheleth's reminiscences as a fictional character is an extension of the implied author's own reflections upon the role of Wisdom in the general pursuit of knowledge.

Of course, such an inter-generational debate is hardly new. As a rule, successive generations read texts differently largely because of the ubiquitous effects of effective historical consciousness.[169] Katherine Dell has cogently argued that subsequent readers have always approached Qoheleth with different interests and reading grids. In her reception-analysis of the book, she concludes that the issues which later readers had with canonizing the book were a result of different reader interests, biases and reading-grids which were strongly influenced by the Torah-

168. Although the Wisdom writers often did this at the level of the individual proverb. See J. Crenshaw, 'Murphy's Axiom: Every Gnomic Saying Needs a Balancing Corrective', in K. Hoglund and E. Huwiler (eds.), *The Listening Heart: Essays in Wisdom and Psalms in Honor of Roland E. Murphy* (JSOTSup, 58; Sheffield: JSOT Press, 1987), pp. 1-17; R. Murphy, 'Wisdom Theses', in J. Armenti (ed.), *The Papin Festschrift: Essays in Honor of Joseph Papin* (Philadelphia: Villanova University Press, 1976), pp. 187-200; *idem*, 'Wisdom: Theses and Hypotheses', in J. Gammie and W. Breuggemann (eds.), *Israelite Wisdom: Theological and Literary Essays in Honor of Samuel Terrien* (Missoula, MT: Scholars Press, 1978), pp. 35-42). The implied author of the book of Ecclesiastes seems to be doing this at the epistemological level.

169. Gadamer, *Truth and Method*, *passim*.

piety of nascent Judaism.[170] In a similar vein, I am simply suggesting that this dynamic was at work even at the level of composition by the implied author regarding the adequacy of personal insight to fully criticize the traditions of one's larger culture, or 'public'. Whoever was responsible for the creation of Qoheleth-the-narrator judged his literary creation by a slightly different standard than the values which are expressed by the narrator, Qoheleth. This creates an underlying and pervasive ironic dimension for the entire discourse. It also creates a sense of unreliability regarding the narrator. The critique of Qoheleth's epistemology by the implied author characterizes the narrator as an unreliable narrator in certain specific ways.

Qoheleth and his implied author therefore occupy rather different posts of observation in this regard. While there was surely a large degree of congruence between the implied author and the narrator simply due to the fact that Qoheleth's oration is given so much rhetorical prominence by the implied author, it is also true that there is a certain level of intellectual distance between the two. As noted in Chapter 2, narrative distance between a narrator and the implied author can take several forms; it may be physical, intellectual, emotional or moral. In Qoheleth's case, narrative distance varies according to the level on which it occurs. At the emotional level, the implied author is quite close to narrator, given the eulogy he receives in 12.9-10. Obviously, the two are separated by the great chasm of death on the physical plane. But at the intellectual level, the implied author has deliberately set a certain distance between himself and his mentor. However, this is probably to be expected. An important stage in the growth of any protégé(e) is the distancing of oneself from the pervasive influence of one's mentor. I would suspect that the sort of intellectual distance the reader infers from the frame-narrator is precisely this sort of distancing. It is probably due to the typical reserve that springs from one generation reflecting upon the achievements of its predecessor and finding room to disagree. This creates yet another level of irony between the two levels of narration. At the level of Qoheleth-the-narrator, subjective insights and private

170. Dell concludes her review of readers' responses to the book: 'in my view the authority of Qoheleth's work comes from its classification as orthodox Wisdom from the time that the text itself was formed... By the time the orthodox of a later generation were considering the book, it was being judged by different standards of orthodoxy which related to harmonization with the Torah' ('Ecclesiastes as Wisdom', p. 328).

knowledge function as a critique for the norms of the larger society. The greater irony of this situation is that, ultimately, it is the larger community which must endorse that critique, making the public the final arbiter in this matter. At the end of this discourse, the Epilogist strongly hints at the limits of such a strictly empirical and subjective approach. Perhaps the most biting irony here is that while part of Qoheleth's function in the Canon is to criticize certain dimensions of public knowledge, that same public knowledge functions to criticize the insights of personal knowledge and its criticisms of orthodoxy.

This creates an epistemological, or perhaps, a rhetorical spiral throughout the discourse in which public knowledge and private insight constantly interact with each other in a never-ending helix of conflict and confirmation. By means of the epistemological spiral, the implied author enables the reader to experience the fundamental rhetorical vanity/absurdity of individual existence. Each of us is caught in a vice whose grips are personal insight and community tradition. The book of Ecclesiastes fully illuminates this aspect of the human equation and enables the reader to come away with a better understanding of the limits of both personal and public knowledge. As I have shown, no one who ventures to stand in for the public can escape this predicament. However, the implied author also hints at the remedy for this situation. By looking at the problem of epistemology as a rhetorical spiral, the implied author suggests to the reader that it is the process of creative dialogue between individualism and corporatism which provides the human community with the means to escape the tyranny of both solipsism and traditionalism. While Ecclesiastes does not suggest that we ever attain absolute certainty in any specific matter, he does communicate quite effectively that the process cannot be subverted by either side without serious ramifications. Radical individualism can result in empty solipsism. Staid tradition can stultify new insights, leading to pessimism and jadedness—something our generation knows only all too well. But somewhere in their exchange, something wonderful happens. Tradition becomes renewed and individualism escapes the limitations of our common mortality. That fundamental insight, which holds these two virtues in creative tension, is the gift of Ecclesiastes' epistemological spiral to readers of Scripture. In this we see where the book of Ecclesiastes is both fundamentally aligned and opposed to postmodern thought. With Postmodernism, it radically insists on the right of the individual to protest and the critical role of the individual in the quest for knowledge. However, the implied author has also seen that if taken too far, this can lead the reader to a dead end in which they are trapped within the inescapable confines of their own experiences. The only way to achieve true knowledge and a modicum of certainty is to travel the rhetorical road which spirals between these epistemological axes. The dialogue between Qoheleth, the narratee and the Epilogist therefore acts as a model, showing the implied reader the sort of intellectual and spiritual honesty it takes to enter into this spiral.

Finally, in true Ricoeurian fashion, one can even say that there is a definite surplus of meaning here when it comes to the effect of the Epilogist on the reader. Admittedly, the insights gained by reading Ecclesiastes' use of a frame-narrator in light of Perelman's concept of the universal audience are patent examples of effective historical consciousness at work in this writer. Nevertheless, because it is the nature of a text to invite interpretations that go beyond their original intended effects, I maintain that such insights, provided they agree with the basic norms of the text, are legitimate realizations of the text's discourse strategy. I, for one, would see this as a real example of how the text has a definite surplus of meaning provided by the insights of modern rhetorical theory. Certainly, the implied author never imagined that his *own* work would also need the validating responses of the later reading community to complete the role he laid down for the frame-narrator. Undoubtedly this creates a fair amount of unstable irony in the book.[171] As a result, the Epilogist offers the reader a role which goes well beyond that which was probably intended.

The ironic conflict between the two levels of narration surely played a part in the book's mixed reception and the hesitancy of the canonizers to fully endorse the book as a Scripture. This ironic bi-functionality is

171. Chatman offers the following definition of irony: 'If the communication is between the narrator and narratee at the expense of a character, we can speak of an ironic narrator. If the communication is between the implied author and the implied reader at the expense of the narrator, we can say that the implied author is ironic and that the narrator is unreliable' (*Story and Discourse*, p. 229). The term 'unstable irony' comes from Booth's now classic analysis of irony. Stable irony is intended, covert, and fixed, while unstable irony is unintended and has a certain undefinable quality about it. Whereas stable irony has fixed boundaries and is limited to specific meanings, the unstable variety 'keeps on going', so to speak. See S. Chatman, *A Rhetoric of Irony* (Chicago: University of Chicago Press, 1974). I have used the term 'unstable' here because it has in all likelihood gone well beyond what the implied author initially proposed.

another aspect of Qoheleth's use of a vain rhetoric. In spite of the persuasiveness of personal testimony and empirical observation, first-person narration is always at risk of being too subjective for common consumption in the trans-generational way that Scripture is meant to function. Brevard Childs has picked up something of this in his analysis of this book from the perspective of its function in the Canon. He argues that the 'authority of the biblical text does not rest on a capacity to match original experiences, rather, on the claim which the canonical text makes on every subsequent generations of hearers'.[172] This insight also applies to the relationship between Qoheleth and the implied author as expressed in the views of the Epilogist. The dynamic interplay between the levels of narration in the book suggests a level of narration which critiques the notions of orthodoxy, but also a level where orthodoxy likewise receives its day before the tribunal of human reason. In that respect, the book in its present literary form possesses a different function than did the original insights and aphorisms which lie behind the literary mask of 'Qoheleth'. 'Private' and 'public' have an *infinitely* ironic relationship to each other in Ecclesiastes. In reaching this conclusion, I concur with Isak Spangenberg's summation of the ironic dimension in Qoheleth's discourse:

> it is important to perceive that the book does not merely contain loose ironic statements but *entirely reflects an ironic tone*. Commentators are able to identify some of the ironic statements, but no one (except Fisch) has ever emphasized the fact that the book *as a whole* has an ironic tone...the confusion which surrounds the concept of irony emanates from the reluctance and inability to distinguish clearly between *primary* and *secondary* forms of irony; namely, between irony as *a disposition* and *the manifestation of it*. The book of Qohelet contains both. The proclivity to be ironic is reflected in scepticism and doubt which abounds in Qohelet while pertinent occurrences can be found in specific sections in the book.[173]

Finally, I should stress the fact that the book retains a different function for successive readers than it did for its authorial audience. The need to validate always includes a broader public than any group can provide. This means that the implied reader of this book must continue the validating roles played by both Qoheleth and the Epilogist. The need for public validation by later generations of readers is also one of

172. Childs, *Introduction*, p. 589.

173. Spangenberg, 'Irony in the Book of Qohelet', p. 62.

the unforeseen results of textuality and distanciation. As Bernard Lategan observes regarding the effects of inscripturation on the reading process:

> inscripturation…not only preserves the message because of its structure, but also makes it transferable insofar as the text is not bound to its situation of origin but free to travel forward in time. Furthermore, the publication of the text means exactly that the text is made public, becomes accessible to others, and forms the basis on which any claim or argument concerning the interpretation of the message must be based. In this sense the text marks out the battlefield on which the struggle for verification and validity of interpretation is to take place.[174]

Due to the effects of textuality and distanciation, every reader indeed becomes the Epilogist for Qoheleth. By the same token, each reader must also enter into dialogue with the broader community just as Qoheleth did. By offering these roles to generations of scriptural readers, Qoheleth and the Epilogist have let every age experience the underlying vanity of human rhetorical existence. Whether we are conscious of it or not, all of us are caught between his or her own limited experiences and the claims of the broader human community. While the specific issues may differ for subsequent generations, the process by which one enters into this debate is still fundamentally the same. Never was such a rhetorical strategy more powerfully enlightening, and yet, so limited. At every level of its rhetorical existence, the book of Ecclesiastes effects a vain rhetoric on the reader. The book's rhetorical strategy is characterized by its stellar strengths and glaring weaknesses which curiously recreate the fundamental vanity/absurdity described by Qoheleth. In the end, that is the reason a reader can both love and hate the book of Ecclesiates, making it the Canon's favourite 'black sheep'.

11. *Summary of Reading Issues in the Book of Ecclesiastes*

Chapters 3 and 4 have attempted to recalibrate traditional historical scholarship for a reader-response/rhetorical model of exegesis, dealing extensively with the implied author's rhetorical utilization of ambiguity and irony respectively as a means to present his argument. Without having fully accented the issues involved in ascertaining the effect of

174. Lategan, 'Reference', pp. 75-76.

Qoheleth's ethos on the reader, these chapters have dealt with the *implied author's* use of first-person discourse and its general effects. With reader-oriented critics like Robert Fowler and Stanley Fish, the various historical and grammatical issues have been gleaned in an attempt to find the sundry reader problems in the book. In Ecclesiastes, these problems can be classified according to whether they exist as textual issues or as persona issues relating to the narrator's character and ethos. The basic overriding tasks which confront the reader revolve around linguistic and structural ambiguities at the textual level and at the persona level, issues pertaining to Qoheleth's use of monologue, the autobiographical quality of the discourse, the fictive nature and effects of the King's/Solomonic Fiction, how first-person discourse affects the use of quotations, the role of the Epilogist as a narrative presenter, and the relationship between private insight and public knowledge. After surveying the options for each general problem, I have attempted to recalibrate each issue in terms of its rhetorical effect on the reader as a vain rhetoric. As such, there is a thorough-going rhetoric of ambiguity at the textual level. On the persona level, the reader is confronted by a rhetoric of reticence which utilizes fiction in order to defamiliarize the reader's understanding of their own existence. Essentially, the major rhetorical strategy of the text is to fully exploit the strengths of first-person discourse by buttressing and critiquing its weaknesses by means of the use of public knowledge, corporate endorsement and public appraisal. In the process, this creates a very strong sense of irony surrounding the narrator's reliability as a critic of society's public knowledge. In Ecclesiastes, the ironist[175] is himself thoroughly ironized by the implied author.

If one looks at the text from a rhetorical perspective, we see that the sort of rhetorical bolstering which Qoheleth needed from a sponsor begins far earlier than the epilogue, the point at which most readers recognize the voice of an advocate. The implied author's task of reinforcing Qoheleth's ethos-related qualities was initiated early in the discourse by his use of the Royal Fiction to color his protagonist's visage with the aura of royal Wisdom and wealth. He continued by establishing a profound sense of intimacy and trust with the reader through the use of a monologic form, the refrain to enjoyment, and the

175. For this characterization of Qoheleth, I am indebted to Fisch who describes Qoheleth as a 'Hebrew ironist' ('Qoheleth: A Hebrew Ironist').

dialogic quality of his 'quotations'. Finally, the implied author finished his task of rhetorical buttressing through the use of third-person commentary in the epilogue and the various external framing techniques which supplement Qoheleth's private knowledge with public knowledge and corporate endorsement.

By separating the text's implied author from the text's narrator, the critical reader comes to understand that Qoheleth is presented to the reader in a favorable, though sometimes ironic fashion. The implied author focalized the narrator's pessimism through the lens of intimacy and authority. Qoheleth's love of ambiguity and irony is presented to the reader through the eyes of the implied author who characterized Qoheleth as a caring and authoritative teacher who nevertheless had exhausted the possibilities of private knowledge. As a result, his views stood in need of a broader, more public perspective. This difference between the narrator and the implied author creates a good deal of subtle irony in the book. In the end, the reader is asked to evaluate the Canon's intimate sceptic, and each generation has continued to do so in a never ending epistemological spiral. The overriding irony in the book of Ecclesiastes is that the effect of Qoheleth's 'I' is so great that even the external points of view that are expressed in the various quotations and the Epilogist's speech are pulled into the gravitational field of Qoheleth's 'I'. Everything is construed as having the limitations of first-person discourse, inviting the reader to argue with both Qoheleth and the Epilogist in a way which is unparalleled in Scripture.

To conclude, the book of Ecclesiastes utilizes a vain rhetoric at every level of its rhetorical existence. The chief effect of a vain rhetoric is to constantly imply its own limitations to the reader. This is the major reason why Qoheleth remains the Canon's most controversial book. It literally invites the reader to argue with it. But it also shows each generation the humble insight that we are all 'Qoheleths' who are equally trapped in the limitations of our own experience. The implied author reminds us through his use of a vain rhetoric not only that our physical existence is fundamentally absurd, but also that our epistemological existence is equally flawed. The book gently prods every age to remember that all human insight is limited and needs the broader perspective of ancient and future generations. Eventually, the book causes the reader to question his or her own hard-fought insights, especially those of us who have experienced the sort of deep-seated pessimism which characterized our text's protagonist. In that respect, the implied author has cast a very powerful light on the private experience of pessimism, disclosing to each generation that the answers to our doubts and railings are not to be found in the experiences which generate a sense of *hebel*. Instead, the subtle message Ecclesiastes gives the reader is that the way to address such radical questioning is by looking outside of one's personal, and limited experiences. As D.W. Hamlyn has so elegantly stated the issue in his overview of empiricism and its various cynical offspring: 'Skepticism is not to be answered by providing absolutely certain truth, but by examining the grounds of skepticism itself'.[176] Ecclesiastes would have heartily agreed with Hamlyn on this point. One responds to the charges of the pessimist not by answering him or her on their own grounds, but by examining the epistemological and rhetorical methods by which they came to such conclusions. In other words, one must do a 'Qoheleth' on Qoheleth, as the implied author so efficiently has done, to adequately respond to the Canon's preeminent pessimist. Ultimately, it is the book's ability to give the reader a narrative encounter with the weaknesses of staid traditionalism, public beliefs, empiricism and personal insight that gives it such a deeply religious character. Perhaps, in the end, that is why such a truly vain rhetoric appears in the Canon—because it so powerfully thrusts upon the reader the need for the transcendental point of view which only faith can provide, however partial that may be.

176. D. Hamlyn, 'Empiricism', in P. Edwards (ed.), *The Encyclopedia of Philosophy* (New York: Macmillan, 1967), II, pp. 499-505.

Chapter 5

ROBUST RETICENCE AND THE RHETORIC OF THE SELF: READER RELATIONSHIPS AND THE USE OF FIRST-PERSON DISCOURSE IN ECCLESIASTES 1.1–6.9

> Statesmen are not only liable to give an account of what they say or do in public, but there is a busy inquiry made into their very meals, beds, marriages, and every other sportive or serious action.[1]

1. *Introduction*

The previous two chapters analyzed the general literary and rhetorical effects of first-person discourse on the reading process. In this chapter and the one that follows I will analyze the specific persuasive and dissuasive effects of the *narrator's* ethos on the reader. As a contribution to the field of reader-response criticism, the various effects of Qoheleth's ethos will be analyzed by means of a linear reading of the book. The basic premise of reader-oriented approaches to literature is that a text unfolds in the mind of a reader in a linear fashion as the textual consumer comes upon succeeding words, sentences, paragraphs and major structural divisions. Because of this fundamental methodological premise, it is best to give a linear accounting of the various effects of Qoheleth's character and ethos in order to show how they develop as the reader progresses through the text. More importantly, a linear discussion of the text will allow the reader-oriented methods utilized by this study to be fully exploited. The lack of such linear readings in the

1. Plutarch, from *Plutarch's Lives. The Translation Called Dryden's*. (5 vols, rev edn. by A.H. Clough; New York: The Athenaeum Society (1905 [orig 1859]), cited by John Barlett in *Familiar Quotations: A Collection of Passages, Phrases, and Proverbs Traced to Their Sources in Ancient and Modern Literature* (10th Edition; revised and enlarged by Nathan Dole, Boston: Little, Brown, and Co, 1930), p. 927.

field is lamentable. Accordingly, I concur with the analysis of T.A. Perry on the current state of Qoheleth studies:

> At any rate, our most usual contact with Kohelet nowadays occurs through the experience of reading, and in the current rage for commentary—which seems for the moment to have replaced theological assertion—one simple literary prerequisite remains on the endangered species list, and that is the naive and linear or sequential reading of the text itself.[2]

Only by following the text as it unfolds, thereby tracking how the text sets up the reader for certain expectancies while arming the reader with specific competencies will the rhetorical impact of Qoheleth's 'I' upon the reader be grasped fully by the critic.

With that in mind, I embark on a linear analysis of the specific rhetorical effects of Qoheleth's character. This study will track three major lines of the reader's response. First, I will carefully track the effect of Qoheleth's ethos on the reader. Specifically, I will analyze the narrator's speech in terms of its attractiveness, trustworthiness and credibility. Second, I must pay careful attention to how the juxtaposition of internal and external posts of observation (that is, private insight vs. public knowledge) influence the reader's evaluation of Qoheleth's radical subjectivity. Third, I will note how the various textual problems, gaps, blanks, incongruities and ambiguities recreate in the reader the fundamental experience of *hebel*. Other reader tasks will also be considered as the text warrants.

2. *I, Qoheleth: The Use of First-Person Discourse in Ecclesiastes 1.1–2.6*

James Crenshaw begins his discussion of Israel's literature of dissent with the observation that 'the question of meaning is more basic than that of God, indeed that biblical man's point of departure was not God but self. In essence, the God question is secondary to self-understanding'.[3] In the case of Qoheleth, this observation has a certain ring of truth to it, even if the scholar does not agree to its general applicability for other First Testament writings. Especially in a book that uses first-

2. T.A. Perry, *Dialogues with Kohelet*, p. xii.

3. J. Crenshaw, 'Popular Questioning of the Justice of God in Ancient Israel', in Crenshaw (ed.), *Studies in Ancient Israelite Wisdom*, pp. 289-304 (291).

person discourse so extensively, the reader can grasp just how important the search for self-understanding, or perhaps more appropriately, world-understanding was for the ancients. When their sacred canopy developed leaks, as it did for Qoheleth, this concern rises to the top of their consciousness. Nowhere is this consciousness of self/world more apparent in the Canon than in the radical 'I' of Qoheleth. In nearly every verse of this text, Qoheleth's consciousness is placed before the reader as a filter through which to view the world. In essence, Qoheleth replaces Israel's sacred canopy not simply with a secular canopy as is often assumed, but with his own peculiar consciousness, a type of radical self-canopy. Some readers, however, have responded to this emphasis on the narrator's peculiar outlook in a negative fashion. So great is the role of Qoheleth's self in this book, that older scholars like Emmanuel Podechard discussed it under the rubric, 'Egoïsme', and named Ernst Renan, Abraham Kuenen and Paul Kleinert as contemporaries who would intersubjectively agree with this characterization of the narrator.[4] To be sure, there is a good degree of 'healthy ego' in the bodaciousness of Qoheleth's radical discourse. However, behind Qoheleth's highly personal and pessimistic outlook there still remains the fundamental vision of the Hebrew Scriptures for a just world. It was not so much that Qoheleth replaced Israel's visionary sacred canopy, but rather, that in his own personal experiences he simply could not 'find' it and went about reporting that fact. In this regard, there is still a good deal of vision behind the narrator's discourse. It is simply that Qoheleth has turned it on its head, so to speak, becoming a form of anti-vision.

3. *Ecclesiastes 1.1-1.11: Prologue and Preparation for Qoheleth's 'I'*

a. *Ecclesiastes 1.1: Qoheleth as Private 'I' and Public Servant*

The problem readers have characterizing Qoheleth begins with the static mimetic statement in the initial verse which names the protagonist.

4. E. Podechard, *L'Ecclésiaste* (Paris: Librairie Lecoffre, 1912), p. 196. He refers to how readers 'have observed the great place which egotism holds in the book of Qoheleth' (p. 196). Nineteenth-century readers who would side with Podechard on this characterization of the narrator include: E. Renan, *L'Ecclésiaste traduit de l'hebréu avec une etude sur l'âge et le caractère du livre* (Paris: Levy, 1882), p. 89; A. Kuenen, *Historische-kritische Einleitung in die Bücher des Alten Testament* (Teil III, vol. 3 of 3 vols; Leipzig: O.R. Reisland, 1897), p. 169; P. Kleinert, *Der Prediger Salomo* (Berlin: G.W.F. Müller, 1864), p. 2.

Ecclesiastes 1.1 informs the reader that the ensuing discourse is indeed 'the words of Qoheleth'. This educates the reader to consume the subsequent discourse as that of the text's protagonist. When a character is given a proper name, the reader naturally begins to attach traits to it.[5] It will also proleptically prepare the reader for the King's Fiction, giving the reader his or her first clue that the characterization in those verses is a mask, or a role-playing by the narrator, Qoheleth.[6] From the very beginning, the implied author arms the reader with the specific interpretative competencies they will need to respond to his literary creation. By beginning the text with a notice of authorship, the text centers the reader's attention squarely on the persona of the narrator. Of course, the critical reader will argue that the superscription is in all likelihood not an original part of the text—a supposition that is entirely reasonable. However, two things should be borne in mind here. The fact that this book has its origins in a late Wisdom setting with definite scribal tendencies should at least raise the possibility that the implied author was wholly capable of imitating the tradition of canonical superscriptions. Furthermore, given the fact that the superscription centers the reader's attention on the protagonist of this text, it could be possible that this verse was consciously constructed by the implied author to specifically introduce his rhetorical aims to the reader, that is, to explore the nature and limits of *individual* insights. By attributing the book to an individual, the superscription immediately begins the process of educating the reader as to the literary aims of the discourse. Conversely, if the superscription originates from the book's later reading community, v. 1 becomes a tacit reading interlude which communicates to the reader the insights of earlier readers who responded to the use of the royal/Solomonic mask, but who had correctly seen through the mask, and thereby attributed the book to the text's protagonist rather than to Solomon. In either case, the primary effect of the superscription is to instruct the reader to direct their full attention on the text's narrator who will be introduced shortly hereafter.

Having instructed the reader to focus their attention on the text's protagonist, a problem immediately raises itself. Most readers have had difficulty understanding the personal nature of this name given its

5. Burnett, 'Characterization and Reader Construction', p. 17.
6. Lux, '"Ich, Kohelet, bin König..."', pp. 335-40.

grammatical form. The name given to a character is often a key element in the characterizing process. Kathleen Farmer notes how names

> have a way of shaping how we feel about the objects they disguise... And the name given to a book has a very real power to shape our expectations of its purpose or its subject matter. We expect a book to give us some clue to the type of material we are going to read.[7]

In Ecclesiastes, the name does indeed shape the competent reader's expectations. The noun 'Qoheleth' is a feminine participle. However, this is hardly what a reader would expect for a masculine narrator. In form, it is similar to the feminine participles found in Ezra 2.55, 57 and Neh. 7.59 which clearly denote various offices in the post-exilic community. Its grammatical form, as is so well rehearsed in the scholarly literature, is thus more appropriate for an office, pen name, acronym or function rather than a specific individual.[8] As a result, Abraham

7. Farmer, *Who Knows What is Good?*, p. 141.

8. J. Crenshaw, 'Ecclesiastes', In J. Crenshaw and J. Willis (eds.), *Harper's Bible Commentary* (San Francisco: Harper & Row, 1989), pp. 518-24 (518). The fact that Qoheleth occurs in 7.7 and 12.8 with the definite article further supports the office interpretation of Qoheleth's name. Bishop notes that in Arabic and other related Semitic languages, the use of the feminine gender is widely used to designate offices like the 'Caliph' (Khalifah), and thus seems to be a 'common Semitic idiom' referring to various public offices. See E. Bishop, 'A Pessimist in Palestine (B.C.)', *PEQ* 100 (1969), pp. 33-41 (33). One can only surmise that perhaps the name is supposed to do double semiotic duty. An analogy in English would be giving the protagonist a personal name like 'Judge', a name which, while not common, is not without some current famous bearers, such as the film-star Judge Reinhold. The name Qoheleth clearly functions with a similar double-entendre as its major effect. This confuses the reader, adding a dimension of mystery to the protagonist's identity. However, the historical background of this office, like all other historical illusions in the book, has been obscured by the passing of time. As Michel has pointed out: 'Unfortunately, the office of a kohelet is otherwise unknown to us' (D. Michel, 'Kohelet und die Krise der Weisheit', *BK* 45 [1990], pp. 2-6 [2] [my translation]). Of the various conjectures on the market, the proposal offered by Crenshaw is probably the closest to having actually uncovered the original office of *haqqōhelet*. Based on the participial form *qᵉhillâ* in Neh. 5.7, which probably means 'harangue', he advocates that the office which the narrator held is that of an 'arguer' or 'haranguer', a type of Devil's advocate. See Crenshaw, *Ecclesiastes*, p. 33. 'Qoheleth' would then be a fictive play on such an office. Although Crenshaw subsequently claims that this interpretation does not fit the way that Qoheleth presents his observations, I fail to understand how the adversarial and disputative quality of the narrator's observations cannot but be consistent with such an office. Of the

Kamenetzky called it the 'Rätselname' or 'mystery/puzzle name'.[9] The irony of this situation should not escape the perceptive reader, given its context in a dramatic monologue which emphasizes personal address. The grammatical oddity of the name creates a good deal of mystery about the character, and adds to the difficulty readers have characterizing Qoheleth.

However, Qoheleth's name also enlightens the reader in some very specific ways. For Rüdiger Lux, the grammatical form is a conscious clue to the reader, forming a part of the text's design to camouflage the narrator behind a theatrical mask. The feminine participial form is a fictive clue to the reader to see the narrator's true identity, and thereby to see through the royal/Solomonic identity which will be offered by the King's Fiction. By giving the narrator such an ambiguous name the text gets the reader to focus on the problem of just who is addressing them, and presents them with their first gap, without providing them a forthright answer. This increases the reader's involvement by tendering

various proposals I have surveyed, this one best accords with the norms of the text in a broad sense. As an example of protest literature, such an office may lie behind the text, although admittedly, we have no external sources to clarify the precise function of this office for the modern reader. However, Whitley attests to such an office in the Qumran community, where we find the office of an 'accuser' or 'plaintiff' in the Community Rule (cf. 1QS 3.23). See C. Whitely, *Koheleth: His Language and Thought* (BZAW, 152; Berlin: W. de Gruyter, 1979), p. 5. He understands the meaning to be something like 'sceptic' based on comparisons with other Semitic languages (p. 6). Given the broad-based questioning of the Wisdom tradition's most cherished tenets by Qoheleth, it is reasonable to understand the function of Qoheleth's public office to have been something like that described in the Manual of Discipline, except on the level of a teacher whose duty it was to critically interpret the tradition (cf. the reference to 'goads' in Eccl. 12.9 would naturally fit into such an office). Such a conjecture seems to be the most natural, given the muted nature of the text's repertoire at this point. However, the reading history of this verse shows that readers have typically understood 'Qoheleth' as an unexpected designation for a personal narrator. The competent reader realizes that the name does not fit the context, creating a gap for anyone who understands its nuances. Other conjectures for this cryptogram have also been adduced, such as 'teacher', 'assembler', 'collector of proverbs', or even possibly an oblique reference to Solomon based on the apposition of 'Solomon' and *qāhal* in 1 Kgs 8.1 (cf. Renan, *L'Ecclésiaste Traduit*, p. 13).

9. A. Kamenetzky, 'Das Koheleth-Rätsel', *ZAW* 29 (1909), pp. 63-69; *idem*, 'Die Rätselname Koheleth', *ZAW* 34 (1914), pp. 225-28; *idem* 'Die ursprünglich beabsichtige Aussprache der Pseudonyms QHLT', *OLZ* (1921), pp. 11-15.

a literary puzzle: 'I am both role and person—who or what am I?' The reader comes away with an answer, but one with which they are not wholly satisfied. The gap raised in the reader's mind begins the process of characterization by presenting a partial identity. This technique of doling out inconclusive answers is a trademark of the implied author's rhetorical design. With Qoheleth's strange name the implied author begins his rhetoric of reticence and ambiguity.

The dual function of Qoheleth's name allows the implied author to emphasize the individuality of the presenting self in an explicit way while exploiting the functional connotations of the eponym to imply the public aspects of the protagonist's role in society. Brevard Childs observes how the professional connotations of Qoheleth's name communicates that the narrator 'had an office or at least a function within the community... His use of wisdom was not just a private affair, hence the name Koheleth.'[10] By giving such a name to his protagonist, the implied author has deliberately chosen a name that would communicate to the reader not only an individual, but also something of the persona's role in the greater community. From the very beginning of the narrative, the implied author subliminally plants in the reader's mind the intent to explore the nature of individual insights within the broader parameters of public knowledge. The name 'Qoheleth', with its personal and public connotations, masterfully fits the text's greater rhetorical context. By hinting at the narrator's role in society it legitimizes the narrator's right to address the public. The narrator's name thereby provides a critical role to play vis-à-vis society. This disarms the reader, allowing them to see the frontal assault of Qoheleth's critical gaze as a societally sanctioned function. However, it also serves to make that critical function accountable to the public for the service that it renders. This prepares the reader for the role played by the Epilogist and creates an expectancy that some sort of public reckoning awaits the narrator. As a result, the name given to the narrator enhances his legitimacy and therefore the trustworthiness of the protagonist in terms of his rhetorical status vis-à-vis the implied reader.

The name 'Qoheleth' occurs seven times in the book of Ecclesiastes (1.1, 2, 12; 7.27; 12.8; 12.9; 12.10). Six of its seven occurrences are located in the speech of the frame-narrator who presents Qoheleth to his reading public. As a result, the name functions as a part of the

10. Childs, *Introduction*, p. 585.

book's external frame, giving the implied reader a post of observation situated outside of the internal perspective of the narrator. By referring to the narrator's personal name the implied author signals to the reader that an external post of observation is being given from which to view and, eventually, to judge the text's protagonist. 'Qoheleth' therefore designates a post of observation aligned with the community's public knowledge, while 'I' denotes a post of observation aligned with the role of private insight. Whenever the reader confronts this third-person perspective—except for 1.12, where the narrator introduces himself—the proper name acts as a signal or conduit for the balancing corrective of the community's broader perspective. This is also true of the lone occurrence of the name outside of the frame in 7.27. The third-person reference to the narrator offers an objectivizing or public post of observation. As a result, 'Qoheleth' and 'I' are dynamically related to each other as a part of the text's structural isomorphism. The appellation functions as more than a personal designation. It performs above all as a focalizing agent that communicates to the reader when an external post of observation is being tendered by the implied author.

To sum up, the name 'Qoheleth' has both a personal and public meaning. It designates at once both an individual identity and a public office whose precise function remains clouded in lexical and historical obscurity. This bi-functionality confuses the reader's first attempts to characterize the narrator. However, it more than makes up for this lack of clarity by bestowing on the narrator a definite aura of legitimacy, trustworthiness and accountability. This insinuates to the reader that the address they are about to hear is both a necessary and socially certified proclamation. In addition, the term functions to intimate the broader rhetorical purposes of the book in terms of its presentation of the ironic relationship between private and public knowledge. 'Qoheleth' therefore acts as a thematizing device in the narrative. Finally, the name Qoheleth functions as a framing mechanism for the narrator's discourse. The third-person designation signals to the reader that the narrator is now being focalized through an exterior post of observation.

b. *Ecclesiastes 1.2-3: Prolegomena to Pessimism—Introducing the Master's Motto*

Ecclesiastes 1.2-3 continues the characterization of Qoheleth by summarizing the narrator's worldview *in nuce*. These verses and the prologue on nature in 1.4-11 defamiliarize the reader, introducing the sort of worldview it takes to adequately understand the narrator's discourse. In the process, a good deal of characterization takes place as the reader must ask what sort of person would see things in this manner. Verse 2 emphasizes the theme of life's absurdity while v. 3 is a rhetorical question that broaches the issue of the overall lack of profit which plagues mankind's collective efforts. Moreover, v. 2 introduces Qoheleth's narrative companion, the frame-narrator, to the reader by way of the phrase 'says Qoheleth'. The book does not begin with the radical I-narration of the text's protagonist. Rather, Qoheleth is introduced to his reader by the community in the guise of the frame-narrator/Epilogist. This softens the shock of Qoheleth's radical emphasis on his self. As a result, Qoheleth is mediated to the reader by the use of quoted speech/monologue. By utilizing quoted monologue, the implied author lends an aura of credibility to the narrator as a duly authorized speaker.[11] However, *ʾāmar qōhelet*, as an indicator of quoted speech, also suggests to the implied reader that what is about to follow should be viewed as the voice of a solitary individual. Having emphasized the discourse's origin in a lone individual, the frame-narrator imbues the narrator's speech with a subtle aura of subjectivity, saying, in effect: 'Dear reader, regard what follows with the same grain of salt you would any individual's insights'. By accentuating the discourse's origin in a single individual, the implied author has marked Qoheleth's speech as an example of private knowledge, insinuating to the reader something about its limitations from the very beginning.

In ascribing these views to the narrator's mental outlook, the frame-narrator offers the reader what Uri Margolin has described as a 'characterizing statement'. It is important to observe that the reader encounters a summary characterization by the frame-narrator. Rather than letting the readers infer this on their own, the implied author offers readers a public perspective on the character of Qoheleth through which their estimation of the narrator is guided. With this summary, the reader begins to characterize the narrator in earnest. Pauline Viviano observes the effect of such a radical opening assertion on the reader:

> Thus the reader is drawn into the text by the extreme nature of the opening statement that all is vanity… The reader wants to know why, to know how, the author came to such a conclusion. The author does not

11. D. Cohn, *Transparent Minds: Narrative Modes for Presenting Consciousness in Fiction* (Princeton: Princeton University Press, 1978), p. 76.

> proceed by answering any of these questions; rather he asks a rhetorical question: 'What profit has anyone from all the labor which one toils at under the sun?' (1.3).[12]

Furthermore, by emphasizing these two themes, the implied author tells the reader exactly what it is that they are to view as the narrator's central point. This gives the reader a specific reading grid or lens through which they are to interpret the narrator's observations. Such a summary characterization of the narrator trains the reader to pay strict attention to the words *hebel* and *yitrôn* in the ensuing discourse. Given the multitude of keywords which litter Qoheleth's monologue (work, do, occupation, trouble, evil, portion, vexation, and so on), the implied author does not allow the reader to wander aimlessly through Qoheleth's semantic universe. These verses act as a compass for the reader, establishing the north (*hebel*) and south (*yitrôn*) poles from which the other terms gain their latitudinal and longitudinal bearings.[13]

However, the semantic opacity of the term *hebel* also adds a great deal of ambiguity to this verse. This compromises its ability to thoroughly direct the reader's understanding of Qoheleth's message and character. Out of the 73 occurrences of *hebel* in the First Testament, 38 occur in Ecclesiastes. It can with some argument be considered the overriding theme of the book—though, admittedly, readers have occasionally argued for other themes, such as the 'what profit?' question,[14]

12. Viviano, 'The Book of Ecclesiastes', p. 81.

13. Of the 28 prominent keywords in the book, the leading candidates for the central theme has to be either the *hebel* or *yitrôn* theme. Although other words or themes are certainly important in the book, such as knowledge, know, death, enjoyment, striving, portion and work, these two words are the ones which seem to influence the reader's overall understanding and orientation of the other terms. The terms *hebel* and *yitrôn* are the two magnetic poles for this semantic universe since each seems to be the antithesis for the other and therefore mark the semantic range for the other terms. In fact, these two terms are so related to each other that Seybold has noted how *yitrôn* 'forces upon *hebel* the special sense of "that which does not count or matter", "null", "vain", "that which yields no result"'. See K. Seybold, '*Hebel*', in *TDOT*, III, pp. 313-20 (319). As a result of their semantic connectedness, *hebel* becomes a shorthand for the implicit conclusion of the 'What profit' question'. See R. Johnson, 'The Rhetorical Question as a Literary Device in Ecclesiastes' (PhD dissertation; Louisville, KY: Southern Baptist Theological Seminary, 1986), pp. 183, 244.

14. Ogden, *Qoheleth*, p. 13.

the call to enjoyment[15] or the labor theme.[16] Nevertheless, my reading suggests that most readers have grasped the *hebel*-theme as the leading concept in the book despite the fact that its precise meaning has been a constant source of frustration. The traditional translation of the term is 'vanity' in the sense of futile, meaningless, vapid, pointless, nugatory and empty. But that traditional understanding has been debated recently, with a plethora of meanings being offered. Edwin Good argued that the term signified the ironic dimension in life and suggested the translation 'incongruous'.[17] Graham Ogden endorses an understanding that is close to Good's, opting to identify the term's referent with 'the enigmatic, the ironic dimension of human experience; it suggests that life is not fully comprehensible… It in no sense carries the meaning "vanity" or "meaningless" '.[18] He concludes that the term simply connotes the mystery of life, and in no manner carries the negative connotations of its usual English translation.

However, other readers do not sense such a positive meaning here. Roland Murphy prefers the translation 'incomprehensible'. Michael Fox, C.B. Peter, Karl Haden, Eben Scheffler and Ardel Caneday translate the term as 'meaningless' or 'absurdity'. Seizo Sekine offers 'nihil' as a possibility. Others have tapped into the latent metaphorical roots of the term to grasp Qoheleth's meaning. Harold Fisch and Rashbam understand its basic meaning to be 'mist' or 'emptiness'. Frank Crüsemann perceives *hebel* to be a 'stirring of the air'. Specifically, it referred to the lack of profit, or economical emptiness that came with the rise of the Ptolemaic state. In that sense, one might even translate it as 'debit' or 'deficit'. He has also suggested the translation of 'shit' to catch Qoheleth's emotive nuances. Kathleen Farmer understands the root metaphor of *hebel* to be 'puff of air', 'breath' or 'vapor', and so translates it as 'breath of breath'. The word is therefore a metaphorical way of communicating something of life's fleetingness and our common

15. Whybray, 'Qoheleth, Preacher of Joy'. Other prominent readers who utilize this reading grid are N. Lohfink, 'Qoheleth 5.17-19: Revelation by Joy', *CBQ* 52 (1990), pp. 625-35; A. Gianto, 'The Theme of Enjoyment in Qoheleth [smch]', *Bib* 73 (1992), pp. 528-32, and the various writings of Gordis.

16. De Jong, 'A Book of Labour'. Another interesting reading which accents the theme of work is Johnston, ' "Confessions of a Workaholic" '.

17. E. Good, *Irony in the Old Testament* (Bible & Literature Series, 3; Sheffield: Almond Press, 2nd edn, 1981), p. 182.

18. Ogden, *Qoheleth*, p. 14.

mortality to the reader. A few obscure meanings have also been offered, such as John McKenna's translation 'contingency' or Karl Knopf's suggestion 'change'.[19]

The problem in understanding *hebel* lies in the terms's underlying metaphor which controls its specific meaning. Metaphor literally means to 'carry across' and thus to 'transfer' meaning from one thing to another. Specifically, a metaphor transfers the properties from thing X to thing Y. It treats one object as if it were another, though usually in some limited and precise manner. Readers process a metaphor by comparing the non-literal elements with the literal elements in a metaphorical statement. I.A. Richards analyzed the components of a metaphor, calling them the tenor, or general idea of the statement, and its vehicle, or the non-literal/pictorial image which imbues the statement with a comparative meaning.[20] According to W. Jordon and W. Adams, the

19. For a fuller discussion of the various proposals reviewed in this discussion see: Murphy, *Ecclesiastes*, p. lix; Fox, *Qoheleth and His Contradictions*, pp. 29-37; C.B. Peter, 'In Defence of Existence: A Comparison between Ecclesiastes and Albert Camus', *BTF* 12 (1980), pp. 26-43; K. Haden, 'Qoheleth and the Problem of Alienation', *CSR* 17 (1987), pp. 52-66; E. Scheffler, 'Qoheleth's Positive Advice', *OTE* 6 (1993), pp. 248-71; Caneday, 'Qoheleth', pp. 21-56; Fisch, *Poetry with a Purpose*, p. 160; Rashbam, *The Commentary of Rabbi Samuel ben Meir Rashbam on Qoheleth* (ed. S. Japhet and R. Salters; Jerusalem: Magness Press; Leiden: E.J. Brill, 1985), p. 90; Crüsemann, 'The Unchangeable World', p. 66; Farmer, *Who Knows What is Good?*, pp. 143-46; J. McKenna, 'The Concept of *Hebel* in the Book of Ecclesiastes', *SJT* 45 (1993), pp. 19-28; K. Knopf, 'The Optimism of Koheleth', *JBL* 49 (1930), pp. 195-99; S. Sekine, 'Qoheleth as Nihilist', in S. Sekine, *Transcendency and Symbols in the Old Testament: A Genealogy of Hermeneutical Experiences* (BZAW, 275; Berlin: W. de Gruyter, 199), pp. 91-128 (99-104). Excellent surveys can be found in Christianson, *A Time to Tell*, pp. 79-91, and D. Miller, 'Qohelet's Symbolic Use of *Hebel*'. Miller's study is extremely insightful. He sees three broad approaches that have used to decipher the meaning of the term. According to Miller, readers have opted for either the abstract sense, the multiple-senses approach, or the single-metaphor interpretation. To this he adds his own—the symbolic approach.

20. I.A. Richards, *The Philosophy of Rhetoric* (Oxford: Oxford University Press, 1936), pp. 95-100. Other useful studies on metaphor from a reader-oriented perspective are: U. Eco, 'The Scandal of Metaphor', *PT* 4 (1983), pp. 217-58 (see also the other contributions in that volume, which is dedicated to exploring the literary dynamics of metaphor); M. Gerhart and A. Russell, 'The Cognitive Effect of Metaphor', *Listening* 25 (1990), pp. 114-26; R. Bontekoe, 'The Function of Metaphor'; F. Brown, *Transfiguration: Poetic Metaphor and the Language of Religious Belief*

reader may choose from four cognitive processes to get at a metaphor's salient meaning. They can either (1) understand the metaphor via its tenor, (2) concentrate on its vehicle, they may (3) 'average' the two, accenting the likenesses of the two components, or they may (4) utilize a 'congruity' model in which both likenesses and differences are compared and contrasted.[21] When Qoheleth says *hakōl hebel* 'Everything is a hebel', *hakōl* ('everything', referring to life in general) is the tenor, while *hebel* ('breath', 'vapor', 'mist') is the vehicle. As a result, readers can either understand the statement via its vehicle or root metaphor, 'breath', or understand the verse via its tenor, that is, its various applications. Alternatively, they could average the meaning between the term and its various case-applications or perhaps delineate and compare the differences and likenesses. Richards underscores the role of contextual information in this process.[22] In situations where contextual clues are missing, the role of the vehicle is increased. In contrast, when a text offers contextual clues, the role of the tenor is magnified. In Ecclesiastes, we see both of these situations at work. Verse 2 is a classic example of a minimal context metaphor.[23] However, as the discourse develops, Qoheleth begins to expand on the meaning of the metaphor by the various cases or illustrations he cites as examples of *hebel*. The *hebel*-metaphor thereby becomes an extended context

(Chapel Hill: University of North Carolina Press, 1983); P. Ricoeur, 'The Metaphorical Process', *Semeia* 4 (1975), pp. 75-106; J. Soskice, *Metaphor and Religious Language*.

21. W. Jordan and W. Adams, 'I.A. Richards' Concept of Tenor-Vehicle Interaction', *CSSJ* 27 (1976), pp. 136-45. In a tenor model, the meaning of the metaphor is reduced to the meaning of the tenor, with the vehicle acting as a 'mere decoration of the tenor' (p. 137). Conversely, in a vehicle model, the meaning of the metaphor is reduced to the meaning of the vehicle. An averaging model attributes the meaning of the metaphor to an equal interaction of both tenor and vehicle. The congruity model asserts that the 'mediating reaction characteristic of each shifts toward congruence with that characteristic of the other, the magnitude of the shift being inversely proportional to the intensities of the interacting reactions...the congruity model takes into consideration the polarity of elements as well as the similarity' (p. 138).

22. I.A. Richards, *The Philosophy of Rhetoric*, p. 120, states: 'an impractical identification, can at once turn into an easy and powerful adjustment if the right hint comes from the rest of the discourse.'

23. A minimal context metaphor is defined by Jordan and Adams as 'one in which the tenor element and the vehicle element consisted of one word each and were paired by an assertion of identity' ('I.A. Richards' Concept', p. 139).

metaphor. This model allows the critic to analyze the semantic debate regarding the meaning of *hebel* with a greater degree of precision. Basically, the various proposals can be analyzed according to which aspect of the cognitive process a given scholar accents. There is no ready-made solution when one field of experience is mapped onto another. Most readers typically accent either the mapped experience (tenor) or the experience being mapped onto (vehicle). The two extremes in this particular debate are best summarized by the analyses of Michael Fox and Kathleen Farmer, who accent the vehicle and tenor respectively.

Farmer underscores the vehicle component of the metaphorical process. She argues that the problems which readers experience understanding the word *hebel* stem from the nature of the root metaphor which underlies the term's usage in Ecclesiastes. The word itself is an example of onomatopoeia—*hebel*'s guttural sounds imitate the act of human exhalation. As a result, the experience of breathing forms the basis of the word's root meaning. However, the function of a metaphor is to map one field of experience onto another, and that is where the problem lies for most readers. While v. 2 is concise and succinct, its simplicity does not allow it to delimit which precise nuance is being mapped onto which dimension of life ('everything' being inclusive, but a bit vague). What the statement gains in summary inclusiveness it loses in terms of specificity. As a result, the verse is not wholly suited to conveying the narrator's message from a semantic point of view. Kathleen Farmer summarizes the problem which lies at the base of understanding Qoheleth's use of *hebel*:

> Metaphors are intentionally provocative figures of speech which can be understood in quite different ways... It is possible then, that *hebel* (meaning a puff of air) might be understood in either a positive or negative sense. If the translation preserves the metaphor...the reader is forced to decide in what sense the comparison should be taken. In my opinion, it is unfortunate that many modern versions of Ecclesiastes have chosen to take the decision away from the reader. Most translations obscure the metaphorical nature of the original statement and replace the concrete, nonjudgmental phrase ('breath' or 'puff of air') with various abstract terms—all of which have decidedly negative connotations in English.[24]

24. Farmer, *Who Knows What is Good?*, p. 143.

The cognitive problem presented in Eccl. 1.2 is for the reader to compare the qualities that 'breath' and life have in common. For Farmer, it is the transitory qualities of a breath which dominates Qoheleth's appropriation of the term's base metaphor. Based on the term's use in Psalms, which describe the brevity of life and the transitory human concerns as compared to the eternity of God, she concludes that *hebel* is essentially mapping the breath's lack of permanence (rather than lack of worth or value) onto the experience of life (cf. Pss. 39.5, 11; 62.9; 78.33; 94.11; 144.4). She states: 'A breath, after all, is of considerable value to the one who breathes. However, it is not something one can hang onto for long. It is airlike, fleeting, transitory, and elusive rather than meaningless.'[25] Furthermore, she observes that frequently the word is paired with *rûaḥ* ('spirit' or 'wind', cf. 1.14; 2.11, 17, 26; 4.4, 16; 6.9), and so argues that the word's meaning is not absurd, but ephemeral or fleeting. The term *hebel*, then, functions to accent the motif of mortality which runs throughout Qoheleth's discourse.

However, Farmer fails to adequately account for the role of contextual information in deciphering a metaphor's salient meaning. The lack of contextual clues creates a gap which, fortunately, is filled in by the text's later appropriations and clarifications. By paying close attention to the term's extended context and, specifically, the tenor of its various applications, Michael Fox translates the term with the word 'absurdity'. His analysis understands the term's meaning not through its base metaphor, but on the basis of the examples which Qoheleth applies to it. While Fox acknowledges that in certain instances the word does possess a primary denotation of 'ephemeral',[26] he argues that the connotation of absurdity carries over even in these cases. In a case-by-case study, he shows how the underlying assumption behind *hebel* is the supposition that the system of rewards in life should be rational, which means for Qoheleth 'that actions should invariably produce appropriate consequences'.[27] Qoheleth illustrates the relationship between tenor and vehicle by discussing toil and its products (2.11, 19, 23, 26; 4.4, 7, 8; 5.9; 6.2), pleasure (2.1; 6.9), wisdom (2.15; 4.16; 7.6), speech (5.6; 6.11), death (11.8), divine justice (8.10, 14), and the totality of life (1.2; 6.4; 9.1; 12.8) as specific examples of how *hebel* as 'breath' or 'vapor' applies to the inner workings of life's general system of rewards. These

25. Farmer, *Who Knows What is Good?*, p. 145.
26. Fox, *Qoheleth and His Contradictions*, p. 43.
27. Fox, *Qoheleth and His Contradictions*, p. 47.

case-studies turn the limited context metaphor in 1.2 into an extended context metaphor, in which the context, that is, the tenor carries the majority of the term's meaning. When interpreted in its extended context, Fox concludes:

> To call something 'absurd' is to claim a certain knowledge of its quality; that it is contrary to reason—perhaps only to human reason, but that is the only reason accessible to humans without appeal to revelation. 'Incomprehensible' indicates that the meaning of a phenomenon is opaque to human intellect, but allows for, and may even suggest, that it is meaningful. 'Absurd' denies meaning, 'incomprehensible' denies only its knowability.

Other readers have also seen the absurd connotations of Qoheleth's appropriation of *hebel*. C.B. Peter compares Qoheleth to the French existentialist Albert Camus, noting that absurdity 'is the failure of the world to satisfy the human demand' and 'the frustration of man's desire for values which will not be transitory and for age which will not be swallowed up by death'.[28] However, one perceives in this definition an appropriation of both the transitory and empty connotations of human breath. Perhaps a better way to handle the semantic opacity of *hebel* is to allow it to have a dominant meaning ('absurd') with a variety of other nuances ('fleetingness', 'meaningless', 'transitory', 'futile').[29] This would accord well with the empirical study by William Jordan and W. Clifton Adams who researched the effects of I.A. Richards's concept of tenor and vehicle relationships on readers' processing of metaphors. Their study showed that with limited context metaphors (such as 1.2, when taken in isolation), the vehicle was a better predictor of readers' responses. However, once that metaphor was placed in a richer literary context, 'the vehicle model was a significantly poorer predictor'.[30] In fact, when contextual cues are added to a metaphor, the

28. Peter, 'In Defense of Existence', p. 37.

29. Caneday has argued that the various nuances of the term should be retained, much as a tapestry might weave different hues of a color to make an artistic impression. He states: 'The theme of evanescence, unsubstantiality, meaninglessness, vanity is carefully carried through the whole book, as a weaver threads this theme color throughout his fabric. It is sufficiently broad in its formulation, for it accurately summarizes the full contents of Qohelet (if one does not restrict the word *hebel* to a rigid or static meaning' ('Qoheleth: ', p. 37).

30. Jordan and Adams, 'I.A. Richards' Concept', p. 142.

tenor, averaging and congruity strategies for solving a metaphor's salient point all increased the reader's ability to solve the metaphor's meaning.[31] However, a good deal of ambiguity always resides in the use of metaphor, 'particularly if the resolution of the metaphor depends on a less salient or more unique dimension of the tenor-vehicle relationship'.[32]

This would be the case with Qoheleth's use of *hebel*. To describe the experience of futility or absurdity is nearly as absurd or futile as the primal experience itself, and one must always account for the inadequacy of language to encapsulate life's awesome presence. Stephen Halloran has aptly summarized the literary experience of absurdity:

> To speak in the face of this experience is always in some sense to move beyond the absurd, for the experience entails that language is grossly inadequate for articulating what is to the absurdist and the mystic the only significant human experience... The writer tries to say what is fundamentally unsayable. In the process, language 'goes on vacation' as Wittgenstein puts it...[33]

As a result, I would argue that one of the reasons that Qoheleth's use of *hebel* is so difficult to understand is that all attempts to describe the absurd encounter this linguistic and semantic difficulty. By choosing a metaphor rather than an abstract term—there were relatively few to choose from given the restrictions of the Hebrew/Aramaic vocabulary—the implied author has chosen a term that would encompass many meanings at one time, and was thus able to cast a wider semantic net in order to capture the various aspects of life's absurdity. This both enlightens and confuses the reader. However, on an emotional level, the metaphor that is latent in the word *hebel*, with its kaleidoscopic ability to describe so many aspects of life's absurdity, has had a tremendous effect on generations of readers. In this respect it has more than adequately served as a summarizing statement for the narrator's ensuing discourse.

In fact, such an effect seems to be purposeful given the rhetoric of ambiguity which rampages throughout the book. Recently, Douglas Miller has proposed that *hebel* functions as a symbol in Qoheleth's discourse. He proposes that Qoheleth builds upon the root metaphor of

31. Jordan and Adams, 'I.A. Richards' Concept', p. 142.
32. Jordan and Adams, 'I.A. Richards' Concept', p. 143.
33. S. Halloran, 'Language and the Absurd', *PR* 6 (1973), pp. 97-108 (98).

hebel as 'vapor' in order to construct 'a symbol by which to represent the entirety of human experience'. He correctly observes that Qoheleth is not consistent in his use of the term and, furthermore, that 'we may do well to consider whether such inconsistency is a part of his purpose'. For Miller, the three primary referents of *hebel* are insubstantiality, transience and foulness.[34] As such, the term is ineluctibly metaphorical, functioning as a tensive symbol which 'holds together a *set* of meanings that can neither be exhausted nor adequately expressed by any single meaning'.[35] That Qoheleth has chosen such a polyvalent and fertile term is quite consistent with the rhetoric of ambiguity which operates thoughout the book. In that, the keyword *hebel* puts Qoheleth's rhetoric of ambiguity in full view of the reader. As Miller surmises: 'Given the extent of Qoheleth's creative use of literary devices, it should not be surprising that his thesis statement (1.2; 12.8) involves *hebel*, a term capable of several senses, which Qoheleth employs with multivalency'.[36] However, it also serves to summarize the complexity of human existence for the protagonist. As Miller correctly adduces: 'None of the three metaphors by itself applies to all of human experience, and yet with this symbol, Qohelet can demonstrate that "all" is *hebel* in one way or another'.[37]

To sum up, the opacity created by the use of *hebel* has created a definite sense of ambiguity regarding the book's overall theme. Specifically, readers have found difficulty understanding the precise ways that the word's root metaphorical meaning summarizes the narrator's overall post of observation. In short, readers have a problem isolating which connotation in such a lively tensive-symbol is the salient aspect that succinctly sums up the overall meaning of the work. The basic problem is that tensive-symbols, which hold a variety of meanings, are poor vehicles for expressing unitary meanings, which is what most readers expect from a summarizing statement. Readers do not understand which of the various nuances of 'breath' is supposed to express the viewpoint of the narrator: the fleeting connotation, the empty connotation, the meaningless or absurd connotation, or some other aspect of the term's root metaphorical meaning. On the surface level of the text,

34. Miller, 'Qohelet's Use of *Hebel*', p. 443.
35. Miller, 'Qohelet's Use of *Hebel*', p. 444.
36. Miller, 'Qohelet's Use of *Hebel*', p. 445.
37. Miller, 'Qohelet's Use of *Hebel*', p. 454.

the implied author has communicated a summarizing statement. However, at a deep level, the verses act more like a statement whose intent is to expand on the subject, possessing lively evocative powers. The reader is confused by the way the surface and deep levels of this verse interact. The average reader does not expect a summarizing statement to complicate matters in the way that a metaphor does. By means of this bewildering interaction, the implied author has constructed a verse that generates in the reader the same sense of complexity and perplexity which underlies the experience of life's '*hebel*-ness'. Once one sees the monumental effects of such a technique, rather than its logical weaknesses, the rhetorical advantage and value of its utilization as a summarizing statement becomes apparent.

Verse 3 continues the assault on the implied reader's cognitive powers with yet another technique from the arsenal of ambiguity. The rhetorical question is used often in Ecclesiastes. There are 32 questions (or 34 if 1.10 and 10.10 are emended) asked of the reader by this text.[38] This accounts for about 12 per cent of the text.[39] Questions literally dot the text with an intensity and regularity that can only characterize the narrator as a questioner of traditional beliefs. Such queries depict the narrator not only as a sceptic, but as a man 'in defiance of school

38. Rhetorical questions are found in verses 1.3, [1.10]; 2.2, 12, 15, 19, 22, 25; 3.9, 21, 22; 4.8, 11; 5.5, 10, 15; 6.6, 8 (2), 11, 12; 7.13, 16, 17, 24; 8.1 (2), 4, 7 [10.10]; 10.14. For a fuller discussion of the rhetorical question in Ecclesiastes, see to the very comprehensive treatment by R. Johnson, 'The Rhetorical Question' and the succinct overview by Loader, *Polar Structures*, pp. 26. Other noteworthy studies of relevance for a reader-oriented approach are: H.A. Brongers, 'Some Remarks on the Biblical Particle *halô*', in *Remembering all the Way* (OTS, 21; Leiden: E.J. Brill, 1981), pp. 177-89; J. Crenshaw, 'Impossible Questions, Sayings, and Tasks', *Semeia* 17 (1980), pp. 19-34; *idem*, 'The Expression *mi yodea'* in the Hebrew Bible', *VT* 36 (1986), pp. 274-88; R. Gordis, 'The Rhetorical Use of Interrogative Sentences in Biblical Hebrew', in J. Burden (ed.), *The Word and the Book: Studies in Biblical Language and Literature* (New York: Ktav, 1976), pp. 152-57; M. Held, 'Rhetorical Questions in Ugaritic and Biblical Hebrew', *ErI* 9 (1969), pp. 71-79; B. Jongeling, 'L'Expression *my ytn* dans L'Ancien Testament', *VT* 24 (1974), pp. 32-40; R. Koops, 'Rhetorical Questions and Implied Meaning in the Book of Job', *BT* 39 (1989), pp. 415-23. Excellent studies of the 'what profit' question in particular are provided by J. Williams, '"What does it profit a man?"'; W.E. Staples, '"Profit" in Ecclesiastes', *JNES* 4 (1945), pp. 87-96; Fox, *Qoheleth and His Contradictions*, pp. 60-62.

39. Christianson, *A Time to Tell*, p. 219.

wisdom's assumptions that much can be known'.[40] However, they do more than simply characterize the narrator as a sceptic. They have a rhetorical effect on the reader at the pathos level as well as the logos level. Few literary techniques have such a powerful ability to defamiliarize reality as the rhetorical question. Although metaphor, analogy and irony are very instrumental in achieving this effect, the rhetorical question is the most direct means to defamiliarize reality and to elicit emotional or psychological responses from the reader (pathos).[41]

Chief among the effects of the rhetorical question is its ability to engage the reader's cognitive abilities and to stimulate the reader's interest in the discourse. The literary effect of such questions is 'to break the monotony of continuous declarative sentences and...invite audience participation'.[42] By asking a question, the reader is called upon to give an answer, thereby allowing the reader to participate in the production of the text's meaning in a most direct way. However, the very act of raising a question creates a gap in the text which cries out for closure. These gaps 'beg to be filled, and as a result hook the audience, the pure psychology of interrogation guarantees the capturing of the reader's attention'.[43] The intellectual 'vacuum' created by the rhetorical question pulls its victim into its circle of influence and drives the reader to solve its intellectual challenge.

In addition, the use of rhetorical questions trains the reader to begin asking questions on their own. By educating the reader to be a questioner, the implied author begins to shape the values of the implied reader, creating the sort of sceptical or pessimistic values that are necessary to accept the text's premises. In this regard, both the narratee and implied reader of the book of Ecclesiastes are characterized by these verses as well. While I would not portray them with the same sceptical qualities that Qoheleth himself possesses, it is clear that the narratee and implied reader are not averse to such questions, and could be cautiously described as the querying sort. This accords well the description of the narratee in 12.12 as 'my son', which obviously has

40. R. Crenshaw, *Old Testament Wisdom: An Introduction* (Atlanta: John Knox Press, 1981), p. 146.

41. R. Johnson, 'The Rhetorical Question', p. 115.

42. R. Johnson, 'The Rhetorical Question', p. 110.

43. Johnson, 'The Rhetorical Question', p. 117. S. Fish also comes to a similar conclusion regarding the literary dynamics of the question (*Self-Consuming Artifacts*, p. 60).

some sort of school scenario implying a youth who is engaged in whatever institutionalized educational process was available at that time, probably a formal setting.[44] However, as Baruch Hochman reminds the critic, it is not the precise setting of the narratee, but the sort of generic human characteristics which matters most in literary characterization. Those generic human values elicited by the characterization of the narratee through the implied author's use of rhetorical questions are openness to new ways of thinking, an ability to question received 'truths' and a strong calling to intellectual honesty.

The rhetorical byproduct of the cognitive gap raised by most questions is that it begins to defamiliarize the reader's worldview. A question 'refocalizes an issue, thereby allowing the reader to see an accepted norm in a new context'.[45] Raymond Johnson summarizes the rhetorical question's power to reorient the reader:

> The dynamics of the interaction between rhetorical question and reader create the potential for reshaping the reader's system of thought in that through the rhetorical question the reader enters into the thought world of the interrogator. Once in the examiner's realm, the reader may be subjected to a new frame of reference, and, as a result, the reader's norms were reoriented.[46]

As a result, the rhetorical question is preeminently suited for an interior monologue like the one found in Ecclesiastes. The question enables the reader to enter directly into the protagonist's mind, allowing them to

44. For a fuller discussion of the various educational settings in the First Testament tradition see E.W. Heaton, *The School Tradition of the Old Testament: The Bampton Lectures for 1994* (New York: Oxford University Press, 1994), esp. p. 137 where he lists Qoheleth's school under that of the 'honest doubters'.

45. J. Resseguie, 'Reader-Response Criticism and the Synoptic Gospels', *JAAR* 52 (1984), pp. 307-24 (310). It should be added that this article provides the best survey of the effects of the rhetorical question, and is foundational for anyone who wants to understand the various techniques of literary defamiliarization. Other techniques of defamiliarization discussed by Resseguie are analogy, contrasting characters, paradox, irony and entrapment (pp. 311-16). Basically, any trope that utilizes the cognitive function of comparison has the power to defamiliarize reality for the reader. By comparing X to Y, the reader is forced to perceive the compared object in a new light with different characteristics. This is the foundational cognitive move that is characteristic of all defamiliarizing techniques. In its most basic cognitive sense, defamiliarization is the taking of a characteristic from one thing and applying it to another so that the new object is perceived to have a different nature or function.

46. R. Johnson, 'The Rhetorical Question', p. 117.

begin to see things through his or her eyes from the inside out. By placing a rhetorical question so early in the text, the implied author is extending an invitation to abide with his protagonist, asking the reader to enter into an intimate relationship inside the mind of the protagonist. The constant asking of questions is a way of offering the reader a post of observation within the mind of the narrator while respecting the reader's own cognitive independence.

As a result, a chief effect of a rhetorical question such as we see in 1.3 is to invite the reader to immerse him or herself into the primal experience of private knowledge, to fully participate in the narrator's worldview and to suspend their own norms in order to more completely understand the world about to be depicted. In this verse the 'now' of the narrator and the reader coalesce, enabling them to share private insights in a very intimate fashion. This brings the reader into the circle of Qoheleth's confidence and begins to engender a sense of trust between reader and narrator. In this sense, the rhetorical question is the literary equivalent of the 'Vulcan mindmeld', to use an analogy from Star Trek. Obviously, such a powerful technique is a primary means by which an implied author consciously shapes the values of a reader to become the text's implied reader. It is a way to gain assent from a dissenting reader.[47] This is accomplished chiefly by inviting the reader to share in the formulation of an argument.[48] By asking questions that have an obvious answer from the narrator's post of observation, the text traps the reader in the 'ironies of faith'.[49] Once trapped by such cognitive gaps, the reader becomes a 'victim to the world view of the text'.[50]

The questions in Ecclesiastes have the following effects on the reader: (1) they convey irony; (2) characterize the narrator with the qualities of pessimism and scepticism; (3) create a prevailing sense of negativity in the reader; (4) create cognitive gaps which create a need for a new premise in the reader; (5) entrap the reader by various attention-getting mechanisms; and (6) defamiliarize the reader's worldview by causing them to question their own assumptions.[51] Questions are also very effec-

47. Gordis, 'The Rhetorical Use', p. 213.
48. R. Johnson, 'The Rhetorical Question', p. 251.
49. R. Johnson, 'The Rhetorical Question', p. 263.
50. R. Johnson, 'The Rhetorical Question', p. 251.
51. R. Johnson, 'The Rhetorical Question', pp. 245, 269.

tive at building rhetorical consensus for controversial positions.[52] They lay bare, in an indirect fashion, the major premises of an argument. At other times they illuminate the issue that is being contended between two parties. By so doing, the rhetorical question 'pulls the weight of the argument to the front'.[53]

In Ecclesiastes questions often build structural frames around pericopes, signaling to the reader the beginning and end of major sections of the discourse (e.g. the questions in 1.3 and 2.22 dofunction to open and close the introduction to Ecclesiastes).[54] Rhetorical questions sometimes announce the theme of the ensuing narrative and anticipate the outcome of the argument.[55] As a result, rhetorical questions often function as reading interludes for the reader, enabling the perceptive reader to determine 'the critical moment or premise of an argument'.[56] In Ecclesiastes they almost constitute a refrain in the text and, as such, provide a major structuring effect in the reader's mind. This is particularly evident in 1.3, where the rhetorical question anticipates an answer that is delayed until 2.11, and which is reintroduced to close the initial section in 2.22.[57] By beginning this section with a question the implied author has set the theme of Qoheleth's interior monologue squarely before the reader's eyes. Ecclesiastes 2.22 utilizes the rhetoric of redundancy to reinforce this effect on the reader. This guides the reader's understanding of the protagonist's essential point, which concerns the loss of proper rewards in life (*yitrôn*, 'profit').

However, while the question in 1.3 surely serves as a structural guide to the reader and invites the reader to see through the narrator's post of observation, its primary function at this time in the linear progression of the text is to defamiliarize reality for the reader so that he or she can be prepared to understand where the narrator 'is coming from'. Thus, 1.3 functions to

> constitute a significant vortex into which the reader of Ecclesiastes is ...unwillingly pulled...the reader's norms are stripped of their context

52. R. Johnson, 'The Rhetorical Question', p. 210.
53. R. Johnson, 'The Rhetorical Question', p. 215.
54. R. Johnson, 'The Rhetorical Question', p. 232.
55. R. Johnson, 'The Rhetorical Question', p. 215.
56. R. Johnson, 'The Rhetorical Question', p. 213.
57. R. Johnson, 'The Rhetorical Question', p. 215. See also vv. 5.9-16.

> and reoriented according to the contexts established by various phenomena, including limited human understanding, the effects of death, and the incomprehensibility of God'.[58]

With the doorway provided by the rhetorical question in 1.3, the reader enters into Qoheleth's world, a universe in which the 'toil-yields-profit' norm does not exist. In this respect the answer to 1.3 has been hinted at in the 'everything is *hebel*' refrain in 1.2. Furthermore, by pairing the word with negative synonyms such as *yitrôn*, Qoheleth insinuates to the reader that *hebel* connotes the sense of 'insubstantiality' as well as 'absurdity'.[59] The reader may begin to surmise the answer at this juncture in the text. However, the closure of that gap must wait until 2.11, at which point the definitative answer is given.

c. *Ecclesiastes 1.4-11: Qoheleth's Private World*

After inviting the reader through the doorway to Qoheleth's consciousness in 1.2-3, the frame-narrator proceeds to give a short guided tour of the narrator's world in 1.4-11. With these verses the reader's initiation into the requisite values needed to appreciate the narrator's counsel becomes complete. In this poem the reader begins to form a *Gestalt* of the narrator's character. Nothing characterizes a person like their worldview. By enabling the reader to see the world through Qoheleth's eyes, the implied author begins to characterize the narrator by employing the strongest means possible. This passage reeks of pathos. Its effect on the reader is more on an affective level than a cognitive one. In this respect the critic can perceive the subtle rhetorical strategy of the narrator, who vacillates between rhetorical strategies that focus on cognitive effects (1.2-3), and those that focus on emotional responses (1.4-11). The prologue on nature balances out the intellectual focus of the opening lines with a poetic introduction whose contours are expressly shaped to elicit an emotional reaction in the reader. As a result, we see where the implied author was not ignorant regarding the subtle power of pathos in a rhetorical situation.

58. R. Johnson, 'The Rhetorical Question', p. 262, see also pp. 255-59.

59. Miller, 'Qohelet's Use of *Hebel*', p. 447. As such, Qoheleth builds lexical competency by pairing words together. Other verses where the model reader is given competency to decipher Qoheleth's dense and metaphorical use of *hebel* via its pairing with other related terms which connotate insubstantiality occur in 1.8, 3.19-20, 5.9, 6.8 and 6.11.

The poem on nature has a 'tired' feeling to it; 'all things are wearisome' to its speaker. It majestic cadences march across the reader's mind with all the fervor of a worn-out old man on the way to visit his own mausoleum. Qoheleth would have heartily agreed with the quip by the poet Edward St Milais: 'Life is not one damn thing after another. It is the same damn thing over and over.' Such a dark, dreary outlook on the wonders of nature powerfully characterizes the one who speaks these words. However, the text is somewhat vague as to who exactly is speaking here, the frame-narrator or Qoheleth. Qoheleth formally introduces himself in 1.2 and in 1.12, the phrase 'says Qoheleth' serving to mark those verses as examples of reported speech. Furthermore, the staccato sound of these verses signal that it is a poetical introduction. Graham Ogden claims that Qoheleth is quoting a poem here, thereby hiding his 'I' underneath the book's poetic introduction.[60] On the other hand, the fact that Qoheleth has not been formally introduced suggests to the reader that either Qoheleth is being quoted by the frame-narrator,[61] or that the frame-narrator is summarizing the protagonist in a poetic fashion. Perhaps we should designate this as an example of ambiguous narration. Whichever it is, the prologue on nature offers a third-person perspective on the protagonist's outlook, and therefore adds a degree of objectivity to this radically different post of observation. Again, the implied author utilizes a rhetoric of reticence, keeping the reader in suspense as to who is speaking.

Several key themes are raised by the prologue. Verse 4 with its emphasis on the passing of generations[62] hints at the spectre of death

60. Ogden, *Qoheleth*, p. 30.

61. A few readers view this passage as an observation by the narrator. Loader argues that this is an observation without an introduction, such as 'I observed', 'I tested' (*Polar Structures*, p. 19). De Jong also reckons this poem is an observation, considering it the frontispiece to the book's first 'observation complex' ('A Book of Labour', p. 108).

62. Nearly all readers have seen a reference to the passing of human generations in this passage. However, Ogden has recently argued that the word *dôr*, here translated 'generations' in its usual sense, refers not to the 'passing of human generations across the stage of an unchanging world, but rather between a cyclic movement within nature which contrasts with earth's permanence' (G. Ogden, 'The Interpretation of *dwr* in Ecclesiastes 1.4', *JSOT* 34 [1986], pp. 91-92 [91]). As such, he concludes that the dominant meaning of the passage does not refer to human transience, but to the 'ebb and flow of nature, its perennial and cyclic movement on the one hand, and on the other, a world-order which remains fixed and immutable'

which looms so large in Qoheleth's thought. The poem on nature and the poem on aging in 12.2-7 envelop the book in a shroud-like fashion. The motif of mortality (note the use of the root *'lm*, 'to endure', in both 1.10 and 12.5) frames Qoheleth's monologue like a casket being prepared for burial. Verses 5-8 turn from the cyclical nature of human existence to the same phenomena in nature. Sun, winds, rain—all these things are found to be a repetitive bore, a *y^egē'îm* ('weariness'). Bo Isaksson observes how the use of participles in 1.4-8 clearly express iteration, further impressing upon the reader the cyclical nature of existence.[63] The force of these participles is to focus the reader's attention on what is the same, thus the narrator 'overcomes his readers with the fatigue of monotony'.[64] In addition, the present tense of these participles occurs on the 'now' level of the reader,[65] suggesting to him or her that this is their life that is being discussed. This overcomes the fatigue

(p. 92). This view has been rebutted by Fox, who notes that *dôr* never means 'cycle' as Ogden argues, and that the key to v. 4 is the meaning of *hā'āreṣ*, which means not 'la terre', but 'le monde'. The verse is concerned not with mortality or transience, but with the fact of the world's permanence. The point is that the passing of generations 'does not change the face of humanity…the persistent, toilsome, movements of natural phenomena of which mankind, taken as a whole, is one, do not really affect anything. All this is meant to show that, by analogy, human toil cannot be expected to do so' (M. Fox, 'Qoheleth 1.4', *JSOT* 40 [1988], p. 109) Still, even though I tend to side with Fox's analysis of the meaning of *dôr* here, the fact remains that the passing of generations does give the reader a veiled insight into the prominence of the theme of mortality for this work. Verses 4-8 of ch. 1 do have the cosmos as their immediate referent rather than humanity per se. However, later in the discourse Qoheleth laments the fact that generations come and go, and that we all die like the rest of the created order (cf. 3.19-21). As such, contra Ogden and Fox, the 'coming and going' of generations in its broader context within the book does have a reference to the 'passing of human generations' and 'human transience'. For the moment, the narrative is content to allow it to function as a part of the cosmological background which is being given to the reader. Later, this aspect of the cosmos's nature will be singled out as a leading contributor to the experience of *hebel* by Qoheleth. The fact that the discourse is content with foregrounding the theme at this juncture should not distract the reader from seeing its thematic importance for the overall text.

63. Isaksson, *Studies in the Language of Qoheleth*, p. 72.

64. L. Alonso Schökel, *A Manual of Hebrew Poetics* (SubBib, 11; Rome: Pontifical Biblical Institute, 1988), p. 71.

65. Isaksson, *Studies in the Language of Qoheleth*, p. 72.

experienced by the reader, and draws him or her further into the text's narrative experience.

However, the particular slant of this post of observation begins to characterize the speaker in a negative fashion. For the scriptural reader who has previously read this book, one can only surmise that the point of view in these verses must somehow be related to Qoheleth's peculiar worldview. With vv. 5-8 the defamiliarization of the reader's world is now complete. This is not the same wonderful orderliness that keeps the horrors of chaos at bay in Genesis 1, nor the majestic creation of Psalm 19. Contrast the two speakers. Psalm 19.4, like Qoheleth, claims 'there is no utterance, there are no words' to describe the glory of God in the cosmos. Qoheleth too is left speechless, but for entirely different reasons. Regularity has become 'daily-ness' for the speaker. Life and the cosmos are simply another burden to bear. Ecclesiastes 1.9-11 builds the sense of boredom to a crescendo of ennui. Not only is life regular, but its cyclical nature strips the universe of anything new, exciting or uplifting. In this respect, there is very little magnificence (a vision that elevates the human spirit) in these verses. The use of *yēš* ('there is') in v. 10 keeps the text focalized from a third-person perspective.[66] However, the use of the second-person imperative, *rᵉʾēh*, is a point of view shifter,[67] and marks the book's first indirect address to the narratee. This provides the reader with a textualized role-model, inviting the reader to see things through the narrator's eyes, but in a more forceful way as the imperative mood gives it a sense a urgency. It should be noted that, generally speaking, for the first part of the discourse the narratee in Ecclesiastes is identical to the implied reader in many respects, particularly in terms of its characterization (but not necessarily with regard to the total role laid down for the implied reader of this text). Logically, they relate to each other much like the overlapping circles in a Venn diagram. Their relationship may be diagrammed as follows.[68]

66. It should be noted that a few readers such as Mitchell see a question in this verse but that the *h*- prefix is lacking. See H.G. Mitchell, 'The Omission of the Interrogative Particle' in R. Harper, F. Brown and G. Moore (eds.), *Old Testament and Semitic Studies in Memory of W.R. Harper* (Festschrift W.R. Harper; Chicago: University of Chicago Press: 1908), pp. 115-29 (119). Loader also takes it in this fashion (*Polar Structures*, p. 26. So too the RSV).

67. Crenshaw, *Ecclesiastes*, p. 68.

68. My reading suggests a rather naive or unsophisticated use of the narratee by

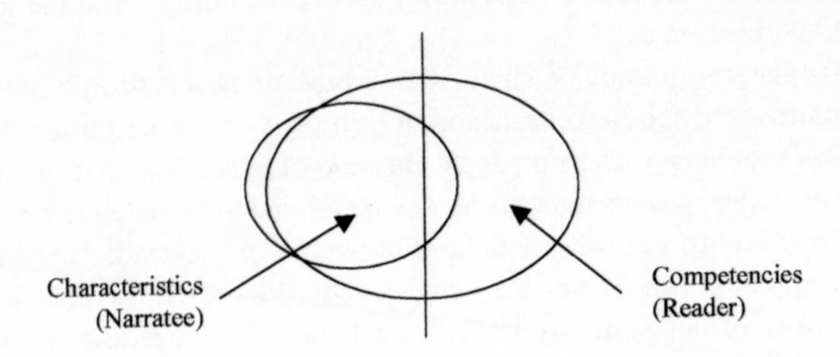

Fig. 5.1: *Narratee/Reader Relationships in Eccl. 1.1–4.16*

The pericope ends in a cascade of negativity with the particles *lō'* and *'ēn* dominating the reader's attention in v. 11. This gives the reader a strong hint that the initial 'what profit?' question posed in v. 3 should be answered in the negative.[69] By arguing that there is no remembrance

the implied author in 1.1–4.16. By this observation, I do not mean to eschew the book's literary sophistication or to criticize its literary techniques. I merely wish to note that unlike modern utilizations of the narratee, Ecclesiastes has virtually collapsed the two into one post of observation in many places during the first observation complex. Not until 4.17–5.8 and also in the Epilogue (12.8-14) do we see a parting of the ways in this regard. As a result, for 1.1–4.16 the reader of this study can read 'narratee/reader' when the term 'implied reader' is being *characterized*, unless a distinction is argued on a particular point. However, it should still be kept in mind that from the perspective of modern literary theory, the two should be considered separate entities. While Qoheleth may have collapsed the two in terms of their mutual characterization, the implied reader has functions which the narratee does not, such as relating the role of the narratee to his or her own role, putting together the perspectives of the implied author, narrator, narratee and implied author into a final *Gestalt* called the meaning of the text, and the like. In these respects the two remain quite separate. However, what Qoheleth asks of the narratee during his first observation complex, he is *generally* (but not always) asking of the implied reader, and whatever characterizes his narratee usually characterizes the implied reader as well. This continues until 4.17, the text's first instruction complex. In that respect, one might consider the narratee a subset of the characteristics laid down for the text's implied reader in these chapters.

69. From the way that Ecclesiastes delays answering his initial question in 1.3, Good argues that the typical strategy in this text is to raise a question, but to delay the answer as long as possible, thus heightening the response of the reader. The delay of gratification for the reader intensifies the affective response of the reader. For instance, Ecclesiastes suspends answering the question posed in 1.3—'What real advantage is there for a man in all the gains he makes beneath the sun?'—until

of a person after his or her death, Qoheleth obliquely suggests that nothing will be 'left over'[70] subsequent to one's death. Because the root meaning of *yitrôn* ('profit') is 'what is left over', the text strongly hints at the answer to the initial question submitted to the reader via this observation. Upon death a person enters into oblivion, with no existential 'plus' remaining (cf. 3.19).

Given the subjective hues of this introduction, the reader surmises that an 'I' lies close beneath the surface of this 'objective' report. Thus far, however, everything is presented in a poetically distanced manner. The form of a man's image begins to appear, but as yet, the face cannot be viewed by the reader. Narratively speaking, what the reader hears in

1.7-8. Qoheleth's style is therefore to 'interpose something else, or what seems like something else, between the expectations and its completion, to give a consequent that is not expected'. (see E. Good, 'The Unfilled Sea: Style and Meaning in Ecclesiastes 1.2-11', in J. Gammie and W. Breuggemann (eds.), *Israelite Wisdom: Theological and Literary Essays in Honor of Samuel Terrien* (Festschrift S. Terrien: Missoula, MT: Scholars Press, 1978), pp. 58-73 (72). How the relations of logical antecedents and consequents carry the load of presentation, set up expectations, delay cognitive fulfillment and 'goad' the reader into wondering what comes next is the key to understanding Qoheleth's discourse. Rhetorically, one might even characterize Ecclesiastes as a 'rhetoric of delay' or a 'rhetoric of frustrated expectations'. In either case, Qoheleth's style of delaying answers serves to undergird his central theme that concrete answers are elusive. Frustration of reader expectations is a powerful means of setting up the reader to actually experience *hebel* in the literary world of this aesthetic work.

70. Perdue has noted that *yitrôn* means 'to survive' in 1 Sam. 25.34 as a part of the semantic range of 'to be left over'. He then speculates that *yitrôn* 'in Qoheleth suggests not so much the idea of "profit" or "advantage" as it does "continuance" or "endurance"' ('"I will make a test of pleasure"', p. 208). While I would not argue against the base meaning of this term being 'profit' as Perdue does, the semantic range of the term is sufficiently broad to cover the idea of death and survival, and so is capable of sending a subtle message to the reader regarding the answer to the programmatic question in v. 3. I would not, however, go so far as Ogden does, who reserves a meaning which includes the 'eternal dimension' for this word (*Qoheleth*, pp. 15-25). There is very little in the rest of the text that would support this line of interpretation given Qoheleth's reference to humans dying like the beasts and other similar statements (cf. 3.19-21). As a result, interpretations such as Ogden's violate the broader norms of the text in quite flagrant ways. The term *yitrôn* hints at the loss of survival after death, and in no way supports such a possibility. With Fox, the word should be understood in two senses: its base sense of 'adequate gain' or 'profit' (cf. 1.3; 2.11; 3.9; 5.15; 10.11) or a comparative sense of 'advantage' (cf. 2.13; 3.19; 5.8; 6.8; 6.11; 7.11, 12; 10.10). See Fox, *Qoheleth and His Contradictions*, p. 60.

this passage is an abstract voice. The enfleshing of this voice must wait for brazen I-narrative which ensues shortly.

Readers have had diverse reactions to these verses, though the predominant response is somewhat reserved. J.A. Loader observes the treadmill-like repetition of Qoheleth's cosmos and notes how his 'pessimistic view of life is so strong it manages to darken even the sun's activities!'[71] But the sun and its seasonal regularity, while expressing a motif in the book, does not really get at the salient issue here.[72] More to the point is the fact that the universe's orderliness has been perceived by Qoheleth as being out of touch with human ambition, particularly concerning the issue of rewards. As Ardel Caneday has astutely commented: 'The earth, methodically plodding along in its routine course,

71. Loader, *Ecclesiastes*, p. 20.

72. The phrase 'under the sun' is a recurrent motif in the book, occurring 29 times in every chapter except ch. 7 (where a similar phrase is utilized in 7.11, 'under the heavens'). See Farmer, *Who Knows What is Good?*, p. 150. However, a minority of readers have turned the phrase into a major reading strategy, whereby they can hold to a dualistic interpretation which preserves the book's claim to revelation while allowing the unorthodox points raised by Qoheleth to function as a description of life without revelation. The 'under the sun' line of interpretation holds that the author 'deliberately concerns himself only with the things of this world… Revelation and the world to come are laid aside for the purpose of argument' (J.S. Wright, 'The Interpretation of Ecclesiastes', p. 20) For another discussion of this reading strategy see S. Holm-Nielsen, 'The Book of Ecclesiastes', p. 80. Nielsen notes how this line of reading is really an attempt to characterize Qoheleth as an orthodox writer. The problem with this reading is that it confuses a motif with a theme. The phrase, 'under the sun', is a coloring agent in the discourse that imbues the discourse with a certain earthly flavor or texture. It interacts with the motif of death which also permeates the text. A theme is more abstract and generally more semantically 'direct' in the way that it appropriates the reader's attention and controls the ideological position(s) of the text. In no way should this phrase be given a higher ideological function than it warrants. It is a texturizing agent, and does not function thematically except in the most vague of senses. See also D. Michel, '"Unter der Sonne": Zur Immanenz bei Qohelet', in Schoors (ed.), *Qohelet in the Context of Wisdom*, pp. 93-111, and H. Grossmann, *taḥat haššamâyim*: Anmerkungen zum Ort des Menschen bei Qohälät', in M. Albani and T. Arndt (eds.), *Gottes Ehre Erzählen* (Festschrift H. Seidel; Leipzig: Thomas Verlag, 1994), pp. 221-23. Grossman concludes that the phrase 'under the sun' has a positive connotation, while the phrase 'under the heavens' is used negatively to denote when humans attempt to go beyond their limitations set by God. If his conclusions are accepted, it may be concluded that the phrase does have an ideological content over and above its role as a texturizing agent.

does not skip a beat of its rhythm to celebrate a man's birth nor to mourn his death'.[73] The world has turned cold and silent for Qoheleth.[74] Johannes Pedersen correctly perceived the problem which Qoheleth laments: 'He has discovered that nature does not exist solely for humanity'.[75] However, both positive and negative readings are possible given the ambiguity of many of the words, phrases and ideas in this passage. Lindsay Wilson concludes: 'This is literary artistry at its best. It is not that a positive or negative reading alone is intended, but that the reader needs to see both the regularity and seemingly pointless repetition are true to life.'[76]

However, the honest reader intuitively senses that Qoheleth is essentially correct in his observations. The universe does not stop for human demands. Such insights do not therefore characterize the narrator in an excessively negative manner. Qoheleth is remarkably similar to many modern existentialists who also lament this common problematic situation. Mihaly Csikszentmihali has poignantly stated the very same thing about modern life:

> The foremost reason that happiness is so hard to achieve is that the universe was not designed with the comfort of human beings in mind. It is almost immeasurably huge, and most of it is hostilely empty and cold... It is not that the universe is random in an abstract mathematical sense. The motions of the stars, the transformations of energy that occur in it might be predicted and explained well enough. But natural processes do not take human desires into account. They are deaf and blind to our needs, and thus they are random in contrast with the order we attempt to establish through goals... A meteorite on a collision course with New York

73. Caneday, 'Qoheleth', p. 38.

74. J. Crenshaw, 'The Eternal Gospel (Eccl. 3.11)', in J. Crenshaw and J. Willis (eds.), *Essays in Old Testament Ethics* (New York: Ktav, 1974), pp. 23-55 (44).

75. J. Pedersen, 'Sceptisme Israelite', *RHPR* 10 (1930), pp. 317-70 (345).

76. L. Wilson, 'Artful Ambiguity in Ecclesiastes 1,1-11', p. 363. For Wilson, the ambiguity discovered in the opening verses of Ecclesiastes is a way to remind the reader that Wisdom observations 'reflect part, not all, of the truth. In other words, *what is being asserted from one viewpoint might need to be qualified by other perspectives*. The effect of this ambiguous opening section is that the reader is warned to tread carefully' (p. 364, my emphasis). In that respect, the artful use of ambiguity dovetails quite nicely with the overarching dialogical quality of Qoheleth's discourse. The ambiguity found in these verses therefore begins to prepare the reader for a rhetorical strategy that will eventually compare and contrast the beliefs of private insight with the hard-won tenets of public knowledge by hinting that all observations are ambiguous and open to qualification.

> might be obeying all the laws of the universe, but it would still be a damn nuisance. The virus that attacks the cells of a Mozart is only doing what comes naturally, even though it inflicts a grave loss on humanity. 'The universe is not hostile, nor yet is it friendly', in the words of J.H. Holmses. 'It is simply indifferent'.[77]

Since Csikszentmihali's book sold over a million copies, this suggests that the average postmodern reader would be quite at home with Qoheleth.

To sum up, although these verses characterize Qoheleth in a grim and pessimistic fashion, such an assessment by the reader does not yet entail any loss of credibility or trustworthiness. The universe presented by the narrator is an opaque world which no longer reflects the glory of God's creation and in which God and the transcendent order are no longer transparent.[78] Qoheleth's 'brave new world' turns out to be a tired old world devoid of wonder and divine presence. The reader correctly surmises that they are standing in the awesome presence of pessimism, and while this paints the narrator in a less than attractive manner for some readers, so far, the discourse has simply given the reader data that would suggest an elderly man who, though somewhat jaded, refuses to pull any punches about reality. Given the right counsel such honesty could benefit the narrator. On the other hand, the wrong counsel could backfire. So, the reader suspends their ultimate judgment on Qoheleth, choosing to wait for more information before they decide to make a clear and definite *Gestalt* regarding Qoheleth's persona. Such austerity demands a longer audience before it can be properly characterized.

4. *Ecclesiastes 1.1–2.26: 'I', Qoheleth —The Search for Self and Knowledge*

a. *Ecclesiastes 1.12-18: Agenda for Private Experience*

As I argued earlier, the book of Ecclesiastes is in some sense a 'theatrical' production captured in a biblical text. Until now, the reader encounters a bodiless voice who expresses an entirely different point of view on the nature of Nature. The stage is darkened. Only a voice is

77. M. Csikszentmihalyi, *Flow: The Psychology of Optimal Experience* (New York: Harper & Row, 1990), p. 9.

78. L. Thompson, *Introducing Biblical Literature: A More Fantastic Country* (Englewood Cliffs, NJ: Prentice–Hall, 1978), p. 171.

heard. Then, the spotlight comes on and a man strolls from behind the curtain, boldly proclaiming to the audience: 'I, Qoheleth, was King over Israel in Jerusalem'.[79] The spotlight centers only on this figure for a moment—a lighted face on a dark stage. The lights then go up, the curtain raises, and the reader is allowed to see the royal palace behind Qoheleth. He or she realizes that the voice we have been hearing is no ordinary old man, but a king! Which is not for the reader to know. But, the palatial setting of the monologue establishes the speaker as one who must be listened to with respect. And for the moment, the reader grants that respect.

So begins the longest observation complex in the book (1.12–4.16).[80] The text's wandering viewpoint changes in these verses. Nature is now placed in the textual background, becoming a part of the text's horizon of values. In its stead, a real life human being is placed in the foreground. To be more precise, the first observation by the narrator in 1.13-15 concerns 'the entirety of human doing'.[81] The private experiences of the protagonist become the theme which engages the reader's attention. From its focus upon the world in general the text now focalizes on the experience of the world by an individual person. In order to accomplish this, the implied author frontloads the narrative use of first-person

79. The reference to Jerusalem is another subtle reference to Qoheleth's identity within the Solomonic guise. As Seow points out, only David and Solomon are said to have ruled 'in Jersusalem' (cf. 1 Sam. 23.17; 2 Sam. 5.2-5 and 1 Kgs 1.34, 3.28; Seow, 'Qohelet's Autobiography', p. 277).

80. De Jong argues that the first observation complex extends from 1.3–4.16 ('A Book of Labour'). However, I am modifying his analysis a little. I would rather see 1.3-11 as an introduction, with the first observation complex beginning with the initial instance of Qoheleth's 'I'-narration. To be sure, 1.3-11 is a disguised observation. However, the words 'I saw', 'I perceived', and other terms implying empirical observation are lacking here. As a result, I would begin Qoheleth's first observation with these verses. Within this observation complex, Krüger sees a chiastic pattern in 1.3–4.12 (1.3 = profit question; 1.4-11 = poem; 1.12–2.26 = reflections with Qoheleth as king; 3.1-8 = poem; 3.9 = profit question). The verses in 3.10–4.12 supposedly function as a 'commentary' on the 'text' of 1.12–2.26 where Qoheleth is not king. However, given the fact that the Solomonic guise is present throughout the discourse for this text's model reader, the latter seems very questionable. See T. Krüger, 'Qoh 2,24-26 und die Frage nach dem "Guten" im Qohelet-Buch', *BN* 72 (1994), pp. 70-84 (80-81). However, his insight that in 1.12–2.26 there is a 'having' or consumer perspective while in 3.10–4.12 there is a 'being' perspective does hold potential for a synchronic reading of this text.

81. Fischer, 'Beobachungen zur Komposition', p. 76.

discourse by the protagonist. As a result, through the constant repetition of the phrases 'I saw', 'I perceived' and 'I said in my heart', the reader comes to understand that private experience has become the center of the text's attention.

Qoheleth begins his monologue by citing the credentials he possesses as a wiseman. For Addison Wright 1.12-18 functions as a double introduction to the book.[82] H.W. Hertzberg calls 1.12-15 'das Programm'.[83] In these verses, Qoheleth clearly sets forth his agenda to the reader. Qoheleth outlines his method, which consists of a strict systematic plan to observe both life and his own observations. The narrator mentions his heart (*lēb*) 12 times in 1.12-26, but only four times after this (7.25; 8.9, 16; 9.1).[84] By focalizing the narrator's 'I' from a radically internal post of observation (the *lēb*), the implied author fully characterizes Qoheleth's knowledge as both empirical and subjective. As a result of his empirical method, Diethelm Michel observes how 'collective experience is no longer available to him'.[85] What we have in

82. A. Wright, 'The Riddle of the Sphinx', p. 322.

83. Hertzberg, *Der Prediger*, p. 81.

84. Fox, 'Qoheleth's Epistemology', p. 143.

85. Michel, *Untersuchungen zur Eigenart*, p. 81. Michel points out how typically, collective knowledge and individual experience mutually support one another. He states: 'Collective knowledge, passed down in the proverbs, reiterates what the individual finds in experience. Collective knowledge confirms through individual experience what is claimed, and individual experience, on the concept brought through the sentence(s) of collective knowledge, mutually support each other and can be thusly inserted argumentatively as a presupposition' (p. 80). However, for Qoheleth, the public knowledge which Wisdom possesses no longer resonates with his soul. There is a cleft between Qoheleth's private experiences and the codified experiences of the public. As a result, Qoheleth has determined to test the community's fund of public knowledge. Michel then argues that the phrase 'I saw'—1.14; 2.13, 24; 3.10, 16, 22(?); 4.4, 15; 5.12, 17(?); 6.1; 7.15; 8.9, 10, 17; 9.13; 10.5, 7(?)—does not refer to the act of physically observing, but means 'I considered'. The term *rā'â* 'comes to expression as a testing analysis which, as collective knowledge is opposed by him regarding the claim of knowledge, the validity of that claim is investigated by him' (p. 81). Again, we see the subtle dialogue in the book regarding the rhetorical validity of public knowledge vis-à-vis private insights. For further review of Michel at this point, see A. Schoors, 'The Verb *rā'âh* in the Book of Qoheleth', in Schoors (ed.), *Qoheleth in the Context of Wisdom*, pp. 227-41. Schoors argues that the word can denote either observation or examination but also realization or conclusion. Once more we perceive the implied author's preference for lexical ambiguity and fluidity.

the narrator's reflections is a type of knowledge that might with good reason be called autobiographical discovery or revelation. The chief effect of v. 12 is to train the reader to view the haphazard reflections that ensue as the result of a single search by a solitary individual.[86]

Refrains of personal experience begin to fill the reader's consciousness.[87] Internalizing phrases such as 'I saw' occur so often in the opening lines of Qoheleth's discourse that personal experience becomes a virtual theme of these chapters. The chief effect of the sheer density of these phrases is to 'pull the reader into the very thought world of the narrator'.[88] In an interesting analogy with the prophets, Robert Johnson astutely observes how v. 12 'serves to function parallel to the call-reports of the prophets: it authorizes Qohelet's right to speak'.[89] Such a radically different call narrative also serves to characterize the narrator. Unlike the Solomon who received Wisdom as a gift of grace in 1 Kings 3, this Solomon only knows Wisdom as a result of arduous effort and intense, comprehensive searching. Qoheleth's internal world is one stripped of grace. It is a world where 'God has disappeared'.[90]

Furthermore, the perceptive reader cannot but help detecting a bit of hubris in Qoheleth's summary of his experiences. In v. 14, Qoheleth claims to have seen everything under the sun (*kol hamma ʿaśîm*). While the probable intent of such a claim was to authorize the narrator's right to speak, its actual effect is to raise suspicion in the reader who correctly surmises that such a claim is beyond any single 'I'. The narrator claims more than he should given the limitations of a first-person reading contract, which strictly limits the speaker's horizon of knowledge and forbids such claims to virtual omniscience. Narratively speaking, he usurps for himself what only a bona fide third-person narrator can possess. Common sense and basic literary competency suggest to the reader that an individual 'I' cannot see everything, and so the narrator claims a type of knowledge which can only be assessed in an ironic and critical light by the reader. One can only wonder why Qoheleth was not cognizant of his own observations in the latter half of this book which fully question the ability of any human being to 'know' in any definitive sense when he claimed this knowledge. By characterizing the narrator

86. Fox, 'Frame-Narrative', p. 90.
87. Further, see Christianson, *A Time to Tell*, pp. 193-215.
88. R. Johnson, 'The Rhetorical Question as a Literary Device', p. 256.
89. R. Johnson, 'A Form Critical Analysis of the Sayings', p. 107.
90. Carr, *From D to Q*, p. 142.

in such a way, the implied author has very discreetly planted in the reader's mind the hint that the ensuing discourse will overstep its rightful place. It is not so much that collective knowledge is not available to Qoheleth, as Michel observes, but that Qoheleth's 'I' actually attempts to substitute for public knowledge. As a result, the reader begins to ask just who is this person who claims so much for his own personal experience. Such raving ego creates a sense of interest for the reader, who desires to understand the broad-based and grandiose claim to knowledge.

Immediately after this Qoheleth quotes, or better, reminiscences upon a proverb to bolster his affirmation in v. 13b that life is an 'unhappy business' (*'inyan rā'*) and his summary conclusion in verse 14b that all is vanity and a chasing after wind (*re'ût rûaḥ*). To make an impact on the reader he utilizes a strong metaphor which evokes the reader's innate sense of emotional exasperation when confronted with futile and worthless efforts.[91] From such global conclusions the reader begins to sense the depth of the narrator's pessimism.[92] J.A. Loader characterizes this judgment as 'fatalism' and suggests that its effect is to 'shock' the reader.[93] The citation of an impossible task (v. 15) further characterizes the narrator as a pessimist in that his first use of the Wisdom tradition

91. Fox, *Qoheleth and His Contradictions*, p. 48. Perdue also discusses the meaning of this term. However, he calls attention to the secondary meaning of *re'ût*, 'to desire'. This raises the possibility that the metaphor of trying to control or harness the wind might also 'be understood to include the God-given breath or spirit that activates and sustains life, [if so,] we have the fundamental, yet tragic, paradox that resides at the heart of human existence and experience; the ephemeral nature of human existence, contrasted with the innate desire to retain the vital spirit that animates human life. If we paraphrase Qoheleth's larger expression, it would be: "all is breath quickly passing and a desire to retain life's animating spirit". Placed within the narrative experience of time in the text of Ecclesiastes this theme emphasizes the unhappy fate of humans who strive for immortality either through accomplishments or through retaining the divine spirit that animates their lives. Neither is possible; they, along with their accomplishments, do not endure' (Perdue, *Wisdom and Creation*, p. 207). Again, the spectre of our common mortality raises itself in some sense for the reader.

92. For the purposes of this study, pessimism will be understood as a person's attitude toward 'the relative evil or goodness of the world or of men's experience of the world' (L. Loemker, 'Optimism and Pessimism', in P. Edwards [ed.], *The Encyclopedia of Philosophy* [New York: Macmillan, 1967], pp. 114-21 [114]). This is quite different from scepticism, which will be discussed in Chapter 6.

93. Loader, *Ecclesiastes*, p. 25.

is to cite its dark side. As Menachem Perry has noted, the first words uttered by a character initiates a powerful primacy effect which can only be undone through great literary effort. Rhetorically, v. 15 functions to build consensus in that it attempts to establish universally acknowledged observations.[94] By assuming such acknowledgment on the part of the implied reader, Qoheleth's discourse presupposes and portrays an implied reader/narratee whose values include a sense of life's skewed nature. Its purpose is to evoke disbelief, incredulity and to give birth to skepticism.[95]

However, Qoheleth's citation surely taxes the gnomic qualities of this particular proverb. Proverbs typically have a limited sphere of application and so refer to various but specific types of instances. Such a broad-based appropriation of a gnomic text to serve the pessimistic ideology of its speaker thoroughly imbues this proverb with a sense of subjectivity, thereby stripping it of its gnomic qualities. It is now thoroughly subsumed into the consciousness of Qoheleth, becoming an example of the narrator's thought-world. This begins to mark him as a subversive sage.[96] Ironically, rather than building consensus, his radical application of it to sum up life in general subverts its rhetorical powers, making it an example not of public knowledge but of his own peculiar worldview, with the end result being that the desired goal of consensus is denied its speaker. The proverb thereby becomes an aphorism.[97] Due to the influence of its setting in an interior discourse, and Qoheleth's emerging ethos, the reader questions the validity of utilizing such a specific insight as a summarizing statement for life in any comprehensive manner.

94. Crenshaw, 'Impossible Questions', p. 19.

95. Crenshaw, 'Impossible Questions', p. 22.

96. Seow has also perceived the subversive nature of this proverb as utilized by Qoheleth. He states: 'Here one sees, again, Qohelet's use of irony. He uses what may have been a wisdom saying to undermine excessive confidence in wisdom. As it turns out, the wise are not really better off than fools' (C.L. Seuw, *Ecclesiastes: A New Translation with Introduction and Commentary* [AB, 18C; New York: Doubleday, 1997], p. 148).

97. Lee also has observed how Qoheleth's fondness for ambiguity again surfaces in this proverb, forcing the reader 'to come to terms with the author's subjective sentiments about a particular issue' (B. Lee, 'A Specific Application of the Proverbs in Ecclesiastes 1.15', *JHStud* 1 [1997], pp. 1-25 [18] [http://www.arts.ualberta.ca/JHS/]).

In addition, this pericope (1.12-15) marks the beginning of a literary and argumentative pattern. Alexander Fischer observes how:

> Three times Qoheleth reports a project with stylistically varying changes, 'I said with my heart', 'I gave my heart', and 'I said in my heart'... Three times he anticipates the result, introduced with 'Behold!', 'I understood', 'it happened', followed again by a stylistically changing *hebel*-saying. The three paragraphs are concluded respectively through the result-concluding proverb, especially through a citation-like claim.[98]

Norbert Lohfink notes that the three initial paragraphs of Qoheleth's discourse function as three overviews for the ensuing monologue.[99] As a result of this pattern, Qoheleth's opening remarks take on the characteristic of something that approaches 'methodological doubt'. Twice in 1.12-18 Qoheleth reports a deed, refers to the result and concludes his point with an adage from proverbial lore. Diegetically, the reader may also perceive a pattern of storytelling, conclusion and reflection which is repeated five times in 1.12–2.3.[100] Furthermore, Qoheleth's own consciousness of himself begins to structure the text, as each pericope (1.13-15, 16-18; 2.1-11) utilizes the pleonastic use of *' 'ănî* to mark out a new section. This further trains the reader to focus their attention on the narrator's 'I'. Beyond that, it also portrays the narrator's discourse as an example of a strictly methodological approach to life's conundrums through its regularity, argumentative structure and use of confirmation by appeal to maxim. By calling upon various proverbs as evidence, Qoheleth attempts to bolster his own observations

98. Fischer, 'Beobachtungen zur Komposition', p. 78. In addition, Fischer analyzes the compositional structure of 1.12–2.21 to have a chiastic structure: A = 1.13-15; B = 1.16-18; C = 2.1-2; C′ = 2.3-11; B′ = 2.12-17; A′ = 2.18-21. Other readers have also called attention to the strategic importance of this literary pattern for understanding the argumentative structure of Qoheleth's discourse. See N. Lohfink, *Kohelet* (DNEB; Würzburg: Echter Verlag, 1980), p. 24; Viviano, 'The Book of Ecclesiastes', pp. 82-83, who refers to an observation/assessment/reason why pattern; Castellino, 'Qoheleth and His Wisdom', p. 18, who classifies his rhetorical strategy as 'confirmation by proverb'. However, I do not understand how Fischer sees a proverb in 2.1-2 since the next proverbial citation (2.14) falls outside the parameters of this passage. The pattern temporarily stops at the end of ch. 1, though 2.1-2 does contain the observation and assessment part of this pattern.

99. Lohfink, *Kohelet*, p. 23.

100. Christianson, *A Time to Tell*, p. 206. He notes that the effect of this pattern is to 'engage the reader in the present of narration/reading. It is to remind us that there is a critical distance being kept'.

through resorting to similar observations made by the public. The net effect of this pattern is to characterize the speaker as a sage with a stringent sense of duty to method but a nefarious predilection for drawing pessimistic conclusions that are more expansive than the evidence warrants.[101]

In v. 16, the narrator draws the reader even further into his consciousness by engaging in interior monologue. This further subjectivizes the narrator's speech. The use of *lēʾmōr* ('saying') is the equivalent of quotation marks. James Crenshaw comments on the rhetorical properties of the passage:

> It functions to dramatize inner thoughts. The author carries on a conversation with himself, thinking aloud, weighing the present situation and his own intellectual accomplishments. This self-critique leaves no room for modesty, particularly since Solomon's legendary wisdom and wealth have inspired the royal fiction.[102]

The narrator's self aggrandizement is inescapable. A man who claims to have experienced all things can hardly be expected to partake of humility. The ethos of Solomon lends a subtle coloring to the text, even though the reference to 'all before me' certainly is a fictive clue to the reader that this cannot be the historical Solomon, since he only had one Israelite predecessor (Jebusites notwithstanding). Again, the author broaches a subject, renders a quick conclusion and then cites a proverbial text in order to confirm it (v. 18). Whereas human activity is the subject of 1.13-15, wisdom becomes the focus of this passage's wandering viewpoint. It too meets the same pessimistic evaluation by the sage. According to Qoheleth, the amassing of wisdom results in vexation/grief (*kāʿas*) and sorrow (*makʾôb*). This is surely another 'shocker' for the reader who is familiar with texts like Prov. 1.2-7 which mentions no such ill effects from the acquisition of wisdom. With Crenshaw, I maintain that such a pessimistic slant on the wisdom tradition characterizes this overview as a 'morbid reflection'.[103] Like the world characterized in the prologue on nature, all things, including the pursuit of Wisdom, are 'weariness'. The gap raised in the prologue as to who is speaking is now closed. The reader surmises that the worldview under-

101. Murphy also observes how Qoheleth generally appropriates proverbial sayings by 'radicalizing them, applying them in the sharpest of ways to the futilities of life' (*Ecclesiastes*, p. 14).

102. Crenshaw, *Ecclesiastes*, p. 74.

103. Crenshaw, *Ecclesiastes*, p. 70.

scored in the prologue is hereby resumed, but in a different garb. So far, however, all we have are conclusions. As yet, the reader has been given no warrants for such dire deductions. The reader is left to wonder why Qoheleth came to these conclusions and how he will defend such grave tenets. The passage thereby raises more questions than it solves.

b. *Ecclesiastes. 2.1-11: The Pleasures of Private Experience*

Just as the reader is about to surmise a total eclipse of all values in Qoheleth's universe, we find out that all is not dark in the narrator's private world. The sun does manage to shine in certain quadrants of this cosmos, most notably those areas associated with pleasure and the enjoyment of those good things which this world has to offer. Ecclesiastes 2.1-2 begins a third observation complex which has an interesting variation on the literary pattern established in ch. 1. Like its predecessors, it begins with *'anî*, observes an event and makes an assessment. But instead of confirming this observation with a reflection upon proverbial lore, the reader's expectations are disappointed by these verses. Rather, the text offers an extended autobiographical narrative (2.3-11). The reader is misled by the literary pattern in ch. 1 and comes to a premature conclusion, which is now corrected by the narrator. Qoheleth boldly substitutes his own private experiences for the confirmation he originally sought by citing public knowledge. The reader's descent into the arena of private insight is now complete. The 'catalog of private pleasures' in 2.3-8 substitutes for the broader experiences of the reading community. By means of this radical exchange Qoheleth challenges the foundations of Wisdom's search for reliable knowledge. However, the reader was prepared for this substitution by the way that Qoheleth's ethos turned the proverbs cited in vv. 15 and 18 of ch. 1 into highly subjective aphorisms, thereby making them examples not of public knowledge but of the narrator's private insights. While the reader was baited by their external form to partake of them as public confirmations, their rhetorical nature prepared the reader for this dramatic substitution.

Qoheleth's speech continues the interior dialogue begun in 1.16 and so keeps the discourse inwardly focalized by debating with himself. The wandering viewpoint changes again, focusing on the pursuit of pleasure and the relative value of *śimḥâ* ('enjoyment').[104] The root *śmḥ*

104. It should also be noted that *śimḥâ* can also refer to pleasures, 'whether

becomes a leitmotif in this chapter, occurring in 2.1, 2, 10 and 26. So important is this theme for understanding the book that many readers, such as Agustinus Gianto, have claimed that the book is consciously structured so as to present the theme of enjoyment as the definitive response to the world's *hebel*-condition.[105] The use of interior monologue and the constant use of *lēb* ('heart', as the seat of intelligence[106]) in this chapter keeps the reader aware of the subjective nature of Qoheleth's discourse. The use of the imperative in 2.1, *r*^{e}*'ēh* ('to see'), while ostensibly acting as an admonition to himself, has the subtle function of addressing the narratee in an oblique fashion. The narratee will then be directed to pursue enjoyment in a more direct manner shortly hereafter via the first call to enjoyment (2.24-26). Since Qoheleth is looking back on his younger self (Level 2b of the narration), the reader can only surmise that an additional function of Qoheleth's 'confession' is to act as an example for his narratee, who is likewise characterized as a youth by the discreet role-modeling provided by the narrator's youthful self, the use of *b*^{e}*nî* in 12.12 and the description of the narratee as 'young man' in 11.9. By introducing the topos of pleasure, Qoheleth completes his thematic trilogy of Wisdom, work and enjoyment.

Ecclesiastes 2.3-8 constitute the heart of the Solomonic fiction. The 'king' tells of how he searched (literally 'spied' in v. 3, cf. 1.13) to find out how various pleasures might contribute to the experience of the good by an individual. It is interesting that Qoheleth repeats the verb *tûr* ('to spy out'), and does not refer to the other word used for searching in 1.12, *dāraš* ('to seek out'). Some of the pleasures he refers to cannot be experienced in public, but rather, have an inherently covert nature to them, especially the sexual ones. When Qoheleth speaks of the pursuit of knowledge via private pleasures, he communicates the private and perhaps sexual nature of that search by the deft use of the

trifling or significant' (M. Fox, 'The Inner Structure of Qohelet's Thought', in Schoors [ed.], *Qoheleth in the Context of Wisdom*, pp. 225-38 [228]; cf. Isa. 22.13). Again, we see Qoheleth's penchant for exploiting words with multivalent meanings and lexical ambiguity when it suits his rhetorical purposes.

105. Gianto, 'The Theme of Enjoyment', pp. 528-29. Gianto observes that the phrase *gam-hû' hābel* ('I said that') in 2.1 forms an inclusio with the similar phrase *šegam-zeh hābel* ('I said...that') in 8.14. He concludes: 'There is thus a strong indication that the two verses, 2.1 and 8.14, mark the beginning and end of a long exposition in which Qohelet tries to present *smḥ* as a response to *hebel*' (p. 529).

106. For an excellent discussion of the intellectual aspects of *lēb* and its role in the process of knowing see Fox, 'Qoheleth's Epistemology', p. 143.

verb *tûr*.[107] This sort of search is not the type of seeking one can do in an academic setting. While it has many of the public trappings which, by necessity, are to be expected in a setting such as the royal palace, its nature is personal. Verses 4-8 provide a catalog of pleasures available to the king—houses, vineyards, gardens, parks, slaves, concubines, herds, silver, gold, and other material possessions. The narrator turns to the motif of sexuality in v. 8 by referring overtly to the 'delight of sons of men, concubine upon concubine'.[108] The word *ta'ᵃnûgôt* means 'daintiness, luxury, exquisite', referring either to the excellent choice of shapely and beautiful women who were available for the king or perhaps to the sexual experience itself. Obviously, the kind of search he reflects upon is typical for a young man who is ascending from adolescence to adulthood, perhaps even to middle-age given the emphasis on material possessions in this passage.

The parallels with Solomon cannot be avoided. Although the name is never mentioned, the inevitable comparison with Solomon is unavoidable as 'the traditional sources make the point that Solomon collected everything (horses, women, wisdom) and displayed a tendency toward

107. The root *tûr* occurs in three strategic places in Ecclesiastes (1.13; 2.3; 7.25), the latter two both dealing with women. The *locus classicus* for the covert and sexual connotations of *tûr* is Josh. 2 where the spies are sent into Jericho to reconnoiter the land. In the process, they conveniently stay at Rahab's house, the local prostitute, who would be privy to much inside knowledge given her profession. The root also occurs in Est. 2.12 and 15, where the nominal form of the root refers to the 'turns' that women in the king's harem would take being with the monarch. The nominal form also occurs in Cant. 1.10, a book renown for its sexual orientation. It should also be noted that the verb form occurs in Num. 15.39, which refers to the going after of 'one's desires', which included the wanton and sexual dimension. Later, Qoheleth seems to consciously make a play on this verse in 11.9b, an admonition which created no small amount of rhetorical dissention among the rabbis (though, it should be noted, Qoheleth was not admonishing the wanton path). (See Salters, 'Exegetical Problems in Qoheleth', pp. 44-59 [57]) Given this semantic background, the root *tûr* would naturally possess the requisite covert and sexual connotations which enabled the implied author to exploit the term for his rhetorical purposes. At the least, its use to refer to abstract qualities here and in ch. 7 'represents a shift from its usual application to tangible qualities like the land that spies explored' (J. Crenshaw, 'Qoheleth's Understanding of Intellectual Inquiry', in Schoors [ed.], *Qoheleth in the Context of Wisdom*, pp. 205-24 [219]).

108. The phrase *šiddâ wᵉšiddôt* may also mean 'women upon women' if *šiddâ* is an Egyptian loan word. See M. Görg, 'Zu einer bekannten Paronomasie', p. 7. Either way, it functions as an illusion to the Solomonic harem in 1 Kgs 11.1-3.

excess and grandeur'.[109] However, the use of reticence has a purpose here. By its use, Qoheleth allows this type of search to function as a role-model, however wise or unwise this particular piece of insinuated advice may be. The bold references to wealth, parks and pleasures hints at the categorical imperative to enjoy life which will shortly be made the focal point of Qoheleth's discourse via the call to enjoyment in 2.24-26. However, those verses will tone down the excesses of this passage.

This passage is a nodal point in the reader's characterization of Qoheleth. If the process of characterization means 'enfleshing' the narrator's voice, as James Lee Marra holds, then this passage is a quintessential example of that process. Much of the reader's sense of the speaker's ethos is generated by the tremendous ways that Qoheleth's 'search' characterizes both the younger and older Qoheleth. As a reader, I cannot escape characterizing the man in a less than positive way. Though Qoheleth tries to defend this search as an experiment in wisdom by noting how even when under the influence of wine that wisdom was still guiding him (v. 3), it seems to me that the juxtaposition of wine and wisdom is an example of satiric characterization by the implied author. A reader could plausibly connect his building projects with wisdom and perhaps the acquisition of goods with that noble pursuit. But to begin this passage with a reference to imbibing as a means to pursue wisdom certainly stretches the reader's sense of credulity. This smacks of lampooning on the part of the implied author.[110] The fact that this juxtaposition is given narrative exposure by its positioning at the beginning of the passage does not escape the careful reader. His other accomplishments could very well have been highlighted.

109. T.A. Perry, *Dialogues with Kohelet*, p. 37.

110. The implied author of our text would not be the first to do so. Newing has analyzed the rhetorical art of the Deuteronomist's treatment of Solomon in 1 Kings, and concluded that the various contradictions in that account are not due to the sloppy work of editors, but are conscious acts of 'lampooning' by the implied author who wished to oppose the ancient, traditional picture of Solomon as 'wise'. He observes how the use of irony, ridicule, satire, sarcasm, and perhaps burlesque has a cumulative effect on the linear reading of the story, rendering a rather negative post of observation regarding Israel's patron sage. I would suggest that the implied author of our text, who was obviously quite familiar with that tradition, continued the rhetorical art of lampooning Solomon in this book. See Newing, 'Rhetorical Art'. Further, see M. Sweeney, 'The Critique of Solomon in the Josianic Edition of the Deuteronomistic History', *JBL* 114 (1995), pp. 607-22.

Again, the implied author resorts to a subtle ironic characterization of the narrator.

Verse 3 has all the hallmark characteristics of a rationalization intended to ease the objections of the implied reader. While one does not dismiss the possibility that *knowledge* was gained as a result of this search, the idea that *Wisdom* was active throughout the process stretches the reader's ability to believe the narrator. What we have in this passage is an example of satire. The implied author's use of the Solomonic mask thereby functions as a critique of Qoheleth's empirical method. In essence, by his subtle depiction of the sage's debauchery and self-centeredness the implied author is asking the implied reader: 'Who else does this type of sage remind you of?' In that respect, the use of the Solomonic masquerade is a technique of irony that lampoons and satirizes the abuses of a radically self-centered approach to knowledge.

What were the results of this search on the young seeker's growing sense of selfhood? The answer is given in 2.9-11. From v. 9 we are told of the narrator's sense of importance and greatness which borders on the egotistic.[111] Verse 10 smacks of self-indulgence and narcissism as

111. It should be noted that Verheij has recently argued based on verbal parallels between Gen. 2 and Eccl. 2.4-6 (*nāṭa'*, 'to plant'; *gan*, 'garden'; *'ēṣ kol-perî*, 'all fruit trees'; *lᵉhašqôt*, 'to drench'; *ṣômēaḥ*, 'to sprout'; *'āśâ*, 'to make') that Qohelet's building of a personal estate parallels that of God's creation in Genesis, particularly the garden of Eden. The 'royal experiment' thereby becomes a 'creation experiment'. See A. Verheij, 'Paradise Retried: On Qohelet 2.4-6', *JSOT* 50 (1991), pp. 113-15. Verheij observes how previous historical reading grids have hampered the reader from seeing this aspect to the narrative: 'Because of this exclusively history-oriented reading, the point is missed that 'Qohelet' not only poses as a king, but even—for a moment—as God' (p. 113). T.A. Perry argues that such an ambition is not only 'anti-Genesis' but is a critique of God's creation as well: 'Kohelet presents a frontal attack against the grounding of creation theology, the notion of the goodness of God's creation; against the statement that "All is good" it counters that "All is vanity" (*Dialogues with Kohelet*, p. 24). Other readers have also seen a connection to the Genesis narratives in Ecclesiastes. See D. Clemens, 'The Law of Sin and Death: Ecclesiastes and Genesis 1–3', *Themelios* 19 (1994), pp. 5-8; C. Forman, 'Koheleth's Use of Genesis', *JSS* 5 (1960), pp. 256-63. The parallels here do suggest a strong intertextual connection at this point in the narrative. I would see this as another example of parody by the implied author, who hints at the inherent gaudiness of such ambitions, and so, portrays the hubris of such pursuits of knowledge. The effect of these parallels is to satirize Qoheleth and so to point out the weaknesses of his approach to knowledge. Again, the subtle negative portrayal of the narrator is characteristic of the implied author's rhetoric of reticence.

Qoheleth relates to his reader that no pleasure was kept from his heart and that whatever his eyes desired, he found a way to enjoy. Finally, we note that the end result (the *ḥēleq*, 'portion' or 'reward') of his intense searching was an increased sense of pessimism as he rather negatively considered the fruits of his toil a pursuit after wind. This is certainly not the sort of *ḥēleq* that is attractive to most readers. Not unimportantly, v. 11 marks the closure of the gap first introduced by the programmatic 'what profit?' question in 1.3. Qoheleth flatly denies that there is anything to be gained under the sun. That such an important question could be answered so negatively and emphatically immediately following the self-absorbed pursuit of pleasure by a young Solomon does not communicate to the reader the sort of mature philosophical reflection that the speaker intended and the reader has come to rightfully expect. This also satirizes the narrator's discourse. In this respect, the reader begins to sense a degree of dissatisfaction with both the narrator's methods and conclusions.

The fact that such a negative and satiric characterization is surmised *at some level* by most readers is demonstrated by the numerous ways that the reading community has attempted to protect the narrator's reputation at precisely this point. Obviously, much of this is typical for a person situated in a social stratification of wealth, power and prestige. But a quick glance at the various readings shows a strong polemical stance taken by many religious readers to defend Qoheleth. T.A. Perry unabashedly refers to this section as 'the collector's greed'.[112] Curiously, he goes on to defend Qoheleth by postulating that 'readers are often willing to overlook the unflattering implications of all this, on the premise that, after all, it was done for the sake of wisdom and as an experiment, and, following Koheleth, pass the blame along and upstairs, to "God the despot", so to speak'.[113] If that were really the case, Perry would not have had to justify this passage, nor would he have characterized the narrator as 'greedy'. Others are similarly guilty of obscuring the ironic characterization of Qoheleth by the implied author. Kathleen Farmer describes the passage as the 'confession of a conspicuous consumer', noting the passage's 'witness to the ultimate lack of satisfaction such things give'.[114] Leo Perdue suggests that its purpose was to portray the failings of Solomon's pursuits as a way of

112. T.A. Perry, *Dialogues with Kohelet*, p. 36.
113. T.A. Perry, *Dialogues with Kohelet*, p. 36.
114. Farmer, *Who Knows What is Good?*, p. 157.

convincing the reader of the inevitable lack of satisfaction in accomplishments and human pleasure.[115] Loader counters that wine here acts metaphorically, functioning as a 'symbol of the pleasure of life in general' (cf. Dt. 14.26; Judg. 9.13, Ps. 104.15; Isa. 5.11.)[116] However, it is interesting to note that Isa. 5.11 speaks not of the good that wine can do, but of the folly of those who 'chase liquor from early in the morning, and till late in the evening are inflamed by wine' (JPSV). Following this line, one could equally say that wine is a symbol for debauchery as well. The honest reader will admit that wine has both meanings and can function either positively or negatively as a symbol. Crenshaw and Whybray simply note how typical such things were of ancient royal dignitaries,[117] as if that eased the characterization of the speaker. To be sure, in the seven-fold call to enjoyment Qoheleth merely admonishes eating and drinking as examples of the good life, which is echoed elsewhere in the First Testament in a positive fashion. But in this specific passage, the narrator's particular pursuit of wisdom lacks positive ethos-related qualities. When commentators go to such great lengths to defend the narrator, it seems to be a classic example of the old adage, 'me thinketh that you protesteth too much'. As someone like Stanley Fish would point out, the various proposals all point to an underlying problem that readers have characterizing the narrator in a positive manner. The point is that if this is a satiric presentation by the implied author the reader is not supposed to defend the narrator, but, rather, should allow the ironic characterization to have its due effects.

To sum up, the characterization of King Qoheleth in 2.1-12 has a less than positive aura about it. First, the intertextuality issues regarding Solomon certainly extend to Qoheleth, with a resulting negative characterization by association. Second, although the text needs to inform the reader that the speaker has the requisite experiences to speak from an empirically-based point of view, the manner of gaining experiences is somewhat disappointing. There is an inescapable sense of egotism, narcissism and self-centeredness that detracts from the speaker's sense of attractiveness and credibility. Certainly one cannot sense the much-needed characteristics of justice or temperance in this passage. While some parts, such as his building accomplishments, project an air of

115. Perdue, '"I will make a test of pleasure"', p. 215.

116. Loader, *Ecclesiastes*, p. 27.

117. Crenshaw, *Ecclesiastes*, p. 79; R. Whybray, *Ecclesiastes* (NCBC; Grand Rapids: Eerdmans, 1989), pp. 53-55.

prudence, this does not overcome the overall effect of the passage, which portrays the narrator's youth as a man obsessed with his own interests and pleasures, however related they may be to the pursuit of wisdom. The confessional increment of 2.1-11 reveals a person absorbed with his own good,[118] which hardly communicates the type of magnificent spirit that would inspire most readers to have confidence in the speaker. As a result, the reader begins to form a *Gestalt* of the narrator's personality, engaging in what Uri Margolin describes as the character-building stage whereby specific traits are joined into a higher, more complex organization. According to Baruch Hochman, the reader relates to this configuration of traits as a personality-type rather than a specific person. However, the reader needs for these characteristics to become a pattern before he or she makes a final decision regarding the character of the narrator. More must be heard to adequately assess Qoheleth. Particularly important will be how the narrator chooses to evaluate his earlier self in retrospect and what exactly Qoheleth will advise on the basis of this search.

c. *Ecclesiastes 2.12-17: Wisdom, Folly and Private Experience*

Qoheleth begins to discuss in greater depth the various insights that have sprung from his lifelong odyssey for wisdom. He returns to the three-fold pattern of argumentation that is characteristic of his first two observations in 1.12-18, citing a proverb in 2.14. This confirms his assessment of the relative value of wisdom and folly that, indeed, wisdom does have an advantage over folly as light is better than darkness. The wandering viewpoint pulls from the text's horizon of values the theme of wisdom and folly which were first discussed in 1.17, thereby resuming his treatment of those issues. Once again, the pronoun *'ᵃnî* functions as a structuring device for the reader, keeping the reader's post of observation focalized from an internal, subjective perspective. He draws the narratee/reader into his circle of contemplation with the rhetorical question in v. 12: 'for what can the person do who comes after the King?' By this the speaker hopes to build a rhetorical consensus with his narratee/implied reader. Having answered his programmatic question in the previous verse, Qoheleth discontinues his rhetoric of delay for the time being. He immediately gives the reader the preferred

118. In an interesting, if not arresting aside on this passage, Christianson wryly observes that 'Qoheleth's younger self is remarkably postmodern for its consumptive nature' (*A Time to Tell*, p. 214).

answer: 'only that which has already been done!' Qoheleth then follows with an observation in v. 13 that conflates the use of 'I saw' (*rā'îtî*) and 'there is' (*yēš*). This educates the reader to view the future utilizations of *yēš* which lack a first-person indicator (cf. 2.24; 5.13, and so on) as examples of the narrator's post of observation, though in ostensive form they *seem* like external posts of observation which simply tell it 'like it is'.

In addition, Qoheleth begins to comment on the proverbs he reflects upon. In v. 14, he notes how it has been observed that 'the wise man has his eyes in his head, but the fool walks in darkness; *yet I know that one fate happens to them all*'. Such comments have several rhetorical effects on the reader. First, they pull these proverbs thoroughly into the gravitational field of Qoheleth's ethos. As a result, they become witnesses to a type of public knowledge that is in reality subordinated to private insight. Because of this, these reminiscences begin to characterize the speaker not only for the values that they espouse, but also for those with which they disagree. Qoheleth's comments mark these reflections as examples of the narrator's own interior thoughts, thereby further stripping these proverbs of their typical gnomic qualities. This trains the reader to view all the ensuing 'citations' as examples of the narrator's peculiar thought-world, so that by the time he or she comes across those in 'series' later, they are consumed not as 'collections', but as the 'recollections' of the sage. The chief effect of Qoheleth's comments is therefore to thoroughly subordinate public knowledge to the evaluation of private experience and its insights. The external post of observation provided by the public's trans-generational observations is thereby reduced to a strictly internal post of observation.

In vv. 12-14 the prudence of the sage is established by his assessment of the value of wisdom over folly. However, his comment in v. 14b negatively characterizes the man, who could not resist the temptation to lend his pessimistic evaluation to this otherwise positive proverb by noting that neither wisdom nor folly will save a person from the eventual clutches of death, obliquely referred to as *miqreh 'eḥād*, 'one fate'. The utilitarian evaluation of the maxim does little to lift the reader's spirit, and so, lacks the rhetorical quality of magnificence which is needed to inspire and, ultimately, suade the reader.

Verses 15-17 begin a new observation regarding wisdom and folly. Qoheleth elaborates on the comment he made in v. 14b. The twofold use of the particle *gam* ('in addition', 'indeed', 'moreover', 'yea') with

ʾanî 'lends the verse an urgent note'.[119] Rhetorically, *gam* functions to emphasize the thought of the entire sentence to the reader, but particularly the word immediately following.[120] The means by which the narrator calls attention to his own private experience reaches new depths in this passage, where the use of *gam* coupled with *b^elibbî* ('in my heart') enables Qoheleth's 'I' to focalize the narrative's post of observation from the most intense of inward perspectives. The narrative has drawn the reader into the innermost citadel of Qoheleth's being. We are now at ground zero in terms of the text's internally-focused aesthetic movement. Now we see what really is at issue with Qoheleth. The protagonist drops all vestiges of objectivity and proclaims in the most self-centered of fashions: 'Like the fortune of the fool, so too will it happen to *me*. For why then have *I* been so excessively wise?' The emotional frustration and anguish of the speaker is evident in his exclamation that this too is *hebel*. Everything is now judged from the point of view of its effect or end result on Qoheleth *himself*.

Verse 16 again raises the issue of humanity's common mortality, which vexes Qoheleth's soul to no end. The prospect that 'when the coming days have passed, everything has been forgotten', including his own remembrance and accomplishments, relativizes everything. In contemplating his own death Qoheleth has experienced what Erich Fromm once described as the 'ego chill' which accompanies the self's realization of its own mortality. All the accomplishments catalogued in 2.3-11 will fade into oblivion. From the heights of pleasure, Qoheleth dips to the lows of pessimism and depression, decrying life's absurdities with the poignant phrase: 'So I hated life'. The cry has all the hallmarks of an elderly man undergoing the final phase of our common march toward ultimate identity, what Erik Erikson called ego-integrity versus despair.[121]

119. R. Johnson, 'The Rhetorical Question', p. 143.

120. BDB, p. 169.

121. Although this verse is filled with despair, we should not too quickly characterize Qoheleth as suicidal or some other unreasonable *Gestalt*. As Gordis has acknowledged: 'In a moment of bitterness or frustration Koheleth may say, "Therefore I hated life" (2.18), but this is not his dominant mood, which is an affirmation of life' (*Koheleth: The Man and His World*, p. 118). We would agree with Caneday that such exclamations are better read as the 'exasperated sentiments of individuals', such as uttered by Job as well in 7.16. See Caneday, 'Qoheleth', p. 36.

The pathos of this cry is an extreme act of intimacy on the part of Qoheleth. The reader is drawn into Qoheleth's circle of friendship, and begins to understand what is 'bugging' the man. Only a reader who has not experienced this side of life will harshly judge him for his honesty and openness. Something new is now added to the narrator's ethos—a sense of openness and frankness that arrests the reader's rising sense of incredulity toward Qoheleth. The implied reader resonates with such pathos. Furthermore, the narratee is hereby characterized as a trusted confidant by means of such personal disclosures. This lends the discourse a tremendous rhetorical advantage when it needs it most.[122] Just as the reader is about to characterize the speaker as fundamentally narcissistic, Qoheleth opens his heart to disclose a person who is essentially concerned with something that concerns all of us as a part of the human community. The intimacy of this passage bolsters the rhetorical standing of the speaker and, for the moment, wards off the increasingly negative characterization of the protagonist. The courage it takes to be this open with the public enhances both the attractiveness and the trustworthiness of the sage. In this passage we see the tell-tale rhetorical persona of Qoheleth. One moment we perceive a man trapped in the confines of his own narcissism and egotism. The next, that same narcissism is utilized to address an issue which concerns all of us. As a result the reader does not yet know what to make of Qoheleth. He or she does surmise, however, that this empiricist is stretching out his heart and his hands in an act of narrative and philosophical friendship. But he or she also begins to realize the limits of such a radically 'I'-centered epistemology. Death has a way of doing that to all grandiose plans and dreams (cf. 5.3, 7), as the narrator disclosed in the early going of this chapter.

d. *Ecclesiastes 2.18-26: Wisdom, Work and Inheritance—The Limits of Private Experience*

The narrative picks up on the emotion of the previous section (*w*[e]*śānē'tî*, 'I hated') while resuming the theme of work found in 2.10-11. On a psychological level, the theme of the wandering viewpoint remains the same. But on the intellectual level, the wandering viewpoint changes again, taking up a theme (*'āmāl*, 'to toil') discussed two passages earlier.[123] In v. 2.18 Qoheleth now realizes the vanity of his earlier

122. Marra, 'The Lifelike "I"', p. 296.

123. The word *'āmāl* has a definite connotation of 'labor' in a negative sense

accomplishments discussed in 2.3-11. Houses, parks and harems do not last. The passage has a definite recency effect which cancels out part of the primacy effect engendered by the egotism expressed in 2.1-12. The elder Qoheleth revises the egotistic estimation he once held regarding the accomplishments of human toil. This raises the reader's sense of the narrator's credibility, as the hubris of the younger Qoheleth receives a more mature re-evaluation. In v. 20, Qoheleth again expresses how he 'let his heart turn to despair over all the accomplishments which I had achieved'. This keeps the passage focused on the narrator's emotional reaction to life's *hebel*, thereby creating a powerful pathos-effect in the reader. David Carr suggests that the effect of the Solomonic masquerade at this point in Qoheleth's confession is to present the '"inside story" of Solomon's real lesson… Unlike 1 Kings 3.2-15 and 2 Chron. 1.1-3, the wisdom of Qohelet's Solomon hardly led to unexpected benefits. Instead, his deeper wisdom about futility led him to despair…'[124] As a result, this passage radically defamiliarizes not only reality and the pursuit of wisdom, but also much about the Solomonic tradition as well. In that, Qoheleth is correct. Not everything a person learns during life's pilgrimage brings joy. The darker side of the universe has a very strong negative existential effect. By voicing such sentiments, Qoheleth shows himself to be very much the realist. This inspires confidence in most readers.

Nevertheless, the pattern of Qoheleth's self-centeredness is evident in this passage. The evaluation of wisdom in vv. 18 and 21 solely in terms of whether it has any lasting effect for oneself is extremely damaging for Qoheleth's ethos. The narrator's words portray him as a greedy old man who resents the fact that he cannot take it with him.[125]

throughout Qoheleth's discourse. Often in Ecclesiastes (particularly 2.22, 24, 3.9; 4.9; 8.15; 10.15) the word refers 'to effort…implies arduous effort, "overdoing" rather than "doing"' (Fox, *Qoheleth and His Contradictions*, p. 54). Another excellent work which deals with the meaning of *'āmāl* in Ecclesiastes is the very insightful article by W. Vogels, 'Performance vaine et performance saine chez Qohelet', *NRT* 113 (1991), pp. 363-85. Vogels's study distinguishes between the person who wants to spend their life in the anxious search for a non-existent materialistic 'more', and the wholesome performance of the one who accepts what God has to give.

124. Carr, *From D to Q*, p. 142.

125. Crenshaw also refers to the 'egocentric perspective' of this passage and comes to a similar characterization of Qoheleth. See Crenshaw, *Ecclesiastes*, p. 88. Others, like Loader, also see the problem, but choose to lessen its effects by

Qoheleth's viewpoint shows the limited viewpoint of his empirical method. Such a method can hardly be understood as sufficient to adequately criticize the stance of the larger public which must reflect a variety of self-interests—something Qoheleth was sadly lacking. In effect, the implied author allows the narrator to hang himself rhetorically.

As a reader, I react viscerally to this passage. The idea that someone who must have inherited most of his wealth (how else could he have been so rich?) now considers it an absurdity when he must do the next generation a similar kindness has an absolutely negative effect on the perceptive reader. Qoheleth's ethos slips to a new low, as the qualities of generosity and magnificence are lacking in their totality. Verses 18 and 21 render an unattractive image for its speaker. Not once in the pursuit of the knowledge of the 'good' does he consider the benefit that he might do for the next generation. He commits a rhetorical error that few readers forget or forgive. This recency effect is unalterable in my opinion. In order to affect a consensus with his reader Qoheleth again resorts to the use of the rhetorical question in v. 22, asking a variation of the 'what profit?' question broached in 1.3. He omits his favorite word *yitrôn* ('profit', 'advantage'), using the ambiguous term *hōweh* ('to happen'), as a means of accenting the fickle and unknowable nature of humanity's fate (*miqreh*) which was introduced in 2.14. This and the use of a 'who knows?' question in v. 19 lends the passage a decided sense of 'resigned inevitability'.[126] The rhetorical effect of these questions is to 'pull the weight of the argument to the front' of

claiming the issue for Qoheleth was that he could not determine whether the inheritor would be wise or foolish. See Loader, *Ecclesiastes*, p. 30. However, in v. 21 no such consideration is given, with the result that egocentricity seems to be Qoheleth's main problem here. There the issue is that someone else might benefit from Qoheleth's possessions 'who did not toil for it'. This is a sad comment upon Qoheleth's sense of family loyalty.

126. Crenshaw, 'The Expression *mi yodea*'', p. 278. Not insignificantly, it should be noted that the 'who knows?' question is extremely rare in the First Testament Canon, occurring only ten times, with nearly half (four times) occurring only in Qoheleth (cf. 2.19; 3.21; 6.12; 8.1). The sceptical characterization of the narrator begins to be felt in this passage. However, it will reside only as a faint theme in the reader's mind until it becomes full-blown in the second half of this book, where it will function as a refrain. The text merely hints at this point that Qoheleth's pessimism has far greater epistemological ramifications. Those ramifications will become the subject of 6.10–12.14.

the reader's consciousness.[127] However, by now, the reader has formed enough of a *Gestalt* of the narrator's character to know the pessimistic answer that Qoheleth is seeking on their part. Without delay, true to form, he answers his own question by calling attention to life's essential pains (*mak'ōbîm*) and vexation (*ka'as*), confessing how he even lost sleep over this realization (v. 23). However, the reader does not typically resonate with this answer. The double-edged rhetorical effect of Qoheleth's vain rhetoric begins to be felt. The reader begins to wonder if this is the ultimate end of life or simply the unavoidable consequences of narcissism and the narrator's self-absorbed worldview. No lampooning is necessary on the part of the implied author. The frame-narrator lets this 'homo salomonicus' do himself in rhetorically.[128]

Having proved the essential lack of worth for material possessions, Qoheleth turns to what is worthy for our existential consideration: the pursuit of pleasure and enjoyment. The first instance of the 'call to enjoyment' signals the end of the passage at hand. His unabashed espousal of this virtue is evident in the fact that Qoheleth essentially coined a new genre, the 'nothing is better' proverb, in order to express its categorical demands upon the reader. Its function is

127. R. Johnson, 'The Rhetorical Question as a Literary Device', p. 215.

128. Höffken refers to Qoheleth in this manner. See Höffken, 'Das Ego des Weisen', p. 127. However, there are some positive results for this sort of lampooning. Höffken argues that the effect of the Royal Fiction in Ecclesiastes is to bring Solomon 'down to earth', so to speak. By utilizing a confessional stance which communicates the struggles Solomon had in his heart with the pursuit of wisdom, Höffken argues that the effect of this passage was to humanize the patron sage of Israel's Wisdom. He states: 'It is also possible, that the reason for this was not without another possible reason: in every instance, the Qoheleth-I presents itself in this ideal autobiographical role in a way, that allows the author to fully play out the possibilities of a King like Solomon, therefore each specific royal activity is deployed, which range from the building activities to the activities in the harem, but no more under the aspect of the fulfilled life, but rather, under that of a failed existence, which serves only more the knowledge that "everything is vain". Thereby, however, Solomon himself knows the failure of wisdom, and the undertaking of that role could prove therefore to be extremely calculated' (p. 126). Perdue also contends that the text demonstrates the problems associated with the amassing of material possessions. See Perdue, ' "I will make a test of wisdom" ', p. 215. Murphy too has also argued for a similar fictive defamiliarization of the life of Solomon. See Murphy, *Ecclesiastes*, p. 26

> to respond in a practical way to the recognition that man has no ultimate 'advantage'...there is a definable relationship between the form and the question of man's *yithrôn*. Having asked what 'advantage' man could anticipate and given the answer 'none', Qoheleth has created the *'ayn tôb* form to address itself to the attitude to life which he wishes to commend.[129]

Although the discourse seemed headed in an utterly nihilistic direction in 2.11, the reader now perceives that there is a silver lining to the cloud that is Qoheleth's radical pessimism. Just when the reader expects pessimism, he or she gets the first of the seven-fold call to enjoyment. In Iserian terms, the reader's expectations are overturned by a negation.[130] Qoheleth first negates the reader's world, then negates his own negation, thereby turning the negative into a quasi-positive admonition. The narrator comes to the realization that the purpose of life is in the living, not in the accomplishments of labor. Life is a process, not a product. And so, at the end of his years, the elder Qoheleth passes on to his narratee what ultimately matters: the enjoyment of a good life which can be summed up in eating, drinking and enjoying one's work (v. 24).

This is the first of seven such calls which, as R.N. Whybray has observed, incrementally increase in direct address as the discourse progresses.[131] As I have observed earlier, this call is in all probability the key structural device which guides the reader's sense of the text's development. The call in 2.24-26 is relatively unadorned, though again, not without its ambiguous moments.[132] Qoheleth chooses to address his narratee in a general way. He simply states his conclusion forcefully and in a straightforward manner. In addition, he fully acknowledges the role that God plays in such enjoyment by describing it as a gift 'from the hand of God' in v. 24c and the stronger rhetorical strategy of

129. G. Ogden, 'Qoheleth's Use of the "Nothing is Better" Form', *JBL* 98 (1979), pp. 339-50 (345).

130. McKenzie, 'Subversive Sages', p. 113.

131. Whybray, 'Qoheleth, Preacher of Joy', p. 87.

132. For instance, *ḥûš* (v. 25) may be read as 'enjoyment' or 'worry'. The passage could therefore mean that 'God is responsible for pleasure and worry'. See Byargeon, 'The Significance of Ambiguity', p. 370. However, for myself, context screens out the polyvalent meanings which gives rise to the latter reading of 'worry'. Once again, Qoheleth's rhetoric of ambiguity is seen in the reception-history of the text. Futher, see Longman, *The Book of Ecclesiastes*, pp. 108-109.

juxtaposing the sinner's amassing of possessions over pleasures in 2.26. At this point in the discourse, Qoheleth seems to have recovered his 'orthodox' personality, dressing himself in an almost traditional garb. Later, the call to enjoyment will be expressed in more direct ways, utilizing the imperative mood in 9.7-9. But for now, the narrator chooses an indirect manner of addressing the narratee. Instead of issuing a direct address, he states his case and then utilizes a rhetorical question, 'for who can eat and rejoice apart from myself?'[133] The use of the rhetorical question assures Qoheleth that he has caught the attention of both the narratee and his implied reader. The rhetorical effect of these questions is to draw 'the reader more tightly into the book's presentation'.[134]

The chief effect of the call to enjoyment in 2.24-26 is to engender a reversal of argumentative direction for the reader. While Qoheleth can find no lasting *yitrôn*, he does seem to have found a *ḥēleq* (or 'portion') in our ability to 'enjoy the ride', so to speak. While we cannot keep it 'all', a portion in the guise of enjoyment is the true reward for our labors.[135] The reader's universe is now thoroughly defamiliarized, but in such a way that Qoheleth's world hovers somewhere between nihilism and traditionalism.[136] Raymond Johnson summarizes how Qoheleth's monologue defamiliarizes wisdom's toil-yields-profit norm:

133. The RSV emends this to read 'for who can eat and rejoice apart from him?', making the verse refer to God. However, the text is quite clear here, with the first person singular suffix (*mimmennî*) demanding a reflexive sense of 'myself'. JPS has the more faithful translation at this point.

134. R. Johnson, 'The Rhetorical Question as a Literary Device', p. 251.

135. T.A. Perry observes how 'early in the narrative, *ḥeleq* is associated with reward or payment (2.10), but later, a second sense emerges, that of inheritance, indicating whatever a man acquires independent of his own efforts or merit, what he is granted as a gift' (*Dialogues with Kohelet*, p. 31). Due to that revision by the text, the reader should not confuse the meaning of *yitrôn* and *ḥēleq*, which stand respectively for the material advantage or profits of labor and the gracious, often non-material by-products of such labors.

136. Part of the polarized aspect of Qoheleth's logic appears at precisely this point. Fox has observed how his advice in 2.26 'goes on to stress what nobody, including Qohelet, would deny: the main thing in life is fear of God and obedience to the commandments. Advice similar to the epilogist's appears within Qoheleth's words, though not in such a simple form…cf. 5.6, 7.18, 2.26, 3.17, 8.12-13… The main difference between Qoheleth and the epilogist is the way the latter asserts the standard religious language' ('Frame-Narrative', p. 103).

> Nevertheless, as each norm is restructured, qualifications are added which prevent the text from becoming truly comic, comedy being simply defined as a 'a movement from bad to good fortune'. Thus while there may be no profit in toil, there is pleasure (2.24-26; 5.17-19; 6.9). Pleasure, however is a (fickle?) gift of God and even then a vanity (2.26; 5.9–6.9). Likewise, wisdom can 'find out' some things, but only in a limited manner and possibly in opposition to God's design for humanity (7.23–8.1). To that extent, the reader of Ecclesiastes is suspended somewhere between tragedy and comedy, pessimism and optimism.[137]

As a result, the portrayal of self-indulgence in 2.3-11 offers a hint to accept the call to enjoyment in 2.24-26. The narrative and confessional aspects of ch. 2 serve the purpose of defamiliarizing the reader's world through the most personal means available to the speaker.

To sum up, most of the reader's characterization of Qoheleth has been achieved by the end of the King's Fiction. As Menachem Perry has noted, the most intensive closure operations by readers occur in the opening chapters of a literary work. The reader now knows that the narrator is a pessimist with strong egocentric tendencies. By means of the Royal Fiction, 'a pronoun had been made flesh'.[138] While the narrator's honesty and openness has created a strong sense of intimacy, some of those intimate secrets shared by Qoheleth have had an adverse effect on his characterization by the reader. More to the point, Qoheleth's radical self-preoccupation has raised certain reservations regarding the sufficiency of his method to duly criticize public knowledge. He has become a classic example of a restricted post of observation.[139] His narcissism has resulted in a loss of attractiveness with a corresponding lack of credibility. This results in a sense of estrangement or alienation within the reader towards the strange worldview of the character.[140]

By the end of ch. 2, the sense of a vain rhetoric is inescapable. Although the reader is aesthetically suaded to accept the narrator's worldview as an accurate presentation of the character's personal worldview, he or she is less positive regarding its ability to summarize the world itself. However, to his credit, Qoheleth manages to rebound from his ethos-related miscues in 2.1-11 through the evaluative reversal he

137. R. Johnson, 'The Rhetorical Question', p. 262.

138. Romberg, *Studies in the Narrative Technique*, p. 85.

139. Prince, *Narratology*, p. 51.

140. Stanzel, 'Towards a "Grammar of Fiction"', *Novel* 11 (1977), pp. 247-64 (260).

gives it in 2.18-26 and especially by the call to enjoyment, which restores a degree of attractiveness to the narrator. However, not all of these negative traits can be easily forgotten by the reader. Qoheleth has managed to duly criticize the wholesale amassing of possessions in a powerful way. But in the process, he has also alerted the reader that such private insights have their limits, and that some of his observations can hardly substitute for the public knowledge which he has so stridently taken to task. Qoheleth has thoroughly engaged in a vain rhetoric which, although it has powerful effects, ultimately fails to suade the reader in a comprehensive and satisfying manner.

5. *Ecclesiastes 3.1-15: Time, Darkness and the Limits of Public Knowledge*

Qoheleth's monologue takes a turn in ch. 3. Although the narrative is still focalized through the eyes of the narrator,[141] the degree of internal focalization lessens. Rather than allowing the reader to peer into the narrator's heart, the monologue turns to consider external reality. Rüdiger Lux observes how at 'the end of the King's Fiction a text-passage now meets us—(2.24-26)… For now the plane of the reality of the heart is left again. The narrator turns to other planes of reality, to that of the world and God.'[142] Although the narrative remains focalized through the eyes of the narrator, the wandering viewpoint changes. The problem of time and its relationship to public knowledge takes the center stage.

Beginning in 3.1-8, the implied author begins to criticize public knowledge via Qoheleth's poem on time and the narrator's subsequent comments. While the implied author has satirized Qoheleth at strategic

141. So too Christianson, *A Time to Tell*, p. 164.

142. Lux, '"Ich, Kohelet, bin König…"', p. 340. Some readers such as Fischer have argued that 3.1-15 belongs to the King's Fiction, with the poems in 3.1-8 and 1.4-11 acting as a poetic framing of the major section in 1.12–2.26, along with two thematically motivated parenthetic remarks in 1.3 and 3.9. The total composition is completed by the reflection in 3.10-15 which, in chiastic order, refers to 1.4-11, 1.12–2.26 and 3.1-8 and reflects the theme of these verses theologically. See Fischer, 'Beobachtungen zur Komposition', *passim*. Although this is an attractive analysis, the structuring effect of the call to enjoyment in 2.24-26 suggests the end of the passage, and the beginning of a new one. The outward focalization of this passages further suggests that 3.1-15 is not intrinsically connected to the inwardly oriented section which precedes it in the tightly composed manner that Fischer suggests.

places, that should not be taken to mean that he totally disagrees with the position being taken by his narrator. In every respect the narrator is his literary creation, expressing a point of view which remains a part of the implied author's own polar thinking. While Qoheleth is clearly characterized in a way that the weaknesses of such an approach to life and its problems are visible to the reader, our protagonist is not a simple foil for the implied author. Rather, he is a literary vehicle for expressing one pole of the epistemological paradox that is held in tension with the other pole known as public knowledge. Although Qoheleth wholeheartedly advises the enjoyment of life, this passage observes that there is a time for everything—including the negative activities of life. Qoheleth does not begin this observation with *'ᵃnî*, but instead utilizes or quotes a poetic text which simulates an objective, external post of observation.[143] This is accomplished by beginning with the phrase *lakōl zᵉmān wᵉ ʿēt lᵉkol-ḥēpeṣ* ('for everything there is an appointed time and a season for everything'). In a manner that is strikingly similar to the beginning of the book, the discourse utilizes a poetic text to defamiliarize the reader's understanding of reality, and so prepares them for the ensuing discourse. This further suggests that Qoheleth's critical gaze has turned toward matters other than the heart. This poem accents the world of human activities whereas the earlier poem in 1.4-11 focalizes on the foundational cosmological environment of human activities. As a result, the wandering viewpoint changes once again. Whereas ch. 2 accented the theme of human accomplishments and pleasures, this chapter concerns the realm of human knowledge which informs our various activities. Qoheleth begins the middle section of his discourse by summarizing human life in general terms as a way to further explore the specific ramifications of the programmatic 'what profit?' question in 1.3.[144]

The principle goal of any Wisdom tract is to educate its readers on the expedient choice in any given situation. This is no less the case with Qoheleth. The poetic nature of 3.1-8 suggests that Qoheleth has taken over a poem from the public treasury of knowledge.[145] While

143. Because of this, Schubert classifies 3.1-15 as a mixed observation. See Schubert, 'Die Selbstbetrachtungen Kohelets', pp. 23-34.

144. Ogden, 'Qoheleth's Use', p. 345.

145. Ogden observes how the poem has the sound of an independent text and contains verbs which are not found elsewhere in the book. See Ogden, *Qoheleth*, p. 52.

some readers have argued that *z^{e}mān* ('appointed time') connotes a sense of theological determinism,[146] the broader context of 3.1-8 gives no confirmation for this reading. The narrator is 'not thinking deterministically… He does not say that man is forced to do his work in the right time, that is, to conform to the work of God. On the contrary, man has the freedom to act as he wants.'[147] The prominence and constant repetition of the word *'ēt* ('time') fatigues the reader, sounding 'like a clock that, inexorably and independent of the wishes of people, keeps ticking and striking'.[148] Taken by itself, the poem in 3.1-8 is entirely devoid of the pessimism which dominates its literary surroundings.

Ecclesiastes 3.1-8 reminds the reader of his or her limits. They are not the lords of time, whose nature is surprisingly resistant to human efforts to domesticate it. The poem celebrates the contradictory nature of life and its propensity to offer various opportunities which demand contrary behaviors on the part of its participants. It simply observes that life presents the sage with a variety of times—a time to be born, to die, to kill, to heal, to weep, to laugh, to mourn, to dance, to love, to hate, to wage war and to wage peace.[149] Life therefore consists of events which are beneficial to the individual, and those which are not. Morally, the poem presents a neutral post of observation on life's various opportunities.[150] These things happen. Precise social settings are

146. Crenshaw, 'The Eternal Gospel', p. 37. However, I would suggest that the idea of an 'appointed time' applies only to the first verset, which deals with the time of dying and being born. These are events which a person cannot choose and are therefore 'appointed'. Elsewhere in the poem, only the neutral term *'ēt* is utilized.

147. Isaksson, *Studies in the Language of Qohelet*, p. 179.

148. Loader, *Ecclesiastes*, p. 35.

149. Wilch describes Qoheleth's use of *'ēt* as expressing the 'occasion of a given opportunity'. In vv. 2-8, the word occurs 28 times. The pairing of occasions does not signify radically opposed events, but rather, to 'include every shade and degree of related occurrence that may be placed between their poles…it is their number that is of greater importance: they form seven double pairs, which in itself symbolizes wholeness. Therefore, the pairs represent all the possibilities that may take place within the range of human activity and experience' (J. Wilch, *Time and Event* [Leiden: E.J. Brill, 1969], p. 120).

150. I am aware that there are readings, such as Brenner's, which would not view this poem as morally neutral. Her reading proposes that this is a 'male text' in which all the parts cohere in relation to the sexual interpretation of 3.5. The Time Poem thereby becomes an 'M Poem of Desire'. See A. Brenner, 'M Text Authority in Biblical Love Lyrics: The Case of Qoheleth 3.1-9 and its Textual Relatives', in Brenner and Van Dijk-Hemmes (eds.), *On Gendering Texts*, pp. 133-63 (150-53).

lacking, perhaps to ensure the ability of the poem to encapsulate all of human existence.[151] No judgment is rendered by the poem itself. It does not prescribe action, but merely describes the various parameters of human choice.[152] According to the principals of text-type theory, it is a

According to this reading, the terms 'uproot' (v. 2d), 'breach' (v. 3), and 'tear' (v. 7) connotate a male point of view on sex (p. 151). However, as Brenner herself admits: 'sexual riddle associations are not easily demonstrable in verses 4, 6, and 7' (p. 148). The reference to the text as 'stereotypically M' (p. 151) causes this reader some concern inasmuch as one wonders what 'stereotypically M' means in any non-sexist way. At the least, I think the poem contains no reading clues which would demand a coherently metaphorical interpretation based on any one verse, let alone one as ambiguous as v. 5.

151. Crenshaw, *Ecclesiastes*, p. 95. In fact, the semantic quality of this passage is an excellent example of the ability of a text to communicate a surplus of meaning to the reader. Take v. 5 as an example. The reference to casting and gathering stones can be taken in a variety of ways denoting a very wide spectrum. For some readers, there is a veiled reference to sexual activity (Loader, *Ecclesiastes*, p. 36). Along that line, Brenner views all the references here as sexual statements, with women being 'metaphorized into stones in Qoheleth 3.5a-b' ('M Text Authority', p. 145). At the other end of the spectrum, some see a reference to war tactics (Crenshaw, *Ecclesiastes*, p. 94). Riffaterre refers to this quality of poetry as semantic overdetermination, whereby the ostensibly lexical denotation of a term takes on a metaphorical meaning due to its terseness and parallelism with adjacent terms. See M. Riffaterre, 'Semantic Overdetermination in Poetry', *PTL* 2 (1977), pp. 1-19. From a reader-oriented perspective, the literal text creates a metaphorical intertext which guides the emergent meaning of the poetic text. This intertext communicates a second level of meaning that surpasses the literal meaning of the words. Readers therefore process a poetic text such as 3.1-8 by comparing the literal and metaphorical levels of a poetic text. Riffaterre summarizes the reading process for poetic texts: 'Only if perceived in relation to that intertext does the poem make sense, or, better, function as a poem rather than a text merely conveying everyday facts or concepts...the intertext forms a network of implied meanings which indirectly guides and controls reading and interpretation, exerting this control as presuppositions. The reader reads two texts at once, as it were, the poem before him and the intertext his linguistic competence enables him to rebuild' (p. 10). Elsewhere he describes the necessity of misreading a poetic text in order to comprehend its meaning: 'The meaning of a poem is discovered only when the failure of a referential interpretation jolts the reader into looking for clues elsewhere, that is, going back over the text in search of help from its own features, since usage will not suffice. Retroactive reading is thus the only means of relating the derivation from its model. A poem is read poetically backwards' (p. 19). This verse is perfect for a rhetoric of ambiguity, whereby the text means ultimately whatever the reader makes of it.

152. However, if Blenkinsopp is correct that *lāmût* in v. 2 reflects the Stoic

descriptive text, not an expository text. The main thrust of the poem is to present a world that is not only humanly indifferent, but is morally detached as well. Such a defamiliarized view on human activities prepares the reader for the pragmatic and utilitarian ethic of the narrator, whose ethics often communicates a sense of the 'golden mean' (cf. 4.17–5.8; 7.16).

However, that objective distance is not retained by Qoheleth, who, true to character, comments on the poem's neutral stance by asking a rhetorical question in v. 9: 'what advantage/profit is there for the worker in which he is toiling?' This gives the passage an 'implicit negativity'.[153] The question presents a blank to the reader, who must labor at understanding the link between the poem and the response by Qoheleth. How the above passage relates to *yitrôn* is not immediately apparent from the aesthetic movement of the poem itself. Accordingly, Qoheleth spells out that relationship in vv. 10-15. The negative character of the narrator is evident in that the query creates an 'ironic twist' to the poem's otherwise dispassionate stance.[154] Viewed from the perspective of individual interest, such a world is hardly a place to invest one's efforts, which requires a sense of the moment's nature in order to maximize its profit for the one undergoing it. The 'what profit?' question serves to further characterize the narrator, who is portrayed as a business person, a 'bottom-line' sort of guy, whose interest in gain surely influences his role as sage or theologian.

Qoheleth's business side is made explicit in vv. 10-15, which highlight the theme of *'inyān*, or the 'business/occupation' that God has given to humanity. He returns to presenting things through his own perceptual grid, again utilizing a first-person observation. Qoheleth omits the pleonastic use of *'ᵃnî* here, principally because, by now, the reader has been trained to understand its rhetorical presence. But just as the reader looks for a negative evaluation, Qoheleth dashes those expectations, referring to everything as *yāpeh* ('beautiful') in its own time. Such optimism only serves to bait the reader, who is pulled into its field

doctrine of the timely putting an end to one's life, the text might indeed prescribe action in a very reticent, or perhaps, veiled manner. See J. Blenkinsopp, 'Ecclesiastes 3.1-15: Another Interpretation', *JSOT* 66 (1995), pp. 55-64.

153. R. Johnson, 'The Rhetorical Question', p. 135.

154. R. Johnson, 'The Rhetorical Question', p. 154.

of influence only to be quickly given a rather pessimistic post of observation. The astute reader understands the delimitations of this evaluation, that is, *in it own time*. An action is advantageous only in its proper setting or time. The discourse vaguely hints at the dark side of such times by carefully noting that beauty and time must coalesce for time's emergent beneficial properties to become available for the individual. Success is therefore described as an emergent property of the universe which depends upon the confluence of proper knowledge and opportunity.

The problem with this arrangement is offered immediately afterward when the caveat to this positive cosmological evaluation is presented: 'but he has also put *'ōlām* in their hearts' (3.11b). No better example of ambiguity and linguistic opacity can be found in the entire Canon. Traditionally, *'ōlām* is translated as 'eternity' or 'duration'. But this translation hardly fits the general thematic context of time and its ability to profit an individual. Given the emphasis on gain in this passage, and the general emphasis on the limits of human knowing in the book, I side with the recent analysis of Francis Holland who argues that the term means 'darkness' and, therefore, 'ignorance'.[155] The problem with public knowledge is that all human knowledge is affected by this limitation. In Qoheleth's evaluation, what we know is surpassed by what we do not know. He shows himself therefore to be the ultimate pessimist by accenting the negative over the positive. This pessimism pushes him over the threshold of scepticism, radically affecting his

155. F. Holland, 'Heart of Darkness: A Study of Qohelet', *PIBA* 17 (1994), pp. 81-101 (92-96). She relies upon Dahood's argument that the root *'lm* means 'to cover' in Ugaritic and that the word phrase *ha'ôrâ 'ôlām* means 'path of ignorance' in Job 22.15 (Dahood, 'Canaanite-Phoenician Influence'). Holland intersubjectively validates this proposed reading by citing G. Barton, *A Critical and Exegetical Commentary on the Book of Ecclesiastes* (ICC; Edinburgh, T. & T. Clark, 1908), pp. 98, 105; O.S. Rankin, 'The Book of Ecclesiastes', in *IB* (12 vols.; Nashville, TN: Abingdon Press, 1956), V, p. 49, and Crenshaw, *Ecclesiastes*, pp. 97-98. Holland's analysis of the contextual meaning of *'ōlām* as 'darkness/ignorance' is extremely insightful. She concludes: 'In the face of this deliberate withholding of ontological knowledge (this heart of darkness) which overshadows the Time Poem, humanity has little choice but to seize the moment (*carpe diem*), which itself is at the disposition of God who may deny it... Qohelet's acknowledgment of the ontological darkness in the human heart and the relative and contingent value of practical wisdom is echoed in his social and anthropological critique' ('Heart of Darkness', pp. 94-95).

epistemology in the latter half of the book (cf. 6.12; 8.1-17). The reader therefore begins to see the deeper sceptical nature of the man coming out in these verses. What really bothers the sage is that we cannot comprehend everything (note the emphasis on *hakōl* in his discourse), especially regarding time and its opportunities. In this Qoheleth deeply laments the limits of both public and private knowledge. For him, 'the tragedy for humans is that God does not reveal to them the direction of cosmos and history'.[156]

By allowing his literary character to speak such thoughts the implied author tacitly acknowledges the general correctness of such insights. Public knowledge does have its limitations. In that regard, it is surely not a cure-all for private insight and its peculiar limitations. While private knowledge is often blinded by self-interest, such as the implied author communicates through Qoheleth's pursuit of pleasure in ch. 2, public knowledge suffers from the finite nature of humanity's corporate powers of knowing. No matter how much we try, there is a 'heart of darkness' in humanity which clouds even the best of public knowledge. As a result, Qoheleth again invokes the existential portion that is available to us given the lack of profit in our efforts. In 3.12-13 he reiterates the call to enjoyment, utilizing a 'nothing is better' proverb to make his point more dramatic. Its importance is apparent in that it occurs a second time only a few scant verses after its first appearance (2.24-26). This time he heightens its force by attaching an asseverative phrase, *yāda'tî kî* ('I realized that'). He concludes with an observation regarding the determinative nature of God's decrees in 3.14-15 which sounds more like a confession than a proof. However, this is no believer's confession of faith. He strongly hints that the limited nature of humanity's epistemological abilities is ultimately dependent upon a decision by God himself, though for the purpose of invoking respect (v. 14b). The first vestiges of Qoheleth's aloof God begins to surface in this confession. Qoheleth finishes his epistemological lament by underscoring the cyclical nature of humanity's existence, returning to a theme first broached in the prologue (1.9).

In these verses, we see where polarized thought is characteristic of both the narrator and the implied author. While the implied author fully acknowledges the limits of private knowledge via the satirical and ironic characterization of the narrator, he also understands that public

156. Perdue, '"I will make a test of pleasure"', p. 218.

knowledge is equally limited in an ontological sense and, indeed, partakes of an even more difficult limitation. While self-interest can be trained or curbed in the individual, what is humanity to do with the absolute limitations placed upon our faculties by God himself? In the end, the implied author has no sure answer to this vexing question, though a quasi-solution comes to him in the form of enjoyment and obedience to God's will (cf. 12.12-14). Ecclesiastes 3.1-15 introduces a critical theme for the entire book. As J.S. Mulder concludes, '3.1–4.6 and 8.1-17 really hit the heart of all Qoheleth: nobody can understand God'.[157] The subtle message to the implied reader from the implied author is to not judge the narrator too harshly for, ultimately, no one knows it 'all'. Ignorance and finite epistemological powers lie at the base of the *hebel*-condition lamented by Qoheleth.

Ecclesiastes 3.1-15 characterizes both Qoheleth and the implied author as sceptics. The difference is that while this causes Qoheleth to keep a distinct distance from God (note the use of *'ᵉlōhîm* here, a word for God that can be translated as 'the Deity'), for the implied author, this is all the more reason to follow the commands of God (cf. 12.13). In this we see where the difference between Qoheleth and the implied author is, as Von Rad pointed out, ultimately it is a matter of faith—a commodity which is sorely lacking in Qoheleth's characterization by the implied author.[158] But even so, the reader is not too negatively predisposed toward Qoheleth here. Most of what he says is simple honesty and openness, which surely does the narrator a lot of rhetorical good. From this passage, the reader gains a sense of Qoheleth's prudence, which marks him as a capable member of the guild of sages.

6. *Ecclesiastes 3.16–4.6: Immorality, Mortality and the Limits of Public Knowledge*

Qoheleth continues to comment on humanity's heart of darkness by noting that there is more than ignorance to consider in the final estimation of public knowledge. The general community is rife with its own corporate kinds of self-interest. Whereas the implied author has satirized

157. J.S. Mulder, 'Qoheleth's Division and also its Main Point', in W.C. Delsman *et al.* (eds.), *Von Kanaan bis Kerala* (Festschrift J. Van der Ploeg; Neukirchen–Vluyn: Neukirchener Verlag, 1982), pp. 140-59 (158).

158. G. von Rad, *Wisdom in Israel* (trans. J. Martin; Nashville: Abingdon Press), p. 236.

Qoheleth for his individual pursuit of self-interest, the narrator now turns the tables on the implied author and says, in effect, 'those who live in glass houses should throw no stones'. Peter Höffken refers to this as 'the destructive role of the I and its observations and reflections opposed to tradition'.[159] By means of the critical role played by Qoheleth's 'I', the implied author obliquely acknowledges that his position also has inherent weaknesses. Not even public knowledge is ultimately reliable in any absolute sense. The specific public of any given historical community has too many pockets of corporate self-interest to generate a truly universal audience.

Qoheleth's discourse discards the initial use of pleonastic *'ᵃnî* at this point, as he no longer wants to emphasize the internal perspective of his observations. Rather, Qoheleth desires to accent the publicly observable nature of the events he is about to describe and so omits the highly visible pronoun, choosing to delay its introduction until vv. 17 and 18. He first selects the justice system from the text's repertoire for critique and defamiliarization. Ecclesiastes 3.16-17 introduces a new theme, observing the role of wickedness (*rāšā'*) even in the courts (*mᵉqôm hammišpāṭ*). Qoheleth notes that there will be a 'time' for such behavior to be duly judged by God (v. 17). Again, we see where reliance upon orthodox beliefs is not beyond the narrator when it serves his rhetorical purposes. Life is a test for public officials—God examines them to see if they are humans or predictors (beasts) disguised as homo sapiens.

In v. 18 Qoheleth again resorts to internal dialogue (*bᵉlibbî*), thereby resuming or perhaps reinforcing the subjective nature of his observations precisely at a point where his deductions are most personal in nature. Such a coincidence must be considered to be a touch of irony by which the implied author, who thus communicates to the reader that Qoheleth's deductions here are perhaps less reliable than others he has made. The theme of v. 19 selects from the horizon of values the idea of our common mortality first introduced in 2.14, noting that our common fate (*miqreh*) is the same as that of the rest of the created order. From this Qoheleth makes a conclusion by reminiscing upon a proverbial text: 'Everyone goes to one place, everyone comes from the dust and everyone returns to the dust'. Immediately following this, Qoheleth again utilizes a rhetorical question: 'Who knows whether the spirit of humanity goes upwards and the spirit of the beasts goes downward to

159. Höffken, 'Das Ego des Weisen', p. 125.

the earth?' This puts the question back upon the reader to answer, forcing the reader to participate in the text's production of meaning. Assuming that the narratee and reader will likewise answer this in the negative (given the fact that we have been duly warned about our ignorance in such matters in 3.11), Qoheleth again admonishes the reader to enjoy life. The seriousness of Qoheleth's call to enjoyment is communicated by the use of asseverative *kî* ('indeed, surely'), which was already employed in 3.12. The use of three such admonitions in the space of just over 20 verses is a way of front-loading the theme for the central part of Qoheleth's discourse, making sure that the reader has been duly trained to answer Qoheleth's rhetorical questions with the appropriate *carpe diem* answer.

Ecclesiastes 4.1-3 continues the theme of the community's self-interested nature. Qoheleth labors the point of 3.16 in 4.1, noting the 'exploitations (*hā'ᵃšuqîm*) that are done under the sun'. The root *'šq* denotes a variety of socially abusive practices: extortions, oppressions and general wrongs. 'King Qoheleth' sounds like a prophet here, criticizing his own government in a way that is reminiscent of the prophet Isaiah (cf. Isa. 3.8-15). In a verse filled with compassion and caring, Qoheleth demonstrates that a you-centered person does indeed reside in his soul, admonishing the narratee to behold 'the tears of the oppressed, yet there was no one to comfort them, but on the side of their oppressors there was power, and not one to advocate for them' (Eccl. 4.1). Qoheleth is at his rhetorical best here, rising to new heights of character. The recency effect created by this passage portrays a man rich in rhetorical attractiveness, trustworthiness and credibility as he exudes the traits of compassion, justice, courage and magnificence in a way which, admittedly, is a bit out of character for the man.[160] The use of *hinnēh* heightens the shift in narrative focalization, functioning as a

160. For instance, Crenshaw sees a negative characterization in these verses, noting that Qoheleth does not urge the reader to correct these injustices. See Crenshaw, *Ecclesiastes*, p. 106. Loader advocates a similar characterization of the narrator. See Loader, *Ecclesiastes*, p. 43. However, this judgment seems a little harsh at this point in the narrative. Such readings are classic examples of how Qoheleth's rhetorical miscues in ch. 2 come back to haunt him later in the discourse. As I argued earlier, most readers neither forgive nor forget the selfish characterization established by his actions at the begining of his discourse. For Crenshaw, Loader and readers like them, the primacy effect of those chapters cannot be overturned by the recency effect of these verses.

virtual command to the narratee/reader. However, the stellar characteristics of the narrator quickly turn to glaring weaknesses as Qoheleth's tendencies toward pessimism cannot be restrained. He summarily proclaims the dead better off than those still living. Of course, the irony exists that such a statement hardly seems to be logical given the way that mortality creates such a vexatious problem for Qoheleth.[161]

Qoheleth ends his critique of the public with yet another observation beginning with the pleonastic use of *'ᵃnî*. In 4.4-6 he returns to the theme of work and toil, observing how our competitive nature as a species is the driving force behind human efforts. Qoheleth quite perceptively calls attention to how envy (*qin'â*) pushes each person to be more skilled than his competitor. This professional rat race he rightfully calls 'an absurdity and a striving after wind'. He then reflects upon two 'dueling proverbs' whose meanings seem at first to be contradictory. Verse 5 calls attention to the fruits of not enough work in a person's life, while v. 6 condemns too much work. The phrase in 4.6b, 'Better is a handful of quietness than two hands full of toil *and a striving after wind*', is surely the 'amen' response of Qoheleth. By citing such proverbial incongruities, Qoheleth accomplishes two things. First, he 'expresses his wisdom in the same manner as the older wisdom teachers'.[162] This further establishes his credentials as a sage and characterizes him as an astute master of the Wisdom tradition. Second, he demonstrates 'the inconsistencies within the wisdom tradition itself'.[163] In a very deft move, he has subtly indicted the community's *general* fund of knowledge for its inability to properly summarize any *specific* situation. By placing these two proverbs adjacent to each other, a blank is opened up in the reader's mind. Such a strategy of incongruity increases the ideational chores of the reader, who must guess at the meaning implied by their juxtaposition. This creates a

161. However, it may be that Qoheleth was simply calling attention to the fact that human oppression often renders death preferable to life, as LaVoie has recently argued. See J. Lavoie, 'De l'inconvénient d'être né: Etude de Qohélet 4,1-3', *SR* 24 (1995), pp. 297-308 (308). Should that be the case, the 'inconvenience of having been born' is an instance of his caring, promoting a sense of good ethos to the reader for Qoheleth.

162. J. Spangenberg, 'Quotations in Ecclesiastes: An Appraisal', *OTE* 4 (1991), pp. 19-35 (24).

163. K. Dell, 'The Reuse and Misuse of Forms in Ecclesiastes', in K. Dell, *The Book of Job as Sceptical Literature* (BZAW, 197; Berlin: W. de Gruyter, 1991), p. 140.

certain level of hermeneutical openness for the passage at hand. The two proverbs do not easily stand next to each other. Their juxtaposition heightens the reader's growing sense that while wisdom addresses general situations, it lacks a great deal when it comes to individual appropriation for specific life-instances. In that it functions as a 'warning to a careless reader' not to read the Wisdom tradition in an inattentive or simplistic manner.[164]

The problem with public knowledge from Qoheleth's post of observation is seen *in nuce* with these proverbs. While public knowledge addresses transgenerational issues, those issues must be worked out in a variety of individual instances to have any lasting gain for the reader. Which of these two gnomic statements applies to any given situation? The answer is determined by the specific nature of the time at hand. Unfortunately, the knowledge of those specific times is missing, or perhaps withheld from humanity. As a result, the effectiveness of public knowledge is a great deal less than claimed. Aarre Lauha has summarized the problem with such proverbs and the public knowledge:

> All exertions of humanity are vain, since its success does not depend upon the activity of the worker...rather it depends upon the conditions of time and of its variations which are independent of him and which are incomprehensible for him.[165]

Wisdom can only offer a partial advantage to its practitioners (and thus the availability of *ḥēleq*, 'portion', but not a *yitrôn*, 'profit'). That limitation is inherent and inescapable given the finite horizons of our knowledge. This existential and epistemological problem Qoheleth laments as *hebel*. Following Addison Wright, I note how the thematic phrase *r*ᵉ*'ût rûaḥ* in 4.6b marks the end of the unit which extends from 3.1–4.6.[166]

To sum up, while 1.1–2.24 accented the limits of private knowledge, the section from 3.1–4.6 highlights its polar opposite, the limitations of public knowledge. In these verses the implied author utilizes Qoheleth's speech to acknowledge that both individual and corporate insights possess only a partial claim to validity. This situation is inherent to the human condition (3.11) and the implied author can offer no solution. The latent scepticism broached in 3.1–4.6 proleptically prepares the

164. Murphy, *Ecclesiastes*, p. 39.
165. Lauha, *Kohelet*, pp. 67-68.
166. A. Wright, 'The Riddle of the Sphinx', p. 321.

reader for the onslaught of scepticism that is forthcoming in ch. 8. Via the critique of society offered by the narrator in 3.16–4.6, the implied author deftly communicates in an almost ironic fashion that while private insight is often too self-centered for public use, the public also partakes of that same weakness, and therefore provides no panacea for humanity's epistemological woes. Both Qoheleth and the implied author are therefore characterized as pessimists with certain sceptical leanings. In this section, Qoheleth strides forth as a bona fide sage who can back up his claims for sagacity. Regarding his general sense of ethos, he is characterized as a caring, sensitive sage who in the best First Testament visionary tradition condemns the social abuse of the poor who are exploited by those who lead the public. This constitutes a recency effect which partially neutralizes the negative effects created by the narrator's opening discourse. In an unexpected turnaround, the protagonist rebounds a great deal from his rhetorical miscues in ch. 2. The reader begins to feel that, indeed, here is a man who might be trusted and whose honesty certainly demands a hearing.

7. *Ecclesiastes 4.7-16: Knowledge and Communal Living*

The wandering viewpoint changes once again as Qoheleth begins to reflect upon the value of living in community. After having indicted the community for its negative side, Qoheleth attempts to balance out his monologue, turning towards the positive benefits of communal living. Qoheleth continues the pattern of beginning his observation with the pleonastic use of *'ᵃnî*, again accenting the personal nature of the ensuing discussion. He begins by concluding in summary fashion that the ensuing observation is a *hebel* rather than waiting to make his judgment at the end as he has typically done. Qoheleth assumes that the reader's world has been adequately defamiliarized by now. As a result, he can begin an observation with his conclusions now acting as premises for his developing argument. Ecclesiastes 4.7 is evidence that by now Qoheleth has trained his model reader with the appropriate reflexes. The discourse begins to actively count on those reflexes as he continues his speech.

Having established and trained his model reader, it is no accident that Qoheleth's discourse takes a turn in ch. 4. The presentation takes on a more impersonal tone in these observations.[167] The quotation or

167. Ogden, 'Qoheleth's Use', p. 346.

reflection upon gnomic texts begins to accelerate, further adding to the impersonal stance of the text. Alyce McKenzie observes the rhetorical advantages of arguing by means of proverbs:

> Proverbs are always indirect speech, they are always quoted. This serves to psychologically remove the protagonist from personal involvement with his arguments, de-emphasizing potential interpersonal conflict, ensuring the greatest degree of stability for continuing conversation. The proverb provides the protagonist (proverb user) with the kind of conflict protection that poetic language affords.[168]

The increased use of proverbial matter rhetorically balances the book, restoring a measure of equilibrium to the reader who has just been bombarded by the intense subjectivity of Qoheleth's interior monologue during the first three chapters. By using proverbs it partakes of the structural isomorphism of art which must balance a work's internal and external perspectives. Since the reader now sees things through his eyes, Qoheleth can dispense with a presentation that focalizes on inward matters. He has now readied the reader, and is prepared to take on the Wisdom tradition in a wholesale, more 'objective' manner.

Verses 8-12 broach the theme of 'two are better than one'. Verse 8 again interrogates the idea of leaving the fruits of one's labor to another (cf. 2.18-26). However, Qoheleth comes to a different conclusion, in essence, questioning the stance he arrived at earlier. Here, unlike in ch. 2, the protagonist questions the wisdom of not leaving something for one's family. This presents another argumentative reversal to the reader. Unlike the former discussion, there is no mention of whether the person has wisdom or not (cf. 2.18-19). In this passage Qoheleth builds upon the observation made in 4.4 that workaholism is the result of our overly competitive natures. He observes in his typically honest fashion how overwork has made many a person a loner. Readers have noted that Qoheleth has contradicted himself in these verses. And, in fact, he has. Having just offered the reader two 'dueling proverbs' in 4.4-6, Qoheleth presents, in the fashion of a wise person, a 'dueling observation' vis-à-vis his earlier observation on this matter. In 2.18-19, nothing was said about the futility of dying alone. By utilizing a strategy of incongruity the text offers an ideological gap to the reader, who cannot understand how the two passages connect in a logical fashion. Qoheleth's private

168. McKenzie, 'Subversive Sages', p. 57.

insights fair no better in that regard than public knowledge. As a result, the reader has a problem with consistency building.

The lasting effect of this inconsistency on Qoheleth's part is to leave the reader with a characterization and meaning *Gestalt* that must remain open. However, by means of this contradiction, the implied author allows Qoheleth to communicate to the reader something about the nature of human observation. All wisdom, both public (4.4-6) and private (2.18-19; 4.8) is subject to the laws of 'time'. Sometimes one observation is correct, while at other times its opposite is the more expedient choice. Qoheleth pulls no punches with his reader on the nature of human existence. For him, the world is inherently contradictory from the individual's point of view due to the corporate heart of darkness/ignorance that exists within humanity. By allowing Qoheleth to contradict himself, the implied author shows the reader how both private insights and public observations are not exempt from this cosmological and existential problem. In presenting this observation about the lonely workaholic, the text presents a wisdom Rubik's Cube to the reader. This bids the reader to think more deeply on the subjects of inheritance, work and human wisdom. The meaning of this text is the same as its effect, that is, to present an ambiguous world to the reader.

The use of the rhetorical question in v. 8 has the further effect of 'pulling the reader into the mind of the narrator's hypothetical loner. Asked in the form of a quotation, the question is placed innocuously in the consciousness of the reader. To that extent, the question involves the reader in the formulation of the argument.'[169] In a manner similar to Baruch Hochman, Hertzberg points out that the reader identifies 'not with a specific person, but rather, with a type of person'.[170] This creates a level of pathos in the text as Qoheleth attempts to attain an emotive response on the part of the reader. Due to the narrator's rhetoric of indirection, Qoheleth himself becomes an ambiguous character to the

169. R. Johnson, 'The Rhetorical Question', p. 160. J. Barton makes a similar observation about this question in v. 8. He states: 'Qoheleth suddenly drops the indirect discourse and transfers us to the soul of the miser, perhaps to his own soul, for this may be a bit of personal experience' (*A Critical and Exegetical Commentary*, p. 115). Such a pathetic description invites the reader to see a confessional increment in these verses even though the observation is couched in a third-person form.

170. Hertzberg, *Der Prediger*, pp. 114-15.

reader, a man about whom we are quite uncertain. Is he the selfish old tycoon in 2.18-19, or the older gentleman who knows that money is not everything in life? Qoheleth's discourse never tells us exactly which one is the true character of the man. However, one thing is for sure here. Again, Qoheleth's discourse characterizes him as a person motivated by self-interest, as the express reasons for his actions have nothing to do with the good he might do for the next generation. The motivation for leaving an inheritance is so that Qoheleth might not 'deprive *myself* of goodness (*miṭṭôbâ*)'. From this the reader again surmises that the narrator is neither altruistic nor humanitarian, but a person who understands the art of self-preservation and how the community facilitates that goal. While the text begins to lend Qoheleth a positive aura, his final comments render a less than attractive personality to the reader.[171]

Qoheleth's comments on the proverbs he reflects upon also characterize the narrator in a very emphatic manner. This habit of commenting and evaluating becomes his primary method of argumentation in the latter chapters of the book.[172] The values these remarks espouse provide the reader with a sense of the character's inner motivations. Such motivational comments provide strong evidence from which to judge and to characterize the narrator. In this instance, the robust sense of self-centeredness detracts from the narrator's ethos. Again, Qoheleth has managed to pull defeat from the jaw's of rhetorical victory by giving voice to such narcissistic tendencies.

Thereafter, Qoheleth again reminisces upon a proverb (*Tob-Spruch*) which confirms his observation.[173] He acknowledges that two are better

171. Jasper has noted the motive of self-interest in these verses, and concludes, as do many readers: 'It is still more a matter of what is advantageous than of what is right' (F.N. Jasper, 'Ecclesiastes: A Note for Our Time', *Int* 21 [1967], pp. 259-73 [266]. Blank also makes a similar observation. See S. Blank, 'Ecclesiastes', *IDB* (4 vols.; Nashville, TN: Abingdon Press, 1962), II, pp. 7-13 (12). Qoheleth's utilitarian ethic is not attractive to most readers.

172. R.F. Johnson, 'A Form Critical Analysis', p. 84.

173. There are 16 'better-than' proverbs in Ecclesiastes: 4.3, 6, 9, 13; 5.4; 6.3b, 9; 7.1a, 2, 3, 5, 8a; 9.4, 16, 18. In this distribution, it should be noted that their appearance dominates the central portion of Qoheleth's discourse. Their rhetorical appropriation by Qoheleth has been thoroughly discussed by G. Ogden, 'The "Better-Proverb" (*Tob-Spruch*), Rhetorical Criticism, and Qoheleth', *JBL* 96 (1977), pp. 489-505. The *Tob-Spruch* is an extremely important tool for the first-person narrator. Most importantly, it is a critical means for presenting the individual's point of view

than one. Qoheleth then comments on the proverb, noting, true to his 'bottom line' persona, that they have a better reward for their productivity. Two also stand against adversity better than one (vv. 11-12). Again, Qoheleth utilizes a rhetorical question to draw in his reader. The *'êk*-question ('how') in v. 11b 'is especially suited for obtaining audience participation, for the form requires more than just a simple "Yes" or "No" answer'.[174] Rhetorically, the observations in 4.10-12 offer 'empirical evidence (commentary) in support of the values implicit in the sayings'.[175] The subtle effect of this is again to make public knowledge dependent upon private observation and insight. What confirms public knowledge for Qoheleth is not its 'publicness', but whether the self can determine its correctness via personal observation.

Again, he reflects upon a proverb in 4.12b: 'A threefold cord will not be broken quickly'. From a Ricoeurian perspective, this verse contains a fair amount of unstable irony. Given the radical emphasis on the individual self as a means of knowing by Qoheleth, this verse also hints at the limits of the narrator's epistemology. Two knowers are also better than one, though this application seemed to escape Qoheleth. Obviously, with his emphasis on the rightful place of public knowledge, this insight did not escape his student, the Epilogist/implied author. And a

on a subject. Ogden notes how they often introduce an emphasis change or summarize a writer's argumentative point (p. 491). They can either conclude or summarize, but when they introduce a passage, as in this verse, they set up the values that are to be explicated in the remainder of the passage (p. 504). However, Ogden also demonstrates that Qoheleth has modified the *Tob-Spruch* by 'appending to it a clause which provides grounds for validating the values proposed' (p. 495; cf. 4.9, 17; 6.3-4; 7.2, 3, 5-6; 9.4). By means of these subtle comments, 'Qoheleth has taken the basic *Tob-Spruch* form with its acknowledged function within the tradition and made personal application of it as a medium for his own unique viewpoint' (p. 504). As a result, the better-than proverbs will have a foundational effect on how the reader characterizes the narrator over and beyond the other proverbial genres utilized by Qoheleth. Both the proverb and Qoheleth's comments will provide key nodal points on which the reader's response will hang. The critic will also observe that even when Qoheleth does not utilize 'I', his style is still to present material though the focalization of his own unique point of view. Again, we see where the proverb and its public knowledge are always subordinated to the all-pervasive influence of Qoheleth's 'I'. So dominant is Qoheleth's 'I', that even when he utilizes third-person forms, he is speaking in a first-person way. This goes to show how little a truly public knowledge existed from Qoheleth's post of observation.

174. R. Johnson, 'The Rhetorical Question', p. 161.

175. R.F. Johnson, 'A Form Critical Analysis', p. 83.

threefold cord, which includes private, public and transgenerational/ reader knowledge, will be even yet stronger. I doubt whether this is a reading based on authorial intention. But, if understood within an epistemological framework similar to that which my study is espousing, the verse does contain a surplus of meaning by taking a certain ironic stance toward the narrator's methodology.

Ecclesiastes 4.13-16 begins a new observation, contrasting youth and old age and their relationship to wisdom. The story told contains numerous ambiguous details and remains hermeneutically open.[176] The narrator notes that wisdom and old age are not synonymous. In order to make his case, Qoheleth resorts to the use of an example story in which he narrates a short story. He proves his point 'by means of a recognizable historical motif' upon which everyone would agree.[177] This example story also serves to characterize the narrator since in all likelihood it disguises the personal experience of the narrator. Gerard Hauser observes that such examples 'are best suited to audiences that have not yet formed general rules from which to reason, such as youths and novices'.[178] As a result, we see where the narratee is a youth, as the use of 'my son' in 12.12 implies. Qoheleth also notes, in a somewhat cynical fashion, that those who would usurp the foolish king are quite numerous, as if the desire for private power is the only motivation for individuals in a society, and that such a king will not be remembered favorably. Again, he concludes that such observations are examples of

176. Both Fox and Wright have noted the numerous details in this story which partake of Qoheleth's consistent use of a rhetoric of ambiguity: Fox, 'What Happens in Qohelet 4.13-16'; A. Wright, 'The Poor But Wise Youth', pp. 142-54. The story also contains some very prominent ironic moments as well. See: Spangenberg, 'Irony in the Book of Qohelet'. A final reading which is worth noting is that of D. Rudman, 'A Contextual Reading of Ecclesiastes 4.13-16', *JBL* 116 (1997), pp. 57-73. Rudman argues that 'the youth who emerges from prison is not an usurper but a counselor in the general tradition of Joseph or Daniel' (p. 62).

177. G. Ogden, 'Historical Allusion in Qoheleth IV.13-16', *VT* 30 (1980), pp. 309-15 (309). Although Ogden argues that Qoheleth is alluding to Joseph here, most readers have concurred that the book affords no sure historical allusions. See Bréton, 'Qoheleth Studies', p. 46. We have here what Alter would describe as a 'type character' set in an imaginative story for heuristic purposes. For both Qoheleth and his narratee, history would have afforded numerous examples, Joseph being only one of many. See R. Alter, *The Art of Biblical Narration*, pp. 47-62.

178. Hauser, *Introduction to Rhetorical Theory*, p. 75.

the absurdity which characterizes this world. In these verses one only detects an honest man with a pessimistic point of view.

8. *Ecclesiastes 4.17–5.8: The Knowledge of Divine Duties*

A qualitative shift occurs in Qoheleth's discourse in 4.17. Even the non-critical reader notices how his speech is suddenly peppered with an abundance of second-person grammatical forms. Some form of second-person address, typically the imperative or jussive, is used in 4.17, 5.1a, 1b, 3a, 4, 5 and 6. This is similar to the front-loading of the first-person pronoun in 1.12-18. The implied author is fond of literary over-kill and redundance when he wants to accent something. As George Castellino observed, Qoheleth is no longer engaging in interior mono-logue, but:

> has turned to the reader, or listener, and is imparting to him admonitions and instructions. The direct speech in second person, however, is not exclusive and consistent, and occasionally Qohelet falls back again into the narrative style. This does not obscure the fact that from 4.17 on we observe a different kind of discourse.[179]

Following Stephan de Jong, I note that the book's first observation complex has ended, and an instruction complex has begun.[180] Eric Christianson observes that such a change constitutes a 'shift from experience to advice and an overall strategy in which the reader is in-vited to partake more and more in the text's story-world'.[181] As a result, the discourse now focuses not on the narrator, but on his narratee. In 4.17–5.8 the reader gets a brief look at Qoheleth's narratee. Just as the earlier passages configured the narrator, so these verses configure the textualized 'listener' for the reader. Most important in this regard is the anonymity of the narratee. The fact that the narratee remains nameless is strategic. The anonymity of the narratee 'creates a gap the reader is invited to fill with her/his own identity, entering into the narrative and confronting the circumstances and situation of the character in the

179. Castellino, 'Qoheleth and His Wisdom', p. 16.

180. De Jong, 'A Book of Labour', p. 108. Although de Jong limits the section to verses 4.17–5.6 (Eng. 5.1-7), I have broadened this first instruction complex to include verses 5.7-8 (Eng. 5.8-9) because of the continued use of second-person address in these verses which continue the trend initiated in 4.17–5.6.

181. Christianson, *A Time to Tell*, p. 245.

text'.[182] An anonymous narratee therefore helps to engender the transformation of the reader's own identity.[183] In addition, anonymous second-person address builds a sense of textual inclusiveness. As Christianson states:

> the element of second-person narration...still does not provide a name. A name would have meant a barrier in this regard and its absence suggests that Qoheleth desired a wide audience to identify as much as possible with this constructed narratee.[184]

Qoheleth begins his direct address to the narratee in as direct a way as possible by utilizing the imperative form *šᵉmōr raglᵉykā* ('watch your steps'). This command summarizes the instructions that follow. Having established his credentials as a sage, he now draws on that notoriety to instruct his narratee. The admonitions in 4.17–5.8 all deal with typical Wisdom themes regarding the problems associated with worship and proper sacrifice, rash speaking, religious vows and obligations, and governmental corruption. Once again, his discourse is judiciously sprinkled with proverbial reflections in 5.1 and 5.5 which express only traditional values. Very little of the advice given here distinguishes Qoheleth from other sages. In fact, if one were to characterize Qoheleth on the basis of this passage alone, a fitting word to describe him would be 'nondescript'. He ascribes to a type of religious devotion that sounds like civil religion in that there is a sense of duty which conspicuously lacks allegiance to a personal God. The deity one meets here is 'not the God of Abraham or the God of Israel, but the God of heavens'.[185] Qoheleth knows his Torah, as he quotes Deut. 23.22a in 5.3.[186] He admonishes his narratee to take the prudent path. The fewer one's words, the better (5.1). In addition, there also seem to be a few more plays on the Solomonic guise in 4.17–5.6. Hubert Tita has convincingly argued that the

182. Beck, 'The Narrative Function', p. 147.

183. Beck, 'The Narrative Function', p. 148.

184. Christianson, *A Time to Tell*, p. 245.

185. A. Tsukimoto, 'The Background of Qoh 11.1-6 and Qoheleth's Agnosticism', *AJBI* 19 (1993), pp. 34-52 (46).

186. Crenshaw, *Ecclesiastes*, p. 116. See also the treatment of this verse by Y. Hoffman, 'The Technique of Quotation and Citation as an Interpretative Device', in B. Uffenheimer and H. Reventlow (eds.), *Creative Biblical Exegesis: Christian and Jewish Hermeneutics through the Centuries*, (JSOTSup, 59; Sheffield: JSOT Press, 1995), pp. 71-79 (76-79).

references to dreams, prayers and the hearing heart all constitute allusions to the story of Solomon in 1 Kgs. 3–11.[187] Once more, we grasp the importance of the King's Fiction for understanding the text's model reader. As Christianson has noted regarding the persistence of the Solomonic guise in Ecclesiastes:

> the guise continually reasserts itself... The Solomonic guise is more complex than that [i.e., a mere rhetorical device]. It provides for the reader an ever-present, if sometimes elusive, sometimes insinuated context in which to grasp the experiments of Qoheleth.[188]

When it comes to divine knowledge Qoheleth knows his place, speaking nothing but the 'party line'. Not wanting to offend the Deity, he merely quotes the status quo, and admonishes the path of least resistance. Not once does he criticize, ironize or defamiliarize the world of divine obligations as he has other parts of the narratee's worldview. Following his own advice, everything seems predicated on not offending the Deity. Nothing is said that would characterize Qoheleth as a devoted follower of the Covenant God. In fact, in verse 5.4, God seems more like a creditor whose bills must be paid. Everything is contoured to keeping the narratee's feet on the ground. As a result, twice Qoheleth evaluates 'dreams' (*ḥalomôt*) in a negative manner (5.2 and 6). While this may have a cultic meaning,[189] the discourse is admittedly vague at this point. I surmise that it refers to the visionary characteristics of youthful fantasies and goals. The point is that one should neither offend God (v. 5) nor the king (v. 7), even if one does object in a youthful fashion to the cries of the oppressed by the system. After all, so long as the economy is kept going, a king has served his purpose (v. 8).

In the space of a few short verses, Qoheleth has moved a great distance from his highly empathic advocation for the oppressed in 4.1. Rhetorically, the reader perceives yet another ethos-related reversal for the narrator. Again, one senses a strategy of incongruity here by comparing this advice with what has just been given in ch. 4. Such inconsistency is rarely dealt with graciously by readers. Given the incessant tendency of the narrator towards self-centered evaluations, the reader begins to realize that although Qoheleth can occasionally escape the

187. H. Tita, 'Ist die thematische Einheit Koh 4,1–5,6 eine Anspielung auf die Salomoerzählung?', *BN* 84 (1996), pp. 87-102.

188. Christianson, *A Time to Tell*, p. 148.

189. Whybray, *Ecclesiastes*, p. 93.

confines of the self, his normal posture is a man obsessed with what is good for himself. By now the reader has seen enough of this pattern to come to a *Gestalt*. The characterizing stage is over. The reader begins to engage in character-building, a process which takes note of persistent patterns and makes due judgments regarding the overall type of personality or character who would possess such traits. Although Qoheleth is seen as possessing the quality of prudence, even that quality is undermined due to the lack of caring, justice, courage, magnanimity and magnificence which are sorely missing in these admonitions. This type of prudence is quickly recognized as mere self-preservation.

For the narratee these admonitions are perceived as authoritative and caring advice given their common situation with the narrator. However, this is not the case for the implied reader who has been 'let in on' the epistemological weaknesses of the narrator and the irony which surrounds this figure. Because of the ironic and satiric characterization offered to the reader by the implied author, while this counsel might be expedient and self-preserving, the narrator comes across as a man who possesses a limited perspective on life and as one who often lacks a visionary perspective on life. That lack of vision (magnificence) creates a rhetorical ethos which suffers a great deal in fundamental attractiveness. This is especially the case for the postmodern reader who has been exposed to the hermeneutics of suspicion—a reading strategy which often informs modern reading habits.

Recalling that the narratee mirrors the narrator, these admonitions presuppose a narratee who is similar to the narrator himself—a young man who is looking to climb the social ladder and who wants to know how to succeed in a world of obligations. Qoheleth warns against youthful flights of fantasy and encourages the narratee to keep both feet on the ground and their mouth shut. The narratee is a social conservative, like Qoheleth himself. Unlike the implied reader who is characterized as a more perceptive person capable of critical judgment and ironic evaluation, the narratee is characterized along lines that suggest a younger version of Qoheleth himself. The narratee is staid and uncommitted to social change. To that extent, there is little difference between Qoheleth and his narratee except for the age factor. As Norman Holland once observed, style seeks itself.[190]

190. N. Holland, *Five Readers Reading* (New Haven: Yale University Press, 1975), p. 114.

A cleft between the narratee and the implied reader erupts in this passage. While at the beginning of the discourse the narratee and the implied reader are virtually the same, in this passage the implied reader begins to perceive a difference between themselves and Qoheleth's narratee. While the narratee is unreflectively conservative, the implied reader is ironically critical. As a result of the implied author's critical and satiric handling of Qoheleth's 'I'-centered epistemology, the implied reader has been trained to sense the deficiencies of an ethic based on such a self-centered epistemology. As a result, the implied reader is given a horizon of ironic knowledge which the young narratee does not yet possess. In that respect, the reader enjoys an elevated position vis-à-vis the narratee. Because of this exalted level of philosophical maturity and vision, I would surmise that the implied reader of the book of Ecclesiastes is slightly older than the narratee. He or she is a person who is more reflective and mature about what constitutes Wisdom and its foundations. Although the narratee is constantly bombarded with rhetorical questions and asked to question a great many tenets of Wisdom, the narratee's essential social stance remains unaffected by these queries. On the other hand, the implied reader is capable of self-transcendence and critique and, more importantly, a post of observation outside the confines of self-interest. While the implied reader is a seasoned sage capable of true Wisdom, the narratee is a debutante who excels only in the naive espousal of the self-expedient path and unreflective questioning. This ironizes the narratee along with the narrator. As a result, it is possible to represent the narratee/implied reader relations after 4.17 as follows:[191]

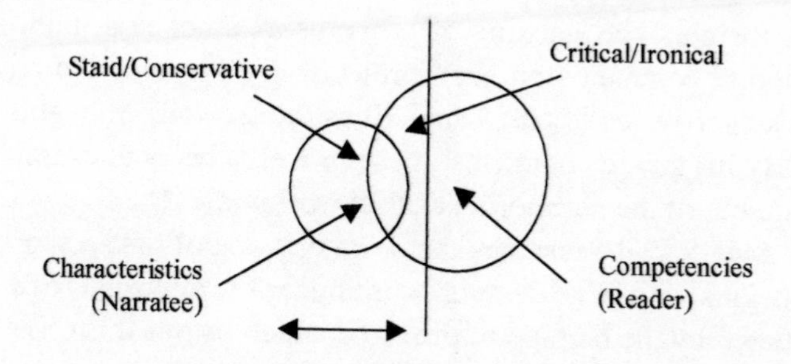

Fig. 5.2: *Narratee/Reader Relations*

191. The horizontal arrows going both ways below the narratee circle suggests the ideological cleft which opens up in these verses.

Due to the cleft which opens up between the narratee and the implied reader in (4.17-5.8), the narratee can function only as an unreliable narratee in terms of providing a textualized role-model for the reader. As a result, we see that the real addressee of the book of Ecclesiastes is not the narratee, but the astute and critical implied reader who is asked to weigh rhetorically both Qoheleth and his narratee in the epistemological balances of public and trans-generational knowledge.

9. *Ecclesiastes 5.9–6.9: Possessions and the Possession of Joyful Knowledge*

Qoheleth ends the first half of his monologue with a critique of the pursuit of wealth. The emphasis again shifts to Qoheleth's own observations regarding the 'meaninglessness of wealth'.[192] He begins his discourse by reflecting on three proverbs which condemn the love of money (v. 5.9), the increase of possessions (v. 10), and greed of the rich (v. 11). His own confirmation of these proverbs is given in 5.9b, 'this also is an absurdity'; v. 10b, 'but what profit has its owner except to see them with his own eyes'; and v. 11b, 'but the greed of the rich will not allow him to sleep'. Qoheleth underlines the emptiness of acquisitions if a person finds no joy in one's work.[193] In each case, as he has done in ch. 4, Qoheleth quotes a proverb in a feigned attempt to establish an impersonal point of view and then subordinates them to his own perceptions by commenting upon them. In these verses he confirms public knowledge by attaching to each proverb a disguised observation couched in a third-person form. Again, we see his tell-tale literary and epistemological method of subordinating public knowledge to private insight. The stacking of proverbs together in a 'mini-series' and his subtle comments on each proleptically prepares the reader to view the longer *series* in chs. 7 and 10 as the '*serious*' thoughts of the narrator. By his subtle comments, these proverbs become examples not of wisdom, but of the narrator's peculiar worldview.[194]

Ecclesiastes 5.12-16 resumes the narrator's use of first-person narration. He begins with *yēš*, 'there is', continuing the atmosphere of objectivity initiated by the barrage of proverbs which begins this observation complex. The use of *yēš* and *hinnēh* in vv. 12 and 17 respectively move

192. Longman, *Ecclesiastes*, pp. 159-60.
193. Whybray, 'Qoheleth, Preacher of Joy', pp. 87-98.
194. R.F. Johnson, 'A Form Critical Analysis', p. 140.

the argumentation to the 'now' level of the reader, further pulling the reader into the flow of the argument.[195] Just in case the narratee assumes that material success is everything as they attempt to climb the social ladder, Qoheleth points out the problems with such an unmitigated pursuit of wealth and possession. Verse 12 observes how possessions can hurt its possessor. Juxtaposed to this proverb and recalling 2.18-19 and 4.8, 5.13 again returns to the theme of leaving wealth to the next generation, noting the horrors of lost inheritance due to bad business decisions on the part of the father. The use of 'grievous/sickening' evil in 5.12a communicates to the reader something of Qoheleth's emotional horror at the thought of having it all and not having any personal peace. However, the emphasis of this observation falls upon the dread of losing one's wealth, as Qoheleth offers an extended reflection on those horrors in vv. 13-16. Winding up like the oppressed in 4.1 is the ultimate nightmare for Qoheleth. Pondering that situation results in one of the most emotion-laden outbursts in the entire book, as Qoheleth lists the results of such a social fall with words that abound in pathotic qualities: darkness (*ḥōšek*), great vexation (*kāʿas harbēh*), disease (*ḥolyô*) and resentment (*qāṣep*). The cascade of emotion in this verse draws the reader into this poor soul's torment, recreating in the reader a sense of terror at the prospect of such a condition. The confessional increment of this emotional outburst portrays a man motivated by fear of failure and abject terror of losing his vaunted social status.

However, the passage does achieve a positive effect by duly criticizing the unabashed pursuit of wealth for its own sake. Via this desolate depiction, Qoheleth begins to 'reconstruct the normative by setting opposites over against one another'.[196] From the entire passage (5.9–6.9), the reader surmises that the new norm espoused by Qoheleth consists of balancing economic stability with an acknowledgment of wealth's limitations and pitfalls. In these verses, Qoheleth asks a rhetorical question five times (cf. 5.10, 15-16; 6.6, 8 [twice]) in an attempt to re-orient the reader's sense of the work-yields-profit norm.[197] The

195. Isaksson, 'The Autobiographical Thread', p. 45.

196. R. Johnson, 'The Rhetorical Question', p. 257.

197. Johnson observes how the 'what profit?' question in vv. 5.9–6.9 undergoes a change in reference. Before this, it refers to the profit of humanity in general. Here, it refers to a single individual, much like the other First Testament usages. See

new norm established by these admonitions and rhetorical questions fully coincides with the *carpe diem* ethic advocated by the seven-fold call to enjoyment. Qoheleth communicates in a negative fashion the categorical obligation to enjoy the fruits of one's labor by portraying in graphic emotional images the futility of acting otherwise. True to his polarized character, Qoheleth wants to 'have his cake and eat it' by advising the narratee to avoid economic excesses in either direction. In admonishing this, he plots a social and economic 'golden mean' ethic for his narratee.

In 5.17-19 the narrative makes another unexpected reversal. Just as the emotional landscape of Qoheleth's discourse darkens into an emotional nightmare, the sun rises again in the fourth call to enjoyment (vv. 17-19). This call is the strongest yet encountered by the reader. It begins with a more 'solemn introduction', *hinnēh ʾašer-rāʾîtî* ('behold, I discovered').[198] The use of *hinnēh* has the effect of aligning the narratee's point of view with the narrator's. It gently commands the narratee to see things his way. Eating, drinking and enjoying one's labor encapsulates Qoheleth's idea of the abundant life. Again, the motif of mortality is raised by his reference to the 'few days of his life which the Deity gives him' (5.18). Given the brevity of human life, enjoyment of those fleeting years is the only logical response to this absurdity. Coming on the heels of Qoheleth's critique of the amassing of wealth, these verses point out how human achievement 'is canceled by the fact that one's shroud has no pockets'.[199]

Verse 18 is the only place in the book of Ecclesiastes where Qoheleth can sense God's grace. The one gift of God in this entire universe is the ability to enjoy one's possessions and to find meaning in work. Verse 19 adds Qoheleth's own 'amen' observation regarding the efficacy of his advice. The person who follows his counsel 'will not brood much over the days of his life, because God keeps him busy enjoying himself' (JPS). However, it may be that the word which is usually translated as 'keep busy' (*maʿaneh*), might be something of a double entendre. Norbert Lohfink has called attention to the two meanings this hiphil participle might have in this verse. The root *ʿnh* can mean either

R. Johnson, 'The Rhetorical Question', p. 167. This change in reference is a consequence of its inclusion in an instruction complex whose focus remains upon the narratee.

198. Whybray, 'Qoheleth, Preacher of Joy', p. 87.

199. Scheffler, 'Qohelet's Positive Advice', p. 256.

'to be occupied, be busy with something' or 'to answer, speak publicly, reveal'.[200] Lohfink takes *'nh* in the latter sense, arguing that

> The joy of the heart must be something like divine revelation. When we experience joy at least in one small moment, we come in touch with that sense of things which normally God alone sees. That could well be the message of 5.19b... Of necessity, he would now understand the verse in the changed sense that 'God answers, speaks, reveals himself by the joy in the heart.[201]

Either way, the reader understands the absolute necessity of enjoying life as a categorical imperative. Unlike the poor loner described in 5.12-16, the reader who puts this advice into practice will be so busy enjoying himself or sensing God's presence that life's wearisome days (cf. 1.8) will hardly be noticed. This is the person described by the proverb in 5.11a, the man who can sleep at night 'whether he eats little or much'. In a moment of magnanimity, Qoheleth expresses a nobility of thought that again raises his 'stock' with the reader. There is also a strong sense of magnificence in these verses as they definitely lift the spirit of anyone who reads them. Never was private insight more beneficial to the public than what the reader encounters here. Again, Qoheleth rises after having fallen rhetorically. To track his ethos is like following the futures market of the Dow Jones Stock Exchange—Qoheleth's ethos is a commodity that has constant spikes and troughs. In this we see the radical effects of the ethos-related pendulum swings which characterize the narrator's vain rhetoric.

Ecclesiastes 6.1-9 functions as a sort of interlude for the two halves of the book. It 'notes the abject misery of a life devoid of joy'.[202] The passage also functions as an extended meditation on what Qoheleth himself has just argued in 4.17–5.19. The call to enjoyment in 5.17-19 is sandwiched between two extended reflections which prepare and debrief the reader regarding the absolute necessity for enjoying one's life regardless of economic status. Such careful argumentation emphasizes the importance of the call to enjoyment for understanding the essential message of Qoheleth's discourse. Verses 1-9 open up a blank which 'exacerbates the audience's cognitive dissonance in that it begins

200. Lohfink, 'Revelation by Joy', p. 627.
201. Lohfink, 'Revelation by Joy', p. 634.
202. Perdue, '"I will make a test of pleasure"', p. 205.

with assertions that diametrically oppose the preceding three verses'.[203] Ecclesiastes 5.17-19, with its emphasis on the enjoyment that 'God gave to him', is now juxtaposed to an equal emphasis in 6.1-2 regarding the enjoyment that 'God did not give to him'.[204] As a result, 'the restructured [toil-yields-profit] norm is defined by the deity; toil does not produce profit, but God produces enjoyment'.[205] Why Qoheleth emphasizes this theological contradiction at this point in his argument is not readily apparent since such determinism essentially undermines the goal of influencing the narratee to choose a *carpe diem* lifestyle. While this creates yet another example of ambiguity and hermeneutical openness for the text, this much is clear: Qoheleth's God is an aloof and distant deity who has a strong streak of capriciousness. Such a God lends a further negative characterization to the narrator.

Qoheleth begins the observation in 6.1-6 by calling attention to a great evil (*rā'āh*)—a person has everything but God does not give him the power to enjoy them so that a stranger enjoys his possessions. He uses the *yēš* ('there is') formula to add a degree of objectivity and formality to his observation. The term *yēš* also moves the narration to the reader's 'now' level.[206] Furthermore, Qoheleth utilizes a *zeh...hû'* construction in 6.2b to emphasize the degree of such an absurdity: 'this *really* is an absurdity'. To forfeit a life in this manner is tantamount to never having existed; in fact, it would be better to have died at birth than to live such an abomination (v. 3). Verses 4-5 observe how the stillborn child finds more rest than such a person. Even length of days, should one live two millennia, cannot compensate for such futility since death will eventually bring an end to this person's existence (v. 6). Qoheleth ends the observation with a rhetorical question so as to force the reader to answer whether a life without enjoyment makes sense.

In vv. 7-9, Qoheleth's mood turns cynical. Again, he returns to the theme of the relative value of wisdom over folly (cf. 1.17; 2.12-13). The sage again picks up the theme of human insatiability that was initiated in 6.1-2.[207] He then reminisces upon a proverb whose referent is difficult to ascertain. The ambiguous use of *l^epîhû* ('for it/him') could

203. R. Johnson, 'The Rhetorical Question', p. 258.
204. R. Johnson, 'The Rhetorical Question', p. 258.
205. R. Johnson, 'The Rhetorical Question', p. 258.
206. Isaksson, 'The Autobiographical Thread', p. 45.
207. Seow, *Ecclesiastes*, p. 226.

refer to the person in 6.2, or perhaps to Sheol (the 'one place' mentioned in 6.6).[208] The proverb simply notes that although we are driven to feed ourselves, humanity is never satiated. Its juxtaposition to 6.1-6 opens up a blank for the reader who must work at understanding its connection to what precedes. Given the lack of satisfaction with humanity's accomplishments and efforts, Qoheleth then asks the double rhetorical question, insinuating that the wise have no advantage over fools. He finishes with a convoluted question regarding the advantage of the poor person who knows how to conduct oneself before the king. Raymond Johnson suggests that the use of the double rhetorical question is to 'heighten reader awareness, and signal both the passage and approach of a significant moment in the argument of the text'.[209] The function of the double question is to generate consensus.[210]

Verse 9 ends the first half of the book with another proverbial reflection and comment. 'Better is the sight of the eyes than the wandering of the appetite' remains hermeneutically open. The use of vague proverbs is yet another technique from Qoheleth's arsenal of ambiguity. This strategy of citing opaque maxims increases over the next few chapters, becoming almost typical of the sage's discourse strategy. I surmise that the proverb in 6.9 refers to the relative value of seeing, that is, wisdom over wandering desire. But how this relates to being an absurdity escapes me. However, perhaps Charles Whitley is correct when he translates this verse as 'Better the pleasure of the moment than the departing of life'.[211] Saying that and concluding it to be an absurdity would make better sense in this context. As a result of the proverb's ambiguity, Qoheleth ends the first half of his monologue with a great crescendo of argumentation, the last note of which sounds decidedly off-key. The use of ambiguity creates another gap which remains hermeneutically open.[212] However, the negativity expressed in these verses seems to suggest that the juxtaposition of 6.1-6 with 6.7-9 is 'to emphasize that joy is not to be identified with the satiation of appetites and the fulfillment of desires'.[213] Such verses portray a sage given to obtuse thought and speech.

208. Crenshaw, *Ecclesiastes*, p. 128.

209. R. Johnson, 'The Rhetorical Question', p. 179.

210. R. Johnson, 'The Rhetorical Question', p. 210.

211. Whitley, *Koheleth*, p. 60.

212. R.F. Johnson, 'A Form Critical Analysis', p. 95.

213. Perdue, 'I will make a test of pleasure', p. 226.

Still, his main point is clear. Ecclesiastes 5.9–6.9 in particular draw the reader into the 'what profit?/*carpe diem*' framework of thinking.[214] Better to enjoy life with less than to amass wealth and miss the point of having lived, which is the real profit or portion available to humanity. In this carefully argued section (4.17–6.9), the reader senses a person who is concerned about their well-being. In spite of his jadedness and self-centered ethic, Qoheleth sincerely cares about the lives of his students. That is established beyond reasonable doubt in these verses. That tender attitude and wise counsel does a lot to influence a positive characterization by the reader. Although the sage is still seen as narcissistic, such passages make him at least palatable and respectable as far as sages go. By ending his discourse in the first half as he does, Qoheleth is characterized as a caring, compassionate pessimist with some rather glaring weaknesses. Unlike other passages, Qoheleth shines with characteristics of temperance, prudence, magnificence and magnanimity. At the moment, Qoheleth's stock has risen to a respectable level. In these verses, the reader senses a sage they can trust.

10. *Ecclesiastes 1.1-6.9 Summarized: A Rhetoric of Robust Reticence*

Ecclesiastes 1.1–6.9 consists of three aesthetic and ideological movements. Ecclesiastes 1.1–2.24 provides a criticism of private insight, whereby the implied author subtly hints at the limits of a self-centered and private epistemology, thereby ironizing and satirizing the protagonist. However, in 3.1–4.16 the implied author criticizes the knowledge of the community, calling attention to its inherent darkness and limitations through Qoheleth's honest observations and subtle comments. The last section (4.17–6.9) offers advice based on this epistemological situation. Given the limits of both private and public knowledge, the text admonishes the narratee/reader to enjoy life as best he or she can as a way to redeem a *ḥēleq* ('portion') from life's overal lack of *yitrôn* ('profit'). In addition, the first half is characterized by what I am calling a 'rhetoric of robust reticence'. By using the word 'robust', I am calling attention to the bodacious use of the self as an epistemological tool by the narrator. Reticence refers to the implied author's tendency to subtly ironize both the narrator and narratee. Juxtaposed to the glaring and bold use of 'I' as a rhetorical technique is the use of literary strategies

214. R. Johnson, 'The Rhetorical Question', p. 257.

which work on a principle of reticence, that is, saying something by innuendo, implication or subtle ironization. The two work in tandem to produce a finely tuned literary character whose rhetorical strategy can, in the final analysis, only be termed a vain rhetoric with its constant pendulum swings between attractive and unattractive ethos.

Chapter 6

A Rhetoric of Subversive Subtlety: The Effect of Qoheleth's First-Person Discourse On Reader Relationships in Ecclesiastes 6.10–12.14

> Of course, I know that the best critics scorn the demand among novel readers for 'the happy ending'. Now, in really great books-in an epic like Milton's, in dramas like those of Æschylus and Sophocles-I am entirely willing to accept and even demand tragedy, and also in some poetry that cannot be called great, but not in good, readable novels, of sufficient length to enable me to get interested in the hero and heroine![1]

1. *Introduction*

Chapter 5 dealt with the characterization of Qoheleth as an intimate pessimist. This chapter will deal with the ethos of scepticism and its suasive and dissuasive effects on the reader. In the book of Ecclesiastes, there is a movement between the two halves from an ethos of pessimism to an ethos of radical scepticism. Beginning with v. 6.11, the book begins to emphasize the root *yd'* ('to know') in a variety of ways. Ecclesiastes 7 and 8 emphasize the phrase 'do not know' and 'no knowledge', with the triple repetition of the phrase 'not find out' in 8.17 marking a major structural division. Following Addison Wright, I note that 7.1–8.17 centers on the theme of 'humanity cannot find out what is good to do' while 9.1–11.6 focuses on the theme 'humanity does not know what will come after them'.[2] The subject of the second half is introduced by the thematizing question in 6.11: 'Who knows what is good for humanity while he lives the few days of his absurd life?'[3]

1. Theodore Roosevelt, *A Book-Lover's Holidays in the Open* (New York: Charles Scribners Sons, 1916), pp. 263-64.
2. A. Wright, 'The Riddle of the Sphinx, p. 108.
3. Farmer views this question as the essence of the book's message. See Farmer, *Who Knows What is Good?*

2. *The Difference a Sceptic Makes*

It is important to note that scepticism and pessimism should not be confused, as they involve two separate and distinct attitudes. Scepticism denotes an epistemological stance toward reality. Under the rubric of scepticism one often finds the broad philosophical problem of how one knows. Specifically, scepticism expresses grave doubts about the ability of humans to adequately know in any absolute and certain sense. Richard Popkin defines scepticism as an extreme questioning of 'all knowledge claims that go beyond immediate experience, except perhaps those of logic and mathematics'.[4] Questions relating to perception and understanding underlie this form of agnosticism. Scepticism therefore arises because of the broad underlying conviction that human apperception and interpretations based on those apperceptions are limited. This is quite a different issue than that which informs the pessimist's convictions, which relate to the broad issue of the relative balance between good and evil in the world. Rather than using the term 'scepticism' to delineate this type of questioning, it is better to differentiate between scepticism and pessimism by having pessimism refer to attitudes about the relative evil or goodness of the world or of people's experience of the world.[5] The matter of questioning the relative strength of good and evil is not the same as that which asks about the sufficiency of human faculties of knowing. One must differentiate between the epistemological and the ontological issues involved in the two attitudes.[6]

4. R. Popkin, 'Skepticism', in Edwards (ed.), *The Encyclopedia of Philosophy*, pp. 449-61 (499).

5. Loemker, 'Optimism and Pessimism', p. 114. For a discussion of the growth of skepticism on Jewish soil, see J. Crenshaw's classic article, 'The Birth of Skepticism in Ancient Israel', in J. Crenshaw and S. Sandmel (eds.), *The Divine Helmsman* (New York: Ktav, 1980), pp. 1-20.

6. For instance, some differentiate scepticism from pessimism based on the whether or not pessimism can be maintained as a distinct philosophy versus a 'way of viewing the world or an attitude towards life based in one's life experiences' (W. Anderson, 'Philosophical Considerations in a Genre Analysis of Qoheleth', *VT* 48 (1998), pp. 289-300 [290-91]). For Anderson, who also relies upon Loemker, scepticism is considered to be a more 'formal investigation and critical analysis of good and evil in the world' whereas pessimism seems to be a more personal, 'attitudinal "response" to proper philosophical constructs and analyses' (p. 291). He concludes that scepticism is the 'intellectual counterpart to pessimism' (p. 291). However, as we have seen, the two may overlap in places, yet pertain essentially to

In the book of Ecclesiastes, this distinction is firmly held at the literary level by the thematic distinction between the first and second halves of this book. Ecclesiastes 1.1–6.9 characterizes the narrator as a pessimist. But that pessimism turns toward scepticism in 6.10–12.7 as the entire epistemological enterprise is subjected to radical questioning. The movement from pessimism to scepticism in the second half of the book is seen above all in the change of overriding questions which controls the aesthetic movement of the two halves. In chs. 1–6, the 'what profit?' question dominates the ideology of the narrator. However, in chs. 7–12 this question wanes in significance, with the 'who knows?' question gaining rhetorical prominence. Raymond Johnson observes:

> Another possible relation between theme questions and structure surfaces when one considers the shift in theme questions that seems to occur at the midpoint of the book (6.9). On the other hand, prior to Ecclesiastes 6.9, seven 'What profit' questions are found as opposed to three questions which inquire after the possibility of knowledge (Eccl. 2.19, 3.21, 3.22). On the other hand, after Ecclesiastes 6.9, the frequency is inverted: there is only one 'What profit' question (Eccl. 6.11) for five questions pertaining to knowledge (Eccl. 6.12a/b; 7.24; 8.7; 10.14).[7]

This shift in thematic question initiates a commensurate modification of the reader's characterization of the narrator. Qoheleth begins to take on a sceptical ethos. In these chapters, the 'quest' takes a back seat. In its place the protagonist's deeper, more philosophical issues come to the forefront as Qoheleth begins to 'speak less of the story of his youth and more of his 'present' concern'.[8] Like the initial "what profit?" question (1.3), the introductory 'who knows?' question in 6.12 also opens a blank for the reader. Once again, the narrator resorts to a rhetoric of frustrated expectations, as it will not be answered until 8.17.

different philosophical issues. There is a larger formal issue at stake here besides whether a statement or ideology springs from a process of formal philosophical deduction as opposed to personal response. Anderson's definition fails to understand that scepticism relates to epistemological issues, while pessimism relates to one's perception of natural ontology, that is, the relative evil or goodness of the world. Like the two circles in a Venn diagram, they may overlap in many strategic places, yet remain distinct realms of thought.

7. R. Johnson, 'The Rhetorical Question', p. 230.
8. Christianson, *A Time to Tell*, p. 243.

Such a degree of honesty and openness is possible because by now Qoheleth has established a trusting and intimate relationship with his reader. As a result, the level of self-disclosure increases. In 6.10–12.7 the reader begins to see Qoheleth as he 'really is'. Trapped within the epistemological confines of empiricism and his own heart (*lēb*), Qoheleth is unable to find a sure means of knowing. The result of his radical scepticism is that he becomes a 'subversive sage', to borrow a characterization from Alyce McKenzie. That subversiveness characterizes the second half of Qoheleth's discourse. It becomes the major ethos-related characteristic around which the reader's response revolves.

3. *The Emergence of the Model Reader*

With the conclusion of the narrator's discourse in 6.9, all the interpretive reflexes needed to read Qoheleth's discourse have been established. The model reader has been trained to understand the various proverbs as examples of the narrator's own reflections through the various 'citations' that occur in 1.15, 18; 2.14; 3.20; 4.4-6, 9, 12, 13, 17; 5.1, 5, 9-11; 6.7, 9. Particularly instrumental in the training of the reader were the dueling proverbs in 4.4-6 which educated the reader to look for incongruities and contradictions in the Wisdom tradition's fund of public knowledge. In addition, the 'mini-series' in 6.1-2, with its prominent use of comments following each proverb, further instructed the reader to view such reflections as examples of Qoheleth's private wisdom and to expect some sort of evaluation following the narrator's utilization of public knowledge.

Because the model reader has been fully equipped with the proper hermeneutic reflexes and specific literary competencies to properly understand the discourse, Qoheleth no longer needs to emphasize his personal 'I'. Nor is it necessary to continue training the reader to question Wisdom's tenets. As a result, emphasis upon first-person discourse proper radically decreases. The narrator's 'I' now punctuates the discourse in a sporadic fashion. There is a decided shift from personal address to a marked impersonal tone, creating a 'shrinking narrator', so to speak. At this point in the discourse, the wholesale reflection upon proverbs/public knowledge becomes Qoheleth's *modus operandi*. By now, the reader knows that the proper stance to take toward both public knowledge and private insight is to critically evaluate their essential claim to universal validity. Like Melville's *Moby Dick*, the model reader of Ecclesiastes has been trained to make contradictions of everything.[9] The reader also understands that the appropriate response to this situation is to lay claim to the only reward that life offers, that is, to wholly embrace a *carpe diem* lifestyle. Having trained the model reader to take such a critical stance and to perceive, regardless of the grammatical orientation of the discourse, the narrator's own underlying post of observation, the implied author can now offer a discourse whose qualities are substantially different from the first half of the book. Even the most unsophisticated of readers notices that a radical shift occurs after 6.9.

While chs. 1–6 are characterized by a rhetoric of gapping, the latter half of Qoheleth's monologue consistently utilizes a rhetoric of subversive blanking. The text consistently juxtaposes various proverbs, aphorisms, reflections and observations in order to increase the reader's role in making sense of the discourse. Earlier I referred to this as the colliding or interacting topics approach to understanding the structure of the text. By constantly invoking a rhetoric of incongruity, Qoheleth offers a discourse which is more rambling in nature. There is an unmistakable granularity to the texture of his monologue. The protagonist's oration no longer flows from point to point, but has a more disjointed quality about it. This juxtaposition of seemingly unrelated topics forces the reader to work harder at consistency building. This results in the reading process slowing down considerably, as the reader labors at 'proverb crunching', to adapt a metaphor from computer technology. In order to read productively the reader compares each statement with what precedes and follows it in the discourse. The text's horizon of topics and values becomes instrumental for helping the reader make sense of Qoheleth's latter discourse. To come to a meaningful *Gestalt*, the reader must infer the relationship between the juxtaposed topics by using the norms established in the earlier discourse as a guide. Given the rhetorical strategy of these chapters, the textual consumer must labor at finding a logical or even meaningful flow to the narrative. While some of this is a bias of a Western reading grid, much of this is natural and quite unavoidable.

The chief effect of the disjointed quality of the discourse and its consistent use of blanks is to constantly generate a sense of uncertainty

9. Mailloux makes a similar point about the discourse strategy of Herman Melville's *Moby Dick* ('Learning to Read', p. 98).

in the reader. Many of Qoheleth's reflections focus upon obtuse proverbs or aphorisms whose individual meaning is opaque to say the least. Given the difficulty of understanding the precise meaning of some of these proverbs, such as the gnomic statement in 6.10, I have come to the conclusion that the effect is the meaning here. The diffuse nature of Qoheleth's monologue suggests a discourse strategy aimed at 'stumping' the narratee/reader. By constructing a narrator-elevated discourse, the implied author allows both the narratee and the implied reader to feel and experience the full effect of wisdom's limitations. Except for the remark by the implied author in 7.27, the reader is no longer given an ironic horizon of knowledge which distinguishes his or her knowledge from that of the narratee. Both are now at the same level of knowledge, or perhaps better, ignorance. Through the utilization of this discourse strategy, the narratee/reader is given a full lesson in the heart of darkness/ignorance (3.11) which limits their mental faculties of perception. In some respects, the text functions like a test for the narratee. As a final exam of sorts, the monologue stresses the correct answer of 'who knows?' over any particular *Gestalt* that a reader may make of any specific passage.

As a result, we see where the argumentative strategy of the latter half of the book is designed to simulate the epistemological problems dealt with in 3.1-21. Having established his credentials as a sage, Qoheleth now demonstrates to his narratee/reader the difficulty of the pursuit of knowledge/wisdom. By means of obtuse and opaque proverbs, contradictory juxtapositions, the continued use of rhetorical questions and the like, the narrator weaves a discourse whose principal effect is to impress upon the narratee/implied reader the fundamental challenge of achieving wisdom. This gives both the narratee and the reader a narrative experience of wisdom's limitations. Via this strategy, the implied author has found a way to impress upon the reader a sense of wisdom's opaqueness by constructing a monologue whose illocutionary force recreates the fundamental experience of *hebel*. The discourse therefore stresses the agnostic and sceptical stance of the narrator not only by constantly repeating the theme of 'not knowing' at the surface level of the discourse, but also at its deep level in the way that 'Professor Qoheleth' consistently argues in a manner that 'stumps the student'. Its overall effect is to produce a very powerful sense of wisdom's essential and unavoidable limitations at both a cognitive and emotional level in the reader. Again, it is possible to see the tell-tale rhetorical trademark of the implied author who prefers to supplement the logos level of the discourse with a complementary pathos dimension designed to reproduce a narrative experience of what is being argued at the cognitive level.

4. *Ecclesiastes 6.10-12: Epistemological Nihilism—Who Knows What is Good?*

Unlike the famous passage in Mic. 6.8, 'He has shown you, O man, what is good', Qoheleth's worldview sees an opaque universe which refuses to reveal ultimate values. The use of a rhetorical question stimulates the reader's questioning. This verse refuses to give an answer to this question, although the model reader knows by now that questions are rarely answered in the affirmative by this sage. Qoheleth again begins to focus on his narratee, engaging in a second instruction complex (6.10–7.22) aimed at educating his young protege on the problems associated with ascertaining the 'good life'. Verses 10-12 begin this complex by acting both as a summary of what precedes and follows. In this regard the passage has a certain 'Janus' function. It faces 'backwards and forwards, recalling certain themes (determinism, powerlessness, and unknown future) that have already surfaced and pointing forward to future treatment'.[10]

Verse 10 recalls those values from the text's horizon of values that are needed to understand the ensuing discourse. Wolfgang Iser refers to this aspect of the reading process as protension and retention. The reader processes such a text by holding the horizon of the text and the immediate theme in creative tension. He states:

> Each sentence correlate contains what one might call a hollow section, which looks forward to the next correlate, and a retrospective section, which answers the expectations of the preceding sentence...every moment of reading is a dialectic of protension and retention, conveying a future horizon yet to be occupied, along with a past (and continually fading) horizon already filled...[11]

The text reminds the reader of the cyclical and determinative nature of the world (v. 10a recalling 1.4-11), the nature of humanity (v. 10b, 'and what mankind is is known to him', presumably referring to the heart of

10. Crenshaw, 'The Expression *mi yodea*'', p. 282.
11. Iser, *The Act of Reading*, p. 112.

ignorance in 3.11), and the nature of God, 'the one stronger than he' (v. 10c). Verse 10 is grammatically ambiguous and opaque,[12] but seems to obliquely summarize the defamiliarized view of nature, both cosmological and human, tendered by the two poems in 1.4-11 and 3.1-8 (including the exposition in 3.10-15).[13] This defamiliarized view on things will now act as the premise for Qoheleth's ensuing argument. The reader therefore must engage in both protension and retention. This creates a sense of expectancy that the questions raised in this passage will somehow be answered. However, the reader must wait for that definitive answer until 8.17.

The passage accents humanity's weakness and ignorance which, by inference, Qoheleth blames on the Creator.[14] The vagueness of the passage's meaning is probably intentional. The muffled reference to 'one stronger than he' (*šehattaqqîp mimmennû*)[15] recalls Qoheleth's similar reticence to say anything that would directly implicate and offend God in his previous discourse (cf. 5.3-5). Verse 11 observes that absurdity can increase even with 'careful speech' (*d^ebārîm harbēh marbîm hābel*).[16] There is a subtle sense of resigned futility which permeates this introduction, lamenting as it were, 'unfortunately, this is the way it is'. The latent sense of futility in this evaluation suggests to the reader that Qoheleth's pessimism has moved to a deeper level. Again, the use of a rhetorical question regarding the lack of 'profit' for

12. For a fuller discussion of the text's various translations see Murphy, *Ecclesiastes*, p. 77.

13. Perdue argues that the act of naming, as a part of the text's repertoire, is a common expression for the act of creation in the ancient Near East (cf. Isa. 40.26). As such, the focus of the verse concerns the problems of the created order. See Perdue, ' "I will make a test of pleasure" ', pp. 226-27.

14. Whybray, *Ecclesiastes*, p. 109.

15. The Massoretes pointed this word by ignoring the *h*, but it makes sense to read it as *šehattaqqîp.* This is an obvious Aramaicism that can be taken either as an adjective with the article or as a hiphil form of the verb. This further contributes to the ambiguity of the verse. To compensate for this grammatical opacity, an early authorial reading-community has provided the text with an external reading interlude, suggested by the *qere* to read the classical adjectival form, *taqqîp* ('mighty') instead. See Crenshaw, *Ecclesiastes*, p. 131. However, the change makes very little difference, as the use of a predicate adjective functions adequately to communicate the narrator's meaning in this context regardless of the dialectic qualities of its grammatical form.

16. Crenshaw, *Ecclesiastes*, p. 131.

humanity (*yōtēr* in the sense of sufficient gain) puts the issue back on the reader to contemplate.

To sum up, the sceptical stance of the narrator is intimated by his use of a double 'who knows?' question in v. 12. This resumes from the text's horizon of values a theme first broached in 3.21, which concerns humanity's inability to know what the future holds.[17] The function of a double rhetorical question is to produce consensus between two parties.[18] These two questions announce the twin themes of 7.1–11.6. Verse 12a, 'Who knows what is good for humanity?', summarizes 7.1–8.17, while 12b, 'Who can tell a person what will be after them?', sums up 9.1–11.6.[19] The mood of the narrator in these verses can be summed up as 'one of resigned inevitability'.[20] However, the ironic dimension of his statement does not escape the careful reader. As Michael Fox has observed, this statement 'radically undermines his own—quite serious —series of statements about what is "good" (7.1-12) by first denying the possibility of knowing "what is good for man"'.[21] Again, we see the ironic qualities of the implied author's characterization of the narrator. The problems involved with private insight's epistemological conclusions arise before they even begin! By having the narrator begin his discourse on such an inconsistent and ironic note, the implied author subtly undermines the rhetorical validity of Qoheleth's tenets.

5. *Ecclesiastes 7.1–8.17: The Ethically Blind Public*

a. *Ecclesiastes 7.1-13: Another Look at Public Knowledge—Proverbs and the Good Life*

In an abrupt fashion, the discourse changes texture on the reader. Suddenly a series of reflections on proverbial knowledge erupts from the narrator's monologue. Roland Murphy voices the problem that readers must mull over the question: 'What is the nature of the relationship between the various sayings in these chapters?'[22] In order to understand the meaning of this passage, a reader must begin to look at the total effect of the verses under discussion. Stephen Brown observes how the

17. Ogden, *Qoheleth*, p. 96.
18. R. Johnson, 'The Rhetorical Question', p. 186.
19. A. Wright, 'The Riddle of the Sphinx', p. 320.
20. Crenshaw, 'The Expression *mi yodea*'', p. 278.
21. Fox, 'Qoheleth's Epistemology', p. 96.
22. Murphy, *Ecclesiastes*, p. 62.

'evocative string of proverbs in 7.1-12...functions like the analogic poem in 1.3-11'.[23] In a rapid-fire fashion, the sage quickly ruminates on the various ways that public knowledge construes the 'good life'. The rhetorical advantage of using proverbs to bolster one's argument is seen in the rhy comment quoted by T.A. Perry: 'Whenever you can attach a proverb, do so, for the peasants like to judge according to proverbs'.[24] This passage inundates the reader with a barrage of 'better-than' proverbs (vv. 1a, 2, 3, 5, 8a) and several others which imply a *min...ṭôb* construction (7.1b, perhaps also 7.10).[25] Five of the sixteen occurrences of the 'better-than' proverb occur here, making this passage a critical place wherein the reader infers the private values of the narrator. By frontloading the 'better-than' proverb at the beginning of the second half of his oration, Qoheleth places the issue of ultimate values clearly before the narratee/reader. As a result, his character plays a critical role in determining the reader's response. Although his 'I' is missing in these reflections, the model reader clearly understands its implied presence.[26] The eclipse of the public nature of Qoheleth's proverbs becomes complete in these chapters.

Much of what he says has a 'thoroughly orthodox' ring to it,[27] such as the value of a good name in v. 1, the ubiquitous counsel found among the sages in v. 5 to heed the rebukes of the wise, and the admonition against quick anger in v. 9. The use of the key word 'good' (*ṭôb*) lends Qoheleth's reflections a thematic unity. However, there are several blanks that the reader must work at overcoming in this extended reminiscence. In this complex, optimistic reflections are balanced by reflections on the dark side of life. Ecclesiastes 7.2 jars the reader by arguing that it is 'better to go to the house of mourning than to go to the house of feasting'. The motif of death/mourning joins 7.1-2 together while the juxtaposition of their essential positive/negative stances toward life causes the reader to question their logical relationship to each other. Furthermore, such a pessimistic denial of the value of feasting on the heels of an extended instruction complex that enjoins the pursuit of pleasure (5.9–6.9) confuses the reader. The reader is left to

23. S. Brown, 'The Structure of Ecclesiastes', p. 202.

24. T.A. Perry, *Wisdom Literature*, p. 2, quoting an anonymous fourteenth-century legal document.

25. Ogden, 'The "Better-Proverb"', p. 492.

26. Polk, 'The Wisdom of Irony', p. 5.

27. R.F. Johnson, 'A Form Critical Analysis', p. 156.

wonder why the narrator commends enjoyment if the 'house of mourning' is better and, if so, in what sense is it better? The comment in v. 2b with its postfixed *kî*-clause, '*for* this is the end/fate of all humanity, and the living will lay it to heart', validates the proverb Qoheleth is reflecting upon. However, it does nothing to overcome this blank except to expose Qoheleth's increasingly negative disposition. Qoheleth presents a paradox in these verses whose final *Gestalt* remains open for the reader.[28] The reflections in v. 3 regarding the value of sorrow over laughter and the reference to the house of mourning in v. 4 continue this paradoxical juxtaposition. Much in 7.1-4 has an underlying ironic tone to it.[29]

In v. 5 the theme changes from mourning to the relative value of folly and wisdom. Qoheleth reflects upon a metaphorical proverb in v. 6 and comments again on how the laughter of such fools is an absurdity. Verse 7 notes that even sages can become fools if they resort to oppression (*'ōšeq*) and bribes (*mattānâ*). The reference to oppression alerts the reader that sometimes the wise themselves are guilty of the reprimands given to society by Qoheleth in 4.1. The reader wonders why Qoheleth places the foolishness of both fools and sages in such close ideological proximity. The proverb in v. 8, 'Better is the end of a matter than its beginning' summarizes the narrator's utilitarian worldview in an economical fashion. Only by the outcome of an action can one tell whether there was profit in it. The emphasis upon the narratee is evident in this passage, as Qoheleth turns directly toward them in v. 9, using a jussive (*'al-tᵉbahēl*, 'do not be hasty') to correct youth's typically rash reactions that often result in anger and folly. This is continued in v. 10, where the jussive (*'al tōmar*, 'do not say') is again utilized to admonish the narratee against living in the past. The extended reflection ends with several proverbs that observe the interaction of money and wisdom and how the two provide mutual support for one another.

Verse 12 is a very ambiguous text. Literally, the text utilizes two beth-preformatives to make a comparison between money and wisdom.

28. R.F. Johnson, 'A Form Critical Analysis', pp. 149-50. He argues that the point of this paradox is that 'it is better to face the reality of finite existence than to delude oneself about his accomplishments and destiny' (p. 149). Clearly, as Crenshaw observes, Qoheleth is captivated with the thought of death's finality by his selection of proverbs and subsequent comments in 7.1-4 (*Ecclesiastes*, p. 135).

29. Spangenberg, 'Irony in the Book of Qohelet', pp. 64-66.

It literally translates: 'For to be in the shelter of wisdom is to be in the shelter of money'. However, this seems so crass and materialistic that some readers have opted to emend the beth-preformatives to *kî* ('like', 'as'), making the clause a double simile: 'for the protection of wisdom is like the protection of money' (RSV). Others, including H.L. Ginsberg, have argued that the verb *b*ᵉ*ṣēl* goes back to an Aramaic verbal root (*b*ᵉ*ṭēl*, 'to cease'), and is not in fact a noun here. He translates it as 'when the wisdom goes, the money goes'.[30] Although the grammatical opacity of this verse remains, there is a clear utilitarian trait which surrounds the narrator's character and subsequently influences the reader's understanding of this verse. Were he alive today, one might justifiably characterize Qoheleth as 'a private-sectorite'.[31] Again, the 'bottom-line' character of Qoheleth's appraisal depicts the sage as one whose values are ultimately aligned with financial interests. Given the consumptive nature of his youth, such values do not surprise the astute reader. However, the reduction of wisdom to financial considerations meets with less than a positive reception among readers, as the attempt to emend the verse shows. The literal translation is the one that best fits the characterization of the protagonist given the norms provided by the overall text, which clearly present a person dominated by financial considerations in his total assessment of life, Aramaic considerations notwithstanding.[32] The reflection ends with a rhetorical question in v. 13 that seems to have no logical connection to what precedes it. The passage ends as it began, with a blank. Verse 13b turns the first proverb cited by Qoheleth (cf. 1.15) into a rhetorical question. The verse hints again, without directly saying it, that God is the culprit here, and makes the narratee/reader cognizant of his or her own limited ability to change things by asking 'who is able to make straight that which he has made crooked?' The reader is left to infer the connection here, deducing that the question bolsters the status quo ethic advocated by a man whose interests are vested in money and status.

30. Ginsberg, *Studies in Koheleth*, p. 22. Rowley also agreed with this reading ('The Problems of Ecclesiastes', p. 88).

31. Merkin, 'Ecclesiastes', p. 402.

32. Rainey calls attention to the necessity of understanding a concatenation of proverbs, such as is found in 7.1-13, by paying strict attention to the 'total impact' of the complete series. See Rainey, 'A Study of Ecclesiastes', p. 155.

To sum up, the proverbial reflection in 7.1-13 depicts Qoheleth as a traditional sage who has a decidedly jaded countenance to his rhetorical visage. The astute reader perceives in this characterization that Qoheleth's jadedness dovetails quite nicely with his own financial interests. One wonders if money serves wisdom with Qoheleth, or does wisdom serve money? Again, Qoheleth's penchant for self-serving wisdom detracts from his attractiveness, and certainly affects his credibility, especially in vv. 11-12. Admittedly, much of what this realist says is true, but it lacks the sort of humane vision that a reader expects from a religious text. As the narrative mirror of the narrator, the narratee is likewise characterized in a similar manner. However, unlike the young narratee the implied reader understands that there is something lacking in Qoheleth's advice given his narcissistic and self-interested evaluations. This is due principally because the implied author has ironically and critically evaluated Qoheleth earlier in the discourse, especially 2.1-11. Second, the narrator's recollection of various proverbs provides a micro-model for the implied reader to understand the limited value of any specific gnomic statement. By means of blanks, jagged juxtapositions and comments, Qoheleth communicates to the reader the 'shortcomings' of public knowledge.[33] But in the process, his own ethos exposes the imperfections of private insight as well.

b. *Ecclesiastes 7.14-29: The Mathematics of Private Insight—Summing It All Up*

Nowhere is the reader caught more tantalizingly between 'fact and opinion' than in this passage.[34] Verses 14-29 reek of a restricted post of observation riddled with subjectivism. While Qoheleth calculates 'two plus two', the reader senses that he did not come up with 'four', especially when it comes to women in vv. 26-28. Because of his jadedness, Qoheleth's bottom line often totals 'zero'. In these verses we meet a person whose pessimism is at the brink of nihilism. The transition from pessimism to scepticism is but a short step away. Qoheleth will overstep that threshold shortly in 8.1-17 and 9.1.

Verse 14 continues Qoheleth's discussion of what is good, referring to the 'day of good things' (*b^e^yôm ṭôbā*). The opening line admonishes

33. Dell, 'The Reuse and Misuse of Forms in Ecclesiastes', p. 141.

34. Grant-Davie has characterized the effect of first-person discourse in these terms. See Grant-Davie, 'Between Fact and Opinion'.

the narratee to 'live in goodness/happiness' (*h^e^yēh b^e^ṭôb*), so that the word *ṭôbâ* now summarizes the existential being of the sage. The use of *r^e^'ēh* ('see!') is a marker of present focalization and communicates that Qoheleth 'does not simply relate thoughts had in the past, but speaks out of his present condition of mind, even though as a true sage he refers to observations he has made'.[35] The narrator's pessimism now steps over the epistemological threshold into scepticism, asserting in v. 14b that 'humanity may not find out anything that comes after it'.

Verses 15-22 present yet another observation. The wandering viewpoint shifts from the theme of goodness to righteousness. Qoheleth's jaded mood is evident in his characterization of his own life as another example of *hebel*, with *bîmê heblî* ('my brief days') presumably referring to the fleeting or transitory qualities of breath which underlies the metaphor latent in the term. Douglas Miller observes quite correctly that there are no contextual clues which might determine which of the various nuances are meant here (as there are earlier in the discourse), and so, 'the reader is meant to recognize that any or all dimensions of *hebel* are being alluded to'.[36] Given that Qoheleth is summarizing his own life, casting such a broad semantic net would seem to serve his narrative interests. Later, in 9.7-10 he will move to cast the same broad net around the reader's life, referring to 'your *hebel* life'. The introduction to this observation therefore communicates to the reader that what ensues is both present and personal for the narrator. So subjective is Qoheleth's monologue at this point that Michael Fox describes 7.15-16 as 'testimony'.[37] Qoheleth observes what every person has seen: a righteous person dies without rewards for his upstanding character while the wicked lives a long and full life. In this, he is simply stating a well-documented fact that cannot be disputed. He directly addresses the narratee with the jussive form, *'al-t^e^hî ṣaddîq harbēh* ('do not be excessively righteous'), to communicate the seriousness of his admonition. Verse 17 states its opposite, commending the narratee to be neither overly wicked, since such foolishness often leads to an untimely death.

It is not too much to say that 7.16-17 has greatly influenced the final *Gestalt* which readers have made of Qoheleth's character. Some readers

35. Isaksson, 'The Autobiographiocal Thread', p. 45.
36. Miller, 'Qohelet's Use of *Hebel*', p. 452.
37. Fox, *Qoheleth and His Contradictions*, p. 88.

have perceived another example of Qoheleth's utilitarian ethic, describing this advice as a 'doctrine of the golden mean'. The juxtaposition of these incongruous admonitions presents yet another blank which results in dueling observations, the final meaning of which is left for the reader to infer. This blank produces 'cognitive dissonance and destabilizes the reader's frame of reference'.[38] Some readers, including R.N. Whybray, have called attention to the prominence of clauses that are 'qualified by *harbēh* or *yôtēr*, words functioning as adverbs with a superlative sense: "greatly, very" '.[39] As a result, he concludes that Qoheleth is neither recommending immorality nor teaching the golden mean. Rather, Whybray concludes 'that Qoheleth is...against the state of mind which claims actually to have achieved righteousness or perfection'.[40]

However, such a reading sounds almost as if Qoheleth had read St Paul, and certainly shows evidence of effective historical consciousness on the reading of this text by conservative readers. Such efforts by readers to protect the canonical Qoheleth from his own character is more likely the result of a reading bias stemming from the texture of the reader's religion.[41] Given the broad norms of the text, and the previous characterization of Qoheleth in 3.1-8 and 5.9–6.9 where a golden mean ethic is strongly insinuated, it is more likely that Qoheleth is remaining true to his character and is again advocating a utilitarian ethic.[42] The sense of relativity and moderation espoused by this verse is the most natural reading of the text. The pragmatic and self-oriented direction of the ethic is evident in the two *kî*-clauses which provide the reasons for this ethic. Both over-righteousness and over-wickedness are judged according to the criteria of how they impact the well-being

38. R. Johnson, 'The Rhetorical Question', p. 253.

39. R. Whybray, 'Qoheleth the Immoralist? (Qoh 7.16-17)', in Gammie and Breuggeman (eds.), *Israelite Wisdom*, pp. 191-204 (192).

40. Whybray, 'Qoheleth the Immoralist?', p. 191. For a summary of the reading history of this passage see W. Brindle, 'Righteousness and Wickedness in Ecclesiastes 7.15-18', *AUSS* 23 (1985), pp. 243-57. The passage has four common reading grids: the golden mean, fanaticism and legalism, overreaction to truth, and the self-righteous grid.

41. For an excellent discussion of the influence of religion on the reading process see Miller, 'The Question of the Book'.

42. Crenshaw, *Ecclesiastes*, p. 140, and Murphy, *Ecclesiastes*, p. 70, also have noted the utilitarian ethic of the passage at hand.

of the individual. Given the emphasis on profit (*yitrôn*) in the preceding passage (v. 12), I doubt very much if Qoheleth waxed overly spiritual in this passage. As a result, Qoheleth again lacks the sort of magnificence and magnanimity that produces a positive ethos. His advice is pragmatic, but is not overly attractive to the average pious reader.

In vv. 19-21 Qoheleth reminisces upon three traditional sayings that deal with righteousness. The three proverbs present yet another blank to the reader, whose precise relations to each other are vague and remain hermeneutically open. Verse 20 is a popular saying which occurs in a similar form in Job 15.14-16 and 1 Kgs 8.46.[43] Again, the narrator tends toward pessimism. Verses 21b-22 reminds the narratee of their common social status as a part of the wealthy upper class. Although Qoheleth advises a course of moderation yet again, there is nothing said about treating the servants so that they have no reason to curse the narratee. The socially empathic attitude offered in 4.1 is a long way from Qoheleth's heart in this passage. Both the narrator and narratee are characterized as people whose vested interests in status and power have divested them of social consciousness.

The previous verses pulled the narratee into the realm of his heart by asking for a brief moment of introspection. Now the monologue shifts to the realm of Qoheleth's heart by returning to an autobiographical style.[44] Qoheleth begins to engage himself in interior monologue as he has done earlier (cf. 1.16; 2.1, 15a; 3.17-18).[45] The emphasis on instruction ends with the beginning of a new observation complex in 7.23-29. The text lays bare Qoheleth's consciousness for public viewing. Verse 23 portrays Qoheleth summing up his previous monologue, as the phrase *kol-zōh* ('all this'), probably refers to everything that occurs between his stated intention to seek and search out wisdom in 1.13 and his conclusion here that he has 'tested' (*nissîtî*) wisdom and found it wanting.[46] Verses 23-25 emphasize that fact that wisdom ultimately escaped the master, that he could not find it, and that it was 'far off'. In

43. R. Johnson, 'A Form Critical Analysis of the Sayings', p. 163.

44. Ogden, *Qoheleth*, p. 116.

45. Loader, *Polar Structures*, p. 25.

46. R.F. Johnson argues that *kol-zōh* refers to the immediately preceding verses in 7.19-21 ('A Form Critical Analysis', p. 162). However, the reference to testing and its relation to 1.13-18 demands a larger context than this presupposes.

a graphically emotional passage filled with pathos, the narrator expresses his absolute exasperation at having not found the certainty which he sought in the quest for wisdom. Here again is another instance where the gravity well of Qoheleth's insistence upon private insight swallows up the public dimension which so characterizes the general Israelite Wisdom tradition. As Christianson notes, the verb *ḥākam* ('to be, become wise') occurs 28 times in the First Testament, but only three of these are couched in the first-person (Eccl. 2.15, 19; 7.23). All the instances of this verb being refracted through the lens of first-person rhetoric occur in Qoheleth's discourse. He states:

> Therefore, only in Ecclesiastes is the idea of becoming wise related so reflexively to the speaker. In the *tunc* of Qoheleth's story, becoming wise is within the grasp of the experience of his self. Unlike Job 28 and Proverbs 8, where the poet seeks wisdom itself, Qoheleth seeks to *be* wise—to *become* wise.[47]

Only here in Qoheleth's discourse is the quest for wisdom limited to such a personal and private endeavor. As Mark Sneed observes, Qoheleth begins to act as a 'deconstructionist' in these verses. The meaning of wisdom becomes personalized and private, with the effect that the term quickly 'becomes vacuous, its edges of distinction quickly assimilating with those of folly'.[48] So powerful is the prism of private insight in these verses that it nearly empties the term of its meaning for the speaker.The use of the rhetorical question in v. 24 pulls the reader into the narrative experience of Qoheleth's exasperation. Again, we see the rhetorical strategy of alternating logos and pathos to make an impression on the reader. Just when the narrative begins to make a negative impression on the reader, Qoheleth opens his heart in an act of intimacy, exposing a man who is honest, sincere and very vulnerable. Such honesty raises the level of trustworthiness and attractiveness for the reader, balancing the negative ethos effected by vv. 15-22.

The subjective tone continues in the observation complex in 7.25-29. The reference to Qoheleth's heart (*lēb*) and the double pleonastic use of *'ᵃnî* in vv. 25-26 remind the reader of Qoheleth's confession-like style

47. Christianson, *A Time to Tell*, p. 208.

48. M. Sneed, 'Qoheleth as "Deconstructionist": "It is I, the Lord, your redeemer ...who turns sages back and makes their knowledge nonsense" (Is 44.24-25)', *OTE* 10 (1997), pp. 303-11 (307).

during the early stages of Qoheleth's monologue. That this observation refers back to the pursuit of pleasure catalogued in 2.1-8 is further supported by the use of the verb *tûr* ('to spy out') used here and in 2.3. Qoheleth's search in 7.25-29 functions like a narrative flashback, recalling the pursuit of wisdom that has characterized his life since youth. But what did he find? Quite unexpectedly, Qoheleth surprises the reader by admonishing the narratee to avoid the woman 'whose heart is snares and nets, whose hands are fetters'. This is simply not the sort of conclusion a reader expects from a lifelong search for wisdom. However, the use of *tûr* here suggests that the search under discussion is not the quest for *general* public wisdom, but rather, the *private* search for wisdom which resulted in the pursuit of pleasures narrated in 2.1-8. As was noted above, *tûr* connotes a covert sense of searching that cannot be done in public, such as sexual exploration. The reference to 'not one in a thousand' probably refers to the tradition of Solomon having 700 wives and 300 concubines.[49]

The intrusion of this private search into the public search for wisdom in 7.15-22 shocks the reader, who justifiably expects something a little more substantial and 'professional' from the sage, and certainly a summary that is less personal and more universal.[50] The reader expects facts, but receives a very jaded and misogynist opinion.[51] In v. 27 Qoheleth persists in his mathematical endeavours, and concludes that 'this' is indeed what he found. In v. 28 men do not fair much better, as the pessimist in Qoheleth can find virtually no righteousness among the male members of the human race either. He ends his summary emphasizing the 'radical depravity' of humans. However, unlike his previous pontifications, God is not the culprit. Rather, free will is to blame, as

49. Perdue, ' "I will make a test of pleasure" ', p. 229. Crenshaw also perceives the subtle influence of intertextuality on this text (*Ecclesiastes*, p. 148).

50. Murphy has also observed this reaction among readers. He notes that the 'reader may be caught off guard after the elaborate introduction to Qoheleth's search in v. 25. The discovery seems to be merely an old *topos* celebrated in the wisdom literature...the adulterous woman' (*Ecclesiastes*, p. 76). However, adultery is not mentioned here, though it could possibly be implied (cf. Prov. 2.16-19; 5.1-4; 7.22-23). Whybray observes how this 'unexpected reference to woman...has perplexed commentators from very early times' (*Ecclesiastes*, p. 125).

51. Loader refers to this passage as 'bitter wisdom', accenting the negative and jaded characterization of the narrator (*Ecclesiastes*, p. 91).

humanity 'craves so many devices/accounts'.[52] At the end of Qoheleth's search for wisdom the reader perceives a jaded, pessimistic person devoid of the qualities of magnificence and magnanimity. One wonders even if he is not telling the reader something about his own heart in this critical evaluation of humanity.

However, more important for the reader's characterization of the narrator in these verses is the implied author's divergence from his normal rhetoric of reticence. In 7.27 the implied author interrupts Qoheleth's conclusion with the words, 'says Qoheleth'. The interruption is dramatic, having been inserted into the middle of a sentence. But the perceptive reader, having noticed the extreme reticence by which the implied author typically communicates to the implied reader, questions why the implied author breaks frame so dramatically in this passage. What is the purpose of such an imposition? First, by emphasizing 'says Qoheleth', the implied author stresses that this is the viewpoint of an individual, and so, accentuates the subjectivity of such a conclusion. It is as if the implied author says directly to the reader, 'so according to his opinion'. That such an evaluation occurs in the midst of a passage which advocates a golden mean approach to ethics and a very negative appraisal of women should not surprise us. Earlier, the implied author lampooned Qoheleth for his narcissistic and private pursuit of knowledge. By having the narrator refer back to the verb *tûr* used in 2.3, the implied author deftly resumes that lampooning here.[53] By reaching such a subjective conclusion the narrator oversteps the boundaries of reasoned approach; as a result, the implied author is forced to reciprocate in equally dramatic fashion. Not able to hold his tongue, the frame-narrator states the obvious, that these are the views of Qoheleth, that is, his private insight.

52. The term typically translated as 'devices, inventions' is *ḥiššbônôt*. The term seems to function as a pun, as it has a double meaning here, also signifying 'accounts, reckoning'. Given the emphasis on wisdom and money in 7.11-12, I would rather interpret it with the latter meaning. In that sense, Qoheleth is commenting on the corruptive nature of money. This accords well with the general norms of the text which consistently depict a person interested in financial matters.

53. Rudman also observes that the phraseology of this passage, especially the use of the proper name Qoheleth and the verb *tûr* in 7.25 (cf. 1.13, 2.3) 'echoes the search that Qoheleth undertakes in the so-called Royal Experiment in 1.12–2.26' (D. Rudman, 'Woman as Divine Agent in Ecclesiastes', *JBL* 116 [1997], pp. 411-27 [415]).

The subtle effect of this *verbum dicendi* is to communicate that Qoheleth's conclusion is the natural outcome of a skewed methodological approach.[54] It also heightens the reader's awareness of this

54. Recently, however, several readings have been offered which attempt to save the narrator from his own misogynist failings as a rhetor. Lohfink argues that the word *mar*, 'bitter', has a double entendre of 'stronger', and so concludes that Qoheleth is speaking about the immortality of women. In this case, not one woman in a thousand escapes death. But does that mean that one man in a thousand does? This seems unlikely. See N. Lohfink, 'War Kohelet ein Frauenfeind?', in M. Gilbert (ed.), *La Sagesse de l'Ancien Testament* (Leuven: Leuven University Press, 1979), pp. 259-87. Baltzer contends that the issue here is military obligations (*'elep* referring to military troops), and that Qoheleth is absolving them from such duties. By saying that women are stronger than death, he shifts the emphasis to the effects of war on women. However, the First Testament is filled with instances of how women die from war. This interpretation is more ingenious than the context permits. See K. Baltzer, 'Women and War in Qohelet 7.23-8.1a', *HTR* 80 (1987), pp. 127-32. Another apologetic has been given by Krüger, who suggests that this woman is 'Dame Wisdom'. Qoheleth is thereby describing the anguish of his soul given the failure of his search to master wisdom and gain knowledge. See T. Krüger, ' "Frau Weisheit" in Koh 7,26', *Bib* 73 (1992), pp. 394-403. However, it can hardly be argued that the person who escapes Dame Wisdom would please God, as Qoheleth states in v. 26b. Rudman argues that the woman mentioned in 7.26 is 'the agent of a deterministic force…she cannot be deemed "wicked" as such since her whole *raison d'être* is to perform God's will by punishing those who have sinned'. As such, she is more a 'huntress of the masses than a temptress of the individual… She is an every woman figure who works for rather than against God in her enactment of judgement upon those who have sinned' (Rudman, 'Woman as Divine Agent', pp. 418-19). Basically, Rudman argues for the role of women as a 'fatalistic attraction' within the divine scheme of things. He states: 'God may pull the strings from heaven, but on earth it is Woman who is the master. In a sense, Qoheleth's worldview is one in which Eve has ganged up with God against Adam' (p. 421). However, this hardly seems any less misogynistic, except that we now have a deterministic misogynism. Pahk argues that the term *'ašer* in 7.26 has conditional force. Thus the verse reads, 'More bitter than death is woman, *if* she is a snare'. In other words, Qoheleth's fondness for ambiguous language has backfired on him, causing readers to misinterpret his intent as misogynist. See 'The Significance of *'ašer* in Qoh. 7,26: "More bitter than death is the woman *if* she is a snare" ', in Schoors (ed.), *Qohelet in the Context of Wisdom*, pp. 373-83. In this view, Qoheleth is not generalizing, but is talking about a 'particular kind of woman' (p. 377). Pakh also views 7.28 as a quotation, the truth of which Qoheleth has not 'yet accepted' (p. 379). However, his analysis seems more apologetic than convincing at many points. His attempt to play mind-reader remains problematic. Long also has attempted to clear Qoheleth of the charge of misogyny. He argues that *māṣā'* ('to

passage.[55] The narrator's jadedness is seen as the result of his own personal excesses and utilitarian approach. When people are reduced to objects for pleasure and observation, the natural consequence is to have a less than human encounter with them, resulting in pessimistic evaluations such as we see here. By having his narrator refer to women in such a negative manner, and by tying this judgment to the narcissistic and self-centered pursuit of pleasure in 2.1-8 through the use of *tûr*, the implied author strongly suggests that this is the sort of unreliable conclusions which will result from a radically self-centered epistemology. The implied author again utilizes the Solomonic mask to criticize the narrator, in effect saying to the implied reader: 'This is the real end product that a strictly empirical approach to knowledge will get you.' As a result, we see that Qoheleth's characterization as a misogynist serves to further lampoon the character. In no way does it express the ideological views or values of the general discourse seen from the perspective of the text's implied author. Such anti-women sentiments are presented to make a parody of the narrator. When readers react negatively to Qoheleth here, they are responding in an appropriate manner to the rhetorical design of the implied author. All attempts to lessen the misogynist effects of the text should therefore be seen as misreadings of the text's intentions.The point of the text is to let such sentiments speak for themselves.[56]

comprehend', 'to fathom') in 7.28 is to be taken in an intellectual sense. Qoheleth was merely saying that he only *understood* men slightly better than women, who 'remain a mystery' to Qoheleth. See B.O. Long, 'One Man Among a Thousand, But Not a Woman Among Them All: A Note on the Use of *māṣā'* in Ecclesiastes vii 28', in K. Schunck and M. Augustin (eds.), *Lasset uns Brücken Bauen* (BEATAK, 42; Bern: Peter Lang, 1995), pp. 101-109 (107). This too seems strained. As a result, I contend that the ironic interpretation best saves the text (though not the narrator) from misogynist readings. An ironic reading still leaves Qoheleth in the rhetorical quandary he so richly deserves given his characterization by the implied author. This proposed reading allows the full ironic dimension of the text to have its due effects on the reader while allowing the sexism of the passage to be duly criticized.

55. Christianson observes: 'by heightening this particular observation syntactically and by employing the guise of Solomon possibly more distinctly than usual, this passage is made inevitably memorable. The frame narrator's insertion here supplements the overall strategy of "setting aside", marking its importance for Qoheleth's narrative' (*A Time to Tell*, p. 95).

56. For other insightful studies which allow the text its full range of sexism, see: E. Christianson, 'Qoheleth the "Old Boy" and Qoheleth the "New Man": Misogy-

Again, the implied author utilizes autobiographical discovery to criticize the epistemological use of the *lēb* by again calling attention to the 'inside story' of such a search. In fact, the use of radical subjectivity and the Solomonic aura acts as a satiric envelope around the 'search' which extends from 1.12–7.29. The implied author does not so much argue with the sceptic's conclusions, but rather, uses an argument based on the quality of existence to fully expose the weaknesses of a strictly self-oriented approach to wisdom. From the implied author's point of view, pessimism and scepticism are their own rewards. Only the gift of satire can save a pessimist from such desserts. Thus the implied reader surmises why the horizon of ironic knowledge has been so graciously offered to them.

The narrator's misogyny is costly to his ethos. The pattern of jaded, negative and excessive subjectivism reaches the point here where its effects are insurmountable. Although Qoheleth retains the benefits of his openness and honesty, the reader realizes the extent of Qoheleth's restricted post of observation. Too many times has Qoheleth shown evidence of a man obsessed with his own interests. He becomes a man trapped inside the limits of his empirical method. By this point in the developing discourse, the reader begins to close his or her *Gestalt*, ultimately characterizing the narrator as a person who lacks the necessary traits to be attractive, and therefore, persuasive in any definitive sense.

To sum up, by breaking frame the implied author again satirizes the narrator's methodology. Such parody stresses that, as Qoheleth himself concluded regarding the 'end of the matter is better than its beginning' (7.8), it is the results that a method reaches which ultimately decides its

nism, The Womb and a Paradox in Ecclesiastes', in A. Brenner and C. Fontaine (eds.), *Wisdom and Psalms: A Feminist Companion to the Bible* (The Feminist Companion to the Bible [2nd ser.], 2; Sheffield: Sheffield Academic Press, 1998), pp. 109-36; C. Fontaine, '"Many Devices" (Qoheleth 7.23–8.1): Qoheleth, Misogyny and the *Malleus Maleficarum*', in the same volume (pp. 137-68). Christianson argues that Qoheleth's preoccupation with the womb is due to his obsession with sexual gratification. This results in a reduction of women to their sexuality, or perhaps, genitalia ('Qoheleth the "Old Boy"', p. 131). Fontaine's study is perhaps the best overview of the reading history of this passage to date. Her study demonstrates the insidiousness of ingenious readings which are so textually focused that they fail to see the negative effect that the plain sense of this text has had on real women's lives over the centuries. A quick scansion of her article gives the sensitive reader much to pause over when reviewing some of the reading strategies surveyed above.

validity. The implied author utilizes the master's own observations to criticize the sage. The corruption that Qoheleth observes among the sages (7.7) and humanity in general (7.19-20) is extended by the implied author to include the narrator's heart as well. As a result, the implied author 'does a Qoheleth' on Qoheleth and meets the challenge of his pessimism by examining the grounds of scepticism with its own methods. A better example of ironization cannot be found. By giving one more peek into Qoheleth's heart, and via the strategic utilization of a *verbum dicendi*, the implied author paints Qoheleth as an unreliable narrator. In the process, the reader is returned to the horizon of ironic knowledge by which he or she can adequately evaluate Qoheleth's private insights. This raises the implied reader to an elevated position of knowledge which is not given to the narratee, who remains ironized along with the narrator. By the implied author's insertion, the reader is gently reminded that Qoheleth's monologue, like so many I-discourses, is caught between fact and opinion and cannot substitute for public knowledge. Via this summary, the implied author has fully satirized the narrator. Not until the epilogue will we hear his voice again.

c. *Ecclesiastes 8.1-9: The Unknowing Sage and Public Life*

The tone of the discourse changes in these verses, returning to the impersonal mode that characterized 7.1-13. Ecclesiastes 8.1-9 constitutes an instruction complex which addresses and admonishes the narratee regarding appropriate actions in the king's court. Qoheleth reflects upon two proverbs in 8.1a, b. The opening rhetorical question, 'Who is like the wise man?', is certainly satiric given its juxtaposition to the ironic treatment of Qoheleth as sage in the previous verses. The phrase also lends a subtle sense of hubris to its speaker. The question is framed in a similar manner to those questions in the First Testament which proclaims the 'incomparability' of Yahweh. The reader who is familiar with the question, 'Who is like Yahweh' (Exod. 15.11; Deut. 32.31; Pss. 86.8; 89.7), does not miss the pompous and Titan-like sound of the sage's words. In its normal setting, the question always anticipates an unqualified, negative response.[57] However, as Aarre Laura argues, in its present context a qualified positive response is anticipated, 'whereby this rhetorical question presupposes the answer: "no

57. R. Johnson, 'The Rhetorical Question', p. 195. For an excellent treatment of the text's repertoire at this point see C. Labuschagne, *The Incomparability of Yahweh in the Old Testament* (POS, 5; Leiden: E.J. Brill, 1966), p. 19.

one but a sage" '.[58] This is not the first time the reader has observed an exalted sense of self in Qoheleth, as the comparison of his estate with the Garden of Eden in 2.1-8 also portrays a person of hubris.[59] Given the recurrence of this trait, the reader again engages in character-building, concluding that this is characteristic of the sage. A sense of 'bad press' begins to cling to Qoheleth's persona. Although the question attempts to create consensus, the ethos of its speaker undermines its implied goal in a fashion that is typical of the narrator's vain rhetoric. Again, the implied author's characterization of the narrator portrays him in a less than palatable manner.

Verses 2-7 are quite conservative and orthodox in terms of the values they espouse. There is nothing that does not strike the reader as anything other than sound and prudent advice here. The *kethib* form found in Mt. v. 2 begins with the pleonastic use of *'ᵃnî*, but most readers emend this to *'eth*, the sign of the accusative. Here Qoheleth uses an imperative (*šᵉmōr*, 'keep!'), and a jussive (*'al-tibbāhēl*, 'do not delay'), to address the narratee, commanding them to 'keep the command of the king' in the strongest of terms. The rhetorical question in v. 4 allows the narratee to figure things out for themselves, should the above imperatives fail to hit their mark. Qoheleth then reflects upon a traditional saying to further his point to the upstart narratee who might be tempted to overstep the mark in the presence of royalty.

Verses 6-7 present yet another blank to the reader. Verse 6 reminds the narratee/reader of the limits of time broached in 3.1-8. Verses 6b-8 pull from the text's horizon of values the theme of humanity's epistemological limits (v. 7) and our common mortality (v. 8). The reader

58. Lauha, *Kohelet*, p. 144.

59. This study has opted to view MT's initial *'ᵃnî* in 8.2 as a corruption for an initial *'ēt* (with the LXX). However, if *'ᵃnî* is original, then surely Qoheleth's pompousness is on full display in a most explicit way here. JPSV takes it in this way, translating the verse as, 'I do! 'Obey the King's orders'. If this should prove the more accurate text, then Qoheleth is actually answering the question in 8.1; 'Who is like the wise, and who knows the interpretation of a matter?… I do!' Beentjes also takes the verse in this manner. Given this scenario, he observes that the personal pronoun *'ᵃnî* functions in a similar manner to that of *'ᵃnî* in 1.12. As such, there may be another allusion to the earlier King's Fiction in these verses. Furthermore, 8.2-5 may be considered an allusion to 1 Kgs 2.43. See P. Beentjes, ' "Who is like the wise?": Some Notes on Qohelet 8,1-15', in Schoors (ed.), *Qoheleth in the Context of Wisdom*, pp. 303-15 (306). In both scenarios, however, Qoheleth's Titanism is deftly communicated to the reader.

must infer the relationship of vv. 6-9 to vv. 2-5. Perhaps the sage is reacting to the naivete and optimism of the proverb in v. 5 which claims that 'the one who keeps a command will experience no evil thing'.[60] Furthermore, the text presents an incongruity as it contradicts other passages which form the text's horizon of values, particularly the reference to the wise person knowing the 'time and decision' (*'ēt ûmišpāṭ*) which contradicts the sage's express worldview presented in 6.12. This presents yet another wisdom Rubik's Cube for the narratee/reader to process and figure out.

The text ends with an observation statement (v. 9) regarding the predatorial nature of those in power and how the abuse of individual authority harms the whole community. Qoheleth typically *begins* an observation complex with a statement like this one. However, it could be that this verse begins the observation which extends from 8.10-15.[61] I surmise that it has a Janus-function in this verse (in a manner similar to 6.10-12), functioning to conclude one segment while expressing a point of view that carries over to the next pericope. In a fashion typical of the rhetorical style of this text, there remains a degree of hermeneutical openness to this blank. This continues to recreate the fundamental experience of wisdom's '*hebel*-ness' for the narratee/reader who is presented with yet another test.[62]

d. *Ecclesiastes 8.(9)10-17: Private Insight and the Problem of Human Observation*

The section 8.(9)10–9.12 begins the last observation complex in Qoheleth's monologue. Ecclesiastes 10-17 continue the observation begun

60. Spangenberg, 'Quotations in Ecclesiastes', p. 20 relying upon D. Michel, 'Qohelet Probleme: Uberlegungen zu Qoh 8,2-9 und 7,11-14, *ThVia* 15 (1979–80), pp. 81-103 (102).

61. Schubert connects v. 9 to what follows, though most readers (Crenshaw, Loader, Whybray, Lauha, to name a few) typically divide the text between vv. 9 and 10 ('Die Selbstbetrachtungen Kohelets', pp. 23-34).

62. Dell refers to this use of blanking in 8.1–10.3 as the 'reuse and misuse of forms in Ecclesiastes'. She characterizes the narrator's rhetorical style as 'the use of an existing tradition to criticize it in a radical way…the scepticism of the author of Ecclesiastes is often expressed in reflective passages which show us the weaknesses of wisdom by providing traditional material with a new context' ('The Reuse and Misuse of Forms', p. 147). The major way that Qoheleth accomplishes this is to place incongruities in Israel's Wisdom tradition side by side, allowing the reader to ascertain the point of those blanks by inference.

in v. 9. Qoheleth's thoughts turn to the problem of unrewarded wickedness, or perhaps better, the delay of just desserts. Kathleen Farmer refers to this passage as 'When Reality Contradicts Tradition'.[63] Verses 10-11 make an observation that few readers will disagree with; life takes a while to justly reward the wicked, and often, this delay of consequences actually incites humanity to unrighteous deeds (v. 11). Furthermore, the wicked sometime receive praises for their deeds even by the religious establishment, an act of hypocrisy which richly deserves Qoheleth's condemnation as an 'absurd' thing (v. 10b). Immediately afterward, another blank is presented to the reader, this time in the form of an observation or testimony rather than a proverb. Verses 12-13 pronounce in a bold-faced manner that God does indeed reward the righteous and judges the wicked, even if there is a delay of rewards. However, v. 14 negates this, observing with the *yēš*-construction that there is an absurdity on earth; that there are 'righteous persons to whom it happens like the deeds of the wicked, and there are wicked folk to whom it happens like the deeds of the righteous'.

The reader must labor to understand how such an optimistic outlook (vv. 12-13) can be sandwiched between two verses (vv. 11 and 14) which negate the value of such an orthodox position. Although the text gives no answer (since it is a test), the inclusio provided by vv. 11 and 14 provides a hint to the reader that the type of optimistic testimonies seen in vv. 12-13 are patently wrong.[64] In an ironic twist Qoheleth undermines such personal observations (note the use of 'I know' in v. 12) with an observation based on reality (note the use of 'there is' in v. 14). Although on the surface v. 14 seems like a rhetoric based on reality, both rhetorical strategies are actually based on the narrator's personal observation. It is just that Qoheleth believes his to be the correct one and hides his 'I know' behind the *yēš*-construction.

By now the text has trained the model reader to understand both statements as examples of the narrator's radical subjectivity. In allowing these two observations to stand next to each other, the implied author alludes to the difficulty of wisdom's task, that is, the use of human faculties of observation to come to certain knowledge.[65] The

63. Farmer, *Who Knows What is Good?*, p. 181.

64. R. Murphy, *Wisdom Literature: Job, Proverbs, Ruth, Canticles, Ecclesiastes, and Esther* (FOTL, 12; Grand Rapids: Eerdmans, 1981), p. 130.

65. Again, it should be stressed that Qoheleth is not always expressing viewpoints with which the implied author disagrees. Many ideas in the narrator's post of

epistemological ramifications of this passage nearly deconstructs the entire Wisdom enterprise. By means of these dueling observations, the implied author hints at the problems associated with all forms of human observation, both public and private. The relationship between Qoheleth's observations and the public's observations becomes so blurred that one cannot tell private insight from public knowledge. Therein the reader experiences theological and epistemological 'vertigo'. Qoheleth's 'I' has nearly destroyed the concept of public knowledge, subverting it into a form that almost undermines the entire process.

Ecclesiastes 8.15 presents the fifth call to enjoyment. Given the fact that humanity may or may not receive adequate compensation for its efforts, Qoheleth commends in the strongest of terms the wholesale enjoyment of life. While the preceding occurrences of the call are based on personal experience, this one begins on a more formal and solemn note, stressing its importance with a prefixed phrase, *w^ešibbaḥtî 'ănî 'et-haśśimḥâ* ('so I commend enjoyment'). To stress its importance, Qoheleth again utilizes a 'nothing is better' form to make a rhetorical impact on his narratee. The postfixed *kî*-clause, 'for this will accompany him in his toil all the days of his life which the Deity gives to him under the sun', stresses the *carpe diem* ethic in the most clear and succinct of manners.

Verses 16-17 conclude the discourse of 6.10–8.17. To make his point, Qoheleth finishes with a flurry of sceptical remarks. He reminds his reader of the search that has characterized his life from the beginning (cf. 1.12-18), concluding that although one might try, 'even if the wiseperson says he knows, he is not able to find it'. This closes the gap raised by the 'who knows what is good?' question in 6.12. In this verse, the discourse reaches a crescendo of scepticism, three times accenting the phrase, 'not find out'. This is the proper conclusion to the feigned summary of wisdom that Qoheleth offers in 7.25-29. Here the reader surmises an acceptable academic summation that is both proper and fitting for a sage. Although a reader may or may not agree, this recapitulation is the sort of answer one expects from an academician.

Qoheleth began this section of the discourse by asking the programmatic question: 'Who knows what is good?' He answers that question just as directly here as he did the initial 'what profit?' question in 2.11.

observation are similar to ideas found in the implied author's post of observation. As Perdue observes: 'Indeed, the mind of the sage—the implied author—and the voice of the narrator often merge' (' "I will make a test of pleasure' ", p. 203).

Nevertheless, the narratee/reader must infer that the question has been answered here. The call to enjoyment in v. 15 is juxtaposed to a passage which denies the availability of knowledge, even that of the good. As a reader, I surmise that Qoheleth found value, but he never found the good in the ultimate sense that he craved. Graham Ogden observes that in these verses 'Qoheleth confesses that he was so obtuse, so blind, that he could not see the answer'.[66] The judgment offered by the sage here extends to all sages, and therefore includes both private and public knowledge. In this respect, Qoheleth's point of view on the public knowledge of the good depicts a community which is fundamentally unable to see things and can, with some reservations, be termed an ethically blind public. However, the model reader understands by now that this too is the subjective post of observation of the narrator, and provides no sure guide to the question he poses.

To sum up, although Qoheleth is capable of streaks of orthodoxy in 6.10–8.17, the characteristics of radical subjectivity, jadedness and scepticism overwhelm the reader. Because of this the narrator is depicted as a subversive sage. Although much of what he says is honest and true, the satiric characterization he receives in 7.25-29 makes a lasting impression on the reader. No longer does Qoheleth's ethos-related swings toward orthodoxy fool the reader. The reader closes the *Gestalt* on the reader's character, deducing quite appropriately, that while Qoheleth means well, his judgment is often clouded by his own narcissism. The path that led him to his deductions is too narrow to support the conclusions he makes. His pessimism and self-centered ethic detracts greatly from his attractiveness as a rhetor. In the end, Qoheleth becomes a limited post of observation whom the reader respects as a sage, but does not necessarily feel obliged to hold in the highest esteem. He possesses prudence and honesty, yet lacks greatly in terms of magnanimity and magnificence. In the process, both Qoheleth and his epistemology are characterized in a satirical light by the overall design of the text, which I am here calling the implied author.

6. *Ecclesiastes 9.1-6: The Depths of Scepticism —Who Knows about God?*

The final quarter of Qoheleth's monologue stresses the theme that humanity does 'not know' what will come after it. The phrase 'no

66. Ogden, *Ecclesiastes*, p. 141.

knowledge' or 'not know' occurs in 9.1, 5, 10, 12; 10.14, 15; 11.2, and three times in 11.5-6.[67] This continues the sceptical trend begun in 6.10–8.17. Qoheleth builds on his previous discourse, noting how he has 'laid all this to my heart' (*zeh*, 'this', referring to what precedes it as it does in 8.15 and its related construction, *gam-hû'* in 2.1). However, the narrator's scepticism reaches a new nadir, as he now extends it to include even God's love. The use of *nātattî* ('I laid') indicates present focalization for the narrator.[68]

Ecclesiastes 9.1-6 deals with the problem of humanity's mortality, the 'one fate' as Qoheleth euphemistically calls it. As Qoheleth's self-interested epistemology only knows the confines of the self, the lack of ultimate rewards for righteousness and wickedness imputes an indifferent attitude toward God. He therefore concludes that we do not know whether God loves or hates us. Of course, Qoheleth deduces this from a minor premise which seems to suggest that love and rewards are related in his mind, something not all First Testament rhetors would share (cf. Hab. 3.17-19). The use of asseverative *kî* ('indeed') further emphasizes the narrator's conclusion. The rhetorical exposure of such an emphatically negative and sceptical text depicts the narrator as a functional agnostic. Again, we see the telltale rhetorical ethos of the narrator, who always seems to follow up an episode of good ethos (the call to enjoyment in 8.14-15) with a statement that affects his characterization in a less than attractive manner.

Verses 2-3 continue the lament that humanity's fate is out of its corporate hands. Qoheleth's practical agnosticism continues quite unabated. Verse 2 virtually deconstructs the need for organized religion. Referring to those who are 'clean' and 'unclean', those who sacrifice and those who do not, Qoheleth insinuates that even obedience to the Torah serves no purpose. This is something that the implied author (cf. 12.13) and the implied reader who resonates with those values would surely dispute. Verse 3 bewails the fact that all of this is an evil, and that humanity's hearts are full of evil as well. Still, the eclipse of lasting value does not diminish the value of life itself, as Qoheleth reminds himself of the better-than proverb: 'a living dog is better than a dead lion' (v. 4). Qoheleth's emotional exasperation leads him to resort to a barbarism in v. 5, proclaiming that the 'dead do not know

67. A. Wright, 'The Riddle of the Sphinx', p. 323.
68. Isaksson, 'The Autobiographical Thread', p. 45.

excrement' (*m^e'ûmâ*, 'nothing', often functions as a euphemism for something that is bad). One might even characterize this judgment as an example of 'cosmic irony'.[69] After their death, all traces of a person vanish, even their love and their hate (v. 6). In these verses the emotional distress of the narrator is quite apparent. Qoheleth's agnosticism decimates his world of any lasting value, a characteristic which leaves the narrator lacking in the qualities of magnanimity and magnificence that are sorely needed by this juncture in the discourse.

7. *Ecclesiastes 9.7-10: Reclaiming the Value of Life—Knowing How to Enjoy Life*

Once more, just when the sun threatens to set in Qoheleth's world, the morning comes in the form of a call to enjoyment. The tone takes an imperative mood as Qoheleth uses a string of commands to commend enjoyment to his narratee. Verse 7 begins with the imperatives *lēk* ('Go!') and *'^{e}kōl* ('Eat!'). The use of these independent command forms adds rhetorical prominence to this call. The pleonastic use of the second person-pronoun *'attâ* in v. 10 also 'personalizes his point',[70] making the reference to his narratee quite explicit. In addition, Qoheleth goes into detail for the first time, giving an extended exposition on what constitutes enjoyment: eating bread with joy, drinking with a merry heart, wearing white garments (celebrative attire), lavishing and pampering oneself with oil, and enjoying the love of a wife all the days of one's life (vv. 7-9).[71] The latter represents something of a change of heart for the narrator, given the negative assessment of women which looms in the text's horizon of values (cf. 7.26-28). Qoheleth boldly proclaims that God has already approved of this course of action (9.7b). For someone who just denied whether one knows whether God loves or

69. Spangenberg, 'Irony in the Book of Qohelet', p. 68.

70. Crenshaw, *Ecclesiastes*, p. 163.

71. As Crenshaw points out, this imperative surely characterizes Qoheleth's narratee as male in gender, unlike the broader audience envisioned by the implied author, who utilizes the sexually generic term 'the people' (*hā'ām*) in 12.9. See Crenshaw, *Ecclesiastes*, p. 163. Undoubtedly, such a characterization marginalizes women readers. However, if one keeps in mind that the implied reader is defined by the implied author, and not the ironized narratee of Qoheleth's discourse, this should provide some degree of solace for the postmodern audience. In addition, a little creative hermeneutics, such as understanding the term in the sense of a 'spouse' will also go a long way here.

hates humanity, a comment like this has a certain ironic tone to it.[72] He ends this effusive admonition on an extremely positive note, commending the narratee/reader to enjoy life with all of one's strength. The reference to 'your *hebel* life' in 9.7 refers back to Qoheleth's description of his own vain life in 7.15. Its precise nuance for this passage remains open.[73] However, by tying the narratee/reader's quality of life to his own, Qoheleth endeavors to draw the reader into his circle of intimate dialogue. Rhetorically, the subtle connection between 7.15 and 9.7 functions to bind Qoheleth's narrated life with that of his listeners.

As with the previous calls, these verses do nothing but good for Qoheleth's ethos. Here the reader perceives a man rich in a spirit that lifts the heart (magnificence), prudence, magnanimity, generosity and attractiveness. Qoheleth's stock begins to rise, partially overcoming the rhetorical faux pas he initiated in the previous chapters. However, the verse does not radically change the reader's *Gestalt* of Qoheleth's characterization. By now the implied author's characterization of the narrator has been completed. This verse only tells the reader what he or she already knows, that Qoheleth is one of those persons with stellar strengths and glaring weaknesses, the sort of 'black sheep' with whom you disagree but for whom you also have some fond feelings.

8. *Ecclesiastes 9.11-12: The Unpredictable and Public Knowledge*

Qoheleth continues his assault on public knowledge by further criticizing the toil-yields-rewards ethic. He seems to repeat himself here, varying only the poetic form of his assault. Verses 11-12 present another observation complex which laments the fact that those who are especially gifted do not always take home the prize. The subjective nature

72. Crenshaw wryly observes regarding the status of knowledge in ch. 9: 'The careful reader will have noted that Qoheleth seems to know far more about God than his theology of divine mystery allows. In truth, he frequently makes assertions about God's will and activity despite the protestations about God's hiddenness' (*Old Testament Wisdom*, p. 139).

73. Miller, 'Qohelet's Use of *Hebel*', p. 452. As in 7.15-18, Miller points out that the context here does not provide other associated terms to distinguish whether *hebel* is meant to be taken as insubstantiality, transience, absurdity, foulness, or any of its various meanings. The reader can therefore recognize any or all of the nuances this word brings to the text. In this sense, it remains hermeneutically open, though in a limited manner since the reader by now has firmly in mind a set of variables this word may mean.

of this observation is evident in the prefixed phrase, 'I saw that', which is not necessary unless one wants to emphasize the personal nature of the statement.[74] By means of this subtle introductory remark, the implied author continues to characterize the narrator in such a way that his subjectivism is always kept before the reader. The problem of time and chance (*'ēt wapega'*) are invoked from the horizon of values as the culprit here. Verse 12 picks up on the theme of time. Time and the evil occasion are depicted as a predator who stalks us. The metaphor is a powerful one and certainly draws the reader into its sphere of pathos. In these verses the reader surmises only honesty and realism.

To sum up, in this observation complex (8.[9]10–9.12), Qoheleth sandwiches two calls to enjoyment (8.16 and 9.7-10) between observations which lament the lack of positive rewards for one's actions. The observation complexes alternate in a negative–positive–negative–positive–negative manner. However, to be fair, much of what Qoheleth observes here is simple realism. Still, the practical agnosticism that is evident in 9.1-3 certainly characterizes the narrator in a less than attractive manner. This alternation of good and bad ethos is consistent with what the reader has come to expect from the narrator's persona. In this respect Qoheleth has become a full-fledged character, whose depth of disposition is now well known to the reader. He has become a round character possessing definite patterns of thought and predictable traits. Like an old friend, Qoheleth no longer surprises his reader.

9. *Ecclesiastes 9.13–12.7: Asking the Narratee to Fill in the Blanks*

Just as the book began with a long observation complex that highlighted the narrator, the book ends with an extended instruction complex which accents the narratee (9.13–12.7). This balances the book from an artistic and ideological perspective and shows something of the structural isomorphism of the text. In a broad-based sense, there is a movement from a narrator to a narratee orientation between the two halves. In the second half, observation complexes become shorter while the narratee oriented sections increase both in length and intensity. The final call to enjoyment in 11.9–12.1 marks the end of what I have termed the cascade of the narratee in the book of Ecclesiastes. This final section therefore balances out some of the inward focus that has dominated the

74. Fox makes this astute observation ('Qoheleth's Epistemology', p. 147).

book so far. In chs. 9–12, there are only two *Selbstbetrachtungen* (in 9.13-16 and 10.5-7) to hermeneutically guide the reading process.[75] Instead, the discourse centers on proverbial texts in a manner that gives it a decidedly disjointed texture, much like Proverbs 10–29. Qoheleth reflects upon proverb after proverb from 9.17–11.4 with scarcely a break in thought. His thought appears rambling, with multiple blanks challenging the reader's cognitive powers.

This is the narratee's 'final exam'. Herein the narrator tests the youth's ability to perceive the inherent contradictions in public knowledge. In the process, the model reader gets tested as well. Qoheleth has done all that is needed to equip the narratee and the model reader with the skills they need to make contradictions of wisdom's public knowledge. As a result, the observations virtually cease. The narratee/reader no longer needs Qoheleth's guiding 'I'. The youthful narratee is asked to become a sage and to stand on his own hermeneutical feet. Qoheleth's discourse builds on the aesthetic movement that has been building since ch. 4 where the extensive use of proverbial texts begins. Throughout his monologue Qoheleth has reflected increasingly upon proverbs and other gnomic texts. In ch. 7 the reader encounters a wholesale meditation upon the problems of proverbial wisdom. Yet throughout those reflections, Qoheleth's observations were constantly interspersed to guide the reading process, making sure that the narratee/reader learned Qoheleth's method of making contradictions. After 10.7 this ceases. Having fully equipped the model reader, Qoheleth in essence offers his student a 'textbook' case that closely resembles the book of Proverbs. The narratee/reader is now asked by the discourse strategy to think like the master. No longer does the narratee think along with the older sage. Instead, Qoheleth withdraws the focalizing properties of his all-guiding 'I', allowing the narratee/reader to become an independent focalizer of the wisdom tradition. They are asked to stand on their own wisdom feet, and to deal critically with the text as they have been so ably trained to do.

75. Although Schubert delimits the twenty-second observation as 9.13–10.3, I fail to understand why the proverbs cited in 9.17–10.3 constitute a part of this complex. As a result, I am limiting its extent to include just that part of the text which is couched in a first-person form. See Schubert, 'Die Selbstbetrachtungen Koheleths', pp. 34-35.

10. *Inferring the Model Reader's Competence*

In these insights, we see the utility of a reader-oriented approach that values the effects of the text over the meaning intended by the author. It was argued at the very beginning of this study that meaning is swallowed up in functionality, and that asking the question, 'How does this text function?', is a more productive place to begin the task of interpretation than beginning with the query, 'What did this author intend to mean?' No better example of that basic premise is available than in the book of Ecclesiastes. The question in this proverb 'collection' is not 'What does Qoheleth mean?' but 'What is happening to the model reader?' Qoheleth engages in what I am calling a rhetoric of subversive inference. In these verses there are few explicit guides. The reader is left to infer on his or her own what Qoheleth intends in these reflections. By this strategy, the text 'goads' the reader to think more deeply about the nature and limits of wisdom, as the implied author himself said in 12.11.

Norman Friedman observed that a text frequently 'runs from one extreme to the other: statement to inference, exposition to presentation, narrative to drama, explicit to implicit, idea to image'.[76] Qoheleth's discourse is a classic literary text in this regard. Throughout the monologue the reader has been asked to perform the six major reading activities, each with its own level of intensity as the changing rhetorical designs of the text have demanded. They have been asked to query, observe, infer, predict, evaluate, and compare in a variety of ways. In 9.17–12.7 the activities of inference and evaluation dominate the reader's cognitive activities.

As a result, the major reading activity changes in ch. 10. Before this point the major reading activity which dominated chs. 7–9 was comparison and the activities of protension and retention. Increasingly, however, the text has asked the reader to *infer* things. Here, however, the horizon of values is complete. No major themes are being added. In fact, Qoheleth has already made his point several times for some themes. He repeats ideas here in order to get the narratee/reader to make sense, or perhaps better, nonsense of them. Comparison is not the major activity required of the reader at this point. Instead, inferring the meaning of the individual proverbs consumes the reader's energies. From the

76. Friedman, 'Point of View in Fiction', p. 1169.

reader's post of observation, these maxims present a real challenge. No longer does Qoheleth's 'I' or strategies of obvious juxtaposition provide clues to their defamiliarized meaning. In addition, many of these proverbs are metaphorical, drawing heavily on the repertoire of the text and the native literary competence of the text's authorial audience. A better test for one's wisdom competency cannot be found in the entire Canon. Through this barrage of inferences, Qoheleth tests the reader's defamiliarized world to make sure that their instinctive reflexes have been established. Once again, the effect of the text is a better guide to meaning than an emphasis on the hypothetical intention of its original author.

This is a very powerful rhetorical strategy. Keith Grant-Davie observes that inference causes the reader to 'own' the opinion they are forming. This increases the suasive properties of the discourse by putting the text's meaning back upon the reader. He astutely observes that 'texts become persuasive inasmuch as readers infer an opinion or point of view which the author seems to invite them to share or demands that they share'.[77] By engaging in a rhetoric of inference, Qoheleth concludes his discourse with a very powerful suasive push that relies on the reader making the correct deductions and inferences, however multivariant those inferences may be.

11. *Ecclesiastes 9.13–11.6: Inferring the Wisdom of Wisdom*

T.A. Perry refers to 9.13–10.1 as 'wisdom's self critique'.[78] The observation in 9.13-16 presents another example story. Qoheleth characterizes the ensuing story as 'wisdom'. This surprises the reader, who expects a *hebel* classification for such a tragic tale.[79] The designation of this story as 'wisdom' is probably due to sarcasm on the part of the narrator. Etan

77. Grant-Davie, 'Between Fact and Opinion', p. 144. He also notes how authorial intention has very little to do with the suasive effects of inference: 'My other major conclusion was that persuasion, as a discourse type, is defined neither by the actual intent of the author, which can seldom be known with certainty, nor by the resulting change in readers of the text, but by their inference of the author's intent' (p. 142). This accords well with a Ricoeurian perspective on methods, which places a premium on the abilities of texts to surpass authorial intention.

78. T.A. Perry, *Wisdom Literature*, p. 61.

79. Crenshaw notes how this has caused some readers to emend the text in various ways, so surprising is the use of 'wisdom' here (*Ecclesiastes*, p. 165).

Levine, while noting the humor in Qoheleth, describes the passage as a 'burlesque of governmental "sagicity" '.[80] Jean-Jacques Lavoie has described the tale as 'eminently ironic'.[81] That there is some form of sarcasm in this passage is further suggested by Katherine Dell. She understands this passage as a 'misuse' of forms by the narrator. Dell argues:

> In 9.13-16 an example story is told which closely resembles example stories in Proverbs (e.g. 21.22). The moral of the story is given in the quotation of a wisdom saying in 16a, but then the author gives his own modification of the saying in the light of reality exemplified in the story. He 'spoils' the form of the story by adding his own viewpoint to the end of it and shows that he is not altogether following the traditional line.[82]

The test in this verse is an interesting one. Qoheleth examines the narratee to see if they understand the difference between wisdom and absurdity. Hence the use of sarcasm here. The story is elegant, simple and paradigmatic—a wise person saves a great city from certain destruction, yet no one remembered the deed. By now the model reader has been thoroughly educated to understand that a lack of reward is the primary criterion for ascertaining a *hebel*-condition. In presenting such a flagrant violation of the toil-yields-rewards norm, the author tests the competency of the narratee/reader, making sure that their worldview has been adequately defamiliarized. However, the attached proverb in v. 16a and Qoheleth's comment in v. 16b present a more difficult blank for the narratee/reader. The verse seems to present a 'yes-but' response to the implied *hebel*-condition; although the wise person's wisdom was despised, the value of this wisdom is not to be denied. The original formulation, 'wisdom is better than might', is 'exposed for what it is, a limited and unwarranted generalization'.[83] However, in a rare example of community-oriented values, Qoheleth judges this situation not by its effects on the individual, but on the general good the sage performed for the city. This sense of self-transcendency surprises the attentive reader, though Qoheleth has done that in the past for short intervals (cf. 4.1; 9.9). The element of surprise is basic to developing a rounded

80. E. Levine, 'The Humor in Qohelet', *ZAW* 109 (1997), pp. 71-83 (77).

81. J. Lavoie, 'La philosophie politique de Qo 9,13-16', *ScEs* 49 (1997), pp. 315-28 (327). He notes how this passage is ironic because the story is told by none other than the king, Qohelet who writes under the Solomonic guise.

82. Dell, 'The Reuse and Misuse of Forms', p. 144.

83. T.A. Perry, *Wisdom Literature*, p. 62.

character with real depth. The implied author constantly presents Qoheleth as a real person whose thoughts are a perpetual challenge to understand.

This story incites Qoheleth to reflect further upon various public affirmations about the role of wisdom and folly. What follows is a 'debate in proverbs'.[84] Ecclesiastes 9.17-18 reflects upon two 'better-than' proverbs which praise the relative value of wisdom over fools. The proverb in v. 18 observes that 'wisdom is better than weapons of war', such as can be seen in the example story. But then Qoheleth adds his own negative comment in v. 18b, noting that a sinner can do an equal amount of destruction. This thought continues in 10.1, as Qoheleth reflects upon a proverb which confirms his previous comment. He notes that just as dead flies spoil costly perfume, so a little folly outweighs wisdom and glory. The positive and negative evaluations of wisdom in 9.17-18a and 9.18b–10.1 function as a blank that the reader must process, forcing the reader to infer its meaning in this context. The model reader understands by now that Qoheleth 'is putting different wisdom sayings together to highlight the contradiction between them'.[85] Such contradictions have the effect of increasing the ideational activity of the reader, who must carefully weigh each proverb and comment that Qoheleth makes. By the use of these blanks, the narrator is making a sage of his narratee. Blanks train the narratee/reader to look for the ironic, the contradictory and the incongruous in life. From Qoheleth's post of observation, such an attitude is the only way a sage can approach public knowledge, and so, constitutes the most basic attitude of the wise person. Qoheleth's rhetoric of cognitive blanks helps create this foundational competency for the would-be sage. Such a strategy characterizes the narrator as a competent sage with a mastery over his chosen field. Although these blanks frustrate the reader, in terms of the narrator's general sense of ethos, they create confidence in the speaker's expertise.

As a result of this the major reading problem in ch. 10 is consistency-building. There are so many blanks in 10.2–11.6 that most commentators refer to this passage as miscellaneous insights, sayings, or some other testimony to the reader's inability to come to a coherent *Gestalt* regarding the text's overall structure and meaning. However, I

84. T.A. Perry, *Wisdom Literature*, p. 61.

85. Dell, 'The Reuse and Misuse of Forms', p. 144.

would argue that the overall structure is not to be found in the use of a common theme, but in the common effect brought about by the text's rhetorical design. The challenge to wisdom and the challenge of wisdom's disparate insights constitutes the major 'thematic' issue which underlies this text. Qoheleth exploits wisdom's limited and context-specific nature by juxtaposing proverb after proverb, insinuating to the reader that the bigger picture is missing, as he deduced in 8.17. The use of blanks created by these disparate proverbs recreates in the reader a narrative experience of that fundamental insight. As a result, we see that the rhetorical design of the text is precisely to leave this sort of open-ended, confused experience with the reader.

The proverbs in 10.2-4 discuss the value of wisdom over folly. Verse 4 addresses the narratee ('if the king rises against *you*', *'ālêkā*), commending composure as a prudent course of action when judgments in error are made in governmental circles. From this, Qoheleth's thought turns to other governmental problems as it relates to wisdom, particularly the evil (*rā'â*) that occurs when a fool is given power, or worse still from his social position, when a slave and a prince trade places—a fear Qoheleth has voiced before in dread terror (cf. 5.12-15). Verses 5-7 constitute the last observation in the book. From the narrator's subjective post of observation, such a topsy-turvy world is evil, though presumably not from the point of view of the poor who endured the wealthy person's oppression (cf. 4.1). Such a situation 'subverts the structured world of the sages, where the wise succeed and prosper and the fools fail because of their own stupidity...The absurdity of the present social order demonstrates the impotency of wisdom to steer a rational course toward certainty and well-being.'[86] However, for the reader who has heard Qoheleth's occasional outbursts which decry the social position of the poor, this observation characterizes him as one of the oppressing class. The only value expressed here is the well-being of the economically advantaged, a position that is hardly attractive. The reader also asks: 'Wisdom is a value, but for whom?' Qoheleth voices a class-biased point of view, holding to the premise that wealth and misfortune are earned.[87] This is something his own observations should

86. Perdue, '"I will make a test of pleasure"', p. 231.

87. Crenshaw, *Ecclesiastes*, p. 171. Habel has shown that the book of Proverbs presents five different paradigms regarding the origin and cause of poverty: (1) The Hard Work Paradigm where the origin of poverty is said to be laziness (cf. Prov. 12.24); (2) The Harsh Reality Paradigm which depicts the horrors of poverty; (3)

have corrected. Once more, the implied author depicts the narrator's post of observation as one characterized by self-interest. Obviously, Qoheleth has forgotten his own position on the subject when he commended a wise but poor youth over an old but foolish king (cf. 4.13). Again, we witness a dueling observation within the discourse.

The ensuing proverbs all draw on the text's repertoire to make a point. Verses 8-11 are notoriously opaque and obtuse. Verses 8-9 express a belief in how actions often beget their own rewards. Verses 10-11 continue this line of thought, noting the negative rewards that result from the loss of diligence or carefulness. The necessity of inference is apparent in v. 11, as Robert Johnson has observed: 'The point of the saying is the value of foresight; however, this value is not stated explicitly, but indirectly'.[88] The proverbs in vv. 12-14 deal with the fool and his mouth. Verse 14 utilizes a rhetorical question, again accenting the theme of humanity's inability to know the future. Verse 14b could be Qoheleth's own comment which agrees with the verdict reached about foolish talk in v. 14a.[89] Verse 15 continues Qoheleth's condemnation of the fool. Verses 16 and 17 show the versatility of our sage, who can even utilize the woe oracle and the blessing to make a point about wisdom. Verse 16 condemns a government run by lads, while v. 17 commends a government run by sensible men, who, in a manner consistent with his class-bias, are naturally defined as 'noblemen' (*ben-ḥôrîm*). These verses condemn the leisure of youthful leadership, a value further condemned in v. 18. Juxtaposed to this condemnation of leisure is a proverb that commends its use: 'One makes bread for laughter, and wine gladdens life'. The reference to 'money answers everything' certainly expresses a jaded point of view on life.

By means of the blank opened up by the juxtaposition of vv. 16-18 and v. 19, Qoheleth subtly shows the inherent contradictions that

The Social Order Paradigm, which counsels the rich to refrain from robbing the poor because they are weak; (4) The Trusting Righteous Paradigm which argues for the integrity of being poor over being unrighteously rich; and (5) The Via Media Paradigm that counsels a middle road, suggesting that the sage plot a middle course between the excesses of wealth and poverty. Accordingly, these five paradigms are ultimately rooted in diverse social settings. See N. Habel, 'Wisdom, Wealth and Poverty Paradigms in the Book of Proverbs', *BibBh* 14 (1988), pp 26-49. Qoheleth espouses all of these at some point in his discourse, though here, paradigm one is being stressed.

88. R.F. Johnson, 'A Form Critical Analysis', p. 180.

89. R.F. Johnson, 'A Form Critical Analysis', p. 181.

proverbial wisdom contains. The reader could well imagine the princes in v. 16 quoting just such a proverb to justify their actions. For that matter, they could even quote the master himself given his admonitions on the subject. The passage possesses a certain sense of unstable irony in that regard. Herein Qoheleth insinuates the obvious dangers in proverbial lore/public knowledge if the narratee/reader is wise enough to catch his drift. Verse 20 ends the concatenation of proverbs on government officials by noting the folly of criticizing the king or the wealthy. The narratee/reader infers from its juxtaposition to vv. 16-19 that the appropriate response from the sage regarding the foolish display by the wealthy is to take a prudent course of silence. Qoheleth hints at the connection of v. 20 to what precedes by using the second-person reference to the narratee ('*your* king', v. 16; '*your* thought', v. 20) as a marker of inclusio.[90] In all of this, the critic notes that not once does Qoheleth explicitly say what he means. As a result, the passage remains hermeneutically open.

The proximity of similar but different proverbs strongly suggests that the blanks opened up by the use of juxtaposed proverbs serves a higher ideological function in these verses. All of this characterizes the narrator as a very clever and subtle sage who has definite subversive tendencies. He is a person of prudence, who, though influenced by the self-interest that blinds all social classes, commends a path of temperance. However, the rhetor lacks the traits of justice and courage (v. 20) which would help his cause. Still, these characteristics are well-known by now. The reader is probably too preoccupied with 'proverb crunching' to pay much attention to that. Generally, Qoheleth comes across as a sage who can be trusted, a person with experience and knowledge of how things run in life.

Beginning with 10.16, Qoheleth's discourse begins to emphasize the word 'your' as a way to engage the narratee's attention. This emphasis continues in 11.1-6. The reference to money in 10.19 incites Qoheleth to turn toward financial matters. Ecclesiastes 11.1-6 deals with the issue of investing one's economic resources. No advice to the youthful narratee would be complete without this. The proverbs in 11.1-3 have a decidedly optimistic tone about them, observing how financial planning is rewarded. Verses 1-2 remind the narratee/reader to invest broadly,

90. Brown, 'The Structure of Ecclesiastes', p. 206.

not putting all of one's eggs in one basket.[91] The use of imperatives (*šallaḥ*, 'Cast!'; *ten*, 'Give!') in vv. 1 and 2 directly engages the narratee/reader's attention. Verse 2b begins, however, to inject the element of uncertainty into this admonition: 'for you do not know what misfortune will happen under the sun'. This continues the theme of human ignorance which is consistently highlighted in chs. 9–11. The proverb in v. 3 is obtuse, but seems to further inject a pessimistic tone into the admonition, suggesting the inevitability of life's ways.[92] The proverb in v. 4 commends the wise person to be attentive to the signs that life gives regarding its vagaries, advising observance and carefulness to the young narratee. This passage reminds the narratee that life has inherent risks. In addition, the blank opened up between vv. 2 and 3 further suggests a critique of traditional wisdom, calling attention to the random aspects of the universe's rewards system. Qoheleth is depicted here as the quintessential wise man who covers his bets. Nothing but prudence characterizes the sage's conservative financial advice. But even here, the influence of his past ethos-related miscues often affects the reader's estimation of his attractiveness. As Robert Johnson concludes regarding these verses:

> This careful, deliberate arrangement of these sayings reflects very clearly Qoheleth's stress on the practical morality of life. He can urge diligence, not for any moral or theological reason, but rather because that is the best way to get along in the world as it is. Thus, while particular exhortations in Ecclesiastes may seem similar to those of traditional wisdom, they originate from another world-view than conventional wisdom and a different conception of human existence in God's world.[93]

Verses 5-6 close the section 9.1–11.7 which accents the theme of humanity's epistemological limitations by emphasizing the phrase 'not know' three times.[94] The theme of human ignorance fully criticizes the

91. However, it should be admitted that this passage maintains a hermeneutical openness about it for most readers. Regarding the various ways readers have taken it, see Tsukimoto, 'The Background of Qoh 11.1-6', p. 42, and Fox, *Qoheleth and His Contradictions*, p. 273. Whether it means taking financial chances, doing deeds of charity, or selling merchandise overseas one cannot say precisely, except that it definitely urges financial advice in some sense. I have taken it as advising to invest broadly.

92. Crenshaw, *Ecclesiastes*, p. 179.

93. R.F. Johnson, 'A Form Critical Analysis', p. 192.

94. A. Wright, 'The Riddle of the Sphinx', p. 323.

optimism of v. 1 in particular by insinuating, 'does anybody really know about all of this?' The narratee is equally stressed here, as the reference to 'you' is given a similar triple rhetorical exposure.[95] Verse 6 continues the direct address to the narratee/reader by using yet another imperative (*zᵉra'*, 'Sow!') and the jussive (*'al-tannaḥ*, 'do not withhold'). In all, there are 12 second-person grammatical forms addressed to the narratee in these two verses using a variety of address forms (imperative mood, jussive and second-person suffixial forms). In these verses and the final section which emphasizes the enjoyment of life while one is still young (11.7–12.7), the cascade of the narratee reaches its climax. The test is over. Qoheleth can now turn to more ultimate matters.

12. *Ecclesiastes 11.7–12.7: Youth, Mortality and the Enjoyment of Life*

The focus of the discourse remains upon the narratee in these verses, but the focalization runs through the dying perspective of the narrator. Verses 7-8 are the thematic introduction to the conclusion of Qoheleth's discourse. He cites a traditional saying about the sweetness of light. The point of this metaphorical proverb is that life is good.[96] Juxtaposed to the emphasis on light is the motif of darkness which is introduced in v. 8. The reference to light and darkness recurs as a motif throughout the final passage (11.7 refers to the sun, 11.9 to sight, 12.2 to the sun and stars, while 11.8 and 12.2-3 refer to darkness). The use of this motif adds a certain emotive texture and depth of pathos to the ensuing discourse. The use of pathos-eliciting metaphors recalls the fact that enjoyment is not just a logical choice, but one that claims the whole person. Verse 8 reminds the narratee/reader to rejoice in all of his or her years and to remind him or herself that darkness will also be a part of the life experience.

The most extensive and direct call to enjoyment is saved until now. Ecclesiastes 11.9-10 addresses the young narratee with the imperative mood one more time. Verse 9 directly characterizes the narratee for the first time, referring to him as a 'young man'. Qoheleth waxes effusive in his admonition: 'let your heart cheer you in the days of your youth; walk in the ways of your heart and the sight of your eyes'. Verse 9b seems to be a conscious play on Num. 15.39 which strictly admonished

95. Mulder, 'Qoheleth's Division', p. 152.

96. R.F. Johnson, 'A Form Critical Analysis', p. 101.

against such things. If so, Qoheleth displays his rebellious and secular sides one final time. This caused not a small amount of debate regarding the appropriateness of Ecclesiastes for the First Testament Canon among the Rabbis.[97] However, the caveat in v. 9b shows that Qoheleth was not advising a wanton path. What he advises is closer to the modern existentialist concept of *Sein zum Tode* ('being to death') advanced by Martin Heidegger.[98] Verse 10 continues his exhortation, stressing the removal of negatives from one's life. Again, Qoheleth concludes that life is a *hebel*, used here not in the sense of absurdity, but with the sense of fleeting or transitory. The use of the imperatives (*w*^{e}*hāsēr*, 'Remove!'; *w*^{e}*ha*^{ʿa}*bēr*, 'Put away!) transforms the call to enjoyment into a categorical imperative.

Ecclesiastes 12.1a could be either the culmination of the call to enjoyment, or the beginning of a new passage.[99] It seems better to view it as the beginning of the narrator's final adieu. Qoheleth's last words to his youthful narratee is to remember *bôr'êka* ('your Creator') in the days your youth. The word *bôr* presents a challenge to the reader. The plural form of the word in MT is a problem. Along with many readers, it seems better to emend this word, and to read *bôr*^{e}*ka* ('your vigor') here.[100] Given the contrast between youth and old age in its surrounding context, this is a preferred reading.

Ecclesiastes 12.1b introduces the opaque and metaphorical poem on aging in 12.2-7. Qoheleth finishes his discourse as he opened it, with a flight of poetic enunciation.[101] The text is dense and highly evocative. One might even call it a rhetoric of metaphorical motifs. So fertile are

97. See Salters, 'Exegetical Problems in Qoheleth', pp. 44-59 (57-59).

98. Scheffler makes this astute observation with which I agree. See Scheffler, 'Qoheleth's Positive Advice', p. 259. As Fisch argues about the purpose of death in Ecclesiastes: 'the book ends in death, but it is death with a difference, death as a warning, an incentive to effort' (*Poetry With a Purpose*, p. 177).

99. Van der Wal also sees such a Janus function for these verses. See A.Van der Wal, 'Qohelet 12,1a: A Relatively Unique Statement in Israel's Wisdom Tradition', in Schoors (ed.), *Qohelet in the Context of Wisdom*, pp. 413-18 (416).

100. Salters, 'Exegetical Problems in Qoheleth', p. 57. He also cites Sir. 26.19 as supporting evidence: 'My son, guard your health in the bloom of your youth'.

101. Merkin also has observed how Qoheleth's poetic flights characterize the narrator in a way which balances his business side: 'But being a personality who wears contradictions without discomfort, he has another side, one that suits another realm—the realm of the artist, where a restless spirit of inquiry soars beyond the walls of the status quo' ('Ecclesiastes', p. 402).

its poetic powers that it functions almost like a Rorschach Test for most readers. Readers have seen an allegory on old age, a reference to a coming storm, an apocalyptic vision, a funeral procession, an allusion to a decaying estate in the poem's images, cosmic deterioration, and more.[102] Some readers, like Frank Zimmermann, have seen a phallic interpretation.[103] The text alternates between metaphorical/allegorical descriptions of old age (vv. 2, 3-4a, 5b, 6) and more literal descriptions (vv. 1, 4b-5a, 5c, 7). This constant interchange of the literal and the metaphorical guides the reader's response, suggesting that old age and death are to be kept clearly in mind. With readers such as Michael Fox, Thomas Krüger, C.L. Seow, T. Beal and H.A.J. Kruger, Qoheleth seems be drawing images from a cultural repertoire which stems from prophetic or apocalyptic traditions. Here too we see the subtle effect of his epistemology. It would seem that Qoheleth has taken these stock images from prophetic or proto-apocalyptic traditions which usually relate to the demise of the nation or, perhaps, cosmos, and then radically reinterpreted them in relation to the demise of the individual. This reduction of prophetic/proto-apocalyptic imagery to another instance of private insight is exactly what the reader has come to expect of the

102. The best overview of this debate is offered by M. Fox, 'Aging and Death in Qoheleth 12', *JSOT* 42 (1988), pp. 55-77. Another excellent treatment is the reader-response analysis offered by B. Davis, 'Ecclesiastes 12.1-8: Death, and the Impetus for Life', *BSac* 148 (1991), pp. 298-318. See also J. Jarick, 'An Allegory of Age as Apocalypse (Ecclesiastes 12.1-7)', *Colloquium* 22 (1990), pp. 19-27; J. Sawyer, 'The Ruined House in Ecclesiastes 12: A Reconstruction of the Original Parable', *JBL* 94 (1976), pp. 519-31; R. Youngblood, 'Qoheleth's "Dark House" (Eccl. 12.5)', *JETS* 29 (1986), pp. 397-410; M. Gilbert, 'La description de la vieillesse en Qohelet XII,7: Est-elle allégorique?', in J. Emerton (ed.), *Congress Volume Vienna* (VTSup, 32; Leiden: E.J. Brill, 1981), pp. 96-109; N. Lohfink, '"Freu dich, junger Mann...": Das Schlussgedicht des Koheletsbuches (Koh 11,9–12,8)', *BK* 45 (1990), pp. 12-19; H. Kruger, 'Old Age Frailty Versus Cosmic Deterioration? A Few Remarks on the Interpretation of Qohelet 11,7–12,8', in Schoors (ed.), *Qohelet in the Context of Wisdom*, pp. 399-411; T. Krüger, 'Dekonstruction und Rekonstruction prophetischer Eschatologie im Qohelet-Buch', in Anja Diesel *et al.* (eds.), *'Jedes Ding hat seine Zeit': Studien zur israelitischen und altorientalischen Weisheit* (Festschrift D. Michel; BZAW, 241; Berlin: W. de Gruyter, 1996), pp. 107-29; C. Seow, 'Qohelet's Eschatological Poem', *JBL* 118 (1999), pp. 209-34; T. Beal, 'C(ha)osmopolis: Qohelet's Last Word', in T. Linafelt and T. Beal (eds.), *God in the Fray: A Tribute to Walter Brueggemann* (Festschrift W. Brueggemann; Minneapolis: Fortress Press, 1998), pp. 290-304.

103. Zimmermann, *The Inner World of Qohelet*, pp. 160-62.

sage. In this we perceive that Qoheleth's epistemology really is a philosophy, or perhaps, worldview for him. Qoheleth is an 'equal opportunity employer' when it comes to the various theological traditions available to him. He is quite capable of reducing any corporate-based tradition to another instance of private insight whenever it suits his purposes. In this regard, all public knowledge, whether it be wisdomic, prophetic, apocalyptic or legal, is refracted through the lens of Qoheleth's all-pervasive epistemology. Since Qoheleth is the literary creation of the implied author, Ecclesiastes, this also affects how one perceives the implied author as well. This much is sure. Whoever crafted such a creative hermeneutic clothed in a monologist's garb and still managed to criticize that hermeneutic through satire and irony was a mind capable of great intellectual precision. The implied author's commitment to dialogical-based thought is thoroughly present in this book. He would have been quite at home in the postmodern world. One can well imagine that were the implied author alive today, he would have given thinkers like Mikail Bakhtin a good run for their money.

The use of poetic imagery in this poem creates a collage of various emotion-producing images which arrest the reader,[104] causing him or her to reflect on the eventuality of old age and death. Death as a motivation for enjoying life has been adduced before (cf. 9.10b). It should therefore come as no surprise that he would expand upon that motivation to make a lasting impact on the narratee/reader one final time. The poem completes a gradual exposition in the book. There seems to be something of a 'readerly journey' implied in Qoheleth's discourse as it pertains to death. We are first told that generations come and go (1.4). Then we learn that there is one destination for all (3.20; 6.6). Later, this becomes explicitly named as 'Sheol' (9.10). Finally in this passage, 'we learn about the permanence of this destination, for this place turns out to be for every mortal, *bêt 'ôlāmô*, an eternal domicile'.[105] The use of poetic imagery thus creates a very intense pathos effect and possesses a very suasive power to influence the reader. Again, Qoheleth's typical rhetorical strategy is to wed both logos and pathos producing strategies into his discourse.

Verses 6 and 7 speak of humanity's mortality, hinting at the narrator's death. Then, in a moment of poetic solemnity, Qoheleth passes over

104. For a provocative analysis of the poem's ability to affect readers' emotions, see Christianson, *A Time to Tell*, p. 253.

105. Seow, 'Qohelet's Eschatological Poem', p. 226.

into literary immortality. This passage reminds me of a scene from a recent movie, *With Honors*, in which the aging and dying protagonist, a homeless person played by Joe Peschi, speaks to his youthful friend who is a college freshman at Harvard. He tries to tell this friend something about the need to enjoy life, much as Qoheleth does. At the end of the movie, he proclaims: 'Harvard, you cannot believe how different life looks going out!' In that respect, I concur with Michael Fox that the real referent to this set of images is the reader. He summarizes:

> these images depict the disaster of a nation or the world at large. For Qohelet they represent the demise of the individual. Qohelet is shaping [prophetic] symbolism in a way contrary to its usual direction of signification. Qohelet views the particular through the general, the small writ large. He audaciously invokes images of general disaster to symbolize every death; more precisely—the death of you, the reader, to whom Qohelet is speaking when he addresses the youth, the ostensive audience.[106]

The conclusion of this passage marks the end of a movement in the book from 'cosmos to history to death'.[107]

The final passage characterizes the sage as an older narrator who is on the verge of death. In a mark of tribute, the implied author allows Qoheleth to pass over the threshold of death on a generally positive note. As a reader I see nothing negative in his final words. Qoheleth describes old age as it is and always was—there is nothing pessimistic nor unwarranted about his portrayal of how old age decimates our powers to enjoy life. The call to enjoyment, if read in its context, is wise and judicious. All that the reader encounters here is an old(er) sage who cared for his youthful student, and how they might live their life to the full. One could justifiably call him 'my Rabbi'. Just so, the implied author lets our protagonist slip away and buries him in that mausoleum known as the book of Ecclesiastes.

106. Fox, 'Aging and Death in Qohelet 12', p. 66. Fox argues that Qoheleth has appropriated images which are typically utilized by the prophets to describe national disaster, but usurps their emotive powers to express a deeper level of pathos to the reader regarding the finality of everyone's death. Krüger also argues that apocalyptic symbolism has been applied to the expectation of the individual's death. See Krüger, 'Dekonstruktion und Rekonstruktion', pp. 125-29.

107. Perdue, '"I will make a test of pleasure"', p. 209.

13. *Ecclesiastes 12.8-14: A Public Perspective on a Private Figure*

In v. 8 we hear again the voice of the frame-narrator, that is to say the Epilogist. Here we meet the 'signature' of the implied author.[108] By repeating the summary statement of 1.3, the implied author returns the reader to the doorway through which they entered into Qoheleth's consciousness. The use of third-person discourse functions in the epilogue as an external perspective which frames Qoheleth's 'autobiography'. It gives the book a lasting sense of artistic isomorphism and signals to the reader that not only is an external post of observations about to be tendered, but that Qoheleth is no more. The tone of the passage resembles an obituary.[109]

Ecclesiastes 12.9-14 continues the implied tribute begun in 11.7–12.7 in an explicit manner. The implied author breaks frame again, but this time not to ironize his fellow colleague. As a duly commissioned representative of the reading community, his duty is to lend the narrator the community's endorsement[110]. Of course, there is a level of irony in

108. Lavoie, 'Un Eloge à Qohelet', pp. 145-70. However, Fox has argued that the Epilogist too is a fictional character. See Fox, 'Frame-Narrative', pp. 104-105. Lavoie's analysis suggests that the Epilogist's words lead straight to the implied author. At the least, one may say that the voice of this fictive character seems to be more closely aligned with the overall values of the implied author in some strategic ways.

109. Perdue, '"I will make a test of pleasure"', p. 199.

110. In that respect, as Shedd has pointed out, the outer frame provides the reader with a 'last word'. He correctly observes that the book repeats *dābār* ('word', 'matter') in 1.1 and 12.9-14 in order to balance the book aesthetically and to maintain a certain 'distance' from the text's protagonist. See M. Shedd, 'Ecclesiastes from the Outside In', *RTR* 55 (1996), pp. 24-37 (27). However, when he urges that the reader must employ a 'frame-driven hermeneutic' in order to find a unifying style (p. 28), Shedd advocates a reductionistic reading grid. As important as the frame is for this book, it must not be allowed to replace the portrait which it holds. As Christianson's study shows so well, an artistic/literary frame is not meant to replace its contents. Having been an artist myself, I cannot imagine a frame being more important than any work of art whose beauty a frame is supposed to augment, not eclipse. At the very least, one cannot imagine many people visiting the Louvre, and saying, 'look at those frames' while virtually ignoring the masterpieces they border. Shedd's proposal, if taken too far, could result in such readings. His summary of the book 'that we should fear God precisely because life is vain' (p. 33) borders on such a reading. Here is a case where the frame has eclipsed the portrait—something with which no artist would agree. Rather, I propose, the frame and

this fact, since Qoheleth himself could scarcely have imagined the need for such a thing given his emphasis on the primacy of private knowledge. Yet such a position cannot be held by the community, which must weigh and validate all individual contributions to the fund of private knowledge. Verses 9-10 portray a sage who was deemed to be wise by the community, a person who 'taught the people knowledge, and...diligently weighed and tested and arranged proverbs'.[111] Verse 10 bequeaths upon Qoheleth the highest of First Testament honors—he is remembered for being upright (*yōšer*). From the community's perspective, all ironization aside, Qoheleth was diligent and his 'expertise beyond question'.[112] He is depicted as the consummate professional sage who labored hard for the public. This confirms what the book hints at by giving the narrator a name that communicates both a sense of individual identity and public office. Qoheleth is deemed a public servant worthy of the office he held. As T.A. Perry summarizes: 'One of the outstanding successes of Kohelet is to have developed a perspective wherein the Pessimist's ranting and ravings can be viewed as limited and also valid'.[113]

Following this, the narrative focalization in vv. 11-12 zooms away, looking at the office of the sage from a yet more distant post of observation. Verses 11-12 give public approval to the office that Qoheleth held. It admonishes the general community regarding the critical role that such individual insight plays in the search for valid knowledge by the community. The sayings act like 'goads', stimulating much needed criticism, and as 'nails' which plant the community's knowledge upon solid ground. He refers to the 'collections that have been given by the *one shepherd*' (12.11). Presumably, those collections refer to writings by the one shepherd, which the reader assumes refers to Qoheleth and this book given its literary setting.[114] However, the verse is vague and

the portrait must be allowed to dialogue with other, as they were meant to by the artist who created both.

111. Alternatively, the NJB translates the verse as: 'Qoheleth taught the people what he himself knew, having weighed, studied and emended many proverbs'.

112. Crenshaw, *Ecclesiastes*, p. 190.

113. T.A. Perry, *Dialogues with Kohelet*, p. 6.

114. However, Terry argued that the 'one shepherd' referred to God. In that case, the implied author is lending Qoheleth divine and human approval. See Terry, 'Studies in Koheleth', p. 367. It might also cryptically refer to Solomon as well. Christianson notes that 'shepherd' here may have found its prototype in traditions such as 1 Sam. 25.7, 'in which shepherds are likened to Israelite kings, possibly

provides no answer as to its referent. Whoever it referred to historically is now a moot question. This endorsement gives the book a certain status among the authoritative, or perhaps nascent canonical writings of the sages.[115] Verse 12 addresses the narratee/reader with the intimate term, 'my son'. In a strange twist, the implied author cautions the reader of anything beyond these collections.

Verse 12b is taken a bit differently by this study. I prefer to understand it along the lines argued by Mitchell Dahood and Anson Rainey, who on the basis of comparative Semitics, translate it as: 'Of making many accounts there is no end, and much reckoning (checking ledgers) is a weariness to the flesh'.[116] Recalling how the word *ḥišbōnôt* in 7.29 can also mean 'accounts', this would reiterate one final time the criticism of materialism which Qoheleth often admonished the narratee against. The implied author reminds the implied reader one last time of the supreme vanity of wasting a life in the pursuit of money.

The final verses may or may not be original, but in their present literary setting, they still have an effect upon the reader. James Crenshaw objects that the theme of these verses is 'alien to anything that Qohelet has said thus far'.[117] However, because reference to the 'fear of God' has occurred elsewhere in Qoheleth's discourse it should not be considered all that alien (cf. 5.6; 7.18; 8.12-13). Verse 13 returns the reader to *their* reality, a reality wherein divine duties have supreme importance.

presuming Qoheleth's association with royalty' (*A Time to Tell*, p. 146). If that is the case, then we have hear a final reference to the King's Fiction which began in 1.1.

115. Although I think this passage is referring only to general authority issues here, there are readers who see a definite canon consciousness by the 'editor' here. See Wilson, 'The Words of the Wise', pp. 175-92; Sheppard, 'The Epilogue to Qoheleth', pp. 57-73; and *idem*, *Wisdom*. An excellent study which surveys the pros and cons of the canonical reading grid for 12.9-14 is offered by Dell, 'Ecclesiastes as Wisdom'. Childs also notes that the function of this verse is to 'set Koheleth's work into the larger context of other wisdom teachers' (*Introduction*, p. 585). In that regard, the community acknowledges that Qoheleth's private insights need the balancing corrective of the community's other individual insights. Only by balancing different individual's insights does the community come to a public knowledge that is valid. The implied author seems to suggest something of this process in this verse.

116. Dahood, 'Canaanite-Phoenician Influence; Rainey, 'Study of Ecclesiastes', p. 149.

117. Crenshaw, *Ecclesiastes*, p. 192.

The implied author stresses what Qoheleth himself stressed; that the commandments are important (cf. 5.3-5). Although the tone of this passage is admittedly different, this can be attributed to the fact that these verses function to sum up the *authorial reader's* world, not Qoheleth's. It is simply a vehicle by which the reader is refamiliarized with their world, a world that has been totally defamiliarized by Qoheleth's monologue.[118] Verse 14 continues the refamiliarization of the reader by stressing the judgment of God. With this Torah-oriented debriefing of the reader, the book has returned the reader from the land of Qoheleth's 'I', and so, abruptly ends where it started. The reader is then left to ponder the relationship between private insight and public knowledge within the context of covenant obligations. As Eric Christianson observes regarding the effect that frames have on readers: 'A frame compels the reader to assess and evaluate the work at hand. By presenting his assessment, the frame narrator solicits the reader's own, personal assessment.'[119]

To sum up, the epilogue lends the authority and validation of the community to this lonely rebel. It depicts the book's protagonist in a positive fashion, with scarcely any of the irony that so characterizes the implied author's literary strategy during the monologue. Qoheleth is presented as a trusted sage, immaculately professional, and as one who has rendered the community a great service in the discharge of his public office. There is a sense of respect and warmth that marks the implied author's evaluation of Qoheleth as a sage. On that note, the book ends. Qoheleth takes his rightful place among the canonical sages.

118. Sheppard argues that the function of this verse serves 'to direct these comments away from the exclusive concern with Qohelet to a larger context' ('The Words of the Wise', p. 178). Although Sheppard understands a canonical meaning for 'larger context', I would argue that the broader context is the reader's life in the real world as well.

119. Christianson, *A Time to Tell*, p. 119. He also observes, quite correctly, in this regard that 'it is clear that the frame narrator *did not* agree with Qoheleth's approach to wisdom, God and tradition' (p. 119). However, as I have argued, this is due to the dialogical commitments of the book's implied author who created both fictional entities in order to explore the nature of human knowing. In other words, the frame leaves the reader exactly where a sage would have them—*themselves* pondering the nature of life, Wisdom, and the problems involved in ascertaining reliable knowledge. In other words, the differences presented here are heuristic in nature, acting as a further 'goad' to the reader in order to stimulate dialogical thinking.

14. *Summary of Reader Relationships in the Book of Ecclesiastes*

a. *Narrator–Narratee Relations*

Throughout the first four chapters, Qoheleth's relationship with his narratee is a rather uncomplicated one. One might even call it a naive use of the narratee. However, that relationship takes a turn in the second half of the book, where the narrator utilizes a rhetorical strategy whose effect is to test and even to 'stump the student'. Throughout the book, the narratee is characterized by an attitude of questioning. However, one does not perceive a critical questioning on the part of the narratee in terms of the broad social values of wealth and status which characterizes the self-interests of both Qoheleth and his student. In the book of Ecclesiastes the narratee listens and queries traditional Wisdom tenets along with the protagonist, but not the position of self-interest which plagues both the narrator and the narratee. Beginning in 7.1, however, the narratee is bombarded with a strategy of incongruity that simulates the contradictory nature of public and private knowledge. The use of literary blanks and incongruities also tests the narratee's ability to correctly perceive this aspect of human knowledge. In these chapters, the master–student quality of their relationship is evident in the subtle utilization of proverbial incongruity to demonstrate to the narratee the limitations of public knowledge, which surely includes the narratee's own level of proficiency in such knowledge.

b. *Narrator–Implied Reader Relations*

Qoheleth relates to the implied reader much as he does the narratee. In no sense does the narrator of Ecclesiastes ever break frame to communicate directly to the reader. Instead, what we see during the course of his monologue is a sage who directs his words solely to the narratee. In this respect, there is a certain level of emotional and narrative distance between the two posts of observation. Although the focus of the implied author's efforts is to communicate to the implied reader something about the ironic nature of Qoheleth's epistemological stance, the focus of Qoheleth's discourse is to address and to suade the young narratee to follow his advice to pursue the enjoyment of life. Whereas his discourse often has the same effects on the implied reader, that reader is not the addressee of his oration. The implied reader overhears the narrator's monologue in the manner of a disinterested third-party whose values are aligned with the implied author. Following the implied author's lead, the implied reader takes a satiric and ironic stance toward the narrator's discourse. However, the level of intimacy and care that Qoheleth extends to his narratee certainly is offered to the reader as well.

c. *Implied Author–Narrator Relations*

The implied author relates to the narrator as a second-generation scholar who finds room to disagree with his mentor. Although there is a degree of warmth and intimacy between the two, there is also an ironic and even satiric intellectual distance between Qoheleth and the implied author, otherwise known as the frame-narrator or the Epilogist. The implied author utilizes the character Qoheleth to explore the limits of both private and public knowledge. Sometimes, as in the King's Fiction, there is a subtle satiric evaluation of the methods of private insight. However, the protagonist is no mere foil for the implied author's ideological stance. Qoheleth presents fully one side of the epistemological debate that rages in the book of Ecclesiastes. However, the implied author is the ultimate ironist, who is fully capable of ironizing both Qoheleth and his own position in order to show the limits of all human knowledge, both public and private. As a result, there is both warmth, closeness and irony between the two narrative personas who debate the relative values of public and private knowledge.

d. *Implied Author–Narratee Relations*

Since the narratee is so closely aligned with the narrator's post of observation in this book, much of what applies to Qoheleth applies to the narratee as well. Both Qoheleth and the narratee are ironized by the implied author. At 4.17–5.8 in particular the implied author relates to the narratee as an older sage who perceives the inherent deficiencies of the youth's allegiance to his mentor's methodology. Because the implied author looks over the narratee's shoulder much like an established scholar might look over the shoulder of a college student from the back of a classroom, the relationship here possesses a certain sense of distance. His evaluation of the narratee results in a less than positive characterization for the master's apprentice. However, a degree of closeness between the two is evident in places, particularly at 12.12 where the use of 'my son' reveals that a caring relationship does indeed exist between them.

e. *Implied Author–Implied Reader Relations*
The implied author relates to the implied reader in a much more direct and positive manner. In this we perceive that the implied reader is the true focus of the implied author's efforts. Throughout the discourse, the implied author attempts to give the reader an horizon of ironic knowledge so as to give him or her an elevated post of observation from which to view and assess both Qoheleth and the narratee. Throughout the discourse, we see that the implied reader is capable of the type of critical questioning that escapes the narratee. As a result, the implied author whispers in the implied reader's ear in a manner that resembles a sage speaking to a fellow colleague who is listening to Qoheleth for the first time. There is a level of respect between the two which suggests a peer relationship. Although at times the implied author allows the implied reader to struggle with Qoheleth's rhetoric of ambiguity and incongruity along with the narratee, the overall position of the implied author is to relate to the implied reader more as an associate than a professor figure. The constant allocation of ironic knowledge to the implied reader characterizes this post of observation as a more mature person who understands the critical nature of the implied author's treatment of his mentor, Qoheleth.

15. *Summary of the Effects of Reading Relationships in the Book of Ecclesiastes*

This intricate set of relationships creates a very rich text filled with irony and satire. All the major characters one would expect in a fictionalized wisdom debate are here: the old professor (Qoheleth), the middle-aged colleague (the implied author/Epilogist), the debutante student (the narratee), and the third-party colleague or friend who listens in on this debate (the implied reader). In that respect, it is not so different from the debate we encounter in the book of Job, except that the level of interaction between the parties is much less pronounced. Crafting a text that reflects the subtle interactions of its main characters, the implied author has constructed a very witty text that looks at the epistemological problems associated with gaining knowledge with an efficiency and appreciation for wisdom's ironic limitations that can only be characterized as truly introspective.

Whoever composed this text was indeed a wise person possessing an uncanny knack for perceiving the ironic. The attitude of intellectual honesty which so completely characterizes the book is an absolute must for the religious pilgrim who has the necessary cognitive fortitude to become this text's implied reader. Such rare intellectual qualities are the mark of a great religious mind. Not for naught did this sage's book enter the Canon. In his ability to so fully portray the ironies involved in the pursuit of human knowledge and religious wisdom, the implied author has constructed a text that deserves a place alongside other canonical personalities, who, it might be noted, sometimes possess only a fraction of this person's intellectual prowess, spiritual acuity or literary sophistication. The text's overall effect is to create a discourse which 'goads' every generation to look at the limits of knowledge and the ironies of faith. By so doing, the book of Ecclesiastes instructs the reader of the Canon in the problems and prospects involved in the search for that knowledge which can lead us to a fuller understanding of life, God and ourselves.

Chapter 7

VAIN RHETORIC: SOME CONCLUSIONS

A definition of a proverb which Lord John Russell gave one morning at breakfast at Mardock's—'One man's wit, and all men's wisdom'.[1]

1. *The Need for a New Loom*

The purpose of this study was to provide fresh insights into the sundry problems that readers have had confronting the radical I-narration of Qoheleth. This study has consciously attempted to follow a different path from that mapped out by previous scholarly reading grids of the book of Ecclesiastes. In this respect I have tried not to replicate what Santiago Bréton rued over a decade ago when he complained that 'most commentators limit themselves to problems discussed by their predecessors (Barucq depends on Podechard; Hertzberg and Loretz on Delitzsch).[2] Referring to the commentaries that had appeared at that time, he observed that they were actually new editions of older works, and as for the 'truly new ones it is not always evident that they represent new approaches or offer new solutions worth considering'.[3] He concluded that 'the traditional canons of exegesis today represent an inadequate approach, while the new ways of interpretation are still in search of a secure basis'.[4] The need for a new 'loom' has been apparent for some time now. And yet, in spite of that, very little has been offered

1. Lord John Russell, '*Memoirs of Mackintosh', vol. ii.*, p. 473, cited by John Barlett in *Familiar Quotations: A Collection of Passages, Phrases, and Proverbs Traced to Their Sources in Ancient and Modern Literature* (10th Edition; revised and enlarged by Nathan Dole, Boston: Little, Brown, and Co, 1930), p. 1053.

2. Bréton, 'Qohelet: Recent Studies', p. 149.

3. Bréton, 'Qoheleth Studies', p. 22.

4. Bréton, 'Qoheleth Studies', p. 22.

that proceeds from newer methodological perspectives. As Carol Newsom concluded in 1995: 'it is also striking that scholarly work on Ecclesiastes has remained, with very few exceptions, the province of traditional historical criticism'.[5] Writing in 1998, Spangenberg could still count less than ten authors 'who had written studies which reflect some influence of the new paradigm'.[6] However, with the methodological innovations brought to bear on the text by my study and the one by Eric Christianson,[7] that situation is being addressed in a more comprehensive manner.

2. *Vain Rhetoric and the* Sitz im Leser*: Summary of Conclusions Reached*

Bréton stood at the cusp of the current methodological crisis which began in the late 1960s and early 1970s.[8] Since then, the scholarly guild has undertaken an extensive questioning and subsequent overhaul of its methodological moorings, though only lately has that revolution seen application to the book of Ecclesiastes. This study is a result of those paradigm shifts, and attests to the need for new methods such as Bréton so insightfully called for nearly 25 years ago. During that time, we have seen a gradual shift from an emphasis on the *Sitz im Leben* of a text to the *Sitz im Leser*.[9] A similar move can be seen in rhetorical circles as well in its shift from author to audience-oriented approaches. My study is an example of how the reader has gradually gained hegemony over historical concerns, at least for a significant minority of critics like myself.

Chapter 1 of this study was an attempt to document this paradigm shift. The historical background of the Cartesian 'Quest for Certainty' was viewed as a context delimited set of axioms which formed the

5. Newsom, 'Job and Ecclesiastes', p. 184. Schoors reaches a similar conclusion, noting that 'modern literary criticism has had only a limited impact on the exegesis of Qohelet' ('Introduction', in Schoors [ed.], *Qohelet in the Context of Wisdom*, p. 3).

6. Spangenberg, 'A Century of Wrestling', p. 75.

7. Christianson, *A Time to Tell*.

8. For a fuller discussion of the paradigm shift which occurred at this time see Spangenberg, 'A Century of Wrestling', p. 66.

9. For this term, I am indebted to its usage by R. Johnson, 'The Rhetorical Question', p. 123.

epistemological basis of historically oriented biblical scholarship for the past 200 years. Therein I argued for a non-historical set of assumptions based on the hermeneutical theory of Paul Ricoeur, John Ellis and other textuality studies. I argued that the basic configuration of rhetorical, reader-response, narrative and textuality approaches can be considered a post-canonical perspective in that they are a better way of getting at the issues addressed in canon critical circles during the late 1970s and 1980s. These non-historical approaches most closely match the interpretative interests of scholars interested in the reading of sacred texts as Scripture and are thus more appropriate for interrogating biblical texts than historical approaches which inevitably reduce scriptures to documents and artefacts. As a result, I see this study as a second-generation canon-critical contribution to the field.

Chapter 2 summarized narrative and reader-response approaches. Therein I delved into the problem of first-person narration as illuminated by the disciplines of narratology and reader-response criticism respectively. The roles of the implied reader, implied author, narrator and narratee were viewed as posts of observation which the reader must navigate and ultimately synthesize into one *Gestalt*. These were described as abstract posts of observation which offer the reader various ideological stances that must be woven together in order to produce that *Gestalt* called 'the meaning of the text'. Especially important for my study was the concept of the structural isomorphism of the text wherein external and internal frames of reference function together to provide the text with a sense of artistic and ideological balance. Literary texts achieve their rhetorical effects through the artful manipulation of external and internal points of view. In a first-person text such as Ecclesiastes, the interplay between internal and external points of view forms the foundational aesthetic dynamic of the text and has the greatest influence on the final *Gestalt* that the reader constructs regarding the text's overall meaning and significance. The structural isomorphism of the book of Ecclesiastes is achieved through the opposing viewpoints of Qoheleth and the the implied author/frame-narrator/Epilogist. Rhetorically, this results in a debate regarding the sufficiency of private insight and public knowledge. The critical theories of Seymour Chatman, Gerald Prince, Wolfgang Iser, Umberto Eco and Boris Uspensky formed the basis of this discussion.

Furthermore, I surveyed reader-response criticism as an example of a rhetorical analysis of the text. The call for a genuinely pragmatic approach which analyses the suasive properties of a discourse was advocated as a perspective which is both timely and needed. The focus on the reader necessarily entails paying close attention to how the text's implied reader is suaded to make certain aesthetic and ideological choices. The central aim of a reader-oriented approach is to focus on the interaction between text and reader rather than on the text itself or the history behind the text. Reading is therefore seen as a series of cognitive activities which take place through time as the reader traverses the text in a linear fashion. In order to validate their insights, reader critics substantiate their analyses by reading along with other critics. The reader critic utilizes historical studies of texts not so much for what they specifically argue, but for the literary problems which these studies inadvertently testify to during the reading of any given work. Texts are thereby viewed as problems and puzzles rather than doorways to another age. The critic ceases being a mind-reader who tries to ascertain the original author's intention, becoming instead a maestro who helps orchestrate the various cognitive maneuvers required by the text. The emphasis of the critic becomes settled on the aesthetic experience of the reader being confronted by the text's various gaps, blanks, wandering viewpoints and other assorted reading problems. Relying upon Umberto Eco's concept of the text's model reader, the reading process was seen to rely upon general literary competence even as it builds the distinctive literary competency demanded by the specific text at hand. The critical theories of Stanley Fish, Steven Mailloux and Wolfgang Iser formed the foundation for this pragmatic approach.

Finally, I surveyed the specific reading problems generated by first-person texts. It was found that the basic rhetorical liability of a first-person text is the aura of subjectivity they inherently lend to a discourse, and the rhetorical limitations that are placed on a narrator once they take human, bodily form. First-person narration also possesses very capable suasive powers, especially during the initial stages of a text. The openness of the speaker, coupled with his or her increased sense of humanity, often builds a sense of trust in the reader. However, it was also seen that during the process of characterization this can sometimes backfire on an author, especially if the ethos of a character or narrator should turn out to be less than positive. Relying upon the reading theories of James Lee Marra and Uri Margolin, I argued that the principle characterization process involved in first-person texts was the 'fleshing out' of the speaker. Therein the 'I' of the first-person text becomes the gravitational center for the reader's response and utterly dominates the reception of a first-person work.

As a result, the reading contract for all first-person works implies the rhetorical limitations of the speaker. Nevertheless, they also have stellar strengths, most notably being their ability to simulate a personal relationship and to build a bridge of trust with the reader. A first-person text therefore possesses both intrinsic liabilities and assets. These liabilities and assets interact in different ways given the basic characterization/ethos-related assessment of the character by the reader. Only by paying close attention to how this gravitational center affects the reader's reception of the text does the reader-critic come to understand the rhetorical powers and properties of any given first-person text. In that respect, ethos is a confounding variable or influence in the rhetorical assessment of a first-person text. Working with rhetorical and reader-response methods construed along Ricoeurian lines, this study has endeavored to look at the specific problem of first-person narration in the book of Ecclesiastes (and the various problems associated with it) with a methodology that would limit the reading-grid problems of past generations who worked principally with historical and referential models of exegesis. Commentaries, monographs and various articles were consulted in order to track the specific literary problems that readers have experienced in the book, and to document the responses elicited by the text as well as the sundry solutions which have been offered by the text's reading community.

Chapter 3 isolated the reader problems at the textual level of the discourse. It analyzed the various linguistic and structural problems as an example of a rhetoric of ambiguity. Specifically, the structuring properties of Qoheleth's 'I' were proposed as the key to understanding how the reader construes the book's literary coherence. Although it was accepted that Addison Wright's logical analysis has the greatest claim to intersubjective validation, it was also argued for a 'both-and' paradigm when approaching Qoheleth's discourse. The role of the various key words, the impact of Qoheleth's observations and the role of the narratee as evoked in the seven-fold call to enjoyment were argued as having the greatest impact on the reader's cognitive structuring of the text.

Chapter 4 looked at the problems relating to persona issues and the various characterization techniques utilized by the discourse: the book's relationship to autobiography, the nature and effects of the King's/Solomonic Fiction, the specific ways that readers build a sense of a character's ethos, the understanding and use of Qoheleth's quotations in a monologic setting and the book's use of third-person narrational techniques. Most notably, I argued for a fictive understanding of the character, Qoheleth. Through the use of fiction, the implied author attempted to recontextualize the Wisdom tradition back into the experience of the solitary individual. Qoheleth's use of 'quotations' thereby become examples of reminiscences spoken within the framework of an interior monologue. This serves to strip the proverbs of their gnomic powers, reducing them to yet another instance first-person discourse. Subsequently, it was noted that the use of third-person narration created an ironic dimension regarding the protagonist's reliance upon private knowledge as the sole means of achieving wisdom. The fact that Qoheleth needed so desperately the validating response of the greater community imbues the discourse with an aura of unstable irony. The confirmation provided by the implied author's use of public knowledge bolsters the protagonist's rhetorical standing vis-à-vis the reader. Obviously, the need for public confirmation/validation by a discourse which so heavily depends on private knowledge thoroughly ironizes the ironist who spoke it. As a result, I would view the ironic relationship between private insight and public knowledge as the foundational element for understanding the text's total rhetorical impact on the reader.

This ironic dimension is achieved through the subtle manipulation of first- and third-person narration by the implied author who stood at a considerable ideological distance from the protagonist. Although on an emotional level the implied author was quite close to Qoheleth, on an ideological level the implied author recognized the rhetorical weaknesses of Qoheleth's ethos as well as the epistemological implications of his empirical approach for the acquisition of knowledge and wisdom. For the implied author, such a radically 'I'-centered epistemology needed the balancing corrective of the reading community's public knowledge before it could be considered a valid rhetorical contribution to the community's fund of truth and knowledge. However, it was also argued that the placement of the Epilogist's frame-narrative in an 'I'-discourse ultimately reduces its use of third-person discourse to yet another example of saying 'I'. This subtle deconstruction of third-person narration by Qoheleth's radical 'I' creates a sense of unstable irony which thoroughly permeates the book. Relying upon Chaim Perelman's concept of the universal audience, I stressed the need for a broader validation of this book than that supplied by the limited community which originally verified Qoheleth's discourse. This sets in motion an epistemological spiral in which the modern reading community is asked to validate the book in a never-ending helix of confirmation and contestation whose twin axes are private and public knowledge. As a result, the reader is asked to play the role of the Epilogist for *both* Qoheleth and the Epilogist. This insight was viewed as a function of the text's surplus of meaning and is considered a valid insight into the text's general overall rhetorical effect on the reader. This is possible from a Ricoeurian perspective because an emphasis on textuality induces the critic to value the text's effects on the reader over the original author's intentions.

Chapters 5 and 6 tracked the specific characterization of Qoheleth in a linear fashion, focusing on the development of the character's ethos, its rhetorical effects on the reader (both negative and positive) and the presentation of public and private knowledge by the discourse. Chapter 5 focused on the ethos of pessimism, while Chapter 6 accented the subversive properties of Qoheleth's dialogic treatment of the Wisdom tradition. Chapter 6 analyzed the ethos of skepticism which permeates the later chapters of Ecclesiastes. While pessimism was viewed as an assessment of the relative value of good and evil in the world, skepticism is seen as an epistemological stance which is much more radical than simple pessimism. By taking such an agnostic stance toward the possibilities of knowledge, Qoheleth characterizes himself as a 'subversive sage', to borrow a term from Alyce McKenzie. The move from pessimism to skepticism further exacerbates Qoheleth's rhetorical standing with the reader and ultimately results in an ambivalent response on the part of the reader. In the movement from pessimism to skepticism it was seen that epistemology is a major theme of the book. This sets the reader up to accept the rhetorical role of the Epilogist who supplements the private insights of empirical observations and methods with the much needed public knowledge of the larger reading community and its fund of tradition-based knowledge.

The general conclusion of Chapters 5 and 6 was that the implied author characterized the narrator as having a rhetorical strategy with powerful effects, but whose final persuasive abilities were considerably mixed. On occasion it was noted how the narrator is characterized in a satiric or ironic fashion by the implied author. Specifically, I argued that the protagonist's lack of generosity, magnificence and magnanimity resulted in a loss of attractiveness and credibility. For most readers the radically self-centered ethic and deep-seated pessimism/skepticism of the narrator results in an additional loss of attractiveness. In some ways Qoheleth's monologue reminds me of a cartoon-strip I saw in *The New Yorker* while writing the dissertation upon which this study is based. It had a man knocking on a door, holding a survey, with the apartment's occupant looking incredulously at the surveyor, who was asking: 'Next question. I believe that life is a constant striving for balance, requiring frequent tradeoffs between morality and necessity, within a cyclic pattern of joy and sadness, forging a trail of bittersweet memories until one slips, inevitably, into the jaws of death. Agree or disagree?' The reader of the book of Ecclesiastes is in some sense asked a similar question by the overall discourse of the text. The result is often a sense of incredulousness that life can be reduced to such nihilistic alternatives. Again, it is possible to see the tale-tell rhetorical trademark of the book, and its penchant for creating ambivalent responses in the reader.

Through my linear reading of the book, I have consciously attempted to generate a truly new approach to the rhetorical crisis that exists within the book's reception history. In that goal I hope to have succeeded in some measurable way. Only the reception of this work by my peers over the next decade will answer whether it has succeeded or not. Like all rhetorical works, it will need the validating responses of the broader scholarly community to intersubjectively authenticate it. At the least, if it has not succeeded in generating a valid new reading, I hope to have explained some of the underlying textual and cognitive problems that have generated the issues that have divided readers over the centuries. If that has been satisfactorily achieved, then this study will have served a useful purpose for the scholarly reading community. As Stanley Fish has astutely observed, the goal of a reader-oriented approach is not always to create new readings, but sometimes its purpose is to elucidate the problems upon which all readers can agree.

3. *Vain Rhetoric: The Rhetorical Backlash of Unabated Subjectivity*

Specifically, I have attempted to show how the decision to anchor the book's persuasive abilities in the powers and deficiencies of first-person narration has been the major rhetorical feature to which readers have responded over the generations. David Goldknopf summarized the rhetorical problems of first-person discourse quite adeptly, noting how the implied author who predominantly utilizes an 'I'-narrator 'deliberately goes forth to battle with one hand tied behind his back'.[10] Based on such insights, I have suggested that the suasion problem which readers have consistently encountered in this book is not due to an underlying historical crisis or psycho-personal dynamic, but rather a literary problem that is endemic and inherent to all first-person discourses, regardless of their historical setting. First-person discourse always communicates a sense of subjectivity to the reader. When utilized too extensively this can backfire on an author or rhetor. As a literary problem attached to the intrinsic possibilities and liabilities attendant upon of the speaker's use of 'I', the book's difficulties are first and foremost a synchronic problem,[11] with diachronic issues supplying various complications of a problem which is not essentially anchored in any historical, cultural or personal matrix.

As a result, the situation lamented by Bréton can be located in the use of inappropriate methods to approach the book's rhetorical problems. Historical critics attempted to solve the book's reading problems with a diachronic method that was wholly unable to address the synchronic dimension which generates the book's basic characteristics and rhetorical properties. I have therefore argued that synchronic methods such as rhetorical and reader-response approaches are much better suited for analyzing the book's sundry problems than those utilized by previous generations of scholarly readers.

To sum up, I have argued that the major reading problem in the book of Ecclesiastes is located in the implied author's decision to anchor the book's rhetorical properties almost exclusively in the powers and liabilities of first-person narration, and that it was that literary decision

10. Goldknopf, 'The Confessional Increment', p. 19.

11. For an interesting and insightful discussion of the meaning of diachronic and synchronic methods in biblical studies, the reader is referred to the article by D.J.A. Clines, 'Beyond Synchronic/Diachronic?', in J.C. de Moor (ed.), *Synchronic or Diachronic? A Debate on Method in Old Testament Exegesis* (New York: E.J. Brill, 1995), pp. 52-71. He argues that the terms are often used in a metaphorical sense by guild members, that most critical methods combine the two in some sense (such as some archeological excavations have been known to do), and finally asks whether the two perspectives stand in need of 'deconstructing' and should not be considered binary opposites. Such a position is helpful, as there is a large amount of truth in holding such a 'both-and' paradigm when confronting postmodern methods.

which has intensely affected its reception by the book's reading community. Literary studies of first-person narration suggest that any rhetorical strategy which uses this method will have a 'double-edged' effect on the reader. First-person discourse can be powerfully persuasive. On the other hand, the use of first-person narration can function to dissuade the implied reader in subtle ways. It is the thesis of this study that the book of Ecclesiastes, as an example of first-person discourse, stands in a long line of examples ranging from ancient to modern times that have generated ambivalent responses in their readers. As such, Qoheleth has utilized a 'vain rhetoric' which produces both acceptance and suspicion towards the major positions argued by the author.

While historical-critical based studies have analyzed the form-critical problem of Qoheleth's 'observations' and discussed the prominence of his use of 'I', none have attempted a comprehensive look at the problem of first-person rhetoric in a scriptural reading context. This study offers such an analysis. As Meir Sternberg has so ably documented, the scriptural reading contract is usually predicated upon the powers of third-person narration and its abilities to simulate divine omniscience.[12] In many respects a first-person rhetoric undermines this contract, with the ultimate effect that the willingness of the reader to 'believe' the text is compromised and, in some cases, derailed. To my knowledge, this is the first study to attempt such an analysis, and offers the scholarly community a compendium of resources for looking at the problem of first-person narration in the biblical text with a new lens. It is my hope that the carefulness with which I documented my sojourn in the land of 'I', as well as the rhetoric and comparative literature departments at a major American university, will be a resource for other critics who would like to tread this path. If this study can act as a resource for reader-response and rhetorical approaches to any other biblical text, as well as the general problem of first-person narration in the Canon, then the study of Qoheleth's specific discourse properties will have served the greater purposes I envisioned while writing this work.

4. *What Do We Mean by a Vain Rhetoric?*

a. *Vain Rhetoric: Begging the Reader to Disagree*

This study has come to several conclusions regarding Qoheleth's use of first-person discourse, which I have termed a 'vain rhetoric'. First, the

12. Sternberg, *The Poetics of Biblical Narrative*, pp. 84-128.

use of first-person narration is a vain rhetoric in the sense that it is the nature of all I-discourses to imply their own limitations, and therefore, to invite dialogic dissension with their major premises and conclusions. They are vain in that the one prevalent effect of the use of 'I' is to generate an argumentative stance in the reader. Typically, 'I' begs to be disagreed with and does not consistently create rhetorical consensus between speaker and audience. This is not a situation that is maximally conducive to persuasion. Gerhard von Rad has eloquently drawn attention to the spirit of Qoheleth's vain rhetoric and the resulting dialogue between Qoheleth and the Epilogist/implied author regarding the adequacy of private experience for generating a valid public knowledge. He cautions:

> Anyone who has listened carefully to Koheleth's dialogue with the traditional doctrines should not find it quite so easy to give one-sided approval to the lonely rebel. He will, rather, be deeply preoccupied with the problem of experience to which both partners in the dialogue urgently referred and yet arrived at such different observations. He will realize how narrowly tied man is as he moves within the circle of experiences which is offered from time to time by his understanding of the world.[13]

The litany of readers who have quarreled with Qoheleth's quarrel is ubiquitous in the literature. His reduction of reliable knowledge to the confines of his own personal experience is unpalatable and untenable to most readers. It is not just that there are 'no controls' on the limits of his generalizations as T.A. Perry has lamented,[14] nor that he deducts from too few examples as Michael Fox has observed.[15] It is the funda-

13. Von Rad, *Wisdom in Israel*, p. 235.

14. T.A. Perry, *Dialogues With Kohelet*, p. 34.

15. Fox, 'Qoheleth's Epistemology', p. 145. See also Fox, *Qoheleth and His Contradictions*, pp. 32-37 and 142-46. Fox states: 'When Qohelet considers life he sees it colored by the exceptions rather than the rule. It is a matter of weighing premises. One person might infer from the fact that most babies are born healthy that God is beneficent and life is orderly and meaningful. This is the temperament of the other Wisdom writers—the author of Job included. Another might infer from this fact that babies are occasionally born schizophrenic that God's ways are arbitrary. This is the religious temperament of Qohelet... For Qohelet, the absoluteness of God's control means that each individual case is an ethical microcosm, so that the local absurdities—and there are many—are irreducible. Qohelet generalized from them no less than from acts of divine justice. As a result, no matter how much right order we see, the absurdities undermine the coherence of the entire system' (*Qoheleth and His Contradictions*, p. 143).

mentally private nature of the knowledge advocated by Qoheleth to which most readers object as a basis for public knowledge. The pitfall of such an approach is acknowledged by Harold Fisch, who argues that 'Ecclesiastes shows us what happens when man withdraws into the inwardness of his own consciousness'.[16] With readers like von Rad and Perry, we can only ask whether private experience can be the 'sole and complete basis for wisdom'.[17]

Nevertheless, first-person narration is a powerful technique that, when it works, is extremely effective—more suasive even than third-person discourse. The basic problem is that it fails to be persuasive just as often, if not more often than it succeeds. Still, it must be admitted that this is not always the case. Some utilizations of I-discourse are very suasive. But given a less than stellar ethos on the part of the speaker, this becomes exponentially more and more unreliable as a means to positively influence the reader in a suasive fashion. Putting all of one's rhetorical eggs in such a volatile basket is a vanity in and of itself. While occasionally one hears a sermon based on first-person rhetoric that is surrealistically persuasive, such as Martin Luther King's 'I Had a Dream' sermon, the all-too-common course for many lesser examples of the I-discourse is for the audience to dismiss such testimony as too subjective for private or communal use. This can be seen in the back-lash against the 'personal testimony' of the TV evangelists during the 1980s. Once the greed and corruption of these rhetors was exposed, the power of their message was soon compromised, with a resulting loss of attractiveness for Christianity in general by the viewing public.

16. Fisch, *Poetry with a Purpose*, p. 158.

17. T.A. Perry, *Dialogues with Kohelet*, p. 35. Von Rad, *Wisdom in Israel*, p. 227, astutely observes how 'experience presupposes a prior knowledge of myself; indeed it can become experience only if I can fit it into the existing context of my experience of myself and the world' (p. 3). Terry made a similar evaluation of Qoheleth's argumentative strategy. See Terry, 'Studies in Koheleth', pp. 370-75. Qoheleth failed to see that the problem with 'experience' is its utter dependence on the self which articulates it. Knowledge, in the sense that Qoheleth craved, is simply not available based on the limited epistemological platform of the self. In fact, von Rad goes on to assert that to 'one who is secure in a fundamental position of faith, events can appear differently from what they do to one who is assailed by doubt. One must indeed go further and say that they not only "appear" different, they are and even become different' (p. 236). Crenshaw argues with the narrator because of the egocentricity of his evaluations which are based solely on his own personal 'safety and comfort' (*Ecclesiastes*, p. 25).

Sometimes 'I experienced' is the most powerful rhetorical technique a rhetor can use. This is particularly true if the person has abundant good ethos and the audience agrees with the speaker's premises, experiences and the deductions he or she makes from them. But if any of those three factors slips the rhetorical use of 'I' becomes dicey to say the least. Unfortunately, this is the case in an emphatic way with Qoheleth, whose ethos vacillates from one possessing good ethos to a rhetor plagued by bad ethos in a constantly dissuasive manner. One moment the narrator is expressing insights that only the dishonest voice of a pseudo-orthodoxy can dismiss. The next, he is making unwarranted deductions that are jaded, misogynist, self-centered or simply devoid of the characteristics of magnificence, generosity or magnanimity. To put it succinctly, *Qoheleth suffers a great deal when it comes to attractiveness, performs averagely in the credibility category, and generally succeeds in the trustworthiness department*. This inconsistent and conflicting configuration of rhetorical characteristics is what lies at the heart of the book's vain rhetoric.

Still, it is the general trustworthiness of the narrator, generated by his honesty and openness, that has compelled the reading community to treasure the book. Qoheleth may be jaded, sceptical, pessimistic, and a host of other things, but one thing he excels in is generating a sense of intimacy his monologue generates in the reader. There is an openness and honesty to the character, coupled with a secular kind of prudence, which endears his soul to the reader who has the requisite experiences to become the text's implied reader, that is, one who has weathered the dark side of life. Perhaps there is no greater tribute to the rhetorical persuasiveness and power of Qoheleth's open and honest self-disclosure than the commendation offered by Rabbi Robert Gordis:

> Whoever has dreamt great dreams in his youth and seen the vision flee, or has loved and lost, or has beaten bare handed at the fortress of injustice and come back bleeding and broken, has passed Koheleth's door, and tarried awhile beneath the shadow of his roof.[18]

How does the critic describe such empathic tributes and still account for the massive criticism of the character and the book throughout its reading history, without calling it a vain rhetoric?

This is the pitfall one always encounters whenever a decision is made to employ first-person rhetoric in a comprehensive manner as the

18. Gordis, *Koheleth*, p. 3.

implied author of this text has chosen to do. It is simply the nature of the beast, so to speak, and cannot be avoided. As a result, I conclude that it is the book's radical dependency upon I-discourse that has generated the mixed reception which lead to such a stormy passage into the Canon. In that respect, the book's foundational problem is a literary problem pure and simple. *The suasion problem encountered in the book of Ecclesiastes is a consequence of the inherent powers and liabilities of first-person discourse as a generic literary and rhetorical discourse strategy*. More specifically, it is a characterization problem for the narrator who lacks those traits needed to supply him with the necessary ethos to effectively suade the reader. To my knowledge, no one has argued this position in the entire reading history of the book. All have responded to the book's overall rhetorical strategy, although some have noted the presence of 'I', or even commented on the inadequacies of the narrator's ethos or character. But the idea that the problem lies in first-person narration per se has not been addressed by the reading community. Nevertheless, it has had a powerful subconscious influence upon the various readings of the text.

One can only surmise how such an insight might benefit other such canonical examples. The Pauline writings stand out as a noteworthy area for future study (cf. 1 Cor. 7.10; 'I' occurs a staggering 208 times in 1 Cor. alone!). For Paul as well, the precise rhetorical nuances of the 'I'-saturation of the discourse has never been adequately discussed from a modern literary or rhetorical perspective. The Psalms would present yet another fruitful field of exploration, as well as those 'confessions' of Jeremiah. Ezra too would make for good rhetorical analysis. The 'I am' speeches of the Johannine Jesus would be quite interesting. The Bible is filled with 'thus says the Lord' speeches and other examples of third-person narration. But what is the effect of the numerous places where saying 'I' dominates the discourse strategy of a given text in a canonical/intertextual setting which typically predicates its rhetorical existence upon the abilities of third-person narration to simulate divine omnisciency? To me, that is an unploughed field for the biblical rhetorical critic.

b. *Vain Rhetoric: Emphasizing the Vanity of Human Rhetorical Existence*

The book of Ecclesiastes utilizes a vain rhetoric in a second sense to enact a lively debate on the adequacy of private experience as a means of achieving public knowledge worthy of scriptural or religious consideration. This is an outgrowth of the essential limits of first-person narration. An important insight afforded by modern literary theory's distinction between the implied author and the narrator of Ecclesiastes is that there exists an ironic interaction between private insight and public knowledge in the book. At a purely narrational level, there seems to be a subtle epistemological debate between the implied author/Epilogist and the character Qoheleth on what constitutes valid rhetorical/public knowledge, that is, wisdom. This should not surprise us, given the numerous times that the root *yāda'* occurs in the book (especially the latter half), the dominance of the rhetorical question throughout Qoheleth's discourse, and the role that epistemology seems to play in the book.[19] Whether this was intended or is simply due to the surplus of meaning that is inherent in all literary texts is irrelevant. What matters is that there is an ironic effect generated by the relationship between the two primary textual agents at this level. Perhaps there was some sort of specific trend among Israel's sages at the time of the composition of the book that looked more to the 'I' of the observer to validate Wisdom's tenets, as Peter Höffken has argued.[20] Maybe the implied author did have an inkling of what he was doing rhetorically. And then again, perhaps he did not and all of this is the gift of textuality and the surplus of meaning which resides in literary texts. Regardless, what we do know is that while the specifics behind the text are quite opaque, the rhetorical situation that is 'in front of the text' is quite clear.[21]

19. Fox, 'Qoheleth's Epistemology'.

20. Höffken, 'Das Ego des Weisen', pp. 121-35. Personally, my scholarly intuition suggests that the growth of individualism which accompanied the influence of Hellenism on post-exilic Judaism was probably a contributor here. I merely want to suggest this as a possible underlying factor for the book's present rhetorical shape. However, proving such a tenet awaits a study whose interests are substantially different than those espoused here.

21. The concept of a rhetorical situation 'in front of the text' is not foreign to rhetorical critics, who have traditionally defined that situation in historical terms. Branham and Pierce have reinterpreted the meaning of the 'rhetorical situation' from a Ricoeurian and reader-response perspective (following, most notably, S. Fish). Observing the importance of interpretive communities in the construction of texts, they argue that not all texts should 'fit' their contexts, but rather, some must reconstruct the rhetorical situation in order to speak to it. See R. Branham and W. Pierce, 'Between Text and Context: Toward a Rhetoric of Contextual Recon-

There is a definite ideological distance between Qoheleth and the implied author regarding the sufficiency of private knowledge for public consumption. At that level, the effect of Qoheleth's utilization of a vain rhetoric as a discourse strategy functions as its own corrective from the perspective of the book's implied author. Not only does Qoheleth's discourse display the weaknesses of any first-person discourse, but the implied author has consciously, or perhaps subliminally exploited this weakness, with the effect that it educates the reader regarding the broader epistemological issues involved in the pursuit of wisdom. To push the issue even further, the overall rhetorical effect of the text is to broach the quintessential question: 'What constitutes valid religious knowledge?' Is it located in the experiencing self, as postmodernism would have it, and Qoheleth as well? Or is the Epilogist correct? Is it to be found in the broad-based collective experiences of the human community, however localized that may be for the individual reader? Is it generated in the interaction between private insight and public knowledge? My reading of the book of Ecclesiastes suggests that the latter is its implied answer to the questions raised by Qoheleth's radical centering of knowledge in the private experiences of the individual.

Qoheleth's 'I' therefore serves to sum up not only a literary character, but also functions as an index to a much larger human problem—the problem of how to integrate individual experience into the broader experiences of the human religious community. On the narrational level of the text, a vain rhetoric functions to criticize the specific message and rhetorical means of the narrator (who ironically attempted to criticize the specific tenets of wisdom himself at the surface level of the

struction', *QJS* 71 (1985), pp. 19-36. They redefine rhetorical context to mean 'the perception of it (context) by various interpretive communities, not the features of the historical situation in which it occurs' (p. 20). Due to the influence of textuality, the rhetorical situation of a text can 'surpass' the original rhetorical situation of a text. A Ricoeurian perspective on rhetorical context emphasizes the importance of the world in front of the text, and therefore, the rhetorical situation that is being projected by the text as a subcomponent of the text's projected world. Context is therefore reinterpreted to refer to the world of the interpretive community and that generated by the literary dynamics of the text (p. 21). See also L. Bitzer, 'The Rhetorical Situation', *PR* 1 (1968), pp. 1-14; A. Brinton, 'Situation in the Theory of Rhetoric', *PR* 14 (1981), pp. 234-48; S. Consigny, 'Rhetoric and its Situations', *PR* 7 (1974), pp. 175-76; W. Ong, 'The Writer's Audience is Always a Fiction', *PMLA* 90 (1975), pp. 9-21; R. Vatz, 'The Myth of the Rhetorical Situation', *PR* 6 (1973), pp. 154-61.

text). Therein the ironist is thoroughly ironized by the externally focalized frame-narrative which surrounds his discourse. At a deeper level, a vain rhetoric acts as an open debate regarding what constitutes valid religious knowledge as it relates to both the individual and the community. The 'I' of Qoheleth and the Epilogist are in reality mere symbols for this broader rhetorical problem which plagues all human attempts to speak for God. Qoheleth symbolizes private knowledge while the Epilogist metonymically substitutes for public knowledge. In Qoheleth's discourse we all experience the fundamental rhetorical vanity of the human religious situation. Each of us struggles with the broad-based claims of our own unique experiences and those of the scriptural, or perhaps, human community. The interaction of these creates a never-ending rhetorical and epistemological spiral which is the deep-level message of Qoheleth's vain rhetoric.

c. *Vain Rhetoric: Illocutionary Speech-Acts that Literarily Re-Enact Life's Absurdity*

Finally, by characterizing the text's rhetoric as a vain rhetoric, I hint at a subtle effect of Qoheleth's extensive use of a rhetoric of ambiguity. Through the constant use of strategies of indirection, the implied author has constructed a text which constantly frustrates the reader, and ultimately, allows the reader no closed *Gestalten* or sure answers. It often leaves the reader in a state of perplexity, confusion or indecision. By so doing, the implied author has consciously constructed a text which would recreate the same sense of *hebel* at a literary level which he experienced in real life. The 'Riddle of the Sphinx' is merely a means of recreating in the reader the iterative experience of life's existential conundrums. Vain rhetoric therefore describes the abiding literary experience of reading the book of Ecclesiastes in a performative sense. The illocutionary force of the implied author's various gapping techniques and strategies of indirection is to recreate in the reader life's penchant for absurdity and ambiguity. As such, vain rhetoric is a powerful technique which allows the reader to experience in a narrative fashion something of the absurdist's primal experience of life. When language goes on vacation, as it does when one attempts to express the absurd, the writer is often left to other indirect, or perhaps, non-cognitive means to express what fills his or her heart. Sometimes, this can only be done obliquely, through the utilization of techniques which mimetically simulate life's darker side. The implied author of Ecclesiastes knew this, compensating for the inability of language to say what he meant by finding a way to communicate that primal experience through literary gapping, blanking and opacity. In that regard, vain rhetoric is a performative concept as well. It functions at the illocutionary level of speech-act theory. It's chief effect is to provide the reader with a narrative experience of life's absurdity.[22]

5. *Three Levels of Vain Rhetoric in the Book of Ecclesiastes*

To sum up, vain rhetoric implies three levels of operation. First, on the surface level, it describes the persuasive and dissuasive properties of the narrator's discourse as a function of his own peculiar characterization and subsequent ethos-related attributes. Second, at the text's deep level, it describes how first-person discourse enables the reader to become aware of the general problem of their own rhetorical existence as it relates to communally-based rhetorical systems such as the Scriptures. All knowledge, both individual and communal, has specific limitations. The debate between Qoheleth and the Epilogist illuminates, or perhaps, hints at that greater issue. Third, at the performative/illocutionary level of the text's use of language, it describes the general effects of the implied author's use of a rhetoric of ambiguity to generate a literary experience which partially escapes language's inability to precisely elucidate the absurd dimension of life. In terms of speech-act theory, a vain rhetoric accomplishes at the illocutionary level what language can only vaguely hint at on the locutionary level. What language cannot adequately express, a vain rhetoric can communicate by re-enacting for the reader the narrative experience of life's essential ambiguities, ironies and absurdities. Besides, absurdity is a linguistic commodity that is best left experienced and can never be delineated in any comprehensive way through language. The book of Ecclesiastes,

22. S. Crites, 'The Narrative Quality of Experience', *JAAR* 39 (1971), pp. 291-311. He observes how narrative is one of the essential ways that life is organized from the inchoate mass of experiences which threaten to overwhelm the individual or society (p. 294). The function of a literary text is to reorganize experiences into a meaningful *Gestalt* through the artful manipulation of life's essentially linear qualities. As a result, Crites argues that 'experience is moulded, root and branch, by narrative forms' (p. 308). From this perspective, the implied author re-enacted life's fundamental absurdity, thereby defamiliarizing the reader's essential experience of the absurd.

with its abundant use of rhetorical questions, constant gapping techniques and other strategies from the arsenal of ambiguity is a stunning testimony to the power of the various strategies of indirection to communicate to the reader something of his or her own rhetorical liabilities and limitations.

6. *Qoheleth's Ethos as Mediator Between the Logos and Pathos Dimensions of the Text*

The book of Ecclesiastes primarily accomplishes these general effects not at an intellectual level (logos), but at an emotional level (pathos). The mediator between the logos and the pathos dimensions of the text is the ethos of the narrator. Via the peculiar ethos of Qoheleth, the reader comes to both experience on an emotional level, and to articulate on an intellectual level, something of life's inherent absurdities at both the existential and rhetorical levels. That is the gift of Qoheleth's ethos—it is a doorway through which the reader comes into contact with life's existential absurdities and one's own rhetorical and epistemological limitations.

7. *The Rhetorical Mirror: Qoheleth and the Postmodern Experience*

Ultimately, the book makes us conscious of the our common rhetorical absurdity which is due to the epistemological weaknesses of our species. Humanity is in essence a collection of separate individuals who live, die and 'know' in community, yet who are trapped in the confines of their solitary existences. Rhetorically, this creates a surd which cannot be easily dismissed by the reader who would aspire to answer the question which dominates the latter half of this book: 'How does humanity know?' The implied author of our text implies epistemological issues that have broad philosophical significance in a rather naive and unsophisticated fashion. Again, we see the book's penchant for raising issues which have no sure answers. And that, in the long run, is the quintessential effect of a vain rhetoric. By implying the rhetorical weaknesses of Qoheleth, the implied author, and humanity in general, the book functions in the Canon as a standing witness to the overarching necessity of approaching life and the transcendental order with an attitude of humility and openness. In this respect, Qoheleth's discourse becomes a rhetorical mirror for the postmodern reader who sees something of himself or herself in the protagonist's epistemological use of the self.

It seems appropriate to end this study with an insightful passage from Lloyd Bitzer. Regarding the necessity for public knowledge in a post-modern society, he argues:

> Man [*sic*] alone, so far as we know, has the capacity to find and create truths which can serve as constituents of the art of life. But there are obstacles and tendencies which thwart generation and recognition of this knowledge. The first is limited individual existence. As individuals we are granted but a short and precious span of existence, hardly enough to acquire the wisdom we need... The second is the countervailing forces of false opinion, poverty of sentiment, and bleak physical conditions. The third is our tendency to yield to the claims of present circumstances, needs, and desires which, while valid in themselves, distract us from enduring truths and separate us from the wisdom of tradition. The fourth is our habit of regarding as true—as knowledge—those propositions which issue from accepted scientific procedures of investigation and confirmation. As a result, principles of moral conduct and maxims of political and social life—indeed all of humane wisdom that may guide civilization—have been regarded as opinion found wanting then put to tests of confirmation. A fifth and related cause is the widespread current belief that truth is to be found in *this* slice of time—in the here and now, in this set of experiences, during this present inquiry, this year. We seem unable or unwilling to acknowledge that some truths are not to be found in these kinds of time frames, but rather *become*, over time, and perhaps pass in and out of existence. Why should we not acknowledge that some truths exist as faint rays of light, perceived perhaps dimly in a near-forgotten past, but which light up again and again in the experience of generations? Finally, our general suspicion of tradition cuts us off from a rich fund of knowledge. We lose important wisdom which ought to be brought into the present where it may enrich culture and assist the resolution of problems... The great task of rhetorical theory and criticism, then, is to uncover and make available the public knowledge needed in our time and to give body and voice to the universal public.[23]

Perhaps, more than we know, the book of Ecclesiastes is the most timely of biblical books for a postmodern consciousness. In its subtle dealings with private and public knowledge, Qoheleth and the Epilogist debate an issue which is ever rhetorical, always timely and much needed for our generation. At the least, it is surely more than a mere theological 'note' for our time.[24]

23. Bitzer, 'Rhetoric and Public Knowledge', p. 92.

24. Jasper once minimized the importance of Ecclesiastes by referring to it as a mere 'note' ('Ecclesiastes').

APPENDIX

WISDOM REFLECTIONS (PUBLIC KNOWLEDGE) IN THE BOOK OF ECCLESIASTES

Adapted from J.J. Spangenberg, 'Quotations in Ecclesiastes: An Appraisal', *OTE* 4 (1991), pp. 19-35. Supplemented with R. Johnson, 'A Form Critical Analysis of the Sayings in the Book of Ecclesiastes (PhD dissertation; Atlanta, GA: Emory University, 1973) and A. McKenzie, 'Subversive Sages: Preaching on Proverbial Wisdom in Proverbs, Qohelet and the Synoptic Jesus through the Reader Response Theory of Wolfgang Iser' (PhD dissertation; Princeton, NJ: Princeton Theological Seminary, 1994).

Gordis (1939)	Whybray (1981)	Michel (1979)	Von Löwenclau (1986)	Johnson (1973)	McKenzie (1994)
	1.4(?)				
	1.8b(?)				
	1.15(?)		1.15	1.15 (P)	1.15 (2)
	1.18(?)		1.18 (P)		1.18
2.13					2.13
2.14	2.14a	2.14	2.14a	2.14a (M)	2.14
					2.16 (IP)
					2.24
					2.25 (IP)
					2.26
					3.1
					3.12
				3.20b (P)	
					3.21 (IP)
					3.22
					3.22 (IP)
					4.2
					4.3
4.5	4.5		4.5	4.5 (P) (M)	4.5
4.6	4.6		4.6	4.6 (M)	4.6
[4.8]					
4.9	4.9b			4.9 (M)	4.9
	4.11(?)				4.11

Gordis (1939)	Whybray (1981)	Michel (1989)	Von Löwenclau (1986)	Johnson (1973)	McKenzie (1994)
4.12				4.12 (P)	4.12
	4.13(?)			4.13 (M)	4.13
	4.17(?)			4.17b (M)	
5.1					5.1
5.2	5.2(?)			5.2 (M)	
					5.3
				5.6a (M)	
			5.7		5.7
			5.8		
5.9	5.9a(?)			5.9 (M)	
				5.10 (P)	
				5.11 (P)	5.11 (IP)
					5.12
	6.7(?)		6.7	6.7 (P)	
					6.8 (IP)
	6.9a(?)			6.9 (M)	6.9
					6.11 (IP)
					6.12-13 (IP)
			6.19		
	7.1(?)		7.1b	7.1 (M)	7.1
7.2a		7.2a(?)		7.2 (M)	7.2
7.3	7.3(?)			7.3 (M)	7.3
7.4	7.4(?)			7.4 (M)	7.4
	7.5			7.5 (M)	7.5
	7.6a			7.6 (P)	7.6
7.7	7.7(?)			7.7 (M)	7.7
	7.8(?)			7.8 (M)	7.8
	7.9(?)			7.9 (A)	7.9
				7.10 (A)	
		7.11		7.11 (M)	7.11
		7.12		7.12 (M)	7.12
					7.13 (IP)
				7.19 (M)	7.19
				7.20 (P)	
				7.21 (A)	
					7.24 (IP)
					7.29
				8.1a (M)	8.1
	8.1b(?)			8.1b (M)	8.1 (IP)
8.2		8.2			
8.3		8.3			
8.4	8.4(?)	8.4			

Gordis (1939)	Whybray (1981)	Michel (1989)	Von Löwenclau (1986)	Johnson (1973)	McKenzie (1994)
		8.5		8.5 (M)	
	8.8(?)				
8.12					
8.13					
					8.15
9.4	9.4b(?)		9.4b	9.4b (M)	9.4
9.16a					9.16
	9.17		9.17	9.17 (M)	9.17
9.18a		9.18a(?)	9.18	9.18 (M)	9.18
				10.1 (M)	10.1
10.2(?)	10.2			10.2 (P)	10.2 (2)
				10.3 (P)	10.3 (2)
				10.4 (A)	10.4
					10.6
10.8(?)	10.8	10.8		10.8 (P)	10.8
10.9(?)	10.9(?)	10.9	10.9 (P)	10.9	
		10.10		10.10 (M)	10.10
		10.11(?)		10.11 (P)	
	10.12			10.12 (P)	10.12
				10.13 (P)	
				10.14 (P)	10.14 (IP)
				10.15 (P)	10.15
				10.16 (W)	
10.18			10.18	10.18 (M)	10.18
	10.19(?)			10.19 (M)	10.19
11.1		11.1(?)		11.1 (E)	
				11.2 (E)	
11.3				11.3 (P)	11.3 (2)
11.4(?)	11.4(?)		11.4	11.4 (P)	
11.5					
11.7				11.7 (M)	

Legend:

(A) = Admonition
(E) = Exhortations
(IP) = Impossible Question
(M) = Moral Sentence
(P) = Proverb
(W) = Woe Saying
(?) = uncertain or questionable citations

BIBLIOGRAPHY

Abrams, M.H., *The Mirror and the Lamp: Romantic Theory and the Critical Tradition* (New York: Oxford University Press, 1953).

Adam, A., 'Twisting to Deconstruction: A Memorandum on the Ethics of Interpretation', *PRS* 23 (1996), pp. 215-22.

Aerts, T., 'Two are Better than One (Qohelet 4.10)', *PtS* 14 (1990), pp. 11-48.

Aichele, G. *et al.* (eds), 'Reader-Response Criticism', in *The Postmodern Bible* (New Haven: Yale University Press, 1995), pp. 20-69.

Alonso-Schökel, L., *A Manual of Hebrew Poetics* (SubBib, 11; Rome: Pontifical Biblical Institute, 1988).

Alter, R., *The Art of Biblical Narrative* (New York: Basic Books, 1981).

—*The Art of Biblical Poetry* (New York: Basic Books, 1985).

—*The Pleasures of Reading in an Ideological Age* (New York: Simon & Schuster, 1989).

—*The World of Biblical Literature* (San Francisco: Harper & Row, 1992).

Amante, D., 'The Theory of Ironic Speech Acts', *PT* 2 (1981), pp. 77-96.

Andersen, K., and T. Clevenger, 'A Summary of Experimental Research in Ethos', *SM* 30 (1963), pp. 59-78.

Anderson, H., 'Philosophical Considerations in a Genre Analysis of Qoheleth', *VT* 48 (1998), pp. 289-300.

Anderson, P., 'Credible Promises: Ecclesiastes 11.3-6', *CurTM* 20 (1993), pp. 265-67.

Anderson, W., 'Philosophical Considerations in a Genre Analysis of Qoheleth', *VT* 48 (1998), pp. 289-300.

—'The Poetic Inclusio of Qoheleth in Relation to 1,2 and 12,8', *SJOT* 12 (1998), pp. 202-13.

Aquine, R., 'The Believing Pessimist: A Philosophical Reading of the Qoheleth', *PSac* 16 (1981), pp. 207-61.

Aristotle, 'Rhetoric' in *The Rhetoric and the Poetics of Aristotle* (trans. W. Rhys Roberts; New York: The Modern Library, 1954).

Armstrong, J., 'Ecclesiastes in Old Testament Theology', *PSB* 94 (1983), pp. 16-25.

Auffret, P., '"Rien du tout de nouveau sous le soleil": Étude structure de Qo 1, 4-11', *FO* 26 (1989), pp. 145-66.

Ausejo, S. de, 'El genero literario del Eclesiastes', *EstBíb* 7 (1948), pp. 369-406.

Austin, J.L., *How To Do Things With Words* (Cambridge, MA: Harvard University Press, 1962).

Auwers, J., 'Problèmes d'interprétation de l'épilogue de *Qohèlèt*', in Schoors (ed.), *Qohelet in the Context of Wisdom*, pp. 267-82.

Aversano, C., '*Mishpaṭ* in Qoh. 11.9c', in A. Vivian (ed.), *Biblische und judaistische Studiën* (New York: Peter Lang, 1990), pp. 121-34.

Avni, O., *The Resistance of Reference: Linguistics, Philosophy, and the Literary Text* (Baltimore: The Johns Hopkins University Press, 1990).

Bach, A., 'Signs of the Flesh: Observations on Characterization in the Bible', *Semeia* 63 (1993), pp. 61-79.

Backhaus, F., *Denn Zeit und Zufall trifft sie Alle: Studien zur Komposition und zum Gottesbild im Buch Qohelet* (ed. F. Hossfeld and H. Merklein; BBB, 83; Frankfurt: Anton Hain, 1993).

—'Qohelet und Sirach', *BN* 69 (1993), pp. 32-55.

—'Der Weisheit letzter Schluss! Qoh 12, 9-14 im Kontext von Traditionsgeschichte und beginnender Kanonisierung', *BN* 72 (1994), pp. 28-59.

Bal, M., 'The Narrating and the Focalizing: A Theory of the Agents in Narrative', *Style* 17 (1983), pp. 234-69.

—*Narratology: Introduction to the Theory of Narrative* (trans. C. Van Boheemen; Toronto: University of Toronto Press, 1985).

—'Metaphors He Lives By', *Semeia* 61 (1993), pp. 185-207.

—'First Person, Second Person, Same Person: Narrative as Epistemology (Reconsiderations)', *NLH* 24 (1993), pp. 293-321.

Baltzer, K., 'Women and War in Qohelet 7.23-8.1a', *HTR* 80 (1987), pp. 127-32.

Bannach, H., 'Bemerkungen zu Prediger 3.1-9', in H. Bannach (ed.), *Glaube und öffentliche Meinung* (Stuttgart: Radius Verlag, 1970), pp. 26-32.

Barr, J., 'Reading the Bible as Literature', *BJRL* 59 (1973), pp. 10-33.

Barré, M., '"Fear of God" and the World View of Wisdom', *BTB* 11 (1981), pp. 41-43.

Barré, M. (ed.), *Wisdom, You Are my Sister* (Festschrift R. Murphy; CBQMS, 29; Washington: Catholic Biblical Association, 1997).

Bartelmus, R., 'Haben oder Sein: Anmerkungen zur Anthropologie des Buches Kohelet', *BN* 53 (1990), pp. 38-69.

Barth, H., 'Autonomie, Theonomie und Existenz', *ZEE* 2 (1958), pp. 321-34.

Barthes, R., *S/Z: An Essay* (trans. R. Miller; New York: Hill & Wang, 1974).

—'Theory of the Text', in R. Young (ed.), *Untying the Text: A Post-Structuralist Reader* (London: Routledge & Kegan Paul, 1981), pp. 133-61.

—'The Death of the Author', in Rice and Waugh (eds.), *Modern Literary Theory*, pp. 114-18.

—'From Work To Text', in Rice and Waugh (eds.), *Modern Literary Theory*, pp. 166-72.

Bartlett, J., *Familiar Quotations: A Collection of Passages, Phrases, and Proverbs Traced to Their Sources in Ancient and Modern Literature* (10th edn; revised and enlarged by Nathan Dole, Boston: Little, Brown, and Co, 1930).

Barton, G., *A Critical and Exegetical Commentary on the Book of Ecclesiastes* (ICC; Edinburgh: T. & T. Clark, 1908).

Barton, J., 'Classifying Biblical Criticism', *JSOT* 29 (1984), pp. 19-35.

—*Reading the Old Testament* (Philadelphia: Westminster Press, 1984).

—'Reading the Bible as Literature: Two Questions for Biblical Critics', *L&T* 1 (1987), pp. 135-53.

Barucq, A., *Ecclésiaste* (VS, 3; Paris: Beauchesne, 1968).

—'Dieu chez les sages d'Israël', in J. Coppens (ed.), *La notion biblique de Dieu* (Leuven: Leuven University Press, 1976), pp. 169-89.

—'Qoheleth (ou livre de l'Ecclésiaste)', in *DBSup*, vol. 9, pp. 609-74.

Baumgärtel, F., 'Die Ochstenstacheln und die Nägel in Koheleth 12,11', *ZAW* 81 (1969), p. 98.

Bea, A., *Liber Ecclesiaste qui ab Hebraeis appelatur Qohelet* (SPIB; Rome: Pontifical Biblical Institute, 1950).

Beal, T., 'C(ha)osmopolis: Qohelet's Last Word', in T. Linafelt and T. Beal (eds.), *God in the Fray: A Tribute to Walter Brueggemann* (Festschrift W. Brueggemann; Minneapolis: Fortress Press, 1998), pp. 290-304.

Beale, W., *A Pragmatic Theory of Rhetoric* (Carbondale, IL: Southern Illinois University Press, 1987).

Beck, D., 'The Narrative Function of Anonymity in Forth Gospel Characterization', *Semeia* 63 (1993), pp. 143-58.

Beentjes, B., '"The countries marvelled at you': King Solomon in Ben Sira 47.12-22', *Bijdragen* 45 (1984), pp. 6-14.

—'"Who is like the wise?": Some Notes on Qohelet 8,1-15', in Schoors (ed.), *Qoheleth in the Context of Wisdom*, pp. 303-15.

Benveniste, E., *Problems in General Linguistics* (Miami: University of Miami Press, 1971).

Bergant, D., *Job, Ecclesiastes* (OTM, 18; Wilmington, DE: Michael Glazier, 1982).

Bergen, R., 'Text as a Guide to Authorial Intention: An Introduction to Discourse Criticism', *JETS* 30 (1987), pp. 327-36.

Bergman, J., 'Discours d'adieu—Testament—Discours posthume; Testaments juifs et enseignements egyptiens', in J. Leclant *et al.* (eds.), *Sagesse et Religion* (1979), pp. 21-50.

Berlin, A., *Poetics and Interpretation of Biblical Narrative* (Bible & Literature Series, 9; Sheffield: Almond Press, 1983).

—*The Dynamics of Biblical Parallelism* (Bloomington: Indiana University Press, 1985).

—'The Role of the Text in the Reading Process', *Semeia* 62 (1993), pp. 143-47.

Bertram, G., 'Hebräischer und Griechischer Qohelet', *ZAW* 64 (1952), pp. 436-44.

Berube, M. *et al.* (eds.), *The American Heritage Dictionary* (Boston: Houghton Mifflin Co., Second College Edition, 1982).

Bianchi, F., 'The Language of Qoheleth: A Bibliographical Survey', *ZAW* 105 (1993), pp. 210-23.

Bishop, E.F., 'Pessimism in Palestine (B.C.)', *PEQ* 100 (1969), pp. 33-41.

Bitzer, L., 'The Rhetorical Situation', *PR* 1 (1968), pp. 1-14.

—'The Rhetorical Situation', in R. Johannesen (ed.), *Contemporary Theories of Rhetoric: Selected Readings* (San Francisco: Harper & Row, 1971), pp. 381-94.

—'Rhetoric and Public Knowledge', in D. Burks (ed.), *Rhetoric, Philosophy, and Literature* (Lafayette, IN: Purdue University Press, 1978). pp. 67-93.

—'Functional Communication: A Situational Perspective', in E. White (ed.), *Rhetoric in Transition: Studies in the Nature and Uses of Rhetoric* (University Park: Pennsylvania State University Press, 1980), pp. 21-38.

Blank, S.H., 'Ecclesiastes', in *IDB* (4 vols.; Nashville, TN: Abingdon Press, 1962), II, pp. 7-13.

Blenkinsopp, J., 'Ecclesiastes 3.1-15: Another Interpretation', *JSOT* 66 (1995), pp. 55-64.

Bons, E., 'Zur Gliederung und Kohërenz von Koh 1.12-2.11', *BN* 24 (1984), pp. 73-93.

—'šiddā w =šiddōt: Überlegungen zum Verständnis eines Hapaxlegomenons', *BN* 36 (1987), pp. 12-16.

—'Ausgewählte Literatur zum Buch Kohelet', *BK* 45 (1990), pp. 36-42.

Bontekoe, R., 'The Function of Metaphor', *PR* 20 (1987), pp. 209-26.

Booth, W., 'Distance and Point-of-View: An Essay in Classification', *EC* 11 (1961), pp. 60-79.

—*A Rhetoric of Irony* (Chicago: University of Chicago Press, 1974).

—*The Rhetoric of Fiction* (Chicago: University of Chicago Press, 2nd edn, 1983 [1961]).

Bordo, S., *The Flight to Objectivity: Essays on Cartesianism and Culture* (New York: State University of New York Press, 1987).

Branham, R., and W. Pearce, 'Between Text and Context: Toward a Rhetoric of Contextual Reconstruction', *QJS* 71 (1985), pp. 19-36.

Bratsiotis, H., 'Der monolog im Alten Testament', *ZAW* 73 (1961), pp. 30-70.

Braun, R., *Kohelet und die frühhellenistische Popularphilosophie* (BZAW, 130; Berlin: W. de Gruyter, 1973).

Bream, H., 'Life Without Resurrection: Two Perspectives from Qoheleth', in H. Bream *et al.* (eds.), *A Light Unto My Path* (Festschrift J. Myers; GTS, 14; Temple University Press, Philadelphia, PA, 1974).

Brenner, A., 'Introduction', in Brenner and Van Dijk-Hemmes (eds.), *On Gendering Texts*, pp. 1-13.

—'M Text Authority in Biblical Love Lyrics: The Case of Qoheleth 3.1-9 and its Textual Relatives', in Brenner and Van Dijk-Hemmes (eds.), *On Gendering Texts*, pp. 133-63.

Brenner, A., and C. Fontaine (eds.), *Wisdom and Psalms: A Feminist Companion to the Bible* (The Feminist Companion to the Bible, Second Series 2; Sheffield: Sheffield Academic Press, 1998).

Brenner, A., and F. Van Dijk-Hemmes (eds.), *On Gendering Texts: Female and Male Voices in the Hebrew Bible* (New York: E.J. Brill, 1993).

Bréton, S., 'Qoheleth Studies', *BTB* 3 (1973), pp. 22-50.

—'Qohelet: Recent Studies', *TD* 28 (1980), pp. 147-51.

Brett, M., 'The Future of Reader Criticisms?', in F. Watson (ed.), *The Open Text: New Directions for Biblical Studies?* (London: SCM Press, 1993), pp. 13-31.

Brindle, W., 'Righteousness and Wickedness in Ecclesiastes 7.15-18', *AUSS* 23 (1985), pp. 243-57.

Brinton, A., 'Situation in the Theory of Rhetoric', *PR* 14 (1981), pp. 234-48.

Brongers, H.A., 'Some Remarks on the Biblical Particle *halô*', in B. Albrektson *et al.* (eds.), *Remembering All the Way* (OTS, 21; Leiden: E.J. Brill, 1981), pp. 177-89.

Brooks, C., and R. Warren, *Modern Rhetoric* (New York: Harcourt, Brace & World, 4th edn, 1979 [1949]).

Brown, F., *Transfiguration: Poetic Metaphor and the Language of Religious Belief* (Chapel Hill: University of North Carolina Press, 1983).

Brown, S., 'The Structure of Ecclesiastes', *ERT* 14 (1990), pp. 195-208.

Brown, W., *Character in Crisis: A Fresh Approach to the Wisdom Literature of the Old Testament* (Grand Rapids: Eerdmans, 1996).

Brueggemann, D., 'Brevard Childs' Canon Criticism: An Example of Post-Critical Naiveté', *JETS* 32 (1989), pp. 311-26.

Brueggemann, W., *Texts Under Negotiation: The Bible and Postmodern Imagination* (Minneapolis: Fortress Press, 1993).

—*Theology of the Old Testament: Testimony, Dispute, Advocacy* (Minneapolis: Fortress Press, 1997).

Bruns, J., 'The Imagery of Ecclesiastes 12,6a', *JBL* 84 (1965), pp. 428-30.

Bryce, G.E., '"Better'-Proverbs: An Historical and Structural Study', in L. McGaughy (ed.) (SBLSP, 2; Missoula, MT: Scholars Press, 1972), pp. 333-54.

Burden, J.J., 'Decision by Debate: Examples of Popular Proverb Performance in the Book of Job', *OTE* 4 (1991), pp. 37-65.

Burnett, F., 'Characterization and Reader Construction of Characters in the Gospels', *Semeia* 63 (1993), pp. 3-28.

Bush, B., *Walking in Wisdom: A Woman's Workshop on Ecclesiastes* (Grand Rapids: Eerdmans, 1982).

Butting, K., 'Weibsbilder bei Kafka und Kohelet: Eine Auslegung von Prediger 7,23-29', *TK* 14 (1991), pp. 2-15.

Buzy, D., 'Le portrait de la vieillesse (Ecclésiaste, Xii, 1-7)', *RB* 41 (1932), pp. 329-40.

—'La notion du bonheur dans l'Ecclésiaste', *RB* 43 (1934), pp. 494-511.

—*L'Ecclésiaste* (Paris: Letouzey & Ané, 1946).

—'Les auteurs de l'Ecclésiaste', *L'A Théo* 11 (1950), pp. 317-36.

Byargeon, R., 'The Significance of the Enjoy Life Concept in Qoheleth's Challenge of the Wisdom Tradition' (PhD dissertation; Fort Worth, TX: Southwestern Baptist Theological Seminary, 1991).

—'The Significance of Ambiguity in Ecclesiastes 2,24-26', in Schoors (ed.), *Qohelet in the Context of Wisdom*, pp. 367-72.

Camp, C., *Wisdom and the Feminine in the Book of Proverbs* (Bible & Literature Series, 11; Sheffield: Almond Press, 1985).

—'Metaphor in Feminist Biblical Interpretation: Theoretical Perspectives', in Camp and Fontaine (eds.), *Women, War and Metaphor*, pp. 3-38.

Camp, C., and C. Fontaine (eds.), *Women, War, and Metaphor: Language and Society in the Study of the Hebrew Bible* (Semeia, 61; Atlanta: Scholars Press, 1993).

Caneday, A., 'Qoheleth: Enigmatic Pessimist or Godly Sage?', *GTJ* 7 (1986), pp. 21-56.

Cardinal, R., 'Unlocking The Diary', *CompCrit* 12 (1990), pp. 71-87.

Carny, P., 'Theodicy in the Book of Qoheleth', in H. Reventlow and Y. Hoffman (eds.), *Justice and Righteousness: Biblical Themes and their Influence* (JSOTSup, 137; Sheffield: JSOT Press, 1992), pp. 71-81.

Carr, D., *From D to Q: A Study of Early Jewish Interpretations of Solomon's Dream at Gibeon* (SBLMS, 44; Atlanta, GA: Scholars Press, 1991).

Carrière, J.M., 'Tout est vanité: L'Un des concepts de Qohélet', *EstBib* 55 (1997), pp. 463-77.

Castellino, G., 'Qoheleth and His Wisdom', *CBQ* 30 (1968), pp. 15-28.

Cazelles, H., 'Conjonctions de subordination dans la langue du Qohelet', *CRGLECS* (1957–60), pp. 21-22.

Ceresko, A., 'The Function of Antanaclasis (*ms'* 'to Find'//*ms'* 'to Reach, Overtake, Grasp') in Hebrew Poetry, Especially the Book of Qoheleth', *CBQ* 44 (1982), pp. 551-69.

—'Commerce and Calculation: The Strategy of the Book of Qoheleth (Ecclesiastes)', *ITS* 30 (1993), pp. 205-19.

Chamakkala, J., 'Qoheleth's Reflections on Time', *Jeevadhara* 7 (1977), pp. 117-31.

Chatman, S., *Story and Discourse: Narrative Structure in Fiction and Film* (Ithaca, NY: Cornell University Press, 1978).

—'Characters and Narrators: Filter, Center, Slant, and Interest-Focus', *PT* 7 (1986), pp. 189-204.

—'The "Rhetoric" of "Fiction"', in J. Phelan (ed.), *Reading Narrative: Form, Ethics, Ideology* (Columbus, OH: Ohio State University Press, 1989), pp. 40-56.

—*Coming to Terms: The Rhetoric of Verbal and Cinematic Narrative* (Ithaca, NY: Cornell University Press, 1990).
Chen, C., 'A Study of Ecclesiastes 10.18-19', *TJT* 11 (1989), pp. 117-26.
Cherwitz, R., 'Rhetoric as a 'Way of Knowledge': An Attenuation of the Epistemological Claims of the 'New Rhetoric' ', *SSCJ* 42 (1977), pp. 207-19.
Cherwitz, R., and J. Hikins, 'Toward a Rhetorical Epistemology', *SSCJ* 47 (1982), pp. 135-62.
Childs, B., *Biblical Theology in Crisis* (Philadelphia: Fortress Press, 1970).
—'The Exegetical Significance of Canon for the Study of the Old Testament', *Congress Volume, Göttingen 1977*, ed. J.A. Emerton (VTSup, 29; Leiden: E.J. Brill, 1977), pp. 66-80.
—*Introduction to the Old Testament as Scripture* (Philadelphia: Fortress Press, 1979).
—*The New Testament as Canon: An Introduction* (Philadelphia: Fortress Press, 1984).
Chopineau, J., 'Image de l'homme: Sur Ecclésiaste 1.2', *ETR* 53 (1978), pp. 366-70.
—'L'Image de Qohelet dans l'exégèse contemporaine', *RHPR* 59 (1979), pp. 595-603.
Christianson, E.S., *A Time to Tell: Narrative Strategies in Ecclesiastes* (JSOTSup, 280; Sheffield: Sheffield Academic Press, 1998).
—'Qoheleth the 'Old Boy' and Qoheleth the 'New Man': Misogynism, The Womb and a Paradox in Ecclesiastes', in Brenner and Fontaine (eds.), *Wisdom and Psalms*, pp. 109-36.
Clemens, D., 'The Law of Sin and Death: Ecclesiastes and Genesis 1–3', *Themelios* 19 (1994), pp. 5-8.
Clines, D.J.A., 'Beyond Synchronic/Diachronic?', in J.C. de Moor (ed.), *Synchronic or Diachronic? A Debate on Method in Old Testament Exegesis* (OTS, 34; Leiden: E.J. Brill, 1995), pp. 52-71.
Cochrane, A.C., 'Joy to the World: The Message of Ecclesiastes', *CCent* 85 (1968), pp. 1596-98.
Cohn, D., *Transparent Minds: Narrative Modes for Presenting Consciousness in Fiction* (Princeton: Princeton University Press, 1978).
Conrad, E., *Reading Isaiah* (Minneapolis: Fortress Press, 1991).
Consigny, S., 'Rhetoric and its Situations', *PR* 7 (1974), pp. 175-76.
Cooper, R., 'Textualizing Determinacy/Determining Textuality', *Semeia* 62 (1993), pp. 3-18.
Coppens, J., 'La structure de l'Ecclésiaste', in M. Gilbert (ed.), *La sagesse de l'Ancien Testament* (BETL, 51; Gembloux: Leuven University, 1979), pp. 288-92.
Corré, A.D., 'A Reference to Epispasm in Koheleth', *VT* 4 (1954), pp. 416-18.
Cosser, W., 'The Meaning of 'Life' in Proverbs, Job and Ecclesiastes', *GUOST* 15 (1953–54), pp. 48-53.
Crane, R.S., 'The Concept of Plot', in Scholes (eds.), *Approaches to the Novel*, pp. 232-43.
Crenshaw, J., 'The Eternal Gospel (Eccl. 3.11)', in J. Crenshaw and J. Willis (ed.), *Essays in Old Testament Ethics* (New York: Ktav, 1974), pp. 23-55.
—'Popular Questioning of the Justice of God in Ancient Israel', in J. Crenshaw (ed.), *Studies in Ancient Israelite Wisdom*, pp. 289-304.
—'The Birth of Skepticism in Ancient Israel', in J. Crenshaw and S. Sandmel (eds.), *The Divine Helmsman* (New York: Ktav, 1980), pp. 1-20.
—'Impossible Questions, Sayings, and Tasks', *Semeia* 17 (1980), pp. 19-34.
—*Old Testament Wisdom: An Introduction* (Atlanta: John Knox Press, 1981).
—'Wisdom and Authority: Sapiential Rhetoric and its Warrants', in *Congress Volume, Vienna 1980* (ed. J. Emerton; VTSup, 32; Leiden: E.J. Brill, 1981), pp. 10-29.

—'Qoheleth in Current Research', *HAR* 7 (1983), pp. 41-56.
—'Education in Ancient Israel', *JBL* 104 (1985), pp. 601-15.
—'The Expression *mi yodea'* in the Hebrew Bible', *VT* 36 (1986), pp. 274-88.
—'The Acquisition of Knowledge in Israelite Wisdom Literature', *WW* 7 (1987), pp. 245-52.
—*Ecclesiastes: A Commentary* (OTL; Philadelphia: Fortress Press, 1987).
—'Youth and Old Age in Qoheleth', *HAR* 11 (1987).
—'Murphy's Axiom: Every Gnomic Saying Needs a Balancing Corrective', in K. Hoglund and E. Huwiler (eds.), *The Listening Heart: Essays in Wisdom and Psalms in Honor of Roland E. Murphy* (JSOTSup, 58; Sheffield: JSOT Press, 1987), pp. 1-17.
—'Ecclesiastes', in J. Crenshaw and J. Willis (eds.), *Harper's Bible Commentary* (San Francisco: Harper & Row, 1988), pp. 518-24.
—'Ecclesiastes: Odd Book in the Bible', *BR* 6 (1990), pp. 28-33.
—'Ecclesiastes, Book of' in *ABD*, II, pp. 271-80.
—'Qoheleth's Understanding of Intellectual Inquiry', in Schoors (ed.), *Qoheleth in the Context of Wisdom*, pp. 205-24.
Crenshaw, J. (ed.), *Studies in Ancient Israelite Wisdom* (New York: Ktav, 1976).
Crites, S., 'The Narrative Quality of Experience', *JAAR* 39 (1971), pp. 291-311.
Croatto, J., *Biblical Hermeneutics: Toward a Theory of Reading as the Production of Meaning* (Maryknoll, NY: Orbis Books, 1987).
Crocker, P.T., '"I made gardens and parks…"', *BH* 26 (1990), pp. 20-23.
Crossan, J., 'Aphorism in Discourse and Narrative', *Semeia* 43 (1988), pp. 121-40.
Crüsemann, F., 'The Unchangeable World: The 'Crisis of Wisdom' in Koheleth' in W. Schottroff and W. Stegemann (eds.), *God of the Lowly* (trans. M. O'Connell; Maryknoll, NY: Orbis Books), 1984, pp. 57-77 (first published as 'Die unveranderbare Welt', in W. Schottroff and W. Stegemann [eds.], *Der Gott der kleinen Leute* [Munich: Chr. Kaiser Verlag, 1979], I, pp. 80-104).
Csikszentmihalyi, M., *Flow: The Psychology of Optimal Experience* (New York: Harper & Row, 1990).
Culler, J., *Structuralist Poetics: Structuralism, Linguistics, and the Study of Literature* (Ithaca, NY: Cornell University Press, 1975).
—'Prolegomena to a Theory of Reading', in Suleiman and Crosman (eds.), *The Reader in the Text*, pp. 46-66.
Culley, R., 'Introduction', *Semeia* 62 (1993), pp. vii-xii.
Dahood, M., 'The Language of Qoheleth', *CBQ* 14 (1952), pp. 227-32.
—*Canaanite-Phoenician Influence in Qoheleth* (Rome: Pontifical Biblical Institute Press, 1952).
—'Canaanite-Phoenician Influence in Qoheleth', *Bib* 33 (1952), pp. 30-52, 191-221.
—'Qoheleth and Recent Discoveries', *Bib* 39 (1958), pp. 302-18.
—'Qoheleth and Qumran: A Study in Style', *Bib* 41 (1960), pp. 395-410.
—'Qoheleth and Northwest Semitic Philology', *Bib* 43 (1962), pp. 349-65.
—'Canaanite Words in Qoheleth 10,20', *Bib* 46 (1965), pp. 210-12.
—'The Phoenician Background of Qoheleth', *Bib* 47 (1966), pp. 264-82.
—'Three Parallel Pairs in Ecclesiastes 10.18: A Reply to Robert Gordis', *JQR* 62 (1971–72), pp. 84-87.
Danker, F., 'The Pessimism of Ecclesiastes', *CTM* 22 (1951), pp. 9-32.
Darr, J., 'Narrator as Character: Mapping a Reader-Oriented Approach to Narration in Luke–Acts', *Semeia* 63 (1993), pp. 43-60.

Davidson, R., *Ecclesiastes and the Song of Solomon* (Philadelphia: Westminster Press, 1986).

Davis, B., 'Ecclesiastes 12.1-8: Death, and Impetus for Life', *BSac* 148 (1991), pp. 298-318.

Davilla, J., 'Qoheleth and Northern Hebrew', *Maarav* 5–6 (1990), pp. 69-87.

Dawson, D., *Allegorical Readers and Cultural Revision in Ancient Alexandria* (Berkeley: University of California Press, 1992).

De Boer, P., 'Note on Ecclesiastes 12;12a', in R. Fischer (ed.), *A Tribute to Arthur Vööbus* (Festschrift A. Vööbus; Chicago: Lutheran School of Theology, 1977), pp. 85-88.

De Bruin, G., 'Ecclesiastes 1.12–2.26 According to a Multidisciplinary Method of Reading' (PhD dissertation; Pretoria: University of Pretoria, South Africa, 1990).

De Jong, S., 'A Book of Labour: The Structuring Principles and the Main Theme of the Book of Qohelet', *JSOT* 54 (1992), pp. 107-16.

—'Qohelet and the Ambitious Spirit of the Ptolemaic Period', *JSOT* 61 (1994), pp. 85-96.

De Waard, J., 'The Structure of Qoheleth', in *Procs. 8th World Congress of Jewish Studies* (Jerusalem: Magnes Press, 1982), pp. 57-63.

Delitzsch, F., *Commentary on the Song of Songs and Ecclesiastes* (trans. M. Easton; Grand Rapids: Eerdmans, repr. edn, 1950 [1875]).

Dell, K., 'The Reuse and Misuse of Forms in Ecclesiastes' in K. Dell, *The Book of Job as Sceptical Literature* (BZAW, 197; Berlin: W. de Gruyter, 1991).

—'Ecclesiastes as Wisdom: Consulting Early Interpreters', *VT* 44 (1994), pp. 301-29.

Delsman, W.C., 'Zur Sprache des Buches Koheleth', in W.C. Delsman *et al.* (eds.), *Von Kanaan Bis, Kerala* (AOAT, 211; Neukirchen–Vluyn: Neukirchener Verlag, 1982), pp. 341-65.

—'Die Inkongruenz im Buch Qoheleth', in K. Jongeling (ed.), *Studies in Hebrew and Aramaic Syntax* (N.P.: 1991).

Detweiler, R., 'What is a Sacred Text?', *Semeia* 31 (1985), pp. 213-30.

Dewey, R., 'Qoheleth and Job: Diverse Responses to the Enigma of Evil', *SpTod* 37 (1985), pp. 314-25.

Dietrich, E., 'Das Religiös-emphatische Ich-Wort bei den jüdischen Apokalyptikern, Weisheitslehren und Rabbinen', *ZRGG* 4 (1952), pp. 289-311.

Donahue, J., *Are You the Christ? The Trial Narrative in the Gospel of Mark* (SBLDS, 10; Missoula, MT: Scholars Press, 1973).

Driver, G.R., 'Problems and Solutions', *VT* 4 (1954), pp. 225-45.

Du Plessis, S.J., 'Aspects of Morphological Peculiarities of the Language of Qoheleth', in *De Fructu Oris Sui* (Festschrift A. Van Selms Editor = I.H. Eybers *et al.*; Leiden: E.J. Brill, 1971), pp. 161-80.

Duesberg, H., and I. Fransen, *Les scribes inspirés: Introduction aux livres sapientiaux de la Bible; Proverbs, Job, Ecclesiaste, Sagesse, Ecclesiastique* (Paris: Maredsous, 2nd edn, 1966).

Dundes, A., 'On the Structure of the Proverb', in A. Dundes (ed.), *Analytic Essays in Folklore* (The Hague: Mouton, 1975), pp. 103-18.

Duranti, A., 'The Audience as Co-Author: An Introduction', *Text* 6 (1986), pp. 239-47.

Eco, U., *The Role of the Reader: Explorations in the Semiotics of Texts* (Advances in Semiotics; Bloomington: Indiana University Press, 1979).

—'The Scandal of Metaphor', *PT* 4 (1983), pp. 217-58.

—'*Intentio Lectoris*: The State of the Art', in U. Eco, *The Limits of Interpretation* (Bloomington: Indiana University Press, 1990), pp. 44-63.

Ede, L., and A. Lunsford, 'Audience Addressed/Audience Invoked: The Role of Audience in Composition Theory and Pedagogy', *CCC* 35 (1984), pp. 155-71.

Edwards, P. (ed.), *The Encyclopedia of Philosophy* (New York: Macmillan, 1967).

Egan, K., 'What is a Plot?', *NLH* 9 (1978), pp. 455-73.

Ehninger, D., 'On Systems of Rhetoric', in R. Johannesen (ed.), *Contemporary Theories of Rhetoric: Selected Readings* (San Francisco: Harper & Row, 1971), pp. 327-39.

Eissfeldt, O., *The Old Testament: An Introduction* (trans. P. Ackroyd; New York: Harper & Row, 1965).

Elgin, S., 'Response from the Perspective of a Linguist', *Semeia* 61 (1993), pp. 209-17.

Ellermeier, F., 'Das Verbum *hws* in Koh 2,25: Eine exegetische, auslegungsgeschichtliche und semasiologische Untersuchung', *ZAW* 75 (1963), pp. 197-217.

—'Die Entmachung der Weisheit im denken Qohelets: zu Text und Auslegung von Qoh 6.7-9', *ZTK* 60 (1963), pp. 1-20.

—'Randbemerkung zur Kunst des Zitierens: Welches Buch der Bible nannte Heinrich Heine 'das Hohelied der Skepsis'?', *ZAW* 77 (1965), pp. 93-94.

—*Qohelet. Teil I, Abschnitt 1* (Herzberg: Jungfer, 1967).

—*Qohelet. Teil I, Abschnitt 2,7*. (Herzberg: Jungfer, 1970).

Ellis, J., *The Theory of Literary Criticism: A Logical Analysis* (Berkeley: University of California Press, 1974).

Faigley, L., and P. Meyer, 'Rhetorical Theory and Readers' Classifications of Text Types', *Text* 3 (1983), pp. 305-25.

Farmer, K., *Who Knows What is Good? A Commentary on the Books of Proverbs and Ecclesiastes* (ITC; Grand Rapids: Eerdmans, 1991).

Fisch, H., 'Qoheleth: A Hebrew Ironist', in H. Fisch, *Poetry with a Purpose* (ISBL; Bloomington: Indiana University Press, 1988, pp. 158-78.

Fischer, A., 'Beobachtungen zur Komposition von Kohelet 1,3–3,15', *ZAW* 103 (1991), pp. 72-86.

Fish, S., *Self-Consuming Artifacts: The Experience of Seventeenth-Century Literature* (Berkeley: University of California Press, 1972).

—'Introduction, or How I Stopped Worrying and Learned to Love Interpretation', in S. Fish, *Is There a Text in This Class? The Authority of Interpretive Communities* (Cambridge, MA: Harvard University Press, 1980), pp. 1-17.

—'Literature in the Reader: Affective Stylistics', in S. Fish, *Is There a Text in This Class? The Authority of Interpretive Communities* (Cambridge, MA: Harvard University Press, 1980), pp. 21-67.

—'What is Stylistics and Why Are They Saying Such Terrible Things About It?', in S. Fish, *Is There a Text in This Class? The Authority of Interpretive Communities* (Cambridge, MA: Harvard University Press, 1980), pp. 68-96.

—'Interpreting the *Variorum*', in S. Fish, *Is There a Text in This Class? The Authority of Interpretive Communities* (Cambridge, MA: Harvard University Press, 1980), pp. 147-173.

—'Interpreting "Interpreting the *Variorum*"', in S. Fish, *Is There a Text in This Class? The Authority of Interpretive Communities* (Cambridge, MA: Harvard University Press, 1980), pp. 174-80.

—'Is There A Text in this Class?', in S. Fish, *Is There a Text in This Class? The Authority of Interpretive Communities* (Cambridge, MA: Harvard University Press, 1980), pp. 303-321.

—'What Makes an Interpretation Acceptable?', in S. Fish, *Is There a Text in This Class? The Authority of Interpretive Communities* (Cambridge, MA: Harvard University Press, 1980), pp. 338-55.

—'Literature in the Reader: Affective Stylistics' in J. Tompkins (ed.), *Reader-Response Criticism: From Formalism to Post-Structuralism* (Baltimore: The Johns Hopkins University Press, 1980), pp. 70-99.

Fisher, W., *Human Communication as Narration: Toward a Philosophy of Reason, Value, and Action* (SRC; Columbia: University of South Carolina Press, 1987).

Fohrer, G., *Introduction to the Old Testament* (trans. D. Green; Nashville: Abingdon Press, 1965).

Fontaine, C., *Traditional Sayings in the Old Testament* (Bible & Literature Series, 5; Sheffield: Almond Press, 1982).

—' "Many Devices" (Qoheleth 7.23–8.1): Qoheleth, Misogyny and the *Malleus Maleficarum*', in Brenner and Fontaine (eds.), *Wisdom and Psalms*, pp. 137-68.

Forman, C., 'The Pessimism of Ecclesiastes', *JJS* 3 (1958), pp. 336-43.

—'Koheleth's Use of Genesis', *JSS* 5 (1960), pp. 256-63.

Forster, E.M., 'The Plot', in Scholes (ed.), *Approaches to the Novel*, pp. 219-32.

Fowl, S., 'Texts Don't Have Ideologies', *BibInt* 3 (1995), pp. 15-34.

Fowler, R., 'Who is the Reader in Reader Response Criticism?', *Semeia* 31 (1985), pp. 5-23.

—*Let the Reader Understand* (Philadelphia: Fortress Press, 1992).

—'Reader Response Criticism', in R. Fowler, *Let The Reader Understand* (Philadelphia: Fortress Press, 1992), pp. 1-58.

—'Characterizing Character in Biblical Narrative', *Semeia* 63 (1993), pp. 97-104.

Fox, M., 'Frame-Narrative and Composition in the Book of Qohelet', *HUCA* 48 (1977), pp. 83-106.

—'The Identification of Quotations in Biblical Literature', *ZAW* 92 (1980), pp. 416-31.

—'The Meaning of *Hebel* for Qoheleth', *JBL* 105 (1986), pp. 409-27.

—'Qoheleth's Epistemology', *HUCA* 58 (1987), pp. 137-55.

—'Aging and Death in Qoheleth 12', *JSOT* 42 (1988), pp. 55-77.

—'Qoheleth 1.4', *JSOT* 40 (1988), p. 109.

—*Qoheleth and His Contradictions* (JSOTSup, 71; Sheffield: Almond Press, 1989).

—'Wisdom in Qoheleth', in L. Perdue, B. Scott and W. Wiseman (eds.), *In Search of Wisdom* (Festschrift J. Gammie; Louisville, KY: Westminster/John Knox Press, 1993), pp. 115-31.

—'What Happens in Qohelet 4.13-16', *JHStud* 1 (1997). (http.//www.arts.ualberta.ca/JHS/)

—'The Inner-Structure of Qohelet's Thought', in Schoors (ed.), *Qoheleth in the Context of Wisdom*, pp. 225-38.

Fox, M., and B. Porten, 'Unsought Discoveries: Qoheleth 7.23-8.1a', *HS* 19 (1978), pp. 26-38.

Fredericks, D., *Qoheleth's Language: Re-Evaluating Its Nature and Date* (ANETS, 3; Lewiston, IL: Edwin Mellen Press, 1988).

—'Chiasm and Parallel Structure in Qoheleth 5.9-6.9', *JBL* 108 (1989), pp. 17-35.

—'Life's Storms and Structural Unity in Qoheleth 11.1-12.8', *JSOT* 52 (1991), pp. 95-114.

Frendo, K., 'The 'Broken Construct Chain' in Qoh 10.10b', *Bib* 62 (1981), pp. 544-45.

Freund, E., *The Return of the Reader: Reader-Response Criticism* (New York: Methuen, 1987).

Freyne, S., 'Our Preoccupation with History: Problems and Prospects', *PIBA* 9 (1985), pp. 1-18.

Friedman, N., 'Point of View in Fiction: The Development of a Critical Concept', *PMLA* 70 (1955), pp. 1160-84.
—*Form and Meaning in Fiction* (Athens: University of Georgia Press, 1975).
Frye, N., *The Great Code: The Bible and Literature* (New York: Harvest/Harcourt Brace Jovanovich Books, 1982).
Gadamer, H., *Truth and Method* (trans. G. Barden and J. Cumming; New York: Seabury, 1975).
Galling, K., 'Koheleth-Studien', *ZAW* 50 (1932), pp. 276-99.
—'Stand und Aufgabe der Kohelet-Forschung', *TRu* NS 6 (1934), pp. 355-73.
—'Prediger Salomo', in O. Eissfelt (ed.) *Die Fünf Megillot: Ruth, Hohelied, Klageslied, Esther, Prediger* (HAT, 18; Tübingen, J.C.B. Mohr, 2nd edn, 1940).
—'Das Rätsel der Zeit im Urteil Kohelets (Koh 3,1-15)', *ZTK* 58 (1961), pp. 1-15.
—*Der Prediger* (HAT, 1; Tübingen, 1969 [1940]).
Gammie, J., and W. Breuggemann (eds.), *Israelite Wisdom: Theological and Literary Essays in Honor of Samuel Terrien* (Festscheift S. Terrien: Missoula, MT: Scholars Press, 1978).
Garrett, D., 'Qoheleth on the Use and Abuse of Political Power', *TJ* 8 (1987), pp. 159-77.
—'Ecclesiastes 7.25-29 and the Feminist Hermeneutic', *CTR* 2 (1988), pp. 309-21.
—*Proverbs, Ecclesiastes, Song of Songs* (NAC; Nashville: Broadman Press, 1993).
Geertz, C., *Local Knowledge: Further Essays in Interpretive Anthropology* (New York: Basic Books, 1983).
Genette, G., *Narrative Discourse: An Essay in Method* (with a forward by J. Culler; Ithaca: Cornell University Press, 1981).
—*Narrative Discourse Revisited* (trans. J. Lewin; Ithaca, NY: Cornell University Press, 1988).
Gerhart, M., and A. Russell, 'The Cognitive Effect of Metaphor', *Listening* 25 (1990), pp. 114-26.
Gese, H., *Lehre und Wirklichkeit in der Alten Weisheit* (Tübingen: J.C.B. Mohr, 1958).
—*Les Sagesses du Proche-Orient an aen: Colloque de Strasborg, 17-19 mai 1962* (Paris: Presses universitaires de France, 1963), pp. 139-51.
—'The Crisis of Wisdom in Koheleth', in J. Crenshaw (ed.), *Theodicy in the Old Testament* (trans. L. Grabbe; Philadelphia: Fortress Press, 1983 [1963]), pp. 141-53.
Gianto, A., 'The Theme of Enjoyment in Qoheleth [smch]', *Bib* 73 (1992), pp. 528-32.
Gibson, W., 'Authors, Speakers, Readers, and Mock Readers', in Tompkins (ed.), *Reader-Response Criticism*, pp. 1-6.
Gilbert, M., 'La description de la vieillesse en Qohelet XII,7, est-elle allégorique?', in J. Emerton (ed.), *Congress Volume, Vienna* (VTSup, 32; Leiden: E.J. Brill, 1981), pp. 96-109.
Ginsberg, H.L., 'The Designation *Melek* as Applied to the Author [Qoheleth]', in Ginsberg (ed.), *Studies in Koheleth*, pp. 12-15.
—'Koheleth 12.4 in Light of Ugaritic', *Syria* 33 (1951), pp. 99-101.
—'Supplementary Studies in Kohelet', *PAAJR* (1952), pp. 35-62.
—'The Structure and Contents of the Book of Koheleth', in M. Noth and D.W. Thomas (eds.), *Wisdom in Israel and in the Ancient Near East* (VTSup, 3; Leiden: E.J. Brill, 1955), pp. 138-49.
—'The Quintessence of Koheleth', in A. Altman (ed.), *Biblical and Other Studies* (Cambridge, MA: Harvard University Press, 1963), pp. 47-59.

—*The Five Megilloth and Jonah: A New Translation* (Philadelphia: Jewish Publication Society of America, 1st edn, 1969).

Ginsberg, H.L. (ed.), *Studies in Kohelet* (New York: Jewish Theological Seminary of America, 1950).

Ginsburg, C., *Coheleth (Commonly Called the Book of Ecclesiastes)* (New York: Ktav, 1970 [1861]). (Reprint)

Glasson, T., '"You Never Know': The Message of Ecclesiastes 11.1-6', *EvQ* 60 (1983), pp. 43-48.

Glender, S., 'On the Book of Ecclesiastes: A Collection Containing 'Diverse' Sayings or a Unified and Consistent Worldview', *BethM* 26 (1981), pp. 378-87 (Hebrew).

Glowinski, M., 'On the First-Person Novel', *NLH* 9 (1977), pp. 103-14 (trans. R. Stone).

Goldknopf, D., 'The Confessional Increment: A New Look at the I-Narrator', *JAAC* 28 (1970), pp. 13-21.

Goldingay, J., *Models for Interpretation of Scripture* (Grand Rapids: Eerdmans, 1995).

Golka, F., 'Die israelitische Weisheitsschule oder des Kaisers neue Kleider', *VT* 33 (1983), pp. 257-70.

—'Die Königs- und Hofsprüche und der Ursprung der israelitischen Weisheit', *VT* 36 (1986), pp. 13-36.

Good, E., 'The Unfilled Sea: Style and Meaning in Ecclesiastes 1.2-11', in Gammie and Breuggemann (eds.), *Israelite Wisdom*, pp. 58-73.

—*Irony in the Old Testament* (Bible & Literature Series, 3; Sheffield: Almond Press, 2nd edn, 1981).

Gordis, R., 'Eccles. 1.17: Its Text and Interpretation', *JBL* 56 (1937), pp. 323-30.

—'Quotations in Wisdom Literature', *JQR* 30 (1939–40), pp. 123-47.

—'The Original Language of Qohelet', *JQR* 37 (1946–47), pp. 67-84.

—'Quotations as a Literary Usage in Biblical, Oriental, and Rabbinic Literature', *HUAR* 22 (1949), pp. 157-219.

—'Koheleth: Hebrew or Aramaic?', *JBL* 71 (1952), pp. 93-109.

—'Was Koheleth a Phoenician?', *JBL* 74 (1955), pp. 103-14.

—*Koheleth: The Man and His World* (New York: Schocken Books, 3rd edn, 1968 [1951]).

—'Quotations in Biblical, Oriental, and Rabbinic Literature' in R. Gordis (ed.), *Poets, Prophets, and Sages* (Bloomington: Indiana University Press, 1971), pp. 104-59.

—'The Rhetorical Use of Interrogative Sentences in Biblical Hebrew', in J. Burden (ed.), *The Word and the Book: Studies in Biblical Language and Literature* (New York: Ktav, 1976), pp. 152-57.

—'Virtual Quotations in Job, Sumer, and Qumran [Eccl. 4.8]', *VT* 31 (1981), pp. 410-27.

Gordon, C.H., 'North Israelite Influence on Post-Exilic Hebrew', *IEJ* 5 (1955), pp. 85-88.

Görg, M., 'Zu einer bekannten Paronomasie in Koh 2,8', *BN* 90 (1997), pp. 5-12.

Goethe, J. von, *The Autobiography of Goethe. Truth and Poetry: From My Own Life* (2 vols.; trans A.J.W. Morrison; London: Henry G. Bohn, 1949).

Grant-Davie, K., 'Between Fact and Opinion: Readers' Representations of Writers' Aims in Expository, Persuasive, and Ironic Discourse' (PhD dissertation; San Diego, CA: University of California, 1985).

Grossberg, D., 'Multiple Meaning: Part of a Compound Literary Device in the Hebrew Bible', *EAJT* 4 (1986), pp. 77-86.

Grossmann, H.C., '*taḥat haššâmâyim*: Anmerkungen zum Ort des Menschen bei Qohälät', in M. Albani and T. Arndt (eds.), *Gottes Ehre erzählen* (Festschrift H. Seidel; Leipzig: Thomas Verlag, 1994), pp. 221-23.

Habel, N., 'Appeal to Ancient Tradition as a Literary Form', *ZAW* 88 (1976), pp. 253-72.
—'Wisdom, Wealth and Poverty Paradigms in the Book of Proverbs', *BibBh* 14 (1988), pp. 26-49.
Haden, N., 'Qoheleth and the Problem of Alienation', *CSR* 17 (1987), pp. 52-66.
Halloran, S., 'Language and the Absurd', *PR* 6 (1973), pp. 97-108.
Hamlyn, D., 'Empiricism', in P. Edwards (ed.), *The Encyclopedia of Philosophy* (8 vols.; New York: Macmillan, 1967), II, pp. 499-505.
Handy, L., 'One Problem Involved in Translating to Meaning: An Example of Acknowledging Time and Tradition', *SJOT* 10 (1996), pp. 16-27.
Harrison, C.R., Jr, 'Qoheleth in Social-Historical Perspective' (PhD dissertation; Durham, NC: Duke University, 1991).
—'Qoheleth Among the Sociologists', *BibInt* 5 (1997), pp. 160-80.
Hart, F., 'Notes on the Anatomy of Autobiography', *NLH* 1 (1970), pp. 485-511.
Hauser, G., *Introduction to Rhetorical Theory* (SCS; San Francisco: Harper & Row, 1986).
Hayman, A.P., 'Qoheleth and the Book of Creation', *JSOT* 50 (1991), pp. 93-111.
Heath, J., 'A Problem of Genre: Two Theories of Autobiography', *Semiotica* 64 (1987), pp. 307-17.
Heaton, E.W., *The School Tradition of the Old Testament: The Bampton Lectures for 1994* (New York: Oxford University Press, 1994).
Held, M., 'Rhetorical Questions in Ugaritic and Biblical Hebrew', *ErI* (1969), pp. 71-79.
Hengel, M., *Judaism and Hellenism* (trans. J. Bowden; 2 vols.; London: SCM Press, rev. edn, 1974 [1973]).
—*Jews, Greeks, and Barbarians: Aspects of the Hellenization of Judaism in the Pre-Christian Period* (trans. J. Bowden; Philadelphia: Fortress Press, 1980).
Henry, M., *Commentary on the Whole Bible. Vol. 6, Acts to Revelation* (New York: Fleming H. Revell Co., rev. edn, 1925).
Hernandi, P., 'Literary Theory: A Compass for Critics', *CritInq* 3 (1976), pp. 369-86.
Hertzberg, H.W., 'Palästinische Bezüge im Buche Kohelet', *ZDPV* 73 (1957), pp. 13-24.
—*Der Prediger* (KAT, 17.4; Gütersloh: Gerd Mohn, 1963).
Hikins, J., 'The Epistemological Relevance of Intrapersonal Rhetoric', *SSCJ* 42 (1977), pp. 220-27.
Hirsch E.D., Jr, 'Author as Reader and Reader as Author: Reflections on the Limits of Interpretation', in E. Spolsky (ed.), *The Uses of Adversity: Failure and Accommodation in Reader Response* (London: Associated University Press, 1990), pp. 36-48.
Hochman, B., *Character in Literature* (Ithaca, NY: Cornell University Press, 1985).
Höffken, P., 'Das Ego des Weisen', *TZ* 4 (1985), pp. 121-35.
Hoffman, Y., 'The Technique of Quotation and Citation as an Interpretive Device', in B. Uffenheimer and H. Reventlow (ed.), *Creative Biblical Exegesis: Christian and Jewish Hermeneutics through the Centuries* (JSOTSup, 59; Sheffield: JSOT Press, 1995), pp. 71-79.
Holland, F., 'Heart of Darkness: A Study of Qohelet 3.1-15', *PIBA* 17 (1994), pp. 81-101.
Holland, N., *Five Readers Reading* (New Haven: Yale University Press, 1975).
—*Holland's Guide to Psychoanalytic Psychology and Literature-and-Psychology* (New York: Oxford University Press, 1990).
Holm-Nielsen, S., 'On the Interpretation of Qoheleth in Early Christianity', *VT* 24 (1974), pp. 168-77.
—'The Book of Ecclesiastes and the Interpretation of It in Jewish and Christian Theology', *ASTI* 10 (1975–76), pp. 38-96.

Horton, E., 'Koheleth's Concept of Opposites', *Numen* 19 (1972), pp. 1-21.

Howarth, W., 'Some Principles of Autobiography', *NLH* 5 (1974), pp. 364-81.

Humbert, P., *Recherches sur les sources egyptiennes de la littérature sapientale d'Israël* (Neuchâtel: Delachaux & Niestlé, 1929).

Irwin, W., 'Eccles. 4.13-16', *JNES* 3 (1944), pp. 255-57.

Isaksson, B., 'The Autobiographical Thread: The Trait of Autobiography in Qoheleth', in Isaksson, *Studies in the Language of Qoheleth*, pp. 39-68.

—*Studies in the Language of Qoheleth: With Special Emphasis on the Verbal System* (AUUSSU, 10; Uppsala: Almquist & Wiksell, 1987).

—'The Pronouns in Qoheleth' in Isaksson, *Studies in the Language of Qoheleth*, pp. 142-71.

Iser, W., 'Indeterminacy and the Reader's Response in Prose Fiction', in J. Hillis Miller (ed.), *Aspects of Narrative: Selected Papers from the English Institute* (New York: Columbia University Press, 1971), pp. 1-45.

—'The Reading Process: A Phenomenological Approach', *NLH* 3 (1972), pp. 279-99.

—*The Implied Reader: Patterns of Communication in Prose Fiction from Bunyan To Beckett* (Baltimore: The Johns Hopkins University Press, 1974).

—*The Act of Reading: A Theory of Aesthetic Response* (Baltimore: The Johns Hopkins University Press, 1978).

—'Narrative Strategies as a Means of Communication', in Valdés and Miller (ed.), *Interpretation of Narrative*, pp. 100-17.

—'The Current Situation of Literary Theory: Key Concepts and the Imaginary', *NLH* 11 (1979), pp. 1-20.

—'The Indeterminacy of the Text', *CompCrit* 2 (1980), pp. 27-47 (trans. R. Foster).

—'Interaction Between Text and Reader', in Suleiman and Crosman (ed.), *The Reader in the Text*, pp. 106-19.

—'The Reading Process: A Phenomenological Approach', in Tompkins (ed.), *Reader-Response Criticism*, pp. 50-69.

—*Prospecting: From Reader Response to Literary Anthropology* (Baltimore: The Johns Hopkins University Press, 1989).

Japhet, S., '"Goes to the South and Turns to the North' (Ecclesiastes 1.6): The Sources and Traditions of the Exegetical Traditions', *JSQ* 1 (1993–94), pp. 289-322.

Jarick, J., 'An Allegory of Age as Apocalypse (Ecclesiastes 12.1-7)', *Colloquium* 22 (1990), pp. 19-27.

Jasper, F.N., 'Ecclesiastes: A Note for Our Time', *Int* 21 (1967), pp. 259-73.

Jastrow, M., Jr, *A Gentle Cynic: Being a Translation of the Book of Koheleth Commonly Known as Ecclesiastes Stripped of Luke's Additions; Also its Origin, Growth and Interpretation* (Philadelphia: Lippincott, 1919).

Jauss, H., 'Theses on the Transition from the Aesthetics of Literary Work to a Theory of Aesthetic Experience', in Valdés and Miller (ed.), *Interpretation of Narrative*, pp. 137-47.

Jefferson, T., 'Winter in Washington, 1807', in B.L. Rayner, *Life of Jefferson with Selections from the Most Valuable Portions of His Voluminous and Unrivalled Private Correspondence* (Boston, MA: Lilly, Wait, Colman & Holden, 1834). {Editor: Sorry, I found the bibliographic information but not the book (page number was given by online source), so I cannot verify the exact pages of this chapter.}

Jenni, E., 'Das Wort *'Olam* im Alten Testament, 1', *ZAW* 64 (1952), pp. 197-248.

—'Das Wort *'Olam* im Alten Testament, 2', *ZAW* 65 (1953), pp. 1-35.

Johnson, R., 'The Rhetorical Question as a Literary Device in Ecclesiastes' (PhD dissertation; Louisville, KY: Southern Baptist Theological Seminary, 1986).

Johnston, R.E., 'From an Author-Oriented to a Text-Oriented Hermeneutic: Implications of Paul Ricoeur's Hermeneutical Theory of the Interpretation of the New Testament' (PhD dissertation; Louvain-la-Neuve and Brussels, Belgium: Catholic University of Leuven, 1977).

Johnson, R.F., 'A Form Critical Analysis of the Sayings in the Book of Ecclesiastes' (PhD dissertation; Atlanta, GA: Emory University, 1973).

Johnston, R.K., '"Confessions of a Workaholic": A Reappraisal of Qoheleth', *CBQ* 38 (1976), pp. 14-28.

Jongeling, M., 'L'Expression *my yodea'* dans l'Ancien Testament', *VT* 24 (1974), pp. 32-40.

Jordan, W., and W. Adams, 'I.A. Richards' Concept of Tenor-Vehicle Interaction', *CSSJ* 27 (1976), pp. 136-45.

Kaiser, O., 'Fate, Suffering and God: The Crisis of a Belief in a Moral World Order in the Book of Ecclesiastes', *OTE* 4 (1986), pp. 1-13.

Kallas, E., 'Ecclesiastes: *Traditum et Fides Evangelica*; The Ecclesiastes Commentaries of Martin Luther, Philip Melanchthon, and Johannes Brenz Considered within the History of Interpretation' (PhD dissertation; Berkeley, CA: Graduate Theological Union, 1979).

Kamenetzky, A., 'Das Koheleth-Rätsel', *ZAW* 29 (1909), pp. 63-69.

—'Die Rätselname Koheleth', *ZAW* 34 (1914), pp. 225-28.

—'Die ursprünglich beabsichtige Aussprache der Pseudonyms QHLT', *OLZ* 1 (1921), pp. 11-15.

Kaufer, D., 'Irony and Rhetorical Strategy', *PR* 10 (1977), pp. 90-110.

—'Ironic Evaluations', *CM* 48 (1981), pp. 25-38.

—'Irony, Interpretive Form, and the Theory of Meaning', *PT* 4 (1983), pp. 451-64.

Kayser, W., *Das sprachliche Kuntswerke deutschen Literatur* (Bern: A. Francke, 1961).

—'Die Wahrheit der Dieter: Wandlung eines Begriffes in der deutschen Literatur (Hamburg: Rowohlt, 1961).

Kazin, A., 'Autobiography as Narrative', *NLH* 3 (1964), pp. 210-16.

Keck, L., 'Will the Historical-Critical Method Survive?', in Spencer (ed.), *Orientation by Disorientation*, pp. 115-27.

Keegan, T., 'Biblical Criticism and the Challenge of Postmodernism', *BibInt* 3 (1995), pp. 1-14.

Keifert, P., 'Mind Reader and Maestro: Models for Understanding Biblical Interpreters', in D. McKim (ed.), *A Guide To Contemporary Hermeneutics* (Grand Rapids: Eerdmans, 1986), pp. 220-38.

Kelber, W., *The Oral and Written Gospel: The Hermeneutics of Speaking and Writing in the Synoptic Tradition, Mark, Paul, and Q* (Philadelphia: Fortress Press, 1983).

Kirschenblatt-Gimblett, B., 'Toward a Theory of Proverb Meaning', *Proverbium* 22 (1973), pp. 821-27.

Kleinert, P., *Der Prediger Salomo* (Berlin: G.W.F. Müller, 1864).

Klopfenstein, M., 'Die Skepsis des Qoheleth', *TZ* 28 (1972), pp. 97-109.

—'Kohelet und die Freude am Dasein', *TZ* 47 (1991), pp. 97-107.

Knopf, C., 'The Optimism of Koheleth', *JBL* 49 (1930), pp. 195-99.

Kobayashi, Y., 'The Concept of Limits in Qoheleth' (PhD dissertation; Louisville, KY: Southern Baptist Theological Seminary, 1986).

Könen, K., 'Zu den Epilogen des Buches Qohelet', *BN* 72 (1994), pp. 24-27.

Koops, R., 'Rhetorical Questions and Implied Meaning in the Book of Job', *BT* 39 (1989), pp. 415-23.

Kruger, H., 'Old Age Frailty Versus Cosmic Deterioration? A Few Remarks on the Interpretation of Qohelet 11,7–12,8', in Schoors (ed.), *Qohelet in the Context of Wisdom*, pp. 399-411.

Krüger, T., '"Frau Weisheit' in Koh 7,26?', *Bib* 73 (1992), pp. 394-403.

—'Qoh 2.24-26 und die Frage nach dem 'Guten' im Qohelet-Buch', *BN* 72 (1994), pp. 70-84.

—'Dekonstruction und Rekonstruction prophetischer Eschatologie im Qohelet-Buch', in Anja Diesel *et a*l. (eds.), *'Jedes Ding hat seine Zeit': Studien zur israelitischen und altorientalischen Weisheit* (Festschrift D. Michel; BZAW, 241; Berlin: W. de Gruyter, 1996), pp. 107-29.

Kuenen, A., *Historische-kritische Einleitung in die Bücher des Alten Testament*, III.2 (Leipzig: O.R. Reisland, 1897).

Kugel, J., *The Idea of Biblical Poetry* (New Haven: Yale University Press, 1981).

—'Ecclesiastes', in P. Achtemeier (ed.), *Harper's Bible Dictionary* (San Francisco: Harper and Row, 1985), pp. 236-37.

—'Qoheleth and Money', *CBQ* 51 (1989), pp. 32-49.

Labuschagne, C.J., *The Incomparability of Yahweh in the Old Testament* (POS, 5; Leiden: E.J. Brill, 1966).

Laertius, D., 'Chrysippus', from *Lives of Eminent Philosophers*. (2 vols.; trans. R.D. Hicks, Loeb Classical Library, edited by T.E. Page, Harvard University Press, Cambridge, MA: 1970).

Lakoff, G., and M. Johnson, *Metaphors We Live By* (Chicago: University of Chicago Press, 1980).

Landy, F., 'On Metaphor, Play and Nonsense', *Semeia* 61 (1993), pp. 219-37.

Lang, B., 'Ist der Mensch hilflos?', *TQ* 159 (1979), pp. 109-24.

—*Ist der Mensch hilflos? Zum Buch Kohelet [Qoh 5.9-6.6, 2.1-3.15, 4.7-16, 7.15-22, 8.10-15, 9.13-10.1]* (Zürich: Benzinger Verlag, 1979).

Lategan, B., 'Current Issues in the Hermeneutical Debate', *Neot* 8 (1984), pp. 1-17.

—'Reference: Reception, Redescription, and Realty', in Lategan and Vorster, *Text and Reality*, pp. 67-93.

—'Some Unresolved Methodological Issues in New Testament Hermeneutics', in Lategan and Vorster, *Text and Reality*, pp. 3-25.

—'Coming to Grips with the Reader in Biblical Literature', *Semeia* 48 (1990), pp. 3-20.

—'Some Implications of Reception Theory for the Reading of Old Testament Texts', *OTE* 7 (1994), pp. 105-12.

Lategan, B., and W. Vorster, *Text and Reality: Aspects of Reference in Biblical Texts* (SBLSS; Philadelphia: Fortress Press, 1985).

Lauha, A., 'Die Krise des religiosen Glaubens bei Koheleth', *VTSup* 3 (1955), pp. 183-91.

—*Wisdom in Israel and in the Ancient Near East* (ed. M. Noth and D.W. Thomas (Leiden, E.J. Brill).

—*Kohelet* (BKAT, 19; Neukirchen–Vluyn: Neukirchener Verlag, 1978).

—'*Omnia Vanitas*: Die Bedeutung von *hbl* bei Kohelet', in J. Kiilunen *et al.* (eds.), *Glaube und Gerechtigkeit* (Festschrift R. Gyllenberg; SFEG, 38; Helsinki: Suomen Eksegeettisen Seura, 1983), pp. 19-25.

Lavoie, J., 'Etude exegétique et intertextuelle du Qohélet: Rapports entre Qohelet 1–11 et Gn 1–11 (PhD dissertation; Montreal: University of Montreal, 1989).

—'A quoi sert-il de perdre sa vie la gagner? Le repos dans le Qohelet', *ScEs* 44 (1992), pp. 331-47.

—'Bonheur et finitude humaine: Etude de Qo 9.7-10', *ScEs* 45 (1993), pp. 313-24.

—'Etude de l'expression *bêt 'ôlāmô* dans Qo 12,5 à la lumière des textes du Proche-Orient ancien', in J. Petit (ed.), *Où demeure Tu? (Jn 1,38): La maison depuis le monde biblique* (Saint Laurent, Quebec: Fides, 1994), pp. 213-26.

—'Un éloge à Qohelet (étude de Qo 12,9-10)', *LavTP* 50 (1994), pp. 145-70.

—'De l'inconvénient d'être né: Etude de Qohélet 4,1-3', *SR* 24 (1995), pp. 297-308.

—'La philosophie politique de Qo 9,13-16', *ScEs* 49 (1997), pp. 315-28.

Leahy, M., 'The Meaning of Ecclesiastes 10.15', *ITQ* 18 (1951), pp. 288-95.

Lee, B., 'A Specific Application of the Proverbs in Ecclesiastes 1.15', *JHStud* 1 (1997), pp. 1-25 (18). (http.//www.arts.ualberta.ca/JHS/).

Leff, M., 'In Search of Ariadne's Thread: A Review of the Recent Literature on Rhetorical Theory', *CSSJ* 29 (1978), pp. 73-91.

Lejeune, P., 'Autobiography in the Third Person', *NLH* 9 (1977), pp. 27-50.

—'The Autobiographical Contract', in T. Todorov (ed.), *French Literary Theory Today* (Cambridge: Cambridge University Press, 1982), pp. 192-222.

Levenson, J., 'Historical Criticism and the Enlightenment Project', in J. Levenson, *The Hebrew Bible, the Old Testament, and Historical Criticism: Jews and Christians in Biblical Studies* (Louisville, KY: Westminster/John Knox Press, 1993).

Levine, D., *The Flight from Ambiguity: Essays in Social and Cultural Theory* (Chicago: University of Chicago Press, 1985).

Levine, E., 'Qoheleth's Fool: A Composite Portrait', in Y. Raday and A. Brenner (eds.), *On Humour and the Comic in the Bible* (JSOTSup, 92; Sheffield: Almond Press, 1990), pp. 277-94.

—'The Humor in Qohelet', *ZAW* 109 (1997), pp. 71-83.

Loader, J.A., 'Qohelet 3.2-8: A 'Sonnet' in the Old Testament', *ZAW* 81 (1969), pp. 240-42.

—'Different Reactions of Job and Qoheleth to the Doctrine of Retribution (Eccl 7.15-20; Prov 10-22)' in Wyk (ed.), *Studies in Wisdom Literature*, pp. 43-48.

—'Relativity in Near Eastern Wisdom (Prov 25, 26; Eccl 2, 7.5, 9)', in Wyk (ed.), *Studies in Wisdom Literature*, pp. 49-58.

—'Sunt lacrimae rerum (et mentem mortalia tangunt)—Qoh 4.1-3', in W. Wyk (ed.) *Aspects of the Exegetical Process* (Pretoria: NHW Press, 1977), pp. 83-94.

—*Polar Structures in the Book of Qohelet* (BZAW, 152; Berlin: W. de Gruyter, 1979).

—*Ecclesiastes: A Practical Commentary* (trans. J. Vriend; Grand Rapids: Eerdmans, 1986).

Loemker, L.E., 'Optimism and Pessimism', in Edwards (ed.), *The Encyclopedia of Philosophy*, pp. 114-21.

Lohfink, N., 'Technik und Tod nach Kohelet', in F. Wulf *et al.*, *Strukturen christlicher Existenz* (Würzburg: Echter Verlag, 1968), pp. 27-35.

—'War Kohelet ein Frauenfeind?', in M. Gilbert (ed.), *La sagesse de l'Ancien Testament*, (BETL, 51; Leuven: Leuven University Press, 1979), pp. 259-87.

—'Der Bible skeptische hintertur', *StZ* 198 (1980), pp. 17-31.

—*Kohelet* (DNEB; Würzburg: Echter Verlag, 1980).

—'*Melek*, *Sallit*, und *Mosel* bei Kohelet und die Aufassungszeit des Buchs', *Bib* 62 (1981), pp. 535-43.

—'Warum is der Tor Unfärig, Böse zu Handeln?', *ZDMG* (1983), pp. 113-20.

—'The Present and Eternity: Time in Qoheleth', *TD* 34 (1987), pp. 236-40.

—'Koh 1,2 'Alles ist Windhauch': Universale oder anthropologische Aussage?', in R. Mosis and L. Ruppert (eds.), *Der Weg zum Menschen* (Freiburg: Herder, 1989), pp. 201-16.

—'Koheleth und die Banken: Zur Übersetzung von Koheleth V 12-16', *VT* 39 (1989), pp. 488-95.

—'Das "Poikilometron": Kohelet und Menippos von Gadara', *BK* 45 (1990), p. 19.

—' "Freu dich, junger Mann...": Das Schlussgedicht des Koheletsbuches (Koh 11,9–12,8)', *BK* 45 (1990), pp. 12-19.

—'Qoheleth 5.17-19: Revelation by Joy', *CBQ* 52 (1990), pp. 625-35.

—'Von Windhauch, Gottesfurcht und Gottes Antwort in der Freude', *BK* 45 (1990), pp. 26-32.

—'Zur Philosophie Kohelets: Eine Auslegung vom Kohelet 7,1-10', *BK* 45 (1990), pp. 20-25.

—'Grenzen und Einbildung des Kohelet-Schlussgedichts', in P. Mommer and W. Thiel (eds.), *Altes Testament: Forschung und Wirkung* (Festschrift H. Reventlow; Bern: Peter Lang, 1994).

—'Les épilogues du livre de Qohélet et les débuts du Canon', in *Ouvrir les écritures* (Paris: Cerf, 1995), pp. 77-96.

Long, B.O., 'Textual Determinacy: A Response', *Semeia* 62 (1993), pp. 157-63.

—'One Man Among a Thousand, But Not a Woman Among Them All: A Note on the Use of *māṣā'* in Ecclesiastes vii 28', in K. Schunck and M. Augustin (eds.), *Lasset uns Brücken bauen* (BEATAK, 42; Bern: Peter Lang, 1995), pp. 101-109.

Longman, T., III, 'Comparative Methods in Old Testament Studies: Ecclesiastes Reconsidered', *TSFB* 7 (1983), pp. 5-9.

—*The Book of Ecclesiastes* (NICOT; Grand Rapids: Eerdmans, 1998).

Loretz, O., 'Zur Darbietungsform der 'Ich-Erzählung' des Buche Qohelet', *CBQ* 25 (1963), pp. 46-59.

—*Qohelet und der alte Orient: Untersuchungen zu Stil und theologischer Thematik des Buches Qohelet* (Freiburg: Herder, 1964).

—'Gleiches Los Trifft Alle! Die Antwort des Buches Qohelet', *BK* 20 (1965), pp. 6-8.

—'Altorientalische und kanaanäische Topoi im Buche Kohelet', *UF* 12 (1980), pp. 267-86.

—'Anfänge jüdischer Philosophie nach Qohelet 1,1-11 und 3,1-15', *UF* 23 (1991), pp. 223-44.

—'Poetry and Prose in the Book of Qoheleth (1.1–3.22; 7.23–8.1; 9.6-10; 12.8-14)', in J.C. de Moor and W.G.E. Watson (eds.), *Verse in Ancient Near Eastern Prose* (AOAT, 42; Neukirchen–Vluyn: Neukirchener Verlag, 1993), pp. 155-89.

Lux, R., ' "Ich, Kohelet, bin König..." Die Fiktion als Schlüssel zur Wirklichkeit in Kohelet 1.12–2.26', *EvT* 50 (1990), pp. 331-42.

Mack, B., *Rhetoric and the New Testament* (Philadelphia: Fortress Press, 1990).

Mailloux, S., 'Learning to Read: Interpretation and Reader-Response Criticism', *SLI* 12 (1979), pp. 93-108.

—*Interpretive Conventions: The Reader in the Study of American Fiction* (Ithaca, NY: Cornell University Press, 1982).

—'Rhetorical Hermeneutics', *CritInq* 11 (1985), pp. 620-41.

Makarushka, I., 'Nietzche's Critique of Modernity: The Emergence of Hermeneutical Consciousness', *Semeia* 51 (1990), pp. 193-214.

Malbon, E., 'Text and Contexts: Interpreting the Disciples in Mark', *Semeia* 62 (1993), pp. 81-102.

Margolin, U., 'Characterization in Narrative: Some Theoretical Prolegomena', *Neophilologus* 67 (1983), pp. 1-14.

—'The Doer and the Deed: Action as a Basis for Characterization in Narrative', *PT* 7 (1986), pp. 205-25.

Marra, J., 'The Lifelike 'I': A Theory of Response to First-Person Narrator/Protagonist Fiction (PhD dissertation; Lubboc, TX: Texas Technical University, 1986).

Martens, L., *The Diary Novel* (Cambridge: Cambridge University Press, 1985).

—'Saying 'I' ', *SLR* 2 (1985), pp. 27-46.

Martin, W., *Recent Theories of Narrative* (Ithaca, NY: Cornell University Press, 1986).

McCracken, D., 'Character in the Boundary: Bakhtin's Interdividuality in Biblical Narratives', *Semeia* 63 (1993), pp. 29-42.

McCrosky, J., 'Ethos: A Confounding Element in Communication Research', *SM* 33 (1966), pp. 456-63.

McGuire, M., 'The Structure of Rhetoric', *PR* 153 (1982), pp. 149-69.

McKane, W., *Proverbs: A New Approach* (OTL; Philadelphia: Westminster Press, 1970).

McKenna, J., 'The Concept of *Hebel* in the Book of Ecclesiastes', *SJT* 45 (1992), pp. 19-28.

McKenzie, A., 'Subversive Sages: Preaching on Proverbial Wisdom in Proverbs, Qohelet and the Synoptic Jesus through the Reader Response Theory of Wolfgang Iser' (PhD dissertation; Princeton, NJ: Princeton Theological Seminary, 1994).

McKnight, E., *The Bible and the Reader: An Introduction to Literary Criticism* (Philadelphia: Fortress Press, 1985).

—*Postmodern Use of the Bible: The Emergence of Reader-Oriented Criticism* (Nashville: Abingdon Press, 1988).

Meade, D., *Pseudonymity and Canon* (Grand Rapids: Eerdmans, 1987).

Merkin, D., 'Ecclesiastes', in D. Rosenberg (ed.), *Congregation: Contemporary Writers Read the Jewish Bible* (London: Harcourt Brace Jovanovich, 1987), pp. 393-405.

Mesner, D., 'The Rhetoric of Citations: Paul's Use of Scripture in Romans 9' (PhD dissertation; Evanston, IL: Northwestern University, 1991).

Michel, D., 'Humanität angesichts des Absurden: Qohelet (Prediger) 1,2–3,15', in H. Förster (ed.), *Humanität heute* (Berlin: Lutherisches Verlagshaus, 1970), pp. 22-36.

—'Vom Gott, der im Himmel ist: Reden von Gott bei Qohelet', *ThVia* 12 (1973–74), pp. 87-100.

—'Qohelet-Probleme: Überlegungen zu Qoh 8,2-9 und 7,11-14', *ThVia* 15 (1979–80), pp. 81-103.

—'Ein skeptischer Philosoph: Prediger Salomo (Qohelet)', in *Universität im Rathaus*, 7 (1987), pp. 1-31.

—*Qohelet* (EF, 258; Darmstadt: Wissenshaftliche Buchgesellschaft, 1988).

—*Untersuchungen zur Eigenart des Buches Qohelet* (BZAW, 183; New York: W. de Gruyter, 1989).

—'Gott bei Kohelet: Anmerkungen zu Kohelets Reden von Gott', *BK* 45 (1990), pp. 32-36.

—'Kohelet und die Krise der Weisheit', *BK* 45 (1990), pp. 2-6.

—'Probleme der Koheletauslegung heute', *BK* 45 (1990), pp. 7-11.

—' "Unter der Sonne": Zur Immanenz bei Qohelet', in Schoors (ed.), *Qohelet in the Context of Wisdom*, pp. 93-111.

Miller, D., 'The Question of the Book: Religion as Texture', *Semeia* 40 (1987), pp. 53-64.

—'Qohelet's Symbolic Use of *Hebel*', *JBL* 117 (1998), pp. 437-54.

Mills, C., 'Relationships Among Three Sources of Credibility in the Communication Configuration: Speaker, Message and Experimenter', *SSCJ* 42 (1977), pp. 334-51.

Min, Y., 'How do the Rivers Flow? (Ecclesiastes 1.7)', *BT* 42 (1991), pp. 226-31.
Misch, G., *A History of Autobiography in Antiquity* (trans. E.W. Dickes; London: Routledge & Kegan Paul, 1950 [1907]).
Mitchell, H.G., 'The Omission of the Interrogative Particle', in R. Harper, F. Brown and G. Moore (eds.), *Old Testament and Semitic Studies in Memory of W.R. Harper* (Festschrift W.I. Harper; Chicago: University of Chicago Press, 1908), pp. 115-29.
Moore, S., *Literary Criticism and the Gospels: The Theoretical Challenge* (New Haven: Yale University Press, 1989).
—'Doing Gospel Criticism as/with a Reader', *BTB* 19 (1990), I, pp. 95-93.
Morgan, R., and J. Barton, *Biblical Interpretation* (Oxford: Oxford University Press, 1988).
Mowinckel, S., 'Die vorderasiastischen Königs-Fürstenschriften, eine stilistiche Studie', in H. Schmidt (ed.), *Eucharisterion* (Festschrift H. Gunkel; Göttingen: Vandenhoeck & Ruprecht, 1923), pp. 278-323.
—' "Ich" und "Er" in der Ezrageschichte', in A. Kuschke (ed.), *Verbannung und Heimkehr* (Festschrift W. Rudolph; Tübingen: J.C.B. Mohr, 1961), pp. 211-33.
Mulder, J.S., 'Qoheleth's Division and also its Main Point', in W.C. Delsman *et al.* (eds.), *Von Kanaan bis Kerala* (Festschrift J. Van der Ploeg; Neukirchen-Vluyn: Neukirchener Verlag, 1982), pp. 140-59.
Müller, H., 'Wie Sprach Qohälät von Gott?', *VT* 18 (1968), pp. 507-21.
—'Neige der althebräischen Weisheit: Zum Denken Qohäläts', *ZAW* 90 (1978), pp. 238-64.
—'Theonome, Skepsis und Lebensfreude: Zu Koh 1,12–3,15', *BZ* 30 (1986), pp. 1-19.
—'Der unheimliche Gast: Zum Denken Koheleths', *ZTK* 84 (1987), pp. 440-64.
—'Kohelet und Amminadab', in O. Kaiser (ed.), *Jedes Ding hat seine Zeit: Studien zur israelitschen und altorientalischen Weishet* (Festschrift D. Michel; BZAW, 241; Berlin: W. de Gruyter, 1996), pp. 149-65.
—'Travestien und geistige Landschaften zum Hintergrund einiger Motive bei Kohelet und im Hohenlied', *ZAW* 109 (1997), pp. 557-74.
Murphy, R., 'The Penseés of Coheleth', *CBQ* 17 (1955), pp. 304-14.
—Assumptions and Problems in Old Testament Wisdom Research', *CBQ* 29 (1967), pp. 407-18.
—'The Interpretation of Old Testament Wisdom Literature', *Int* 23 (1969), pp. 289-301.
—'A Form-Critical Consideration of Ecclesiastes 7', in G. MacRae (ed.) (SBLSP, 1; Missoula, MT: Scholars Press, 1974), pp. 77-85.
—'Wisdom Theses', in J. Armenti (ed.), *The Papin Festschrift: Essays in Honor of Joseph Papin* (Philadelphia: Villanova University Press, 1976), pp. 187-200.
—'Qohélet le Sceptique', *Concordia* 119 (1976), pp. 57-62.
—'Wisdom: Theses and Hypotheses', in Gammie and Breuggemann (eds.), *Israelite Wisdom*, pp. 35-42.
—'Qoheleth's 'Quarrel' with the Fathers', in D. Hadidian (ed.), *From Faith to Faith* (Pittsburgh: Pickwick Press, 1979), pp. 234-44.
—*Wisdom Literature: Job, Proverbs, Ruth, Canticles, Ecclesiastes, and Esther* (FOTL, 12; Grand Rapids: Eerdmans, 1981).
—'Qohelet Interpreted: The Bearing of the Past on the Present', *VT* 32 (1982), pp. 331-36.
—'The Faith of Qoheleth', *WW* 7 (1987), pp. 253-60.
—'On Translating Ecclesiastes', *CBQ* 53 (1991), pp. 571-79.
—'Qoheleth and Theology?', *BTB* 21 (1991), pp. 30-33.
—*Ecclesiastes* (WBC, 23A; Waco, TX: Word Books, 1992).

Newing, E., 'Rhetorical Art of the Deuteronomist: Lampooning Solomon in First Kings', *OTE* 7 (1994), pp. 247-60.

Newsom, C., 'Job and Ecclesiastes', in J. Mays, D. Petersen and K. Richards (eds.), *Old Testament Interpretation: Past, Present, and Future* (Festschrift G. Tucker; Nashville: Abingdon Press, 1995), pp. 177-94

Niehoff, M., 'Do Biblical Characters Talk to Themselves?', *JBL* 111 (1992), pp. 577-95.

Nietzche, F., 'History in the Service and Disservice of Life', in F. Nietzche, *Unmodern Observations* (ed. W. Arrowsmith; trans. G. Brown; New Haven: Yale University Press, 1990), pp. 75-145.

No Author, 'Introduction: Darwin and Us' (http.//www.clark.net/pub/wright/introduc.htm).

Noble, P., 'Hermeneutics and Postmodernism: Can We Have a Radical Reader-Response Theory? (Part 2)', *RS* 31 (1995).

Ogden, G., 'The 'Tob-Spruch' in Qoheleth: Its Function and Significance as a Criterion for Isolating and Identifying Aspects of Qoheleth's Thought' (PhD dissertation; Princeton: Princeton Theological Seminary, 1975).

—'"Better-Proverb" (*Tob-Spruch*), Rhetorical Criticism, and Qoheleth', *JBL* 96 (1977), pp. 489-505.

—'Qoheleth's Use of the 'Nothing Is Better' Form', *JBL* 98 (1979), pp. 339-50.

—'Historical Allusion in Qoheleth IV: 13-16?', *VT* 30 (1980), pp. 309-15.

—'Qoheleth IX.17–X.20: Variations on the Theme of Wisdom's Strength and Vulnerability', *VT* 30 (1980), pp. 27-37.

—'Qoheleth IX.1-16', *VT* 32 (1982), pp. 158-69.

—'Qoheleth XI.1-6', *VT* 33 (1983), pp. 222-30.

—'The Mathematics of Wisdom: Qoheleth IV: 1-12', *VT* 34 (1984), pp. 446-53.

—'Qoheleth XI.7–XII.8: Qoheleth's Summons to Enjoyment and Reflection', *VT* 34 (1984), pp. 27-38.

—'The Interpretation of *dwr* in Ecclesiastes 1.4', *JSOT* 34 (1986), pp. 91-92.

—*Qoheleth* (Readings: A New Biblical Commentary; Sheffield: JSOT Press, 1987).

—'"Vanity" It Certainly Is Not', *BT* 38 (1987), pp. 301-307.

—'Translation Problems in Ecclesiastes 5.13-17', *BT* 39 (1989), pp. 423-28.

Ong, W., 'The Writer's Audience is Always a Fiction', *PMLA* 90 (1975), pp. 9-21.

—*Orality and Literacy: The Technologizing of the Word* (New York: Methuen, 1982).

Pakh, J., 'The Significance of *'ašer* in Qoh. 7,26: 'More bitter than death is the woman *if* she is a snare' ', in Schoors (ed.), *Qohelet in the Context of Wisdom*, pp. 373-83.

Paterson, J., 'The Intimate Journal of an Old-Time Humanist', *RL* 19 (1950), pp. 245-54.

Pedersen, J., 'Scepticisme Israelite', *RHPR* 10 (1930), pp. 317-70.

Perdue, L., *The Collapse of History: Reconstructing Old Testament Theology* (OBT: Minneapolis, MN: Fortress Press, 1994).

—'"I will make a test of pleasure": The Tyranny of God and Qoheleth's Quest for the Good', in L. Perdue, *Wisdom and Creation* (Nashville: Abingdon Press, 1994), pp. 193-242.

Perelman, C., *The Realm of Rhetoric* (trans. W. Kluback; with an introduction by C. Arnold; Notre Dame: University of Notre Dame Press, 1982).

Perelman, C., and L. Olbrechts-Tyteca, *The New Rhetoric: A Treatise on Argumentation* (trans. J. Wilkinson and P. Weaver; Notre Dame: University of Notre Dame Press, 1969).

Perry, M., 'Literary Dynamics: How the Order of a Text Creates Its Meanings (Part One)', *PT* 1 (1979), pp. 35-64.

—'Literary Dynamics: How the Order of a Text Creates Its Meanings (Part Two)', *PT* 1 (1979), pp. 311-61.

Perry, T.A., *Dialogues with Kohelet: The Book of Ecclesiastes, Translation and Commentary* (University Park: The Pennsylvania State University Press, 1993).

—*Wisdom Literature and the Structure of Proverbs* (University Park: Pennsylvania State University Press, 1993).

Peter, C.B., 'In Defence of Existence: A Comparison between Ecclesiastes and Albert Camus', *BTF* 12 (1980), pp. 26-43.

Petersen, N., 'Literary Criticism in Biblical Studies', in Spencer (ed.), *Orientation by Disorientation*, pp. 25-52.

—'The Reader in the Gospel', *Neot* 18 (1984), pp. 38-51.

Phelan, J., 'Narrative Discourse, Literary Character, and Ideology', in J. Phelan (ed.), *Reading Narrative: Form, Ethics, Ideology* (Columbus: Ohio State University Press, 1989), pp. 132-46.

—*People Reading, Reading Plots: Character, Progression, and the Interpretation of Narrative* (Chicago: University of Chicago Press, 1989).

Pick, B., 'Ecclesiastes or the Sphinx of Hebrew Literature', *Open Court* 17 (1903), pp. 361-71.

Piwowarczyk, M., 'The Narratee and the Situation of Enunciation: A Reconsideration of Prince's Theory', *Genre* 9 (1976), pp. 161-77.

Plumptre, E., *Ecclesiastes* (CB, 23; Cambridge: Cambridge University Press, 1890).

Plutarch, *Plutarch's Lives. The Translation Called Dryden's*. (5 vols., rev. edn. by A.H. Clough; New York: The Athenaeum Society (1905 [orig 1859]),cited by J. Barlett, *Familiar Quotations: A Collection of Passages, Phrases, and Proverbs Traced to Their Sources in Ancient and Modern Literature* (10th edn; revised and enlarged by Nathan Dole, Boston: Little, Brown, and Co, 1930).

Podechard, E., 'La composition du livre de l'Ecclésiaste', *RB* 21 (1912), pp. 161-91.

—*L'Ecclésiaste* (Paris: Librairie Lecoffre, 1912).

Polk, T., 'The Wisdom of Irony: A Study of *Hebel* and its Relation to Joy and the Fear of God in Ecclesiastes', *SBTh* 6 (1976), pp. 3-17.

Popkin, R., 'Skepticism', in P. Edwards (ed.), *The Encyclopedia of Philosophy* (New York: Macmillan, 1967), VII, pp. 449-61.

Prince, G., 'Notes Toward a Categorization of Fictional "Narratees"', *Genre* 4 (1971), pp. 100-106.

—'On Readers and Listeners in Narrative', *Neophilologus* 55 (1971), pp. 117-22.

—'Introduction to the Study of the Narratee', in Tompkins (ed.), *Reader-Response Criticism*, pp. 7-25.

—'Notes on the Text as Reader', in Suleiman and Inge (eds.), *The Reader in the Text*, pp. 225-40.

—'Reading and Narrative Competence', *L'Esprit créateur* 21 (1981), pp. 81-88.

—*Narratology: The Form and Functioning of Narrative* (Berlin: Mouton de Gruyter, 1982).

—*A Dictionary of Narratology* (Lincoln, NB: University of Nebraska Press, 1987).

Rabinowitz, P., 'Truth in Fiction: A Re-Examination of Audiences', *CritInq* 4 (1977), pp. 121-41.

Rad, G. von., *Wisdom in Israel* (trans. J. Martin; Nashville: Abingdon Press, 1971).

Rainey, A., 'A Study of Ecclesiastes', *Concordia* 35 (1964), pp. 148-57.

—'A Second Look at *Amal'* in Qoheleth', *Concordia* 36 (1965), p. 805.

Rankin, O.S., 'The Book of Ecclesiastes', in *IB* (12 vols.; Nashville, TN: Abingdon Press, 1956), V, pp. 3-88.
Ranston, H., 'Koheleth and the Early Greeks', *JTS* 24 (1923), pp. 160-69.
—*Ecclesiastes and the Early Greek Wisdom Literature* (London: Epworth Press, 1925).
Raschke, C., 'From Textuality to Scripture: The End of Theology as 'Writing' ', *Semeia* 40 (1987), pp. 39-52.
Rashbam, *The Commentary of Rabbi Samuel ben Meir Rashbam on Qoheleth* (ed. S. Japhet and R. Salters; Jerusalem: Magnes Press; Leiden: E.J. Brill, 1985).
Rashkow, I., 'In our image we create him, male and female we create them': The E/Affect of Biblical Characterization', *Semeia* 63 (1993), pp. 105-13.
Reed, E., 'Whither Postmodernism and Feminist Theology?', *FemTh* 6 (1994), pp. 15-29.
Reed, W., *Dialogues of the Word: The Bible as Literature According to Bakhtin* (New York: Oxford University Press, 1993).
Renan, E., *L'Ecclésiaste traduit de l'hebréu avec une étude sur l'âge et le caractère du livre* (Paris: Levy, 1882).
Rendtorff, R., *The Old Testament: An Introduction* (trans. J. Bowden; Philadelphia: Fortress Press, 1986).
Renza, L., 'The Veto of the Imagination: A Theory of Autobiography', *NLH* 9 (1977), pp. 1-26.
Resseguie, J., 'Reader-Response Criticism and the Synoptic Gospels', *JAAR* 52 (1984), pp. 307-24.
Rice, P. and P. Waugh, 'The Postmodern Perspective', in Rice and Waugh (eds.), *Modern Literary Theory: A Reader*, pp. 428-40.
Rice, P. and P. Waugh (eds.), *Modern Literary Theory: A Reader* (London: Edward Arnold, 1989).
Richards, I.A., *The Philosophy of Rhetoric* (Oxford: Oxford University Press, 1936).
Richards, K., 'From Scripture to Textuality', *Semeia* 40 (1987), pp. 119-24.
Ricoeur, P., 'The Metaphorical Process', *Semeia* 4 (1975), pp. 75-106.
—'Philosophical Hermeneutics and Theological Hermeneutics', *SR* 5 (1975), pp. 14-33.
—'The Specificity of Religious Langauge', *Semeia* 4 (1975), pp. 107-39.
—*Interpretation Theory: Discourse and the Surplus of Meaning* (Fort Worth: Texas Christian University Press, 1976).
—*Essays on Biblical Interpretation* (with an introduction by L. Mudge; Philadelphia: Fortress Press, 1980).
—*Hermeneutics and the Human Sciences: Essays on Language, Action and Interpretation* (ed. and trans. J. Thompson; Cambridge: Cambridge University Press, 1981).
Rideout, P., 'Narrator/Narratee/Reader Relationships in First Person Narrative: John Barth's *The Floating Opera*, Albert Camus' *The Fall*, and Gunter Grass' *Cat and Mouse*' (PhD dissertation; Tallahassee, FL: Florida State University, 1981).
Riffaterre, M., 'Interpretation and Descriptive Poetry', *NLH* 4 (1972–73), pp. 229-56.
—'Semantic Overdetermination in Poetry', *PTL* 2 (1977), pp. 1-19.
—*Semiotics in Poetry* (Advances in Semiotics; Bloomington: Indiana University Press, 1978).
Rimmon-Kenan, S., *Narrative Fiction: Contemporary Poetics* (New York: Methuen, 1983).
Rogers, R., 'Amazing Reader in the Labyrinth of Literature', *PT* 3 (1982), pp. 31-46.
Romberg, B., *Studies in the Narrative Technique of the First-Person Novel* (Lund: Almqvist and Wilksell, 1962).

Roosevelt, T., *History as Literature and Other Essays* (New York: Charles Scribners and Sons, 1913).
—*A Book-Lover's Holidays in the Open* (New York: Charles Scribners Sons, 1916).
Rosendal, B., 'Popular Wisdom in Qohelet', in K. Jeppensen, K. Nielsen and B. Rosendal (eds.), *In the Last Days: On Jewish and Christian Apocalyptic and its Period* (Aarhus: Aarhus University Press, 1994), pp. 121-27.
Rosenthal, P., 'The Concept of Ethos and the Structure of Persuasion', *SM* 33 (1966), pp. 114-25.
Rousseau, F., 'Structure de Qohelet I 4-11 et plan du livre', *VT* 31 (1981), pp. 200-17.
Routledge, D., 'Faithful Reading: Poststructuralism and the Sacred', *BibInt* 4 (1996), pp. 271-87.
Rowley, H.H., 'The Problems of Ecclesiastes', *JQR* 42 (1951–52), pp. 87-90.
Rudman, D., 'A Contextual Reading of Ecclesiastes 4.13-16', *JBL* 116 (1997), pp. 57-73.
—'Woman as Divine Agent in Ecclesiastes', *JBL* 116 (1997), pp. 411-27.
Russell, J., '*Memoirs of Mackintosh*', in J. Barlett, *Familiar Quotations: A Collection of Passages, Phrases, and Proverbs Traced to Their Sources in Ancient and Modern Literature* (10th Edition; revised and enlarged by Nathan Dole, Boston: Little, Brown, and Co, 1930). {Or, just use the Bartlett reference cited above and omit this reference to Russell in lieu of citing Bartlett, see note in Footnotes addendum.}
Sacken, J., *'A Certain Slant of Light': Aesthetics of First-Person Narration in Gide and Cather* (New York: Garland, 1985).
Salters, R., 'The Word for 'God' in the Peshitta of Koheleth', *VT* 21 (1971), pp. 251-54.
—'Qoheleth and the Canon', *ExpTim* 86 (1975), pp. 339-42.
—'A Note on the Exegesis of Ecclesiastes 3,15b', *ZAW* 88 (1976), pp. 419-20.
—'Text and Exegesis in Koh 10,19', *ZAW* 89 (1977), pp. 423-26.
—'Notes on the History of the Interpretation of Koh 5,5', *ZAW* 90 (1978), pp. 95-101.
—'Notes on the Interpretation of Qoh 6,2', *ZAW* 91 (1979), pp. 282-89.
—'Exegetical Problems in Qoheleth', *IBS* 10 (1988), pp. 44-59.
Salyer, G., 'Vain Rhetoric: Implied Author/Narrator/Narratee/Implied Reader Relationships in Ecclesiastes' Use of First-Person Discourse' (PhD dissertation; Berkely, CA: Graduate Theological Union, 1997).
Sanders, J., *From Sacred Story to Sacred Text* (Philadelphia: Fortress Press, 1987).
Savignac, J., de, 'La sagesse du Qohéléth et l'épopée de Gilgamesh', *VT* 28 (1978), pp. 318-23.
Savran, G., *Telling and Retelling: Quotation in Biblical Narrative* (ISBL; Indianapolis: Indiana University, 1988).
Sawyer, J., 'The Ruined House in Ecclesiastes 12: A Reconstruction of the Original Parable', *JBL* 94 (1976), pp. 519-31.
Scharlemann, R., 'Theological Text', *Semeia* 40 (1987), pp. 5-20.
—'The Measure of Meaning in Reading Texts', *Dialog* 28 (1989), pp. 247-50.
Scheffler, E., 'Qohelet's Positive Advice', *OTE* 6 (1993), pp. 248-71.
Schneiders, S., *The Revelatory Text: Interpreting the New Testament as Scripture* (San Francisco: HarperSanFrancisco, 1991).
Scholes, R. (ed.), *Approaches to the Novel: Materials for Poetics* (San Francisco: Chandler, rev. edn, 1966).
Schoors, A., 'La structure littéraire de Qoheleth', *OLP* 13 (1982), pp. 91-116.
—'Koheleth: A Perspective of Life after Death?', *ETL* 61 (1985), pp. 295-303.

—'Emphatic and Asseverative *kî* in Koheleth', in H. Vanstiphout *et al.* (eds.) *Scripta Signa Vocis* (Groningen: Egbert Forster, 1986), pp. 209-15.

—'The Pronouns in Qoheleth', *HS* 30 (1989), pp. 71-90.

—*The Preacher Sought to Find Pleasing Words: A Study of the Language of Qoheleth* (OLA, 41; Leuven: Peeters, 1992).

—'Bitterder dan de dood is de vrouw (Koh 7,26)', *Bijdragen* 54 (1993), pp. 121-40.

—'Qoheleth: A Book in a Changing Society', *OTE* 9 (1996), pp. 68-87.

—'Introduction' in Schoors (ed.), *Qoheleth in the Context of Wisdom*, pp. 1-13.

—'The Verb *rā'âh* in the Book of Qoheleth', in Schoors (ed.), *Qoheleth in the Context of Wisdom*, pp. 227-41.

Schoors, A. (ed.), *Qoheleth in the Context of Wisdom* (BETL, 136; Leuven: Leuven University Press, 1998).

Schubert, M., 'Die Selbstbetrachtungen Kohelets: Ein Beitrag zur Gattungsforschung', in *Theologische Versuche* 24 (1989), pp. 23-34.

—*Schöpfungstheologie bei Kohelet* (ed. M. Augustin and M. Mach; BEATAK 15, Bern: Peter Lang, 1989).

Schwarzschild, R., 'The Syntax of *'shr* in Biblical Hebrew with Special Reference to Qoheleth', *HS* 31 (1990), pp. 7-39.

Scott, R.B.Y., *Proverbs, Ecclesiastes* (AB, 18; Garden City, NY: Doubleday, 1965).

Sekine, S., 'Qoheleth as Nihilist', in S. Sekine, *Transcendency and Symbols in the Old Testament: A Genealogy of Hermeneutical Experiences* (BZAW, 275; Berlin: W. de Gruyter, 1999), pp. 91-128.

Seow, C.L., 'Qohelet's Autobiography', in A. Beck *et al.* (eds.), *Fortunate the Eyes that See* (Festschrift D.N. Freedman; Grand Rapids: Eerdmans, 1995), pp. 275-87.

—'Linguistic Evidence and the Dating of Qoheleth', *JBL* 115 (1996), pp. 643-66.

—' "Beyond Them, My Son, Be Warned": The Epilogue of Qoheleth Revisited', in Barré (ed.), *Wisdom, You Are My Sister*, pp. 125-41.

—*Ecclesiastes: A New Translation with Introduction and Commentary* (AB, 18C; New York: Doubleday, 1997).

—'Qohelet's Eschatological Poem', *JBL* 118 (1999), pp. 209-34.

Serrano, J.J., 'I Saw the Wicked Buried (Ecc 8,10)', *CBQ* 16 (1954), pp. 168-70.

Seybold, K., '*Hebel*', in *TDOT*, III, pp. 313-20.

Shank, H.C., 'Qoheleth's World and Lifeview as Seen in His Recurring Phrases', *WTJ* 37 (1974), pp. 57-73.

Shedd, M., 'Ecclesiastes from the Outside In', *RTR* 55 (1996), pp. 24-37.

Sheppard, G., 'The Epilogue to Qoheleth as Theological Commentary', *CBQ* 39 (1977), pp. 182-89.

—*Wisdom as a Hermeneutical Construct: A Study in the Sapientializing of the Old Testament* (BZAW, 151; Berlin: W. de Gruyter, 1980).

Siegfried, C.G., *Prediger und Hoheslied* (Göttingen: Vandenhoeck & Ruprecht, 1898).

Smith, M., *Persuasion and Human Action: A Review and Critique of Social Influence Theories* (Belmont, CA: Wadsworth, 1982).

Sneed, M., 'The Social Location of Qoheleth's Thought: Anomie and Alienation in Ptolemaic Jerusalem (Israel)' (PhD dissertation; Madison, NJ Drew University, 1990).

—'Qoheleth as 'Deconstructionist': 'It is I, the Lord, your redeemer…who turns sages back and makes their knowledge nonsense' (Is 44.24-25)', *OTE* 10 (1997), pp. 303-11.

Soskice, J., *Metaphor and Religious Language* (Oxford: Oxford University Press, 1985).

Staples, W., ' "Profit' in Ecclesiastes', *JNES* 4 (1945), pp. 87-96.

Spangenberg, I.J.J., 'Quotations in Ecclesiastes: An Appraisal', *OTE* 4 (1991), pp. 19-35.
—'Irony in the Book of Qohelet', *JSOT* 72 (1996), pp. 57-69.
—'A Century of Wrestling with Qohelet: The Research History of the Book Illustrated with a Discussion of Qoh 4,17-5,6', in Schoors (ed.), *Qoheleth in the Context of Wisdom*, pp. 61-91.
Spencer, R. (ed.), *Orientation by Disorientation: Studies in Literary and Biblical Criticism* (Pittsburgh: Pickwick Press, 1980).
Spengemann, W., *The Forms of Autobiography* (New Haven: Yale University Press, 1980).
Spolsky, E., 'The Uses of Adversity: The Literary Text and the Audience that Doesn't Understand', in E. Spolsky, *The Uses of Adversity: Failure and Accommodation in Reader Response* (London: Associated University Press, 1990), pp. 17-35.
Spriggs, W., 'Evolutionary Psychology for the Common Person'. (http.//www.evoyage.com/index.html).
Staley, J., *The Print's First Kiss: A Rhetorical Investigation of the Implied Author in the Fourth Gospel* (SBLDS, 82; Missoula, MT: Scholars Press, 1988).
Stanzel F., 'Towards a 'Grammar of Fiction' ', *Novel* 11 (1977), pp. 247-64.
—*A Theory of Narrative* (trans. C. Goedsche; with a preface by P. Hernadi; Cambridge: Cambridge University Press, 1984 [1979]).
Staples, W.E., ' "Profit' in Ecclesiastes', *JNES* 4 (1945), pp. 87-96.
Steiner, G., ' "Critic'/'Reader' ', *NLH* 10 (1979), pp. 423-52.
Sternberg, M., *Expositional Modes and Temporal Ordering in Fiction* (Baltimore: The Johns Hopkins University Press, 1978).
—*The Poetics of Biblical Narrative: Ideological Literature and the Drama of Reading* (ISBL; Bloomington: Indiana University Press, 1985).
Stout, J., 'What is the Meaning of a Text?', *NLH* 14 (1982), pp. 1-12.
Strauss, H., 'Erwagungen zur seelsorgerlichen Dimension von Kohelet 12,1-7', *ZTK* 78 (1981), pp. 267-75.
Suleiman, S., 'Introduction: Varieties of Audience-Oriented Criticism', in Suleiman and Crosman (ed.), *The Reader in the Text*, pp. 3-45.
—'Of Readers and Narratees: The Experience of *Pamela*', *L'Esprit Créatur* 21 (1981), pp. 89-97'.
Suleiman, S., and I. Crosman, (eds.), *The Reader in the Text* (Princeton, NJ: Princeton University Press, 1980).
Sweeney, M., 'The Critique of Solomon in the Josianic Edition of the Deuteronomistic History', *JBL* 114 (1995), pp. 607-22.
Tadmor, H., 'Autobiographical Apology in the Royal Assyrian Literature', in H. Tadmor and M. Weinfeld (eds.), *History, Historiography and Interpretation: Studies in Biblical and Cuneiform Literatures* (Jerusalem: Magnes Press, 1984), pp. 36-57.
Tamir, N., 'Personal Narration and Its Linguistic Foundation', *PTL* 1 (1976), pp. 403-29.
Tannen, D., 'Oral and Literate Strategies in Spoken and Written Narratives', *LJLSA* 8 (1982), pp. 1-21.
—'The Oral/Literate Continuum in Discourse', in D. Tannen (ed.), *Spoken and Written Language: Exploring Orality and Literacy* (ADP, 9; Norwood, NJ: Ablex, 1982), pp. 1-21.
Tate, W.R., *Biblical Interpretation: An Integrated Approach* (Peabody, MA: Hendrickson, rev. edn, 1997 [1991]).
Terry, M., 'Studies in Koheleth', *MR* 70 (1888), pp. 365-75.

Thompson, J., *The Form and Function of Proverbs in Ancient Israel* (The Hague: Mouton, 1974).

—'Preface to Ricoeur', in J. Thompson (ed.), *Hermeneutics and the Human Sciences: Essays on Language, Action and Interpretation* (trans. J. Thompson; Cambridge: Cambridge University Press, 1981), pp. 1-26.

Thompson, L., *Introducing Biblical Literature: A More Fantastic Country* (Englewood Cliffs, NS: Prentice–Hall, 1978).

Thompson, M., '"God's voice you have never heard, God's form you have never seen": The Characterization of God in the Gospel of John', *Semeia* 63 (1993), pp. 177-204.

Tita, H., 'Ist die thematische Einheit Koh 4,17–5,6 eine Anspielung auf die Salomoerzählung?', *BN* 84 (1996), pp. 87-102.

Todorov, T., 'Reading as Construction', in Suleiman and Crosman (ed.), *The Reader in the Text*, pp. 67-82.

—'Reading as Construction', in C. Porter (ed.), *Genres in Discourse* (Cambridge: Cambridge University Press, 1990), pp. 39-49.

Tompkins, J., 'An Introduction to Reader-Response Criticism', in Tompkins (ed.), *Reader-Response Criticism*, pp. ix-xxvi.

Tompkins, J. (ed.), *Reader-Response Criticism: From Formalism to Post-Structualism* (Balitimore: The Johns Hopkins University Press, 1980).

Torrey, C., 'The Question of the Original Language of Qohelet', *JQR* 39 (1948–49), pp. 151-60.

—'The Problem of Ecclesiastes IV,13-16', *VT* 2 (1952), pp. 175-77.

Toulmin, S., *Cosmopolis: The Hidden Agenda of Modernity* (Chicago: University of Chicago Press, 1993).

Tsukimoto, A., 'The Background of Qoh 11.1-6 and Qoheleth's Agnosticism', *AJBI* 19 (1993), pp. 34-52.

Urban, G., 'The 'I' of Discourse', in B. Lee, G. Urban and T. Sebeok (eds.), *Semiotics, Self, and Society* (Berlin: Mouton de Gruyter, 1989), pp. 27-51.

Uspensky, B., *A Poetics of Composition: The Structure of the Artistic Text and Typology of a Compositional Form* (trans. V. Zavarin and S. Wittig; Berkeley: University of California Press, 1973).

Utzchneider, H., 'Text—Reader—Author: Towards a Theory of Exegesis; Some European Views', *JHStud* 1 (1996), pp. 1-22 (http.//www.arts.ualberta.ca/JHS/).

Valdés, M., and O. Miller (eds.), *Interpretation of Narrative* (Toronto: University of Toronto Press, 1978).

Van der Wal, A., 'Qohelet 12,1a: A Relatively Unique Statement in Israel's Wisdom Tradition', in Schoors (ed.), *Qohelet in the Context of Wisdom*, pp. 413-18.

Van Wolde, E., 'Texts in Dialogue with Texts: Intertextuality in the Ruth and Tamor Narratives', *BibInt* 5 (1997), pp. 1-28.

Vatz, R., 'The Myth of the Rhetorical Situation', *PR* 6 (1973), pp. 154-61.

Verheij, A., 'Paradise Retried: On Qoheleth 2.4-6', *JSOT* 50 (1991), pp. 113-15.

Viviano, P., 'The Book of Ecclesiastes: A Literary Approach', *TBT* 22 (1984), pp. 79-84.

Vogels, W., 'Performance vaine et performance saine chez Qohélet', *NRT* 113 (1991), pp. 363-85.

Vorster, W., 'The Reader in the Text: Narrative Material', *Semeia* 48 (1990), pp. 21-40.

Vorster, W. and B. Lategan, *Text and Reality: Aspects of Reference in Biblical Texts* (SBLSS; Philadelphia: Fortress Press, 1985).

Wallace, M., *Recent Theories of Narrative* (Ithaca, NY: Cornell University Press, 1986).

White, H., 'A Theory of the Surface Structure of the Biblical Narrative', *USQR* 34 (1979), pp. 159-73.

Whitley, C., 'Koheleth and Ugaritic Parallels', *UF* 11 (1979), pp. 811-24.

—*Koheleth: His Language and Thought* (BZAW, 148; Berlin: W. de Gruyter, 1979).

Whybray, R.N., *The Succession Narrative: A Study of II Samuel 9–20; I Kings 1 and 2* (SBTheo, Second Series, 9; Naperville, IL: Alec R. Allenson, 1968).

—*The Intellectual Tradition in the Old Testament* (BZAW, 135; Berlin: W. de Gruyter, 1974).

—'Qoheleth the Immoralist? (Qoh 7.16-17)', in Gammie and Brueggemann (eds.), *Israelite Wisdom*, pp. 191-204.

—'The Identification and Use of Quotations in Ecclesiastes', in *Congress Volume, Vienna 1980*, ed. J.A. Emerton (VTSup, 32; Leiden: E.J. Brill, 1981), pp. 435-51.

—'Qoheleth, Preacher of Joy', *JSOT* 23 (1982), pp. 87-98.

—'Ecclesiastes 1.5-7 and the Wonders of Nature', *JSOT* 41 (1988), pp. 105-12.

—*Ecclesiastes* (NCBC; Grand Rapids: Eerdmans, 1989).

Wilch, J., *Time and Event* (Leiden: E.J. Brill, , 1969).

—'Qoheleth as a Theologian', in Schoors (ed.), *Qoheleth in the Contect of Wisdom*, pp. 234-65.

Williams, J., ' "What does it profit a man?": The Wisdom of Qoheleth', in Crenshaw (ed.), *Studies in Ancient Israelite Wisdom*, pp. 375-89.

—'The Power of Form: A Study of Biblical Proverbs', *Semeia* 17 (1980), pp. 35-58.

—*Those Who Ponder Proverbs: Aphoristic Thinking and Biblical Literature* (Bible & Literature Studies, 2; Sheffield: Almond Press, 1981).

Wilson, G., 'The Words of the Wise: The Intent and Significance of Qoheleth 12.9-14', *JBL* 103 (1984), pp. 175-92.

Wilson, L., 'Artful Ambiguity in Ecclesiastes 1,1-11', in Schoors (ed.), *Qohelet in the Context of Wisdom*, pp. 357-65.

Wilson, P., *Man, the Promising Primate: The Conditions of Human Evolution* (New Haven: Yale University Press, 1980).

Wimsatt, W.K., Jr, 'Intention', in J. Shipley (ed.), *World Dictionary of Literature* (New York: The Philosophical Library, 1942), pp. 326-29 (reprinted in W.K. Wimsatt Jr and M. Beardsley, *The Verbal Icon* [Lexington: The University of Kentucky Press, 1954], pp. 3-18).

Winquist, C., 'Preface', *Semeia* 40 (1987), pp. i-iii

Wise, M., 'A Calque from Aramaic in Qoheleth 6.12; 7.12; and 8.13', *JBL* 109 (1990), pp. 249-57.

Worster, W., 'The Reader in the Text: Narrative Material', *Semeia* 48 (1990), pp. 21-40.

Wright, A., 'The Riddle of the Sphinx: The Structure of the Book of Qoheleth', *CBQ* 30 (1968), pp. 313-34.

—'The Riddle of the Sphinx Revisited: Numerical Patterns in the Book of Qoheleth', *CBQ* 42 (1980), pp. 35-51.

—' "For everything there is a season": The Structure and Meaning of the Fourteen Opposites (Ecclesiastes 3,2-8)', in J. Dore *et al.* (eds.), *De la Tôrah au Messie: Mélanges Henri Cazelles* (Paris: Desclée de Brouwer, 1981), pp. 321-28.

—'Additional Numerical Patterns in Qoheleth', *CBQ* 45 (1983), pp. 32-43.

—'The Poor But Wise Youth and the Old But Foolish King (Qoh 4.13-16)', in Barré (ed.), *Wisdom, You Are My Sister*, pp. 142-54.

Wright, C.H., *The Book of Koheleth* (London: Hodder & Stoughton, 1883).

Wright, J.S., 'The Interpretation of Ecclesiastes', *EvQ* 18 (1946), pp. 18-34.
Wright, R., *The Moral Animal: Why We Are the Way We Are; The New Science of Evolutionary Psychology* (New York: Vintage Books, 1994).
—*The Moral Animal: Evolutionary Psychology and Everyday Life* (New York: Peter Smith, 1997).
Wuellner, W., 'Where is Rhetorical Criticism Taking Us?', *CBQ* 49 (1987), pp. 448-63.
Wyk, W. (ed.), *Studies in Wisdom Literature* (OTSWA, 15-16; Pretoria: NHW Press, 1972–73).
Youngblood, R., 'Qoheleth's 'Dark House' (Eccl. 12.5)', *JETS* 29 (1986), pp. 397-410.
Zimmerli, W., 'Das Buch Kohelet: Traktat oder Sentenzensammlung?', *VT* 24 (1974), pp. 221-30.
—' "Unveranderbare Welt" oder "Gott ist Gott?": Ein Plädoyer für die Unaufgebbarkeit des Predigerbuches in der Bibel', in H. Geyer *et al.* (eds.), *Wenn nicht jetzt, wann dann?* (Festschrift H. Kraus; Neukirchen–Vluyn: Neukirchen Verlag, 1983), pp. 165-78.
Zimmermann, F., 'The Aramaic Provenance of Qohelet', *JQR* 36 (1945–46), pp. 17-45.
—'The Question of Hebrew in Qohelet', *JQR* 40 (1949–50), pp. 17-45.
—*The Inner World of Qoheleth* (New York: Ktav, 1973).
Zuck, R. (ed.), *Reflecting with Solomon: Selected Studies on the Book of Ecclesiastes* (Grand Rapids: Baker Books, 1995).

INDEX

INDEX OF REFERENCES

OLD TESTAMENT

Genesis
1 265
2 282

Exodus
15.11 348

Numbers
15.39 227, 280, 367

Deuteronomy
14.26 284
23.22 314
32.31 348

Joshua
2 280

Judges
9.13 284

1 Samuel
23.17 271
25.7 373
25.34 267

2 Samuel
5.2-5 271

1 Kings
1.34 271
2.43 349
3–11 188, 315
3 273
3.2-15 220, 289
3.28 271
4.21-28 192
4.29-34 192
8.1 244
8.46 341
10.5 192
11.1-3 280

1 Chronicles
5.12-13 192
27.27-31 192

2 Chronicles
1.1-13 220
1.1-3 289

Ezra
2.55 243
2.57 243

Nehemiah
5.7 243
7.59 243

Esther
2.12 280
2.15 280

Job
15.14-16 341
22.15 300

Psalms
19 265
19.4 265
39.5 253
39.11 253
62.9 253
78.33 253
82.1 55
86.8 348
89.7 348
94.11 253
104.15 284
144.4 253

Proverbs
10–29 172, 358
1–9 183, 229
1.2-7 277
1.1 191
1.12 188
1.16 188
2.4-9 188
2.16-19 343
5.1-4 343
7.22-23 343
8.15 187
12.24 363
21.22 361

22.17 191
24.23 191
26.4-5 135, 203
30.1 191

Ecclesiastes
1.1–12.14 215
1–6 328, 330
1.1–6.9 324, 328
1.1–4.16 152, 158, 266
1.1–2.24 306, 324
1 276, 278
1.1-11 135, 213
1.1 144, 155, 158, 190, 205, 242, 245, 372
1.2–12.8 144
1.2-18 313
1.2-11 82, 90, 158, 213, 267
1.2-3 217, 246, 262
1.2 155, 211-13, 245, 247, 252-54, 256, 262, 263
1.3–12.7 156
1.3–4.16 155, 271
1.3–4.12 271
1.3–3.15 146, 204
1.3–3.9 146
1.3-11 152, 271, 335
1.3 72, 141, 146, 152, 207, 247, 248, 257, 260-62, 266, 267, 271, 283, 290, 295, 296, 328, 372
1.4-11 143, 146, 161, 217, 246, 262, 271, 295, 296, 332, 333
1.4-8 264
1.4 204, 263, 264, 370
1.5-8 264, 265
1.7-8 267
1.7 225
1.8 204, 262, 321
1.9-11 265
1.9 301
1.10 257, 264, 265
1.11–2.11 169
1.11 266
1.12–12.7 213, 216
1.12–7.29 347
1.12–6.9 159
1.12–4.16 175, 271
1.12–2.26 142, 146, 162, 185, 186, 189, 271, 295
1.12–2.3 276
1.12-26 272
1.12-18 272, 276, 285, 352
1.12-15 158, 272, 276
1.12 90, 141, 142, 158, 169, 170, 172, 190, 192, 245, 246, 263, 273, 279, 349
1.13-17 192
1.13-18 341
1.13-15 181, 204, 271, 276, 277
1.13 150, 274, 279, 280, 341, 344
1.14 253, 272, 273, 274
1.15 204, 205, 207, 274, 275, 278, 329, 337
1.16-18 131, 158, 181, 204, 276
1.16 175, 176, 225, 277, 278, 341
1.17 285, 322
1.18 204, 207, 277, 278, 329
2 46, 85, 103, 294, 296, 301, 304, 307, 308
2.1–2.17 213
2.1-12 284, 289
2.1-11 105, 106, 124, 142, 158, 276, 285, 294, 338
2.1-8 343, 346, 349
2.4-7 192
2.7-10 192
2.1-2 181, 276, 278
2.1 175, 225, 253, 279, 341, 354
2.2 72, 152, 207, 257, 279
2.3-11 131, 181, 276, 278, 287, 289, 294

Ecclesiastes (cont.)
2.3-8 278, 279
2.3 279-82, 343, 344
2.4-8 280
2.4-6 282
2.4 189
2.7 192
2.8 192, 280
2.9-11 282
2.9 282
2.10-11 160, 288
2.10 279, 282, 293
2.11 189, 253, 261, 262, 267, 283, 292, 352
2.12-17 158, 276
2.12-14 181, 286
2.12-13 322
2.12 72, 152, 207, 257, 285
2.13-15 131
2.13 267, 272, 286
2.14 276, 285, 286, 290, 303, 329
2.15-17 181, 286
2.15 72, 152, 207, 253, 257, 342
2.16 287
2.17 253
2.18–6.9 158
2.18-26 158, 295, 308
2.18-21 276
2.18-19 308-10, 319
2.18 287-90
2.19 72, 152, 207, 253, 257, 290, 328, 342
2.20-23 181
2.20 289
2.21 289, 290
2.22 72, 152, 207, 257, 261, 289, 290
2.23 253, 291
2.24-26 157, 187, 279, 281, 292-95, 301
2.24 158, 162, 272, 286, 289, 292
2.25 72, 152, 257
2.26 253, 279, 293, 294
3 85, 207, 295
3.1–4.16 324
3.1–4.6 306
3.1-22 160
3.1-21 331
3.1-15 143, 158, 181, 295, 296, 302
3.1-13 162
3.1-8 146, 195, 271, 295-98, 333, 340, 349
3.2-8 297
3.2 298
3.3 298
3.5 297, 298
3.7 298
3.9 72, 146, 152, 207, 257, 267, 271, 289, 295, 299
3.10–4.12 271
3.10-15 295, 299, 333
3.10 272
3.11 104, 127, 131, 210, 304, 306, 331, 333
3.12-13 157, 158, 187, 301
3.12 160, 162, 163, 304
3.14-22 162
3.14-15 301
3.16–4.6 307
3.16–4.3 152
3.16-22 181
3.16-17 303
3.16 272, 304
3.17-18 131, 341
3.17 175, 293, 303
3.18 175, 303
3.19-21 264, 267
3.19-20 262
3.19 267, 303
3.20 329, 370
3.21 72, 152, 207, 257, 290, 328, 334
3.22 72, 152, 157, 162, 187, 257, 272, 328
4–5 203
4 207, 307, 315, 318, 358
4.1–5.19 162
4.1-3 181, 304
4.1 304, 315, 319, 336, 341, 361, 363
4.3 310
4.4-6 104, 181, 305, 308, 309, 329
4.4-5 229
4.4 253, 272, 308
4.5-6 203
4.5 305

4.6	305, 306, 310
4.7-12	181
4.7-9	158
4.7	253, 307
4.8-12	308
4.8	72, 152, 207, 253, 257, 308, 309, 319
4.9-12	203
4.9	289, 310, 311, 329
4.10-12	311
4.11-12	311
4.11	72, 152, 207, 257
4.12	329
4.13-16	131, 134, 181, 312
4.13	310, 329, 364
4.15	272
4.16	155, 253
4.17–12.8	152
4.17–6.9	324
4.17–5.19	321
4.17–5.8	155, 299, 313, 314, 318, 377
4.17–5.6	152, 153, 314
4.17	152-55, 158, 175, 205, 207, 266, 311, 313, 317, 329
5	155
5.1	152, 205, 207, 314, 329
5.2-3	160
5.2	315
5.3-5	333, 375
5.3	152, 229, 288, 314
5.4-5	207
5.4	152, 310, 313, 315
5.5	152, 207, 257, 313-15, 329
5.6	138, 207, 253, 293, 313, 315, 374
5.7–6.12	152
5.7-8	313
5.7	152, 207, 288, 315
5.8	267, 315
5.9–6.9	155, 294, 319, 324, 335, 340
5.9-12	203
5.9-11	329
5.9	253, 262, 318
5.10	257, 318, 319
5.11	318
5.12–6.9	158
5.12-16	181, 318, 321
5.12-15	363
5.12	272, 318, 319
5.13-17	181
5.13-16	319
5.13	286, 319
5.15-16	319
5.15	257, 267
5.17-19	157, 158, 181, 187, 294, 320-22
5.17	162, 163, 272, 318
5.18	320
5.19	320
5.1a	313
5.1b	313
5.3	313
6.1–8.15	162
6.1-9	181, 321
6.1-6	322, 323
6.1-2	322, 329
6.1	272
6.2	253, 323
6.3-4	311
6.3	322
6.4-5	322
6.4	253
6.6	207, 257, 319, 322, 323, 370
6.7-9	322, 323
6.7	329
6.8	138, 207, 257, 262, 267, 319
6.9	158, 159, 253, 294, 310, 323, 328-30
6.10–12.14	290
6.10–12.7	328, 329
6.10–8.17	352-54
6.10–7.22	155, 332
6.10–7.14	146
6.10-12	158, 332, 350
6.10	158, 331-33
6.11	253, 257, 262, 267, 326, 328
6.12	207, 257, 290, 301, 328, 334, 350, 352
6.2	322
6.3	310
7–12	328
7–10	203
7–9	359
7	158, 268, 280, 318, 326, 358
7.1–12.8	152
7.1–11.6	334
7.1–8.17	159, 326, 334

Ecclesiastes (cont.)
7.1-14 158
7.1-13 146, 337, 338, 348
7.1-12 334, 335
7.1-10 203
7.1-4 336
7.1-2 335
7.1 205, 310, 335, 376
7.2 310, 311, 335
7.3 310, 311, 335, 336
7.4-5 138
7.4 336
7.5-6 311
7.5 310, 335, 336
7.6 156, 253, 336
7.7 131, 138, 243, 336, 348
7.8 210, 336, 347
7.9-10 205
7.9 335, 336
7.10 207, 335, 336
7.11-14 203
7.11-12 131, 338, 344
7.11 267, 268
7.12-13 203
7.12 267, 336, 341
7.13 205, 257, 337
7.14-29 338
7.14 207, 338
7.15-24 158
7.15-22 181, 339, 342, 343
7.15-18 136, 356
7.15-16 339
7.15 272, 356
7.16-17 207, 339
7.16 257, 287, 299
7.17 207, 339
7.18 293, 374
7.19-21 341
7.19-20 348
7.20 138, 341
7.21-22 207
7.21 70, 207
7.23–8.1 294
7.23-29 155, 341
7.23-25 341
7.23 175, 341, 342
7.24 257, 328, 342
7.25-29 158, 181, 342, 343, 352, 353
7.25-27 113
7.25-26 160, 342
7.25 272, 280, 343, 344
7.26-28 338, 355
7.26 345
7.27 211, 212, 245, 246, 331, 343, 344
7.28 343, 345, 346
7.29 374
7.2b 336
7.8 310, 335
8 158, 307, 326
8.1–10.3 350
8.1-17 158, 301, 338
8.1-9 348
8.1-8 155
8.1 257, 290, 348, 349
8.2-7 349
8.2-5 349, 350
8.2-4 203
8.2 207, 349
8.4 257, 349
8.5-6 203
8.5 350
8.6-9 350
8.6-7 349
8.6 138
8.7 158, 257, 328, 349
8.8 349
8.9–9.12 155
8.9-15 181
8.9 272, 350, 351
8.10-15 350
8.10-11 351
8.10 253, 272
8.11-14 203
8.11-12 139
8.11 351
8.12-13 293, 351, 374
8.12 351
8.14-15 354
8.14 253, 279, 351
8.15–9.10 162
8.15 157, 162, 187, 289, 352-54
8.16–9.10 181
8.16-17 352
8.16 272, 357
8.17 159, 272, 326, 328, 333, 363
8.25 205
8.67 138
9–12 358
9–11 366
9 158, 356
9.1–11.7 366
9.1–11.6 159, 326, 334
9.1-12 158, 160
9.1-6 159, 354
9.1-3 357
9.1 253, 272, 338, 354
9.2-3 354

9.2	354	10.2-4	363	11.7-8	367
9.3	354	10.2-3	131	11.7	367
9.4-6	203	10.4	207, 363	11.8	253, 367
9.4-5	131, 138	10.5-7	131, 181, 358, 363	11.9–12.1	357
9.4	310, 311, 354	10.5	272	11.9-10	367
9.5	354	10.7	272, 358	11.9	74, 162, 163, 227, 279, 280, 367
9.6	355	10.8-11	364	11.10	163, 207, 368
9.7-10	136, 157, 159, 187, 339, 357	10.8-9	364	12	162, 185
9.7-9	162, 205, 293, 355	10.10-17	205	12.1-7	217
9.7	160, 207, 355, 356	10.10-11	364	12.1	163, 207, 368, 369
9.9-10	207	10.10	257, 267	12.2-7	264, 368
9.9	361	10.11	267, 364	12.2-3	367
9.10	354, 355, 370	10.12-15	203	12.2	367, 369
9.11–11.10	162	10.12-14	364	12.3-4	369
9.11-12	159, 181, 356	10.12-13	203	12.5	264
9.11	351	10.14	328, 354, 364	12.6	369, 370
9.12	354, 357	10.15	289, 354, 364	12.7	369, 370
9.13–12.7	155, 357	10.16–11.2	159	12.8-14	212, 213, 266
9.13–10.15	158	10.16-19	365	12.8	155, 156, 189, 211, 213, 217, 243, 245, 253, 256, 372
9.13–10.3	181, 358	10.16-18	364	12.9-14	144, 152, 155, 158, 213, 219, 372, 374
9.13–10.1	360	10.16	364, 365	12.9-10	120, 218, 231, 373
9.13-16	85, 358, 360, 361	10.17-18	160	12.9	148, 190, 219, 229, 244, 245, 355
9.13	272	10.17	364	12.10	245, 373
9.16-17	203	10.18	364	12.11-12	373
9.16	131, 310	10.19	364, 365	12.11	144, 146, 154, 359, 373
9.17–12.7	359	10.20	207, 365	12.12-14	74, 144, 302
9.17–11.4	358	11.1-6	365		
9.17–10.3	358	11.1-3	365		
9.17–10.1	131	11.1-2	365		
9.17-18	362	11.1	207, 366, 367		
9.18	310, 362	11.2	207, 354, 366		
10–17	350	11.3-6	159		
10	85, 158, 318, 359, 362	11.3	366		
10.1-20	146	11.4	366		
10.1-3	203	11.5-6	159, 207, 354, 366		
10.1	362	11.6	207, 367		
10.2–11.6	362	11.7–12.8	143, 158		
		11.7–12.7	146, 367, 372		
		11.7–12.1	162		
		11.7-10	157, 187		

Ecclesiastes (cont.)
12.12-13 144
12.12 74, 140-42, 207, 217, 258, 279, 312, 374, 377
12.13-14 218, 229
12.13 229, 302, 354, 374
12.14 375
12.4-5 369
12.5 369
2.14 286
2.15 175, 341
2.17-19 194
2.24 163
2.24-26 138, 292
3.11 300
3.14 301
3.22 163
4.11 311
4.12 311
5.10 318
5.11 318, 321
5.18-20 181
6.10 332, 333
6.12 328, 334
7.13 337
7.14 339
7.21-22 341
7.26 345
717 257
8.(9)10-9.12 350, 357
8.10 351
8.12-15 131
8.15 163
8.6-8 349
9.10 370
9.16 361
9.17-18 362
9.18-10.1 362
9.7-9 163

Song (Cant.)
1.1 191
1.10 280
8.11-12 191

Isaiah
3.8-15 304
5.11 284
22.13 279
40.26 333
44.24-25 342
5.11 284

Ezekiel
1.1 172

Micah
6.8 332

Habakkuk
3.17-19 354

OTHER ANCIENT REFERENCES

Apocrypha
Wisdom of Solomon
6–9 191

New Testament
Matthew
23 46, 47

1 Corinthians
7.10 393

Pseudepigrapha
Pss. Sol.
17–18 191

Qumran
1QS
3.23 244

Unknown/Other
Bar.
3.9–4.4 144

Keret
90–91 140

Sirach
16.24–17.14 144
24.3-29 144
26.19 368
44.1–50.24 220
47.12-22 220
47.19 220
2.7 190

INDEX OF MODERN AUTHORS

Adam, A.K.M. 98, 99
Adams, W. 250, 251, 254, 255
Aichele, G. 91, 108, 110, 111, 115
Albright, W.F. 137, 139
Alonso-Schökel, L. 264
Alter, R. 58, 83, 109, 215, 227, 312
Andersen, K. 195
Anderson, W. 217, 327, 328
Armstrong, J. 191
Avni, O. 118

Baltzer, K. 345
Barthes, R. 38, 58, 179
Bartlett, J. 239, 380
Barton, G. 300
Barton, J. 37, 93, 309
Bea, A. 145
Beal, T. 369
Beck, D. 74, 75, 314
Beentjes, P. 221, 349
Benveniste, E. 118
Bergant, D. 141
Berlin, A. 113, 114
Bianchi, F. 139
Bishop, E. 243
Bitzer, L. 222, 395, 399
Blank, S. 310
Blenkinsopp, J. 298, 299
Bontekoe, R. 130, 250
Booth, W. 63, 75, 78, 92, 118
Bordo, S. 33
Branham, R. 394
Bratsiotis, N. 174-76, 206
Braun, R. 228
Brenner, A. 58, 297, 298
Bréton, S. 61, 132, 133, 148, 187, 312, 380, 381, 388
Brett, M. 91
Brindle, W. 340
Brinton, A. 395
Brongers, H.A. 257
Brooks, C. 83
Brown, F. 250
Brown, S. 145, 159, 160, 335, 365
Brown, W. 116, 151
Brueggemann, D. 36, 45
Brueggemann, W. 32, 33, 222
Burden, J. 197
Burnett, F. 60, 112, 124, 242
Buzy, D. 196
Byargeon, R.W. 127, 135, 137, 292

Camp, C. 129, 208-10
Camus, A. 254
Caneday, A. 131, 148, 156, 249, 250, 254, 269, 287
Carr, D. 221, 273, 289
Carriére, J.M. 135
Castellino, G. 145, 151-53, 155, 157, 158, 276, 313
Ceresko, A. 141
Chatman, S. 57, 63-66, 73, 78-80, 83, 84, 92, 93, 109, 119, 233, 382
Chesterton, G.K. 213
Childs, B. 37, 41, 145, 190, 234, 245, 374
Christianson, E.S. 83-85, 115, 170, 172, 177, 187, 189, 194, 212, 214, 215, 250, 257, 273, 276, 285, 295, 313-15, 328, 342, 346, 347, 370, 372, 373, 375, 381
Clemens, D. 282
Clevenger, T. 195
Clines, D.J.A. 388
Cohn, D. 247
Conrad, E. 51
Consigny, S. 395

Cooper, R. 41
Crane, R.S. 84
Crenshaw, J. 132, 133, 138, 141, 144, 149, 150, 156, 174, 230, 240, 243, 257, 258, 265, 269, 275, 277, 280, 284, 289, 290, 297, 298, 300, 304, 314, 323, 327, 332-34, 336, 340, 355, 356, 360, 363, 366, 373, 374, 391
Crites, S. 397
Croatto, J. 51
Crüsemann, F. 140, 141, 168, 249, 250
Csikszentmihalyi, M. 269, 270
Culley, R. 36

Dahood, M. 137, 139-41, 300, 374
Davis, B. 369
Dawson, D. 50
De Bruin, G. 115
De Jong, S. 145, 151, 155-57, 161, 168, 249, 263, 271, 313
Delitzsch, F. 138, 143, 169
Dijk-Hemmes, F. van 58
Dell, K. 229-31, 305, 338, 350, 361, 362, 374
Delsman, W.C. 137, 138
Derrida, J. 109
Detweiler, R. 51
Dickens, C. 57
Dietrich, E. 173
Donaue, J. 35
Duesberg, H. 178
Duranti, A. 54, 55, 106

Eco, U. 81, 82, 91, 99, 250, 382, 383
Ehninger, D. 93
Eichorn, D. 146
Elgin, S. 129
Ellermeier, F. 145
Ellis, J. 47, 48, 50, 56, 382
Eslinger, L. 227

Faigley, L. 83
Farmer, K. 149, 170, 193, 197, 243, 249, 250, 252, 253, 268, 283, 326, 351
Fisch, H. 172, 174, 236, 249, 250, 368, 391
Fischer, A. 146, 169, 204, 271, 276, 295
Fish, S. 36, 59, 82, 94-97, 101, 110, 236, 258, 383, 387, 394
Fohrer. G. 145
Fontaine, C. 129, 200, 208, 347
Forman, C. 282
Forster, E.S. 84
Foucault, M. 109
Fowl, S. 50
Fowler, R. 51-53, 66, 68, 73, 79, 91, 94, 96-98, 110, 123, 128, 130, 131, 236
Fox, M. 131, 133, 134, 141, 144, 145, 148, 150, 164, 171, 181, 182, 186, 197, 204, 211, 212, 217, 219, 228, 249, 250, 252, 253, 264, 267, 272-74, 279, 312, 334, 339, 357, 366, 369, 371, 372, 390, 394
Fransen, I. 178
Fredericks, D. 138
Freund, E. 91
Freyne, S. 46
Friedman, N. 87, 125, 359
Frye, N. 29

Gadamer, H. 54, 97, 230
Galling, K. 132, 145, 149
Geertz, C. 224
Gerhart, M. 250
Gese, H. 168
Gianto, A. 249, 279
Gibson, W. 63, 75
Gilbert, M. 369
Ginsberg, C.D. 145
Ginsberg, H.L. 133, 142, 143, 337
Goethe, J. von 62
Goldingay, J. 40, 44, 45, 48
Goldknopf, D. 117, 118, 388
Good, E. 249
Gordis, R. 133, 137, 139, 145, 168, 197, 257, 260, 287, 392
Gordon, C.H. 138, 139
Görg, M. 186, 280
Grant-Davie, K. 93, 106, 338, 360
Grossmann, H. 268

Habel, N. 363, 364
Haden, K. 249, 250
Halloran, S. 255
Hamlyn, D. 238

Handy, L. 55
Harrison, C.R., Jr 168
Hart, F. 150, 166
Hauser, G. 93, 195, 312
Heaton, E.W. 259
Heidegger, M. 368
Held, M. 257
Hengel, M. 168
Henry, M. 11
Hertzberg, H.W. 131, 145, 272, 309
Hochman, B. 176, 177, 182, 285, 309
Höffken, P. 169, 205, 291, 303, 394
Hoffman, Y. 314
Holland, F. 300
Holland, N. 111, 316
Holm-Nielsen, S. 138, 196, 268

Isaksson, B. 138, 139, 145, 163, 172, 173, 179, 193, 264, 297, 319, 322, 339, 354
Iser, W. 78-80, 82, 91, 94, 95, 101-105, 107, 146, 147, 193, 332, 382, 383

James, H. 52
Jarick, J. 369
Jasper, F.N. 310, 399
Jastrow, M., Jr 172
Jefferson, T. 167
Johnson, M. 129
Johnson, R. 115, 141, 203, 248, 257-62, 273, 287, 291, 293, 294, 299, 309, 311, 319, 320, 322-24, 328, 334, 340, 341, 348, 381, 400
Johnson, R.F. 146, 178, 197, 203, 204, 310, 311, 318, 323, 335, 336, 341, 364, 366, 367
Johnston, R. 203
Johnston, R.K. 141, 156, 249
Jongeling, B. 257
Jordan, W. 250, 251, 254, 255

Kaiser, O. 168
Kamenetzky, A. 244
Kayser, W. 179
Keegan, T. 33
Kleinert, P. 241
Knopf, K. 250
Koops, R. 257
Kruger, H.A.J. 369
Krüger, T. 271, 345, 369, 371
Kuenen, A. 241
Kugel, J. 141

Labuschagne, C. 348
Laertius, D. 21
Lakoff, G. 129
Lang, B. 145
Lategan, B. 46, 51, 99, 235
Lauha, A. 147, 148, 306, 349
Lavoie, J. 144, 219, 305, 361, 372
Lee, B. 275
Leff, M. 224
Lejeune, P. 38, 117
Levenson, J. 33, 40
Levine, D. 131
Levine, E. 361
Loader, J.A. 141, 142, 149, 168, 176, 185, 257, 265, 268, 274, 284, 289, 290, 297, 298, 304, 341, 343
Loemker, L. 274, 327
Lohfink, N. 156, 162, 186, 249, 276, 321, 345, 369
Long, B. 109
Long, B.O. 346
Longman, T., III 186, 292, 318
Loretz, O. 148-50, 152, 170, 179-81, 184, 185, 187, 188, 191, 205, 228
Lux, R. 44, 169, 170, 184-86, 189, 190, 192, 193, 242, 244, 295

Mailloux, S. 91, 94, 98-101, 330, 383
Makarushka, I. 34
Margolin, U. 119-21, 247, 285, 383
Marra, J. 76, 77, 117, 122-24, 154, 281, 288, 383
Martens, L. 123
Martin, W. 80
McCroskey, D. 87
McKane, W. 208
McKenna, J. 250
McKenzie, A. 102, 115, 129, 130, 195, 197, 200, 201, 203, 292, 308, 329, 386, 400
McKnight, E. 42, 49, 51, 57
Meade, D. 142, 186, 191, 192
Melville, H. 329, 330

Merkin, D. 186, 337, 368
Messner, D. 197
Meyer, P. 83
Michel, D. 138, 170, 177, 178, 243, 268, 272, 350
Miller, D. 51, 52, 135, 136, 250, 255, 256, 262, 339, 340, 356
Mills, C. 220, 221
Misch, G. 150, 151, 180, 183
Mitchell, H.G. 265
Mowinckel, S. 174
Mulder, J.S. 151, 159, 302, 367
Müller, H. 169, 186
Murphy, R. 100, 132, 133, 138, 139, 148, 151, 156, 159, 196, 197, 205, 230, 249, 250, 277, 291, 306, 333, 334, 340, 343, 351

Newing, E. 221, 281
Newsom, C. 134, 381
Niehoff, M. 174
Nietzche, F. 34, 56

Ogden, G. 141, 144, 248, 249, 263, 264, 292, 296, 307, 310-12, 334, 335, 341, 353
Olbrechts-Tyteca, L. 223, 225
Ong, W. 53, 395

Pahk, J. 345
Paterson, J. 154
Pearce, W. 394
Pederson, J. 269
Perdue, L. 43, 186, 267, 274, 283, 284, 291, 301, 321, 323, 333, 343, 352, 363, 371, 372
Perelman, C. 222-25, 386
Perry, M. 95, 96, 101, 104-106, 153, 275, 294
Perry, T.A. 71, 148, 163, 196, 207, 240, 281-83, 293, 335, 360-62, 373, 390, 391
Peter, C.B. 249, 250, 254
Peterson, N. 78
Pick, B. 143, 177
Piwowarczyk, M. 70, 71
Plumptre, E. 143, 177
Podechard, E. 196, 241
Polk, T. 141, 181, 202, 335
Popkin, R. 327
Prince, G. 67, 69, 70, 72-74, 78, 79, 118, 123, 294, 382

Rabinowitz, P. 74, 76
Rad, G., von 150, 302, 390, 391
Rainey, A. 140, 141, 337, 374
Rankin, O.S. 300
Rashbam, S. 249, 250
Rashkow, I. 124, 125
Rayner, B.L. 167
Reed, W. 54
Renan, E. 241, 244
Rendtorff, R. 151, 159
Renza, L. 171, 184
Resseguie, J. 259
Rice, P. 31
Richards, I.A. 250, 251
Ricoeur, P. 41, 42, 44-46, 51, 56, 57, 59, 66, 97, 180, 208, 382
Rideout, P. 70, 72-74, 117, 123
Riffaterre, M. 298
Rimmon-Kennan, S. 63, 64, 79
Rodgers, R. 101
Romberg, B. 218, 294
Roosevelt, T. 126, 326
Rosendal, B. 197, 198
Rousseau, F. 151, 161, 162
Rowley, H.H. 132, 133, 142, 337
Rudman, D. 312, 344, 345
Russell, A. 250
Russell, J. 380

Salters, R. 227, 280, 368
Sanders, J. 37
Savran, G. 201
Sawyer, J. 369
Scheffler, E. 249, 250, 320, 368
Schneiders, S. 51
Schoors, A. 135, 138, 139, 150, 151, 159, 168, 169, 272
Schubert, M. 178, 181, 206, 296, 350, 358
Sekine, S. 249, 250
Seow, C.L. 138, 186, 189, 271, 275, 322, 369, 370
Seybold, K. 248

Shank, H. 149, 165
Shedd, M. 372
Sheppard, G. 144, 229, 374, 375
Siegfried, C.G. 138
Smith, M. 93
Sneed, M. 168, 342
Soskice, J. 129
Spangenberg, I.J.J. 33, 129, 234, 305, 312, 336, 350, 355, 381, 400
Spolsky, E. 127
Spriggs, W. 77
Staley, J. 78
Stanzel, F. 122, 216, 294
Staples, W.E. 257
Steiner, G. 96
Sternberg, M. 58, 64, 104, 105, 126, 226, 389
Suleiman, S. 91
Sweeney, M. 281

Tadmor, H. 189
Tamir, N. 119, 201-203
Tate, W.R. 39
Terry, M. 169
Thompson, J. 203
Thompson, L. 128, 270
Thompson, M. 108, 122, 124
Tita, H. 315
Todorov, T. 78, 81
Tompkins, J. 91
Torrey, C. 138, 139
Toulmin, S. 31-37, 56, 223
Tsukimoto, A. 314, 366

Uspensky, B. 72, 88-90, 176, 184, 213, 216-19, 382
Utzchneider, H. 91, 92

Van Wolde, E. 198-200
Van der Wal, A. 368
Vatz, R. 395
Verheij, A. 282
Viviano, P. 148, 207, 247, 248, 276
Vogels, W. 289
Vorster, W. 51

Warren, R. 83
Waugh, P. 31
White, H. 118, 119
Whitley, C. 138, 139, 244, 323
Whybray, R.N. 151, 156, 161-63, 168, 183, 186, 197, 249, 284, 292, 315, 318, 320, 333, 340, 343
Wilch, J. 297
Williams, J. 141, 257
Wilson, G. 144, 229, 374
Wilson, L. 135, 269
Wilson, P. 77
Winquist, C. 51
Worster, W. 78
Wright, A. 89, 134, 143-45, 151, 157-60, 272, 306, 312, 326, 334, 354, 366, 384
Wright, C.H.H. 138
Wright, J.S. 128, 149, 196, 268
Wright, R. 77
Wuellner, W. 176

Youngblood, R. 369

Zimmerli, W. 145, 146
Zimmermann, F. 138, 139, 168, 180, 369

JOURNAL FOR THE STUDY OF THE OLD TESTAMENT
SUPPLEMENT SERIES

200 M. Daniel Carroll R., David J.A. Clines and Philip R. Davies (eds.), *The Bible in Human Society: Essays in Honour of John Rogerson*
201 John W. Rogerson, *The Bible and Criticism in Victorian Britain: Profiles of F.D. Maurice and William Robertson Smith*
202 Nanette Stahl, *Law and Liminality in the Bible*
203 Jill M. Munro, *Spikenard and Saffron: The Imagery of the Song of Songs*
204 Philip R. Davies, *Whose Bible Is It Anyway?*
205 David J.A. Clines, *Interested Parties: The Ideology of Writers and Readers of the Hebrew Bible*
206 Møgens Müller, *The First Bible of the Church: A Plea for the Septuagint*
207 John W. Rogerson, Margaret Davies and M. Daniel Carroll R. (eds.), *The Bible in Ethics: The Second Sheffield Colloquium*
208 Beverly J. Stratton, *Out of Eden: Reading, Rhetoric, and Ideology in Genesis 2–3*
209 Patricia Dutcher-Walls, *Narrative Art, Political Rhetoric: The Case of Athaliah and Joash*
210 Jacques Berlinerblau, *The Vow and the 'Popular Religious Groups' of Ancient Israel: A Philological and Sociological Inquiry*
211 Brian E. Kelly, *Retribution and Eschatology in Chronicles*
212 Yvonne Sherwood, *The Prostitute and the Prophet: Hosea's Marriage in Literary-Theoretical Perspective*
213 Yair Hoffman, *A Blemished Perfection: The Book of Job in Context*
214 Roy F. Melugin and Marvin A. Sweeney (eds.), *New Visions of Isaiah*
215 J. Cheryl Exum, *Plotted, Shot and Painted: Cultural Representations of Biblical Women*
216 Judith E. McKinlay, *Gendering Wisdom the Host: Biblical Invitations to Eat and Drink*
217 Jerome F.D. Creach, *Yahweh as Refuge and the Editing of the Hebrew Psalter*
218 Harry P. Nasuti, *Defining the Sacred Songs: Genre, Tradition, and the Post-Critical Interpretation of the Psalms*
219 Gerald Morris, *Prophecy, Poetry and Hosea*
220 Raymond F. Person, Jr, *In Conversation with Jonah: Conversation Analysis, Literary Criticism, and the Book of Jonah*
221 Gillian Keys, *The Wages of Sin: A Reappraisal of the 'Succession Narrative'*
222 R.N. Whybray, *Reading the Psalms as a Book*
223 Scott B. Noegel, *Janus Parallelism in the Book of Job*
224 Paul J. Kissling, *Reliable Characters in the Primary History: Profiles of Moses, Joshua, Elijah and Elisha*
225 Richard D. Weis and David M. Carr (eds.), *A Gift of God in Due Season: Essays on Scripture and Community in Honor of James A. Sanders*

226 Lori L. Rowlett, *Joshua and the Rhetoric of Violence: A New Historicist Analysis*
227 John F.A. Sawyer (ed.), *Reading Leviticus: Responses to Mary Douglas*
228 Volkmar Fritz and Philip R. Davies (eds.), *The Origins of the Ancient Israelite States*
229 Stephen Breck Reid (ed.), *Prophets and Paradigms: Essays in Honor of Gene M. Tucker*
230 Kevin J. Cathcart and Michael Maher (eds.), *Targumic and Cognate Studies: Essays in Honour of Martin McNamara*
231 Weston W. Fields, *Sodom and Gomorrah: History and Motif in Biblical Narrative*
232 Tilde Binger, *Asherah: Goddesses in Ugarit, Israel and the Old Testament*
233 Michael D. Goulder, *The Psalms of Asaph and the Pentateuch: Studies in the Psalter, III*
234 Ken Stone, *Sex, Honor, and Power in the Deuteronomistic History*
235 James W. Watts and Paul House (eds.), *Forming Prophetic Literature: Essays on Isaiah and the Twelve in Honor of John D.W. Watts*
236 Thomas M. Bolin, *Freedom beyond Forgiveness: The Book of Jonah Re-examined*
237 Neil Asher Silberman and David B. Small (eds.), *The Archaeology of Israel: Constructing the Past, Interpreting the Present*
238 M. Patrick Graham, Kenneth G. Hoglund and Steven L. McKenzie (eds.), *The Chronicler as Historian*
239 Mark S. Smith, *The Pilgrimage Pattern in Exodus*
240 Eugene E. Carpenter (ed.), *A Biblical Itinerary: In Search of Method, Form and Content. Essays in Honor of George W. Coats*
241 Robert Karl Gnuse, *No Other Gods: Emergent Monotheism in Israel*
242 K.L. Noll, *The Faces of David*
243 Henning Graf Reventlow (ed.), *Eschatology in the Bible and in Jewish and Christian Tradition*
244 Walter E. Aufrecht, Neil A. Mirau and Steven W. Gauley (eds.), *Urbanism in Antiquity: From Mesopotamia to Crete*
245 Lester L. Grabbe (ed.), *Can a 'History of Israel' Be Written?*
246 Gillian M. Bediako, *Primal Religion and the Bible: William Robertson Smith and his Heritage*
247 Nathan Klaus, *Pivot Patterns in the Former Prophets*
248 Etienne Nodet, *A Search for the Origins of Judaism: From Joshua to the Mishnah*
249 William Paul Griffin, *The God of the Prophets: An Analysis of Divine Action*
250 Josette Elayi and Jean Sapin, *Beyond the River: New Perspectives on Transeuphratene*
251 Flemming A.J. Nielsen, *The Tragedy in History: Herodotus and the Deuteronomistic History*
252 David C. Mitchell, *The Message of the Psalter: An Eschatological Programme in the Book of Psalms*

253 William Johnstone, *1 and 2 Chronicles, Volume 1: 1 Chronicles 1–2 Chronicles 9: Israel's Place among the Nations*
254 William Johnstone, *1 and 2 Chronicles, Volume 2: 2 Chronicles 10–36: Guilt and Atonement*
255 Larry L. Lyke, *King David with the Wise Woman of Tekoa: The Resonance of Tradition in Parabolic Narrative*
256 Roland Meynet, *Rhetorical Analysis: An Introduction to Biblical Rhetoric*
257 Philip R. Davies and David J.A. Clines (eds.), *The World of Genesis: Persons, Places, Perspectives*
258 Michael D. Goulder, *The Psalms of the Return (Book V, Psalms 107–150): Studies in the Psalter, IV*
259 Allen Rosengren Petersen, *The Royal God: Enthronement Festivals in Ancient Israel and Ugarit?*
260 A.R. Pete Diamond, Kathleen M. O'Connor and Louis Stulman (eds.), *Troubling Jeremiah*
261 Othmar Keel, *Goddesses and Trees, New Moon and Yahweh: Ancient Near Eastern Art and the Hebrew Bible*
262 Victor H. Matthews, Bernard M. Levinson and Tikva Frymer-Kensky (eds.), *Gender and Law in the Hebrew Bible and the Ancient Near East*
263 M. Patrick Graham and Steven L. McKenzie, *The Chronicler as Author: Studies in Text and Texture*
264 Donald F. Murray, *Divine Prerogative and Royal Pretension: Pragmatics, Poetics, and Polemics in a Narrative Sequence about David (2 Samuel 5.17–7.29)*
265 John Day, *Yahweh and the Gods and Goddesses of Canaan*
266 J. Cheryl Exum and Stephen D. Moore (eds.), *Biblical Studies/Cultural Studies: The Third Sheffield Colloquium*
267 Patrick D. Miller, Jr, *Israelite Religion and Biblical Theology: Collected Essays*
268 Linda S. Schearing and Steven L. McKenzie (eds.), *Those Elusive Deuteronomists: 'Pandeuteronomism' and Scholarship in the Nineties*
269 David J.A. Clines and Stephen D. Moore (eds.), *Auguries: The Jubilee Volume of the Sheffield Department of Biblical Studies*
270 John Day (ed.), *King and Messiah in Israel and the Ancient Near East: Proceedings of the Oxford Old Testament Seminar*
271 Wonsuk Ma, *Until the Spirit Comes: The Spirit of God in the Book of Isaiah*
272 James Richard Linville, *Israel in the Book of Kings: The Past as a Project of Social Identity*
273 Meir Lubetski, Claire Gottlieb and Sharon Keller (eds.), *Boundaries of the Ancient Near Eastern World: A Tribute to Cyrus H. Gordon*
274 Martin J. Buss, *Biblical Form Criticism in its Context*
275 William Johnstone, *Chronicles and Exodus: An Analogy and its Application*
276 Raz Kletter, *Economic Keystones: The Weight System of the Kingdom of Judah*
277 Augustine Pagolu, *The Religion of the Patriarchs*

278 Lester L. Grabbe (ed.), *Leading Captivity Captive: 'The Exile' as History and Ideology*
279 Kari Latvus, *God, Anger and Ideology: The Anger of God in Joshua and Judges in Relation to Deuteronomy and the Priestly Writings*
280 Eric S. Christianson, *A Time to Tell: Narrative Strategies in Ecclesiastes*
281 Peter D. Miscall, *Isaiah 34–35: A Nightmare/A Dream*
282 Joan E. Cook, *Hannah's Desire, God's Design: Early Interpretations in the Story of Hannah*
283 Kelvin Friebel, *Jeremiah's and Ezekiel's Sign-Acts: Rhetorical Nonverbal Communication*
284 M. Patrick Graham, Rick R. Marrs and Steven L. McKenzie (eds.), *Worship and the Hebrew Bible: Essays in Honor of John T. Willis*
285 Paolo Sacchi, *History of the Second Temple*
286 Wesley J. Bergen, *Elisha and the End of Prophetism*
287 Anne Fitzpatrick-McKinley, *The Transformation of Torah from Scribal Advice to Law*
288 Diana Lipton, *Revisions of the Night: Politics and Promises in the Patriarchal Dreams of Genesis*
289 Jose Krazovec (ed.), *The Interpretation of the Bible: The International Symposium in Slovenia*
290 Frederick H. Cryer and Thomas L. Thompson (eds.), *Qumran between the Old and New Testaments*
291 Christine Schams, *Jewish Scribes in the Second-Temple Period*
292 David J.A. Clines, *On the Way to the Postmodern: Old Testament Essays, 1967–1998 Volume 1*
293 David J.A. Clines, *On the Way to the Postmodern: Old Testament Essays, 1967–1998 Volume 2*
294 Charles E. Carter, *The Emergence of Yehud in the Persian Period: A Social and Demographic Study*
295 Jean-Marc Heimerdinger, *Topic, Focus and Foreground in Ancient Hebrew Narratives*
296 Mark Cameron Love, *The Evasive Text: Zechariah 1–8 and the Frustrated Reader*
297 Paul S. Ash, *David, Solomon and Egypt: A Reassessment*
298 John D. Baildam, *Paradisal Love: Johann Gottfried Herder and the Song of Songs*
299 M. Daniel Carroll R., *Rethinking Contexts, Rereading Texts: Contributions from the Social Sciences to Biblical Interpretation*
300 Edward Ball (ed.), *In Search of True Wisdom: Essays in Old Testament Interpretation in Honour of Ronald E. Clements*
301 Carolyn S. Leeb, *Away from the Father's House: The Social Location of na'ar and na'arah in Ancient Israel*
302 Xuan Huong Thi Pham, *Mourning in the Ancient Near East and the Hebrew Bible*
303 Ingrid Hjelm, *The Samaritans and Early Judaism: A Literary Analysis*

304 Wolter H. Rose, *Zemah and Zerubabbel: Messianic Expectations in the Early Postexilic Period*
305 Jo Bailey Wells, *God's Holy People: A Theme in Biblical Theology*
306 Albert de Pury, Thomas Römer and Jean-Daniel Macchi (eds.), *Israel Constructs its History: Deuteronomistic Historiography in Recent Research*
307 Robert L. Cole, *The Shape and Message of Book III (Psalms 73–89)*
308 Yiu-Wing Fung, *Victim and Victimizer: Joseph's Interpretation of his Destiny*
309 George Aichele (ed.), *Culture, Entertainment and the Bible*
310 Esther Fuchs, *Sexual Politics in the Biblical Narrative: Reading the Hebrew Bible as a Woman*
311 Gregory Glazov, *The Bridling of the Tongue and the Opening of the Mouth in Biblical Prophecy*
312 Francis Landy, *Beauty and the Enigma: And Other Essays on the Hebrew Bible*
314 Bernard S. Jackson, *Studies in the Semiotics of Biblical Law*
315 Paul R. Williamson, *Abraham, Israel and the Nations: The Patriarchal Promise and its Covenantal Development in Genesis*
317 Lester L. Grabbe (ed.), *Did Moses Speak Attic? Jewish Historiography and Scripture in the Hellenistic Period*
320 Claudia V. Camp, *Wise, Strange and Holy: The Strange Woman and the Making of the Bible*
321 Varese Layzer, *Signs of Weakness: Juxtaposing Irish Tales and the Bible*
322 Mignon R. Jacobs, *The Conceptual Coherence of the Book of Micah*
323 Martin Ravndal Hauge, *The Descent from the Mountain: Narrative Patterns in Exodus 19–40*
324 P.M. Michèle Daviau, John W. Wevers and Michael Weigl (eds.), *The World of the Aramaeans: Studies in Honour of Paul-Eugène Dion*, Volume 1
325 P.M. Michèle Daviau, John W. Wevers and Michael Weigl (eds.), *The World of the Aramaeans: Studies in Honour of Paul-Eugène Dion*, Volume 2
326 P.M. Michèle Daviau, John W. Wevers and Michael Weigl (eds.), *The World of the Aramaeans: Studies in Honour of Paul-Eugène Dion*, Volume 3
327 Gary D. Salyer, *Vain Rhetoric: Private Insight and Public Debate in Ecclesiastes*
332 Robert J.V. Hiebert, Claude E. Cox and Peter J. Gentry (eds.), *The Old Greek Psalter: Studies in Honour of Albert Pietersma*